ADVANCED ACCELERATOR CONCEPTS

ADVANCED ACCELERATOR CONCEPTS

Ninth Workshop

Santa Fe, New Mexico   10–16 June 2000

EDITORS
Patrick L. Colestock
Sandra Kelley
Los Alamos National Laboratory

◎ CD-ROM INCLUDED

Melville, New York, 2001
AIP CONFERENCE PROCEEDINGS ■ VOLUME 569

Editors:

Patrick L. Colestock
MS H851
Los Alamos National Laboratory
Los Alamos, NM 87545
USA

E-mail: colestock@lanl.gov

Sandra Kelley
MS H831
Los Alamos National Laboratory
Los Alamos, NM 87545
USA

The articles on pp. 258–265, 287–293, 335–339, 582–589, 604–614, 678–684, 694–700, 848–855, 888–900, and 901–911 were authored by U. S. Government employees and are not covered by the below mentioned copyright.

Authorization to photocopy items for internal or personal use, beyond the free copying permitted under the 1978 U.S. Copyright Law (see statement below), is granted by the American Institute of Physics for users registered with the Copyright Clearance Center (CCC) Transactional Reporting Service, provided that the base fee of $18.00 per copy is paid directly to CCC, 222 Rosewood Drive, Danvers, MA 01923. For those organizations that have been granted a photocopy license by CCC, a separate system of payment has been arranged. The fee code for users of the Transactional Reporting Service is: 0-7354-0005-9/01/$18.00.

© 2001 American Institute of Physics

Individual readers of this volume and nonprofit libraries, acting for them, are permitted to make fair use of the material in it, such as copying an article for use in teaching or research. Permission is granted to quote from this volume in scientific work with the customary acknowledgment of the source. To reprint a figure, table, or other excerpt requires the consent of one of the original authors and notification to AIP. Republication or systematic or multiple reproduction of any material in this volume is permitted only under license from AIP. Address inquiries to Office of Rights and Permissions, Suite 1NO1, 2 Huntington Quadrangle, Melville, N.Y. 11747-4502; phone: 516-576-2268; fax: 516-576-2450; e-mail: rights@aip.org.

L.C. Catalog Card No. 2001089399
ISBN 0-7354-0005-9 Set
ISBN 0-7354-0006-7 CD-ROM (not available separately)
ISSN 0094-243X
Printed in the United States of America

CONTENTS

Preface xi
Scientific Advisory and Organizing Committee xiii

TRIBUTE TO PROFESSOR JOHN DAWSON

Personal Recollections on the Development of Plasma Accelerators
and Light Sources 3
    J. M. Dawson
Tribute to John M. Dawson: "Pre-historic Days" of the
Dawson-Tajima 1979 Paper 23
    T. Tajima
John Dawson's Advanced Accelerator Years 26
    C. Joshi

WORKING GROUP SUMMARIES

Report of Working Group 1: The Physics of High Energy Density 33
    J. B. Rosenzweig
Summary Report of Working Group 2 on Laser-Plasma
Acceleration Concepts 35
    W. B. Mori and M. C. Downer
Summary Report of the Working Group on Electromagnetic
Structure-Based Acceleration Concepts 47
    E. R. Colby and W. Gai
Beam Generation, Monitoring, Conditioning and Control
at High Frequencies and Ultra-Fast Time Scales 57
    D. T. Palmer and J. G. Wang
Summary Report: Working Group 5 on "Electron Beam-Driven
Plasma and Structure Based Acceleration Concepts" 61
    M. E. Conde and T. Katsouleas
Summary Report of Working Group 6: Millimeter-Wave Sources 65
    G. S. Nusinovich and S. H. Gold
Summary of Working Group 7 on "Exotic Acceleration Schemes" 77
    T. Tajima

CONTRIBUTED AND INVITED PAPERS

Laser-Plasma Accelerators: A Status Report 85
    C. Joshi
Relativistic Ion Acceleration by Ultraintense Laser Interactions 97
    K. Nakajima, J. K. Koga, and K. Nakagawa

Production and Characterization of a Fully-Ionized He Plasma Channel 105
    E. W. Gaul, S. P. Le Blanc, A. R. Rundquist, R. Zgadzaj,
    N. H. Matlis, H. Langhoff, and M. C. Downer

GeV Laser Acceleration Research at JAERI-APR 112
    K. Nakajima, T. Hosokai, S. Kanazawa, M. Kando,
    S. Kondo, H. Kotaki, and T. Yokoyama

Numerical Simulation for Plasma Electron Acceleration by 12TW 50 Fs Laser Pulse 122
    N. Hafz, J. Koga, R. Hemker, and M. Uesaka

Status of the LILAC Experiment 127
    N. Saleh, P. Han, C. Keppel, P. Gueye, V. Yanovsky,
    and D. Umstadter

Laser Wakefield Accelerator Experiments at LBNL 136
    W. P. Leemans, D. Rodgers, P. E. Catravas, G. Fubiani,
    C. G. R. Geddes, E. Esarey, B. A. Shadwick,
    G. J. H. Brussaard, J. van Tilborg, S. Chattopadhyay,
    J. S. Wurtele, L. Archambault, M. R. Dickinson, S. DiMaggio,
    R. Short, K. L. Barat, R. Donahue, J. Floyd, A. Smith,
    and E. Wong

Demonstration of a Laser-Driven Prebuncher Staged with a Laser Accelerator—The STELLA Program 146
    M. Babzien, I. Ben-Zvi, L. P. Campbell, C. E. Dilley,
    D. B. Cline, J. C. Gallardo, S. C. Gottschalk, W. D. Kimura,
    P. He, K. P. Kusche, Y. Liu, R. H. Pantell, I. V. Pogorelsky,
    D. C. Quimby, J. Skaritka, A. van Steenbergen, L. C. Steinhauer,
    and V. Yakimenko

Fluid Modeling of Intense Laser-Plasma Interactions 154
    B. A. Shadwick, G. M. Tarkenton, E. H. Esarey,
    and W. P. Leemans

Emittance Control in Laser Wakefield Accelerator 163
    S. Cheshkov, T. Tajima, C. Chiu, and F. Breitling

Optimization of Laser Wakefield Acceleration 177
    A. R. Rundquist, S. P. LeBlanc, E. W. Gaul, S. Cheshkov,
    F. B. Grigsby, T. T. Tajima, and M. C. Downer

Laser Shaping and Optimization of the Laser-Plasma Interaction 183
    A. Spitkovsky and P. Chen

Parametric Excitation of Plasma Waves by Counter-Propagating Laser Beams Detuned by $2\omega_p$ 195
    G. Shvets

Suppression of Raman Forward Scattering in Plasma Channels 204
    G. Shvets and X. Li

Nonparaxial Propagation of Intense Laser Pulses in Plasmas 214
    E. Esarey, B. A. Shadwick, C. B. Schroeder, J. S. Wurtele,
    and W. P. Leemans

Beam-Quality Simulations for Channel-Guided LWFA 223
    V. V. Goloviznin, A. J. W. Reitsma, L. P. J. Kamp,
    and T. J. Schep

Resonant and Hollow Beam Generation of Plasma Channels 231
    I. Alexeev, K. Y. Kim, J. Fan, E. Parra, H. M. Milchberg,
    L. Y. Margolin, and L. N. Pyatnitskii

GeV Energy Gain in a Channel Guided Laser Wakefield Accelerator 242
    P. Sprangle, B. Hafizi, J. R. Peñano, R. F. Hubbard,
    A. Ting, A. Zigler, and T. M. Antonsen, Jr.

IFEL Experiment at the Neptune Lab 249
    P. Musumeci and C. Pellegrini

**Experimental Measurements of Wakefields in a Multimode,
Dielectric Structure Driven by a Train of Electron Bunches.** 258
    J. G. Power, M. E. Conde, W. Gai, A. Kanareyken,
    R. Konecny, and P. Schoessow

Direct Laser Acceleration in a Capillary Channel 266
    L. C. Steinhauer

S-Band and X-Band Integrated PWT Photoelectron Linacs 274
    D. Yu, D. Newsham, and J. Zeng

**Experimental Demonstration of Dielectric Structure Based
Two Beam Acceleration.** 287
    W. Gai, M. E. Conde, R. Konecny, J. G. Power, P. Schoessow,
    X. Sun, and P. Zou

Acceleration Results from the Microwave Inverse FEL Experiment 294
    R. B. Yoder, T. C. Marshall, and J. L. Hirshfield

**Bunch Stability during Wave Field Generation in a
Dielectric-Lined Waveguide** 305
    S. Y. Park and J. L. Hirshfield

Multi-mode, Multi-bunch Dielectric Wake Field Resonator Accelerator 316
    T. C. Marshall, J.-M. Fang, J. L. Hirshfield, and S.-Y. Park

Laser-Driven Clyclotron Autoresonance Accelerator 326
    J. L. Hirshfield and C. Wang

Twisted Waveguide Accelerating Structure 335
    Y. W. Kang

A Possible Plasma Source for a SLAC Afterburner 340
    C. Joshi

Acceleration Concepts Based on Electromagnetic Structures 342
    L. Schächter

**Parameters of a 2×200 GeV Linear Collider with Microstructures
Excited by Laser Radiation.** 365
    A. Mikhailichenko

**Production and Synchronization of Electron Beams from
RF Photoinjector/Compressor Systems for Ultra-Fast Applications.** 374
    M. C. Thompson and J. B. Rosenzweig

**RF Photoinjector Development for a Short-Pulse, Hard X-Ray
Thomson Scattering Source** 391
    G. P. Le Sage, S. G. Anderson, T. E. Cowan, J. K. Crane,
    T. Ditmire, and J. B. Rosenzweig

UMER: The University Maryland Electron Ring 405
 P. G. O'Shea, M. Reiser, R. A. Kishek, S. Bernal, H. Li,
 M. Pruessner, M. Virgo, V. Yun, W. Zhang, T. Godlove,
 D. Kehne, P. Haldemann, and I. Haber

Confinement of Bunched Beams. 415
 M. Hess and C. Chen

Studies of Space-Charge Effects in Ultrashort Electron Bunches 423
 G. Fubiani, W. Leemans, and E. Esarey

Low Emittance Electron Beam Formation with a 17 GHz RF Gun 436
 W. J. Brown, S. E. Korbly, K. E. Kreischer, M. A. Shapiro,
 and R. J. Temkin

Three-Dimensional Theory of Emittance in Compton Scattering 450
 F. V. Hartemann, A. Le Foll, A. K. Kerman, B. Rupp,
 D. J. Gibson, E. C. Landahl, A. L. Troha, N. C. Luhmann, Jr.,
 and H. A. Baldis

**A Ferroelectric Cathode Electron Gun for Use in High Power
Microwave Sources.** 465
 Y. Hayashi, X. Song, J. D. Ivers, D. Flechtner, J. A. Nation,
 and L. Schächter

Betatron Radiation from Electron Beams in Plasma Focusing Channels 473
 E. Esarey, P. Catravas, and W. P. Leemans

Commissioning and Measurements of the Neptune Photo-injector. 487
 S. G. Anderson, M. Loh, P. Musumeci, J. B. Rosenzweig,
 H. Suk, and M. C. Thompson

Summary of Japanese Advanced Accelerator Work 500
 M. Uesaka

Plasma Focusing of High Energy Density Electron and Positron Beams. 518
 J. S. T. Ng, P. Chen, H. A. Baldis, P. Bolton, D. Cline,
 W. Craddock, C. Crawford, F. J. Decker, R. C. Field,
 Y. Fukui, V. Kumar, M. J. Hogan, R. Iverson, F. King,
 R. E. Kirby, T. Kotseroglou, K. Nakajima, R. Noble,
 A. Ogata, P. Raimondi, D. Walz, and A. W. Weidemann

High-Brightness Electron Beam Production, Transport, and Measurement 529
 B. E. Carlsten

Aperture Effects in Intense Beams 544
 S. Bernal, P. G. O'Shea, R. Kishek, and M. Reiser

Laser Acceleration of Protons from Thin Film Targets 553
 K. Flippo, S. Banerjee, V. Y. Bychenkov, S. Gu,
 A. Maksimchuk, G. Mourou, K. Nemoto,
 and D. Umstadter

Recent Advances in Electron and Positron Sources 563
 J. E. Clendenin

**Development of High-Brightness Laser Synchrotron Source
at BNL ATF.** 571
 I. V. Pogorelsky, I. Ben-Zvi, T. Hirose, S. Kashiwagi, K. Kusche,
 T. Kumita, T. Omori, V. Yakimenko, K. Yokoya, J. Urakawa,
 and M. Washio

Muon Cooling—Emittance Exchange 583
 Z. Parsa

Modeling Beam-Driven and Laser-Driven Plasma Wakefield Accelerators with XOOPIC...............591
    D. L. Bruhwiler, R. Giacone, J. R. Cary, J. P. Verboncoeur,
    P. Mardahl, E. Esarey, and W. Leemans

Transformer Ratio Enhancement Using a Ramped Bunch Train in a Collinear Wakefield Accelerator...............605
    J. G. Power, W. Gai, and A. Kanareykin

Particle Beam Stability in the Hollow Plasma Channel Wake Field Accelerator...............616
    C. B. Schroeder and J. S. Wurtele

Trapping of Background Plasma Electrons in a Beam-Driven Plasma Wake Field Using a Downward Density Transition...............630
    H. Suk, N. Barov, J. B. Rosenzweig, and E. Esarey

Development of Multiple Beam Guns for High Power RF Sources...............640
    L. Ives and G. Miram

Symmetric and Asymmetric Mode Interaction in High-Power Traveling Wave Amplifier...............652
    P. Wang, Z. Xu, C. Grabowski, J. A. Nation, S. Banna, and L. Schächter

Development of a 10 MW, 91 GHz Gyroklystron for Accelerator Applications...............663
    R. L. Ives, W. Lawson, J. M. Neilson, and M. Read

X-Band Dielectric Loaded Traveling-wave Acceleration Structure...............679
    P. Zou, W. Gai, R. Konecny, X. Sun, and T. Wong

RTA Beam Dynamics Experiments: Limiting Cumulative Transverse Instability Growth in a Linear Periodic System...............686
    T. Houck, S. Lidia, and G. Westenskow

CARM-Klystron Amplifier for Accelerator Applications...............695
    S. H. Gold and A. W. Fliflet

Multi-Moded Passive RF Pulse Compression Development at SLAC...............702
    C. D. Nantista and S. G. Tantawi

Design and Fabrication of a 94 GHz Klystron...............712
    G. Scheitrum, G. Caryotakis, A. Haase, L. Song, B. Arfin, Y. Cheng, B. Shew, and B. James

Comparison of Discrete Klystron Produced RF to Two-Beam Produced RF for Large Accelerator Systems...............725
    R. Pitthan

Two-Channel Active High-Power X-Band Pulse Compressor...............741
    A. L. Vikharev, A. M. Gorbachev, O. A. Ivanov, V. A. Isaev,
    V. A. Koldanov, S. V. Kuzikov, A. G. Litvak, M. I. Petelin,
    J. L. Hirshfield, and O. A. Nezhevenko

Some Thoughts about Millimeter-Wave Drivers for Future Linear Colliders...............751
    G. S. Nusinovich

10-MW, W-Band RF Source for Advanced Accelerator Research 765
    J. L. Hirshfield, O. A. Nezhevenko, C. Wang, V. P. Yakovlev,
    A. A. Bogdashov, V. L. Bratman, A. V. Chirkov, G. G. Denisov,
    A. N. Kuftin, S. V. Samsonov, and A. V. Savilov

Design of Flat-Field, High-Aspect Ratio RF Structures 775
    D. Yu and A. V. Smirnov

34 GHz Pulsed Magnicon Project 786
    O. A. Nezhevenko, M. A. LaPointe, S. V. Schelkunoff,
    V. P. Yakovlev, J. L. Hirshfield, E. V. Kozyrev, G. I. Kuznetsov,
    B. Z. Persov, and A. Fix

3-D Space Charge Simulations of High-Power Magnicon Amplifiers 797
    V. P. Yakovlev, O. V. Danilov, B. Hafizi, and O. A. Nezhevenko

Designs of Three-cavity Frequency Quadrupling Coaxial Gyroklystrons 809
    I. Yovchev, G. S. Nusinovich, W. Lawson, V. L. Granatstein,
    and E. S. Gouveia

**Optically Driven Emitter of Neutrinos for Testing of
Neutrino Oscillations** 825
    A. V. Pakhomov and Y. Takahashi

**Multi-Stage, High-Gradient, Cyclotron Resonance Proton
Accelerator Concept** 833
    J. L. Hirshfield, C. Wang, and R. Symons

W-Band Accelerator Study in KEK 844
    X. Zhu and K. Nakajima

Principle of Alternating Gradient Acceleration 850
    M. Xie

**The Status of Ionization Cooling Tests for Muon Colliders and
Neutrino Factories** 858
    D. B. Cline

Wake-Amplification by a Solid-State Active Medium 863
    L. Schächter, E. Colby, and R. H. Siemann

**Resonant Absorption Instability: Acceleration and Radiation
Amplification** 873
    L. Schächter

Table Top Accelerator with Extremely Bright Beam 881
    A. Mikhailichenko

Muon Sources—ν Factory to $\mu^{\pm}$ Colliders 890
    Z. Parsa

An Ultra-High Gradient Plasma Wakefield Booster 903
    P. Chen, S. Cheshkov, R. Ruth, and T. Tajima

Author Index 915

Preface

The Ninth Workshop on Advanced Accelerator Concepts, AAC2000, was held at the Santa Fe Hilton from June 10 – 16, 2000 in Santa Fe, NM. This workshop was sponsored by the Los Alamos National Laboratory, and locally organized by members of LANL. This was another in a series of workshops on this topic held since the early 1980's, which incidentally began with a workshop sponsored by Los Alamos. This workshop was also sponsored by the US Department of Energy High Energy Physics Division, Advanced Technology Branch, under the leadership and encouragement of Dr. David Sutter.

This is the only such workshop regularly sponsored to permit a cross-disciplinary discussion of the innovative and evolving area Advanced Accelerators. By its nature, the field and the workshop cover a broad range of topics ranging from laser and beam-driven acceleration methods, novel particle sources, high-power rf and mm-wave generation methods, beam diagnostics and control, to exotic acceleration schemes. In recent years, both theoretical and experimental progress has been rapid, and this workshop saw the revelation of new and encouraging results in a number of areas.

The format of the week-long workshop was a series of invited papers held in common plenary sessions in the mornings, with afternoons open for individual working group discussions. Invited papers were presented chiefly early in the week to foster discussions in the working groups. Within the working groups, contributed papers were given and discussions guided by the working group leaders and co-leaders. In addition, a continuous poster session was held during the week to permit extended topical discussions. To further encourage participant interaction, receptions were held at the Santa Fe Museum of Indian Arts and Culture and the Contemporary Southwest Gallery. The informal outdoor banquet was held at the Bishop's Lodge in the foothills of the Sangre de Cristo range, just north of Santa Fe.

The workshop was attended by 155 participants, by invitation only, that represented a broad spectrum of research interests. Particular attention was given to ensuring participation by a significant number of junior members of the advanced accelerator community, and graduate students. To encourage student participation, 28 scholarships were granted.

An overall theme of this particular workshop was the celebration of the seventieth birthday of Prof. John Dawson, a man who has done so much to create and develop the concept of advanced accelerators over the span of his productive career. John was in attendance to receive the accolades and gratitude of his fellow colleagues and friends. A number of tributes to his many and varied contributions to the field were given, and these are included as an introduction to these proceedings, along with Prof. Dawson's own personal recollections on how some of the ideas evolved. We are all grateful for Prof. Dawson and his work, and for the opportunity to celebrate with him during the workshop.

I would like to take the opportunity to acknowledge the generous support of this field over the last two decades by the US Department of Energy High Energy Physics Division, in particular Dr. David Sutter, Chief of the Advanced Technology Branch. I would also like to thank the Scientific Advisory and Organizing Committee for their hard work and support. A special thanks goes to the local organizing committee and dedicated staff including Jaime McDonald, Christopher Webster, Marianna Martinez and Marion Hutton. Thanks also goes to Robert Wheat for computer support, and to Sandra Kelley for help in editing these proceedings. Thanks also to the professional staff at the Santa Fe Hilton.

It was a privilege to host the Advanced Accelerator Workshop this year, both for the honor of celebrating Prof. John Dawson, and for the significance of the technical progress recorded here. We hope you will find these proceedings informative and inspirational.

Michael Fazio
Los Alamos National Laboratory

Scientific Advisory and Organizing Committee

Ilan Ben-Zvi
Bruce Carlsten
Pisin Chen
Eric Colby
Patrick Colestock
Michael Fazio
Wei Gai
Victor Granatstein
Chan Joshi
Tom Katsouleas
Warren Mori
Wim Leemans
K. Nakajima
Gregory Nusinovich
John Nation
Patrick O'Shea
Jamie Rosenzweig
Linda Spentzouris

TRIBUTE TO
PROFESSOR JOHN DAWSON

Personal Recollections On The Development of Plasma Accelerators And Light Sources

John M. Dawson
UCLA

This **Story**, of my involvement with Plasma Accelerators, is one that goes back more than forty years; the work of many hands in many places with a great deal of serendipity and good fortune on my part. I will tell you how it evolved as best I can remember it. It is impossible for me to cover all the important aspects and developments by so many people, so I apologize in advance for leaving out some critical contributions, hence, I hope you will simply put these up to the failing memory of an ancient physicist.

In many ways, this path began soon after I joined Project Matterhorn at Princeton in 1956. In those days, work on Controlled Fusion was classified, and I could not work directly on Fusion until my clearance came through which took about four months. They put me to learning Plasma Physics, a virtually unknown science at the time, from Lyman Spitzer's little book. One of the things I studied was electrostatic plasma (Langmuir) waves. I soon found that for a Cold Plasma, by going to a Lagrangian Coordinate System, waves of an arbitrary amplitude became linear, and I could obtain exact large amplitude solutions as long as there was no crossing of orbits[1]. You can't imagine what problems I had with referees in getting this result published. For plane waves there was a critical amplitude ($\delta X \sim 1/k, \delta X$ is the electron displacement and k is the wave number) at which orbit crossing took place. I found that the technique worked even if the motion was relativistic, but in this latter case the amplitude had to be constant everywhere and, of course, the motions had to be properly phased. If these conditions were not met then breaking always developed in time. Breaking also tend to develop for non-planar wave forms, or if the plasma has a non-uniform density. I wondered what would happen to a wave when it broke? Of course, the crest fell into the trough (Fig. 1), and it appeared that the result would be pretty chaotic and destructive to the wave.

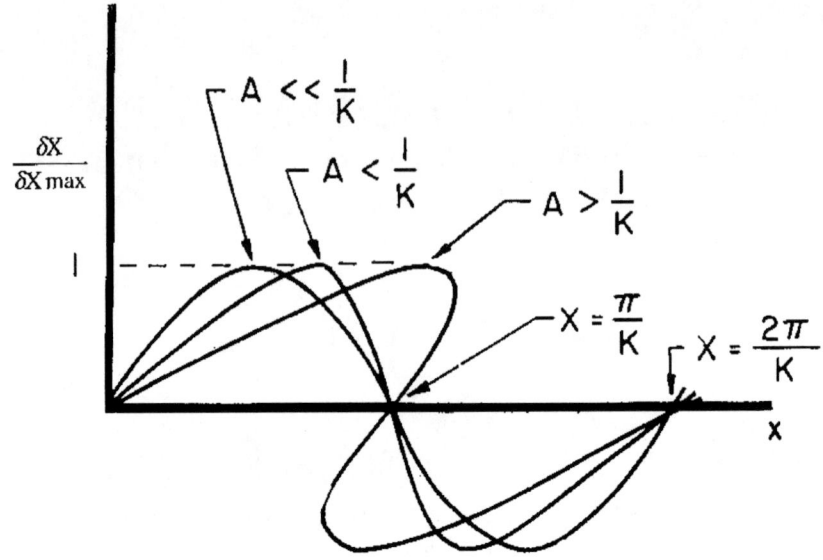

FIGURE 1. Wave form vs. amplitude

I was sure electrons would be accelerated to high energy, but how high and how many of them? To do anything analytically meaningful seemed almost impossible. Fortunately for me, the high speed electronic computer was just becoming available. Princeton got an IBM 650 about this time. This machine had all of 8000 words of memory for program and data and could do about 100 arithmetic operations a sec. Nevertheless, I realized that this machine could give me real information on the Breaking Wave Problem. I realized that we could treat the problem as the motion of a large number of charged sheets embedded in a neutralizing background[2] (Fig. 2).

The electron sheets were started out with a sinusoidal wave form of sufficient amplitude that wave breaking took place. The wave amplitude and the particle energies could be followed as a function of time and also as a function of the initial amplitude. We found that breaking was quite destructive to the wave, and that some particles were pushed to twice the phase velocity of the initial wave[3]. I will return to this shortly, as it is quite important to plasma accelerators. Figure 3, shows some early results from these breaking wave calculations.

Soon after I discovered this model, I recognized that it was a wonderful and powerful

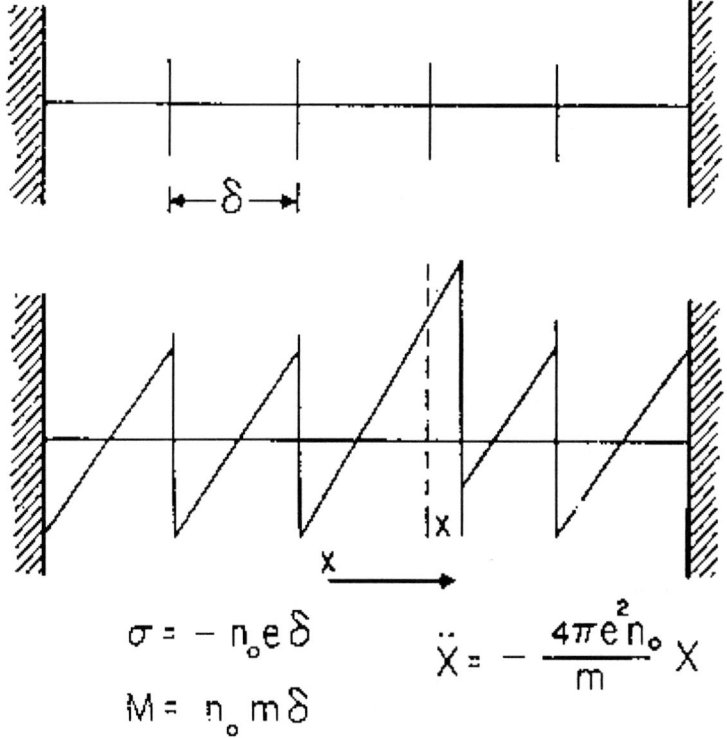

FIGURE 2. One-dimensional plasma model

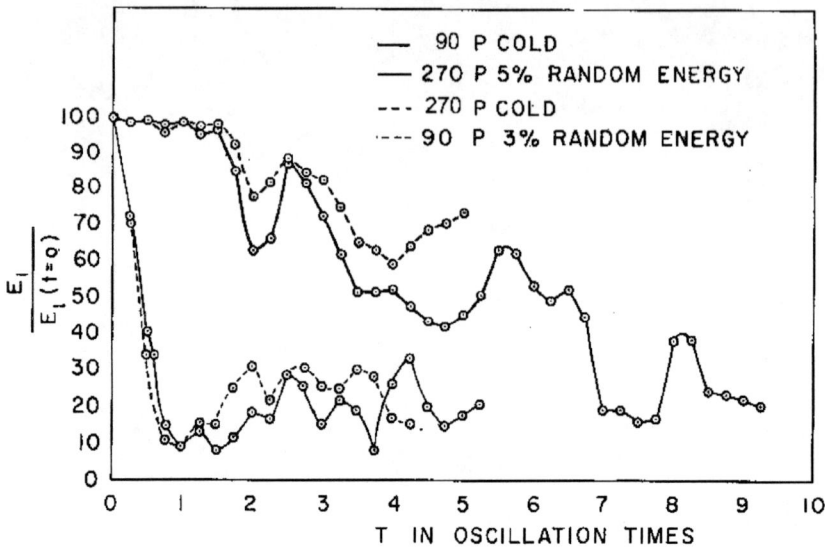

FIGURE 3. Reseults from Breaking Wave calculations

tool for studying the physics of plasmas. At the time there was much work going on the kinetic theory of plasma[4,5], and I realized that many tests of this theoretical work could be made using this model. It was also clear that many nonlinear processes could be looked at, and I could see that this was a great liberating idea that allowed us to look at theory, check our intuition, explore new regimes, verify correct ideas, and correct false ideas. It was a great learning tool for Plasma Physics.

The first work on one-dimensional plasma models, although quite limited, demonstrated Debye Shielding, Landau Damping (Figs. 4 and 5), particle drag and diffusion in velocity space, and electric field fluctuations. This work clearly showed the power of the method and led to a great number of student theses.

At this point, I would like to pay tribute to my first student, Craig Smith, who I recently learned passed away. Craig was great with computers and worked under some very adverse conditions. He developed a model that was composed of both positive and negative sheets. This system could be solved exactly between crossings, and this is what he programmed the computer to do. His code conserved energy and momentum to one part in 10^{12}; a feat I believe no one else has matched[6].

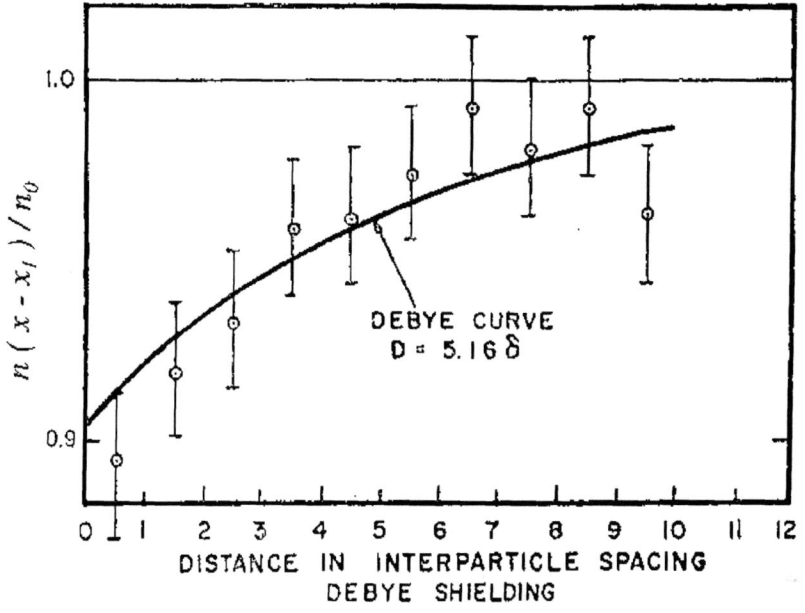

FIGURE 4. Debye cloud

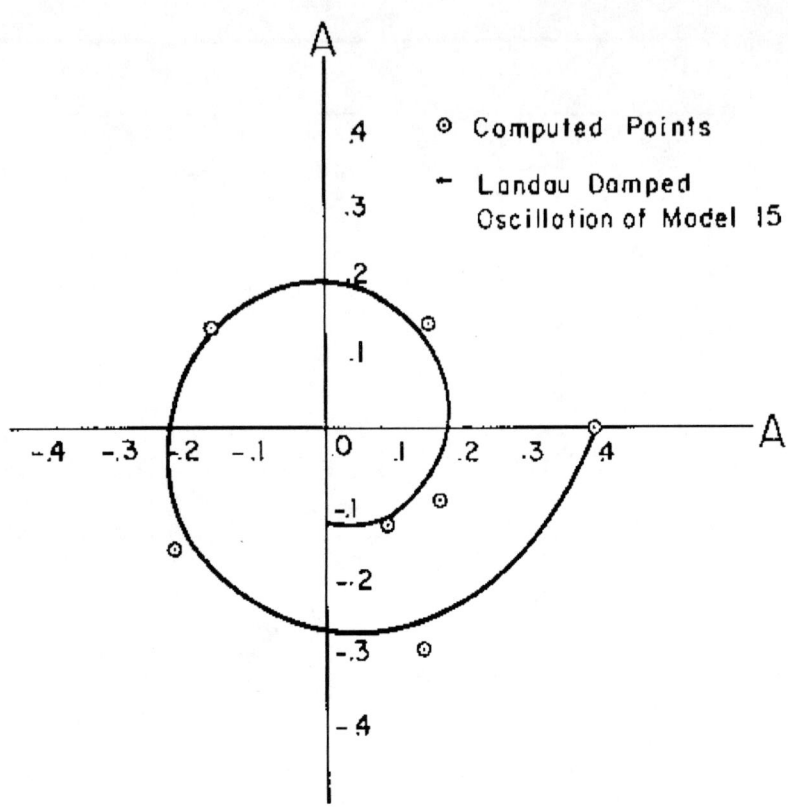

FIGURE 5. The Landau damping of Fourier mode

There were a lot of students in the 1960's, and unfortunately I cannot cover them all. The next work I want to mention is that of Bill Kruer on parametric instabilities of a plasma subject to an intense AC electric field. People had developed Q switched laser and were making plasmas with them. The most powerful of such lasers produced Gigawatts for nano secs. , and could produce plasmas in the multi hundred ev. range. The idea of using these to achieve fusion of course came up. It was clear that at the intensities achieved, the conventional theory of collisional absorption would break down. This was particularly so for the most powerful and efficient laser of the day, the 10μ CO_2 laser which would only penetrate into regions of relatively low density where collisions are ineffective. However, it was also clear that at the electric fields involved, the plasma was being treated very roughly; it appeared that the plasma might be subject to instabilities that would give it an anomalous absorption coefficient. This is the problem that Bill set out to address and the Numerical Model seemed the best and most straight forward way to attack it. Bill's investigations, indeed showed that parametric instabilities did occur and that they lead to anomalous absorption; the absorption increased by many orders of magnitude. The instability produced hot electrons, ~100s of Kev. Figure 6, shows the time development of the anomalous collision frequency as a function of time for a plasma subjected to an intense AC field near the plasma frequency.

At first we thought this might be OK; the hot electrons could heat the outer shell of the pellet. However, they proved too penetrating and heated the core and prevented its compression. Even in these early days the solution was clear, go to higher frequency and reduce the coherence of the laser. However, the technology was not available for quite sometime.

Although these early experiments did not achieve their primary goal, however, they did illustrate that parametric instabilities in plasmas could generate intense plasma waves with fields of Mev to 10s of Mev per cm. It was natural to ask, "can't we build a laser plasma accelerator?"

From our studies of plasma waves we knew that they could have any phase velocity, $v_p = \omega_p/k$, this could in fact be the speed of light. From our work on large amplitude waves, we knew that the electric field that could be produced by a speed of light wave could be extremely large ($E_{Max} \sim [n]^{1/2}$ V/cm or 10^9 V/cm for a density of 10^{18}). It was also clear that for a wave with phase velocity close to C, there would be very little gain in velocity, but a enormous gain in energy as the particle has difficulty in out running the wave; the limit on energy gain turns out to be $\delta W = 2(1 - v_p^2/C^2)^{1/2} \, mC^2$.

I was thinking about these ideas as early as 1970. However, there were many, many questions. How could we create such a wave? I knew such waves had a tendency

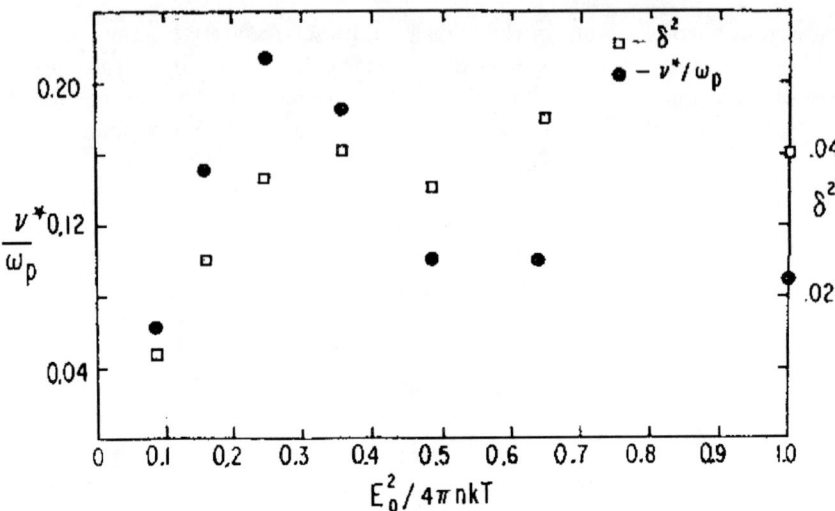

FIGURE 6. The effective collision frequency and the ion density saturation vs. the energy in the driving field.

to break and Bill Kruer's work had showed that they would be unstable. Even if I could create them, would they hang together long enough to make a useful accelerator?

Laser experiments were starting to be done which showed stimulated Raman instabilities where the reflected light was down shifted by the plasma frequency, and thus indicating that plasma oscillations were being excited. The theoretical work of Rosenbluth and Liu[7] predicted that the forward Raman instability could excite a plasma wave with a phase velocity equal to the group velocity of the laser, i.e., close to C. Thus, the possibility of generating the desired wave seemed to be there. This was the situation when I arrived at UCLA in 1973. Again, the ideal way to carry the investigation forward seemed to be through computer modeling. I simply needed someone to carry out the calculations. Bruce Cohen of LLNL visited me in 1974. I suggested that he look at the problem. He had access to electromagnetic relativistic codes, and LLNL had the computing resources that could do the job. I thought LLNL would be interested in the result, however, Bruce declined. The problem had to wait until 1976, when Toshi Tajima joined me. Toshi picked it up and carried out a very successful investigation which showed that substantial electron acceleration could be obtained[8]. Figure 7, shows particle acceleration found in that work. In the process, the concepts of the laser wake field accelerator, the laser beat wave accelerator, and the spontaneous Raman instability accelerator were developed. The expected acceleration was seen and, moreover, we found that the accelerated electrons closely followed the laser pulse so that an extended stable wake of plasma waves was not needed.

At this point, a piece of rare good fortune bestowed upon me. It was totally unforeseen and was a high point of my scientific career — Chand Joshi came to UCLA. Chand was a believer in our work and wanted to test our ideas experimentally. However, Chand was in the Electrical Engineering Department, and not yet a faculty member, and I was in the Physics Department. We needed a laboratory to carry out his experiments. At that time, Dean Harold Ticho was gracious enough in granting our space request in the Physics Department. Chand also got funding for the project, and we were off and running. Chand successfully demonstrated the generation of Beat Waves with phase velocities close to C in 1983[9]. Other experiments followed around the world.

Next came the problem of accelerating electrons by such waves. This was a problem orders of magnitude more difficult than simply generating the waves. For this Chand set up a lab in Engineering, the Mars Lab. He obtained one of the amplifiers from LANL CO_2 Fusion program which had been discontinued. Chand and his group engineered it to put out an intense short pulse containing two frequencies suitable for doing beat wave experiments. The experiment required pre-accelerating some electrons and injecting them into the focus of two beating laser beams. The size of the confocal spots was perhaps 100μ. A plethora of nitty-gritty problems had to be overcome which had nothing to do with the basic physics of the experiment. For example, the first plasma source we tried was a Θ pinch, which no matter how hard Chand tried, it produced a **B** field that could not be made zero or reproducible. It deflected the e-beam sufficiently that it missed the focal spot. It took Chand about 10 years to solve these problems, but he had faith. It took a lot of courage for a young scientist to stake his career on this. Other places around the world had essentially given up by the time Chand got it to work. Then in 1992, everything came together. In short order Chand was accelerating electrons to 30 Mev in a cm or two[10].

During the 1980's, and up to the present day, the Modeling Group at UCLA has attracted a truly remarkable set of students, postdocs, and visitors. To name a few students: Tom Katsouleas, Warren Mori, Yi-To Yan, J.J. Su, Scott Wilks; and postdocs, Pisin Chen, Jean-Noel LeBeouf, Luis Silva. On the visitor side, we were particularly lucky to establish collaboration with Bob Bingham of the Rutherford Appleton Laboratory in England, where the flow of ideas from this collaboration has been boundless. I should not overlook the wonderful contributions of our research

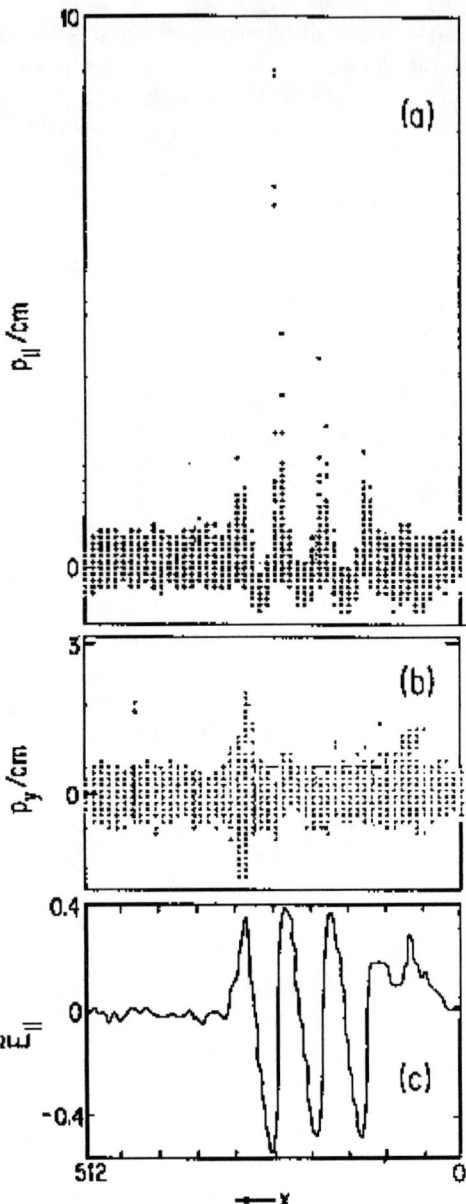

FIGURE 7. Wake-plasmon excitation and trapping of electrons. The head of the photon packet has proceeded forward to $x = 310$ at $t = 24\omega_p^{-1}$. $\omega/\omega_p = 4.3$. (a) The longitudinal momentum ($p_x = p_\parallel$) vs position ($p_x - x$ phase space) of electrons. (b) $p_y - x$ phase space. (c) The longitudinal field $E_L = E_\parallel$ vs position.

staff, particularly those of Viktor Decyk who has made simulation a world of three dimensional models with all the vast richness that comes with it.

All this talent brought with what I call an "explosion of ideas." Suddenly, many things started to come together, and we could solve problems we never could before. Many of the concepts that arose during this era have yet to be investigated, and are just beginning to be looked at. They touch on things from fusion, to particle acceleration, to tuned light, to acceleration of cosmic rays, to the dynamics of the explosion of supernova and, possibly to the development of the Big Bang.

I will list a few of the things that happened during this period.

1. The realization that electron bunches can replace laser beams as generators of accelerating waves. In the mid-80's, it was proposed that SLAC use a plasma booster to enhance their energy which is now being investigated in the E-157 experiment.

2. The realization that a plasma can act as a lens of unprecedented strength for focusing electrons or positrons. First tested at Wisconsin, later at UCLA, and most recently at SLAC on E-150.

3. The realization that laser focusing and guiding can be achieved by a plasma.

4. The realization that plasma waves contain regions of focusing and defocusing of high energy particles, and that the accelerating and focusing regions overlap.

5. The invention of the Surfatron to prevent phase slip between accelerated particles and a wave.

6. The realization that there are different regimes for wave generation, small perturbations, and plasma blow out regimes which have different properties and advantages.

7. The realization that the laser wake field can be reversed, a plasma wave can be used to add energy to a light pulse (upshift its frequency); the photon accelerator.

8. The realization that through the interactions of electron beams, plasmas and lasers, tunable sources of radiation can be created.

9. The realization that the enormously intense Neutrino fluxes produced in Supernova explosions should cause parametric instabilities similar to those produced by intense lasers, and that these could be responsible for the deposition of a significant amount of the Neutrino energy in the Stellar envelope. This would effect the dynamics of supernova explosions.

10. The realization that there existed a unified method for treating the interactions of intense electron beams, intense laser beams, and intense Neutrino beams

with plasma. A wave kinetic approach where only the strength of the coupling coefficients need be adjusted.

11. The development of particle methods for treating photons and Neutrinos for generating numerical solutions of real problems.

This was an extremely exciting time with new ideas coming along every few weeks. Just to give you a better idea of the excitement and how many people were involved, I have attached an addendum of a list of 50 papers and reports that were generated. The list is incomplete, but it gives some feel of the excitement and how many people were involved.

During this time the phenomenal growth of computer power and our ability to use it has played an important and critical role. Warren Mori, in the mid-80's, took on the Toshi problem in two dimensions. His results demonstrated laser self focusing, focusing of accelerated particles and the production of currents of Mev electrons in the 30,000 amp range[11]. Some results from his work are shown in Figures 8, 9 and 10. These early results gave us confidence that we were on the right track.

Recently, the development of parallel computing and particularly the work of Viktor Decyk in making very large calculations feasible and cheap has moved us into the three dimensional regime. The challenge of understanding results here are very great, but that is where real world problems are.

This has been a *Thumb Nail Sketch* of the path my group has taken. Of necessity, much has had to be left out and credits go to many, many people in many places; many of the ideas were in the air and they simply came together with the tools that opened them to solutions at UCLA. I apologize that I am unable to give proper credit to all who deserve it.

This is the story of **Science as a Living Thing** taking unexpected turns in directions that were never foreseen. Science must have goals, but it must also have the freedom to follow up interesting and unexpected results when they turn up. This is what excites the good young researcher and it is in their hands that our future rests.

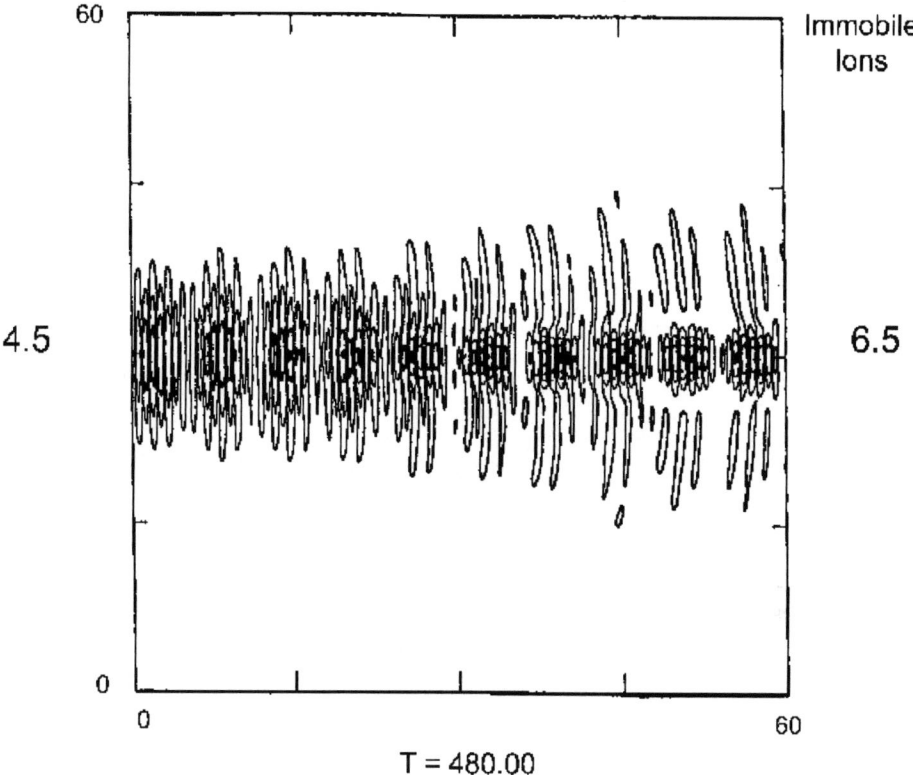

FIGURE 8. Contour plot of laser Electric field.

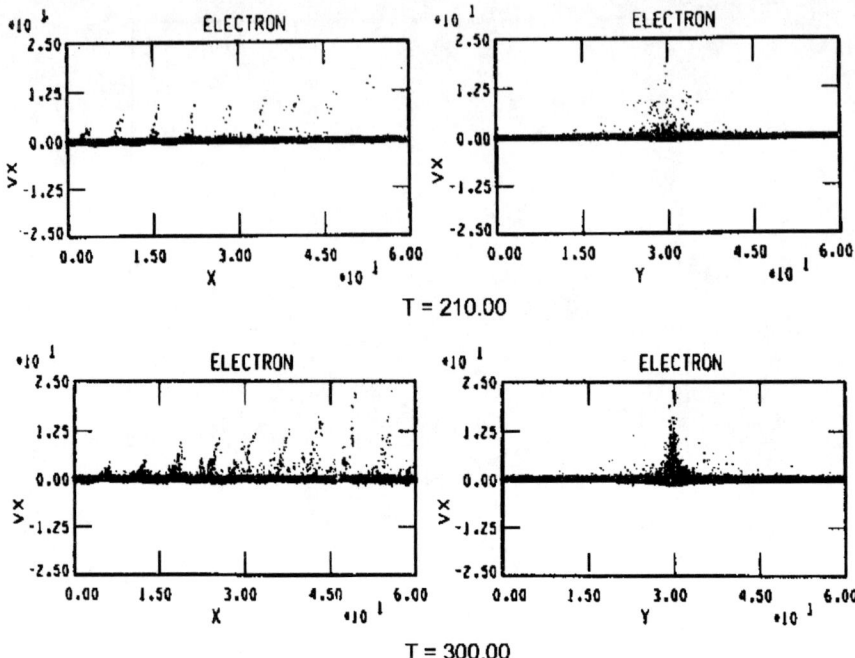

FIGURE 9. Electron Acceleration and focusing from 2-D Beat Wave

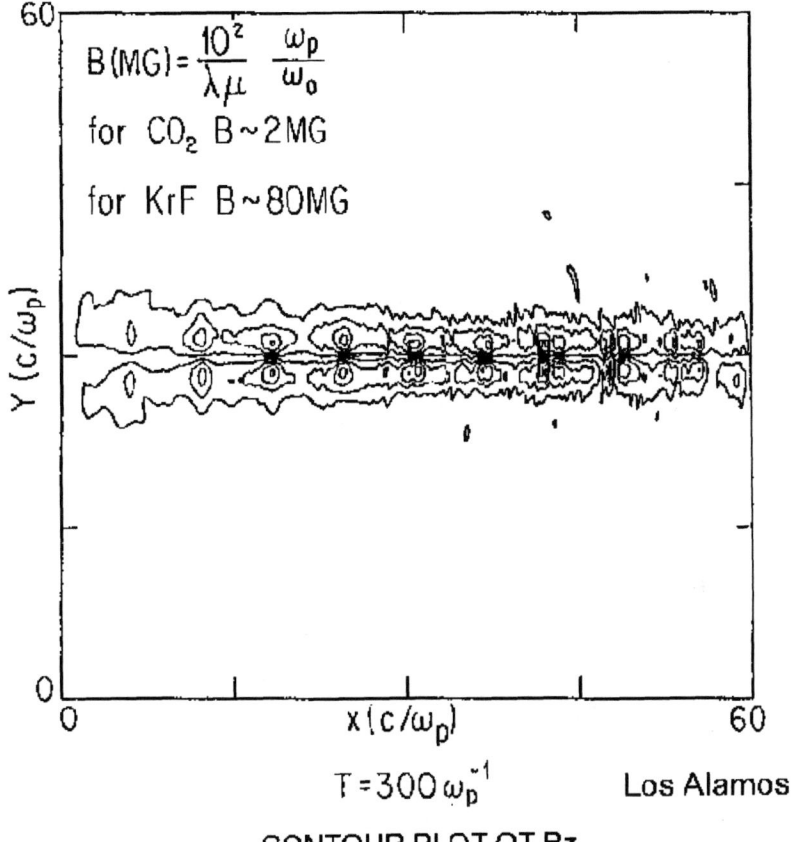

FIGURE 10. Magnetic field produced by Accelerated electrons.

REFERENCES

1. J.M. Dawson, "Nonlinear Electron Oscillations in a Cold Plasma," Phys. Rev., 113, 383 (1959).
2. J.M. Dawson, "A One-Dimensional Plasma Model," Phys. Fluids, 5, 445 (1962).
3. J.M. Dawson, "The Breaking of Finite Amplitude Plasma Oscillations," Project Matterhorn Report, Matt-4 (1959).
4. N. Rostoker and M.N. Rosenbluth, Phys. Fluids 3, 1 (1960).
5. A. Lenard, Ann. Phys. (N.Y.) 3, 390 (1960).
6. C. Smith and J.M. Dawson, "Some Computer Experiments With a One-Dimensional Plasma Model," Princeton University Plasma Physics Laboratory, Report Matt-151 (1963).
7. M.N. Rosenbluth and C.S. Liu, Physical Review Letters 29, 701 (1972).
8. T. Tajima and J.M. Dawson, "Laser Electron Accelerator," Phys. Rev. Lett., 43, 267 (1979).
9. C.E. Clayton, C. Joshi, C. Darrow, and D. Umstadter, "Relativistic Plasma-Wave Excitation by Collinear Optical Mixing," Phys. Rev. Lett., 54, 2343 (1985).
10. M. Everett, A. Lal, D. Gordon, C.E. Clayton, K.A. Marsh, and C. Joshi, Nature, 368, 527 (1994).
11. C.E. Clayton, K.A. Marsh, A. Dyson, M. Everett, A. Lal, W.P. Leemans, R. Williams, and C. Joshi, Phys. Rev. Lett., 70, 37 (1993).

Partial List of Publications and Reports from 1979 to 1998:

1. T. Tajima and J.M. Dawson, "Laser Electron Accelerator," Phys. Rev. Lett., 43, 267 (1979).

2. C. Joshi, T. Tajima, J.M. Dawson, H.A. Baldis, and N.A. Ebrahim, "Forward Raman Instability and Electron Acceleration," Phys. Rev. Lett., 47, 1285 (1981).

3. T. Katsouleas and J.M. Dawson, "Unlimited Electron Acceleration in Laser-Driven Plasma Waves," Phys. Rev. Lett., 51, 392 (1983).

T. Katsouleas, C. Joshi, W. Mori, and J.M. Dawson, "Prospects of the Surfatron Laser Plasma Accelerators," Proc. 12th Int'l Conf. on High Energy Accelerators, pg. 460, ed. by F.T. Cole and R. Donaldson, August 1983.

5. T. Katsouleas and J.M. Dawson, "A Plasma Wave Accelerator–Surfatron I," IEEE Trans. on Nucl. Sci., NS-30, 3241 (1983).

6. W.B. Mori, C. Joshi, and J.M. Dawson, "A Plasma Wave Accelerator – Surfatron II," IEEE Trans. on Nucl. Sci., NS-30, 3244 (1983).

7. J.M. Dawson, V.K. Decyk, R.W. Huff, I. Jechart, T. Katsouleas, J.-N. Leboeuf, B. Lembege, R.M. Martinez, Y. Ohsawa, and S.T. Ratliff, "Damping of Large-

Amplitude Plasma Waves Propagating Perpendicular to the Magnetic Field," Phys. Rev. Lett., 50, 1455 (1983).

8. P. Chen, R.W. Huff, and J.M. Dawson, "A Plasma Booster for Linear Accelerators", UCLA Report PPG-802, 1984, and Bull. Am. Phys. Soc. vol. 29, 1355 (1984).

9. P.K. Shukla and J.M. Dawson, "Stimulated Compton Scattering of Hydromagnetic Waves in the Interstellar Medium," Astrophys. J., 276, L49 (1984).

10. C. Joshi, W.B. Mori, T. Katsouleas, J.M. Dawson, J.M. Kindel, and D.W. Forslund, "Ultrahigh Gradient Particle Acceleration by Intense Laser-Driven Plasma Density Waves," Nature, 311, 525 (1984).

11. J.M. Dawson, "Beat Wave and Surfatron Accelerator of Particles," Proc. of Int'l Conf. on Plasma Physics, Lausanne, Switzerland, p. 837, ed. by M.Q. Tran and R.J. Verbak, June 27-July 3, 1984.

12. P. Chen, J.M. Dawson, Robert W. Huff, and T. Katsouleas, "Acceleration of Electrons by the Interaction of a Bunched Electron Beam with a Plasma," Phys. Rev. Lett., 54, 693 (1985).

13. W.B. Mori, C. Joshi, J.M. Dawson, K. Lee, D.W. Forslund, and J.M. Kindel, "Studies of the Plasma Droplet Accelerator Scheme," IEEE Trans on Nucl. Sci., NS-32, 3503 (1985).

14. P. Chen and J.M. Dawson, "The Plasma Wakefield Accelerator," Proc. Second Workshop on Laser Acceleration of Particles, Malibu, CA, p. 201, Jan 7-18, 1985.

15. T. Katsouleas, J.M. Dawson, D. Sultana, and Y.T. Yan, "A Side-Injected-Laser Plasma Accelerator," IEEE Trans on Nucl. Sci., NS-32, 3554 (1985).

16. R. Bingham, W.B. Mori, and J.M. Dawson, "Some Nonlinear Processes Relevant to the Beat Wave Accelerator," Proc. Second Workshop on Laser Acceleration of Particles, Malibu, CA, pg. 138, Jan. 7-18, 1985.

17. P. Chen, J.J. Su, J.M. Dawson, K. Bane, and P. Wilson, "On Energy Transfer in the Plasma Wake Field Accelerator", Phys. Rev. Lett. vol. 56, 1252 (1986).

18. Y.T. Yan and J.M. Dawson, "ac Free-Electron Laser," Phys. Rev. Lett., 57, 1599 (1986).

19. S. Wilks, T. Katsouleas, J.M. Dawson, P. Chen, and J.J. Su, "Beam Loading in Plasma Waves", IEEE Trans. Plasma Sci. PS-15, 210 (1987).

20. T. Katsouleas, S. Wilks, P. Chen, J.M. Dawson and J.J. Su, "Beam Loading in Plasma Accelerators," Particle Accel., 22, 81 (1987).

21. P. Chen, J.J. Su, T. Katsouleas, S. Wilks, and J.M. Dawson, "Plasma Focusing for High Energy Beams," IEEE Trans. on Plasma Science, PS-15, 218 (1987).

22. S. Wilks, J.M. Dawson and T. Katsouleas, "The Focusing of a Relativistic Electron Beam Using a Performed Ion Channel," submitted to Phys. Rev. A., 1987.

23. J.J. Su, J.M. Dawson, T. Katsouleas, S. Wilks, P. Chen, M. Jones, and R. Keinigs, "Stability of the Driving Bunch in the Plasma Wakefield Accelerator," Proc. of 1987 IEEE Particle Accelerator Conference, ed. E.R. Lindstrom and L.S. Taylor, Vol. 1, Washington, D.C., pg. 127, March 1987.

24. C. Joshi, T. Katsouleas, J.M. Dawson, Y.T. Yan, and F.F. Chen, "Plasma Wave Wigglers for Free Electron Lasers," Proc. of 1987 IEEE Particle Accelerator Conference, ed. E.R. Lindstrom and L.S. Taylor, Vol. 1, Washington, D.C., pg. 199, March 1987.

25. S. Wilks, T. Katsouleas, J.M. Dawson, and J.J. Su, "Beam Loading Efficiency in Plasma Accelerators," Proc. of 1987 IEEE Particle Accelerator Conference, ed. E.R. Lindstrom and L.S. Taylor, Vol. 1, Washington, D.C., pg. 100, March 1987.

26. F.F. Chen, C.E. Clayton, C. Darrow, J.M. Dawson, C. Joshi, T. Katsouleas, W. Leemans, K. Marsh, W.B. Mori, J. Su, D. Umstadter, and S. Wilks, "Particle Acceleration by Plasma Waves," Proc. of Int'l. Conf. on Plasma Physics, Kiev, USSR, p. 797, (1987).

27. W.B. Mori, C. Joshi, J.M. Dawson, D.W. Forslund, and J.M. Kindel, "The Evolution of Self-Focusing of Intense Electromagnetic Waves in Plasma," Phys. Rev. Lett., 60, 1298 (1988).

28. S.C. Wilks, J.M. Dawson, and W.B. Mori, "Frequency Up-Conversion of Electromagnetic Radiation with Use of an Overdense Plasma," Phys. Rev. Lett., 61, 337, 1988.

29. T. Katsouleas, J.J. Su, and J.M. Dawson, "Underdense Plasma Lenses for Focucing Particle Beams," Proc. 1988 Linear Accelerator Conference, CEBAF, Williamsburg, VA.

30. S.C. Wilks, J.M. Dawson, W.B. Mori, T. Katsouleas, and M.E. Jones, "A Photon Accelerator," Phys. Rev. Lett., 62, 2600, 1989.

31. J.M. Dawson, "Plasma Particle Accelerators," Scientific American, 260, 54, 1989.

32. T. Katsouleas, J.J. Su, C. Joshi, W.B. Mori, J.M. Dawson, and S. Wilks, "A Compact 100 MeV Accelerator Based on Plasma Wakefields," SPIE Conf. Proc. OE/LASE '89, Vol. 1061, Los Angeles, CA, p. 428.

33. T. Katsouleas, J.J. Su, S. Wilks, W.B. Mori, and J.M. Dawson, "The Role of Plasmas in Future Accelerators," Proc. of HEACC-89, Tsukuba, Japan, August 22-26, 1989; Particle Accelerators, 32, 185 (1990).

34. S.C. Wilks, T. Katsouleas, and J.M. Dawson, "The Photon Accelerator: A Novel Method of Frequency Upshifting Sub-Picosecond Laser Pulses," AIP Conference Proceedings 193, Advanced Accelerator Concepts, ed. by C. Joshi, p. 448, 1989.

35. D.H. Whittum, A.M. Sessler, and J.M. Dawson, "An Ion-Channel Laser," Phys. Rev. Lett., 64, 2511 (1990).

36. J.J. Su, T. Katsouleas, J.M. Dawson, and R. Fedele, "Plasma Lenses for Focusing Particle Beams," Phys. Rev. A., 41, 3321 (1990).

37. T. Katsouleas, J.J. Su, W.B. Mori, and J.M. Dawson, "Plasma Physics at the Final Focus of High Energy Colliders," Phys. Fluids B, 2, 1384 (1990).

38. K.R. Chen, T. Katsouleas, and J.M. Dawson, "On the Amplification Mechanism of the Ion-Channel Laser," IEEE Trans. on Plasma Sci., 5, 837 (1990).

39. K.R. Chen, and J.M. Dawson, "Ion-Ripple Laser as an Advanced Coherent Radiation Source," sub. to SPIE 1991 Int'l Symp. on Optical Applied Science and Engineering, San Diego, CA; July 21-26, 1991.

40. K.R. Chen, K.R. and J.M. Dawson, "Theory and Simulation of High Gain Ion-Ripple Lasers," Phys. Rev. A, 45, 4077 (1992).

41. C. Joshi, C.E. Clayton, W.B. Mori, J.M. Dawson, and T. Katsouleas, "The Prospects for a GeV Plasma Beat Wave Accelerator," Comments Plasma Phys. Controlled Fusion, 16, No2, 65-77 (1994).

42. J.J. Su, R. Bingham, J.M. Dawson, and H.A. Bethe, "Nonlinear Neutrino Plasma Interactions," Physica Scripta, T52, 132-134, (1994).

43. R. Bingham, J.M. Dawson, J.J. Su, and H.A. Bethe, "Collective Interactions Between Neutrinos and Dense Plasmas," Phys. Lett. A, 193, 279-284, (1994).

44. R. Bingham, H.A. Bethe, J.M. Dawson, P.K. Shukla, and J.J. Su, "Nonlinear Scattering of Neutrinos by Plasma Waves: A Ponderomotive Force Description," Phys. Lett. A., 220, 107-110, (1996).

45. J.M. Dawson, R. Bingham, and V.D. Shapiro, "X-rays from Comet Hyakutake," Plasma Phys. Control. Fusion, 39, No.5A SISI, 185- 193, (1997).

46. P.K. Shukla, R. Bingham, H.A. Bethe, J.M. Dawson, and L. Stenflo, "Nonlinear Coupling between Intense Neutrino Fluxes and Dense Magnetoplasmas," Physical Scripta, Vol.55, 96-98, (1997).

47. J.T. Mendonsa, L. Oliveira e Silva, R. Bingham, N.L. Tsintsadze, P.K. Shukla, and J.M. Dawson, "Equivalent Charge of Photons and Neutrinos in a Plasma," Phys. Lett. A, 239, 373-377, (1998).

48. B.J. Kellett, H. Negoro, F. Nagase, R. Bingham, J.M. Dawson, D.A. Mendis, and V.D. Shapiro, "ASCA X-ray Observations of Comet Hale-Bopp," in Press Monthly Notices of Royal Astrominical Society, (1998).

49. L.O. Silva, R. Bingham, and J.M. Dawson, "Pondermotive Force of Quasi-Particles in a Plasma," submitted to Physics of Plasmas, 1998.

50. L.O. Silva, R. Bingham, J.M. Dawson, J.T. Mendoná, and P.K. Shukla, "Anomalous Scattering of Neutrinos in Dense Supernovae Plasma, (1998).

Tribute to John M. Dawson: "Pre-historic Days" of the Dawson-Tajima 1979 Paper

T Tajima[*][†]

[*]*Lawrence Livermore National Laboratory, Livermore, CA 94551*
[†]*Department of Physics and Institute for Fusion Studies
The University of Texas at Austin, Austin, TX 78712*

It is a great honor to warmly reminisce in my fortune to work with John some quarter of a century ago. Sunday's keynote speaker George Cowan discussed plasticity of infant's brain in the first three years. If I acquired any iota of creativity, it is a testimony that John was able to plastically alter a 28-year old underdeveloped primate brain even after a Ph.D. cast. I am forever grateful for that. Life makes sense, sometimes, only after a long time. For example, my encounter of February 1976 may have contributed to a root of today's meeting. The vision of the original laser acceleration conception is closer to a reality.

Since it is nearly impossible to second-guess the creativity of John's and later he will give his own historic account, let me outline how I view the background of our paper, Dawson-Tajima 1979. In the 1950s Budker and Veksler, independently, suggested the notion of "collective acceleration." The idea was a new one: instead of accelerating each charged particle by a given external electromagnetic field, one may want to accelerate such a particle by "collective fields" created by some means. Their suggestion was to use intermediate energy electrons (in the range of MeV) to excite "collective fields" behind or around them. These fields may be a precursor of the nowadays "wakefields." They try to accelerate ions in this field, with the rationale that the energy gain of ions is enhanced by the mass ratio M/m. In order to capture sluggish ions, it was implied that the phase velocity of the accelerating field is much less than the speed of light c. I observed that because of this feature the accelerating structure is "plastic" or easily deformable or unstable.

In the 1960s Basov and Krokhin published a paper on laser (1963). Dawson immediately followed it, suggesting laser fusion in 1964. Laser has entered in the physics of light-matter interaction. Much exciting research ensued and many conferences have been organized, including the series "Anomalous Absorption Conference," as it was soon recognized that intense laser interacts with matter "anomalously," i.e. in ways more than the classical collisional absorption. John played a leading intellectual role in this.

Meanwhile, the allure of "collective acceleration" attracted many physicists' attention ever since Budker and Veksler's ideas. These include the distinguished participants of this conference, Martin Reiser and John Nation, and my former advisor Norman Rostoker and my departmental colleague Bill Drummond. In the 1960s and 1970s they erected projects to test or explore these ideas in a laboratory. In fact, Rick Mako, who is in this conference back there, and I spent a great deal of time in Professor Rostoker's laboratory for collective ion acceleration, by poking my neck into Rick's experiment when we were graduate students at Irvine in the early 1970s. What I learned through this experience is that, as I said above, the accelerating structure caused by the electron beam with low-phase velocity seems plastic and unable to grab ions in its pocket for a sufficiently long time. I felt that a rigid structure, as opposed to a plastic one, is needed. I further observed that the conventional accelerators are good and efficient enough to accelerate particles (particularly electrons) to relativistic energies. Thus there was no need to pick up particles in a nonrelativistic speed and vary the phase velocity, which compounds the above problem. Instead, if one starts with a phase velocity close to c, since no particles can exceed c, this structure remains (relatively) rigid and coherent.

Soon after I joined John's group at UCLA, probably in March or April of 1976 we had a general discussion on collective acceleration. When I mentioned my experience at Irvine, John responded to use laser right away. We did not start on this problem immediately, though. In part because I was given a project, advanced fuel in magnetic fusion, another pioneering subject John spearheaded. In May of 1976, if I am not mistaken, however, John came back from the Anomalous Absorption Conference in Canada, cutting his trip short, as he suddenly fell ill. Sometime in 1977 and 1978 John Dawson and I worked intensively on this subject. The years between 1976 and 1979 are remarkable in many ways, but let me mention one very personal experience below. By the way, my son was born in April 1977, named Yuhki JOHN, carrying a legacy of the man I was working so passionately for.

Let me indulge in reading my own article [Fusion Tech. **19**, 409 (1991)] written on the occasion of the international Dawson Symposium honoring John Dawson's 60th birthday (September 30, 1990). At this symposium Tudor Johnston dedicated his poem to John:

A plasma magician named Dawson
Is fertile with ideas that blossom,
Though his concepts are wild,
Each latest brainchild
Is backed by code runs that are awesome.
Breaking waves he found was so fine,
Near a surf beach he never would pine,
"Surf's up in the hypercube,
Comes John in a supertube.

> *Riding waves till the end of the line!"*
> *Catalina's the place to say, "John,*
> *Though phase-space is maybe a con,*
> *And flows and vortexes*
> *Just sent to perplex us,*
> *May you always find waves to ride on!"*

However, I shall refrain from reciting it here for the benefit of the audience. Instead I recount my personal observation of Dawson as scientist and man from this article: "The second example is more personal....The medicines must have been extremely strong, as he invariably became very sick and lost most of his blonde hair. What was most awesome and inspirational to me was his courageous attitude. In spite of the major life-threatening illness and the equally severe medication, he never stopped working on physics! Dawson used to be a chubby man, but after the surgery and medication he became quite thin. When he was too sick to come to the office, we were summoned to his Pacific Palisades home to discuss our results. Without fail, he fully discussed the subject at hand when I visited him. A more surprising thing was that he was even more creative, or at least it seemed to me, during this serious period than the previous time. Charlie Kennel jokingly said that because he was free from all the daily chores, he was more creative. Charlie was right. During this period, he worked on advanced fuel fusion, various ideas on fusion reactors, MHD particle codes, isotope separation, free electron lasers, initiation of space plasma simulation, and *laser acceleration*, among other topics, in addition to the more mundane duties of teaching students. This, I believe, more than anything else is the testimony of what kind of man John Dawson is. Nancy Dawson made an emotional speech on his contribution to isotope separation, which produced the ^{102}Pd isotope used to help therapy of prostate cancer."

Little did I know that 20 years later I had to "emulate" and "follow his footsteps" again, but this time a similar affliction he had suffered and to seek counsel and to find tremendous comfort and encouragement from him. His example and counsel gave me courage to cope and to get free from the fear. It gives one a greater sense and satisfaction of life. Today I feel fortunate to be able to give this tribute and celebrate his contribution. I am wishing for your strong recovery from your recent ailment again, to continue to inspire us how a courageous man can contribute to mankind's well being.

Thank you, John, for all of us!

John Dawson's Advanced Accelerator Years

C. Joshi

UCLA, Los Angeles, CA 90095

It's a great pleasure for me to recount the excitement and creativity of John Dawson's so called "Advanced Accelerator Years" which span the period 1979-1992.

In 1979, the now famous Tajima and Dawson paper,[1] "Laser-Electron Accelerator," appeared in the *Physical Review Letters*. In it, Tajima and Dawson proposed the use of a short but intense pulse to accelerate electrons by surfing on a wake produced by such a pulse in a plasma. This scheme later became known as the Laser-Wake Field Accelerator. I was a postdoctoral fellow at the National Research Council in Ottawa, Canada at the time and read it with great interest. I and my colleagues were experimentally looking at fast electron generation from high frequency (Raman, two-plasmon decay) instabilities[2] and resonant absorption[3] in CO_2 laser-plasma interactions. We had some intriguing results on high energy electron emission from thin foils (which generated sub-critical density plasmas) that suggested that perhaps the Raman forward instability which excites a relativistic plasma wave ($v_{ph} \sim c$), may have been excited in the plasma.

In April 1980, I came, as a Research Engineer, to the Electrical Engineering Department of UCLA to work on Laser-Plasma Interactions with Frank Chen. Our laboratory was rather modest and I was looking for something exciting to do with our limited resources. Within a very short time after arriving at UCLA, I went over to the Physics Department to meet Toshi Tajima about his 1979 paper and talk to him about the electron emission data we had on thin foils. Toshi and I walked over to John Dawson's office and he introduced me to John. I was struck by how unassuming, friendly and interested he was. He and Toshi agreed that they could simulate the Raman Forward Scattering problem using a 1D PIC code. It was at that time that I first really understood the Beat-Wave Acceleration scheme, which utilized the beating pattern produced by two different frequency laser beams to resonantly excite a plasma wave.

Unfortunately for me, Toshi soon left UCLA for U. T. Austin to take up a faculty position and the simulations he promised me would have to wait for almost another year. In the meantime, I went to work in our laboratory with Chris Clayton, who was a student at the time, and got our CO_2 laser to work on multiple frequencies using an intracavity saturable absorber cell. Within a week we were exciting a "beat-wave" using an arc-plasma source. We observed that the self-focusing effect was enchanced when a two frequency laser beam was sent into the plasma at the resonant density.[4] We were baffled by the data but John Dawson had an idea of why this may have been the case. The ponderomotive force of the finite diameter longitudinal

CP569, *Advanced Accelerator Concepts: Ninth Workshop*, edited by P. L. Colestock and S. Kelley
© 2001 American Institute of Physics 0-7354-0005-9/01/$18.00

plasma wave was greater than that of the light waves leading to self-focusing of light below the usual single frequency threshold.[4]

By Dec. 1981 both the Raman Forward Instability[5] and the Resonant Self Focusing[4] papers had been published and were received with some interest by the laser-fusion community. When I came back from my "marriage leave" in February 1982, there was a letter waiting for me from Andy Sessler in which he invited me to a workshop at Los Alamos on Laser Acceleration of Particles. Actually, he wanted John Dawson to go but John couldn't go and suggested that I might go instead. I arrived at Los Alamos a day after the conference had started. Toshi Tajima had already given a talk on "A Laser Electron Accelerator" the opening day. I found that the audience of about 80 rather influential physicists was very interested in my experimental talk that showed both beat-wave excitation and relativistic electron acceleration by the Raman Forward Instability. By the end of the Workshop the Plasma Accelerator field was born and endorsed by the Workshop Organizing Committee in their summary as an area worth pursuing. Andy Sessler encouraged me to submit a proposal to Dr. David Sutter's office at the DOE to do dedicated plasma accelerator work at UCLA.

When I came back from the Los Alamos Workshop I talked with John. As I was not a faculty member I had no prospect for obtaining space to build a lab. John arranged a meeting with Dean Ticho of the Physical Sciences and he agreed to lend me his personal laboratory in the Physics Department. I wrote a 3-year proposal to the DOE to demonstrate beat-wave acceleration of electrons which, in retrospect was incredibly naïve on my part. At about the same time, Warren Mori joined our group as a graduate student in Electrical Engineering and I asked him to reproduce Toshi's laser wake-field results using the same 1D PIC code. Warren turned out to be a very talented student and we very quickly realized that we could and should do 2D PIC simulations of the beat wave scheme. Serendipitously, during a conversation with me, Joe Kindel and David Forslund of Los Alamos mentioned that they were interested in having a graduate student work in the X1 group over the summer. Warren went off to LANL which had just taken delivery of the CRAY-XMP machines and Dave Forslund made sure Warren had enough time to run first large scale 2D-PIC simulations using the code WAVE. This collaboration was so successful that we wrote up a full length article in *Nature*[6] on, "Ultra-High Gradient Acceleration by Intense Laser-Driven Plasma Density Waves" and two papers in *Physical Review Letters*.[7]

John had a large group of students and postdocs at the time. Among them was one particularly bright and enthusiastic student named Tom Katsouleas who picked up the topic of overcoming the dephasing limited energy gain in a plasma accelerating structure. The solution, which came to be known as the surfatron,[8] was deceptively simple. By applying a perpendicular magnetic field of the appropriate strength, particles could be phase-locked in the plasma wave while "surfing" transversely or parallel to the plasma wave front. The solution while elegant was very difficult to implement in practice because it required rather large cross-section plasma waves. To my knowledge no one has successfully demonstrated the Surfatron in a laser-plasma experiment. The Surfatron concept clearly illustrates John's creativity and his ability to nurture talented students. At about the same time John and his collaborators were working on collisionless dissipation mechanisms of electrostatic waves moving across magnetic fields. This problem was of great interest to space-plasma physics and would appear to be completely unrelated to accelerators. However, John's intuition told him that new and interesting physics would result if the phase velocity of these electrostatic waves approached the speed of light. He told Tom about it, who quickly worked out the specifics. Although many famous plasma physicists, most notably R. Sagdeev, had studied this collisionless dissipation mechanism, none had appreciated the profound effects when waves became relativistic.

It so happened that Tom Katsouleas was sharing an office with Warren Mori. This is how I met Tom and eventually, John, Tom, Warren and I began to work together; a collaboration and friendship that has continued for the past fifteen years. This is also common when working with John. His demeanor and personality is such that it is easy to work together and collaborators become close friends

After Tom Katsouleas and Chris Clayton graduated I persuaded both to stay on as postdocs at UCLA supported by the newly arrived 3-year DOE grant. We were furiously building a new lab in the Physics Department. We had just about a year before the Second Laser Acceleration of Particles Workshop in Malibu in January 1985, which we were hosting. I would not have agreed to host this were it not for Tom's enthusiasm and John's encouragement. We were fortunate to have some talented students join our groups; Chris Darrow and Don Umstadter on the experimental side and Y. T. Yan, J. J. Su and Scott Wilks on the theory side working with John. At about the same time Pisin Chen joined John as a postdoc and John asked him to look into accelerating electrons in a plasma using an electron bunch which came to be known as the Plasma Wake Field Accelerator.[9] John Dawson has a fantastic intuition about physics and even as early as in 1984 he had already proposed the PWFA scheme as a booster or energy doubler for the SLAC linac. After the first paper on the Plasma Wake Field Accelerator was published in *Physical Review Letters* in 1985, Pisin Chen expressed an interest in working at SLAC. John personally went to SLAC to meet with Richter and arranged for Pisin to spend a year at SLAC. This was very fortunate because Pisin developed many fruitful collaborations with scientists at SLAC and subsequently was offered a staff position at SLAC. Pisin Chen, in collaboration with K. Bane and P. Wilson of SLAC, John Dawson and J. J. Su of UCLA wrote an important paper[11] in 1986 showing how an asymmetric current distribution of the drive beam had some advantages in increasing the transformer ratio, and therefore ultimately the maximum energy, that a trailing particle could gain in the Plasma Wake Field Accelerator. The idea of a plasma lens was also developed at about this time.[10] Plasma focusing was first studied in conjunction with betatron motion of particles in the relativistic plasma wave as a result of strong radial focusing fields, and subsequently as a result of partial or complete charge neutralization of an electron beam as it propagates through a plasma[10] even without necessarily exciting a longitudinal wake.

In 1986, Y. T. Yan and John proposed the idea of using a purely oscillating electric field as a wiggler to obtain free electron laser radiation.[12] At this time the "Star-Wars" program on directed beam weapons was in full swing. John was more interested in making cheap and compact FELs for medical and commercial applications such as x-ray crystallography of proteins. In 1987, John and his disciples, myself included, T. Katsouleas and others[13] proposed a relativistic plasma wave as a wiggler for obtaining FEL radiation. This technique was a practical realization of the

AC FEL scheme which, instead of using a capacitor field, used the oscillating field of a propagating relativistic plasma wave to wiggle an electron beam fired transverse to the plasma wave.

John began to think about how one might generate tunable radiation using a plasma but without using an electron beam as a source of energy. He straightaway realized that self-phase modulation of an e.m. wave in a plasma would lead to large frequency upshifts. In a scheme which came to be known as "flash-ionization,"[14] the medium surrounding the e.m. wave was rapidly ionized everywhere before the wave could leak-out of the medium. Since the wavenumber of the wave was "frozen," the frequency had to upshift to accommodate an index of less than 1. The work on flash-ionization was followed up by a now classic paper authored with Scott Wilks and others titled, "Photon Accelerator."[15] Of course the title itself is an oxymoron if one is talking about vacuum propagation of light but it cleverly referred to the speeding up of a photon pulse as it surfed the plasma density wake left behind by an electron or a photon pulse. In this process the frequency of the surfing photon pulse is upshifted.

It is revealing to see how John came up with the photon accelerator idea. Rather than working with a complicated set of equations, he realized that if one laser pulse could give energy by making a wake in a plasma, then another photon pulse that was placed π out of phase in this wake could extract energy from this wake. In analogy to a pulse of electrons, the second photon pulse takes this energy by being "accelerated." Since the number of photons stays fixed, then each photon needs to have a higher frequency. In plasma, high frequency photons have a higher group velocity than the lower frequency photons and it is in this sense that photons are accelerated.

The work on photon accelerator also lead John, J. J. Su and R. Bingham to ask the following question. If photons can be accelerated in a plasma, why can't other particles such as neutrinos be accelerated? Once again John's intuition said yes implying that neutrinos-plasma interactions would lead to a slew of instabilities in analogy with laser-plasma and electron beam-plasma interactions. It is of interest to note that John's collaboration with R. Bingham began in 1985 at the Malibu AAC Workshop and continues to-date, as do many of John's other collaborations.

In 1989 John spent a sabbatical at LBL. He interacted closely with David Whittum and Andy Sessler who were working out the concept of an ion channel laser.[16] In an ion-channel laser, a relativistic electron beam propagating through an ion-channel undergoes betatron oscillations and emits betatron radiation which in turn bunches the electron beam and produces coherent radiation with a large frequency upshift. Upon coming back to UCLA, John with his student, K. R. Chen worked out the theory of an ion ripple laser.[17]

John's Advanced Accelerator Years were incredibly exciting for those of us who had the fortune to participate in many of his inventions. During these years, he invented or co-invented the Plasma Beat Wave Accelerator,[1,6] Plasma Wake Field Accelerator,[9] Laser Wake Field Accelerator,[1] Plasma Lens,[10] Plasma Wiggler,[12,13] Photon Accelerator,[14,15] Ion Channel (Ripple) Laser[16,17] and other concepts. Many of his students and postdocs have followed up his ideas and indeed established themselves as leaders in the Advanced Accelerator Concepts community. This

September John Dawson turns 70 and we would like to recount his leadership, vision, and support of this field as a way to say, "Thank-you John."

ACKNOWLEDGMENTS

I thank Warren Mori, Tom Katsouleas, Pisin Chen and Chris Clayton for their accounts of interactions with John Dawson.

REFERENCES

1. T. Tajima and J. M. Dawson, *Phys. Rev. Lett.* **43**, 267 (1979).
2. N A. Ebrahim et al., *Phys. Rev. Lett.* **45**,1179 (1980).
3. N. A. Ebrahim and C. Joshi, *Phys. Fluids* **24**, 138 (1981).
4. C. Joshi et al., *Phys. Rev. Lett.* **47**, 1285 (1981).
5. C. Joshi et al., *Phys. Rev. Lett.* **48**, 874 (1982).
6. C. Joshi et al., *Nature* **311**, 525 (1984).
7. D. W. Forslund et al., *Phys. Rev. Lett.* **54**, 558 (1985); W. B. Mori et al., *Phys. Rev. Lett.* **60**, 1298 (1988).
8. T. Katsouleas et al., *Phys. Rev. Lett.* **51**,392 (1983).
9. P. Chen et al., *Phys. Rev. Lett.* **54**, 693 (1985).
10. P. Chen et al., *IEEE Trans. Plasma Sci.* **PS-15**, 218 (1987); J. J. Su et al., *Phys. Rev. A* **41**, 3321 (1990).
11. P. Chen et al., *Phys. Rev. Lett.* **50**, 1252 (1986)
12. Y. T. Yan et al., *Phys. Rev. Lett.* **57**, 1599 (1986).
13. C. Joshi et al., *IEEE J. Quant. Elec.* **QE-23**, 1571 (1987).
14. S. C. Wilks et al., *Phys. Rev. Lett.* **61**, 337 (1988).
15. S. C. Wilks et al., *Phys. Rev. Lett.* **52**, 2600 (1989).
16. D. Whittum et al., *Phys. Rev. Lett.* **64**, 2511 (1990).
17. K. R. Chen et al., *Phys. Rev. Lett.* **68**, 29 (1992).

WORKING GROUP SUMMARIES

Report of Working Group 1: The Physics of High Energy Density

J.B. Rosenzweig, for the working group

UCLA Dept. of Physics and Astronomy
405 Hilgard Ave., Los Angeles, CA 90095

Abstract. The working group on the physics of high energy density at the 2000 Advanced Accelerator Conference (AAC2K) was envisioned to encompass many aspects of high phase-space density beam interactions with radiation, plasmas, solids, and their own collective fields. These include advanced beam diagnostics, cooling concepts such as ionization cooling, and acceleration and focusing of particle beams in plasmas. As the achieving of high energy density or intensity often implies that energy pulses in laser or particle beams are very short, this working group was also concerned with ultra-fast phenomena. These phenomena include the generation of femtosecond x-ray pulses from high-intensity laser-electron beam collisions (Compton light sources), and proton or other heavy ion acceleration in laser-target interactions.

While the AAC Workshop traditionally is focused on acceleration and manipulation of electrons and positrons, at AAC2K there was much discussion of advances in heavy particle accelerator physics. The recent emphasis of laser-plasma based sources of electrons has been recently paralleled by the unexpected discovery of ion beam generation from multi-TW to PW, sub-psec laser-solid interactions. At AAC2K, results from the UM group (in the ten TW regime), as well as from LLNL (PW class laser) were reported. In the lower power experiments, reported by K. Flippo, it has been shown that the energetic (several MeV) ions are generated on the upstream sided of the thin-foil targets. In the LLNL experiments (discussed by T. Cowan), on the other hand, the accelerated ions (>30 MeV) are found to be derived from the back side of the target, ejected along the surface normal. In both experiments, the emitted ions showed small transverse phase space extent. Even more remarkably, in the LLNL case the energy spectrum showed a notable peak near the maximum observed energy — the emitted ions formed a true narrow band (and initially sub-picosecond) beam. While such observation of such unprecedentedly high brightness ion beam sources capture the imagination, it is not yet clear how one captures such a beam into a more conventional linac to use it in applications. Such applications were discussed in the working group by Pakhomov, who proposed a novel high-flux neutrino source based on laser-produced proton beams, and by Nakajima, who discussed general aspects of high-flux proton beam acceleration in laser-produced solitons. Also, implications for scaling both the UM and LLNL results to different parameter regimes of interest for applications were discussed by Flippo and Cowan, respectively.

The recent emergence of muon colliders, and the related concept of neutrino factories, was recognized in working group 1 at AAC2K, with overview talks by D. Cline and Z.

Parsa. There are many challenging research problems which need to be investigated before such projects are undertaken, such as ionization cooling, and manipulation of beam phase spaces under conditions where the angular momentum of the beam is non-zero, as discussed by G. Penn. It was apparent from the presentations at AAC2K he muon/neutrino factory field is still in its youth, and that with the onset of active experiments to prove the relevant principles, that this proposed high-energy physics instrument will grow in importance in the near future.

Ionization cooling is one area in which the interaction of intense particle beams with matter was discussed in the working group. In a related subject, J. Rosenzweig discussed, in a joint session on diagnostics with WG3 and WG4, the role of collective effects in beam diagnostics, and experience in experiments at UCLA, LANL and BNL. V. Yakimenko added many illustrative examples of measurements in which these effects were found important at the BNL ATF. A similar review of ultra-fast measurement systems at the Univ. of Tokyo twin-linac facility was given by M. Uesaka. The need to extend ultra-fast, ultra-small emittance beam measurements to unprecedented degrees of precision, driven by direct laser acceleration experiments, was discussed by E. Colby (in the context of the Stanford LEAP experiment), and A. Mikhailachenko. As the systems at different labs were seen to be complementary, a very informative discussion resulted from these sessions.

Some of the highest electromagnetic energy densities which can be created in the laboratory will be found in the final focus interaction region of a linear collider. The physics of creating such small spot sizes is limited by the strength and compactness of the final focus lenses. Recent measurements of the performance of plasma lenses as final focusing elements were reported by J. Ng. These measurements impressively showed both electron and positron focusing of beams, further reducing few micron spots. In addition K. Nakajima discussed extensions of this idea to beam collisions inside of solids.

The use of high intensity, ultra-fast laser pulses colliding with high brightness electron beam pulses, to create sub-picosecond, high flux, nearly monochromatic x-ray pulses by Compton scattering was reported by a number of groups at AAC2K. High quality experimental measurements of such effects with a long wavelength (10 micron), long pulse (10 psec laser and electron beams) at the ATF were reported by I. Pogorelsky. Ambitious proposals for using TW to PW level, femtosecond lasers at LLNL for producing LCLS-like photon spectra from Compton scattering off of photoinjector-derived electron beam (not coherent, however) were discussed by Cowan and F. Hartemann. The applications of Compton sources discussed included biomedical imaging, ultra-fast shock physics, and photon production for gamma-gamma colliders. This last high energy physics application was discussed by Cline, who also proposed the use of a plasma accelerator and lens in the final acceleration/focusing region of the collider.

Summary Report of Working Group 2 on Laser-Plasma Acceleration Concepts

W. B. Mori
University of California at Los Angeles
Department of Physics
Los Angeles, CA 90095

M. C. Downer
University of Texas at Austin
Department of Physics
Austin, TX 78712-1081

I INTRODUCTION

The charge of Working Group 2 was to review the status of, and explore new directions in, experiments, theory and simulation related to laser-driven plasma accelerators. The working group began with status reports from key experimental facilities. These reports were from: U. of Texas (Downer); LBL (Leemans); NRL (Ting); UCLA (Tochitsky); U. of Maryland (Milchberg); Japan (Uesaka, Nakajima); BNL (Pogorelsky); and U. of Michigan (Umstadter). Following these status reports from key experimental facilities, discussion of experiments focused on three issues:

1. *Laser guiding experiments.* Questions included: What are the prospects for controlled guiding of a $v_{group} \sim c$ pulse over ~ 3 cm at intensity 10^{18} W/cm^2? How can intense laser pulses be coupled into waveguides without severe distortion or loss? What are the prospects for propagating intense laser pulses through hollow channels?

2. *Particle injection into plasma accelerators.* Questions included: Are there new ideas for phased particle injection into short-wavelength plasma waves? What experimental progress has occurred in demonstrating laser injection schemes proposed at previous workshops?

3. *New laser-plasma acceleration results.* Areas of interest included new experiments with self-modulated laser wakefield accelerators (LWFA), standard LWFAs,

and second generation plasma beat wave accelerators (PBWA).

Discussion of theory and simulation focused on the status of, and new directions in, techniques that can be used to determine the overall efficiency for producing a useful wake when an intense lase pulse propagates through uniform plasma or preformed channels.

II PLASMA ACCELERATION EXPERIMENTS

A Guiding

Plasma waveguides capable of guiding intense femtosecond laser pulse over multiple Rayleigh lengths without optical distortion are essential to developing practical laser-driven plasma accelerators that operate near the dephasing limit. [1] Most WG2 reports focused on transient plasma fiber waveguides, which possess a density minimum on axis. Contrasting methods of producing such channels are being developed in different laboratories, including shock waves induced by a line-focused laser, [2] ablation of the inner wall of ceramic or organic capillaries heated by microsecond-scale electrical discharge, [6,7] and nanosecond-scale z-pinches. [8,9] In addition, there was continuing discussion of passive solid capillary waveguides for intense pulses. [10]

Progress with development of channels based on laser-induced cylindrical shocks was reported by H. M. Milchberg (U. Maryland), C. G. R. Geddes and W. M. Leemans (LBL), and E. W. Gaul and M. C. Downer (U. Texas). Milchberg's group originally developed this method by focusing a ~ 100 ps, ~ 300 mJ Nd:YAG laser pulse with an axicon lens to a ~ 1 cm long line focus in a several hundred Torr backfill of N_2, Ar, and other gases. [2] The YAG pulse creates a seed electron population by field ionization, then further avalanche-ionizes the gas and heats the plasma by inverse bremsstrahlung, thus launching a cylindrical shock wave. The resulting channel is typically 10 to 50 μm in radius, depending on timing. Such plasma channels successfully guided low intensity pulses over cm path lengths. However, high intensity pulses coupled poorly in a backfill because of ionization-induced distortions in the incompletely ionized end regions. To solve this problem, the Maryland and LBL groups now form their channels in elongated, high repetition rate gas jets. [3,4] The truncated end regions permit pulses with peak intensity as high as 10^{17} W/cm^2 to couple efficiently into the channel. Instead of the cylindrically symmetric axicon lens used by the Maryland group, the LBL uses two intersecting pulses, each line-focused with a separate cylindrical lens: a 75 fs, 20 mJ "ignitor" pulse that initiates field ionization, and a 250 ps, 200 mJ Nd:YAG "heater" pulse. [4] The channel's shape is controlled by adjusting the intersection angle of the two pulses. Coplanar intersection yields slab channels, while perpendicular intersection yields cylindrical channels. Both planar and cylindrical modes have been guided at high intensity. Potential disadvantages of the elongated gas jet are limited length scalability, because they are difficult to operate stably at lengths greater than 1 to

2 cm, and density nonuniformity along the length of the channel, which can lead to radial and phase nonuniformities in a channeled wakefield.

Milchberg reported a new resonant coupling mechanism, in which the pulse to be guided enters the plasma channel through its sides, rather than its end, after being focused onto the channel's axis with an axicon lens similar to the one that focused the channel-forming pulse. A resonance in coupling efficiency is predicted and observed when the component $k_\parallel$ of the wavevector of the injected pulse parallel to the channel axis becomes equal to the wavevector β of the quasi-bound mode inside the channel. This novel coupling mechanism circumvents end distortions altogether. It may be of interest for wakefield generation if sufficiently high coupling efficiencies can be achieved.

The Texas group reported a different approach to eliminating ionization-induced distortions by forming a Milchberg-type channel in a backfill of pure He gas. Previous attempts to form such channels in He failed because of the high field ionization threshold ($\sim 10^{15}$ W/cm^2) of He, which is not reached by line-focused 100 ps, 300 mJ channel-forming pulses. Gaul et al. solved this problem by striking a pulsed electrical discharge in the He to create a seed electron population $n_e \sim 10^{16}$ cm^{-3} prior to arrival of the channel-forming laser pulse. This plays a role similar to the LBL "ignitor" laser pulse, but is simpler and cheaper. The Texas group also found that 400 ps, > 0.6 J Nd:YAG pulses could form high quality channels in He with without a pre-discharge, which they presumed to start up from field ionization of trace impurities. Transverse interferometry showed full ionization in the He channel, and pulses of peak guided intensity 2×10^{17} W/cm^2 (50 % throughput) were guided over 1.5 cm without any spectral distortion (e.g. blueshifting) caused by ionization. Evidently end distortions were suppressed by the high field ionization threshold of He, while internal distortions were absent because of full ionization of the channel. Working group participants pointed out that the threshold for end region ionization could be raised further by using circularly-polarized pulses. The He backfill method is scalable to 10 cm lengths, and yields very uniform axial density.

A group at Hebrew University [6,7] developed an alternative method for forming a plasma channel by ablating the plastic inner wall of a $\sim 300\mu$m diameter capillary by passing a pulsed ($\sim 1 - 100\mu$s) electrical discharge (typically 1 kA at 1-2 kV) through it. A temporary cylindrical channel with a density minimum on axis forms as plasma from the heated walls diffuses inward. Using this method, the Naval Research Laboratory (NRL) group reported guiding of intense ($\sim 10^{17}$ W/cm^2) laser pulses ($\lambda = 1\mu$m) over 2 cm (more than 20 Rayleigh lengths) in a 35μm-diameter mode with over 70 % transmission efficiency. The Brookhaven National Laboratory (BNL) group reported development of a two-stage capillary discharge (1-4 cm long, 0.3 to 1.5 mm diameter) system for guiding intense CO$_2$ ($\lambda = 10\mu$m) laser pulses. This illustrates the general principle that the guiding condition for plasma channels is wavelength-independent. The ablating capillary approach avoids the use of expensive channel-forming lasers altogether. However, some working group members suggested that laser ablation of the inner capillary wall could be substituted, in order to avoid magnetic fields from the strong discharge current that could potentially distort charged particle beams in a channelled accelerator. Discussions also emphasized the need for improved characterization of the transverse channel structure. Drawbacks of the ablating capillary approach at present are the limited capillary lifetime ($\sim$ 100 shots), and limited repetition rate.

Groups at Japan Atomic Energy Research Institute (JAERI) [8] and University of Würzburg [9] have reported success with a third method of forming a plasma channel using a fast ($\sim$ 15 ns) z-pinch in a preionized gas. A ceramic capillary (typically 1 mm diameter) supports the discharge electrodes. However, the discharge current (30 to 60 kA at 30 - 300 kV) passes through preionized gas inside the capillary rather than through its walls. Thus the ceramic tube is long-lived, and can operate at 10 Hz repetition rate. A large (tens of Tesla) azimuthal magnetic field drives plasma implosion by the $J \times B$ force. A temporary channel with density minium on axis forms as the shock wave reaches the axis, and again after the reflection of the shock wave on the axis. K. Nakajima (JAERI) reported guiding of intense Ti:S laser pulses (λ = 790 nm, 90 fs, 10^{17} W/cm^2) over 2 cm with energy transmission between 30 and 64%. Fauser and Langhoff at Würzburg [9] used alumina capillaries of 2 mm radius and 14 cm length filled with 100 Pa of helium or hydrogen, and guided a low intensity dye laser pulse over 180 Rayleigh lengths. For accelerator applications, some working group members expressed concern that the high magnetic fields needed for the z-pinch would distort charged particle beams, which need to be injected with high spatial and temporal precision into a channelled wakefield.

Finally, the working group discussed guiding of intense laser pulses in a passive dielectric capillary. Last year, Dorchies et al. [10] reported monomode guiding of intense pulses (10^{16} W/cm^2, 120 fs) over 100 Rayleigh lengths (10 cm) in hollow dielectric capillary tubes (45-70 μm inner diameter) without inner wall damage. M. Xie (LBL) argued that oversized open dielectric waveguides provide the best long-term figure-of-merit as a guiding approach for accelerator applications, and should therefore be vigorously pursued. M. Murnane (U. Colorado), whose group has used similar capillaries at lower intensity for phase-matched high harmonic generation, [11] however, cautioned that transmission at intensities higher than 10^{16} W/cm^2 would be extremely difficult in practice because of wall damage. T. Antonsen (U. Maryland) presented simulations which reproduced the high quality transmission observed by Dorchies et al. for hollow capillaries, but showed significant loss over several cm and poorer mode quality when it contained a plasma of $n_e \sim 10^{18}$ cm^{-3}. Clearly experiments at intensities (10^{18} W/cm^2) and plasma densities ($n_e \sim 10^{18}$ cm^{-3}) of interest for laser-driven accelerators are needed to resolve these issues.

B Injection

Phased injection of charged particles into laser-plasma accelerators requires unusually high spatial and temporal precision because, for accelerating gradients of interest, plasma waves have much short wavelengths than conventional RF accelerators. Experimental approaches to the injection problem were discussed in a joint session with working group 5. Several all-optical approaches to injection have been proposed in recent years, [12–14] and were actively discussed in working group 2. We consider each of these in turn.

Umstadter et al. [12] proposed the laser-injected linear accelerator (LILAC) scheme several years ago. In this scheme, an intense pump pulse resonantly excites a wakefield. A second temporally-synchronized injection pulse is focused into the wakefield, where it perturbs background plasma electrons by the ponderomotive force. Some of these perturbed electrons are raised above the injection capture threshold. Temporal and spatial alignment of pump and injection laser pulses determines the acceleration phase. N. Saleh and D. Umstadter reported on the status of experiments underway at U. Michigan to demonstrate this concept. A chirped pulse amplifier providing high contrast 25 fs, 150 TW laser pulses, expected to be fully operational this year, will provide both pump and injection pulses. Preliminary studies of ultrashort pulse propagation in the resonant regime of wakefield generation were reported. [15] These experiments revealed the existence of a regime of pump powers ($P_{crit} < P_{pump} < 5P_{crit}$) in which resonant-wakefield-producing laser pulses propagate without breakup. Furthermore no measurable dark current out of the accelerator was detected, demonstrating the need for an injection pulse. At $P_{pump} > 5P_{crit}$, relativistic filamentation correlated with production of 1°-divergence multi-MeV electron beam was observed, signalling onset of self-modulated wakefield generation.

Esarey et al. [13] proposed the colliding pulse injector. In this scheme an intense pump pulse again resonantly excites a wakefield. Two additional laser pulses, counterpropagating along the same path as the pump, collide within the wakefield to form a low-phase-velocity beatwave that traps background cold plasma electrons, thereby pre-accelerating some of them to the capture threshold. Temporal synchronization between the pump and one of the colliding pulses determines the injection phase. Experiments to demonstrate this concept are underway at LBL.

Moore et al. [14] proposed and demonstrated the laser ionization and ponderomotive acceleration (LIPA) scheme for all-optical injection. A wakefield is again resonantly driven by a pump pulse, but in LIPA the injection pulse intersects its path in neutral gas prior to the wakefield, rather than inside the wakefield as in LILAC. Ionization and subsequent ponderomotive acceleration of initially bound electrons from the gas provide the injected electrons, which have a pulsewidth shorter than the ionizing pulse. Timing of the arrival of the injection pulse determines injection phase. By spatially separating the injection pulse from the wakefield, wakefield disruption which may occur in the previous schemes is avoided. C. Moore presented measurements performed at NRL of the electrons from the LIPA region, the first experimental demonstration of an all-optical acceleration concept. More than 10^7 electrons with energy peaked at 500 keV were observed. This is somewhat below the required capture energy for wakefields of interest, and the electron number is lower than desired for full energy extraction from the wakefield. Some anomalies in the observed angular distribution of LIPA electrons were also discussed, and need to be explained. Simulations by the NRL group, however, show that considerable improvement in electron output is possible with other laser parameters. Further experiments are underway at NRL.

S. Tochitsky (UCLA) described a double-beat wave injection concept, similar in concept to the LILAC and colliding beam injectors, that is being implemented for the Neptune beatwave accelerator experiment. The main plasma wave is generated by a loosely focused 100 ps CO_2 pulse that contains two phased frequency components whose difference-frequency beat is resonant with the plasma frequency. A portion of this CO_2 pulse is split off and focused more tightly inside the plasma beatwave of the main pulse, creating a more localized plasma beatwave that serves as an injector. Injection phase is again determined by adjustment of the timing between pump and injector pulses.

Other injection schemes were also presented, and are described more completely in the reports of working group 5. We briefly mention two important results. First, the Brookhaven group achieved the first demonstration of staging between two laser-driven accelerator stages (inverse free electron lasers (IFELs)), known as STELLA. Second, K. Nakajima (JAERI) presented preliminary results of bunch slicing, in which a femtosecond laser pulse, split from the wakefield driving pulse, energy-modulates a small portion of a long electron pulse in an IFEL. The modulated portion is then separated by a bunch slicer, or chicane, and emerges synchronized on a femtosecond time scale with the wakefield driving pulse.

C New laser-plasma acceleration results

Several significant new results from self-modulated laser wakefield acceleration (SM-LWFA) experiments were reported. The LBL group reported nuclear activation of Cu and Pb by the output of a SM-LWFA. Since these activation processes require a threshold electron energy of 25 MeV, they demonstrate unambiguously the production of electrons at > 25 MeV. The Cu activation $Cu^{63}(\gamma,2n) \rightarrow Cu^{61}$ is initiated when electrons from the SM-LWFA produce γ rays in the target, and produced neutron yield as high as 3mR/hr. Maximum neutron yield was confirmed to correlate with maximum electron yield from the SM-LWFA, optimum laser overlap with the gas jet, and approximately with the shortest laser pulse duration. The SM-LWFA thus appears to be an efficient table-top source for nuclear activation studies.

Two recent SM-LWFA experiments show evidence that a significant fraction of the electrons can be accelerated directly by the laser laser field, in addition to conventional electrostatic acceleration by the plasma wave. A recent experiment at Garching [16] used 200 fs, 1.2 TW pulses to drive a SM-LWF in a plasma of density $n_e = 4 \times 10^{20}$ cm^{-3}, resulting in electrons from <1 to more than 10 MeV. A 3D PIC simulation of the experiment showed that the high energy tail could be explained only by direct lateral laser acceleration acting in conjunction with a betatron resonance to produce forward accelerated electrons. Conventional electrostatic acceleration accounted for only the lower energy electrons. A second experiment at Rutherford-Appleton Laboratory used 400 fs, circularly polarized pump pulses focused into a plasma of density $n_e = 10^{19}$ cm^{-3}. Faraday rotation of a co-propagating probe pulse was observed when it temporally coincided with the pump. The rotation was consistent with an axial magnetic field of more than 1 MG from azimuthal currents driven directly by the circularly polarized pump. This effect has not been seen in any simulations to date

A. Ting (NRL) reported production of up to 100 MeV electrons from a SM-LWFA along with narrow Forward Raman Scattering (FRS) peaks, indicating a coherent plasma wave. This contrasted with earlier SM-LWFA results which showed breakup of the FRS peaks when 100 MeV electrons were produced, indicating a noisy or chaotic plasma wave.

Outside of wakefield acceleration, Umstadter (U. Michigan) reported acceleration of a collimated beam of $\sim 10^9$ protons to energies as high as 1.5 MeV from excitation of a thin film solid target by an intense high-contrast laser pulse. [17] The protons, which appear to originate from impurities on the front side of the target, are accelerated over a region extending into the target and exit out the back side in a direction normal to the target surface. Acceleration field gradients 10 GeV/cm were inferred. The maximum proton energy can be explained by the charge-separation electrostatic-field acceleration due to "vacuum heating." [18]

III THEORY AND SIMULATION

The feasibility of any high-energy laser plasma accelerator scheme hinges on being able to propagate intense laser beams for pump depletion distances while controlling the amplitude and phase velocity of the accelerating structure, i.e., the wake. To this end there has been much recent work within the community on developing the theoretical and computational foundation for understanding the evolution of intense lasers in plasmas. The quality of the WG2 reports indicated the tremendous progress that has been made. Furthermore, the computational tools now enable many of the experiments to be modeled with confidence. Full-scale modeling will become an even more important tool during the next decade.

A Theory

The greatest theoretical challenge is to model the evolution of true three-dimensional laser pulses. Techniques for studying the stability of plane waves have existed for decades. However, this community has pioneered the study of finite width (i.e., Gaussian) pulses [20] [21], [22], [23], [31], [32]. There are now two major theoretical formalisms being used to study the evolution of Gaussian pulses. The LBL (Esarey *et al.*) and NRL groups (P. Sprangle *et al.*) are using the source dependent expansion (SDE) approach [21], while the UCLA group (B. J. Duda, C. Ren *et al.*) is using a variational approach [22]. While the two approaches are similar in some respects, they are not identical. The existence of two independent methods is useful for comparisons.

E. Esarey reported on the LBL's efforts. They have spent considerable effort including the "dispersion" term (or "non-paraxial" or "mixed derivative" term) in the nonlinear wave equation using the SDE method. It has long been recognized that this term is needed to properly include direct forward Raman scattering [30], that it is required to include the nonlinear corrections to the group velocity, and that it leads to pulse steepening. They studied how this term modifies the evolution of the laser for "short" pulse drivers [20]. In addition, they studied how this term couples direct forward scattering to the spot-size modulational instability for "longer" pulses. He also described how this term modifies the behavior of unmatched beams in preformed plasma channels [26].

B.J. Duda reported on the UCLA efforts to include the dispersion term using the variational principle formalism. This formalism was new since the last workshop so he spent some time describing the approach [22]. Using his approach he showed that in the absence of dispersion, the stability of a Gaussian laser could be decomposed into three modes. These are hosing, symmetric spot-size self-modulation, and asymmetric spot-size self-modulation. The asymmetric mode is new. He then showed how the dispersive term can be included in a straight-forward yet tedious manner. His results showed that this term not only couples direct forward scatter to the symmetric mode, but that it also modifies the growth rate of the hosing and asymmetric spot-size modes. He also showed that the variational principle approach can be used to study the stability of pulses even when all the nonlinearities are included. During the workshop he calculated how the growth rates are modified in a preformed density channel.

P. Sprangle and B. Hafizi reported on the NRL efforts. They have also been studying the effects of the dispersion term. Much of their recent effort has been directed on studying the optimum pulse length. Very short pulses (pulse lengths much less than a plasma period) do not make a large wakes, but by the same token they do not distort. So an interesting question is whether the overall energy gain is larger when very short pulses are used. The NRL group is studying this. They also proposed that dephasing might be overcome by tapering the radius of the channel. This is a twist on T. Katsouleas' so-called accordion effect in density gradients [28]. They also presented results on self-focusing when both the relativistic and ponderomotive blowout terms are kept to all orders.

T. Antonsen, Jr. presented some recent theoretical and simulation results on how lasers evolve as they form a channel or propagate down a gas filled capillary. He and his graduate students found that a new modulational instability results as a channel is formed by the heater beam. Modulations in the heating rate lead to modulations of the channel parameters which in turn cause the axicon fields to scatter into a channel mode. The beating of the channel mode and the axicon field reinforces the modulation of the heating rate leading to an instability. They also found that when the capillary is filled with a gas that ionization induced refraction scatters the laser to the walls at a larger angle causing the laser to leak out at a much faster rate.

Anatoly Spitkovsky presented results on the optimum pulse shape. Part of his talk was devoted to defining what optimum means and this elicited a lively discussion. Based on his criterion he showed that the optimum pulse shape was a ultra short precursor followed by a "wedge shape" (it is a wedge shape in the weakly nonlinear limit). He also used the insight obtained from tranformer ratios when particle beam drivers are considered. While his results are limited to only one-dimension they are an important first step in the ultimate design of a high efficiency laser-plasma accelerator stage.

C. Ren presented an interesting talk on how laser beams can mutually interact in a plasma. While not of direct relevance to particle acceleration, this talk served to add to the understanding of how short pulse lasers evolve in a plasma. In particular, he showed that two partially overlapping beams in a plasma will be attracted to each other [23]. He showed that if two lasers enter the plasma with some initial angular momentum about the axis that joins their centers, then this attractive force will cause them to spiral around each other. Furthermore, he showed that the plasma wake causes different axial cross sections to spiral at different rates leading to a braided pattern. This work could have implications for beat wave excitation when the two lasers do not exactly overlap.

Vladimir Golovizmin presented results on beam loading. He and his collaborators studied emittance growth and energy spread that arises when non-tailored bunches of electrons are accelerated. In particular they extended the 1D results of Chiou and Katsouleas [27] to 3D. They showed that the 1D conclusion that phase slippage leads to lower than expected energy spread still holds in 3D.

Sergey Cheshkov presented results on the emittance growth if 100's to 1000's of GeV LWFA stages are combined to make a TeV collider. He and his coauthors used nonlinear dynamics maps which indicate that weak focusing has advantages to strong focusing in terms of emittance growth. To obtain weak focusing while still maintaining high beam loading efficiency requires the use of very narrow bunches or hollow channels.

C. Ren also presented a talk on curious effects in beat wave excitation. He considered what happens if the beat frequency is not ω_p but $2\omega_p$ or $3\omega_p$. He showed that a plasma wave is still excited even when it is driven at these subharmonic resonances. He showed that in contrast to the secular growth that results when driven resonantly, when driven at $2\omega_p$ the wave grows exponentially while when driven at $3\omega_p$ it grows explosively. In both of these cases the growth eventually saturates due to relativistic detuning.

Last, there was a series of talks on interesting and related physics by G. Shvets, G. Fubiani, C. Philip, and M. Xie. Shvets talked about some curious and perhaps useful features of exciting wakes using colliding pulses. He showed that large wakes could be excited by rather modest laser intensities and this mode of operation could be used to overcome dephasing. G. Fubiani, described some results on space charge effects of the intense electron beams that result in the wavebreaking regime of the self-modulated LWFA concept. C. Philip presented a experimental design for phase locking an electron beam with the plasma wave in the beat wave experiment in the NEPTUNE lab at UCLA. Ming Xie gave a talk entitled "Two pillars for a new landscape of laser accelerators" in which he basically gave his ideas on ways to overcome diffraction and phase slippage.

B Simulation

The rapid improvement in computer modeling capability is impressive. The hierarchy of codes and techniques now available were elegantly described by T. Antonsen Jr. in a plenary talk. In the working group several talks were give which described the codes and approximations in more detail.

F. Tsung gave a presentation on the UCLA fully explicit PIC code called OSIRIS [19]. This code is a fully explicit, fully parallelized, fully object oriented (in Fortran90), and fully relativistic three-dimensional particle-in-cell (PIC) code. It also has a moving window and an advanced diagnostic and visualization package. The code has now been extensively used and tested. Results from the code were presented by C. Ren.

Richard Hubbard (substituting for Dan Gordon) described a new code called turboWAVE [24]and presented some recent results obtained from it. This code is a 2 1/2 dimensional fully explicit code written in C++. It can also be run in parallel using 1D domain decomposition and it also has a moving window. It has an ionization package and the option of using the ponderomotive guiding center approximation to model the laser and the motion of the plasma particles. Using turboWAVE, Gordon and co-workers showed the differences between the full PIC and guiding center PIC (both with and without ionization) when modeling the self-modulated concept. The results showed that including Raman backscatter is important. This code is an important new addition to the modeling infrastructure for laser-plasma acceleration. This code was used by C.Ren when he studied subharmonic beat wave excitation.

There were also several presentations which used WAKE. WAKE was written by P. Mora and T. Antonsen, Jr. [25]. This code was the first to successfully incorporate the pondoromotive guiding center approximation into a particle model. The key difference between this code and turboWAVE is that WAKE makes the quasistatic approximation so that in principle it can make bigger time steps in the laser evolution equation (currently, there are also differences in how the dispersive term is included into the laser evolution equations). However, the quasi-static approximation means that trapped particles cannot be self-consistently included and it prevents WAKE from being used to study the longer pulse beat wave experiments where the pulse length greatly exceeds the plasma length.

There were also presentations on the development of new fluid codes. J. Penano from NRL described a code he recently wrote called SIMILAC which uses the weakly nonlinear fluid equations. The code is fully 3D and it uses the wave equation for the electric field rather than the vector potential. In so doing his dispersion term is more complicated, but he is able to include nonlinear optical effects from the neutral gas. He presented results in which they studied the pulse compression which results when a laser with a pulse length roughly one plasma wavelength long propagates through many Rayleigh lengths of a preformed plasma channel. Finally, Brad Shadwick described his efforts to develop a fully explicit fluid code. This code would use the fluid equations but it would fully resolve the laser wavelength and hence Raman backscatter. He also described LBNL codes which make the quasi-static approximations which they use to study the predictions of their group's theoretical predictions.

Finally, N. Havz reported on simulations of the LWFA concept. These simulations indicated that wavebreaking might be possible with their existing laser. Uesaka described their plans for an upcoming experiment using their 50fs 12TW laser in his status report talk.

IV CONCLUDING REMARKS

It was clear that this field has matured tremendously since the previous workshop. Experimental efforts have shifted from acceleration via wave breaking to controlled acceleration of injected electrons. Furthermore, the community is now actively studying optical guiding of intense lasers. The theory and computational efforts have begun to address the need for high fidelity predictive capabilities for long term laser evolution. All in all the community seems to have a coherent and directed vision for demonstrating high quality 100 MeV acceleration of a beam and perhaps GeV peak acceleration of single electrons by the next workshop.

REFERENCES

1. P. Sprangle and B. Hafizi, Phys. of Plasmas **6**, 1683 (1999) and references therein.
2. C.G. Durfee and H.M. Milchberg, Phys. Rev. Lett. **71**, 2409 (1993); C.G. Durfee, J. Lynch, and H.M. Milchberg, Phys. Rev. E **51**, 2368 (1995).
3. J. Fan, T.R. Clark, and H.M. Milchberg, Appl. Phys. Lett. **73**, 3064 (1998); S. P. Nikitin, I. Alexeev, J. Fan, and H.M. Milchberg, Phys. Rev. E **59**, R3839 (1999).
4. P. Volfbeyn, E. Esarey and W.P. Leemans, Phys. of Plasmas **6**, 2269 (1999).
5. E.W. Gaul, S.P. LeBlanc, A.R. Rundquist, R. Zgadzaj, H. Langhoff, and M.C. Downer, Appl. Phys. Lett. (2000).
6. Ehrlich *et al.*, Phys. Rev. Lett. **77**, 4186 (1996).
7. D. Kaganovich *et al.*, Appl. Phys. Lett.**75**, 772 (1999); Phys. Rev. E **59**, R4769 (1999); Appl. Phys. Lett. **71**, 2925 (1997).
8. T. Hosokai *et al.*, Opt. Lett. **25**, 10 (2000).
9. C. Fauser, H. Langhoff, unpublished (2000).
10. F. Dorchies *et al.*, Phys. Rev. Lett. **82**, 4655 (1999).
11. A.R. Rundquist *et al.*, Science **280**, 1412 (1998).
12. D. Umstadter *et al.*, Phys. Rev. Lett. **76**, 2073 (1996).

13. E. Esarey *et al.*, Phys. Rev. Lett. **79**, 2682 (1997).
14. C.I. Moore *et al.*, Phys. Rev. Lett. **82**, 1688 (1999).
15. D. Umstadter *et al.*, Phys. Rev. Lett. **84**, 5324 (2000).
16. C. Gahn *et al.*, Phys. Rev. Lett. **83**, 4772 (1999).
17. A. Maksimchuk *et al.*, Phys. Rev. Lett. **84**, 4108 (2000).
18. F. Brunel, Phys. Rev. Lett. **59**, 52 (1987); M. K. Grimes *et al.*, Phys. Rev. Lett. **82**, 4010 (1999).]
19. R.G. Hemker *et al.*, Proc. of the 1999 Particle Accelerator Conference, (1999); R.G. Hemker Ph.D. Dissertation, UCLA, 2000.
20. E. Esarey *et al.*, Phys. Rev. Lett. **84**, 3081 (2000).
21. P. Sprangle *et al.*, Phys. Rev. A **36**, 2773 (1987).
22. B.J. Duda and W.B. Mori, Phys. Rev. E **61**, 1925 (2000).
23. C. Ren *et al.*, Phys. Rev. Lett. **85**, 2124 (2000).
24. D. Gordon *et al.*, to appear in IEEE Trans. on Plasma Sci.
25. P. Mora and T. Antonsen, Jr., Phys. Plasmas **4**, 217 (1997).
26. E. Esarey and W.P. Leeman, Phys. Rev. E, **59**, 1082 (1999).
27. T.C. Chiou and T. Katsouleas, Phys. Rev. Lett. **881**, 3411 (1998).
28. T. Katsouleas, Phys. Rev. A **33**, 2056 (1986).
29. P. Sprangle *et al.*, Phys. Rev. Lett. **82**, 1173 (1999).
30. W.B. Mori *et al.*, Phys. Rev. Lett. **72**, 1482 (1994); C.D. Decker *et al.*, Phys. Rev. E **50**, R338 (1994); W.B. Mori, IEEE, J. Quant. Elec., **33**, 1942 (1997).
31. G. Shvets and J. Wurtele, Phys. Rev. Lett. **73**, 3540 (1994).
32. T. Antonsen, Jr. and P. Mora, Phys. Rev. Lett. **69**, 2204 (1992).

Summary Report of the Working Group on Electromagnetic Structure-Based Acceleration Concepts

E. R. Colby[1] and W. Gai[2]

[1]*Stanford Linear Accelerator Center, Mail Stop 07, 2575 Sand Hill Road, Menlo Park, CA, 94025*
[2]*Argonne National Laboratory, Mail Stop 362, 9500 Cass Road, Argonne, IL, 60439*

Abstract. Significant progress in both the study of new acceleration structures and in the production and experimental testing of a variety of structures has taken place since the last Conference. Staging of laser accelerator sections and acceleration of particles by a microwave inverse FEL was demonstrated for the first time, fabrication of 92 GHz structures compatible with high power was demonstrated, new experimental results on dielectric based acceleration concept were presented, and detailed analysis of a new accelerator scheme, the inverted medium accelerator, was discussed. Novel structure development continued apace, with twisted waveguide structures, resonantly driven dielectric wakefield accelerators, and evanescent-mode capillary guide laser accelerators all being introduced. Power sources present and future and their impact on structure development were examined together with the need for renewed vigor in studying rf breakdown at higher frequencies. The needs and potential applications for table top radiation sources were discussed, with the expectation that development efforts in this area would directly impact the design of a future collider.

INTRODUCTION

Structure-based acceleration concepts, as a category, was defined rather liberally at this workshop. Acceleration structures coupling either external power sources or drive beams in either slow-wave or fast-wave structures were considered as well as schemes in which negative resistivity media were employed. Schemes involving plasmas in any (intentional) form were not considered in this working group. L. Schächter gave an excellent and comprehensive tutorial on structure-based acceleration methods to kick off the working group[1].

Acceleration methods are divided below into far-field and near-field, with the latter case subdivided according to the method by which the synchronism condition is met: with dielectrics, or with metal iris loading.

INVERSE PROCESS ACCELERATORS

Lorentz reciprocity provides that all radiative processes have "inverse" processes that may be used to accelerate charged particles, with the instructive corollary that all successful accelerating structures will cause the beam to radiate power into the accelerating mode. Indeed all structure-based acceleration methods can be viewed as the inverse of a radiative process, so the definition of some methods as "inverse processes" is artificial, and here is restricted to methods for which field-altering boundaries are in the far field.

Proof-of-principle experiments have successfully demonstrated the inverse free electron laser (IFEL) mechanism[2] and the Inverse Cerenkov Acceleration mechanism[3]. The key issues of timing stability of the test electron bunches relative to the laser pulses used to power the inverse accelerative process have been demonstrated to be difficult but soluble, as is discussed below. The next step was to demonstrate that two successive acceleration stages could be made to work in synchronism, despite significantly more stringent requirements imposed by maintaining timings good to a few degrees of the optical wavelength, typically 1-10 µm. STELLA has conclusively done so, maintaining timing on the scale of $\sim\lambda/4$ at 10.6 µm over a distance of nearly 2 meters, almost 200,000 wavelengths of drift. By contrast, the two-mile accelerator at SLAC is approximately 30,000 wavelengths of the 10 cm microwaves used to power the structures.

W. Kimura reported the recent successes of the Staged Electron Laser Acceleration (STELLA) experiment in observing staged acceleration of electrons by the inverse FEL mechanism. Earlier operational experience with the Inverse Cerenkov Accelerator (ICA) had proved that the laser mode converter required to produce the radially polarized beam was very time-intensive to set up, an Inverse FEL was substituted for the ICA and the staging experiment carried out. As mentioned earlier, the relative timing between the bunching and acceleration stages was observed to be stable, which is all the more remarkable as no exotic preparations were made to eliminate thermal or microphonic effects, nor was exotic electron beam focussing required to prevent path length differences from "washing-out" the optical bunching imposed by the initial IFEL.

R. Yoder reported on the first success of the Microwave Inverse Free Electron Laser Accelerator (MIFELA) in demonstrating acceleration by a microwave FEL. Using short bunches derived from an s-band photoinjector, a helical undulator was used to couple the beam to s-band microwave power. Acceleration was seen, with the 5-6 MeV electron beam gaining 380-400 keV on passing through the MIFEL. Clear change in the energy spectrum was observed for different phases of the drive power to the FEL relative to the photoinjector.

DIELECTRIC WAKEFIELD ACCELERATORS

Recent theoretical work by J. Hirshfield and S. Park to describe the TM modes of an infinitely long axisymmetric dielectric-loaded waveguide modes as a generalized HEM decomposition was presented. Dipole wakefield-driven beam breakup was pointed out to be potentially as serious a problem for dielectric structures as it is with conventional iris-loaded metal structures. Some additional discussion resulted from Hirshfield and Park's conclusion that the Poynting vector associated with the Čerenkov wakefield was oppositely directed to the beam momentum in the laboratory frame. Schächter subsequently demonstrated that in fact the energy flow was in the same direction as the beam momentum in the laboratory frame, as is the observed case in Čerenkov radiation-based diagnostics.

T. Marshall proposed driving an over-moded dielectric resonator with a pulse train, with the current harmonics set to match the cavity TM_{0nm} resonances. Use of a low dispersion dielectric such as alumina would allow high-Q structures to maintain good phase coherence of the harmonics and therefore high accelerating gradients.

Work at the Argonne Wakefield Accelerator facility (AWA) was summarized, covering experimental demonstration of deceleration of a bunch train in a multimode dielectric wakefield accelerator, and experimental plans for testing an x-band dielectric accelerator at high gradient using both an external power source and using pulse trains. W. Gai presented new experimental results of two beam acceleration using dielectrics as both the power extraction and acceleration structures. A witness beam was accelerated and decelerated in this experiment. P. Zou presented details of an instrumented dielectric wakefield accelerator section that will be powered by a SLAC 75MW x-band klystron to test issues of ceramic breakdown and dark current. Gai described plans to use the AWA injector to produce short bunch trains of 4 pulses to resonantly excite wakefields in a multimode structure in the manner described by Marshall. Gai expects test that gradients of several hundred MeV/m are achievable at 30 GHz with a structure Q of 1000. J. Power described experimental measurement of resonant wakefield generation in a multimode $CaTiO_3$-$LaAlO_3$ ceramic structure. Energy loss measurements for the 4 successive pulses in the drive train were in good agreement with predictions from theory.

Experimental results on wakefields in a 92 GHz dielectric structure were presented by M. Hill. Beam from the x-band NLCTA injector was used to regeneratively excite a ceramic structure. Power production of nearly 200 kW at 92 GHz was observed with structure deceleration gradients reaching 20 MV/m with no signs of breakdown, and no evidence of dielectric charging.

METALLIC STRUCTURES

Work on metallic structures continues at a vibrant pace, despite these being the oldest of the coupling methods used in rf acceleration. Work to develop new slow-wave metal structure geometries that are inexpensive to produce, can support power densities typical of laser-driven acceleration, or that operate at unusually high frequencies was presented.

Planar accelerating structures for millimeter-wave acceleration and power production were presented by D. Yu, including a modification of the Miller barbell cavity[4] that used dielectric loads at the ends of the cutoff waveguide section. Results of beam tests at the CLIC Test Facility on a 30 GHz planar accelerating structure produced in collaboration with H. Henke of T. U. Berlin by precision CNC milling techniques were also presented. The structure was powered from the CLIC transfer structure with up to 80 MW for 4 ns and up to 40 MW for 16 ns with no evidence of breakdown. D. Palmer showed loss and bead pull measurement results on the Zipper structure[5], a 25-cell constant impedance travelling-wave $2\pi/3$ planar accelerating structure, that were in very good agreement with 3-dimensional GdfidL simulations, providing the first example of a complex 92 GHz planar accelerating structure with construction consistent with high power usage.

Lithographic fabrication techniques for producing very high frequency structures (w-band and beyond) were discussed by Y. Kang and E. Colby. Kang described work at Argonne which has centered on LIGA fabrication of a 32-cell 108 GHz constant-impedance accelerator cavity and a 66-cell 94 GHz constant-gradient accelerator cavity. They are preparing a 26 GHz vector network analyzer with up- and down-mixers for use as a w-band vector network analyzer, and expect measurements shortly. Colby described work at SLAC in collaboration with Sandia National Laboratories of California to make a w-band sheet beam klystron by LIGA, and of plans to study issues of dimensional accuracy by optical and quasi-optical techniques.

Kang described efforts to look for very simple metal structures with good electrical properties, and gave as an example an s-band helical ridged waveguide structure. The structure supports a rectangular HEM_{11}-mode with potentially interesting electrical and beam dynamical properties with a phase velocity directly related to the pitch angle of the helix. It is believed such a constant cross-section structure can be produced cheaply by extrusion.

Metallic structures for laser acceleration received considerable attention, with L. Steinhauer presenting a capillary waveguide for use at 10.6 µm wavelength. The 300 µm diameter smooth-wall capillary guide supports an evanescent surface-wave mode with zero field variation across the vacuum aperture of the capillary. For 0.3 TW input power, Steinhauer estimates acceleration gradients of 1.3 GeV/m are obtainable without material damage. With attenuation lengths approaching 10 cm, reasonable energy gains per structure are possible. Efficiently coupling a to the surface wave mode will be challenging.

General features of fast-wave laser acceleration structures were discussed by M. Xie. As an alternative to slow-wave structures, he argued for "Alternating Gradient Acceleration": fast-wave structures in which the energy loss incurred while the particle phase is in the decelerative half-cycle is minimized by one or more of i) using plasmas or other means to hasten the phase slip, ii) lessening the field strength in these regions, or iii) lessening the transit-time factor in these regions. Several examples were given, with either $\varepsilon<1$ media used to speed up the EM wave, or particle path-length increasing mechanisms (solenoid, wiggler, etc.) invoked to slow down the forward motion of the particle beam. In effect, this scheme raises the phase velocity of all the space harmonics until again a synchronism condition exists, but with a space harmonic other than the fundamental. Xie also advocated the use of open oversize structures[6] for their potentially better peak-to-acceleration field ratio and relative lack of higher order modes. This interesting combination of schemes allows for fairly simple, open structures to be used for laser acceleration. The impact of the strongly excited non-synchronous fundamental space harmonic on beam quality remains an important issue for this scheme.

GRADIENT LIMITATIONS: JOINT SESSION ON RF BREAKDOWN

Given the level of effort directed at producing accelerating structures operating in the millimeter-wave frequency range, the natural question of interest to both the power source and accelerator communities is whether the empirical scaling ($f^{7/8}$) for rf breakdown[7,8] is accurate much beyond present measurements in the 1-10 GHz range. At sufficiently high frequencies surface effects will dominate breakdown and the empirical scaling will no longer be valid. Additional work is needed to understand whether this point occurs within the millimeter-wave band presently under active development.

A discussion on rf breakdown research, held in joint session with the Millimeter-Wave Power Sources Working Group, was led by D. Sprehn. Given that present understanding of rf breakdown (e.g. [9]) involves material properties of the cavity surfaces in considerable detail, the expansive sweep of experiments required to adequately understand rf breakdown requires significant surface science capabilities, rigor, and expense, and consequently the coordinated efforts of a number of laboratories. Further, it is clear that although single-cell cavities are an excellent test vehicle for material property issues, they do not begin to address issues related to geometry and dark current properties (e.g. beam energy, current distribution) in initiating breakdown in multicell cavities. Also of issue is the significant discrepancy between power tube output circuit and accelerator structure breakdown levels, with the latter generally holding off significantly higher gradients than the former. Here again issues of beam interception and x-rays from nearby beam strike in the collector are markedly different than those reproduced in single-cell tests.

The importance of rf breakdown to accelerator and power tube development is such that a conference on the subject, as suggested by Sprehn, would be indispensable.

EXOTIC ACCELERATION MECHANISMS

All of the acceleration methods described thus far have relied on a lossy coupling structure to convert fast waves from a power source or slow waves from another beam to slow waves of suitable field geometry to accelerate charged particle beams. L. Schächter proposed modifying the coupling structure to include an active medium (a medium for which the resistivity is negative) to supply the energy for acceleration. Lasers employ active media to amplify optical signals, which are typically Gaussian wave packets. Used in an accelerator, the Cerenkov radiation of the "trigger" bunch replaces the "seed pulse" and is amplified, resulting in a large amplitude wakefield some distance behind the trigger bunch, where a suitably placed "witness" bunch will be accelerated. As only modes which (1) are synchronous with the trigger bunch, (2) lie on the dispersion curve of the media, and (3) have a frequency within the bandwidth of the media can be amplified, the amplified wakefield will be generally composed of only one or very few modes, and will be synchronous with the trigger bunch. This latter property provides a natural mechanism for synchronizing multiple stages of acceleration by using a single trigger bunch / witness bunch pair throughout the accelerator. Schächter estimates that the ~2 µW Cerenkov radiation signal from a 3 nC bunch focussed to 75 µ traversing a properly pumped 3 cm long Nd:YAG rod will undergo almost 120 dB amplification, resulting in an amplified Cerenkov wake power of nearly 3 MW a short distance behind the trigger pulse. He estimates that amplified spontaneous emission will be some -90dB below the Cerenkov wake, hence wholly negligible. Work is underway to design a proof-of-principle experiment at SLAC.

JOINT SESSION ON TABLE TOP RADIATION SOURCES

As the proposed table top radiation sources all demand high brightness electron beams from a machine with a compact footprint, many of the technical requirements are similar to those imposed by a linear collider: namely, a compact accelerator producing bunches with excellent transverse emittances and short bunch lengths, good diagnostics and control systems, and excellent stability of beam properties[10]. A joint session of the Diagnostics and Sources, High Density Beams, and Structure-Based Acceleration groups was called to discuss the problems inherent in such systems, with an eye towards their probable impact on linear collider design.

The discussion on diagnostics began with V. Yakimenko who gave an overview of beam diagnostics at the Brookhaven ATF and described recent developments towards performing tomographic reconstructions of beam profiles. He also described comparison measurements of fluorescing and OTR screens, revealing significant space-charge induced[11] blooming of the image on YAG fluorescent screens for incident charge densities above 1 nC/100 μ^2, a potentially serious resolution limit.

J. Rosenzweig surveyed present beam diagnosis techniques and described the difficulties in making reliable measurements of beams of the densities and emittances that are being produced by the best sources and accelerators. X-ray backgrounds expected in the LCLS will limit the resolution of wire scanners, making indirect methods, such as the BPM-based technique demonstrated by Russell[12], more attractive. He also pointed out that the antiquated argument that OTR resolution worsens with increasing beam energy is invalid because the angular distribution of OTR is not Gaussian. Rather, the RMS angular spread of OTR is largely independent of beam energy, giving no degradation of resolution at higher beam energies.

Electron/photon beam synchronization at the picosecond level and beyond was the subject of three talks. State-of-the-art streak camera-based and spectral methods of timing measurement with picosecond resolution were described by M. Uesaka. As with all spectral techniques, the key issues in successfully Fourier transforming the frequency-domain data are the ansatz for the low-frequency portion of the spectrum and the presumed functional form of the distribution. Fluctuation-measurement based bunch length measurements using Cerenkov radiation were also presented.

High harmonic phase detection methods for sub-picosecond resolution relative timing measurements for the Laser Electron Acceleration Project were described by

E. Colby. Signal from an rf cavity tuned to the 238th harmonic of the laser is mixed against the commensurate comb-generated harmonic from photodiode detection of the laser to produce a phase. A benchtop precision of 50 fs was demonstrated, with a long-term measurement stability of ±100 fs over 12 hours.

Synchronization by optical mixing was described by M. Thomson (using a Kerr cell as the mixing element) as proposed for use in the synchronization of a CO_2 laser used to excite plasma waves, and a UV laser used to produce photoelectrons in an rf photoinjector. In this way one laser "gates" a short pulse out of the other, longer pulse, guaranteeing synchronism of the induced plasma wave and photo-emitted electron bunch. With a dispersive dipole chicane in the injector set to give pulse compression, additional suppression of phase jitter of the electron beam is possible.

Integration of a laser-based accelerator, a novel "super-tip" field emission-based electron source, and diagnostics was discussed by A. Mikhailichenko, who presented an adaptation of his table-top accelerator[13] scheme for a 100 GeV×100 GeV table top linear collider, a subsection of which could easily serve as the electron source for a table top light source. The accelerator is based on a scanned, 10.6 µm, 100 ps laser pulse illuminating a "fox-hole" structure. He estimates gradients of 1.2 GeV/m are possible in silicon microstructures without damage.

Present research efforts to develop Thomson scattering and Compton scattering-based tunable x-ray sources formed the final part of the discussion on table-top radiation sources.

I. Pogorelski gave a summary presentation of work at Brookhaven to construct a Laser Synchrotron Source (LSS) for bright, incoherent x-rays. Thomson scattering signals ~100 times the brem■trahlung background and totaling some 3×10^6 x-rays have been recorded in crossings between 1 nC, 60 MeV, 3.5 ps FWHM electron bunches and 0.2 J, 180 ps CO_2 laser pulses at the ATF, in good agreement with prediction. Plans to upgrade the laser to obtain much higher peak and average brightnesses were also discussed, as well as plans for a plasma channel guided LSS experiment to further enhance the electron and photon densities at the crossing.

T. Cowan described progress in commissioning the Falcon multi-terawatt laser and a new HIP copper photocathode rf gun at Livermore, in preparation for Compton light source development there. Plans were discussed to use 5 MeV, 100 A peak current electron bunches and 400 fs, 50 mJ laser pulses to generate sub-picosecond x-ray pulses for pump-probe and ultrafast chemistry experiments. The narrow-band tunable radiation from this source also facilitates microradiography.

F. Hartemann discussed detailed 3D computations of Compton scattering spectral yields, and plans for a joint UC Davis-SLAC Advanced Compton Light Source, presently under construction in the klystron test facility at SLAC. Using 50 MeV, 0.5 nC electron bunches and 50 mJ, 100 fs laser pulses at 1 kHz, they estimate average brightnesses of 4.1×10^{10} γ/mm^2 mr^2 0.1%BW are obtainable at 12.77 keV.

The distinguishing characteristics of Compton and Thomson scattering-based radiation sources are the narrow-band tunability of the electron beam energy, and consequently the ray energy, and the short electron and photon pulses. Synchrotron- and rotating-anode-based ray sources produce high average brightness, broadband, long pulses of x-rays. Scattering sources have the potential to provide highly flexible, ultrafast, tunable x-ray sources. Additionally, some discussion as to the potential applicability of Thomson-scattering sources (the UCD source in particular) to protein crystallography occurred. W. Leemans and H. Padmore pointed out that a typical crystallographic data scan at the ALS requires some 1.2×10^{17} photons in a 0.1 mm aperture and 1.5 eV bandwidth, resulting in an estimated ~333 hr exposure time using the UCD brightness numbers, while for Multi-wavelength Anomalous Diffraction (MAD), Hartemann and B. Rupp[14] estimate that the exposure requirement is some 1.4×10^{15} photons in a 1 eV bandwidth, resulting in a more tenable ~10 hr exposure time per wavelength scanned.

It is clear that the success of scattering-based table top light sources will rest on many of the same issues important to e^+e^- colliders, namely: smallest spot sizes achievable at the interaction point, timing and pointing stability, energy spread, and beam divergence, and that progress made in this field will benefit linear collider efforts. It is also clear that if such table top sources are to be truly compact and affordable, inexpensive means of high gradient acceleration will be a key enabling technology.

JOINT SESSION ON MILLIMETER-WAVE SOURCE CAPABILITIES

The reciprocal interests of the power source and accelerator communities were addressed in two separate discussions on power source capabilities in comparison with accelerator structure requirements. Power requirements for collider, medical, and industrial application accelerators were developed in the structures working group and presented to the millimeter-wave power sources group early in the conference. Subsequently, in the last working group session, the power sources and structures groups met in joint session to discuss what is potentially achievable in the near- and intermediate-term future.

Table 1 below summarizes power source requirements adapted from P. Wilson[15] to 11.4, 30, 60, and 94 GHz, together with predictions and measurements of near-term micro- and millimeter wave power sources. Numbers in parenthesis are the present measured quantities shown next to design values.

TABLE 1. Near-Term Micro- and Millimeter-Wave Power Sources					
	Frequency [GHz]	Peak Power [MW/m]	Pulse Length [µs]	Efficiency [%]	Rep Rate [Hz]
P. Wilson 2x500 GeV, 80 MV/m	11.4	130	0.25		180
L.B.N.L. T.B.A.	11.4	200	0.25	30-35	180
U.M.D. Gyroklystron	17	80/tube	1.0	34	1
P. Wilson 2x500 GeV, 180 MV/m	30	440	0.6		180
C.L.I.C. T.B.A.	30	1000(180)	0.13(0.015)		
OmegaP Magnicon	34	45/tube	1.5	35	10
P. Wilson 2x500 GeV, 340 MV/m	60	1000	0.02		180
A.W.A. T.B.A.	60	1000	0.004		10
P. Wilson 2x500 GeV, 500 MV/m	94	1800	0.01		180
U.M.D./C.C.R. Gyroklystron	94	10/tube	1	37	180

There was lively discussion about the difficulties inherent in making high power sources at high frequencies, about the potential severity of pulsed heating problems, and of the need for continued work to understand rf breakdown scalings to millimeter-wave frequencies. Substantial optimism was expressed that mm-wave sources capable of power levels approaching 100 MW (short pulse) at w-band would be possible in the coming decade, which combined with pulse compression reaches a significant way towards achieving even the extraordinary requirements for w-band high gradient acceleration.

CONCLUSION

Significant progress in structure-based acceleration was made since the last conference, both in the development of new structure-based acceleration concepts and theoretical frameworks for describing them, and in the experimental demonstration of staged IFEL acceleration, the IFEL mechanism at microwave frequencies, and of fabrication techniques for metallic mm-wave structures. Significant progress in developing compact radiation sources has begun to offer promise of a new set of diagnostic tools for chemistry and ultrafast pump-probe experiments, and has brought renewed attention to beam physics issues of fundamental importance to linear colliders.

For most of the acceleration methods discussed here, key questions of beam quality preservation and diagnosis of the exotic beams typically called for by laser acceleration schemes remain, as does integration of these technologies into a working linear collider.

ACKNOWLEDGMENTS

The authors would like to thank the members of the Electromagnetic Structure-Based Acceleration Concepts working group for making this conference productive and stimulating. Thanks also to J. Rosenzweig/T. Cowan (High Density Beams), J.G. Wang/D. Palmer (Sources & Diagnostics), and G. Nusinovich/S. Gold (Millimeter Wave Sources) for their efforts in making the joint sessions possible and productive. Lastly, the author is indebted to L. Schächter for valuable guidance in preparing the summary talk presented at the workshop. This work was supported in part under DOE contract DE-AC03-76SF00515.

REFERENCES

1. L. Schächter, "Acceleration Concepts Based on Electromagnetic Structures", in these proceedings.
2. A. van Steenbergen *et al*, "Status of the BNL IFEL Accelerator", in Proc. AAC1996, ed. S. Chattopadhyay, AIP Conf. Proc. 398, New York: AIP, 1998, p.591.

3. W. D. Kimura et al, "STELLA Experiment: Design and Model Predictions", in Proc. AAC1998, ed. W. Lawson, AIP Conf. Proc. 472, New York: AIP, 1998, p.563.
4. K. Eppley, W. Hermannsfeldt, R. Miller, "Design of a Wiggler-Focussed, Sheet Beam X-Band Klystron", SLAC-PUB-4221, Feb. 1987.
5. N. Kroll et al, "Planar Accelerator Structures for Millimeter Wavelengths", in Proc. IEEE Part. Accel. Conf., New York, N.Y., pp. 3612-5, (1999).
6. N. Kroll, "General Features of the Accelerating Modes of Open Structures", in AIP Conf. Proc. V. 130, p.253, (1985).
7. G. Loew and J. W. Wang, "RF Breakdown Studies in Room Temperature Electron Linac Structures", SLAC-PUB-4647, (1988).
8. A. Vlieks et al, "Breakdown Phenomena in High Power Klystrons", SLAC-PUB-4546, (1988).
9. R. V. Latham, "Prebreakdown Electron Emission", IEEE Trans. Elect. Ins. EI-18, No. 3, June 1983.
10. E. Weihreter, *Compact Synchrotron Light Sources*, Singapore: World Scientific, 1996.
11. J. Rosenzweig, private communication.
12. S. Russell, "Emittance Measurements of the Sub-picosecond Accelerator Electron Beam Using Beam Position Monitors", Rev. Sci. Inst., **70**, 1362, (1999).
13. Mikhailichenko, "Laser Linear Collider with a Travelling Laser Focus Supply", in. proc. Part. Accel. Conf, New York, NY, pp. 3633-5, (1999).
14. F. Hartemann et al, "Three-Dimensional Theory of Emittance in Compton Scattering", in these proceedings.
15. P. Wilson, "RF Power Sources for 5-15 TeV Linear Colliders", SLAC-PUB-2774, (1997).

Beam Generation, Monitoring, Conditioning and Control at High Frequencies and Ultra-Fast Time Scales

D. T. Palmer (SLAC) and J. G. Wang (ORNL)

Working Group 4 Charge

"The goal of WG 4 is to study the generation of electron/positron bunches and the associated diagnostics on an ultra fast time scale, including conditioning and control issues. The work will focus about the design of an injector for small phase space acceptance systems such as plasma, laser or W-band accelerators."

A total of twenty-one talks were given in working group 4. Two of the talks were invited contributions to give an overview of the progress and status in the field. Some of the talks were presented in joint sessions with working groups 1, 3 and 5. All the presentations in working group 4 are classified in three research areas, namely, Particles Sources (9 talks), Diagnostics (8 talks), and Beam Conditioning and Control (4 talks). A detailed listing of the working group 4 authors, their affiliation, along with the title of their talks is included in this summary. Most presentation are discussed in the following text, but not all. This is due to many factors beyond the individual authors control.

Particle Sources

Author	Lab	Title
J. Clendenin *	SLAC	Recent Advances in Electron and Positron Sources
L. Serafini	INFN	Photon gated electron source at W-band
W. Leemans	LBNL	Laser wakefield acceleration, - Electron sources
W. Brown	MIT	MIT 17 GHz RF gun experiment
D. Yu	Duly	PWT photoinjector
K. Nakajima	KEK	PAMELA simulation of W-band RF gun
T. Tajima	LLNL	Attosecond electron source
T. Cowan	LLNL	RF photoinjector at LLNL
A. Mikhailichenko	Cornell	Micro RF gun

* Invited talk.

J. Clendenin, SLAC, was invited to give a review talk in WG4 on recent advances in electron and positron sources. Three major topics were covered during the working group review talk. These included the present state of the art in electron and positron sources along with a discuss of Simultaneous of laser damping and polarization. The LCLS injector was discussed in detail. This injector, using a new operating paradigm developed by M. Ferrario et al. along with HOMDYN simulation indicate that the

LCLS injector will be able to produce a 0.4 mm mrad normalized rms emittance, of $\varepsilon_{n,rms} = 0.4\,\pi$ mm mrad, for a 1 nC 10 psec electron bunch.

A full W-band injector lay-out, delivering a 8 MeV beam was presented by L. Serafini (INFN), which uses a 90 GHz booster and an 13.8 mm period undulator driving a FEL-CSE process to filter the head of the long electron pulse produced from the trailing part, achieving in this way a production of synchronized short (w.r.t. the RF period) electron bunches. This technique eliminates the need to mode lock the laser pulse to the rf with a stability of $\delta\tau = 33$ fsec. ITACA simulation show that a peak current of 600 A with a beam quality of $\varepsilon_{n,rms} = 1.5\,\pi$ mm mrad is possible. An initial w-band rf gun has been produced by the authors and is now undergoing low power rf testing.

W. Brown, MIT, presented experimental results from a 17 GHz rf gun. Accelerating field gradients of 200 MV/m have been attained in the gun which produces an electron beam of $\gamma = 1$ MeV with an energy spread of $\delta\gamma/\gamma = 2.5$ %. This injector has produce electron bunches with a total charge of 10 –100 pC and a normalized rms emittance of $\varepsilon_{n,rms} = 3\,\pi$ mm mrad. The production of a symmetries 2.4 cell rf gun is planned for the future.

D. Yu, Duly Research, presented design parameters for S-band PWT rf gun. This gun is presently installed at PEGAUS, which is located in the Department of Physics at UCLA. PARMELA simulation show that 1 nC, 20 MeV electron beam can be produced with a beam quality of $\varepsilon_{n,rms} = 1\,\pi$ mm mrad. An X-band PWT design was also presented.

A. Mikhailichenko, Cornell, presented work on a micro rf gun. This gun is design to be the electron sources for a Laser-Driven Tabletop Linac. This source has been design to produce 100 MeV electrons with a bunch population of 10^5 electrons. Questions of maximum charge density threshold where raised and are still under investigation.

During the group discussion we tried to specify beam parameters for a high brightness electron injector for advanced accelerators. The number of electron particles is in the order of 10E10. The beam energy is usually in many MeV, depending specific accelerator schemes. The beam emittance is typically 1 mm mrad per nC charge. The beam length is usually about 5% of acceleration wavelength, which is around 3 mm for w-band accelerator, 100 μm for plasma accelerator, and 10 μm for CO_2 laser accelerator. The bunch jitter should not exceed 10% of acceleration wavelength. For other advanced applications such as GPT, the injector should produce 100 pC charge with a bunch length of 70 fs.

Positron source, which is an important topic for the AAC, was not presented in the workshop, except that J. Clendenin reviewed the recent activity in the development of polarized electron beams for future colliders.

Diagnostics

Author	Lab	Title
V. Yakimenko *	BNL	ATF diagnostics
M. Hogan	SLAC	E-157 optical diagnostics
M. Uesaka	U. Tokyo	Ultrashort e-bunch diagnostics
E. Colby	SLAC	Picosecond resolution timing measurement
P. Catravas	LBNL	Measurement of emission near an atomic spectral resonance in E-157
S. Russel	LANL	Emittance measurement with a BPM
S. Anderson	UCLA	Commissioning and status of the Neptune photoinjector
M. Zolotoreva	LBNL	Pondermotive focusing,

* Invited talk.

Based on the experimental results from ATF at BNL, V. Yakimenko presented an invited overview talk in WG4 on various beam diagnostics techniques for high brightness photoinjectors. These include charge measurement by Faraday cup, beam size measurement by phosphor screen, transition radiation, YAG crystal, wire scanner, etc, beam position measurement by ps Compton x-ray, Smith-Purcell radiation, IFEL & ICR, bunch density measurement from phase space tomography, and ps slice emittance measurement. M. Uesaka of University of Tokyo talked performance and comparison of bunch size and shape measurement with different methodologies: streak camera, far-infrared polychromatic with two gratings, CTR intererometers, and fluctuation method. He also helped greatly during the group discussion on ultra short electron bunch diagnostics by providing very useful information about the status and directions in this field. There were two talks from the E-157 experiment: M. Hogan of SLAC presented beam transverse profile (beam size and distribution) measurement by optical transition radiation diagnostics; P. Catravas talked about plasma density and neutral density measurement by Cerenkov radiation diagnostics. E. Colby's talk was about ps resolution timing (both RF and beam) and monitoring by streak camera in a laser electron accelerator project at SLAC. In the Neptune experiment at UCLA, there are beam emittance measurement by slits, pulse length measurement by CTR, and quantum efficiency measurement, all of which were presented by S. Anderson. An emittance measurement by BPM was performed at LANL, and its result was presented by S. Russel.

The diagnostics for a high brightness photoinjector requires the measurement of beam bunch charge, length, size, shape, distribution, time jitter, emittance, etc. During the past two years, significant progress has been made in this area. Bunch charge in the range of 50 pC to 1 nC can be measured accurately with destructive and non-destructive methods. For the bunch size measurement, streak camera works well above 1 ps bunch length, and has reached to 100 fs, a possible limit to this technique, by reflective optics. CR polychromatic has been developed to work in the range of 1 ps to 100 fs, and probably can reach to 10 fs in future. It is always important to apply two different methodologies for the same measurement in order to compare and to get an agreement. Great challenges ahead is to extend the bunch size measurement to around 1 fs, or even shorter. CR interferometer looks most promising to achieve this goal because of laser "autocorrelation". We expect more accomplishment in this area will be achieved during the next two years.

Beam Conditioning and Control

Author	Lab	Title
J. Hirshfield	Yale	Laser-driven cyclotron auto resonance acceleration - LACARA
M. Thompson	UCLA	Photoinjector beam timing and synchronization
M. Hess	MIT	Confinement of bunched beams
S. Bernal	UMD	Effects of an aperture on the phase-space particle distribution

In the area of beam conditioning and control, there were only four contributed papers, which covers rather diversified research. J. Hirshfield presented a new mechanism of gyroresonance acceleration of electrons in vacuum by high power CO-2 laser to achieve an acceleration gradient of the order of 100 MeV/m. An experiment to verify the prediction is also planned at BNL/ATF. The talk of M. Thompson was about an experiment at UCLA where a chicane compressor is employed to achieve good photoinjector beam timing, which is not sensitive to laser jitter. The longitudinal dynamics such as space-charge defocusing and beam compression by the chicane were also analyzed. M. Hess, who worked with Chiping Chen at MIT, presented their theoretical investigation of the confinement criteria for the space charge, which is important to controlling high-intensity bunched beams in high-gradient accelerators and intense microwave sources. An experiment performed by S. Bernal at University of Maryland produced very clear transverse wave-like phenomena in highly space-charge dominated electron beams. His presentation generated great interest in the audience and led broad discussions on space-charge dynamics in intense beams.

Overall, the beam conditioning and control is under-presented in the workshop. We believe there are many important, unresolved research issues in this area, which are important to the development of a high brightness electron/positron injector suitable for advanced accelerator schemes. In fact, B. Carlston presented an excellent tutorial talk in the WG4 area. He mentioned many challenging beam dynamics problems for very high brightness particle sources. A few of his examples include emittance oscillation/growth and phase space dilution, wave breaking, space-charge effects, etc. We hope that there will be more contributions in this area in the next AAC workshop.

Acknowledgements

We would like to thank all the participants in WG4 for their contributions to the presentations and discussions in the Workshop. Especially, we thank J. Clendenin and V. Yakimenko for giving overview talks on particle sources and diagnostics and for leading the group discussions.

Summary Report: Working Group 5 on "Electron Beam-Driven Plasma and Structure Based Acceleration Concepts"

Manoel E. Conde[†] and Thomas Katsouleas[*]

[†]*Argonne National Laboratory*
High Energy Physics Division
9700 S. Cass Ave. Bldg. 362
Argonne, IL 60439

[*]*University of Southern California*
Department of Electrical Engineering-Electrophysics
Los Angeles, CA 90089-0484

Abstract. The talks presented and the work performed on electron beam-driven accelerators in plasmas and structures are summarized. Highlights of the working group include new experimental results from the E-157 Plasma Wakefield Experiment, the E-150 Plasma Lens Experiment and the Argonne Dielectric Structure Wakefield experiments. The presentations inspired discussion and analysis of three working topics: electron hose instability, ion channel lasers and the plasma afterburner.

INTRODUCTION

This summary of the activities of the Working Group 5 is divided into two main parts: the first one giving a brief summary of the talks that were presented in the working group sessions, and the second part covering the material that was generated by small sub-groups that further explored the subject of some talks.

The talks were classified among the five following topics: Transverse Beam Dynamics and Radiation in Plasmas, PWFA (plasma wakefield accelerators) and Lens Experiments, Injection via Plasma Trapping (joint session with Working Group 2), Structure Based Acceleration, and Ionization and Positron Acceleration. The topics that were discussed by the small sub-groups were: Hose Instability, Ion Channel Radiation, and Afterburner Design.

SUMMARY OF TALKS

The following tables list the talks that were presented at the working group sessions. Each table is followed by a brief summary of some of the talks. We apologize for not covering extensively all the talks and also for not listing the speaker's co-workers (this was done for the sake of keeping this paper as just a summary; more information can, of course, be found in these proceedings). The talks that led to the discussions by small sub-groups are covered in the second part of this paper.

TABLE 1. Talks on Transverse Beam Dynamics and Radiation in Plasmas.

Presenter and affiliation	Title
Patrick Muggli (USC)	Betatron Dynamics in E157
Evan Dodd (UCLA)	Hosing Simulations
Brent Blue (UCLA)	Tail Oscillations in E157
Sho Wang (UCLA)	Betatron X-Radiation in E157 and ICLs
Eric Esarey (LBNL)	Spontaneous Ion Channel Radiation Spectra and limits to ICL gain

All the talks above are related to two of the working sub-groups and are discussed later.

TABLE 2. Talks on PWFA and Lens Experiments.

Presenter and affiliation	Title
Mark Hogan (SLAC)	Optical Diagnostics in E157 (joint w/ WG4)
Palma Catravas (LBNL)	Cerenkov Diagnosis of Plasma Density
Steve Russel (LANL)	LANL PWFA Observation (60 MeV/m deceleration)
Nick Barov (UCLA / FNAL)	PWFA Results at ANL and FNAL Plans
Patrick Muggli (USC)	σ_z^2 scaling of PWFA: proposed experiment at ATF
Johnny Ng (SLAC)	E150 Plasma Lens Results
Pisin Chen (SLAC)	100 GeV PWFA

Hogan presented an extensive discussion of the optical diagnostics used in the E-157 experiment at SLAC, including the motivation for reflective optical elements used in pairs to cancel out aberrations. Catravas showed the measurements of Cerenkov radiation cones as a means to diagnose the plasma density and neutral gas density in the E-157 experiment. Russel presented preliminary results of the PWFA experiment at LANL that indicated a relatively high decelerating gradient. Barov showed acceleration of the electron bunch tail in his PWFA experiments at Argonne and the plans for a new series of experiments at Fermilab. Muggli presented the plans for experiments at Brookhaven to study the dependence of PWFA on bunch length.

J. Ng showed results from the plasma lens experiment at SLAC (E-150), indicating a reduction in the beam waist of about a factor two. This experiment was the first to realize the very high focusing gradients promised by plasma lens theory (order GG/m) and showed that plasma lenses also focus positrons. Chen showed alternative plans for the possibility of an Afterburner experiment at SLAC. The Afterburner concept aims at doubling the energy of a high energy beam in a single plasma wakefield accelerator stage and decreasing the spot size by a factor two.

TABLE 3. Talks on Injection via plasma trapping (w/ WG 2).

Presenter and affiliation	Title
Christopher Moore (NRL)	Short pulse high-energy e- production from laser ionization
Ned Saleh (U. Mich.)	Recent results from U of Michigan
Hyyong Suk (UCLA)	Trapping at a density gradient in PWFA

Suk presented results from PIC simulations showing the trapping of electrons due to phase mixing at the sharp boundary of two plasma regions with different densities.

TABLE 4. Talks on Structure Based Acceleration

Presenter and affiliation	Title
Wei Gai (ANL)	Two Beam Acceleration at ANL
John Power (ANL)	Transformer Ratio Enhancement in PWFAs
Paul Schoessow (ANL)	30 GHz Power via Dielectric Structures
Ilan Ben-Zvi (BNL)	Electron bunching at 3 fs

W. Gai presented the first results of two beam acceleration using dielectric loaded waveguides as both the power extraction and the accelerating structures. Principal results were the generation of 4 MW of RF power from a single bunch and the coupling of that power to a second bunch in a parallel dielectric tube. The transformer ratio in the first dielectric structure was 1.8, followed by an enhancement of the accelerating gradient by a factor 2.5 as the RF power was coupled into the second dielectric structure, yielding an effective transformer ratio of 4.5. Acceleration and deceleration of the second bunch at up to 7.5 MeV/m was demonstrated. J. Power showed a novel scheme to achieve a transformer ratio greater than two by using a bunch train with increasingly higher charge per bunch. Schoessow presented numerical calculations for the design of a 15 GHz and a 30 GHz dielectric loaded waveguide to be used as power extraction structures. Ben-Zvi presented measurements from the IFEL experiment (with a second IFEL as a pre-buncher) that showed bunch lengths of the order of 3 fs.

TABLE 5. Talks on Ionization and Positron Acceleration

Presenter and affiliation	Title
Dave Bruhwiler (Tech-X)	PIC Simulations with Ionization
Seung Lee (USC)	Positron Wakes

Bruhwiler showed PIC simulations of electron-impact ionization. The work suggests that impact ionization can be a candidate plasma formation mechanism for Afterburner parameters. Lee showed positron wake simulations in the nonlinear regime. Positron wakes are smaller in homogeneous plasma but can be comparable to electron wakes in hollow plasmas.

WORKING SUB-GROUP TOPICS

The discussions that followed some of the working group talks led to the formation of sub-groups that further explored some of the issues under debate. The three main topics studied by the sub-groups were: Hose Instability, Ion Channel Radiation, and the Afterburner Design

The Hose Instability sub-group was motivated by the talks given by Muggli, Dodd and Blue. Muggli and Blue presented E-157 data on bunch tail motion, which could be either hose instability or simply bunch tail amplification, as pointed out by Katsouleas. Dodd showed results from PIC simulations for a wide range of parameters, including cases with and without hose instability. The working sub-group generated a table with parameters for the electron beam and plasma for various experiments, and used an asymptotic expression for the hose instability growth rate to predict whether hosing should be observed in the given experiments.

The sub-group also examined the parameters used in Dodd's simulations and calculated the expected growth rate with the asymptotic expression. Surprisingly, the PIC simulations did not show hosing for some cases in which the asymptotic expression would predict instability. Apparently the PIC simulations indicated that hosing occurred when the asymptotic expression predicted it, but only if the beam density was sufficiently higher than the plasma density. This fact raises some concerns about the growth of this instability in a possible afterburner experiment, where the beam density may be approximately a thousand times higher than the plasma density.

The Ion Channel Radiation sub-group resulted from the discussion that followed the talks given by Wang and Esarey. The main topic under debate was the spectrum of the emitted radiation. Particularly, the consequences of the radial dependence of the betatron strength parameter. Some new ideas were brought up during the discussions: (1) to use a beam with small transverse dimension, minimizing the effect of the radial dependence of the betatron strength parameter; (2) to use a precursor beam with considerably larger transverse dimension, therefore creating a channel that would seem uniform to the smaller trailing beam; (3) to use a beam with an asymmetry in the transverse plane (slanted beam), therefore making use of the radial dependence of the radiation spectrum to create a chirped pulse.

The Afterburner Design sub-group considered the PWFA Afterburner design presented by Katsouleas in the plenary session discussed several issues related to the design of such an experiment. These included: (1) the possibility of increasing the total number of particles in the beams, in order to avoid using half of the present SLC beam to accelerate the other half, resulting in a smaller luminosity that would have to be compensated by smaller transverse beam dimensions; (2) the need of a careful parameter study in terms of energies of drive and accelerated beams, plasma density and length, absolute gradients, and transformer ratio; (3) the background level at the detector due to beam – plasma collisions; (4) the comparison between photoionization and collisional ionization to form the plasma; (5) the necessity of a unified treatment for the alignment, corrections and beam – plasma instabilities in the afterburners and interaction region.

ACKNOWLEDGEMENTS

We would like to thank the working group for their many excellent presentations and collegial interactions that made the duty of chairing this session a pleasure. Work supported by USDOE.

Summary Report of Working Group 6 Millimeter-Wave Sources

Gregory S. Nusinovich

Institute for Plasma Research
University of Maryland
College Park, MD 20742

Steven H. Gold

Beam Physics Branch, Plasma Physics Division
Naval Research Laboratory
Washington, DC 20375-5346

Abstract. The focus of the Working Group on Millimeter-Wave Sources was the development of high-power amplifiers operating at Ka-band and above, including both discrete sources and two-beam accelerator concepts. A comparative review was made of the present status of millimeter-wave source research as well as of the potential for future progress. In addition, recent progress in the development of electron guns and rf pulse compression technology was reviewed.

INTRODUCTION

The focus of Working Group 6 was to carry out a comparative review of the present status of high-power millimeter-wave amplifier tube development, as well as the potential for future advances. Consideration was given both to the development of discrete rf sources, and to two-beam accelerator concepts. In addition, the development of active and passive rf pulse compressors was considered, as well as advances in electron gun technology. In order to consider these areas, a number of sessions were devoted to the presentation of research results by members of the Group, while other sessions were devoted to discussions of important issues related to high-power millimeter-wave source development. The efforts of WG6 were coordinated with those of WG3 (RF Structures). At the start of the group sessions, WG3 gave WG6 some guidance on the requirements for future millimeter-wave colliders as well as other accelerators for medical or industrial applications. At midweek, a joint session was held dealing with two-beam accelerator concepts, and a second joint session dealing with rf breakdown issues. The first of the joint sessions will be summarized in this report, and the second in the report of WG3. In the last working group session, WG6 provided feedback to WG3 on the status of ongoing projects and the potential for future development of millimeter-wave sources.

"Millimeter-Wave Sources" is one of the Working Groups of this Workshop because it is believed that structure-based accelerators can be developed with higher accelerating gradients at higher frequencies. For instance, Wilson has considered colliders at center-of-mass energies of 1, 5 and 10 TeV that operate at frequencies of 11.4, 34.3, and 91.4 GHz with gradients of 77, 225, and 500 MV/m [1]. Such colliders would require increasing large rf powers per unit length (96, 530, 1700 MW/m), but for progressively shorter pulse lengths (360, 80, 16 ns). This optimistic scaling with frequency ignores other effects that may limit achievable gradients, such as pulse heating of the structures. Nevertheless, it seems likely that the future of colliders will be at higher frequencies. Developing sources, perhaps combined with rf pulse compressors, to power such colliders is a major challenge.

Working Group 6 began with an introductory talk by Nusinovich. It had sessions dealing with source development (Ka-band and W-band), rf pulse compressors, two-beam accelerators (jointly with WG3), advances in electron guns, and miscellaneous subjects, including beam halos in relativistic klystrons, photonic band-gap cavities, and the resonant absorption instability. There were also discussions dealing with fast-wave device scaling, slow-wave device scaling, the comparison of two-beam accelerators (TBAs) to discrete sources, guns and cathodes, and a summary discussion: "Where are we today, and where should we go?"

PRESENTATION SUMMARIES

Rf Sources

The development of millimeter-wave sources for future accelerators is just beginning, as shown in Table I. Demonstrated results range from 7 to 17 GHz, programs in progress range from 11.4 to 91 GHz, and designs and projections cover the range of 34 to 91 GHz at higher power or efficiency. The presentations covered principally the middle category, progress reports on currently funded projects.

Nation of Cornell University has carried out experiments on X-band relativistic traveling-wave tube (TWT) amplifiers for a number of years, and reported that his best results were 60-120 MW at 9 GHz with 54% electronic efficiency. An important conclusion was that the HEM_{11} mode was not significantly excited in his devices, even when the TM_{01} and HEM_{11} dispersion curves overlapped. He recently changed his focus to 35 GHz, and presented some preliminary results from a relativistic TWT driven by a 900-keV, 100-A electron beam. This TWT has a short dielectric stage and 80-cell disk-loaded $\pi/2$-slow-wave structure separated by a sever of microwave absorber. It has thus far achieved 2 MW with 30-dB gain, and in simulations produces 35-70 MW at 40% efficiency.

Lawson of the University of Maryland (UMD) has carried out a number of gyroklystron experiments at ~8.5 and 17 GHz. He has previously reported 75 MW at 8.6 GHz with 32% efficiency from a fundamental-harmonic 3-cavity coaxial gyroklystron, and an experiment in progress is designed to produce 80 MW at 17.14 GHz with 32% efficiency from a second-harmonic version of the device. At this workshop, Lawson

TABLE 1. Source Development.

Demonstrated:	
INP 7-GHz magnicon	55 MW @ 56% efficiency
UMD 8.5-GHz gyroklystron	80 MW @ 33% efficiency
Cornell 9-GHz relativistic TWT	~100 MW, 54% efficiency (single shot)
SLAC 11.424-GHz PPM klystron	75 MW @ 55% efficiency
UMD 17-GHz gyroklystron	30 MW @ 32% efficiency
HRC 17-GHz klystron	26 MW @ 49% efficiency
Experiments in progress (design values):	
NRL/Omega-P Inc. 11.4-GHz magnicon	60 MW @ 60% efficiency
UMD 17-GHz gyroklystron	80 MW @ 30% efficiency
Omega-P Inc. 34-GHz magnicon	45 MW @ 45% efficiency
CCR 91-GHz gyroklystron	10 MW @ 37% efficiency
Cornell 35-GHz relativistic TWT	35-70 MW @ 40% efficiency
FMT 34-GHz MPG klystron	150 MW @ 53% efficiency
SLAC 94-GHz "klystrino"	100 kW @ 38% efficiency
SLAC 91-GHz sheet-beam klystron	1 MW @ 49% efficiency
Designs and Projections:	
UMD 34-GHz gyroklystron	55 MW @ 37% efficiency
Omega-P Inc. 34-GHz magnicon	100 MW @ 55% efficiency
MRC 34-GHz coaxial RKA	160 MW @ 40% efficiency
UMD 91-GHz gyroklystron	100 MW @ 35% efficiency

reported on the design of a 34.272-GHz, 4-cavity second-harmonic coaxial gyroklystron designed to produce 55 MW at 37% efficiency, using a 500-keV, 300-A electron beam with a velocity pitch ratio $\alpha \sim 1.5$. The design is zero-drive stable, with a gain of 47 dB. Lawson has designed a double-anode magnetron injection gun (MIG) to produce the required electron beam, and carried out microwave circuit simulations.

Nusinovich of the University of Maryland reported on a study of the design of a fourth-harmonic, frequency-quadrupling gyroklystron at 34.3 GHz. Its advantage is the possibility of using the existing Maryland electron gun, solenoid (<6 kG), and X-band driver to produce multi-MW Ka-band radiation. Its design values are 40 MW at 16% efficiency with 40-dB gain. The low efficiency is a result of operating in the fourth cyclotron harmonic.

Nezhevenko of Omega-P, Inc. reported on the development of a frequency-tripling magnicon amplifier that is designed to produce 45 MW at 45% efficiency with 55-dB gain and 1.5-µs pulse length. It consists of a drive cavity, three gain cavities, and two penultimate cavities, operating in rotating TM_{110} modes at 11.424 GHz, followed by a 34.272-GHz output cavity operating in a rotating TM_{310} mode. The electron gun is designed to produce a 500-keV, 200-A electron beam with a diameter of less than 1 mm from a 4.4-cm-diam. cathode, for an area compression ratio of 3000. A superconducting solenoid will provide a magnetic field of 13 kG in the deflection system, increasing to 23 kG at the output cavity. The modulator has been delivered and tested, the electron gun has been delivered, the superconducting magnet is expected in July, 2000, and the collector and rf circuit are in fabrication. This experiment is scheduled for assembly and initial operation early in 2001.

Yakovlev of Omega-P, Inc. described the development of a new physical model and computer code for magnicon simulation. Previous 3-D models and codes included realistic rf field distributions and DC magnetic field profile, as well as finite beam size effects. These codes accurately modeled previous magnicon experiments, including the performance of the 7-GHz Budker Institute of Nuclear Physics (INP) magnicon experiment. The new magnicon simulation code models the electron beam dynamics in the presence of space charge.

The micropulse gun (MPG), developed by Mako of FM Technologies (FMT), is an rf cavity that uses secondary emission to produce a beam bunched at the rf frequency. It operates by a resonant build-up of charge between the rear surface of the cavity and a semi-transparent emitting screen at the output of the cavity. The MPG offers long lifetime, high current density, and immunity to poisoning by exposure to air. The MPG beam can be post-accelerated and magnetically compressed to drive a klystron-like interaction in an output cavity operating at an integer multiple of the MPG frequency. Mako discussed a frequency-tripling 34-GHz MPG klystron. It is designed to produce 150 MW at 34.2 GHz from a post-accelerated 727-keV electron beam at 60-dB gain and 53% system efficiency, but would require 50 MW of rf power at 11.424 GHz from a SLAC klystron to drive the MPG. Mako reported that the X-band TM_{020}-mode MPG cold testing was complete, and the prototype MPG was ready for brazing. Len of FMT reported on a 91-GHz "Gatling" MPG-driven rf source. The 91-GHz design employs an 11.424-GHz MPG operating in a rotating TM_{110} mode that emits sequentially from 8 equally-spaced azimuthal locations to produce a beam bunched at 91 GHz. The beam would be post-accelerated to ~650 keV before injecting into the output cavity to excite the TM_{020} mode at 91.4 GHz. The point design produces 1 MW of output power at 51% efficiency.

Ives of Calabazas Creek Research (CCR) reported on the design of a 91-GHz frequency-doubling gyroklystron amplifier that uses a 500-keV, 55-A, α=1.6 electron to produce 10 MW at 37% efficiency with 55-dB gain. The design builds on the work at the Naval Research Laboratory (NRL) to develop W-band gyroklystrons for radar applications, as well as on the work at UMD to develop high peak power gyroklystrons at lower frequencies for accelerator applications. The 5-cavity circuit operates at a magnetic field of 27 kG and uses the TE_{021} mode in the second harmonic buncher and output cavities. Ives hopes to have initial results within a year.

Scheitrum of SLAC reported on his work on "klystrinos," modular PPM-focused klystrons that could be stacked in blocks to produce multiples of the single-tube output power. These klystrinos are designed to use 110-keV, 2.4-A electron beams passing through 6 cavities separated by 800-µm-diam. drift tubes, with a 5-gap output cavity, and would produce 100-kW peak power, 1-kW average power, and 40-dB gain. The circuit will be manufactured by a LIGA process, which produces a ±2-µm accuracy combined with good surface finish. (A more ambitious 91-GHz sheet-beam PPM-focused klystron variant is also under investigation at SLAC by Colby and coworkers. It would produce 1 MW at 63-dB gain with 49% efficiency. The 15-A, 140-keV beam would be transported through a 0.8x7.2-mm channel. Colby noted that the fabrication tolerances are ~1 µm RMS.)

Hirshfield of Omega-P, Inc. described the design of a novel 8th-harmonic gyroharmonic converter. In this device, 28 MW of X-band power at 11.424 GHz would be used to accelerate and spin up an injected 480-keV, 40-A beam in a TE_{111} cavity. The device would produce more than 10 MW of power at 91.392 GHz using a TE_{811}

output cavity. The design included the electron gun, the magnetic system, the rf-cavity system, the mode converter to convert the output wave to a Gaussian beam, and the beam collector.

Pulse Compressors

As accelerators are pushed up in frequency to achieve higher sustainable accelerating gradients, the required power per unit length is increased, while the corresponding rf pulse lengths are reduced. The increasing need for microwave pulse compressors grows out of these twin facts. Either higher power, shorter pulse rf sources must be developed, or lower power, longer pulse sources combined with rf pulse compression. In general, with the maximum desirable voltage of modulator-driven sources ~500 kV, the maximum beam perveance for most types of device ~1-2x10^{-6} A/V$^{3/2}$, and the maximum efficiency ~50%, the single-tube power is limited to ~100 MW. Moreover, the energy lost in modulator rise and fall times limits the efficiency of sources with pulse lengths much below 1 µs. However, even at 11.4 GHz, ~1 GW is needed per accelerator rf feed, with pulse length ~300 ns.

Nantista of SLAC described the work under way on the passive pulse-compressing power distribution system for the Next Linear Collider (NLC), called the multimode delay-line distribution system (DLDS). In one version of this concept, one solid-state modulator would power 8 pairs of 75-MW, 1.5-µs klystrons. The multimode DLDS system would then sequentially provide 600-MW, 375-ns pulses to four separate accelerator feeds, for an effective power multiplication of 8x. These four accelerator feeds might each be separated by eight other feeds powered by other DLDS sections, so that a cluster of 9 DLDS systems containing 72 klystrons would power 36 sequential sections of the accelerator. The power distribution to the four feeds of a single DLDS section via TE_{01} and TE_{12} modes is controlled by the phase of the 8 klystrons, and makes use of a variety of newly developed overmoded waveguide components and delay lines, including 8-port devices such as the "cross-potent super hybrid," to combine the power at the separate accelerator feeds. These complicated components are being designed on computer codes, fabricated, and tested.

Active pulse compressors offer the possibility of achieving higher compression ratios and higher efficiencies. For instance, a version of the present SLED2 pulse compressor at SLAC could achieve 84% efficiency with active coupling irises versus the present 64% with fixed coupling irises. This active version of SLED is known as SWIRL (switched resonant delay lines). Higher compression ratios, e.g. 16-20x, might also be possible, making possible the use of fewer, longer-pulse klystrons or other rf tubes. Moreover, active pulse compressors should make possible shorter compressor lengths (i.e., fewer 100's of km of waveguide for a full-scale collider). Thus, the potential is to significantly lower the cost of a collider.

Two approaches to high-power active pulse compression are under investigation. The experimental work on these pulse compressors is being carried out at 11.424 GHz. Tantawi of SLAC described his work on multi-MW semiconductor "active window" TE_{01}-mode microwave switches that are switched by applying an electric pulse to a PIN/NIP-diode-array active window. The metallic terminals on the window are radial, that is perpendicular to the electric field of the TE_{01} mode. With forward biasing, carriers are injected into the I region, which then becomes a conductor, reflecting the rf signal. As an example, a test switch changed from 600-kW transmission and 1-MW reflection at no bias, to 50-kW transmission and 1.7-MW reflection at 130-V bias. The switching speed is thus far limited by the external biasing circuit. Test switches have been able to produce up to ~15 MW in 150-ns pulses; however, higher power tests resulted in damage to the switch element due to breakdown across the aluminum lines used for biasing. Tantawi is investigating the failure mode for these switches and hopes to extend their range to significantly higher peak powers, higher average powers (perhaps by means of thermally-conductive dielectric coatings, such as diamond), and higher switching speeds.

Vikharev of the Institute of Applied Physics in Nizhny Novgorod and Omega-P, Inc. reported on his work on active Bragg compressors using gas-discharge switches. In this approach, a large TE_{01}-mode rf storage cavity is formed by a Bragg reflector at the input end and a reflector cavity at the output end. The storage cavity is Q-switched in about 10 ns by an electrical discharge through a gas-filled tube that is located in the output reflector cavity. This discharge changes the center frequency of the reflector cavity, allowing the storage cavity to empty through it. In low power tests, this pulse compressor demonstrated a power gain of 11-12x, with a 1-µs drive pulse and a 55-ns output pulse, at an efficiency of 50%. Planned high-power tests using the NRL/Omega-P X-band magnicon as a driver will attempt to compress 10 MW at 1 µs into a 100-MW output pulse. Cold-test versions of a two-channel active Bragg compressor have also been built, with a 9.4-GHz device producing a 3.4-MW, 52-ns output pulse at 21x power gain and 44% efficiency, using a 160-kW, 2.5-µs drive pulse from a magnetron. In these tests, phase control was demonstrated by showing that the output of the two arms added constructively to produce four times the power of a single arm. A high-power 11.4-GHz version of the two-channel compressor is also planned, which would produce a 500-600 MW, 50-ns output pulse at 10-15x power gain, using a 50-MW, 1-µs drive. Efficiencies as high as 60-65% are predicted with a modified plasma switch. A high-power quasioptical 3-dB hybrid coupler is also under development, to connect to the two-channel compressor.

Two-Beam Accelerators

A session on two-beam accelerators was moderated by Gold. He pointed out that the history of the current two-beam accelerator concepts could be traced back to the work of Sessler, who presented his proposal at the first Workshop on Laser Acceleration of Particles in 1982. Versions of this concept are under development by CERN (the CLIC concept), by Lawrence Berkeley National Laboratory and Lawrence Livermore National Laboratory (LBL/LLNL), and by Argonne National Laboratory (ANL).

In the two-beam accelerator concept, a lower voltage, high-current electron beam runs parallel to the main accelerated beam, and is used to generate rf to feed the high-gradient accelerator cavities. The first beam transits many extraction cavities, producing rf at the feed of successive stages of the collider, thus eliminating the need for large numbers of discrete sources to drive the collider cavities. Moreover, matching the pulse length of the first beam to the requirements of the collider also eliminates the necessity of rf pulse compression.

Pitthan of SLAC reported on the CERN 30-GHz two-beam accelerator concept, known as CLIC. In this concept, a drive-beam accelerator powered by 250 50-MW, 937-MHz klystrons produces a 7.6-A, 1.24-GeV, 92-µs bunched electron beam. The electron bunch spacing is decreased by a factor of 32 through pulse stacking employing a 2x delay ring and two 4x combiner rings, ultimately producing 22 244-A, 1.24 GeV, 130-ns drive beams. The resulting train of pulses is transported along the main accelerator beam in a counterflow pattern. Each of the 22 beams sequentially powers ~625 m of the main accelerator, consisting of 1000 30-GHz accelerating structures, at 150 MV/m by means of 500 30-GHz transfer structures that progressively decelerate the 1.24 GeV beam down to a final energy of 124 MeV while generating the required rf power.

Pitthan also carried out a cost comparison of an 11.424-GHz two-beam collider (of the basic CLIC design) with a similar collider built with discrete sources. While the cost elements are different for the two approaches, the overall capital costs are very similar, with no clear cost advantage to the two-beam approach. Since overall collider wall-plug efficiencies are also comparable (~8%), predicted operating costs would also be similar. However, the two-beam approach seems more favorable at higher frequencies, where discrete sources and components, including pulse compressors, are increasingly difficult to develop.

The CLIC test facility has reported some results from a 1-m-long, 15-mm-diam. power extraction and transfer structure (PETS). The PETS produced 90 MW at 30 GHz in a 15-nsec pulse, using 48 5-nC bunches at 62 MeV. Using the 30-GHz rf power, they achieved a 290 MV/m accelerating gradient in their acceleration cavities without breakdown; however, breakdown was observed at higher gradients.

Houck of LLNL reported on the LBL/LLNL two-beam accelerator concept, in which a 10-MeV drive beam would undergo successive rf extraction and reacceleration using induction modules, in order to power a 300-m section of a collider. A 3-TeV e^+e^- collider would employ 76 units on a side, each of which would provide 360-MW, ~360-ns pulses of rf power to 150 rf output structures. This would produce an unloaded gradient of 100 MV/m in the main linac. This program has previously reported the generation of collider-scale drive beams in induction linacs, the extraction of more than 250 MW at 11.4 GHz from a single structure, the reacceleration of a bunched beam, and power extraction from multiple structures. Recent work has focused on suppressing the exponential growth of the beam-breakup (BBU) instability by such means as phase-mix damping and control of the betatron phase advance between cavities. The long-range goal is to build a smaller prototype TBA at LBNL. It would operate at lower energy (2.5-4 MeV), use a shorter extraction section (10-20 structures), but test all pertinent elements of the TBA concept.

There is also a two-beam concept using dielectric cavities that is under development by Gai of ANL. They produced ~4 MW at ~7.8 GHz, and transferred >90% of the power into the second (accelerating) cavity, with an inferred gradient of ~10 MV/m.

Haimson of Haimson Research Corporation (HRC) described a 17-GHz linac 4x power amplifier project, in which the 17-GHz accelerating structure is part of a resonant ring circuit. It is planned to use this power amplifier in conjunction with a 75-MW, 17-GHz second-harmonic gyroklystron which is currently under development at UMD. This will produce a circulating power of 280 MW, and will allow one to reach accelerating gradients of up to 200 MV/m, which is important for studying the issues associated with rf breakdown at 17 GHz as well as for particle acceleration.

Guns and Cathodes

True of Litton Electron Devices discussed recent advances in Pierce gun and magnetron injection gun design. One important advance has been the realization that emission from the edge of cathodes, both in Pierce guns and MIGs, can substantially degrade beam quality, and the development of new means to suppress this emission. As a result, the discrepancy between the beam quality predicted by electron optics codes for MIGs and the lower measured beam quality has been reduced, and recent gyroklystron experiments with ultralow velocity spreads are now performing in good agreement with theoretical predictions. True also discussed the Hobetron, a novel current-regulating switch tube for modulator applications that uses a hollow-beam electron gun. True also described the recent development of advanced gun design codes that are capable of designing 3-dimensional structures that lack cylindrical symmetry, such as multi-beam klystron guns.

Fortgang of LANL discussed his characterization of an 8" thermionic dispenser cathode. An important goal was to minimize the heater power and maximize the cathode lifetime by having a uniform temperature profile. The cathode had two separate heater elements, and temperature profiles were measured for various combinations of filament power. In addition, the cathode emission was studied as a function of space, time, and temperature.

Schächter discussed the work at Cornell University on ferroelectric guns. A gun was developed that produced 200-A electron beam at 450 kV for ~250 ns with no sign of impedance collapse. The beam was used for amplification in X band in a single stage TWT amplifier. It appears that this cathode has a reasonable lifetime, when used at modest current densities (~30 A/cm^2), but was only tested to ~20,000 shots.

Ives reported on his work on multiple-beam klystron guns, in which the cathode emits a number of separate beamlets that traverse the microwave tube in separate beam tunnels. This reduces the space-charge forces, leading to higher-current, lower-voltage designs that offer higher efficiency than conventional klystrons. Singly and doubly-convergent designs have been studied, that take into account beam area convergence, magnetic compression, beam location, and cathode loading.

Miscellaneous Topics

Shapiro of MIT discussed the use of photonic bandgap (PBG) cavities in place of conventional cavities for both accelerators and rf tubes. (This concept was initially proposed by Kroll of Stanford several years ago [2].) A TM_{01}-like mode can be formed by a two-dimensional triangular lattice of metal rods with a defect in the center, and coupling can be achieved by removal or partial withdrawal of some of the rods. Such cavities are substantially larger than conventional cavities, for a particular frequency, and may offer lower pulsed heating effects by proper selection of the metal rod dimensions. Measurements were reported for a 17-GHz PBG cavity.

Chen of MIT described his theoretical study of halo formation in high-power electron beams. His simulation results were in good agreement with the experimental observations of electron beam loss in the SLAC 11.424-GHz PPM klystron. These halos may contribute to breakdown effects in the klystrons. Chen derived electron confinement criteria that may lead to new means to control beam loss in high-power microwave sources.

Schächter of the Technion discussed a new instability, called the resonant-absorption instability, that occurs when a space-change wave oscillates at a frequency close to a resonance of the medium in which the electron beam propagates. The growth rate is directly related to the attenuation of a pure electromagnetic mode in the medium. For an electron beam propagating through a gas, resonant absorption in the gas can cause the spatial growth of a space-charge wave on the electron beam. Schächter proposes that this instability can be used for the amplification of millimeter-wave radiation or for particle acceleration.

DISCUSSION SUMMARIES

<u>Fast-Wave Device Scaling, moderated by Lawson.</u> The present focus of fast-wave devices for accelerator applications is on gyroklystrons. Lawson discussed the results of scaling studies of gyroklystron design as a function of frequency. This scaling begins with the electron gun, typically a double-anode MIG with adiabatic magnetic compression. Scaling laws for this type of gun have been previously published [3]. So long as limits on magnetic compression, peak electric field, and cathode loading are not reached, MIG peak-power scaling at constant voltage is proportional to wavelength. The beam power can be increased by increasing the voltage, increasing the electric field, increasing the cathode size, increasing the cathode current density (which may require new cathode materials), or changing the cathode geometry, such as by using an inverted MIG. To scale the rf circuit, all cavity dimensions are scaled with wavelength to scale the resonant frequencies, drift tube dimensions are scaled with wavelength to allow passage of the electron beam, and cavity quality factors are scaled with frequency to maintain the relative start-oscillation operating point. If the drive power is also scaled with wavelength, the circuit gain and efficiency are maintained. The increased Q required at higher frequencies also indicates that the output-cavity shape must be modified, and at some point the diffraction Q may approach the ohmic Q, reducing the overall efficiency. Gyrotron oscillator scaling has traditionally taken advantage of very high-order cavity modes to scale to high peak powers, but multicavity gyroklystrons are limited by the requirement to isolate the bunching cavities. Gyroklystron scaling to high power has taken advantage of coaxial cavities to decrease the mode density. Following these approximate scaling laws, a 105-MW, 17-GHz frequency-doubling gyroklystron would scale to 55 MW at 34 GHz and 20 MW at 91 GHz. In each case, the efficiency would be ~35%.

During this discussion, Gold pointed out that cyclotron autoresonance masers (CARMs) have some potential advantages compared to gyroklystrons, since they can achieve comparable efficiencies (~40%) at much lower values of beam α (~1 vs 1.5), provided that good bunching can be achieved. The necessity to operate at high α decreases the threshold for parasitic mode excitation throughout a gyroklystron, and also degrades the beam quality. However CARMs are sensitive to axial velocity spread, and can only achieve high efficiency for very low values of spread (~1-3%). It is not yet clear whether an effective CARM-klystron configuration can be designed.

Another fast-wave device that has potential relevance to accelerator applications is the ubitron, or low-voltage free-electron laser. A design was published for an X-band ubitron [4], but extending this upwards in frequency to 34 GHz may be problematic at 500 keV because of the smaller wiggler period required.

Slow-Wave Device Scaling, moderated by Symons. Symons of Litton Electron Devices presented some scaling laws for klystrons. These laws, based on wavelength scaling, transit angle scaling, and cavity scaling yield quite different predictions for the dependence of the output power on the wavelength. For instance, assuming that a klystron is scaled in wavelength at constant perveance and constant electron beam current density, with the product of beam radius and electron wavenumber held constant, leads to the relationship $P \propto \lambda^{10}$ for constant-transit-angle scaling (in the non-relativistic limit). However, keeping the ratio of beam radius to wavelength constant at constant perveance and beam current density leads to the relationship $P \propto \lambda^{10/3}$. An empirical relation governing the limiting efficiency of conventional klystrons is given by Symon's Law: $\eta = 0.9 - 0.2 k_\mu$, where k_μ is the electron-beam microperveance. Symons also discussed some possible configurations for coupling the klystrons to accelerating structures.

Gold pointed out that one way to beat the typical scaling of klystron power with wavelength, and klystron efficiency with perveance, is the intense-beam relativistic klystron amplifier (RKA), using a mildly relativistic annular beam. This approach was pioneered by Friedman at NRL [5]. In his work, operation close to the space-charge limit enhances the bunching effects. These devices have produced gigawatts of power at ~1 GHz. Pasour of Mission Research Corporation (MRC) is now developing a coaxial version of this device in X-band using a carbon-fiber cathode. (This device is referred to as "triaxial," since the annular electron beam is spaced between center and outer conductors.) It is designed to operate in the range of 300-400 keV at 3-4 kA, in order to produce in the powers in the range of 300 to 800 MW at 40-50% efficiency in a 1-µs pulse. Simulations of a Ka-band version of the device suggest powers of 160 MW at 40% efficiency from a 1-kA, 400-keV device.

Guns and cathodes, moderated by Nusinovich. Ferroelectric-cathode experiments are apparently dominated by emission from triple points in the mesh electrode deposited on the ferroelectric material, and long-pulse operation suffers from non-constant impedance that indicates the presence of plasmas. Field-emitter arrays (Spindt cathodes) seem to be practical only at low current densities, and high-current density field-emission cathodes also suffer from plasma formation. It appears that the best path to increasing current densities may be the use of oxide cathodes, which have been replaced in most applications by tungsten-matrix dispenser cathodes. According to Scheitrum, "old-fashioned" oxide cathodes can furnish current densities inversely proportional to the pulse length, for pulses on the order of nanoseconds up to a few microseconds, and can produce ~100 A/cm² for 1-µs pulses. According to Ives, field-emission cathodes are being used in a prototype low-power C-band traveling-wave tube at Northrop-Grumman, in place of a thermionic cathode (90 mA at 3.5 keV demonstrated from 1-mm-diam. surface containing 50,000 molybdenum tips, with maximum design value of 160 mA at 5 keV (20 A/cm² current density) [6]). However, the use of field-emitter arrays in high-peak-power devices is generally accompanied by plasma generation.

Pulse Compressors, moderated by Tantawi. Tantawi of SLAC described the various rf pulse-compression techniques, including both passive and active approaches. The passive approach uses rf-phase control, at the feed to the high-power klystrons, to control the pulse compression, while active approaches directly change the Q of rf pulse-storage elements. Three passive approaches have been investigated at SLAC, binary pulse compressors, the SLED approach with resonant cavities or resonant delay lines, and the Delay Line Distribution System (DLDS). Binary pulse compression has 100% intrinsic efficiency, but involves very long delays lines, e.g. ~1000 km of waveguide delay lines for a TeV collider. Compared to this, the SLED approach is relatively compact (only 100's of km of delay lines; however it has lower intrinsic efficiency, and appears to be impractical for power multiplication beyond a factor of 6-8. The DLDS is also relatively compact and has high intrinsic efficiency, but employs a complicated topology, involving overmoded components. Its rigid design is almost impossible to upgrade (for instance, to retrofit a collider for higher center-of-mass energies.) A fourth approach that is frequently used for short-pulse lasers, involving compression of chirped pulses, does not seem applicable to accelerators.

Active compressors can potentially produce power multiplication factors of 15-20x at high efficiency. A question was raised concerning the potential efficiency of active compressors using gas-filled discharge tubes to Q-switch the rf storage cavities, due to the power loss in the discharge tube. It was pointed out that the efficiency reported by Vikharev for these compressors was dominated by reflection loss, in filling the cavity, and wall losses, and that the losses in the switch region were only several percent. There seems to be no fundamental reason that the efficiency of discharge-switched active pulse compressors, such as the active Bragg compressor, cannot exceed 80%.

Panel Discussion: "Where are we and where should we go?", moderated by Gold. The invited members of this panel were Symons, Scheitrum, Schächter, Lawson, Nezhevenko, and Tantawi. However, other members of the group were also active participants in the discussion. These discussions covered the recent progress in the development of specific microwave sources, in the development of guns and cathodes, and in the development of microwave pulse compressors, as well as the future potential to develop millimeter-wave drivers for TeV-scale linear colliders. Some details of this discussion are summarized in the Conclusions.

CONCLUSIONS

Substantial progress has been made in the past two years. New high-frequency source designs and experiments, including klystrons, gyroklystrons, and magnicons, routinely exceed the naive Pf^2=constant scaling that might be expected based on the decreasing size of rf circuits for a given operating mode. A number of techniques have been used or are available to continue to push the high-power limit of high-frequency amplifier tubes. These means include use of multiple beams, annular or sheet electron beams, coaxial structures, overmoded output circuits, quasioptical circuits, photonic band-gap cavities, higher current density cathodes, diamond surface coatings, high-conductivity dispersion hardened copper cavities to minimize pulse-heating effects, and possibly Doppler-shifted devices, including CARMs and ubitrons. However, some fundamental limits remain, including pulse heating damage to circuits (particularly accelerating cavities as well as output stages of millimeter-wave sources), rf breakdown, mode competition, cathode loading, and problems with beam transport in increasingly tiny structures. The next two years should produce new results from a variety of promising millimeter-wave amplifier experiments, as well as new results from pulse-compressor research.

ACKNOWLEDGMENTS

We are grateful to the Department of Energy for their support of this work, and to the Working Group 6 participants for generously sharing their research results.

REFERENCES

1. Wilson, P.B., "Scaling Linear Colliders to 5 TeV and Above," SLAC-PUB-7499 (1997).
2. Kroll, N., Smith, D.R., and Schultz, S., in *Advanced Accelerator Concepts 1992*, edited by J. Wurtele, AIP Conference Proceedings 279, New York: American Institute of Physics, 1993, p. 197.
3. Lawson, W., *IEEE Trans. Plasma Sci.* **16**, 290-295 (1988).
4. Balkum, A.J., McDermott, D.B., Phillips, R.M., Lin, A.T., and Luhmann, N.C., Jr., *IEEE Trans. Plasma Sci.*, **24**, 802-807 (1996).
5. Friedman, M., Krall, J., Lau, Y.Y., and Serlin, V., *Rev. Sci. Instrum.* **61**, 171-181 (1990).
6. Whaley, D. R., Gannon, B., Smith, C.R., Armstrong, C.M., and Spindt, C.A., *IEEE Trans. Plasma Sci.*, June, 2000 (in press), and D. Whaley, private communication.

Summary of Working Group 7 on "Exotic Acceleration Schemes"

T Tajima

Lawrence Livermore National Laboratory, Livermore, CA 94551

Abstract.
Exotic concepts of advanced acceleration technologies have been explored by Group 7 under the leadership of T. Tajima and T. Smith (who could not attend) at the AAC. Explored concepts are: (1) proton (ion) acceleration by laser, (2) additional ion acceleration methods, (3) crystal x-rays and acceleration, (4) vacuum acceleration, (5) active medium acceleration, and (6) some advanced methods in laser wakefield. The first subject of laser photon acceleration was discussed jointly with Group 1 and in the end the participants came to an agreement on the mechanism of proton acceleration by laser irradiation.

Participants:

K. Flippo, A. Pakhomov, R. Carrigan, R. Garrett, P. Schoeschow, G. Shuets, D. Umstadter, A. Bojacz, T. Cowan, M. Zolotorev, S. Chattopadhay, J. Rosensweig, D. Cline, Z. Pensa, R. Noble, P. Chen, K. Nakajima, N. Saleh, I. Pogorelsky, J. Hirshfield, P. Channell, P. Musumeci, C. Pellegrini, L. Steinhouse, A. Spitokovtsky, P. Catravas

    I. Proton (Ion) Acceleration (w/Gr. 1)

    Experiments, their analysis, mechanisms, scaling(s), and applications

    II. Additional Ion Acceleration

       (a) Soliton acceleration

       (b) Compact high-current multistage cyclotron

       (c) Neutrino beams

    III. Crystal Acceleration/ x-rays

(a) FNAL A0 Experiment/Channeling

(b) Parametric X-rays

(c) Future issues

IV. Vacuum Acceleration

(a) Ponderomotive acceleration

(b) IFEL

V. Active Media

(a) PASER

VI. LWFA

(a) Colliding laser

I LASER PROTON ACCELERATION

The experimental results of proton acceleration by Lawrence Livermore National Laboratory (LLNL) [by Tom Cowan], Rutherford Appleton Laboratory (RAL) [represented by Cowan], and the University of Michigan [by Don Umstadter and K. Flippo], have been reviewed. In these experiments the parameters of laser and targets, the way the irradiation was conducted, quality and energy of laser, etc. vary widely and the phenomena and effects observed as well as the energy flux, efficiency, the direction of protons etc. also widely differ. It should be noted that in spite of varied parameters the accelerator parameters are already quite impressive: the emittance 2.2π mmmrad (normalized one 0.5π mmmrad), the longitudinal emittance $\Delta E \Delta t \sim$ MeV· ps, based on LLNL results (Cowan). After closer scrutiny at the conference, however, we came to recognize that the underlying mechanism and physics are the same. Leaving the details of observed phenomena and experimental setups, here we focus on the common elements and mechanism, from which the above wide variabilities of observed phenomena may be explained from a unified point of view. The basic mechanism of proton acceleration is as follows. The laser irradiation causes the heating of electrons of the solid target. Though the heating mechanism and efficiency vary depending on the laser intensity etc., once the energy of hot electrons is given, the rest of physics is common. These hot electrons form strong electrostatic fields away from the solid surface, which pull protons to high energies.

In the LLNL experiment as well as in RAL, ion acceleration takes place primarily on the non-irradiated side of the solid and in the direction perpendicular to that surface, indicating the electrostatic nature of the field. The energy gain of ions is

$$\mathcal{E}_i = q\ell\, E_0,$$

where q is the ion charge, ℓ the acceleration distance, and E_0 the induced electrostatic field. These take the following values

$$\ell \sim 5 - 10\lambda_D,$$

$$E_0 \sim \frac{T_e}{\lambda_D},$$

where λ_D and τ_e are the Debye length and the temperature/or energy of hot electrons. In the intense laser irradiation (when $I > I_{cr} \sim 10^{18}\,\text{W}(\text{cm}^2)$) such as for the LLNL experiments, the heated electron energy scales as $T_e \propto \sqrt{I}$. Thus, the ion energy gain is

$$\mathcal{E}_i \sim (5-10)T_e \propto \sqrt{I}.$$

On the other hand, the Michigan experiment, which also shows the acceleration from mainly the non-irradiated side, scales differently. When the laser is in the fundamental frequency ($1\,\omega$) with precursor, the electron heating precedes the density pedestal, which leads to $T_e \propto I$ (in their non-relativistic intensity), yielding

$$\mathcal{E}_i = \alpha T_e \propto I.$$

Meanwhile when 2ω is used with much sharper contrast, the electron heating mechanism is that of Brunel, to make $\ell \sim \ell_0$, where ℓ_0 is the wavelength of laser. This leads to the scaling as

$$\mathcal{E}_i = q\ell_0 E_0 = \frac{\lambda_0}{\lambda_D} T_e \propto \sqrt{T_e} \propto I^{1/4},$$

as $T_e \propto \sqrt{I}$ for the Brunel heating.

In earlier French experiments (Fews et al., for example), the ion acceleration takes place from the irradiated side, when a low contrast, short pulse laser was used. This is because the laser energy is small and hot electrons cannot penetrate the target, giving rise to the opposite direction of ion acceleration and the scaling of energy as

$$\mathcal{E}_i \alpha T_e \propto I^\beta \quad \text{with} \quad \beta = 1 \text{ or } 1/2$$

depending on the intensity below or above I_{cr}.

Applications such as the injection to spallation neutron source drift tube, the fast ignition spark, and neutrino sources among others have been suggested.

II ADDITIONAL ION ACCELERATION

A Soliton acceleration driven by laser

Nakajima (JAERI/KEK) introduced a hadron collider scheme based on a laser driven soliton-like wake. According to James Koga's (JAERI) simulation, a short pulse (20 fs) intense ($I = 5\times 10^{19}\,\text{W/cm}^2$) laser injection on a plasma ($6\times 10^{19}\,\text{cm}^{-3}$) induces a soliton-like structure that possesses an accelerating electrostatic field

$\sim 3\,\text{TV/m}$. Nakajima suggests to use denser plasma ($10^{21}\,\text{cm}^{-3}$) targets, from which two opposing colliding ion beams with energy $1\,\text{GeV/u}$ are generated. he estimates the luminosity of $10^{35}\,\text{cm}^{-2}$ per shot is achievable. He further suggests that if the intensity if $10^{23}\,\text{W/cm}^2$, a $1\,\text{TeV} \times 1\,\text{TeV}$ collider at the luminosity of $10^{36}\,\text{cm}^{-2}\,\text{s}^{-1}$ using a 10 Hz laser is possible.

B Multistage, high-gradient proton cavity cyclotron

Hirshfield (Yale) introduced a concept that employs multistages of cyclotron cavities that successively reduce the cyclotron frequency as the beam moves onto an adjacent cavity. He claims that this way much more efficient high-current ion acceleration can be achieved. His case study shows that protons with 1 GeV energy are accelerated with large current ($\sim 100\,\text{mA}$) efficiently ($> 75\%$) in a cw fashion. Good energy applications which require large current should exist.

C Neutrino beams

Pakhomov (Univ. of Alabama, Huntsville) introduced a possible generation of neutrino beams with a very short pulse (40 fs) high fluence, using hypothetically strong laser ($10^{23}\,\text{W/cm}^2$) driven proton beams ($10^{13}\,\text{GeV}$ protons/pulse). As a laser might exist, if, for example, a large energy laser such as an NIF laser arm is employed as a driver for short laser pulses. He estimates about $10^{11}\nu_\mu\,\text{sr}^{-1}$ per pulse. Using these neutrinos, he suggests a possibility of neutrino oscillation experiments.

III CRYSTAL ACCELERATION/X-RAYS

With the motivation that a high-density plasma can sustain proportionally high wakefield (Tajima-Dawson, 1979), Chen and Noble suggested a design of crystal acceleration where the electron density is $10^{22} - 10^{24}$/cc. Carrigan (Fermi) reported the experiment (A0 Channeling Radiation experiment at FANL), where high-energy electrons (20 MeV) are injected (10^{13} electrons/cm^2) in Si crystal channels. The FANL collaboration was able to have electrons channel through the crystal. They reported strong x-ray signal (detector saturated) coming from electron channeling radiation. This is a spectacular experiment, pushing the crystal usage for accelerator physics by an important step. Parsa (BNL) followed this to theoretically discuss pulsed parametric x-ray radiation from a crystal.

IV VACUUM ACCELERATION

Zoloforev (LBL) introduced a concept to use 1 J short pulse (20 fs) laser to accelerate electrons in vacuum. This method produces very short bunches ($\sim 0.03\,\mu\text{m}$)

with excellent emittance (~ 0.01 mmmrad). About $10^5 \sim 10^6$ electrons are accelerated to 5 MeV. This can be a good candidate for an injector.

Musemeci (UCLA) discussed an Inverse Free Electron Laser.

V ACTIVE MEDIA

Bogacz (Jefferson) discussed the concept of PASER scheme via relativistic ion beams pumping. The plenary speaker, Schaechter, discussed the usage of active media (such as He/Ne) to accelerate electrons. Bogacz suggests L_i^{+2} ions that are streaming relativistically, encountering the counterstreaming lower frequency laser that pumps the Li ions, emitting forward up-converted photons that drive electrons. By the way, the utilization of active media for superluminous laser light (for the purpose of accelerator) was suggested by Fisher and Tajima in 1993, which was recently picked up in the renewed interest in superluminous light propagation (Wang et al. in Nature, 2000).

VI LWFA

Shvets introduced colliding laser pulses, one a forward propagating short pulse, the other the train of longer pulse pump lasers (with slightly higher and slightly lower frequencies than the short pulse laser). With this method the wakefield phase can be flipped by π after the short pulse laser exist the first longer pulse laser and when it enters the second longer pulse laser. This way, it may be possible for electrons and the wakefield to adjust their phase slippage. This may become an important technique for a multistaged laser wakefield accelerator, in which the dephasing and pump depletion are crucial issues.

Finally, unrelated to the above issued, Cline (UCLA) discussed the SLC afterburner and mentioned the possibility of making it a $\gamma + \gamma$ collider.

CONTRIBUTED AND INVITED PAPERS

Laser-Plasma Accelerators: A Status Report

C. Joshi

UCLA, Los Angeles, CA 90095

INTRODUCTION

In this paper, the status of the Laser-Plasma Accelerator field within the Advanced Accelerator Concepts area of research is reviewed. In particular, I review the status of the Plasma Beat Wave Acceleration[1] (PBWA) scheme, the Self-Modulated, Laser-Wake Field Acceleration[2] (SMLWFA) scheme and the Laser-Wake Field Acceleration[3] (LWFA) scheme. In all these three schemes, charged particles are accelerated by a relativistic (phase velocity ~ c), space-charge wave excited in a plasma by a photon beam. The wave potential becomes large enough to trap either the background plasma electrons, in a process known as self-trapping,[4] or electrons (positrons) must be externally injected with a certain minimum energy[5] to be trapped and accelerated by this potential. There is a simple scaling law that gives the accelerating electric field of such waves as $E_z(V/cm) \sim \varepsilon(n_e)^{1/2}$ (cm^{-3}) where $\varepsilon = \tilde{n}/n_o$ is the density perturbation associated with the wave and n_e is the initial unperturbed plasma density. The maximum energy gain limited by dephasing is given by $\sim 2\gamma_{ph}^2$ (MeV) where $\gamma_{ph} = (\omega_o/\omega_p)$ is the relativistic Lorentz factor associated with the phase velocity of the wave, ω_o is the laser frequency and ω_p is the plasma frequency.

STATUS REPORT ON THE PLASMA BEAT WAVE ACCELERATION (PBWA)

In the PBWA, two co-propagating laser beams, with slightly different frequencies and wave numbers, beat-excite the relativistic plasma wave if the frequency difference $\Delta\omega = \omega_1 - \omega_2$ is equal to the plasma frequency ω_p. In this case the wave number of the plasma wave is $k_p = \Delta k = k_1 - k_2$ and $\omega_p/k_p \cong c$ with $\gamma_{ph} = (\omega_1 + \omega_2)/2\omega_p$. The physical mechanism for displacing the plasma electrons from their initial position is the, so-called, ponderomotive force, which is proportional to the gradient of the dot product of the electric fields of the two lasers. Since beat excitation is a resonant process, the laser intensity required excite a large amplitude ($\varepsilon > 0.1$) plasma oscillation is relatively modest and the laser pulse can be fairly long.

CP569, *Advanced Accelerator Concepts: Ninth Workshop*, edited by P. L. Colestock and S. Kelley
© 2001 American Institute of Physics 0-7354-0005-9/01/$18.00

Characterizing the laser intensities and the pulse length in the normalized units of the vector potential a = eE/mcω and collisionless skin depth c/ω_p respectively, the PBWA typically requires $a_{1,2} = 0.1$ and $\tau \cong 10$–100 c/ω_p. This translates to $I_{1,2} = 10^{14}$ W/cm^2 and $\tau \cong 100$ ps for a CO_2 laser driver and $\gamma_{ph} \cong 30$ plasma wave.

The PBWA scheme was anticipated in the original Laser Electron Accelerator[3] paper of Tajima and J. M. Dawson. However, many scientists doubted whether such a scheme would work in practice because of the relatively long and intense pulses used to excite the plasma wave. There was worry that competing laser-plasma instabilities would "kill" the scheme. It was not until 1987 when first 2D, PIC code simulations of the PBWA scheme were carried out in the context of a particle accelerator that the potential of this method as an ultrahigh gradient particle accelerator became widely accepted.[1] Figure 1 below shows one result from Reference (1) showing that there was a quantitative agreement between the growth rate of the plasma wave seen using PIC simulations and that predicted by the fluid theory developed by M. N. Rosenbluth and C. S. Liu.[6]

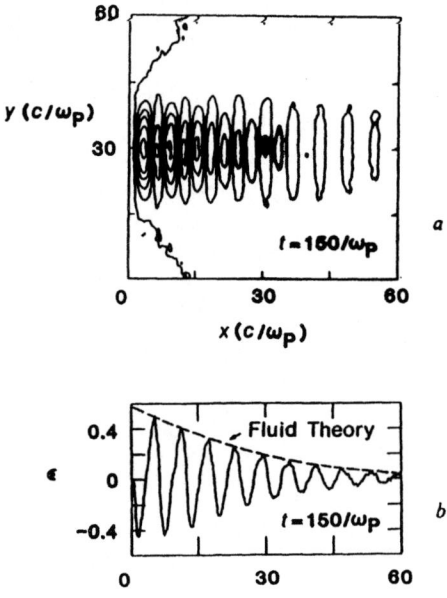

Figure 1. (a) 2D contours of the wave potential in space and (b) the growth of the plasma wave in time as predcted by 2D—PIC simulations of the PBWA (ref. 1).

Reference (1) was followed shortly by an experimental verification of the excitation of relativistic plasma waves by collinear optical mixing[7] by the UCLA group. Figure 2 shows key results from this paper which demonstrated the frequency ($\omega_p = \Delta\omega$) and wave number ($k_p = \Delta k$) relationship for the excited plasma wave; the excitation of the Stokes and anti-Stokes side bands and a proportionate relationship between the amplitude of the first Stokes side band and the amount of Thomson scattering from the plasma wave. It was this paper that gave credibility to the whole Plasma Accelerator field and resulted in many experiments being funded around the world on actual acceleration of electrons using such waves.

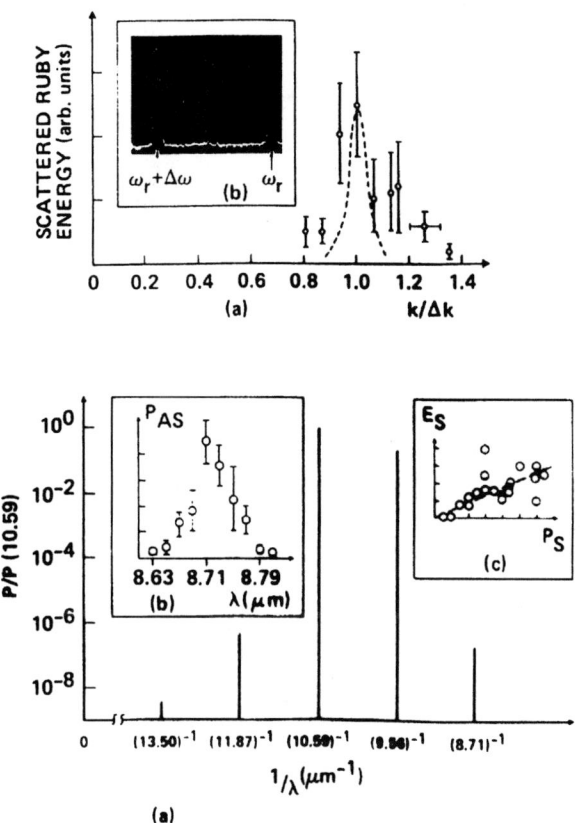

Figure 2. (a) The frequency and wavenumber spectrum of the beat-excited relativistic plasma wave, (b) the stokes and anti-stokes side bands including the anti-stokes spectrum (inset) and (c) linear relationship between thomson scattered light (E_s) and stokes light P_s (Ref. 7).

It was 1993 when first conclusive results were presented by the UCLA group of acceleration and substantial energy gain of externally injected electrons using the PBWA technique.[8] Figure 3 shows key results from that publication. Many plasma physicists doubted that one could conclusively demonstrate acceleration of test particles in a violent plasma environment. So the UCLA group used first a monochromer to momentum select electrons of a certain energy and then an auxiliary cloud chamber to visualize the particles. A secondary orthogonal B field was used to bend the electrons in the cloud chamber and measure their relativistic

Larmor radius to confirm the energy gain. Futhermore, both the electron energy gain and the number of accelerated electrons were found to maximize at the resonant density. (In Fig. 3(b) the horizontal axis is pressure but later this was shown to be directly proportional to electron density).

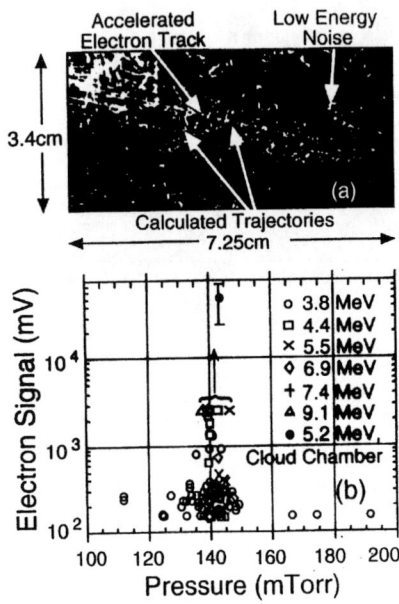

Figure 3. (a) Cloud chamber tracks produced by accelerated electrons and (b) the number of accelerated electron as the gas pressure (plasma density) is varied showing the expected resonance (Ref. 8).

In 1994, the UCLA group showed trapping and acceleration of externally injected electrons by the PBWA.[9] The injected electron energy was 2 MeV while γ_{ph} was 33 so the observation of electron energies of greater than 16.5 MeV was conclusive proof that electrons had been trapped by the wave potential (Fig. 4). Maximum electron energies of 30 MeV were observed implying an acceleration gradient of ~ 2.8 GeV/m as the acceleration length was measured to be less than 1 cm long. Furthermore, loss of electron energy was seen as well as gain. This was to be expected as the wavelength of the plasma oscillation was 300 μm (or 1 ps duration) whereas the injected electron microbunch was 10 ps long. The data shown in Fig. 4 was obtained over many laser shots and as there was a ± 50 ps jitter between the laser and the electron beam, the electron beam sampled plasma waves of greatly different amplitudes from one shot to the next.

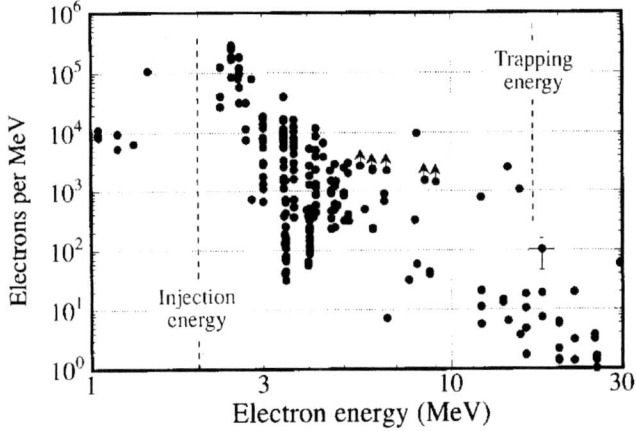

Figure 4. Trapping of externally injected electrons in the beat-excited relativistic plasma wave. (Ref.9)

It had taken 9 years from the time a relativistic plasma wave was first excited using the beat wave technique[7] to do a convincing acceleration experiment[8] mainly because the new Laser Acceleration field turned out to be cross-disciplinary requiring state-of-the-art expertise in lasers, particle accelerators and plasmas. In fact in 1994, the UCLA group proposed a 1 GeV PBWA experiment[10] based on then available 1 μm CPA laser technology, but unfortunately such an experiment was never funded due to lack of resources.

The UCLA group subsequently proposed a more modest 100 MeV PBWA experiment whose goal is to demonstrate acceleration of a significant number of electrons while maintaining a reasonable emittance (< 10π mm mrad) and energy spread (< 10% $\Delta\gamma/\gamma$). This is known as the Neptune project. This meant that a photoinjector linac with an extremely good beam quality had to be built as an injector[11] and the CO_2 laser had to be upgraded to give 1 TW of peak power.[12] The Neptune facility recently successfully finished its construction phase and the experimental team has begun the experiments on the 100 MeV PBWA.

To complete the status report on the PBWA, other notable groups that have contributed in this field are those from U. Osaka (Japan),[13] Imperial College (U.K.),[14] Ecole Polytechnique (France)[15] and Chalk River Laboratory of AEC, Canada.[16] These experiments were carried out using both 1 μm and 10 μm laser pulses and have demonstrated acceleration of self-trapped[13] as well as externally injected electrons.[15,16]

STATUS REPORT ON THE SELF-MODULATED, LASER-WAKE FIELD ACCELERATION (SMLWFA)

Now I will discuss the SMLWFA scheme.[2] In the 1D manifestation of this scheme, the Raman Forward Instability (RFI)[17] plays a critical role in establishing the relativistic plasma wave. In the RFI, the laser beam decays into a forward propagating Stokes wave, an anti-Stokes wave and a relativistic plasma wave. Once the Stokes and the anti-Stokes waves become sufficiently intense they beat with the pump wave to produce a deeply amplitude modulated envelope of the electric field. As in the PBWA scheme the modulation has a frequency of ω_p and wave number $k_p = k_o - k_s$ where "o" denotes the pump and "s" denotes the Stokes wave, respectively.

When the laser pulse is short compared to Z_R/c where Z_R is the Rayleigh length of the focused laser beam, the RFI is in the so-called spatio-temporal regime[18] where the spatio-temporal gain $G = e^g/(2\pi g)^{1/2}$ with

$$g = (a_o/\sqrt{2})(\omega_p/\omega_o)^2(\omega_o/c)\sqrt{(x-\phi)/\phi} \qquad (1)$$

Figure 5 shows the number of exponentiations of the growth G from an initial noise level of $\varepsilon_{noise} \sim 10^{-5}$ as the laser intensity (expressed in units of a_o) is varied for different densities assuming a 1 μm laser. One can see that the number of e-foldings of growth increases first with a_o but subsequently remains rather constant or even decreases for $a_o > 1$). On the other hand, Ln(G) is a strong function of plasma density. There are only 2 e-foldings of growth when $n = 5 \times 10^{18}$ cm^{-3} but this number approaches 12 when the density increases to 1.5×10^{19} cm^{-3}. Thus, at these high densities we expect to see self-trapping of background plasma electrons and indeed wave breaking[19] of the plasma oscillation.

The first paper that pointed out the RFI's role in electron acceleration was by C. Joshi et al.[20] who used a relatively long but intense CO_2 laser pulse ($a_o \sim 0.3$, $\omega/\omega_p \sim 2.2$) to form a thin foil carbon plasma and measured the spectrum of forward and

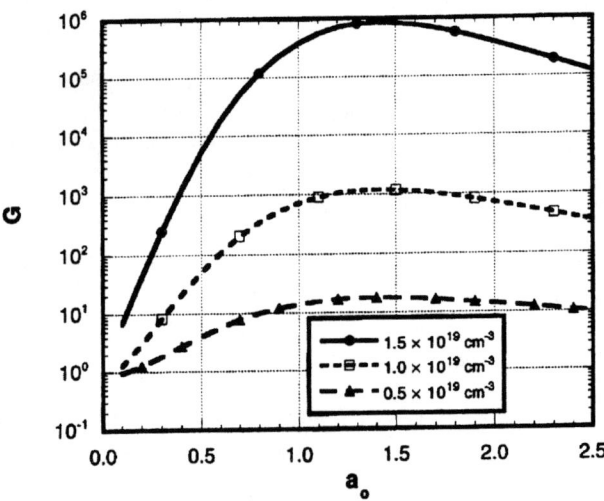

Figure 5. Spatio-temporal growth rate for RFI as a function density and laser power for a 1 μm laser, nominally 800 fs long (Ref. 19).

backward emitted electrons. (Figure 6) Electrons with higher energies (up to 1.4 MeV) were emitted in the forward direction compared with up to 0.8 MeV in the backward direction and were attributed to RFI. Electrons were observed without external injection which means that they came from self-trapping of plasma electrons.

It was not until 1995 that the research on SMLWFA scheme began in earnest. The reason for this being very simple. The CPA technique[21] had by then enabled many groups to have access to TW class 1 μm lasers which were essential for this scheme. In 1993, the LLNL and the UCLA group jointly showed acceleration of electrons via the RFI conclusively using a 1 μm laser.[22] In the RFI, the single frequency pump laser decays into a comb of Stokes and anti-Stokes side bands each frequency shifted by ω_p. In this experiment, the first two anti-Stokes side bands were seen. The amplitude of the first anti-Stokes wave was found to correlate well with the number of energetic electrons observed in the forward direction. Electrons up to 2 MeV energy were seen when an $a_o = 0.9$ laser was used to produce and interact with a 10^{19} cm^{-3} gas jet plasma.

The Livermore-UCLA experiment[22] was soon followed by a much more sophisticated experiment at the Rutherford Appleton Laboratory[23] in the U.K. by the Imperial College, UCLA and Ecole Polytechnique groups. The 1 μm Vulcan laser at the laser facility had the capability of delivering up to 30 TW of power in an 800 fs pulse. Using this laser ($a_o \gtrsim 2$) and a gas-jet plasma target this group was able to see copious fast electron generation via breaking of the Raman Forward plasma wave. The evidence for wave breaking came from the sudden broadening of the comb of satellites in the forward direction (See Fig. 7) as the plasma density was increased

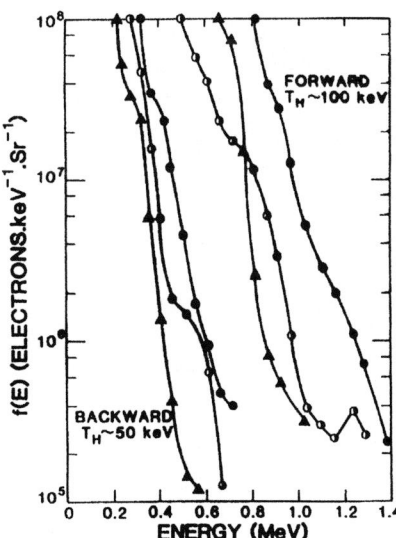

Figure 6. Forward and backward emitted electron spectrum from thin carbon foil plasma irradiated by an intense CO_2 laser pulse (Ref. 20).

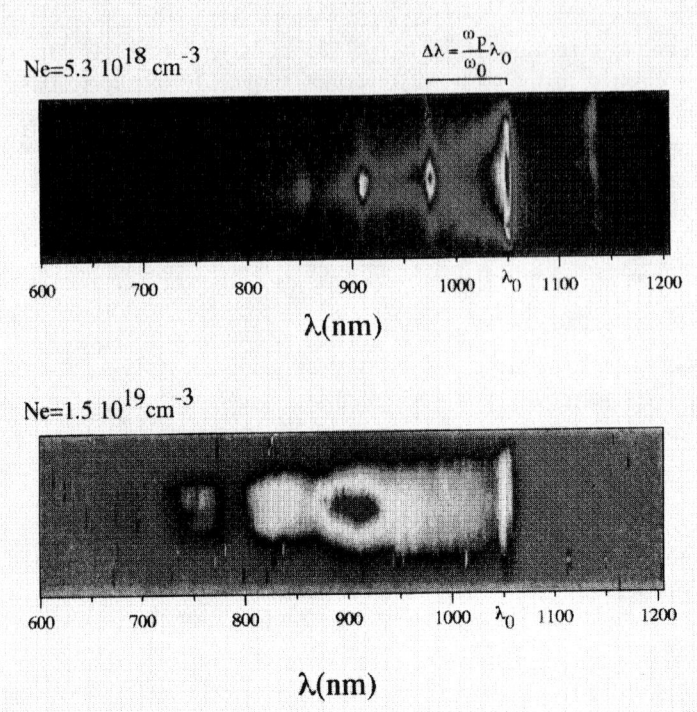

Figure 7. Spectrum of satellites to the pump laser in the forward direction at two different densities in the Rutherford experiment (Ref. 23).

from 5.3×10^{18} cm^{-3} to 1.5×10^{19} cm^{-3}. The spatial extent of the relativistic plasma wave was also directly measured using Thomson scattering of a probe beam[24] as was the spectrum of the relativistic electrons that were escaping the plasma in the forward direction (Fig. 8). From the maximum observed energy of 94 MeV the accelerating gradient was deduced to be greater than 150 GeV/m which represents a record of sorts for the highest gradient terrestrial acceleration of charged particles. Another interesting aspect of this experiment is that the maximum energies observed were greater than those expected from the phase slippage between the electrons and the accelerating electric field of the plasma wave as given by the linear theory for pre-injected electrons.

Many other groups around the world soon experimented with the SMLWFA scheme. Most notably the U. Michigan group[25] and the NRL group[26] both observed the anti-Stokes side bands and electrons. The Michigan group showed that the electrons were emitted in a well-defined beam[25] in the same direction as the laser. A result from the NRL group is shown in Fig. 9 which qualitatively looks similar to the Rutherford results however, what is surprising about the NRL data is that the electron spectrum was found to extend out to 100 MeV at a much lower peak laser power (~ 2 TW), whereas Rutherford experiments were carried out at around 25 TW. The NRL

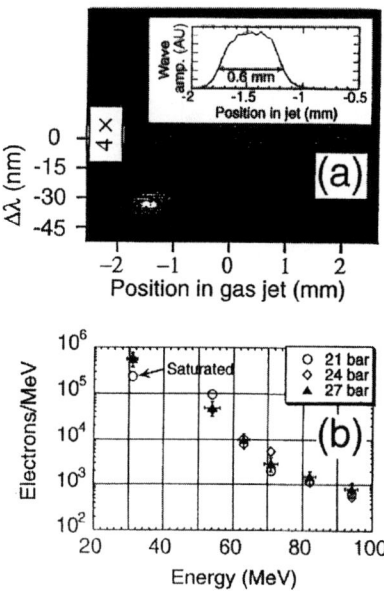

Figure 8. (a) transverse, small angle Thomson scattering showing the spatial extent of the RFS plasma wave and (b) the electron spectrum emitted in the forward direction in the Rutherford experiments (Ref. 20).

group did coherent Thomson scattering measurements on the plasma wave and determined that it lasts for about 30 ps or roughly 100 oscillations before decaying into ion acoustic waves.

An interesting byproduct of the SMLWFA experiments is the observation of relativistic self-focusing and filamentation of the laser beam in the plasma. This is so because the threshold for both Raman Forward and relativistic self-focusing are about the same for n_c/n_e of about 100. In the Rutherford experiments[27] where the ratio of P/P_c was about 20 where P_c is the critical power for whole beam relativistic guiding, a relativistic plasma wave that was about 24 Rayleigh lengths long was observed and presumed to be inside a filament of similar length observed by imaging sidescattered light. The observation of the plasma wave puts a lower bound on the intensity of light inside the filament (Fig. 10) to be around 10^{18} W/cm^{-2}. Experiments at U. Michigan[28] observed the onset of relativistic guiding very close to the theoretically predicted threshold. Furthermore, when a preformed channel with a density minimum on axis was produced using a radially expanding plasma column, this group showed that a second probe pulse could be guided by this plasma fiber.[29] Experiments on preformed plasma fibers have been pioneered by the Maryland group[30] and reproduced in various forms by others such as at U. Texas,[31] LBL,[32] NRL[33] and elsewhere and peak laser intensities of ~ 2×10^{17} W/cm^2 have been transported over ~ 20 Rayleigh lengths.

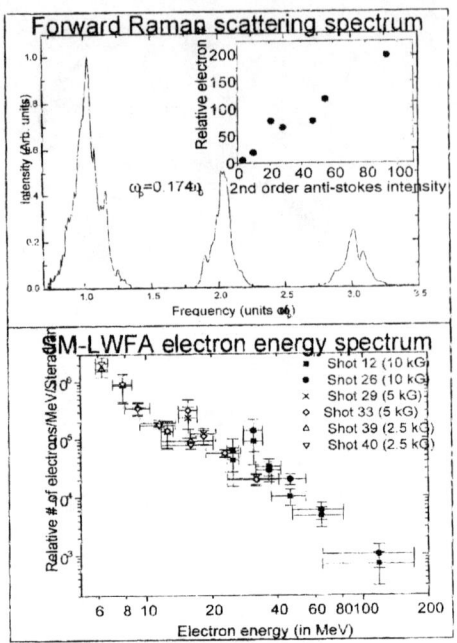

Figure 9.(a) Forward emitted anti-Sokes side bands in the NRL experiments and the linear relationship between number of accelerated electrons with the amplitude of the second anti-Stokes (inset; b) the accumulated electron energy spectrum in the forward direction in the NRL-SMLWFA experiments.

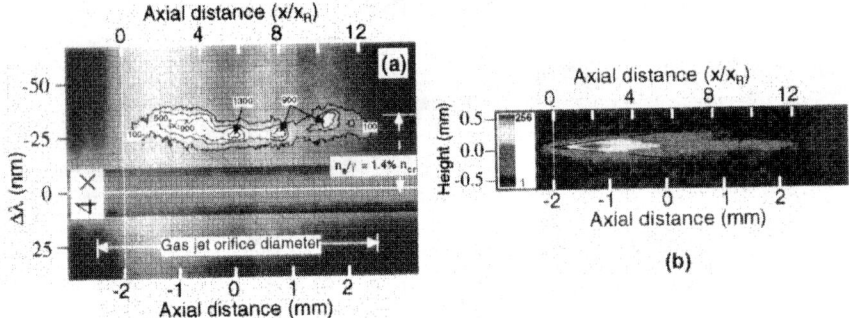

Figure 10. (a) Frequency-resolved image of EPW amplitude along the laser propagation axis. Contours are of constant scattered probe energy and are artificially suppressed at the edges relative to x = 0 due to the temporal profile of the probe pulse. Spatial-modulated Bremsstrahlung continuum is also apparent. (b) Side-view sidescatter near 1 µm for the same shot (color).

An interesting variation on the laser guiding experiments is being attempted at NRL and Ecole Polytechnique. This is laser pulse propagation through thin capillary tubes.[34] In the experiment at NRL for instance, an electrical discharge through a 2 cm long capillary generates a density minimum on axis. A laser beam with a diameter of ~ 35 μm and a peak intensity of 10^{17} W/cm^2 is found to be guided over 22 Rayleigh lengths with an energy transfer efficiency of greater than 70% with an excellent mode structure. These experiments are significant and important because if the laser intensity transported could be increased to ~10^{18} W/cm^2 over about 1 cm then a 1 GeV SMLWFA suddenly becomes realizable.

STATUS REPORT ON THE LASER WAKE FIELD ACCELERATION (LWFA) EXPERIMENTS

In the LWFA a short ($\tau \sim \omega_p/c$) but intense ($a_o \sim 1$) laser pulse sent through a plasma leaves behind it a "wake" of plasma oscillation which has a phase velocity v_ϕ which is equal to the group velocity v_g of the laser pulse in the plasma. For $\omega_o \gg \omega_p$, $v_g \cong c$ and the resultant plasma wake is relativistic. This was the original scheme proposed by T. Tajima and J. Dawson.[2] The scheme was however not shown to work as the required laser pulses (<100 fs in duration with focused intensities of > 10^{18} W/cm^2 for $\lambda < 1$ μm) did not exist until the Ti-saphire revolution in the late 1990's.

The first observation of a wake produced by a single short laser pulse was in 1996 by the Ecole Polytechnique[35,36] and by the U.T. Austin groups.[37] In both of these experiments the laser was focused to a spot size much smaller than wavelength of the plasma oscillation and consequently, the oscillation was dominated mainly by the radial motion of the electrons. Such cylindrical electron wakes were measured, with a temporal resolution much better than ω_p/c by frequency domain interferometry.

In the Ecole Polytechnique experiments, the cylindrical plasma wake field excited by a 130 fs Ti-saphire laser was seen to have a nonlinear increase in oscillation frequency as the plasma density was decreased below the optimum density. The plasma wave was also seen to damp in a few plasma periods.[35] These experiments were followed up by a proof-of-principle acceleration experiment by the same group.[38] By injecting a 3 MeV electron beam into the wake a maximum energy gain of 1.6 MeV was measured, corresponding to a maximum longitudinal field of 1.5 GeV/m. Experiments on LWFA are in progress at JERI/KEK laboratories with published data on acceleration that has caused much excitement and controversy.

Finally, the IC group,[39] the LBL group,[40] and the U. Michigan group[41] are all beginning to use nuclear activation technique as a method for characterizing the flux of electrons above a certain threshold energy in their respective experiments. It is quite possible that the first use of these laser electron accelerators may well turn out to be for generating rare radioactive isotopes for specialized uses.

ACKNOWLEDGMENTS

I am grateful to Mike Downer, Tony Ting, J. Wurtele, D. Umstadter, W. Leemans, and Zulfikar Najmudin for providing me with information on their latest research.

REFERENCES

1. C. Joshi et al., *Nature* **311**, 525 (1984).
2. P. Sprangle et al., *Phys. Rev. Lett.* **72**,2887 (1994).
3. T. Tajima and J. M. Dawson, *Phys. Rev. Lett.* **43**, 267 (1979).
4. T. Coffey, *Phys. Fluids* **14**, 1402 (1971).
5. R. Williams et al., *Laser and Particle Beams* **8**(3), 427 (1990).
6. M. N. Rosenbluth and C. S. Liu, *Phys. Rev. Lett.* **29**, 701 (1972).
7. C. Clayton et al., *Phys. Rev. Lett.* **54**,2343(1985).
8. C. Clayton et al., *Phys. Rev. Lett.* **71**, 37(1993).
9. M. Everett et al., *Nature* **368**, 527 (1994).
10. C. Joshi et al., *Comments on Plasma Phys. and Nucl. Fusion*, **16**, 65 (1994).
11. J. Rosenzweig et al., *Nucl. Intsr. Meth. A*. **410**, 437 (1998).
12. S. Tochitsky et al., *Optics Lett.* **24**, 1717 (1999).
13. Y. Kitagawa et al., *Phys. Rev. Lett.* **68**, 48 (1992).
14. A. Dyson et al., *Plasma Phys. Controlled Fusion*, **38**, 509 (1996).
15. F. Amiranoff et al., *Phys. Rev. Lett.* **68**, 3710 (1992).
16. N. A. Ebrahim, *J. Appl. Phys.* **76**, 7645 (1994).
17. D. W. Forslund et al., *Physics Fluids* **18**, 1002 (1075).
18. W. Mori et al., *Phys. Rev. Lett.* **72**, 1482 (1994).
19. J. M. Dawson, *Phys. Fluids* **113**, 383 (1959).
20. C. Joshi et al., *Phys. Rev. Lett.* **47**, 1285 (1981).
21. P. Maine et al., *IEEE J. Q. E.* **24**, 398 (1988).
22. C. Coverdale et al., *Phys. Rev. Lett.* **74**, 4659 (1998).
23. A Modena et al., *Nature* **377**,606 (1995).
24. D. Gordon et al., *Phys. Rev. Lett.* 80, 2133 (1998).
25. D. Umstadter et al., *Science* **273**,472 (1996).
26. A. Ting et al., *Phys. Rev. Lett.***77**, 5377 (1996).
27. C. Clayton et al., *Phys. Rev. Lett.* **81**, 100 (1998).
28. R. Wagner et al., *Phys. Rev. Lett.* **78**, 3125 (1997).
29. S. Y. Chen et al., *Phys. Rev. Lett.* **80**, 2610 (1998).
30. C. Durfee et al., *Phys. Rev. Lett.* **71**, 2409 (1993).
31. G. Gaul et al., Proc. Int'l. Conf. LASERS'99, Quebe City, 13-16 Dec. (1999).
32. Volfbeyn et al., *Phys. Plasmas* **4**, 3403 (1997).
33. C. Krushelnick et al., *Phys. Rev. Lett.* **78**, 4047 (1997).
34. Y. Ehrlich et al., *Phys. Rev. Lett.* **77**, 4186 (1996); F. Dorchies et al., *Phys. Rev. Lett.* **82**, 4655 (1999).
35. J. R. Marqués et al., *Phys. Rev. Lett.***78**, 3463 (1997).
36. J. R. Marqués et al., *Phys. Rev. Lett.* **76**, 3570 (1996).
37. C. W. Siders et al., *Phys. Rev. Lett.* **76**, 3370 (1996).
38. F. Amiranoff et al., *Phys. Rev. Lett.* **81**, 995 (1998).
39. Z. Najmudin, Private Communication.
40. W. Leemans, at this meeting, see proceedings.
41. D. Umstadter, at this meeting, see proceedings.

Relativistic Ion Acceleration by Ultraintense Laser Interactions

K. Nakajima[1,2], J. K. Koga[1], and K. Nakagawa[1]

[1] *Advanced Photon Research Center, JAERI, Kyoto-fu, 619-0215, Japan*
[2] *High Energy Accelerator Research Organization, Tsukuba, 305-0801, Japan*

Abstract. There has been a great interest in relativistic particle generation by ultraintense laser interactions with matter. We propose the use of relativistically self-focused laser pulses for the acceleration of ions. Two dimensional PIC simulations are performed, which show the formation of a large positive electrostatic field near the front of a relativisitically self-focused laser pulse. Several factors contribute to the acceleration including self-focusing distance, pulse depletion, and plasma density. Ultraintense laser-plasma interactions are capable of generating enormous electrostatic fields of ~ 3 TV/m for acceleration of protons with relativistic energies exceeding 1 GeV.

INTRODUCTION

Ultraintense laser interactions with matter can generate enormous number of highly energetic electrons, photons and ions that induce nuclear interactions. Recent laser-matter interaction experiments have revealed acceleration of $\sim 10^{10}$ electrons with a maximum energy up to ~ 100 MeV and the production of $\sim 10^{12}$ ions up to several ten MeV [1]. High energy electron acceleration can be elucidated by the self-modulated laser wakefields with accelerating gradients of ~ 100 GV/m excited due to strong interactions of a relativistically self-focused laser pulse with highly nonlinear plasma waves. In the meantime, a few 100 keV ions are driven due to explosions from strong space charge forces, the so called Coulomb explosion, excited by the electron density depression in clusters of atoms or a plasma channel. It may be considered that several MeV ions are accelerated in electrostatic fields induced by a charge separation between fast thermal electrons and cold ions. However, acceleration mechanisms of energetic ions are not well understood.

In the two-dimensional PIC simulations for propagation of ultraintense laser pulses of the order of 10^{20}W/cm^2 in underdense plasmas of $\sim 10^{20}$cm^{-3}, we found that a large amplitude of positive electrostatic fields of the order of a few TV/m is generated over ~ 1 mm scale in the front of a relativistically self-focused laser pulse. These enormous fields implies the capablility of accelerating protons up to ~ 1GeV within matter. A solitary potential with a spacial and temporal size of the order of μm can produce a femtosecond ion pulse. We propose a new acceleration mechanism for ions due to enormous accelerating fields generated by relativistically self-focused laser pulses in plasmas. This mechanism may open up a new regime of ultraintense laser-matter interactions and new fields of high energy particle physics.

STRONG ELECTROMAGNETIC FIELDS

The peak amplitude of the transverse electric field of a linearly polarized laser pulse is given by

$$E_L[\text{TV/m}] \simeq 2.7 \times 10^{-9} I^{1/2}[\text{W/cm}^2] \cong 3.2 a_0/\lambda_0[\mu\text{m}], \qquad (1)$$

where I is the laser intensity, λ_0 is the laser wavelength, and a_0 is the laser strength parameter defined by $a_0 \equiv eA_0/m_e c^2$ in terms of the peak amplitude of the laser vetor potential A_0 and the electron rest energy $m_e c^2$. Using the laser peak intensity $I = cE_L^2/8\pi = ck^2 A_0^2/8\pi$, the laser strength parameter is given by

$$a_0 = (2e^2 \lambda_0^2 I/\pi m_e^2 c^5)^{1/2} \cong 0.85 \times 10^{-9} \lambda_0[\mu\text{m}] I^{1/2}[\text{W/cm}^2]. \qquad (2)$$

Physically a_0 is equal to the normalized momentum of the electron quiver motion in the laser field. The corresponding magnetic field is given by $B_L[\text{T}] = E_L/c$ and the radiation pressure exerted by the laser intensity I is given by $P_L[\text{Bar}] = 0.1 I/c[\text{J/cm}^3]$. In fact, advances in laser technology provide us with ultraintense ultrashort lasers capable of generating high intensities more than 10^{20} W/cm^2 and short pulses less than 1 ps. Such laser pulses can generate the electric field more than $E_L \sim 27$ TV/m and the relativistic electron motion with $a_0 \simeq 10$ for $\lambda_0 = 1 \mu$m. At these intensities, the magnetic field is $B_L \sim 10^5$ T and the radiation pressure exceeds $P_L \sim 30$ GBar.

In laser-matter interactions, an energy spectrum of electron quiver motion may be expressed by a Maxwellian distribution with an effective temperature given by an average energy of the electron quiver motion in the laser field, $\varepsilon_{avg} = mc^2[1 + a_0^2/2]^{1/2}$. As the effective electron temperature is $\varepsilon_{avg} \sim 4$ MeV at 10^{20} W/cm^2, the maximum electron energy will reach up to more than 30 MeV. It is inferred that those fast heated electrons are ejected from plasma region to create a charge separation with a positive high gradient field.

SIMULATION PARAMETERS

To study the self-focusing of a high intensity short pulse laser in a plasma we use a 2 dimensional fully relativistic particle-in-cell(PIC) code. The code has been parallelized to run on the Intel Paragon S120MP. The simulation box is 1234μm (16384 cells) by 38.4μm (512 cells) in the x and y directions respectively. The boundary conditions are periodic in the y direction and outgoing in the x direction.

There is a vacuum region at both ends of the simulation box of length $22.6\mu m$. The plasma density is chosen to be $5 \times 10^{19} cm^{-3}$ which corresponds roughly to doubly ionized Helium gas at atmospheric pressure. There is one electron and one ion in each simulation cell, setting the ion mass ratio to electron to be 1836. The linearly s-polarized laser pulse (E_z, B_y) starts in the vacuum region on the left and propagates to the right. The parameters of the laser are that of the newly developed sub 20fs 100 TW Ti:sapphire laser at the Japan Atomic Energy Research Institute (JAERI) [2]. The pulse length is 19 fs with a spot size of $10\mu m$. The wavelength is $0.8\mu m$. The corresponding unitless laser strength parameter $a_0 = 7.4$ where $a_0 = eE_0/m_0\omega_0 c$, E_0 is the peak electric field, m_0 is the electron mass, and ω_0 is the laser frequency.

Using the formula for the critical power P_{cr} for the relativistic self-focusing of a Gaussian laser pulse [5]

$$P_{cr}[GW] = 17(\frac{\omega_0}{\omega_p})^2, \qquad (3)$$

where ω_0 is the laser frequency and ω_p is the plasma frequency. We find that $P/P_{cr} = 160$, where P is the laser power. Thus, the laser pulse should relativistically self-focus in the plasma. Also given the condition for length of the laser pulse L_t necessary for the optimum generation of a wake field [6]: $L_t = \pi c/\omega_p$, we find that $L_t = 2.4\mu m$ whereas the laser pulse length is $5\mu m$. Under these conditions a large wake field behind the pulse should not occur.

SIMULATION RESULTS

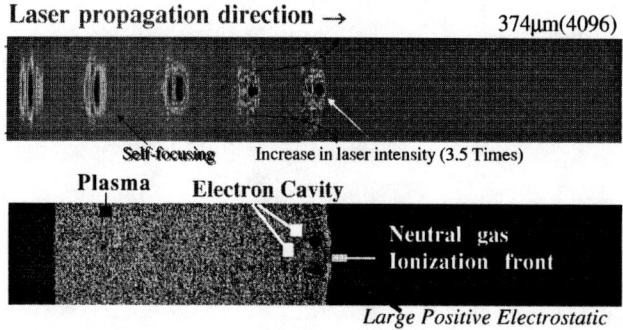

FIGURE 1. (a) Sequence of frames showing the E_z field of the self-focusing laser pulse, (b) electron density, and (c) ion density at the final frame of the E_z field plot after the laser has propagated $200c/\omega_p$.

Figure 1(a) shows a sequence of frames of the laser pulse after it has propagated about $140\mu m$ ($200c/\omega_p$) into the plasma. It can be seen that the pulse has relativistically self-focused and has filamented. Figure 1(b) and (c)show the electron and ion density at the same propagation distance, respectively. In the central portion of the pulse the electrons have been completely ejected, whereas, the ions still remain. Figure 2 shows the maximum intensity of the laser pulse and the electrostatic field as a function of propagation distance into the plasma. Due to the self-focusing of the pulse the maximum laser intensity increases to be more than 3.5 times higher than the initial pulse intensity after propagating $150c/\omega_p$.

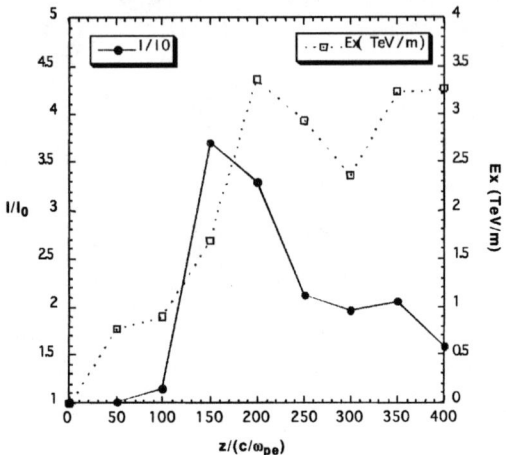

FIGURE 2. Maximum laser intensity (solid line with solid circles) and x component of the electrostatic field (dotted line with empty squares) as a function of propagation distance of the laser pulse.

Figure 3 shows the structure of the light bullet after it has propagated $321\mu m(350c/\omega_p)$ into the plasma. In Figure 3(a) we can see that the laser pulse

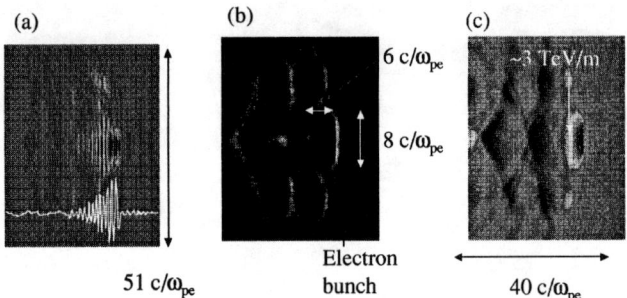

FIGURE 3. The structure of the light bullet after it has propagated $321\mu m(350c/\omega_p m)$ into the plasma of (a) the laser pulse, (b) electron density, and (c) E_x electrostatic field

has filamented and the central portion has narrowed. From the line profile taken down the center of the pulse we can see that the front of the laser pulse has steepened. Figure 3(b) shows the electron density at the same time. Many cavities have formed from filaments of the laser pulse. An electron cavity from the main part of the pulse has formed which is $8c/\omega_p$ wide and $6c/\omega_p$ long. In the front of this cavity the electrons have built up. The maximum density is about 10 times higher than the initial density. Due to this large buildup of electrons at the front of the pulse, there is a large positive electrostatic field in the propagation direction of the laser pulse created there (Figure 3(c)). It reaches a maximum of greater than 3 TeV/m. The peak in the electrostatic field comes after the peak in the laser field intensity. The source of this huge field is due to the evacuation of electrons from the laser self-focus region and the pile up of electrons at the front of the pulse. This electrostatic field remains high in spite of the decrease in the intensity of the laser pulse.

ION ACCELERATION CONCEPT

We propose to use this large field created at the front of the pulse to accelerate protons or heavy ions. Figure 4 shows a surface plot of the electrostatic potential at the front of the pulse at the same propagation distance. It can be seen that

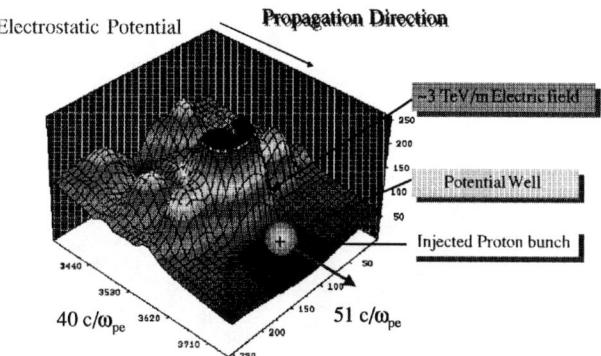

FIGURE 4. Surface plot of the electrostatic potential after the laser pulse has propagated $321\mu m (350 c/\omega_p m)$ into the plasma.

there is a potential well which has been created. If ions are appropriately injected there, it will be trapped and accelerated. Also it can be seen that there is a transverse potential well which exists there. This will help to confine the ion bunch transversely. Upon closer examination of the electron density the source of this potential well is a small bunch of electrons which exist in front of the main wall of electrons (see arrow indicating the location of the electron bunch in Figure 3(b)). The width of the well is approximately $5.7\mu m$ ($8c/\omega_p$) with the length being approximately $7.2\mu m$ ($10c/\omega_p$).

There are several factors which affect or limit the acceleration of ions. One is the electric field itself. In Figure 5 we show the electric field as a function of propagation distance. The electric field grows to its maximum value approximately

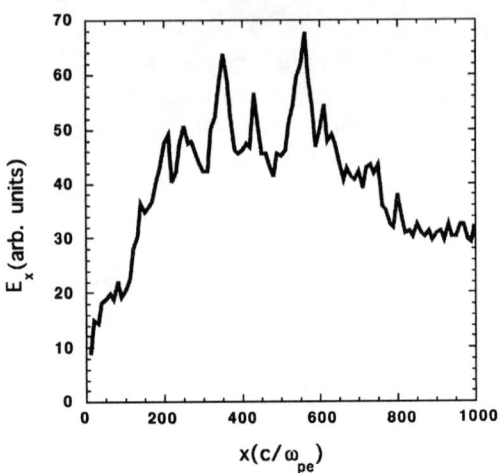

FIGURE 5. Electric field in the propagation direction as a function of propagation distance

over the distance it takes the laser pulse to relativisticly self-focus. This distance can be adjusted by the introduction of various density profiles which affect the location of the self-focus point of the laser pulse [7]. After the laser pulse self-focuses the electric field remains fairly constant over a distance of 180μm. After which the laser pulse begins to deplete as seen in Figure 6, where the z component of the laser field is plotted. It can be seen that as the laser pulse propagates its

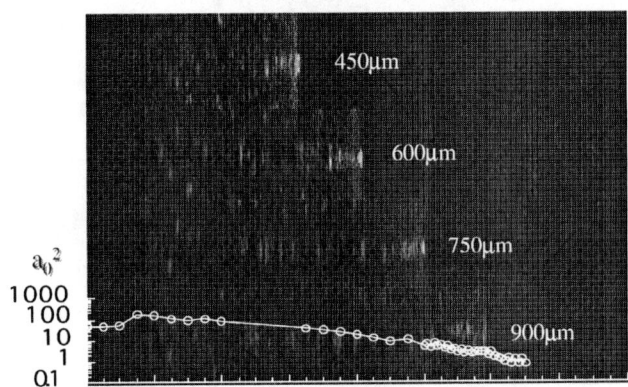

FIGURE 6. Laser field depletion as a function of propagation distance

intensity is diminishing. At the bottom of the figure is shown the unitless laser intensity, a_0^2, as a function of propagation distance (white line with empty circles). After the initial increase in the intensity after self-focusing of the laser pulse, the intensity gradually decreases with propagation. This depletion may be due in part to the conversion of the pulse energy to the electron plasma wave created at the front of the pulse [8]. In the one dimensional approximation the depletion time t_{dep}

from this is theoretically given by [8]:

$$t_{dep} \approx \tau \omega_0^2/\omega_p^2, \quad (4)$$

where τ is the pulse length. Using the parameters from the simulation we get by assuming that the pulse propagates at nearly the speed of light that the distance over which the pulse should deplete $l_{dep} = ct_{dep}$ is approximately 159μm ($218c/\omega_p$). This is shorter than the observed depletion length seen in the simulation. This is probably due to the two dimensional self-focusing of the laser pulse. This depletion leads to the eventual disappearance of the large electrostatic field at the front of the pulse.

In order to trap and accelerate ions in the electrostatic potential moving at the velocity nearly equal to the group velocity $v_g = c(1 - \omega_p^2/\omega_0^2)^{1/2}$ of a laser pulse in a plasma, the potential velocity must coincide with a particle velocity v. In case stationary ions trapped in the moving potential can be accelerated in the effective accelerating field E_{acc} over a distance L_{acc}, the energy gain of an ion is given by

$$W_f = m_i c^2 (\gamma_f - 1) = ZeE_{acc}L_{acc}, \quad (5)$$

where W_f is the final kinetic energy, m_i is an ion mass, $\gamma_f = (1 - v_f^2/c^2)^{-1/2}$ is the Lorentz factor for the final ion velocity v_f, and Z is the atomic number. In order to accelerate initially stationary ions up to the final energy W_f, the required plasma density gradient for accelerating the group velocity of the laser pulse should be

$$\frac{n_{ef} - n_c}{L_{acc}} = -\beta_f^2 \frac{n_c}{L_{acc}}, \quad (6)$$

where n_{ef} is the final electron plasma density, $n_c = \pi/(r_e \lambda_0^2)$ is the critical plasma density, r_e is the classical electron radius, and $\beta_f = v_f/c$ is the final normalized ion velocity. Laser-solid interactions with a pre-pulse on a thin foil could produce inhomogeneous plasmas with a density gradient from a solid density to vacuum in its thermal expansion process. When a main high peak power pulse is injected into the critical density region at an appropriate timing, the moving electrostatic potential could trap stational ions to accelerate them until the potential is depleted due to loading. Consider proton acceleration by the proposed mechanism. According to the simulation results, letting the acceleration length to be $L_{acc} \sim 1$mm, the effective accelerating gradient may be $eE_{acc} = 1.5$ TeV/m. This gives $W_f = 1.5$ GeV, $\gamma_f = 2.6$, $\beta_f = 0.92$. For this proton acceleration, the required plasma density distribution is given by varying the electron density from $n_c = 1.7 \times 10^{21}$cm^{-3} to $n_{ef} = 2.6 \times 10^{20}$cm^{-3} over ~ 1 mm.

CONCLUSIONS

In this paper we have proposed using the large electrostatic field created in the front of a relativistically self-focused laser pulse to accelerate ions. We have shown from 2 dimensional PIC simulations that the accelerating field of the order of 3 TeV/m can be excited by a 100 TW laser pulse propagating in a plasma at a density of 5×10^{19} cm^{-3}. Ions trapped in the front of such a self-focused pulse can be accelerated in coincidence of the ion velocity with the potential velocity nearly equal to the laser group velocity in plasmas. The ions can be transversly confined by the transverse potential well which exists in the front of the pulse. The acceleration will be limited by the depletion of the laser pulse. The kinetic energy gained by initially stationary ions could reach 1.5 GeV in an inhomogeneous plasma with density gradient over 1 mm for the laser intensity of the order of 10^{20}cm^{-3} with the pulse duration of $\sim$ 20 fs. The laser-matter interactions will open up a new regime of high energy beam science.

ACKNOWLEDGMENTS

We would like to acknowledge useful discussions with T. Tajima and support from the Advanced Photon Research Center of JAERI-KANSAI.

REFERENCES

1. M. H. Key et al., "Studies of the Relativistic Electron Source and Related Phenomena in Petawatt Laser Matter Interactions" in Proceedings of *Inertial Fusion Sciences and Applications 99*, edited by C. Labaune et al., ELSEVIER, Paris, 2000, pp.392-400.
2. K. Yamakawa, M. Aoyama, S. Matsuoka, T. Kase, Y. Akahane, and H. Takuma, Opt. Lett. **23**, 1468 (1998).
3. M. Kando, H. Ahn, H. Kotaki, K. Tani, T. Watanabe, T. Ueda, M. Uesaka, Y. Kishimoto, J. Koga, H. Watanabe, K. Nakajima, M. Arinaga, T. Kawakubo, H. Nakanishi, and A. Ogata, in Advanced Accelerator Concepts, S. Chattopadhyay, J. McCullough, and P. Dahl, eds., AIP Press, New York (1997).
4. S. V. Bulanov, I. N. Inovenkov, V. I. Kirsanov, N. M. Naumova, and A. S. Sakharov, *Phys. Fluids B* **4**, 1935 (1992).
5. G. Z. Sun, E. Ott, Y. C. Lee, and P. Guzdar, Phys. Fluids **30**, 526 (1987). D. C. Barnes, T. Kurki-Suonio, and T. Tajima, IEEE Trans. Plasma Sci. **PS-15**, 154 (1987). P. Sprangle, C. M. Tang, and E. Esarey, IEEE Trans. Plasma Sci. **PS-15**, 145 (1987).
6. T. Tajima and J. M.Dawson, Phys. Rev. Lett. **43**, 267 (1979).
7. D. P. Garuchava, Z. I. Rostomashvili, and N. L. Tsintsadze, Sov. J. Plasma Physics **12**, 776 (1986).
8. S. V. Bulanov, V. I. Kirsanov, N. M. Naumova, A. S. Sakharov, and H. A. Shah, Physica Scripta **47**, 209 (1993).

Production and characterization of a fully-ionized He plasma channel

E. W. Gaul, S. P. Le Blanc, A. R. Rundquist, R. Zgadzaj, N. H. Matlis, H. Langhoff*, and M. C. Downer

University of Texas at Austin, Department of Physics, Austin, TX 78712
permanent address: Physikalisches Institut der Universität Würzburg, 97074 Würzburg, Germany

Abstract. We report guiding of intense ($I = 1.3 \pm 0.7 \times 10^{17} \text{W/cm}^2$) 80 fs laser pulses with negligible spectral distortion through 1.5-cm-long preformed helium plasma channels. Channels were formed by axicon-focused Nd:YAG laser pulses of either 0.3 J energy, 100 ps duration, after pre-ionizing a 200-700 Torr backfill of He gas to $n_e \sim 10^{16} \text{cm}^{-3}$ with a pulsed electrical discharge; or 0.6-1.1 J energy, 400 ps duration, which required neither pre-ionization nor intentional impurities for seeding. Transverse interferometry showed that He was fully-ionized on the channel axis in both cases. Identical fs pulses suffered substantial ionization-induced blueshifts after propagating through Ar and Ne channels of similar dimensions.

INTRODUCTION

Plasma waveguides capable of guiding intense fs laser pulses over multiple Rayleigh lengths without optical distortion are essential to developing practical laser-driven plasma accelerators and coherent short-wavelength light sources [1]. Although sufficiently intense pulses can self-guide by relativistic self-focusing [2-4], preformed plasma channels [5-8] offer superior control over the mode quality of the guided pulse [9]. Milchberg and co-workers [7-9] have developed a method for producing 2-cm-long, 10-50 μm radius channels of excellent optical quality by ionizing and heating heavy noble gases or N_2 with axicon-focused 0.3 J, 100 ps Nd:YAG laser pulses to induce a radially expanding shock. However, these pulses generally leave the channel only partially ionized, and its entrance/exit regions nearly un-ionized. Consequently, ionization-induced phase modulation can distort intense fs pulses propagating through the channel and end regions. Solutions to this problem so far have been either expensive (e.g. generating, in addition to channel-forming and guided pulses, a third intense pulse to completely pre-ionize the channel), incomplete (e.g. finding conditions for which most channel ions are stripped to noble gas cores, thus leaving an ionization-free intensity "window"), or limited in length scalability (truncating end regions by using a gas jet) [10,11]. In this Letter, we demonstrate distortion-free propagation of 0.3 TW laser pulses through 1.5 cm-long, *fully*-ionized He plasma channels *and* their un-ionized end regions. Past channel-forming pulses lacked the peak intensity to field-ionize enough He atoms to seed avalanche ionization (AI) - the main breakdown mechanism - by the same pulse. We have overcome this problem by either: 1) striking a 2-cm-long pre-plasma ($n_e \sim 10^{16}$ cm^{-3}) in a He backfill with an inexpensive pulsed electrical discharge, thus seeding AI by our synchronized 100 ps, 0.3 J Nd:YAG channel-forming pulse; or 2) lengthening the Nd:YAG pulse to 400 ps and amplifying it to $\simeq 0.6$ J so that AI and inverse bremsstrahlung (IB) heating have sufficient time to build up even from field ionized trace contaminants ($n_e^{con} \ll 10^{16}$ cm^{-3}) without a pre-plasma. In addition the high field-ionization threshold ($I_P = 24.6$ eV) of He suppresses ionization by the guided pulse in the end regions. Thus we can form a distortion-free channel in a backfill, which is more length-scalable and uniform in density than a gas jet. This important advantage is lacking with other fully-ionizable, but low I_P gases such as H_2 ($I_P = 15.4$ eV).

I EXPERIMENT

Fig. 1 shows a schematic of our experiment. The 100 ps, 0.3 J Nd:YAG laser pulse (P1) from a self-filtering regenerative amplifier [12] is line-focused by an axicon lens to generate a plasma channel. A $\sim$30 mJ, 80 fs Ti:S laser pulse (P2) is focused with an $f/8$ off-axis parabolic mirror, reflected from a dichroic mirror into a 2 mm diameter axial hole in the axicon and coupled into the channel. Its transverse mode entering or exiting the channel is imaged onto CCD1 by L1 or into a spectrometer by L2. The channel profile is probed transversely by a Mach-Zehnder interferometer using probe pulse (P3), which is split from P1 and delayed -3ns< $\Delta T2$ <+1ns. The interferogram is imaged onto CCD2 by L3, evaluated by a Fourier-transform method [13] to extract the phase shift profile, then Abel transformed [14] to yield an electron density profile $n_e(r)$. Jitter in the electronically-adjusted delay $\Delta T1$ between P1 and P2 is held to ±200 ps accuracy by controlling the Ti:S oscillator length with a piezo-electric transducer.

The insert shows the electrical discharge circuit used to pre-ionize the He. The charge on a 0.5 nF, $\simeq$30 kV storage capacitor C1 is held off by a spark gap (SG), which is enclosed in a sealed chamber that is pressurized with 2-3 atm N_2 and triggered by a synchronized 10 mJ, 8 ns laser pulse. When SG conducts, charge flows onto a second capacitor C2, thereby building up voltage across the 2 cm gap between two knife-edge electrodes that protrude into the He backfill near the axicon. The low inductance (< 100 nH) circuit builds up overvoltage quickly, triggering rapid ($\simeq$30 ns) breakdown of the He and depositing 100 mJ into a plasma column (conductivity $\sigma \sim 0.1 \Omega^{-1}$) of initial radius $r_i \simeq .1$ mm, which expands later to $r_f \sim 1$ mm, as seen in the time integrated luminescence image (Fig. 1 inset). From σ and r_i we estimate $n_e \simeq 10^{16}$ cm^{-3} and electron temperature $kT_e \simeq 1$ eV,

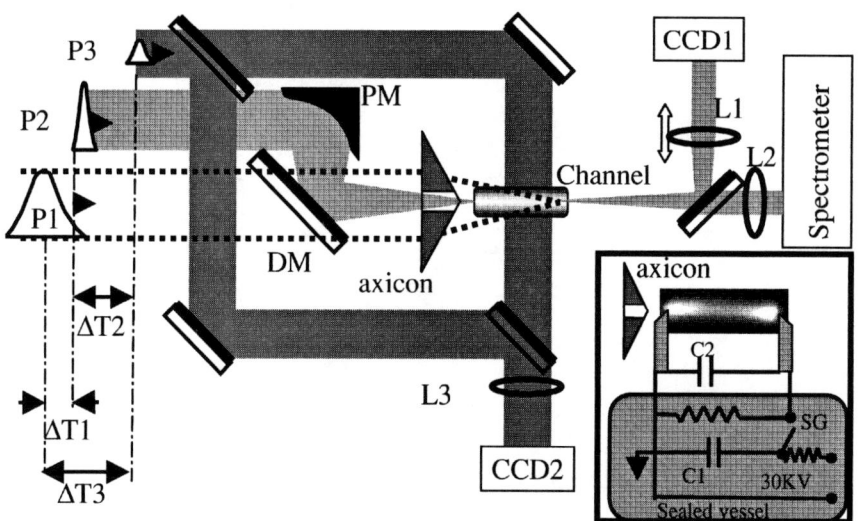

FIGURE 1. Experimental layout. P1: Nd:YAG channel-forming pulse. P2: 80 fs, 0.4 TW Ti:S pulse coupled into channel by parabolic mirror PM and dichroic mirror DM through hole in axicon. P3: weak 80 fs pulse to probe $n_e(r)$ in channel interferometrically. CCD1, 2 record guided mode and interferograms imaged by lenses L1, L3, respectively. Inset: discharge circuit used to pre-ionize He; shaded area is pressurized to 2-3 atm N_2.

sufficient to absorb $> 1\%$ of the energy of P1 at $I = 10^{13}$ W/cm^2 in 200 – 750 Torr He, which generates a plasma filament over ~ 1.5 cm of the pre-ionized region. The laser pulse formed no detectable filament without the discharge. Precise spatial ($\pm 20\mu$m) and temporal (± 5 ns) overlap of discharge with the laser pulse were found to be essential.

Longer heating pulses of equal peak intensity should produce equivalent plasma filaments from lower seed n_e [15]. To produce such a pulse we inserted a 3 mm solid fused silica etalon before the Nd:YAG regenerative amplifier to stretch the pulse to 400 ps and added one amplifier stage to obtain up to 1.1 J. We observed that 0.6-1.1 J pulses create plasma channels in He without any additional pre-seeding device. Only $n_e \sim 10^{10}$ cm^{-3} is required to seed efficient avalanche ionization of our interaction region by this laser pulse. Nevertheless, our 400 ps pulse cannot field ionize [16] even this density in pure He. Since our chamber base pressure is ~ 10 mTorr, field ionization of low I_P trace contaminants evidently supplies the seed electrons. For 100 ps pulses much higher partial pressure of low I_P (but not fully ionizable) contaminants must be introduced [7,8].

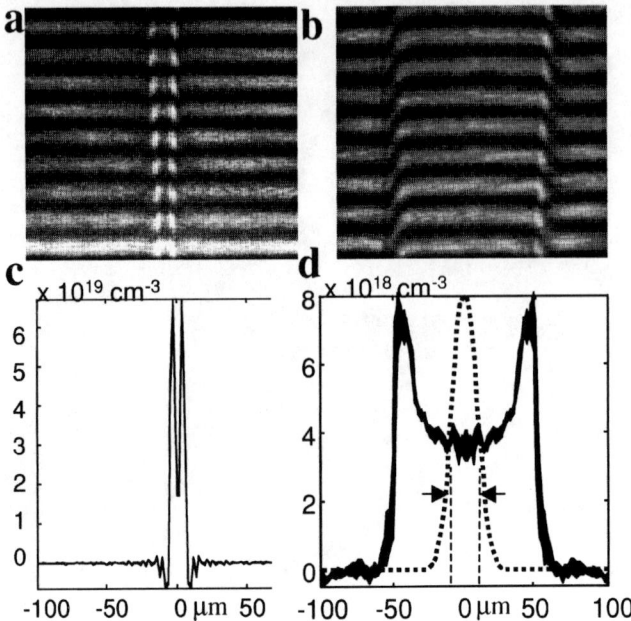

FIGURE 2. Typical interferograms (a, b) of He channels and corresponding $n_e(r)$ in cm^{-3} (c, d) for $\Delta T3 = 200$ ps and $\Delta T3 = 6$ ns. Note the different scales on panel c versus d. Dotted curve in d - predicted guided mode in parabolic approximation ($w_0 = 14\mu$m), dashed lines - e^{-2} width of the measured mode ($w_0 = 8\mu$m).

II RESULT AND DISCUSSION

Fig. 2a, b show interferograms, and Fig. 2c, d corresponding $n_e(r)$ profiles, of filaments formed in 750 Torr He at $\Delta T3 = 200$ ps and 6 ns after the peak of a 400 ps pulse. The 200 ps profile shows a 10 μm radius plasma column of average $n_e = 6 \times 10^{19}$ cm^{-3}, i.e. *fully-ionized* He, immediately following the Nd:YAG pulse. A shockwave then spreads out radially, creating the $n_e(r)$ profile in Fig. 2d, which contains nearly the same integrated number $\int 2\pi r n_e(r) dr$ of electrons per unit length as Fig. 2c. When this profile is approximated by a parabola, the dashed curve in Fig. 2d shows the intensity profile of the lowest-order Gaussian mode (e^{-2} radius $w_0 = 14\mu$m) expected to be stably guided.

Fig. 3a shows images of the transverse mode of P2 as the object plane of L1 was translated 0-4 mm from the focus of P2 with no channel present, and thus shows the expanding mode of a normal vacuum focus. The data in Fig. 3b shows the mode near the channel exit, at a location $\sim$12 mm from the focus, which is a typical exiting guided mode of radius $w_0 \sim 8\mu$m which we measure to contain 50% of the energy of the entering pulse, implying guided intensity $I_{guided} = 2 \times 10^{17}$ W/cm^2. The discrepancy with the expected 14μm mode diameter (which implies $I_{guided} =$

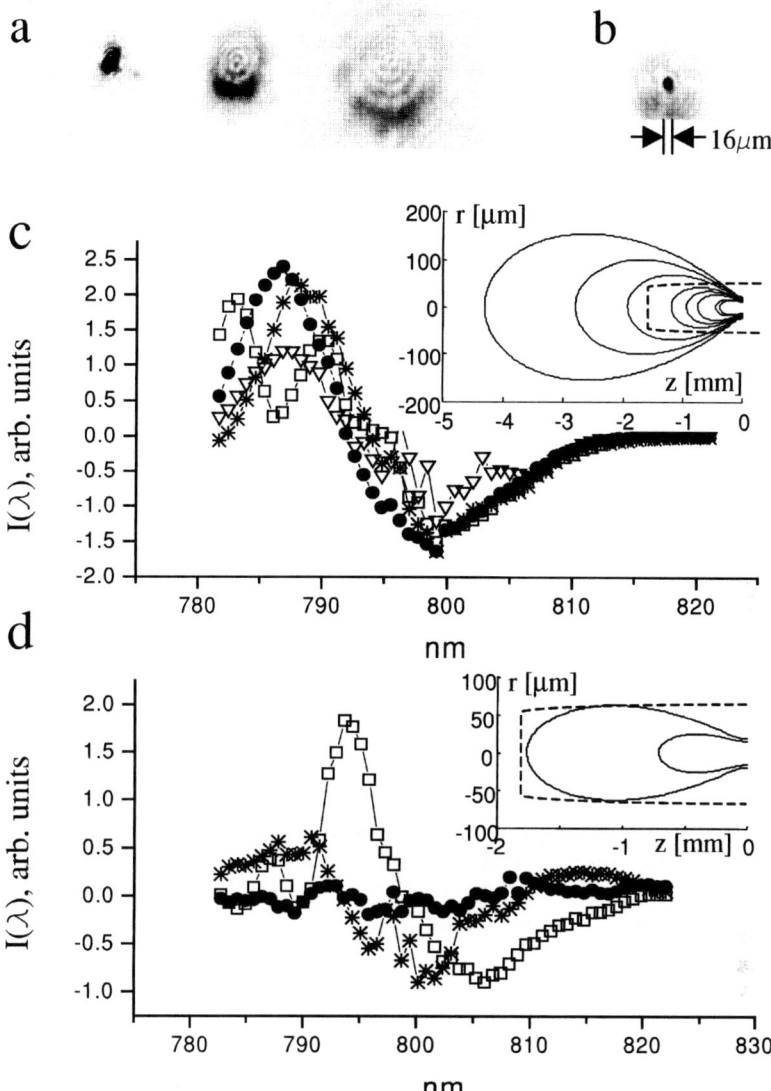

FIGURE 3. Images of a) expanding unguided and b) exiting guided transverse laser modes at various positions 0-12 mm after the vacuum focus in He. Panel b and c: difference spectra of Ti:S laser pulse after channel in Ar c) $-\bullet-$ 200 Torr, $-*-$ 100 Torr, $-\nabla-$ 50 Torr) and He d) $-\bullet-$ 750 Torr, $-*-$ 630 Torr) backfill pressures. Squares: unchanneled pulses in 200 Torr Ar, 750 Torr He. Inserts in c, d: isointensity contours corresponding to ionization levels of Ar, He, respectively, for Ti:S pulses focused to $I_{guided} = 2 \times 10^{17}$ W/cm^2. Dashed curve: boundaries of plasma channel.

0.6×10^{17} W/cm^2) may originate from uncertainty in the Abel-transformed $n_e(r)$, but is not quantitatively understood. Thus we estimate $I_{guided} = 1.3 \pm 0.7 \times 10^{17}$ W/cm^2.

In order to evaluate ionization-induced distortion, we compared the spectral blueshifts of intense ($I = 2 \times 10^{17}$ W/cm^2) pulses exiting Ar, Ne, N$_2$, and He channels of varying density. The results for Ar and He are shown by the differential power spectra $\Delta I(\lambda) = I_{exit}(\lambda) - I_{in}(\lambda)$ in Fig. 3c and d, respectively. We compare He pressures four times larger than corresponding Ar pressures to keep the electron number approximately the same. When the incident pulses were focused into neutral gas without a channel, we observed substantial ionization-induced blue-shifts for all gases, as shown by the open squares in Fig. 3c and d. Intense guided pulses exiting 1.5 cm long (or 60 Rayleigh lengths for $w_0 = 8\mu$m) Ar channels were also strongly blue-shifted throughout the investigated density range, as shown by the remaining data sets in Fig. 3c. Very similar results were obtained for Ne and N$_2$ channels. On the other hand, identical pulses exiting He channels showed significantly weaker blueshifts (stars, Fig. 3d) or, in some cases, no detectable blueshift whatsoever (filled circles, Fig. 3d).

The persistent blueshifts observed with Ar (Ne, N$_2$) channels can be attributed in part to ionization within the channel. Since the average energy per electron needed to strip the outer shell is nearly twice that required to fully ionize He, the Nd:YAG pulse evidently leaves the channel only partially ionized. In addition, the focusing pulse ionizes neutral Ar while approaching the channel entrance. The inset of Fig. 3c shows that the iso-intensity contours (solid curves) corresponding to various field-ionization levels of Ar extend as much as 17 Rayleigh lengths into the neutral Ar preceeding the plasma column (dashed curve) for $I_{guided} = 2 \times 10^{17}$ W/cm^2. Ionizing propagation through this distance (mirrored at the exit) creates blueshifts approaching those of the unchanneled pulse. Both sources contribute to the observed blueshifts, although the present results do not distinguish their relative importance quantitatively.

For He channels, on the other hand, the full ionization evident from interferometry eliminates blue-shifting within the channel. Equally important, the iso-intensity contours corresponding to field-ionization levels are much smaller than for Ar. In fact, as shown by the inset of Fig. 3d, they fit approximately within the 50 μm radius plasma column present at $\Delta T_3 = 6$ ns. Thus, with strategic adjustment of the focus of P2, ionization-induced distortions in the entrance region can be almost completely eliminated, as suggested by the result (filled circles) in Fig. 3d. For the results obtained at $p_{He} = 630$ Torr ($-*-$), although the interferometer still indicated full ionization in the middle of the channel, a shorter channel formed. However, the focus of P2 was not moved, and thus no longer coupled to a pre-ionized position. The weak spectral distortion observed in this case can therefore be attributed to ionization in the end region. Similar distortions were observed as the focus of P2 was translated along the channel axis. We estimate that end distortions in He will remain acceptable for intensities up to 10^{18} W/cm^2. At higher intensites, relativisitic plasma nonlinearities also begin to distort the pulse [17].

III CONCLUSION

In summary, we have demonstrated that intense ultrashort pulses couple into, guide at $I = 1.3 \pm 0.7 \times 10^{17}$ W/cm^2 and couple out of a fully-ionized, 1.5 cm long plasma channel formed in a He backfill without spectral distortion. The channel-forming techniques used here are scalable to channel lengths of $\sim$10 cm. This will enable studies of laser wakefield acceleration into and beyond the dephasing limit [18] for $n_e \leq 10^{18}$ cm^{-3}. This research was supported by U.S. Department of Energy grant DEFG03-96-ER-40954.

REFERENCES

1. P. Sprangle and B. Hafizi, Phys. of Plasmas **6**, 1683, (1999) and references therein.
2. K. Krushelnick, A. Ting, C. I. Moore, H. R. Burris, E. Esarey, P. Sprangle, and M. Baine, Phys. Rev. Lett.**78**, 4047 (1997).
3. R. Wagner, S.-Y. Chen, A. Maksimchuk, and D. Umstadter, Phys. Rev. Lett. **78**, 3125 (1997).
4. J. J. Fuchs, G. Malka, J. C. Adam, F. Amiranoff, S. D. Baton, N. Blanchot, A. Hron, G. Laval, J. L. Miquel, P. Mora, H. Ppin, and C. Rousseaux, Phys. Rev. Lett. **80**, 1658 (1998).
5. Y. Ehrlich, C. Cohen, A. Zigler, J. Krall, P. Sprangle, and E. Esarey, Phys. Rev. Lett.**77**, 4186 (1996).
6. T. Hosokai, M. Kando, H. Dewa, H. Kotaki, S. Kondo, N. Hasegawa, K. Nakajima, and K. Horioka, Opt. Lett. **25**, 10(2000) and references therein.
7. C. G. Durfee and H. M. Milchberg, Phys. Rev. Lett.**71**, 2409 (1993).
8. C. G. Durfee, J. Lynch and H. M. Milchberg, Phys. Rev. E **51**, 2368 (1995).
9. T. R. Clark and H. M. Milchberg, Phys. Rev. Lett. **81**, 357 (1998).
10. S. P. Nikitin, I. Alexeev, J. Fan, and H. M. Milchberg, Phys. Rev. E **59**, R3839 (1999).
11. P. Volfbeyn, E. Esarey, and W. P. Leemans , Phys. of Plasmas **6**, 2269, (1999)
12. P. G. Gobbi and G. C. Reali, Opt. Comm., **52**, 195 (1984).
13. M. Takeda, H. Ina, and S. Kobayashi, J. Opt. Soc. Am **72**, 156 (1980).
14. G. Pretzler Z. Naturforschung, **46 a**, 639 (1991).
15. N. E. Andreev, L. Y. Margolin, I. V. Pleshanov, L. N. Pyatnitskii, Quantum Electronics **28**, 910 (1998)
16. M. V. Ammosov, N. B. Delone, and V. P. Kranov, Soviet Physics - JETP, **64**, 1191 (1986)
17. P. Sprangle, B. Hafizi, and J. R. Peano, Phys. Rev. E. **61**, 4381 (2000).
18. W. Leemans, C. W. Siders, E. Esarey, N.E. Andreev, G. Shvets, and W. B. Mori, IEEE Trans. Plasma Sci.**24**, 331 (1996).

GeV Laser Acceleration Research at JAERI-APR

K. Nakajima[†*], T. Hosokai[*], S. Kanazawa[*],
M. Kando[*], S. Kondo[*], H. Kotaki[*], T. Yokoyama[*]

[*] *Japan Atomic Energy Research Institute, Kizu, Kyoto, 619-0215, Japan*

[†] *High Energy Accelerator Research Organization, Tsukuba, Ibaraki, 305-0801, Japan*

Abstract. Recent experiments on ultrahigh field particle acceleration driven by ultraintense laser pulses in plasmas have succeeded in electron acceleration up to the order of 100 MeV with large energy spreads. The next step of laser acceleration research has been focused on the injection of ultrashort electron bunches into a correct acceleration phase of laser wakefields and on the optical guiding of ultraintense ultrashort laser pulses in underdense plasmas. A proposal of laser wakefield accelerator (LWFA) experiments has been presented to accomplish high energy gains of the order of GeV with high quality electron beam injection. The recent status of the project proceeded at JAERI-APR is reported on the developments of high quality electron beam injectors and the capillary plasma waveguide for optical guiding of ultrashort intense laser pulses.

INTRODUCTION

A novel particle acceleration concept was proposed by Tajima and Dawson [1], which utilizes plasma waves excited by intense laser beam interactions with plasmas for particle acceleration, known as laser-plasma accelerators. In particular recently there has been a great experimental progress on the laser wakefield acceleration of electrons since the first ultrahigh gradient acceleration experiment made by Nakajima et al. [2]. Recent world-wide experiments have successfully demonstrated that the self-modulated LWFA mechanism is capable of generating ultrahigh accelerating gradient of ~ 100 GeV/m. In the self-modulated LWFA, however, the maximum energy gain has been limited at most to 100 MeV with energy spread of ~ 100 % because of dephasing and wavebreaking effects in highly dense plasmas where thermal plasma electrons are accelerated. The first high energy gain acceleration exceeding 200 MeV has been observed with the injection of an electron beam at an energy matched to the wakefield phase velocity in a fairly underdense plasma [3]. Hence the second-generation research has dealt with the injection of ultrashort electron bunches into a correct acceleration phase of laser wakefields and the optical guiding of ultraintense ultrashort laser pulses in underdense plasmas in order to accomplish high energy gains of more than 1 GeV and high quality beam acceleration with a small energy spread. Here the conceptual designs of GeV laser wakefield accelerator are discussed from the points of view on the ultrashort pulse beam injection and the optical guiding. The recent achievements of the laser acceleration research at JAERI-APR are reported.

CONCEPTUAL DESIGNS OF GEV LWFA

Plasmas provide some advantages as an accelerating medium in laser-driven accelerators. Plasmas can sustain ultrahigh electric fields, and can optically guide the laser beam and the particle beam as well under appropriate conditions. For a nonrelativistic plasma wave, the acceleration gradients are limited to the order of the wave-breaking field given by $eE_{WB}[\text{eV/cm}] = m_e c \omega_p \simeq 0.96 n_0^{1/2}[\text{cm}^{-3}]$, where $\omega_p = (4\pi n_0 e^2/m_e)^{1/2}$ is the electron plasma frequency and n_0 is the ambient electron plasma density. It means that the plasma density of $n_e = 10^{18}$ cm^{-3} can sustain the acceleration gradient of 100 GeV/m.

As an intense laser pulse propagates through an underdense plasma, the ponderomotive force expels electrons from the region of the laser pulse. This effect excites a large amplitude plasma wave (wakefield) with phase velocity approximately equal to the group velocity of laser pulse, given by $v_p = c(1 - \omega_p^2/\omega_0^2)^{1/2}$, where ω_0 is the laser frequency. The maximum axial wakefield occurs at the plasma wavelength, $\lambda_p[\mu\text{m}] \simeq 0.57\tau$ in a plasma with the resonant electron density, $n_0[\text{cm}^{-3}] \simeq 3.5 \times 10^{21}/\tau^2$ in terms of a FWHM pulse duration τ [fs]. When a Gaussian driving laser pulse with the peak power P [TW] is focused on the spot size r_0 [μm], the maximum axial wakefield yields

$$(eE_z)_{max}[\text{GeV/m}] \simeq 8.6 \times 10^4 P \lambda_0^2 / (\tau r_0^2 \gamma_0), \tag{1}$$

where $\gamma_0 = (1 + a_0^2/2)^{1/2}$ takes account of nonlinear relativistic effects, and $a_0 = 6.8\lambda_0 P^{1/2}/r_0$ is the laser strength parameter for the linear polarization [4].

Several effects limit the energy gain in a single-stage of laser-plasma accelerators; laser diffraction, electron dephasing, pump depletion and laser-plasma instabilities. For a Gaussian beam propagation of the laser pulse with the peak power P in an underdense plasma, the effective acceleration length can be limited to a diffraction length $L_{dif} = \pi Z_R$, where $Z_R = \pi r_0^2/\lambda_0$ is the Rayleigh length. For a properly phased electron, the maximum energy gain limited by diffraction effects is given by

$$\Delta W_{dif}[\text{GeV}] \simeq 0.85 P[\text{TW}] \lambda_0[\mu\text{m}]/(\gamma_0 \tau[\text{fs}]). \tag{2}$$

As the electron is accelerated, its velocity v_z will increase and approach the speed of light, $v_z \to c$. If the phase velocity of the plasma wave is constant with $v_p < c$, the electrons will eventually outrun the accelerating phase and move into the

TABLE 1. Parameters of the GeV channel-guided laser wakefield accelerators.

Energy gain [GeV]	0.5	1	5
Pulse duratuon τ [fs]	20	50	100
Peak power [TW]	100	40	20
Spot radius [μm]	30	20	10
Laser strength parameter	1.8	1.7	2.4
Plasma density [10^{18} cm^{-3}]	8.8	1.4	0.35
Accelerating gradient [GeV/cm]	1.9	0.7	0.55
Diffraction length [cm]	1.1	0.5	0.12
Dephasing length [cm]	0.4	5.5	56
Channel length [cm]	No	1.5	10
$N_{\max}[10^9]$	7	1.1	0.2

decelerating phase. For a highly relativistic electron with $v_z \simeq c$, the dephasing length is given by $L_d = \lambda_p \gamma_p^2$, where $\gamma_p = (1 - v_p^2/c^2)^{-1/2} = \omega_0/\omega_p$ is the relativistic factor associated with the phase velocity of the plasma wave. The maximum energy gain limited by dephasing effects is

$$\Delta W_d[\text{GeV}] = 0.01 P[\text{TW}] \tau^2[\text{fs}]/r_0^2[\mu\text{m}]. \qquad (3)$$

As the laser pulse excites a plasma wave, it loses its energy. The pump depletion length L_{pd}, in which the laser pulse loses a half of its total energy to excite plasma waves, is given by $L_{pd} \simeq 2.65 \lambda_p \gamma_p^2 a_0^{-2}$. The maximum energy gain limited by pump depletion is

$$\Delta W_{pd}[\text{GeV}] = 0.91 \times 10^{-3} \tau^2[\text{fs}] \gamma_0^2 / \lambda_0^2[\mu\text{m}]. \qquad (4)$$

The maximum number of electrons capable of accelerating with 100% energy spread is estimated to be $N_{\max} \sim 3.55 \times 10^9 P \lambda_0^2 / (\tau \gamma_0)$.

In order to achieve the acceleration energy gains of higher than 1 GeV in a single stage of cm-scale, it is necessary to extend the acceleration length limited by diffraction effects of laser beams. We propose the channel-guided LWFA in which both the driving laser pulses and particle beams can be guided through Z-pinch capillary discharge plasmas of cm-scale. The parameters to test electron acceleration of GeV energies are shown in Table 1. The designs of the LWFA are based on availability of the 10 Hz table-top ultrashort, ultrahigh peak power Ti:Sapphire laser with 20 fs and 100 TW developed at JAERI-APR.

THE ELECTRON BEAM INJECTORS

In order to produce a high quality electron beam with low momentum spread and small pulse-to-pulse energy stability, it is required that femtosecond electron bunches should be injected with the energy higher than trapping threshold and femtosecond synchronization with respect to a wakefield accelerating phase space which is typically less than 100 fs in a longitudinal scale and 10 μm in a transverse size. For the second-generation LWFA experiments, we have developed an electron injection system consisting of a photocathode RF gun and a compact race-track microtron shown in Fig. 1.

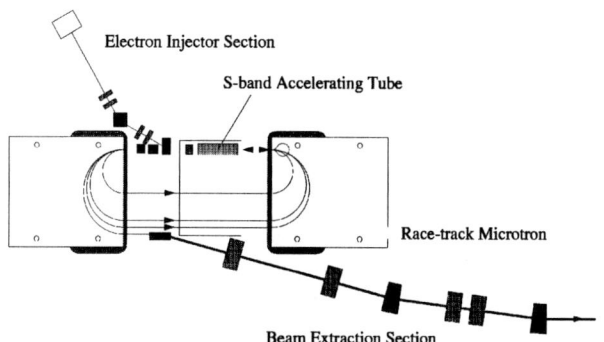

FIGURE 1. A schematic of the electron injection system.

The photocathode RF gun

The S-band RF gun has been developed by the collaboration with BNL, KEK and Sumitomo Heavy Industries, Ltd (SHI). This cavity was based on the gun developed by the BNL / SLAC / UCLA collaboration [5] and was improved for high duty operation at 50 Hz. The gun can produce $\sim 1\ \pi$ mm·mrad electron pulses owing to an emittance compensation solenoid magnet. A copper cathode is illuminated by an UV light of 263 nm with an incident angle of 68°, delivered from a compact all solid-state Nd:YLF laser system (SHI PULRISE II). This laser can generate the output pulse energy of 200 μJ at 263 nm with fluctuation of 0.5 % and the pulse width of approximately 6 ps FWHM.

We have performed tests of the photocathode RF gun and the driving laser at the Nuclear Engineering Research Laboratory, the University of Tokyo. The results of beam tests are summarized in Table 2. The quantum efficiency has been increased up to about 3 times higher values than that at the installation by a vacuum improvement.

The Microtron booster

The race-track microtron (RTM) manufactured by SHI is utilized as a booster accelerator. The original RTM accelerates electrons emitted by a thermionic gun from 120 keV to 150 MeV after 25 turns. We replaced the thermionic gun with the

TABLE 2. Results of the beam tests on the photocathode RF gun.

Maximum beam charge	2.7 nC (1.7×10^{10}e)
Quantum efficiency	1.4×10^{-4}
Output energy	3.7 MeV
Energy spread, $\Delta E/E$	4.7 % (FWHM)
Pulse width	< 7.6 ps (FWHM)
Normalized emittance	< 10π mm·mrad
Beam stability	1.1 % (rms)

photocathode RF gun to inject a low emittance electron beam at 4.5 MeV. According to the tracking simulation for an injection beam to the RTM, the transmission efficiency of the beam through the RTM has been improved up to 50 %, which was much better than the RTM with the thermionic gun [6].

The RTM has two straight sections; the one is non-dispersive where the accelerating tube is placed, and the other is dispersive due to the 180° bending magnet of the RTM. Because of the space limitation, beam extraction magnet is placed in the dispersive straight section. We designed a dispersion-compensated beam extraction section in order to avoid emittance growth caused by dispersion. The section was designed by employing transfer matrix technique to satisfy doubly achromatic conditions. Finally, we found that three bending magnets and four quadrupole magnets system can efficiently work with transverse focusing.

The RTM combined with the photocathode RF gun was installed to JAERI-APR in August 1999. Their commissioning has been started in March 2000, and we succeeded in generating a 150 MeV single electron bunch with a charge of 91 pC at 10Hz. Fig. 2 shows the synchrotron light images observed at each lap of the electron paths.

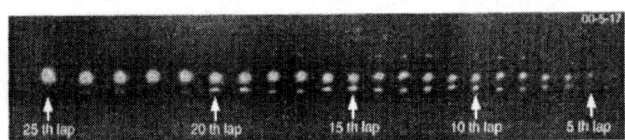

FIGURE 2. The synchrotron light images observed at each lap of the electron paths in RTM.

Bunch Slicing and Femtosecond Synchronization

The current performance of photocathode RF electron guns and the bunch compression technology has achieved a few 100 fs femtosecond bunch length and timing accuracy. These achievements could not necessarily satisfy requirements for high quality beam acceleration of GeV LWFA. Although novel schemes which use laser triggered injection of plasma electrons into the wakefield have been proposed as a plasma cathode, there has been no experimental proof so far to demonstrate generation of a femtosecond electron beam with a narrow energy spread. We have conceived a method capable of generating a femtosecond electron pulse injected into a correct wakefield phase within a few femtoseconds. Generation of femtosecond electron pulses with femtosecond synchronization is based on slicing a bunch through a process of energy modulation created in the interaction of electrons with a femtosecond laser pulse splitted from a main pump pulse. This technique of slicing a bunch of energy modulated electrons has been demonstrated to produce femtosecond synchrotron radiation pulses in a electron storage ring at the Advanced Light Source of Lawrence Berkeley National Laboratory [7].

We apply the energy modulation technique to production of an ultrashort slice of a few 10 femtoseconds duration from a electron bunch of a few picoseconds duration delivered by the RTM. The mechanism of energy modulation is based on the inverse free electron laser, which generates the efficient energy exchange between electrons and laser fields in an undulator when the laser wavelength λ_L satisfies the resonance condition of free electron lasers, given by $\lambda_L = \lambda_u(1+K^2/2)/(2\gamma^2)$, where λ_u is the undulator period, γ is the Lorentz factor, and $K = eB_0\lambda_u/(2\pi m_e c) = 0.934\lambda_u[\text{cm}]B_0[\text{T}]$ is the deflection parameter of the undulator with the peak magnetic field of B_0. In addition to the resonance condition, when the transverse mode and the spectral bandwidth matching between the laser and the undulator spontaneous radiation can be satisfied, an amplitude of energy modulation is given by [8]

$$(\Delta E)^2 \simeq 4\pi\alpha A_L \hbar \omega_L \frac{K^2/2}{1+K^2/2}\frac{M_u}{M_L} \qquad (5)$$

for $M_u \leq M_L$ where A_L is the laser pulse energy, $\hbar$ is the Plank's constant, α is the fine structure constant, $\omega_L = 2\pi c/\lambda_L$, M_u is the number of undulator periods, and M_L is the laser pulse length in optical cycles. We consider the energy modulation for the microtron beam at $E = 150$ MeV with an expected energy spread of 0.1 % using the undulator with $\lambda_u = 3.3$ cm, $M_u = 61.5$ and the maximum K value of 1.8. The resonance condition is satisfied for $K = 1.6$, provided by adjusting the gap. Assuming a femotosecond laser pulse with $\lambda_L = 400$ nm, the second harmonics of a 50 fs Ti:sapphire laser pulse splitted from a high peak power pump pulse, the pulse energy required to produce an energy modulation ΔE can be estimated as $A_L[\mu\text{J}] \sim \Delta E^2[\text{MeV}]$. The energy modulation $\Delta E = 15$ MeV can be produced by a laser pulse with $A_L = 225\mu\text{J}$.

OPTICAL GUIDING IN PLASMA CHANNELS

In order to increase the energy gain beyond the diffraction limitation, it is essential to propagate a laser pulse in an underdense plasma beyond the vacuum Rayleigh length. A promising method is the relativistic self-guiding induced by relativistic quiver motion of the plasma electrons for the laser power exceeding a critical power, given by $P_c = 17(\omega_0^2/\omega_p^2)$ GW. Since the index of refraction, however, becomes modified by the laser pulse on the plasma frequency time scale, $\sim 1/\omega_p$, relativistic optical guiding is ineffective in preventing diffraction of ultrashort pulses, $L_L \leq \lambda_p/\gamma_0$ [9]. It is known that the relativistic self-guiding is associated with instabilities induced by ultraintense laser interactions with plasmas, such as filamentations and hose instabilities.

Optical guiding of a Gaussian laser pulse with a focal spot radius of r_0 can be made through the plasma density channel with a parabolic electron-density profile of the form $n(r) = n_0 + \Delta n r^2/r_0^2$, where $\Delta n \geq 0$. For a low power, $P \ll P_c$, low intensity $a^2 \ll 1$ laser pulse, the index of refarction is given by

$$\eta \simeq 1 - \frac{\omega_p^2}{2\omega_0^2}\left(1 + \frac{\Delta n}{n_0}\frac{r^2}{r_0^2}\right). \tag{6}$$

An envelope equation for a Gaussian laser beam of the form $a^2 = a_0^2 \exp(-2r^2/r_s^2)$ is

$$\frac{d^2 R}{dz^2} = \frac{1}{Z_R^2 R^3}\left(1 - \frac{\Delta n}{\Delta n_c}R^4\right), \tag{7}$$

where $R = r_s/r_0$, $\Delta n_c = 1/(\pi r_e r_0^2)$ is the critical channel depth, and $r_e = e^2/(m_e c^2)$ is the classical electron radius [10]. This indicates that a parabolic channel can guide a Gaussian beam with $r_s = r_0$ provided that the density channel depth is equal to the critical depth, $\Delta n = \Delta n_c$, which is $\Delta n_c[\text{cm}^{-3}] = 1.1 \times 10^{20}/r_0^2[\mu\text{m}]$.

The Fast Z-pinch Capillary Discharge Plasma Waveguide

The optical guiding experiments have been made by using a discharge capillary in a vacuum [11]. A plasma channel is formed in a capillary due to diffusion of a wall ablation plasma. We have developed a stable cm-scale plasma channel produced by an imploding phase of fast Z-pinch discharge in a gas-filled capillary without wall ablation. A high current fast Z-pinch discharge generates strong azimuthal magnetic field, which contracts the plasma radially inward down to ~ 100 μm in diameter. The imploding current sheet drives the converging shock wave ahead of it, producing a concave electron density profile in the radial direction just before the stagnation phase. The concave profile is approximately parabolic to out a radius of ~ 50 μm, after which the density falls off.

The typical experimental setup is shown in Fig. 3 [12]. We have used a capillary with an inner diameter of 1 mm and a length of up to 2 cm. With this configuration, the discharge current generated a peak of 4.8 kA with a rise time of about 15 ns and a duration of 70 ns (FWMH). The capillary was filled with helium, under differential pumping at an initial pressure which was varied from 0.5 to 5 Torr. A DC discharge circuit was used to form an uniformly preionized helium gas.

A high intensity Ti:Sapphire laser pulse ($\lambda = 790$ nm, 90 fs, $> 1 \times 10^{17}$ W/cm^2) was focused on the front edge of the capillary to a spot size of 40 μm in diameter.

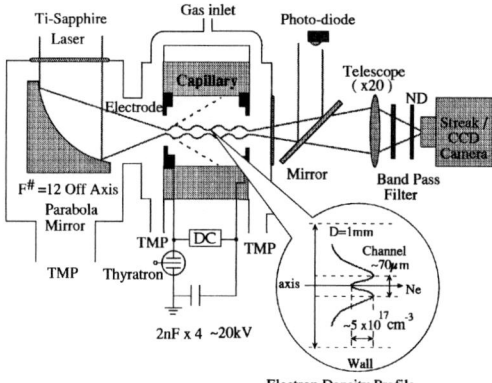

FIGURE 3. Experimental setup of a fast capillary discharge for a plasma waveguide. A typical electron density profile of the implosion phase of the fast capillary discharge in the radial direction is illustrated in the circle.

The transmitted laser beam profile at the exit of the capillary was observed through a band pass filter ($\Delta \lambda = 10$ nm) with a CCD camera. Fig. 4 shows typical CCD images of the transmitted high intensity Ti:sapphire laser pulse profile through the capillary discharge plasma. These show clearly that a high intensity laser pulse could be guided through the channel over a distance of 2 cm corresponding to $\sim 12.5\, Z_R$, where $Z_R \sim 1.6$ mm is the vacuum Rayleigh length. The electron density profile in the channel was estimated by corroborating the transmitted laser images with the results of ray trace calculation. We assumed that the electron density profile of the channel in the radial direction was parabolic and the channel was fully ionized with the electron density of 6.0×10^{16}cm^{-3} on the axis, which was twice the initial gas density. We also assumed that the incident laser had a radial Gaussian profile, the channel was uniform over the channel length and the channel radius r_{ch} was 35 μm according to the observed value. Under these assumptions, we can estimate the electron density on the peaks of the channel to be 1.5×10^{18} cm^{-3} and the density gradient to be $\sim 4 \times 10^{20}$ cm^{-4}. The matched beam radius r_m was given by $r_m = [r_{ch}^2/(\pi r_e \Delta n)]^{1/4}$, where Δn was the channel depth. With $r_{ch} = 35$ μm and $\Delta n \sim 1.5 \times 10^{18}$ cm^{-3}, the matched beam radius is $r_m \sim 17$ μm. The observed spot radius of ~ 20 μm was consistent with this value.

Electron Beam Dynamics in the Z-pinch Plasma Waveguide

In the Z-pinch discharge plasma, a strong Z-pinch force is exerted on the electron beam by the discharge driving current of several kA in a plasma column. The magnetic field gradient G [T/m] induced by the Z-pinch plasma current will be the order of 10^4 T/m inside the uniform plasma column with $\sim 100\mu$m radius for the

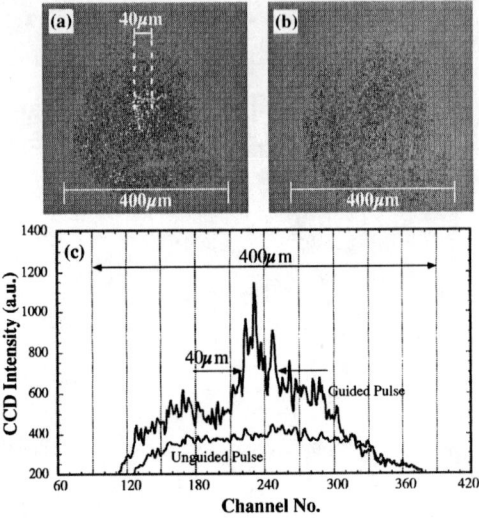

FIGURE 4. Typical CCD images of the transmitted a high intensity Ti:Sapphire laser pulse ($\sim 1 \times 10^{17}$ W/cm^2) through the capillary at an initial pressure of 0.9 Torr He; (a) t = 8.5 ns, (b) no discharge, and (c) their intensity profiles in the radial direction.

discharge current of 5 kA. The beam envelope equation on the r.m.s. beam radius, σ_{rb}, along the z-axis can be given by

$$\frac{d^2\sigma_{rb}}{dz^2} + K_F\sigma_{rb} - \frac{\varepsilon_n^2}{\beta^2\gamma^2\sigma_{rb}^3} = 0, \qquad (8)$$

where ε_n is the normalized emittance of the beam and $\gamma = (1-\beta^2)^{-1/2}$ is the relativistic factor of the electron beam and K_F [m^{-2}] $\simeq 600G/\gamma$ is the focusing strength induced by the discharge current. The equilibrium beam radius can be obtained from equating the focusing force and the transverse thermal spreading (emittance) force as σ_{rb}[mm] $\simeq 0.2\varepsilon_n^{1/2}$[mm · mrad]$\gamma^{-1/4}G^{-1/4}$[T/m]. The equilibrium radius of accelerated electron beams will be $\sim 3\mu m$ for the 1 GeV LWFA with an injection beam of $\varepsilon_n \sim 1$ mm·mrad. The channel guiding of the accelerated beam will occur by controlling the discharge current and the injection phase of the discharge.

CONCLUSIONS

The laser wakefield acceleration experiments have demonstrated ultrahigh gradient acceleration exceeding 100 MeV with the same order of energy spread. In the next step of LWFA research, two major problems must be overcome to accomplish so called controlled acceleration. The one is the injection matched to wakefields in their phase space to accelerate a high quality beam with a low emittance and a small energy spread. The other is the optical guiding to achieve high energy gains exceeding 1 GeV in a single stage. A design of GeV range LWFA indicates that the energy gain of 1 GeV will be accomplished by means of the channel-guided scheme over 1.5 cm pumped by 50 TW, 40 fs laser pulses.

From these points of view, our researches have concentrated on developments of the high quality beam injector and the plasma waveguides. We have developed the compact microtron with the photocathode RF gun as a high quality beam injector for the laser wakefield acceleration experiments. Recently commissioning of the injector system has been successfully conducted to produce a 150 MeV beam with a single bunch of 100 pC electrons at 10 Hz. The optical guiding development have resulted in a cm-scale plasma waveguide using the fast Z-pinch capillary discharge. We have succeeded in demonstrating propagation of 2 TW, 90 fs laser pulses over 2 cm in the Z-pinch plasma waveguide. In addition to these experimental achievements, we will construct the femtosecond bunch slicing stage as a part of the laser acceleration test facility to generate a femtosecond electron pulse injected into laser wakefields with femtosecond accuracy. Finally the facility including the 100 TW laser system constructed at JAERI-APR will enable us to conduct the laser wakefield acceleration experiments achieving high energy gains more than 1 GeV as well as high quality beam acceleration.

REFERENCES

1. T. Tajima and J. M. Dawson, Phy. Rev. Lett. **43**, 267 (1979).
2. K. Nakajima et al., Rhy. Rev. Lett. **74**, 4428 (1995).
3. H. Dewa et al., Nucl. Instr. and Meth. in Phys. Res. **A410**, 357 (1998); M. Kando et al., Jpn. J. Appl. Phys. **38**, L967 (1999).
4. K. Nakajima, Nucl. Instr. and Meth. in Phys. Res. **A410**, 514 (1998).
5. D. T. Palmer et al., Proc. of Part. Accel. Conf., 982(1995).
6. M. Kando et al., Proc. of the 1999 Part. Accel. Conf. 5, 3704 (1999).
7. R. W. Schoenlein et al., Science, **287**, 2237-2240 (2000).
8. A. A. Zholents and M. S. Zolotorev, Phys. Rev. Lett. **76**, 912-915 (1996).
9. P. Sprangle et al., Phys. Rev. Lett. **69**, 2200 (1992).
10. E. Esarey et al., IEEE Trans. Plasma Sci. **24**, 252 (1996).
11. Y. Ehrlich et al., Phys. Rev. Lett. **77**, 4186 (1996).
12. T. Hosokai et al., Optics Letters, **25**,10-12 (2000).

Numerical Simulation for Plasma Electron Acceleration by 12TW 50 Fs Laser Pulse

N. Hafz[1], J. Koga*, R. Hemker, and M. Uesaka

Nuclear Engineering Research Laboratory, Graduate School of Engineering, University of Tokyo, Tokai, Naka, Ibaraki, 319-1106, JAPAN
*Japan Atomic Energy Research Institute, Kizu, Kyoto, 619-0215 JAPAN

Abstract. Particle-in-cell simulation in two dimensions (PIC-2D) have been performed in order to study the trapping and acceleration of plasma electrons in a plasma wave excited resonantly by a 12TW 50fs laser pulse. The phase-space characteristics of the electrons are presented for the resonant and non-resonant densities. The effect of the ramp thickness of the plasma-vacuum interface is also studied. The present work is relevant to the experimental work on the advanced accelerator research at the University of Tokyo. The experiment uses 12TW 50fs Titanium-Sapphire laser (λ=800nm) driver beam and gas-jet target.

I. INTRODUCTION

There has been a remarkable progress over the last two decades in the field of advanced accelerator research aiming to understand the detailed physics of various schemes (such as Beat Wave, Laser Wake Field, Plasma Wake Field Accelerators, etc.) and hence overcoming the difficulties (such as problems of laser guiding in plasma, electron beam injection and dynamics in the plasma wave, exact synchronization between the laser and electron beams, etc.) that face the realization of advanced accelerators. Recently, various proof-of-principle experiments[1-5] (in the self-modulated laser wake field accelerator scheme, SM-LWFA) have demonstrated electron acceleration up to very high energies in distances as short as few millimeters from gas-jet targets (acceleration gradients of the order of 1-100 GeV/m) by large amplitude plasma waves excited by intense (1-30 terawatt) laser pluses. In the SM-LWFA, the ambient plasma density $(n_e \sim 10^{19} cm^{-3})$ is high enough that the laser pulse is long compared with the plasma wavelength, $L > \lambda_p$ and the laser power is larger than the critical power for relativistic self-focusing[10], $P \geq P_{cr} = 17.4 \times \omega^2 / \omega_p^2$ [GW] where ω and ω_p are the laser and plasma frequencies respectively. In this regime, a large amplitude plasma wave (wakefield) is generated via a self-modulation or Raman forward scattered instability. The plasma-wave amplitude rapidly grows to extreme values (called wavebreaking limit[10], $E_0 [V/m] = cm_0 \omega_p / e \approx 97 n_e^{1/2} [cm^{-3}]$ where $\omega_p = (4\pi n_e e^2 / m_0)^{1/2}$ and n_e is the ambient electron plasma density) such that

---

[1] Corresponding author. E.mail<nasr@tokai.t.u-tokyo.ac.jp>

the background electrons can get trapped and accelerated by the wave which result in sub-ps electron bunch generation. It has been suggested that the wavebreaking is the responsible mechanism for self-trapping. A detailed beam characteristics form laser-driven wavebreaking by PIC simulations are found in Refs. [6,7]. The difficulties of the SM-LWFA are that it produces electron bunches with about 100% energy spread and the fact that self-modulation relies on instabilities, i.e. trapping and acceleration occur in an uncontrolled manner.

II. 2D-PIC SIMULATION

In the present work the laser pulse duration is 50 fs which is about one order of magnitude shorter than pulses used in some previous simulations and experiments[1-6]. The plasma density is chosen such that the laser pulse length L $(=c\tau_L)$ and the plasma wavelength λ_p are approximately equal, and the excitation of the plasma wave is similar to the standard LWFA i.e. in the resonance regime and no longer by the self-modulation. In the present simulation a fully relativistic 2D-PIC code has been used, with computational box of 2500×300(corresponds to $250\lambda \times 30\lambda$) Cartisian grid in x-y plane and the simulation follows 750,000 particles for about 5000 timesteps. The Gaussian laser pulse is plane polarized in z-direction, propagating along the x-axis and the spot size along the y-axis is r_0 ~7.5 μm. The laser pulse starts in vacuum then interacts with the plasma slab. The plasma temperature is $1 keV$ and the density varies linearly from zero at $x_1 = 50\lambda$ to $n_{e1}=3\times10^{18}$ cm^{-3} at some distance x_2 where $\lambda \sim 0.8\mu m$ is the laser wavelength. At the distance $x_3 = 240\lambda$ the plasma density drops to zero again with a scale length of 10λ. In a series of different simulations the linear vacuum-plasma ramp x_2-x_1 and the value of density n_{e1} were changed to see the effects. Figure 1 shows the plasma wave excited by laser pulse with $a_0 = 2.5$, $\tau_L= 50fs$, r_0 ~7.5 μm, where $a_0 = eE/m_0\omega c = 8.5\times10^{-10}\lambda I^{1/2}$ is the laser normalized vector potential, E is the laser electric field, and I $[W/cm^2] = cE^2/8\pi$. Due to beam loading, the oscillations have low amplitude in the region x=500-2000. The peak of the plasma wave is $E_{0x} \approx -97$ GV/m at $x \approx 2150$ and the position of the laser pulse is located at x ≈ 2400. In this simulation run we chose the approximate resonance $(c\tau_L \approx \lambda_p)$ plasma density $n_e=3\times10^{18} cm^{-3}$ at the flat top of the profile shown by dashed lines in Fig.1, and the vacuum-plasma ramp $x_2-x_1= 35\mu m \approx 2\lambda_p$.

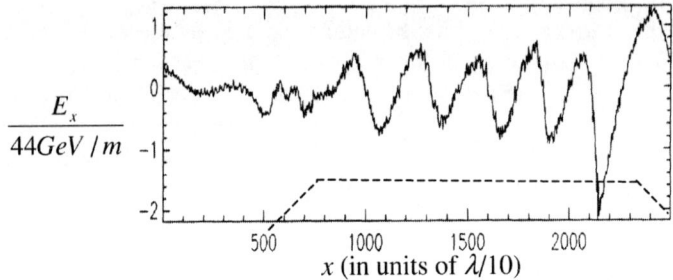

FIGURE 1. The longitudinal plasma wave (wakefield) excited by a laser pulse with $a_0 = 2.5$, $\tau_L = 50\text{fs}$, $r_0 \sim 7.5$ µm. The laser pulse propagates from left to right. The lowering in the amplitude of plasma wave in the region $x=500\text{-}2000$ is due to beam loading. The peak value of the wakefield at $x \approx 2150$ is $E_{0x} \approx -97$ GV/m. The laser pulse is located at $x \approx 2400$. The dashed lines shows the plasma density profile (shown for illustration and not to scale).

Figure 2a shows the longitudinal phase space (P_x-x) for the accelerated electrons which are trapped in the plasma wave given in Fig.1. The maximum energy of the accelerated electrons is about 35MeV. There are also 4 groups of low energy (1-5MeV) electrons trapped at each plasma wavelength's accelerating phase, the total number of accelerated particles is about 10^{10}, a large number of electrons that might be accelerated due to wavebreaking. The y-P_x phase space is shown is Fig.2b which shows the collimation of the accelerated electron beam near the y-axis (y=150), and there are also electrons (having relatively low energy) that are thrown out by the transverse wakefield.

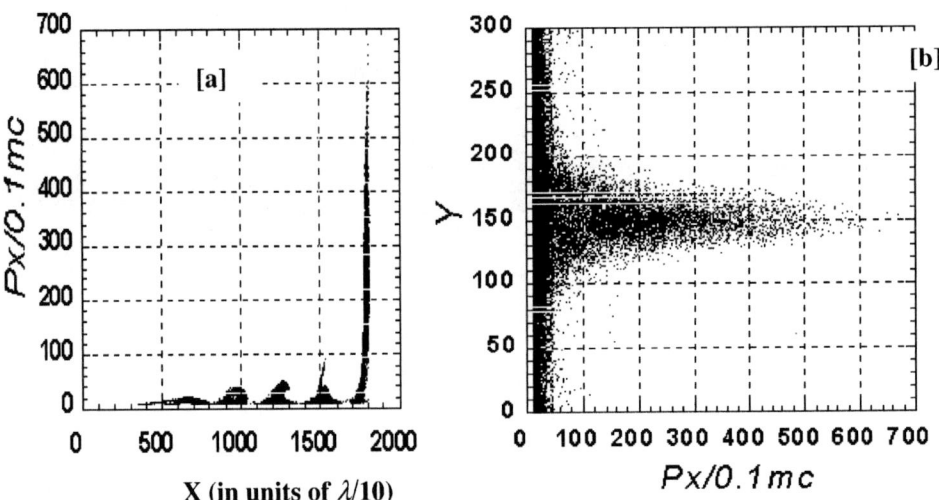

FIGURE 2. [a] The longitudinal phase space (P_x-x) of trapped plasma electrons in the plasma wave shown in Fig.1. [b] The phase space (y-P_x). These results are for sharp vacuum–plasma density ramp ($\approx 2\lambda_p$).

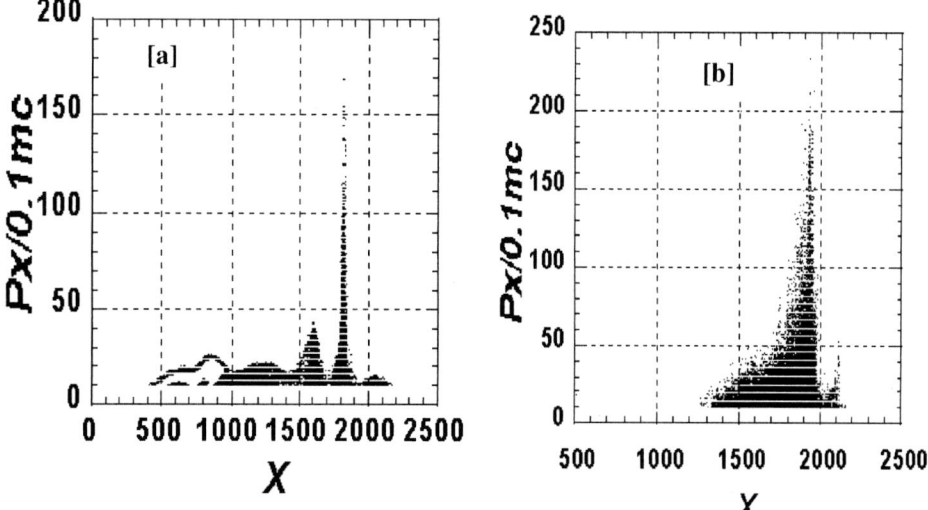

FIGURE 3. [a] The longitudinal phase space (P_x-x) of trapped plasma electrons in the case of ~$4\lambda_p$ density ramp and $n_{el}=3\times10^{18}\,cm^{-3}$. [b] The phase-space for ~$37\lambda_p$ ramp and $n_{el}=3\times10^{20}\,cm^{-3}$. X is in units of $\lambda/10$.

Bulanov et al.[7] interpreted that this low energy electron component appears because of the 2-D structure of the breaking of the plasma wave, during the break, those electrons at first move toward the axis, where their trajectory intersects, and then leave the plasma wave and propagate outward. By increasing the ramp thickness within the range 72μm-500μm ($\approx 4\lambda_p$-$27\lambda_p$ at $n_{el}=3\times10^{18}cm^{-3}$ where $\lambda_p \cong 3.3\times10^{10}n_e^{-1/2}[cm^{-3}]\mu m$), the maximum energy of the accelerated electrons is about 8MeV, this result is shown in Fig.3a. By increasing the ramp thickness up to $37\lambda_p$ (for $n_{el}=3\times10^{20}cm^{-3}= 0.17n_c$, where $n_c=1.7\times10^{21}\,cm^{-3}$) the phase space became as shown in Fig.3b. We still see trapped electrons with relativistic energies up to ~ 12MeV but the structure of the accelerated electrons are different in this case. Furthermore, when looking at the phase-space y-P_x in Fig.4 for those electrons of Fig.3b, we see that the electrons are not collimated along the y-axis which means that the wave's focusing force is less effective than the case of Fig. 2b.

We conclude by saying that, far from the SM-LWFA regime, by resonating the plasma density with the 50 fs laser pulse, it is possible to trap and accelerate plasma electrons up to 35MeV in a well collimated beam. By increasing the density to about $0.1n_c$ there still trapped particles but not in a single beam-like shape (nearly filamentary structure). The trapping and acceleration mechanisms in both cases seem to be very different. In the resonant case, wavebreaking may take place due to both the excitation of plasma wave up to very large amplitude resonantly with the laser and due to the sharpness of the vacuum-plasma ramp[8]. In the higher density non-resonant case and for long ramps, the direct laser acceleration[9] (DLA) is responsible for the appearance of relativistic electrons.

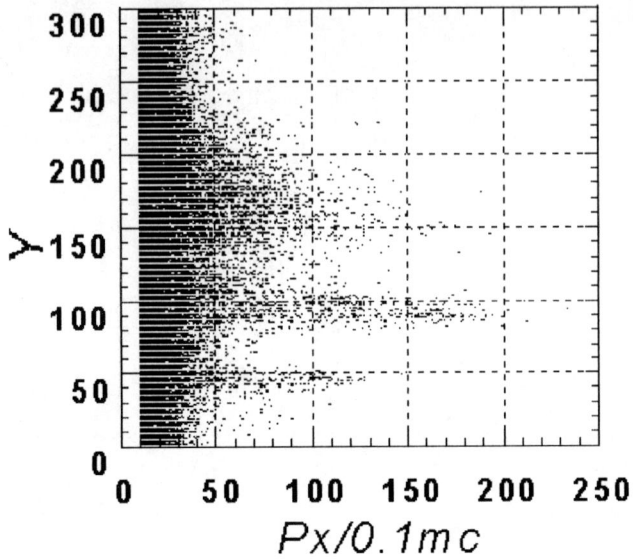

FIGURE 4. The phase space (y-P_x) for electrons in case of $\sim 37\lambda_p$ thick ramp and $n_{el}=3\times10^{20}cm^{-3}$. Y is in units of $\lambda/10$.

ACKNOWLEDGMENTS

The authors are grateful for Prof. Y. Katsumura, and Dr. K. Nakajima, for useful discussions and for continuous encouragement.

REFERENCES

1. K. Nakajima *et al.* Phys. Rev. Lett. **74**, 4428 (1995).
2. D. Umstadter *et al.* Science **273**, 472 (1996).
3. D. Gordon *et al.* Phys. Rev. Lett. **80**, 2133 (1998).
4. C. Coverdale *et al.* Phys. Rev. Lett. **74**, 4659 (1995).
5. A. Modena *et al.* Nature **377**, 606 (1995).
6. K. Tzeng et al. Phys. Rev. Lett. **79**, 5258 (1997).
7. S. Bulanov et al. Plas. Phys. Rep. **25**, 468 (1999).
8. S. Bulanov *et al.* Phys. Rev. E. **58**, R5257 (1998).
9. Pukhov et al. Phys. of Plasmas **6**, 2847 (1999).
10. E. Esarey et al IEEE Trans. Plas. Sci. **24**, 252 (1996).

Status of the LILAC Experiment

N. Saleh[†], P. Han[†], C. Keppel,[*] P. Gueye,[*] V. Yanovsky[†] and D. Umstadter[†]

[†]*Center for Ultrafast Optical Science*
University of Michigan, Ann Arbor, MI 48109, USA
[*]*Dept. of Physics, Hampton University, Hampton, VA 23668*

Abstract. We present the status of the LILAC experiment [1], including results on the propagation of 30-fs duration laser pulses in plasmas of the requisite density, and measurements of the dark current [2]. We also discuss the status of a laser upgrade, an electron beam line and plans for the future.

INTRODUCTION

The concept of optical injection of electrons in laser-produced plasma waves has generated much recent interest. We have proposed an acceleration schemes based on this concept [1], and embarked upon implementing it at the University of Michigan's Center for Ultrafast Optical Science. Variants of this concept were also proposed [3]. We have explored the parameter space necessary to implement this concept and are in the process of building the requisite laser system and diagnostics.

Description of the Laser System

The technology developed at our Center, along with the specific laser system we used, is described, respectively, in [4][5]. Fig. 1 and Fig. 2 depict a schematic of the laser system and experimental chambers.

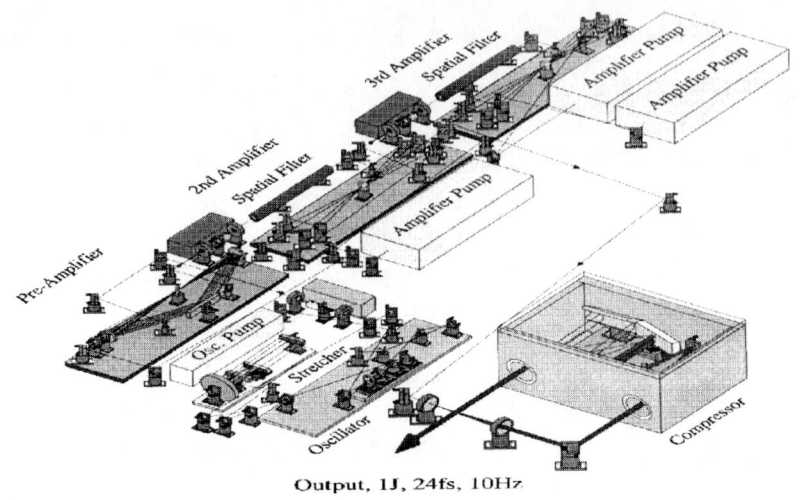

FIGURE 1. High peak power CPA laser system used in first run of the LILAC Experiment.

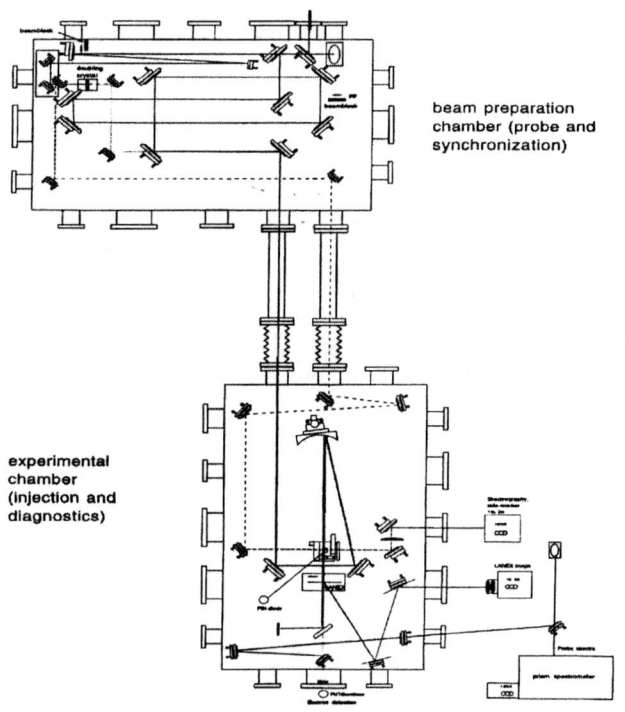

FIGURE 2. Experimental chambers for the LILAC experiment first run.

Experimental Procedure

The LILAC (Laser Injected Laser ACcelerator) concept utilizes a loosely focused (1-D) short laser pulse, which resonantly produces a wake field in under-dense plasma, and forms the acceleration structure. Another short pulse, transversely overlapped, tightly focused, and properly synchronized with the first pulse, suitably injects electrons into this acceleration structure so that they get trapped and accelerated. The resulting accelerated bunch is predicted to have novel characteristics [1]. Some issues within the experimental dynamics may cause fatally serious problems to the LILAC schemes; among these issues are, filamentation and laser beam break up, strong dark current from the resonant wakefield itself, and the failure of the short pulse to self focus in the resonant regime. We found out in our first run of the experiment that the former two points are not going to fail the LILAC scheme. The last point, however, remains to be investigated in the second run of the experiment. The experimental parameters are listed in Table1. We proceeded along with our experiment and scanned the density of the plasma by changing the backing pressure on the gas. By doing so, we were actually varying the plasma wave (wakefield) period and sweeping from the sub-resonant regime, passing through the resonant regime, and inching along towards the self-modulated regime. We had initial evidence that we were actually generating a resonant wakefiled (plasma wave) from our weak frequency-doubled probe that was collected in the forward direction from the collective collinear Thompson scattering of the anti-Stokes satellites, this is depicted in Fig. 3. We were interested in experimental observations near the resonant regime; the favorable regime for the LILAC scheme. The resonant density, n_r, is related to the critical density, n_c, of the plasma by the relation

$$n_r = (\lambda_{laser}/ c\tau_{laser})^2 n_c \quad (1)$$

with λ_{laser} being the laser wavelength, and τ_{laser} being the laser pulse duration. Fig. 4, shows that the laser pulse can be sustained before, and suitably after the resonant regime without breaking up. This range extends beyond the critical power for relativistic self-focusing, and up to five times the critical power where relativistic filamentation takes place. Only then, we observe the generation of narrow (1-degree)

TABLE 1. Experimental parameters of the first run of LILAC experiment

Laser pulse duration	29 fs
Laser wavelength	810 nm
Maximum power	35 TW
Maximum intensity	3×10^{18} W/cm^2
Plasma medium	He-like N_2 @ $n_e \leq 10^{20}$ cm^{-3}, supersonic gas jet
Laser beam diameter	4 cm
Focusing parabola	f/4.5 (13°).

MeV-electron bursts correlated with emission of red-shifted laser light, unlikely from forward Raman scattering [6]. We also show in [2] how the filaments' characteristics closely compare to what was theoretically predicted.

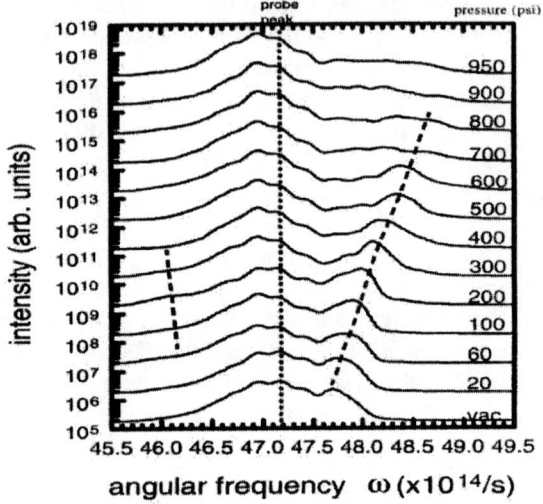

FIGURE 3. Initial evidence of collective collinear Thompson scattering of the anti-Stoles satellites from a resonant wakefield; satellite position varies with gas backing pressure (plasma period).

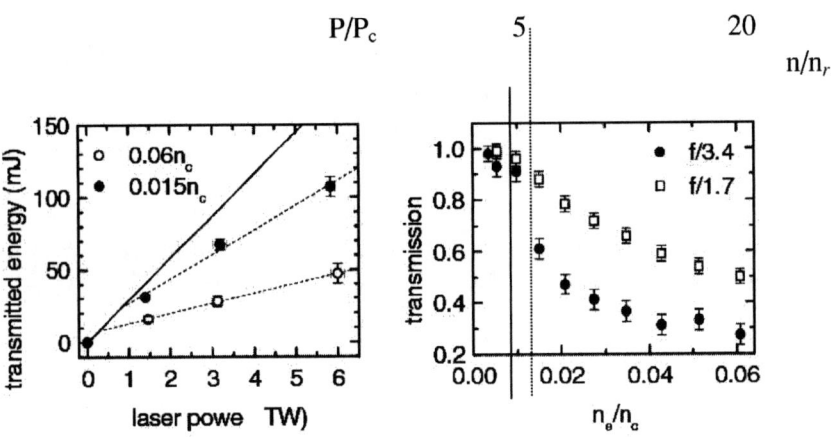

FIGURE 4. Data collected from the transmitted laser light when focused onto the supersonic gas jet. When using the f/3.4 collecting optic, and at five times the critical power, we observe strong filamentation, and the diffracted light exiting the gas jet significantly misses the collecting optic. The solid line represents the resonant regime, when both ratios; n/n_r and τ_{laser}/τ_{pl} are unity (τ_{pl} is the plasma wave duration). The dotted line represents the onset of relativistic filamentation, and MeV-electron generation.

We emphasize here that electrons were observed only when filamentation took place, but prior to that, including the vicinity of the resonant regime, no significant electron signal was measured. This brings us to the conclusion that there was no dark current in the resonant regime, and that the LILAC operational power parameters are: $P_C < P < 5\,P_C$, with P_C being the critical power for relativistic self-focusing. This concludes the data set for the first experimental run. The significance of the absence of dark current, i.e., pre-injection self-trapped electrons, is that the resonant wakefield will not be distorted or beam-loaded, so that the predicted novel characteristics (monochromaticity, duration, contrast, transverse emittance, etc.) of the accelerated bunch, after injection, will not be compromised.

Laser System Upgrade

We are developing and new laser system at the Center for Ultrafast Optical Science of the University of Michigan [8]. The new system is a high contrast, CPA, 150 TW on-target, with a dual compressor system, which permits the independent control of the optical properties of two laser pulses to optimize them for production of the wakefiled, and injecting electrons in it. A simple schematic is shown below. We anticipate the system to be completed no later than the end of the current year.

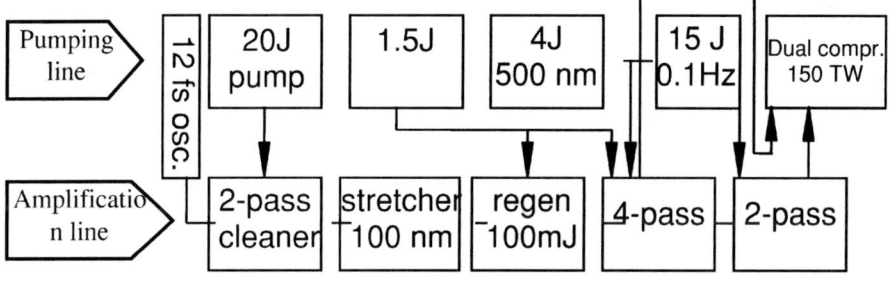

QQQDD Electron Spectrometer System

We have also developed a 500-MeV electron spectrometer at the University of Michigan's Center for Ultrafast Optical Science with collaborators from Hampton University [7]. The QQQD part of the spectrometer has been installed in the experimental area at Michigan, and ready to receive beam. We await the last dipole to arrive at some later time. This system was designed and simulated using two beam transport codes; TRANSPORT and OPTIM. Another code, GEANT, was also used to investigate the generation of secondary particles from the beam pipe. Fig. 5 to Fig. 8 show some details of this spectrometer design. The magnets were manufactured by Danfysik, and tested, calibrated, and characterized by Hampton University using a rotating coil probe with %0.1 and %1 accuracy for the dipoles and quadrupoles, respectively.

FIGURE 5. Spectrometer under construction, three quadrupoles, and one dipole.

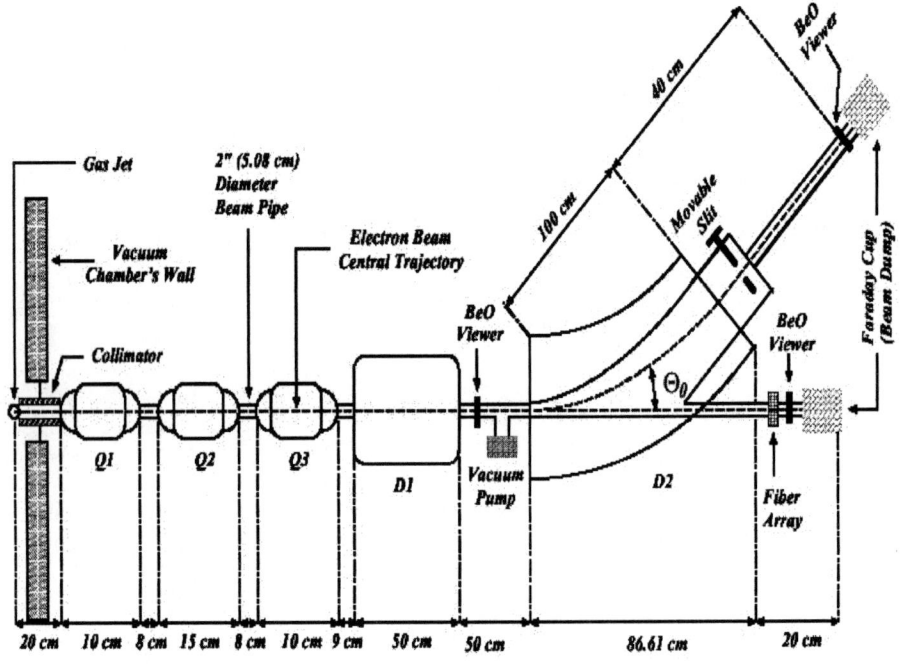

FIGURE 6. Schematic of the electron spectrometer at CUOS, Michigan.

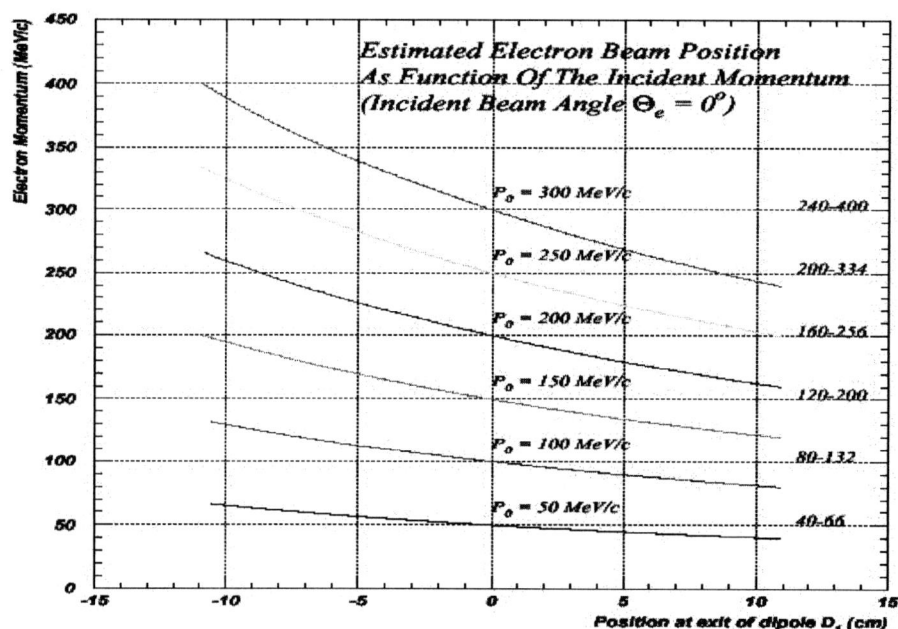

FIGURE 7. Dispersion curves generated by OPTIM at the exit of the first dipole magnet for various electron momenta.

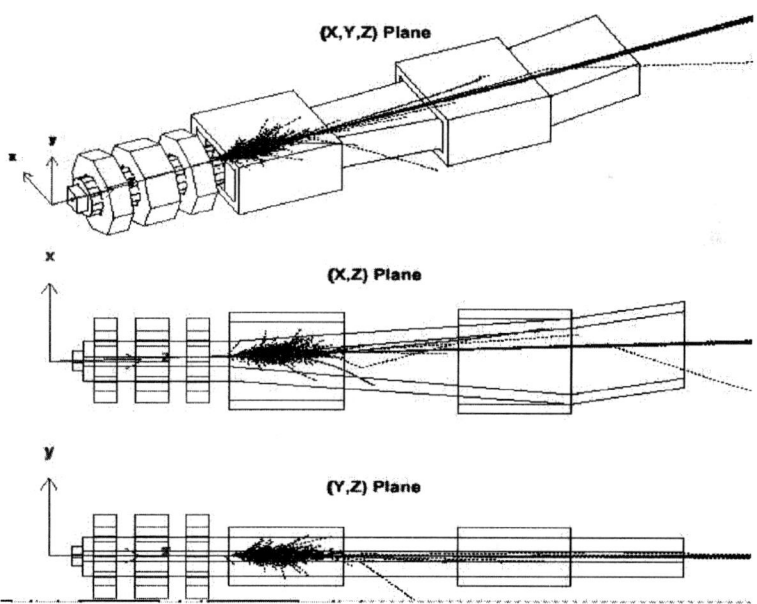

FIGURE 8. GEANT simulation of secondary particles production for a worst-case-scenario of 50 GeV incident electron beam.

In-Chamber Coherent Transition Radiation (CTR) Detector

We also plan on installing a coherent transition radiation detector inside the vacuum chamber where the experiment optics are set up. The reason for doing so is that the LILAC scheme, predicts the productions of an electron bunch as short as 10 fs, this corresponds to infrared CTR of 3 microns and longer, and we are not aware of any window material with transmission properties that cover such a broad band of radiation. The setup for this detector consists of a generic Michelson interferometer, with a broadband pyroelectric IR detector from Molectron. See Fig. 9.

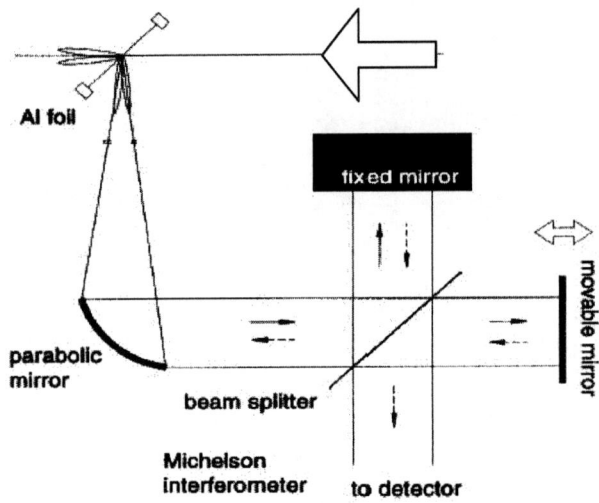

FIGURE 9. Schematic for CTR detector intended for the second run of the LILAC experiment.

CONCLUSION

The first run of the LILAC experiment produced forthcoming results in terms of the feasibility of the LILAC scheme. The resonant-wake-producing laser pulse was sustained in the plasma above the critical power for relativistic self-focusing, and the produced wake did not self-trap and accelerate any significant charge (no dark current). The Laser system is being upgraded to operational specifications closer to the optimum experimental conditions than before. We also designed an electron spectrometer that will be used to analyze the produced electron bunch. Other diagnostics are also being developed to provide full characterization of the anticipated accelerated bunch. This promotes the LILAC scheme to a promising run in the near future.

ACKNOWLEDGEMENT

We acknowledge the support of the High Energy Physics Division of the US Department of Energy, with laser facilities supported by the NSF. We would also like to thank X. Wang, G. Mourou, H. Kapteyn, M. Murnane and S. Backus for their many contributions.

REFERENCES

1. D. Umstadter *et al.*, *Physical Review Letters*, **76**, 2073 (1996).
2. X. Wang *et al.*, *Physical Review Letters*, **84**, 5324 (2000).
3. E. Esarey *et al.*, *Physical Review Letters,* **79**, 2682 (1997), B. Rau *et al., Physical Review Letter,* **78**, 3310 (1997), R. G. Hemker *et al., Phys. Rev. E,* **57**, 5920 (1998), S. Bulanov, *Plasma Phys. Rep.* **25**, 468 (1999), C.I. Moore *et al., Physical Review Letters.* **82**, 1688 (1999).
4. S. Backus *et al.*, *Review of Scientific Instruments*, **69**, 1207 (1998).
5. H. Wang *et al.*, *JOSA B*, **16**, 1790 (1999).
6. See details in reference [1] on this point.
7. P. Gueye *et al.*, proposal to be published.
8. V. Yanovsky *et al.*, *Digest of CLEO Technical Papers 2000*, 288 (2000).

Laser Wakefield Accelerator Experiments at LBNL

W.P. Leemans, D. Rodgers, P.E. Catravas, G. Fubiani, C.G.R. Geddes, E. Esarey, B.A. Shadwick, G.J.H. Brussaard,[†] J. van Tilborg,[†] S. Chattopadhyay, J.S. Wurtele, L. Archambault, M.R. Dickinson, S. DiMaggio, R. Short, K.L. Barat, R. Donahue, J. Floyd, A. Smith, and E. Wong

Ernest Orlando Lawrence Berkeley National Laboratory, University of California
1 Cyclotron Road, Berkeley CA 94720

Abstract. The status is presented of the laser wakefield acceleration research at the l'OASIS laboratory of the Center for Beam Physics at LBNL. Experiments have been performed on laser driven production of relativistic electron beams from plasmas using a high repetition rate (10 Hz), high power (10 TW) Ti:sapphire (0.8 μm) laser system. Large amplitude plasma waves have been excited in the self-modulated laser wakefield regime by tightly focusing (spot diameter 8 μm) a single high power (≤ 10 TW), ultra-short (≥ 50 fs) laser pulse onto a high density ($> 10^{19}$ cm^{-3}) pulsed gasjet (length 1.2 mm). Nuclear activation measurements in lead and copper targets indicate the production of electrons with energy in excess of 25 MeV. This result was confirmed by electron distribution measurements using a bending magnet spectrometer. Progress on implementing the colliding pulse laser injection method is also presented. This method is expected to produce low emittance ($< 1\pi$ mm-mrad), low energy spread ($< 1\%$), ultrashort (fs), 40 MeV electron bunches containing 10^7 electrons/bunch.

INTRODUCTION

Plasma-based accelerators [1], such as the laser wakefield accelerator (LWFA), offer the potential of developing ultra-compact accelerators capable of producing high quality relativistic electron beams. Acceleration of electrons to energies as high as 100 MeV over mm-size distances has been demonstrated in several experiments [2] - [8]. These energy gains correspond to accelerating electric fields in plasmas greater than 30 GV/m. The excitation of these large amplitude plasma waves was done by operating in the so-called self-modulated laser wakefield acceleration (SM-LWFA) regime [1], [2] - [5].

[†] On leave from Technische Universiteit Eindhoven

In the SM-LWFA [1], a single, long laser pulse with duration $L > \lambda_p$ breaks up (self-modulates) into a train of short pulses, each of which has a width on the order of the plasma wavelength λ_p. Strong self-modulation occurs when $L > \lambda_p$ and for pulse powers $P > P_c$, where $P_c = 17\omega^2/\omega_p^2$ GW is the critical power for relativistic self-focusing. Since $\lambda_p \sim n_0^{-1/2}$ and $P_c \sim n_0^{-1}$, for fixed laser parameters, the conditions $L > \lambda_p$ and $P > P_c$ can usually be satisfied by operating at a sufficiently high plasma density n_0. Associated with the break up of the long pulse is a large amplitude plasma wave that can self-trap and accelerate electrons from the background plasma. This results is an electron beam with a large energy spread. To improve the electron beam quality, several schemes are currently being pursued using the standard LWFA [1], [6] - [8] (in which $L \approx \lambda_p$) that use additional laser pulses to inject electrons directly into the wakefield [9] - [14].

In this paper we describe experiments performed at the l'OASIS laboratory of LBNL [15] - [16] on the SM-LWFA and progress on implementing the LWFA colliding pulse injection method [11] - [14]. The SM-LWFA phase of the experiment has served two purposes: (i) The development and commissioning of the laser system, target chamber and various laser beam, plasma and electron beam diagnostics; and (ii) the production of relativistic electron beams from the SM-LWFA regime at high repetition rate which, in turn, has allowed the first demonstration of radio-isotope production in a lead and copper target. The next phase will aim at producing electron beams in the standard LWFA regime by relying on optical injection using one or two additional laser beams.

EXPERIMENTAL ARRANGEMENT

The layout of the experiment is shown in Fig. 1 and consists of the high power Ti:Al$_2$O$_3$ laser, a pulsed gasjet for the plasma source, laser and plasma diagnostics, and electron beam diagnostics. Pulses from a Kerr lens mode-locked Ti:Al$_2$O$_3$ oscillator, lasing at about 0.8 μm, were first stretched by a grating stretcher with all-reflective optics to a length of up to 300 ps, controllable through the bandwidth of the injected oscillator pulses. The stretched pulses were amplified in a regenerative amplifier, pumped with a 1 kHz intra-cavity doubled Nd:YLF laser.

The output of the regenerative amplifier, 1.0 - 1.2 mJ per pulse, was sent to a three-pass pre-amplifier, producing about 40 mJ per pulse at a repetition rate of 10 Hz. A fraction of the pulse (8%) was split off and sent to a large aperture five pass main amplifier (AMP1). The rest of the beam was injected into separate amplifier used for laser plasma channeling experiments which are discussed elsewhere [15,16]. AMP1 brings the beam to an energy of up to 1 J per pulse. This high energy 200-300 ps chirped pulse was propagated into a shielded cave below the laser lab through an evacuated beam pipe. The pulse was then compressed in a vacuum compressor to peak powers of 8-10 TW in a pulse as short as 50 fs. This high power pulse served as the main drive laser pulse for the self-modulated LWFA experiment and will also be the main drive pulse for the upcoming colliding laser wakefield experiments. The

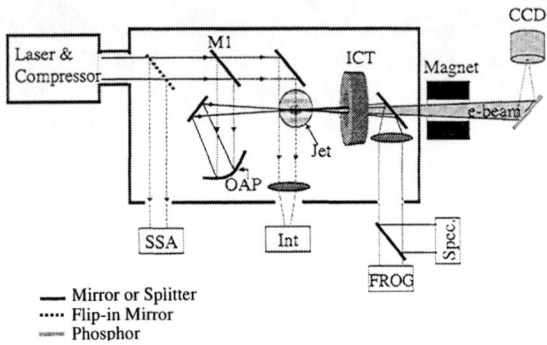

FIGURE 1. Lay-out of experiment showing the laser beam exiting the compressor, being reflected by mirror M1 onto the off-axis parabola (OAP), which focuses it onto the gasjet. The resulting electron beam is measured using the integrated current transformer (ICT) and is dispersed in the magnetic spectrometer onto a phosphor screen. The screen is imaged with the CCD. Plasma densities are measured with the interferometer (INT) and the laser beam is analyzed using the single-shot autocorrelator (SSA), the frequency resolved optical gating system (FROG) and an imaging optical spectrometer (Spec.).

peak power of the laser was varied using the pulse duration and laser energy.

The amount and sign of the chirp and, consequently, laser pulse duration, was varied by changing the grating distance in the vacuum compressor. Measurement of the laser pulse duration and laser chirp was done with a commercial single shot autocorrelator (SSA) and a frequency resolved optical gating (FROG) system, respectively. Both systems are located outside the vacuum chamber. To avoid linear and non-linear dispersion effects, the compressor chambers and beam transport tubes were evacuated. A typical compressor scan is shown in Fig. 2(a) and accompanying FROG images in Fig. 2(b).

After compression, the laser beam was reflected with mirror M1 onto an F/4, 30 cm focal length off-axis parabola (OAP), which focused the beam onto a high pressure pulsed gasjet. The gasjet was operated with hydrogen, helium and nitrogen at backing pressures up to 72 bar. OAP alignment was optimized for minimum aberrations, providing a spot size of approximately 8 μm. A final steering mirror after the OAP was used to provide independent control of the pointing direction. After the interaction region, the main laser beam was reflected by a gold or silver coated 5 μm nitrocellulose pellicle. This material and thickness was chosen to minimize Coulomb scattering of electrons propagating through the pellicle, while maintaining optical flatness. After appropriate attenuation, the spectral properties and pulse duration of the exiting laser beam were then analyzed on either a FROG system or an imaging spectrometer.

The density profile of the laser produced plasmas was measured using side-on interferometry of the folded-wave type (figure 3). Laser radiation leaking through M1 was reflected onto a variable optical delay line and sent through the interaction

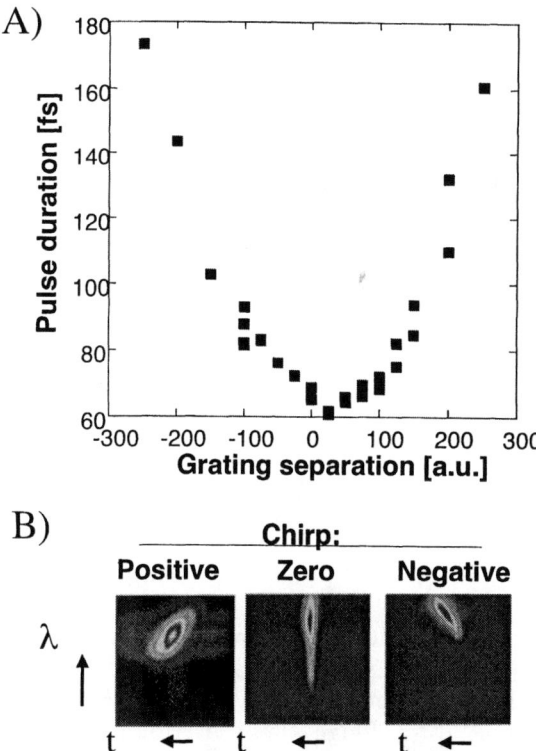

FIGURE 2. (a) Laser pulse duration vs. compressor grating separation. (b) Spectrum vs. time measured using a frequency resolved optical gating system. The laser chirp sign changes as the pulse compression crosses its minimum value.

region above the gasjet at right angles to the main beam. After exiting the chamber the probe beam was split and recombined, forming two identical interferograms at the CCD camera. The interaction region was imaged onto the camera using an achromatic lens. Phase changes imparted to the beam by the plasma were extracted from the interferograms, and density profiles were obtained by Abel inverting the two dimensional phase profiles.

The total charge per bunch in the electron beam was measured using a commercial integrating current transformer (ICT). This ICT had been calibrated against a Faraday cup and found to be in very good agreement. The spatial profile was measured with a phosphor screen that was imaged onto a 16 bit CCD camera. The energy distribution of the electron beam was measured by placing the same phosphor and camera downstream of a dipole spectrometer magnet. The ICT as well as an identical magnetic dipole had been previously used at the Beam Test Facility [17], located at the Advanced Light Source of LBNL, with 30 ps long bunches at

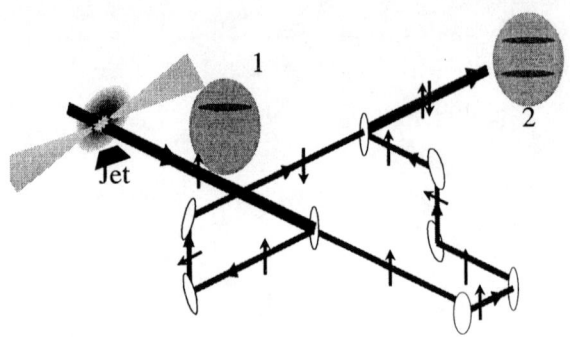

FIGURE 3. Lay-out of the folded wave interferometer used for measuring the plasma density profile. A single image of the interaction region at (1) is split, folded, and recombined to form two interfering images at the detector (2).

50 MeV containing typically 1-1.5 nC.

Neutrons and gamma rays produced during operation of the experiment were monitored with a variety of different detectors, allowing both use of this radiation as a beam diagnostic and the evaluation of various detectors' performance for ultra short radiation pulses [18]. Most of the gamma radiation was produced from the acceleration and deceleration of the electron beam, while neutrons were produced by interactions of high energy gammas with the target. Neutron production therefor served as a rough diagnostic of high energy electron production (Fig. 4).

The high repetition rate and high power levels sustainable by the l'OASIS laser system produce high energy beams with doses on target sufficient to perform nuclear activation experiments. Nuclear activation through (γ,n) reactions was chosen to provide a lower bound to the electron beam maximum energy. The target material was designed to maximize the high energy Bremsstrahlung yield, generate reaction

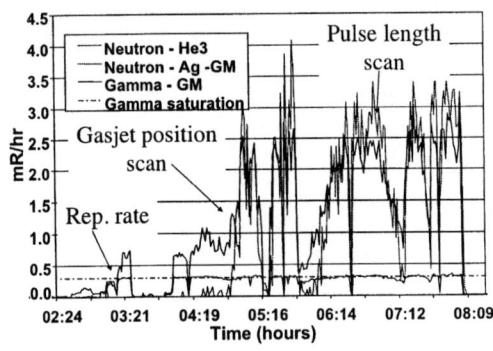

FIGURE 4. Neutron and Gamma production as a function of time illustrates repeatable, controllable, high energy electron beam production.

products with half-life time greater than 5 minutes but shorter than 2 days that emit detectable quantities of characteristic gamma rays, provide incremental indicators over a gamma energy range from 8 MeV to 30 MeV, and be practical to use (available and inexpensive). Candidate elements and reaction products were determined using Refs. [19] - [21].

The electron beam was stopped in a lead/copper target and the Bremsstrahlung gamma rays activated the target material. After the target was removed, the reaction products were analyzed by gamma spectroscopy for identification. Transportation to the spectrometer facility meant that counting began fifteen minutes after the beam shut off time. Each reaction, (γ,n), (γ,2n), and (γ,3n), has a threshold for the γ energies below which the reaction cannot occur, yielding an unambiguous lower bound on the electron beam energy.

In our experiments, a multi section target constructed of 13 two-piece blocks of various sizes arranged in a bulls eye pattern centered on the beam path approximately 60cm downstream of the gasjet. Each piece was composed of 6.3 mm of Pb at the front and 12.7 mm of Cu at the back. The Cu was selected, because it had all three reactions detectable with a gamma ray energy spread of 10.8 MeV to 31.4 MeV. Pb was chosen to generate Bremsstrahlung photons as well as for the complimentary (γ,n) indicators at 8 and 15 MeV. The choice of thickness of the Pb was a compromise between maximum yield of high energy Bremsstrahlung photons, and minimal absorption before entering the Cu.

EXPERIMENTAL RESULTS

Detailed studies of the dependence of electron and neutron production on such parameters as plasma density, laser power, pulse length, chirp, and focal position with respect to the gas jet were made, along with nuclear activation experiments in lead and copper. A typical electron density density profile is shown in Fig. 5. Plasma densities on the order of $1-5\times 10^{19}$ cm^{-3} were produced, which for multi-TW powers is in the SM-LWFA regime. Generating wakefields in the standard LWFA configuration with such pulses requires a density of $n_0 \simeq 5 \times 10^{18}$ cm^{-3}. Hence, for this laser pulse, the LWFA will be reached by decreasing the plasma density by a factor of 10 compared to the SM-LWFA configuration.

Figure 6 shows laser pulse width measured by the SSA along with the blue shifting of the main drive pulse (caused by the interaction of the laser beam with the rapidly ionizing gas jet plume) as a function compressor grating position. From a one-dimensional ionization blue shifting model it can be seen that the maximum blue shifting occurs at the minimum pulse width due to the fact that the ionization rate, and hence blue shift magnitude, increases with peak laser intensity. Hence the minimum of the blue shift curve indicates the minimum pulse width at the interaction point. Note that this minimum occurs at a slightly different position from that measured with the SSA,due to finite temporal dispersion of the exit BK7 window on the vacuum chamber, as well as from the optics of the SSA .

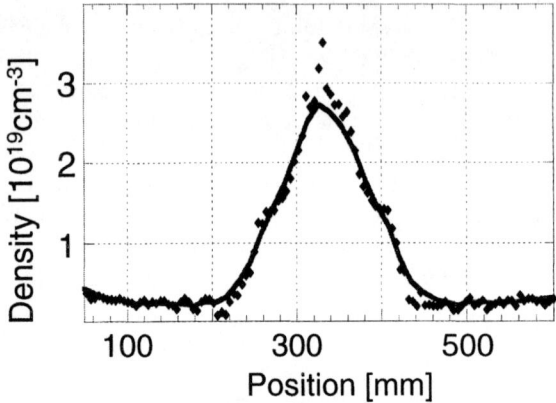

FIGURE 5. Extracted density profile obtained with the folded wave interferometer. The helium plasma was produced by laser ionization of the gasjet plume.

As is evident from Fig. 7, an asymmetry is observed in electron yield measured with the ICT and laser pulse length as a function of compressor grating position. Using the optical imaging spectrometer, spectral sidebands around the center laser wavelength have been observed which also exhibit a similar asymmetric behavior with grating position. As discussed above, the amount and sign of the laser chirp changes while scanning through the compressor minimum. Details of these observations will be discussed in a later paper [22].

Electron yield and neutron yields were found to be very well correlated and large increases in yield were observed by adjusting the position of the gasjet edge with respect to the location of the vacuum focus. The yield in electrons and neutrons

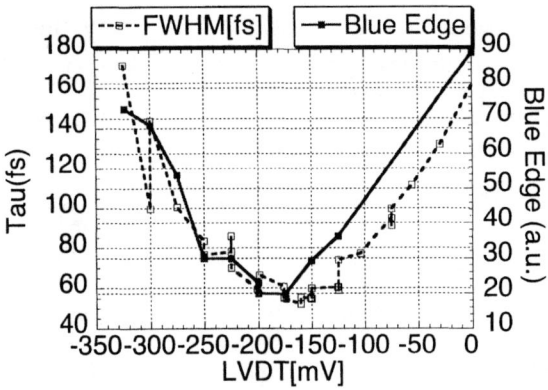

FIGURE 6. Laser wavelength ionization blue shift and laser pulse duration vs. compressor grating separation.

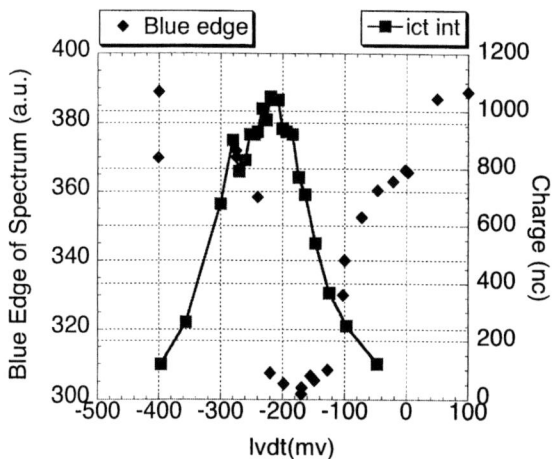

FIGURE 7. Electron yield in nC and laser pulse duration vs. compressor grating separation.

was also found to scale with increased laser power which will be discussed in a later paper [22] (see Fig. 8).

For nuclear activation experiments, the target plate was removed from the vacuum chamber after irradiation for 3.5 hours, transferred to the remote counting facility, and individual blocks were removed from the plate for counting. An example gamma spectrum from the counting is shown in Fig. 9. Initial surveying of the target with a Geiger survey meter revealed significant radioactivity on the order of 0.5 μ Ci. The distribution of relative activity on the target was indicative of a well collimated relativistic electron beam emerging from the gasjet, with the majority of all of the activity being from the central 1" diameter block. We identified gamma rays for the ^{63}Cu (γ,n) and (γ,2n), ^{65}Cu (γ,n), Pb204 (γ,n), and Pb206 (γ,2n)

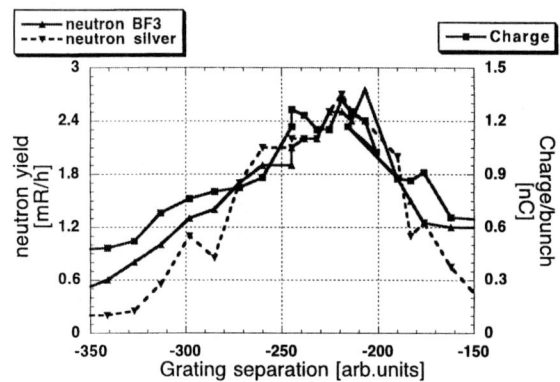

FIGURE 8. Electron and neutron yield versus compressor position.

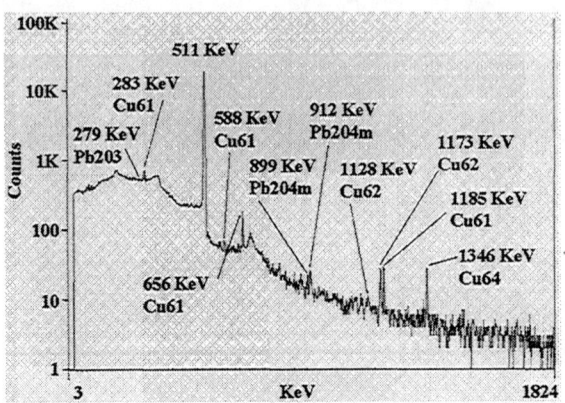

FIGURE 9. Example gamma ray spectrum from the nuclear activation measurements, showing the peaks corresponding to each isotope produced.

reactions. Successful observation of the 3.3 hr ^{61}Cu from the ^{63}Cu(γ,2n) reaction confirmed that the γ-ray (electron) energy distribution had a significant component above 19.7 (25) MeV.

SUMMARY AND FUTURE WORK

Recent experiments in the SM-LWFA regime at the l'OASIS laser facility have produced repeatable, high repetition rate electron beams with charge over 1nC and relativistic peak energies over 25MeV. Beam dependence on plasma and laser parameters has been studied, and these measurements will be refined in the near future.

To significantly reduce the energy spread and increase the mean energy, injection of two additional laser pulses is being implemented. In this method, referred to as the colliding pulse injection method [11] - [14], the ponderomotive force of the high-power drive pulse excites a large amplitude wakefield via the standard LWFA mechanism. The two lower power injection pulses collide behind the drive pulse and provide a time-gated electron trapping mechanism by shifting the momentum and relative phase of the plasma electrons. Electrons are injected at a very specific phase into the wakefield for acceleration to high energy. This method allows control of the injection process through the injection phase (position of the forward injection pulse), beat wave velocity (frequencies of the injection pulses), and the beat wave amplitude parameter (injection pulse intensities). Simulations with a drive pulse power of 5 TW and injection pulses of 1 TW each indicate the production of ultrashort ($\sim$ 1 fs), relativistic electron bunches (40 MeV in 1 mm) with low fractional energy spread ($\sim$ 1 %) and low normalized transverse emittance ($\sim$ 1 mm mrad).

At the present time (Aug 2000), new target chambers have been manufactured and installation of optics is in progress. The system is expected to become fully operational during Fall 2000. Since the energy distribution of the electron beam produced with the colliding pulse method is expected to be significantly narrower than what was produced in the SM-LWFA regime, two different magnetic spectrometers have been designed: a low dispersion electromagnet with round poles and a high dispersion square pole magnet using Sm-Co magnets with a surface field strength of 1.1 T. The round and rectangular spectrometers offer broad energy range and good energy resolution, respectively.

ACKNOWLEDGMENTS

This work was supported by the Department of Energy under Contract No. DE-AC-03-76SF0098.

REFERENCES

1. For a review see, E. Esarey et al., IEEE Trans. Plasma Sci. **PS-24**, 252 (1996).
2. A. Modena et al., Nature **377**, 606 (1995); D. Gordon et al., Phys. Rev. Lett. **80**, 2133 (1998).
3. K. Nakajima et al., Phys. Rev. Lett. **74**, 4428 (1995).
4. D. Umstadter et al., Science **273**, 472 (1996); R. Wagner et al., Phys. Rev. Lett. **78**, 3125 (1997).
5. A. Ting et al., Phys. Plasmas **4**, 1889 (1997); C.I. Moore et al., Phys. Rev. Lett. **79**, 3909 (1997).
6. H. Dewa et al., Nucl. Instr. Meth. A **410**, 357 (1998).
7. F. Amiranoff et al., Phys. Rev. Lett. **81** , 995 (1998).
8. D. Bernard et al., Nucl. Instr. Meth. A **432**, 227-231 (1999).
9. D. Umstadter et al., Phys. Rev. Lett. **76**, 2073 (1996).
10. R. G. Hemker et al., Phys. Rev. E **57**, 5920 (1998).
11. E. Esarey et al., Phys. Rev. Lett. **79**, 2682 (1997).
12. W. P. Leemans et al., SPIE Conf. Proc. **3451**, 41-50 (1998).
13. C.B. Schroeder et al., Phys. Rev. E **59**, 6037 (1999).
14. E. Esarey et al., Phys. Plasmas **6**, 2262 (1999).
15. W. P. Leemans et al., Phys. Plasmas **5**, 1615 (1998).
16. P.Volfbeyn et al., Phys. Plasmas **6**, 2269 (1999).
17. W. Leemans et al., Proc. 1993 Part. Acc. Conf., 83 - 85. (1993).
18. W.P. Leemans et al, in preparation.
19. see http://ie.lbl.gov/education/isotopes.htm
20. R.E. Sund et al., Phys. Rev. **176**, 1366 (1968).
21. S.C. Fultz et al., Phys. Rev. B **133**, 1149 (1964).
22. W.P. Leemans et al., in preparation.

Demonstration of a Laser-Driven Prebuncher Staged With a Laser Accelerator – The STELLA Program

M. Babzien,[†] I. Ben-Zvi,[†] L. P. Campbell,[*] C. E. Dilley,[*] D. B. Cline,[‡]
J. C. Gallardo,[†] S. C. Gottschalk,[*] W. D. Kimura,[*] P. He,[‡] K. P. Kusche,[†]
Y. Liu,[‡] R. H. Pantell,[§] I. V. Pogorelsky,[†] D. C. Quimby,[*] J. Skaritka,[†]
A. van Steenbergen,[†] L.C. Steinhauer,[¶] and V. Yakimenko[†]

[*]*STI Optronics, Inc., Bellevue, Washington 98004*
[†]*Brookhaven National Laboratory, Upton, New York 11973*
[‡]*University of California, Los Angeles, Los Angeles, California 90095*
[§]*Stanford University, Stanford, CA 94305*
[¶]*University of Washington, Redmond Plasma Physics Laboratory, Redmond, Washington 98052*

Abstract. During the Staged Electron Laser Acceleration (STELLA) program we demonstrated for the first time the staging together of laser accelerator modules. We also demonstrated the capture and acceleration of laser-generated microbunches, which are 1-2 µm in length and have <2% energy spread. The modules consist of an inverse free electron laser (IFEL) prebuncher with an IFEL accelerator. The two IFELs use identical permanent-magnet wiggler arrays (wiggler period = 3.3 cm, wiggler length = 33 cm) and are separated by 2.3 m with a triplet located between them. The first IFEL is used to modulate the *e*-beam energy so that microbunches are formed at the entrance to the second IFEL. This second IFEL is designed to capture and accelerate the microbunches. These devices are driven by a CO_2 laser beam, which has been split into two beams that are sent to the IFELs. A trombone delay line in the laser beam transport system is located before the second IFEL to allow phase delay adjustment between the microbunches entering the accelerator and the laser beam driving the second IFEL. Typical energy spectra of the staged IFELs as a function of this phase delay will be shown. The data agree well with model simulations. Reliable and reproducible control of the phase delay is demonstrated.

INTRODUCTION

Routine laser acceleration of electrons has been performed at the Brookhaven National Laboratory Accelerator Test Facility (ATF)[1,2] and at facilities around the world.[3] However, the acceleration is typically achieved using a single pass of the laser beam with the electrons either through direct coupling or through indirect coupling, such as generation of a plasma wave. To achieve high net energy gain requires staging the process whereby the electrons are repeatedly accelerated by the laser field. This implies the need to create well-formed microbunches, whose lengths are a fraction of the accelerated wave, and to rephase these microbunches with the laser field.

The goal of the Staged Electron Laser Acceleration (STELLA) experiment is to demonstrate staging of the laser acceleration process. There are two important steps to this demonstration. The first is creating the aforementioned microbunches using a laser-driven prebuncher (modulator) and sending these microbunches into a laser-driven accelerator. The second is capturing and accelerating the microbunches in the accelerator. Achieving the first step does not guarantee achieving the second step. This is because effects such as bunch smearing and rephasing jitter can interfere with the ability to capture the microbunches effectively in the accelerator.

In this paper we report the successful completion of both steps during the STELLA effort.

DESCRIPTION OF EXPERIMENT

Two identical inverse free electron lasers (IFEL) are used as the prebuncher (IFEL1) and accelerator (IFEL2). Each has a permanent magnet, planar, nontapered, fixed-gap wiggler ($L = 33$ cm, $\lambda_w = 3.3$ cm, $K = 2.9$), manufactured by STI Optronics, which is designed for a resonant energy of 45.6 MeV. The wigglers can be easily raised manually off the beamline in order to remove the wigglers when other users need the beamline. Kinematic feet on the wiggler ensure accurate realignment of the device when it is lowered back to the beamline. Figure 1 shows one of the wigglers in the raised position.

FIGURE 1. Photograph of permanent-magnet wiggler in raised position above the beamline.

The prebuncher and accelerator are arranged in series along the beamline as depicted in Fig. 2. The electron beam (*e*-beam) is modulated by the prebuncher and travels through a drift region (2.3 m) where a triplet focuses the *e*-beam into the accelerator. Downstream of the accelerator is an energy spectrometer. A beamsplitter divides the CO_2 laser beam ($\lambda = 10.6$ μm) into two beams, one drives the prebuncher and the other is sent through a trombone delay line before entering the accelerator. The delay line permits adjusting the phase of the laser field inside the accelerator relative to the field inside the prebuncher. Since the phase of the microbunches produced by the prebuncher is directly related to the phase of the laser field inside the prebuncher, this delay line controls whether the microbunches are accelerated or decelerated in the accelerator.

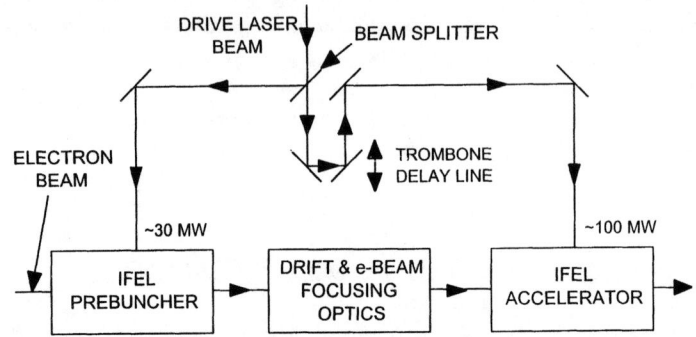

FIGURE 2. Conceptual layout for the STELLA staging experiment.

COMPUTER MODEL FOR STELLA EXPERIMENT

A simulation model was developed to aid in performance predictions and sensitivity studies for the STELLA experiment. The code simulates the free-electron laser (FEL) interaction in the prebuncher and accelerator stages, and includes a particle tracking model for the electron paths through the intervening drift region. It includes a full 3-D implementation permitting the sensitivity to emittance and misalignment effects to be determined. A simple 1-D longitudinal space charge model was added to study possible bunch smearing, which may occur in the drift region due to this effect. (See Ref. 4 for more details concerning possible space charge effects.)

As shown schematically in Fig. 3, the integrated model consists of two modules: an FEL model and a drift region model. In the simulation results shown in this paper, the e-beam is modeled using 5000 simulation particles. The electron parameters in longitudinal and transverse phase space are passed between modules when needed. The modules are run repeatedly as needed to simulate staged systems. Initiated with the electron distribution provided by the linear accelerator, the FEL module calculates the energy modulation induced by the FEL interaction. The drift region model then computes the microbunching resulting from the induced energy spread, including bunch smearing effects resulting from longitudinal space charge as well as finite emittance and pathlength differences through the quadrupole system used for refocusing the beam into the second IFEL. The FEL module is then reused to simulate the acceleration of the prebunched beam.

When comparing the model with the data, the e-beam Twiss parameters and emittance for the model are chosen until the best agreement with the measured e-beam sizes on the various beam position monitors (BPMs) along the beamline are obtained. All other parameters, such as the quadrupole field gradients and the separation distances between all beamline components, are the actual values during the experiment.

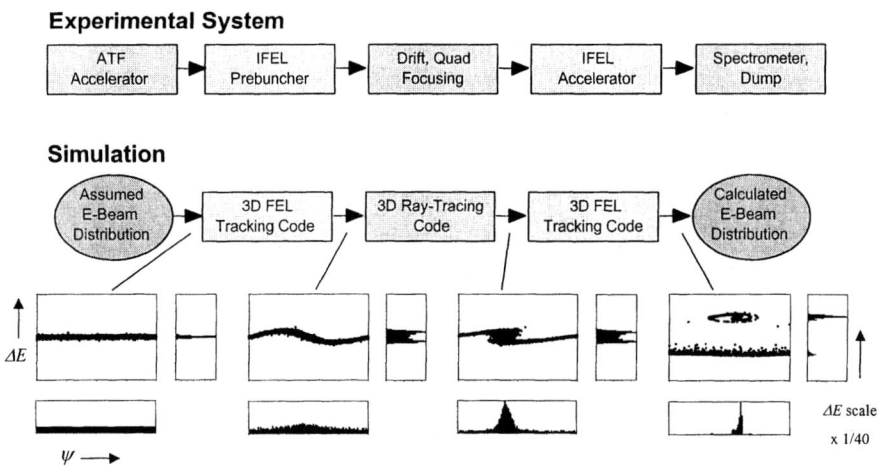

FIGURE 3. Flow chart for STELLA computer model.

Table 1 lists the basic parameters used in the model simulation comparisons with the experimental data given in this paper.

TABLE 1. Parameters used in model for comparison with experimental data.

Parameter	Value	Comments
E-beam energy	45.6 MeV	Resonance for both wigglers
E-beam intrinsic $\Delta E/E$ (1σ)	0.04%	From spectrometer measurement
Spectrometer spread (1σ)	0.07%	From zero dispersion measurement
E-beam charge	0.2 nC	Corresponds to 67 A peak current
E-beam pulse length	3 ps	Estimated value
Emittance (normalized)	1.5 mm-mrad	From matching BPM sizes with model
β_x (m), α_x (in wiggle plane)	0.07, 16.0	From matching BPM sizes with model
β_y (m), α_y	0.016, -5.3	From matching BPM sizes with model
Triplet #4 quad settings (kG/m)	-13.8, 33.2, -28.3	Use actual experimental values
Laser pulse length	≈180 ps	Leading peak of laser pulse
Laser power to IFEL1	30 MW	Typical value
IFEL1 laser beam size (1σ)	1.7 mm	Measured value in middle of wiggler
Laser power to IFEL2	100-300 MW	Typical value
IFEL2 laser beam size (1σ)	0.62 mm	Measured value in middle of wiggler

EXPERIMENTAL RESULTS

Figure 4 is an example of the energy spectrum produced by the accelerator (IFEL2) when it is operated alone, i.e., the prebuncher is not used. In this situation the electrons enter the wiggler over all phases, thereby resulting in a symmetric energy gain/loss distribution as shown in Fig. 4(b), which compares a line profile in the center of the raw data image [Fig. 4(a)] with the computer model histogram. (In all the results in this paper the model and data curves are scaled to have equal areas.) We see the agreement with the model is very good.

A similar, but narrower, double-peaked energy spectrum is observed when operating the prebuncher (IFEL1) since it is intentionally driven with less laser power.

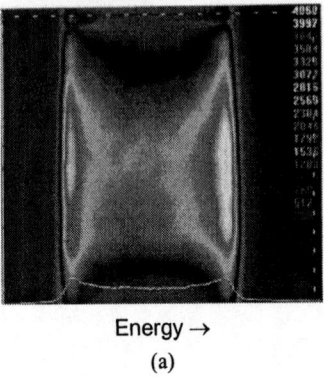
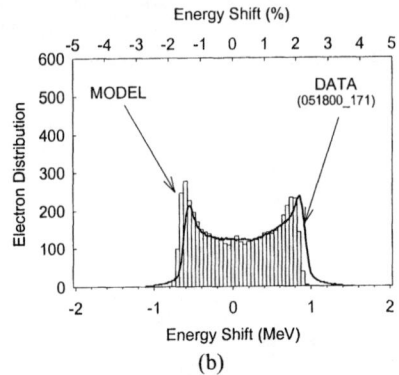

FIGURE 4. Second wiggler (IFEL2) only results. (a) False-color raw video image from energy spectrometer. Energy dispersion is in the horizontal direction. Laser power ≈100 MW. (b) Model prediction with no space charge for energy spectrum from (a) overlaid on model histogram.

The energy histogram changes dramatically when the two stages are operated together. Typical staging results are shown in Fig. 5 for three different phase delays. The left-column figures are the raw video output from the spectrometer. The middle-column figures are energy spectrum profiles overlaid on the model predictions. And, the right-column figures are the corresponding energy-phase plots from the model.

In this particular set of data the phase delay used in the model has been chosen to yield the best fit with the data. The model includes space charge effects and 0.2 mrad of angular misalignment exiting the prebuncher in both the x and y directions. This misalignment is an estimate based upon comparisons between the model and data. This angular error is well within the accuracy with which the e-beam can be tuned through the triplet between the prebuncher and accelerator.

The data clearly shows a grouping of electrons, which moves across the energy spectrum as a function of phase delay. The model energy-phase plots indicate this is the microbunch moving across the background of uncaptured electrons. These energy-phase plots also show evidence of some bunch smearing caused by space-charge effects and angular misalignment.

Good agreement with the model is also obtained when intentionally overmodulating the e-beam in the prebuncher, thereby causing partial debunching to occur at the entrance to the accelerator.

Due to gradual phase drift that occurs during the measurements, it was not possible to obtain a detailed phase delay history of the energy spectra. Nevertheless, within a series of back-to-back shots taken several minutes apart, the data exhibits very good repeatability and control. This is illustrated in Fig. 6, which depicts the raw spectrometer output for four consecutive shots. The first shot was at some arbitrary phase delay point we shall call 0°. During the second shot the delay was moved by 180° from the initial phase delay point. Then for the third shot, the delay was moved back to 0°, and for the fourth shot it was moved back to 180°. We see that the overall profiles return to their original distributions at each of the two phase delay points.

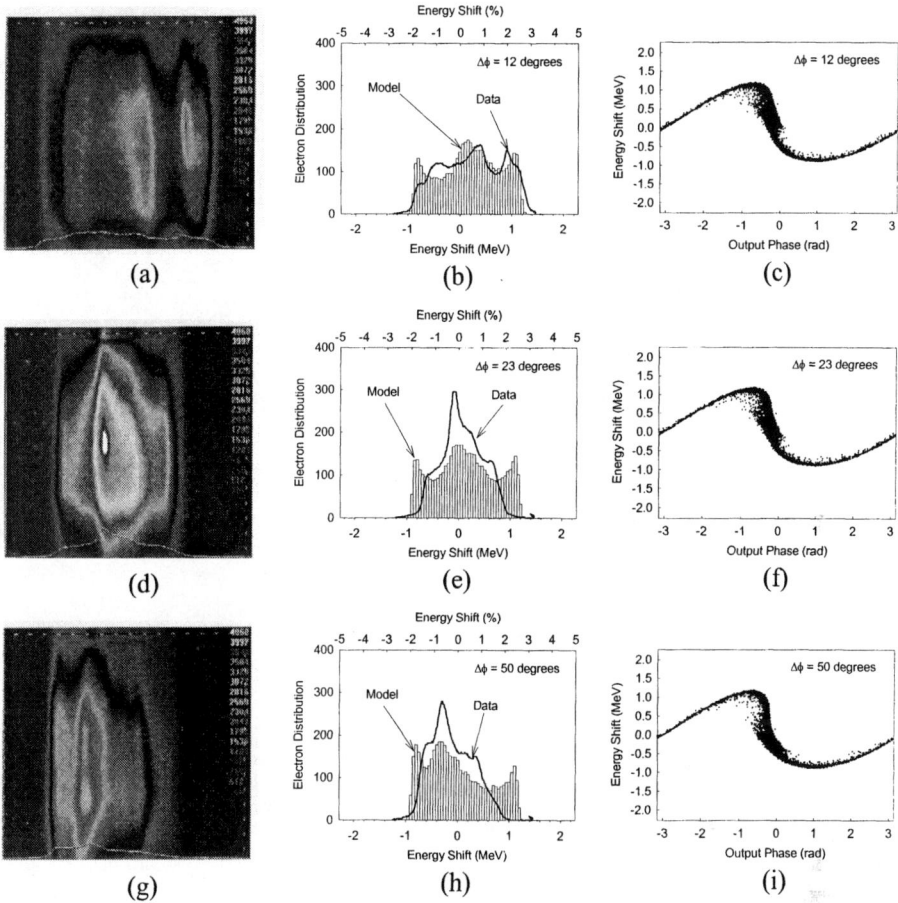

FIGURE 5. Typical staging results at three different phase delays. False-color raw video images from the energy spectrometer are given in (a), (d), and (g). Energy dispersion is in the horizontal direction. Laser power to the accelerator in the model is 150 MW. Model prediction comparisons of the energy spectra with the data are shown in (b), (e), and (h). Red curves are line profiles through the video images in (a), (d), and (g), respectively. Corresponding model predictions for the energy-phase plots are given in (c), (f), and (i).

Overall the model agrees well with the data. It also indicates that the microbunch length is 1-2 μm (FWHM) with an energy spread of <2% (FWHM). Thus, we can confidently claim we are capturing and accelerating the microbunch in a controlled and reproducible manner.

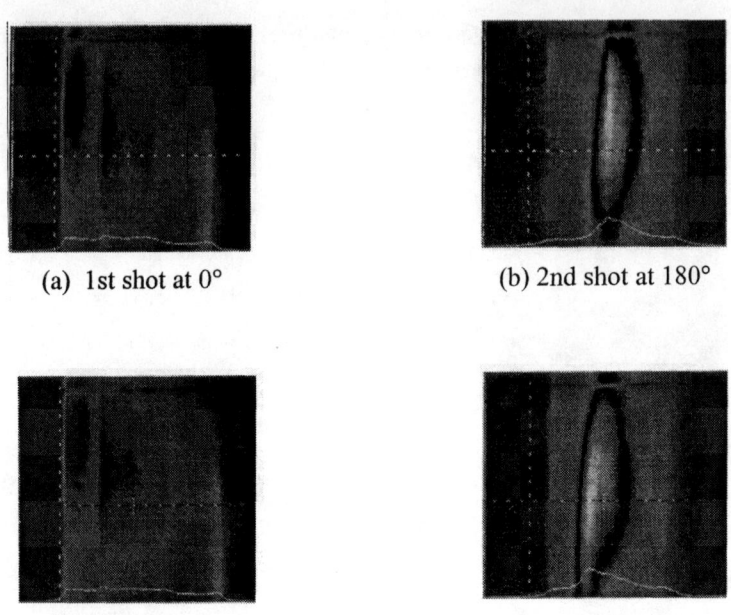

FIGURE 6. Data taken of four consecutive shots with the phase delay varied back and forth showing good repeatability. (a) Arbitrarily assigned 0°. (b) 180° from 0°. (c) Return to 0°. (d) Return to 180°.

DISCUSSION AND CONCLUSIONS

The accomplishments of the STELLA program are listed below.

1) First demonstration of an IFEL prebuncher staged together with an IFEL accelerator.
2) First demonstration of an IFEL prebuncher staged together with an inverse Cerenkov accelerator.[5]
3) First generation of 3-4 fs microbunches using an external laser driver. In addition these were created in a reliable and reproducible fashion.
4) First demonstration of capturing and acceleration of microbunches. We demonstrated the ability to control the phase of the microbunches within the accelerating field over periods of many minutes.
5) First demonstration of laser-accelerated microbunches with energy spreads of 1-2% (FWHM).
6) Demonstration of peak energy gains in an untapered IFEL accelerator of $\approx 5\%$.
7) Demonstration of good agreement between data and staging model.
8) Observation of the strongest coherent transition radiation signal from microbunches to date.

In conclusion, the STELLA experiment successfully accomplished its primary goal. No major "show stoppers" were encountered. We believe the technology is ready to move on to the next challenge, which is to demonstrate monoenergetic acceleration, whereby the accelerated microbunches in the accelerator are separated from the noncaptured electrons while maintaining a small energy spread, e.g., ~1 %. Our model indicates this will be possible by delivering higher laser power (~70 GW). The ATF is currently upgrading their CO_2 laser system to deliver over several hundred gigawatts.

ACKNOWLEDGMENTS

The authors wish to acknowledge the helpful support of Dr. X. J. Wang and the staff at the ATF. This work was supported by the U.S. Department of Energy, Grant Nos. DE-FG03-98ER41061, DE-AC02-98CH10886, DE-FG03-92ER40695.

REFERENCES

1. W. D. Kimura, G. H. Kim, R. D. Romea, L. C. Steinhauer, I. V. Pogorelsky, K. P. Kusche, R. C. Fernow, X. Wang, and Y. Liu, Phys. Rev. Lett. **74**, 546-549 (1995).
2. A. van Steenbergen, J. Gallardo, J. Sandweiss, J.-M. Fang, M. Babzien, X. Qiu, J. Skaritka, and X.J. Wang, Phys. Rev. Lett. **77**, 2690 (1996).
3. See for example *Advanced Accelerator Concepts*, Baltimore, MD, AIP Conference Proceedings No. 472, W. Lawson, C. Bellamy, and D. Brosius, Eds., (American Institute of Physics, New York, 1999).
4. L.C. Steinhauer and W.D. Kimura, Phys. Rev. Special Topics-Accelerators and Beams **2**, 081301 (1999).
5. L. P. Campbell, C. E. Dilley, S. C. Gottschalk, W. D. Kimura, L. C. Steinhauer, M. Babzien, I. Ben-Zvi, J. C. Gallardo, K. P. Kusche, I. V. Pogorelsky, J. Skaritka, A. van Steenbergen, V. Yakimenko, D. B. Cline, P. He, Y. Liu, and R. H. Pantell, "Inverse Cerenkov Acceleration and Inverse Free Electron Laser Experimental Results for Staged Electron Laser Acceleration," to be published in IEEE Transactions on Plasma Science Special Issue on Second Generation Plasma and Laser Accelerators.

Fluid Modeling of Intense Laser-Plasma Interactions

B. A. Shadwick,[1,2] G. M Tarkenton,[1] E. H. Esarey,[2] and W. P. Leemans[2]

[1] *Institute for Advanced Physics, 10875 U.S. Hwy. 285, Suite 199, Conifer, CO 80433*
[2] *Center for Beam Physics, Ernest Orlando Lawrence Berkeley National Laboratory, University of California, Berkeley, CA 94720*

Abstract. We discuss various aspects of implementing numerical solutions to cold fluid models of laser-plasma interactions. Using the conservative formulation of the fluid model allows us to apply standard computational fluid dynamics methods. We discuss some of the details of this process showing the complications that can arise, and the trade-offs between performance and accuracy involved. We also discuss some results in one- and two-dimensions, showing pump-depletion effects and short-pulse generated nonlinear wake fields. Additionally, we describe work-in-progress on the numerical analysis of the algorithms and the different forms for the model equations.

INTRODUCTION

Much of the physics relevant to understanding the interaction of intense laser pulses with underdense plasmas is contained in the Vlasov equation coupled to Maxwell's equations. In limited, specialized cases, analytical progress can be made towards solving the Vlasov equation, but in full generality, the Vlasov–Maxwell system is considered to be intractable to analytic solution. Fluid models (both warm and cold) represent a significant simplification over full kinetic treatments of plasma dynamics; of course, this simplification comes at the price of having discarded most of the kinetic behavior (such as particle trapping), but retains enough physics to be qualitatively and quantitatively useful.

Consider the typical cold fluid model where the relativistic fluid momentum, p is driven by the Lorentz force:

$$\frac{\partial p}{\partial t} + v \cdot \nabla p = q\left(E + \frac{v}{c} \times B\right), \tag{1}$$

where $\gamma m v = p$, $\gamma = \sqrt{1 + |p|^2}$, q is the electron charge, E and B are the self-consistent fields obtained from Maxwell's equations with the current given $j = q n v$, n being the fluid electron density which satisfies the continuity equation:

$$\frac{\partial n}{\partial t} + \boldsymbol{\nabla} \cdot n\boldsymbol{v} = 0\,. \qquad (2)$$

While much interesting physics has been obtained from various simplifications of this model (typically involving some averaging procedure to remove the fast time-scale associated with the laser), we are going to concern ourselves with the direct solution of Eqs. (1) and (2) along with Maxwell's equations. The most obvious cost of this approach is that the longitudinal resolution will have to increase over that used in "envelope" simulations, by approximately ω_0/ω_p. For the current generation of laser-plasma acceleration experiments, this ratio tends to be in range of 5 and 10. This cost is, however, offset by the fact that averaging tends to lead to quite complex equations, and thus the true cost of solving the primitive equations compared to an envelope model is significantly less than the ratio of ω_0/ω_p would suggest. This notwithstanding, it turns out that the solution of this system is not entirely straightforward.

SOME REPRESENTATIVE RESULTS

Before moving to a detailed discussion of the numerical subtleties involved in solving these equations, we will describe some representative results from our preliminary one- and two-dimensional implementations of this model [1] that execute on workstation class machines, taking anywhere from a few minutes for one-dimensional runs to several hours for two-dimensional runs. Our codes are modular, fully object-oriented implementations in C++, allowing us to adjust the model's physics content very quickly. To minimize the size of the computational domain, we make use of the standard moving window by changing coordinates from (x, z, t) to $(x, \xi = t - z/c, t)$ in two-dimensions and from (z, t) to (ξ, t) in one-dimension. We use slab geometry and assume a linearly polarized laser field; hence we only must follow n, p_x, p_z, E_x, E_z, and B_y.

One-Dimensional Example: Wake Excitation and Pump Depletion

Consider a one-dimensional (z, t) configuration. The initial laser pulse has a Gaussian envelope of width $L = 1/k_p$, central wavenumber $k = 5k_p$ and amplitude $a_0 = 1$, such that the laser pulse excites a large amplitude wake field [2]. For these parameters, the depletion length [2] $L_{\rm pd} \approx 160/k_p$. As the laser pulse propagates in the plasma, we expect the amplitude will decrease and the spectrum will be redshifted due to pump depletion, *i.e.* laser field energy is lost to plasma wake excitation. Figures 1–3 show space-time diagrams for the resulting wake field, plasma density modulation, and laser magnetic field respectively. The sloped phase fronts seen in Figures 1 and 2 are a result of the laser pulse sliding backwards in the window. In one dimension the linear group velocity of the laser is, for $\omega_0^2 \gg$

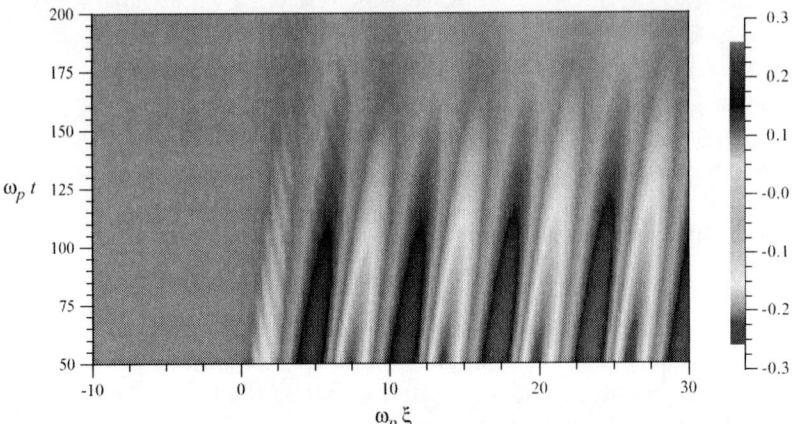

FIGURE 1. Longitudinal wake field of the laser pulse, normalized to cold waved breaking value.

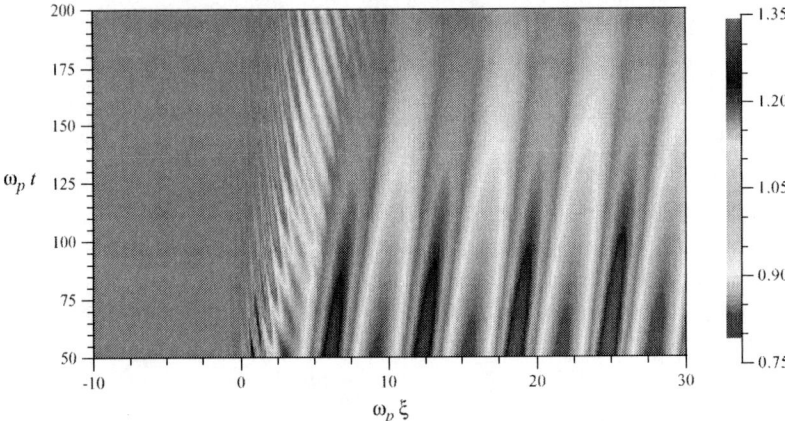

FIGURE 2. Relative density modulation.

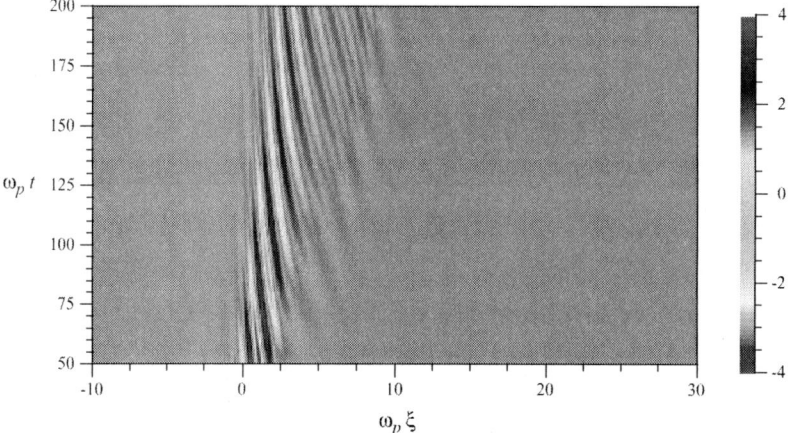

FIGURE 3. Magnetic field of the laser pulse as it propagates through the plasma. Same normalization as in Figure 1.

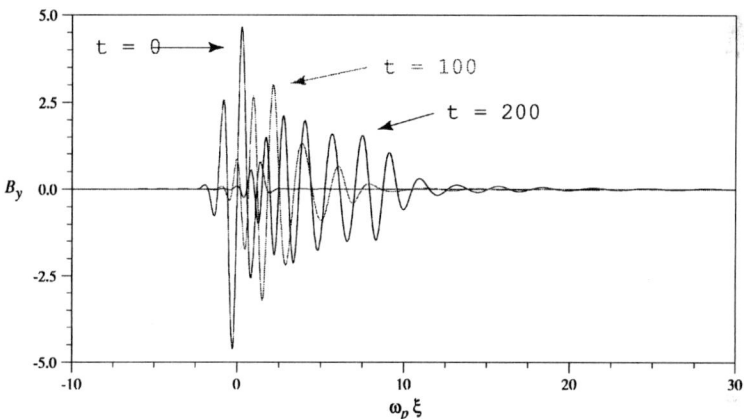

FIGURE 4. Laser field at $\omega_p t = 0$ (red), $\omega_p t = 100$ (green), and $\omega_p t = 200$ (blue).

ω_p^2, $\beta_g = 1 - \omega_p^2/2\omega_0^2$. For our parameters, $\beta_g = 49/50$ and so we expect to see the laser slide backwards in the window at $c/50$. The slope of the wake phase fronts agrees very well with this value of β_g. We see that the wake amplitude diminishes rather rapidly starting at approximately $\omega_p t = 150$ which agrees with the estimated depletion length. The short wavelength oscillations at the left of the figures are modulations driven by the fast component of the ponderomotive force of the laser at half the laser wavelength. Figure 4 shows the laser at $\omega_p t = 0$, $\omega_p t = 100$, and $\omega_p t = 200$. The slipping, lengthening and redshifting of the pulse is clearly evident.

Two-Dimensional Example: Short Pulse LWFA

In two dimensions, one is faced with implementing transverse boundary conditions. While periodic boundaries are by far the easiest to implement, one has to be very careful to avoid introducing numerical artifacts. While computationally more difficult, we are currently using absorbing boundaries and are investigating transmitting boundary conditions. Consider, for example, a laser wake field accelerator (LWFA) driven by an initial laser pulse with a Gaussian envelope of width $L = 2/k_p$, central wavenumber $k = 7.5 k_p$, amplitude $a_0 = 1$, and Rayleigh length $Z_R = 300/k_p$. Shown in Figures 5 and 6 are the resulting longitudinal wake field and plasma density at $\omega_p t = 32$, respectively. Notice that plasma oscillations are highly nonlinear, with the wake exhibiting significant steepening in the longitudinal direction. The curved phased fronts are a consequence of the relativistic shift of the plasma frequency; the plasma momentum, and hence γ, has a maximum on axis leading to lower plasma frequency. The transverse decrease in γ then results in a larger plasma frequency off axis and to the curved phase fronts. As part of our benchmarking process, this example is being carefully compared to PIC simulations with identical physical parameters. Preliminary results indicate that the agreement is remarkably good [3].

NUMERICAL CHALLENGES

The cold fluid equations are a nonlinear, hyperbolic system[1] which presents some significant numerical challenges. This partly explains our interest in one-dimensional models; many of the technical hurdles associated with numerical schemes for hyperbolic systems are independent of the dimensionality of the system. The leaves us in the happy circumstance where the techniques developed for the one-dimensional case can also be applied to higher-dimensional models [4].

Perhaps the most difficult aspect of the numerical solution of hyperbolic systems has to do with dispersion. One generally finds that the numerical dispersion relation

---

[1] As we will see below, these equations are not *strictly hyperbolic* but this does not affect the current discussion

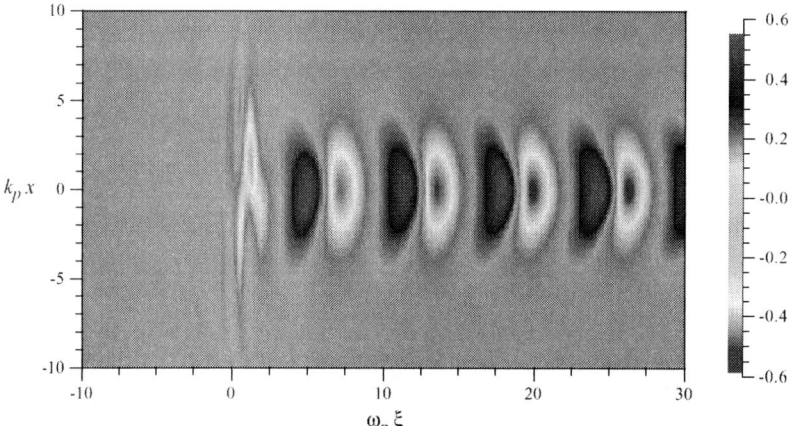

FIGURE 5. Longitudinal wake field at $\omega_p t = 32$.

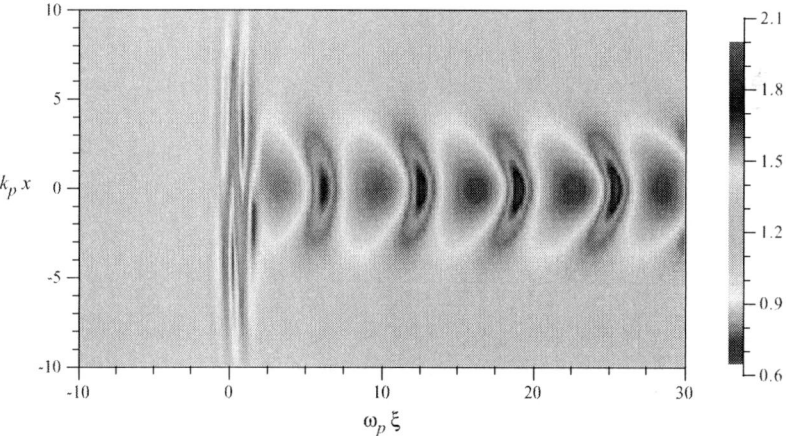

FIGURE 6. Relative density modulation at $\omega_p t = 32$.

differs from the analytical dispersion relation for large wavenumber modes.[2] For *linear* hyperbolic systems, this is not a serious limitation. One can always ensure that the spatial grid has adequate resolution so that the spatial modes present in the initial condition occupy that the portion of the dispersion curve where the analytical and numerical dispersion relations are sufficiently similar. For nonlinear systems, however, the spectrum typically expands due to nonlinear mixing. Even if the initial condition has a compact spectrum, this may well not be true at a later time. In the present case, the convective derivative leads to the well-known unending propagation of the initial disturbance to ever-larger wavenumbers, *i.e.* to ever-decreasing spatial scales. Thus, for nonlinear hyperbolic systems, one generally has to accept that the numerical dispersion relation will differ from the true dispersion relation. The question then shifts to how does one minimize the distortion of the physical content of the solution.

An example of the difficulty posed by this phenomena can be found in a vacuum-plasma interface. At sufficiently high power, as the laser pulse enters the plasma, it tends to steepen the density gradient. When this process is modeled using typical second-order differencing methods (the Lax-Wendroff method for instance), one finds that ripples develop at the base of the density pedestal, leading to *negative* values for the density. This unphysical behavior is accompanied by an additional error, namely a density overshoot at the top of the pedestal. The overshoot occurs because the total electron density is conserved[3]; thus the overshoot is required to compensate the negative contribution from the ripples.

There exists a class of numerical methods for hyperbolic equations know as Total Variation Diminishing (TVD) [6], whose dispersion properties are guaranteed not to introduce spurious ripples into the solution. Unfortunately, the simple TVD methods (*e.g.* upwinding) are only first order accurate in space and time, and consequently suffer large errors due to numerical diffusion. Moreover, it can be proven that all TVD methods have an accuracy *less* than second order. In the mid 1970'2, two novel approaches to constructing high accuracy TVD methods appeared: flux-corrected transport [7-10] and flux-limiting [11]. While these methods differ in implementation, they are quite similar in spirit. In both cases, the numerical solution is taken to be a linear combination of a high accuracy solution, which may suffer from unphysical ripples, and a low order ripple-free solution, which may suffer from unphysical diffusion. This linear combination is adjusted to favor the high-accuracy solution as much as possible, stoping just short of introducing ripples. The net result is that the high-accuracy solution is used wherever possible, only falling back to the low accuracy (but ripple-free) solution when absolutely necessary. Empirically, these methods tend to have a spatial order between 1.5 and 1.9. We have applied Zalesak's method [10] to a case with a density ramp tapering to zero and found that the unphysical density ripples are completely eliminated [4].

---

[2] Of course, the dispersion relations must match for small wavenumbers, lest the numerical method not be consistent with the differential equations.

[3] The total density being the spatial integral of n, is a *linear invariant* of the system and is, therefore, exactly conserved by virtually all numerical methods. See Ref. [5].

Lastly, we make a brief comment on the singular nature of the cold fluid limit. A system of partial differential equations is *hyperbolic* if the eigenvalues of the Jacobian of the flux function (also called the speed matrix) are all real. A *strictly hyperbolic* system has *distinct* real eigenvalues. If the eigenvalues are not distinct, the system is degenerate. All of the numerical analysis for hyperbolic systems rests, in one form or another, on the property of strict hyperbolicity [6]. The cold-fluid system is degenerate because the sound speed vanishes in this limit, leading to a triple degeneracy of eigenvalues and an incomplete eignevector specturm. Physically, this couples the density variations exclusively to the electromagnetic field. Applying the standard numerical techniques in this case becomes a delicate matter, as the characteristic curves can have higher-order contact, complicating the analysis of discontinuities. We have done a preliminary analysis and have observed the existence of discontinuities associated with the crossing of density characteristics [4].

FLUX CONSERVATIVE FORM

The numerical methods discussed in the previous section require that the dynamical equations be cast as flux conservation laws. The conservative formulation of the fluid model is most conveniently obtained starting from the relativistically covariant expression for energy-momentum conservation and the cold approximation to the ideal-fluid energy-momentum tensor [1]. Reducing these expressions leads to the momentum density equation:

$$\frac{\partial n\boldsymbol{p}}{\partial t} + \boldsymbol{\nabla} \cdot (n\boldsymbol{v}\boldsymbol{p}) = q\,n\left(\boldsymbol{E} + \frac{\boldsymbol{v}}{c} \times \boldsymbol{B}\right). \tag{3}$$

To obtain the non-conservative form, Eq. (1), of this equation, we use the continuity equation to eliminate $\partial n/\partial t$ and divide by n. The covariant formulation is also of value in that it allows recourse to relativistic thermodynamicsc [12] to correctly incorporate finite temperature effects into our fluid models [4].

When the plasma density is zero, Eqs. (1) and (3) give very different results. In particular, Eq. (1) predicts, for a laser pulse propagating in vacuum, non-zero longitudinal and transverse plasma momentum whereas, Eq. (4) predicts vanishing momentum density. When this pulse subsequently enters a region of non-zero density, the fields in the two cases are markedly different [4]. This a consequence of the fact that the division of Eq. (3) by n to obtain Eq. (1) is simply not valid when $n = 0$. Ultimately, Eq. (3) is the fundamental equation, *i.e.* only the currents have physical significance since only the currents couple the plasma motion to the electromagnetic fields.

CONCLUSIONS

Although there are a surprising number of technical issues surrounding a direct solution of the cold fluid equations, we have made significant progress towards the implementation of these models in one and two dimensions. An immediate application of this modeling will be to aid in the interpretation of the laser-plasma experiments at LBNL [13]. In addition to continuing the work described here, we are also beginning the extension to three dimensions. Much of the experience gained for the one- and two-dimensional cases is readily applied to three-dimensional models. In three dimensions we are faced with the added complexity of moving to a parallel computing environment, since the memory requirements for all but the simplest cases well exceed the resources of single workstation-class systems.

ACKNOWLEDGEMENTS

The authors acknowledge useful conversations with David Bruhwiler, John Cary, and Rodolfo Giacone, This work was supported in part by the Institute for Advanced Physics and by the Department of Energy under Contract No. DE-AC-03-76SF0098.

REFERENCES

1. G. M. Tarkenton and B. A. Shadwick, Bull. Am. Phys. Soc. **44**, 243 (1999).
2. E. Esarey, P. Sprangle, J. Krall, and A. Ting, IEEE Trans. Plasma Sci. **24**, 252 (1996).
3. R. Giacone and J. R. Cary, private communication.
4. B. A. Shadwick and G. M. Tarkenton, in preparation.
5. B. A. Shadwick, J. C. Bowman, and P. J. Morrison, SIAM J. Appl. Math **59**, 1112 (1999).
6. E. Godlewski and P.-A. Raviart, *Numerical Approximation of Hyperbolic Systems of Conservation Laws, Applied Mathematical Sciences Volume 118* (Springer, New York, 1996).
7. J. P. Boris and D. L. Book, J. Comput. Phys. **11**, 38 (1973).
8. D. L. Book, J. P. Boris, and K. Hain, J. Comput. Phys. **18**, 248 (1975).
9. J. P. Boris and D. L. Book, J. Comput. Phys. **20**, 397 (1976).
10. S. T. Zalesak, J. Comput. Phys. **31**, 335 (1979).
11. B. Van Leer, J. Comput. Phys **14**, 361 (1974).
12. D. Mihalas and B. Weibel-Mihalas, *Foundations of radiation hydrodynamics* (Oxford University Press, New York, 1984).
13. W. P. Leemans *et al.*, Laser Wakefield Accelerator Experiments at LBNL, this proceedings.

Emittance Control in Laser Wakefield Accelerator

S. Cheshkov[1], T. Tajima[1,2], C. Chiu[1], F. Breitling[1]

[1] *Department of Physics, University of Texas at Austin, Austin, TX 78712*
[2] *Lawrence Livermore National Laboratory, CA 94550*

Abstract. In this paper we summarize our recent effort and results in theoretical study of the emittance issues of multistaged Laser Wakefield Accelerator (LWFA) in TeV energy range. In such an energy regime the luminosity and therefore the emittance requirements become very stringent and tantamount to the success or failure of such an accelerator. The system of such a machine is very sensitive to jitters due to misalignment between the beam and the wakefield. In particular, the effect of jitters in the presence of a strong focusing wakefield and initial longitudinal phase space spread of the beam leads to severe transverse emittance degradation of the beam. To improve the emittance we introduce several methods: a mitigated wakefield focusing by working with a plasma channel, an approximately synchronous acceleration in a superunit setup, the "horn" model based on exactly synchronous acceleration achieved through plasma density variation and lastly an algorithm based on minimization of the final beam emittance to actively control the stage displacement of such an accelerator.

INTRODUCTION

The concept of LWFA was originally proposed by Tajima and Dawson [1]. Since then there has been significant advance in this area. For a review, see Esarey et al [2].

In pursuit of the next energy frontier, a laser-based wakefield linear e^+e^- collider at high energies (such as 5 TeV) has been considered for which many wakefield units are needed to reach the desired energy. However, the high energy is not the only requirement, such a collider demands an extremely small beam emittance and thus extremely precise beam handling. To identify the crucial physical and technological problems associated with this, a systems approach through a dynamical map has been introduced [3–5]. There was also an earlier study on a 5 TeV laser-wakefield collider [6]. Emittance degradation in TeV-accelerators for the case of a full filamentation in the transverse phase space was considered in ref. [7].

In [5], the study of emittance degradation in the presence of jitters, associated with stochastic misalignment between the beam and the wakefields was carried out where the plasma medium is uniform and the beam is accelerated over a full quarter-wave region. One finds that the system is sensitive to transverse offsets due to the wakefield averaging over the entire accelerating phase and thus typically providing a very strong focusing.

A possible way to decrease the strong focusing wakefield is to use the hollow channel design [8]. A drawback is that due to the finite density gradient near the wall of the cavity, there is a local plasma frequency which would match the wakefield frequency and lead to resonance absorption [9]. In [5], numerical models with beam acceleration over a full quarter-wave-region were considered. These models are for both without involving the plasma channel and with the plasma channel ignoring the resonance absorption effect. The former will be referred to as the CTHY model and the latter the CTHY1 model and are briefly described in Sec.2.

From a general consideration, one expects that the emittance degradation should depend on the phase-range through which the acceleration occurs. Using two different approaches [10,11] we explore ways to improve the resilience against jitters through variations of the loading phase and also of the phase interval of acceleration. Computer simulation indicates that when the acceleration phase is approximately fixed (the phase slip is small) there is an inverse power behavior (for a fixed final energy of the particles), in particular, the emittance degradation decreases like $1/N_T$, where N_T is the total number of acceleration units. This confirms the theoretical expectation of CTHY deduced from a statistical theory [5]. The inverse power law suggested that through the use of small acceleration intervals one may be able to achieve high resilience against jitters. The second approach is to work with a synchronous acceleration model, where there is no phase slip at all. It was pointed out by Katsouleas [12] over a decade ago that synchronous acceleration can be achieved by varying the plasma density. More specifically, consider the case where the local density along the beam direction is gradually increasing. Then the wavelength of the plasma waves, on which the beam electrons are riding, becomes shorter and shorter. If the rate of the phase-slip of the beam electrons exactly matches the rate of the phase advance due to the shrinkage of the plasma waves, a continuous acceleration without any phase-slip may be achieved. From a study on the hydrodynamics of a nozzle flow [13], we find that if there is a steady flow opposite to the direction of the beam propagation, by fine tuning the increase of the nozzle cross section along the beam, one can control the corresponding increase of the plasma density and in turn achieve a synchronous acceleration. Here the acceleration unit has a horn shape and we refer to this model as the "horn model" [11]. Based on the Katsouleas's matching condition, we have derived a set of analytic expressions which have been incorporated in the dynamical map. Our work [11] also takes into account the conservation of energy in the context of the pump-depletion effect [14] and the adiabatic invariance property throughout the acceleration process [15]. The computer simulations for the horn model with a small loading phase show a definite improvement over CTHY model. Lastly we propose an active alignment control which can significantly reduce the final emittance of the beam as illustrated by our preliminary numerical results. The analysis in done by introducing a feedback in our multistage systems code which adjusts the accelerator stages based on the calculation of the final emittance only and minimization criteria.

MULTISTAGE ACCELERATION, MAP APPROACH AND EMITTANCE DEGRADATION

In general, TeV center of mass energies of colliding particles require multistage acceleration even if we assume large acceleration gradients typical for the plasma based accelerators. To study such an accelerator system we introduced a map approach described in [3–5]. The longitudinal phase space transformations (for the differential phase $\delta\Psi$ and the differential Lorenz factor $\delta\gamma$, respectively) stage by stage are described by the longitudinal map:

$$\delta\Psi_{n+1} = \delta\Psi_n \tag{1}$$

$$\delta\gamma_{n+1} = 2\gamma_p^2 \Phi_0 (\cos(\Psi_s + \Delta) - \cos\Psi_s)\delta\Psi_n + \delta\gamma_n, \tag{2}$$

where n enumerates the stage, Ψ_s is the "synchronous" or the loading phase, Δ is the phase slippage, γ_p is the Lorenz factor of the plasma wave, and $\Phi_0 \approx a_0^2$, where a_0 is the normalized vector potential of the laser. Characteristic for these equations is that the particles are extremely relativistic ($\gamma \sim 10^5 - 10^7$) so that the synchrotron oscillation frequency approaches zero. Transverse motion is described by the following equation (for each stage, z is defined with respect to the beginning point of this stage along the beam propagation direction)

$$\ddot{\tilde{x}} + \omega_\beta^2 \sin(\omega_s z + \Psi_s + \delta\Psi_n)\tilde{x} = 0, \tag{3}$$

where $\ddot{\tilde{x}} = \frac{d^2\tilde{x}}{dz^2}$,

$$\omega_s = \frac{k_p}{2\gamma_p^2}, \quad \omega_\beta = \frac{2}{r_s}\left(\frac{\Phi_0}{\gamma}\right)^{1/2}, \tag{4}$$

and $\tilde{x} = \sqrt{\gamma}\, x$. In Eq. (4), r_s is the laser spot size and k_p is the plasma wavenumber. For analytic solution additional approximations are needed. The analysis here is based on harmonic oscillator model and free drift of the particles between the stages. So the transverse map becomes

$$\begin{pmatrix} \tilde{x}_{n+1} \\ \dot{\tilde{x}}_{n+1} \end{pmatrix} = M_n \begin{pmatrix} \tilde{x}_n \\ \dot{\tilde{x}}_n \end{pmatrix}, \tag{5}$$

$$M_n = \begin{pmatrix} \cos(\omega l), & \frac{1}{\omega}\sin(\omega l) \\ -\omega\sin(\omega l), & \cos(\omega l) \end{pmatrix} \cdot M_{\text{gap}}, \quad M_{\text{gap}} = \begin{pmatrix} 1 & L \\ 0 & 1 \end{pmatrix}, \tag{6}$$

where $l = \Delta/\omega_s$ is the stage length. In Eq. (6), ω is the betatron frequency[1] (in units of 1/m), and M_{gap} is for a free drift space. We have also considered M_{gap} in

---

[1] In Eq. (6) the betatron frequency ω is defined by $\omega^2 = \omega_\beta^2 <\sin\Psi>$, for details, see [11].

which magnets are included [10,11]. Even though our simplified notation in Eq. (6) does not explicitly show it, it is important to remember that the transverse matrix depends on the stage number (since particles are being accelerated and $\omega \propto \frac{1}{\sqrt{\gamma}}$) and also is different for different particles due to the spread in ω which in turn is caused by the particle spread in the longitudinal phase space ($\delta\Psi, \delta\gamma$).

We now introduce errors [5] in the accelerating structure, namely jitter of the aligned wakefield (by whatever mechanism) stage by stage. The result of this is a phase space mixing, which depends on absolute spread of the betatron frequencies and jitter magnitude. It degrades the transverse emittance of the beam. Typical strength of the focusing force is of great importance for the rate at which mixing and correspondingly the emittance degradation takes place. In our model the dislocation of the aligned position of each stage is a stochastic variable with Gaussian distribution of standard deviation $\sigma_\mathcal{D}$. The transverse map is modified according to

$$\begin{pmatrix} \tilde{x}_{n+1} \\ \dot{\tilde{x}}_{n+1} \end{pmatrix} = M_n \begin{pmatrix} \tilde{x}_n - \tilde{\mathcal{D}}_n \\ \dot{\tilde{x}}_n \end{pmatrix} + \begin{pmatrix} \tilde{\mathcal{D}}_n \\ 0 \end{pmatrix} \qquad (7)$$

where $\mathcal{D}_n$ is the misalignment of stage n. The total transverse map after N stages becomes

$$\begin{pmatrix} \tilde{x}_{N+1} \\ \dot{\tilde{x}}_{N+1} \end{pmatrix} = M_N M_{N-1} \ldots M_2(1 - M_1) \begin{pmatrix} \tilde{\mathcal{D}}_1 \\ 0 \end{pmatrix} + \ldots (1 - M_N) \begin{pmatrix} \tilde{\mathcal{D}}_N \\ 0 \end{pmatrix}$$

$$+ M_N M_{N-1} \ldots M_1 \begin{pmatrix} \tilde{x}_1 \\ \dot{\tilde{x}}_1 \end{pmatrix} \qquad (8)$$

The stochastic map (8) leads to a transverse emittance degradation. Computer simulation with small random dislocations of magnitude $\sigma_\mathcal{D} = 1 \cdot 10^{-7}$ m is presented in Figure 1a and Figure 1c [4]. We see that in this case (corresponds to design I in [6]) we have a severe emittance growth (the initial normalized emittance is 2.2 nm). Additional results can be found in [3–5]. In general, the problem can be cured by decreasing the focusing of the accelerator system. One possible way is to use a plasma channel [8,9]. It provides a linear weak focusing and we showed in [4] that its performance in a collider application is promising.

A detailed study of the map and emittance degradation properties can be found in [5]. In the limit of small betatron frequency ω, namely $\omega\, l < 1$ and small distance between the stages and in fixed energy approximation the emittance growth after N stages as shown in [5] is:

$$\Delta\epsilon \approx \frac{1}{2}\gamma\omega(\omega\, l)^2 \sigma_\mathcal{D}^2 N \ . \qquad (9)$$

The alignment errors introduce randomness in the phase space particle positions upon reentry to the next stage, the differential betatron oscillations mix these positions causing an emittance growth. This is valid in the case of a small drift

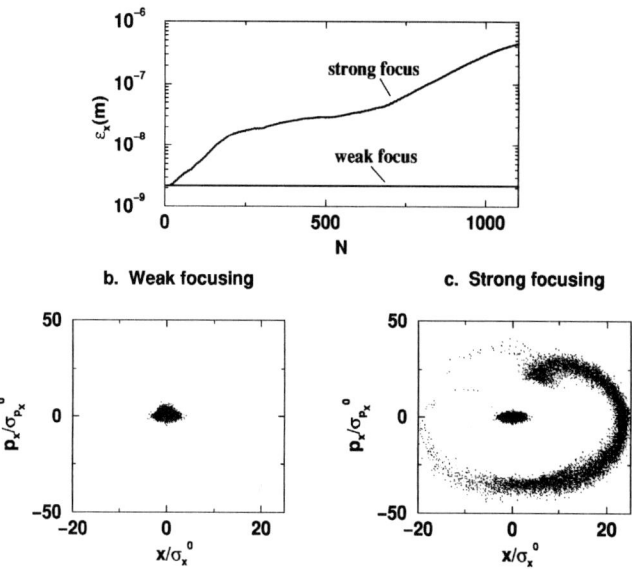

FIGURE 1. The normalized x-emittance vs. stage number and the transverse phase space before and after the acceleration to 2.5 TeV for weak and strong wakefield properties. Parameters for the CTHY model, i.e. the strong focusing case (uniform plasma) are $\gamma_p = 100$, $L = 15$ cm, $\epsilon_x^0 = 2.2$nm, $r_s = 0.5$mm, $a_0 = 0.5$, dislocation size=0.1μm, $\sigma_\gamma/\gamma=0.01$, $\sigma_\Psi = 0.01$. Parameters used for the CTHY1 model, i.e. the weak focusing (plasma channel) are given in Sec. 2

$L \ll l$. If this is not the case, but $L\omega \ll 1$ is still satisfied we can modify the above equation by introducing $\omega' = \omega\sqrt{l/(l+L)}$. Then the betatron phase advance per stage (wakefield and drift) is given by $\omega'(l+L)$ and in all formulas ω should be replaced by ω' and l by $l+L$.

Since the energy increases Eq. (9) needs to be further modified. Denote the Lorenz factor increase per stage by $\Delta\gamma$. In the adiabatic limit we obtain

$$\Delta\epsilon \approx \frac{1}{2}\gamma\omega(\omega\, l)^2\sigma_\mathcal{D}^2\left(\frac{\gamma}{\Delta\gamma}\right)^{1/2}\sqrt{N\ln\left(1+\frac{\Delta\gamma N}{\gamma}\right)}, \qquad (10)$$

where now γ is the initial particle Lorenz factor and ω is the initial betatron frequency. Typically $\Delta\gamma \approx \frac{ea_0^2 E_0 l}{mc^2} = a_0^2 k_p l$, where E_0 is the nonrelativistic wavebreaking field, and $\omega \propto \frac{a_0}{r_s\sqrt{\gamma}} <\sin\Psi>^{1/2}$, so we obtain

$$\Delta\epsilon \propto \frac{l^{3/2}a_0^2\sigma_\mathcal{D}^2}{r_s^3 k_p^{1/2}}\sqrt{N\ln\left(1+\frac{\Delta\gamma N}{\gamma}\right)} <\sin\Psi>^{3/2}. \qquad (11)$$

A very important result is the strong dependence of the emittance growth on the magnitude of the betatron frequency (or wakefield curvature). Of course, better control of the errors reduces the emittance degradation. We can also see from Eq. (11) that for a fixed final energy reducing the length of a single stage decreases the emittance growth. This point is exploited in the next section. Scalings of the emittance growth with various parameters can be found in [5]. When the number of stages is relatively small and the phase space mixing is not complete, numerical results appear to be the only reliable way to analyze the properties of the map, analytical estimations are rather difficult. Analytical estimations of emittance growth due to stage misalignment valid in the case of full filamentation (phase space mixing) in a single stage can be found in [7]. In this limit (corresponds to a very strong focusing of the wakefield), control over the emittance growth can be achieved only by precise handling of the beam (namely error control better than the beam size). The results in this limit are reproduced in our theory by replacing the factor ωl in Eq. (9) by unity.

From the computer simulations for the small emittance design [6] for a multi TeV collider we see that in the case of an initially homogeneous plasma it is difficult to avoid a severe emittance growth of the accelerated beam in the presence of small alignment errors stage-by-stage based on reasonable parameters (laser spot size, dislocation size and number of stages). The difficulty is primarily due to the fact that the wakefield focusing too strong in this case. The above considerations do not include the transverse nonlinear effects which also contribute to the emittance increase.

A possible way to decrease the focusing is the "hollow channel" design proposed by [8] in which a preformed vacuum channel in an underdense plasma is discussed. This case offers several important advantages: the focusing force is almost exactly (because the phase velocity of the wakemode is very close to the speed of light) linear and weak in the channel (the weak focusing is a very important improvement over that of a uniform plasma case); there exists a stable propagation solution for the laser mode; the acceleration gradient is very uniform in transverse coordinates within the channel. A drawback is the loss in the magnitude of the accelerating field. The equations for the wakefield in the channel can be found in [8]. There are no major changes to our previous map scheme, there is a reduction in Φ_0 and the magnitude of the focusing changes:

$$\omega_\beta = \frac{k_{ch}}{\sqrt{2}\gamma_p}\left(\frac{\Phi_0}{\gamma}\right)^{1/2},$$

where k_{ch} is the wake wavenumber. Since the γ_p factor is usually large the magnitude of the focusing force decreases significantly. Run shown on Figure 1a and Figure 1b [4] indicates a significant improvement over the previous design. Here we are able to preserve even design I [6] normalized emittance of 2.2 nm. The parameters of the CTHY1 model are: $\gamma_p = 150$, the channel radius $a = 30\mu$m, the laser spot size $r_s = 50\mu$m, the plasma density (outside the channel) $n = 5 \cdot 10^{16}$cm^{-3}, the laser wavelength $\lambda \approx 1\mu$m, and the drift space of 0.3 m. The magnitude of the stage dislocations is increased to $\sigma_\mathcal{D} = 0.5\mu$m. The remaining parameters are the same as in the CTHY case. From the graphs we see that the emittance growth of the accelerated beam is now practically negligible and the design is more promising. Unfortunately, there is an additional effect: because in reality we have a finite density gradient it leads to a resonant absorption where the local plasma frequency matches the wakefield frequency. This effect has been studied in [9], where an expression for the quality factor of the hollow channel is derived. Possible low values of this factor limit the acceleration of multiple bunches in a single shot created wakefield. Another way to decrease the wakefield curvature is through the use of transversely shaped laser pulses. A "flat top" laser pulse would produce a small curvature wakefield and correspondingly small focusing force. Creation and propagation of such pulses needs to be studied. In the case of PWFA the density shaping of the driver electron bunch can be achieved by using octuple magnets. Lastly we note that in the weak focusing cases achieved in plasma the collision-induced emittance degradation becomes important, since it is inversely proportional to the betatron frequency. Correspondingly, there is an optimal wakefield focal strength. We will present the results on this in a follow-up paper.

DESIGN ISSUES, APPROXIMATELY SYNCHRONOUS ACCELERATION, HORN MODEL

Accelerator with superunits, chips and magnets

For a fixed final energy it is advantageous [10,11] to increase the number of acceleration stages to have better control of the emittance. See Figure 2. In such scenario of having a very large number of stages, each stage becomes very short (e.g. of the order of 1 cm) and we are led to consider a superunit which is made out of many short tubes or chips, the wakefield within each chip is created by an independent laser pulse. We consider distances of the order of 1m between adjacent superunits to allow the experimental set up needed to maintain superunits including magnets placed over a certain period of length to ensure the quality of the beam. We have considered an illustrative system:

- Total energy: 2.5 TeV, which is used as each of the two arms of the 5 TeV collider. The acceleration is from 0.5 TeV to 2.5 TeV.

- Total number of superunits (SU): 500

- Within one super-unit (SU) there are:
    - 100 stages per SU, and
    - gap = tube = 0.83 cm.

- There is a large-gap between two adjacent super-units: 1m

- Length of the accelerator: about 1300 m.

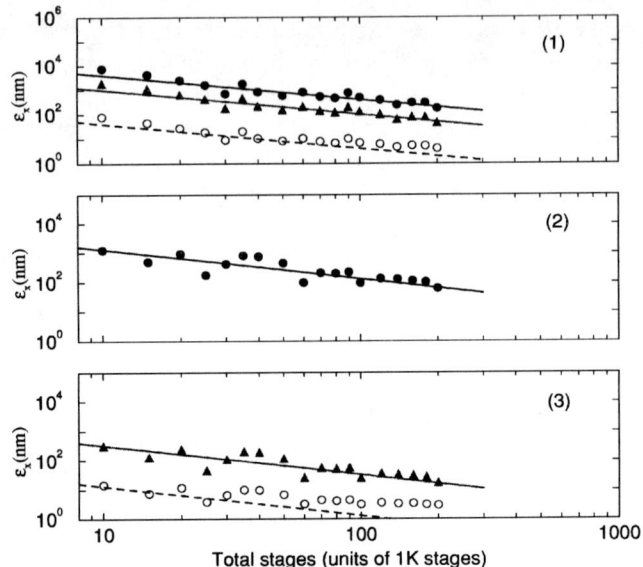

FIGURE 2. Multistages [11]: (1) Gap=10 tubes, σ_D=1, 0.5, & 0.1 μm, (2) Gap=tube, σ_D=1 μm, (3) Gap=tube, σ_D=0.5 & 0.1 μm. Each case is compared with the inverse power law $1/N_T$. N_T is the total number of stages.

We proceed to look at how emittance degradation varies as a function of the loading phase for the system of superunits with chips. From Eq. (9) one expects in some average sense

$$\epsilon \propto \omega^3 \propto (\sin \Psi_m)^{3/2}, \tag{12}$$

where Ψ_m is the mean phase of the beam, taken to be $\Psi_m = \Psi_s + 0.5\Delta$. Here Ψ_s is the loading phase and Δ the total phase slip. This implies that that the resilience of the present system against jitters can be further improved, at least in the small loading phase region, by lowering the loading phase value. We consider two loading phases, i.e. $\Psi_s = 0.15$ rad and $\Psi_s = 0.05$ rad. Figure 3 (taken from [11]) shows the interim emittance degradation for three cases. There are 50K stages and all cases are at the final energy of 2.5 TeV. The stochastic theory if applicable implies that the intermediate emittance should grow approximately[2] linearly with the number of stages. Approximate mean linear behavior is observed for curves a and b. For curve c, there is a rapid rise up to about 20% of the total stages, which is followed by an approximately linear mean behavior. To conclude, within the present chip-model the final emittance has been reduced to say less than

---

[2] This approximate linearity is valid as is Eq. (9) only under assumption of a constant typical energy beam transport. Obviously, the linearity is not a good approximation for very large N, when the asymptotic behavior is $\Delta\epsilon \propto \sqrt{N} \ln N$ as seen from Eq. (11)

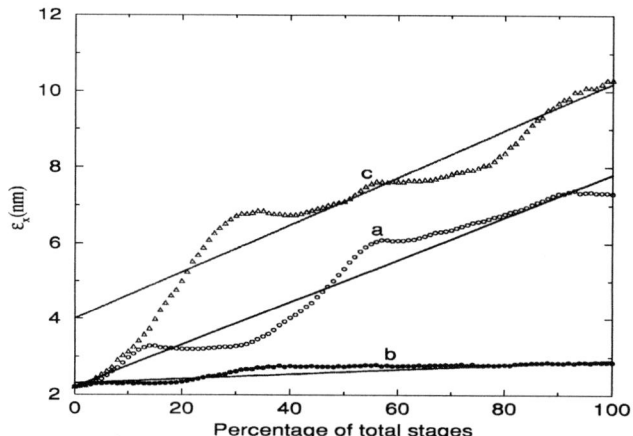

FIGURE 3. The interim emittance degradation behavior as the beam particles traverse through the system of a chip model for $\sigma_D = 0.1\mu$m. For each case, a solid line of a linear behavior is included to guide the eyes.
Curve a: Total stages: 50K, $\Psi_s = 0.15$ rad.
Curve b: Total stages: 50K, $\Psi_s = 0.05$ rad.
Curve c: Total stages: 20K, $\Psi_s = 0.05$ rad.

$2\epsilon_0$, where ϵ_0 is the initial normalized transverse emittance of 2.2nm. This is to be compared to the situation in the CTHY model, where the final emittance is beyond $100\epsilon_0$. This, however, is at the expense of introducing 50 times more laser pulses, which increases the power consumption by many folds. Thus it has severe practical limitations. These limitations might be ameliorated by adopting a technique to flip a phase by π by introducing two counterpropagating lasers with slightly different colors (G. Shvets' method [16]).

Synchronous Acceleration

As mentioned earlier for the "horn model" [11], synchronous acceleration may be achieved through a specific variation of the plasma density [12]. Consider a steady adiabatic flow of a fluid from a reservoir through a nozzle say in the z direction. Let the static fluid density of the fluid in the reservoir be ρ_0, which will be referred to as the quiescent density. Denote the fluid density at z along the nozzle be $\rho(z)$. In the [11] we showed that based on fluid dynamics [13] the following relation is valid:

$$A(z) = const \left(\frac{\rho(z)}{\rho_o}\right) \sqrt{1 - \left(\frac{\rho(z)}{\rho_o}\right)^{\gamma-1}}, \qquad (13)$$

where γ is the usual ratio of the specific heat at a constant pressure to that at a constant volume. The region of interest is characterized by a subsonic fluid flow

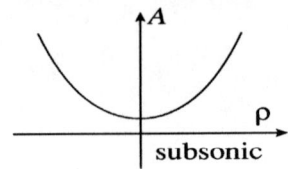

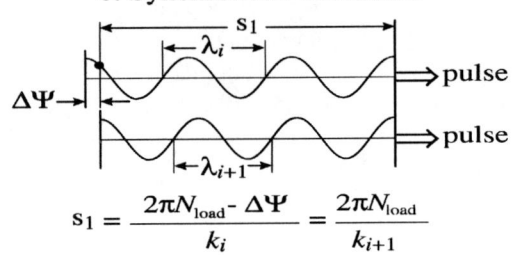

FIGURE 4. The Horn model and the matching condition

(see Figure 4a), where there is a one-to-one relationship between the cross sectional area A, and the plasma density ρ. By increasing the cross section along the beam direction in a specified way one may achieve the required density function. Looking down the stream of the beam, the accelerator consists of a system of aligned horns, although in some cases the increase in radius may be slight. This is the reason why we refer to the present model as the "horn model". See Figure 4b. We now come to Katsouleas's matching condition, see Figure 4c. Consider the wakefield acceleration of a beam electron which is located at the center of the beam. Let the "loading number" N_{load} be the number of wave crests the electron is lagging behind the laser pulse. If the initial electron phase relative to the local wakefield, as defined earlier, is Ψ_s, then the electron phase relative to the laser pulse defined by the local plasma wave number k_p is

$$k_p s_1 = 2\pi N_{load} - \Psi_s,$$

where s_1 is the distance from the electron to the pulse measured in the rest frame of the pulse. To motivate the matching condition, for the time being imagine the horn has been divided into many segments. We will assume that the density is constant within each segment. For the ith and the i+1th segments, the wave numbers are k_{pi} and k_{pi+1}, respectively. The ith segment has a width Δz and has a phase-slip $\Delta \Psi$. Here its phase relative to the pulse measured by the wave number of the ith segment is $k_{pi}s_1 = 2\pi N_{load} - \Psi_s - \Delta\Psi$. The synchronous condition requires the recovery of the initial phase at the start of the i+1th segment, i.e. $k_{pi+1}s_1 = 2\pi N_{load} - \Psi_s$. In other words, the matching condition is given by

$$\frac{2\pi N_{load} - \Psi_s - \Delta\Psi}{k_{pi}} = \frac{2\pi N_{load} - \Psi_s}{k_{pi+1}}. \tag{14}$$

We write $k_{pi+1} = k_{pi} + \frac{dk}{dz}\Delta z$, where Δz is the width of the ith segment. In the continuum limit, after some algebra it leads to

$$\frac{1}{k_p} \cdot \frac{dk_p}{dz} = \frac{1}{(2\pi N_{load} - \Psi_s)} \cdot \frac{d\Psi}{dz} = \frac{1}{2(2\pi N_{load} - \Psi_s)c} \cdot \frac{\omega_p^3}{\omega_0^2}. \tag{15}$$

The first equality is the Katsouleas' condition for synchronous acceleration. The frequency of the laser pulse is denoted by ω_0. To evaluate the number density variation within the horn, we first recall that the frequency of plasma waves is proportional to the square-root of the number density. Thus the z-dependence of all three quantities: the number density of the plasma medium, the frequency and the wave number of plasma waves, may be specified by a single z-dependent function $\zeta(z)$. In particular, one may write

$$n(z) = n_0\zeta(z)^2, \ \omega_p(z) = \omega_{p0}\zeta(z), \text{ and } k_p(z) = k_{p0}\zeta(z), \tag{16}$$

Substituting Eq.(16) into Eq.(15) gives

$$\frac{1}{k} \cdot \frac{dk}{dz} = \frac{1}{\zeta} \cdot \frac{d\zeta}{dz} = \frac{1}{2(2\pi N_{load} - \Psi_s)c} \cdot \frac{\omega_{p0}^3}{\omega_0^2}\zeta^3, \tag{17}$$

To the extent one neglects the pump-depletion effect [14], i.e. the loss of laser pulse energy as it traverses through the horn, the intensity and the frequency of the laser pulse is assumed to be constant. Integrating over Eq. (17) leads to

$$\zeta(z) = \frac{1}{(1 - z/z_0)^{1/3}}, \text{ with } z_0 = \frac{2(2\pi N_{load} - \Psi_s)c}{3} \cdot \frac{\omega_0^2}{\omega_{p0}^3}. \tag{18}$$

In a similar fashion pump depletion effect and the adiabatic invariance can be included in our scheme, however it involves significant amount of algebra and is left for [11]. All this leads to modifications in the longitudinal and transverse transfer map. Here we discuss only the numerical results. For the present synchronous acceleration case, there is no quarter-wavelength restriction, so the tube length can *a priori* vary over a range of values. We have verified that the emittance degradation is also not too sensitive to the loading number. Here is an illustrative case [11]. The tube length is 0.35m and the loading number is 5. The density variation per horn is 7%, with the acceleration energy per stage 2.08 GeV, which is comparable to that of the CTHY model. In Figure 5 [11], curve-a corresponds to the case where

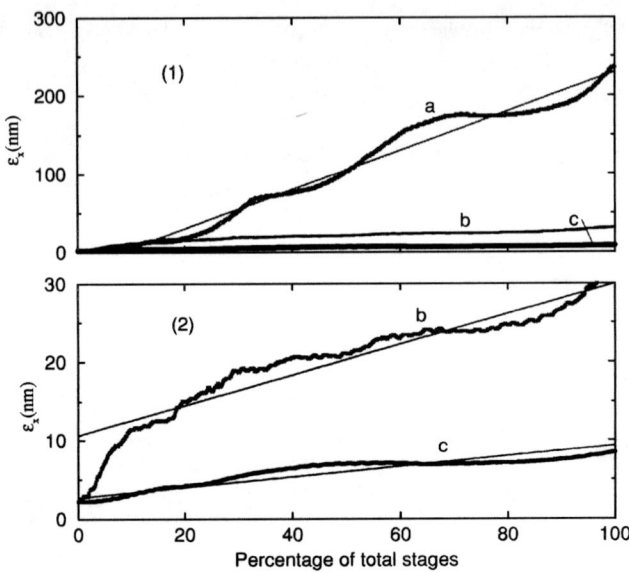

FIGURE 5. The emittance degradation for three cases of the horn model.
Curve-a: $\Psi = 0.15$ rad
Curve-b: $\Psi = 0.04$ rad
Curve-c: $\Psi = 0.04$ rad and $\sigma_\psi = 0.0001$ rad

$\Psi_s = 0.15$ rad. Here the final emittance is $\epsilon = 237$nm $\sim 108\epsilon_0$, which is in the same ball park as that of the CTHY model. So far we have not gained much ground. The important case is curve-b, which is the case where $\Psi_s = 0.04$ rad. It has a final emittance $\epsilon = 31.7$nm $\sim 14.5\epsilon_0$, which is about an order of magnitude reduction compared to that of the CTHY model. The interim emittance for this case [11] is shown in Figure 5-1 and with an amplified scale in Figure 5-2. The emittance degradation is sensitive to the longitudinal phase spread of the beam which for all cases considered up to now has been taken to be $\sigma_\psi = 0.01$ rad. Curve-c illustrates the case for a negligibly small value of the spread, i.e. $\sigma_\psi = 0.0001$ rad. Here the final emittance is given by $\epsilon = 8.4$nm $\sim 3.8\epsilon_0$. See [11] for more details.

EMITTANCE MINIMIZATION CONTROL OF LWFA

In this section we describe preliminary results of our studies on active feedback [17] (and feed forward) control of beams of the LWFA based collider. In the past we introduced the feedforward control of laser optics by the neural net in order to minimize the jitter of the mirror positions [18]. The idea here is to correct the stage positions based on the measurements of the final emittance only rather than to "measure" more difficult quantities of the beam. This is the entropy minimization strategy. We implemented this strategy in our model CTHY. In our computer code, each transverse stage displacement consists of two parts: constant (in time), with a magnitude in the micron range and random (in time), with a magnitude in the submicron range. After each shot a stage is moved transversely by a certain fraction of a micron. If the emittance is decreased the new position is accepted otherwise the previous position is reset. As a result the emittance can be significantly reduced if the stochastic (in time) jitter is not very large. Typical runs are shown in Figure 6. However, there are several problems: after adjustments the stages are still misaligned (the algorithm finds local minima of the emittance) and correspondingly the beam centroid is usually kicked too much, this method reduces the emittance by a large factor when the emittance growth is large but does not work that well for smaller emittance growth. As a future plan we want to incorporate the beam centroid position in the algorithm and also study the efficacy of the algorithm in different accelerator scenarios.

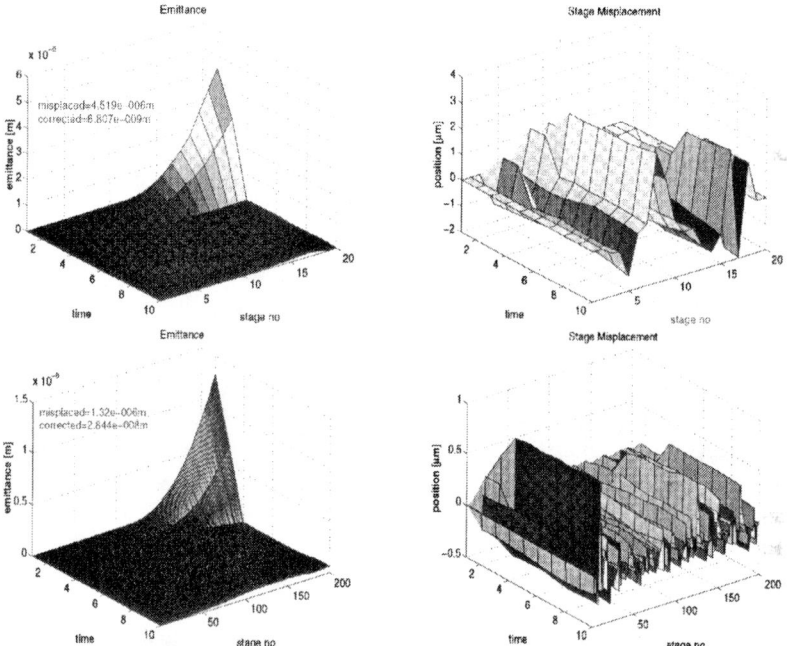

FIGURE 6. The improved control of emittance by feedback control to minimize beam entropy (final emittance). The transverse normalized emittance and stage positions in 20 (upper row) and 200 (lower row) stage units. Magnitude of the constant in time misalignment is $2\mu m$ and of the stochastic one is $0.1\mu m$

CONCLUSIONS

Emittance control in a high energy accelerator is of crucial importance. In previous work we identified main effects that degrade the emittance of the beam in plasma wakefield based collider. In this paper we considered various methods for emittance control and discussed their efficacy and applicability.

ACKNOWLEDGMENT

We thank Mike Downer and also his Femtosecond Spectroscopy Group, especially Andy Rundquist and Erhard Gaul for valuable discussions. We also thank Boris Breizman for discussions on fluid dynamics related issues. This work is supported in part by the US Department of Energy (DOE) and Japan Atomic Energy Research Institute (JAERI). One of us (TT) is also supported in part through a US DOE contract to LLNL, W-7405-Eng.48.

REFERENCES

1. Tajima, T., and Dawson, J., *Phys. Rev. Lett.* **43**, 267 (1979).
2. Esarey, E., Sprangle, P., Krall J., and Ting, A., *IEEE Trans. Plasma Sci.* **24**, 252 (1996).
3. Tajima, T., Cheshkov, S., Horton, W., and Yokoya, K., in *Advanced Accelerator Concepts 8*, edited by W. Lawson, (AIP, New York, 1999), p.153.
4. Cheshkov, S., Tajima, T., Horton, W., and Yokoya, K., in *Advanced Accelerator Concepts 8*, edited by W. Lawson, (AIP, New York 1999), p.343.
5. Cheshkov, S., Tajima, T., Horton, W., and Yokoya, K. accepted to *Phys. Rev. ST Accel. Beams*.
6. Xie, M., Tajima, T., Yokoya, K., and Chattopadhyay, S., in *Advanced Accelerator Concepts 7*, edited by S. Chattopadhyay, (AIP, New York, 1997), p.233.
7. Assmann, R., and Yokoya, K., *Nucl. Instrum. & Methods. A* **410**, 544 (1998).
8. Chiou, T., Katsouleas, T., Decker, C., Mori, W., Wurtele, J., Shvets, G., and Su, J., *Phys. Plasmas* **2**, 310 (1995).
9. Shvets, G., Wurtele, J., Chiou, T., and Katsouleas, T., *IEEE Trans. Plasma Science* **24**, 351 (1996).
10. Chiu, C., Cheshkov, S., and Tajima, T., *Beam Dynamics Newsletter*, **21**, 110 (2000).
11. Chiu, C., Cheshkov, S., and Tajima, T., submitted to *Phys. Rev. ST Accel. Beams*.
12. Katsouleas, T., *Phys. Rev. A* **33**, 2056 (1986).
13. Landau, L., and Lifshitz, E.,*Fluid Mechanics*, Pergamon Press,2d Ed., 1987, Sec. 83 and Sec. 97.
14. Horton, W., and Tajima, T., *Phys. Rev. A* **34**, 4110 (1986).
15. Kuehl, H., Zhang, C., and Katsouleas, T., *Phys. Rev. E* **47**, 1249 (1993).
16. Shvets, G., Fisch, N., Pukhov, A., and Meyer-ter-Vehn, J., *Phys. Rev. E.* **60**, 2218 (1999).
17. Seeman, J. et al. *SLAC-PUB-5439* (1991), and references therein.
18. Breitling, F., Weigel, R., Downer, M., and Tajima, T., submitted to *Rev. Sci. Instrum.*

Optimization of Laser Wakefield Acceleration

A. R. Rundquist, S. P. LeBlanc, E. W. Gaul, S. Cheshkov, F. B. Grigsby, T. T. Tajima, and M. C. Downer

University of Texas at Austin, Department of Physics
Austin, TX 78712

Abstract. Using an evolutionary strategy algorithm, we optimize the generalized transformer ration of a laser wakefield accelerator. The algorithm tests several realistic pulse shapes by integrating the fluid wakefield differential equation and it converges to the shape that most efficiently produces a strong accelerating gradient while experiencing minimal distortion.

INTRODUCTION

With acceleration fields orders of magnitude greater than standard linear accelerators, laser wakefield accelerators (LWFA's) have drawn considerable attention. While several groups have achieved the generation and characterization of laser-produced wakefields [1-10], the need for multi-parameter optimization has arisen so that LWFA's can become the next generation particle accelerators. The ideal LWFA will make the most efficient use of the laser energy by allowing the laser pulse to propagate with a minimum of distortion so that the wakefield accelerating structure will be as long as possible. While high intensity guiding is being pursued by several groups to overcome the Rayleigh length limitation of LWFA's [11-16] and tailored plasma densities are being considered to overcome the dephasing length limitation [17], it is also necessary to ensure that the accelerating structure remains as stable as possible throughout the interaction length. A standard resonant LWFA with a gaussian pulse produces a portion of a plasma oscillation *during* the pulse that will lead to pulse distortions and to a reduced wake after propagation. The minimization of this distortion is the subject of this work. For the case of a plasma beam based wakefield accelerator, there has been theoretical work done by Bane et al which produced an analytical plasma beam shape to optimize the so-called transformer ratio of the accelerator [18]. The transformer ratio is defined to be the ratio of the maximum accelerating field after the beam to the maximum decelerating field during the beam. Clearly, maximizing the transformer ratio in this case will produce the maximum wakefield after the beam while minimizing the distortion of the beam. Chen et al extended this result to the analogous case of the LWFA [19]. Their optimized result produced no pulse distortion and an arbitrarily large wakefield after the pulse. Unfortunately, the pulse derived by Chen et al is not possible to produce in the lab since it requires a half-delta function and an instantaneous falling edge of the pulse. Neither of these effects is possible with a laser of limited bandwidth. Our approach to this problem for the LWFA is to use an evolutionary strategy computer code to "evolve" towards a pulse shape that maximizes a generalized transformer ratio [20]. We perform numeric simulations with realistic limitations built-in (fixed bandwidth lasers and pulse shaping that is achievable with state-of-the-art pulse shapers) and with fully relativistic, 1D fluid codes. This paper is organized as follows: we will first describe the equations we use along with a careful description of the generalized transformer ratio we use; we will then describe our optimization procedure and finally we will discuss the results and conclude.

THEORY

The coupled, 1D fluid equations which govern wakefield production and pulse propagation in a plasma are [21]

$$\frac{\partial^2 a}{\partial \zeta^2} = -\frac{\omega_p^2}{c^2(\beta_{ph}^2 - 1)} \frac{a}{1+\phi} \tag{1}$$

$$\frac{\partial^2 \phi}{\partial \zeta^2} = \frac{\omega_p^2}{2c^2}\left(\frac{1+a^2}{(1+\phi)^2} - 1\right) \tag{2}$$

where a is the normalized vector potential of the laser, ϕ is the electrostatic potential of the plasma, ζ is the space/time coordinate moving in the frame of the pulse, β_{ph} is given by $(1-v_{ph}/c)^{-1}$, and ω_p is the plasma frequency. We concentrate on the second equation which describes the production of a wakefield from a given pulse envelope $a(\zeta)$. The first equation describes the self-consistent evolution of the pulse as it propagates in the plasma. Our approach is to find a pulse envelope that, while maximizing the wakefield accelerating gradient after the pulse, would experience negligible distortion as it propagates by finding a solution of (2) with minimal plasma disturbance *during* the pulse. We solve for ϕ and then take up to three derivatives: the first gives the accelerating field of the plasma wave, the second gives the plasma density variation (and hence the plasma's refractive index), and the third derivative gives the ζ-dependence of the phase of the pulse. If the laser pulse phase changes with space or time, pulse instabilities arise and grow which diminish the pulse's ability to produce efficient wakefield accelerating structures. It is well known, for example, that the frequency of the pulse will change when it experiences a time-dependent refractive index [22,23]. These new frequencies enable the linear dispersion of the plasma to disrupt the pulse more quickly. Thus by monitoring the derivative of the phase, we can predict which pulses will produce the best results after propagation.

Rather than using the transformer ratio as defined by Chen et al for the LWFA, we parameterize the pulse distortion by considering the weighted effect of all the possible frequency shifts during the pulse. Our "fitness" function is therefore given by the integral of the product of the pulse envelope and the third derivative of the electrostatic potential subtracted from the maximum of the wakefield accelerating field at any point:

$$fitness = w_1 \left.\frac{\partial \phi}{\partial \zeta}\right|_{max} - w_2 \int a^2(\zeta) \frac{\partial^3 \phi}{\partial \zeta^3} d\zeta \tag{3}$$

where w_i are the weights for each term. The optimum pulse shape will be orthogonal (in the general sense) to the frequency shift term. An addition to the fitness function for numeric sampling reasons is a strong negative weight to the value of a^2 at the first point in the array (ζ_{min}). This ensures that the wakefield does not have an unphysical beginning before the pulse. The pulse shaping is accomplished by varying the spectral phases and amplitudes with the limitations present in real world shapers (e.g. a fixed total bandwidth, a fixed sampling rate etc). It should be noted that temporal shaping of these pulses is generally not possible. When spectral amplitude shaping is considered, the fitness function needs to be augmented to protect against extra nonlinear phase in the amplifier (since most pulse shapers for experiments like that described here are placed before the amplifier). We subtract a term that is essentially the "B-integral" of the amplifier chain for the newly shaped spectral amplitude [24]. The B-integral is affected if there is amplitude shaping of the spectrum if one assumes a fixed total energy from the amplifier. If the peak of the spectrum contains more energy, the B-integral is larger.

Procedure

The genes in our evolutionary strategy are the values of the spectral phase and amplitude of an originally transform limited pulse of fixed duration (set to be resonant with the plasma density). These values mutate at a variable (and evolving) mutation rate to produce the children of each new generation. The children are tested according to the fitness (3) and the best are chosen to mutate into the next generation. This process is repeated until the increase in the fitness from one generation to the next goes to zero (typically 1000 iterations).

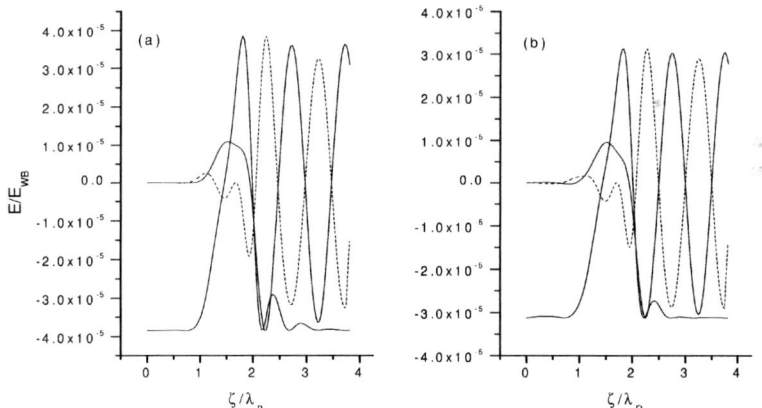

FIGURE 1. Pulse shaping results for $a^2_{max}=0.01$ for phase-only (a), and amplitude and phase shaping (b). Shown are the pulse envelope (solid), electric field (solid), and the frequency shift (dashed). All plots are normalized to the field which is given in units of the wave-breaking field. The x-axis is the space/time coordinate normalized to the plasma wavelength.

RESULTS

Figure 1a shows the result of phase-only shaping (used for maximum energy throughput and no change in the B-integral) for the case where the maximum a^2 of the original transform limited pulse was 0.01 (in the linear regime [21]). Figure 1b shows the result at the same intensity for both amplitude and phase shaping. In both cases the dominant feature of the pulse is a slowly rising triangle with an abrupt falling edge where the wakefield amplitude dramatically increases. This pulse is reminiscent of the prediction of Chen et al but without the half delta function in front [19]. The best fitness values achieved in the linear regime are nearly twice that of the original transform limited gaussian. The cost of reducing the maximum intensity by lengthening the pulse is overcome in this case by greatly reducing the amount of frequency shifting occurring during the energetic portions of the pulse. It is clear that the greatest amount of frequency shifting occurs during the fast falling edge and hence the evolutionary strategy tries to converge to something with as fast a fall as possible. The frequency shifting is minimized during the pulse because the wakefield produced at the beginning of the pulse (either by a small prepulse or by the first appreciable slope of the envelope – it can be shown that the electron density fluctuations are caused by the gradient of the pulse envelope [21]) is out of phase with the wakefield produced by the main slope of the triangle. Thus the wakefields add destructively until the final falling edge when the electron oscillations attain much larger values. Finally, the largest plasma density gradient for electron acceleration occurs just after the pulse where there is a small postpulse that reduces the wakefield amplitude (i.e. it produces a wake that is out of phase with that of the falling edge of the main pulse). We feel that the postpulse is a limitation of the constraints on the evolutionary strategy as the number of postpulses is reduced when amplitude shaping is allowed.

Figure 2 shows the best result of phase shaping in the nonlinear regime (a^2_{max} = 0.1 and 1) [21]. At a^2=0.1 (figure 2a), the result is quite similar to the linear regime in figure 1 but with a pronounced prepulse that sets up the destructive interference inside the pulse mentioned above. As the pulse becomes more and more nonlinear, the best pulse shape begins to converge to the original gaussian. We have found that there is a smooth transition from the triangular shape of the linear regime to the gaussian shape. In the nonlinear regime, the cost of reducing the peak intensity to produce a pulse with less frequency shifting during the pulse is too great. Thus the original transform limited gaussian is the best pulse in the nonlinear regime since it has by definition the greatest peak intensity [24]. We have found that the results of phase-only and amplitude and phase shaping are very similar for a^2_{max}>0.1.

DISCUSSION

There are two methods of experimental implementation for this work. The first is to produce the pulse that the simulation converges to and then test it experimentally. There are several pulse shapers (most notably those based on an acousto-optic crystal) that can take a given amplitude and phase mask and produce a desired pulse [25-30].

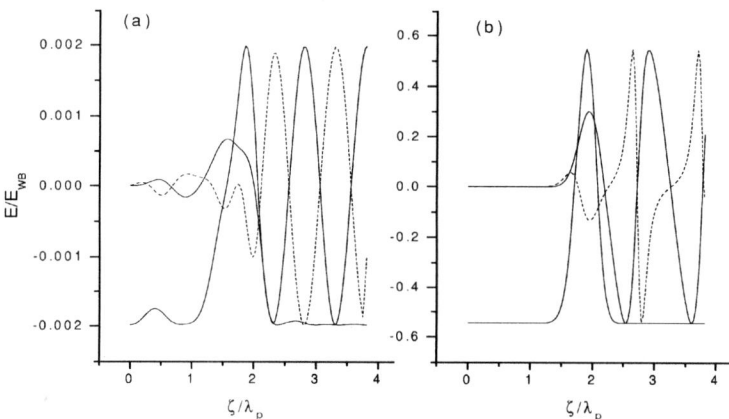

FIGURE 2. Pulse shaping results for $a^2_{max}=0.1$ (a) and $a^2_{max}=1.0$ (b). Shown are the pulse envelope (solid), electric field (solid), and the frequency shift (dashed). All plots are normalized to the field which is given in units of the wave-breaking field. The x-axis is the space/time coordinate normalized to the plasma wavelength.

We feel, however, that a more efficient experimental procedure would be to allow all real-world limitations to be taken into account in the evolutionary strategy algorithm. This is accomplished by replacing the numeric integration of equation (2) with an actual measurement of a wakefield produced by a given pulse. Physical realities such as a focusing geometry which is not fully 1D and pulse distorting optics between the pulse shaper and the experiment (most notably the amplifier) can be included in the optimization routine to produce the most efficient LWFA possible with a real laser. Just a few years ago such an experiment would not have been considered because the full results available by integrating equation (2) could not be measured quickly (typically a multi-shot experiment was required that could take hours [4]). Since the evolutionary strategy requires testing up to thousands of different pulse shapes, a single-shot feedback mechanism is necessary for experimental implementation. Such a single-shot diagnostic of wakefields has been proposed and is being tested by the authors [31]. With a fast feedback diagnostic available, the physical implementation of the evolutionary strategy approach should be feasible.

We have determined the optimum, physically realizable laser pulse shape for LWFA. By extending the concept of the transformer ratio and concentrating on the actual distortion of the pulse as it propagates through the plasma we have found pulse shapes that experience minimal frequency shift while still producing large accelerating gradients. In the linear regime of LWFA the optimum shape is an asymmetric triangle with the fast edge on the trailing side while in the nonlinear regime the best pulse is the original transform limited gaussian. We feel that the evolutionary strategy we have employed can be extended to the real-world experiment in order to best optimize LWFA even with non-ideal experimental conditions.

ACKNOWLEDGMENTS

We would like to acknowledge Erik Zeek for guidance in constructing our evolutionary strategy algorithm. This research was supported by U.S. Department of Energy grant DEFG03-96-ER-40954.

REFERENCES

1. Chen, S.Y., Krishnan, M., Maksimchuk, A. et al., *Physics of Plasmas* **6**, 4739-49 (1999).
2. Le Blanc, S.P., Downer, M.C., Wagner, R. et al., *Physical Review Letters* **77**, 5381-4 (1996).
3. Moore, C.I., Ting, A., Krushelnick, K. et al., *Physical Review Letters* **79**, 3909-3912 (1997).
4. Siders, C.W., Le Blanc, S.P., Fisher, D. et al., *Physical Review Letters* **76**, 3570-3 (1996).
5. Ting, A., Krushelnick, K., Moore, C.I. et al., *Physical Review Letters* **77**, 5377-5380 (1996).
6. Ting, A., Moore, C.I., Krushelnick, K. et al., *AIP. Physics of Plasmas* **4**, 1889-99 (1997).
7. Dorchies, F., Amiranoff, F., Malka, V. et al., *Physics of Plasmas* **6**, 2903-13 (1999).
8. Amiranoff, F., Baton, S., Bernard, D. et al., *Physical Review Letters* **81**, 995-8 (1998).
9. Clayton, C.E., Tzeng, K.C., Gordon, D. et al., *Physical Review Letters* **81**, 100-3 (1998).
10. Gordon, D., Tzeng, K.C., Clayton, C.E. et al., *Physical Review Letters* **80**, 2133-6 (1998).
11. Hosokai, T., Kando, M., Dewa, H. et al., *Proceedings of the 1999 Particle Accelerator Conference (Cat. No.99CH36366). IEEE. Part vol.5, 1999*, 3690-2 vol.
12. Ehrlich, Y., Cohen, C., Kaganovich, D. et al., *Journal of the Optical Society of America B-Optical Physics* **15**, 2416-23 (1998).
13. Durfee, C.G., III, Lynch, J., and Milchberg, H.M., *Physical Review E. Statistical Physics, Plasmas, Fluids, & Related Interdisciplinary Topics* **51**, 2368-89 (1995).
14. Gaul, E.W., Le Blanc, S.P., Rundquist, A.R. et al., *these proceedings*, (2000).
15. Kaganovich, D., Ting, A., Moore, C.I. et al., *Physical Review E. Statistical Physics, Plasmas, Fluids, & Related Interdisciplinary Topics* **59**, R4769-72 (1999).
16. Kaganovich, D., Sasorov, P.V., Erlich, Y. et al., *Applied Physics Letters* **71**, 2925-7 (1997).
17. Kaganovich, D., Cohen, C., Zigler, A. et al., *Applied Physics Letters* **75**, 772-4 (1999).
18. Bane, K.L.F., Chen, P., and Wilson, P.B., *IEEE Transactions on Nuclear Science* **NS-32**, 3524-6 (1985).
19. Chen, P., Spitkovsky, A., Katsouleas, T. et al., *Elsevier. Nuclear Instruments & Methods in Physics Research Section A-Accelerators Spectrometers Detectors & Associated Equipment* **410**, 488-92 (1998).
20. Haupt, R.L. and Haupt, S.E., *Practical genetic algorithms*, Wiley, New York, 1998.
21. Esarey, E., Sprangle, P., Krall, J. et al., *IEEE Transactions on Plasma Science* **24**, 252-88 (1996).
22. Wood, W.M., Siders, C.Q., and Downer, M.C., *Physical Review Letters* **67**, 3523-6 (1991).
23. Le Blanc, S.P., Sauerbrey, R., Rae, S.C. et al., *Journal of the Optical Society of America B-Optical Physics* **10**, 1801-9 (1993).
24. Siegman, A.E., *Lasers*, University Science Books, Mill Valley, CA, 1986.
25. Dugan, M.A., Tull, J.X., and Warren, W.S., *Journal of the Optical Society of America B-Optical Physics* **14**, 2348-58 (1997).
26. Hillegas, C.W., Tull, J.X., Goswami, D. et al., *Optics Letters* **19**, 737-9 (1994).
27. Meshulach, D., Yelin, D., and Silberberg, Y., *Journal of the Optical Society of America B-Optical Physics* **15**, 1615-19 (1998).
28. Zeek, E., Maginnis, K., Backus, S. et al., *Optics Letters* **24**, 493-495 (1999).
29. Weiner, A.M., Leaird, D.E., Patel, J.S. et al., *IEEE Journal of Quantum Electronics* **28**, 908-920 (1992).
30. Wefers, M.M. and Nelson, K.A., *Journal of the Optical Society of America B-Optical Physics* **12**, 1343-1362 (1995).
31. Le Blanc, S.P., Gaul, E.W., Matlis, N.H. et al., *Optics Letters* **25**, 764 (2000).

Laser Shaping and Optimization of the Laser-Plasma Interaction

Anatoly Spitkovsky* and Pisin Chen[†]

Department of Physics, University of California at Berkeley, Berkeley, CA 94720
[†] *Stanford Linear Accelerator Center, Stanford University, Stanford, CA 94305*

Abstract. The physics of energy transfer between the laser and the plasma in laser wakefield accelerators is studied. We find that wake excitation by arbitrary laser shapes can be parameterized using the total pulse energy and pulse depletion length. A technique for determining laser profiles that produce the required plasma excitation is developed. We show that by properly shaping the longitudinal profile of the driving laser pulse, it is possible to maximize both the transformer ratio and the wake amplitude, achieving optimal laser-plasma coupling. The corresponding family of laser pulse shapes is derived in the nonlinear regime of laser-plasma interaction. Such shapes provide theoretical upper limit on the magnitude of the wakefield and efficiency of the accelerating stage by allowing for uniform photon deceleration inside the laser pulse. We also construct realistic optimal pulse shapes that can be produced in finite-bandwidth laser systems and propose a two-pulse wake amplification scheme using the optimal solution.

INTRODUCTION

Recent advances in laser technology allow one to create laser pulses with virtually arbitrary temporal intensity profiles using amplitude and phase shapers [1–3]. Such laser pulses with non-Gaussian axial intensity are now being considered for applications as drivers in Laser Wakefield Accelerators (LWFA). Shaped lasers provide the means of controlling the generation of plasma wake and thus offer the possibility of optimization of wake excitation and accelerating efficiency. However, progress in finding "the optimal" shape has been hindered by the apparent complexity of the problem. Not only is the parameter space of possible shape functions huge, but also the generated wakefield is a nonlinear function of laser intensity, requiring numerical solution of differential equations in a variational calculation. As a result, several groups turned to trial and error methods such as genetic algorithms for optimization [2,3]. Still, even these methods require consistent classification of laser shapes so that different pulses can be meaningfully cross-compared while desired properties such as wake amplitude or efficiency are optimized. In this paper we reanalyze the process of wake generation and argue that the only two physical parameters that describe a laser shape from the stand point of wake excitation are the total pulse energy and its depletion length. Using these parameters we find the *analytic* expression for the family of optimal laser shapes that maximize both the wakefield *and* the accelerating efficiency. We also develop a method for determining the shape of a laser that produces a required value of wakefield *without* explicitly solving the wake equation. This opens the way for obtaining laser shapes that satisfy other optimization criteria specific to given experimental conditions.

ENERGY TRANSFER IN LWFA

Wakefield accelerators such as the laser-driven LWFA [4] or electron beam driven Plasma Wakefield Accelerator (PWFA) [5] can be viewed as two-step energy transfer systems: in the first step the driver deposits energy into wake excitation of the plasma, and in the second step the energy is taken from the wake by the accelerating beam. While the second step is the same for both accelerating schemes, the physics of driver energy deposition is quite different between them. In PWFA the electron beam loses energy to the plasma through interaction with the induced electrostatic field, while in the LWFA laser energy loss occurs via photon red-shift or deceleration [6]. This process can be understood as follows. Poderomotive force of the laser modifies both the density n_e and the Lorentz factor γ of the plasma electrons. This produces modulations in the nonlinear index of the refraction $\eta \equiv [1 - (\omega_p/\omega)^2 n_e/\gamma n_p]^{1/2}$, where $\omega_p \equiv \sqrt{4\pi e^2 n_p/m_e}$ is the unperturbed plasma frequency, and ω is the frequency of the laser. The wake-induced modulations of the refractive index appear stationary in the reference frame co-moving with the laser and cause laser photons to red- or blue-shift depending on the sign of refractive index gradient [7]. Due to negligible scattering in the setting of laser accelerators, the photon number in the laser is essentially constant, and the energy deposition into the plasma is therefore determined by the photon deceleration. To address this quantitatively we consider a laser propagating along the z axis with initial frequency $\omega_0 \gg \omega_p$. In the laser comoving frame, the plasma response can be written in terms of the independent dimensionless variables $\zeta = k_p(z - v_g t)$ and $\tau = k_p c t$, where k_p is the plasma wavenumber, and $v_g \approx -c$ is the laser group velocity (for convenience, the laser is moving in the negative z direction). Introducing dimensionless normalized scalar and vector potentials $\phi(\zeta)$ and $a(\zeta)$, the parallel and perpendicular electric fields are $E_\parallel = -(mc^2 k_p/e)\partial\phi/\partial\zeta$ and $E_\perp = -(mc/e)\partial a/\partial t = -(mc^2 k_p/e)\partial a/\partial\zeta$. The wakefield generation equation can then be written as [8,9]:

$$\frac{d^2 x}{d\zeta^2} = \frac{n_e}{n_p} - 1 = \frac{1}{2}\Big(\frac{1 + a^2(\zeta)}{x^2} - 1\Big), \tag{1}$$

where $x \equiv 1+\phi$ is the modified electrostatic potential, and $a^2(\zeta)$ is the dimensionless laser intensity envelope averaged over fast oscillations. Prior to the arrival of the laser the normalized wakefield $\mathcal{E} \equiv eE_\parallel/mc\omega_p = -dx/d\zeta$ is zero. A formal solution for the electric field outside the laser can be written as the first integral of (1): $[\mathcal{E}^{out}(\zeta)]^2 = -(x-1)^2/x + \int_{-\infty}^{\infty} a^2 x'/x^2 d\zeta$, which reaches a maximum value at $x = 1$:

$$[\mathcal{E}_{max}^{out}]^2 = -\int_{-\infty}^{\infty} a^2(\zeta)\Big(\frac{\partial}{\partial \zeta}\frac{1}{x}\Big) d\zeta. \quad (2)$$

This expression can be understood in terms of the deposition of laser energy into plasma. For this we use the formula for local frequency shift of laser photons obtained from the analysis of laser evolution equation [10,11]:

$$\frac{\partial \omega}{\partial z} = -\frac{1}{2}\frac{\omega_p^2}{\omega}k_p\frac{\partial}{\partial \zeta}\frac{n_e}{\gamma n_p} = -\frac{\omega_p^2}{2\omega}k_p\Big(\frac{\partial}{\partial \zeta}\frac{1}{x}\Big). \quad (3)$$

The energy density in the wake from (2) can then be interpreted as the intensity-weighted integral of the photon deceleration throughout the pulse. Let's denote the wake-dependent part of the photon deceleration function as $\kappa(\zeta) \equiv x'/x^2$. The value of the peak wakefield in (2) is then bounded from above by the total laser energy (the integral of a^2) and the maximum photon deceleration κ_{max}:

$$[\mathcal{E}_{max}^{out}]^2 = \int_{-\infty}^{\infty} a^2(\zeta)\kappa(\zeta)d\zeta \leq \kappa_{max}\int_{-\infty}^{\infty} a^2(\zeta)d\zeta, \quad (4)$$

where κ_{max} is the maximum of $\kappa(\zeta)$ inside the laser. Maximum photon deceleration κ_{max} actually has a simple physical interpretation. It is closely related to the characteristic laser depletion length l_d, or the distance in which the maximally decelerated laser slice red-shifts down to ω_p (assuming no evolution of the wakefield). From (3) this characteristic depletion length is:

$$l_d = [(\omega_0/\omega_p)^2 - 1]/k_p\kappa_{max}. \quad (5)$$

The peak wakefield outside the laser then scales with depletion length l_d and dimensionless pulse energy $\varepsilon_0 \equiv \int_{-\infty}^{\infty} a^2(\zeta)d\zeta$ as: $\mathcal{E}_{max}^{out} \leq \sqrt{[(\omega_0/\omega_p)^2 - 1]\varepsilon_0/k_p l_d}$. The range of achievable wakefields is therefore set by the total pulse energy and its depletion length. For pulses of fixed energy and depletion length the actual value of the wakefield within this range will depend only on particular laser shape, and can be optimized by varying the shape subject to constraints.

WAKEFIELD OPTIMIZATION

One possible optimization problem can be formulated as follows: given some fixed laser energy ε_0 and the desired depletion length ($\kappa_{max} = \kappa_0$), what laser shape would produce the largest possible wakefield? According to (4) this amounts to finding a shape that maintains the largest $\kappa(\zeta)$ for the duration of the pulse. Since the maximum $\kappa(\zeta)$ is fixed by the depletion length, such a shape should have

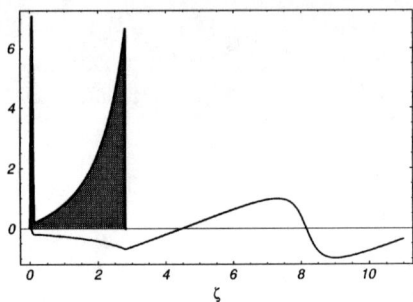

FIGURE 1. General shape of the nonlinear optimal laser intensity profile and its corresponding wakefield (arbitrary units)

a constant photon deceleration throughout the pulse, $\kappa(\zeta) = \kappa_0$. If the laser is present for $\zeta > 0$, then in order to satisfy the boundary condition of quiescent plasma before the laser, the photon deceleration should rise from 0 to value κ_0 at the very beginning of the pulse, e.g., like a step-function: $\kappa(\zeta) = \kappa_0 \theta(\zeta^+)$. Here, $\zeta^+ \equiv \zeta - 0^+$ in order to avoid ambiguities with the values of step-function at 0. The corresponding laser profile is then found from the wake equation (1):

$$a_l^2(\zeta) = \frac{2\kappa_0 \delta(\zeta^+)}{(1-\kappa_0\zeta)^4} + \frac{4\kappa_0^2 \theta^2(\zeta^+)}{(1-\kappa_0\zeta)^5} + \frac{1}{(1-\kappa_0\zeta)^2} - 1, \qquad (6)$$

where $\zeta \in [0, \zeta_f < 1/\kappa_0]$, and $\delta(\zeta^+)$ is a delta-function such that $\int_0^{\zeta>0} \delta(y^+) dy = 1$. A schematic drawing of the optimal laser intensity variation and its associated plasma wakefield are shown in Fig. 1. Generally, the shape consists of a δ-function at the front followed by a ramp in intensity which is cut off at ζ_f. In the linear regime, when $a^2 \ll 1$, $\kappa_0 \to 0$, the ramp reduces to a triangular shape found in [11,12]: $a^2 = 2\kappa_0(\delta(\zeta^+) + \zeta)$. We note that (6) describes a family of shapes, rather than a fixed shape. The actual profile of the optimal pulse depends on the deceleration parameter κ_0 set by the desired depletion length and the pulse length ζ_f, which is determined from the available total energy:

$$\varepsilon_0 = 2\kappa_0 + \frac{\zeta_f[\kappa_0^2 + (1-\kappa_0\zeta_f)^3]}{(1-\kappa_0\zeta_f)^4}. \qquad (7)$$

Although the pulse length cannot exceed $\zeta_c \equiv 1/\kappa_0$, the rise of a^2 towards the end of the pulse guarantees that any finite laser energy can be accommodated for $\zeta_f < \zeta_c$. The two terms in (7) represent the energy contained in the δ-function precursor and the main pulse. It is clear that for a fixed total energy there exists a maximum value of $\kappa_0 = \varepsilon_0/2$ which is achieved when $\zeta_f \to 0$, i.e., all of the energy is concentrated in the δ-function. This shape, which is a particular case of the general optimal shape (6), excites the largest possible wakefield and has the smallest depletion length among all pulses of fixed energy. For circularly polarized pulses with cylindrical transverse crossection of radius r_0 and wavelength λ, the maximum achievable wake is then given by:

$$E_{\max} = 6.54 E_{wb} \Big[\frac{U_0}{1\text{J}}\Big]\Big[\frac{\lambda}{1\mu\text{m}}\Big]^2 \Big[\frac{10\mu\text{m}}{r_0}\Big]^2 \Big[\frac{n_p}{10^{18}\text{cm}^{-3}}\Big]^{1/2} \quad (8)$$

where U_0 is the total pulse energy (in Joules) and $E_{wb} = 96[n_p/10^{18}\text{cm}^{-3}]\text{GV/m}$ is the nonrelativistic wavebreaking field.

EFFICIENCY OPTIMIZATION

While generation of large accelerating gradients is a prerequisite for a successful accelerating scheme, the efficiency of acceleration should also be considered. For an accelerating scheme that involves transfer of energy from the driver beam to the accelerating beam, efficiency is measured in terms of the *transformer ratio*, or the ratio of the maximum rate of energy gain per particle of accelerating beam to the maximum rate of energy loss per particle of the driving beam. In the case of laser-plasma accelerators, where the driving and accelerating beams consist of particles of different species, the following kinematic definition is more useful:

$$R \equiv \frac{|\partial \gamma_a/\partial z|_{max}}{|\partial \gamma_d/\partial z|_{max}}, \quad (9)$$

where γ_d and γ_a are Lorentz factors for the driving and accelerating beams. In LWFA the particles in the trailing electron bunch are accelerated via electrostatic interaction with the wake, so $|\partial \gamma_a/\partial z|_{max} = |eE_{\parallel}^{max}|/m_e c^2 = k_p |\mathcal{E}_{max}^{out}|$. For the laser propagating in plasma $\gamma_d \approx \omega/\omega_p$, so $|\partial \gamma_d/\partial z|$ is the photon frequency shift given by (3). The transformer ratio for LWFA is then:

$$R^{\text{LWFA}} = \frac{2\omega}{\omega_p} \frac{|\partial x/\partial \zeta|_{max}^{out}}{|\partial (1/x)/\partial \zeta|_{max}^{in}} \propto |\mathcal{E}|_{max}^{out} k_p l_d. \quad (10)$$

Defined this way the transformer ratio can have several interpretations. On the one hand, it is a measure of local accelerating efficiency, or the amount of increase in γ of the accelerating electron per unit loss of γ of the laser. On the other hand, transformer ratio is proportional to the maximum energy that can be transferred to the accelerating beam particle over the laser depletion length (assuming no evolution of the wake during propagation). Therefore, an efficient accelerating scheme should attempt to maximize the transformer ratio.

There are several ways to find the laser shape that maximizes R. Among the pulses of fixed energy and depletion length, R is maximized by a pulse that produces the largest wakefield as can be seen from (10). The optimal shape found in Eq. (6) satisfies this requirement. Alternatively, one can relax the energy constraint and instead look for a laser profile that has the largest depletion length among all the shapes that produce a given maximum wakefield behind the laser. Although this reasoning leads to the same resulting shape, we include the proof for completeness as it demonstrates a useful technique for determining laser shapes that satisfy constraints on the values of the wakefield.

In order to find the shape that maximizes the transformer ratio, we vary the photon deceleration function $\kappa(\zeta)$ inside the laser. We require that $\kappa(\zeta)$ be positive definite, i.e., laser photons only *lose* energy to the plasma and do not reabsorb energy from the wake. The advantage of varying $\kappa(\zeta)$ rather than $a^2(\zeta)$ directly is that one can immediately write down the solution for the wakefield potential $x(\zeta)$ in terms of the photon energy deposition function $\psi(\zeta) \equiv \int_{-\infty}^{\zeta} \kappa(\zeta_1) d\zeta_1$, i.e., $x(\zeta) = 1/(1 - \psi(\zeta))$. The corresponding laser shape is then determined from the wakefield equation (1):

$$a^2(\zeta) = (2x''(\zeta) + 1)x(\zeta)^2 - 1 = \frac{2\psi''(\zeta)}{[1 - \psi(\zeta)]^4} + \frac{4\psi'(\zeta)^2}{[1 - \psi(\zeta)]^5} + \frac{1}{[1 - \psi(\zeta)]^2} - 1. \quad (11)$$

Note that not all functions $\psi(\zeta)$ should produce physical, i.e., positive $a^2(\zeta)$, and the validity of a given $\kappa(\zeta)$ should be checked through (11). By considering photon energy deposition in the pulse all possible laser shapes that produce a given wakefield can be mapped onto a bounded space of monotonically increasing functions $\psi(\zeta)$, whose end values on the interval $[0, \zeta_f]$ and derivatives at ζ_f are constrained by the required maximum value of the wakefield. From the first integral of the wakefield equation we can relate the modified potential $x_f \equiv x(\zeta_f)$ and the electric field $x'_f \equiv x'(\zeta_f)$ evaluated at the end of the pulse:

$$(\mathcal{E}_{max}^{out})^2 = \frac{(x_f - 1)^2}{x_f} + (x'_f)^2 = \frac{\psi(\zeta_f)^2}{(1 - \psi(\zeta_f))} + \frac{\psi'(\zeta_f)^2}{(1 - \psi(\zeta_f))^4}. \quad (12)$$

Monotonicity of $\psi(\zeta)$ follows from the requirement $\kappa(\zeta) \geq 0$, and the bounds on $\psi(\zeta)$ are $0 = \psi(0) \leq \psi(\zeta) \leq \psi_{max} < 1$. A few sample solutions for $\psi(\zeta)$ and corresponding photon deceleration and laser shapes are plotted in Fig. 2. The function $\psi(\zeta)$ that results in the largest transformer ratio should possess the smallest maximal slope in the interval $[0, \zeta_f]$ – this will maximize the depletion length for a fixed $\mathcal{E}_{max}^{out}$. Such curve is unique and is represented by curve 2 in figure 2. It is a straight line with slope $\psi'(\zeta_f) = \psi(\zeta_f)_{crit}/\zeta_f$, where the value of $\psi(\zeta_f)_{crit}$ is determined from substituting $\psi'(\zeta_f)$ into eq. (12). Let's show that this line has the smallest maximum slope. Since $\psi'(\zeta_f)$ is a decreasing function of $\psi(\zeta_f)$ for a fixed $\mathcal{E}_{max}^{out}$ (eq. (12)), all curves $\psi(\zeta)$ with $\psi(\zeta_f) < \psi(\zeta_f)_{crit}$ (such as curve 1 in figure 2) will automatically have larger slope at ζ_f: $\psi'(\zeta) > \psi(\zeta_f)_{crit}/\zeta_f$. On the other hand, the curves with $\psi(\zeta_f) > \psi(\zeta_f)_{crit}$ (such as curves 3 and 4) should have a slope larger than $\psi(\zeta_f)_{crit}$ somewhere between 0 and ζ_f in order to be larger than $\psi(\zeta_f)_{crit}$ at ζ_f. We therefore prove that the function $\psi(\zeta) = \kappa_0 \zeta$, where $\kappa_0 \equiv \psi(\zeta_f)_{crit}/\zeta_f$, is an integrated photon deceleration profile that maximizes the transformer ratio.

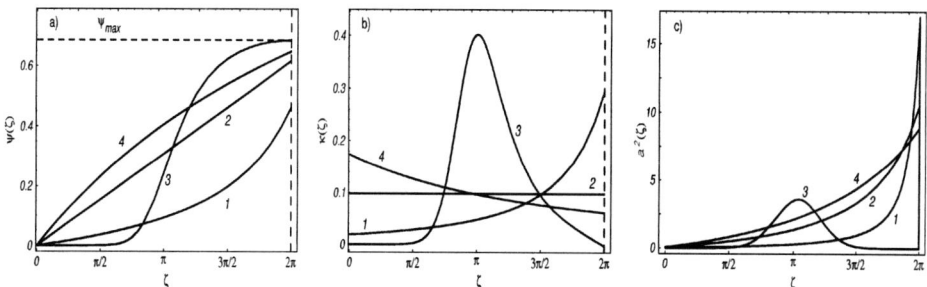

FIGURE 2. a) Sample photon energy deposition $\psi(\zeta)$ for pulses of length not exceeding $\zeta_f = 2\pi$; b) corresponding photon deceleration functions; c) resulting laser intensity profiles. All shapes produce the same maximum wakefield.

The photon deceleration function associated with this $\psi(\zeta)$ is a constant $\kappa(\zeta) = \kappa_0$ and the resulting laser shape is the same as given by (6). The optimal transformer ratio associated with this shape can be found from (10):

$$R^{\text{LWFA}} = \frac{2\omega}{\omega_p}\sqrt{\frac{1 + (k_p L_p)^2[1 - \kappa_0(k_p L_p)]^3}{[1 - \kappa_0(k_p L_p)]^4}}, \quad (13)$$

where $L_p = \zeta_f/k_p$ is the pulse length. In the linear regime optimal transformer ratios for both LWFA and PWFA schemes scale identically with the pulse/beam length: $R^{\text{LWFA}} \to (2\omega/\omega_p)\sqrt{1 + (k_p L_{pulse})^2}$, $R^{\text{PWFA}} \to \sqrt{1 + (k_p L_{beam})^2}$ [13]. The LWFA scheme is intrinsically more efficient by a factor of $2\omega/\omega_p$, which is needed for viability of LWFA since lasers are typically "slower" drivers than electron beams.

UTILITY OF PULSE SHAPING

The advantage of using the optimal pulse shape is best seen in comparison with the unshaped (Gaussian) pulse. For a given Gaussian pulse (or any other non-optimal shape) one can always construct a corresponding optimally shaped pulse with the same laser energy such that the photon deceleration across the optimal pulse equals to the peak photon deceleration in the unshaped one (i.e., both pulses have equal depletion lengths). Unshaped pulses deplete first in the region where photon deceleration is the largest, whereas a laser with the optimal shape loses *all* its energy in a depletion length due to uniform photon deceleration, thus enhancing instantaneous energy deposition and wakefield. For a numerical example, we consider the optimal and Gaussian pulses of total energy 0.5J, wavelength 1μm and cylindrical radius 10μm in a plasma with $n_p = 10^{18}$cm^{-3}. The transformer ratio, the maximum wakefield, the required pulse length, and the corresponding peak a_0 are shown in Fig. 3 as a function of depletion length.

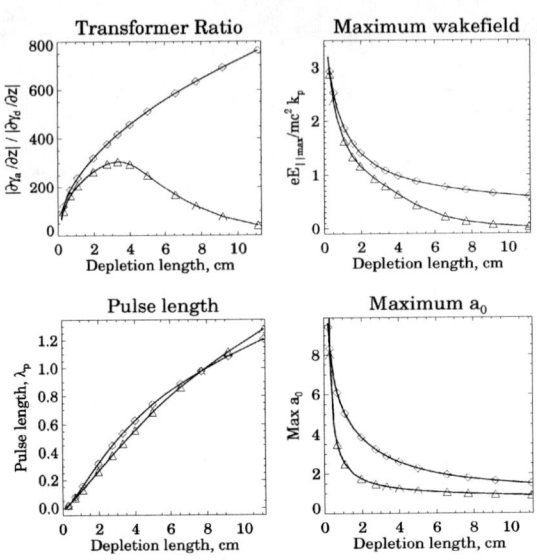

FIGURE 3. Comparison of the transformer ratio, maximum wakefield, pulse length, and maximum normalized vector potential in shaped (diamonds) and Gaussian (triangles) pulses of equal depletion lengths and constant pulse energy of 0.5J.

From Fig. 3 we see that the transformer ratio and the maximum wakefield are consistently larger for shaped pulses. In fact, the lines for optimal pulse wakefield and transformer ratio represent the theoretical upper limits for all pulses of given energy. The Gaussian pulse achieves a maximum transformer ratio when its length (measured here as FWHM) equals 1/2 of the relativistic plasma wavelength. The effects of shaping are especially prominent for longer pulses, where Gaussian pulse yields almost no wake excitation due to plasma oscillations inside the pulse that cause part of the laser photons to absorb energy from the wake. On the other hand, a shaped laser postpones plasma oscillation until the end of the pulse, and all photons decelerate uniformly. For very short pulses, the differences between the two shapes are minimal. This is due to the fact that very short Gaussian pulses of fixed energy asymptotically approach the delta-function limit of the short optimal shape. For these short pulses the wakefield attains the maximum value given by (8) as the depletion length reaches the minimal value for given pulse energy.

Although short pulses generally produce the largest wakefields, their efficiency is close to minimal possible, as the depletion length decreases faster than increase in the wake. Therefore, the choice of the appropriate pulse shape for LWFA stage will depend on specific experimental conditions. If the laser-plasma interaction distance is limited by instabilities, diffraction or dephasing, then in order to maximize the electron energy gain one should try to achieve the largest accelerating gradient, which can be accomplished with ultrashort pulses. For some regimes of plasma density and laser energy available laser systems may be unable to produce pulses short enough so that the pump depletion length is longer than the characteristic instability or dephasing length. In this case shaping the laser will increase the wakefield over the interaction distance, even though it will be below the maximum possible if a shorter pulse were used. If the interaction length is less constrained, such as the case for propagation in plasma channels [14], then using a finite-length shaped pulse will result in a greatly improved overall energy gain per stage as can be seen from Fig. 3. An added benefit of pulse shaping is the suppression of modulational instability that affects unshaped pulses that are longer than plasma wavelength. When all photons red-shift, or "slow down", at the same rate, different slices of the laser do not overrun each other, and the 1D laser self-modulation is suppressed.

REALISTIC PULSE SHAPING

As the optimal pulse shape is associated with a delta-function precursor, the feasibility of such a structure may be a concern. We note that the purpose of this precursor is to bring the photon deceleration from zero in the quiescent plasma before the laser to a finite value κ_0 at the beginning of the main pulse. This can also be achieved with a more physical prepulse, whose shape can be found from the wake equation once a smooth function $\kappa(\zeta)$ is chosen.

For our example we choose a photon deceleration function that varies as a hyperbolic tangent: $\kappa(\zeta) = \kappa_0[1 + \tanh(\alpha(\zeta - \zeta_0))]/2$, where α is a steepness parameter and ζ_0 is an arbitrary offset. The photon energy deposition is then $\psi(\zeta) = \kappa_0[\zeta + \ln(\cosh(\alpha(\zeta - \zeta_0)))/\alpha]/2$, and the corresponding laser shape is found from equation (11):

$$a^2(\zeta) = \frac{\kappa_0 \alpha \operatorname{sech}^2(\alpha(\zeta - \zeta_0))}{\chi^4(\zeta)} + \frac{\kappa_0^2[1 + \tanh(\alpha(\zeta - \zeta_0))]^2}{\chi^5(\zeta)} + \frac{1}{\chi^2(\zeta)} - 1, \quad (14)$$

where $\zeta \leq \zeta_f$ and the function in the denominator is $\chi(\zeta) = 1 + (\kappa_0/2\alpha)\ln\frac{1}{2} + \zeta_0\kappa_0/2 - \psi(\zeta)$. As before, the pulse length ζ_f can be found from the total available energy of the pulse. By varying α we can change the slope of $\kappa(\zeta)$ as it rises from 0 to κ_0 and construct a pulse shape that satisfies experimental constraints yet retains essential physics of the optimal shape. For a step-function photon deceleration ($\alpha \to \infty$) expression (14) asymptotes to equation (6). However, for finite values of α the delta-function precursor spreads out and can even disappear as shown in Fig. 4. The family of shapes given by (14) is better suited for the finite-bandwidth laser systems that have a lower limit on achievable feature size. The values of maximum wakefield for pulses in Fig. 4 is within few percent of the value for a delta-function optimal pulse of the same energy and depletion length. This is due to the fact that the bulk of the laser pulse still experiences constant maximal photon deceleration. The wakefield further degrades with longer rise times of $\kappa(\zeta)$.

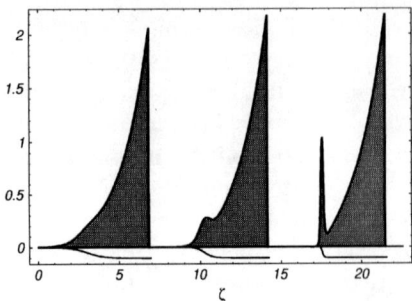

FIGURE 4. Laser intensity (shaded) and associated photon deceleration $(-\kappa(\zeta))$ for pulses of the same total energy and characteristic depletion length in the order of increasing α.

The pulse shaping techniques described so far have assumed that the laser is incident on an unperturbed plasma. However, this does not have to be the case, and we can construct an optimally-shaped laser that enters the plasma at some phase of a pre-existing plasma oscillation. Such oscillation could be left from a precursor laser pulse or electron beam as shown in Fig. 5. When there is an existing plasma wave, the value x_0 of the modified electrostatic potential at the beginning of the optimal pulse will generally be different from unity. In this case the expression for the optimal pulse without the delta-function precursor is modified into:

$$a_l^2(\zeta) = \frac{4\kappa_0^2}{[x_0^{-1} - \kappa_0(\zeta - \zeta_0)]^5} + \frac{1}{[x_0^{-1} - \kappa_0(\zeta - \zeta_0)]^2} - 1, \qquad (15)$$

where we assume that the main pulse lies between ζ_0 and ζ_f, and $\kappa_0 = x'(\zeta_0)/x_0^2$. If this shape is placed in a correct phase of the oscillation (so that $a_l^2(\zeta_0)$ from (15) is positive), it acts as an amplifier of the existing wakefield. The ratio of maximum wakefield behind the optimal pulse to the field in front of it scales as $(R/x_0^2)(\omega_p/2\omega)$ which for pulse lengths around λ_p from Fig. 3 can be of order 10. A detailed discussion of this scheme and a comparison to the resonant laser-plasma accelerator concept [15] will be reported elsewhere.

DISCUSSION

As we have shown, the huge phase space involved in shaping laser drivers for applications in laser wakefield accelerators can be described using only two parameters: total pulse energy and characteristic depletion length. The shape of photon energy deposition (photon deceleration) inside the pulse plays a crucial role for both the wake excitation and the evolution of the laser driver. By varying the shape of the photon deceleration function for pulses of fixed energy and depletion length we were able to optimize both the generated wakefield and the efficiency of

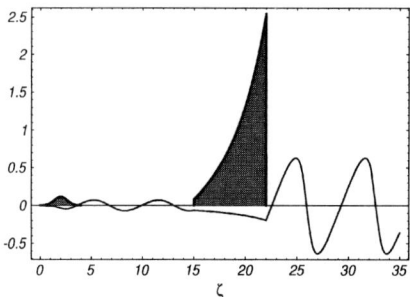

FIGURE 5. Laser intensity profiles ($a^2(\zeta)$, shaded) and normalized electric field for optimally shaped main pulse following a Gaussian precursor.

the accelerating scheme. The method used for obtaining the optimal shapes (6) and (14) is actually more general and can be used to determine laser shapes that generate other variations in the nonlinear index of refraction. Having a physical requirement for the refractive index, which in this case is the requirement of uniformity of photon deceleration, provides a constraint on the functional form of the wakefield, which can then be used to find the required laser shape. Alas, such a "reverse" solution is not always guaranteed to yield a physical (i.e., positive) $a^2(\zeta)$, so, in general, caution is advised.

Several issues should be addressed before the laser pulse shaping concept can be fully utilized. Even without the delta-function precursor, the finite laser bandwidth will necessarily smooth out steep rises and falls of the optimal pulse shape. Although we do not anticipate adverse effects when the feature size is much smaller than the plasma wavelength, the 1D self-consistent laser evolution and stability of realistic optimal shapes are currently under investigation. Another consideration is the influence of the laser-plasma interaction in the transverse dimension on the evolution of the pulse. Many of the laser-plasma instabilities are seeded by the wakefield-induced perturbations of the index of refraction. As we have demonstrated in this paper, the nonlinear index of refraction can be effectively controlled through laser shaping, thus suggesting the method of delaying the onset of these instabilities. Whether this approach increases the growth rates of other instabilities, particularly in the transverse dimension, remains to be investigated.

We would like to thank J. Arons, A. Charman, T. Katsouleas, W. B. Mori, and J. Wurtele for fruitful discussions and suggestions.

REFERENCES

1. F. Verluise, V. Laude, Z. Cheng, *et al.*, Optics Lett. **25**, 575 (2000).
2. M. Murnane, these proceedings.
3. M. Downer, these proceedings.

4. T. Tajima, J. M. Dawson, Phys. Rev. Lett. **43**, 267 (1979).
5. P. Chen, J. M. Dawson, R. Huff, T. Katsouleas, Phys. Rev. Lett. **54**, 693 (1985).
6. S. Wilks, J. M. Dawson, W. B. Mori, T. Katsouleas, M. Jones, Phys. Rev. Lett. **62**, 2600 (1989).
7. W. B. Mori, IEEE J. Quant. Elec. **33**, 1942 (1997)
8. P. Sprangle, E. Esarey, J. Krall, and G. Joyce, Phys. Rev. Lett. **69**, 2200 (1992).
9. E. Esarey, P. Sprangle, J. Krall, A. Ting, IEEE Trans. Plasma Sci. **24**, 252 (1996).
10. E. Esarey, A. Ting, and P. Sprangle, Phys. Rev. A **42**, 3526, (1990).
11. P. Chen, A. Spitkovsky, AIP Conf. Proc. **472**, 321 (1999).
12. P. Chen, A. Spitkovsky, T. Katsouleas, W. B. Mori, Nucl. Instr. Meth. **410**, 488 (1998).
13. P. Chen, J. J. Su, J. M. Dawson, K. Bane, and P. Wilson, Phys. Rev. Lett. **56**, 1252 (1986).
14. E. Esarey, P. Sprangle, J. Krall, A. Ting, G. Joyce, Phys. Fluids B **5**, 2690 (1993).
15. D. Umstadter, E. Esarey, J. Kim, Phys. Rev. Lett **72**, 1224 (1994)

Parametric excitation of plasma waves by counter-propagating laser beams detuned by $2\omega_p$

Gennady Shvets

Princeton Plasma Physics Laboratory[1]
Princeton NJ 08543

Abstract. Short and long wavelength plasma waves can become strongly coupled in the presence of two counter-propagating laser pulses detuned by twice the cold plasma frequency ω_p. This coupling leads to the exponential amplification of the plasma waves at a rate proportional to the product of the laser amplitudes. A minimum product of the laser intensities is required for the instability when the laser detuning differs from $2\omega_p$. Even for significant departures from the optimal detuning, this finite intensity threshold remains modest and achievable using modern lasers.

INTRODUCTION

Rosenbluth and Liu (RL) pointed out in 1972 [1] that electrostatic plasma waves can be resonantly driven by two laser beams detuned in frequency by one or two plasma frequencies $\omega_p = \sqrt{4\pi e^2 n_0/m}$, where $-e$, m, and n_0 are, respectively, the electron charge, mass, and density. Two distinct excitation geometries were considered by RL: co- and counter-propagating laser beams. In a tenuous plasma, the phase velocity of thus driven plasma waves is sub-relativistic in the counter-propagating case and close to the speed of light in the co-propagating case. The utility of such fast plasma waves for particle acceleration was later realized by Tajima and Dawson [2], sparking an enormous interest in plasma acceleration over the past decade. Particle acceleration using a plasma wave excited by two co-propagating laser pulses detuned by ω_p (known as Plasma Beat Wave Accelerator), has been experimentally observed [3]. Counter-propagating laser beams were not considered relevant for accelerator applications.

More recently [4], the combination of a short and long counter-propagating laser beams was shown to produce anomalously high (up to 10 GeV/m) accelerating plasma waves at the modest ($I \ll 10^{18}$ W/cm^2) laser intensities. The physical explanation of this effect is that the slow (short-wavelength) plasma waves, driven

---
[1] This work was supported by the US DOE Division of High-Energy and Nuclear Physics

by the interference between the counter-propagating laser beams, can nonlinearly drive the fast plasma wave. Since the slow plasma waves are easily driven to large nonlinear amplitudes even at modest laser intensities, the fast plasma waves are produced as their natural by-product. Generation of the slow plasma waves is caused by backscattering of the laser light, and it is the property of the cold electron plasma that all parametric processes which involve backscattering occur faster than the ones which involve forward scattering. For example, for the same laser intensity, the rate of the Raman backscattering is higher than that of the Raman forward scattering by a large factor $(\omega_0/\omega_p)^{3/2}$, where ω_0 is the laser frequency.

In this paper I demonstrate that *fast* plasma waves can be driven unstable and amplified by two *counter-propagating* laser beams detuned by $\Delta\omega = |\omega_0 - \omega_1| = 2\omega_p$. The beatwave due to the counter-propagating optical mixing couples the fast and the slow plasma waves, driving them unstable. This possibility has been overlooked by RL who only considered the nonlinear coupling of the slow (fast) wave to itself, but not the coupling between the two. Using the counter-propagating beams is advantageous because of the very high growth rate of the instability, $\gamma = 2\vec{a}_0 \cdot \vec{a}_1 \omega_0^2/\omega_p$, where $\vec{a}_0$ and $\vec{a}_1$ are the normalized vector potentials of the forward and backward moving laser beams. This growth rate is a factor of ω_0^2/ω_p^2 higher than in the co-propagating geometry originally calculated by RL. Curiously, about the same growth rate was predicted by RL for the slow wave (unsuitable for particle acceleration) in the counter-propagating geometry. Therefore, by extending the RL calculation, I find that the "best of both worlds" is achieved in the counter-propagating geometry: (a) the growth rate is high, characteristically of the slow-wave instability, and (b) the fast (accelerating) wave is produced along side with the slow wave.

LAGRANGIAN DESCRIPTION OF THE INSTABILITY

Consider two counter-propagating circularly polarized laser beams with their normalized vector potentials given by $\vec{a}_0 = a_0 \times (\vec{e}_+ \exp(i\theta_0) + \text{c. c.})$ and $\vec{a}_1 = a_1(\vec{e}_- \exp(i\theta_1) + \text{c. c.})$, where $\vec{e}_{+(-)} = (\vec{e}_x \pm i\vec{e}_y)/2$, $\theta_0 = k_0 z - \omega_0 t$, and $\theta_1 = k_1 z + \omega_1 t$. The forward-propagating pulse labeled by 0 is assumed to be shorter than its long counter-propagating counterpart, which is assumed to have the duration $2L/c$, where L is the length of the plasma. The assumption of the tenuous plasma $\omega_p \ll \omega_0$ justifies taking the group velocity of the short pulse $v_{g0} = c^2 k_0/\omega_0 \approx c$, and $|\vec{k}_0 - \vec{k}_1| \approx 2k_0$.

Interference between the two beams creates a modulation in the total laser intensity $|\vec{a}|^2 = |\vec{a}_0|^2 + |\vec{a}_1|^2 + 2\vec{a}_0 \cdot \vec{a}_1$. This intensity modulation is the source of the ponderomotive force $\vec{F} = -0.5mc^2 \nabla |\vec{a}^2|$ acting on the electron plasma. The ponderomotive force has the spatial periodicity of $2\pi/(k_0 + k_1) \approx \lambda_0/2$ and the frequency $\Delta\omega = \omega_0 - \omega_1 \approx 2\omega_p$. Besides the ponderomotive force, plasma electrons experience the restoring electrostatic force of the stationary ions. This force is easily calculated using the Lagrangian displacement of the plasma electrons $\xi \equiv z - z_0$

[5]: $\ddot{\xi} + \omega_p^2 \xi = F_z/m$. At this point, to study the excitation of the plasma waves, an appropriate assumption about the electron motion has to be made.

Rosenbluth and Liu assumed that a *single* spatial harmonic of the plasma wave is excited: $\xi = A(t)\sin[k_0 z - \omega_p t + \phi(t)]$. Inserting this expression into the formula for the ponderomotive force and assuming that $\Delta\omega = 2\omega_p$ and $\dot{A} \ll \omega_p A$ yields [1]

$$\dot{A} = \frac{k_0 c^2}{2\omega_p} a_0 a_1 J_1(k_0 A), \tag{1}$$

where the dot stands for a time derivative. In the limit of $k_0 A < 1$ Eq. (1) predicts an instability with a temporal growth rate $\omega_0^2 a_0 a_1 / 4\omega_p$. The phase velocity of the plasma wave is $\omega_p/k_0 \ll c$ for a tenuous plasma. Therefore, despite its relatively high growth rate, this wave cannot be used for accelerating relativistic particles.

Two-wave ansatz

It turns out that the single-wave ansatz used by RL is not the only self-consistent assumption which leads to an instability. Consider the following two-wave ansatz:

$$\xi = A_f \sin[k_p z - \omega_p t + \phi_f] + A_s \sin[(2k_0 - k_p)z - \omega_p t + \phi_s], \tag{2}$$

where A_f (ϕ_f) and A_s (ϕ_s) are the amplitudes (phases) of the fast and slow plasma waves. Substituting ξ into the equation of motion yields

$$\frac{\partial^2 \xi}{\partial t^2} + \omega_p^2 \xi = ik_0 c^2 a_0 a_1 \sum_{k,l} (-1)^{k+l} J_k(2k_0 A_f) J_l(2k_0 A_s)$$

$$e^{ik[k_p z_0 - \omega_p t + \phi_f]} e^{il[(2k_0 - k_p)z_0 - \omega_p t + \phi_s]} e^{i[\Delta\omega t - 2k_0 z_0]} + c.c., \tag{3}$$

where $J_{k,l}$ are the Bessel functions, and $\Delta\omega \equiv \omega_0 - \omega_1 = 2\omega_p + \delta\omega$. Retaining only the $(k=0, l=1)$ and $(k=1, l=0)$ terms and neglecting the second derivatives of the amplitudes and phases, obtain:

$$\frac{\partial \phi}{\partial t} = \delta\omega - \frac{\Omega_B^2}{4} \omega_p G(A_f, A_s) \sin\phi \tag{4}$$

$$\frac{\partial (k_0 A_f)}{\partial (\omega_p t)} = \frac{\Omega_B^2}{4} J_0(2k_0 A_f) J_1(2k_0 A_s) \cos\phi \tag{5}$$

$$\frac{\partial (k_0 A_s)}{\partial (\omega_p t)} = \frac{\Omega_B^2}{4} J_1(2k_0 A_f) J_0(2k_0 A_s) \cos\phi, \tag{6}$$

where $\phi = \phi_s + \phi_f + \pi/2 + \delta\omega t$, $\Omega_B^2 = 4a_0 a_1 \omega_0^2/\omega_p^2$ is the square of the electron bounce frequency in the optical lattice created by the interference of the counter-propagating lasers (normalized to the plasma frequency), and

$$G(A_f, A_s) = \frac{J_0(2k_0 A_f) J_1(2k_0 A_s)}{k_0 A_f} + \frac{J_1(2k_0 A_f) J_0(2k_0 A_s)}{k_0 A_s}.$$

If $\delta\omega = 0$ phase-locking at $\phi = 0$ occurs, resulting in an instability with *both* the fast and the slow waves growing with the growth rate $\Omega_i = \omega_0^2 a_1 a_0/\omega_p$. In a nutshell, this is the principal result of this work: fast plasma wave capable of accelerating relativistic particles can be produced with a high temporal growth rate Ω_i. It is useful to recall that in the case of the co-propagating laser beams (also considered by RL) the growth rate of the fast plasma wave $\gamma_{RL} \approx \omega_p a_0 a_1/2$ is too small for the nonrelativistic intensity lasers to be of practical interest. In the counter-propagating case a much higher growth rate $\Omega_i/\gamma_{RL} \approx 2\omega_0^2/\omega_p^2$ is achieved. Such a high growth rate is more typical for the slow plasma waves. In this case, it is achieved for the fast wave A_f only because the fast wave is parametrically coupled to the rapidly growing slow wave A_s. The coupling mechanism is the ponderomotive force due to the counter-propagating optical mixing of the laser beams.

Another practical issue is the sensitivity of the instability on the deviation from the exact two-plasmon resonance $\delta\omega$. For the finite frequency detuning from resonance $\delta\omega \neq 0$ there is an intensity threshold: phase-locking takes place only if $\Omega_B^2/2 > \delta\omega/\omega_p$. Here, again, the counter-propagating geometry offer an advantage over the co-propagating case: the intensity threshold for the finite detuning case is fairly modest, given by

$$\sqrt{I_0 I_1}[\text{W}/\text{cm}^2] = 1.4 \times 10^{-3}(\delta\omega/\omega_p)n_0[\text{cm}^{-3}] \qquad (7)$$

For example, if the laser wavelengths $\lambda_0 = 0.8\mu m$ and $\lambda_1 = 1.0\mu m$, and plasma density $n_0 = 10^{19}$ cm^{-3} (corresponding to $\omega_0 - \omega_p = 2.5\omega_p$), the geometric mean of the laser intensities should exceed the threshold value of 8.0×10^{15} W/cm^2. Since this threshold is not too high, the instability is quite robust to plasma inhomogeneity and frequency mismatch error.

Reduction to single-mode equation

The system of the nonlinearly coupled equations (4,5, 6) can be simplified by noting that $J_0(2k_0 A_f)/J_0(2k_0 A_s) = $ const. If both waves start out negligibly small, the constant is equal to unity, and one can assume that $A_f = A_s$ at all times. This assumption is, of course, meaningful only when the instability takes place, resulting in a significant amplification of both A_s and A_f, so that the small differences in the initial amplitude values are not important. The equations for the phase and the normalized amplitude $u = 2k_0 A_s = 2k_0 A_f$ become

$$\dot\phi = (\delta\omega/\omega_p) - \Omega_B^2 \frac{J_0(u)J_1(u)}{u}\sin\phi \qquad (8)$$

$$\dot u = \frac{\Omega_B^2}{2} J_0(u)J_1(u)\cos\phi, \qquad (9)$$

where the dot indicates a derivative with respect to $\omega_p t$. The conserved invariant of Eqs. (8,9) is $\mathcal{H} = \Omega_B^2 u^2 \sin\phi - 2(\delta\omega/\omega_p)F(u)$, where

$$F(u) = \int_0^u dx \frac{x^2}{J_0(x)J_1(x)}.$$

Note is that $F(u)$ diverges for $u \to \mu_0$, where $\mu_0 = 2.405$ is the first zero of J_0.

For the excitation which starts out infinitesimally small $\mathcal{H} \approx 0$, and the $\sin\phi$ can be expressed in terms of the amplitude u. The expression for the $\cos\phi$ is then substituted into Eq. (9):

$$\dot{u} = \pm \frac{J_0(u)J_1(u)}{2}\left[\Omega_B^4 - \frac{4F^2(u)(\delta\omega)^2}{u^4\omega_p^2}\right]^{1/2}. \quad (10)$$

Equation (10) gives the trajectory of the wave amplitude as a function of time. The plus (minus) sign corresponds to the increasing (decaying) portions of the trajectory. For a finite detuning $\delta\omega$, the "motion" of u is periodic between its initial starting value u_0 and the maximum value $u_{\max}$.

For a perfect laser detuning $\delta\omega = 0$, the mode amplitude has a stable attractor at $u = \mu_0$. Since $\mu_0 > 1$, Eq. (3) no longer holds because of the breaking of the slow wave [5]. For $0 < (\delta\omega/\omega_p) < \Omega_B^2/2$ the amplitude u oscillates periodically between its initially small value u_0 and $u_{\max} < \mu_0$ which is found by solving the equation $F(u_{\max})/u_{\max} = \Omega_B^2\omega_p/2\delta\omega$. This equation has no solutions for $(\delta\omega/\omega_p) < \Omega_B^2/2$, i. e. there is no instability. Temporal evolution of u is shown in Fig. 1 for a fixed $\Omega_B = 1$ and three different detunings corresponding to $\epsilon = 0.2, 0.1, 0.05$. Analytic progress can be made in the limit of $u < 1$, which is, in any case, the applicability limit of Eq. (3). This limit corresponds to the laser intensities slightly above the instability threshold, $\Omega_B^2/2 = (1+\epsilon)(\delta\omega/\omega_p)$, where ϵ is small. Then expanding $F(u) = u^2 + 3/16 u^4 + ...$, the maximum amplitude can be evaluated as $u_{\max} = 4\sqrt{\epsilon/3}$. The oscillation period is given by

$$(\delta\omega)T = \frac{8\sqrt{6}}{3u_{\max}}\ln[2u_{\max}/u_0]. \quad (11)$$

Figure 1 confirms that the smaller is the peak amplitude of the wave, the longer is the oscillation period.

The physics of the amplitude oscillation can be understood as follows. Initially, u is very small, and since the ratio $F(u)/u^2$ is approximately a constant, the relative phase is locked at a constant $\phi = \sin^{-1} 2(\delta\omega/\omega_p)/\Omega_B^2$. As u undergoes an exponential growth, the phase "unlocks" and drifts towards $\phi = \pi/2$, at which time the amplitude peaks at $u = u_{\max}$ and starts dropping. After the amplitude drops to its initial value u_0, the phase locks again, and the process repeats.

It is instructive to qualitatively understand how two plasma oscillations (fast and slow) can become strongly coupled by a beatwave which has a frequency

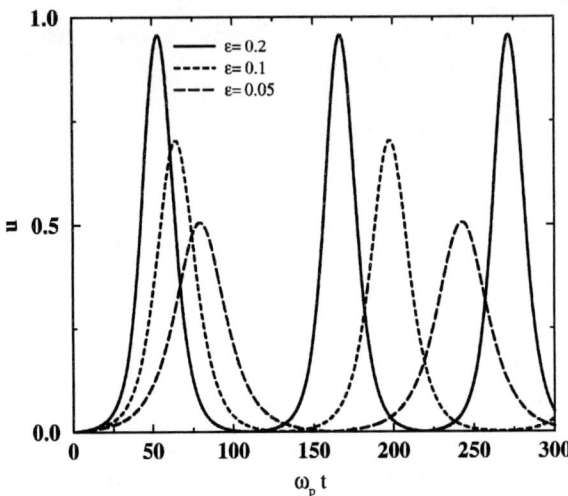

FIGURE 1. Fast and slow wave amplitude $u = 2k_0 A_f$ as a function of time for three detunings: $\Omega_B^2/2 = (1+\epsilon)(\delta\omega/\omega_p)$. Initial excitation $u_0 = 10^{-3}$, $\Omega_B^2 = 1$

$\Delta\omega \neq \omega_s + \omega_f$. First consider the small-intensity regime $(\delta\omega/\omega_p) \gg \Omega_B^2/2$. Then using $\phi \approx (\delta\omega)t$ and expanding Bessel functions to the lowest order in $A_{s,f}$, it can be shown from Eqs. (4-6) that both ϕ_s and ϕ_f acquire a time-averaged drift $\dot\phi_{s,f} = -\delta\Omega_{s,f}$, where $\delta\Omega_s = \delta\Omega_f = \Omega_B^4/(32\delta\omega/\omega_p)$. Therefore, in the presence of the nonresonant beatwave the frequencies of both modes are shifted in the direction of $(\delta\omega)$. A rough estimate of the instability threshold can be obtained by requiring that $\delta\Omega_s + \delta\Omega_f = (\delta\omega/\omega_p)$. This results in $\Omega_B^2 = 4(\delta\omega/\omega_p)$, overestimating the earlier obtained expression for the intensity threshold by a factor 2. As shown below, there is an additional mechanism of shifting the frequency of the slow plasma wave via backscattering the short laser pulse. This frequency shifting can significantly modify the threshold intensity. This effect is missed by the single plasma wave approximation used in Ref. [1].

TWO-SCALE ANALYSIS AND PARTICLE SIMULATIONS

The above analysis is only qualitative because it neglects several potentially important effects: (i) the nonresonant plasma oscillation driven at frequency $\Delta\omega$; (ii) modification of a_1 by the backscattering of a_0 off this driven density perturbation; (iii) the renormalization of the slow wave frequency due to its interaction with the short laser pulse. Therefore, we supplement the above calculation by a more rigorous two-scale particle simulation, which takes advantage of the scale separation between the short period of the slow plasma wave and a much longer period of the fast wave.

The small-scale dynamics of the plasma electrons is characterized by their location (or phase) $\theta_j = \theta_0 + \theta_1 \approx 2k_0 z_j$ inside the optical lattice produced by the interference of the two lasers. Equations of motion for the $j's$ electron in a reference frame moving with the short pulse are described in Refs. [6,7]:

$$\ddot{\theta}_j + \Omega_B^2 \sin(\theta_j - \Delta_0 \zeta) = -\sum_{l=1}^{\infty} \hat{n}_l e^{il\theta_j} - \tilde{e}_z + \text{c. c.}, \qquad (12)$$

where a dot denotes a derivative with respect to $\zeta = \omega_p(t - z/c)$, $\hat{n}_l = i \left\langle e^{-il\theta_j}/l \right\rangle_{\lambda_0/2}$ is the l-th harmonic of the small-scale electron plasma wave averaged over one lattice period, and $\Delta_0 = \Delta \omega_0 / \omega_p$. The global electric field $\tilde{e}_z = 2\omega_0 e E_z / mc\omega_p^2$ is generated owing to the average momentum deposition from the lasers into the plasma: As the photons are exchanged between the lasers, the recoil momentum is deposited into the plasma, generating electron current. In 1-D this current is balanced by the displacement current, which drives the electric field according to the Ampere's law $\partial E_z / \partial t = -4\pi < J_z >$, where $< J_z >$ is the current averaged over the period of the slow wave. In normalized units this equation and the Maxwell's equation for the pump can be written as

$$\frac{\partial \tilde{e}_z}{\partial \zeta} = \left\langle \dot{\theta}_j \right\rangle_{\lambda_0/2}, \quad \frac{\partial a_1}{\partial \zeta} = -i \frac{\omega_p a_0^*}{4\omega_0} \left\langle e^{-i\theta_j} \right\rangle_{\lambda_0/2} \qquad (13)$$

Equations (12,13), complemented by the initial conditions at $\zeta = -\infty$, can be numerically solved using macro-particles. Scale separation between the slow and fast plasma waves significantly simplifies the problem, justifying the assumption that the particles in different ponderomotive buckets interact with particles in other buckets only via the global electric field $\tilde{e}_z$ and the backward radiation a_1. Therefore, Eqs. (12,13) are simulated using the particle-in-cell method on a large scale, and the molecular dynamics method on a small scale. As an initial condition, we assume that at $\zeta = -\infty$ the plasma is uniform ($\hat{n}_l = 0$ for all l) and stationary ($\dot{\theta}_j = 0$ for all j), and that a small initial fast plasma wave is present ($\tilde{e}_z = \tilde{e}_0$). The presence of a much larger plasma wave inside the short pulse (taken here in the form $a_0 = 0.5\bar{a}_0[\tanh(-\zeta/\tau_L) + 1]$) indicates an instability.

The fast electric field E_z obtained by integrating Eqs. (12,13) is shown in Fig. 2 for two sets of laser field amplitudes a_0 and a_1. Simulation parameters are $\omega_0/\omega_p = 10$, $\omega_0 - \omega_1 = 2.5\omega_p$, and $\tilde{e}_0 = 10^{-3}$. In Fig. 2(a) $a_0 = a_1 = 0.06$ were assumed fixed. Evolving a_1 according to the second Eq. (13) did not result in any significant change of E_z. We also simulated the case of the fixed $a_0 = 0.19$ and $a_1 = 0.015$, which did not show any instability since in this case Ω_B^2 is smaller then in Fig. 2(a). However, when a_1 was self-consistently evolved, large electric field was excited, as shown in Fig. 2(b). This result is a manifestation of the physics which was not included in the above two-wave analysis which predicted that the threshold for the instability is determined by the frequency detuning $\delta\omega$ and $\Omega_B^2 = 4a_0 a_1 \omega_0^2/\omega_p^2$

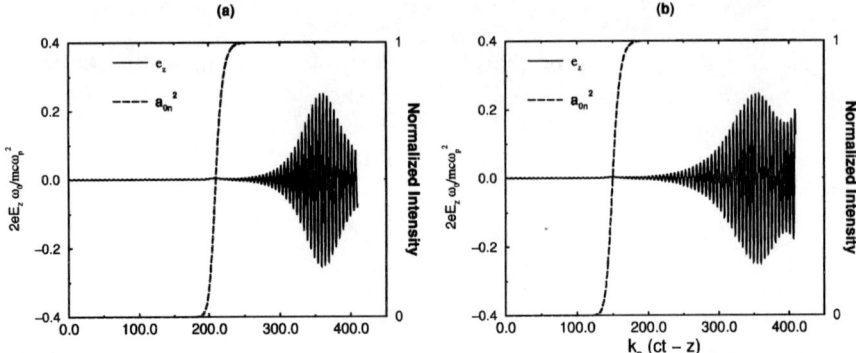

FIGURE 2. Solid line: fast electric field $\tilde{e}_z$, dashed line: normalized intensity of short pulse a_{0n}^2. (a) $a_0 = a_1 = 0.06$, fixed a_1; (b) $a_0 = 0.19$, $a_1(\zeta = 0) = 0.015$, and a_1 is solved for from Eq. (13)

which depends only on the *product* of the laser amplitudes, not on the individual amplitudes. As was explained earlier, the instability threshold arises because the finite Ω_B^2 is needed to shift the frequencies of the fast and slow plasma waves to compensate for the frequency detuning $\delta\omega$. However, there may be other mechanisms of frequency shifting unaccounted by the two-wave treatment. In particular, it follows from Eq. (13) that a slow wave with amplitude $\hat{n}_1 \sim e^{-i\zeta}$ excites a backward wave $\delta a_1 = \omega_p a_0^* \hat{n}/4\omega_0(\Delta_0 - 1)$, which then forms a beatwave with a_0 and acts back on the plasma electrons. Substituting δa_1 into Eq. (12), obtain an additional frequency shift of the slow plasma wave $\delta\Omega_s^+ = \omega_p^2|a_0|^2/4\omega_0(\Delta\omega - \omega_p)$. Therefore, for the simulation parameters of Fig. 2(b) an additional frequency shift *independent* of a_1 is produced, effectively reducing the $\delta\omega = 0.5\omega_p$ frequency mismatch, thereby reducing the Ω_B^2 required to bridge the remaining gap. For the simulation parameters of Fig. 2(a) this reduction was negligible because of the smallness of a_0^2. The relatively modest intensity threshold, given by Eq. (7), ensures that the amplification process is robust with respect to small variations of the plasma density. This threshold is somewhat lowered due to the additional frequency shift $\delta\Omega_s^+$, and can be lowered even further by employing chirped laser pulses. Frequency chirp (which results in a ζ-dependence of $\delta\omega$) also provides the benefit of suppressing the Raman backscattering of the more intense short pulse which can evolve from noise [8]. In Fig. 3 we plotted the amplitudes of the fast and slow plasma waves, $\tilde{e}_z$ and $\langle\cos\theta_j\rangle$, for a linearly-chirped Gaussian pulse $a_0 = 0.15\exp-\zeta^2/2\tau_L^2$ with $\tau_L = 25$, $d\delta\omega/d\zeta = -9.5 \times 10^{-3}\omega_p$, and the central frequency $\omega_0 = \omega_1 + 2.35\omega_p$ ($\omega_0/\omega_p = 10$). The initial fast plasma wave $\tilde{e}_0 = 10^{-3}$ and $a_1 = 0.0165$ have been assumed. In this example an accelerating plasma field of up to 9 GeV/m is generated.

CONCLUSIONS

In conclusion, we uncovered a novel mechanism of parametrically exciting large-amplitude fast plasma waves by combining two counter-propagating laser pulses

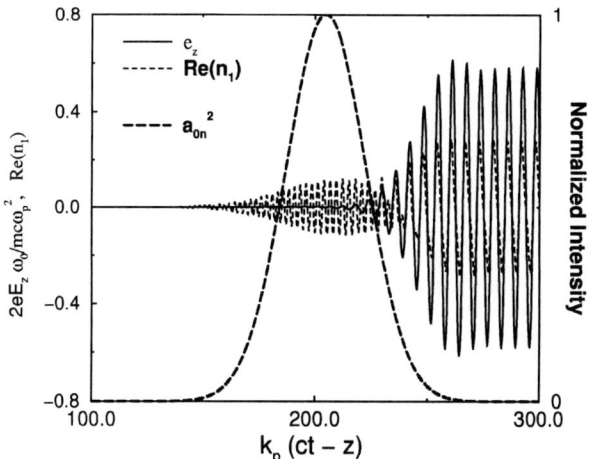

FIGURE 3. Solid line: fast electric field $\tilde{e}_z$, long-dashed line: normalized intensity of short pulse a_{0n}^2, dashed line: density bunching of the slow plasma wave $\text{Re}(\hat{n}_1) = \langle \cos\theta_j \rangle$. Rapidly-varying part part of $\hat{n}_1$ is the driven plasma response inside the laser pulse.

detuned by approximately two plasma frequencies. The short-wavelength ponderomotive force, driven by the counter-propagating optical mixing, couples the fast and the slow plasma waves. The high growth rate and low intensity threshold of this instability makes it an interesting candidate for exciting large-amplitude plasma waves for particle acceleration using modest-intensity laser beams.

REFERENCES

1. M. N. Rosenbluth, C. S. Liu, Phys. Rev. Lett. **29,** 701 (1972).
2. T. Tajima and J. M. Dawson, Phys. Rev. Lett. **43,** 267 (1979).
3. C. Clayton *et. al.*, Phys. Rev. Lett. **70,** 37 (1993); F. Amiranoff *et. al.*, Phys. Rev. Lett. **74,** 5220 (1995).
4. G. Shvets, N. J. Fisch, A. Pukhov, and J. Meyer-ter-Vehn, Phys. Rev. E **60,** 2218 (1999).
5. J. Dawson, Phys. Rev. **113,** 383 (1959).
6. G. Shvets, J. S. Wurtele, and B. A. Shadwick, Phys. Plasmas, **4,** 1872 (1997).
7. G. Shvets, N. J. Fisch, A. Pukhov, and J. Meyer-ter-Vehn, Phys. Rev. Lett. **81,** 4879 (1998).
8. V. M. Malkin, G. Shvets and N. J. Fisch, Phys. Rev. Lett. **84,** 1208 (2000).

Suppression of Raman Forward Scattering in Plasma Channels

Gennady Shvets and Xiaohu Li

Princeton Plasma Physics Laboratory[1]
Princeton NJ 08543

Abstract. Raman Forward Scattering (RFS) instability of an intense laser pulse in a single-moded plasma channel is studied in the slab geometry. For a particular class of channels the growth rate is found to be significantly smaller than in the homogeneous plasma. This reduction, appreciable even for sub-relativistic laser intensities and shallow plasma channels, is caused by the radial shear of the plasma frequency and the existence of the collisionlessly damped hybrid (electrostatic/electromagnetic) modes of the transversely inhomogeneous plasma.

INTRODUCTION

Plasma channels [1-3] have been utilized for guiding intense laser pulses over extended distances. Plasma density minimum on the channel axis produces a necessary for guiding maximum of the refraction index $n = \sqrt{\epsilon} = (1 - \omega_p^2/\omega_0^2)^{1/2}$, where $\omega_p^2 = 4\pi e^2 n_0(r)/m$, $n_0(r)$ is the electron density, $-e$ and m are the electron charge and mass. Guided laser pulse can still suffer from various parametric instabilities, such as the small-angle or forward Raman scattering [4] or laser-hosing [5]. Recently, the RFS instability was studied in two limiting cases of a hollow [7] and infinitely shallow [8] channels. In the latter case RFS instability proceeds at the same rate as in a homogeneous plasma.

In this paper, we concentrate on the Raman Forward Scattering (RFS) in a single-mode plasma channel. The physical mechanisms underlying the RFS in the channel and in the homogeneous plasma are similar: the ponderomotive beatwave between the injected laser pulse (pump) and its Stokes/anti-Stokes sidebands drives a plasma wave, which then acts as a grating, scattering the pump and re-enforcing the sidebands. Despite this conceptual similarity, two new effects come into play in a channel, significantly modifying the instability growth rate.

First, the shear of the plasma frequency across the channel modifies the excitation of the plasma wave responsible for RFS. Indeed, the low-frequency beatwave with

---

[1] This work was supported by the US DOE Division of High-Energy and Nuclear Physics

$\omega \ll \omega_0$ efficiently excites a plasma wave only in the vicinity of a specific transverse location x inside the channel where $\omega_p(x) = \omega$. As a result, the overlap between the lasers and the plasma wave is reduced, and so is the instability growth rate. The quantitative measure of the channel inhomogeneity is $\eta = \omega_0 \Delta \omega_p^2 / u_0 \omega_{p1}^3$, where u_0 is the dimensionless laser vector potential on axis.

Second, the low-frequency plasma wave, which is purely electrostatic in a homogeneous plasma, acquires an electromagnetic component in a channel, becoming a weakly-damped quasi-mode [9,10]. This quasi-mode also participates in the RFS instability further modifying its growth rate. Capturing these two effects requires an accurate description of the plasma response to the beatwave of the pump and its sidebands. Both the phase-mixing and the quasi-mode excitation are missed if the plasma wave is modeled as a simple harmonic oscillator with the natural frequency equal to the plasma frequency at the center of the channel [8]. In this paper, we demonstrate that in a realistic plasma channel, in which the plasma frequency monotonically varies between $\omega_{p1} = \sqrt{\omega_{p2}^2 - \Delta \omega_p^2/2}$ at $x = 0$ and ω_{p2} at $|x| \to \infty$, these two effects significantly modify the growth rate of the RFS even for relatively shallow channels and weakly-relativistic pump beams.

BASIC FORMALISM AND ASSUMPTIONS

To distill the physics associated with the RFS instability, we assume that the plasma channel supports a single bound laser mode. This assumption enables neglecting side-scattering instabilities, concentrating instead on direct forward scattering. In order to quantitatively describe the phase-mixing of plasma oscillations in a channel (which is responsible for the modification of RFS), we neglect the self-modulation instability (SMI) [11,12] which, undoubtedly, is also modified by phase-mixing. To enable analytic progress, we restrict ourselves to the slab geometry assuming that $n_0(x)$ is a function of a single transverse coordinate. Dispersion relation for a circularly polarized signal $\vec{a}_0 = a_0/2(\vec{e}_x + i\vec{e}_y)e^{i\theta_0}$ + c.c., with $\theta_0 = (k_0 z - \omega_0 t)$ and $a_0(x) = eA_0/mc^2$, can be derived by solving the eigenvalue equation $\mathcal{L}_0 a_0 = \lambda_0 a_0$, where $\lambda_0 = \omega_0^2/c^2 - k_0^2$ and

$$\mathcal{L}_0 = -\frac{\partial^2}{\partial x^2} + \frac{\omega_p^2(x)}{c^2}\left(1 - \frac{|a_0^2|}{2}\right). \tag{1}$$

In deriving Eq. (1) $|a_0^2| < 1$ is assumed (weakly-relativistic pump), and the density depression created by the ponderomotive pressure of the laser pulse is neglected. Relativistically modified plasma density $U_0(x) = k_p^2(1 - |a_0^2|/2)$ plays the role of the self-consistent confining potential with a minimum at $x = 0$ (here and elsewhere $k_p = \omega_p/c$). To enable analytic progress, a particular plasma density profile is assumed:

$$\omega_p^2 = \omega_{p2}^2 - \frac{\Delta \omega_p^2}{2 \cosh^2(x/\sigma)}, \tag{2}$$

where $\Delta\omega_p^2 < \omega_{p2}^2$ is assumed. Neglecting the second-order $\Delta\omega_p^2 u_0^2/\omega_{p2}^2$ term, it can be shown that the amplitude u_0 of a single bound (fundamental) mode $a_0(x) = u_0\psi_0 \equiv u_0\cosh^{-1}(x/\sigma)$ is related to the laser spotsize through $(\Delta k_p^2 + k_{p2}^2 u_0^2)\sigma^2 = 4$, resulting in the weakly-nonlinear pump dispersion relation $\omega_0^2/c^2 - k_0^2 = k_{p2}^2 - \sigma^{-2}$. Confining potential now becomes $U_0 \approx k_{p2}^2 - (\Delta k_p^2 + k_{p0}^2 u_0^2)/2\cosh^2(x/\sigma)$. Using $y = \tanh(x/\sigma)$ as a transverse variable, the eigenmodes ψ_q of the transverse operator $\mathcal{L}_0$ are found by solving the eigenvalue equation:

$$\frac{\partial}{\partial y}\left[(1-y^2)\frac{\partial \psi_q}{\partial y}\right] + \left[s(s+1) - \frac{\mu^2}{1-y^2}\right]\psi_q = 0, \qquad (3)$$

where ψ_q's are defined inside the $-1 < y < 1$ interval, $2s(s+1) = (\Delta k_p^2 + k_{p0}^2 u_0^2)\sigma^2$, and $\mu^2 = (k_{p2}^2 - \lambda_q)\sigma^2$. The solutions of this equation are the associated Legendre functions $P_s^\mu(y)$ [13], and the spectrum contains s discrete energy levels with $\mu^2 > 0$ (in our case $s = 1$) and a continuum of modes with $\mu^2 = -q^2\sigma^2 < 0$. Here q is used to label the continuum modes which behave $\propto \exp(\pm iqx)$ at infinity. In the remainder of this paper we assume that focusing is primarily provided by the pre-formed channel, $k_{p2}^2 u_0^2 \sigma^2 \ll 4$, which is equivalent to assuming that the laser power is below the relativistic focusing threshold in slab geometry. Modes of the continuum are orthogonal to the bound mode and to each other:

$$<\psi_0,\psi_q> \equiv \int_{-\infty}^{+\infty} dx\, \psi_0\psi_q = 0$$
$$<\psi_q,\psi_{q'}> \equiv \int_{-\infty}^{+\infty} dx\, \psi_q\psi_{q'} = \delta(q-q') \qquad (4)$$

We proceed by Fourier-Laplace transforming the envelope of the perturbed laser field and separating the Stokes/anti-Stokes components $\tilde{a}_\pm$ according to

$$\delta\tilde{a} = \sum_{\omega,k}\left(\tilde{a}_+ e^{i(kz-\omega t)} + \tilde{a}_- e^{-i(kz-\omega^* t)}\right) + \text{c. c.} \qquad (5)$$

Wave equation for for a_+ (tildes are dropped for compactness) becomes

$$(\mathcal{L}_0 - \Delta_+)a_+ = \frac{\omega_p^2(x)}{2c^2}a_0^2(x)(a_+ + a_-^*) - \frac{\omega_p^2(x)}{c^2}\left(\frac{\delta n}{n_0}\right)a_0, \qquad (6)$$

where $\Delta_\pm = (\omega_0 \pm \omega)^2/c^2 - (k_0 \pm k)^2$. The first term in the RHS of Eq. (6) is due to the relativistic mass increase. For ω close to ω_p (Raman process) we assume it to be smaller than the second term in the RHS of Eq. (6) which is due to the resonant excitation of a plasma wave. Plasma wave is driven by the intensity modulation of the laser field given by

$$|a|^2 = a_0^2 + a_0(a_+ + a_-^*)e^{i(kz-\omega t)} + \text{c. c.} \qquad (7)$$

To calculate the ponderomotively-driven perturbation $\delta n/n_0$, we expand the perturbed laser field in the eigenfunctions of $\mathcal{L}_0$:

$$a_\pm = a_\pm^{(0)} \psi_0(x) + \sum_q a_\pm^{(q)} \psi_q(x), \qquad (8)$$

multiply Eq. (6) by $\psi_0(x)$, and integrate over x. Using the orthogonality conditions (4), obtain

$$(\lambda_0 - \Delta_+) a_+^{(0)} = -a_0 \frac{\langle \delta k_p^2, \psi_0^2 \rangle}{\langle \psi_0, \psi_0 \rangle} + \sum_q N_q, \qquad (9)$$

where $\delta k_p^2 = k_p^2 \delta n / n_0$ is proportional to the density variation due to the perturbation of the bound mode $a_\pm^{(0)}$, and $N_q = a_0 \delta k_p^{2(q)}/2\langle \psi_0, \psi_0 \rangle$ is the partial contribution of the continuum mode q to the $\delta n/n_0$ term in the RHS of the wave equation.

Continuum and bound modes are treated separately because we assume that only the bound mode becomes unstable. The continuum modes are not independently unstable because they describe diffracting radiation which does not overlap with the pump a_0 for a sufficient time to get significantly amplified by the small-angle Raman scattering. Some nonlinear coupling to the continuum modes does, however, take place: they are driven by the exponentially growing bound mode $a_\pm^{(0)}$ [14]. This drive is described by the analog of Eq. (9):

$$(\lambda_q - \Delta_+^{(q)}) a_+^{(q)} = -a_0 \langle \delta k_p^2, \psi_0 \psi_q \rangle, \qquad (10)$$

where the $q - q'$ coupling between the continuum modes is dropped. If the growth rate is small, the continuum modes can, in effect, be eliminated by assuming that $a_\pm^{(q)}$ adiabatically follow the bound mode (i. e. $\Delta_+^{(q)} = \Delta_+^{(0)} \approx \lambda_0$):

$$a_+^{(q)} = -\frac{a_0 \langle \delta k_p^2, \psi_0 \psi_q \rangle}{(\lambda_q - \lambda_0)}.$$

This expression for $a_+^{(q)}$ can be used for calculating $\delta k_p^{2(q)}$ and inserted into N_q, resulting in the modified dispersion of the bound mode. In particular, balancing $\sum_q N_q$ against the first term in the RHS of Eq. (9) provides the rate of the self-modulation instability in the paraxial approximation [8,11,12] which is neglected in this paper in order to elucidate the physics of RFS in the regime where phase-mixing of the plasma wave is important.

ANALYSIS OF THE DISPERSION RELATION

Neglecting the continuum modes' correction $\sum N_q$ in Eq. (9) results in a familiar from Refs. [15,16] expression

$$\frac{D_+ D_-}{D_+ + D_-}(a_+ + a_-^*) = a_0 \frac{\langle \delta k_p^2, \psi_0^2 \rangle}{\langle \psi_0, \psi_0 \rangle}, \qquad (11)$$

where $D_\pm = \pm 2(\omega_0 \omega/c^2 - k_0 k) + (\omega^2/c^2 - k^2)$. Ponderomotively driven density perturbation δk_p^2 in a channel is given by

$$\delta k_p^2 = \frac{1}{2}\left[\frac{\omega_p^2(x)}{\omega_p^2 - \omega^2}\nabla^2 + \frac{\epsilon'}{\epsilon^2}\frac{\partial}{\partial x}\right]a_0(a_+ + a_-^*) - \frac{\epsilon'}{\epsilon^2}\frac{eB_y}{mc^2}, \qquad (12)$$

where $\epsilon(x,\omega) = 1 - \omega_p^2(x)/\omega^2$, the prime denotes a derivative with respect to x, and $\tilde{B} = e_y B_y$ is the magnetic field derived in Ref. [10]:

$$B_y'' - \frac{\epsilon'}{\epsilon}B_y' - \frac{\omega_p^2(x)}{c^2}B_y = -\frac{m\omega^2 \epsilon'}{2e\epsilon}a_0(a_+ + a_-^*). \qquad (13)$$

Note from Eq. (12) that the density perturbation can be broken up into two parts: $\delta k_p^2 = \delta k_p^{2(L)} + \delta k_p^{2(B)}$, where the first contribution (in square brackets) is locally-driven (i. e. $\delta k_p^{2(L)}(x)$ is determined by the ponderomotive force at the same x), while the second contribution $\delta k_p^{2(B)}$ is related to the magnetic field B_y and is manifestly non-local.

Local Density Perturbation

First, neglecting $\delta k_p^{2(B)}$ and substituting $\delta k_p^{2(L)}$ into Eq. (11), obtain a dispersion relation

$$D_+ D_-/(D_+ + D_-) = u_0^2 Q, \qquad (14)$$

where Q is proportional to the overlap integral $\langle \delta k_p^2, \psi_0^2 \rangle$:

$$Q = \int_{-1}^{1} dy(1-y^2)\left[\frac{y^2}{\sigma^2} + \frac{\omega^2}{4c^2}\right]\frac{\omega_{p1}^2 + \Delta\omega_p^2 y^2/2}{(\omega^2 - \omega_{p1}^2) - \Delta\omega_p^2 y^2/2},$$

where we assumed $\partial^2/\partial z^2 \approx -\omega^2/c^2$. Note that this integral becomes singular for the real $\omega_{p1} < \omega < \omega_{p2}$, where $\omega_{p1}^2 = \omega_{p2}^2 - \Delta\omega_p^2/2$. However, for $\mathrm{Im}\,\omega > 0$ the integral is well-defined, and can be calculated analytically. For a broad shallow channel $\omega^2 \sigma^2/c^2 \gg 1$ the expression for Q is particularly simple:

$$Q = \frac{\omega^2}{4c^2}\frac{\omega_{p1}^2}{\omega^2 - \omega_{p1}^2}\left[\frac{2}{B^2}\left(1 - \frac{2C^2}{3} + \frac{C^2}{B^2}\right) - \frac{(B^2+C^2)(1-B^2)}{B^5}\ln\left(\frac{1+B}{1-B}\right)\right], \qquad (15)$$

with $C^2 = \Delta\omega_p^2/2\omega_{p1}^2$ and $B^2 = \Delta\omega_p^2/2(\omega^2 - \omega_{p1}^2)$. Logarithm in Eq. (15) is analytically continued to satisfy causality: Q is real for $|B| < 1$ on the real axis. To the best of our knowledge, Eqs. (14,15) represent the first analytic expression for the growth rate of the RFS which takes into account phase mixing of the plasma oscillations.

The almost-homogeneous plasma limit is recovered by expanding Eq. (15) in powers of small B and $\Delta\omega_p/\omega_{p1}$: $Q \approx k_{p1}^2\omega^2/3(\omega^2 - \omega_{p1}^2)$. For $\omega = \omega_{p1} + i\gamma$ and $k = \omega_{p1}/v_{g0}$ the peak temporal growth rate $\gamma_{\text{hom}} \approx u_0\omega_{p1}^2/\sqrt{6}\omega_0$ (where we used $D_+D_-/(D_+ + D_-) \approx -i\omega_0^2\gamma/\omega_{p1}c^2$, and $v_{g0} = c^2k_0/\omega_0$ is the group velocity of the pump). This growth rate is almost identical to the well-known γ_{rfs} in a homogeneous plasma [15,16]. The above estimate of γ relies on $|B^2| \ll 1$, or $u_0 > (\sqrt{6}/4)\Delta\omega_p^2\omega_0/\omega_{p1}^3$. Even for a very shallow $\Delta\omega_p^2/\omega_{p1}^2 = 0.2$ (10% density depression) channel with $\omega_0/\omega_{p1} = 10$, the homogeneous plasma result is valid for $u_0 > 2.5$. Physically, this implies that the phase-mixing of plasma oscillations in a channel is so severe that it can only be neglected if the RFS growth rate is high, $\gamma > \Delta\omega_p^2/2\omega_{p1}$.

We can now introduce the measure of the plasma inhomogeneity $\eta = \Delta\omega_p^2\omega_0/u_0\omega_{p1}^3$. For $\eta < 1$ phase mixing is negligible, homogeneous plasma response [8] is valid, and the growth rate of the RFS is almost identical to that in the homogeneous plasma. In the opposite limit of $\eta > 1$ phase mixing is important. Taking the $|B^2| \gg 1$ limit of Q, obtain

$$Q = \frac{\omega^2}{c^2}\frac{\omega_{p1}^2}{\Delta\omega_p^2}\left[\frac{-i\pi\Delta\omega_p}{\sqrt{8}(\omega^2 - \omega_{p1}^2)^{1/2}} + \left(2 - \frac{\Delta\omega_p^2}{3\omega_{p1}^2}\right)\right] \quad (16)$$

The leading term ensures an instability for all $\omega_{p1} < \omega_r < \omega_{p2}$: in a plasma channel there is always a resonant location x where $\omega_p(x) = \omega_r$. Note that the commonly used [8] simplified equation $(\partial_t^2 + \omega_{p1}^2)\delta n/n = -k_p^2|a|^2/2$ for the density perturbation cannot correctly describe the phase mixing since it implies that the plasma wave has a fixed response frequency ω_{p1}. In reality, there is a continuum of plasma frequencies between ω_{p1} and ω_{p2}, and the phase-mixing of various plasma oscillations from this continuum leads to the overlap integral (16).

As we see below, the growth rate is reduced for $|B^2| \gg 1$. The physical meaning of $|B|$ is the transverse extent $\delta x \approx \sigma/|B|$ of the plasma wave, which becomes more localized than the driving beatwave. This sharp localization reduces the overlap between the plasma wave and the pump – hence, the reduced growth rate. Analytic estimates of the growth rate are obtained in two limits: gain-dominated (γ_g) and dispersion-dominated (γ_d). In the gain-dominated case $\gamma > \omega_{p1}^3/2\omega_0^2$ (which corresponds to a strong pump with $u_0^2 > 0.8\Delta\omega_p/\omega_0$) we obtain

$$\omega - \omega_{p1} = e^{i\pi/3}\left(\frac{u_0\omega_{p1}^2}{\omega_0}\right)\left[\frac{\pi^2}{16\eta}\right]^{1/3} \Rightarrow \gamma_g \approx \frac{1.8\gamma_{\text{hom}}}{\eta^{1/3}} \quad (17)$$

while in the opposite dispersion-dominated (weak-pump) limit

$$\gamma_d \approx \gamma_{\text{hom}}\sqrt{6\omega_{p1}/\omega_0\Delta\omega_p}. \quad (18)$$

It is instructive to compare the laser intensity scaling of the growth rate in a channel and in a homogeneous plasma. From Eq. (17), $\gamma \sim I_0^{2/3}$, and $\gamma_{\text{rfs}} \sim I_0^{1/2}$.

The stronger scaling with intensity in the case of the channel is related to the fact that the instability is "less" resonant than in the homogeneous case: the spread in the plasma frequencies exceeds the peak growth rate. Something similar occurs in silica fibers, where molecular vibrational bands spread out into bands which overlap and create a continuum. As a result, the Raman gain in silica fibers extends continuously over a broad range and scales $\propto I_0$ [17]. The scaling in plasma channels is weaker because plasma frequencies are not randomly spread out: instead, there is a well-defined region of almost-uniform plasma near the axis.

In Fig. 1(a) we show the dependence of the peak growth rate on u_0, which is obtained by solving the dispersion relation with exact value of Q for a plasma channel with a 40% density depression. The dotted line shows the almost-homogeneous plasma growth rate γ_{hom}, which over-estimates the growth rate for all values of u_0. The dashed-dotted and dashed lines are the growth rates in the dispersion and gain-dominated regimes, respectively. Note that for large u_0 Eq. (17) approximates the growth rate better than γ_d while the opposite is true for small u_0.

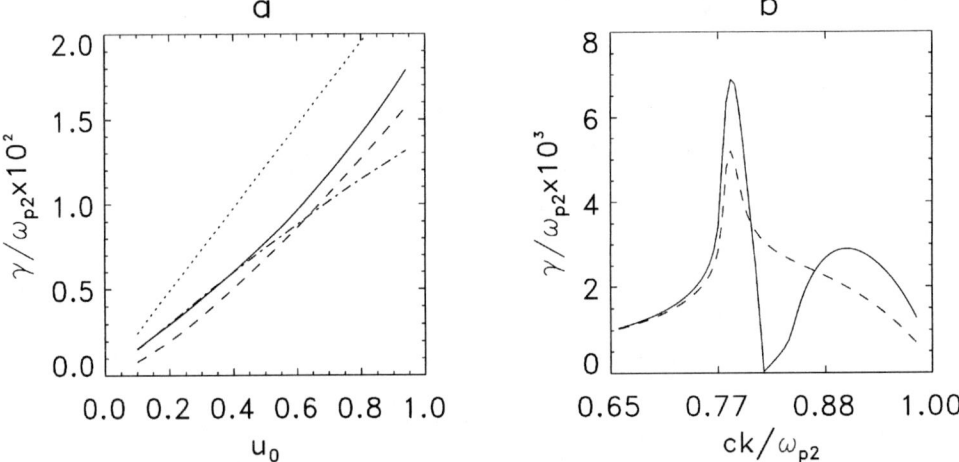

FIGURE 1. (a) Peak growth rate γ of RFS near the on-axis plasma frequency ω_{p1} as a function of normalized laser amplitude u_0. Only local density perturbation $\delta k_p^{2(L)}$ is taken into account. Growth rate is numerically calculated from: dispersion relation $D_+D_-/(D_+ + D_-) = u_0^2 Q$ (solid line); homogeneous plasma estimate $\gamma = \gamma_{\text{hom}}$ (dotted line); Eq. (17) in gain-dominated regime (dashed line); $\gamma = \gamma_d$ in dispersion-dominated regime (dot-dashed line). (b) Instability spectrum with (solid line) and without (dashed line) non-local density perturbation due to magnetic field. Channel parameters: $\omega_{p1}/\omega_{p2} = 0.775$ and $\omega_0/\omega_{p2} = 10$.

For a fixed laser amplitude $u_0 = 0.4$, the scan of the peak growth rate in the vicinity of $\omega = \omega_{p1}$ is plotted in Fig.2(a) as a function of the fractional channel depth $\Delta\omega^2/2\omega_{p2}^2$. As in Fig. 1(a), the solid line is growth rate obtained by numerically solving the dispersion relation $D_+D_-/(D_+ + D_-) = u_0^2 Q$. The numerical solution is very well approximated by the γ_d (dot-dashed line) and slightly under-estimated by the Eq. (17) (dashed line). On the other hand, the almost-homogeneous plasma growth rate γ_{hom} (dotted line) significantly over-estimates the growth rate. The inaccuracy of the γ_{hom} is explained by Fig. 2(b), where the dependence of the overlap coefficient $|B|$ is plotted. For the parameters of Fig. 2, $|B|$ always exceeds unity, invalidating the homogeneous plasma approximation: even for a relatively shallow plasma channel with a 20% density depression $|B| = 3.2$.

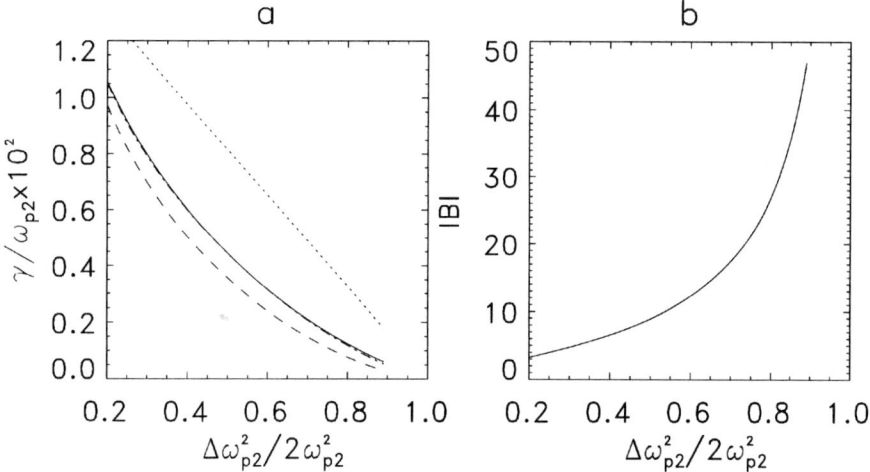

FIGURE 2. (a) Same as in Fig. 1(a) as a function of the channel depth $\Delta\omega^2/2\omega_{p2}^2$ for a fixed laser amplitude $u_0 = 0.4$ and $\omega_0/\omega_{p2} = 10$. (b) Overlap coefficient $|B|$ for the same laser and channel parameters.

Density perturbation due to magnetic field

The local density $\delta k_p^{2(L)}$ perturbation, assumed up to this point, is only one part of the total plasma response to the ponderomotive force. The total $\delta k_p^2 = \delta k_p^{2(L)} + \delta k_p^{2(B)}$ contains the contribution of the magnetic field calculated from Eq. (13). The overlap integral $\langle \delta k_p^{2(B)}, \psi_0^2 \rangle$ can only be calculated numerically:

$$u_0 \frac{\langle \delta k_p^{2(B)}, \psi_0^2 \rangle}{\langle \psi_0, \psi_0 \rangle} = \frac{u_0^2}{2c^2} \frac{\omega^4 \Delta\omega_p^4}{(\omega^2 - \omega_{p1}^2)^3} \int_0^1 dy \frac{y(1-y^2)H_1}{(1-B^2y^2)^2}$$

where the normalized magnetic field H_1 satisfies Eq. (13), re-written using $y = \tanh(x/\sigma)$ instead of x:

$$\frac{\partial}{\partial y}\left[\frac{1-y^2}{1-B^2y^2} \frac{\partial H_1}{\partial y}\right] + \sigma^2 \left(\frac{\Delta\omega_p^2}{2c^2} - \frac{k_{p2}^2}{1-y^2}\right) \frac{H_1}{1-B^2y^2} = -\frac{y(1-y^2)}{(1-B^2y^2)^2}, \quad (19)$$

with the boundary conditions $H_1(y=0,\omega) = H_1(y=1,\omega) = 0$. With vanishing RHS, Eq. (19) describes the collisionlessly-damped global quasi-mode [10] of a plasma channel. Boundary conditions for the quasi-mode can be satisfied for a single value of B which determines its frequency and damping rate. For example, for a plasma channel with a 40% density depression they are equal to $\omega_D = 0.82\omega_{p2}$ and $\gamma_D = 0.045\omega_{p2}$. The plot of the dependence of the real and imaginary parts of the quasi-mode frequency ω_D on the relative channel depth $\Delta\omega_p^2/2\omega_{p2}^2$ is shown in Fig. 3 for $u_0 = 0.4$ ($I_0 = 4\cdot 10^{17}$W/cm^2).

For relatively deep plasma channels, the density perturbation $\delta k_p^{2(B)}$ arising from the non-electrostatic nature of the plasma wave can be appreciable. In Fig. 1(b) the normalized growth rate γ/ω_{p2} is plotted v. s. the wavenumber ck/ω_{p2} for a channel with $\omega_{p1} = 0.775\omega_{p2}$, $\omega_0/\omega_{p2} = 10$, and $u_0 = 0.35$. Since the peak growth rate corresponds to $k \approx \omega_r/v_{g0}$, Fig. 1(b) is the spectrum of RFS in a channel. Dashed line is the growth rate due to the local density excitation $\delta k_p^{2(L)}$ only. When magnetic field contribution $\delta k_p^{2(B)}$ is added (solid line), the peak growth rate at ω_{p1} is modified, and a distinctive broad amplification band peaked at $\omega_r = 0.91\omega_{p2}$ develops. It is slightly shifted from the quasi-mode frequency $\omega_Q = 0.82\omega_{p2}$. Although the growth rate at ω_r is only half of that at ω_{p1}, large transverse electric field E_x associated with the off-axis mode may result in an early wavebreaking off the channel axis, turning the channel wall into a source of hot electrons.

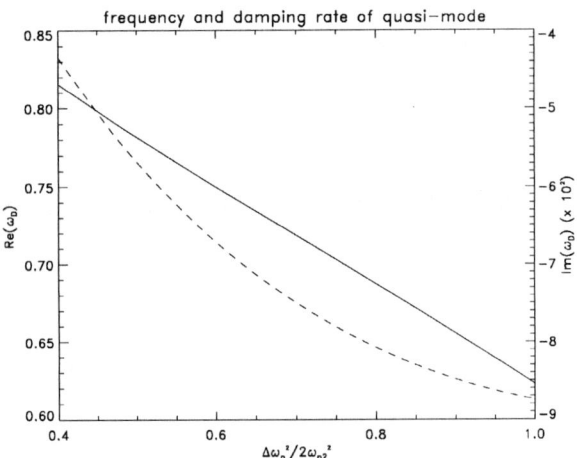

FIGURE 3. The real frequency Re(ω_D) (solid line) and the damping rate Im(ω_D) (dashed line) of the quasi-mode as a function of the normalized channel depth $\Delta\omega^2/2\omega_{p2}^2$. Laser amplitude is $u_0 = 0.4$

CONCLUSIONS

In conclusion, the growth rate of the Raman Forward Scattering is significantly modified in a plasma channel as a result of two effects: (a) plasma wave localization caused by phase-mixing of plasma oscillations in a channel, and (b) non-electrostatic nature of plasma waves in a strongly-inhomogeneous channel. The importance of channel inhomogeneity is characterized by a single dimensionless parameter $\eta = \Delta\omega_p^2 \omega_0 / u_0 \omega_{p1}^3$. These newly uncovered phenomena might also affect other laser instabilities in plasma channels, such as self-modulation instability.

REFERENCES

1. C. Clark et. al., J. Opt. Soc. Am. B **3**, 371 (1986).
2. C. Durfee and H. Milchberg, Phys. Rev. Lett. **71**, 2409 (1993).
3. A. Ziggler et. al., J. Opt. Soc. Am. B **13**, 68 (1996).
4. T. M. Antonsen, Jr.,and P. Mora, *Phys. Rev. Lett.* **69**, 2204 (1992); T. M. Antonsen, Jr., and P. Mora, Phys. Fluids B **5**, 1440 (1993).
5. G. Shvets and J. S. Wurtele, Phys. Rev. Lett. **73**, 3540 (1994); P. Sprangle, J. Krall, and E. Esarey, Phys. Rev. Lett. **73**, 3544 (1994).
6. T. M. Antonsen, Jr. and P. Mora, Phys. Rev. Lett. **744**, 4440 (1995).
7. T. C. Chiou et. al., AIP Conference Proceedings, **398**, 357 (1997).
8. E. Esarey et. al., Phys. Rev. Lett. **84**, 3081 (2000).
9. G. Shvets et. al., IEEE Trans. Plasma Sci. **24**, 351 (1996).
10. G. Shvets and X. Li, Phys. Plasmas **6**, 591 (1999).
11. E. Esarey et. al., Phys. Rev. Lett. **72**, 2887 (1994).
12. N. E. Andreev et. al., Phys. Plasmas **2**, 2573 (1995).
13. I. S. Gradshteyn and I. M. Ryzhik, *Table of Integrals, Series, and Products*, Academic Press, New York (1965).
14. G. Tempea and T. Brabec, Opt. Lett. **23**, 762 (1998).
15. W. L. Kruer, *The Physics of Laser Plasma Interactions* (Addison-Wesley, Reading, MA, 1988).
16. C. J. McKinstrie and R. Bingham, Phys. Fluids B **4**, 2626 (1992).
17. G. P. Agarwal, *Nonlinear Fiber Optics*, Academic Press, 2-nd edition, New York (1998).

Nonparaxial Propagation of Intense Laser Pulses in Plasmas

E. Esarey, B.A. Shadwick, C.B. Schroeder,[1] J.S. Wurtele, and W.P. Leemans

Center for Beam Physics, Ernest Orlando Lawrence Berkeley National Laboratory, University of California, Berkeley CA 94720

Abstract. Non-paraxial propagation of ultrashort, high power laser pulses in plasma channels is examined. In the adiabatic limit, pulse energy conservation, nonlinear group velocity, damped betatron oscillations, self-steepening, self-phase modulation, and shock formation are analyzed. In the non-adiabatic limit, the coupling of forward Raman scattering (FRS) and the self-modulation instability (SMI) is analyzed and growth rates are derived, including regimes of reduced growth. The SMI is found to dominate FRS in most regimes of interest.

INTRODUCTION

Guiding of intense laser pulses in plasma channels [1] is beneficial to various applications, including harmonic generation [2], x-ray lasers [3], advanced laser-fusion schemes [4], and plasma-based accelerators [5]. A laser pulse in vacuum diffracts after a distance on the order of a Rayleigh length $Z_R = \pi r_0^2/\lambda$, where r_0 is the spot size at focus, $\lambda = 2\pi c/\omega$, and ω is the frequency. A preformed plasma density channel can prevent diffraction, e.g., a channel with a radially parabolic density profile $n(r) = n_0 + \Delta n r^2/r_0^2$ can guide a laser pulse of spot size r_0 provided $\Delta n = \Delta n_c$, where $\Delta n_c = 1/\pi r_e r_0^2$ is the critical channel depth and $r_e = e^2/m_e c^2$ [6]. Plasma channels have been created experimentally by various methods and have been used to guide laser pulses over distances $\lesssim 100 Z_R$ [1], [7], [8].

Conventional theories of intense, finite-radius pulse propagation in plasmas have assumed the paraxial approximation (PA) [1], which assumes a fixed group velocity and neglects many important finite pulse length effects. In the PA, axial transport of energy within the pulse is not permitted. Hence the PA is incapable of describing many phenomena, e.g., forward Raman scattering (FRS) [9], [10], in which intensity modulations arise from an axial transport of energy. The PA does describe the self-modulation instability (SMI) [5], [11], [12], i.e., intensity modulations from a radial

---

[1] Present address: Department of Physics, University of California, Los Angeles CA 90095

transport of energy. There has been debate within the community [5], [9]- [13] as to which of these instabilities is responsible for intense pulse modulation observed in experiments [13]. A comprehensive theory of FRS and SMI is currently lacking.

Recently, a nonlinear theory of non-paraxial pulse propagation was formulated that is valid for ultrashort, high power $P \leq P_c$ pulses in plasmas with or without a parabolic channel [14]. Here $P_c[\text{GW}] = 17(\lambda_p/\lambda)^2$ is the critical power for relativistic self-focusing [1], $\lambda_p = 2\pi c/\omega_p$, and $\omega_p = ck_p = (4\pi n_0 e^2/m_e)^{1/2}$ is the plasma frequency. This paper discusses several results of that formulation. First, pulse propagation in the adiabatic limit is discussed, e.g., pulse energy conservation, nonlinear group velocity, damped betatron oscillations, pulse self-steepening, self-phase modulation, and shock formation. In the adiabatic limit the plasma response reduces to a standard third-order nonlinearity in the field. Hence, the adiabatic wave equation typifies a general class of problems in nonlinear media. In the non-adiabatic limit, which includes time dependent coupling to plasma waves, instabilities are analyzed. Next, the explicit coupling and interplay between SMI and FRS is discussed and analytic expressions for the growth rates are presented, including regimes of reduced growth. The SMI is found to dominate FRS in most regimes of interest.

LASER PULSE ENVELOPE EQUATIONS

The wave equation for the transverse component of the normalized vector potential $a_\perp = eA_\perp/m_e c^2$ of the laser field, in terms of the independent variables $\zeta = z - \beta_{g0}ct$ and z, is [1]

$$\left[\nabla_\perp^2 + 2\left(ik + \frac{\partial}{\partial \zeta}\right)\frac{\partial}{\partial z} + \gamma_{g0}^{-2}\frac{\partial^2}{\partial \zeta^2} + \frac{\partial^2}{\partial z^2}\right]\hat{a} = K^2\hat{a}, \tag{1}$$

where $a_\perp = (\hat{a}/2)\exp(ikz - i\omega t) + \text{c.c.}$ (c.c. denotes the complex conjugate), ω and k are the central frequency and wavenumber, $v_{g0} = c\beta_{g0}$ is the linear group velocity of a matched fundamental Gaussian pulse in a channel [15], i.e., $\gamma_{g0}^{-2} = 1 - \beta_{g0}^2 = \omega_p^2/\omega^2 + 4c^2/r_0^2\omega^2$, and $\omega\beta_{g0}/ck = 1$. Here $K^2 \equiv k_p^2(\rho_0 + \delta\rho) - \gamma_{g0}^{-2}\omega^2/c^2$, $\rho_0 = 1 + \Delta n r^2/n_0 r_0^2$, and $\delta\rho$ is the nonlinear plasma response which, in the limits $\hat{a}^2 \ll 1$ and $k_p^2 r_0^2 \gg 1$, is given by [9]- [12] $\left(\partial^2/\partial\zeta^2 + k_p^2\right)\delta\rho \simeq -k_p^2\hat{a}^2/2$, assuming circular polarization such that $a_\perp^2 = \hat{a}^2$.

For a short pulse of length L propagating in a plasma channel, the operators on the left of Eq. (1) scale as $\nabla_\perp \sim 1/r_0$, $\partial/\partial\zeta \sim 1/L$, and $\partial/\partial z \sim 1/Z_R$. In the following analysis, the last two terms on the left of Eq. (1) are small in the parameter regime of interest (underdense plasmas $k_p^2/k^2 \ll 1$) and will be neglected. This is valid provided $|\partial^2\hat{a}/\partial z^2| \ll 2|\partial^2\hat{a}/\partial\zeta\partial z|$, which implies $L \ll 2Z_R$, and $\gamma_{g0}^{-2}|\partial^2\hat{a}/\partial\zeta^2| \ll 2|\partial^2\hat{a}/\partial\zeta\partial z|$, which implies $2L/Z_R \gg (1 + 4/k_p^2 r_0^2)k_p^2/k^2$. These two conditions, along with $k_p^2 r_0^2 \gg 1$, imply $k^2 r_0^2/4 \gg kL \gg k_p^2 r_0^2/4 > 1$. For an underdense plasma $\gamma_{g0}^{-2} \ll 1$, and the $\partial^2/\partial\zeta\partial z$ term dominates. At high densities (e.g., $k_p/k \sim 1$), the $\partial^2/\partial\zeta^2$ term dominates, as in conventional nonlinear optics.

In Eq. (1), the term $2\partial^2/\partial\zeta\partial z$ represents the leading-order correction to the paraxial wave equation. It proves convenient to further approximate this operator by using the paraxial expression for the operator $\partial/\partial z$, i.e, $\partial \hat{a}/\partial z \simeq (-i/2k)(K^2 - \nabla_\perp^2)\hat{a}$. Using this approximation in the term $2\partial^2/\partial\zeta\partial z$, Eq. (1) becomes [14]

$$\left(\nabla_\perp^2 + 2ik\frac{\partial}{\partial z}\right)\hat{a} \simeq \left[K^2 + \frac{i}{k}\frac{\partial}{\partial\zeta}\left(K^2 - \nabla_\perp^2\right)\right]\hat{a}. \qquad (2)$$

The second and third terms on the right represent the lowest order (first order in $1/kL$) contributions of the $2\partial^2\hat{a}/\partial\zeta\partial z$ term.

Equation (2) can be solved using the source-dependent expansion method [1], [11], wherein $\hat{a}$ is expanded in a series of Laguerre-Gaussian source-dependent modes, $\hat{a} = \sum_m \hat{a}_m L_m(\chi) \exp[-(1 - i\alpha)\chi/2]$, where $m = 0, 1, 2, ...$, $\hat{a}_m(\zeta, z)$ is the complex amplitude, $\chi = 2r^2/r_s^2$, $r_s(\zeta, z)$ is the spot size, $\alpha(\zeta, z)$ is related to the curvature, $L_m(\chi)$ is a Laguerre polynomial of order m, and axisymmetry has been assumed, i.e., $\hat{a} = \hat{a}(r, \zeta, z)$. Assuming that $\hat{a}$ is adequately described by the lowest order mode ($m = 0$), the evolution of the real parameters r_s, α, a_r, and θ, where $\hat{a}_0 = a_r \exp(i\theta)$, is given by [14]

$$\dot{r}_s/r_s = 2\alpha/kr_s^2 - H_I \qquad (3)$$

$$\dot{a}_r/a_r = -2\alpha/kr_s^2 + G_I + H_I \qquad (4)$$

$$\dot{\alpha} = 2(1+\alpha^2)/kr_s^2 + 2H_R - 2\alpha H_I \qquad (5)$$

$$\dot{\theta} = -2/kr_s^2 - G_R - H_R \qquad (6)$$

where $\dot{Q} = \partial Q/\partial z$ (for a function Q), and the subscripts R and I denote the real and imaginary parts. Also, $(G, H) = \sum (G, H)_j$ with $j = a, b$ and c:

$$k^2 r_0^2 G_a = \left(2 - \Delta_c r_s^2/r_0^2\right)(T_A - k) + \left(1 - \Delta_c r_s^2/r_0^2\right) T_B \qquad (7)$$

$$k^2 r_0^2 H_a = (\Delta_c r_s^2/r_0^2)(T_A - k) - \left(1 - 2\Delta_c r_s^2/r_0^2\right) T_B \qquad (8)$$

$$k^2 r_s^2 G_b = -(1+\alpha^2)T_A + (i-\alpha)\alpha T_B \qquad (9)$$

$$k^2 r_s^2 H_b = -(1-i\alpha)^2 T_A + (1 + 2\alpha^2 + i\alpha)T_B \qquad (10)$$

$$G_c = -\frac{4k_p}{k^2} \int_{\zeta_0}^{\zeta} d\zeta_1 S(\zeta,\zeta_1) \left[T_C + \frac{r_{s1}^2 T_D}{2(r_s^2 + r_{s1}^2)} \right] \quad (11)$$

$$H_c = -\frac{4k_p}{k^2} \int_{\zeta_0}^{\zeta} d\zeta_1 \frac{r_s^2 S(\zeta,\zeta_1)}{(r_s^2 + r_{s1}^2)} \left[T_C + \frac{r_{s1}^2(r_s^2 - r_{s1}^2) T_D}{2r_s^2(r_s^2 + r_{s1}^2)} \right] \quad (12)$$

where $\Delta_c = \Delta n/\Delta n_c$, $\hat{P} = P/P_c = k_p^2 a_r^2 r_s^2/16$, $T_A = \theta' - ia_r'/a_r$, $T_B = \alpha' - 2(\alpha + i)r_s'/r_s$, $T_C = k - T_A + 2ia_{r1}'/a_{r1}$, $T_D = -T_B + 4ir_s^2 r_{s1}'/r_{s1}^3$, $S = (r_s^2 + r_{s1}^2)^{-1} \hat{P}_1 \sin k_p(\zeta - \zeta_1)$, $Q' = \partial Q/\partial \zeta$, $Q_1 = Q(\zeta_1)$, and ζ_0 is chosen before the pulse ($\zeta \leq \zeta_0$). Notice that Eqs. (3) and (4) imply $\partial \hat{P}/\partial z = 2\hat{P} G_I$. When $Q' = 0$, Eqs. (3)-(12) reduce to paraxial limit [1] and $H = G = 0$ describes paraxial vacuum diffraction of a Gaussian beam.

ADIABATIC LIMIT

Consider the adiabatic limit in which the pulse length is long compared to the plasma wavelength ($k_p^2 L^2 \gg 1$) and coupling to the plasma wave (e.g., FRS) is neglected, i.e., $\delta \rho \simeq -\hat{a}^2/2$. The wave equation then contains a cubic nonlinearity. In this limit, Eqs. (11) and (12) reduce to $k^2 r_s^2 G_c = 2k^2 r_s^2 H_c + \hat{P}\left[\alpha'/2 - (\alpha + 3i) r_s'/r_s\right]$ and $k^2 r_s^2 H_c = \hat{P}(\theta' - k - 3ia_r'/a_r)$. This implies $\partial \hat{P}/\partial z + \partial(\delta\beta_g \hat{P})/\partial \zeta = 0$, i.e., the local group velocity is given by $\beta_g \simeq \beta_{g0} + \delta\beta_g(\zeta, z)$, where [14]

$$k^2 \delta\beta_g = 2/r_0^2 - (1 + \alpha^2)/r_s^2 - \Delta_c r_s^2/r_0^4 + 3\hat{P}/r_s^2. \quad (13)$$

Furthermore, the total pulse energy $W = \int d\zeta \hat{P}$ is conserved, i.e., $\partial W/\partial z = 0$. This is not true for the general non-adiabatic case, since pulse energy is lost to the generation of plasma waves.

In the low power ($\hat{P} \ll 1$) adiabatic limit with $\Delta_c = 1$, $r_s = r_0 + \delta r$, and $\alpha = \delta \alpha$ (where $\delta Q/Q \sim \hat{P}$), we obtain $\delta\beta_g \simeq 3\hat{P}/k^2 r_0^2$ and the power evolution is given by $\hat{P} = f(\zeta - 6\hat{P}z/k^2 r_0^2)$ where f is a function, e.g., $f(\zeta) = \hat{P}_0 \exp(-2\zeta^2/L^2)$ for a Gaussian with a peak power $\hat{P}_0$. This describes self-steepening of the pulse power profile, i.e., the higher the local power, the higher the local group velocity, $\delta\beta_g$, and power is shifted forward within the pulse. The pulse peak moves at a velocity $\beta_{peak} = \beta_{g0} + \delta\beta_{peak}$ with $\delta\beta_{peak} = 6\hat{P}_0/k^2 r_0^2$. In the absence of dispersive pulse broadening [from the term $\gamma_{g0}^{-2} \partial^2/\partial \zeta^2$ in Eq. (1)], steepening continues until a shock is formed ($\partial \hat{P}/\partial \zeta \to \infty$). For a Gaussian $f(\zeta)$, shock formation occurs after a distance $z = Z_S$, where $Z_S = (e^{1/2}/6) k L Z_R/\hat{P}_0$.

Spot size evolution in the low-power adiabatic limit can be examined by perturbing about the zero-power, matched-pulse equilibrium with $\Delta_c = 1$, i.e, $r_s = r_0 + \delta r_s$, $\alpha = \delta \alpha$, $a_r = a_{r0}(\zeta) + \delta a_r$, etc. In particular, Eqs. (3) and (5) imply [14]

$$\left[\left(\frac{\partial}{\partial z} - \frac{2}{kZ_R}\frac{\partial}{\partial \zeta}\right)^2 + \frac{4}{Z_R^2}\right]\frac{a_{r0}\delta r}{r_0} \simeq -\frac{\hat{P}a_{r0}}{Z_R^2}. \tag{14}$$

For the initial conditions $\delta r_s = \delta r_0$, $\delta r_s' = 0$, and $\hat{P} = \hat{P}_0 \exp(-2\zeta^2/L^2)$,

$$\delta r_s/r_0 = \left(F_\beta \delta r_0/r_0 + F_\beta^3 \hat{P}/4\right)\cos(k_\beta z) - \hat{P}/4, \tag{15}$$

where $F_\beta = \exp(-2z\zeta/Z_\beta L - z^2/Z_\beta^2)$, $k_\beta = 2/Z_R$ is the betatron wavenumber, and $Z_\beta = kLZ_R/2$ is the betatron damping distance. In the linear limit ($\hat{P} = 0$), Eq. (15) describes damped betatron oscillations of a pulse mismatched ($\delta r_0 \neq 0$) in a channel [15]. Asymptotically, these oscillations damp via $\delta r_s \sim \exp(-z^2/Z_\beta^2)$ for fixed ζ, with a head-tail asymmetry. For finite powers, however, betatron oscillations arise even when $\delta r_0 = 0$, only now with an enhanced damping rate, i.e., $\exp(-3z^2/Z_\beta^2)$. This is the case since a pulse with $\hat{P} > 0$ is no longer matched when $r_s = r_0$ in a channel with $\Delta_c = 1$. Recall that paraxial theory [1] gives a matching condition $r_s^4/r_0^4 = (1 - \hat{P})/\Delta_c$. For $\Delta_c = 1$ and $\hat{P} \ll 1$, this gives $r_s/r_0 \simeq 1 - \hat{P}/4$, precisely the asymptotic ($z \gg Z_\beta$) behavior given by Eq. (15).

Phase distortions (self-phase modulation) also develop. In the limit $\hat{P} \ll 1$ and $\Delta_c = 1$, Eq. (6) implies $\delta\dot{\theta} \simeq (4\delta r/r_0 - 3\hat{P})/kr_0^2$. This results in local frequency shifts via $\delta\omega/\omega = \delta\theta'/k$. Asymptotically, for $z \gg Z_\beta$ (neglecting betatron oscillations), the self-phase modulation due to self-steepening is given by $\delta\dot{\theta} \simeq -4\hat{P}/kr_0^2$, which implies $\delta\omega/\omega \simeq (2/3)\ln[P/P(z=0)]$.

Numerical solutions to Eqs. (3)-(12) in the adiabatic limit are shown in Figs. 1-3 for the parameters $\lambda = 1$ µm, $r_0 = 10$ µm ($Z_R = 310$ µm), $\lambda_p = 15$ µm ($\Delta n = \Delta n_c = 1.1 \times 10^{18}$ cm^{-3} and $n_0 = 4.9 \times 10^{18}$ cm^{-3}), $a_0 = 0.4$ ($\hat{P}_0 = 0.18$) and $L = 5$ µm (FWHM 20 fs) with an initially Gaussian profile, $P(0) = P_0\exp(-2\zeta^2/L^2)$. The spot size evolution $r_s(z)$ is shown in Fig. 1 near the (a) front $\zeta = L$, (b) center $\zeta = 0$, and (c) back $\zeta = -L$ of the pulse. The numerical (solid curve) and analytical (dashed curve), Eq. (15), solutions show good agreement in Figs. 1 (a) and (b). At the back of the pulse, discrepancies arise, e.g., a nonlinear betatron wavenumber shift, however, excellent agreement is obtained for smaller $\hat{P}_0$. Self-steepening of the power profile $\hat{P}(\zeta)$ is shown in Fig. 2 at $z = 0$ (solid curve), $z = 20Z_R$ (dashed curve), and $z = 40Z_R$ (dotted curve). The velocity of the peak is in good agreement with theory ($\delta\beta_{peak} = 2.7 \times 10^{-4}$), as is the position of shock formation $Z_s = 0.55Z_\beta/\hat{P} = 48Z_R = 1.5$ cm. The evolution of the intensity profile $a_r^2(\zeta, z)$ is shown in Fig. 3 with the effects of the damped betatron oscillations and self-steepening clearly evident.

A recent paper [16] has proposed using the quasi-paraxial approximation (QPA) to analyze the adiabatic limit, in which the $\partial/\partial\zeta$ term in Eq. (1) is replaced by a term proportional to ζ. We note that in the QPA the pulse energy increases via $W \simeq W_0\exp(z^2/2Z_\beta^2)$. Hence, to approximately conserve energy, the QPA is restricted to $z \ll Z_\beta$ and, thus, it is incapable of describing the phenomena

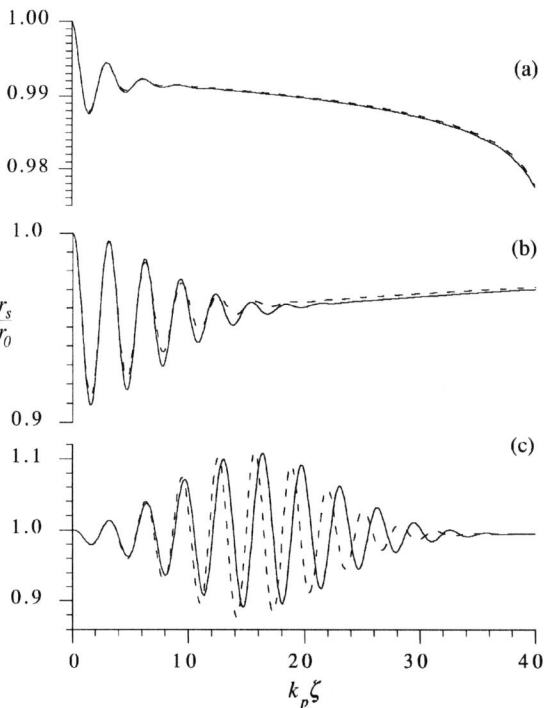

FIGURE 1. Spot size $r_s(z)$ at (a) front $\zeta = L$, (b) center $\zeta = 0$, and (c) back $\zeta = -L$ of pulse, from simulation (solid curve) and theory (dashed curve), for $\lambda = 1$ μm, $r_0 = 10$ μm, $\lambda_p = 15$ μm, $\Delta_c = 1$, $\hat{P}_0 = 0.18$, and $L = 5$ μm, with an initially Gaussian profile.

analyzed in the present work. Also, we find no evidence for the "enhanced" self-focusing discussed in [16].

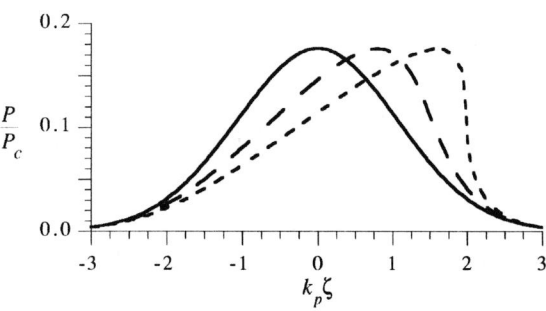

FIGURE 2. Power profile $\hat{P}(\zeta)$ at $z = 0$ (solid curve), $z = 20Z_R$ (dashed curve), and $z = 40Z_R$ (dotted curve) for the parameters of Fig. 1.

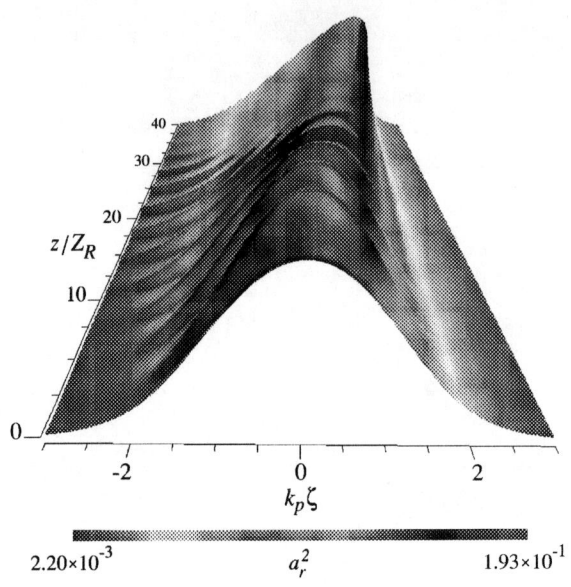

FIGURE 3. Intensity profile $a_r^2(\zeta, z)$ for the parameters of Fig. 1.

INSTABILITIES

Laser-plasma instabilities of finite-radius pulses (as opposed to plane waves) can be examined using the full equations, Eqs. (3)-(12), including coupling to the plasma wave, as in FRS and SMI. Analytically, this is done by expanding Eqs. (3)-(12) about the optically-guided, matched-beam equilibrium given by $r_s = r_0$, $a_r = a_0$, $\alpha = 0$, and $\theta' = 0$, where a_0 and r_0 are constants (a flat-top axial profile) and $\Delta_c + \hat{P} = 1$ is assumed. Letting $Q = Q_0 + \delta Q$ and $\delta Q = \delta \hat{Q} \exp(ik_p\zeta)$ with $|\partial \delta \hat{Q}/\partial \zeta| \ll |k_p \delta \hat{Q}|$ (modes resonant with the plasma wave), give [14]

$$\mathcal{L}_1 \mathcal{L}_2 \delta \hat{r} = i C_c \delta \hat{r}, \tag{16}$$

where $\mathcal{L}_1 = \partial^2/\partial \hat{\zeta} \partial \hat{z} + \hat{k}_p \hat{P}$, $\mathcal{L}_2 = (\partial^2/\partial \hat{z}^2 + \hat{k}_\beta^2)\partial/\partial \hat{\zeta} + i\hat{P}$, $C_c = \hat{k}_p \hat{P}^2/2$, $\hat{k}_\beta = k_\beta Z_R = (4 - 2\hat{P})^{1/2}$, $\hat{k}_p = k_p/k$, $\hat{\zeta} = k_p \zeta$ and $\hat{z} = z/Z_R$. Notice that $\mathcal{L}_1 \delta \hat{r} = 0$ describes conventional 1D FRS [9], [10] and $\mathcal{L}_2 \delta \hat{r} = 0$ describes conventional 2D SMI [11], [12]. In general, Eq. (16) describes the nonlinear coupling of these two instabilities.

Using Eq. (16), asymptotic expressions for the number of e-folds N_e, $\delta \hat{r} \sim \exp(N_e)$, have been obtained in the appropriate spatial-temporal regimes. Typically, two branches to Eq. (16) are identified, associated with SMI and FRS, with either conventional (C) or reduced (R) growth rates. For the SMI branch, $N_e = (2\hat{P}|\hat{\zeta}|\hat{z}/\hat{k}_\beta)^{1/2}$ (C) is found to be valid in the short-pulse regime $\hat{P}/2\hat{k}_\beta \ll |\hat{\zeta}|/\hat{z} \ll 2\hat{k}_\beta^3/\hat{P}$; $N_e = c_0(\hat{P}|\hat{\zeta}|\hat{z}^2)^{1/3}$ (C) is valid in the intermediate regime $\hat{k}_\beta^3/2\hat{P} \ll$

$|\hat{\zeta}|/\hat{z} \ll 1/2\hat{P}\hat{k}_p^3$, where $c_0 = (1 + i/3^{1/2})3^{3/2}/2^{5/2}$; and $N_e = c_0(\hat{P}|\hat{\zeta}|\hat{z}^2/2)^{1/3}$ (R) is valid in the long-pulse regime $1/\hat{P}\hat{k}_p^3 \ll |\hat{\zeta}|/\hat{z}$. For the FRS branch, $N_e = (4\hat{k}_p\hat{P}|\hat{\zeta}|\hat{z})^{1/2}$ (C) is found to be valid in the short-pulse regime $\hat{k}_p\hat{P} \ll |\hat{\zeta}|/\hat{z} \ll \hat{k}_p\hat{k}_\beta^4/\hat{P}$; $N_e = (2\hat{k}_p\hat{P}|\hat{\zeta}|\hat{z})^{1/2}$ (R) is valid in the intermediate regime $2\hat{k}_p\hat{k}_\beta^4/\hat{P} \ll |\hat{\zeta}|/\hat{z} \ll 2/\hat{P}\hat{k}_p^3$; and $N_e = (4\hat{k}_p\hat{P}|\hat{\zeta}|\hat{z})^{1/2}$ (C) is valid in the long-pulse regime $1/\hat{P}\hat{k}_p^3 \ll |\hat{\zeta}|/\hat{z}$. Note that SMI dominates FRS in the short-pulse (assuming $\hat{k}_\beta\hat{k}_p < 1/2$) and intermediate regimes. FRS dominates SMI in the long-pulse regime, however, here growth is significant only in the tail of long pulses, i.e., $\hat{\zeta} \gg 1/2\hat{k}_p^2\hat{P}$.

As an example, consider parameters relevant to recent experiments on self-modulated laser wakefield acceleration [13]: $\lambda = 1$ μm, $L = 100$ μm (400 fs FWHM), $\lambda_p = 10$ μm ($n_0 \sim 10^{19}$ cm^{-3}), $\Delta_c = 0$, $P \simeq P_c \simeq 2$ TW, and a plasma of length $25Z_R \sim 2$ mm. Near the end of the pulse, $|\zeta| = L$, FRS can occur in the long-pulse regime if $\hat{z} \ll \hat{k}_p^3\hat{P}\hat{\zeta}$ (before transitioning to the intermediate regime at larger z). Letting $\hat{z} = \epsilon^2\hat{k}_p^3\hat{P}\hat{L}$ (with $\epsilon < 1$) gives $N_e \simeq 1.3\epsilon$, i.e., FRS will not undergo significant growth. On the other hand, near the front of the pulse $|\zeta| = L/4$, SMI will reach saturation in the intermediate regime, e.g., $N_e \simeq 12$ after $z = 5Z_R$.

SUMMARY

In summary, a nonlinear theory [14] of finite-radius pulse propagation has been discussed that includes finite pulse length and group velocity effects. In the adiabatic limit, effects such as the nonlinear group velocity, damped betatron oscillations, and self-steepening were analyzed. In the non-adiabatic limit, the nonlinear coupling of FRS and SMI was described and asymptotic growth rates were derived in various regimes. For sub-ps pulses, SMI dominates in typical regimes. The validity of this theory has been restricted to underdense plasmas ($k_p/k \ll 1$) with $z < Z_S$, but these constraints can be relaxed by a straightforward extension of this theory to include the $\gamma_{g0}^{-2}\partial^2/\partial\zeta^2$ term in Eq. (1).

ACKNOWLEDGMENTS

This work was supported by the Department of Energy under contract No. DE-AC-03-76SF0098. The authors acknowledge useful conversations with W.B. Mori.

REFERENCES

1. For a review see, E. Esarey et al., IEEE J. Quantum Electron. **33**, 1879 (1997).
2. H.M. Milchberg et al., Phys. Rev. Lett. **75**, 2494 (1995); A. Rundquist et al., Sci. **280**, 1412 (1998).

3. D.C. Eder et al., Phys. Plasmas **1**, 1744 (1994); B.R. Benware et al., Phys. Rev. Lett. **81**, 5804 (1998).
4. M. Tabak et al., Phys. Plasmas **1**, 1626 (1994); S.C. Wilks and W.L. Kruer, IEEE J. Quantum Electron. **33**, 1954 (1997).
5. For a review see, E. Esarey et al, IEEE Trans. Plasma Sci. **24**, 252 (1996).
6. P. Sprangle et al., Phys. Rev. Lett. **69**, 2200 (1992).
7. S.P. Nikitin et al., Phys. Rev. E **59**, R3839 (1999); D. Kaganovich et al., Phys. Rev. E **59**, R4769 (1999); P. Volfbeyn et al., Phys. Plasmas, **6**, 2269 (1999).
8. K. Krushelnick et al., Phys. Rev. Lett. **78**, 4047 (1997); S.Y. Chen et al., Phys. Rev. Lett. **80**, 2610 (1998); J. Fuchs et al., Phys. Rev. Lett. **80**, 1658 (1998); C.E. Clayton et al., Phys. Rev. Lett. **81**, 100 (1998).
9. T.M. Antonsen and P. Mora, Phys. Fluids B **5**, 1440 (1993); P. Mora and T.M. Antonsen, Phys. Plasmas **4**, 217 (1997).
10. W.B. Mori et al., Phys. Rev. Lett. **72**, 1482 (1994); C.D. Decker et al., Phys. Plasmas **3**, 1360 (1996); W.B. Mori, IEEE J. Quantum Electron. **33**, 1942 (1997).
11. E. Esarey et al., Phys. Rev. Lett. **72**, 2887 (1994); P. Sprangle et al., Phys. Rev. Lett. **73**, 3544 (1994).
12. N.E. Andreev et al., Phys. Plasmas **2**, 2573 (1995); N.E. Andreev et al., IEEE Trans. Plasma Sci. **24**, 363 (1996).
13. R. Wagner et al., Phys. Rev. Lett. **78**, 3125 (1997); C.I. Moore et al., Phys. Rev. Lett. **79**, 3909 (1997); D. Gordon et al., Phys. Rev. Lett. **80**, 2133 (1998).
14. E. Esarey et al., Phys. Rev. Lett. **84**, 3081 (2000).
15. E. Esarey and W.P. Leemans, Phys. Rev. E **59**, 1082 (1999).
16. P. Sprangle et al., Phys. Rev. Lett. **82**, 1173 (1999); Phys. Rev. E **59**, 3614 (1999).

Beam-Quality Simulations for Channel-Guided LWFA

V.V. Goloviznin[1,2], A.J.W. Reitsma[1], L.P.J. Kamp[1], T.J. Schep[2]

[1] *Technische Universiteit Eindhoven,*
P.O. Box 513, 5600 MB Eindhoven, The Netherlands

[2] *FOM Instituut voor Plasmafysica Rijnhuizen,*
P.O. Box 1207, 3430 BE Nieuwegein, The Netherlands

Abstract. The report presents our first results on two-dimensional dynamics of a short electron bunch in a strong plasma wave excited by a channel-guided laser pulse. The problem is studied both analytically and numerically in slab geometry. Collective self-interactions within the bunch are fully taken into account. Similarly to the one-dimensional case, the natural evolution of the bunch is shown to lead, under proper initial conditions, to a minimum in the relative energy spread. The existence of adiabatic invariants of motion is of crucial importance for the final beam quality.

INTRODUCTION

In the last decade, plasma-based methods of particle acceleration have made a spectacular progress. Several successful proof-of-principle experiments [1–4] demonstrated enormous longitudinal fields existing in plasma: up to 100-200 GeV/m, which is three to four orders of magnitude higher than accelerating gradients achievable with conventional linacs. Jets of accelerated electrons with an energy of up to 100 MeV, but also with a large energy spread, have been observed [2–5]. Now worldwide research efforts in the field of plasma acceleration enter a new stage when the demonstration of a table-top electron accelerator with a final particle energy of several hunderd MeV to one GeV becomes possible. At this new stage, final beam quality becomes a matter of major concern. This is what constitutes the subject of our paper: self-consistent bunch dynamics in a channel-guided plasma wave, and, in particular, the energy spread and transverse emittance achievable with the LWFA technique.

The present theoretical study is part of a larger project run at the Technical University of Eindhoven in collaboration with FOM-Institute for Plasma Physics 'Rijnhuizen' in The Netherlands. The project is aimed at the demonstration of fully controllable plasma-based acceleration of an injected bunch of 10-MeV electrons up to an energy of 200-300 MeV. Specifically, it includes the creation of a plasma channel to provide optical guiding of the laser pulse through the acceleration region, the generation of an ultrashort electron bunch with a duration of about 50–100 fs, and its pre-acceleration and precise injection in the laser-excited plasma wave at a correct phase.

As is well known, the energy acquired by an electron in a plasma wave depends on the injection phase and, due to the short plasma wavelength, phase control is a severe problem. Controlled acceleration is possible for short electron bunches, either directly injected from an RF gun [4], or optically injected by means of additional laser pulses [6–8]. To minimize final energy spread, a phasing strategy for the acceleration of such a short bunch in a one-dimensional plasma wave has been developed in [9]. In this paper, we discuss the generalization of this strategy to the case of a two-dimensional setting, including the effects of transverse motion and bunch self-interaction.

PLASMA WAKEFIELDS

As mentioned before, we consider a channel-guided laser wakefield acceleration scheme. The pre-formed plasma channel is assumed to have a stationary density profile that depends only on the transverse coordinate. To be specific, we use the following expression

$$n_0(x) = n_{00}\left(1 - \Delta\left(1 - x^2/W^2\right)e^{-x^2/2W^2}\right), \qquad (1)$$

where n_{00} is the ambient plasma density, x is the transverse coordinate, Δ is the density modulation and W the channel width. This form of plasma channel is close to what has been found in recent experiments [10,11].

The excited wakefield is defined by a set of fully nonlinear hydrodynamic equations [12,13] for the motion of plasma electrons (plasma ions are taken to be immobile). The quasistatic approximation is applied, so that all fields depend on the longitudinal coordinate z and time t only through the combination $\zeta = z - v_\varphi t$, where v_φ is the group velocity of the laser pulse. The corresponding Lorentz factor γ_φ is taken to be large ($\gamma_\varphi \gg 1$), so that in calculating the wakefields v_φ may be approximated with c.

The ponderomotive potential I of the laser pulse is considered to be a given function of x and ζ. This approximation can be justified in the case of a relatively short pulse of weakly-relativistic intensity. We also assume the laser pulse to be perfectly matched to the channel: for the transverse size of the pulse we use the value that gives steady-state propagation without intensity oscillations in the case of a parabolic channel [14].

For convenience, throughout the paper dimensionless variables are used: time $t \to \omega_p t$, coordinates $(x, z) \to k_p(x, z)$, ion (background) density $n_0 \to n_0/n_{00}$, plasma and bunch electron density $n \to n_e/n_{00}$, $n_b \to n_b/n_{00}$, electron momentum $(p_x, p_z) \to (p_x, p_z)/(m_e c)$, velocity $(v_x, v_z) \to (v_x, v_z)/c$, wakefield components

$(E_x, B_y, E_z) \to e(E_x, B_y, E_z)/(m_e \omega_p c)$. Here ω_p and k_p are the electron plasma frequency and the wavenumber associated with the ambient plasma density n_{00}: $\omega_p^2 = 4\pi n_{00} e^2/m_e$, $k_p = \omega_p/c$. To be specific, in all numerical examples throughout the paper, we take $n_{00} = 10^{17}$ cm^{-3} and an optical wavelength of 1 μm, which correponds to a plasma wavelength $\lambda_p = 100\mu$m and $\gamma_\varphi = 100$.

As usual, the (averaged) Lorentz factor of plasma electrons is defined as $\gamma_p = \sqrt{1 + p_x^2 + p_z^2 + 2I}$ and the wakefield potential $\Phi = \gamma_p - p_z$ such that

$$F_x = -E_x + B_y = \frac{\partial \Phi}{\partial x}$$

$$F_z = -E_z = \frac{\partial \Phi}{\partial \zeta},$$

where F_x, F_z denote components of the Lorentz force acting on an ultra-relativistic (bunch) electron. The wakefield equations are combined to

$$\frac{\partial^2 \Phi}{\partial \zeta^2} - \frac{\partial^2 \Phi}{\partial x^2} - \frac{\partial^2}{\partial x \partial \zeta}\left(\frac{1}{\eta}\frac{\partial^2 \Phi}{\partial x \partial \zeta}\right) + \eta \Phi =$$

$$= \eta \gamma_p - \frac{\partial^2 \gamma_p}{\partial x^2} + n_b, \qquad (2)$$

where we introduce the relativistic density $\eta = n/\gamma_p$. Eq. (2) is highly nonlinear through the dependence of γ_p and η on Φ:

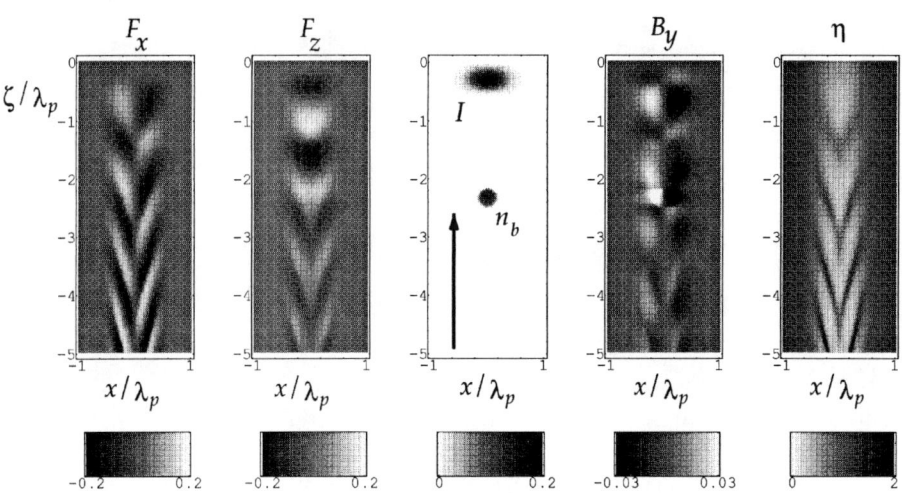

FIGURE 1. Contour plots of F_x, F_z, B_y, η, I and n_b as functions of x and ζ.

$$\eta = \frac{1}{\Phi}\left(n_0 + \frac{\partial^2 \Phi}{\partial x^2}\right) \tag{3}$$

$$\gamma_p = \frac{1}{2\Phi}\left(1 + \Phi^2 + \left(\frac{1}{\eta}\frac{\partial^2 \Phi}{\partial x \partial \zeta}\right)^2 + 2I\right). \tag{4}$$

The above equations are solved numerically under the conditions that the plasma is at rest ahead of the laser pulse ($\Phi = 1$ for $\zeta \geq 0$) and that the fields fall off exponentially at large $|x|$. For illustration, a typical field and density distribution is shown in Fig. 1, which demonstrates contour plots of F_x, F_z, B_y, η and a combined plot of I and n_b. The parameters are: channel width $W = 0.5\ \lambda_p$, channel modulation $\Delta = 0.5$, laser spot size $0.3\ \lambda_p$, laser pulse length $0.16\ \lambda_p$. The (dimensionless) peak electron bunch density is 0.143, and the laser peak power $I_0 = 0.2$. The arrow indicates the direction of propagation.

In Fig. 1 it can be seen that the amplitude of the focusing force increases with the distance behind the pulse and the amplitude of the accelerating force decreases with the distance behind the pulse. The overlap of focusing and accelerating regions behind the laser pulse is clearly visible. These features are in accordance with the results of ref. [15]. Also visible in Fig. 1 is the influence of the electron bunch on the wakefields, caused by *beam loading*. Inside and directly behind the bunch there is enhanced focusing and diminished acceleration. The magnetic field around the bunch is strong, even outside the given range. The cutoff was used in order not to spoil fine details of the field structure behind the bunch. The plot of η exhibits narrow regions with high plasma-electron density, corresponding to sharp gradients in F_x, F_z.

PARTICLE DYNAMICS

Our numerical model combines fluid description of the background plasma with Particle-In-Cell methods for the description of the accelerated bunch. To study bunch dynamics self-consistently, numerical integration of the equations of motion

$$\frac{dx}{dt} = v_x \qquad \frac{dP_x}{dt} = F_x$$

$$\frac{d\zeta}{dt} = v_z - v_\varphi \qquad \frac{dP_z}{dt} = F_z$$

has been performed for a set of macroparticles that model the bunch. The bunch density as a source term is included in the calculation of F_x, F_z, which allows to study the influence of beam loading on the acceleration process.

An important feature of particle dynamics is the existence of adabatic invariants of motion. As is well known, the motion of a single electron in a wakefield is governed by the Hamiltonian

$$H = \gamma - v_\varphi P_z - \Phi, \tag{5}$$

where $\gamma = \sqrt{1 + P_z^2 + P_x^2}$ is the electron's Lorentz factor and $\vec{P} = (P_x, P_z)$ its (dimensionless) momentum. Assuming the shape of the wakefield potential to be close to parabolic in the vicinity of the axis, one can easily estimate typical timescales for longitudinal and transverse oscillations in the small-angle approximation:

$$\tau_\| \simeq \gamma_\varphi^2, \quad \tau_\perp \simeq \gamma^{1/2} |\Phi_{xx}^{(0)}|^{-1/2},$$

where $\Phi_{xx}^{(0)}$ denotes the second derivative of Φ at $x = 0$. Apparently, the transverse oscillations are much faster than the longitudinal ones: the condition $\tau_\perp \ll \tau_\|$ is satisfied for a large part of all trajectories unless the particle's energy is extremely high (of order 100 TeV for $\gamma_\varphi = 100$) or the electron slips too close to a defocusing region (i.e. near a point where $\Phi_{xx}^{(0)} = 0$). Particle evolution in the longitudinal phase space (ζ, P_z) may thus be considered adiabatically slow as compared to its evolution in the transverse phase space (x, P_x).

The existence of adiabatic invariance for the transverse motion means the conservation of the area enclosed in the transverse phase space

$$\oint P_x \, dx = \text{Constant}. \tag{6}$$

From this we conclude that the transverse emittance is conserved insofar as the bunch distribution is *matched* to the phase space orbits of the transverse Hamiltonian [16]. Combining this with the approximate equations of motion results in

$$x_0 \propto \gamma^{-1/4} |\Phi_{xx}^{(0)}|^{-1/4}, \tag{7}$$

where x_0 is the amplitude of the transverse oscillations.

Since during acceleration an electron remains inside the focusing region of the plasma wave, $\Phi_{xx}^{(0)}$ changes only slightly, but the electron's γ may go up from about

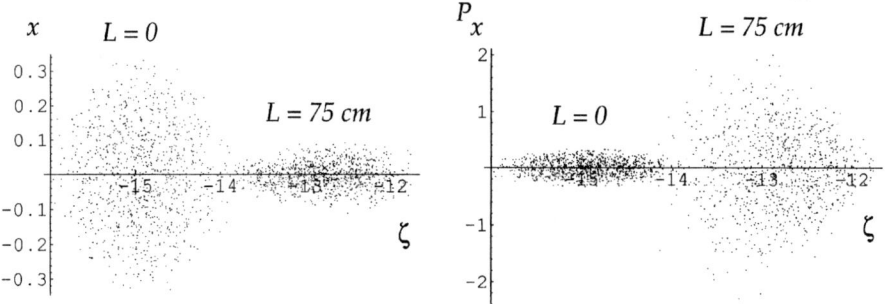

FIGURE 2. Phase space snapshots at the entrance of the plasma channel and at an acceleration distance of 75 cm. Transverse focusing is clearly seen.

10 to about 3000. In that case, Eq. (7) indicates that the amplitude of the transverse oscillations goes down rapidly by a factor of about 4. One must note that the contribution to the energy spread due to the transverse oscillations of bunch electrons is approximately quadratic in x_0. As a result of the transverse focusing, this contribution becomes negligibly small as compared to other sources of the energy spread (see Eqs. (9)-(12) below).

The focusing effect of the accelerated electrons, as a consequence of adiabatic invariance, is clearly observed in our simulations. Fig. 2 demonstrates two snapshots of (ζ, x), (ζ, P_x)-phase spaces that are depicted at the entrance and at the exit of the plasma channel. One can see that the transverse extent of the bunch decreases and a typical transverse momentum of electrons increases with the particle energy, while the normalized emittance remains practically constant.

Note that the effect of beam loading on transverse emittance is relatively unimportant. The bunch wakefields introduce extra focusing in the rear part of the bunch, which leads to a change in $\Phi_{xx}^{(0)}$. The bunch adjusts itself to the potential on the short timescale $\tau_\perp$, whereas the evolution of bunch wakefields takes place on the long timescale $\tau_\parallel$ of longitudinal motion. Consequently, the adiabatic invariance is not influenced by collective effects and there is no additional emittance growth. By a similar reasoning one finds that effects of finite bunch length on transverse emittance are unimportant.

Now let us discuss the effect of finite bunch length on the energy spread. Since $\gamma \gg \gamma_\varphi$ holds for a large part of the acceleration, one may approximate the energy gain ΔP_z for a particle injected at ζ_{in} and extracted at ζ_{ex} with:

$$\Delta P_z = 2\gamma_\varphi^2 \left(\Phi^{(0)}(\zeta_{ex}) - \Phi^{(0)}(\zeta_{in}) \right) \qquad (8)$$

For a short bunch injected with typical phase spread $\delta\zeta \ll \lambda_p$ around the injection

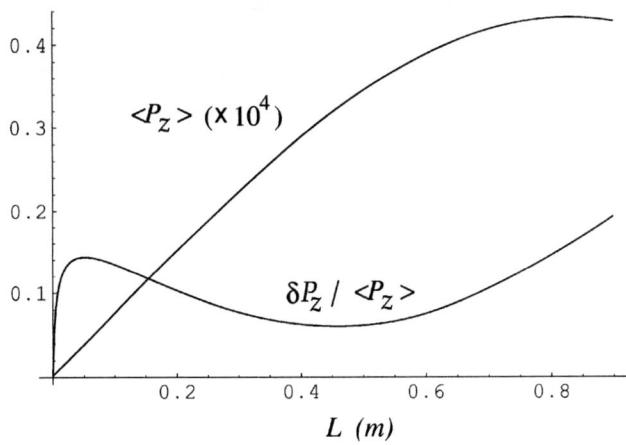

FIGURE 3. Bunch energy $<P_z>$ and the relative energy spread $\delta P_z / <P_z>$ as functions of the acceleration distance L.

phase, the cumulative energy spread $\delta P_z^{(1)}$ can be estimated from eq. (8) as

$$\delta P_z^{(1)} = 2\gamma_\varphi^2 \, \delta\zeta \, \left(\Phi_\zeta^{(0)}(\zeta_{ex}) - \Phi_\zeta^{(0)}(\zeta_{in})\right) \tag{9}$$

This equation suggests that there are two strategies to minimize energy spread: either to use a very short bunch to minimize $\delta\zeta$, or to arrange injection and extraction phases such that

$$\Phi_\zeta^{(0)}(\zeta_{ex}) - \Phi_\zeta^{(0)}(\zeta_{in}) = 0 \tag{10}$$

Note that this can also be written as

$$(\zeta_{ex} - \zeta_{in}) \, \Phi_{\zeta\zeta}^{(0)}(\zeta_M) = 0 \tag{11}$$

where ζ_M denotes a certain point in the interval $[\zeta_{in}, \zeta_{ex}]$. Therefore, this second strategy requires that the accelerating gradient has its maximum somewhere between ζ_{in} and ζ_{ex}. At the same time, the transverse stability requires that the particles remain inside the focusing region. For two-dimensional wakefields in a homogeneous plasma, the above conditions cannot be satisfied simultaneously, since typically the accelerating gradient has its maximum at the edge of the focusing region. However, the overlap of focusing and accelerating regions provides such an opportunity in the case of a plasma channel, which makes it possible to apply the second optimizing strategy.

The effect of beam loading on energy spread is determined by the modification of the accelerating force inside the bunch due its own wakefields. The cumulative energy difference $\delta P_z^{(2)}$ due to beam loading can be estimated as

$$\delta P_z^{(2)} = 2\chi \gamma_\varphi^2 \left(\zeta_{ex} - \zeta_{in}\right) \tag{12}$$

where χ denotes the beam loading efficiency (which is approximately linear in the total charge of the bunch). The condition for minimum energy spread is that the terms in Eq. (9) and Eq. (12) cancel each other:

$$\delta\zeta \left(\Phi_\zeta^{(0)}(\zeta_{ex}) - \Phi_\zeta^{(0)}(\zeta_{in})\right) + \chi(\zeta_{ex} - \zeta_{in}) = 0 \tag{13}$$

This condition can be satisfied if a point ζ_M in the interval $[\zeta_{in}, \zeta_{ex}]$ exists such that

$$\Phi_{\zeta\zeta}^{(0)}(\zeta_M) = -\frac{\chi}{\delta\zeta} \tag{14}$$

The above condition coincides with the one derived for one-dimensional acceleration [9,17], but it turns out to be applicable in the two-dimensional case as well, because the contribution due to the transverse oscillations is suppressed by the focusing effect.

As an example, Fig. 3 shows the bunch energy $< P_z >$ and relative energy spread $\delta P_z / < P_z >$ as functions of the acceleration length L. A minimum in the energy spread of about 6.5 % is seen to occur after accelerating over a length $L \simeq 40$ cm. At this point, $< P_z >$ is 2750 (1.4 GeV).

CONCLUSIONS

We have thus considered the dynamics of an accelerated electron bunch in a channel-guided laser wake-field accelerator, with particular attention to beam quality. The influence of transverse bunch dynamics on its normalized emittance and energy spread was studied with analytical and numerical methods. An important feature of the bunch dynamics is the large difference in timescales of the longitudinal and transverse motion, which leads to the existence of an adiabatic invariant for the transverse motion. This adiabaticity results in conservation of transverse emittance for matched beams, even with beam loading effects taken into account.

As for the relative energy spread, we have shown that the strategy of optimal phasing developed earlier for one-dimensional acceleration can be successfully applied to two-dimensional settings as well. The main reason for this is the transverse focusing of bunch electrons, which strongly reduces the influence of transverse oscillations on the final energy spread. Under proper initial conditions we find that the two main contributions to the energy spread, namely due to finite bunch length and due to beam loading, may cancel each other, resulting in a minimized relative energy spread.

REFERENCES

1. Clayton, C.E. et.al., *Phys. Rev. Lett.* **70** 37 (1993).
2. Modena, A. et.al., *Nature* **377**, 606 (1995).
3. Nakajima, K., *Phys. Rev. Lett* **74**, 4428 (1995).
4. Dorchies, F. et. al., *Phys. Plasmas* **6**, 2903 (1999).
5. Chen, S.-Y. et.al., *Phys. Plasmas* **6**, 4739 (1999).
6. Umstadter, D., Kim, J.K., and Dodd, E., *Phys. Rev. Lett.* **76**, 2073 (1996).
7. Esarey, E. et.al., Phys. Rev. Lett. **97**, 2682 (1997).
8. Esarey, E., Schroeder, C.B., Leemans, W.P., and Hafizi, B., *Phys. Plasmas* **6**, 2262 (1999).
9. Chiou, T.C., and Katsouleas, T., *Phys. Rev. Lett.* **81**, 3411 (1998).
10. Clark, T.R., and Milchberg, H.M., *Phys. Rev. E* **61**, 1954 (2000).
11. De Wispelaere, E. et.al., *Phys. Rev. E* **59**, 7110 (1999).
12. Khachatryan, A.G., *Phys. Rev. E* **60**, 6210 (1999).
13. Lotov, K.V., *Phys. Plasmas* **5**, 785 (1998).
14. Esarey, E., Krall, J., and Sprangle, P., *Phys. Rev. Lett.* **72**, 2887 (1994).
15. Andreev, N.E. et.al., *Physics of Plasmas* **4**, 1145 (1997).
16. Katsouleas, T. et.al., in *Nonlinear and collective phenomena in beam physics, Arcidosso 1996*, AIP Conf. Proc. **395**, p. 75, ed. S. Chattopadhyay, M. Cornacchia, C. Pellegrini.
17. A. Reitsma, R. Trines, V. Goloviznin, to be published in *IEEE Trans. Plasma Sci.* **26-2** (2000).

Resonant and Hollow Beam Generation of Plasma Channels

I. Alexeev, K.Y. Kim, J. Fan, E. Parra, and H.M. Milchberg

Institute for Physical Science and Technology
University of Maryland
College Park, MD 20742

L.Ya. Margolin and L.N. Pyatnitskii

Institute for High Temperatures
Russian Academy of Sciences, Izhorskaya ul. 13/19, 127412 Moscow

Abstract. We report two variations on plasma channel generation using the propagation of intense Bessel beams. In the first experiment, the propagation of a high intensity Bessel beam in neutral gas is observed to give rise to resonantly enhanced plasma channel generation, resulting from resonant self-trapping of the beam and enhanced laser-plasma heating. In the second experiment, a high power, hollow Bessel beam (J_5) is produced and the optical breakdown of a gas target and the generation of a tubular plasma channel with such a beam is realized for the first time. Hydrodynamic simulations of the laser-plasma interaction of are in good agreement with the results of both experiments.

I. RESONANT GENERATION OF PLASMA CHANNELS USING BESSEL BEAMS

Over the past few years, we have pioneered the generation of laser-produced plasma waveguides using the elongated focus of Bessel beams [1]. Here we demonstrate a new type of resonant self-trapping, correlated with enhanced absorption, which occurs in the propagation of high intensity Bessel beams. This leads to the resonant generation of plasma channels. We consider a *generalized* Bessel beam $E(\vec{r}_\perp, z, \omega) = e^{i\beta z} u(\vec{r}_\perp, \omega)$ in an isotropic medium invariant along the optical axis z to be a solution to the wave equation

$$\nabla_\perp^2 u + \kappa^2 u = 0 \ , \qquad (1)$$

where $\kappa^2(\vec{r}_\perp, \omega) = k^2(1 - \beta^2/k^2 + \delta_{plasma}(\vec{r}_\perp, \omega) + 4\pi\chi(\vec{r}_\perp, \omega))$ is the square of the local transverse wave number accounting for transverse variation of the dielectric medium which may include plasma contributions δ_{plasma} , $\nabla_\perp^2$ is the Laplacian in the transverse coordinate $\vec{r}_\perp$, $k = \omega/c$ is the vacuum wavenumber of the laser, β is the wavenumber along the propagation axis, and χ is the total atomic and ionic susceptibility. For a uniform, non-absorbing medium, the transverse wavenumber $\kappa = \kappa_0$ is a real constant, and Eq. (1) yields the standard Bessel beam solution $E(r,\phi,z,\omega) = E_i e^{i\beta z} J_m(\kappa_0 r) e^{im\phi}$, where E_i is the peak electric field, and m is an azimuthal index. Such uniform medium Bessel beam solutions have been considered in previous work [2,3].

Bessel beams of zero order ($m=0$) have been produced using phase masks [2], or using special conical lenses called axicons [1,4]. More recently, $m>0$ beams have been produced using a phase mask and an axicon in combination [5]. This will be discussed in more detail in part II of this paper. Using moderate pulse energies and pulse widths, the peak intensity of a Bessel beam can be sufficient for breakdown of neutral gases, which is of great interest for generating elongated plasma channels. Such channels have been used for optical guiding [1], and have applications to laser-driven plasma accelerators [6], short wavelength generation [7], and high current, high speed switching [4]. Plasma channels produced by Bessel beam pulses shorter than a few hundred picoseconds and at gas pressures below ~1 atm have negligible axial structure and are cylindrically symmetric [8].

To our knowledge, the self-consistent Bessel beam-plasma interaction has never been considered. Recent pump-probe experiments, however, show that weak, 1 ps Bessel beam probe pulses focused onto preformed plasma channels produced by 100 ps pump pulses in low pressure gases (<150 torr) can be resonantly trapped and quasi-guided after a few hundred picoseconds of channel evolution, when a local electron density minimum appears on axis as a result of the plasma's radial expansion [9]. This suggests that at higher initial gas pressures, increased laser- plasma heating and more rapid formation of such a channel structure could result in trapping of the 100 ps heating pulse itself.

Insight into the dynamical interaction of a Bessel beam with an evolving plasma can be obtained by examining the solution to Eq. (1) for a stationary plasma. Well outside the plasma boundary ($r>>r_b$), the solution gives $E(\vec{r}_\perp, z, \omega) \approx E_i e^{i\beta z}[e^{-i\kappa_0 r} + \eta e^{-i(m+1/2)\pi} e^{i\kappa_0 r}]e^{im\varphi} / (2\pi\kappa_0 r)^{1/2}$, a sum of conical waves incident upon and scattered by the plasma at the angle $\gamma = \tan^{-1} \kappa_0/\beta$ with respect to the plasma axis, where $\kappa_0 = (k^2 - \beta^2 + 4\pi\omega^2 \chi_0/c^2)^{1/2}$, $\chi_0 = \chi(r>r_b)$, and η is the complex scattering coefficient of the outgoing wave, which depends on the specific plasma structure. In the stationary limit, the fractional plasma absorption of the Bessel beam is given by $1-|\eta|^2$. We note that the transit time for light across a plasma channel of maximum diameter ~100 μm [8] is < 1 ps, which is much shorter than either the 100 ps laser pulse used here or the plasma hydrodynamics time scales. Indeed, measurements have confirmed that these time scales are ~50 ps for avalanche-driven electron density growth, ~100ps for shock development, and ~1 ns for radial evolution of the plasma column [8]. Therefore, in our numerical simulations of the time-dependent interaction (see below) we solve Eq. (1) for an evolving sequence of plasma profiles; our simulations are quasi-stationary.

Figure 1 shows the experimental setup. Pulses (100ps, 1064nm, 350 mJ) from a Nd:YAG laser system were focused by an axicon with base angle 25° (J_0 beam with ray approach angle $\gamma=15°$, where $\beta=k\cos\gamma$), making a 1.5 cm long plasma column in an ambient gas of 20 torr N_2O and variable pressure argon. The N_2O component, which field ionizes at ~10^{13} W/cm^2, provides seed electrons for the uniform avalanche breakdown of Ar [8]. The relative self- trapping of the Bessel beam by its self-generated plasma was measured by integrating CCD camera images of the mode intensity profile at the exit plane of the plasma channel. For the measurement of plasma absorption, all axicon rays passing through the plasma (including the trapped portion of the beam) were collected and focused by a second axicon through a 1064 nm interference filter onto an energy meter. The channel's hydrodynamic evolution was measured by interferometry using a variably delayed probe beam (70 ps, 532 nm, ~100 µJ) which was directed transversely through the plasma [8].

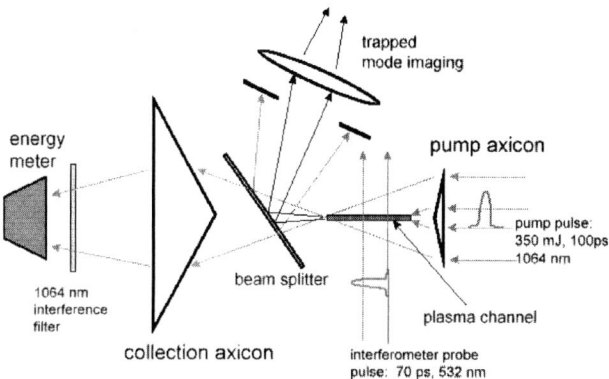

FIGURE 1. Experimental setup to measure Bessel beam trapping (via mode imaging), absorption (with collection axicon and energy meter) and plasma heating (from plasma expansion speed determined by time-resolved interferometry).

The self-trapping of the Bessel beam vs. gas fill pressure is plotted in Fig. 2. As the pressure increases, strongly increased coupling occurs to the $m=0$, $p=0$ lowest order mode (p is the radial mode index) with maximum coupling at ~300 torr, with the mode at that pressure shown in the inset. Just beyond the peak, the coupling drops sharply, consistent with a coupling resonance at 300 torr. Beyond 400 torr, the coupling increases again, now to $m>0$ modes. We note that even though the generated channel is cylindrically symmetric [8], slight azimuthal variation in the input beam is sufficient for such $m>0$ coupling [9]. At ~500 torr, the $m=1$ mode is optimally coupled (shown in inset), followed by the onset of closely spaced $m>1$ modes at even higher pressure, with fluctuations appearing in m from shot-to-shot channel variations. Figure 3 shows absorption of the pulse as a function of pressure. Note that the non-trapped or transmitted part of the beam collected by the second axicon was a well-defined ring; no off-forward scattering was observed from the plasma, which is axially locally uniform. The pressure dependence of the absorption corresponds well to the coupling resonances of Fig. 2.

Direct evidence of the enhanced plasma heating from these coupling resonances is obtained from measuring the channel shock expansion velocity. As the expansion is adiabatic after the first few hundred picoseconds [8], the square of the channel expansion velocity at later times is a relative measure of the initial heating by the laser pulse. The channel radius R_s, taken as the position of peak electron density in the expanding shock, was measured as a function of time by interferometry and the velocity was obtained as the time derivative of the best fit curves for the radius, which to excellent accuracy follow cylindrical blast wave behaviour $R_s \sim (\varepsilon/\rho_0)^{1/4} t^{1/2}$ [8,10,11], where ε is the energy per unit length available to drive the expansion and ρ_0 is the initial mass density.

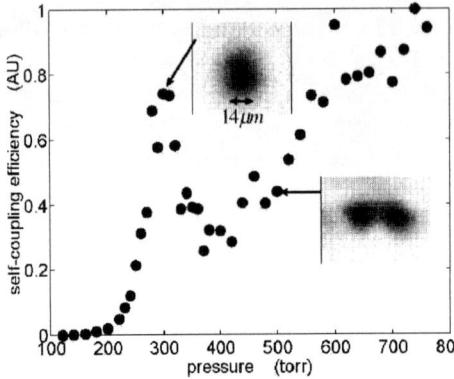

FIGURE 2. Self-coupled energy measured at channel exit versus pressure (20 torr N$_2$O, variable pressure Ar, laser energy 350 mJ, γ=15°). Each point is a 10 shot average. Trapped mode images for m=0 (14μm FWHM) and m=1 are also shown.

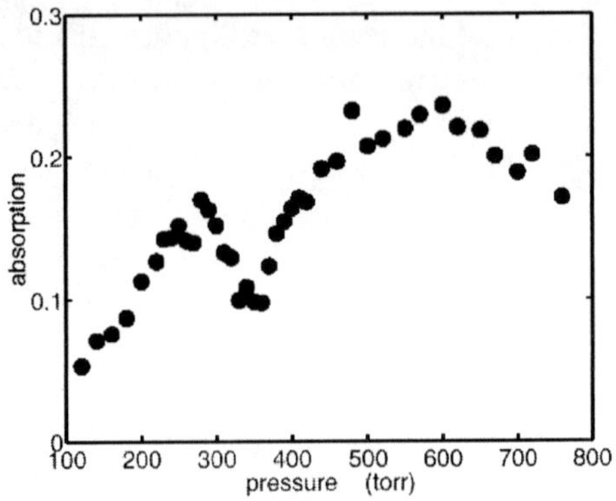

FIGURE 3. Absorption versus pressure (for same conditions as in Fig. 2). Each point is a 500 shot average.

Figure 4(a) is a plot of $v_s^2 = (dR_s/dt)^2$ versus pressure at t=1.6 ns after the pump, and a clear peak is seen at ~300 torr, coincident with the m=0 peaks in both the relative trapping and in the absorption, direct evidence for enhanced heating at a trapping resonance. Just past 300 torr there is a dip, followed by an increase in heating at higher pressure where coupling occurs to m>0 modes. Figure 4(b) is a plot of $v_s^4 P$ versus pressure (P), where $v_s^4 P$ is proportional to ε from the blast wave expression above. The similarity to Fig. 3 is striking, showing the direct impact of the coupling resonances on the plasma hydrodynamics.

We have simulated the Bessel beam-plasma interaction using an axially and cylindrically symmetric, quasi-stationary model in which Eq. (1) is coupled to a 1D Lagrangian hydrocode. Starting with neutral gas, at each time step Eq. (1) is solved using the density profile of the previous time step, yielding the complex scattering coefficient $\eta(t)$ as a function of time. The updated electric field then heats and ionizes the plasma, advancing the hydrodynamics. We assume a Drude model for the plasma, for which $\delta_{plasma}(r,\omega) = -\xi + i\xi \nu/\omega$, where $\xi=N_e(r)/N_{cr} (1+\nu^2/\omega^2)^{-1}$, $N_{cr} = m\omega^2/4\pi e^2$ is the critical density, and $\nu=\nu(r)$ is the collision frequency which includes electron-ion and electron-neutral collisions. The calculation includes field ionization [12], collisional ionization, gradient-based and flux-limited thermal conduction, and a collisional-radiative ionization package.

For a peak vacuum focus intensity of 5x10^{13} W/cm^2, a 100 ps FWHM gaussian pulse envelope, λ=1064 nm, γ=15°, and m=0 (corresponding to our experimental parameters), Fig. 5(a) shows the time-dependent transmission $|\eta(t)|^2$ calculated for initial pressures of 250, 300, and 350 torr of argon. Consistent with our coupling and absorption measurements of Figs. 2 and 3, optimal coupling takes place at 300 torr, where the strong dips in $|\eta(t)|^2$ result from dynamic resonant coupling during the pulse. The dips correspond to successive 'attempts' at self-trapping to the lowest order quasi-bound mode. At the first dip, the density profile supports a lowest order quasi-bound mode whose central spot radius extends beyond the radial electron density peak at $r = r_s$, outside of which the solution is oscillatory (~15% of peak amplitude). At the second dip, the mode is far more tightly bound, with the central spot well within r_s with very small amplitude oscillation (~1% of peak) beyond this radius. In between the dips, the field is more excluded from the channel centre since the depth and wall thickness have increased without a sufficiently large increase in r_s. For 300 torr, Fig. 5(b) shows the electron temperature $T_e(t)$ at the channel centre, and the relative incident and absorbed powers, $P_{inc} = E_i^2(t)$ and $P_{abs} = (1-|\eta(t)|^2)E_i^2(t)$, illustrating the effect of the resonant heating. For these conditions, the resonances occur past the midpoint of the pulse, after a channel confining structure has formed. Figure 5(c) shows a time sequence of laser electric field profiles in the channel near the time of the second dip in $|\eta(t)|^2$, showing the dynamic trapping which occurs during the heating pulse. The enhanced heating is seen to be associated with the enhanced amplitude of the trapped field. An intensity weighted average of calculated intensity profiles over the pulse history agrees well with the $m=0$ mode size in Fig. 2.

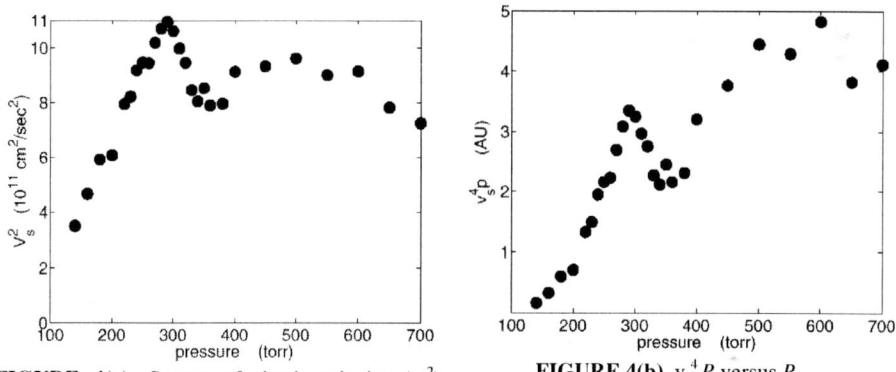

FIGURE 4(a). Square of shock velocity (v_s^2) versus pressure (P), for the conditions of Fig. 1.

FIGURE 4(b). $v_s^4 P$ versus P.

A necessary but not sufficient requirement for resonant self-trapping of a Bessel beam is that $\kappa^2(r_s)<0$, ensuring some exponential damping in the channel wall. This is equivalent to $N_{es} > N_{cr}\sin^2\gamma = N_{cr}^{eff}$, where $N_{es}=N_e(r_s)$. This criterion can be used to roughly estimate the minimum neutral density required for self-trapping as $N_0 \approx N_{cr}\sin^2\gamma/Z$, where Z is the average ionization level. Using $\lambda=1064$nm ($N_{cr}\sim10^{21}$cm^{-3}), $\gamma=15°$, and argon average ionization of $Z\sim8$ [8] gives a minimum equivalent pressure of ~240 torr, in reasonable agreement with our measurements and simulations. However, even if $\kappa^2(r_s)<0$, but the combination of λ and γ do not correspond to a channel transverse quasi-resonance [9], much of the wave will be reflected from the outside of the channel. In Fig. 5(a), the regions of the 300 torr $|\eta(t)|^2$ curve between the two dips and after the second dip correspond to this situation.

We note that in using shallow axicons with small γ in order to make long channels, the electron density can easily exceed N_{cr}^{eff} very early in the pulse. In those cases, *direct* laser heating at the channel centre can only proceed through dynamical coupling to channel resonances. Beams with some azimuthal variation will assist in this process since the $m \neq 0$ resonances are much more closely spaced than the $m=0$ resonances for varying p [9]. This is borne out by the second broad peaks of Figs. 2 and 3, which are for resonant coupling to closely spaced $m>1$ modes.

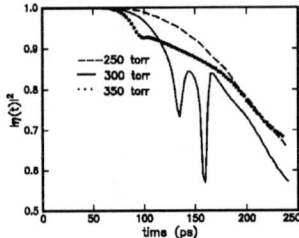 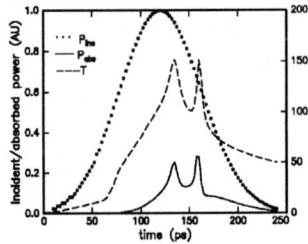 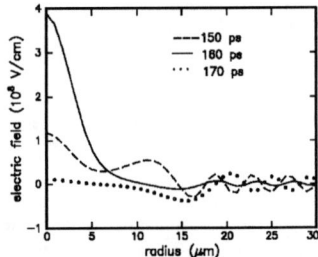

FIGURE 5(a). Calculation of time-dependent transmission $|\eta(t)|^2$ during the laser pulse for argon initial pressures of 250 torr, 300 torr, and 350 torr.

FIGURE 5(b). Relative incident and absorbed power (P_{inc} and P_{abs}), and channel-centre electron temperature (T_e) versus time for the 300 torr case.

FIGURE 5(c). Electric field profiles just before (150 ps), during (160 ps), and after (170 ps) the second dip in $|\eta(t)|^2$, for the 300 torr case.

We have demonstrated a new type of resonant self-trapping and enhanced heating in plasmas which is an integral feature of intense Bessel beam propagation, and which makes possible the more efficient generation of plasma channels. For much shorter and more intense Bessel beam pulses than those considered here, the plasma hydrodynamics may be largely driven by ponderomotive rather than thermal forces. For cases where the quasi-static analysis is appropriate, similar criteria for self-trapping will apply.

II. GENERATION OF PLASMA CHANNELS USING A HOLLOW BESSEL BEAM

In this experiment, we use an intense hollow laser beam to break down a gas to form a tubular plasma, with maximum electron density initially located radially away from the optical axis. In our previous work, laser-produced plasma waveguides have been generated using zero order Bessel beams (J_0) with an intensity maximum on axis, which produces a plasma with an initial electron density maximum on axis. Subsequent shock wave generation at the radially expanding plasma periphery results in a hollow density profile [8]. The time scale for the initial generation of the shock wave is ~100ps, consistent with the time over which ion-ion collisions can cause a radial mass density buildup in a sub-atmospheric density gas target. A hollow density profile sufficiently deep for strongly bound optical guiding develops several nanoseconds after the laser pulse, as the shock propagates radially outward [1,9].

Here, the plasma is generated with a 5th order Bessel beam (J_5), which has an intensity maximum radially off-axis. We measure prompt generation of electron density in the high intensity locations of the beam, at times before significant ion mass motion can occur. This makes possible the generation of plasma waveguides with much smaller effective core diameters than those produced through shock expansion. The results shown here constitute the smallest core diameter plasma waveguides produced to date by this method. Earlier work by some of us using J_5 beams produced with 5 ns pulses resulted in plasmas in which it was difficult to observe hollow electron density profiles [13]. For such long pump pulses, plasma hydrodynamic evolution ensures that the on-axis density depression will not survive beyond the earliest times in the pulse.

There are two main methods for generating hollow Bessel beams [14, 15]. The most straightforward method uses a phase screw followed by an axicon. The phase screw is an optical plate whose thickness is constant with radius but increases linearly with azimuthal angle φ about the optical axis. A planar phase front passing through a phase screw picks up an azimuthally dependent phase Φ $(\varphi)=k(n_p-1)(d/2\pi)\varphi =s\varphi$, where k is the wavenumber in the medium outside the phase screw, d is the maximum thickness of the screw, n_p is the refractive index of the screw plate material, and s is called the phase screw parameter. If $s=m$, where m is a positive integer, then this parameter plays the role of an azimuthal mode index. An axicon is a cone shaped lens, which converts a plane wave into a conical beam, where the incident beam's rays are redirected toward the optical axis at an angle γ and an extended focus is formed [1,4]. Following the phase screw plate by an axicon adds a radial phase shift, so that the total transverse phase factor imposed on the beam immediately after these elements is $\exp(i\Phi)$, where $\Phi(\rho,\varphi)= -k_{\perp,\text{axicon}}\rho+s\varphi$, where ρ is the radial coordinate with respect to optical axis, $k_{\perp,\text{axicon}}= k\sin\gamma$, $\gamma = \sin^{-1}(n\sin\alpha)-\alpha$ is the angle between the focused axicon rays and the optical axis, α is the axicon cone base angle, and n is the refractive index of the axicon material. A hollow Bessel beam of order m (J_m), for $s=m$, results from propagating this beam to the focal region of the axicon, as will be discussed below.

In the second method, a discrete approximation of the transverse phase Φ, combining the functions of both the phase screw and axicon, can be attained by using a transparent plate with an etched surface microrelief [15]. For use with high power laser pulses, the discrete phase plate was easier to fabricate than the continuous phase screw, so this method was used in the experiments described here. Here, the plate consisted of N×N radial and azimuthal segments ($i=1,...,N$ and $j=1,...N$, where N=3000). For $2\pi n < -k_{\perp,\text{plate}}\rho_{ij}+m\varphi_{ij} <2\pi n+\pi$, the depth of the microrelief features was set to give $\Phi_{ij}=0$, and for $2\pi n+\pi < -k_{\perp,\text{plate}}\rho_{ij}+m\varphi_{ij}<2\pi n+2\pi$, it was set to give $\Phi_{ij}=\pi$ (where n is a positive or negative integer). Here $k_{\perp,\text{plate}}=k\sin\gamma_{\text{plate}}$, where γ_{plate} is the angle of the rays with respect to the optical axis focused by the discrete phase plate. The hollow Bessel beam was formed by the phase plate (with $m=5$, and $\gamma_{\text{plate}}=1°$) closely followed by an axicon (with $\alpha=30°$), giving $\gamma=19°$. Since $k_{\perp,\text{plate}}/k_{\perp,\text{axicon}} <<1$, the phase plate acted mainly as an azimuthal phase screw, with most of the radial phase shift provided by the axicon.

Consider a linearly polarized laser beam $E_i(\rho)\hat{x}$ normally incident on an $m=5$ phase plate followed by an axicon, where $\hat{x}$ is the polarization unit vector, ρ is the radial coordinate in the incident beam, and the unit vector $\hat{z}$ is along the optical axis. Applying Kirchhoff's integral to the source amplitude and phase distribution just beyond the axicon, $E_i(\rho)\exp(i\Phi)\hat{x}$, and using the stationary phase approximation gives the following squared magnitudes of the field components in the paraxial region ($r<z\sin\gamma$, $kr^2<z$) of the focus:

$$|E_x(r,z)|^2 = \frac{\pi}{32} \frac{kz\sin^2\gamma}{\cos\gamma - \sin\gamma\tan\alpha}(1+\frac{1}{\cos(\alpha+\gamma)})^2 E_i^2(\rho)\left\{(1+\cos\gamma)J_5(x)+\frac{1}{2}(1-\cos\gamma)(J_7(x)+J_3(x))\right\}^2$$

$$|E_z(r,z)|^2 = \frac{\pi}{32} \frac{kz\sin^4\gamma}{\cos\gamma - \sin\gamma\tan\alpha}(1+\frac{1}{\cos(\alpha+\gamma)})^2 E_i^2(\rho)(J_4(x)-J_6(x))^2$$

Here r is the radius in the focal plane, and $J_m(x)$ is the mth order Bessel function, where $x=kr\sin\gamma$. If $z=0$ is located at the vertex of the axicon, then radial locations ρ of the incident beam are directed to axial locations z given by $\rho = z\tan\gamma /(1-\tan\alpha\tan\gamma)$. For $r=0$, the intensity is zero so that a hollow beam is formed along the optical axis over a maximum length $z_0=\rho_0(1-\tan\alpha\tan\gamma)/\tan\gamma$, where ρ_0 is the incident beam radius. For $\sin^2\gamma <<1$ and $(1-\cos\gamma)<<1$, as is the case here, $|E_z(r,z)| \sim 0$ and the focal intensity distribution is approximately

$$I(r,z) = \frac{\pi}{32} \frac{kz\sin^2\gamma}{\cos\gamma - \sin\gamma\tan\alpha}(1+\frac{1}{\cos(\alpha+\gamma)})^2 I_0(\rho)\left[(1+\cos\gamma)J_5(x)\right]^2$$

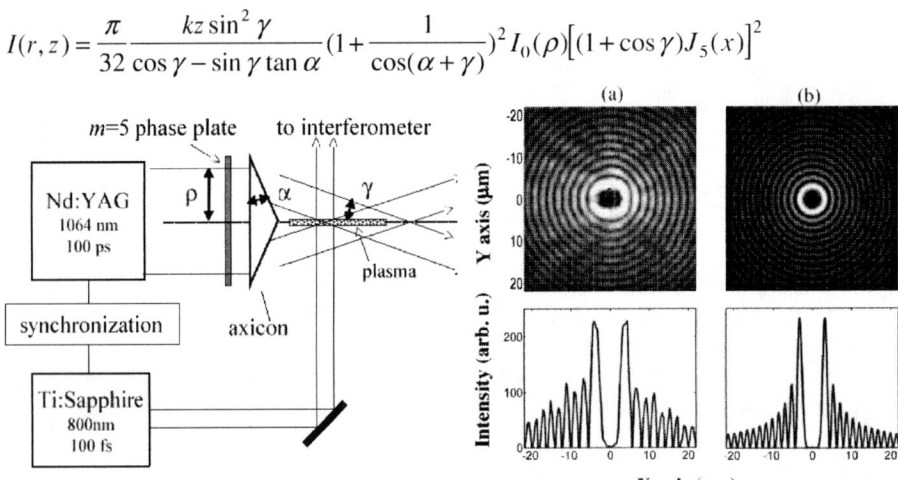

FIGURE 6. Experimental arrangement, showing synchronization of Nd:YAG and Ti:Sapphire laser systems, hollow beam optics, and interferometer setup.

FIGURE 7(a). Image and lineout of focal region of $m=5$ phase plate/ axicon combination for $\lambda=1064$nm; **(b).** Theoretical plot and lineout for J_5^2 for $\gamma=19°$ and $\lambda=1064$nm.

where $I_0(\rho)$ is the radial intensity distribution of incident laser beam. The first maximum of $J_5(x)$ occurs at $x \approx 6.43$, giving $r_0=6.43/k\sin\gamma$ as the radius of the hollow beam. For $\lambda=1064$nm and $\gamma=19°$, $r_0 \sim 3.3\mu m$.

The experimental setup is shown in Figure 6. Pulses (500 mJ, 100ps, 1064nm) from a 10 Hz Nd:YAG laser system [16] were directed to the $m=5$ phase plate and axicon combination to generate a plasma in an ambient gas of 700 torr argon. A focal plane image of the hollow beam and a centered lineout is shown in Fig. 7(a), taken at low energy using a 60X microscope objective. The measured radius for the first maximum is ~4μm, somewhat larger than for the $|J_5|^2$ theoretical result, for which an image and lineout is shown in Fig. 7(b) for $\lambda=1064$ nm and $\gamma=19°$. There was little distortion of the focal image when translating the microscope ~1 cm along the optical axis, and this was used as a test of the optical alignment accuracy. The peak intensity in the first ring was ~ 2.5×10^{12} W/cm^2, more than an order of magnitude weaker than the peak intensity from the axicon without the phase plate. This results from the large area of the first ring compared to the central maximum of a J_0 beam.

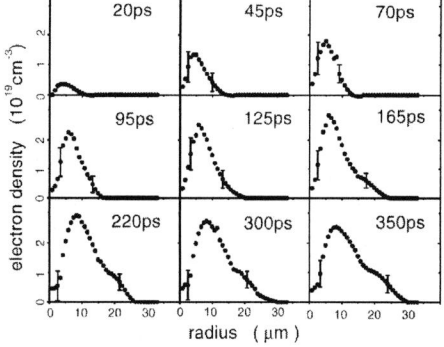

FIGURE 8. Electron density profiles determined from interferometry. Pump pulse $\lambda=1064$nm, 100ps, peak intensity 2.5×10^{12}W/cm^2. The time $t=0$ corresponds to a peak electron density of 5×10^{17} cm^{-3}, representing the minimum detectable phase shift.

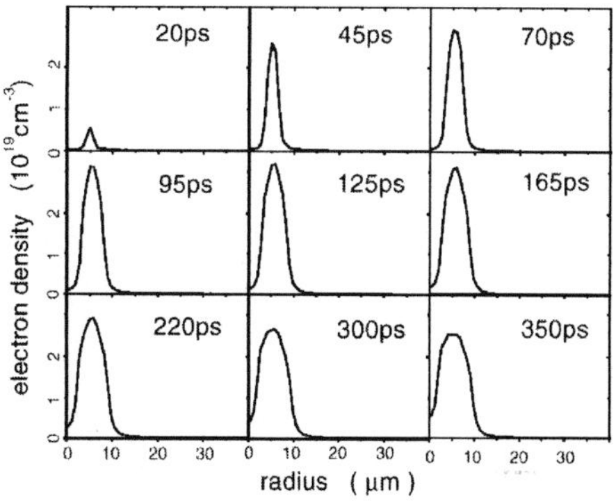

FIGURE 9. Electron density profiles from simulation. The time $t=0$ is set from the appearance of a peak electron density of 5×10^{17} cm^{-3}.

The plasma generated by the hollow beam was ~0.8cm long and uniform along the axis, with some tapering over ~ 1 mm at the ends. This was determined by scanning the interferometer viewing field along the full length of the plasma. To measure the time evolution of the plasma, a 100 fs, 800 nm, 1 mJ interferometer probe pulse was obtained from a separate Ti:Sapphire laser synchronized to the Nd:YAG system with less than 10 ps of jitter. This synchronization is described in reference [17]. The probe pulse passed through an optical delay line (−200 ps to +500 ps with respect to the peak of the 1064nm pump pulse) and was directed transversely across the plasma into an imaging Michelson interferometer. The interferogram phase was extracted using a fast fourier transform technique and the electron density was determined through Abel inversion [8]. Electron density profiles from the central region of the plasma column for a number of delays are shown in Fig. 8. The zero time reference was assigned by the first appearance of a phase disturbance in the interferogram, at an electron density threshold of ~5×10^{17} cm^{-3}. Each profile was obtained from averaging the extracted interferogram phase for ~200μm along the optical axis, which was the extent of the field of view. The vertical bars represent the statistical standard deviation of profiles derived from phase lineouts from individual columns of CCD camera pixels. It is seen that a hollow electron density profile clearly appears by 20ps. The off-axis peak electron density continues to rise until it saturates by 170ps at a level of ~3×10^{19} cm^{-3}, at which point the pump pulse is well past its peak. From saturation out to 350ps, the central part of the profile changes little except for a slight increase in the central density and a slight broadening at the radial periphery. Although we have not yet attempted to inject pulses into this waveguide, we calculate numerically [9] that the fundamental modes (which are wavelength independent [7]) are strongly bound with a nearly constant intensity FWHM of 3.1μm for the electron density profiles between 95ps and 350ps. Compared to plasma waveguides generated with a J_0 Bessel beam, where the earliest time for very leaky quasi-bound guiding is ~120ps after the channel peak density is generated (see part I above and ref. [9]), the waveguide generated with a J_5 Bessel beam can guide strongly bound modes at earlier times and at smaller mode diameters.

The hollow Bessel beam-plasma interaction was simulated self-consistently as in part I above by coupling the radial wave equation for the hollow beam electric field, $d^2E/dr^2 + (1/r)dE/dr + (\kappa^2(r) - m^2/r^2)E = 0$, to the 1D Lagrangian hydrocode, using the same model for the plasma refractive index. (The theoretical profile of Fig. 7(b) was calculated by solving the radial wave equation for the vacuum case for $\gamma=19°$). Figure 9 shows the electron density evolution calculated for the interaction with 700 torr of argon of a pulse with peak intensity of 2.5×10^{12} W/cm^2, λ=1064nm, pulsewidth 100ps, and m=5. An approach angle of γ=15° rather than 19° was used so that in vacuum the first ring would be positioned to agree with the measurement of Fig. 7(a). The time t=0 was set by appearance of a peak density of 5×10^{17} cm^{-3}. By 20ps, a hollow profile with peak electron density of $\sim 5 \times 10^{18}$ cm^{-3} has developed, and by 70ps the peak off-axis density has saturated at $\sim 3 \times 10^{19}$ cm^{-3}. At 700 torr, the ionization proceeds primarily through collisions (avalanche).

There are several differences between the measurements and the simulation. While the peak saturated densities and the central densities are in agreement, the saturation occurs faster in the simulation. This cannot be explained by a maximum probe pulse jitter of 10ps, and at present we have no explanation. In addition, the simulation profiles are narrower in outer extent, and the location of the density peak remains almost stationary compared to the experimental profiles. This is likely a result of the wider main ring and the broader radial distribution of energy in the subsidiary rings of the experimental hollow beam (compare Figs. 7(a) and 7(b)), which would heat the outer region of the density profile to a higher temperature. The simulation predicts a peak temperature at the ring location of ~5eV and a peak ionization of Z=1. This explains the long time persistence of the central hole in the density profile, which fills in at about ~800ps in the simulation. The temperature is too low for radially inward thermal conduction to quickly raise the ionization yield of the weakly ionized gas on axis or for rapid outward expansion of the channel. Even though the breakdown is in the avalanche regime, the low temperature and slow hydrodynamics is made possible by the modest peak intensity of 2.5×10^{12} W/cm^2 of the 100ps pulse. A similar result could be obtained using ultrashort pulses at much higher intensity in the field ionization regime [18], where the electron density would be strongly localized at the first ring and the residual electron temperature would be low [19].

In conclusion, we have demonstrated the generation of elongated, tubular plasmas using a high order Bessel beam. For modest peak intensity pump pulses, some applications of this channel, such as guiding for laser-plasma accelerators, might be complicated by the remaining neutral and weakly ionized gas on axis, while others such as harmonic generation might benefit from its presence. Higher intensity pump pulses might generate high enough temperature for additional ionization of the guide core. In any case, the prompt generation of hollow, deep electron density profiles, without the need for hydrodynamic evolution to establish them, will be beneficial for preformed channel guiding of high intensity laser pulses of unprecedentedly small spot size.

ACKNOWLEDGEMENTS

This work is supported by the US Dept. of Energy (DEF G0297 ER 41039), the National Science Foundation (PHY-9515509), and the Civilian Research and Development Foundation (CRDF grant RP2-130).

REFERENCES

1. H.M. Milchberg et al., Phys. Plasmas **3**, 2149 (1996); S. P. Nikitin, I. Alexeev, J. Fan, and H.M. Milchberg, Phys. Rev. E **59**, R3839 (1999).
2. J. Durnin et al., Phys. Rev. Lett. **58**, 1499 (1987); T. Wulle and S. Herminghaus, Phys. Rev. Lett. **70**, 1401 (1993).
3. P. Sprangle and B. Hafizi, Phys. Rev. Lett. **66**, 837 (1991); B. Hafizi, E. Esarey, and P. Sprangle, Phys. Rev. E **55**, 3539 (1997).
4. F.V. Bunkin, V.V. Korobkin, Yu. A. Kurinyi, L. Ya Polonsky, and L.N. Pyatnitsky, Kvantovaya Elektron. 10, 443 (1983) [Sov. J. Quant. Electron **13**, 254 (1983)]; L.Ya Polonsky and L.N. Pyatnitsky, Optica Atmosphery **1**, 86 (1988).
5. N.E. Andreev, S.S. Bychkov, V.V. Kotlyar, L. Ya Margolin, L.N. Pyatnitsky and P.G. Serafimovich, Quant. Electron. **26**, 126 (1996).
6. E. Esarey, P. Sprangle, and A. Ting, IEEE Trans. Plasma Sci. **24**, 252 (1996).
7. H.M. Milchberg, C.G. Durfee III, and T.J. McIlrath, Phys. Rev. Lett. **75**, 2494 (1995); H.M. Milchberg, C.G. Durfee, and J. Lynch, J. Opt. Soc. Am. B **12**, 731 (1995).
8. T.R. Clark and H.M. Milchberg, Phys. Rev. Lett. **78**, 2373 (1997).
9. T.R. Clark and H.M. Milchberg, Phys. Rev. Lett. **81**, 357 (1998); T.R. Clark and H.M. Milchberg, Phys. Rev. E **61**, 1954 (2000).
10. T.R. Clark and H.M. Milchberg, Phys. Rev. E **57**, 3417 (1998).
11. L. Sedov, *Similarity and Dimensional Methods in Mechanics* (Academic, New York, 1959).
12. M. V. Ammosov, N. B. Delone, and V. P. Krainov, Sov. Phys JETP **64**, 1191 (1987) [Zh. Eksp. Teor. Fiz. **91**, 2008 (1986)].
13. S.S. Bychkov, S.V. Gorlov, L. Ya. Margolin, L.N. Pyatnitskii, A.D. Tal'virskii, and G.V. Shpatakovskaya, Quant. Electron. **29**, 229 (1999).
14. N.E. Andreev, L. Ya. Margolin, I. V. Pleshanov, and L. N. Pyatnitskii, JETP **78**, 663(1994).
15. N.E. Andreev, S.S. Bychkov, V.V. Kotlyar, L Ya Margolin, L.N. Pyatnitskii, and P.G. Serafimovitch, Quantum Electronics **26**, 126(1996).
16. T.R. Clark, Ph.D. Thesis, University of Maryland, College Park (1998).
17. S.P. Nikitin, T.M. Antonsen, T.R. Clark, Yulein Li, and H.M. Milchberg, Opt. Lett. **22**, 1787 (1997).
18. M.V. Ammosov, N.B. Delone, and V.P. Krainov, Sov. Phys. JETP **64**, 1191 (1987).
19. B. M. Penetrante and J. N. Bardsley, Phys. Rev. A **43**, 3100 (1991).

GeV Energy Gain in a Channel Guided Laser Wakefield Accelerator

P. Sprangle,[1] B. Hafizi,[2] J.R. Peñano,[3] R.F. Hubbard,[1] A. Ting,[1] A. Zigler,[4] and T.M. Antonsen, Jr.[5]

[1] *Plasma Physics Division, Naval Research Laboratory, Washington, DC*
[2] *Icarus Research, Inc., Bethesda, MD*
[3] *LET Corporation, Washington, DC*
[4] *Hebrew University, Israel*
[5] *University of Maryland, College Park, MD*

Abstract. A 3-D envelope equation for a laser pulse in a tapered plasma channel is derived, which includes wakefields, relativistic and non-paraxial effects such as finite pulse length and group velocity dispersion. It is shown that electron energies of ~ GeV in a channel-guided LWFA can be achieved by using short pulses where the forward Raman and modulation nonlinearities tend to cancel. Further energy gain can be achieved by tapering the plasma density to reduce electron dephasing.

1. INTRODUCTION

In the standard laser wakefield accelerator (LWFA) a short laser pulse, on the order of a plasma wavelength long, excites a trailing plasma wave that can trap and accelerate electrons to high energy [1-3]. Raman, modulational, and hose instabilities can disrupt the acceleration process [4-11]. Extended propagation of the laser pulse is necessary since, in the absence of guiding, the acceleration distance is limited to a few Rayleigh lengths which is far below what is necessary to reach GeV electron energies [1,12]. The physics of laser beams propagating in plasmas has been studied in great detail [4,13-18]. Besides laser beam propagation issues, dephasing of electrons in the wakefield can limit the energy gain [19-22].

This article addresses the guiding and stability of an intense laser pulse in a uniform plasma channel and wakefield acceleration in a tapered plasma density. A coupled pair of laser and wakefield equations that include wakefields, relativistic and non-paraxial effects, such as finite pulse length and group velocity dispersion (GVD) is derived for pulses propagating in a tapered plasma channel. It is shown that the Raman and modulation instabilities tend to cancel in the short pulse, broad beam limit. Using short pulses in a uniform channel a dephasing-limited electron energy gain of $\sim GeV$ is predicted. In addition, the wakefields in tapered plasma are obtained and the plasma density taper necessary for the wakefield phase velocity to equal the speed of light is determined.

2. FORMULATION

The linearly polarized laser electric field $\mathbf{E}(x,y,z,t)$ in a tapered channel, correct to order $\mathbf{E}^3$, is described by [4]

$$\left(\nabla^2 - c^{-2}\partial^2/\partial t^2\right)\mathbf{E} = \omega_p^2(z)/c^2(1+r^2/R_{ch}^2(z) + \delta n/n_o(z) - |a|^2/4)\mathbf{E},$$

where $\omega_p(z) = \left(4\pi e^2 n_o(z)/m\right)^{1/2}$ is the plasma frequency, $n_o(z)$ is the non-uniform plasma channel density, $R_{ch}(z)$ is the channel radius, δn is the plasma density perturbation associated with the wakefield, $|a| = \sqrt{2}(|e|/mc\omega_o)\langle\mathbf{E}\cdot\mathbf{E}\rangle^{1/2}$ is the magnitude of the electron oscillation momentum normalized to mc, and the brackets denote a time average. In the equation for $\mathbf{E}$, the last three terms on the right hand side represent, respectively, a parabolic plasma density channel, plasma wakefields and the relativistic mass correction. The field can be represented as

$$\mathbf{E}(x,y,z,t) = (1/2) E(x,y,z,t) \exp(i(\int^z k_o(z)dz - \omega_o t))\hat{\mathbf{e}}_\perp + c.c.,$$

where $E(x,y,z,t)$ is the complex laser envelope, $k_o(z)$ is the spatially varying wavenumber, ω_o is the frequency, $\hat{\mathbf{e}}_\perp$ is a transverse unit vector, and c.c. denotes the complex conjugate. Substituting the above field representation into the wave equation and changing independent variables from (z, t) to (z, τ) where $\tau = t - \int^z dz'/v_g(z')$, the envelope equation becomes

$$\left[\nabla_\perp^2 + \frac{4}{r_o^2} - \frac{\omega_p^2(z)}{c^2}\frac{r^2}{R_{ch}^2(z)} + 2ik_o(z)\left(1 + \frac{i}{k_o(z)v_g(z)}\frac{\partial}{\partial\tau}\right)\frac{\partial}{\partial z}\right.$$
$$\left. + i\frac{\partial k_o}{\partial z}\left(1 - \frac{i}{k_o(z)v_g(z)}\frac{\partial}{\partial\tau}\right) + v_g^{-2}(z)\gamma_g^{-2}(z)\frac{\partial^2}{\partial\tau^2}\right]a(x,y,z,\tau) \quad (1)$$
$$= \frac{\omega_p^2(z)}{c^2}\left(\frac{\delta n}{n_o} - \frac{|a|^2}{4}\right)a(x,y,z,\tau)$$

where $a=|e|E/(mc\omega_o)$, r_o is the initial laser spot size, $v_g(z) = c^2 k_o(z)/\omega_o$ is the group velocity, $k_o(z) = c^{-1}(\omega_o^2 - \omega_p^2(z) - 4c^2/r_o^2)^{1/2}$, $\gamma_g(z) = (1 - \beta_g^2(z))^{-1/2}$ and $\beta_g = v_g/c$. The Helmholtz operator in the wave equation has not been approximated (except for neglecting the $\partial^2/\partial z^2$ term) and therefore finite pulse length and group velocity dispersion (GVD) effects, to all orders, are contained in Eq. (1). In Eq. (1), wakefields (δn) and relativistic ($|a|^2$) effects can lead to the Raman and modulation instabilities and GVD is represented by terms proportional to $\partial^2/\partial\tau^2$ and higher order τ derivatives introduced through the $\partial^2/\partial\tau\partial z$ term.

The electric field $\mathbf{E}_p(x,y,z,\tau)$ associated with the wakefield in a tapered plasma, correct to order in a^2, is given by

$$\left(\partial^2/\partial\tau^2+\omega_p^2(z)\right)\mathbf{E}_p = -(mc^2/|e|)\omega_p^2(z)\nabla|a|^2/4,$$

and the perturbed wakefield density is $\delta n/n_o(z) = -(|e|/m)\omega_p^{-2}(z)\nabla\cdot\mathbf{E}_p$ and $\nabla = \nabla_\perp + (\partial/\partial z - v_g^{-1}\partial/\partial\tau)\hat{e}_z$.

3. PULSE PROPAGATION

To utilize laser pulses for electron acceleration or radiation generation it is necessary to propagate intense pulses many Rayleigh lengths in plasma without disruption. This can be accomplished by propagating a short pulse in a plasma channel. In the short pulse, $L \ll \lambda_p$, broad beam, $r_o \gg L$, limit the wakefield density perturbation becomes $\delta n/n_o \cong |a|^2/4$ and the right hand side of Eq. (1) vanishes, i.e., the Raman and modulation nonlinearities cancel. The possibility of suppressing the Raman instability in the short pulse LWFA was noted in [1]. The cancellation of the right hand side of Eq. (1) also implies that short pulses do not undergo relativistic focusing, as first discussed in [23]. This indicates that the long-pulse self-modulated LWFA [19,21,22,24,25], which undergoes disruptive instabilities, may not be the optimal configuration for achieving high energies.

Additional limitations on the acceleration distance in the LWFA include i) the diffraction (Rayleigh) length, $Z_R = \pi r_o^2/\lambda$, ii) the dephasing length, $L_d \cong \gamma_g^2 \lambda_p = \lambda_p \omega_o^2/(\omega_p^2 + 4c^2/r_o^2)$, iii) pulse energy depletion length, $L_e \cong |a|^2(\omega_o/\omega_p)^2(E_{wb}/E_{p,z})^2 L$, and iv) the pulse dispersion length, $Z_{GVD} \cong \pi(\omega_o^2/(\omega_p^2 + 4c^2/r_o^2))L^2/\lambda$. Here, $E_{wb} = \omega_p mc/|e|$ is the wavebreaking field, $E_{p,z}$ is the axial component of the wakefield and L is the laser pulse length. In an untapered channel the acceleration distance is limited by the dephasing length and the electron energy gain is $\Delta W \cong \alpha|eE_{zo}|L_d$, where $E_{zo} \cong |\delta n/n_o|E_{wb}/(1+8c^2/r_o^2\omega_p^2)$, and $\alpha \cong 1/3$ which accounts for dephasing (slippage) and transverse focusing requirements. In the short pulse, broad beam limit, $\delta n/n_o \cong |a|^2/4$, and the energy gain is

$$\Delta W \cong (\pi/2)\alpha mc^2|a|^2(\omega_o/\omega_p)^2(1+8c^2/r_o^2\omega_p^2)^{-1}(1+4c^2/r_o^2\omega_p^2)^{-1}.$$

An example of extended pulse propagation and wakefield generation in a plasma channel is shown in Fig. 1. The parameters are, $n_o = 1.1 \times 10^{17} cm^{-3}$ ($\lambda_p = 100\mu m$), $\lambda = 2\pi c/\omega_o = 1\mu m$, $r_o = 70\mu m$, $a_o = 0.6$, $c\tau_o = L = 37\mu m$ is the pulse length, $Z_R = 1.5 cm$, $P_p = 174 TW$ is the critical power for relativistic focusing [12], $P = 38 TW$ is the peak laser pulse power, $L_d = 53 Z_R$, $Z_{GVD} \cong 2400 Z_R$ and $L_e = 890 Z_R$. Figure 1 shows the intensity and density on a planar cut through the center of the pulse at $z = 0$ and at $z = 53 Z_R$. In this example the average peak wakefield is $|E_{zo}/E_{wb}| \approx 0.1$, the peak perturbed density is $|\delta n/n_o| \approx 0.15$ and the estimated energy gain is $\Delta W = \alpha |eE_{zo}|L_d \cong 0.9 GeV$.

4. WAKEFIELDS IN TAPERED CHANNELS

For a tapered plasma channel the wakefield associated with a matched laser pulse, of the form $|a| = a_o(k_o(0)/k_o(z))^{1/2}\exp(-r^2/r_o^2)\sin(\pi\tau/\tau_o)$, for $0 \leq \tau \leq \tau_o$ and zero otherwise, can be obtained in the broad pulse limit ($r_o \gg L$). Note that for a matched pulse of constant spot size the channel radius is given by $R_{ch}(z) = r_0^2 \omega_p(z)/2c$. The axial component of the wakefield behind the pulse, $\tau \geq \tau_o$, on axis, is $E_{p,z}(0,z,\tau) = -E_o(z)\sin(\omega_p(z)\tau_0/2)\cos(\omega_p(z)(\tau - \tau_o/2))$, where

$$E_o(z) = E_{wb}(c/v_g(z))(\omega_p(z)/\omega_p(0))(k_o(0)/k_o(z))(\pi a_o/\tau_o)^2/(\omega_p^2(z) - (2\pi/\tau_o)^2).$$

The phase velocity of the wakefield is found to be

$$v_{ph}(z,\tau) = v_g(z)/\bigl(1 - (\partial\omega_p(z)/\partial z)(v_g(z)/\omega_p(z))(\tau - \tau_o/2)\bigr). \qquad (2)$$

The phase velocity of the wakefield increases (decreases) with distance from behind the pulse for an increasing (decreasing) plasma density. For an increasing plasma density, the phase velocity can equal the speed of light at some point behind the pulse. In general this point moves relative to the back of the pulse. The existence of a luminous point behind the laser pulse has been noted earlier in wakefield simulations

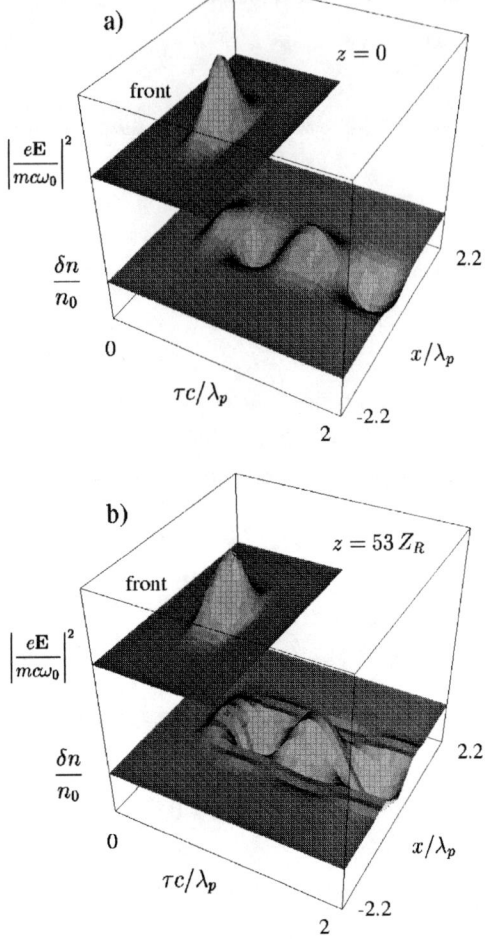

FIGURE 1. Surface plots of the normalized intensity $|e\mathbf{E}/mc\omega_o|^2$ and density perturbation $\delta n/n_o$ on a planar cut through the center of the pulse for $\lambda=1\mu m$, $\lambda_p=100\mu m$ and $r_o=70\mu m$, at a) $z=0$ and b) $z=53Z_R$ where $|e\mathrm{E}(z=0)/mc\omega_o|^2_{max}=0.36$ and $|\delta n(z=0)/n_o|_{max}=0.15$.

[26]. For the luminous point to remain fixed relative to the wakefield, say N plasma wavelengths behind the pulse, the plasma density taper must satisfy

$$\partial \hat{\omega}_p / \partial \hat{z} = (\hat{\omega}_p^2 / 2\pi N)\left((\pi r_o / \lambda)^2 \hat{\omega}_p^2 + 1\right), \tag{3}$$

where $\hat{\omega}_p = \omega_p(\hat{z})/\omega_o$, $\lambda = 2\pi c/\omega_o$ and $\hat{z} = z/Z_R$. Figure 2 shows both the solution of Eq. (3) with $N=5$ and the wakefield amplitude as a function of

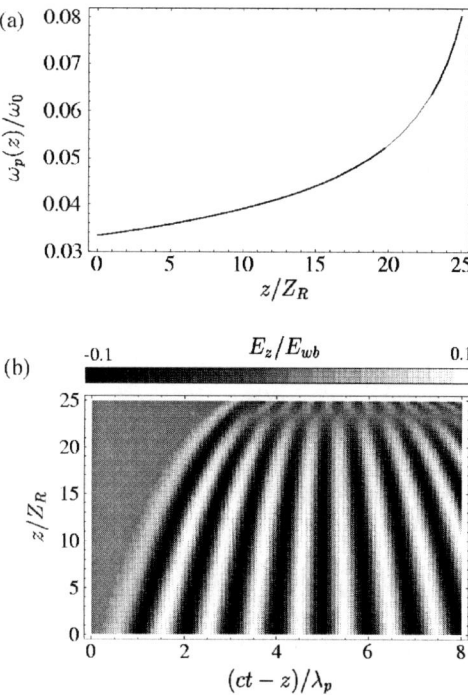

FIGURE 2. a) Solution of Eq. (3) showing the normalized plasma frequency $\omega_p(z)/\omega_{po}$ as a function of z/Z_R, for $N=5$, $r_o = \lambda_p$, $\omega_o/\omega_{po} = 30$, $a_o = 0.5$, and $c\tau_o = \lambda_p$. **b)** Plot of the axial electric field as a function of $(ct-z)/\lambda_p$ and z/Z_R for a laser pulse propagating in plasma with density taper shown in a).

$(ct-z)/\lambda_p$ (speed of light frame) and z. The dark bands represent regions of accelerating axial electric field while the bright bands correspond to decelerating field. The vertical band at $(ct-z)/\lambda_p = 5$ represents the wakefield bucket with phase velocity equal to c.

5. CONCLUSIONS

In summary, a 3-D envelope equation for a laser pulse propagating in a plasma channel has been derived and applied to LWFA. This equation includes wakefields, finite pulse length, GVD and relativistic effects. Numerical solutions demonstrate long-range short pulse propagation and wakefield amplitudes sufficient to accelerate electrons to $\sim GeV$. The dephasing length can be increased and higher energy obtained in a tapered plasma channel where wakefield phase velocities equal to the speed of light can be achieved over an extended distance.

ACKNOWLEDGMENTS

This work was supported by the Department of Energy and the Office of Naval Research. The authors acknowledge useful discussions with D.F. Gordon.

REFERENCES

1. P. Sprangle et al., Appl. Phys. Lett. 53, 2146 (1988); A. Ting et al., AIP Conference Proceedings, 193, 398 (1989).
2. E. Esarey et al., IEEE Trans. Plasma Sci. 24, 252 (1996).
3. F. Amiranoff et al., Phys. Rev. Lett. 81, 995 (1998); F. Dorchies, et al., Phys. Plasma 6, 2903 (1999)
4. E. Esarey et al., IEEE J. Quantum Electron. 33, 1879 (1997).
5. P. Sprangle et al., Phys. Rev. Lett. 79, 1046 (1997), Phys. Rev. E 56, 5894 (1997).
6. T.M. Antonsen and P. Mora, Phys. Rev. Lett. 69, 2204 (1992); Phys. Fluids, B5, 1440 (1993).
7. C.D. Decker and W. B. Mori, Phys. Rev. Lett. 72, 490 (1994); Phys. Rev. E 51, 1364 (1995).
8. W.B. Mori, IEEE J. Quantum Electron. 33, 1942 (1997).
9. P. Sprangle et al., Phys. Rev. E 61, 4381 (2000).
10. P. Sprangle, J. Krall, and E. Esarey, Phys. Rev. Lett. 73, 3544 (1994).
11. W.B. Mori et al., Phys. Rev. Lett. 72, 1482 (1994).
12. P. Sprangle and B. Hafizi, Phys. Plasmas 6, 1683 (1999).
13. G.Z. Sun et al., Phys. Fluids 30, 526 (1987); A.B. Borisov et al., Phys. Rev. A 45, 5830 (1992); B. Hafizi et al., Phys. Rev. E 62, September (2000).
14. P. Sprangle et al., Phys. Rev. Lett. 82, 1173 (1999); Phys. Rev. E. 59, 3614 (1999).
15. E. Esarey et al., Phys. Rev. Lett. 84, 3081 (2000).
16. D. Umstadter et al., Science 273, 472 (1996).
17. Y. Ehrlich et al., J. Opt. Soc. Am. B 15, 2416, (1998).
18. H.M. Milchberg et al., Phys. Plasmas 3, 2149 (1996).
19. K.C. Tzeng et al., Phys. Rev. Lett. 79, 5258 (1997)
20. R.F. Hubbard et al., IEEE Trans. Plasma Sci. (to be published); R.F. Hubbard et al., Phys. Rev. E (to be published)
21. D. Gordon et al., Phys. Rev. Lett. 80, 2133 (1998).
22. E. Esarey et al., Phys. Rev. Lett. 80, 5552 (1998).
23. P. Sprangle et al., Phys, Phys. Rev. A 41, 4463 (1990).
24. J. Krall et al., Phys. Rev. E, 48, 2157 (1993).
25. C.I. Moore et al., Phys. Rev. Lett. 79, 3909 (1997); S.P. LeBlanc, et al., Phys. Rev. Lett. 77 (1996)
26. T. Katsouleas, Phys. Rev. A 33, 2056 (1986).

IFEL Experiment at the Neptune Lab

P. Musumeci and C. Pellegrini

Department of Physics and Astronomy, University of California at Los Angeles,
405 Hilgard Avenue, Los Angeles, CA, 90095-1547

Abstract. We present a two stage Inverse Free Electron Laser accelerator proposed for construction at the UCLA Neptune Lab. Proof-of-principle experiments on the IFEL scheme have been carried out successfully. This experiment is intended to achieve a 100 MeV energy gain, staging two IFEL modules. It will use a 16 MeV electron beam, a 1 TW CO_2 laser and two different tapered helical undulators. The problem of refocusing both laser and electron beam is analyzed in detail. A preliminary beam-line layout and numerical simulation are presented.

INTRODUCTION

One of the most appealing possibilities for the acceleration of charged particle is to make them interact with the very large high electric fields easily available in today's high power lasers. One important advantage of far field accelerator with respect to other advanced accelerator scheme, is that the acceleration takes place in vacuum and the interaction does not require the presence of a plasma or other media at a wavelength distance from the beam, thus avoiding problems of electrical breakdown, beam intensity limitations due to electromagnetic interaction with material boundary, and beam quality degradation due to the interaction with a plasma. In principle every reverse process of a charged particle radiation can be used for acceleration. In this paper we study the inverse process of the Free Electron Laser, namely the interaction of a quasi monochromatic electromagnetic wave, with a relativistic electron beam inside an oscillating static magnetic field.

This idea has been proposed initially by Palmer [1] and then extensively explored by Courant, Pellegrini and Zakowicz [2] and others [3-4]. Proof-of-principle Inverse Free Electron Laser experiments have already been carried out successfully and recently also the possibility of staging of different IFEL modules has been proved [5]. In particular a system with many accelerating regions can be obtained either by using a number of laser beams each focused only once, or by multiple focusing of one laser beam. In the first case the main problem is to keep the phase coherence of the different laser beams so that the particles remain in step with the accelerating field [6]. We explore the second case, where the main problem is the transport and focusing of a high power laser beam.

The goal of the proposed experiment is to realize an IFEL accelerator raising the beam energy from about 14 MeV, to about 100 MeV, and to test the feasibility of a staging scheme using only one laser beam.

The Neptune Laboratory at UCLA has already a high brightness split photoinjector [6], and the high power MARS laser. The electron beam and laser parameters are given in Table 1.

Table 1. Initial parameters	
Electron beam energy	14 MeV
Electron beam charge	1 nC
Electron beam pulse length	6 ps
Electron beam normalized emittance	5 mm-mrad
Laser wavelength	10.6 μ
Laser energy	100 J
Laser pulse duration	100 ps

In the first part of this paper we propose a solution to the problem of focusing and transporting a laser pulse with 3-4 order of magnitude more energy respect to other IFEL experiments. A particular study of the IFEL interaction including the effect of the laser diffraction is also presented. The Guoy phase shift that a Gaussian beam experiences going through a waist is compensated by a gap between two half-undulators to allow re-phasing of electrons and photons. With this new scheme it is particularly important to control the effect of the wigglers on the transverse beam dynamics. At the end we present the results of 3 dimensional simulations of the beam phase space dynamics.

DEALING WITH TERAWATT LASER

We describe the laser beam with a Gaussian approximation:

$$El, Bl \propto \frac{e^{-\frac{(x^2+y^2)}{w(z)^2} + i\left(kz - \omega t + \varphi_0 + \frac{k(x^2+y^2)}{2R(z)} - arctg\left(\frac{z-zw}{zr}\right)\right)}}{\sqrt{1+\left(\frac{z-zw}{zr}\right)^2}} \quad (1)$$

The best possible optical configuration for an IFEL application would be a laser beam focused at the center of the undulator to a spot size such that the Raleigh range is comparable with the length of the interaction region, that is the undulator length. To reach this optimum situation is complicated by the limit set by the damage threshold of the materials used in the transport system (2J/cm^2) [6]. In fact the spot size on the focusing lens cannot be smaller than 50 cm^2 and the focal distance is limited by the fact that for practical space reasons, the lens cannot be more that 2-3 m away from the waist point. Using these numbers in the relation valid for Gaussian beams:

$$f = \pi w_0 w_f / \lambda \qquad (2)$$

where w_0 is the initial spot size, we obtain a final spot size w_f of about 0.25 mm, and the associated Raleigh range of about 2 cm. Focusing a 1 TW CO2 laser beam to this small spot size will give an electric field at the waist as high as 60 GV/m. Because the Raleigh range is much shorter than the undulator length, it is important to include the effect of diffraction in the analysis of the Inverse Free Electron Laser interaction.

A DIFFRACTION-DOMINATED IFEL INTERACTION

The Resonant Acceleration

To describe a diffraction-dominated Inverse Free Electron Laser interaction we modify the classical IFEL equations [12] to include the diffraction effects, in particular the dependence of the electric field from the spot size, and the Guoy phase shift effect.

$$\begin{cases} \dfrac{\partial \gamma}{\partial z} = \dfrac{eE_0}{mc^2} K \dfrac{1}{\sqrt{1+\dfrac{(z-z_w)^2}{z_r^2}}} \sin(\psi) \\ \dfrac{\partial \psi}{\partial z} = k_w + k - \dfrac{k}{\beta_z} - \dfrac{1}{z_r(1+\dfrac{(z-z_w)^2}{z_r^2})} \end{cases} \qquad (3)$$

valid for helical geometry with constant undulator parameter K. We assume that the laser wave amplitude and phase are given, and do not change during the interaction with the electron beam. If the undulator is properly tapered electrons and photons can maintain a definite phase relationship and there can be an energy transfer from the wave to the electrons [2], as shown in Fig. 1.

Stability Of Acceleration

Fig.2 shows the longitudinal phase space of the electrons. It is evident that going through the laser waist, the change in parameters, in particular the fast 180° phase shift is not adiabatic, and the accelerating bucket concept, useful in describing the dynamics for slowly changing Hamiltonian [12] is not valid anymore. The accelerating bucket disappears near the end of the second half-undulator.

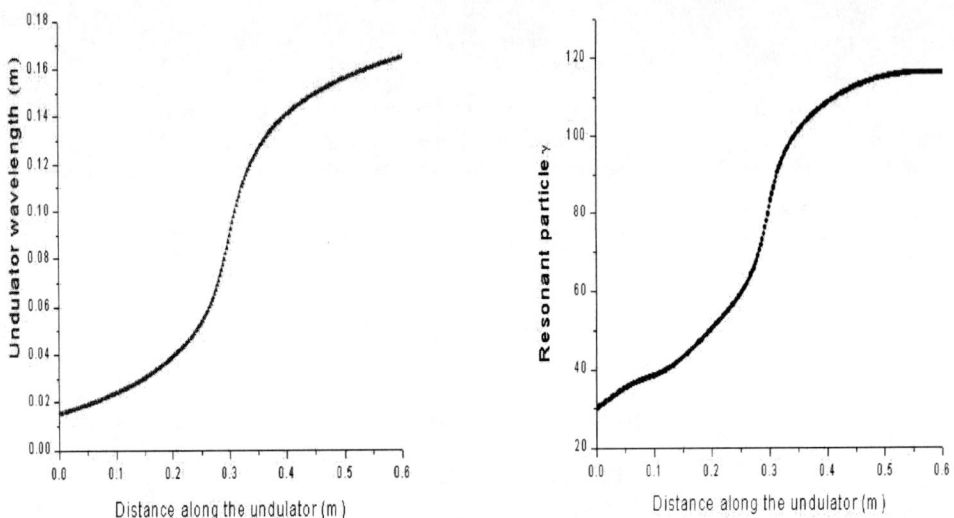

FIGURE 1. Energy and wavelength along a constant K optimally tapered undulator

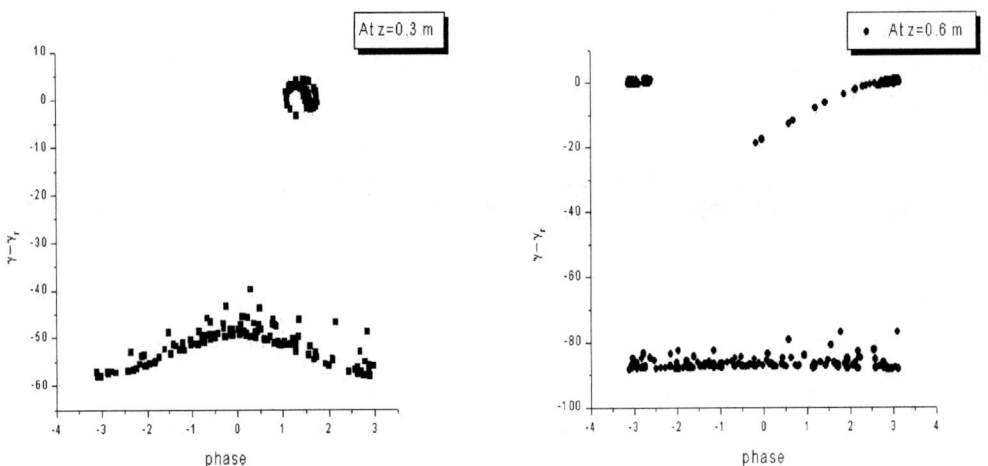

FIGURE 2. Longitudinal phase space

Solution Of The Guoy Phase Shift Problem

To avoid this problem we can insert in the region around the laser waist, a gap in the undulator magnetic field such that electrons and photons have the right accelerating phase at the entrance of the second undulator section. The laser phase shift is 180°, and if the length of the gap is given by:

$$\Delta z = \lambda \gamma^2 \approx 4\,cm \tag{4}$$

the electrons slip another 180° respect of the electromagnetic wave and the resonant phase is preserved. A 1-dimensional simulation confirms that with this scheme, the bucket is preserved at the end of the accelerating region.

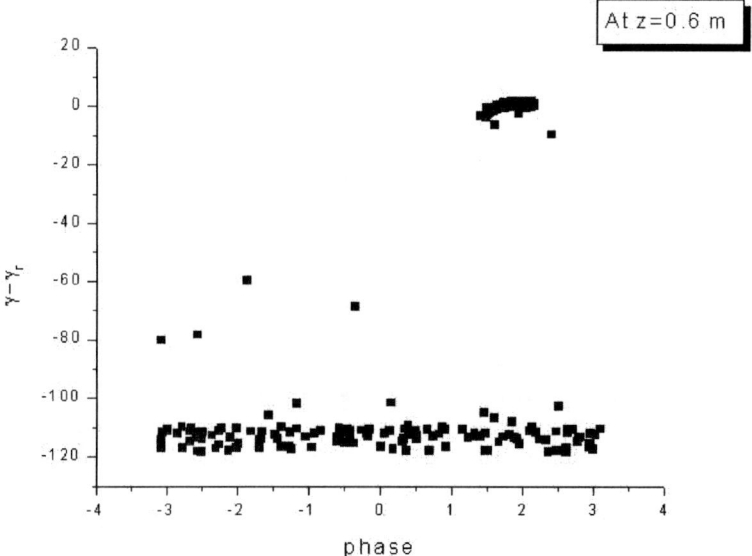

FIGURE 3. Phase space at the end of the second half-undulator with a gap around the laser waist

Fig.4 clearly shows what happens in the critical region: the resonant phase slips 2π at the laser waist and the energy starts to grow again when the particle enter the second half-undulator.

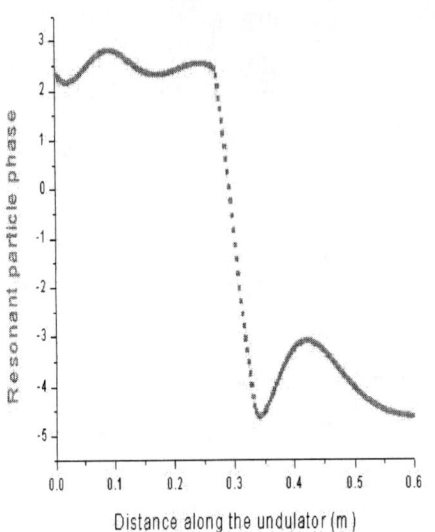
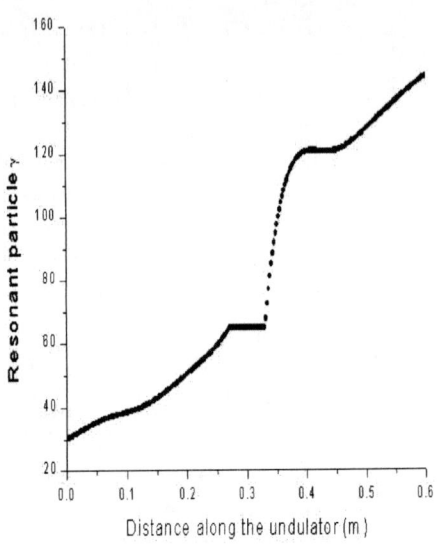

FIGURE 4. Resonant phase and resonant energy

UNDULATOR DESIGN

The undulator parameters are in Table 2. The helical geometry is convenient because the Inverse Free Electron laser interaction is always "turned on". The choice of keeping constant K is made for convenience. A crucial concern for the undulator design is the period tapering. The best tapering function is shown in Fig.5. To simplify the undulator design though, a linear approximation is made to this function in each half-undulator. The value of K is also increased going from the first interaction region to the next one. In the first half-undulator K has to be as low as possible to meet the resonant condition for a 14 MeV electron beam. Then, when the electrons have been already accelerated, the value of the undulator parameter can be raised to give even bigger accelerating gradient.

Table 2. Undulator parameters

Parameters	1st half-undulator	2nd half-undulator
Initial λ_w	1.5 cm	1.5 cm
K	0.5	1.5
B	0.2 T	0.6 T
Undulator length	0.27 m	0.27 m
Linear tapering coefficient	0.08	0.14

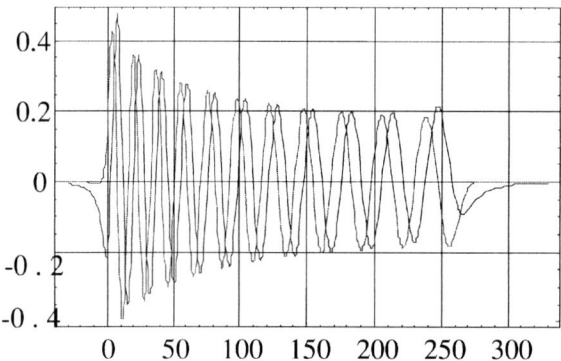

FIGURE 5. Field map for the first half undulator. B(T) vs. z(mm). Note the linear period tapering. Note that in the bifilar design the radius of the winding helices was increased to decrease the field amplitude to maintain constant K.

The undulator parameters can be achieved in different ways. Either hybrid design with permanent magnet and iron, or an electromagnetic undulator appear to have satisfying performances. As a first step towards undulator design, in order to study the particle evolution, a 3 dimensional magnetic field map form RADIA [10] was generated for two bifilar helical undulators with dipole kickers at the entrance and exit to compensate for the transverse kick due to the undulator magnetic fields.

3D SIMULATION

TREDI [11], a Lienard-Wiechert based, particle tracking code, using 4th order Runge-Kutta, was used to follow the particles in the RADIA 3d map, and the gaussian laser field (1). The results are compatible with the 1-d simulations.

The variation of the percentage of captured particles with electron beam size and transverse initial displacement of the position of the bunch centroid (fig. 6) can be explained observing that because of the gap around the waist is about 2 Raleigh range, the smallest laser beam size that the electrons see inside the undulator is about 0.4 mm.

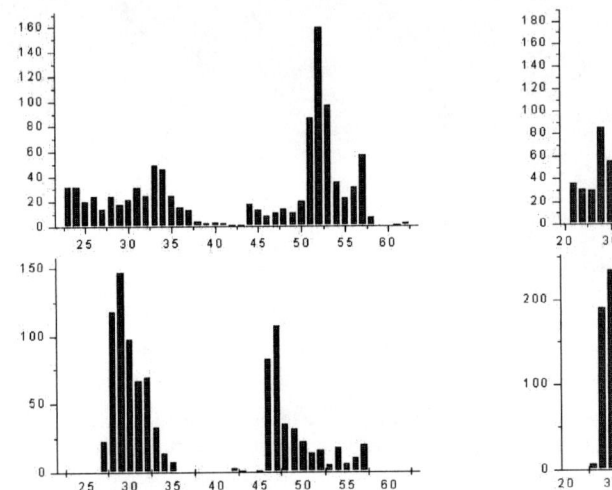
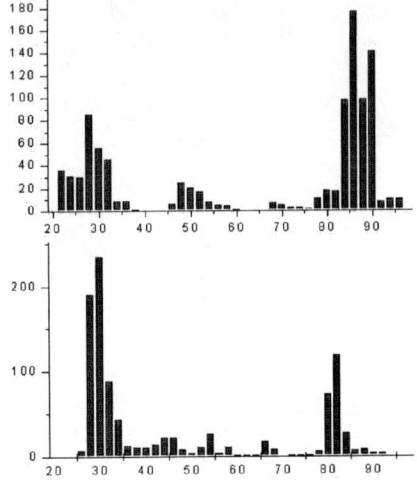

FIGURE 6. Histogram plots of the particle energy -in rest mass unit- at the end of first half-undulator (on the left) and at the end of second half-undulator (right). The top pictures show the 1-dimensional results.

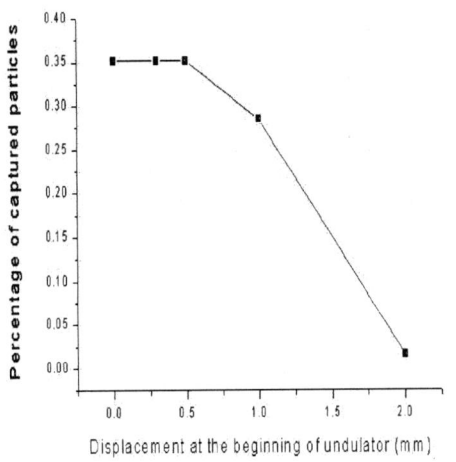
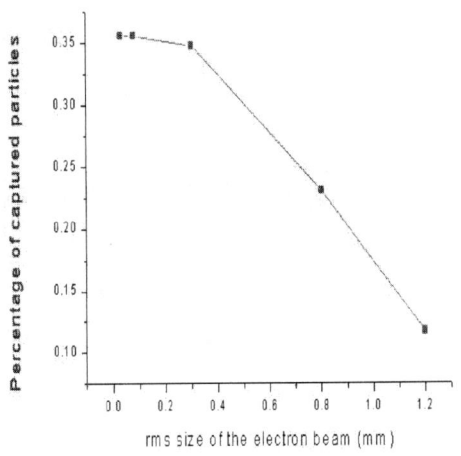

FIGURE 7. Variation of the percentage of captured particle with transverse parameters. Initial displacement and initial beam size.

CONCLUSION

The results of the initial study of the Inverse Free Electron Laser Accelerator at the Neptune Lab at UCLA are summarized in Table 3.

Table 3. IFEL parameters	
Initial energy	15 MeV
Final energy	75 MeV
Averaged energy gradient	100 MeV/m
Microbunch length	10 fs
Fraction of particle within a 10% width at the final energy	0.40

The limit on the charge that can be accelerated by this system is set evaluating when the longitudinal space charge force becomes comparable with the bunching force. This limit is about 2000 A. The Neptune beam has an average current of 500 A. The power efficiency of the accelerator is 10 %. Finally, the calculations including the synchrotron radiation effects show that this effect at these low electron energies is negligible. The proposed solution to the problem of focusing and transporting the high power laser is not the only one possible. Laser waveguides, or the optical properties of an already ionized medium can also solve the problem and we will study them in the future. The initial calculations and the simulation result show, however, that interesting results can be obtained in this diffraction-dominated configuration.

REFERENCES

1. R.Palmer, *J. Appl. Phys.* **43**, 3014 (1972)
2. E. Courant, C. Pellegrini, W. Zakowicz, *Phys Rew A*, **32**, 2813 (1985)
3. A.Fischer, J. Gallardo, J. Sandweiss, A. Van Steenbergen, *Advanced Accelerator Concepts, AIP Conference Proceedings* **219**, Port Jefferson, NY, 299 (1993)
4. I. Wernick, T.C.Marshall, *Phys. Rev. A,* **46**, 3566 (1992)
5. A. Van Steenbergen, J. Gallardo, J. Sandweiss, J. Fang, M. Babzien, X. Qiu, J. Skaritka, X-J.Wang, *Phys.Rev.Letters* **77**, 2690 (1996)
6. W.Kimura *These Proceedings*
7. C.Clayton, C.Joshi, K.Marsh, C.Pellegrini, J. Rosenzweig, *Proceedings of Particle Accelerator Conference 97*, 678 (1997)
8. R. Wood *Laser Damage in optical materials,* Adam Hilger, Briston and Boston 1996
9. A. Siegman *Lasers,* University Science Books, 1986
10. P. Elleaume, et al. , *Proceedings of Particle Accelerator Conference 97*, 3509 (1997)
11. L. Giannessi, et al. *Nucl. Instr. Meth.* **393**, 434, 1997

Experimental Measurements of Wakefields in a Multimode, Dielectric Structure Driven by a Train of Electron Bunches

J.G. Power, M.E. Conde, W. Gai, A. Kanareyken[†], R. Konecny, and P. Schoessow

Argonne National Laboratory, 9700 S. Cass Ave. Argonne, IL 60439
[†] St. Petersburg Electrical Engineering University, 5 Prof. Popov St., St. Petersburg 197376, Russia

Abstract. We report on the experimental results of a new wakefield acceleration scheme. The multibunch driven, multimode, dielectric wakefield accelerator was demonstrated at the Argonne Wakefield Accelerator (AWA). In this experiment, a bunch train of 4, 5 nC electron bunches, separated by 760 ps, was passed through a 60 cm long dielectric-lined cylindrical waveguide. The separation was chosen to match the net acceleration wavelength of the multimode structure. By carefully measuring the energy spectrum of the 4 beams after they passed through the waveguide, we demonstrated that the wakefield is indeed enhanced by a train of periodically spaced electron bunches. The analysis of the multimode structure driven by a bunch train was done by a trivial extension to the existing theory since we are operating in the linear regime. This work represents the first experimental demonstration of this concept and also shows that multibunch operation of wakefield accelerators is worthy of further investigation.

INTRODUCTION

We report on a new collinear acceleration scheme [1,2] in which a multimode, dielectric structure is driven by a bunch train. - i.e. a multimode dielectric wakefield accelerator (multimode DWFA) driven by a bunch train. This scheme has two distinguishing features compared with conventional collinear wakefield devices: the use of closely spaced, multiple drive bunches and the intentional excitation of multiple cavity modes to create the net wakefield. That is not to say that higher order modes (HOM's) are not excited in all structure wakefield accelerators, that is a well-established fact taken into account by all modern analysis. Rather, it is that the HOM's in most structures are regarded as parasitic. The dimensions of the multimode DWFA structure are chosen so the excited cavity modes are nearly harmonic (i.e. equally spaced in frequency) and thus superimpose (linearly) periodically to produce sharply peaked fields (approximately equal to the electron pulse length). (In retrospect, a more appropriate name would have been the 'harmonic' DWFA since all structures have multiple modes, but the 'multimode' name has already been established.) The wakefield left behind the beam looks like an alternating series of delta functions rather than the sine-wave like wakefield of the single-mode device.

This scheme has two particularly interesting characteristics: relaxed drive beam requirements and (possibly) improved *rf* breakdown properties. The former characteristic comes about by dividing the single high charge bunch into a train of low charge bunches, thus making it easier to produce low emittance, short pulse bunches. (Note that the use of pulse trains can benefit both multimode and single mode accelerators such as the CLIC decelerator.) The later feature arises since the short duration of the individual wakefield pulses may slightly increase the *rf* breakdown voltage. This is simply because the dielectric is exposed to the high field for a shorter time than in the single-mode case.

In the first section of this paper, we show how the analytic solution for the usual DWFA excited by a single drive bunch [3] can be extended to the calculation of multimode DWFA driven by a bunch train. In the remainder of the paper we describe multimode structure design, the experimental setup, and the experimental results compared to numerical simulations. We end with some concluding remarks about the merits of this scheme.

MULTIBUNCH, MULTIMODE DEVICES

In this section we show how to extend the analytic solution for the usual DWFA excited by a single drive bunch [3] to the calculation of a multimode DWFA driven by a bunch train.

Direct Solution of the Wave Equation

One can determine the wakefield excitation of any cylindrically symmetric structure [3] by first finding the point charged particle solution (Green Function) of the wave equation in cylindrical coordinates. For a particle of charge q located at position z_0 and moving with axial velocity v, the point charge density and current are written as, $\rho(r, z_0, t) = q \frac{\delta(r)}{r} \delta(z_0 - vt)$ and $J_z = v\rho$ respectively. Defining $z = z_0 - vt$, one can now solve the wave equation and write the Green function for the axial electric field as,

$$G_z(r,z) = \sum_{n=1}^{\infty} G_n(r) \cos(k_n z) \tag{1}$$

where $G_n(r)$ are the coefficients of the Fourier expansion and k_n are the wave numbers of the nth mode, which are determined by the boundary conditions. Since all bunches travel along the cylinder axis, for the rest of the paper we need only consider fields at $r = 0$ and will drop the explicit dependence on r. Using (1), one can solve for the longitudinal wakefield, $W_z(z)$, for any arbitrary charge distribution $f(z)$ by taking the convolution of the Green function over the axial charge distribution,

$$W_z(z) = \int_{-\infty}^{z} dz' f(z') \{ \sum_{n=1}^{\infty} G_n \cos[k_n(z-z')] \} \quad (2)$$

This is a general solution for the longitudinal wakefield of any DWFA since all modes are accounted for. The only difference between the multimode device and the single mode device is that in the former many modes ($n \sim 50$) are excited while in the later only a few modes ($n \sim 3$) are appreciably excited. It is *trivial* to extend this solution to a "bunch train" of M bunches separated by a distance, λ, by writing the multibunch charge distribution as, $f(z) = \sum_{m=1}^{M} f_m(z - m\lambda)$. Finally, the wakefield excited by the bunch train at any point z (which may be inside a bunch) is obtained by substituting this multibunch charge distribution into (2),

$$W_z(z) = \int_{-\infty}^{z} dz' \sum_{m=1}^{M} f_m(z' - m\lambda) \{ \sum_{n=1}^{\infty} G_n \cos[k_n(z-z')] \}$$
$$\equiv \sum_{m=1}^{M} W_{zm}(z) \quad (3)$$

where W_{zm} is the longitudinal wakefield due to the mth bunch (located at $z = m\lambda$) acting alone. This form of writing makes our assumption of linear superposition explicit and we now see that absolutely no extension to the existing theory is required other than invoking linear superposition.

DESIGN OF THE MULTIMODE STRUCTURE

In this section we discuss how the multimode, dielectric-lined, cylindrical wakefield structure (Fig. 1) was designed using the analytic work derived in [3].

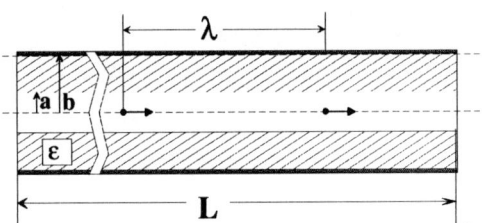

FIGURE 1. The Multimode Waveguide driven by a Bunch Train. The thick walled dielectric waveguide used in this experiment has inner radius $a = 0.5$ cm, outer radius $b = 1.44$ cm, length $L = 60$ cm and dielectric constant $\varepsilon = 38.1$. Two bunches from the pulse train, separated by λ, are shown - the experiment uses four.

The AWA Drive Beam

Using the existing multibunch capabilities of the AWA, we have designed and constructed an experiment to demonstrate the multimode concept. The AWA can provide a train of electron bunches spaced at multiples of 760 ps, with energy 15 MeV, emittance 20 mm mrad, and bunch length of 3-5mm. For this experiment we passed 4 electron bunches, separated by our minimum achievable distance of 760 ps, though a dielectric-lined tube.

The Dielectric Waveguide

There are four parameters that must be chosen for the waveguide: (see Fig. 1) $a, b, \varepsilon,$ and L. The choice of inner radius a is constrained, primarily, by the emittance and energy of the AWA drive beam. Due to the relatively large emittance and low energy, the inner radius of the tube must be kept large to maximize the charge transported through the tube. The downside of a large inner radius is that a relatively low gradient is excited, which makes it difficult to obtain a large total energy gain (loss) unless the tube is long. An inner radius of 0.5 cm and tube length of 60 cm was found to be capable of transmitting about 5 nC which yields energy changes that are large enough to measure with a spectrometer. The choice of dielectric constant was based on finite group velocity effects and availability of materials. In order for the *rf* packets of the 4 bunches to interact, the energy of the first bunch must not have propagated away by the time the fourth bunch enters the tube, therefore we want to keep the group velocity (v_g) low. Since $v_g \sim c/\varepsilon$, we want ε high. We chose $\varepsilon = 38.1$ which gives $v_g \sim 0.026c$. The dielectric material [5] is $CaTiO_3 - LaAlO_3$ with a perovscit structure. With a and ε fixed, the outer radius b was adjusted until the net wavelength of the wakefield in the multimode tube was equal to the distance between bunches. For the multiple beam experiment at AWA, the drive bunch spacing is 23.05 cm since λ_{rf} is 23.05 cm at 1.3 GHz; thus the outer radius b is solved to be 1.44 cm.

EXPERIMENTAL SETUP

The AWA drive linac was configured to provide a collinear bunch train of 4 bunches, each with charge $Q = 5$ nC, bunch length $\sigma_Z = 4.5$ mm and energy spread $(\Delta E/E)_{FWHM} = 0.3\%$. Each bunch must be spaced so that the individual wakes left by each bunch adds constructively to the wake of the preceding bunch. For the multimode experiment, simulation shows that the spacing between bunches must be accurate to within σ_Z out of λ (or 4.5 mm out of 230.5 mm) for appreciable constructive interference, and hence large wakefields, to obtain.

Optical Splitter

A train of 4 laser pulses is made by optically splitting a single laser pulse into 4 separate pulses as shown in Fig. 2. The distance between laser pulses is adjusted

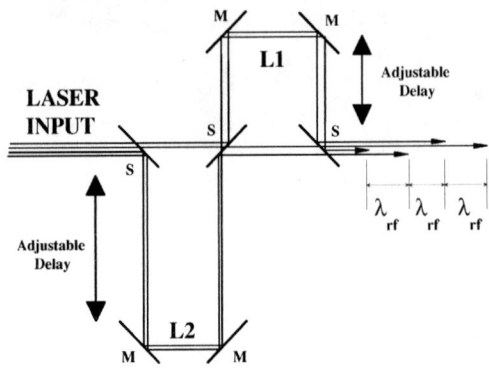

FIGURE 2. Laser Beam Splitter Optics. Three 50/50 beam splitters (S) in combination with four mirrors (M) are used to produce four laser pulses of variable separation. For this experiment, legs L1 and L2 are adjusted until the four pulses are separated by one wavelength of the rf (λ_{rf}) driving the photoinjector. (Lines separated for clarity.)

optically by moving mirrors on translation stages in the delay line. Initially, the distance between bunches is crudely measured with a streak camera, using the 10 ns sweep rate, giving timing resolution of ~10 ps between bunches. Final bunch spacing must be done during the experiment by maximizing the deceleration of trailing bunches after they emerge from the multimode structure.

Spectrometer

A central part of the experimental effort is to study the wakefield enhancement due to the bunch train; thus measurement of the energy spectrum of each drive beam is crucial. Also, because the 4 drive beams must be spaced to within one σ_z to achieve significant acceleration, a high demand is placed on the AWA's energy measurement system. The energy measurement system must have a resolution of 0.5% or better and be insensitive to position jitter since bunches within the train may not exactly follow the same path. To this end, a new imaging spectrometer was designed and built (Fig. 3) to provide improved momentum resolution over a wider range while reducing the sensitivity to beam jitter. Setting the tungsten slit to 300 μm, the resolution of the spectrometer is calculated to be 0.2%.

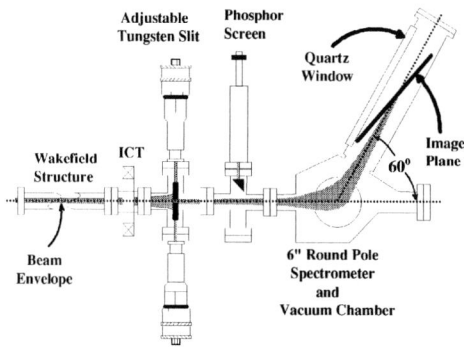

FIGURE 3. Energy Measurement System. The decelerated drive bunch train passes through a tungsten slit of adjustable width, which is imaged through a 60° dipole (diameter = 6") onto a phosphor screen located at the image plane. The phosphor screen is viewed with an intensified camera through the quartz window.

EXPERIMENTAL RESULTS

In this section we first calculate the expected energy spectrum of the bunch train based on the beam parameters used during the run with our analytic model. We then compare this expected energy spectrum to the experimental data.

Using (3), the computed wakefield as a function of z (the distance behind the first bunch) for our specific experimental parameters is plotted in Fig. 4. The parameters used to create the plot were a charge of 4.8 nC, a rms. bunch length of 4.5 mm, a FWHM energy spread of 0.3% and an initial energy of 15.46 MeV. From Fig. 4, one can see that peak accelerating wakefield is only about 1.2 MeV/m.

Using the computed wakefields (Fig. 4) one may easily extract the monochromatic energy spectrum generated by the 4 drive beams. Projecting the wake experienced by each particle within the bunch onto the energy axis and scaling by the appropriate

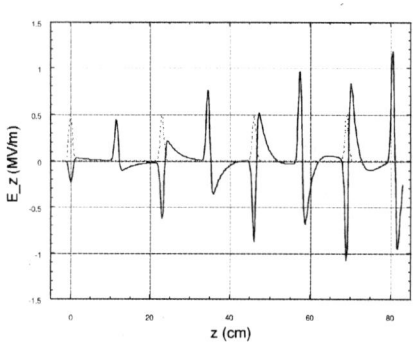

FIGURE 4. The wake excited (solid line) by 4 monochromatic drive beams (charge = 4.8 nC/each and bunch length = 4.5 mm) spaced one wavelength (23.05 cm) apart. The beam distribution is shown as the dashed line and is moving right to left.

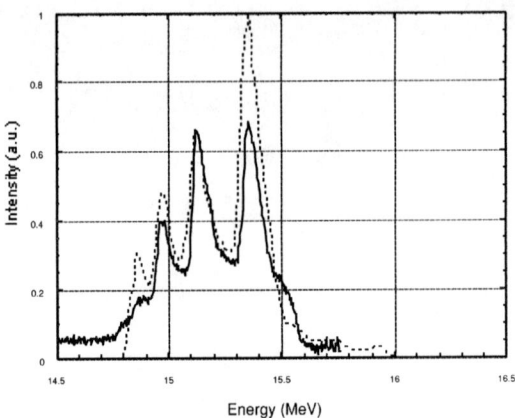

FIGURE 5. Energy spectrum of the 4 x 4.8 nC bunch train. The solid line is the measured energy spectrum and the dotted line is the analytic model.

value of L gives the monochromatic energy spectrum. The total energy spectrum is then computed by taking the convolution of the monochromatic energy spectrum over the energy distribution of the drive bunch, which is measured to be approximately Gaussian in shape.

The energy spectrum of the decelerated bunch train was measured at the image plane (Fig. 3) and acquired with a frame grabber for off-line analysis. For tube length L = 59.87 cm and the parameters given earlier in this section, we plot the measured energy spectrum (Fig. 5, solid line) and compare it to the computed energy spectrum (Fig. 5, dotted line). For the purposes of this experiment, the figure of merit is the location of the energy peaks in Fig. 5. The intensity of the individual energy peaks (here, intensity is proportional to the charge) is meaningless for a simple reason. Namely, the transverse position of the individual bunches striking the 300 µm wide tungsten slit of Fig. 3 fluctuate causing the amount of charge transported to the image plane to fluctuate. The energy peaks (in MeV) from the data are located at 15.35, 15.12, 14.98 and 14.88 while the fits peaks are located at 15.35, 15.12, 14.98 and 14.86. As is readily seen from this data, the agreement between theory and experiment is excellent.

To ensure that the measured energy spectrum was indeed due to multimode wakefields, we sent a low charge pulse train (< 1 nC each) through the multimode structure. Since the wakefield excitation amplitude is proportional to the total charge we expect the total width of the energy spectrum to be narrower in this case than in the high charge case. The projected energy spectrum of the low charge case is shown in Fig. 6. The narrow spread of the energy spectrum in Fig. 6 confirms that we are indeed seeing the effects of the multimode wakefield deceleration in the data of Fig. 4.

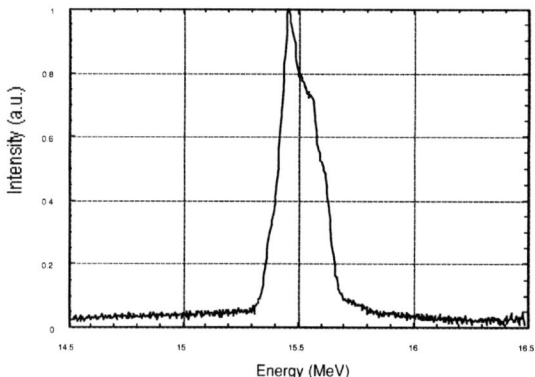

FIGURE 6. Energy Spectrum of low charge pulse train (4 x ~0.5 nC) with an average energy of E_0 = 15.46 MeV. Total energy spread is observed to be much less than in the high charge case.

CONCLUSION

We have experimentally demonstrated the concept of multibunch excited wakefields in a multimode dielectric structure. We found that the wakefield is indeed reinforced by a pulse train and the measurement is in complete agreement with the existing theory [3] from the conventional single mode DWFA based on linear superposition. Due to the fact that we used a dielectric tube with a relatively large inner radius, the observed gradient is low (1.2 MV/m). However, if we use the upgraded AWA gun (under construction) with Q = 50nC and σ_z =1mm, and modify the dielectric structure to have inner radius of 2 mm and outer radius of 11.5 mm, a pulse train of 4 bunches can produce 145 MV/m accelerating gradient.

ACKNOWLEDGEMENTS

We appreciate the interesting ideas put forth by Dr. T-B. Zhang, Prof. J. L. Hirshfield, and Prof. T. C. Marshall and the useful discussions with Prof. J. Rosenzweig and Dr. J. Simpson. This work was supported by the US Department of Energy, Division of High Energy Physics, under contract W-31-109-ENG-38.

REFERENCES

[1] T-B. Zhang, J. L. Hirshfield, T. C. Marshall, B. Hafizi, Phys. Rev. **E56** 4647 (1997).
[2] J.G. Power, W. Gai, P. Schoessow, Phys. Rev. **E60** 6061 (1999).
[3] see M. Rosing, W. Gai, Phys. Rev. **D42** 1829 (1990) or K. Ng, Phys. Rev. **D42** 1819 (1990).
[4] P. Schoessow, M. E. Conde, W. Gai, R. Konecny, J. Power, J. Simpson, J. Appl. Phys **84** 663 (1998).
[5] St. Petersburg Electrical Engineering University, Physics Dept.

Direct Laser Acceleration in a Capillary Channel

Loren C. Steinhauer

Redmond Plasma Physics Laboratory
University of Washington

Abstract. A simple channel with a conducting boundary acts as a slow-wave structure under certain conditions. For infrared frequencies and "capillary" channels in a metal the wave moves at the speed of light and has a forward electric field component. For a diameter of ~600 μm the *channel wave* can be directly driven by a long-wavelength laser (λ = 10 μm). Acceleration gradients in the GeV/m range over many cm distances are projected using state-of-the-art high power CO_2 lasers. Projected accelerations and possible limitations of this concept are discussed.

INTRODUCTION

The goal of laser-particle acceleration is to create a large forward electric field with a phase velocity approximately synchronous with a relativistic particle beam. A familiar approach is to employ some sort of "slow wave" structure to slow down the laser wave. This is conventional in radio-frequency acceleration, but may also be done at optical frequencies. The slow wave structure may be a periodically-loaded structure or a dielectric medium in the beam path or a surrounding waveguide. The simplest of all structures is a simple vacuum channel with a conducting wall; these are well known to support waveguide modes with phase velocities *exceeding* the speed of light c and as such are not suitable for acceleration. However, under certain conditions, namely infrared frequencies and small "capillary" dimensions, such channels can support a *synchronous* slow wave (phase velocity = c). These can be TM modes with a substantial forward electric field. Moreover, the wave can be excited directly by injecting an infrared laser at one end of the channel. As such this is a method of *direct* laser acceleration.

The direct acceleration concept is illustrated in Fig. 1. A radially-polarized laser (TM wave) is focused into the end of the capillary channel; it then propagates along it as a traveling channel wave. Laser guiding is achieved automatically since the channel acts like an optical fiber.[1,2] This concept is ideally suited for long wavelength lasers. In particular at the CO_2 laser wavelength (~10 μm) the capillary diameter is about 600 μm. Focusing the laser into such channels (diameter ~60 wavelengths) is well within the capabilities of state-of-the-art optics.

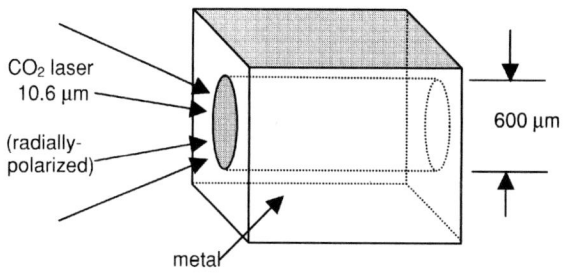

FIGURE 1. Schematic of laser injection into a capillary channel.

Direct laser acceleration in a capillary is akin to other concepts that have been proposed, although there are important differences. The *plasma-fiber beat wave* concept uses a narrow plasma channel with low density inside the channel and high density outside.[1] The *rippled plasma fiber* concept also uses a narrow plasma channel, but with a wavy wall to slow down the traveling wave.[1] The *hollow-channel wakefield* accelerator[2] uses a vacuum channel with (ideally) a uniform plasma outside the channel. The plasma-fiber beat wave and the hollow-channel wakefield concepts rely on nonlinear excitation of a traveling channel wave. The laser and channel waves are two completely different waves, the latter with somewhat lower frequency. In the plasma fiber beat wave[1] concept the coupling is through a laser-beat wave interaction. In the hollow-channel wakefield concept the coupling is through the pondermotive potential of the laser.[2] On the other hand the rippled plasma fiber concept is direct drive, *i.e.* the laser wave becomes a channel wave that is slow enough to cause acceleration because of the rippled channel.[1]

The outline of the paper is as follows. Section II examines the propagation characteristics of slow channel waves. Section III makes projections about the acceleration gradient. Section IV summarizes the attractive features of direct laser acceleration in a capillary and discusses possible liabilities.

SLOW CHANNEL WAVE PROPAGATION

Dispersion Relation

In traveling wave accelerators the "traditional" method to slow down the wave is to introduce a periodic structure (ripples, irises). This is necessary because the phase velocity exceeds c in a standard, longitudinally-uniform waveguide. However if the conditions (frequency, channel dimensions) are such that the wall cannot be treated as a perfect conductor then the results may be different. This case is analyzed here using a metal wall with finite conductivity and some damping. These channel waves are related to surface waves[2] associated with the jump in the dielectric constant at the channel boundary. Surface waves are evanescent in both directions from the surface.

TM modes are investigated since they have a forward component of the electric field suitable for accelerating a beam.

Consider a cylindrical vacuum channel with radius b surrounded by a metal wall. The surface wave limit has $b \to \infty$. These waves are transverse ($\nabla \cdot \mathbf{E} = 0$). Then the nonzero electric field components in the channel ($r \leq b$) are

$$E_z = G I_0(\alpha_1 r) \cdot e^{-i\phi}; \quad E_r = G(k/\alpha_1) I_1(\alpha_1 r) \cdot i e^{-i\phi}, \quad (1,2)$$

and in the wall ($r > b$) are

$$E_z = H K_0(\alpha_2 r) \cdot e^{-i\phi}; \quad E_r = -H(k/\alpha_2) K_1(\alpha_2 r) \cdot i e^{-i\phi}, \quad (3,4)$$

where G, H are constants, and the subscripts 1, 2 denote channel and wall, respectively. Note that G is the acceleration gradient for a perfectly phased particle traveling along the axis. Here $\phi = \omega t - kz$ is the phase, ω is the wave frequency, k is the longitudinal propagation constant, and the transverse propagation constants in the two regions are

$$\alpha_{1,2} = \sqrt{k^2 - (\omega/c)^2 \varepsilon_{1,2}(\omega)}, \quad (5)$$

where $\varepsilon_{1,2}(\omega)$ is the dielectric function. Note that in ideal waveguide modes this constant in the channel is imaginary $\alpha_1 = i|\alpha_1|$, and $E_z \sim J_0(|\alpha_1|r)$, $E_r \sim J_1(|\alpha_1|r)$.

The fields in the channel and the wall match according to the continuity conditions from Maxwell's equations, $[\varepsilon E_\perp] = 0$ and $[E_z] = 0$ where the square brackets denote the jump at the boundary. These connect G and H and give the dispersion relation

$$\varepsilon_1 I_1(\alpha_1 b)/[\alpha_1 b I_0(\alpha_1 b)] = -\varepsilon_2 K_1(\alpha_2 b)/[\alpha_2 b K_0(\alpha_2 b)]. \quad (6)$$

This is a dispersion relation because $\varepsilon_{1,2} = \varepsilon_{1,2}(\omega)$, $\alpha_{1,2} = \alpha_{1,2}(\omega, k)$. Note that $\varepsilon_1 = 1$ for a vacuum channel. In the surface wave limit $b \to \infty$ with a cold-electron plasma ($\varepsilon_{1,2} = 1 - \omega_{p1,2}^2/\omega^2$, $\omega_p = (4\pi e^2 n/m_e)^{1/2}$ = plasma frequency, n = electron density, m_e = electron mass), the dispersion relation takes the simple form

$$c^2 k^2 = \frac{(\omega^2 - \omega_{p1}^2)(\omega^2 - \omega_{p2}^2)}{2\omega^2 - \omega_{p1}^2 - \omega_{p2}^2}. \quad (7)$$

Inspection of this shows that a synchronous slow wave ($ck = \omega$) can only occur if the plasma is *underdense* on one side, $\omega_{p1} < \omega$, and *overdense* on the other $\omega_{p2} < \omega$. The same property carries over to the case of channel waves (finite b).

Ideal Metal Wall

Figure 2 shows the slow-wave dispersion curve for a vacuum channel and an ideal metal wall, i.e. ε_2 is that of a cold-electron plasma, and n_2 is the free electron density. This example has a channel with $b = 10c/\omega_{p2}$. Also shown is the surface wave ($b \to \infty$). The straight line shows the conditions where the phase velocity ω/k is exactly c. The surface wave is asymptotic to $\omega/k = c$ for $k \to 0$ and is asymptotic to $\omega = \omega_{p2}/\sqrt{2}$ for $k \to \infty$.[3] The channel wave has a nonzero frequency as $k \to 0$. It crosses the $\omega/k = c$ line at a particular point, $k = 0.4164 \omega_p/c$.

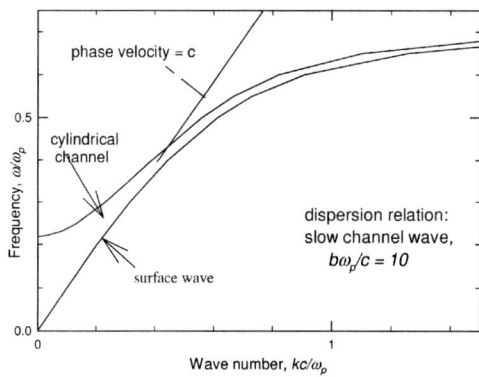

FIGURE 2. Dispersion relation for slow waves.

It is instructive to find the channel sizes for the slow wave as a function of the frequency. Assume a vacuum channel and the highly overdense limit ($\omega_{p2} \gg \omega$). Then at the synchronous condition $\omega/k = c$, the dispersion relation Eq. (6) leads to an expression for the channel radius

$$b \approx (e\lambda^2/\pi c)\sqrt{n_2/\pi n_c} \qquad (8)$$

Although ideal waveguides are unsuitable for acceleration since the phase velocity always exceeds c, they are a familiar yardstick with which to compare slow-wave channels. Refer to a cylindrical waveguide with an infinitely conducting wall, $\omega/k = \sqrt{2}c$ in the fundamental mode, which has radius $b = 0.54\lambda$ ($\lambda = 2\pi c/\omega$ is the vacuum wavelength). Table 1 shows the channel sizes for the synchronous slow-wave and waveguide at several common RF and optical wavelengths. The slow-wave examples assume a free electron density of $n_2 = 8.5 \times 10^{22}$ cm^{-3} (copper). The slow-wave channels for RF frequencies are impractically large. The channel radius for a CO_2 laser can be readily fabricated and is consistent with the transverse dimensions of high-quality relativistic beams. The channel radius for a Ti-Sapphire laser is possible with nano-technologies, but requires an exceedingly narrow particle beam. This illustrates the suitability of a CO_2 laser for directly driving a slow channel wave.

TABLE 1. Channel diameter for slow and fast waves		
	Slow-Wave	Waveguide
X-band RF, $\nu = c/\lambda = 9$ GHz	4.7 km	3.6 cm
W-band RF, $\nu = 90$ GHz	47 m	3.6 mm
CO_2 laser, $\lambda = 10.6$ μm	600 μm	11 μm
Ti-Sapphire laser, $\lambda = 0.8$ μm	2.7 μm	0.9 μm

Resistive Metal Wall

Consider a realistic metal wall with a finite damping constant. Then the dielectric function in the wall is $\varepsilon_2 \approx i\omega_p^2/\omega(\gamma - i\omega)$ where the contribution due to bound electrons is ignored and γ is the damping constant of the metal.[4] For complex ε_2 the propagation constant is also complex, $k = k_R + ik_I$: the real part gives the phase velocity ω/k_R, and the imaginary part k_I is the damping constant ($1/k_I$ is the damping length). Assume weak damping ($k_I \ll k_R$) and the synchronous condition $\omega/k_R = c$. Then the *real* part of Eq. (6) determines the channel width. In a cylindrical channel

$$b = \frac{e}{\pi c}\sqrt{\frac{n_2}{\pi m_e}}\lambda^2 \sqrt{\frac{1+\sqrt{1+\gamma^2/\omega^2}}{2(1+\gamma^2/\omega^2)}}. \tag{9}$$

For $\gamma/\omega < 1$, the second radical is $(1 - 3\gamma^2/8\omega^2 ...)$, thus the ideal expression, Eq. (8), is recovered for $\gamma \to 0$. The *imaginary* part of Eq. (6) gives the damping constant

$$k_I = \frac{2\lambda}{\pi b^2}\sqrt{\frac{-1+\sqrt{1+\gamma^2/\omega^2}}{1+\sqrt{1+\gamma^2/\omega^2}}}. \tag{10}$$

For $\gamma/\omega < 1$, the radical can be expanded, giving

$$k_I = \frac{\gamma\lambda^2}{2\pi^2 cb^2}\left(1 - \gamma^2/4\omega^2 ...\right) \tag{11}$$

For a CO_2 laser ($\lambda = 10.6$ μm) and a room-temperature copper wall ($n_2 = 8.5\times 10^{22}$ cm^{-3}, $\gamma = 4\times 10^{13}$ sec^{-1}) the channel diameter is $2b \approx 610$ μm, and the damping length is $1/k_I \approx 12.4$ cm. For this example the damping correction on channel size [from the second radical in Eq. (9)] is less than 2%. The skin depth in the wall is extremely short; for the same example, $1/\alpha_2 = 0.018$ μm (18 nm).

Basic Properties of Slow Channel Waves

Slow channel waves differ enough from more familiar wave types that it is helpful to review their basic properties. Slow channel waves are similar to waveguide modes in that both employ a simple channel (no periodic structures), both have a *forward* electric field (TM modes), and both are traveling waves. As traveling waves they differ from plasma oscillations, which have zero group velocity (cold plasma); although, the phase velocity may be near c. However, waveguide modes have a "radial" propagation component in that the fields are expressed using *cyclic* functions (J_0, J_1 Bessel functions in cylindrical geometry), while slow channel waves are *evanescent* inward and outward from the wall (I_0, I_1, K_0, K_1 modified Bessel functions). Thus, the field in the channel is actually the *near-field* of a surface wave running along the wall. Even so the evanescence length *inward* from the wall, $1/\alpha_1$, can be much larger than the wavelength, and in fact is infinite for a vacuum channel and a synchronous wave ($\omega/k = c$). For particle acceleration a slow channel wave is a structure-based concept by virtue of the channel wall; however, the featureless channel is the simplest of all slow-wave structures.

ACCELERATION PERFORMANCE

The acceleration gradient for direct laser acceleration in the capillary channel is

$$G \approx (1/\pi)E_L \lambda/b, \qquad (12)$$

where E_L is the *maximum* electric field in the channel (E_r at $r = b$). This compares favorably with laser wakefield acceleration (LWFA),[5]

$$G \approx 0.4 E_L (\lambda/\lambda_p) \cdot a_L / \sqrt{1+a_L^2}, \qquad (13)$$

where $a_L = eE_L/\omega m_e c$ is the normalized laser field, and $\lambda_p = 2\pi c/\omega_p$ is the plasma wavelength. In comparing these, the power-field relationship for a Gaussian profile with $1/e$ radius r_L is $P_L = (cE_L^2/8)\pi r_L^2$, while in the channel wave $P_L = (cE_L^2/16)\pi b^2$. The acceleration gradients of these are compared in Table 2.

TABLE 2. Comparison of acceleration gradients.

	Capillary Channel	LWFA Long Wavelength	LWFA Short Wavelength
Laser power, P_L	1 TW	1 TW	1 TW
Wavelength, λ	10.6 μm	10.6 μm	0.8 μm
Density, n	—	2×10^{16}(cm^{-3})	7×10^{17}(cm^{-3})
Radius, b or r_L	305 μm	224 μm	40 μm
Laser pulse length, τ_L	—	370 fsec	67 fsec
Acceleration gradient, G	0.79 GV/m	0.29 GV/m	0.30 GV/m

In the LWFA examples the density is determined by requiring the dephasing length $L_{ph} = \lambda_p^3/\lambda^2$ to exceed 10 cm. The density is zero in the capillary channel. In the LWFA examples, the laser radius is set by the requirement of quasi-one-dimensionality of the wakefields, $r_L > \lambda_p$, and the laser pulse lengths are set by the resonance condition $\tau_L = \pi/\omega_p$. The capillary channel has no laser pulse length requirement. Comparable acceleration gradients can be achieved by the two methods of acceleration.

DISCUSSION

The direct-drive capillary channel concept described here has several attractive features. (1) *Simplicity*. The smooth capillary channel is the simplest of all slow-wave structures. A byproduct of this may be insensitivity to wall damage by laser-ablation. (2) *Competitive gradient*. The acceleration gradient is competitive with that predicted for LWFA. (3) *Laser guiding*. The capillary automatically channels the laser beam for the full channel length. (4) *Dephasing*. Perfect synchronism ($L_{ph} \to \infty$) is achieved for proper choice of channel radius and wall density. (5) *Direct drive*.

The channel wave is driven directly by the laser. As such it does not rely on nonlinear coupling processes from the laser wave to another wave, which may be difficult to control or have deleterious byproducts. Neither is there a resonance condition requiring a very short laser pulse. (6) *Beam quality*. The "flat" accelerating field E_z in the channel minimizes energy spread introduced by the acceleration fields, and the linear focusing field E_r minimized emittance growth. Further, there is no beam scattering since the channel is a vacuum.

Several liabilities and issues of this concept need investigation. (1) *Acceptance limitations*. Capillary channels raise the question of how well the particle beam fits into the narrow channel over significant lengths (say 10 cm). (2) *Channel wave damping*. The resistivity of the metal damps the propagation of the channel wave. It needs to be determined whether the damping lengths (~10 cm) are tolerable. (3) *Resonance absorption*. The density passes through cutoff at the wall. Resonance absorption at the overdense channel edge might degrade the laser as seen in simulations of laser acceleration schemes involving overdense plasmas.[6] This might be mitigated in the capillary channel since the density jumps sharply through the resonance. (4) *Channel degradation*. The intensity at the wall is large, 7×10^{14} W/cm^2 in the example in Table 2. This will cause cyclic vaporization and recondensation of the wall during successive shots. The simple geometry may make this a benign effect. Experiments with high-voltage discharges in polyethylene capillaries as small as 300 μm diameter showed that ablation was minor and the shape of the capillary remained almost the same after 1000 shots.[7] If wall damage is mitigated in this way, it ceases to be a limitation on the allowed electric field in the channel. Wall damage might also be mitigated by using a suitable liquid metal surface layer on a dielectric substrate. (5) *Channel smoothness requirements*. It is unclear how sensitive the slow channel wave is to wall roughness such as might be introduced during fabrication. Surface quality may actually improve because of the repeated ablation. (6) *Plasma effect on channel wave propagation*. The intensity is large enough to break down the ablated wall material and form a plasma. This affects the channel wave propagation. However the amount of blowoff is extremely small during a short pulse. The blowoff distance in 10 psec is only $0.1 - 0.2$ μm for a copper plasma temperature of 100-500 eV, which is a tiny fraction of the channel diameter. (7) *Nonlinear propagation effects*. While direct laser acceleration in a capillary channel does not rely on ponderomotive effects to excite the wave, nonlinear effects will be present at some level and may affect the propagation and synchronism. (8) *Coupling efficiency*. Ideally a radially-polarized laser wave turns into the channel wave once it enters the capillary. However some loss by reflection or into parasitic modes may occur.

ACKNOWLEDGMENTS

This work was supported by U.S. Department of Energy Grant No. DE-FG03-98ER41061. The author acknowledges useful discussions with Dr. W.D. Kimura.

REFERENCES

1. T. Tajima, *Laser Accelerator for Ultra-High Energies*, in Proc. 12th Int. Conf. On High-Energy Accelerators, F.T. Cole and R. Donaldson, eds., Fermi National Accelerator Laboratory, Batavia, Illinois, 1983, p. 470.
2. T.C. Chiou, T. Katsouleas, C. Decker, W.B. Mori, J.S. Wurtele, G. Shvets, J.J. Su, Phys. Plasmas **2**, 310 (1995).
3. P.K. Kaw and J.B. McBride, Phys. Fluids **13**, 1784 (1970).
4. J.D. Jackson, *Classical Electrodynamics*, 2nd Ed., Wiley, New York, 1975, p284ff.
5. P. Sprangle, E. Esarey, and A. Ting, Phys. Rev. Lett. **64**, 2011 (1990).
6. T. Katsouleas, J.M. Dawson, D. Sultana, and Y.T. Yan, IEEE Trans. Nucl. Sci. **NS-32**, 3555 (1985).
7. D. Kaganovich, P.V. Sasorov, Y. Ehrlich, C. Cohen, and A. Zigler, Appl. Phys. Lett. **71**, 2925 (1997).

S-Band and X-Band Integrated PWT Photoelectron Linacs

D. Yu, D. Newsham, J. Zeng

DULY Research, Inc., Rancho Palos Verdes, CA 90275
E-mail: duly@technologist.com

J. Rosenzweig

Department of Physics and Astronomy, UCLA, Los Angeles, CA 90095-1547

Abstract. A compact high-energy injector, which has been developed by DULY Research Inc., will have wide scientific, industrial and medical applications. The new photoelectron injector integrates the photocathode directly into a multicell linear accelerator. By focusing the beam with solenoids or permanent magnets, and producing high current with low emittance, high brightness and low energy spread are achieved. In addition to providing a small footprint and improved beam quality in an integrated structure, the compact system considerably simplifies external subsystems required to operate the photoelectron linac, including rf power transport, beam focusing, vacuum and cooling. The photoelectron linac employs an innovative Plane-Wave-Transformer (PWT) design, which provides strong cell-to-cell coupling, relaxes manufacturing tolerances and facilitates the attachment of external ports to the compact structure with minimal field interference. DULY Research Inc. under the support of the DOE Small Business Innovation Research (SBIR) program, has developed, constructed and installed a 20-MeV, S-band compact electron source at UCLA. Cold test results for this device are presented. DULY Research is also actively engaged in the development of an X-band photoelectron linear accelerator in a SBIR project. When completed, the higher frequency structure will be approximately three times smaller. Design considerations for this device are discussed following the S-band cold test results.

INTRODUCTION

We report the progress in the development of integrated photoelectron linear accelerators designed to operate in S- and X-band. In this integrated design, the photocathode is included in the accelerating structure using the Plane-Wave-Transformer (PWT) design [1][2]. The S-band prototype structure has been completed by DULY Research and is in the commissioning process at UCLA. The PWT design is illustrated in Figure 1, in which a standing wave, iris-loaded open disk assembly is suspended in a large tank, providing easy access to vacuum and rf ports. The photocathode is located at the center of one of the end plates of the linac. The open cavities provide very strong cell-to-cell rf coupling. Electrons are accelerated in a TM02-mode along the axis of the disk assembly. The PWT design is named after the fact that the TEM-like mode in the annular region between the disk assembly and the tank has the appearance of a plane wave, when the region is energized by rf power transmitted via an external waveguide coupled to the tank. The standing wave structure is designed to transform the TEM-like mode in the annular region into a TM-like mode along the axis of the structure [4][5].

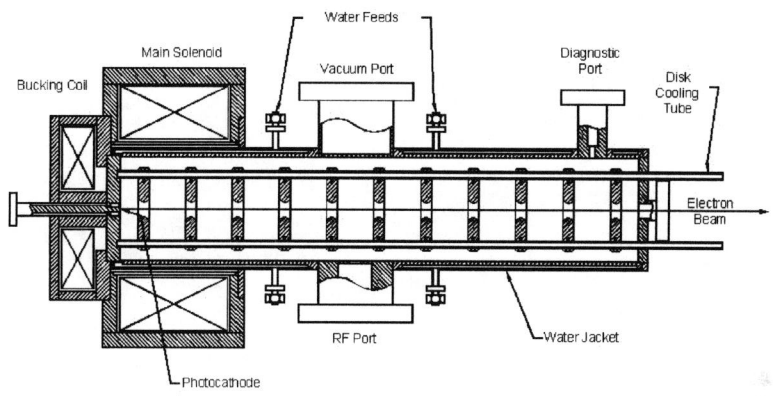

FIGURE 1. Schematic of the S-Band PWT Photoelectron Linac.

The features and benefits of the DULY PWT photoelectron linac include:
- High brightness – The electron beam has high charge, a short bunch length, and low emittance, resulting in a very bright beam.
- Low energy spread – Short laser pulse length results in a small beam length and low energy spread.
- Integrated photocathode and linac – The removal of the long drift space between a conventional photocathode and linac system simplifies the rf coupling, electron beam focusing, and vacuum and cooling requirements.
- PWT design – The strong cell-to-cell coupling reduces manufacturing tolerances while increasing accelerating gradient.
- Unique focusing design – Emittance compensation is provided by using compact solenoidal magnets in the S-band design, and hybrid permanent magnets in the X-band design.
- Compact – The design has a very small footprint.

S-BAND PWT PHOTOELECTRON LINAC COLD TEST

Figure 2 shows the interior and exterior of the S-band PWT hardware. A prototype S-band PWT photoelectron linac (see Ref [1][2][3]) has been developed, fabricated, and cold tested by a team of DULY Research Inc. and UCLA researchers. The design parameters and expected performance from simulations are shown in Table 1. The

20 MeV linac has been installed and is being beam tested in the UCLA PEGASAS Laboratory (see Figure 3).

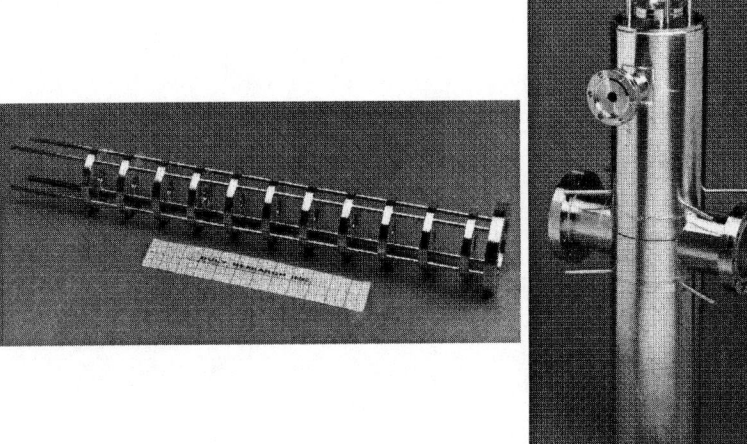

FIGURE 2. Interior accelerating structure and exterior tank of the S-band PWT photoelectron linac.

FIGURE 3. Installation of the S-band PWT photoelectron linac in the PEGASUS Laboratory at UCLA.

TABLE 1. S-Band PWT Design Parameters.

Parameter	Value
Frequency	2856 MHz
Energy	20 MeV
Charge per Bunch	1 nC
Normalized Emittance	1π mm-mrad
Energy Spread	< 0.1%
Bunch Length (rms)	2 psec
Rep Rate	5 Hz
Peak Current	100 A
Linac Length	58 cm
Beam Radius	< 1mm
Peak B Field	1.8 kG
Peak Gradient	60 MV/m
Peak Brightness	2×10^{14} A/(m-rad)^2

The results of the cold test of the S-band PWT have been compared with simulations. Figure 4 shows a 10-cell GdfidL model that includes the cooling/support rods, rf ports, and end plates. Figure 5 shows the S_{11} resonance curve for the π-mode, measured during the cold test after all of the brazing cycles were complete and the PWT was ready for installation at UCLA. The full measured spectrum, as measured by a network analyzer, and the field amplitude inside the S-band PWT as measured by a bead pull, are shown in Figure 6. Using a waveguide that is coupled at the center cell of the 10-cell GdfidL model and the Kroll-Yu [6] method, the simulated value of the coupling constant was calculated to be $\beta_{K-Y} = 1.08$. There was excellent agreement between the resonant frequency of the GdfidL model (2856.0 MHz) and the measured value from the cold test (2856.2 MHz). In addition, despite the small overcoupling of the rf port, Figure 6 shows that the electric field is quite flat, with only a slight dip at the ends and in the center where the rf port is attached.

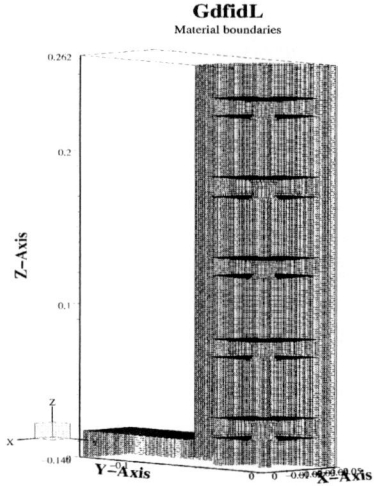

Figure 4. 10-cell GdfidL model.

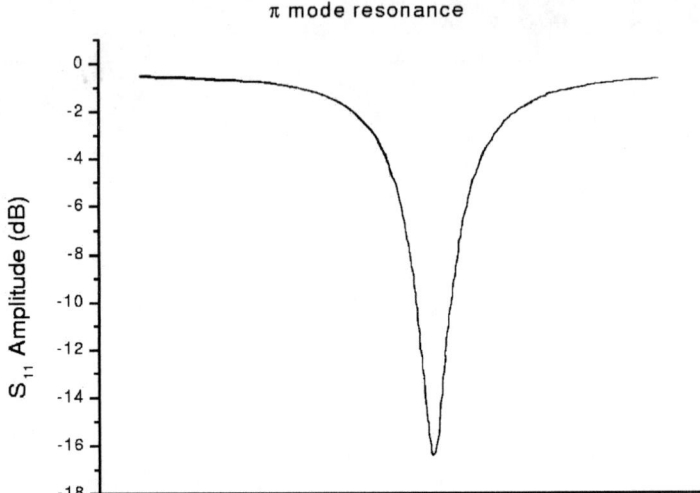

FIGURE 5. S_{11} curve taken from the S-band PWT cold test centered around the π-mode.

There was an approximately 50% discrepancy between measured value of Q_o (12,450) and the value determined from GdfidL simulation (24,305). This difference between the observed value and the calculated value of Q_o can have a significant effect on the efficiency of the linac. A 50% drop in Q_o will result in a factor of $\sqrt{2}$ drop in the accelerating gradient.

Two possible mechanisms to explain this discrepancy were investigated. In the PWT design, the current that supports the magnetic field is carried on the surface of the cooling/support rods, as opposed to a disk-loaded design where the current is carried on the outer wall. The large number of braze joints in the S-band PWT results in an increased chance for an imperfect braze. Such a braze joint, with increased local resistivity, could impede the necessary flow of the current and result in greater losses in the rods. The greater loss in the rods results in a lower value of Q_o. In order to try to model this possibility, a single-cell GdfidL model was used. To simulate the braze joint, a different value of the material resistivity was given to a small region between the cooling rod and the copper disk. Figure 7 shows the relative drop in the value of Q_o as the resistivity of the braze material is increased. A 50% reduction in Q_o, if attributed entirely to the braze defects, would require that the resistivity be increased by an overwhelming factor of 40,000. The braze material used in these joints was 40Au-60Cu which has a resistivity approximately 6 times larger than OFHC copper. The additional factor of ~7,000 required would have to be a result of inclusions, "bubbles", or other imperfections in the braze material.

The tank walls and the cooling rods were constructed out of stainless steel, then copper plated prior to brazing. It is possible that some unknown contamination during

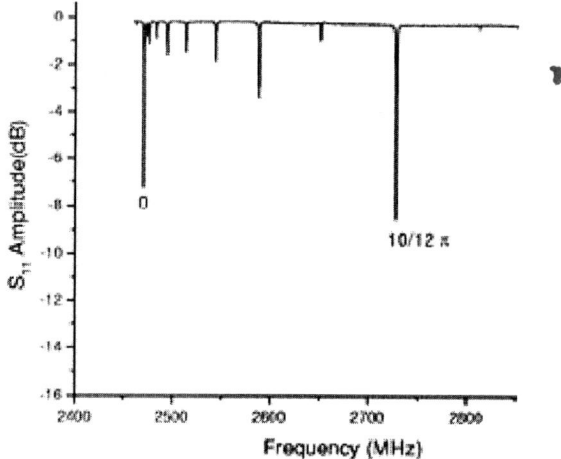

the plating process and/or possible surface damage at the high brazing temperatures may cause the resistivity of these surfaces to be greater than was used in the GdfidL

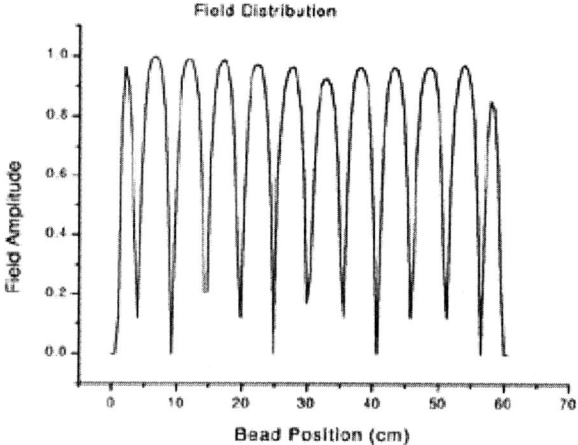

FIGURE 6. Full spectrum S_{11} curve and results of the bead pull measurement from the S-band PWT cold test.

model. This is the second mechanism investigated to explain the decreased value of Q_o. In the GdfidL model, the resistivity of the wall and rod material was increased from the original value corresponding to OFHC copper. The results are shown in Figure 8. To achieve the 50% reduction in Q_o that is observed, the resistivity of the copper surfaces on the tank wall and rods must increase by a factor of 10. As a reference point, the resistivity of 304 stainless steel is approximately 40 times larger than that of OFHC copper, and as mentioned before, the resistivity of the braze material is 7 times larger. If surface contamination or degradation is the source of the decreased value of Q_o, then these issues should be reduced in the X-band PWT, where the smaller size allows for the cooling/support rods to be made from copper tubing instead of copper plated stainless steel.

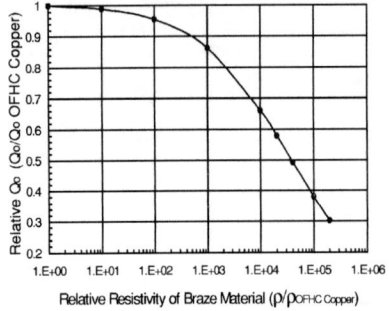

FIGURE 7. Degradation of Q_o as the relative resistivity of the braze material is increased.

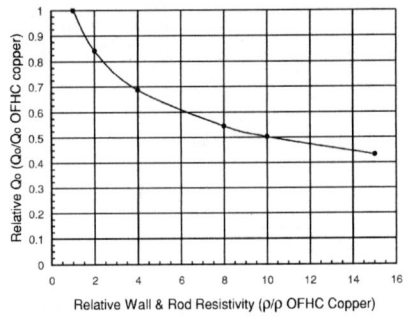

FIGURE 8. Degradation of Q_o as the resistance of the walls and cooling rods is increased.

X-BAND PWT PHOTOELECTRON LINAC

Spurred on by the success of the S-band PWT, DULY Research Inc. is now developing an X-band version of the integrated photoelectron linac. The X-band PWT photoelectron linac promises to produce an even brighter beam than the S-band PWT, due in part to the smaller bunch length and lower emittance of the beam. While in principle the physics of the PWT photoinjector linac is scaleable in frequency [7], the daunting engineering tasks of a high frequency integrated photoinjector linac are mandated by the requirements of a much higher magnetic field, higher accelerating gradient, and higher peak electric field – all in a smaller package. The development of this photoelectron linac is supported by an SBIR grant from the Department of Energy, and is conducted in collaboration with UCLA and ILSA/UCD at LLNL, where an X-band (8.547 GHz) klystron is available for beam tests.

A schematic of the X-band PWT is shown in Figure 9. A simple 1/3 scaling from the S-band PWT indicates that the required magnetic field should be on the order of 4-5.5 kG. This would require a solenoid magnet significantly larger than the one currently used on the S-band PWT and operating at much higher current. Such a magnet would dwarf the X-band structure and add potentially unnecessary bulk and expense to the final product. DULY has designed a hybrid magnet system, which is primarily comprised of permanent magnets, but uses solenoidal windings for precise trim. This hybrid-focusing magnet is shown in Figure 10 and has a footprint which is acceptable relative to the size of the linac. The primary disadvantage of a permanent magnet focusing system is the inability to make fine tune adjustment in the strength of the magnetic field. Currently, DULY is working on a hybrid magnet design in which the solenoidal windings play a more significant role than in the present design, shown in Figure 10, where the windings provide minor trim control of the position of the magnetic null. A pair of pancake solenoids such as shown in Figure 11, may provide greater flexibility during operation so that the focusing fields can be adjusted over a wider range to provide better measured emittance, energy spread, and/or beam size.

X-Band PWT Photoinjector

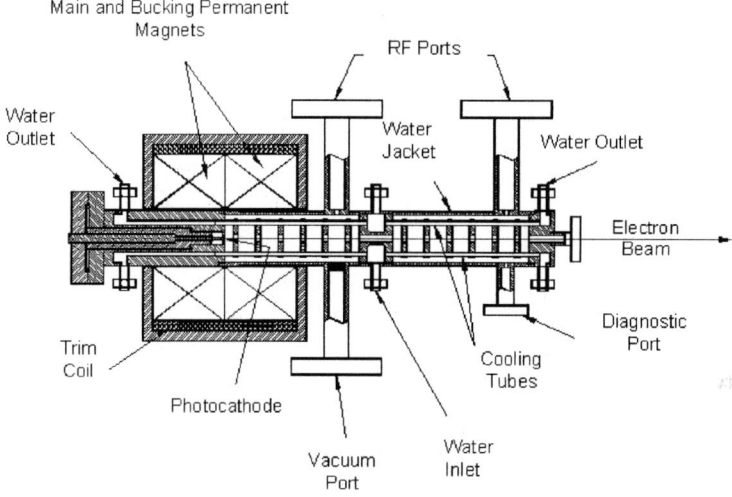

FIGURE 9. Schematic of the X-Band PWT Photoelectron Linac.

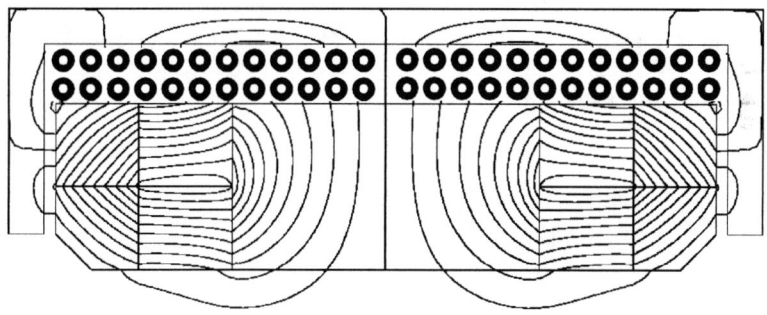

FIGURE 10. Sketch of the proposed hybrid focusing magnet for the X-band PWT.

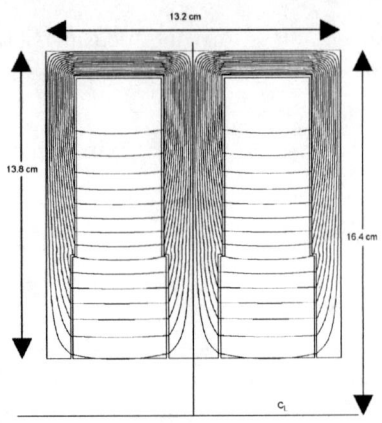

FIGURE 11. Solenoidal magnet design for the X-band PWT.

Because the PWT is a standing wave structure, some of the rf power from the klystron may be reflected during startup and conditioning. This issue is resolved in the S-band PWT by use of an rf isolator; however, at X-band, rf isolators are expensive and difficult to produce. To prevent the reflected power from damaging the klystron, we will use a magic-T power splitter to divide the main rf feed into two feeds of equal phase. Differences in the waveguide length will provide a relative 90° phase shift between the two feeds. The PWT linac is also split, and the sections are powered by the separate feeds from the power splitter. If the two linac sections have the same coupling coefficients and are both tuned to resonance, any reflected rf power from the 90° shifted feed will receive an additional 90° phase shift from the longer waveguide on the return trip. The reflected power from this feed will arrive at the power splitter 180° out of phase with the power from the other feed, and all the power will be delivered into the load. Under these conditions, there is no reflected power back to the klystron and no high-power rf isolator is required. To compensate the phase difference, the two linac sections are connected by a short drift tube. Several other conditions may affect the amount of reflected power at the klystron. Some of these situations are summarized in Table 2.

TABLE 2. Efficiency of the Magic-T Isolator under Several Mismatch Conditions.

Mismatch Condition	Reflected Power at Klystron
Equal reflection from both sections	0%
Total reflection from only one section (none from the other)	25%
Coupling constant (β) ≠ 1 in one section	< 3% for $\Delta\beta = 0.5$
Waveguide phase mismatch of 10° (½ mm error)	< 1%
Cavity detuning ($\Delta f = 200$ kHz)	~3%

One of the issues associated with splitting the linac into two sections and separating them with a drift tube is the tuning of the last cell of the 1st section and the first cell of the 2nd section. Ideally, these cells would be tuned so that at the correct resonant frequency, the position of the electric field maximum would be in phase with the electron beam. In this way, these cells could be used for active acceleration. For a closed half-cell, like the first half-cell with the photocathode, the electric field maximum is at the metal wall. When an aperture is opened for the drift tube, the electric field is "pulled" into the drift tube and the electric field maximum shifts away from the wall. The main effect of this in the PWT structure is to add some spread in energy. Figure 12 shows a design of the 6^+-cell X-band structure. The protrusion near the outer wall combined with the position of the end plate provides the ability to tune the frequency and the position of the peak electric field simultaneously.

When the drift tube transition cell is properly tuned, the electric field in the 6^+-cell structure is flat, except for the transition cell. In this particular design, the transition cell has an electric field that is approximately 10% larger than the field in the main portion of the linac section (Figure 13). With other design choices, the field in the transition cell can be larger than, smaller than, or equal to the field in the other cells. When the tuning ring was shifted by 20 mil, SUPERFISH gave a frequency shift in the entire structure of $\delta f/f_o = 0.1\%$ (9.2 MHz). The field flatness in the "normal" region was $\delta E/E = 9\%$, and the electric field in the final cell was 2% lower than the average field in other ("normal") cells. Because SUPERFISH is a 2D axisymmetric code, the cooling rods cannot be modeled; and these results are indicative of the order of magnitude of the effects. In a final detailed design of the X-band PWT integrated photoinjector, GdfidL will be used to perform a fully 3D simulation of the drift tube transition cell.

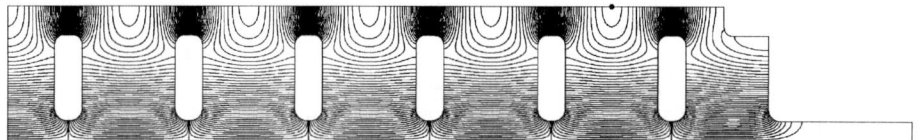

FIGURE 12. SUPERFISH model of the 6^+-cell linac section for the X-band PWT.

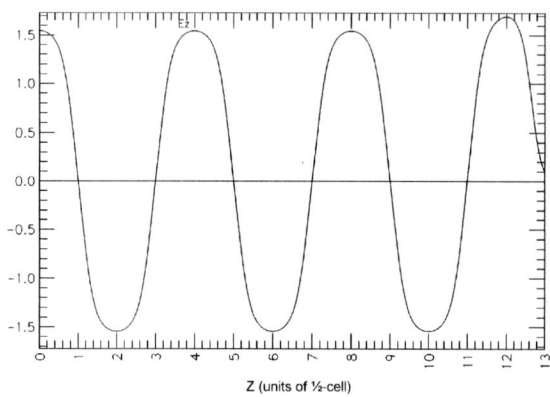

FIGURE 13. Axial electric field plot for X-band PWT.

From the results of the cold test measurement on the S-band PWT discussed above, we should anticipate a reduction in the value of Q_o for the X-band PWT as well. In order to determine the effect this reduction in Q_o has on the parameters of the X-band PWT, 4 cases are being studied using PARMELA. These cases are based on 19 MW output power from the klystron. Half of the cases use the ideal value of Q_o as calculated by GdfidL, and the other half assume a value of $\frac{1}{2}Q_o$. Additionally, SLED pulse compression can provide additional power that is dependent upon Q_o. The presence or absence of the SLED pulse compression combined with the ideal or reduced Q_o give 4 different cases. The parameters for these cases are given in Table 3. SLED provides a 71% (144%) increase in rf power that translates into a 30% (11%) increase in the accelerating gradient for Case 2 (4) from the nominal X-band PWT design of Case 1. At this point, installation of a SLED pulse compression system is not a part of the project, but is intended as an upgrade after additional funding is secured. The result is that Case 3 should most closely reflect the initial operating regime of the X-band PWT.

TABLE 3. Four Operating Cases for X-Band PWT.

Case	Q_o	Power (MW)	E_o (MV/m)	SLED	Comment
1	15,717	19.0	137.2	NO	Ideal Q_o
2	15,717	32.5	179.4	YES	Ideal Q_o
3	7,859	19.0	97.0	NO	½ Ideal Q_o
4	7,859	46.4	151.6	YES	½ Ideal Q_o

The normalized emittance curves for each of the 4 cases are shown in Figure 14, the energy spreads are in Figure 15, the rms beam radii are in Figure 16, and a summary of the simulation results appears in Table 4. Currently, the input parameters are being studied to optimize each of these cases with a particular emphasis on Case 3. The parameter space for the input conditions is vast, and the search is a long process. As is clear from the data, Case 3 results in the worst beam parameters, making optimization critical. All of the PARMELA data presented in Table 3 used a charge per bunch of 1 nC. One option under strong consideration is to operate the X-band PWT at a total charge lower than the nominal 1 nC. When the system is upgraded with the SLED pulse compression, this limitation on the charge should not be necessary.

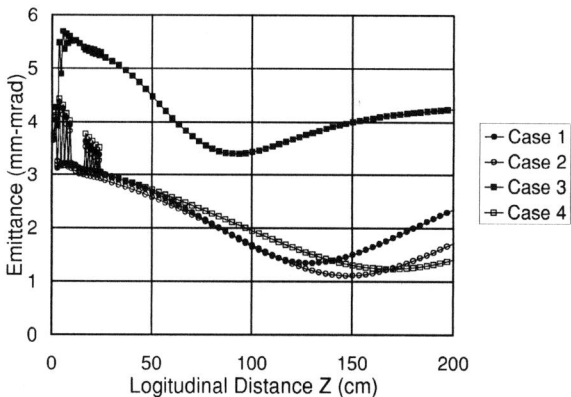

FIGURE 14. Normalized rms emittance for the four operating cases in the X-band PWT obtained from PARMELA simulation.

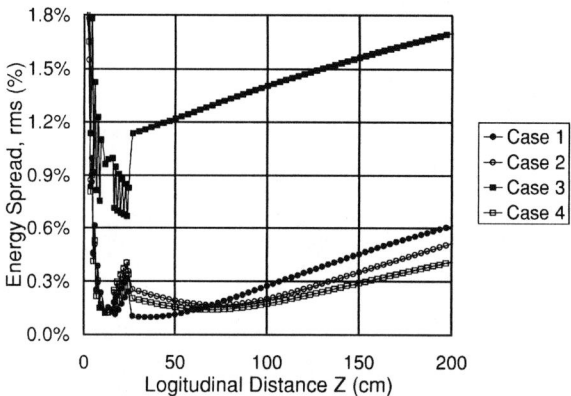

FIGURE 15. rms energy spread for the four operating cases in the X-band PWT obtained from PARMELA simulation.

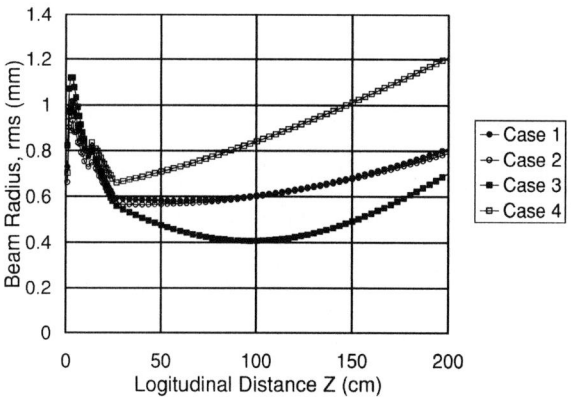

FIGURE 16. rms beam radius for the four operating cases in the X-band PWT obtained from PARMELA simulation.

TABLE 4. PARMELA Results for the Operating Cases (data at point of minimum emittance).

Case	Beam Energy (MeV)	Energy Spread	Emittance (mm-mrad)	Beam Radius (rms mm)	Brightness (10^{14} A/(m-rad)^2)
1	23.30	0.37%	1.35	0.63	5.40
2	29.20	0.34%	1.11	< 0.70	11.9
3	16.59	1.38%	3.40	0.41	0.50
4	25.40	0.34%	1.24	1.1	7.40

ACKNOWLEDGMENTS

The authors would like to thank C. Pellegrini, P. Wilson, F. Hartemann, X. Ding, and S. Telfer for useful discussions, comments and technical assistance. This work is supported by the U.S. Department of Energy SBIR Grants DE-FG03-96ER82159 and DE-FG03-98ER82566.

REFERENCES

1. D. Yu et al., Proc. of Particle Accelerator Conf., Vancouver, B.C. Canada, May 1997, p.2802.
2. D. Yu et al., Proc. of Particle Accelerator Conf., New York, NY, March 1999, p.2203.
3. D. Yu, Proc of 2nd ICFA Adv. Accel. Workshop, UCLA, November 1999, *to be published*.
4. V.B. Andreev, Soviet Physics — Technical Physics, 13, 1070 (1969).
5. D.A. Swenson., European Particle Accel. Conf., Rome, Italy, ed. S. Atzzari, 2 (1988).
6. N.M. Kroll and D. Yu, *Particle Accelerators* 34, 231-250 (1990).
7. J. Rosenzweig, N. Barov, and E. Colby, Proc. Adv. Accel. Concepts, p.724, AIP 335, 1995.

Experimental Demonstration of Dielectric Structure Based Two Beam Acceleration

Wei Gai, M. E. Conde, R. Konecny, J. G. Power, P. Schoessow, X. Sun and P. Zou

*High Energy Physics Division, Argonne National Laboratory
Argonne, IL 60439*

Abstract. We report on the experimental results of the dielectric based two beam accelerator (step-up transformer). By using a single high charge beam, we have generated and extracted a high power RF pulse from a 7.8 GHz primary dielectric structure and then subsequently transferred to a second accelerating structure with higher dielectric constant and smaller transverse dimensions. We have measured the energy change of a second (witness) beam passing through the acceleration stage. The measured gradient is > 4 times the deceleration gradient. The detailed experiment of set-up and results of the measurements are discussed. Future plans for the development of a 100 MeV demonstration accelerator based on this technique is presented.

INTRODUCTION

The development of robust accelerating structures is critical for future high energy accelerators. The concept of using dielectric loaded structure as acceleration structure has been around for many years [1], and in particular there have been a number of proposals to use dielectric loaded structures as collinear wakefield accelerators [2,3]. Recently, there have also been proposals to also use dielectric structures as power extraction devices for high frequency rf generation [4]. The simplicity of this method, as well as the relative ease with which parasitic higher order modes can be damped compared to conventional structures[5] operating at comparable frequencies makes this technology an attractive option for future high energy linear colliders.

Another important issue for linear collider development is RF sources. Using bunched train driven dielectric based technology for RF extraction directly from an intense relativistic bunched electron beam has significant advantages since the radiation frequency only depends on the dielectric device geometry. The RF source frequency can be easily tuned to a harmonic of the linac RF frequency with some tuning range (by adjusting the laser spacing for the RF photocathode gun). In fact, at very high frequencies, it can be viewed as a continuous RF source. This RF power is then directly transferred to a second dielectric tube with much higher dielectric constant, compressing the pulse and hence enhancing the acceleration field. This scheme is also called a step-up transformer; the use of separate beam paths allow transformer ratios> 2 to achieved [4]. By using a multiple drive beam, longer acceleration distance can also be achieved, thus obtain higher gradient and sustained accelerations than the collinear schemes.

In this paper, we report on the proof of principle experimental results on dielectric loaded two beam acceleration experiment. In the past year, we have completed an upgrade of the Argonne Wakefield Accelerator [6,7] with a parallel witness beam line with a magnetic spectrometer for energy measurement of the witness beam. The typical witness beam energy is about 3.5 MeV with charge of 0.5 nC. The step-up transformer experiment is shown in Figure 1 schematically.

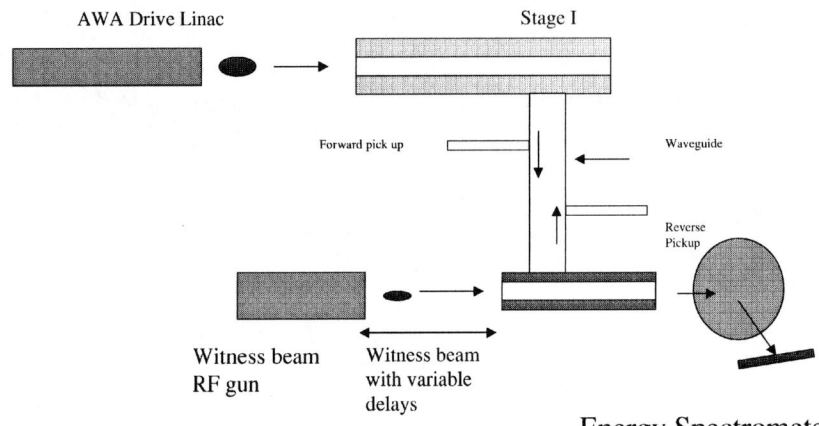

FIGURE 1. Schematic diagram of the two beam acceleration (step-up transformer) experiment. The wakefield generated in Stage I structure is taken out and transferred to a second structure to establish higher acceleration gradient. A less intense witness is injected in the second structure for acceleration experiment. Its energy change depends on the relative delay between the drive and witness beam.

Dielectric Structure Design and Construction

The structures used for these experiments were designed to demonstrate the physics of dielectric based two beam acceleration. The choice of operating frequency is 7.8 GHz which is compatible with the drive bunch length available at AWA while also a harmonic of the 1.3 GHz frequency. The device parameters are summarized in Table 1.

TABLE 1. Parameters for staged dielectric step-up transformer

	Stage I	Stage II
Inner Radius a	6 mm	3 mm
Outer radius b	11.15 mm	5.41 mm
Dielectric constant ε	4.6	20
Group velocity β_g	0.25	0.05
E_z (Max)	3 MV/m	8 MV/m
Designed field step up		2.5
Interaction length for a single beam	30 cm	3 cm

The structures were constructed in house and cold tested using a HP8510C network analyzer. It was very challenging to achieve good coupling between the stages because of the wave impedance mismatch between the dielectric loaded waveguide and standard rectangular waveguide. This problem was solved by tapering the dielectric inner radius near the coupling slot so it can act as an impedance transformer. The measured transmission from stage I to stage II was >96%, sufficient for the proof of principle experiment.

Beam Measurement Results

After the structures were constructed, they were installed in the beamline for experiment using electron beam from the drive gun. In this section we discuss the experimental results.

RF Power Spectrum and Transfer Measurements

The first experiment conducted was to measure the RF power generated from the stage I dielectric tube. We used a HP spectrum analyzer capable of measurements up to 50 GHz. Figure 2 shows the measured spectrum from the forward power flow. It shows that majority of the energy is concentrated near 7.8 GHz as expected. However, a small peak around 8.3 GHz is not expected and further investigation is needed. Expanding the frequency range showed the second deflection mode at 11 GHz as expected. Due to the finite coupling iris band width, we could not observe the first deflection mode.

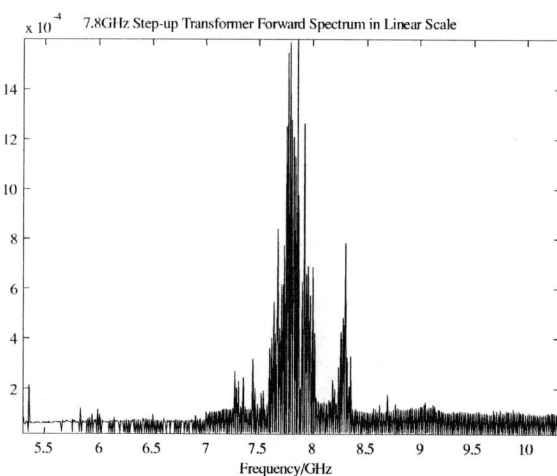

Figure 2. Measured spectrum of the beam induced RF signal from the stage I tube. The energy is concentrated around the TM_{01} fundamental frequency 7.8 GHz as expected.

The RF power flow was also measured between the two stages. In order to do this, a directional coupler was installed in the transfer waveguide. The forward and reflected RF power signal envelope is detected using a high frequency diode as shown in FIG 3. The measured RF pulse width is 2.5 ns, which agrees very well with the prediction based on the dielectric property and structure length. The timing between the forward and reflected power indicates that the small reflection is from the coupling iris of the stage II tube. The estimated reflection coefficient is <5%, as expected from the bench coupling measurements.

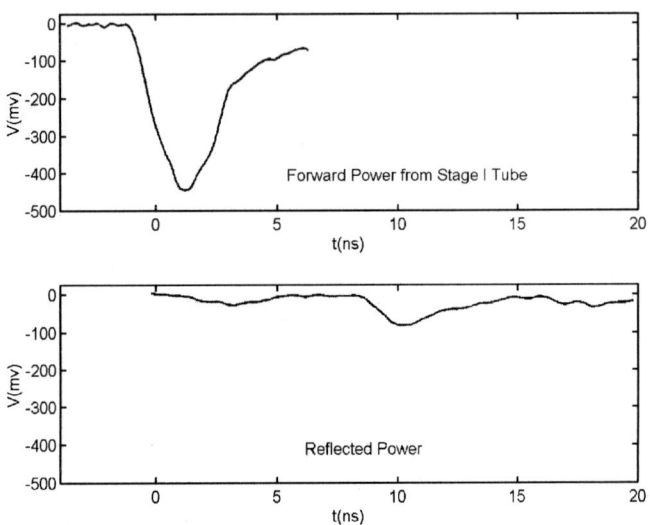

Figure 3. Measured forward and reflected RF power envelope from the stage I to stage II tube. The peak power is 4 MW. The estimated reflected RF power is <5%, corresponding to the bench measurement of the stage I-II coupling.

Acceleration of the Witness Beam in Stage II

When the witness beam injected into the stage II structure, its energy will change depending on its phase relative to the drive beam. At the AWA, this phase is adjusted by varying simultaneously the laser delay and RF phase to the witness gun [7]. Figure 4 shows the results of a typical measurement. The horizontal axis is the energy (bend) plane of the spectrometer. The top picture is the witness beam imaged on a phosphor screen in the spectrometer focal plane with no drive beam passing through the stage 1 tube. In this experiment, the drive beam charge is ~20nC, which would be expected to generate 1.6 MV/m decelerating gradient in the stage 1 tube (and ~ 3 MV/m peak field) as measured indirectly from the forward RF power (4 MW, in Fig 3). The middle picture shows the witness energy at 56 ps nominal delay. (Note the nominal drive-witness delays given below are measured with repect to an arbitrary zero-delay point). The maximum energy gain is about 250 KeV which is corresponding to about 7.5 MeV/m, agreeing well with the prediction as shown in table 1. Therefore, the field step-up ratio is > 2 as predicted. Moreover, This also mean the transformer ratio (Maximum accelerating field/maximum decelerating field in driving tube) exceeds 4. This demonstrates experimentally that the wakefield theorem can be violated in a noncollinear geometry devices. When the delay is changed to − 8 ps, the witness beam experienced maximum decelerating phase of the wakefields. Thus the wavelength in the second stage is ~ 132 ps, which corresponds to 7.8 GHz excitation in stage II with phase velocity c.

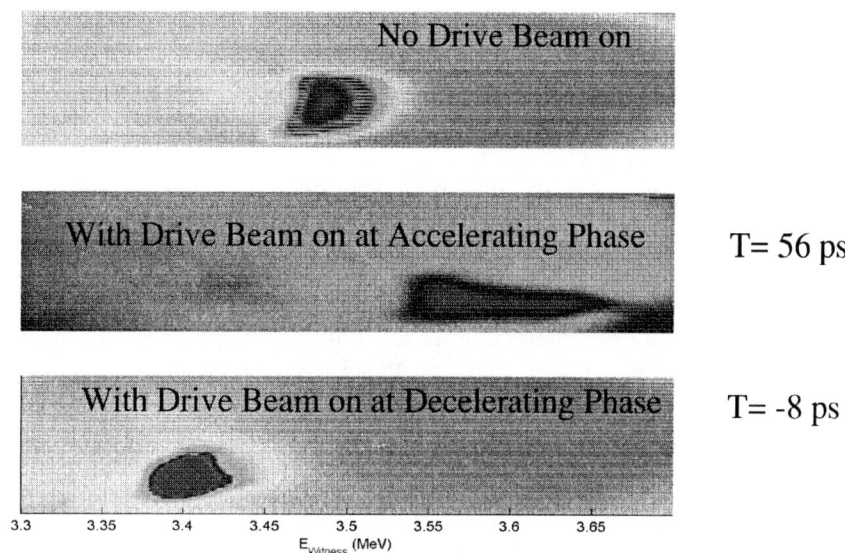

Figure 4. Measured energy change of the witness beam. The top is with drive beam off and middle is with drive beam on and with acceleration phase of beam delay at 56 ps. The bottom shows the deceleration.

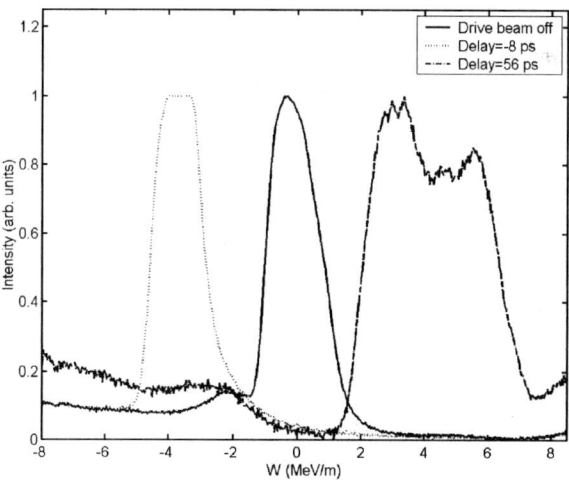

Figure 5. Bend view projections of the data in Figure 4.

Projections of the beam profile in the energy plane are shown in Figure 5. The measured energy change of the witness beam in the stage II dielectric tube demonstrated the following results: the RF generated is efficiently coupled to the TM_{01} accelerating mode between the stages. The field gradient in stage II is stepped-up by a factor > 2 with transformer ratio of >4. Finally, the efficient coupling technique developed using bench measurements is found to couple to the desired phase velocity for TM_{01} acceleration mode.

Future Plans

Although we have successfully tested the step-up transformer concept using the two beam acceleration method, there are still some detailed experiments to be done. In the short term, we will map in detail the accelerating field by adjusting the pulse delay between the drive and witness beams in small increments and observing the change in witness beam energy. Then we will use the multiple drive beam to increase the acceleration distance from 3 cm to 12 cm using 4 drive pulses.

The limiting factor in accessing higher acceleration gradients is the drive beam properties, in particular that the current drive beam has relative high emittance and longer bunch length. A further limiting factor is the number of drive bunches produced due to the quantum efficiency of the photocathode. There is major ongoing effort to improve the drive beam properties by constructing a third generation drive gun[8]. The new gun is a 1 ½ cell RF photocathode gun with axial electric field of 80 –100 MV/m. Based on PARMELA[9] simulations, it will produce much lower emittance (by a factor of 20) and shorter beams (3 – 4ps) for intensities of 40 - 100 nC. Another improvement will come from the operating vacuum and cathode upgrades. We intend to replace the current Mg cathode with a CsTe type high QE cathode. Based on the current available AWA laser system (5 mJ, 4ps @ 248 nm), we could produce a pulse train up to 4 pulses with 100 nC each. By using this pulse train we could easily achieve 100 MV/m acceleration gradient using the same two beam acceleration (step-up transformer) structures used in this experiment. This gun would also produce up to 64 pulses with 40 nC each. By using the same step up transformer, we would not only achieve 100 MV/m, but also could accelerate the beam to 100 MeV in less than a meter. This is the equivalent of powering the second stage tube with external 500 MW RF power source with a 50 ns pulse length. The new electron gun is expected to be in operation in the fall of year 2000.

SUMMARY

Considerable progress has been made towards a demonstration of the dielectric loaded two beam accelerator concept. A proof of principle experiment clarified the associated physics and engineering issues such as RF coupling and acceleration in the correct acceleration mode. With the new AWA electron gun, we will further extend this experiment by using the high current pulse train to provide > 100 MV/m gradient over an acceleration distance of 1 meter.

ACKNOWLEDGMENTS

This work is supported by the US Department of Energy, Division of High Energy Physics, under contract W-31-109-ENG-38.

REFERENCES

1. G. Flesher and G. Cohn, AIEE Trans. **70**, 887 (1951)
2. R. Keinigs, M. Jones, and W. Gai, Part. Accel. **24**, 223 (1989)
3. T. B. Zhang, J. Hirshfield, T. Marshall, and N. Hafizi, Phys. Rev. E 56, 4647 (1997)
4. E. Chjonacki, W. Gai, P. Schoessow, and J. Simpson, in Proceedings of Particle Accelerator Conference, edited by L. Lizama and J. Chew (IEEE, San Francisco, 1991), pp. 2557-2559.
5. E. Chjonacki *et al*, J. Appl. Phys. **69**
6. M. E. Conde et al, Phys. Rev. ST Accel. Beams **1**, 041302, 1998.
7. M. E. Conde et al, "A High-Charge High-Brightness L-Band Photocathode RF Gun", Submitted to the Proceedings of the ICFA Advanced Accelerator Workshop (Los Angeles, Nov. 1999) edited by J. Rosenzweig, World Scientific Publisher, 2000
8. J. G. Power and M. E. Conde, Rev. Sci. Instrum. 69, 1295 (1998).
9. SUPERFISH and PARMELA, Los Alamos National Laboratory Report No. LA-UR-96-1834, 1997; Report No. LA-UR-96-1835, 1996.

Acceleration Results from the Microwave Inverse FEL Experiment

R. B. Yoder,[a] T. C. Marshall,[b] and J. L. Hirshfield[a,c]

[a]*Physics Dept., Yale University, PO Box 208120, New Haven, CT 06520-8120*
[b]*Dept. of Applied Physics, Columbia University, New York, NY 10027*
[c]*Omega-P, Inc., 202008 Yale Station, New Haven, CT 06520-2008*

Abstract. An inverse free-electron-laser accelerator has been developed, built, and operated in the microwave regime. Development of this device has been described at previous Workshops; the accelerator is driven by RF power at 2.8 GHz propagating in a smooth-walled circular waveguide surrounded by a pulsed bifilar helical undulator with tapered pitch, while an array of solenoid coils provides an axial guide magnetic field. In low-power experiments, injected electron beams at energies between 5 and 6 MeV have gained up to 0.35 MeV with minimal energy spread, and the phase sensitivity of the IFEL mechanism has been clearly demonstrated for the first time. Agreement with simulation is very good for accelerating phases, though less exact otherwise. Scaling the device to high power and high frequency is discussed.

INTRODUCTION

A proof-of-principle experiment to accelerate electrons via the inverse FEL interaction at microwave frequencies has recently been completed at Yale University and Omega-P, Inc., after several years of development [1,2,3]. While IFEL devices have been discussed since 1972 [4] and demonstrated in the millimeter-wave [5] and optical [6] regimes, the results presented here are the first to show IFEL acceleration in an RF structure; we have also demonstrated the variation in energy change with the initial phase at which the electrons are injected into the accelerator, a measurement that has not previously been attempted.

The free-electron-laser interaction is commonly approximated by the one-dimensional relation (for constant wiggler period)

$$\frac{d\gamma}{dz} = \frac{k_s a_s a_w}{\gamma} \sin \Phi \qquad (1)$$

where $k_s = 2\pi/\lambda_s$ is the free-space wavenumber of the RF driving wave, γ is the relativistic energy factor $(1-\beta)^{-1/2}$, $a_w = eB_w/k_w mc$ is the undulator parameter, and $a_s = eE_0/\omega_s mc$ is the normalized electric field of the RF wave. The so-called ponderomotive phase Φ represents the position of the electron in the beat wave formed by the wiggler and electromagnetic fields, and the sine term shows the reciprocal

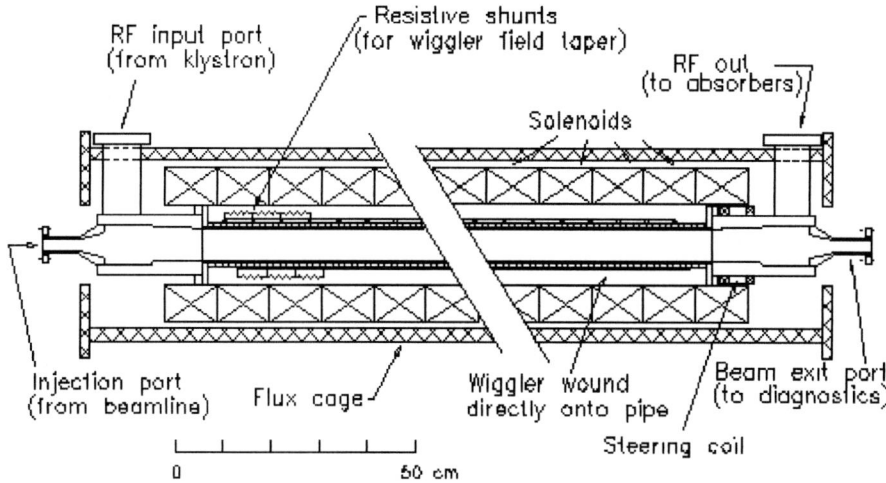

FIGURE 1. Drawing of the MIFELA structure, with the beam path from left to right along the axis of the waveguide. The wiggler and solenoid magnets are shown, as well as RF input and output couplers and an iron flux cage to contain the axial field.

nature of the FEL and IFEL mechanisms, since positive values of $\sin\Phi$ denote accelerating phases (IFEL) and negative values decelerating phases (FEL radiation). Injection at a single value of this phase thus enables clear measurement of energy gain or loss for an entire beam, avoiding the spread in output energies that results from injection over a broad range of phases.

In the Yale device, known as the Microwave IFEL Accelerator or MIFELA, the acceleration structure is simply a cylindrical waveguide of prescribed radius, into which RF power at 2.856 GHz is fed through a specially designed input coupler which sets up a circularly polarized traveling wave. The helical wiggler field is tapered in pitch for maximum acceleration gradient; the undulator parameter $a_W = eB_W/k_W mc$ increases from 2.4 to 2.75 at an initial period of 11.75 cm. A second, axial, magnetic field is used for orbital stability and guiding. The injected MIFELA beam is produced by an RF gun, which operates at the same microwave frequency as the accelerator itself. Up to 23 MW of RF power from the laboratory klystron is thus divided between gun and accelerator, with appropriate phase delays. The beam travels through an 11-element achromatic beamline which performs energy selection and focusing before injection into MIFELA. Current and position monitors allow fine-tuning of the dipole and quadrupole fields, and focusing and energy selection are carried out by a 19-element beamline before injection into the accelerator. After its exit from MIFELA, the energy of the beam is analyzed by a magnetic dipole spectrometer, in which the beam is dumped into a Faraday cup and the return current measured.

The wiggler consists of a pair of helical conductors, which are wound directly on the exterior wall of the waveguide and pulsed at high current via a capacitor discharge bank. Up-tapering the wiggler period accomplishes the increase in the wiggler field over the length of the structure, while the nonlinear field shape in the beam entry region is brought about by reducing the current using resistive shunts between the two windings. The details of the construction of MIFELA are described in the next section, its parameters are summarized in Table I, and a schematic drawing of MIFELA is shown in Figure 1.

TABLE 1. Parameters for MIFELA during experiments reported in this paper.

Dimensions		
	Total length	180 cm
	Adiabatic entry region	58.75 cm
	Acceleration region	105 cm
RF parameters		
	Frequency	2.856 GHz
	Drive power	3–7 MW
	Normalized RF field $a_s = eE_0/mc\omega_s$	0.07–0.09
	Pulse length	2 µs
	Polarization	Circular
	Waveguide radius	3.14 cm
	Waveguide refractive index $n = \omega/ck$	0.2
	Waveguide mode	TE_{11}
Magnet parameters		
	Axial guiding field	1.58 kG, flat
	Wiggler winding radius	3.84 cm
	Wiggler period	11.75–12.33 cm, linear taper
	Wiggler field on axis	1.15–1.3 kG
	Wiggler strength parameter $a_w = eB_w/mck_w$	2.4–2.75
Beam parameters		
	Initial energy	5.1–6.1 MeV
	Peak current	• 0.3 A
	Micropulse length	5 ps

CONSTRUCTION OF THE EXPERIMENT

Acceleration Structure and Magnets

The circular waveguide which makes up the MIFELA serves the functions of maintaining the correct waveguide mode, acting as the vacuum vessel, and supporting the windings for the wiggler magnet, which is wound on its exterior wall. The MIFELA operates in the TE_{11} waveguide mode with circular polarization; in the interests of maintaining high power density, the guide radius is near cutoff for the driving frequency of 2.856 GHz, with a waveguide refractive index $n = \omega/ck = 0.2$.

With a refractive index near zero, small changes in radius have a large effect on RF propagation velocity, so that the construction tolerance on the inner radius was very tight; this was a machining challenge which limited the MIFELA structure to less than 180 cm in length. The construction material was stainless steel with the minimum possible wall resistance (4 mm); this design was used in order to minimize eddy-current shielding of the pulsed wiggler field.

The wiggler windings are heavy-gauge copper, with an initial period of 11.75 cm—on the order of an RF wavelength—which is increased to 12.3 cm over the length of the MIFELA with the use of a precise insulating spacer between the filament windings. Such a taper maintains the FEL resonance as the particle energy increases, and also keeps the acceleration gradient maximal, as Eqn. 1 implies. For acceleration to take place, the beam must have transverse velocity components, and at these fields, the radius of the transverse orbits is relatively large; hence the beam, injected on axis, must be 'spun up' until it attains its final gyration radius. This adiabatic entry is accomplished in an injection region in which the wiggler and guiding fields are gradually ramped up from zero to their acceleration values, so that the beam's momentum is transferred from axial to transverse, with little energy change [3]. This injection region occupies the first 58.75 cm, in which the wiggler period remains constant at 11.75 cm. The up-taper in the wiggler field follows a nonlinear taper [$B(z) = B_0\sin^2(2kz/\bullet)$, $0<z<\bullet/2k$, where k is the wiggler wavenumber], which proved to give the best axis-encircling orbits; this ramp is produced by a series of 23 resistive shunts linking the two filaments of the wiggler, so that the current in the filaments falls to zero in small decrements. Figure 2 shows the circuit of the wiggler and the measured field on axis. On exit from the accelerator, the beam must again be steered back to the axis; this is accomplished by a long-pulse steering magnet, there being insufficient space for an adiabatic exit region.

The axial guide field is generated by a series of 18 independently-controlled solenoids, in order that a range of field profiles may be produced. For acceleration, a constant 1.58 kG field was used over the middle of the structure. Plots of the fields in the structure are shown in Figure 3, along with the calculated evolution in energy of an electron bunch in the accelerating phase, showing that acceleration does not begin until the fields have reached their resonant values.

RF System, Beamline, and Diagnostics

The beam accelerated by MIFELA is produced by a 2-1/2 cell RF gun operating at 2.856 GHz and consists of a train of 5 ps microbunches of about 10^8 particles each, with an energy near 6 MeV. With an RF wavelength of 10.5 cm, this bunch size represents a phase spread of roughly $\bullet/30$ radians—so small as to approximate injection at a single phase. RF power from an XK-5 klystron is split between the gun and the accelerator structure using a variable splitter and phase shifter, so that phase synchronism between the beam pulse and the accelerating bucket is assured. After exiting the gun, the beam travels through an achromatic beamline containing an energy selection slit, which reduces the energy spread on the injected beam to the 1%

level. Diagnostics on the beamline include toroidal current monitors and phosphor screens.

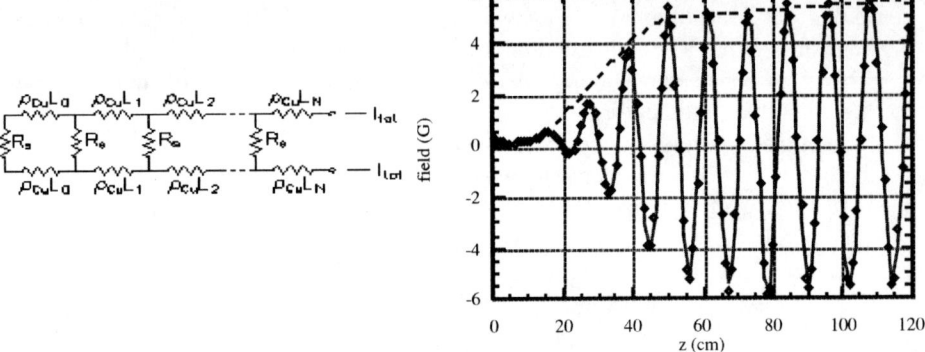

FIGURE 2. The left side shows the wiggler shunt circuit, where R_s is the resistance of a shunt, ρ_{Cu} the resistance/unit length of copper, and L_i the length of filament between two shunts. On the right side is plotted the x-projection of the wiggler magnetic field (DC benchtop measurement); the solid line is interpolated from the data points and the dashed line shows total field amplitude.

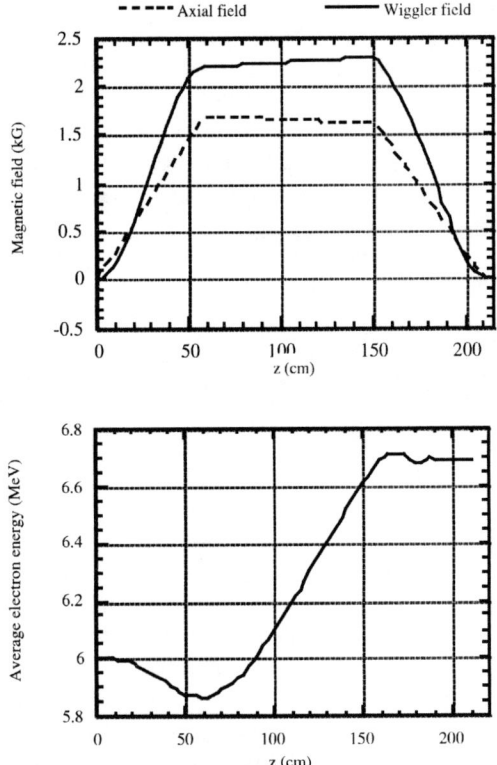

FIGURE 3. Left side: ideal calculated field profiles for MIFELA, showing the field up-taper before acceleration begins. Right side: energy evolution with those fields and 15 MW of RF power; acceleration does not begin until fields are resonant.

With 7 MW of RF used to power the electron gun, up to 15 MW of 2.856 GHz power is left for use in the accelerator itself, and is injected via an input coupler having two arms 90 degrees apart; this sets up the required circularly polarized travelling wave in the waveguide. Beam loading is minimal, and most of the RF power is coupled out of the guide at the end of the acceleration section and absorbed in matched loads.

The energy distribution of the electron beam is measured with a magnetic dipole spectrometer, the beam being collected in a Faraday cup after passing through a 3 mm slit. Calculations and comparison with known attributes of the beamline indicate a resolving power (for this energy range) of 1.5%, sufficient to detect energy changes on the order of 100 keV or less. The beam may also be allowed to continue through a series of quadrupoles and steering magnets and viewed on a series of phosphor screens.

EXPERIMENTAL RESULTS

Data Collection Procedures

For each experimental run, an initial measurement of beam energy, giving an unaccelerated-beam spectrum at a particular phase value, was obtained using a beam passing through the MIFELA in the absence of RF power and wiggler field (although the axial field was retained in order to stabilize the beam as it traveled through the acceleration tube). The steering magnet field required for the run was calibrated on this unaccelerated beam with the wiggler energized. Once the steering current had been found, both steering and wiggler currents were left unchanged for the rest of the run. At this point, the overall RF power was increased and the power split adjusted to bring power to MIFELA while maintaining the energy of the injected beam. Data was taken by operating the MIFELA in single-shot mode and building up a beam spectrum from a series of shots as the spectrometer was scanned through the energy range. The energy spectrum with RF in MIFELA in the absence of wiggler field proved to be unchanged from the null result, so that an unaccelerated spectrum could be obtained by scanning the spectrometer without pulsing the wiggler. This provided a check against slow drift in the beam energy.

Because of the pulsed, single-shot nature of the experiment, each individual energy spectrum consisted of a series of collected current signals vs. time, each obtained at a different spectrometer setting. A spectrum was assembled by assigning a single peak value to each current trace, after baseline subtraction to eliminate voltage offset and noise in the electronics. Sources of uncertainty in this reduction come both from indefinite peak heights and from the uncertainty in the spectrometer current/energy calibration. To compensate for shot-to-shot variation in injected beam current, the peak current measured in the current monitor nearest to the gun was also recorded for each shot, and the collected currents were scaled to correspond to an identical injected beam current. Shot-to-shot variation in the wiggler current was monitored throughout; it was negligible compared to other sources of uncertainty for data presented here.

Acceleration Data

The data presented here are in the form of spectrum comparisons for beam energies with and without the wiggler field operating. (No energy deviation was seen if only RF power was present.) Figure 4 shows output spectra obtained at two different phase values, 155° apart, taken with an input beam energy of 5.62 MeV and RF power of about 6 MW. The wiggler current was 32 kA, leading to an initial field of 1.3 ± 0.1 kG, and beam current at injection was equal to 80 mA. The plot for 6° shows an acceleration of 0.34 MeV, with an uncertainty of roughly 40 keV; there is clear separation between the unaccelerated and accelerated spectra, since both of them have widths of 0.15 MeV. There are no measurable unaccelerated particles. While there is beam loss on the order of 30%, the spectral shape is nevertheless quite consistent between the two, with a suggestion of structure on the high-energy side appearing in both curves. The effective acceleration gradient is 0.43 MV/m; the energy gain of 6% is the highest reported to date for an IFEL. The plot for 161°, with a distorted spectral shape in which the average and peak energies differ, is harder to interpret. If average energies are compared in this case, the beam has gained about 0.07 MeV. While Eqn. (1) predicts that the FEL interaction should be symmetric with respect to phase—that is, two phase values nearly 180° apart should lead to equal and opposite energy changes—the use of a tapered-period wiggler breaks this symmetry, so we do not expect the two spectra here to be opposites, or MIFELA performance to be similar at different phases. We believe that the unoptimal conditions of injection into the decelerating phase lead to greater beam instability and sensitive dependence on conditions during the interaction in that case.

Figure 5 shows a more complete investigation of the IFEL phase response with a different set of initial conditions. For this data set, input power was about 3 MW, with wiggler current of 30 kA producing an initial field of 1.1 ± 0.1 kG and axial field of 1.58 kG. The beam energy at input was 5.24 MeV, with a peak injected current near 40 mA.

The maximum energy gain in this case is 0.20 ± 0.02 MeV, occurring at a phase value of 40°. Again, there is an asymmetric result for the energy change vs. phase, with results for "decelerating" phases being both mostly positive (although near zero) and less clean than results for accelerating phases. The error bars on this graph give an indication of the uncertainty involved in assigning a single number to the energy in each case, with larger bars corresponding to multiply-peaked or widely spread output spectra.

On the same figure, for comparison, are the results of simulation for these conditions. The simulated beam was monoenergetic but was Gaussian in divergence angle; v_z/v values were taken from a distribution with width $\sigma=0.01$.

Figure 5 shows an excellent fit between modeling and theory for acceleration phases. The only free parameter in use was an overall additive constant on the energy data, representing additive uncertainty in the spectrometer calibration with the steering correction included; this constant has a best-fit value of –70 keV. Data in the deceleration phases is much less clean; agreement is imprecise, and one of the data

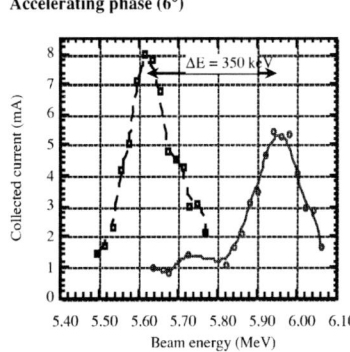
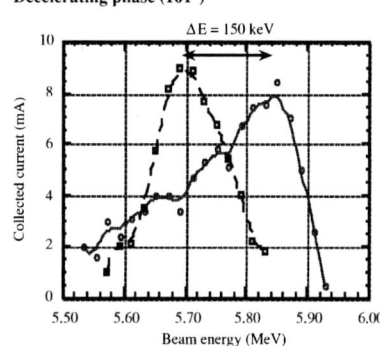

FIGURE 4. Experimental energy spectra for two phases, separated by 155°, showing results for accelerating and decelerating phases. The exiting beam spectrum under MIFELA operation (solid line, circles) is compared with the spectrum in the absence of wiggler fields (dashed line, squares).

points appears to be well above the expected energy. This discrepancy is partially resolved below.

In Figure 6, the acceleration spectra from Fig. 4 are superimposed with the results of model calculations for those cases, using the same best-fit additive constant of −70 keV as above. Use of a physical beam with energy spread on the order of the measured spectral width results in an accelerated-beam spectrum that very closely matches the experimental data. Again, the decelerating phase match was not as good, although part of the distribution does overlap with that of the injected beam. The simulation, however, does predict a higher degree of beam loss, a greater beam spread, and a somewhat distorted spectral shape for the decelerating phase as compared to the accelerating phase; asymmetry in the gain curve is also present. It is probable that these characteristics indicate overall instability in the decelerating-phase case, when the wiggler taper no longer matches the energy change, and that the resulting sensitivity of the beam to exact field and injection conditions would make its simulation less reliable.

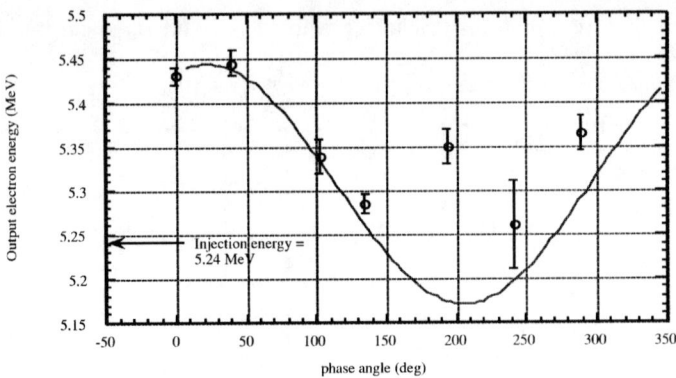

FIGURE 5. Experimental plot of output beam energy for an input beam energy of 5.24 MeV and 3 MW of RF power, as a function of injection phase, compared with simulation (solid line). Error bars denote the uncertainty in the identification of output energy values.

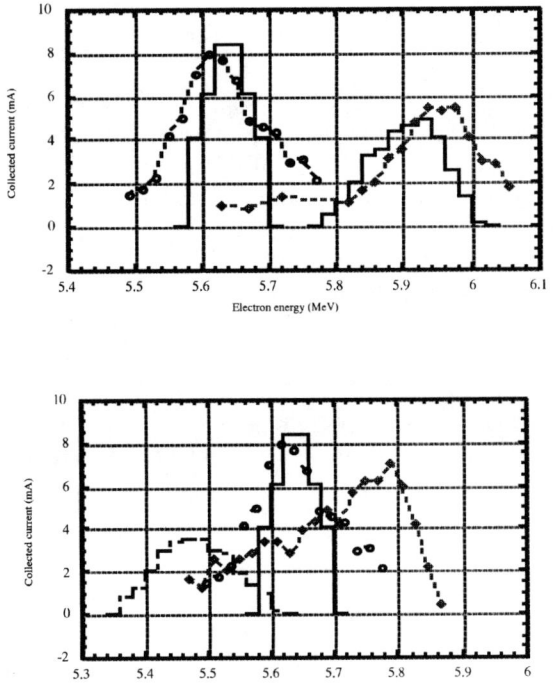

FIGURE 6. Simulation results compared with data from Fig. 4. Upper: accelerating phase, with injected beam (circles) compared with exiting beam (diamonds) and simulation of each. Lower: decelerating phase, showing injected beam (circles) and simulation (solid line) compared with exiting beam (diamonds) and simulation (dashed line.)

CONCLUSION

We have constructed and operated an IFEL accelerator using microwave power, designed to accelerate a 6 MeV electron beam produced by an RF gun. Engineering solutions and features include a pulsed bifilar helical winding to create the wiggler field, shunt resistors to ensure adiabatic beam entry, and continuously variable injection phase, with a very small effective "phase window." Automatic synchronization between RF fields and electron beam is enforced by splitting a single RF pulse between the RF gun and the accelerator structure. When operated at the appropriate phase and using 6 MW of RF power, the MIFELA exhibited energy gain of 0.35 MeV, with virtually unchanged energy spread, and an accelerated fraction of nearly 70%. We emphasize that no one to date has reported IFEL acceleration results of this kind: unambiguous acceleration of the entire exiting beam, with no untrapped particles, and with energy gain more than twice the width of the spectral peak. The percentage energy change of 6% for all exiting particles currently exceeds others in the literature. The importance of injection phase was demonstrated repeatedly; for accelerating-phase values, simulation results agree well with experiment, while off-phase injection results in degraded beam quality, energy spread, and insignificant energy change, and leads to less exact agreement with simulation. These findings imply that in high-gradient IFELs, energy spreads created by large phase width at injection could be a concern, and efficient pre-bunching would then become a priority [7].

The magnitude of the observed energy change is a function of the frequency and power of the driving radiation. Acceleration gradients can be shown to scale roughly linearly with the RF frequency and with the square root of drive power; use of a cavity rather than a traveling-wave device would increase the effective power levels considerably, so that the gradient of the MIFELA could increase by a factor of 3 to 4, and using 150 MW of 34 GHz radiation, gradients well above 30 MeV/m appear likely. An RF-powered IFEL could thus be of interest as a pre-injector for other high-energy machines.

ACKNOWLEDGEMENTS

The authors acknowledge the collaboration of M. A. LaPointe, Mei Wang, and T. B. Zhang, of Yale and Omega-P, helpful discussions with A. K. Ganguly (Omega-P) and S. Y. Park (POSTECH, Pohang, Korea), and technical assistance from S. Gold (NRL) and M. Shapiro (MIT). This work was supported by the U.S. Dept. of Energy, High Energy Physics Division.

REFERENCES

1. Hirshfield, J. L., *et al.*, *Nucl. Instr. Meth. Phys. Res.* A **358,** 129–130 (1995).
2. Yoder, R. B., Zhang, T. B., Marshall, T. C., and Hirshfield, J. L., "Simulation Results and Experimental Design for the Microwave Inverse FEL Accelerator" in *Advanced Accelerator Concepts, Seventh Workshop* (Lake Tahoe, 1996), edited by S. Chattopadhyay et al., AIP Conference Proceedings 398, New York: American Institute of Physics, 1997, pp. 629–637.
3. Yoder, R. B., Marshall, T. C., Wang, M., and Hirshfield, J. L., "Status of the Microwave Inverse FEL Experiment" in *Advanced Accelerator Concepts, Eighth Workshop* (Baltimore, 1998), edited by W. Lawson et al., AIP Conference Proceedings 472, New York: American Institute of Physics, 1999, pp. 635–643.
4. Palmer, R. B., J. Appl. Phys. **43**, 3014–3023 (1972).
5. Wernick, I., and Marshall, T. C., Phys. Rev. A **46**, 3566–3568 (1992).
6. van Steenbergen, A., Gallardo, J., Sandweiss, J., and Fang, J.-M., *Phys. Rev. Lett.* **77**, 2690–2693 (1996).
7. Liu, Y., *et al.*, *Phys. Rev. Lett.* **80**, 4418–4421 (1998).

Bunch Stability During Wake Field Generation in a Dielectric-Lined Waveguide

S. Y. Park[*,†,¶] and J. L. Hirshfield[*,†]

*Beam Physics Laboratory, Yale University, 272 Whitney Ave., New Haven, CT 06511
†Omega-P, Inc., 345 Whitney Ave., New Haven, CT 06511
¶Department of Physics, POSTECH, Pohang 790-784, Korea

Abstract. A recently-developed analytic theory for wake fields generated when a charge bunch, or train of bunches, passes along a dielectric-lined waveguide is applied to examine stability issues for this system. In particular, examples are calculated of longitudinal spreading of an initially tight drive bunch, due to non-uniform drag forces along the bunch. The effect of this spreading on the acceleration of a following test bunch is also determined. Furthermore, when the drive and test bunches are injected parallel to, but displaced from, the waveguide axis, transverse (mainly dipole) forces cause the tail of the bunch or bunches to swerve sharply towards the waveguide wall, and to intersect the wall after a relatively short time. This effect is stronger for a test bunch that feels transverse forces from the drive bunch together with its own non-uniform drag forces. These results suggest that successful exploitation of strong wake fields generated by one or more drive bunches will require short high-gradient acceleration modules, together with some means of transverse focusing, such as a FODO.

INTRODUCTION

An analytical theory has been developed recently for wake field generation by one or more charge bunches moving parallel to the axis of a dielectric-lined cylindrical waveguide [1]. This configuration is of interest because of the potential it holds as the basis for a high-gradient two-beam accelerator for electrons and positrons [2]. Attributes of the new theory include the following:

- An explicit field solution has been found for the wake fields in terms of an orthonormal set of HEM_{mn} eigenfunctions that, for a bunch moving on the axis, reduce to an orthonormal set of TM_{0n} (monopole) modes.
- Orthonormalization constants are found (for the first time), both for a stationary source (i.e., continuous beam) and for a localized source (one or more bunches); these constants are shown to be different.
- Explicit solutions are found for forces within a distributed bunch, and on one bunch from bunches that precede it; longitudinal and transverse forces are found to satisfy the Panofsky-Wenzel theorem [3], as of course they must.

- Poynting's theorem for this system dictates that convected Coulomb energy must be subtracted from the Poynting flux to obtain the radiation power, which in this case is shown to propagate in a direction opposite to the bunch motion.
- Short bunches are shown to excite many (>100) waveguide modes, that can coalesce spatiotemporally to produce intense localized periodic wake fields.
- A dielectric-lined waveguide can be designed so that the period of the localized wake fields is equal to the period of a train of drive bunches, thereby allowing coherent build-up of the wake fields.
- It is shown that kA short bunches can produce >100 MV/m accelerating fields in a model dielectric-lined waveguide, but that dipole (and higher multipole) forces can lead to serious instability in the absence of strong focusing.
- Extensions to this theory are required to account for axial boundaries that are obviously part of any realistic dielectric wake field module [4]. Boundaries can allow buildup of wake fields due to multiple internal reflections.

The idea of arranging coherent superposition of wake fields from a succession of drive bunches to produce a strong localized accelerating field for a trailing test bunch was first put forward in 1997 [5]. Since, further analysis has appeared [6], and experiments have been proposed and carried out [7,8]. Growing interest in this approach for achievement of a high-gradient lepton accelerator has also prompted a necessary study of stability issues [9]. However, until now, no analytical theory for multipole longitudinal and transverse forces on one or more distributed bunches in a dielectric-lined waveguide has been available. Prior stability analyses rested upon particle simulation codes [9] which, while of enormous power, cannot compete with an analytical theory for providing insight and scaling estimates for the underlying physics. The purpose of the present paper is to show examples of the use of the new analytical theory in computing the motion of charges within a distributed bunch, either due to non-uniform drag forces within a bunch, and/or due to fields of a preceding bunch. Such analysis can provide the basis for evaluation of means of stabilization, chiefly among which ranks externally applied focusing and defocusing (FODO) [9]. However, a significant caveat deserves mention: Present theory is for an infinitely-long waveguide, so effects of boundaries for a (realistic) finite-length accelerating module are not included. This circumstance is in the process of being remedied [10].

BACKGROUND

In ref. [1], closed-form analytical solutions were found for the wake fields generated by a charge bunch moving parallel to (but not necessarily along) the axis of a dielectric-lined cylindrical waveguide. To illustrate, solutions for the wake fields were used to generate a plot of electric field flux lines emanating from a 50 MeV drive bunch moving along the axis; this plot is shown in Fig. 1. For this example, the inner and outer radii of the dielectric liner are $R_1 = 0.05$ cm and $R_2 = 0.15$ cm; the relative dielectric constant is 9.5; and the bunch length is $\Delta = 0.02$ cm. The Cerenkov cone is clearly discernable, as it suffers multiple reflections from the outer cylindrical conducting boundary. For a 2 nC drive bunch, the peak axial accelerating field at $z =$

1.85 cm is 155 MV/m. Fig. 2 shows the longitudinal electric field for two such drive bunches, with an accelerating field at the location of a trailing test bunch of about 280

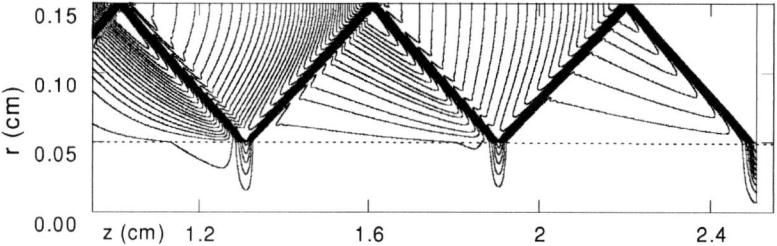

Figure 1. Plot of electric field flux lines for wake fields emanating from a test bunch moving to the right in a dielectric-lined waveguide. See text for parameters

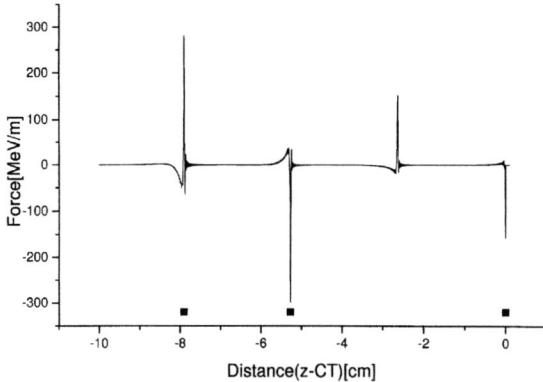

Figure 2. Longitudinal electric field due to two 2 nC, 30 MeV, 0.2 mm drive bunches in a dielectric waveguide. Black dots signify bunch locations. Accelerating field at the test bunch is ~280 MV/m.

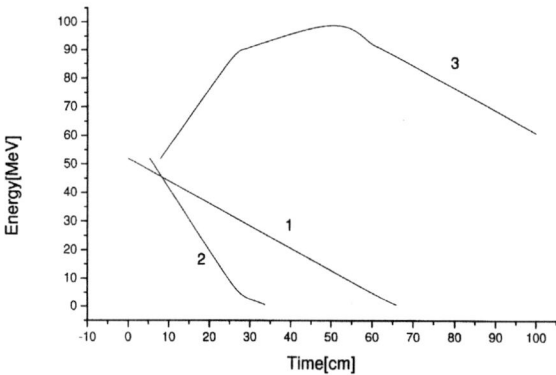

Figure 3. Energy loss for drive bunches (1 and 2), and energy gain for test bunch (3).

MV/m; coherent superposition of fields is evident. Fig. 3 shows the history of energy loss for the two drive bunches, and energy gain for the test bunch; an energy gain for the test bunch of about 48 MeV is seen, after traveling a distance of about 40 cm, for an average gradient of 120 MV/m. This is less than what one might expect from the peak accelerating field, but can be explained as due to (i) the sharp diminution in gradient after each drive bunch loses nearly all its energy, and (ii) averaging over the finite-length test bunches. Nevertheless, this significant magnitude of energy gain by a test bunch from modest drive bunches provides motivation to seek deeper understanding of issues that surround this approach to high-gradient acceleration.

STABILITY

An expression derived in ref. [1] for the forces on a test charge q due to the wake fields generated by a drive bunch charge q_o, valid in the limit as the beam energy factor for the moving charges $\gamma \to \infty$, is

$$\begin{pmatrix} F_z(\mathbf{r},t) \\ F_r(\mathbf{r},t) \\ F_\theta(\mathbf{r},t) \end{pmatrix} = -2qq_o \sum_{\ell=-\infty}^{\infty} \sum_{n=1}^{\infty} \frac{1}{C_n}\left(\frac{r_o}{R_1}\right)^\ell \begin{pmatrix} \left(\dfrac{r}{R_1}\right)^\ell \cos\ell\theta\, g_{z,n}(s) \\ \ell\left(\dfrac{r}{R_1}\right)^{\ell-1} \cos\ell\theta\, g_{\perp,n}(s) \\ -\ell\left(\dfrac{r}{R_1}\right)^{\ell-1} \sin\ell\theta\, g_{\perp,n}(s) \end{pmatrix}. \quad (1)$$

In Eq. 1, ℓ and n are azimuthal and radial mode indices; C_n is the orthonormalization constant; R_1 is the inner radius of the dielectric liner; r_o is the radial displacement of the charges from the axis; $s = z - vt$ with v the velocity of the drive bunch; and, for a rectangular drive bunch of length Δ,

$$\begin{pmatrix} g_{z,n}(s) \\ g_{\perp,n}(s) \end{pmatrix} = \frac{1}{k_n\Delta} \begin{pmatrix} \sin k_n s' \\ \cos k_n s' \end{pmatrix} \Theta(s') \Bigg|_{s'=-s-\Delta/2}^{s'=-s+\Delta/2}, \quad (2)$$

where k_n is the axial wavenumber and $\Theta(x)$ is the Heaviside function (1 for $x > 0$, and 0 for $x < 0$). When more than one drive bunch is present, Eq. 1 is generalized by summing over bunches, with s for each bunch suitably displaced in z. It is straightforward to show for the components in Eq. 1 that $\nabla_\perp F_z = (\partial/\partial z)\mathbf{F}_\perp$, in conformity with the Panofsky-Wenzel theorem [3]. Close examination of Eq. 1 also reveals that the forces for $|\ell| > 0$ enjoy no favorable inverse-γ scaling that might provide a qualitative advantage in stability, as compared with that for a conventional rf linac structure. It is helpful in interpreting Eq. 1 to give explicit forms for the g-functions. In front of the bunch $(s > \Delta/2)$, one finds $g_{z,n}(s) = g_{\perp,n}(s) = 0$, reflecting causality. In the bunch $(\Delta/2 > s > -\Delta/2)$, one finds $g_{z,n}(s) = (k_n\Delta)^{-1}\sin k_n(-s+\Delta/2)$ and $g_{\perp,n}(s) = (k_n\Delta)^{-1}[\cos k_n(-s+\Delta/2) - 1]$. Finally, behind the bunch $(s < -\Delta/2)$, one finds $g_{z,n}(s) = \alpha_n(\Delta)\cos k_n s$ and $g_{\perp,n}(s) = \alpha_n(\Delta)\sin k_n s$, with $\alpha_n(\Delta) \equiv \sin(k_n\Delta/2)/(k_n\Delta/2)$. These forms indicate that the longitudinal and transverse fields are oscillatory within the bunch, but in a conjugate relationship. Transverse wake fields can lead to shear displacements along the bunch, while longitudinal wake fields can lead to spreading in length as the bunch propagates. The forms valid behind the bunch show that the field solutions for a finite rectangular bunch are the same as those for a point bunch, except for the form factor $\alpha_n(\Delta)$. This factor has the effect of reducing the mode amplitudes $1/C_n$ that apply for a point charge to α_n/C_n for a rectangular bunch of width Δ. From the familiar properties of $\alpha_n(\Delta)$, it is seen that higher-order modes for which $k_n\Delta > \pi$ are reduced in amplitude. Those are modes with half-wavelengths less than the bunch length. This demonstrates the rather obvious point that excitation of short wavelength higher-order wake field modes requires short drive bunches.

The monopole $(\ell = 0)$ and dipole $(\ell = \pm 1)$ components are of greatest interest in determining stability. The only significant monopole component is

$$F_z|_{\ell=0} = -2qq_o \sum_{n=1}^{\infty} \frac{1}{C_n} g_{z,n}(s) \tag{3}$$

which is independent of γ, r_o, and r. The monopole radial force $F_r|_{\ell=0}$ is zero to order γ^{-1}: the lowest order contribution is of order γ^{-2}, as a result of the near cancellation between electric and magnetic forces. The monopole azimuthal force $F_\theta|_{\ell=0}$ is identically zero to all orders, by symmetry. The dipole force is the most significant deflecting force that arises from slight displacements of the bunch off axis, on account of the $(r_o/R_1)^\ell$ factors in Eq. 1. Knowledge of the dipole force is essential in analyzing stability of both the drive bunch or bunches, and the test bunch. The longitudinal component of the dipole force is given by

$$F_z(x, y, s)|_{\ell=1} = -4qq_o \frac{r_o}{R_1} \frac{x}{R_1} \sum_{n=1}^{\infty} \frac{1}{C_n} g_{z,n}(s), \tag{4}$$

since the displacement r_o is taken to be in the x-direction. A factor of two in Eq. 4 comes from summing contributions from the $\ell = \pm 1$ terms. The dipole portion of the axial force is seen to be either decelerating or accelerating, depending upon the sign of

x. Such an effect can contribute to energy spread within the bunch. The transverse components of the dipole force are given by

$$F_x(x,y,s)\big|_{\ell=1} = -4qq_o \frac{r_o}{R_1} \sum_{n=1}^{\infty} \frac{1}{C_n} \frac{g_{\perp,n}(s)}{k_n R_1}, \text{ and } F_y(x,y,s)\big|_{\ell=1} = 0. \quad (5)$$

One notes that the transverse dipole force is proportional to r_o, the displacement of the drive bunch off axis. Furthermore, the dipole force is zero in the direction normal to the direction of the displacement, it is independent of the position of the test charge, and it is independent of γ. Plots of exact evaluations of transverse and longitudinal forces for monopole, dipole, and quadrupole modes are given in ref. [1].

Determining bunch dynamics is straightforward once the forces are known. Examples are given in Figs. 4-11. Figs. 4-6 are for the forces within a single drive bunch moving along the axis. The bunch is divided for simplicity into five equal point

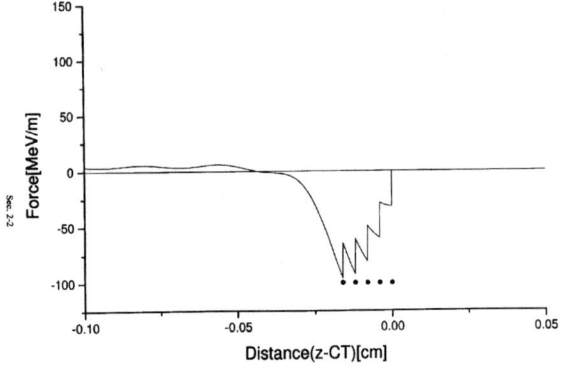

Figure 4. Initial axial drag force within a 2 nC bunch of initial length 0.02 cm. The bunch is divided into five equal beamlets, whose initial locations are shown by the black dots.

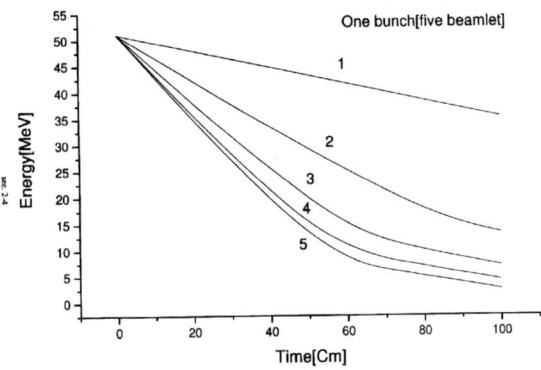

Figure 5. Energy histories for the five beamlets of Fig. 4, as the beam progresses along the dielectric waveguide.

charge beamlets. The position of each beamlet is shown by the black dots in Fig. 4, which depicts the drag force acting on each beamlet. Fig. 5 shows the resulting energy decrease of each beamlet due to drag forces, and Fig. 6 shows the relative spreading-in-z between beamlets as the bunch propagates.

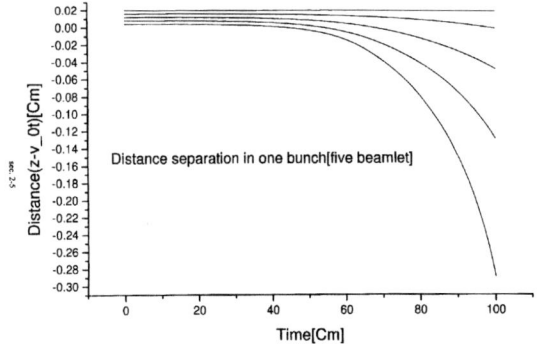

Figure 6. Relative separation between beamlets as the bunch propagates along the dielectric-lined waveguide, as in Figs. 4 and 5.

Figs. 7 and 8 are for a 2 nC, 0.02 cm test bunch that trails by 2.63 cm a 2 nC drive bunch. The test bunch is divided into nine equal beamlets, shown as black dots in Fig. 7. For these two figures, the drive bunch is taken to be a point charge. Axial accelerating fields (reduced by the self-fields of the test bunch) that act on the nine beamlets in the test bunch are shown in Fig. 7, with a variation of essentially 100%. This large variation in accelerating fields arises since the test bunch has a width comparable to the width of the accelerating peak, similar to that shown in Fig. 2. Fig. 8 shows the history of energy gain and loss for the drive bunch and the 9 beamlets in the test bunch, labeled 2-1 through 2-9. As is seen, the drive bunch loses its total energy through drag in traveling a distance of 65 cm. The nine beamlets gain and lose energy at quite different rates, but all beamlets lose energy once the accelerating field

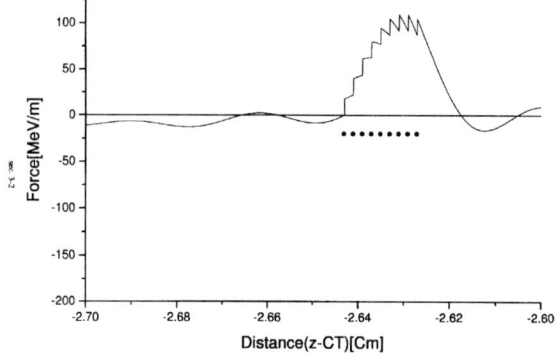

Figure 7. Axial accelerating field, in the wake of a 2 nC drive bunch located at $s = 0$, that acts on nine beamlets within a 0.02 cm, 2 nC test bunch. Note the strong variation in accelerating field.

due to the drive bunch ceases and only drag on the beamlets remains. A final energy spread of about 2:1 is seen for the test bunch. Most of this energy spread originates with the strong variation in accelerating fields across the test bunch.

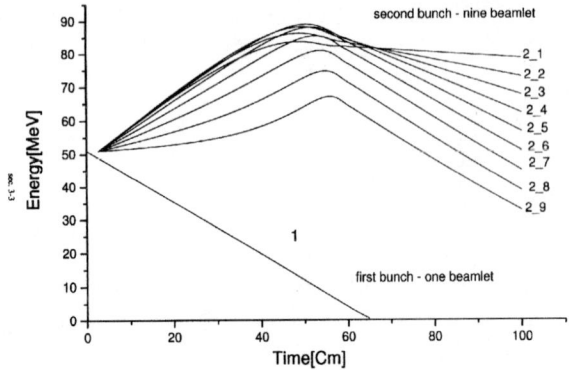

Figure 8. Energy histories for the drive bunch, and for the nine beamlets in the test bunch, as in Fig. 9.

Fig. 9 is for a situation similar to that in Figs. 7 and 8, except that the drive bunch is here divided into five beamlets, labeled 1-1 through 1-5. Fig. 9 shows the energy histories of all 14 beamlets. The main difference between this case and that of Figs. 7 and 8 is that the drive bunch here is dispersing as it travels (see Figs. 5 and 6), thereby giving rise to a wake differing from that for Figs. 7 and 8. After the test bunch has traveled 100 cm, energies in the test and drive beamlets are dispersed essentially uniformly from zero up to 120 MeV.

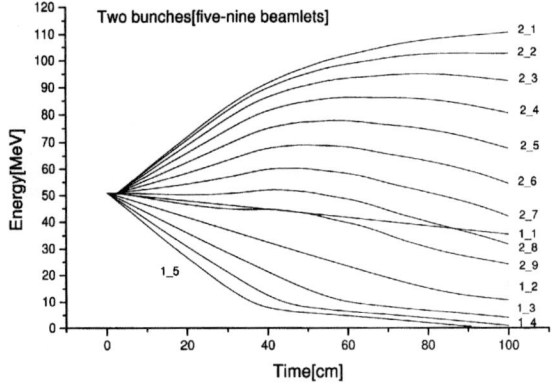

Figure 9. Energy histories for 5 beamlets in the drive bunch, and 9 beamlets in the test bunch.

For the examples shown so far, the bunches have been moving on the axis of the dielectric-lined waveguide, so that transverse forces are absent. But in Figs. 10 and 11 results are shown for the trajectories of beamlets when the bunch or bunches are moving parallel to the axis, but initially displaced from it by 0.005 cm, i.e. by 10% of the radius of the vacuum channel in the dielectric-lined waveguide. Fig. 10 shows the radial trajectory for the five beamlets in a 2 nC, 0.02 cm drive bunch, with axial forces essentially as in Fig. 4. The 1st beamlet moves without transverse acceleration, but the 5th beamlet hits the vacuum channel wall after traveling less than 20 cm. Clearly, a head-to-tail instability is in evidence here, wherein fields due to prior beamlets within the bunch create strong transverse forces on the tail of the bunch that cause it to self-destruct.

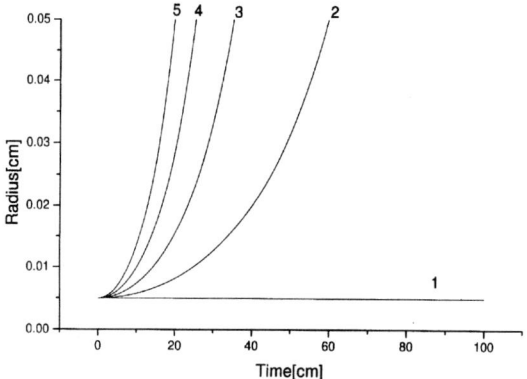

Figure 10. Radial trajectories for 5 beamlets in a 2 nC, 0.02 cm bunch moving parallel to the axis, but initially displaced off-axis by 0.005 cm.

Fig. 11 shows radial trajectories for the nine beamlets in a 2 nC test bunch that trails a 2 nC point charge drive bunch, when both bunches are initially displaced off axis by 0.005 cm. In this case, the 9th beamlet in the test bunch is seen to hit the vacuum channel wall after traveling only 12 cm. This example shows the cumulative power of the head-to-tail instability when the transverse fields add constructively.

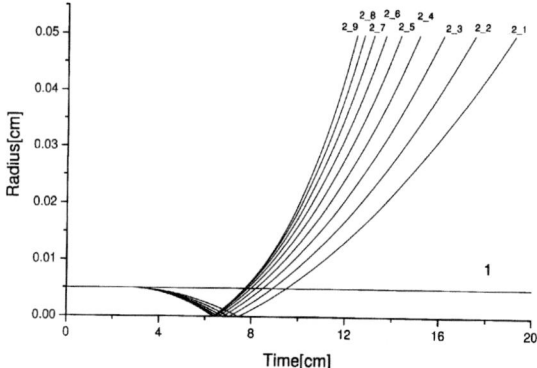

Figure 11. Trajectories for nine beamlets in a 2 nC test bunch that follows a 2 nC point charge drive bunch, when both bunches are displaced off-axis by 0.005 cm. Severe distortion of the test bunch is seen, with intersection at the wall after traveling only 12-20 cm.

DISCUSSION

Results for bunch dynamics have been obtained from an analytic theory [1] for wake fields in a cylindrical dielectric-lined waveguide. Examples were presented for a structure in which a single 2 nC, 0.02 cm bunch moving along the waveguide axis produces an initial peak accelerating axial field of 150 MV/m, and in which two such bunches when properly spaced, produce a peak initial accelerating field of 280 MV/m (see Figs. 1 and 2). Self-forces within a single such bunch have been shown to lead to longitudinal dispersion of the bunch, due to the strong variation in drag force between its head and tail. After traveling about 70 cm, the bunch length was shown (see Fig. 6) to double, and thereafter to spread more rapidly. The accelerating force on a similar test bunch has been shown to vary considerably over its 0.02 cm length (Fig 7). As a result, the acceleration history of the test bunch shows strong dispersion (Fig. 8), even when the drive bunch is taken to be a point charge; the dispersion is even stronger when the drive bunch is distributed (Fig. 9). These results suggest that acceleration of a test bunch in such a configuration should be at the highest possible rate, to minimize effects due to inherent self-spreading of the drive bunches. Acceleration gradient can be enhanced by superposition of fields from successive bunches, and/or by reflections from ends of a short accelerating module. Of course, when drive bunches are longer (for the same total charge), the inherent self-spreading will proceed more slowly, but the accelerating gradient will be smaller as well.

When a bunch, or train of bunches, is injected parallel to but displaced from the axis of the waveguide, significant transverse dipole (and higher multipole) acceleration results. Results in Fig. 10, for a 2 nC, 0.02 cm bunch displaced 0.005 cm off-axis show self-destruction by intersection with the vacuum wall at 0.05 cm radius after 20 cm travel for the tail of the beam. The tail of a similar test bunch trailing a point charge drive bunch is shown in Fig. 11 to intersect the wall after traveling only 12 cm. These results indicate the need for a means of transverse stabilization, such as a FODO channel [9]. In any case, high gradient acceleration in a short module is preferable to lower gradient acceleration in a longer module, on account of the roughly quadratic rate-with-z of transverse displacement.

One can conclude that use of the high wake field acceleration gradients that can be generated in a dielectric waveguide by one or more short, moderate charge bunches, will involve serious issues of longitudinal and transverse stability. These issues are reminiscent of those encountered with traditional rf accelerating structures that are driven using external sources. It can thus be anticipated that corresponding measures to provide the needed stabilization will be required as well in a dielectric wake field accelerator.

ACKNOWLEDGMENT

The authors acknowledge fruitful discussions with T. C. Marshall. This work was sponsored by the US Department of Energy.

REFERENCES

[1] S. Y. Park and J. L. Hirshfield, "Theory of wake fields in a dielectric-lined waveguide," *Phys. Rev. E* **62**, pp. 1266-1283 (2000).
[2] See, for example, E. Chojnacki, *Proc. 1991 Particle Accelerator Conf. (San Francisco)* vol. 1, pp. 2557-2559 (IEEE, 1991).
[3] W. K. H. Panofsky and W. A. Wenzel, *Rev. Sci. Instrum.* **27**, 967 (1956).
[4] T. C. Marshall, J.-Y. Fang, J. L. Hirshfield, and S. Y. Park, *this volume*.
[5] T.-B. Zhang, J. L. Hirshfield, T. C. Marshall, and B. Hafizi, *Phys. Rev.* E **56**, 4647 (1997).
[6] J. G. Power, W. Gai, and P. Schoessow, *Phys. Rev.* E **60**, 6061 (1999).
[7] J.-M. Fang, T. C. Marshall, J. L. Hirshfield, M. A. LaPointe, and X. J. Wang, "Experimental test of the theory of the stimulated dielectric wake-field accelerator," in *Proc. 1999 Particle Accelerator Conf.*, A. Luccio and W. MacKay, eds., p. 3630- (IEEE 1999).
[8] J. G. Power *et al*, *this volume*.
[9] W. Gai, A. D. Kanareykin, A. L. Kustov, and J. Simpson, "Numerical simulations of intense charged particle beam propagaion in a dielectric wakefield accelerator," in *Advanced Accelerator Concepts - Fontana, WI 1994*, P. Schoessow, ed. pp. 463- 473 (AIP Conf. Proc **335**, AIP Press, Woodbury, NY, 1995).
[10] S. Y. Park, *personal communication*.

Multi-mode, Multi-bunch Dielectric Wake Field Resonator Accelerator

T.C. Marshall[1], J-M. Fang[1], J.L. Hirshfield[2] and S-Y. Park[2,3]

[1] *Department of Applied Physics, Columbia University, New York City*
[2] *Omega-P, Inc., and Physics Department, Yale University, New Haven CT 06520*
[3] *POSTECH, Pohang, Korea*

Abstract We describe a multi-mode, dielectric-lined cylindrical resonator equipped with end reflectors in which wake fields are built up by a sequence of compact drive bunches. The parameters of the resonator are chosen such that the period of the wake fields is the same as the spacing of the drive bunches, and the length of the resonator is taken to be a half-integer multiple of the wake field period. Thus the wake field of a passing charge bunch will travel down the resonator and back so as to arrive at the front reflector just as the next bunch enters. Wake fields remain well defined because, excepting the lowest frequency TM mode, the resonator length is very nearly an integer multiple of the individual mode half-wavelengths. The device thus resembles a mode-locked laser resonator equipped with an "optical switch" (the passing bunches here). By numerical simulation, we find that the wake field amplitudes will increase with additional bunches, and show an example for an experiment to be done at the Yale Beam Physics Laboratory. For the first time we show how wake fields are reflected from boundary surfaces, an effect that should occur in every dielectric wake field apparatus and which we now exploit to advantage. The resonator concept permits a staged accelerator system, and could reduce the severity of beam bunch breakup due to charge asymmetries.

INTRODUCTION

In this we describe a new concept that allows generation of very strong wake fields in a short dielectric-lined resonator driven by a train of passing, compact drive bunches containing modest charge. Each resonator can be designed to allow localized buildup in amplitude of multi-mode wake fields [1] from successive drive bunches, leading to a higher acceleration gradient in a short module than would otherwise prevail. Staging of several such resonator accelerating modules should permit very high energies to be reached in a linear, two beam system. Wake field accelerators are attractive because no external source of rf is used in the structure itself, other than the source which is used to drive the conventional rf linac that generates the bunches. Thus neither new rf sources, high intensity lasers, nor plasmas are required in this new system. Of course, all wake field devices are limited in the energy available for acceleration in one module, and staging is required to reach high energy. A distinctive advantage in the proposed system is the possibility that the short modules and high accelerating fields will permit energy to be extracted from the drive bunches before higher-order mode instabilities that distort the bunches can develop appreciably. The resonator has reflecting surfaces at each end: prior calculations have not included wake field reflections at both faces of the dielectric module, so in this paper for the first time we account for such reflections using a PIC code [2] KARAT. An analytic theory has yet to be developed.

In the conventional dielectric wake field accelerator, a dielectric-lined cylindrical waveguide supports wake fields with longitudinal electric fields induced by the passage of an electron bunch containing high charge (the "drive bunch"). Phase velocities for the modes of dielectric-lined waveguide can be less than the speed of light [3], so that Cerenkov radiation occurs [4], manifesting itself as a wake field that reflects periodically from the conducting lateral wall and fills the waveguide behind the drive bunch. The dielectric constant and the dimensions will determine the periodicity of the wake fields, which can be chosen equal to the spacing of a train of drive bunches. If a "test bunch" of smaller charge is injected at a suitable interval after the drive bunch, it can move synchronously with the wake fields and experience net acceleration [5-7]. Recently, we have pointed out [1] the advantages of using very short bunch lengths, which excite a wide spectrum of TM_{0n} modes: these set up very short high E_z field pulses located on the axis of the device, and much larger radiation levels are produced by such short bunches than for the longer ones. Of course, field gradients in all dielectric-lined waveguides must be below the breakdown limit of the dielectric [8]. Tunneling ionization, the most likely breakdown mechanism sets a limit, ~ 1GeV/m for dielectrics which are exposed to psec pulses of high gradient. The remarkably high accelerating gradients that can apparently be induced by short, moderate charge drive bunches have spurred activity at Argonne National Laboratory [9], and within the Yale/Columbia/Omega-P beam physics collaboration[10], aimed at definitive experiments.

Our new approach [1,12] exploits the excitation of a large number of synchronous wake field modes by a train of short drive bunches. However, a variety of theoretical oversights, since corrected [11], render erroneous the wake field magnitudes computed in earlier publications. The corrected theory has a number of internal checks, such as consistancy with Gauss's law and the Panofsky-Wenzel theorem. Moreover, computations based on the new theory have been benchmarked against a powerful relativistic particle-in-cell electrodynamic simulation code KARAT. Simulated and calculated wake fields now agree with each other to better than 10%; this small difference could be ascribable to the finite cell size taken in KARAT.

One complication that might interfere with the linear buildup of wake fields from several bunches might be dispersion in the material dielectric. In the case of alumina, it has been demonstrated that the dielectric constant is essentially constant up to at least 200 GHz [13]. This range includes a very large number of wake field modes in most practical situations. (Alumina is also attractive because of its high dielectric constant $\kappa = 9.6$, its good vacuum properties, and its amenability to precision grinding.) Another source of dispersion is intrinsic to cylindrical geometry itself, and arises from the non-periodicity of Bessel functions which enter into the radial eigenfunctions for the fields in the dielectric. However, a short drive bunch will excite lower frequency wake field modes more weakly than higher frequency modes. The latter have nearly-uniform spacing: the mode frequency spacing $\Delta\omega$ can be shown to approach asymptotically $\Delta\omega = \pi\beta c[(R - a)\sqrt{(\kappa\beta^2 - 1)}]^{-1}$ as the mode index tends to infinity. Perfect uniformity in mode spacing leads to ideal constructive superposition of mode amplitudes. Short drive bunches thus excite a wake field with recurring highly-localized features over a number of periods trailing the initial drive bunch, and radiate more power than longer bunches having the same charge.

THE WAKE FIELD RESONATOR

We now turn our attention to the wake field resonator itself. Each resonator resembles a laser resonator: it is of length L and has a reflector at each end, and the electrons enter and exit through a small hole in each reflector. The length L is chosen to be an integer multiple of one-half the wake field period; the wake field period is also the spacing between drive bunches in a train. Thus the wake field from the leading bunch will travel down and back inside the resonator and reinforce the wakefield emitted from the next bunch as it enters. An example, where the wake field period is 21.6 cm, is given in Table I. The first column identifies the TM_{0n} ("monopole") mode. The second column provides the axial wavelength of the eigenfrequency (observed in [10] and computed using the dispersion relation from [1], with good agreement). The third column is the axial wavelength computed using the asymptotic relationship where the frequency-spacings of the modes are equal. Taking $2L = 21.6$ cm in this example, the fourth and fifth columns show the number of half-radiation wavelengths contained in the resonator for the actual frequencies, and using the asymptotic relationship, respectively. Apart from the first mode, the resonator is nearly an integer multiple of radiation half-wavelengths in length for each mode in the spectrum, within a few percent.

TABLE I. Dielectric resonator properties.

$(n-1)$	λ_z (actual)	λ_z (asympt.)	$2L/\lambda_z$ (actual)	$2L/\lambda_z$ (asympt.)
0	15.6 cm	21.6 cm	1.38	1
1	6.8	7.2	3.18	3
2	4.4	4.3	4.9	5
3	3.15	3.1	6.8	7
4	2.48	2.4	8.7	9
5	2.02	1.96	10.7	11

This device resembles a mode-locked laser. In the latter, the atoms provide radiation, which is pulsed by an optical switch at an interval corresponding to the round trip travel time of light waves in the resonator. In the wake field resonator, the radiating electron bunches are timed to the same interval as the optical switch in the atomic laser. The important point in each case is that identical radiation waveforms are injected into the resonator at the same interval, and this should lead to a sharpening of the "mode-locked" pulses. Losses in the system eventually will remove waves which do not correspond to the axial mode frequencies.

Stability is a serious issue for propagation of charge bunches along a narrow dielectric channel. This could limit the useful channel radius, channel length, bunch charge, or a combination of all three. This would be important since achievement of a strong wake field has been shown to require a small channel radius. Moreover, superposition of wakes from a bunch train requires that the channel length be no shorter than the train.

Calculations of transverse wakes that accompany a non-axisymmetric drive bunch have recently been carried out [11]. The transverse forces are typically a factor of 10 less than the axial force (caused by the "monopole" TM modes) for the dipole modes, and a factor of 100 less for the quadrupole modes. Since transverse deflection is approximately quadratic in time, instability for a shorter module with higher fields is more tolerable than for a long module with proportionately weaker fields.

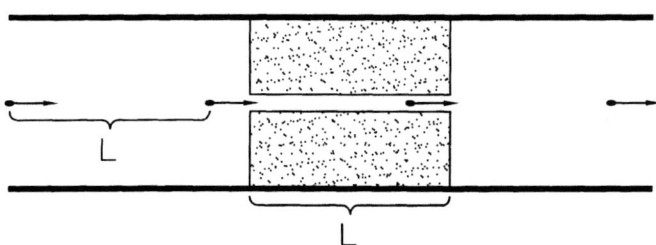

Figure 1. The wake field resonator, showing an entering train of drive bunches.

Our new approach for realization of high-gradient wake fields could mitigate the effects of instability, since a much shorter wake field module is employed. It is thus expected that head-to-tail bunch distortion and/or bunch motion into the channel wall will have less time to develop before the bunch leaves the module. A resonator module and a segment from a train of drive bunches that are key to this new concept are shown in Fig. 1. The key feature here is to exploit the strong reflections of wake fields from both end faces of the dielectric material; these reflections can be made equal to essentially 100% by placing metal disks with small beam holes at each face. The module length is set equal to a multiple of the wake field's half-period, as in Table I. Thus a bunch ceases to radiate once it has deposited a one-wavelength slug of wake energy in the resonator, and a following bunch enters the module just on top of the first decelerating peak of the preceding bunch. The energy lost by each drive bunch is stored in the resonator, apart from ohmic losses, until it can be imparted to one or more test bunches.

In an ideal scenario, we suppose that a long train of drive bunches establishes a steady-state inside the resonator such that their energy loss would be nearly complete in passage through the module; then the energy gained by test bunches in a train interleaved between the drive bunches would be nearly equal to the initial energy of the drive bunches. For example, consider a train of 105 MeV bunches each having charge 1 nC, spaced by 10.5 cm. We take the dielectric resonator to be 10.5 cm in length, with parameters adjusted so that the wake field periodicity is also 10.5 cm. Once a steady-state is reached, each drive bunch is decelerated by ~100 MeV, due to a composite decelerating wake of 0.95 GeV/m. The energy input to the resonator from each drive bunch is balanced by the energy gained by each test bunch that will have been injected mid-way between the drive bunches. In an illustrative example, 1 nC test bunches would each absorb the 0.10 J, for a net increase of 100 MeV. The often-discussed "ratio of two" ("transformer ratio") between the first accelerating wake and the decelerating (drag) field felt by the drive bunch (Wilson's theorem) does not apply here, because of the finite length of the drive bunches and as a consequence of the superimposition of many wakes. About 10^4 modules are needed to accelerate the test beam up to 1 TeV; but many fewer would be needed if the driver beam energy were 1GeV. The drive trains can be obtained from a long train, chopped into shorter trains and deflected into each dielectric resonator *en passant*. As to staging, a series of wake field resonators can be assembled in which the drive and test bunches move co-linearly. The drive bunches may be separated from the accelerated test bunches by magnetic dipoles located between the resonators, since the energy of the test bunches is much higher than the drive bunches. It is also conceivable that the drive and test bunch beamlines can be separated using a *bicylindrical* waveguide [14] structure which would make it easier to deflect the drive bunches without disturbing the test bunches.

SIMULATIONS

Instructive insights into the effects of reflections of wake field waves inside a dielectric-filled resonator can be obtained using the PIC code KARAT. We describe some results obtained using KARAT, for excitation of monopole wake fields set up by a train of several on-axis 6 MeV, $Q = 0.1$ nC drive bunches, each 10 psec long, that traverse a 10.5 cm long resonator having $a = 1.5$ mm (radius of vacuum channel) and $R = 11.5$ mm, lined with alumina, and bounded by perfectly conducting surfaces. There are no losses in this simulation. The module length equals the wake field period for these parameters, which are appropriate for a planned experiment to be done at the Yale Beam Physics Laboratory. A short module helps limit the run time of a dedicated 600 MHz PC, and a solution up to about 3.25 nsec was sufficient to see the accelerating wake after several bunches have passed and a composite wake was obtained. Results are shown in Figs. 2-5. It takes 0.35 nsec for a bunch to pass through the resonator; drive bunches enter from the left on the axis and exit on the right.

Fig. 2 illustrates how the wake field of a 10 psec rectangular profile bunch of charge is modified by proximity to the end conducting wall of the resonator. Parts a, b and c of the figure show $E_z(r = 0, z)$ (the on-axis component of the axial electric field) set up by the bunch at 20, 40 and 100 psec, measured from the entrance of the front of the first drive bunch at the left side of the device. In Fig. 2a, the first drive bunch has just moved away from the conducting end wall; the sharp positive pulse is at the location of the bunch, but there is a comparable disturbance which trails the wake to the left of the bunch at the wall, which may be caused by transient currents on the wall surface. In Fig. 2b, the field at the wall has reversed sign. In Fig. 2c, where the bunch is 3 cm (100 psec) from the left wall, the wake field is well-localized and the transients following the bunch have largely disappeared; but observe that the wake field pulse amplitude is not the same in parts a, b, and c. From this, we conclude that there are important effects that remain to be understood about the effect of the end wall on the bunch's wake field.

In Fig. 3a, at $t = 0.50$ nsec, the first bunch has passed out of the device and its accelerating wake (followed by the field of the second bunch) is nearing the exit reflector. The two features are the self-field of the #2 following bunch (left) and the

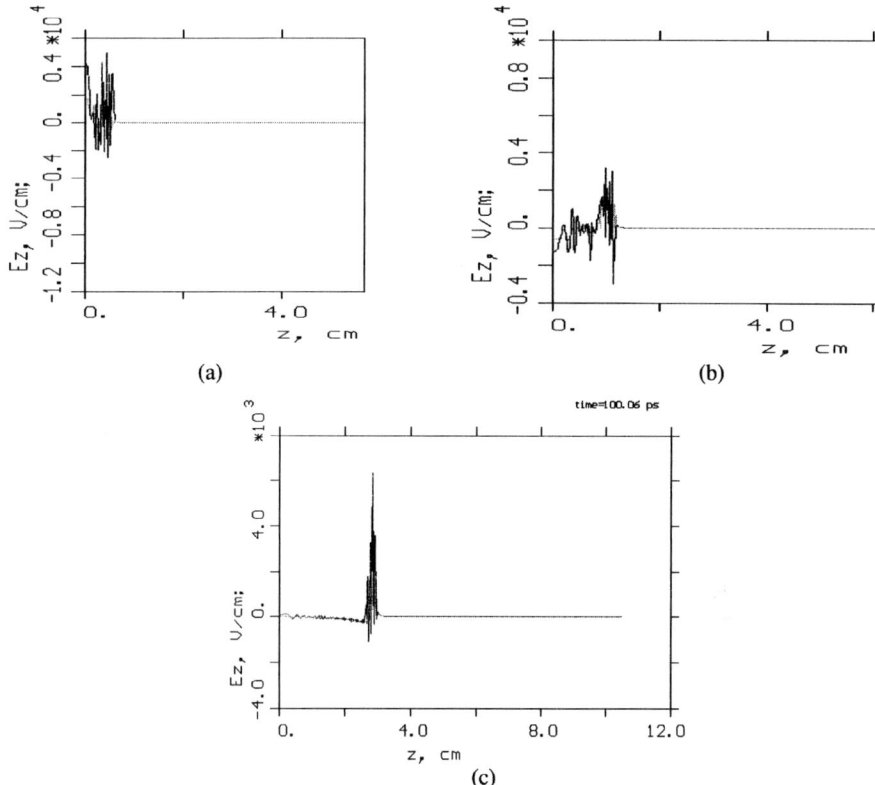

Figure 2. Wake field E_z at three different times, measured from the time of entry of the first bunch. The dielectric ($\kappa = 9.6$) fills the structure between $r = 0.15$ cm and $r = 1.05$ cm, so that the length of the structure and the wake field period are the same (10.5 cm). Parts (a), (b), and (c) show $E_z(r = 0, z)$ at 20, 40 and 100 psec. Notice the wake-field structure extending behind the bunch and "attaching" to the end wall in (a) and (b), a feature which does not appear in (c).

accelerating field of the leading #1 bunch (right). In Fig. 3b is shown $E_z(r = 0, z)$ at time 1.5 nsec after the first bunch enters; the accelerating field has accumulated from the first, second, third, and fourth bunches respectively (compare amplitude with Fig. 3a) and wake fields have had the opportunity to make over two complete round trips inside the device. Fig. 4a shows the synchronism of the Cerenkov wakes at 0.5 nsec. Fig. 4b shows a comprehensive view of the wake fields at 2.55 nsec: the 8th bunch has just entered from the left, and the accelerating wake pulse (to its right) has been built up from the seven preceding drive bunches. The wake field structure still maintains considerable integrity. In progressing from the accelerating wake of bunch #1 to bunch #7, we find that the accelerating gradient builds in magnitude from 2.2 to 10.5 kV/cm. The composite wake field is found to increase linearly, including the contributions of seven bunches, as shown in Fig. 5.

The wake fields excited here are not large because of the small bunch charge and large bunch length, but are indicative of what we might obtain using the apparatus available to us at the Yale Beam Physics Laboratory. However, one may increase the wake fields by increasing the bunch charge and reducing the dimensions of the bunch [1,11].

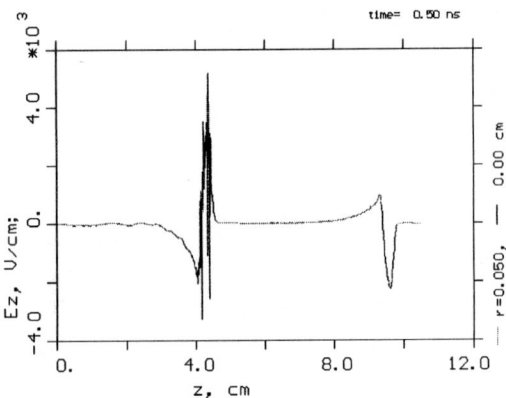

Figure 3a. $E_z(r = 0, z)$ in the wake field structure, at 0.5 nsec after the injection of the first bunch. At this time, the first bunch has exited to the right, the second bunch is seen at the left, and the accelerating wake field from the first bunch is at $z = 9.5$cm.

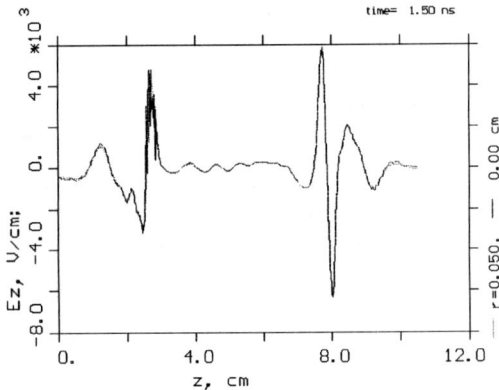

Figure 3b. $E_z(r = 0, z)$ at time 1.5 nsec after the first bunch enters the resonator. The accelerating field (at $z = 8$cm) has built up from the superposition of four bunches. The fifth bunch is to the left of the accelerating wake field pulse.

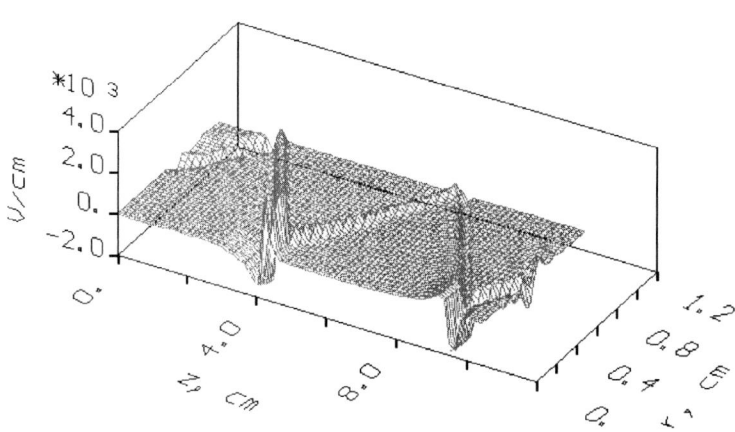

Figure 4a. The composite wake field E_z at $t = 0.5$ nsec after injection of the first bunch, which has now left the structure. The second bunch's wake field is set up following the accelerating wake of the first bunch to its right. Notice the synchronization of the wakes.

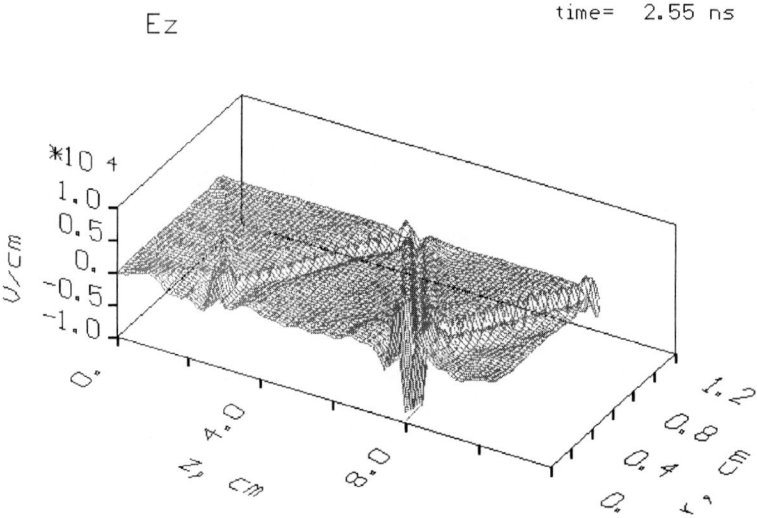

Figure 4b. The composite wake field E_z at 2.55 nsec set up by seven drive bunches in the resonator; bunch #8 has just entered to the left.

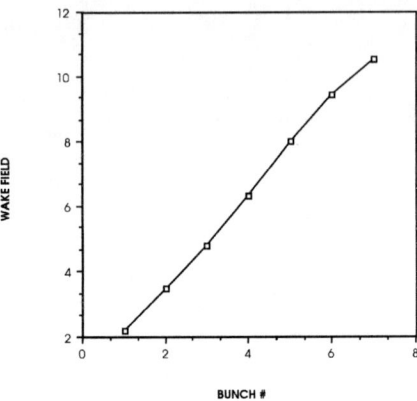

Figure 5. The accumulated accelerating wake field amplitude E_z from seven input drive bunches in the resonator.

Increasing the charge to 2 nC would increase the wake field a factor of 20, and decreasing the bunch length to 1.3 psec would increase the wake field by an additional factor ~4 [1]. Whereas the actual device has cavity modes with finite Q and bandwidth, the code has no dissipation and therefore does not damp away waves that do not fit into the resonator as an integer number of half-wavelengths.

While the gross features of the simulated results bear some resemblance to wake field phenomena computed in unbounded waveguides [1,9], the details are different. Notable is the presence of counter-moving reflected fields along the path of the test bunch, and a more complicated distribution of fields in the resonator, particularly when a bunch is near an end wall. The latter might be due to transients in the dielectric and end wall as the bunch enters/exits the chamber. One may expect that the case of the fields leading to transverse instabilities of the bunch would also be more complicated. It could be that the use of a PIC code, might introduce computational artifacts. We hope that a rigorous analytical theory (yet to be formulated) may be of value in sorting through these issues; nevertheless, thus far our studies show that predictions about wake fields in finite-length structures are suspect if they are based on theory that does not take into account reflections.

Any technique which offers the possibility to secure high gradient acceleration in a staged way that does not require the development of new rf sources or lasers should be pursued. We intend to do simple experiments in the coming year, using the Yale beamline, which provides 6 MeV bunches of charge obtained from a thermionic rf gun (driven by a 2.85 GHz klystron). The idea is to inject a train of drive bunches into the wake field resonator, and to measure the spectrum of electron energies that emerge. Given that the bunch charge is not high (~100 pC), we shall use a resonator that will be one wake field period (10.5 cm, equal to the bunch spacing) in length to secure a measurable energy loss. By varying the macropulse length, we shall explore how the energy lost by the bunches depends on the number of bunches injected into the resonator. One can also monitor the resonator time constant and filling time, by monitoring the millimeter radiations that escape from the end wall holes (3 mm dia.) in the resonator. This proposed experiment resembles one reported by Onishchenko *et al* [15], conducted at the Kharkov Instute of Physics and Technology, Ukraine. Although this group used a dielectric resonator traversed by energetic charge bunches, the device was not the correct length or radius to set up the effects we have described; yet they did nevertheless report some interesting effects regarding substantial energy loss of the electron bunches, together with appreciable microwave radiation emitted from the apparatus.

ACKNOWLEDGMENTS

This research was supported by the Department of Energy, High Energy Physics Division.

REFERENCES

[1] Marshall, T.C., Zhang, T-B., and Hirshfield, J.L., p.27, Eight Advanced Accelerator Workshop (AIP Conference Proceedings **472**), Lawson, Bellamy, and Brosius, eds. (1999)
[2] Tarakanoff, V.P., User's Manual for Code KARAT, BRA Inc., VA, USA (1992)
[3] Chang, C.T.M., and Dawson, J.W., J. Appl. Phys. **41**, 4493 (1970)
[4] Bolotovski, B.M., Sov. Phys. Uspekhi, **4**, 781 (1962)
[5] Ng, K-Y., Phys. Rev. **D42**, 1819 (1990)
[6] Rosing, M., and Gai, W., Phys. Rev. **D42**, 1829 (1990)
[7] Gai, W., et al., Phys. Rev. Lett. **61**, 2756 (1988)
[8] Sprangle, P., Hafizi, B., and Hubbard, R.F., Phys. Rev. **E55**, 5964 (1997)
[9] Power, J.G., Gai, W., and Schoessow, Paul, Phys. Rev. **E60**, 6061 (1999)
[10] Fang, J-M., et al., p.3627, **5**, Proc. 1999 Particle Accelerator Conf.,, Luccia and MacKay, eds. (IEEE, 1999)
[11] Park,S-Y., and Hirshfield, J.L., Phys. Rev. **E62,** July (2000)
[12] Zhang, T-B., Hirshfield, J.L., Marshall, T.C., and Hafizi, B., Phys. Rev. **E56**, 4647, (1997)
[13] Ibarra, A. et al., J. Nucl. Materials **191**, 530 (1992); **212**, 1029 (1994); **212**, 1113 (1994); **219**, 182 (1995); **253**, 141 (1998)
[14] Ivanyan, M.I., Radiotekhnika i Elecktronika **44**, 401 (1999)
[15] Onishchenko, I.N., et al., p. 782, **2**, Proc. 1995 Particle Accelerator Conf. (IEEE 1996)

Laser-Driven Cyclotron Autoresonance Accelerator

J. L. Hirshfield[*,†] and Changbiao Wang[*]

Beam Physics Laboratory, Yale University, 272 Whitney Ave., New Haven, CT 06511
†*Omega-P, Inc., 345 Whitney Ave., New Haven, CT 06511*

Abstract. Simulations are presented for a laser-driven cyclotron autoresonance accelerator with a practical guide magnetic field, as produced by a single coil. Effects that are investigated include acceleration in the non-resonant magnetic field profile, reduced acceleration due to misalignment between the magnetic field axis and the laser beam axis, acceleration with finite beam loading, and reduced acceleration due to slippage for a finite length radiation pulse and finite length bunch.

INTRODUCTION

Electron acceleration using intense lasers has engendered significant attention within the accelerator research community. This interest stems from the enormous optical electrical field strengths E that can be obtained with a focused laser, i.e. of the order of $E = 3 \times 10^{-9} \sqrt{I}$ TV/m, where the intensity I is in W/cm^2. Since compact terawatt focused lasers can have $I > 10^{18}$ W/cm^2, field strengths of the order of TV/m are possible. Of course, since this field is transversely polarized, it cannot give much net acceleration to a charged particle directly, so an indirect means must be employed to achieve net acceleration, such as the laser-driven cyclotron autoresonance acceleration (LACARA) [1-6]. The basis upon which LACARA rests is cyclotron resonance, using an axial static magnetic field. The magnetic field can be adjusted to allow transverse deflections of electrons that move along a helical path to be synchronous with the rotating transverse electric field of a circularly-polarized laser beam, thereby allowing the field to do work on the electrons.

LACARA is a laser-driven accelerator that operates in vacuum. It does not require a pre-bunched beam; nevertheless all injected electrons can enjoy nearly the same acceleration history. LACARA is operated without a tight laser focus, so the Rayleigh length can be 10's of cm for a 10.6 μm laser wavelength, and continuous acceleration in vacuum over several Rayleigh lengths can take place. Phase bunching—but not spatial bunching—occurs in LACARA, which explains how all injected electrons can experience nearly the same accelerating fields, since circularly-polarized laser radiation is used. Furthermore, the effective group velocity in LACARA exceeds the particle's axial velocity, so operation with strong pump depletion is possible without causing undue energy spread for the accelerated beam.

Colson and Ride [1] analyzed LACARA with a uniform applied magnetic field and a circularly polarized plane wave. In this idealized model, the wave refractive index is equal to unity and cyclotron resonance can be maintained during acceleration through autoresonance. For a practical laser beam, the refractive index and the wave intensity vary along the acceleration path. Thus, a uniform magnetic field cannot maintain resonance between the particles and the laser field. Sprangle, Vlahos, and Tang [2] studied *LACARA* with a resonant magnetic field and a Gaussian laser beam, finding that the acceleration energy is considerably increased compared to the uniform magnetic field case. Radiation loss in an accelerator with a gyrating beam is a concern as well. Loeb and Friedland analyzed *LACARA* with a plane wave model and indicated that the radiation losses in a 1-TeV accelerator are negligible. Recently, using computation, Chen obtained scaling laws for acceleration energy gain, distance, and gradient based on a set of self-consistent equations of motion governing the inverse cyclotron autoresonance maser (*CARM*) interaction. The empirical scaling law for energy gain is confirmed by the analytic formula [7]. However, analysis has not appeared prior to 1999 that included realistic profiles for the applied magnetic field and the focused laser beam. Those deficiencies are remedied in [5,6], and in the present work.

Using Maxwell's equations, a Gaussian laser beam can easily be shown to have longitudinal components of electric and magnetic field. Based on a model that includes these longitudinal components with longitudinal and transverse gradients, but a resonant magnetic field, *LACARA* was studied by computer simulation for electron beams with finite initial emittance [6]. The simulations have shown that beam stalling can be avoided as beam energy increases during acceleration.

In this paper, *LACARA* is analyzed by simulation using a practical magnetic field profile produced by a single coil. This one-coil magnet is affordable for an experimental *LACARA* prototype that is under design. A sketch of the prototype *LACARA* is shown in Fig. 1. Copper mirrors direct a mildly-focused laser beam to pass along the axis of a 6 T solenoid in the surrounding cryomagnet. For this prototype, the CO_2 laser power is 2 TW at a wavelength $\lambda = 10.6$ μm, the minimum laser spot radius is $w_o = 1.0$ mm, with a Rayleigh length $z_R = \pi w_o^2/\lambda = 29.6$ cm. The electron beam and the laser interact over a length of $6z_R = 178$ cm, but the uniform portion of the magnetic field only extends for about 150 cm. The (nominal) 1 A, 50 MeV beam injected at $z = 0$ has a normalized emittance of 2.0 mm-mrad. Compromises were made in selecting these parameters to be the basis for design of the prototype *LACARA*, mainly on account of the high cost of a magnet with a field profile that better approximates the resonant profile. The mirror spacing is 230 cm, and the 8-cm i.d. coil length of 180 cm provides a nearly uniform field region of 150 cm in length. Electron orbits are computed in the region between mirrors, all through the fringing fields at the ends of the coil.

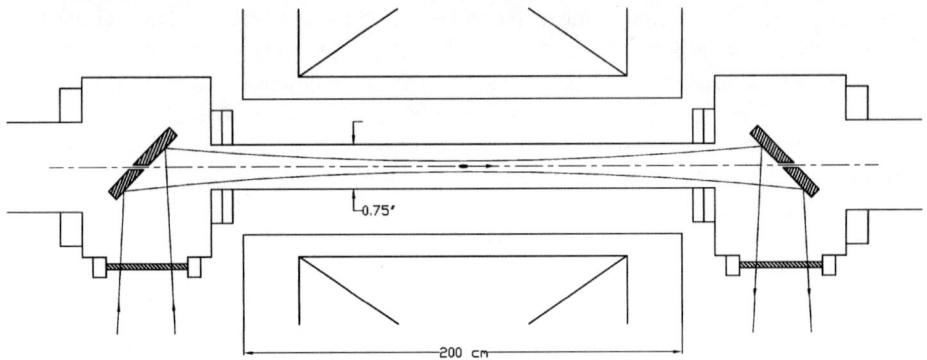

Figure 1. Sketch of *LACARA* prototype, not to scale. Accelerating charge bunch is shown at center.

RESULTS OF COMPUTATIONS

Simulation results are presented below to show the effects on acceleration due to (*i*) the non-resonant one-coil magnetic field profile, (*ii*) misalignment between the magnetic field axis and the laser beam axis, and (*iii*) beam loading. In the computation, a total of 904 computational particles were injected, uniformly distributed in optical phase and transverse phase spaces within emittance ellipses having major and minor axes r_b and $\beta_{\perp max}$, where r_b is the beam radius and $\beta_{\perp max}$ is the maximum normalized transverse velocity. The parameters used in the simulation are summarized in Table I.

Table I. Parameters in simulation.

Initial beam energy	50 MV
Beam current	1 A
Normalized rms emittance	2 mm-mrad
Beam radius r_b	0.3 mm
Laser power	2 TW
Laser wavelength λ	10.6 μm
Minimum laser spot size w_o	1.0 mm
Rayleigh length z_R	29.6 cm
Mirror separation L	230 cm
Coil length	180 cm
Coil inner diameter	8 cm
Magnetic field (uniform portion)	60.6 kG

Fig. 2 shows, with the solid lines, the magnetic field profile $B_o(z)$ and the average relativistic energy factor $\langle \gamma(z) \rangle$ as they vary along the axis of *LACARA*. Using dashed lines, the same quantities are shown for the ideal resonance magnetic field profile. Table II compares results for the actual and ideal magnetic field profiles.

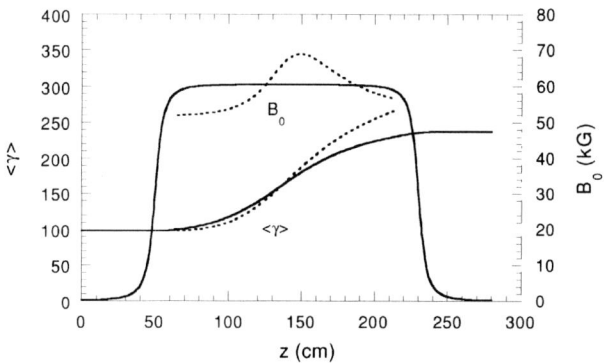

Figure 2. Energy gain and magnetic field profile produced by one coil for the prototype *LACARA* (solid lines). Dashed lines show these parameters for the idealized resonant magnetic field profile.

Table II. Comparison of *LACARA* performance for the proposed affordable one-coil magnet system with that for the ideal resonant magnetic field profile.

	one-coil B-field profile	ideal B-field profile
final beam energy	120.9 MeV	135.5 MeV
average accel. gradient	47.2 MeV/m	57.0 MeV/m
maximum accel. gradient	75.7 MeV/m	100.6 MeV/m

Fig. 3 shows the acceleration along *LACARA* for three different peak magnetic field values, the optimum value of 60.6 kG and two other values 58.5 kG and 62.7 kG that are 3.5% higher and lower. This degree of mistuning of the magnetic field strength is seen to degrade the final beam energy from 121 MeV to 113 MeV, a 7% decrease in energy gain. This calculation shows that it is important to set the magnetic field strength to the desired value within a fraction of a percent, so as not to cause a diminution in energy gain. This requirement is not particularly demanding.

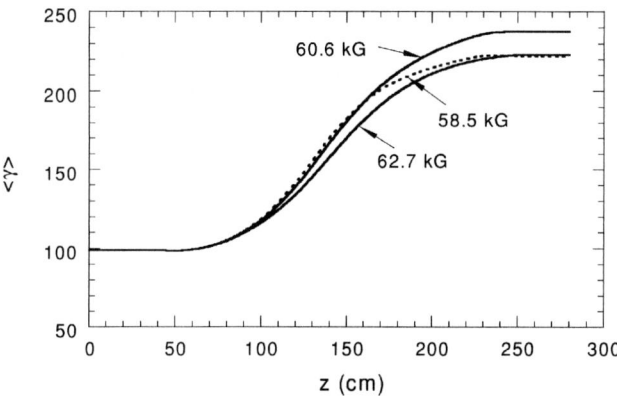

Figure 3. Acceleration in a one-coil magnetic profile *LACARA* for three values of peak magnetic field. The diminution in energy gain is 7% for a 3.5% change in peak magnetic field.

A practical issue is the question of alignment between the magnetic axis and the laser beam axis. The effect of misalignment has been computed, for the *LACARA* as shown in Fig. 2, except that the magnetic field is taken to be tilted about the center point of the system by an angle θ. Fig. 4 shows the mean energy gain profiles, for three different tilt angles. For comparison, the result with perfect alignment $(\theta = 0)$ is also shown (case 1). Cases 2, 3 and 4 are for tilt angles θ = 0.675 mr, 1.35 mr, and 2.70 mr, respectively. These correspond to transverse displacements $d = L \sin \theta$ of $d = w_0$, $2w_0$, and $4w_0$, respectively. Not surprisingly, when misalignment causes the two axes to be displaced by more than about two laser waist radii, the acceleration diminishes significantly. This is shown clearly in Fig. 4. The lesson taught by this exercise is that provision is required to align the magnetic and laser beam axes to within about 1 mr. This is a degree of precision not uncommon in many laser-acceleration schemes.

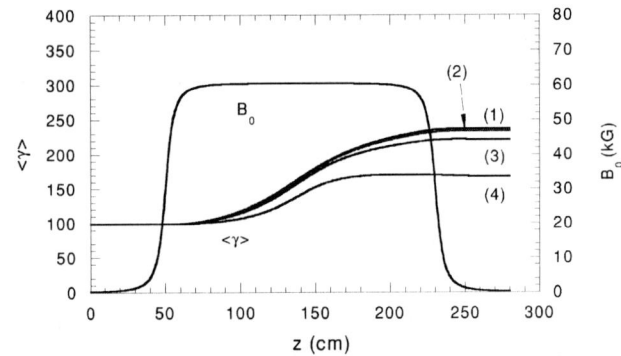

Figure 4. Effect upon acceleration of a misalignment between the magnetic axis and the laser beam axis. For cases 1-4, the misalignment angles are 0, 0.675, 1.35, and 2.7 mr.

All the examples of *LACARA* performance discussed so far are for a 1 A beam. If the average energy gain per electron is 71 MeV (as for the one-coil magnetic profile), then the power added to the beam is 71 MW, far less than the 2 TW in the laser beam. Fig. 5a shows the power profiles of the beam and laser. Here, the efficiency η for transfer of laser power to electron beam power is $\eta = 3.5 \times 10^{-3}$ %, and beam loading is negligible. Thus the laser power is seen to not diminish measurably along the acceleration path. To examine performance of *LACARA* when beam loading is not negligible, higher beam currents were introduced, namely 10 kA, 20 kA, and 40 kA. Power profiles are shown for these currents in Figs. 5b, 5c, and 5d, respectively.

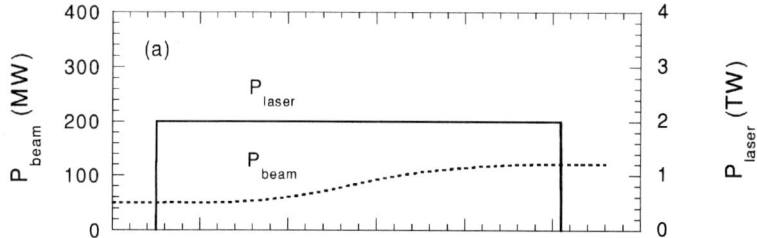

Figure 5a. Laser and beam power profiles along *LACARA* for a 1 A beam, where beam loading is negligible. Final beam energy is 120.86 MeV. $\eta = 3.5 \times 10^{-3}$ %.

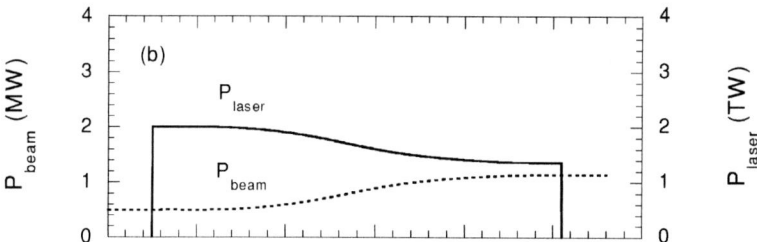

Figure 5b. Laser and beam power profiles along *LACARA* for a 10 kA beam, where beam loading is significant. Final beam energy is 114.05 MeV. $\eta = 32.0$ %.

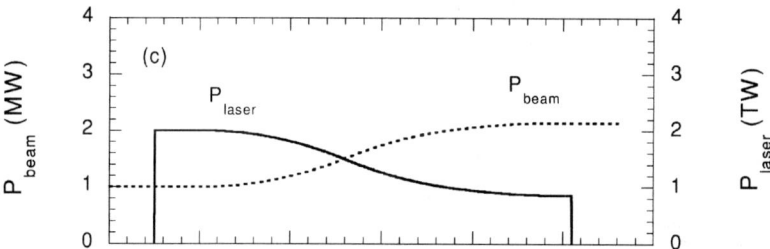

Figure 5c. Laser and beam power profiles along *LACARA* for a 20 kA beam, where beam loading is more significant. Final beam energy is 107.24 MeV. $\eta = 57.2$ %.

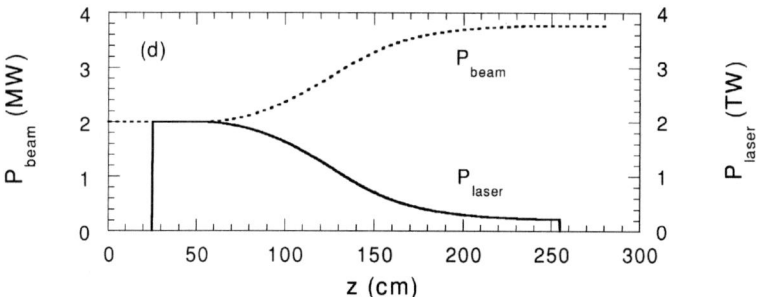

Figure 5d. Laser and beam power profiles along *LACARA* for a 40 kA beam, where beam loading is even more significant. Final beam energy is 94.2 MeV. $\eta = 88.4$ %.

Several messages are conveyed by the results shown in Figs. 5b-5d. *First, LACARA* is a laser-based accelerator that can have very high efficiency for transfer of laser power to electron beam power. *Second*, the efficiency is evidently not critically dependent upon the choice of magnetic field profile, since the results shown here are for a profile with significant deviations from the ideal resonance magnetic field profile. *Third*, operation with higher beam loading and higher efficiency carries a penalty of lower net acceleration; this is a feature that is natural and is common with other traveling-wave accelerators, where the accelerating field is reduced as the particle beam soaks up power from the radiation beam. The high efficiency values that are shown here serve to distinguish *LACARA* from other laser-based accelerators, where efficiency is so small as to be hardly ever discussed.

FINITE BUNCH-LENGTH EFFECTS AND SLIPPAGE FOR A FINITE LENGTH RADIATION PULSE

All of the computations presented so far in this paper assume that the laser field is at a steady-state level during passage of the electrons. Invariably, however, both the laser pulse and the electron beam bunch length are finite in duration and comparable, typically in the psec range. Operation of *LACARA* then requires precise synchronism in timing between injection of the electron bunch and the laser pulse. In this case, it is possible for there to be an excess energy spread that arises during acceleration from two sources: (*i*) laser amplitude variations if the laser pulse is not wider than the electron bunch, even with good synchronization; and (*ii*) slippage of the laser pulse over the electron bunch, since the two do not move at exactly the same speeds. In the experiments that are planned at Brookhaven National Laboratory Accelerator Test Facility, it is intended to operate *LACARA* with an electron bunch width of less than 1 ps, if possible, and with a laser pulse of width of greater than 1 psec. Thus, if the slip between the pulse and the bunch is rather less than 1 psec, slippage should not cause significant non-uniformity in the electron acceleration. This is shown as follows. To estimate slippage, the transit times for electron bunches (t_b) and for laser pulses (t_L) were computed, using the one-coil magnetic field profile with all other parameters identical to Fig. 2. The transit times were defined as

$$t_b = \frac{1}{c}\int_0^L \frac{dz}{\langle \beta_z \rangle}, \quad \text{and} \quad t_L = \frac{1}{c}\int_0^L \frac{dz}{\beta_g},$$

where $<\beta_z>$ is the electron's normalized average axial velocity, normalized group and phase velocities for the laser pulse are related by $\beta_g \beta_p = 1$, and $\beta_p = ck_z/\omega$ with k_z the axial wave number and ω the laser angular frequency. The group velocity is used here since it is motion of a wave packet (laser pulse) that influences the field amplitudes that act on the electrons; phase variations between the laser fields and electron orbits are already included in the dynamics as analyzed for a continuous laser beam. The slip time is given by

$$\Delta t = t_b - t_L = \frac{1}{c}\int_0^L dz \left[\frac{1}{\langle\beta_z\rangle} - \frac{1}{\beta_g}\right] \approx \frac{1}{c}\int_0^L dz \left[\beta_g - \langle\beta_z\rangle\right],$$

since $\beta_g \approx \langle\beta_z\rangle \approx 1$. Fig. 6 is a plot of $\beta_g - \langle\beta_z\rangle$ along *LACARA*. Numerical integration of this curve from $z = 25$ cm to $z = 255$ cm gives $\Delta t = 0.27$ ps. This suggests, for example, that a 1 ps electron bunch can remain within a 1.3 psec laser pulse without slipping out of the accelerating field, when the laser pulse and electron bunch start off in perfect synchronism. The relatively small margin for slip without a strong effect on acceleration indicated by this estimate suggests that care must be taken in the experiments. Moreoever, it appears that actual time-dependent simulations of acceleration history should be carried out in future for a finite laser pulse width and a finite electron beam width, in order to better model this important phenomenon.

CONCLUSIONS

Simulations have been carried out for a *LACARA* with a practical magnetic field profile produced by a single coil. Compared to the ideal resonant magnetic field profile, the acceleration energy is reduced by 11%. The effect of the misalignment between the magnetic field axis and the laser beam axis is examined. To get effective acceleration, the displacement of the two axes should be less than two laser waist radii. The acceleration efficiency is not critically dependent on the choice of magnetic field profile, but operation with higher beam loading and higher efficiency carries a penalty of lower net acceleration. Slippage between laser and beam pulses can cause diminution in energy gain, if the laser pulse is too short.

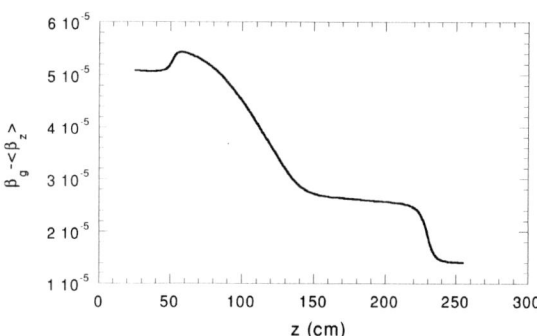

Figure 6. Slip between laser pulse and electron bunch along *LACARA*.

ACKNOWLEDGMENT

Constructive discussions with B. Hafizi, T. C. Marshall, M. A. LaPointe, and V. L. Bratman are acknowledged. This research was sponsored by US Department of Energy.

REFERENCES

[1] W. B. Colson and S. K. Ride, "A laser accelerator," *Appl. Phys.* **20**, 61 (1979).
[2] P. Sprangle, L. Vlahos, and C. M. Tang, "A cyclotron autoresonance accelerator," *IEEE Trans. Nucl. Sci.* **NS-30**, 3177 (1983).
[3] A. Loeb and L. Friedland, "Autoresonance laser accelerator," *Phys. Rev. A* **33**, 1828 (1986).
[4] C. Chen, "Theory of electron-cyclotron-resonance laser accelerators," *Phys. Rev. A* **46**, 6654 (1992).
[5] C. Wang and J. L. Hirshfield, "Laser-driven cyclotron autoresonance accelerator," in *Proc. of the 1999 Particle Accelerator Conference*, edited by A. Luccio and W. MacKay, 1999, p. 3630.
[6] J. L. Hirshfield and C. Wang, "Laser-driven cyclotron autoresonance accelerator with production of an optically chopped electron beam," *Phys. Rev. E* **61**, 7252 (2000).
[7] C. Wang and J. L. Hirshfield, "Energy limit in cyclotron autoresonance acceleration," *Phys. Rev. E* **51**, 2456 (1995).

Twisted Waveguide Accelerating Structure

Yoon W. Kang

Advanced Photon Source, Argonne National Laboratory

Abstract. Regular straight hollow waveguides have phase velocities of propagating electromagnetic waves greater than the free-space speed of light. However, it has been found that, if the waveguide is twisted, the phase velocities of the waveguide modes become slower. By choosing optimum shape of the cross section and pitch angle of a twisted waveguide, a desired phase velocity of the accelerating mode can be obtained for accelerating charged particles. This type of accelerating structure may have advantages over the conventional iris-loaded structures. The twisted waveguide structure has been modeled and computer simulated in 3-D electromagnetic solvers to show the slow-wave properties of the accelerating mode.

INTRODUCTION

Slow-wave structures are used in many radio frequency (rf) applications including charged-particle accelerators. A slow-wave structure employs reactive loadings in a hollow waveguide to slow down the phase velocity of electromagnetic fields in a specific mode to be used. However, a question can be raised: "Is it possible to have a slow-wave hollow waveguide structure with a uniform cross section?". There exists interest for finding accelerating rf structures that can be inexpensive to manufacture especially for high-energy linear accelerators. The cost reduction can be important for large-scale accelerators, such as the next-generation linear colliders.

The above question sounds useless if one recalls the fundamental dispersion relations of waves in hollow waveguides. However, the above question could be answered in an unexpected way that is discussed and shown in the following. In a twisted noncircular hollow waveguide, the phase velocities of propagating modes can be slower than the velocities in a straight waveguide. A twisted waveguide has been usually a twisted rectangular waveguide section with a slow pitch angle. This type of waveguide has been implemented in waveguide circuits for simple plumbing purposes with no attention to their phase properties. In this paper, a slow-wave TM_{01}-like mode in a twisted waveguide that is useful for acceleration of charged particles is discussed.

Since the twisted waveguide structure has a uniform cross section, it may be built without welding or brazing many parts unlike the conventional disk-loaded accelerating cavity structures. A special extrusion technique or electroforming may be used for mass production of the waveguide and can lower the manufacturing cost significantly. The uniform smooth cross section along the direction of propagation in the structure means that higher-order modes may have lower impedances and can exit more easily. The reduction of the higher-order mode power harmful to the particle beam may deliver better beam properties.

CP569, *Advanced Accelerator Concepts: Ninth Workshop,* edited by P. L. Colestock and S. Kelley
2001 American Institute of Physics 0-7354-0005-9

TWISTED WAVEGUIDE

Figure 1 shows a section of twisted rectangular waveguide. Intuitively, the fields in such a waveguide are seen as twisted along the guide. For computer simulation and any accurate analysis of the structure, a short section of the twisted waveguide shown in the figure must be used. A short section of a long twisted waveguide structure must have end walls that can satisfy boundary conditions for the twisted electric and magnetic fields. That means the end walls are orthogonal to the twisted surfaces; the end-wall surfaces must also be twisted, thus they are no longer flat 2-D planes.

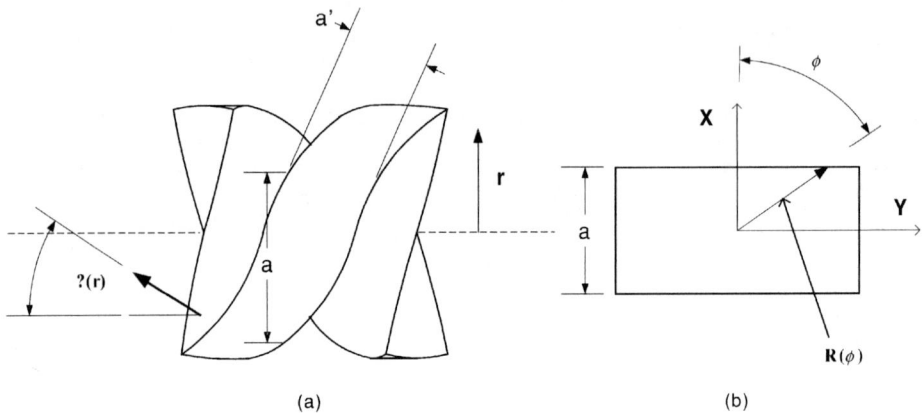

FIGURE 1. Geometry of a section of twisted waveguide. (a) a waveguide section with twisted port surfaces, (b) cross section of the waveguide. (A rectangular shape is shown, but it can be arbitrary.)

The twisted field vectors of a mode in the waveguide are assumed normal to the conductor surface for E-field, and tangential to the surface for H-field. These vectors will have a tilt angle, $\theta(r)$, comparing with the fields in a straight waveguide, which is a function of the radial distance r. Effective height of the waveguide can be approximated as

$$a'(r) = a \sin \theta(r) . \qquad (1)$$

Therefore, if a rectangular waveguide is twisted, the twisting effectively squeezes the waveguide height down to a lower height as r increases. This suggests that the cross section defined in the *x-y plane* of a regular rectangular coordinate system must be modified to have an effective cross section similar to that of a rectangular waveguide. This goal may be achieved if the cross section in the *x-y plane* has a bow-tie shape as shown in Figure 2. The fields near the narrow walls of the waveguide are stretched as the waveguide is twisted. Therefore, the volume of the waveguide is increased when the waveguide height is increased with the angle $\theta(r)$.

The twisted waveguide has greater volume by expanded space along the narrow walls. This structure may be treated as an equivalent loaded waveguide with either dielectric or ferromagnetic materials with radially nonuniform weighted permittivity

$\varepsilon_r(y)$ or permeability $\mu_r(y)$, respectively, considering the stretched path with the on-axis length [1]. An approximate expression for the change of propagation constant of a waveguide with material perturbation is given as [2]

$$\beta - \beta_o \approx \omega \frac{\iint\limits_S (\Delta\mu |H_o|^2 + \Delta\varepsilon |E_o|^2) ds}{\iint\limits_S (\mathbf{E}_o^* \times \mathbf{H}_o + \mathbf{E}_o \times \mathbf{H}_o^*) \cdot \mathbf{u}_z ds}, \qquad (2)$$

where β and β_o are the perturbed and unperturbed phase constants respectively, E_o and H_o are the unperturbed electric and magnetic fields, and * denotes a complex conjugate. In a hollow waveguide, the stored magnetic energy is greater near the narrow walls of the waveguide. The above expression suggests that a decrease in the phase velocity may result in a loaded waveguide described above and equivalently in a twisted waveguide with a properly modified cross section. The envelope of the longitudinal cross section can be determined by transforming the radial distance $R(\phi, z=z_o)$ into $R'(z)$, where ϕ is the azimuth angle in a circular cylindrical coordinate system [1].

SIMULATION

Because existing commercial electromagnetic codes use orthogonal curvilinear coordinate systems, the twisted structure can not be modeled accurately. The twisted waveguide used here as an example has a cross section resembling a bow tie. The twisted waveguide has been modeled in the MAFIA code [3] using stacked waveguide slices and is shown in Figure 2. As mentioned in the previous section, the end walls used in the simulations do not satisfy the boundary conditions of the accelerating mode fields of a long twisted waveguide structure. A constant displacement angle between the slices determines the pitch of the twist. The displacement angle of the slices is adjusted until the free-space half wavelength of the TM_{01}-like resonance frequency becomes identical to the length of the twisted structure. This is the case for relativistic particles, $\beta_b = 1$.

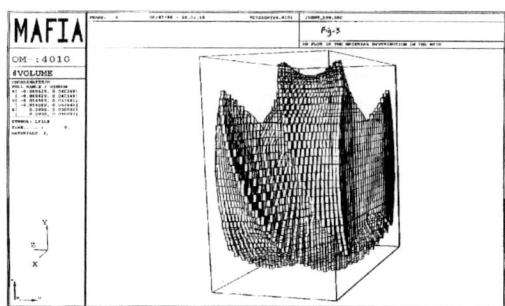

FIGURE 2. A twisted waveguide modeled in MAFIA with a stack of slices turned by a constant angle.

For a structure with a length of 5 cm, the angle was varied until a resonant frequency of 3 GHz was obtained from the desired TM_{01}-like mode. The fields in the simulated waveguide do not look like the fields in a regular straight hollow waveguide. In a regular rectangular waveguide, the magnetic field vectors of the TM_{11} mode (comparable to TM_{01} mode in a circular cylindrical waveguide) are lying on the transverse plane. However, the magnetic field of the present TM_{01}-like mode is twisted to conform to the true orthogonal walls of the twisted waveguide. That is, the H_z portion of the magnetic field vector increases as radial distance r increases.

The computed properties of the waveguide modes are also shown in the following. Figure 3 shows the magnetic field of the TM_{01}-like mode in the bow-tie shaped twisted waveguide with $v_p=c$.

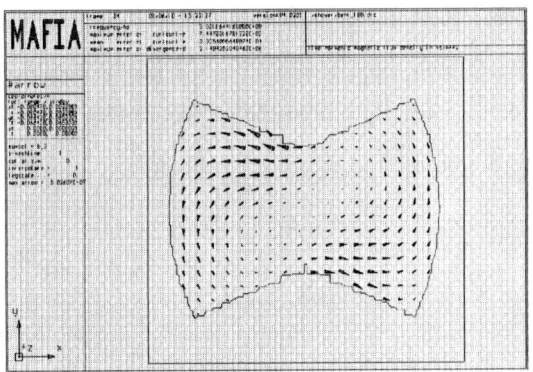

FIGURE 3. Magnetic field of the TM_{01}-like mode in the bow-tie twisted waveguide.

Figure 4 shows the electric field of the TM_{01}-like mode in the twisted waveguide with $v_p=c$. The electric field will have E_z only on the beam axis but will have E_ϕ if $r \neq 0$. It is interesting to note that the cavity envelope and the field configurations are similar to those of the conventional iris-loaded rotationally symmetrical structure.

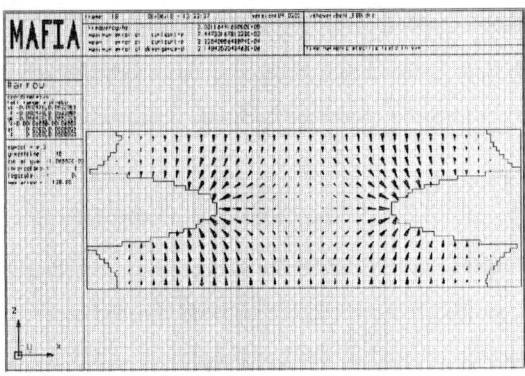

FIGURE 4. Electric field of the TM_{01}-like mode in the bow-tie twisted waveguide.

CONCLUSION

Computer simulations have been made to show that a twisted waveguide can support a slow-wave accelerating mode. Commercially available MAFIA and Agilent HFSS codes have been used, and the results of the MAFIA simulations have been presented in this paper. Result of HFSS also showed similar slow-wave properties. Although approximations are made in the modeling due to nonorthogonal end walls, it has been shown that such twisted waveguide structures can support an accelerating mode with $\beta_h \leq 1$. A slow wave with a specific β_h at a frequency can be excited in a twisted waveguide by choosing an optimum shape of the cross section and a pitch angle.

The simulations verified that the phase velocity in the waveguide is equal to the free-space speed of light when the angle of twist is 180 degrees. For the specific example for 3 GHz, the waveguide made of copper can support a TM_{01}-like mode with Q = 12,620 and the shunt impedance R_s = 258 MΩ / m.

The inner surface of the structure is smooth and free from any sharp corners. This means that, in the accelerating structure, damping the higher-order modes is easier and the particle beam properties can be enhanced. This also means that the maximum field inside the waveguide may be greater so that the field gradient limit could be raised. Prototype fabrication and bench measurements of the twisted waveguide are in progress. Investigation on the beam properties in the waveguide is also in progress. Development of new code for the specific coordinate system, which is conformal to the twisted structure, may be needed. Using this type of new code may aid in making more accurate assessments of rf and beam properties in the structure.

ACKNOWLEDGEMENT

This work was supported by the U. S. Department of Energy, Office of Basic Energy Sciences, under Contract No. W-31-109-ENG-38.

REFERENCES

1. Y. Kang, "Twisted waveguide – A slow-wave structure," to be published in *IEEE Trans. on Microwave Theory and Techniques*.
2. R. Harrington, *Time Harmonic Electromagnetic Fields*, McGraw Hill, 1961, Chapter 7.
3. *MAFIA Release 4 Users Manual*, Darmstadt: Computer Simulations Technology, 1997.

A Possible Plasma Source For a SLAC Afterburner

C. Joshi

UCLA, Los Angeles, CA 90095

In the proposed SLAC afterburner[1] experiment a 7 meter long plasma source with a density of 2×10^{16} cm^{-3} is required. Since such a plasma source has never been operated at the required repetition rate of at least 1 Hz, we outline one possible way to make such a source.

In this idea a low energy relativistic electron beam collisionally ionizes a column of low z gas. The propagation of the electron beam through this gas is achieved by propagating this beam through a low density plasma produced by single photon photo-ionization of an additive gas with a low ionization energy.

For the sake of illustration we consider the ETA/ATA accelerator parameters as the characteristics for the ionizing beam.[2,3]

	ETA	ATA
Beam γ	10	100
Current	10 kA	10 kA
Pulse Length	30 nsec	70 ns
Total Charge	1.8×10^{15}	4.3×10^{15}

Table 1: Ionizing Beam Parameters

The primary gas is hydrogen with a fill pressure of about 10 Torr corresponding to a peak neutral density of 3×10^{17} cm^{-3}. The electron beam is focussed to a radius of 1 mm at the entrance of the plasma column. The beam density is typically 6×10^{13} cm^{-3}. To transport this beam without spreading one will have to rely on plasma column focusing. In other words the beam will have to see plasma focusing force that will balance its spreading due to emittance and space-charge. Thus $n_e \sim 10^{13}$ cm^{-3} density plasma will have to be preformed using ionization of an easily ionizable organic molecule such as TMAE using a laser beam.

Now we will consider the ionization of the hydrogen by the beam. The ionization cross section is [4]

$$\sigma_i = 1.87 \times 10^{-20} A_1 \left[\ln(7.52 \times 10^4 A_2 \gamma^2) - 1\right] cm^2 \qquad (1)$$

Where $A_1 = 0.695$, $Z = 2$ and $A_2 = 1.57$ and for $\gamma = 10$, $\sigma_I = 5 \times 10^{-19}$ cm^2.

Where n_p is the plasma density, n_g is the neutral gas density, n_b is the beam density and

$$\frac{dn_p}{dt} = n_g n_b \sigma_i c \qquad (2)$$

c the speed of light.

Assuming that the ionizing beam is gaussian of the form

$$n_b = \frac{N}{\pi^{3/2} \sigma_r^2 \sigma_z} \exp\left(-\frac{r^2}{\sigma_r^2}\right) \exp\left(-\frac{\xi^2}{\sigma_z^2}\right) \qquad (3)$$

where $\xi = z - ct$ and N = total number of electrons in the beam, we obtain

$$n_p = \frac{\sqrt{\pi}}{z} n_b n_g \sigma_i \sigma_z [1 - \exp(\xi)] \exp\left(-\frac{r}{\sigma_r}\right) \qquad (4)$$

And the maximum density

$$n_{p\,max} = \frac{1}{\pi} \frac{N}{\sigma_r^2} n_g \sigma_i \qquad (5)$$

Substituting for $N = 1.8 \times 10^{15}$, $\sigma_r = 1$ mm, $n_g = 3 \times 10^{17}$ cm^{-3} and $\sigma_i = 5 \times 10^{-19}$ cm^2, one obtain $n_{p\,max} = 2.7 \times 10^{16}$ cm^{-3}. This is in the ballpark of the desired density. Of course in order to produce a uniform 1 mm radius plasma column one should spoil the emittance of the electron beam such that the equilibrium spot size is 1 mm.

ACKNOWLEDGEMENTS

Thanks to T. Katsouleas and D. Gordon for useful discussions. This work was carried out in the Working Group section on " Beam Driven Accelerator Schemes."

REFERENCES

1. T. Katsouleas, see these proceedings.
2. T. Fessenden et al., Proceedings of the Fourth International Topical Conference on High-Power Electron and Ion Beam Research Technology, Palaiseau, France. ed. by J. H. Doucet and J. M. Buzzi, p. 813 (1988).
3. L. L. Reginato, IEEE Trans. Nucl. Sc.**NS-30**, 2970 (1983).
4. Theory and Design of Charged Particle Beams, M. Reiser, J. Wiley (1994).

Acceleration Concepts Based on Electromagnetic Structures

Levi Schächter

Department of Electrical Engineering
Technion – Israel Institute of Technology
Haifa 32000, ISRAEL

Abstract. In the last two decades a variety of novel concepts of electrons acceleration have been suggested and investigated. According to the interaction mechanism, these can be classified into three main groups: *inverse radiation processes*, *space-charge wakes* and *electromagnetic wakes*. In the first case, electrons are injected in structures that in "normal" situations would generate radiation. Electromagnetic radiation is actually injected into the system and it accelerates a bunch of electrons. In the second case, a space-charge wake is generated in plasma by an intense laser pulse or short bunch of electrons. This space-charge wake may also accelerate a bunch of electrons. In the third case, an intense electron beam (bunch) generates radiation in a structure. This radiation is guided into an acceleration structure. Based on this classification, we shall review the basic concepts behind the first and the third group.

INTRODUCTION

When the goal is the largest energy possible, subject to severe budget constraints, the accelerator community put forward an impressive list of innovative ideas that may facilitate gradients on the order of 100MV/m and higher in a feasible way. The goal here is to give a tutorial overview of the basic concepts and to present some of the experimental data that has accumulated during recent years. For the purpose of this presentation it is instructive to determine a set of criteria that will help to present the various new acceleration schemes in a logical and coherent way. The three groups are: *(a)* inverse radiation processes, *(b)* space-charge wakes and *(c)* electromagnetic wakes. This classification is not unique (e.g. near or far field schemes) and the reason it is adopted here is since in my opinion it is more adequate for a tutorial presentation.

In the framework of the first group, *Inverse Radiation Processes*, the schemes rely on the fact that electrons when injected in various structures may generate radiation and therefore lose energy. With proper design, radiation can be injected into the same structure and electrons may actually gain energy from the injected electromagnetic energy flux. Clearly the applied field has to exceed the deceleration in order to have net acceleration consequently, each such process has a threshold intensity. In most cases, the radiation is in the form of a laser pulse and since the latter is a vacuum eigen-mode it does not posses a longitudinal component of the electric field that may accelerate electrons. As a result, the specific structure has to generate it or cause the electrons to experience such an effective field.

In the second group of acceleration schemes, *Space-Charge Wakes*, the longitudinal field that accelerates the electrons is part of a space-charge wake generated by either laser pulse(s) or intense electron bunches. This group is discussed separately therefore we shall consider here only topics that have direct relevance to electromagnetic structures. The third group, *Electromagnetic Wakes*, relies on the intense radiation generated by electron bunches in order to accelerate a different bunch.

With a few exceptions, in the first and second group the initial energy is almost always stored in the radiation field (laser pulse) whereas in the third, the driving electron bunch carries the initial energy. With this regard, in all the concepts to be presented the accelerator is a transformer whose primary is either a laser or a bunch of electrons of high current and relatively low voltage whereas the secondary is a bunch carrying low current at high voltage. The goal is always to generate the highest *transformer ratio* possible, subject to the highest efficiency.

Let us examine now the typical requirements from such a transformer. From the energy perspective, bringing 10^8 electrons to an energy of 1TeV at a 200Hz repetition rate implies a total energy of 3.2kJ. If we assume a one kilometer long accelerator, then each meter the bunch will need about 3.2 Joules. However, if we are optimistic and assume 10% overall efficiency then the *average power* per unit module (1 meter) is 32W. The space-charge force does not allow to keep 10^8 electrons in a bunch that is significantly shorter than a 1μm wavelength (say 1/36) therefore these will need to be split into micro-bunches of about 10^6 electrons and a total of 100 macro-bunches. The instantaneous current associated with a single micro-bunch is 1.7kA and a 1m long module of acceleration requires $(1.7 \times 10^3) \times [(1.0 \times 10^9) \times 1.0] = 1.7$TW. It will be necessary to have a pulse that is at least 300fsec for the acceleration of the entire macro-bunch and the typical impedance is $[(1.0 \times 10^9) \times 1.0]/1.7 \times 10^3 = 0.6$MΩ. For comparison, 2×10^9 electrons in 0.7mm long bunch correspond to 0.13kA of current therefore assuming a gradient of 100MV/m in one meter, as is the basic design of the NLC, the interaction impedance is 0.7MΩ.

The remainder of this tutorial has three additional main sections according to each one of the groups indicated above: next we discuss inverse radiation processes, followed by a short section on space-charge wakes and finally a section that details the electromagnetic wakes alternatives.

INVERSE RADIATION PROCESSES

Inverse Cerenkov Accelerator. When an electron exceeds the characteristic speed of light in the medium ($v_{ch} \equiv c/\sqrt{\varepsilon_r}$), it generates the so-called Cerenkov radiation that propagates at an angle θ relative to the electron trajectory; this angle is given by $\cos(\theta) = v_{ch}/v$ where v is the velocity of the electron. The electromagnetic power that is generated in this process comes at the expense of the kinetic energy of the electron. Imagine now the reverse situation, in which an electron that traverses a dielectric medium is illuminated by an intense laser beam at the Cerenkov angle θ. If the phase is properly chosen, the bunch of electrons will be accelerated. The concept was first demonstrated in 1981 [1] and it is being currently investigated at ATF(BNL)

[2-3]. Figure 1 illustrates schematically the basic configuration of the setup and the experimental parameters are presented in Table 1. An axicon lens focuses a laser beam along an extended length in a pressurized (1.8atm) hydrogen. An early experiment with a similar set-up but with 580MW laser pulse, a 40MeV bunch of electrons gained up to 3.7MeV. An improvement of the performance may be anticipated by operating close to one of the resonances of the background gas. In such a case the effective dielectric coefficient is significantly larger and the pressure reduced [4]. It should be pointed out that the bunch was long on the scale of the laser wavelength and therefore only a fraction of its electrons were actually accelerated.

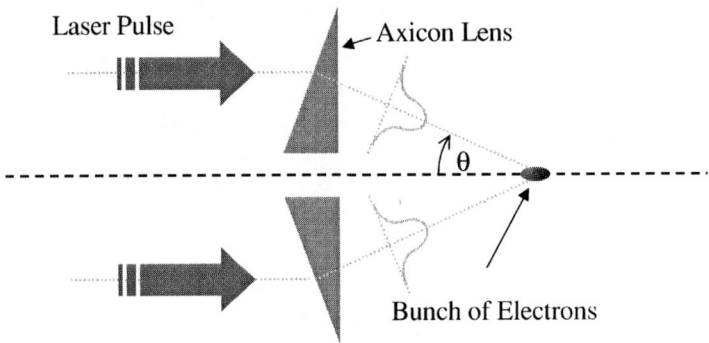

FIGURE 1. Illustration of the Inverse Cerenkov Acceleration scheme. A laser beam is focused by an axicon lens so that on axis the incident angle is identical with the Cerenkov angle for the medium and the electrons.

TABLE 1. ICA parameters [Ref. 3]

Gas Cell		e-Beam Parameters		Laser Pulse Parameters (CO_2)	
Pressure (atm)	1.8	Energy (MeV)	45	Wavelength (μm)	10.6
θ (mrad)	20	Charge (nC)	>0.1	Power (GW)	>10
Interaction length (cm)	6.5	Emittance (mm-mrad)	0.85π	Focal width (μm) FWHM	200
Window (μm)	1	Pulse duration (psec)	10	Pulse duration (psec)	200
		Diameter (μm)	60		

Inverse FEL Accelerator. Another source of electromagnetic radiation is a free-electron laser. The concept was proposed by R. Palmer [5] in 1972 to accelerate electrons in a periodic magnetic field. During the seventies and eighties it was thoroughly investigated as high-power, high-frequency radiation source. However only recently [6], electrons acceleration was actually demonstrated experimentally.

In the case of a magneto-static wiggler, electrons oscillate in a periodic magnetic field and generate or amplify radiation. A bunch of electrons that is small compared to the radiation wavelength, can be accelerated by a radiation field propagating in parallel, if the phase is adjusted properly. In case of a long bunch, only a fraction of the electrons will be accelerated. Schematic of the basic configuration is illustrated in Figure 2.

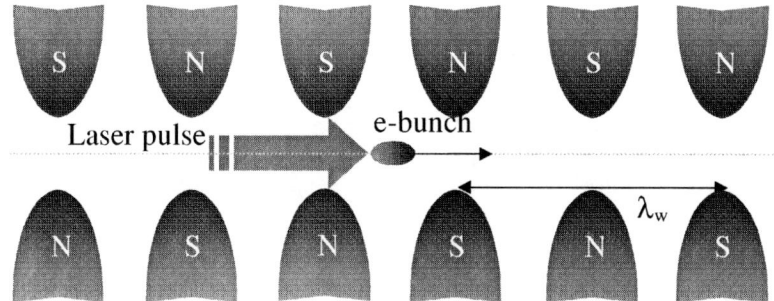

FIGURE 2. Illustration of the inverse free electron laser (IFEL).

In the framework of this scheme, the electromagnetic wave does not posses a longitudinal electric field parallel to the main momentum component of the electrons. These oscillate in the transverse direction under the effect of a periodic magnetic field [e.g. $B_x(z) = B_w \cos(2\pi z/\lambda_w)$] and a transverse electric field (E) and magnetic field (H) associated with the laser beam. The transverse motion combined with the transverse field components may lead to a gradient that its maximum value is given by $E_{acc} = E(ecB_w\lambda_w/2\pi\gamma mc^2)$. Clearly, this gradient is proportional to the square root of the intensity of the laser field and the wiggler's amplitude but it is inverse proportional to the energy of the accelerated bunch. Regardless of the laser field, when an electron is injected in such a wiggler it emits radiation therefore it losses energy. The decelerating field associated with this loss (for $\beta \approx 1$) is

$$E_{dec} = (\gamma^2/2)(e/4\pi\varepsilon_0\lambda_w^2)(ecB_w\lambda_w/mc^2)^2 . \qquad (1)$$

Such a decelerating field introduces a threshold for the radiation energy flux required in order to get net acceleration namely $E_{acc} > E_{dec}$ that can be formulated in terms of the radiation intensity as

$$I > I_{th} \equiv \left[\frac{1}{2\eta_0}\left(\frac{e}{4\pi\varepsilon_0\lambda_w^2}\right)^2\right]\left(\frac{ecB_w\lambda_w}{mc^2}\right)^2 \pi^2\gamma^6 . \qquad (2)$$

This reveals the main limitation of the IFEL at high energies namely, the necessary laser flux increases as γ^6. As an example, consider a wiggler with $B_w = 1T$ and $\lambda_w = 2cm$; this leads to $I_{th} = 2.4 \times 10^{-13} \gamma^6$ [W/cm^2]. At 1TeV the required laser intensity would be 1.5×10^{25}[W/cm^2] and this is only to balance the decelerating force! However for the acceleration of an 1GeV electron the threshold intensity is only 1.5×10^7[W/cm^2] that is orders of magnitude below the available today.

The experiment that demonstrated the operation of the IFEL was performed at BNL based on a 1GW CO_2 laser pulse and using 40MeV electron pulses. The total wiggler length was 47cm and the energy increase was 2.3MeV corresponding to an average gradient of 4.6MV/m. The complete list of parameters of this experiment is presented in Table 2.

TABLE 2: IFEL parameters [Ref. 6]

Wiggler			e-Beam Parameters		Laser Pulse Parameters	
Length (cm)	47		Energy (MeV)	40	Wavelength (μm)	10.6
Period (cm)	3		Charge (nC)	0.1	Power (GW)	1
Gap (mm)	4		Emittance (mm-mrad)	0.88π	E_{max} (GV/m)	0.78
B_{max} (T)	1		Pulse duration (psec)	5	Pulse (psec)	220
			Diameter (mm)	0.6	Radius (mm)	1

Two IFEL's were used recently to demonstrate *staging* of laser driven acceleration modules (STELLA). Although the two wigglers are almost two meters apart, micron size bunches generated in the first section were preserved and accelerated or decelerated by the second IFEL according to the phase of the driving laser field; technical details about this experiment are presented in Ref. 7.

Conceptually, free electron lasers are not limited to a periodic *magneto-static* field (wiggler) and it is possible to generate or amplify radiation also with an *electromagnetic* or *electro-static* field. If we replace the magneto-static wiggler with an *electromagnetic wiggler*, that is a wave propagating anti-parallel to the beam with a phase velocity $c\beta_{ph}$, the wavelength of the radiation emitted by a relativistic electron is $\lambda=\lambda_w/2\gamma^2(1+\beta_{ph})$. The outcome is a shorter effective wavelength of the wiggler that may drop by a factor of 2 (when $\beta_{ph}=1$) or even shorter in the case of a guided wave. But the significant advantage of this scheme is the possibility to generate for short periods of time an intense wiggler using high-power microwaves. This concept was demonstrated [8], as a radiation source, when a high-power (500MW @12.5GHz) microwave pulse from a backward wave oscillator was used as a wiggler to generate a few hundreds of kilowatts of millimeter waves (140GHz); the electrons energy in this case was 1MeV. For a similar microwave field (3cm) as a wiggler and an accelerating laser operating at 1μm, the electrons' typical energy will have to be of the order of 44MeV for an interaction to occur.

Another radiation mechanism that, in principle, can be used for particles acceleration is the *channeling radiation*. The lattice of a crystal forms a natural periodic electrostatic wiggler. If a beam of relativistic electrons is injected parallel into one of the symmetry planes of a lattice, then in the frame of reference attached to the electrons, the structure has two very distinct periods. One in the longitudinal direction that has a negligible effect since the potential is much smaller than the beam's energy. The other periodicity is in the transverse direction. Associated with this transverse periodic potential there is a discrete set of eigen-states similar to a harmonic oscillator. The moving electrons "populate" one of these states. If they are not in the ground state, which is the case when the beam is not ideally parallel to the symmetry plane, then they may emit a photon and jump to a lower state; the photon's energy depends on the atomic potential and the energy of the beam. In classical terms, if the electrons have a small momentum in the transverse direction, they are ``reflected" by the atoms on the lattice-plane and therefore they undergo an oscillatory motion [9-10] - as a result, radiation is emitted. Since there are no appropriate radiation sources at the relevant wavelengths (X-ray), this mechanism does not seem to become feasible as an acceleration scheme in the near future. This, in spite the fact that it has significant advantages in what regards the emittance [11] since, as is the case of a harmonic oscillator, there is a minimum transverse energy-state.

Inverse Cyclotron Radiation. In the case of the magneto-static IFEL the laser field does not have a longitudinal electric field and the acceleration occurs due to transverse motion and the transverse electric field components. A similar process may occur in a uniform magnetic field. In "normal" situation a gyrating electron emits radiation. In principle, the opposite situation is also possible namely, a circularly polarized electromagnetic wave can accelerate electrons in the presence of a magnetic field[12].

Inverse Smith-Purcell. When a charged particle moves at a velocity $v=c\beta$ above a metallic grating of periodicity λ_g, it generates radiation at frequencies that satisfy

$$\omega = \frac{2\pi n}{\lambda_g} \frac{c}{\beta^{-1} - \cos(\theta)} \qquad (3)$$

where θ is the angle this radiation propagates and n is an integer [13]. For a relativistic particle the, highest frequency is generated close to the axis ($\theta \ll \pi$) which is parallel to the motion of the particle. Explicitly, in the case of n=1 the angular frequency is given by $\omega = (2\pi c/\lambda_g) 2\gamma^2/[1+(\gamma\theta)^2]$. Conceptually, the inverse process is feasible in the sense that a laser pulse is injected at a small angle θ, and particles at the correct phase are accelerated – the basic configuration is illustrated in Figure 3.

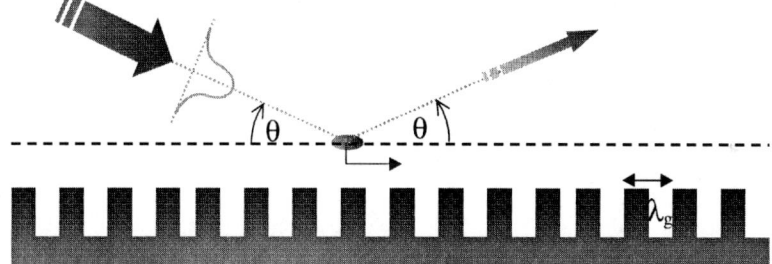

FIGURE 3. Illustration of the Inverse Smith-Purcell Acceleration. The bunch emits radiation at an angle θ and the laser required for acceleration is schematically shown in the left.

Contrary to the magneto-static wiggler, the decelerating force for relativistic particles is independent of their energy and it depends primarily on the height of the charged particle from the top of the grating [14]. Although this is a significant advantage, it imposes a significant constraint on the distance between the tip of the grating and the bunch.

For continuous acceleration over many periods it is necessary that the injected wave will excite an eigen-mode of the structure. This mode needs to have a phase velocity c on one hand and it is strictly necessary that the wave decays perpendicular to the grating, otherwise the energy carried by this mode will decay exponentially parallel to the electron motion. This implies that in the second transverse direction (parallel to the grating) there must be some boundary conditions that impose a periodic solution [15]. To the best of our knowledge, no acceleration experiment based on Smith-Purcell effect was performed. However, preliminary steps along this line were made at BNL when a 45MeV electron beam was injected parallel to a 126mm long grating of 1mm periodicity and 10μm radiation was measured [16].

Inverse Transition Radiation. An electromagnetic plane-wave does not interact with an electron over an extended length in vacuum. However, an interaction may occur if the interaction is over a finite region and/or some boundary conditions are imposed. A simple example comes in mind: an electron that moves in free space. It obviously does not emit any radiation however, if it traverses a grounded metallic membrane, it emits transition radiation. Similar radiation is generated when the electron traverses two metallic parallel plates that form a transmission-line. Now we can conceive the opposite situation. An electromagnetic wave propagates along the transmission-line as a charged-particle traverses the *confined* plane-wave. The separation of the two plates is w and θ is the angle of the particle's trajectory and the direction the wave propagates - see Figure 4.

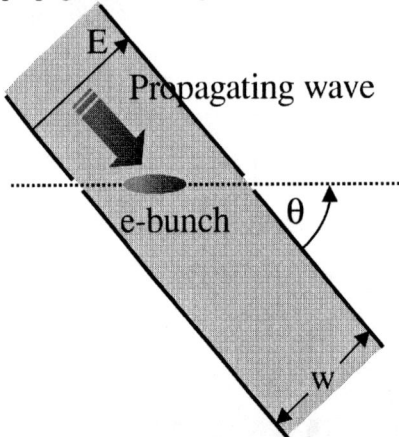

FIGURE 4. A point-charge traversing a confined TEM wave that propagates at angle θ. The effective electric field experienced by the electron is Esin(θ) and the interaction length is w/sin(θ) therefore the maximum energy it may gain is eEw.

The TEM wave has a transverse electric field component E, that that has a projection, E sin(θ), on the direction parallel to the particle's trajectory. The total distance this particle is exposed to the wave is D=w/sin(θ), therefore ignoring the deceleration associated with the transition through the confining plates, the maximum possible acceleration is $\Delta\gamma=eEw/mc^2$. For this to happen it is necessary that the overall phase-slip of the electron relative to the wave will be *smaller* than π. In fact, three conditions have to be satisfied: θ<<π, (γθ)^2>>1 and θ<<λ/w. In addition, it is necessary to set the correct phase for acceleration. Assuming an effective area A_{eff}, the total electromagnetic power is P=(E^2/2η)A_{eff} and with this relation the maximum change in the particle's energy is $\Delta\gamma = e(2\eta P w^2 / A_{eff})^{1/2}/mc^2$ hence a 1TW electromagnetic pulse can change the γ of a particle by a factor of 50 or explicitly,

$$\Delta\gamma = 53.7 \, P^{1/2}[TW] \qquad (4)$$

where the ratio $w/A_{eff}^{1/2}$ is assumed to be unity.

An experiment is underway at Stanford University in collaboration with SLAC for a conceptual test of the scheme. The experimental setup relies on a 34MeV accelerator that can deliver 1psec bunch consisting of 10^8 electrons with an energy spread of 16keV (FWHM). The electron bunch is intersected at an angle of 20mrad by two laser beams generated by a 1psec, 1mJ and 0.8µm Ti: sapphire laser. Its conceptual setup is illustrated in Figure 5 and Table 3 provides additional information regarding this experiment.

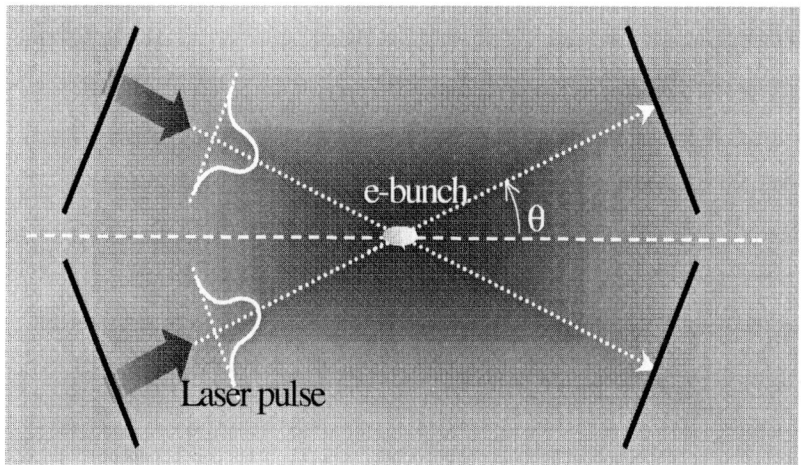

FIGURE 5. Schematic of the LEAP experiment at Stanford University. Two laser beams generate a region in space that electrons may be accelerated if injected when the wave has the proper phase. The dimension of the aperture is a tradeoff between small size that may generate problems in the beam transfer or large aperture that entails broad termination of the acceleration field that in turn reduces the energy gain.

A bunch of electrons traverses the space between two apertures[17]. In the absence of an external laser field, radiation is emitted at both ends and the bunch is decelerated. The laser pulse matches to the characteristic radiation except that it propagates in the opposite direction. As in the case of an inverse free electron laser, for net acceleration this deceleration has to be smaller than the acceleration associated with the external laser field. It is important to point out that the details of the aperture size and curvature are critical for the operation of the system. Its geometry is a trade-off between several constraints: sharp termination (small aperture) entails maximum energy transfer but also high gradients that may cause breakdown as well limited bunch transmission. On the other hand, large aperture (and curvature) reduces the surface gradients facilitates better transmission of the electron bunch but the energy gain is smaller. In the experiment the bunch is much longer than one optical wavelength therefore a fraction of the electrons are decelerated and another fraction is accelerated. According to simulation, an increase of 150keV in the energy of a fraction of the electrons in the bunch is anticipated [18].

TABLE 3: LEAP parameters [Ref. 18]

e-Beam Parameters		Laser Pulse Parameters	
Energy (MeV)	32	Wavelength (μm)	0.8
Electrons in bunch	10^7	Power (GW)	1
Emittance (mm-mrad)	8π	FWHM (μm)	130
Bunch length (psec)	2	Pulse (psec)	0.1-30
Rep. Rate (Hz)	10		
Energy spread - FWHM (keV)	16		

Inverse Laser Effect: Particle Acceleration by Stimulated Emission of Radiation (PASER). In gas lasers pumped by an electron beam, atoms are excited by transferring energy to bound electrons via collisions of the first kind. When the excited atom is stimulated by an adequate photon, two identical photons are emitted – this is the essence of light amplification by stimulated emission of radiation (LASER). We can now conceive the inverse process: consider an excited atom and a free electron moving in its vicinity. Attached to the moving electron, there is a broad spectrum of virtual photons part of which corresponds to the resonance of the atom. The latter is stimulated emitting two *identical* photons that can be absorbed by the initial electron. Thus energy stored in the excited atom can be transferred to the free electron – see Figure 6.

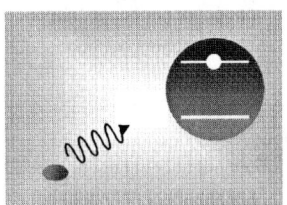  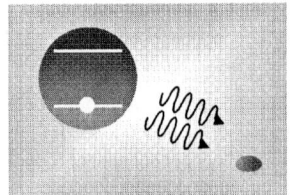

FIGURE 6. Particle Acceleration by Stimulated Emission of Radiation (PASER). An excited atom and a free electron moving in its vicinity. Attached to the moving electron, there is a broad spectrum of virtual photons one of which corresponds to the resonance of the atom. The latter is stimulated emitting two identical photons that are absorbed by the initial electron.

From the perspective of a conventional accelerator, in this case electromagnetic energy is stored in *microscopic* rather than macroscopic cavities and for effective acceleration it is necessary that the size of the bunch will be much smaller than the radiation wavelength. Subject to this condition it was shown that a point-charge (q) injected in an active medium described by a frequency dependent dielectric coefficient

$$\varepsilon(\omega) = \varepsilon_r + \frac{\omega_p^2}{\omega_0^2 - \omega^2 + 2j\omega\omega_1} \quad , \quad \omega_p^2 = (N_1 - N_2)\frac{2\omega_0 \mu_{12}^2}{\hbar\varepsilon_0} \tag{5}$$

can be accelerated [19-20]. In this last expression ω_0 is the resonance angular frequency, ω_1 is the resonance width, the term in the right defines "plasma" frequency in terms of the microscopic parameters of the constituents; $N_1[N_2]$ is the density of atoms in which the bound electron is in the low [high] state; μ_{12} is the dipole-moment of the transition from one energy state to the other. Clearly this quantity is negative if the population is inverted and since when ignoring saturation the reaction-field of the medium to the presence of the relativistic particle on the particle itself is

$$E_z = \frac{q}{4\pi\varepsilon_0\varepsilon_r} \frac{2\omega_p^2}{c^2\varepsilon_r} = \frac{q}{4\pi\varepsilon_0\varepsilon_r} \frac{\omega_0}{c} \frac{4\mu_{12}^2(N_1 - N_2)}{c\hbar\varepsilon_0\varepsilon_r} \quad , \tag{6}$$

this corresponds to an accelerating field; saturation was neglected here. In order to envision the potential of this scheme, consider a 10μm radiation, assuming $\mu_{12}=$ $(1.6\times10^{-19}) \times (2\times10^{-10})$, a point charge that consists of 10^7 electrons and a large population inversion $N_2-N_1 = 10^{25} m^{-3}$, the accelerating gradient is of the order of 1.3GV/m. The advantage of this scheme is the fact that the energy is stored in microscopic "cavities" namely atoms and there is only one bunch involved comparing to two in other cases to be discussed below. However, we have to bear in mind that it is difficult to maintain 10^7 electrons in a bunch that that is significantly smaller than 10μm. A solution to this problem was suggested and it will be discussed in the electromagnetic wakes section.

SPACE-CHARGE WAKES

Acceleration schemes that rely on the generation of space-charge wakes in plasma are reviewed elsewhere [21] therefore we shall limit the present discussion to the strict necessary for the introduction of a concept that combines microscopic electromagnetic cavities namely, resonant atoms. Plasma based schemes may be divided according to the energy source into either laser or electron bunch driven accelerator. In the case of laser wake field acceleration (LWFA) a short and intense laser pulse is injected in the plasma. The space-charge wake generated behind is used to accelerate electrons. A similar process occurs in a plasma wake field accelerator (PWFA), when the driver is an intense electron pulse. The plasma beat wave accelerator (PBWA) uses the resonance between two different laser pulses and the plasma frequency to generate intense space-charge wave.

Another resonant mechanism that can be used is resonant absorption of the background plasma. Consider a gas of neutral or partially stripped atoms that its dielectric coefficient is given by (5) except that the population is not inverted[22]. A beam of electrons is injected in this medium and it is slightly modulated at the resonance of the latter. The absorption associated with the medium leads to an instability that is similar to the well-known resistive wall instability. The space-charge wave that develops along the beam, *grows* in space according to

$$\text{Im}(k) = \frac{\omega_{p,beam}}{v} \text{Im}\left[\frac{1}{\sqrt{\varepsilon(\omega)}}\right] = \frac{1}{R_b}\sqrt{\frac{\eta I}{\pi V_e \varepsilon_r}}\frac{1}{(\gamma\beta)^3}\begin{cases} 1/\Omega_p\sqrt{2} & \Omega_p^2 \gg 1 \\ \Omega_p^2/2 & \Omega_p^2 \ll 1 \end{cases} \quad (7)$$

In the right hand side of Eq.(7) is the estimate of Im(k) at resonance, $V_e = mc^2/e$, R_b is the beam radius and $\Omega_p^2 \equiv \omega_{p,gas}^2 / 2\omega_0\omega_1\sqrt{\varepsilon_r}$. For an illustration of the process see Figure 7.

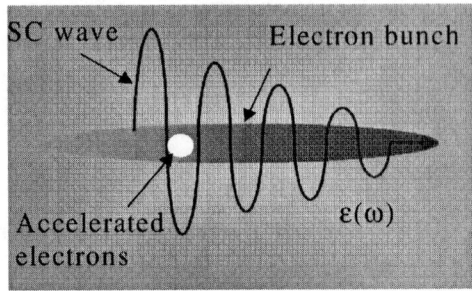

FIGURE 7. Resonant Absorption Acceleration. A long electron bunch propagates in a resonant medium. The bunch is modulated at the frequency of the medium. Due to loss associated with the medium, a space-charge wave grows in a way similar to the well know resistive wall instability.

The energy source is the electron beam itself that transfers part of its (dc) kinetic energy to the build-up of the oscillatory wave. As an example one may conceive an Ar^+ plasma in which a relatively long bunch of electrons is injected. This bunch has a small modulation generated by an Ar^+ laser. As the bunch propagates in the resonant plasma, a space-charge wave grows exponentially along the bunch. A witness electron bunch located towards the tail of the main bunch, can be accelerated by this space-charge wave.

ELECTROMAGNETIC WAKES

The electromagnetic wakes group is conceived here as comprising all the schemes in which an electron beam (or bunch) generates radiation in one structure and subsequently this radiation accelerates a different bunch in the same or a different structure. As such, we can describe in the same framework conventional accelerating structures fed by 250nsec long pulses as in the basic NLC design and a few picoseconds long pulses as is the case in dielectric wake-field accelerator.

Periodic Acceleration Structures – Damped Detuned Structure (DDS). Acceleration structures are designed to operate with a symmetric (TM_{01}) mode and in principle, both the structure and the electron bunch are symmetric. In practice however there are asymmetries both in the structure and in the bunch. These may cause the excitation of non-symmetric modes that in turn may deflect the beam to the wall due to non-zero transverse magnetic field at the bunch location. A substantial effort was and still is directed towards suppression of these hybrid electromagnetic (HEM) modes. The basic idea is to design a structure that looks symmetric and periodic to the TM_{01} but non-periodic to the HEM modes such that asymmetric modes do not develop or at least their power level is kept minimal. Moreover, in order to avoid trapping of these modes in the tapered structures, the HEM modes are damped. A variety of damped-detuned structures (DDS) were designed and tested at SLAC during the years[23]. Typically, these structures are driven by power levels < 100MW generating gradients as high as 80MV/m.

In order to have a more quantitative measure of the detuning process in such a quasi-periodic structure let us consider a simple model of a waveguide filled periodically with 200 dielectric layers. The phase-advance per cell of the TM_{01} mode at 11.424GHz is assumed to be $2\pi/3$ thus the periodicity of the structure is set to be L=8.75mm. For a given radius (R=1cm) the dispersion relation of a periodic structure determines the relation between the thickness (d) and the dielectric coefficient (ε); between the dielectric layers, $\varepsilon=1$. We calculated three pairs of values that correspond to this phase-advance per cell but to different group velocities (d=2,3,5mm and ε=39.851, 7.815, 3.514 correspondingly); the structure terminates with dielectric material of the same type. Since in such a structure the TE and TM modes are de-coupled no hybrid modes can exist therefore, we shall examine the closest "relative" namely, the asymmetric TM mode. The transmission coefficient was calculated for two different cases: in the first case all 200 cells were *identical* (d=3mm and ε=7.815) therefore we shall refer to it as a uniform structure. In the second case there are 200 cells of three types that are distributed *randomly*; this structure will be referred to as the random structure.

Figure 8 illustrates the transmission coefficient of the first symmetric and asymmetric modes for these two cases. The top left frame illustrates the transmission coefficient of TM_{01} for a uniform periodic structure whereas the top-right frame shows the first asymmetric mode (TM_{11}). The lower frames illustrate this quantity for the case of a random structure. When comparing the corresponding frames we observe that this choice of non-periodic structure filters significantly all the frequencies with the exception of the one that was preset (11.424GHz) In fact the transmission of the second pass-band of the TM_{01} is negligible in the random structure as is the first pass-band of the TM_{11}. Although this simple model illustrates the "filtering" of homogeneous waves that is injected from outside it does not describe mode trapping in a fraction of this structure. This can happen when a bunch of electrons propagates in the system. These may excite a wave that will not propagate along the entire structure but rather be confined to a relatively small number of cells. In order to suppress these trapped modes and further reduce the excitation of regular modes, damping of the hybrid modes is also introduced. In fact the damping of the HEM modes has to be very aggressive since not only that this kind of instability may occur on a scale of a single acceleration section but also on the scale of many segments [24-25].

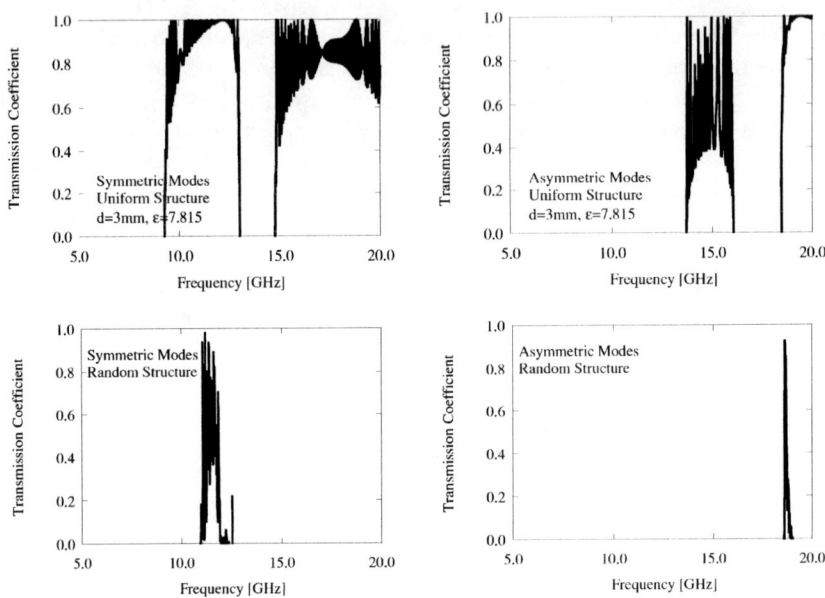

FIGURE 8. The transmission coefficient of the symmetric and asymmetric TM modes for a radius of 1cm and 200 cells. Top frames illustrate the case of 200 identical cells and at the bottom we observe the mode suppression when the 200 cells are randomly chosen to have one of the three characteristics: $(d,\varepsilon)=(2mm,39.851)$, $(3mm,7.815)$ or $(5mm,3.514)$.

Periodic Acceleration Structures – W-band Structure. In zero order, for a given gradient the power in an acceleration structure scales as $1/f^2$ therefore if at X-band the requirement is for a total of 250MW for about 250nsec, operation at W-band entails an rf power of less than 4MW. As a result, in recent years there is an undergoing effort at SLAC to design [26] and manufacture [27] an acceleration structure at $32 \times 2.856GHz = 91.392GHz$. A schematic of the structure is presented in Figure 9. Manufacturing constraints basically rule out the possibility of a circular structure and therefore a planar acceleration structure is being developed. The structure can be conceived as a muffin-tin with the sides removed and replaced by a pair of side chambers which act as side terminations. When viewed from the side, the structure presents two periodic arrays of vanes facing one another from above and below the beam plane. They extend towards the beam plane from upper and lower plane metallic surfaces. The vane pairs are the analogs of the beam iris in cylindrical structures, and the space between the irises and terminated by the upper and lower plane metallic surfaces correspond to the cavities. The geometric parameters are illustrated also in Figure 9; the quality factor as calculated with MAFIA is 1605 and the group velocity is 0.098c. For comparison the typical quality factor of the DDS is 7000 whereas the group velocity may vary from 0.03c to 0.1c.

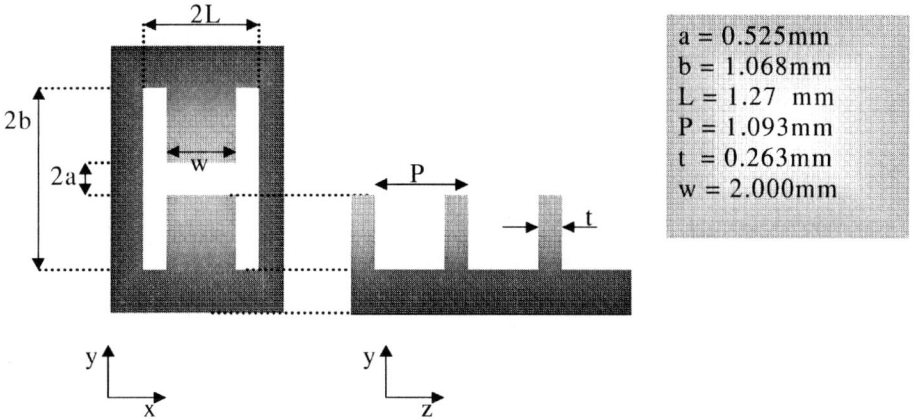

FIGURE 9. Schematic of the W-band acceleration structure that is being developed at SLAC.

Periodic Acceleration Structures – Photon Band-Gap (PBG). A different concept was developed by Kroll, Schultz and their collaborators at the University of California at San Diego (UCSD). It replaces the basic pill-box cavity with an open and periodic structure that consists of a series of metallic cylindrical posts – it is called photonic band gap (PBG) cavity [28]. The posts are confined by two metallic plates whose separation is smaller than the wavelength – see Figure 10. A double periodic structure of this type may be designed such that at the frequency of interest, it will not transmit any electromagnetic waves. Based on such a design, the central post may be removed and as a result a discrete eigen-frequency is formed in the forbidden band thus the corresponding eigen-mode may be confined by the posts. In this way, a cavity similar to the regular pill-box cavity was generated, except that from the perspective of HEM modes this is an open cavity therefore the corresponding quality factor is low.

In order to envision the process consider a series of very *thin* (on the scale of one wavelength) posts organized in a 7×7 matrix with one post in the center removed. Each post can be electromagnetically represented by a current I_n located at (X_n, Y_n) and since they are made of metal the tangential electric field vanishes. The entire system is assumed to be excited by a dipole that carries a current I_0, located at (X_0, Y_0) in the vicinity of the missing post in the center – see left frame of Figure 10. The electromagnetic field is a superposition of the contributions from all the currents thus,

$$A_z(x,y) = \frac{1}{2\pi\mu_0} \sum_{n=0}^{N} I_n H_0^{(2)}\left[\frac{\omega}{c}\sqrt{(x-X_n)^2 + (y-Y_n)^2}\right] \quad (8)$$

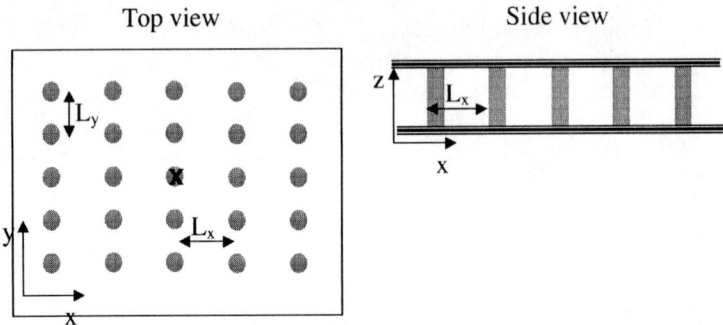

FIGURE 10. Schematic of a Photonic Band Gap (PBG) cavity used for an acceleration structure. The removal of the central post creates an eigen-frequency inside the forbidden band. This is the equivalent of the TM_{01} mode in a pill-box cavity. Higher modes have low quality factor since the cavity is open.

The currents, I_n, are determined from the condition that all the other dipoles contribute a total zero electric field at the location of any given post. The total power emitted normalized to the power that I_0 would emit in the absence of the posts is given by $\overline{P} = \left|\sum_{n=0}^{N} I_n\right|^2 |I_0|^{-2}$. This quantity is reproduced in Figure 11. The period in both directions is L=4.3cm and we observe that there is a very distinct peak around 1GHz; the normalized power in this case is almost -30dB. This can be understood in terms of a disctructive interference namely, the contribution to the far-field due to the currents associated with all the posts combined, almost cancels the far-field contribution of the driving dipole (I_0). It is also evident that at least up to 10GHz there is no other frequency at which the electromagnetic energy is well confined. It should be emphasized that in this simple model, the posts are assumed to be very thin; their finite (radial) dimension may increase the frequency where the field confinement occurs.

The UCSD group developed a periodic acceleration structure made of a series of photonic band gap cavities. According to calculations performed at S-band, the wake-field contribution in the range 0-10GHz is half the corresponding value of a structure based on a pill-box cavity. Table 4 presents some generic parameters for an acceleration structure that operates at 2.856GHz and their equivalent at (32x2.856GHz=) 91.392GHz.

TABLE 4: Photonic Band Gap parameters [Ref. 28]

	S-Band (2.856GHz)	W-Band (91.392GHz)
Plates separation (mm)	34.99	1.093
Post diameter (mm)	16.09	0.504
Periodicity (mm)	42.9	1.344
Thickness of the plates (mm)	6.0268	0.188
Phase advance per cell	$2\pi/3$	$2\pi/3$

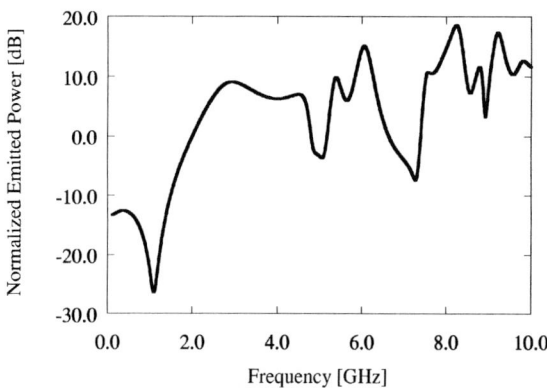

FIGURE 11. Normalized power emitted by an ideal dipole located in the center of a photonic band gap (PBG) cavity. Close to 1GHz the contribution of all posts cancels out the driving dipole.

Two Beam Accelerator (TBA). In conventional linear accelerators multiple rf sources feed one acceleration structure. Each such source has one electron beam therefore the input power to this large transformer is split into many primary ports. The two beam accelerator uses only two beams: the primary is a high current and relatively low energy electron beam and the secondary, is a regular acceleration structure that each one of its modules is fed from a "twin" module in the primary- for a schematic illustration see Figure 12. The concept was proposed by Sessler [29] in 1982 and at present there are two different approaches: at CERN, the driving beam is planned to have an energy in the GeV range and there will be no need for re-acceleration units. Each extraction module consists of a traveling-wave structure that already generated almost 80MW of power at 30GHz corresponding to accelerating gradients of 125 MV/m. In the framework of the current optimized design [30] a 3 TeV accelerator will be 35 km long assuming gradients of 100-200 MV/m.

An alternative design is pursued at LBNL/LLNL. It is planned to use a beam of more modest energy (only a few MeV's) and re-accelerate the bunches after they leave each extraction unit where they generate power at 11.424GHz. For performance similar to the CERN design namely, a similar gradient ($P \times f^2$ = constant), each module will need to generate order of 520MW of power.

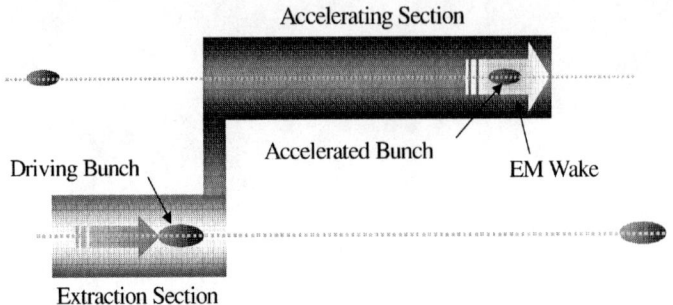

FIGURE 12. Schematic of a Two-Beam Accelerator. An electromagnetic wake is generated in the extraction section and then guided into the acceleration module.

Dielectric Wake Field Accelerator (DWFA). The idea in this case is similar to the TBA except that both the extraction and accelerating module are basically *dielectric loaded waveguide*. A proof of principle when both beams were on the same axis was performed back in 1988[31] at Argonne National Laboratory. A top gradient of 15MV/m was demonstrated experimentally; the details are presented in Table 5. Although limited by surface breakdown of the dielectric materials, the advantage of this configuration is its simplicity. Consider a simplified version of this structure: a waveguide of radius R filled with dielectric (ε_r). A charged-loop of radius $R_L<R$ and infinitesimal thickness moves along the axis of the waveguide (ignoring collisions of the charged-loop with the dielectric). It generates a longitudinal electric field that for a relativistic velocity ($\beta \approx 1$) is given by

$$E_z(r,\tau) = \frac{q}{4\pi\varepsilon_0 \varepsilon_r R^2} \sum_{s=1}^{\infty} \left[\frac{2}{J_1(p_s)}\right]^2 J_0\left(p_s \frac{R_L}{R}\right) J_0\left(p_s \frac{r}{R}\right) \cos[\Omega_s \tau] h(\tau) \quad (9)$$

where $J_0(p_s)=0$, $\Omega_s = p_s c/R(\varepsilon_r-1)^{1/2}$, h(t) is the step function and $\tau=t-z/v$. Here we observe an important property of wakes generated by electrons and this is: the phase velocity of *all* the eigen-modes that represent this wake, is always the same (v). Different phase velocities may develop only due to discontinuities along the structure that may excite homogeneous waves.

TABLE 5: Dielectric Wake-Field Accelerator

	Driver Bunch	**Witness Bunch**
Charge (nC)	20-50	1
Pulse duration (psec)	15-30	3
Radius (mm)	3	1-3
Energy (MeV)	15	3.8
Emittance (mm-mrad)		5
Gradient (MV/m)		15
Energy gain (MeV)		1.5

A second result that can be readily shown here based on (9) is the so-called *transformer ratio* defined as the ratio between the accelerating field on the witness bunch and the average field on the drive bunch. Assuming that the latter is uniformly distributed radially (R_b), azimuthally and longitudinally (Δ), then the average decelerating field is

$$E_{dec,av} = \frac{q}{4\pi\varepsilon_0\varepsilon_r R^2} \sum_{s=1}^{\infty} \left[\frac{2}{J_1(p_s)}\right]^2 \left[\frac{2J_1(p_s R_b/R)}{p_s R_b/R}\right]^2 \left[\frac{1}{2}\left(\frac{\sin(\Omega_s\Delta/2c)}{\Omega_s\Delta/2c}\right)^2\right] \quad (10)$$

whereas the accelerating field on axis is

$$E_{acc} = \frac{-q}{4\pi\varepsilon_0\varepsilon_r R^2} \left[\frac{2}{J_1(p_1)}\right]^2 \left[\frac{2J_1(p_1 R_b/R)}{p_1 R_b/R}\right]^2 \left[\frac{\sin(\Omega_1\Delta/2c)}{\Omega_1\Delta/2c}\right]. \quad (11)$$

In case the first Bessel harmonic is dominant the transformer ratio is 2. However when this is not the case the wake is more complex and it is illustrated in Figure 13. The solid line represents a point-charge whereas the dashed-line a Gaussian distribution that its width is 1/6 of the wavelength of the first mode. The interference peaks are evident as well as the regions of destructive interference. This may have the advantage that the metallic surface that guides the wake is exposed to an intense field for a shorter period of time comparing to the case of a dominant single harmonic.

The distance between two peaks may be readily estimated using geometrical optics arguments. Cerenkov radiation leaves the particle at an angle $\theta = \cos^{-1}(1/\beta\sqrt{\varepsilon_r})$, it is reflected by the waveguide that is at a distance R away from the particle and intersects the axis at a distance $2R/\sqrt{\varepsilon_r\beta^2 - 1}$. It is evident that for a given radius the separation of the peaks is inverse proportional to the square root of ε_r thus the larger the latter, the shortest the separation. In addition, close to the Cerenkov condition, the distance between peaks may become significant.

In principle, it is possible to take advantage of this periodic pattern by injecting instead of a single bunch a train of bunches distributed in a way that by virtue of constructive interference, the wake will be significantly larger although the total amount of charge is the same. Experiments along similar lines are underway at the Argonne National Laboratory [32].

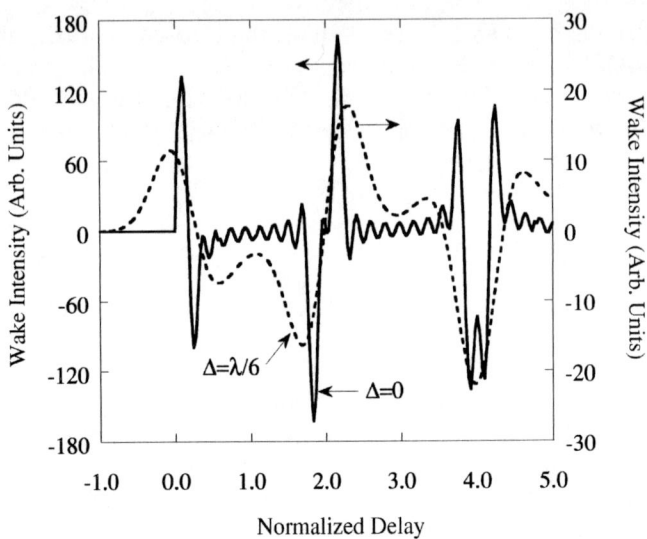

FIGURE 13. Wake generated by a point-charge (solid-line) and by a Gaussian distribution of charge (dashed-line) which its width is 1/6 of the first Bessel harmonic wavelength.

Amplified Cerenkov Wake. In principle, it is possible to generate a transformer ratio larger than 2 by changing the longitudinal distribution of the pulse, however this may not be a trivial task both practically and theoretically (having non-symmetric modes in mind). One option that we suggested [33-34], is to use energy stored in active medium in order to enhance significantly the transformer ratio. In fact, rather than having a *driving* bunch that carries all the initial energy, it is necessary to have only a *triggering* bunch that triggers the adequate mode in the system. This mode is then amplified by the active medium since it is the latter that stores the energy required for acceleration. A conceptual set-up is illustrated in Figure 14. Two periodic dielectric mirrors are designed such that at the wavelength of interest, the reflection coefficient for an evanescent wave is virtually unity. In this way, at resonance most of the electromagnetic energy is either in the vacuum region or in the active medium region that is attached to each one of the dielectric mirrors.

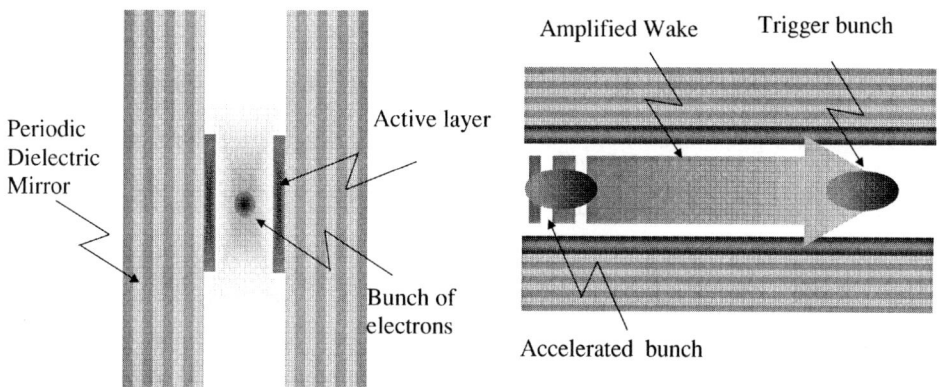

FIGURE 14. Cross-section (left) and side (right) view of a Wake-Amplified Accelerator. A couple of dielectric mirrors confine the radiation to the vacuum and active medium regions. The bunch excites a wake that is amplified by the medium. The wake moves at the speed of the trigger bunch thus it can accelerate a trailing bunch.

The triggering bunch generates Cerenkov radiation in the active dielectric layers and only one of the excited eigen-modes will be amplified by the resonant medium i.e.

$$E_z(t,z) = E_0 \cos[\omega_0(t - z/v_T)] e^{\text{Im}(\omega)(t - z/v_T)} h(t - z/v_T). \qquad (12)$$

As all the other waves attached to the propagating bunch, this mode has the same phase-velocity as the triggering bunch. Many wavelengths behind the triggering bunch, a different bunch of electrons may be accelerated by the amplified wake – see right frame of Figure 14. This mechanism has the advantage that the *longitudinal* electric field is inherent to the excitation mode contrary to the regular "vacuum" mode of a laser that its longitudinal electric field is limited by the Rayleigh length. It also eliminates the need for the optical system that is part of any laser based setup. And finally, since the energy is stored in the medium, it releases many constraints from the driving bunch that in comparison with the dielectric wake-field accelerator it carries the entire energy.

CONCLUDING REMARKS

An adequate comparison of the various acceleration schemes will not be appropriate since the concepts are different and the available means differ substantially from one scheme to another. For example, if we compare either one of the microwave (or millimeter wave) schemes with the either one of the optical schemes there is a significant advantage in favor of the longer wavelength schemes regarding the availability of small electron bunches. While millimeter long or even sub-millimeter long bunches are available, there are no single electron bunch that is even of the same order of magnitude as optical wavelength – except if it is part of a train of bunches. This directly affects the flexibility of experiments.

Another aspect that the optical schemes endure is alignment and synchronization problems. While at X-band, Ka-band or even W-band the methods are still scalable from the vast experience at S-band, in the optical range most of the related questions are still open. For this reason, the STELLA experiment has provided confidence for experiments to come regarding the feasibility of staging laser-driven acceleration modules. The investigators presented results that indicate good control of the relative phase between the laser field and micron size bunches train although the latter drifted almost 2 meters between two acceleration stages.

Operation at millimeter wave (W-band) is a trade-off between the present operation regime (S and X-band) and the optical regime. It has the advantage that many of the design concepts are very similar to present ones yet it benefits from the high frequency operation and planar manufacturing. In fact, the design is in process of being adapted to the *manufacturing* and mechanical constraints of the structure. Sources are not readily available but in the foreseeable future, klytrinos developed at SLAC or gyro-klytrons developed at University of Maryland may fill this gap. Operation at optical wavelength benefits from availability of high-power lasers and, in the context of vacuum acceleration (e.g. LEAP), it may also benefit from the vast experience in the manufacturing of planar structures using photo-lithographic techniques developed for micro-electronics.

S and X-band experience indicates that symmetry of the acceleration structure is very important. In fact, even when the structure is symmetric, an electron bunch may excite *asymmetric modes* that in turn can deflect the electrons. In laser-driven schemes these asymmetric modes may prove to be a significant impediment since in many cases, the electromagnetic structure itself is asymmetric.

There is little information about *breakdown* and *dark-current* beyond what is known in the S-band. However, from the dark-current perspective, vacuum systems are to be preferred since in a gas filled system, electrons may be extracted and accelerated by the laser field. Yet we do not know how the dark-current in laser-driven schemes compares with rf-driven structures. Breakdown on the other hand, seems to impose less stringent constraints in optical schemes comparing to rf-driven structures. For example, assuming a gradient of 1GV/m then a surface field of 5GV/m is probably a reasonable estimate. Therefore the energy flux that impinges upon the surface is of the order of 0.25 TW/cm^2 (assuming a *local* impedance similar to that of the vacuum). Consequently, for a 1psec pulse duration, the energy is by a factor of 4 below a conservative *surface damage threshold* of 1J/cm^2.

In the context of wake-field schemes where the wake is generated by electron bunch(es) in an electromagnetic structure, it should be possible to *tailor a train of bunches* for the generation of the highest gradient subject to a given amount of charge for a given structure configuration i.e. boundary conditions.

It was emphasized here that using *microscopic cavities*, namely resonances of atoms or molecules, may reduce or eliminate some inherent problems associated with the various acceleration schemes. They do not impose any machining constraints as in the case of W-band systems. The medium can be gaseous such as CO_2 or Argon or it can be solid-state such as Nd:YAG or Ti:Sapphire. Gaseous medium has the advantage that it does not require a minimum energy for the electron bunch but collisions and breakdown are significant drawbacks. On the other hand, in the vicinity of an active solid-state the last two problems are reduced dramatically, but because of the exponential decay of an evanescent wave, the distance of the bunch from the surface of the solid material has to be on the order of $\lambda\gamma\beta/2\pi$ and for most practical purposes this is relevant only for highly relativistic particles. Another advantage associated with microscopic cavities is the way energy is stored. This is *distributed* along the interaction region thus instead of compressing all the energy in a bunch that is one millimeter long and several microns in radius, the energy is stored in a hollowed cylinder that is many centimeters long and order of one micron in thickness. Since the energy is stored in the resonant medium, the requirements from the triggering bunch, are significantly reduced. In fact, as indicated earlier, in the case of the amplified wake-field acceleration scheme, the role of the trigger bunch is only to excite the adequate mode which eventually is amplified by the medium.

This work was supported by the United States Department of Energy.

REFERENCES

1. J.A. Edighoffer et.al., Phys. Rev. A, **23**, p. 1848 (1981).
2. W.D. Kimura et. al., *Phys. Rev. Lett.* **74**, p.546 (1995).
3. W.D.Kimura et. al., "STELLA Experiment: Design and Model Predictions", in *Advanced Accelerator Concepts -1998*, edited by W. Lawson, AIP Conference Proceedings 472, New York: American Institute of Physics, 1998, p. 563.
4. L. Steinhauer, J. Appl. Phys. 68, p. 4929(1990).
5. R. Palmer, *J. Appl. Phys.*, **43**, p. 3014 (1972).
6. A. van Steenbergen et.al., "Status of the BNL IFEL Accelerator" in *Advance Accelerator Concepts - 1996*, Edited by S. Chattopadhyay, AIP Conference Proceedings 398, New York: American Institute of Physics, 1998, p. 591.
7. See report by W.D. Kimura in this Proceedings.
8. Y. Carmel et.al., *Phys. Rev. Lett.* **51**, p.566 (1983).
9. M.A. Kumachov; *Phys. Lett A.* **57**, p.17 (1976).
10. R.W. Terhune and R.H. Pantell; *Appl. Phys. Lett.*, **30**, p.265 (1977).
11. Z. Huang, P. Chen and R. Ruth;. *Phys. Rev. Lett.* **74**, p.1759 (1995).
12. C. Chen; *Phys. Rev. A.* **46**, p. 6654 (1992).
13. S.J. Smith and E.M. Purcell; *Phys. Rev.* **92**, p.1069 (1953).
14. L. Schächter and D. Schieber; *Nucl. Instr. Meth. Phys. Res. A* **440**, pp.1-4 (2000).
15. N.M. Kroll, in *Laser Acceleration of Particles* edited by C. Joshi and T. Katsouleas, AIP Conference Proceedings 130, New York: American Institute of Physics, 1985, p. 253.
16. R.C. Fernow et. al., "Observation of 10μm Smith-Purcell Radiation from 45MeV Electrons" in *Advanced Accelerator Concepts - 1996*, edited by S. Chattopadhyay, AIP Conference Proceedings 398, p. 601.
17. E. Esarey, P. Sprangle and J. Krall; *Phys. Rev. E*, **52**, p.5443 (1995).
18. R.H. Siemann and R.L. Byer, Private communication (2000).
19. L. Schächter, *Phys. Lett. A.*, **205**, p. 355 (1995).
20. L. Schächter, *Phys. Rev. E.*, **53**, p. 6427 (1996).
21. See the three invited talks by C. Joshi, T. Katsouleas as well J. Rosenzweig on Plasma Based Accelerators in this Proceedings.
22. L. Schächter, "Resonant Absorption Instability", this Proceedings.
23. J.W. Wang et.al. "Accelerator Structure R&D for Linear Colliders" in *the Proceedings of the 1999 Particle Accelerator Conference*, New York 1999, Editor A. Luccio, W. MacKay, p.3423.
24. R.H. Helm and G.A. Loew, "Beam Break-up", Ch. B.1.4, *Linear Accelerator*, Eds. P. Lapostolle and A. Septier, pp. 173-221, North Holland Publishing Co., Amsterdam, 1970.

25. J.Haimson and B.Mecklenburg; "Suppression of Beam-induced Pulse-shortening Modes in High Power RF Generator TW Output Structures", SPIE Proceedings Vol. 1629, *Intense Microwave and Particle Beams III*, Edited by H.E. Brandt, Los Angeles California, p.209 (1992).
26. N.M. Kroll et.al., " Planar Accelerator Structures for Millimeter Wavelengths", " in *the Proc. of the 1999 Particle Accelerator Conference*, New York 1999, Editor A. Luccio, W. MacKay, p.3612.
27. D.T. Palmer. et.al., "Material Research related to W-band Cavity Construction" in *the Proc. of the 1999 Particle Accelerator Conference*, New York 1999, Editor A. Luccio, W. MacKay, p.545 .
28. N.M. Kroll et.al., "Photonic Band Gap Accelerator Cavity at 90GHz" in *the Proceedings of the 1999 Particle Accelerator Conference*, New York 1999, Editor A. Luccio, W. MacKay, p.830.
29. A.M. Sessler, in *Laser Acceleration of Particles*, AIP Conference Proceeding, 91, Editor P.J. Channel p.154 (1982). A.M. Sessler and S.S. Yu, *Phys. Rev. Lett.*, **58**, p.2439 (1987).
30. Jean-Pierre Delahaye et.al., "The CLIC Study of a multi-TeV e+-e- linear collider", in *the Proc. of the 1999 Particle Accelerator Conference*, New York 1999, Editor A. Luccio, W. MacKay, p. 250
31. W. Gai et. al.; *Phys. Rev. Lett.*, **61**, p.2756 (1988).
32. W. Gai this Proceedings
33. L. Schächter, *Phys. Rev. Lett.*, **83**, p.92 (1999).
34. L. Schächter, "Wake Amplification by a Solid-State Active Medium" in this Proceedings.

Parameters of 2 ×200 GeV Linear Collider with Microstructures Excited by Laser Radiation

A. Mikhailichenko

Cornell University, Wilson Laboratory, Ithaca, NY 14850

Abstract. Parameters of 2×200 GeV linear collider represented here. Accelerating structures of this collider are scaled down to meet the requirements of a micrometer wavelength laser as driver. The length of this collider is 2×150 m, and luminosity is 10^{33}cm^{-2}sec^{-1}. Energy required to feed the microstructures is 0.1J/m in 100 ps laser pulse at repetition rate 160 Hz.

1. INTRODUCTION

As the discovery of the Higgs boson with no doubt will be done at LHC (if not at LEP), the direct production of these bosons in electron-positron colliders will be a top priority task for nearest future. The energy of e^+e^- collider required, according to latest result from LEP, could be not higher, than 200 GeV in any case. Reach this level was planned using conventional technology, what is basically an upgraded version of equipment, developed at the times of radar invention (what is not true, however for TESLA, if the accelerating structure only is taken into account). A lot of job done convinced everyone now, that 2×200 GeV linear collider could be made with this technology. The cost of this 10 km long (total) machine estimated to be in multi-billion ranges however.

In contrast, we pointed out, that investment even a small fraction of this amount to implementation of *existing* micro-technology into accelerating business, might give a rise for very fast development much more compact machine [1] for this energy scale. In case of success this will give a green light for developing a multi-TeV machines with accelerating gradients ~10 GeV/m.

The method of acceleration of particles in microstructures described in [1] (see also [2]). In this scheme the accelerating structures scaled down to a micrometer level with the help of technology available from micro fabrication. In *this case* they could be feed by laser radiation. A special feeding procedure proposed, what was called Travelling Laser Focus (TLF), lowers the probability of destruction of the structure under extreme laser radiation exposure. The key element of the TLF method is a *sweeping device*, which allows local excitation of the microstructure in accordance with instant position of accelerated bunch. The microstructures have independent feeding of each cell from the open side orthogonal to the beam trajectory. The excitation is going through these openings *locally* by sweeping. Special sweeping device, which prepares the laser beam, as it shown in Fig.1, allows very short illumination of the every point on the surface of the microstructure. This helps to concentrate all instant power of the laser onto small part of the microstructure. In a transverse direction (transverse to the motion), a cylindrical lens provides a local focus of the laser radiation. That is why this arrangement was called a Travelling Laser Focus (TLF).

In this publication we scale down parameters of accelerator [1] so that they are reflecting now the requirements for 2×200 GeV machine (instead of 2×3 TeV).

We estimate, that collider described here could be made faster, than any of others types for this energy interval. This estimation takes into account the time required for R&D stage.

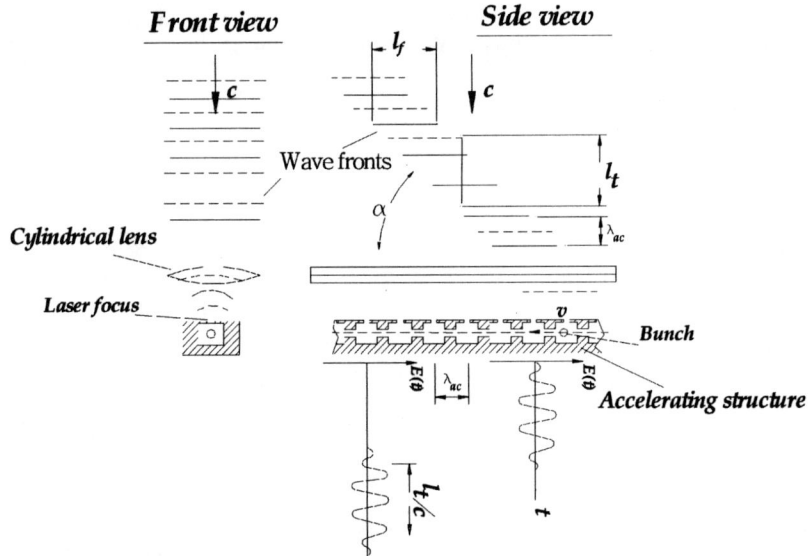

FIGURE 1.1: This configuration of the feeding field prepared by *sweeping* the laser radiation along the trajectory of the bunch. Time dependence of the field in two different points along the structure is at the bottom graphs. Beam has a speed $\vec{v} \cong c$ to the left, so $\tan\alpha = c/v \cong 1$. Cylindrical lens focuses the laser beam in transverse direction, see Fig 2.1. It made on low dispersion material. At left side of the picture the beam is going rectangular to the plane of this picture.

2. LINEAR ACCLERATOR

Time duration of the radiation at each point on the structure is $\sim l_t/c$, where l_t is the instant height of the sloped laser bunch along the perpendicular to the plane of the structure, see Fig. 1.1. The number of the cells excited simultaneously defined by $\sim l_f/\lambda_{ac}$. Due to this arrangement, all pulsed laser power is acting for generation of accelerating field at the momentary particle's location only. By this way total power, required from the laser, is much less, than if the structure illuminated as whole. The illuminating time for any point on the structure is lowered by factor $\approx L/l_f \cong L/l_t$.

The device for sweeping the laser beam made preferably on electro-optical material, which controllably changes its refraction index by application of electric field [1]. Fragment of accelerating structure is represented in Fig. 2.2. Sweeping device works for accelerating structure, which is 3 cm long. Accelerating structures and sweeping devices installed in series, so that after passage one of the structure particles goes into another. Synchronization of modules is going through common power laser source. There is no necessity in exact following the bunch by the focused laser spot so. Accelerator structure [2] serves as housing for accelerating field.

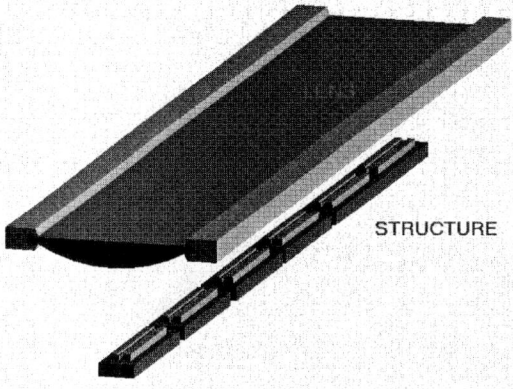

FIGURE 2.1: The accelerating structure and the short focusing lens. Structures installed on piezoelectric movers.

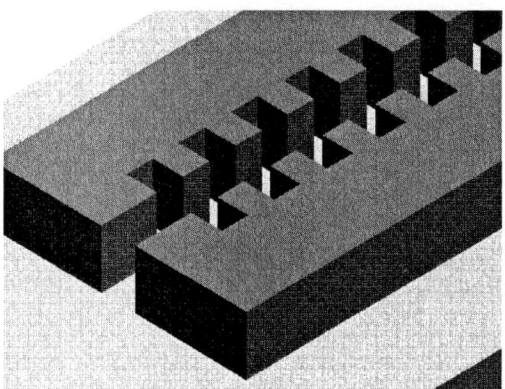

FIGURE 2.2: The accelerating structure fragment. Beam is going at the half height of the structure.

Simplified view of such an assembly is represented in Fig. 2.3. All elements are placed in vacuumed volume, not shown in this Figure.

Primary laser beam 1 goes to the end of accelerator. Mirrors 2 redirect it back, pos.3, trough the sequence of splitters. In the similar way the particle's beam 5, goes trough bending system 6 and further trough structures to next modules, 4. 7 and 8 –are the focusing elements for the laser and particle's beam respectively. Optical platform 9 is standing on legs 10 with active damping system to minimize vibrations. 13– cylindrical lenses, 14–are the accelerating structures. All elements on the table are located in a vacuumed volume, not shown here.

The direction of the accelerating structures as indicated in Fig. 2.3 helps to run the elecron/positron bunch through the structures, as in horizontal plane the corresponding emittance is bigger.

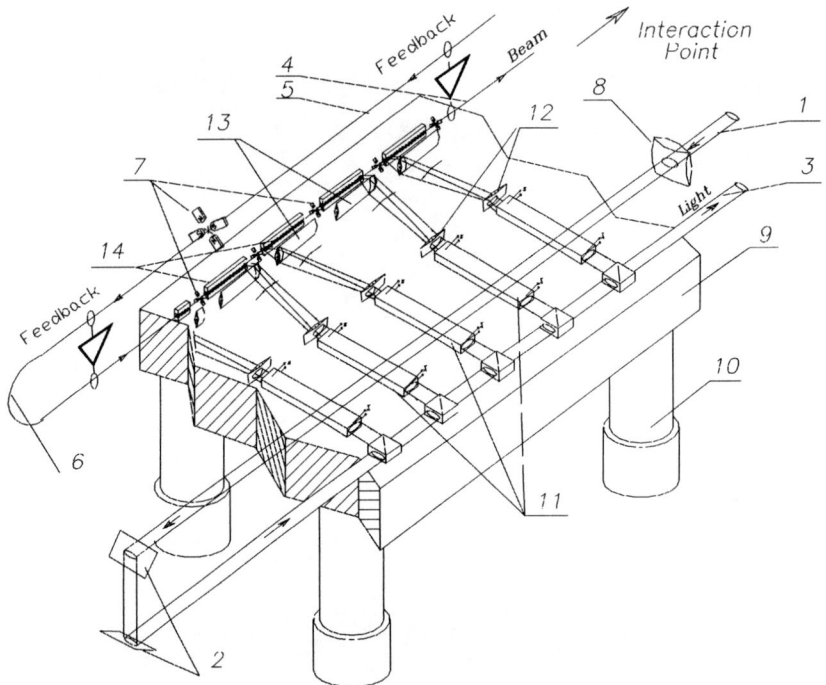

FIGURE 2.3: Direction of largest beam size in damping ring (horizontal, if wigglers have vertical magnetic field) is going in the structures along the passing slots. This helps to keep emittance low during acceleration. Light –means laser beam. Beams are covered by numerous feedback loops, schematically shown in this figure. The laser beams also have feedback loops arranged with the help of electro-optical elements. Other comments are in the text.

3. INJECTOR

Injector must generate the beam with minimal emittance. Positron production requires a damping ring in any case. The only restriction from the lower side on emittance is (see Ref. In [1])

$$(\gamma\varepsilon_x)(\gamma\varepsilon_y)(\gamma\varepsilon_s) = (\gamma\varepsilon_x)(\gamma\varepsilon_y)(\gamma l_b(\Delta p/p_0)) \geq \tfrac{1}{2}(2\pi\lambda_c)^3 N , \qquad (3.1)$$

where $\lambdabar_C = r_0/\alpha$ –is a Compton wavelength, $r_0 = e^2/mc^2$ –is a classical electron radius, $\alpha = e^2/\hbar c$ –is a fine structure constant, $\gamma = E/mc^2$, N–is a bunch population, l_b –is a bunch length, $\Delta p/p_0$ –is a relative momentum spread in the bunch, $(\gamma\varepsilon_s) = \lambdabar_b(\Delta p/p_0)$ –is an invariant longitudinal emittance.

A Kayak-Paddle Cooler (KPC) [1], Fig. 3.1, is a racetrack, which made as a sequence of wigglers and accelerating structures, installed along straight line and having the bends at the end.

Bends made with many short period divisions to prevent emittance dilution. In first approximation it could be considered as a *very* short-period *planar* wiggler with lot of bending elements. Each element of this bend could be considered as an achromatic three-magnet bend with the focusing quads in between. One of the three-magnet bending elements is represented in scale at the right top. It has three magnets and two quads.

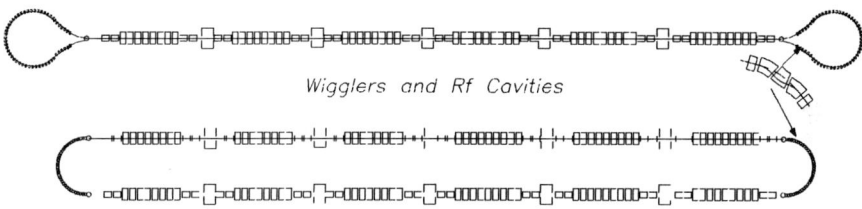

FIGURE 3.1: Top: A Kayak-paddle cooler [1]. At the lower part the sketch of more conservative design is represented. Here the straight sections are the same as at the top part. Two of these sections joint by semi-arcs. This injector is more expensive, but, probably, is easy in tuning. It provides less input in emittance dilution from the bends also.

The straight sections squeezed together for a compact size, so the back and forward trajectories *might be* congruent. The longer the straight section is–the smaller fraction the bends give to the cooling process. The bends itself could be made to give a small input into cooling dynamics. We used separate short quadrupoles for transverse focusing. These quadrupoles are positioned between three-magnet achromatic bend.

In this cooler a particle needs to re-radiate its full energy few times, like in ordinary damping ring. The cooling time associated with these losses is

$$\tau_{cool} \cong -\frac{\gamma}{cd\gamma/ds} = \frac{3}{2}\frac{\lambdabar^2}{cr_0 K^2 \gamma}, \qquad (3.2)$$

where $K = eH_\perp \lambdabar/mc^2$, $H_\perp$ –is a magnetic field value in the wiggler, $2\pi\lambdabar$ –is the wiggler period. One can see from here, that the cooling (damping) time *does not depend on the wiggler period*. So that is valid even when the wiggling arranged by laser radiation. This cooling time obviously does not depend on the length of the straight section, as the influence of the bends was neglected –shorter the length, faster the revolution.

Emittance dynamics defined by well-known equations (averaging over the period)

$$\frac{d\varepsilon_{x,y}}{ds} \cong \left\langle \left(H_{x,y} + \frac{\beta_{x,y}}{\gamma^2} \right) \frac{d(\Delta E/E)^2_{tot}}{ds} \right\rangle - 2\alpha_{x,y}\varepsilon_{x,y}, \qquad (3.3,3.4)$$

where $H_{x,y}$ defined

$$H_{x,y} = \frac{1}{\beta_{x,y}} \left(\eta^2_{x,y} + (\beta_{x,y}\eta'_{x,y} - \frac{1}{2}\beta'_{x,y}\eta_{x,y})^2 \right), \qquad (3.5,3.6)$$

$\eta_{x,y}$ –Are dispersion functions. Derivatives are taken over longitudinal direction, which is s. One can suggest any orientation of the wiggler field polarization. The orientation chosen, however, gives the direction of largest emittance along the narrow side of the structure. Partial decrements $\alpha_{x,y,s}$ are defined as $\alpha_i = J_i / 2 I_s$, where $J_x \cong 1, J_y = 1, J_s \cong 2, J_x + J_s = 3$. Partial decrement for energy spread is the same as for emittance.

For a dipole wiggler, with vertical polarization of magnetic field, the periodic solution for η_x can be expressed as*

$$\eta_x = \frac{K_x \lambdabar}{\gamma} Sin\frac{s}{\lambdabar} = \frac{\lambdabar^2}{\rho_x} Sin\frac{s}{\lambdabar}, \qquad (3.7)$$

where $\rho_x = \lambdabar \gamma / K_x$ is the bending radius in magnetic field of the wiggler. The length of formation of radiation is $\rho_x / \gamma = \lambdabar / K_x$. For the function H_x we can estimate $H_x \cong \beta_x \eta'^2_x$. As the $\eta'_x \cong K_x / \gamma \cdot Cos(s/\lambdabar)$ one can obtain

$$H_x \cong \beta_x K_x^2 / \gamma^2 \cdot Cos^2(s/\lambdabar). \qquad (3.8)$$

So formulas (3.4)-(3.5) yield

$$\frac{d\varepsilon_x}{ds} = \left\langle \left(1 + K_x^2 Cos^2(s/\lambdabar)\right) \frac{\beta_x}{\gamma^2} \frac{d(\Delta E/E)^2_{tot}}{ds} \right\rangle - 2\alpha_x \varepsilon_x \qquad (3.9)$$

For vertical emittance $K_x = 0$ (if the wiggler field has no horizontal polarization). Equilibrium emittances defined by condition $d\varepsilon_{x,y} / ds = 0$. Combining (3.3)–(3.9) for quantum excitation *only* one can find

$$(\gamma \varepsilon_x) \cong \tfrac{1}{2} \cdot \lambdabar_c \bar{\beta}_x (1 + K_x^2/2)\gamma / \rho_x \cong \tfrac{1}{2} \cdot \lambdabar_c \bar{\beta}_x (1 + K_x^2/2) K_x / \lambdabar \qquad (3.10)$$

$$(\gamma \varepsilon_y) \cong \tfrac{1}{2} \cdot \lambdabar_c \bar{\beta}_y \gamma / \rho_x \cong \tfrac{1}{2} \cdot \lambdabar_c \bar{\beta}_y K_x / \lambdabar, \qquad (3.11)$$

where $\bar{\beta}_{x,y}$ – are averaged envelope functions in the wiggler. The last formulas together with (3.2), $\tau_{cool} \cong \tfrac{3}{2} \cdot (\lambdabar^2 / r_0 c K^2 \gamma)$, define the cooling dynamics under SR. One can see that equilibrium invariant emittances *do not depend on energy*. In addition, quantum equilibrium *vertical emittance* and the cooling time do not depend on the wiggler period at all. The same is true for radial emittance if $K \le 1$. So the wiggling in laser field does not give any advantage.

---

* The wiggler must have the ending field in poles relation as ¼, -¾, 1, n(-1,1), - ¾ , ¼ . n=0,1,... This will give ~zero averaged displacement of trajectory, see Ref. in [1].

Substitute for estimation $\overline{\beta}_{x,y} \approx 1m$, $\lambda \cong 5cm$, $K \cong 5$, one can obtain for *quantum emittances* the following estimations

$$(\gamma\varepsilon_x) \cong 2.5 \cdot 10^{-8} \, cm \cdot rad, \tag{3.10a}$$

$$(\gamma\varepsilon_y) \cong 9.5 \cdot 10^{-10} cm \cdot rad. \tag{3.11a}$$

On expense of damping time these last values could be reduced by the factor 5, putting $K \cong 1$. The bunch number in the damping ring needs to be increased to satisfy the repetition rate. Additional decrease of emittance could be obtained by reducing $\overline{\beta}_{x,y}$. So a decrease of the order of 10 for each emittance could be expected here. For IBS scattering we have

$$\varepsilon_x^{5/2} \cong \frac{3}{2} \cdot \frac{\left(1+K_x^2/2\right)^{3/2}}{K^2} \frac{N}{\gamma^6} \frac{\beta_x^{3/2} \cdot r_0^2 \cdot \lambda^2 \ln_C}{l_b \beta_y^2}. \tag{3.12}$$

One can see that there is a strong dependence on the beam energy $\varepsilon_x \sim \gamma^{-12/5}$. For $N \cong 10^9$, $\gamma \cong 4 \cdot 10^3$, $l_b \cong 1cm$, $\overline{\beta}_{x,y} \approx 1m$, $\lambda \cong 5cm$, $K \cong 5$, we obtain for *IBS emittances*

$$\gamma\varepsilon_x \cong 4 \cdot 10^{-6} cm \cdot rad,$$

$$\gamma\varepsilon_y \cong 3 \cdot 10^{-7} cm \cdot rad.$$

So the only way to come to emittances as required compared with ones defined by quantum process only– is increasing the energy of the beam and following *scrapping* the extra particles. For successful operation of Laser Linear Collider the only $N \approx 10^6$ particles are required.

As the cooler described above keeps the dispersion invariant (3.5,3.6) as small as possible, the emittance (and brilliance) is the smallest one and could not be made smaller (as quantum limitations define it). Summarizing, we can say that the source of electrons and positrons with emittance required could be build with no doubt.

4. PARAMETERS

General view of Linear Laser Collider (LLC) complex is represented in Fig.4.1. In figure's capture all sufficient elements are mentioned. A lot of feedback elements pursuit the beam's fluctuations. For this purposes, both bunches- the laser and particle's one go first to the end of all linac apart from the central station- this was mentioned in second Chapter. On the way there the bunch's (both, optical and particle's) parameters picked up, processed with appropriate algorithms locally and applied to correcting elements [1]. We reduce the accelerating gradient and (hence) a power of laser to relax the operation. In optimistic case the length could be shrink to 2×20 meters with 3 *mJ/cm* of laser energy density.

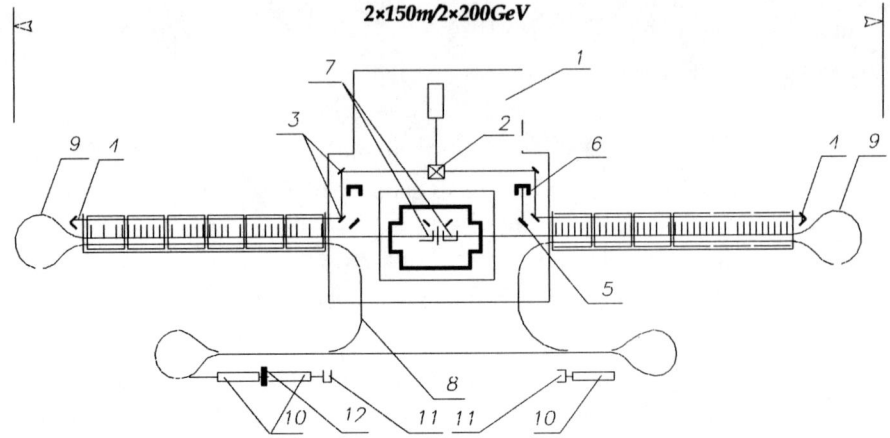

FIGURE 4.1: Laser Linear Collider (LLC) complex. 1–is a laser master oscillator platform, 2 –is an optical splitter, 3,4–are the mirrors, 5–is a semi-transparent mirror, 6–is an absorber of laser radiation. 7–are the Final Focus Systems. 8–are the damping systems for preparing particle's beams with small emittances, 9–are the bends for particle's beam. 10–are the accelerating X-band structures, 11–is an electron gun, 12–is a positron converter. The scheme with the damping rings as sources are shown here.

A small *crossing angle* required for preventing illumination of the final lenses by used beam. This angle is absolutely necessary for multi-bunch operation mode. Tiny dimensions of the beam can help to push the beams through optics. Repetition rate, even few *kHz*, allows easily manipulation with the beams. In Table 1 we summarized parameters of 200×200 *GeV* machine.

TABLE 1. Parameters of the Laser Linear Collider.

Wavelength	$\lambda_{ac} \cong 1 \mu m$
Energy of $e^{\pm}$ beam	200×200 GeV
Total two-linac length	2×200 m
Main linac gradient	1.2 GeV/m
Luminosity/bunch	$10^{32} cm^{-2} s^{-1}$
No. of bunches/pulse	10 (≤ 100)*
Laser flash energy/Linac	3J
Repetition rate	160 Hz
Beam power/Linac	5 W
Bunch population	10^5
Bunch length	0.1 μm
$\gamma \varepsilon_x / \gamma \varepsilon_y$	$\approx 10^{-8} / 10^{-9} cm \cdot rad$
Damping ring energy	2 GeV
Disruption parameter	1.4
Length of section/Module	3cm
Wall plug power**	$2 \times 5kW$

*–Maximal possible number.
**–The power for supplemental electronics and for feeding the damping ring is not included.

Disruption parameter is small due to short bunch length and small bunch population. The beams of electrons and positrons can be *polarized* what gives the effective gain in luminosity and reduces the background [1].

5. REFERENCES

[1] A. Mikhailichenko, *Particle Acceleration in Microstructures Excited by Laser Radiation*, CLNS 00/1662, Cornell 2000.
[2] A. Mikhailichenko, *TableTop Accelerator with extremely Bright Beam*, this Workshop.

Production and synchronization of electron beams from RF photoinjector/compressor systems for ultra-fast applications

M.C. Thompson and J.B. Rosenzweig

UCLA Dept. of Physics and Astronomy
405 Hilgard Ave., Los Angeles, CA 90095

Abstract. The RF photoinjector, when coupled with a magnetic pulse compression system, is now a ubiquitous tool for production of sub-picosecond electron beam pulses which are to be used in advanced accelerator and light source experiments. As the time-scale for both pulse lengths and synchronization to external systems approaches the femtosecond level, a clear understanding of the longitudunal dynamics of the electron injector is required. This paper presents an analysis of the longitudinal beam dynamics of electron bunches in the photoinjector/compressor system from birth at the photocathode, through their initial violent acceleration in the RF gun, and subsequent phase space manipulation in the post-acceleration linac and magnetic chicane. The phenomena of phase focusing due to RF forces, and defocusing due to longitudinal space-charge, are discussed, as is the process of magnetic pulse compression. The issues relevant to synchronization of electron pulses with external lasers are examined, using the examples of beat-wave acceleration and Compton light sources, and solutions involving appropriate compressor configurations are proposed. Diagnosis of the relevant physical effects in such schemes is discussed.

INTRODUCTION

The advanced accelerator (AA) and free-electron laser (FEL) communities are now reliant on the RF photoinjector as the electron source which can provide needed short pulse, high current and high brightness beams required by the demanding experimental state-of-the-art. As the obtaining of low emittances requires limiting the beam density in a way that the beam plasma frequency is matched to the applied betatron frequency, there are limits on the peak current (minimum pulse length and/or maximum charge) which can be extracted from these devices[1,2]. In order to break these constraints, as well as to relax the requirements on the photocathode drive laser pulse length, many laboratories now employing magnetic pulse compression systems[3]. An example of such a system is displayed in Fig. 1, which shows the Neptune Advanced Accelerator Laboratory photoinjector[4] at UCLA. This system is typical of the present-day RF photoinjector; it has a split accelerator configuration (high gradient gun[5], with an adjustable phase, post-acceleration linac), and a magnetic chicane. Its applications include investigation of beat-wave (two-frequency laser-driven) and wake-field (electron beam-driven) acceleration in plasma.

The most demanding of the applications which photoinjector/compressor systems must serve are in fact found in the field of advanced accelerators, where plasma-based acceleration schemes with femtosecond beams are envisioned[6,7], as well as in Compton light sources[8], where sub-picosecond electron beams are collided with ultra-short laser pulses. In both cases, the electron beam is required not only to be ultra-short, but also synchronized at well below the sub-picosecond level to the external system — the short-wavelength accelerating wave in the plasma accelerator case, and the laser pulse intensity envelope in the Compton light source scenario. In order to understand the pulse length and synchronization issues in such applications, we must first develop a clear understanding of the longitudinal beam dynamics in the injector. The purpose of this paper is to provide an analysis of these dynamics, and to show how they can be manipulated to properly synchronize the electron beam pulse to the external systems in question.

Our analysis of the electron beam longitudinal dynamics proceeds as follows. First, we develop an analytical model based on an approximate Hamiltonian treatment, in order to gain insight into the relevant physical processes in these systems. This analysis is intended to allow study of initial acceleration dynamics, including the backward wave component of a standing wave RF gun, and longitudinal space-charge. The results of the model are then compared to exact solutions of the equations of motion and to simulation of multiparticle systems with PARMELA. The various competing effects — phase focusing due to RF acceleration, and defocusing due to repulsive longitudinal space-charge forces can then be quantitatively discussed.

The fact that the beams are longitudinally focused during initial RF acceleration implies that they become partially locked to the RF "clock", in that the electrons are attracted (focused) towards a certain optimum RF phase. Thus any photocathode laser injection jitter is partially removed during initial acceleration. This effect is desirable from the viewpoint of pulse length, but also causes problems in synchronization. This is because the photocathode laser itself cannot be considered a good "clock", as it does not produce a beam which arrives at the end of the injector with a constant time delay from injection. The presence of the magnetic compression system allows one to choose which good "clock" one wishes to lock to, however. In the case of optimum compression, and neglecting defocusing beam self-forces, the compressor takes all initial injection timing errors, and maps them to a final state in which the injection error disappears to lowest order, at the expense of enhanced momentum error. Thus the beam timing is, to lowest order approximation, perfectly locked to the RF "clock" at the compressor exit. To use this beam in an application, one must lock all other (external) components of the experimental system to this clock as well. This occurs naturally in the case of a multi-stage electron-beam plasma wake-field accelerator, and methods for locking a short pulse laser to the RF have also been proposed for Compton scattering experiments as well.

On the other hand, in the case of a plasma beat-wave accelerator, it is difficult to lock the beat envelope (which is sensitive to the relative phase of the two laser lines) to the RF clock. As a way of injecting the electron beam into the beat-wave with the correct time structure, it has been proposed that the photocathode drive laser be sliced into a periodic array of micropulses, by modulating the laser intensity using nonlinear effects based on the beat-wave intensity itself. The RF phase focusing in the gun destroy this initially perfectly locked (to the beat-wave) photoelectron time structure, however. As is discussed below, this problem can be removed by simply using the magnetic compressor to reverse the compression in the gun — to slightly decompress the beam. With this procedure, the timing of the electron beam micropulses at the compressor exit is, to lowest order approximation, exactly locked to the initial laser micropulse train.

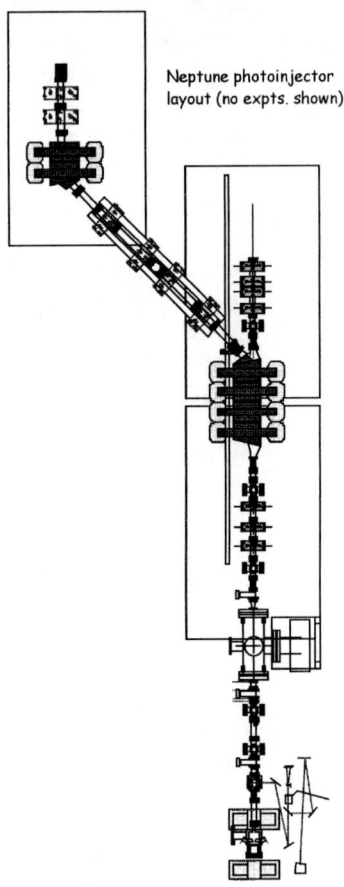

FIGURE 1. UCLA Neptune photoinjector (no experiments shown).

AN ANALYTICAL MODEL OF LONGITUDINAL BEAM DYNAMICS IN AN RF PHOTOINJECTOR

Longitudinal beam dynamics in a RF photoinjector[9] are dominated by the evolution the beam undergoes in the first few millimeters of propagation when the beam is not yet highly relativistic. Once the beam is relativistic its longitudinal profile is essential frozen in and any further significant longitudinal evolution requires outside manipulation of the beam phase space, typically using the enhanced longitudinal dispersion introduced by bend magnets. If we wish to describe the longitudinal evolution of the beam correctly during the early non-relativistic period the effects of the forward traveling RF wave must be considered along with those of the backward wave, space-charge, and the elongation of the resonance cavity in 1.625 cell type photoinjectors.

The travelling wave approximation

The longitudinal evolution of electrons emitted in photoinjectors is dominated by the resonant forward wave. As electrons are accelerated from rest they slip with respect to the forward wave until they achieve relativistic velocity. Since the RF field is typical rising when the beam is launched later electrons see a higher initial field and do not slip as far as the earlier electrons, which tends to compress the beam. This process is typically referred to as velocity bunching, or RF focusing. To begin to understand this process, we start by including only the effects of the forward wave. The longitudinal field of the forward traveling RF wave in a photocathode gun can be written simply as

$$E_z = E_0 \sin(\omega t - k_z z) = -\frac{\partial A_z}{\partial t}, \quad (1)$$

where,

$$A_z = \frac{E_o}{k_z v_\varphi} \cos(k_z(v_\varphi t - z)). \quad (2)$$

Inserting this expressing for the vector potential into the general form of the one dimensional Hamiltonian of a charged particle in an electromagnetic field

$$H = \sqrt{(p_{z,c} - qA_z)^2 c^2 + (m_0 c^2)^2}, \quad (3)$$

gives a Hamiltonian that is explicitly dependant on time and can not be used as a constant of the motion.

$$H = \sqrt{\left(p_{z,c} - \frac{qE_o}{k_z v_\varphi} \cos(k_z(v_\varphi t - z))\right)^2 c^2 + (m_0 c^2)^2}. \quad (4)$$

In order to use the Hamiltonian as a constant of the motion we need to make a canonical transformation that eliminates time as an explicit variable. Ultimately we would like to describe the position of the particle by the phase it occupies in the forward wave. The logical choice for the new coordinate is

$$\zeta = v_\varphi t - z, \quad (5)$$

which is equivalent to a Galillean (purely mathemetical, not physical *a la* Lorentz) tranformation of z into the frame of the forward traveling wave. Taking this new coordinate and equating the new momentum to the old $p_\zeta = p_z$ we can write down the transformed Hamiltonian[10] so as to preserve the proper equations of motion,

$$\tilde{H}(\zeta, p_{\zeta,c}) = H(\zeta, p_{\zeta,c}) - v_\varphi p_{\zeta,c},$$

$$\tilde{H}(\zeta, p_{\zeta,c}) = \sqrt{\left(p_{\zeta,c} - \frac{qE_o}{k_z v_\varphi}\cos(k_z\zeta)\right)^2 c^2 + (m_0 c^2)^2} - v_\varphi p_{\zeta,c}, \quad (6)$$

$$\frac{d\zeta}{dt} = \frac{\partial \tilde{H}}{\partial p_{\zeta,c}} = \frac{p_\zeta}{\gamma m_0} - v_\varphi = v_z - v_\varphi, \quad (7)$$

$$\frac{dp_{\zeta,c}}{dt} = -\frac{\partial \tilde{H}}{\partial \zeta} = \frac{qE_o v_z}{v_\varphi}\sin(\omega t - k_z z). \quad (8)$$

Now we can simplify the transformed Hamiltonian using the following definitions and approximations

$$\alpha_{RF} = \frac{\gamma'_{max}}{k_z}, \quad \chi = \sqrt{\frac{1+\beta_z}{1-\beta_z}}, \quad \beta_\varphi \approx 1, \quad \phi = k_z \zeta, \quad (9)$$

$$\tilde{H} = m_0 c^2 (\chi^{-1} - \alpha_{RF}\cos\phi), \quad (10)$$

where ϕ is the phase location of the particle in the forward RF wave. Since this Hamiltonian is a constant of the motion we can use it to derive the mapping function $\phi_{final}(\phi_{initial}) \equiv \phi_f(\phi_i)$ which gives each particle's final position in the forward wave from knowledge of its initial position,

$$\tilde{H}_i = m_0 c^2 (1 - \alpha_{RF}\cos\phi_i), \quad \tilde{H}_f = -m_0 c^2 \alpha_{RF}\cos\phi_f, \quad (11)$$

$$\tilde{H}_i = \tilde{H}_f, \text{ and} \quad (12)$$

$$\phi_f = \cos^{-1}\left(\cos\phi_i - \frac{1}{\alpha_{RF}}\right). \quad (13)$$

We can also derive a simple compression function for beams short compared to the RF wavelength. This is the quantity we are really interested in since it conveys the degree of longitudinal focusing/defocusing that the particles experience. The simplest way to accomplish this is to differentiate Eq. 12 with respect to ϕ_i and solve for $\frac{d\phi_f}{d\phi_i}$

$$\frac{d\tilde{H}_i}{d\phi_i} = \frac{d\tilde{H}_f}{d\phi_i} \Rightarrow m_0 c^2 \alpha_{RF}\sin\phi_i = m_0 c^2 \alpha_{RF}\sin\phi_f \frac{d\phi_f}{d\phi_i}. \quad (14)$$

The desire to create a beam with good transverse emittance implies that $\phi_f = \pi/2$, which further implies a certain choice of ϕ_i. Making this substitution we are left with a compression function dependant only on the initial launch coordinate,

$$\frac{d\phi_f}{d\phi_i} = \sin\phi_i. \quad (15)$$

This quantity, which is manifestly less than unity, is mapping factor for compressing the final beam phase extent with respect to its initial extent. It can also be interpreted as the degree of "jitter suppression" that the RF focusing supplies, since errors in injection timing (jitter) will be ameliorated by this factor.

Standing wave acceleration: inclusion of the backward wave

In reality, photoinjector RF guns are standing wave devices and the backward wave has significant influence on the longitudinal evolution of beam before the particles become relativistic[9]. The full RF field in the photoinjector is given by

$$E_z = E_0\left(\sin(\omega t - k_z z) + \sin(\omega t + k_z z)\right), \quad (16)$$

If we transform this equation into the phase variable ϕ [9],

$$E_z = E_0\left(\sin\phi + \sin(\phi + 2k_z z)\right), \quad (17)$$

we see that it is not possible to write the backward wave term solely in terms of the phase variable. In order to include the backward wave term in our Hamiltonian and have it remain a constant of the motion we must make an approximation. Since the electric field that the particles experience at launch dictates the degree of the primary velocity focusing effect, it follows that a good approximation of the backward wave contribution must be correct at launch. In consideration of this fact we choose to make the approximation

$$E_z = E_0\left(\sin\phi + \sin\phi_0\right), \quad (18)$$

where $\phi_0 = \phi_i$ so the backward wave term is set at launch and kept constant thereafter. At first this seems drastic, but Eq. 18 is exactly correct at lauch and we will show that it is reasonable later in this paper. Note that Eqs. 16-18 illustrate how the backwave boosts the field at launch, which reduces velocity focusing by ensuring that all the particles will accelerate faster and experience less phase slippage.

Taking the backwave term as a constant and proceeding as in the previous section we arive at the new Hamiltonian

$$\tilde{H} = m_0 c^2\left(\chi^{-1} - \alpha_{RF}(\cos\phi - \phi\sin\phi_0)\right), \quad (19)$$

Rewriting Eq. 11 with the new Hamiltonian (Eq. 19) reveals a slight sublety in the calculation of the compression function

$$\begin{aligned}\tilde{H}_i &= m_0 c^2\left(1 - \alpha_{RF}(\cos\phi_i - \phi_i\sin\phi_i)\right), \\ \tilde{H}_f &= -m_0 c^2 \alpha_{RF}\left(\cos\phi_f - \phi_f\sin\phi_0\right).\end{aligned} \quad (20)$$

Here we see the very important distinction. Intitially the backward wave contribution must be treated as variable since $\phi_i = \phi_0$ is varying. In the equation for the final Hamiltonian, however, the backward wave term is a constant whose derivative is zero. Now if we equate the equations (20) and differentiate as in (14) taking care to follow the rule detailed above we find

$$\frac{d\phi_f}{d\phi_i} = \frac{2\sin\phi_i + (\phi_i - \frac{\pi}{2})\cos\phi_i}{(1+\sin\phi_i)}. \tag{21}$$

A one-dimensional model of longitudinal space-charge effects

In order to stay consistent with the plan of keeping the fields that the electrons experience as accurate as possible at the point of launch, the interaction of the charges with their image in the cathode must be included in the description. This effect is often comparable to the backward wave in size for parameter regimes of interest.

To good approximation the space-charge (SC) field experienced by the electron can be modeled as that of a one-dimensional system varying only in z. Including image charge effects, the electric field at a given point in the bunch is dependent only on the surface charge density of the beam population emitted before the point in question[11],

$$E_{SC}(t) = 4\pi\sigma(t), \tag{22}$$

where

$$\sigma(t) = \frac{4Q}{\pi d^2} = -\frac{4eN_b}{\pi d^2}\left(\frac{t}{\Delta t_p}\right) = -\frac{4eN_b}{\pi d^2}\left(\frac{\phi_i - \phi_{head}}{\omega_{RF}\Delta t_p}\right), \tag{23}$$

N_b is the total beam population, d is the laser spot diameter, Δt_p is the total laser pulse length, and ϕ_{head} is the launch phase of the head of the beam. Note that this model correctly indicates that space-charge is zero for the first electron and then grows linearly reaching a maximum at the tail of the beam. In addition, this treatment has the benefit of depending only on the initial phase ϕ_i. Much like the backward wave approximation we used in the previous section, this model overestimates the field since it does not fall off once the beam is far from the cathode. This exaggeration is actually useful. Since the backward wave and space-charge terms have opposite effects (focusing and defocusing, respectively), the errors made by approximating both terms tend to cancel out.

If we add in the space-charge term to the rest of the Hamiltonian, again keeping the new term as a constant dependent only on initial phase ϕ_0,

$$\tilde{H} = m_0 c^2 \left(\chi^{-1} - \alpha_{RF}(A\phi(\phi_0 - \phi_{head}) + \cos\phi - \phi\sin\phi_0)\right), \tag{24}$$

where

$$A = \left(\frac{16N_b e^2}{m_e c \alpha_{RF}(\omega_{RF}d)^2 \Delta t_p}\right) = \frac{1}{\alpha_{RF}}\left(\frac{\omega_{p0}^2}{\omega_{RF}^2}\right), \tag{25}$$

with the initial phase extent of the injected beam $\Delta\phi = \omega_{RF}\Delta t_p$, and ω_{p0} is the square of the nonrelativistic plasma frequency of a bunch of diameter d and length $c\Delta t_p$.

Note that ϕ_{head} serves merely as an index that records the phase where the beam starts and it is a constant in both the initial and final Hamiltonian equations. Again we can solve for the compression function as described in the previous section,

$$\frac{d\phi_f}{d\phi_i} = \frac{2\sin\phi_i + (\phi_i - \frac{\pi}{2})\cos\phi_i - A(2\phi_i - \phi_{head} - \frac{\pi}{2})}{(1 + \sin\phi_i - A(\phi_i - \phi_{head}))}. \tag{26}$$

Effects of elongated initial cell

Up until now we have neglected the fact that the 1.625 cell geometry of many modern photoinjector guns[5] we are interested in significantly distorts the RF fields from their vacuum wavelength. We simulate this effect by assuming an offset in the initial conditions

$$k_z z_o = -\phi_l, \tag{27}$$

where in our case $\phi_l = \pi/8$, which gives a correct additional length in the initial cavity, but does not quite give the correct spatial dependence of the RF wave. This leads to the following transformation of the RF fields

$$(\omega t - k_z z) \rightarrow (\omega t - k_z z + \phi_l), \quad (\omega t + k_z z) \rightarrow (\omega t + k_z z - \phi_l). \tag{28}$$

We can easily add this correction into the Hamiltonian

$$\tilde{H} = m_0 c^2 \left(\chi^{-1} - \alpha_{RF} \left(A(\phi + \phi_l)(\phi_0 - \phi_{head}) + \cos(\phi + \phi_l) - (\phi + \phi_l)\sin(\phi_0 - \phi_l) \right) \right) \tag{29}$$

There is a convenient simplification that can be made in writing down the initial and final Hamiltonians using Eq. (29). Since we have already state that the electron must arrive at a final phase of $\pi/2$ we are free to absorb the constant phase shift ϕ_l into ϕ_f,

$$\tilde{H}_i = m_0 c^2 \left(1 - \alpha_{RF} \left(A(\phi_i + \phi_l)(\phi_i - \phi_{head}) + \cos(\phi_i + \phi_l) - (\phi_i + \phi_l)\sin(\phi_i - \phi_l) \right) \right)$$
$$\tilde{H}_f = -m_0 c^2 \left(\alpha_{RF} \left(A(\phi_f)(\phi_0 - \phi_{head}) + \cos(\phi_f) - (\phi_f)\sin(\phi_0 - \phi_l) \right) \right) \tag{30}$$

Solving for the compression mapping function as before, we obtain

$$\frac{d\phi_f}{d\phi_i} = \frac{2\sin(\phi_i)\cos(\phi_l) + (\phi_i + \phi_l - \frac{\pi}{2})\cos(\phi_i - \phi_l) - A(2\phi_i + \phi_l - \phi_{head} - \frac{\pi}{2})}{(1 + \sin(\phi_i - \phi_l) - A(\phi_i - \phi_{head}))}. \tag{31}$$

NUMERICAL SOLUTIONS OF THE EQUATIONS OF MOTION: COMPARISON WITH THE ANALYTICAL MODEL

In order to judge the accuracy of our Hamiltonian theory we have solved the equations of motion numerically and compared the conclusions to those given in the previous sections. The results of this comparison for high charge are summarized in Table 1. The charge of 1nC is typical of plasma-based advanced accelerator experiments we are considering. In all calculations in this section we assume parameters typical of the Neptune photoinjector[4]: $\alpha_{RF} = 1.6$, a laser spot diameter on the cathode of 4.3 mm, $\omega_{RF} = 2.856$ GHz, $\Delta t_{pulse} = 6$ psec, and $\phi_f = \pi/2$.

TABLE 1. Comparison af Hamiltonian and numerical solutions for various approximations with and without high charge.

Calculation Method		Force Equation	$d\phi_f/d\phi_i$	$\Delta\phi_f/(\Delta t_{pulse}\omega_{RF})$
Hamiltonian		$\sin\phi$	0.722	0.723
Numerical		$\sin\phi$		0.723
Hamiltonian		$\sin\phi + \sin\phi_0$	0.892	0.901
Numerical		$\sin\phi + \sin(\phi + 2k_zz)$		0.974
Hamiltonian	Q = 1nC	$\sin\phi + \sin\phi_0 - \mathcal{E}(\phi_0 - \phi_{head})$	1.23	1.18
Numerical	Q = 1nC	$\sin\phi + \sin(\phi + 2k_zz) - \mathcal{E}(\phi_0 - \phi_{head})$		1.46
Hamiltonian	Q = 1nC	$\sin(\phi + \phi_l) + \sin(\phi_0 - \phi_l) - \mathcal{E}(\phi_0 - \phi_{head})$	1.20	1.19
Numerical	Q = 1nC	$\sin(\phi + \phi_l) + \sin(\phi + 2k_zz - \phi_l) - \mathcal{E}(\phi_0 - \phi_{head})$		1.28
PARMELA	Q = 1nC	Complete		1.08

As a check on our derivations, we compare the results of the derivative treatment to direct evaluation of the finite ratio ($\Delta\phi_f / (\Delta t_{pulse}\omega_{RF})$) using the Hamiltonian examined at initial and asymptotic final conditions. In all cases we see that there is very good agreement between the derivative method and the difference method for calculating the compression function from the Hamiltonian, which is a good check on our derivations. Since the Hamiltonian treatment is essential exact in the traveling wave only case, the precise agreement with numeric integration is expected. These results show the strength of velocity focusing which accounts for a 28% reduction in beam length for this case.

When the backward wave is added (the third and fourth rows of Table 1) we see a notable reduction in velocity focusing, as expected, due to the increased accelerating field at launch. At this point the approximation of the backward wave made in the Hamiltonian causes its result to diverge from that of exact numerical integration, but only by 7%. Interestingly, even thought the average field of the backward wave is overestimated in the Hamiltonian it still does not reduce velocity focusing as much as in the numerical solution. In the exact force equation the backward wave field $\sin(\phi + 2k_zz)$ rises immediately after launch giving the electrons an extra acceleration. This small extra acceleration from the backward wave is lost in the Hamiltonian description.

Space-charge was added to both the Hamiltonian and numerical force equation using the same approximate term for comparison purposes (see the fifth and sixth rows of Table 1). The strong space-charge term overcomes the velocity focusing and causes the beam to expand longitudinally. This term causes the numerical result to show greater expansion than the Hamiltonian results because the space-charge force experienced by the electrons is greatly overestimated while the backward wave term is treated exactly. In the

Hamiltonian treatment the average field of the backward wave is also greatly overestimated. Since the space-charge and backward wave fields act in opposite directions and are of similar magnitude the errors made by exaggerating each term tend to cancel each other.

Finally, the initial starting position of the electrons is moved back by $\pi/8$ to simulate the elongated geometry of a 1.625 cell gun in both Hamiltonian and numerical treatments (the seventh and eighth rows of Table 1). This has the effect of decreasing the RF field at launch and thus increases the strength of velocity focusing, but not enough to overcome space-charge. Once again the numerical result is larger due to the reasons mentioned above. The Hamiltonian derivative result differs from that of the UCLA PARMELA simulation (the final row of Table 1), which uses a precise modeling of the space-charge and 1.625 cell geometry considerations, by only 10%. This is exceptional agreement considering the rather severe approximations used in the Hamiltonian description and indicates that our treatment successfully captures the majority of relevant physics.

The analysis of the physical effects relevant to longitudinal dynamics in RF photoinjectors that we have presented above gives us a fairly complete picture of the issues one faces in using these beams for AA and other ultra-fast applications. It is clear that velocity focusing is always present during the capture process, and also that it can be opposed and even overcome by the effects of space-charge. The balance of these two effects requires a more a subtle analysis, however, as the longitudinal space-charge force is not constant as a function of radial position — the problem is no longer one-dimensional. In addition, the longitudinal defocusing of the space-charge does not move the *centroid* of the beam in the same way as the longitudinal focusing of the RF capture process. Thus the timing change of the centroid relative to an external event such as photocathode drive laser injection cannot be compensated by space-charge. Schemes which address the resultant timing synchronization problems are discussed in the next section.

It is also clear from our analyses in this section that beams with finite space-charge forces cannot be made arbitrarily short. The introduction of a magnetic compressor is a powerful tool in this regard — beams an order of magnitude shorter than those which can be emitted directly from an rf gun are made possible. In addition, possibilities are introduced by the compressor system which allow synchronization of lasers and photoinjector-derived electron beams. In both cases of interest involving the chicane, ultra-short beam production and beam-laser synchronization, the performance of the system will be limited by collective (*i.e.* space-charge) effects.

PHASE SPACE MANIPULATION AND ITS APPLICATION TO BEAM SYNCHRONIZATION

Our study of longitudinal dynamics in photoinjectors shows that the beam produced in these devices is related to the injected drive laser by a compression factor $d\phi_f/d\phi_i$ which typically differs from unity by up to 20%. This has many implications for synchronizing these beams with down stream experiments. The compression of low charge beams in the photoinjector seen in previous section can be viewed as an attraction of all the particles in the beam to a single RF phase. If this attraction can be enhanced to the point were all the particle collapse to a single phase regardless of their initial phase ($d\phi_f/d\phi_i=0$), concerns over laser jitter are eliminated and the RF can be used as a reliable system clock. Alternatively, if we operate in a regime where $d\phi_f/d\phi_i=1$ the phase attraction is removed and the laser jitter translates directly into jitter in the beam. We will discuss these two regimes of operation as they relate to timing in electron beam plasma wake field accelerators and laser plasma beat-wave accelerators, respectively.

Either mode of operation discussed above requires additional manipulation of the beam beyond the photoinjector. The choice of parameters for photoinjector operation is largely determined by the need to maintain good transverse dynamics. Therefore we need to use phase space manipulation to choose our mode of operation. The effect of the linac and chicane dipole magnets on the phase space of the beam can be modeled to first order by the transformations

$$\delta p_{linac} = \delta p_0 + \alpha E(\sin(\phi + \delta\phi) - \sin\phi),$$
$$\delta\phi_{linac} = \delta\phi_0, \tag{32}$$

$$\delta p_{chicane} = \delta p_0,$$
$$\delta\phi_{chicane} = \delta\phi_0 + R_{56}\left(\frac{\delta p_0}{p_0}\right), \tag{33}$$

where the constant $R_{56} = \partial z_f / \partial(\delta p/p_0)_i$ is the linear dependence (matrix element) of the final particle position on its initial momentum error. To good approximation the linac changes only (in the absence of noticeable velocity differences) the momentum of the particles and the chicane alters only (in the absence of significant longitudinal space-charge and coherent synchrotron radiation) the phase. By passing the beam through the chicane off crest the beam's momentum can be ramped either with either a positive or negative slope. When a ramped beam passes through the chicane magnets it will expand or contract depending on the sign and magnitude of its momentum slope.

Electron wake-field timing

The beam compression and attraction to one phase that we saw during the analysis of longitudinal dynamics in photoinjectors can be greatly increased by the type of manipulation described above. By running the beam ahead of crest in the linac, the tail of the electron beam can be given greater energy than the head, resulting in a negative momentum slope. When the beam passes through the chicane dipole magnets lower energy particles will have a longer path length than higher energy particles. This differential path length compresses the beam. If the linac and chicane parameters are chosen well the process can come close to the ideal of a $d\phi_f/d\phi_i = 0$ system, where to lowest order, all particles regardless of their initial injection phase, are mapped to the same final phase. The PARMELA simulation results shown in Fig. 2 below are an example of this type of operation. It can be seen that final phase spread in the are indeed dominated by the quadratic dependence of momentum error on RF phase.

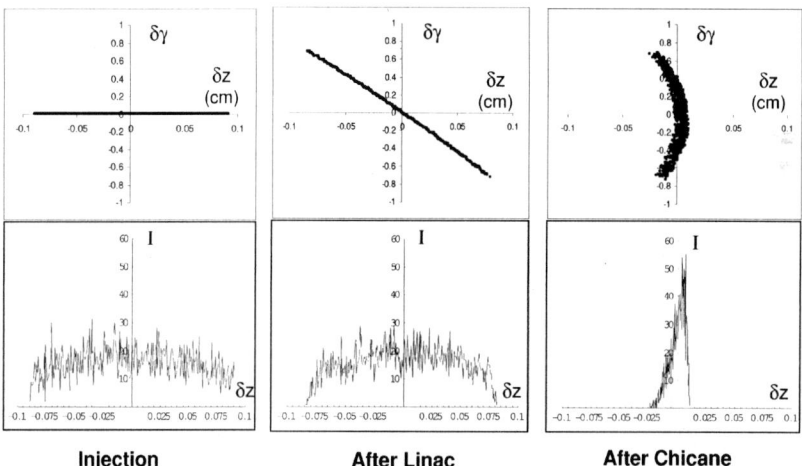

FIGURE 2. Example of $d\phi_f/d\phi_i = 0$ operation using the Neptune photoinjector and chicane, as simulated using PARMELA without space-charge.

The simulation shows how the particles originating in a large range of phases end up compressed into greatly reduced area. We can see from Eqs. 32 and 33 that the compression truly attracts all particles toward the same final point in phase. If the drive laser, and subsequently the beam, jitters relative to the design phase the momentum ramp and compression force grow with the displacement. Thus, beam jitter is suppressed by the same mechanism that compresses the beam and the compressed beam is reliably locked to a single RF phase every shot[3].

The simulation does not at present include space-charge due to difficulties we are having implementing space-charge and related collective forces such as coherent synchrotron radiation[12] (CSR) properly in the compressor. A preliminary analysis of space-charge effects in compressors on the $d\phi_f/d\phi_i = 0$ condition was given in Ref. 4. Our experimental team has recently demonstrated this type of RF and chicane manipulation experimentally at the UCLA Neptune photoinjector[3] by compressing a 0.5 nC beam from 4 psec to less than 1 psec[13] in rms bunch duration, σ_t. This measurement used

Michelson interferometry to obtain of a multi-shot autocorrelation[14,15] of the coherent transition radiation (CTR) pulse emitted from the beam.

Beam compression which results in RF phase locking has application to any experiment that requires precise timing of high current, short pulse electron beams. One important example is electron wake-field plasma accelerators. Driving and accelerating beams must be synchronized precisely to achieve acceptable performance. Using magnetic compression to operate in a $d\phi_f/d\phi_i = 0$ regime eliminates laser jitter concerns and allows the RF to be used as the universal clock. Such a system makes staged wake-field acceleration using multiple beams from one injector feasible[16]. In addition, it has been proposed that an RF feedback scheme be used in an "optical chicane" which uses an electro-optic device to produce time delays that lock the envelope of a short laser pulse to the RF clock[17]. This type of scheme may be used to minimize electron beam-laser timing jitter in sub-picosecond Compton scattering x-ray sources.

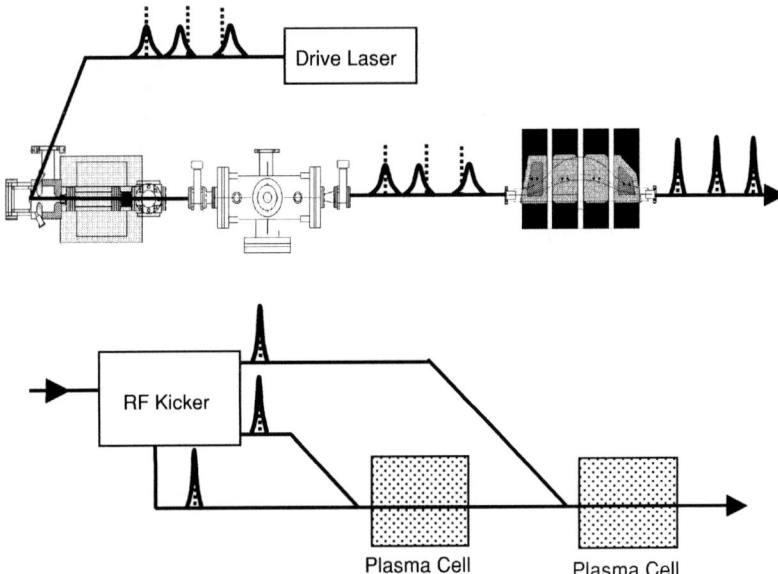

FIGURE 3. Conceptual drawing of multi-stage electron wake-field accelerator where magnetic compress is used to synchronize the electron bunches with RF clock.

Plasma Beat-Wave Injection

Laser beat-wave driven plasma accelerators have difficult timing issues related to laser jitter. While it is quite possible to lock the beam to the RF clock as illustrated in the previous section, the beat-wave envelope, being sensitive to the relative *phase* of the two laser frequencies, is uncorrelated to usual timing signals, and is thus very difficult to lock to the RF clock. In order to circumvent this problem we examine the option to abandon the RF clock, and lock the drive laser to the beat-wave envelope, which then becomes the experimental clock. If we operate the beam line in $d\phi_f/d\phi_i = 1$ mode, where the initial phase with respect to the RF is unchanged at output, synchronization to an external clock, such as the beat-wave, can be preserved in the electron beam. This type of synchronization to the beat-wave can be achieved using a scheme like the one illustrated in Fig 4, which has been proposed for use at the Neptune laboratory[18]. The high power CO_2 laser beat-wave modulates the photocathode drive laser directly using nonlinear optical effects, producing a beating envelope in the drive laser which directly reflects the CO_2 beat envelope. Thus that any shift (jitter, as seen from the RF clock point of view) in the beat-wave envelope shifts the modulation of the drive laser by the same amount. The $d\phi_f/d\phi_i=1$ beam line transfers this shift precisely to the electron beam bunch train which exits the injector. Once the overall phase between the beat-wave and electron beam is set the electron pulses will be properly loaded into beat-wave on every shot regardless of jitter between the beat-wave and RF clocks.

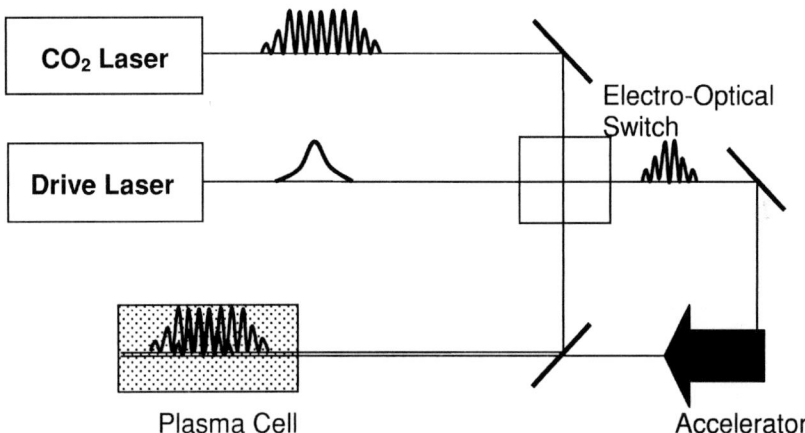

FIGURE 4. Conceptual drawing of purposed plasma beat-wave synchronization scheme.

The proposed beat-wave experiment at Neptune[7] calls for the loading of ~50 pC (chosen in order to make space-charge effects small) into the beat-wave structure. Our analysis of photoinjector longitudinal dynamics tells us that a beam of such low charge will leave the photoinjector compress by about 20%. If not corrected, this compression will destroy the matching to the beat-wave. The compression can be corrected by running the beam slightly behind crest in the linac and thereby giving the phase space the opposite momentum tilt as in the compression case. This makes the chicane an expander rather than a compressor allowing us to reverse the compression from the photoinjector, and gives an overall mapping of $d\phi_f/d\phi_i=1$, as desired.

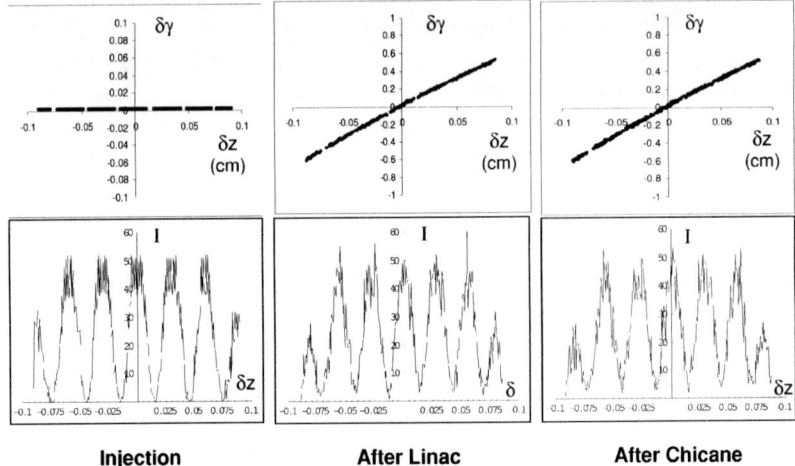

FIGURE 5. Example of $d\phi_f/d\phi_i = 1$ with the Neptune photoinjector and chicane, as simulated using PARMELA with no space-charge, and beam modulated at beat-wave periodicity.

Space-charge is also neglected in this simulation as the beam charge is low enough that its inclusion would not be significant. While the decompression process itself does contribute to slight spreading of the beamlets, due to the nonlinearity of the RF wave (the same effect which fundamentally limits compression), they are on the whole well preserved.

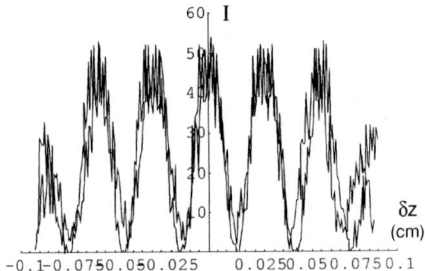

FIGURE 6. Superposition of initial and final beam current showing successful reconstruction of the beat-wave envelope.

As the beat-wave laser system at the Neptune lab fires once per five minute period, it is imperative that we develop a single-shot diagnostic of the beam's longitudinal profile, to see how well the imprinting of the beat-wave structure on the beam current is accomplished. In the case of the UCLA Neptune beat-wave parameters, the two laser lines are at 10.6 and 10.3 micron wavelength, giving a beat period near to one picosecond, as illustrated in Figs. 5 and 6. The current sub-picosecond diagnostic is the CTR interferometer used in recent[14] and present[13] experiments, which produces an autocorrelation using many shots. It is not necessary to autocorrelate the CTR pulse in order to deduce the profile, however, one may reconstruct the profile with good confidence measuring only the frequency spectrum[19]. This is especially true in the case of microbunched distributions, where the signature of microbunching is the appearance of an isolated peak in the spectrum in the vicinity of the modulation frequency[20,21]. In the

Neptune beat-wave case, the need for single-shot diagnosis of the modulation, or microbunching, has pushed us to consider the multi-channel polychromator device recently developed by Shibata and Uesaka[22] for sub-picosecond measurements at the Univ. of Tokyo.

In order to display the expected performance of this diagnostic, we display in Fig. 7 the frequency spectra of the beat-wave modulated beam both after the gun, and after the chicane run in $d\phi_f/d\phi_i=1$ mode, as simulated and shown Fig. 5. It can be seen that the peak in the spectrum is upshifted by velocity compression after the gun, but the decompression in the chicane allows the peak to be placed back at the original modulation frequency. We thus expect this diagnostic to provide a definitive measurement and tuning tool for eventual phase locking of the injected electron beam into the plasma beat-wave accelerator experiment at Neptune.

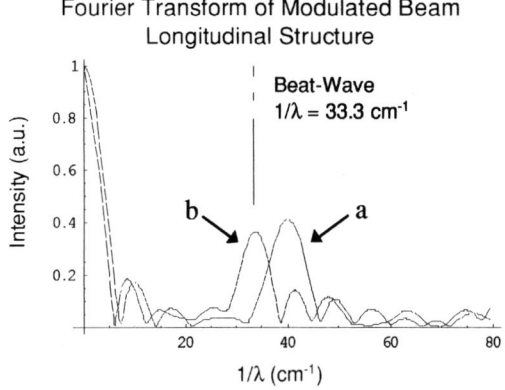

FIGURE 7. Frequency spectrum of modulated beam from simulations in Fig. 5, (a) after the gun, and (b) after the chicane run in of $d\phi_f/d\phi_i = 1$ mode.

ACKNOWLEDGMENTS

This work was supported by the U.S. Department of Energy, under contracts DE-FG03-92ER4069 3 and DE-AC-03-76SF0098.

REFERENCES

1. B. E. Carlsten, *Nucl. Instrum. Methods A* **285**, 313 (1989).
2. Luca Serafini and James Rosenzweig, "Envelope Analysis of Intense Relativistic Quasi-Laminar Beams in RF Photoinjectors: A Theory of Emittance Compensation" *Physical Review E* **55**, 7565 (1997).
3. J.B. Rosenzweig, N. Barov and E. Colby, "Pulse compression of RF photoinjector beams: advanced accelerator applications", *IEEE Trans. Plasma Sci.* **24,** 409 (1996).

4. J.B. Rosenzweig, *et al.,* "The Neptune Photoinjector", *Nucl. Instr. Methods A A* **410**, 437 (1998).
5. D.T. Palmer, *The Next Generation Photoinjector,* PhD. thesis, Stanford University, Dept. of Applied Physics, 1998.
6. W. Kimura, *et al.*, these proceedings.
7. C. Clayton, *et al.,* "Second Generation Beat-wave Experiments at UCLA", *Nucl. Instr. Methods A* **410,** 378 (1998).
8. R. W. Schoenlein, et al., "Femtosecond X-Ray Pulses At 0.4 Angstrom Generated By 90-Degrees Thomson Scattering: a Tool For Probing the Structural Dynamics Of Materials," Science 274, 236 (1996).
9. K.J. Kim, *Nucl. Instruments and Methods A* **275,** 201 (1989).
10. J.B. Rosenzweig, "Trapping, Thermal Effects and Wave Breaking in the Nonlinear Plasma Wake-field Accelerator", *Phys.Rev. A* **38**, 3634 (1988).
11. S. Hartman, *et al.,* "Initial Measurements of the UCLA RF Photoinjector", *Nucl. Instr. Methods A* **340**, 219 (1994).
12. R. Li, "Progress on the study of CSR effects", in *Physics of High Brightness Beams* (Eds. J.B. Rosenzweig and L. Serafini, World Scientific, 2000).
13. S.C. Anderson, *et al.,* "Commissioning results from the Neptune photoinjector and compressor", these proceedings.
14. A. Murokh, *et al.*, "Bunch length measurement of picosecond electron beam from a photoinjector using coherent transition radiation" *Nuclear Instruments and Methods A* **410,** 549 (1998).
15. U. Happek, A.J. Sievers and E.B. Blum, *Phys. Rev. Lett.* **67** (1991) 2962.
16. J. Rosenzweig, *et al.*, "Towards a Plasma Wake-field Acceleration-based Linear Collider", *Nuclear Instruments and Methods A* **410**, 532 (1998).
17. J. B. Rosenzweig, Greg LeSage, "Synchronization of Sub-picosecond Electron and Laser Pulses", *Advanced Accelerator Concepts Eight Workshop Conference Proceedings*, 795 **472**, (AIP, 1999).
18. C. Filip, *et al.*, these proceedings.
19. R.Lai and A.J. Sievers, *Phys. Rev. E* 52 (1995) 4576.
20. J. Rosenzweig, G. Travish and A. Tremaine, "Coherent Transition Radiation Diagnosis of Electon Beam Microbunching", *Nucl. Instr. Methods A* **365** 255 (1995).
21. A. Tremaine, *et al.,* "Observation of Self-Amplified Spontaneous Emission-induced Electron Beam Microbunching Using Coherent Transition Radiation", *Physical Review Letters,* **81** 5816 (1998).
22. M. Uesaka, *et al.*, these proceedings.

RF Photoinjector Development for a Short-Pulse, Hard X-Ray Thomson Scattering Source

G.P. Le Sage[1], S.G. Anderson[2], T.E. Cowan[1], J.K. Crane[1], T. Ditmire[1] and J.B. Rosenzweig[2]

1. University of California, Lawrence Livermore National Laboratory, Livermore CA 94550 USA
2. UCLA Department of Physics and Astronomy, Los Angeles CA 90059 USA

Abstract. An important motivation in the development of the next generation x-ray light sources is to achieve picosecond and sub-ps pulses of hard x-rays for dynamic studies of a variety of physical, chemical and biological processes. Present hard x-ray sources are either pulse-width or intensity limited, which allows ps-scale temporal resolution only for signal averaging of highly repetitive processes. A much faster and brighter hard x-ray source is being developed at LLNL, based on Thomson scattering of fs-laser pulses by a relativistic electron beam, which will enable x-ray characterization of the transient structure of a sample in a single shot. Experimental and diagnostic techniques relevant to the development of next generation sources including the Linac Coherent Light Source can be tested with the Thomson scattering hard x-ray source. This source will combine an RF photoinjector with a 100 MeV S-band linac. The photoinjector and linac also provide an ideal test-bed for examining space-charge induced emittance growth effects. A program of beam dynamics and diagnostic experiments are planned in parallel with Thomson source development. Our experimental progress and future plans will be discussed.

INTRODUCTION

The use of ultrafast laser pulses to generate high brightness, ultrashort pulses of x-rays is a topic of interest to the x-ray user community. Femtosecond-scale pump-probe experiments can in principle be used to temporally resolve structural dynamics of materials on the time scale of atomic motion. A Thomson scattering source is being developed at LLNL to provide ultrashort (ps to fs) x-rays. With this machine we intend to improve on the performance of a previously demonstrated Thomson source at LBNL in which 300 fs, 30 keV pulses were generated [1,2]. Our ultimate goal is to increase the x-ray brightness by four to five orders of magnitude.

Using a short-pulse laser, RF linac, and photoinjector, the LLNL source will provide a means of performing pump-probe experiments on a sub-picosecond time scale, with flux suitable for single-shot measurements. Single-shot measurement enables experiments on samples undergoing irreversible damage: shocks, plasma ablation, or ultrafast melting.

The LLNL Thomson scattering x-ray source combines a 35 fs laser system, currently producing 0.6 J at a 1 Hz repetition rate, with a planned upgrade to 4 J with a 0.1 Hz repetition rate, a photoinjector capable of producing 5 MeV electron pulses with charge up to 1 nC, pulse length of 0.2-10 ps, and normalized, rms emittance of 5

π mm-mrad, and an RF linac with output energy adjustable from 30 to 100 MeV. Each subsystem is described below, including current progress and future plans.

PHOTOINJECTOR

The LLNL Photoinjector uses the BNL/SLAC/UCLA 1.6 cell, standing wave accelerator geometry [3] The photoinjector cavity and diamond-turned cathode were constructed using High Isostatic Pressure annealed OFHC Copper [4]. The photocathode surface has a measured flatness of 79 nm peak-to-valley over a 2 inch diameter ($\lambda/8$ at 633 nm). With the experience of the groups involved in the construction of previous photoinjectors, development of the LLNL photoinjector progressed quickly. First metal cutting for the photoinjector began in December of 1998 at the UCLA Physics department precision machine shop. High power RF was applied to the photoinjector in July of 1999. The first photoelectron beam was produced in January of 2000.

In order to test the field profile of the photoinjector with the actual photocathode in place, a "bead-drop" was performed where a dielectric bead was hung from the end of a thin nylon line and lowered into the central axis of the photoinjector cavity. The usual technique requires a second test cathode plane to be made with a hole in the center for a string to pass through and suspend a bead in the horizontal direction: "bead-pull." The result of the bead-drop measurement is presented in Fig. 1, in comparison to the field profile generated by SUPERFISH, with a background image of the ρH_ϕ profile.

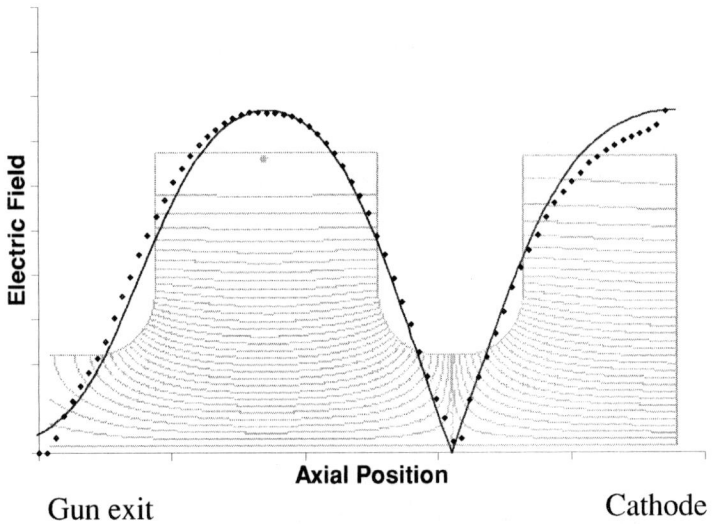

FIGURE 1. Bead-drop characterization of photoinjector field profile.

The photoinjector is currently operated separately from the RF linac, and has been fully characterized using a diagnostic beamline. Characterization of the photoinjector has included measurement of cavity Q and filling time, dark current, charge and quantum efficiency, Schottky scan, energy, and emittance. Emittance has also been characterized as a function of charge and pulse length.

The energy range of field emitted electrons (dark current) in the cavity ranges broadly up to the peak beam energy. Characterization of the total field emission through current collection with a Faraday cup or integrating current transformer is subject to beam dynamics issues as the RF power and beam energy changes. An equivalent measurement of the Fowler-Nordheim macroscopic field enhancement factor β can be made by measuring the radiation produced by the gun as a function of RF drive power [5]. A time-averaging radiation meter was placed at approximately half a meter from the photoinjector and measured radiation levels up to 500 mR/h with RF drive power up to 7 MW. An image of the dark current emission, focused by the photoinjector solenoids, and captured on a scintillator shows the emission from the cathode surface and the iris between the two gun cells in Fig. 2.

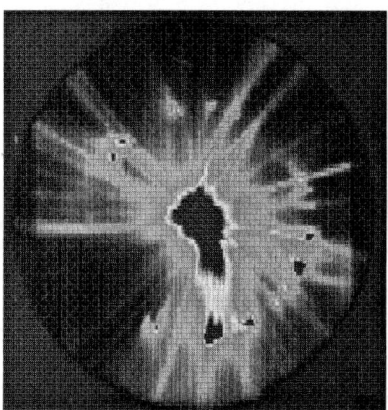

FIGURE 2. Dark current imaged with photoinjector solenoids.

A plot of 1/E versus the logarithm of the dose rate divided by E^4 is shown in Fig. 3, and provides the means of measuring the field enhancement factor β. The β value measured for the LLNL HIP Copper photoinjector, taken from the slope of this plot was 62.

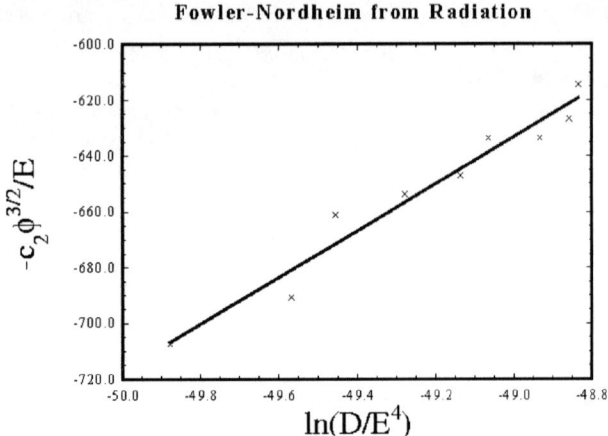

FIGURE 3. Fowler-Nordheim equivalent field emission plot.

The photoinjector cavity Ohmic Q was measured in cold test before and after brazing, and had a final cold-test value of 13,000. The value predicted by SUPERFISH, taking into account the loss of the Copper surface including the cathode plane was 15,878. The Q value of 13,000 was also verified by examining the filling time of the RF cavity with high power RF applied. The cavity fills with the characteristic Q_0 and empties with $Q = \left[\frac{1}{Q_0} + \frac{1}{Q_e} \right]^{-1} = \frac{1}{2} Q_0$ as shown in Fig. 4.

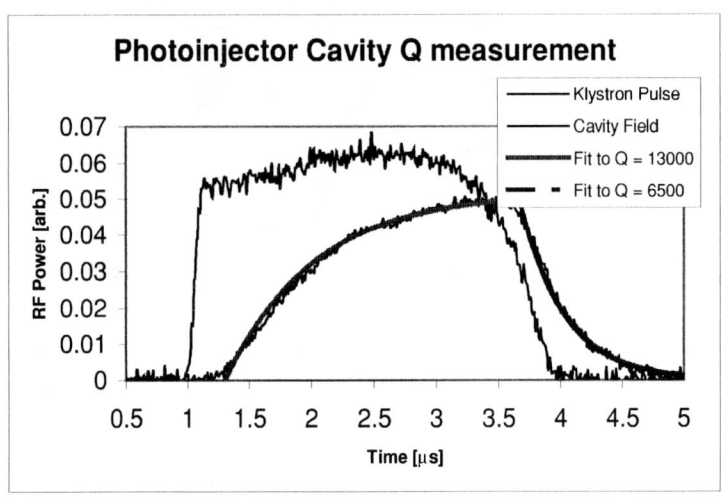

FIGURE 4. High power measurement of photoinjector cavity Q_0.

A summary of the LLNL photoinjector RF characteristics is presented in Table 1. The drive power and cathode peak field gradient correspond to production of a 5 MeV photoelectron beam.

Table 1. Photoinjector cavity parameters.	
Beam parameter	Measured value
Peak Cathode Gradient	112 MV/m
Drive Power	7.1 MW
Ohmic Q	13,000
Cavity $|S_{11}|$	less than –30 dB
Fowler Nordheim β parameter	62

Photoelectron current was first measured on January 21, 2000. The program of beam measurements that followed this first beam production included Schottky scanning, in order to determine the injection phase for optimum emittance, and measurement of the total charge and quantum efficiency. An image of one of the first photoelectron beam profiles measured with the system is shown in Fig. 5.

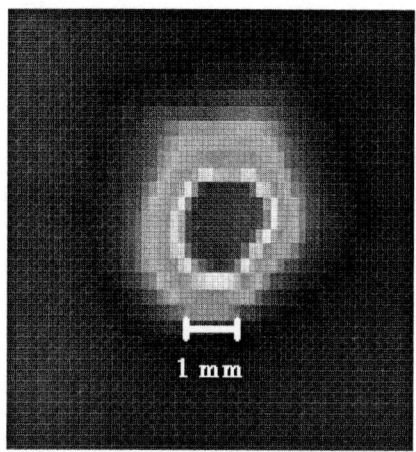

FIGURE 5. Photoelectron beam profile at 1m from cathode.

The photoelectron bunch charge has been measured using both Faraday cups and a fast integrating current transformer. The peak charge per pulse that has been measured to date is 0.4 nC, and was purposely limited due to the small UV spot size on the photocathode. A full nC of charge extraction at high gradient with a small spot size would approach the region where cathode damage can result [6].

The quantum efficiency of the photoinjector operating at 23° phase angle off of the RF crest was measured as 2.2×10^{-5} using an integrating current transformer positioned directly at the photoinjector output, and a calibrated UV energy meter monitoring a split beam from the photocathode laser path. The required UV energy for 1 nC of charge is thus 300 µJ. The UV laser system has produced up to 400 µJ in its present configuration. The IR (805 nm wavelength) pulse length produced by the laser compressor is adjustable from 150 fs to more than 10 ps, giving a UV pulse length in the range from less than 100 fs to more than 9 ps. The electron pulse length has not been verified with a streak camera to date, and the electron pulse length is currently assumed to equal the UV pulse length.

The electron beam energy was measured using the steering magnets that are part of the diagnostic beamline. The steering magnet profile was integrated over a long longitudinal range extending into the fringe field region to give more accurate prediction of steering versus beam energy and magnet current. A linear fit of several magnet deflections, with variation of the RF drive power confirm a beam energy of 4.95 MeV at the current RF drive power level of 7.1 MW.

The photoinjector electron beam emittance has been measured using a slit collimator [7]. This technique removes space charge from the beam measurement, and provides a single-shot means of measuring emittance through computer control and analysis. Launch phase and solenoid settings were optimized in real time using this diagnostic. The emittance as a function of beam launch parameters was characterized over a wide range of charge, pulse length, and beam radius on the photocathode. A schematic of the slit collimator arrangement is shown in Fig. 6. The collimator-slit widths are 50 µm, with collimator thickness is 5 mm, and the drift length is 53 cm.

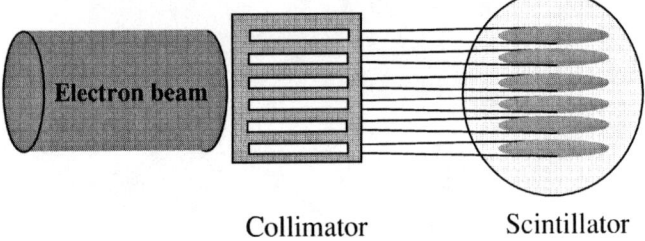

FIGURE 6. Collimator slit projection schematic.

An image of the photoelectron beam collected by the scintillator is shown in Fig. 7, and the virtual instrument panel for the emittance diagnostic is shown in Fig. 8.

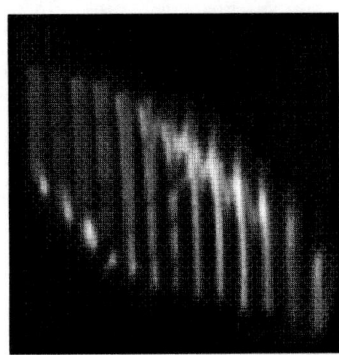

FIGURE 7. Collimator slit projection of photoinjector beam.

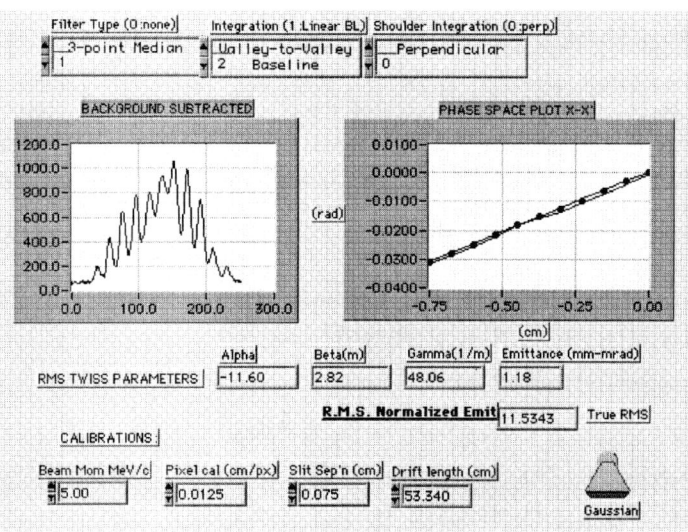

FIGURE 8. Slit emittance measurement virtual instrument panel.

The characteristics of the photoinjector for a specific pulse length are listed in Table 2. The measured emittance should be sufficient for the planned Thomson scattering source at 100 MeV, provided that excessive emittance growth through the linac is not encountered. The design value for the normalized emittance is 10 π mm-mrad, which implies a physical emittance at 100 MeV of ~51 μm-mrad. With a 20 cm focal length final electron optic, and a 1 mm entrance spot radius (5 mr convergence) the emittance limited spot size is ~10 μm. Since the focal spot size of the FALCON laser beam with an F/10 optic is estimated to be 20 μm, the electron focal spot should overlap well with the laser spot.

Table 2. Photoinjector beam parameters.

Beam parameter	Measured value
Beam energy	5 MeV
UV laser pulse length	7.1 ps FWHM
Bunch charge	200 pC
Emittance$_{x, rms, n}$	5.8 π mm-mrad
Quantum efficiency	2.2×10^{-5} at 23° off RF crest

LASER SYSTEM

There are two laser systems included in the Thomson scattering project. The first is the high power, short pulse "FALCON" laser [8,9] which produces 35 fsec, 0.6 J pulses at 1 Hz repetition rate. This laser system is based on Ti:Sapphire, and relies on chirped pulse amplification. The ultimate specification goal for the FALCON laser is 4 J pulse energy (~100 TW) with a repetition rate of 0.1 Hz using two 15 J custom-built Nd:glass pump lasers. In addition to the Thomson scattering project, the FALCON laser is used for a variety of laser-plasma and solid target interaction experiments. A second laser system takes seed pulses from the FALCON master oscillator and amplifies them for conversion to UV and photoelectron production. When the FALCON laser, photoinjector, and linac are integrated, the same laser pulse from the oscillator that is amplified through the FALCON laser amplifier chain will also be amplified by the photoinjector laser system to produce electron pulses.

The FALCON oscillator, which drives both the high power laser amplifiers and the photoinjector UV laser system is a Kerr-lens modelocked, Ti:Sapphire system with a feedback loop controlling the cavity length and thus pulse repetition frequency. The feedback system controls kHz level thermal and mechanical drifts. The laser oscillator is not actively modelocked by an RF master oscillator, but instead provides the master RF signal to the entire linac and photoinjector system.

100 MeV RF LINAC

The LLNL RF linac in its present configuration is composed of five travelling wave accelerator sections, a thermionic injector and a travelling wave buncher. The travelling wave accelerator sections are iris-loaded waveguides, and each section has an overall length of 2.6 m, boosting the beam energy by up to 30 MeV. The thermionic injector is a pulsed DC thermionic source, followed by a travelling wave buncher that accelerates the beam from 150 keV to 3 MeV in a total length of 0.75 m. The original purpose of the linac was to produce high average currents of electrons, positrons, and intense pulse neutrons. The machine is thus capable of very high repetition rates with long macropulses. In "long pulse mode" (up to 2.8 μs macropulses), the linac produces roughly 8000 micropulses per RF macropulse, with a total of 2 μC per pulse, at a repetition rate up to 300 Hz. In this configuration, the linac can produce 0.6 mA of average current, and 0.7 A during each RF macropulse. In "short pulse mode" (2-20 ns macropulses), the linac can operate at a repetition rate of 1440 Hz, producing a macropulse current of 10 A, and an average current of 173 μA. The linac beam energy has been tuned over the range of 13 to 130 MeV, and the emittance at 100 MeV has been measured as 260 π mm-mrad$_{x,rms,n}$ integrated over one macropulse, using the thermionic injector. While the high average current is important for the continuing production of positron beams at the LLNL facility, the Thomson scattering experiment relies on the low beam emittance provided by the photoinjector, and requires a repetition rate of 10 Hz or less, matched to the laser system.

A magnetic beam switchyard has been designed to pass the thermionic injector beam around the photoinjector when high average current operation is desired. In this way, the capabilities of both injector systems will be maintained by the LLNL accelerator facility.

SYNCHRONIZATION

Thomson scattering the FALCON laser pulse with the linac electron beam requires synchronization on the picosecond to sub-picosecond time scale. To date, synchronization between an RF linac and modelocked laser system has been demonstrated with an rms jitter of 2.2 ps by the E-144 collaboration at SLAC [10]. In this case, linac RF was used to actively modelock a laser oscillator, which in turn produced laser pulses for the Thomson scattering interaction. The LLNL system instead uses a laser oscillator that also functions as the master RF oscillator at the 35th subharmonic of the linac RF. The ultimate timing jitter between the laser pulses and electron pulses will be due to long time scale contributions: thermal and mechanical in transport, and to fast time scale sources: jitter inherent to laser master oscillator, and phase noise introduced in the RF system. Assuming that the slow time scale contributions can be remedied by feedback loops with kHz bandwidth, the ultimate limiting factor for final synchronization rests with inherent laser oscillator jitter and RF phase noise. The LLNL system can improve upon previous synchronization results for three key reasons described as follows.

First, the laser-electron jitter that arises from the inherent jitter of the laser master oscillator will be minimized. Using the laser master oscillator as the source of the linac RF allows the linac RF system to track laser oscillator phase changes up to the characteristic frequency of the entire system. The question then becomes how fast the RF system can respond to changes in the laser-produced RF phase. The photoinjector cavity bandwidth is the limiting factor since its Ohmic Q is 13,000 (filling time is 725 ns). The response time is thus on the order of 1 µs, compared to RF amplifier bandwidths, travelling wave accelerator filling times, and transmission line path lengths, each of which add up to 10's to 100's of ns. The phase noise integration bandwidth for the LLNL system corresponds to at most microseconds of reaction time.

The timing jitter of the laser pulses produced by a laser oscillator can be characterized by integrating laser RF harmonic spectral noise [11]. A high-bandwidth photodiode and a spectrum analyzer were used to make the jitter characterization of the FALCON laser oscillator. The integration bandwidth is determined in the case of an actively (externally) modelocked laser slaved to an RF system by the time between application of the modelocking RF signal and the final interaction of the laser and electron pulses. The integration bandwidth is larger, so the jitter is proportionately larger. Additionally, if a master RF oscillator is used to both modelock a laser and drive the linac RF system, jitter inherent to the laser oscillator driven by a perfect RF sine wave has no chance of being tracked by the RF system over any bandwidth.

The LLNL (FALCON) laser oscillator was characterized with an inherent jitter of < 1.5 ps integrating over 1 µs to 1 ms, and a jitter of 5.5 ps integrating over > 1 ms. The noise floor of the detector limited the accuracy of this measurement, and 1.5 ps jitter should be considered a worst case estimate of ultimate system performance. The spectrum of the 22nd harmonic of the FALCON oscillator 81 MHz RF is shown in Fig. 9. The noise floor was –70 dB at frequencies more than 1 kHz from the harmonic peak, and the resolution bandwidth was 100 Hz.

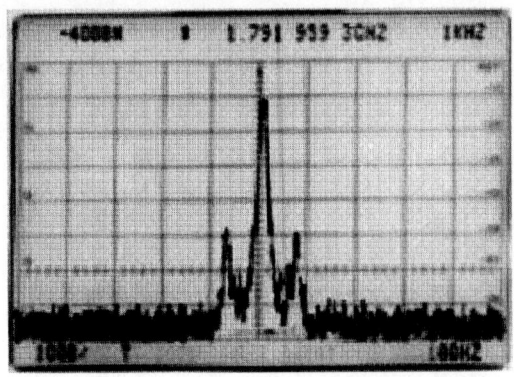

FIGURE 9. Laser oscillator spectrum of 22nd harmonic of 81 MHz RF.

At the SLAC Gun Test Facility, jitter measurements were made using an independent RF source to drive both a modelocked laser and the RF system, and using the laser as the master reference [12]. The laser-electron jitter was measurably lower in the case where the laser provided the master RF signal than in the case where the laser and RF systems were driven by an external reference, demonstrating the principle described here.

The second reason why the LLNL system will achieve low jitter is that phase and amplitude control of each linac section can be accomplished at the mW RF power level, allowing fast electronic feedback control of amplitude and phase. Each linac section and the photoinjector are driven separately by a total of six Klystron amplifiers, each in turn is driven by a booster amplifier with mW level RF reference input. For systems where power is split in waveguide networks between accelerator sections, mechanical phase and amplitude control is implemented, allowing only for < kHz level corrections at. Short feedback loops built around individual Klystrons will allow phase and amplitude flattening on a time scale short compared to the several µs-long RF macropulses.

Finally, the laser transport and RF and beam transport systems are arranged so that the same laser pulse from the FALCON master oscillator that produces the photoelectron pulse also gets amplified for the final Thomson scattering interaction. Shot-to-shot laser pulse jitter thus does not add directly to the jitter of the overall system.

THOMSON SCATTERING AT 5 MeV

The first demonstration of Thomson scattering is planned using the 5 MeV beam directly from the photoinjector interacting with a 30 mJ IR laser pulse produced by the residual energy of the photocathode laser system. The interaction with the 5 MeV beam is expected to produce 10^4 to 10^5 600 eV photons per shot with an interaction angle of 135 degrees between the electron and laser beams [13]. A drawing of the 5 MeV Thomson scattering experiment beamline is shown in Fig. 10.

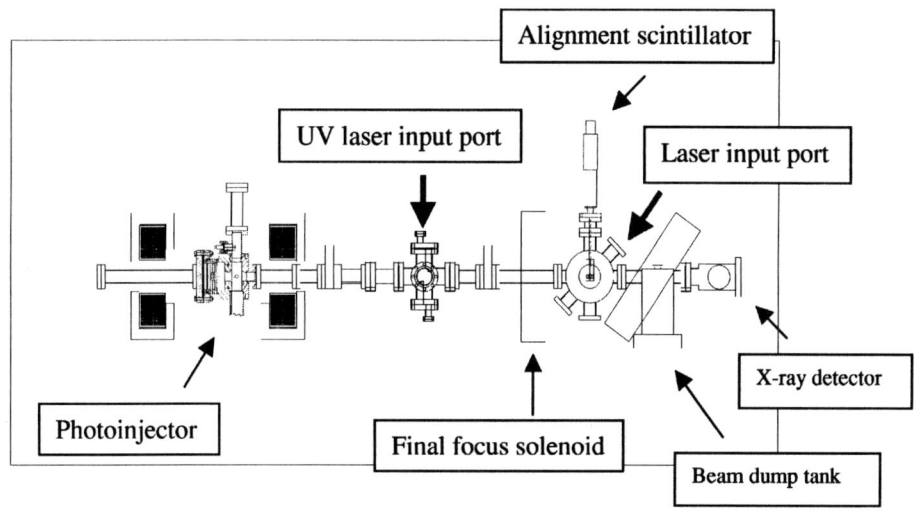

Figure 10. Beamline for Thomson scattering at 5 MeV.

A high field solenoid was chosen for the final focusing of the 5 MeV electron beam for the initial Thomson scattering experiment. The peak field for the final focus magnet is 3 kG, with a yoke radius of 2 cm, and a length of 7.5 cm. The focal length at 5 MeV is 18 cm. Final focus parameter optimization involves a tradeoff between the competing effects of emittance, energy spread, and spherical aberration. The scaling of these effects can be represented by a quadrature addition [14].

$$R_{final} = \left(\frac{\varepsilon}{R_0}f_0\right)^2 + R_0^2\left(\frac{\Delta f}{f_0}\right)^2 + \left(AR_0^3\right)^2 \qquad (1)$$

Where R_{final} depends on the beam radius entering the solenoid R_0, the emittance ε, the lens focal length f_0, the chromatic spread $\Delta f/f_0$ related to the energy spread, and the spherical aberration term A. The spherical aberration term is determined by integrating the magnetic field through the solenoid at different radii. The POISSON simulated fields were utilized for this calculation. Using the parameters of our photoinjector and beamline, the competing effects and the total final radius were plotted as a function of the beam radius entering the final focus solenoid as shown in Fig. 11. This model assumes a continuous beam with uniform characteristics. Because of the photoinjector beam dynamics, entering the final solenoid with the optimum radius based on this model increases the energy spread significantly, which in turn changes the chromatic term. The model does however demonstrate that the optimization for the 5 MeV experiment is achieving the minimum energy spread. Optimizing for this parameter, PARMELA simulations show that the configuration planned for the experiment should focus a 100 pC, 2 ps electron pulse to a radius of approximately 50 μm (~90% of particles).

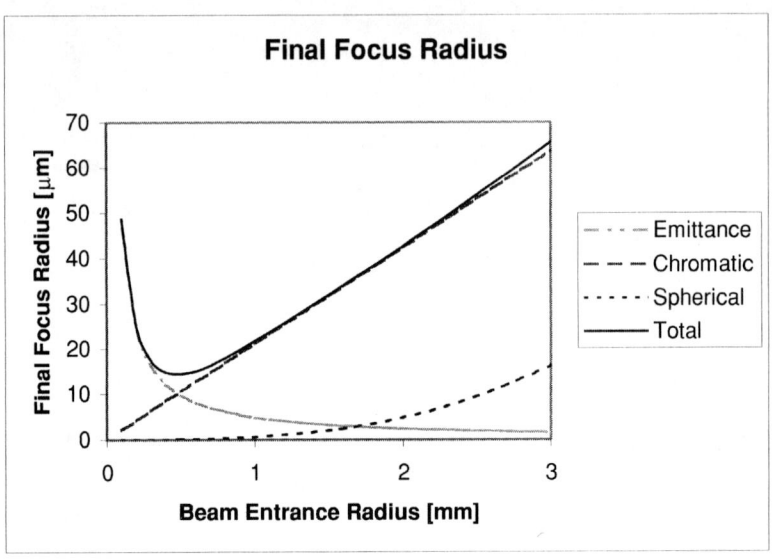

Figure 11. Final Focus calculations at 5 MeV.

With a 3 mm beam radius at the input of the final focus solenoid, the emittance, chromatic, and spherical aberration effects contribute 1.6 µm, 64 µm, and 16 µm respectively to the final spot radius. The energy spread $\Delta\gamma/\gamma$ decreases significantly with acceleration for short electron pulses. Therefore, at the future operating conditions of 100 MeV, the emittance can be expected to have a more dominant effect on the final focus spot size.

BEAM DYNAMICS AND DIAGNOSTIC EXPERIMENTS

In addition to the Thomson scattering experiment, the photoinjector and 100 MeV RF linac also provide an ideal test-bed for examining space-charge induced emittance growth effects. A program of beam dynamics and diagnostic experiments are planned in parallel with Thomson source development for both the photoinjector, and the linac. So far a new emittance measurement technique has been demonstrated on the RF linac in order to characterize the beam produced by the thermionic injector. Based on optical transition radiation interferometry (OTRI), the emittance measurement system images and optically masks the OTRI produced by the electron beam passing through a pair of thin foils [15]. Scanning the optical mask through the electron beam allows the divergence to be measured as a function of position within the beam envelope. A proof of principle experiment was conducted to show the differences in divergence and average direction of the beam particles as a function of transverse position within the beam. Determination of the phase-space ellipse tilt using this technique changed the measured value of emittance by 45% compared to the assumption that the beam was exactly at a waist. This technique also allows real time tuning of the beam convergence at the diagnostic foil plane. This new diagnostic technique will be used to optimize the Thomson scattering interaction at 100 MeV.

Using the photoinjector, measurements of the emittance using the quad scan technique and slit collimator technique are being compared as a function of plasma wavelength at the photocathode. The LLNL system can be used to investigate a wide range of plasma parameters since the IR pulse length is adjustable from 150 fs out to several ps. Since the quad-scan technique of emittance measurement is susceptible to space charge forces at low beam energy, the quad-scan and slit collimator emittance results are expected to diverge as the plasma wavelength decreases. This measurement and its implications are currently being examined.

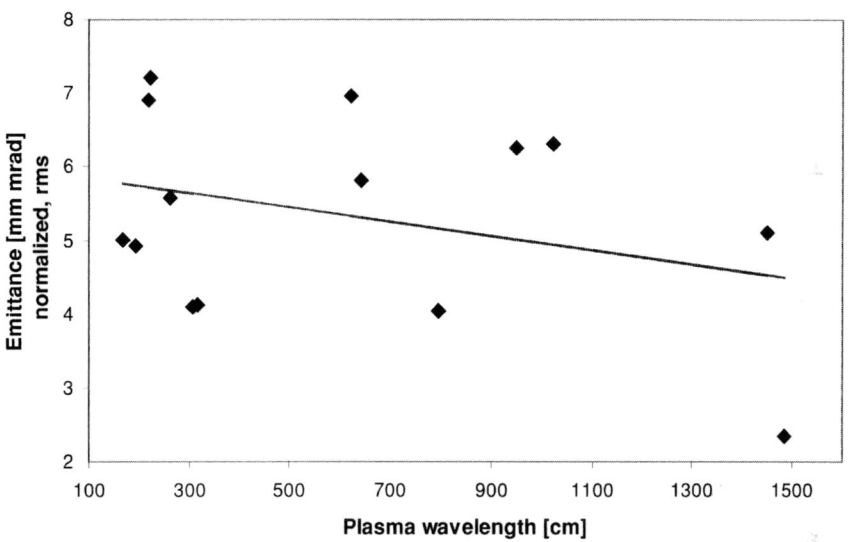

FIGURE 12. Photoinjector beam emittance scaling with plasma wavelength.

SUMMARY

Present hard x-ray sources are either pulse-width or intensity limited, which allows ps-scale temporal resolution only for signal averaging of highly repetitive processes. A Thomson scattering source is being developed at LLNL to provide ultrashort (ps to fs) x-rays. Using a short-pulse laser, RF linac, and photoinjector, the source will provide a means of performing pump-probe experiments on a sub-picosecond time scale, with flux suitable for single-shot measurements. Experimental and diagnostic techniques relevant to the development of next generation sources including LCLS can be tested with the Thomson scattering hard x-ray source. This source will combine an RF photoinjector with a 100 MeV S- band linac and will also provide an ideal test-bed for examining space-charge induced emittance growth effects. A program of beam dynamics and diagnostic experiments are planned in parallel with Thomson source development.

After the Thomson scattering demonstration at 600 eV is complete, the next major step is the installation of the photoinjector on the Linac. Propagating the electron beam through the linac, and characterization of the 100 MeV beam will follow photoinjector installation. Compression of the FALCON beam, and delivery of the laser beam to the linac interaction region will proceed in parallel. First demonstration of Thomson scattered light from the linac beam and FALCON laser pulses represents the key milestone of the project.

ACKNOWLEDGMENTS

Work performed under the auspices of the U.S. Dept. of Energy by the Lawrence Livermore National Laboratory under Contract W-7405-Eng-48.

REFERENCES

1. R.W. Shoenlein, et al., "Femtosecond x-ray pulses at 0.4 angstrom generated by 90 degree Thomson scattering: A tool for probing the structural dynamics of materials," Science **274**, 236 (1996).
2. W.P. Leemans et al., "Interaction of relativistic electrons with ultrashort laser pulses: Generation of femtosecond x-rays and microprobing of electron beams," IEEE Jour. of Quant. Elec. **33**, 1925 (1997).
3. D.T. Palmer et al., "Emittance studies of the BNL/SLAC/UCLA 1.6 cell photocathode rf gun," IEEE Part. Accel. Conf. (1997).
4. H. Matsumoto et al., "Applications of hot isostatic pressing (HIP) for high gradient accelerator structure," Proc. PAC '91 1008 (1992).
5. D. Gooden and J. Rosenzweig, "Modeling of the x-ray radiation dependence on power in high gradient radio-frequency accelerator structures," UCLA Part. Beam Phys. Lab. Internal Report (1999).
6. X.-J. Wang et al., "Intense electron emission due to picosecond laser produced plasmas in high gradient electric fields," Journal of Applied Physics, 72(3), p. 888-894 (1992).
7. J. Rosenzweig and G. Travish, "Design Considerations for the UCLA PBPL Slit-based Phase Space Measurement Systems," Nucl. Instrum. Meth. A **341**, p. 379 (1994).
8. T. Ditmire and M.D. Perry, "High intensity physics with a table-top 20 TW laser system", LLNL Internal Report, UCRL-ID-133293 (1999).
9. V.P. Yanovsky et al., "Multiterawatt Laser-Linac Facility," Conference on Lasers and Electron Optics OSA Technical Digest (OSA, Washington DC), 410 (1999).
10. T. Kotseroglou et al., "Picosecond timing of terawatt laser pulses with the SLAC 46 GeV electron beam," Nucl. Instrum. Meth. A **383**, 309-317 (1996).
11. M.J.W. Rodwell, D.M. Bloom, K.J. Weingarten, "Subpicosecond laser timing stabilization," IEEE J. Quantum Electronics **25**, 817 (1989).
12. J.F. Schmerge et al., "Photocathode rf gun emittance measurements using variable length laser pulses," SPIE Photonics West, LASE 99, Proceedings of SPIE Vol. 3614.
13. E. Esarey et al, "Nonlinear Thomson Scattering of Intense Laser Pulses from Beams and Plasmas," Phys. Rev. E **48**, 3003 (1993).
14. Yu-Jiuan Chen, personal communication.
15. G.P. Le Sage st al., "Transverse phase space mapping of relativistic electron beams using optical transition radiation," Phys. Rev. Spec. Topics – Accel. And Beams **2**, 122802 (1999).

UMER: The University Maryland Electron Ring

P.G. O'Shea, M. Reiser, R.A. Kishek, S. Bernal, H. Li, M. Pruessner, M Virgo, V. Yun, W. Zhang,
Institute for Plasma Research, Univ. of Maryland, College Park, MD 20742, USA

T. Godlove, D. Kehne, P. Haldemann, *FM Technologies, Inc. Fairfax, USA*,

I. Haber, *Naval Research Laboratory, Washington, DC, USA*

Abstract. A detailed understanding of the physics of space-charge dominated beams is vital for many advanced accelerators that desire to achieve high beam intensity. In that regard, low-energy, high-intensity electron beams provide an excellent model system. The University of Maryland Electron ring (UMER), currently under construction, has been designed to study the physics of space-charge dominated beams with extreme intensity in a strong focusing lattice with dispersion. At 10-keV, 100 mA, the UMER beam has a generalized perveance in the range of 0.0015. Though compact (11-m in circumference), UMER is a very complex device. In this paper, the unique design features of this research facility, the beam physics to be investigated, and simulation studies are reviewed.

INTRODUCTION

A long-term goal of beam physics is the development of ever more intense high quality beams. Generally, recirculators and rings have been limited to lower intensities than linear accelerators. One of the most important limitations in beam current, luminosity and brightness in circular machines is the Laslett or space-charge tune shift [1]. Fear of destructive resonances has constrained existing rings to low intensity and relatively modest tune depression. With space charge, the resonances are no longer caused by the interaction between the single particle orbits and the harmonics of the field errors. Instead, a resonance can occur when the frequency of a collective beam mode coincides with one of the harmonics of the error frequency spectrum [2]. Conventional recirculators (i.e. synchrotrons) have been limited in intensity by the necessity for the beam to make a very large number of orbits. Previous work [3] has shown that it may be possible to exceed the Laslett limit in a machine with a small number, perhaps less than 100 turns.

There has never been an opportunity to perform experiments on such recirculating machines in the region of deep tune depression, and extreme beam intensity. Therefore, almost all of our understanding in this region is based on theory, simulation and conjecture. The University of Maryland Electron Ring (UMER) [4], currently under construction, is a low-cost flexible electron model of intense ion recirculators.

In order to put UMER in a proper context, it is important to define what we mean by intensity, and to illustrate how the increased intensity influences the collective beam motion. We use the dimensionless intensity parameter, χ, as the ratio of the space-charge force to the external focusing force at the beam radius $\chi = \dfrac{K}{k_0^2 a^2}$. The space charge term is represented by K/a^2 where $K = \dfrac{2I}{I_0(\beta\gamma)^3}$ is the generalized perveance, a is the 2x rms beam radius, I is the beam current, and $I_0 \approx \dfrac{mc^2}{30q}$ (I_0 = 3.1 x 10^7 (A/Z) Amps for ions with charge state Z and atomic number A, and I_0 = 17 kA for electrons). The external focusing forces are represented by k_0, the zero-current betatron wavenumber. These quantities are related to the 4x rms emittance, ε, through the matched beam envelope equation, in the smooth approximation: $k_0^2 - \dfrac{K}{a^2} = \dfrac{\varepsilon^2}{a^4} = k^2$ (where k is the depressed betatron wavenumber) or $1 - \chi = \dfrac{\varepsilon^2}{k_0^2 a^4} = \dfrac{k^2}{k_0^2}$. The tune depression can then be expressed in terms of χ as $\dfrac{v}{v_0} = \dfrac{k}{k_0} = \sqrt{1-\chi}$, and the plasma wavenumber k_p as $\dfrac{k_p}{k_0} = \sqrt{2\chi}$. For a zero current, fully emittance dominated beam we have χ = 0, while χ =1 for a fully space charge dominated beam, with zero emittance. At the lowest values of χ, the motion is dominated by single particle effects and emittance. As χ approaches unity, collective plasmas oscillations become increasingly dominant. For χ = 0.5, the space-charge and emittance terms in the envelope equation are equal, Thus, for the range 0 <χ < .5, we can say that the beam radius (hence the beam physics) is emittance-dominated, while for 0.5 < χ <1 the beam radius (physics) is space-charge dominated.

For example, heavy ion fusion drivers will likely operate with 0.89 <χ < 0.98. We see that it is possible to achieve such values of χ with electrons at 10 keV, 100 mA, ε = 50 µm and a =1 cm. Figure 1 shows the range of the intensity parameter for UMER. With χ ranging from 0.2 - 0.98 UMER offers a unique opportunity to study intense beam physics in a completely new regime that begins near the upper intensity range of existing machines. In the space-charge dominated regime of UMER, not only will the tune depression be unprecedented, but also the plasma frequency will exceed the zero-current betatron frequency. This implies that, at the high end of the UMER intensity range, collective effects will have an enormous impact on the beam dynamics. Further, the experimental and theoretical study of beam dynamics in UMER will have important applications to high-current circular accelerators other than heavy-ion fusion recirculators; examples of such applications include high-energy physics booster synchrotrons, muon colliders, and spallation neutron sources.

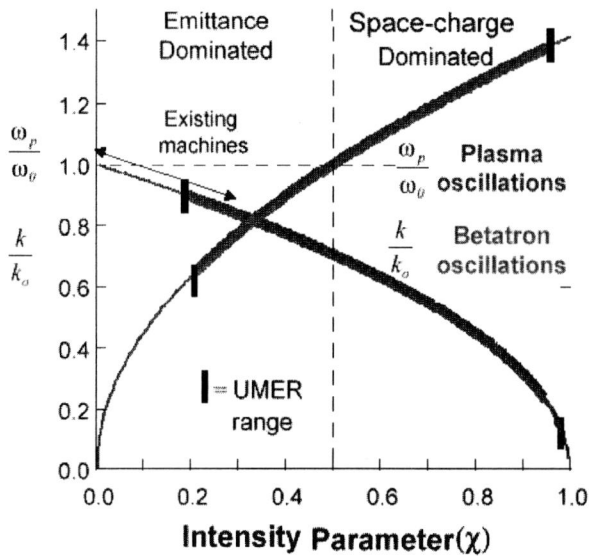

FIGURE 1. Emittance dominated and space-charge dominated regimes, showing the betatron tune depression and plasma tune enhancement with increasing intensity parameter.

The unknown territory in the extreme space-charge dominated regime will be very challenging and should provide a wealth of new phenomena. The UMER facility will allow us to investigate emittance growth due to conversion of free energy, halo formation, and equipartitioning in a circular machine. So far, these effects have only been studied in linear transport lines. In addition, UMER will permit experimental investigations of longitudinal-transverse coupling and beam profile changes resulting from dispersion; the behavior of bunch ends, resonance traversal; the longitudinal resistive wall instability; and other effects in the space-charge dominated regime that is currently inaccessible.

UMER DESIGN FEATURES

In its initial phase UMER will operate at a fixed energy of 10 keV ($\beta = 0.2$). Future upgrades will allow UMER to operate as a fast cycling recirculating accelerator at energies of up to 50 keV ($\beta = 0.4$). Table 1 gives the nominal specifications of UMER at 10 keV. The intensity parameter, tune depression or beam current can be varied in UMER over a wide range by changing to different apertures sizes in the beam collimator at the exit of the gun, by changing the anode–cathode spacing, or by changing the beam energy. Several examples are shown in Table 2 for both 10 and 50 keV operation. A complete description of the UMER design features is beyond the scope of this paper. The reader is referred to our web page for additional details [5].

TABLE 1. UMER design specifications

Energy	10 keV
β (= v/c)	0.2
Current	100 mA
Generalized perveance	1.5×10^{-3}
Emittance, 4x rms, norm	10 µm
Pulse Length	40 ns
Circumference	11.52 m
Lap time	197 ns
Rep. rate	60 Hz
Mean beam radius	< 1 cm
FODO period	0.32 m
Phase advance, σ_o	76°
Betatron tune, ν_o	7.6
Tune Depression	>0.12

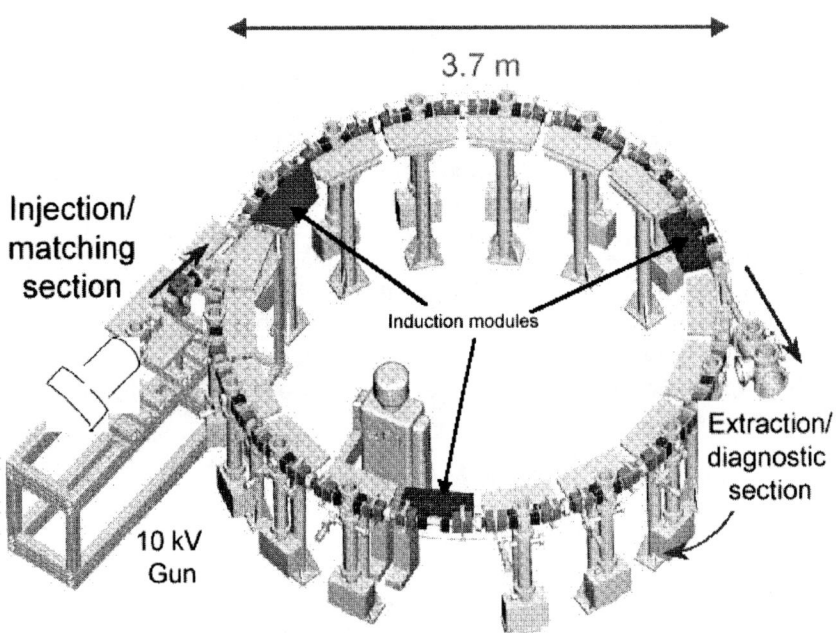

FIGURE 2. UMER layout.

TABLE 2. Some examples of the range of UMER parameters at 10 keV and 50 keV

I (mA)	$\beta\gamma$	ε_f (μm)	σ_{of} (deg)	a (mm)	χ	ν/ν_0	Assumption
100	0.20	50.0	76.0	9.48	0.98	0.13	**Full Current**
14	0.20	18.7	76.0	3.72	0.89	0.33	**14 % aperture**
1	0.20	5.0	76.0	1.31	0.51	0.70	**1 % Aperture**
14	0.20	37.4	76.0	4.14	0.72	0.53	**2* emittance**
100	0.20	250.0	76.0	10.90	0.74	0.51	**5* emittance**
100	0.45	21.9	50.3	4.58	0.86	0.38	**Full Current**
14	0.45	8.2	50.3	2.12	0.56	0.66	**14 % aperture**
1	0.45	2.2	50.3	0.95	0.20	0.89	**1 % Aperture**

A schematic layout of UMER is shown in Fig. 2. The focusing lattice consists of 36 FODO periods of length 0.32 m and the ring circumference is 11.52 m. Each FODO section contains two printed-circuit quadrupole magnets and one printed-circuit dipole. A detailed mechanical layout of a full lattice period is shown in Figure 3. The zero-current phase advance per period is $\sigma_0 = 76°$, corresponding to a tune of $\nu_0 = 7.6$. The maximum tune depression due to space charge is expected to be between 0.12 and 0.2. There are 13 diagnostic ports, and three induction modules that provide fast-rising "ear fields" to prevent expansion of the bunch ends [6] and permit acceleration to 50-keV in a future extension of the ring operation.

The electron bunch is injected into the ring at a repetition rate of 60 Hz or less from the injector system [7] with the help of two pulsed Panofsky quads and a pulsed dipole [8]. The bunch can be extracted within the first turn or after any number of turns with a system that duplicates the features of the injector line except that the electron gun is replaced by a large diagnostic chamber with phosphor screen, emittance meter and energy analyzer [7].

The operating vacuum is determined by gas-scattering-induced emittance growth. The entire vacuum system is designed to be bakeable. All pumping is by ion pumps, one at each diagnostic station, approximately 70 cm apart; our goal is to reach the low 10^{-9} Torr range after bakeout.

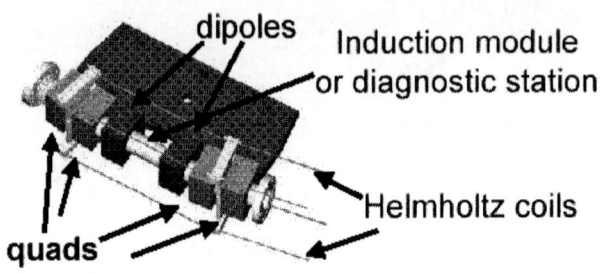

FIGURE 3. Detail of the UMER lattice

The electron gun

The design of the electron gun using the well-known code EGN has been described elsewhere [7]. It employs a commercial 8-mm diameter dispenser cathode with integral filament and grid. The e-gun includes a micrometer-controlled adjustable A/K gap to vary the beam current, a gate valve and an ion pump. A built-in Rogowski-type current monitor is included, as is a rotatable calibrated aperture plate with six masks. The plate includes a pepperpot and a five-beamlet aperture as well as four round apertures ranging from 0.5% to 100% transmission. The grid, cathode and grid pulser all float at 10 kV on a conventional high voltage deck. Pulses of 50-100 ns can be generated using an internal pulse-forming cable that can be changed as desired. The fast switch can be operated to produce either rectangular or parabolic pulses.

Magnet system

The design of the UMER magnetic elements presented a particular challenge. The uniform magnetic field required to make a 10 keV electron orbit at the radius of UMER (1.8 m) is 1.5 Gauss – about three to five times the value of the vertical component of the earth's magnetic field in the UMER laboratory. The earth's field in the region of the ring has been mapped with milligauss accuracy using a 3-axis flux-gate magnetometer. The vertical component will be used to assist the bending in the ring. The horizontal component is sufficient to deflect the beam about 1 mm vertically in a half lattice period. Compensation for the horizontal component will implemented with a series of nine segmented Helmholtz coils placed in a toroidal geometry around the ring. Segmentation of the compensation is necessary because of the sinusoidal variation of the magnetic force from the horizontal with angle around the ring, and because of local variations in the field strength.

When complete the ring magnetic lattice will consist of over 140 quadrupoles, dipoles and steering magnets. The typical focusing gradients and bending fields are on the order of 5 Gauss/cm and 10 Gauss, respectively. The use of iron-based magnets is impractical for such low fields. Therefore, the UMER magnets are based on an iron-free printed circuit (PC) design [9]. These printed circuits are fabricated to a very high tolerance, carefully mounted, measured and then installed and aligned on the beamline. Tables 3 and 4 give the general design characteristics of the PC magnets. The general tolerances on these PC magnets are similar to those of the iron-based magnets of large accelerators. The performance of the magnets on the test bench is summarized in Table 5. While the multipole content meets our design specifications, we are continuing to improve the design, fabrication and testing techniques. The details of the measurement techniques used and the results will be published elsewhere.

TABLE 3. UMER PC dipole design specifications.

Dipole field	15.4 G (5.2 G/A)
Current	3 A
Physical length	4.4 cm
Effective length	3.8 cm
Radius	2.8 cm
Field integral	20 G-cm/A
Resistance	3 Ω
Allowed harmonic content	<1%
Transverse alignment error	< 0.05 mm

TABLE 4. UMER PC quadrupole design specifications.

Field gradient	4.1 G/cm/A
Current	2 A
Physical length	4.4 cm
Effective length	3.6 cm
Radius	2.8 cm
Field integral	15 G/A
Resistance (room temp.)	7 Ω
Allowed harmonic content	<1%
Transverse alignment error	< 0.05 mm

TABLE 5. Measurements of the relative multipole components of the PC magnets.

Multipole components	PC dipole	PC quadrupole
Dipole	1	-
Quadrupole	0.010	1
Sextupole	0.0035	0.0035
Octupole	0.003	0.0041
Decapole	0.0009	0.0009
Duodecapole	0.0011	0.0011

Diagnostics

To allow detailed comparison between theory and experiment, UMER will have a comprehensive set of beam diagnostics. Each of the 13 diagnostic stations around the ring will have a phosphor screen and capacitive beam position monitor [10]. In addition, fast current monitors and resistive beam position monitors will also be installed. A sophisticated diagnostic end-chamber has been fabricated by FM Technologies. It houses emittance meters of the slit-wire and pepper-pot types, a retarding-field energy analyzer with sub-eV resolution for energy and energy spread measurements, a movable phosphor screen with 1.5 meter travel for insertion into the complete transport line, and a Faraday cup for current measurement.

UMER SIMULATIONS

A characteristic feature of the intense beam research at the University of Maryland has been the close coupling between experiment, theory and simulations. To continue this tradition on UMER, we are employing a comprehensive suite of design and simulation codes. These codes are the same as are being used for the design of HIF drivers.

The main code for studying important beam physics is the particle-in-cell (PIC) code WARP [11]. The PIC model, although slow relative to other techniques, has the advantage of a self-consistent calculation of the self-fields, which is often necessary for accurate analysis of certain phenomena such as space charge waves, and instabilities.

We have benchmarked the WARP code against our beam collimation experiment [12,13]. These simulations, which underline the importance of the initial particle distribution to its downstream behavior, exhibit very good agreement with the experiments. We believe that the residual disagreements between our simulation and experiment result from uncertainties in some of the variables (e.g. magnetic fields, initial phase space distributions) in the experiment. Therefore, we plan to ensure that UMER has much improved diagnostics over our previous experiments.

The primary application of the WARP simulations has been to study the ring design and thoroughly analyze its tolerance to various errors. This effort has been fully intermeshed with the magnet measurements, the electronics specifications, and the alterations to the mechanical design. Details of simulations related to UMER can be found in ref [14]

UMER PLANS

The UMER project can be divided into four phases as follows:
- Design/Prototype Phase
    This phase is now complete .
- Construction/Experimental Phase I

Sequential installation of injector and ring section beam physics experiments for about 75% of one turn. The phase is just beginning (August 2000) and will take three years.
- Experimental Phase II
  Ring closure and multi-turn operation.
  Low current operation (10 mA) for 100 turns.
  High current operation (100 mA) for at least 10 turns with $\frac{\Delta \varepsilon}{\varepsilon} \leq 4$
  This phase will begin toward the end of the current grant period
- Upgrade Phase
  Upgrade of UMER to a fast cycling synchrotron to accelerate the beam to 50 keV over 50-100 turns to study resonance crossing.

At the time of writing, August 2000, the UMER electron gun is under test, and the design current and emittance have been achieved..

CONCLUSION

UMER is a low-cost flexible electron model of intense ion recirculators. It offers the opportunity to study the physics of space-charge dominated beams at unprecedented intensity with strong focusing and dispersion. It facilitates the development and benchmarking of HIF driver codes, and is adaptable to many problems in beam physics because of its low cost and modular nature. We anticipate that UMER will uncover a considerable amount of new physics in the unexplored high-intensity regime.

ACKNOWLEDGMENTS

We would like to acknowledge our colleagues at Michigan State University, led by Richard York, who have been responsible for the design and construction of the main mechanical component on the ring.

This work is supported by the U.S. Department of Energy grant numbers DE-FG02-94ER40855 and DE-FG02-92ER54178.

REFERENCES

[1] Martin Reiser, "*Theory and Design of Charged Particle Beams*", (New York: John Wiley & Sons, Inc., 1994), p. 262 ff.
[2] Richard Baartman, "*Resonances with Space Charge*," Proceedings of Workshop on Space-Charge Physics in High Intensity Hadron Ring at Shelter Island, NY, 1998, (New York: AIP Press, no. 448, 1998), p. 56.
[3] I. Hofmann and K. Beckert, IEEE Trans. Nucl. Sci., 32 (1985) 2264.
[4] M. Reiser et al. Proc.1999 IEEE Particle Accelerator Conference, (2000) 234.
[5] UMER web site: www.ipr.umd.edu/umer

[6] Y. Li et al., Proc. 1999 IEEE Particle Accelerator Conference (2000) 1656.
[7] D. Kehne, et al "Injector for the University of Maryland Electron Ring (UMER)" to appear in Nuclr. Instr. Meth.
[8] Y. Li et al., Proc. 1999 IEEE Particle Accelerator Conference (2000) 3369.
[9] T. F. Godlove, S. Bernal, and M. Reiser, Proc. 1995 Particle Accelerator Conference (1995) 2117P
[10] Y. Zou, et al., "Development of a Prototype Capacitive BPM," Proc.1999 IEEE Particle Accelerator Conference (2000) 2102.
[11] D. P. Grote, *et. al.*, Fusion Eng. & Des. **32-33**, (1996)193.
[12] S. Bernal, P. Chin, R. A. Kishek, Y. Li, M. Reiser, J. G. Wang, T. Godlove, and I. Haber, Phys. Rev. ST Accel. Beams **1** (1998) 044202
[13] S. Bernal, R. A. Kishek, M. Reiser, and I. Haber, Phys. Rev. Lett., **82**, 20 (1999) 4002
[14] R.A. Kishek, M. Reiser, P. O'Shea, M. Venturini, W.W. Zhang, "Errors, Resonances, and Corrections in the Space-Charge-Dominated Beam of the University of Maryland Electron Ring (UMER)", to appear in the proceedings of the 2nd ICFA Workshop on High Brightness Beams, UCLA, November 1999 (http://pbpl.physics.ucla.edu/papers/index.html#ICFA99)

Confinement of Bunched Beams

Mark Hess and Chiping Chen

Plasma Science and Fusion Center, Massachusetts Institute of Technology, Cambridge, MA 02139, USA

ABSTRACT. The non-relativistic motion is analyzed for a highly bunched beam propagating through a perfectly conducting cylindrical pipe confined radially by a constant magnetic field parallel to the conductor axis, using a Green's function technique and Hamiltonian dynamics analysis. It is shown that for the confinement of beams with the same charge per unit length, the maximum value of the effective self-field parameter for a highly bunched beam is significantly lower than the Brillouin density limit for an unbunched beam.

INTRODUCTION

Confinement and transport of high-intensity charged-particle beams are important subjects in both plasma physics and beam physics [1,2]. It is well known that for a continuous, non-neutral, charged-particle beam propagating in a uniform magnetic field, the maximum beam density is determined by the so-called Brillouin density limit [3,4]. For nonrelativistic beams, the Brillouin density limit corresponds to the condition $2\omega_p^2/\omega_c^2 = 1$, where $\omega_p = (4\pi n q^2/m)^{1/2}$ is the nonrelativistic plasma frequency, and $\omega_c = qB/mc$ is the nonrelativistic cyclotron frequency. Although there is a large body of literature on the equilibrium and stability properties of high-intensity continuous non-neutral charged-particle beams, high-intensity bunched beams are rarely discussed in the literature [5-7].

There is a need to gain a fundamental understanding of high-intensity bunched beams because they are widely employed in high-power microwave (HPM) sources, such as klystrons and traveling wave tubes, as well as in high-intensity particle accelerators such as high-intensity linacs. In both HPM sources and high-intensity particle accelerators, an important problem associated with lack of full beam confinement caused by the bunching of the electron and ion beam in the direction of beam propagation is beam loss, through such mechanisms as beam halo formation [8-10].

In this paper, we analyze the nonrelativistic motion of a highly bunched beam propagating through a perfectly conducting cylindrical pipe confined radially by a constant magnetic field parallel to the conductor axis. In the present analysis, the beam is treated as either a thin rod distribution representing a continuous (unbunched) beam or periodic collinear point charges representing a highly bunched beam. Use is made of the Green's function to compute the electrostatic force on the beam due to the induced surface charge in the conductor wall. From Hamilton's equations, the radial phase space is studied for both unbunched and bunched beams. In general, the radial phase space contains both closed orbits (i.e., trapped particle orbits) and untrapped orbits (i.e., orbits which intersect the conductor wall) at sufficiently low beam densities, whereas only untrapped orbits exist at sufficiently high beam densities. By determining the conditions for the disappearance of trapped particle orbits, a criterion for the confinement of a highly bunched beam is derived. It is shown that for the confinement of beams with the same charge per unit length, the maximum value of the effective self-field parameter is $2\omega_p^2/\omega_c^2 \cong 2a/L$ for a highly bunched beam with $a \ll L$, where a is the radius of the conducting cylinder and L is the periodic spacing of the bunches. This result is significantly lower than the Brillouin density limit $2\omega_p^2/\omega_c^2 = 1$ for an unbunched beam.

HAMILTONIAN DYNAMICS

We analyze the radial dynamics of one rod of charge with line charge, λ, (2-D) as in Fig. 1 (a) and one string of charges with charge q (3-D) as in Fig. 1(b) interacting with its self-field and a constant applied magnetic field, $\vec{B} = B\hat{e}_z$. In such systems, there are no forces in the longitudinal direction. Therefore, we may describe all of the dynamics using a Hamiltonian in the radial and azimuthal directions and set $v_z = 0$ without loss of generality. In particular, the Hamiltonian for transverse motion is given by

$$H = \frac{1}{2m}\left[\left(P_r - \frac{qA_r}{c}\right)^2 + \frac{1}{r^2}\left(P_\theta - \frac{rqA_\theta}{c}\right)^2\right] + q\phi^{self} \quad (1)$$

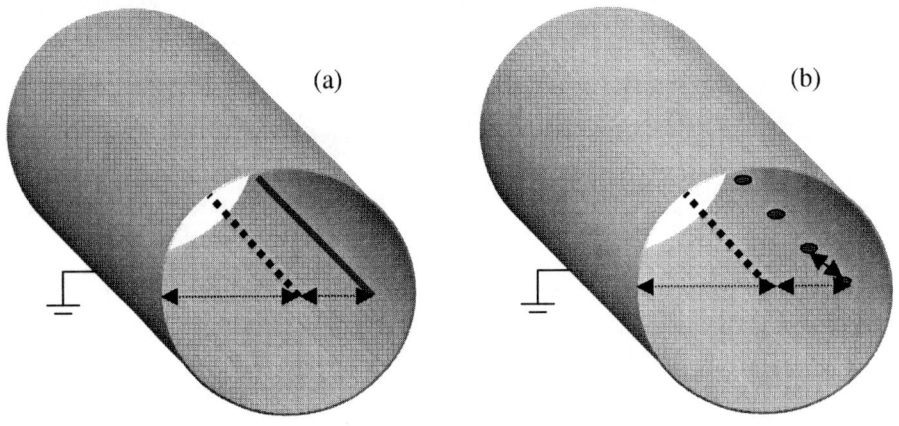

Figure 1. Schematics of (a) line charge and (b) periodic array of charges in a perfectly conducting cylinder.

where $\vec{P}$ is the canonical momentum, $\vec{A} = rB/2\hat{e}_\theta$ is the vector potential, $q = \lambda L$ and $m = \rho L$ for the 2-D system (ρ is the mass density of the rod), and ϕ^{self} is the electrostatic potential associated with the electric field produced by the induced charge at the conducting wall. It is readily shown from the Green's function that ϕ^{self} is given by [11,12]

$$\phi^{self} = \begin{cases} \lambda \ln(1 - \hat{r}^2/\alpha^2) & (2-D) \\ q/L\left[\ln(1-\hat{r}^2/\alpha^2) - 2\sum_{n=1}^{\infty}\sum_{l=-\infty}^{\infty} I_l^2(n\hat{r})K_l(n\alpha)/I_l(n\alpha)\right] & (3-D) \end{cases} \quad (2)$$

where $\hat{r} = 2\pi r/L, \alpha = 2\pi a/L$. $I_l(x)$ and $K_l(x)$ are the order l modified Bessel functions of the first and second kind, respectively.

Applying Hamilton's equations to (1), gives the following set of normalized equations:

$$\frac{d\hat{r}}{d\tau} = \hat{P}_r \quad , \quad \frac{d\hat{P}_r}{d\tau} = \frac{\hat{P}_\theta^2}{\hat{r}^3} - \hat{r} - \xi\frac{d\hat{\phi}^{self}}{d\hat{r}}$$

$$\frac{d\theta}{d\tau} = \frac{\hat{P}^\theta}{\hat{r}^2} - 1 \quad , \quad \frac{d\hat{P}_\theta}{d\tau} = 0 \quad (3)$$

where normalized variables and parameters are defined $\tau = \omega_L t$, $\hat{P}_r = 2\pi P_r/mL\omega_L$, $\hat{P}_\theta = 4\pi^2 P_\theta/m\omega_L$, $\hat{\phi}^{self} = L\phi^{self}/2q$, $\xi = 32\pi^2 mc^2/L^3 B^2$, $\omega_L = qB/2mc$ and ω_L represents the Larmor frequency. From (3), it is obvious that the canonical angular momentum is conserved. Combining the first two equations in (3), and denoting initial conditions with a subscript 0, we can find an expression relating the canonical radial momentum with the radial position,

$$\hat{P}_r = \pm\sqrt{\hat{P}_{r0}^2 + F(\hat{r}_0) - F(\hat{r})} \quad (4)$$

where F represents an effective potential energy, and is given by,

$$F(\hat{r}) = \frac{\hat{P}_\theta^2}{\hat{r}^2} + \hat{r}^2 + 2\xi\hat{\phi}^{self}. \quad (5)$$

Depending on the parameters $(\alpha, \xi, \hat{P}_\theta)$, the function $F(\hat{r})$ for the 3-D case may have two possible behaviors; $F(\hat{r})$ is either monotonically decreasing or has a kink. A similar property occurs in the 2-D case when we vary $(\xi, \hat{P}_\theta)$. When $F(\hat{r})$ is monotonic, the phase space $(\hat{r}, \hat{P}_r)$ only contains untrapped orbits, which result in particle loss to the wall. However, when $F(\hat{r})$ has a kink, both trapped and untrapped orbits. Fig. 2(a) and 2(b) illustrate the two phase spaces for the 3-D case.

CONDITIONS FOR CONFINEMENT

The complete criterion for trapped particle orbits is threefold, a) $F(\hat{r})$ must have a kink, b) the initial particle radius must be chosen between the local maximum of $F(\hat{r})$ and the other point on $F(\hat{r})$ corresponding to the same value, and c) the initial radial momentum must be sufficiently small, such that

$$\hat{P}_{r0}^2 \leq F(\hat{r}_0) - F(\hat{r})\big|_{min}. \qquad (6)$$

The most important of the three criteria for trapped particle orbits is the first. We therefore determine the region in parameter space $(\alpha, \xi, \hat{P}_\theta)$ space for both the 2-D and 3-D systems, such that $F(\hat{r})$ has a kink. In order to find this criterion for $F(\hat{r})$, i.e. that trapped particle orbits may exist, we must look for the conditions such that $F'(\hat{r}) = F''(\hat{r}) = 0$, where $F'(\hat{r}) = dF(\hat{r})/d\hat{r}$ and $F''(\hat{r}) = d^2 F(\hat{r})/d\hat{r}^2$. This represents that transition point between $F(\hat{r})$ being monotonic and non-monotonic.

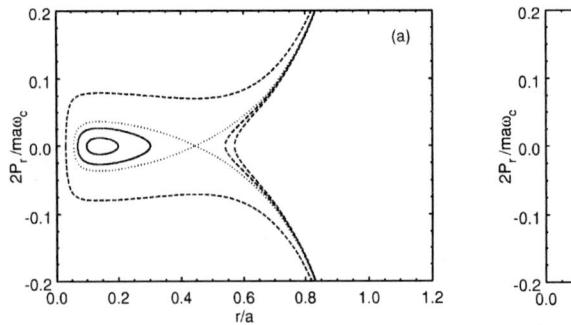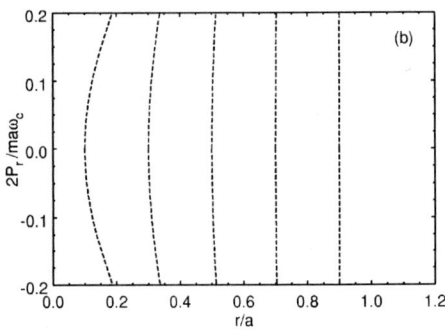

Figure 2. Plots of the radial phase space in the 3-D system for the choices of system parameters corresponding to: (a) $\xi/\alpha^2 = 0.7$, $\hat{P}_\theta/\alpha^2 = 0.01$ and $\alpha = 4.0$, and (b) $\xi/\alpha^2 = 0.7$, $\hat{P}_\theta/\alpha^2 = 0.01$ and $\alpha = 4.0$.

When we apply the transition condition to (5), we find that for the 2-D system, the following inequality must be satisfied for trapped particle orbits to occur [12],

$$8\mu^2 \leq 27 - 18(1-v) - (1-v)^2 - \sqrt{(27 - 18(1-v) - (1-v)^2)^2 - 64(1-v)^3} \qquad (7)$$

where $\mu \equiv \hat{P}_\theta/\alpha^2$ and $\nu \equiv \xi/\alpha^2$. Note that since μ and ν are both independent of L, (7) is also independent of L. Equation (7) is plotted later in Fig. 3 in terms of normalized P_θ and the effective plasma frequency, as we compare the 2-D case with the 3-D case.

Since the effective density of particles for both systems is given by $n = (\pi a^2 L)^{-1}$, we can relate ξ/α^2 to the effective plasma frequency $\omega_p = (4\pi n q^2/m)^{1/2}$ (where $q = \lambda L$), and the cyclotron frequency $\omega_c = qB/mc$ by $\xi/\alpha^2 = 2\omega_p^2/\omega_c^2$, which is the familiar self-field parameter. As shown in Fig. 3, the maximum of the self-field parameter occurs at $|\hat{P}_\theta| = 0$, and the maximum value is $2\omega_p^2/\omega_c^2 = 1$. Therefore, the criterion for the confinement is:

$$2\omega_p^2/\omega_c^2 \leq 1. \tag{8}$$

Note that $\omega_p^2 = \omega_c^2/2$ corresponds to the Brillouin density limit [3,4].

For the 3-D system, it can readily be shown that the maximum density limit for that the lowest order non-constant term, the quadratic term, will be positive when [12]

$$1 - \xi/\alpha^2 - \xi \sum_{n=1}^{\infty} n^2 \left[\frac{K_0(n\alpha)}{I_0(n\alpha)} + \frac{K_1(n\alpha)}{I_1(n\alpha)} \right] \geq 0. \tag{9}$$

By utilizing a formula related to the Wronskian, $I_m(z)K_{m+1}(z) + I_{m+1}(z)K_m(z) = 1/z$, we can simplify (9) to

$$\xi/\alpha^2 = 2\omega_p^2/\omega_c^2 \leq \left(1 + \sum_{n=1}^{\infty} n\alpha/I_0(n\alpha)I_1(n\alpha) \right)^{-1} \tag{10}$$

Fig. 3 illustrates a few of the critical transition curves in a normalized P_θ and $2\omega_p^2/\omega_c^2$ space. In obtaining the results in Fig. 3, we use Newton's method to simultaneously solve the equations, $F'(\hat{r}) = F''(\hat{r}) = 0$ for fixed values of $\hat{r}$ and α. Seed values are given to ξ and $\hat{P}_\theta$, and convergence of these values typically occurs within five iterations. Because the 2-D system corresponds to the limit $a/L \to \infty$, the transition curve for $a/L = \infty$ is identical to the results predicted by (7).

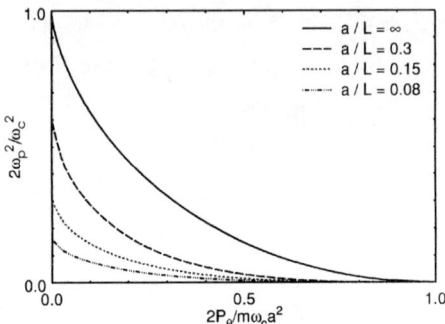

Figure 3. Plots of the maximum value of the self-field parameter $2\omega_p^2/\omega_c^2$ for confinement as a function of normalized canonical angular momentum $2P_\theta/m\omega_c a^2$ for several values of the aspect ratio a/L in the 3-D system. Note that the 2-D system corresponds to the limit $a/L = \infty$, and the curve with $a/L = \infty$ is obtained from (7).

Fig. 4 shows a plot of the upper bounds for transition to occur in the 2-D and 3-D systems. The upper bounds are precisely the intersections of the curves in Fig. 3 with the $P_\theta = 0$ axis. Two limits are worth mentioning in (10). First, expanding (10) in the limit $\alpha \gg 1$ (i.e. a nearly unbunched beam) and $I_0(n\alpha) \approx I_1(n\alpha) \approx e^{n\alpha}/(2\pi n\alpha)^{1/2}$, we obtain $2\omega_p^2/\omega_c^2 \cong 1 - 2\pi\alpha^2 e^{-2\alpha} = 1 - 8\pi^3 a^2 e^{-4\pi a/L}/L^2$ which shows that the system asymptotically approaches the 2-D system's Brillouin flow limit for large a/L. The other important limit of (10), $\alpha \ll 1$ (i.e. a strongly bunched beam), may be solved numerically, and yields $2\omega_p^2/\omega_c^2 \cong 2a/L$, which is significantly lower than the Brillouin density limit.

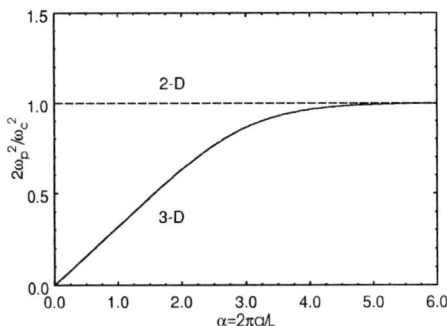

Figure 4. Plots of the maximum value of the self-field parameter $2\omega_p^2/\omega_c^2$ for confinement as a function of the aspect ratio a/L for $P_\theta = 0$ in both the 2-D and 3-D systems.

SUMMARY

To summarize, we have derived confinement criteria for a highly bunched beam and an unbunched beam propagating down a perfectly conducting cylinder with an applied magnetic field. We have modeled these two systems by approximating the unbunched beam as a rod of charge and the bunched beam as collinear periodic charges. For these two models, we have derived the equations of transverse motion from the Hamiltonian.

The criteria have been obtained by examining the properties of the beam's radial phase space. There are two possible phase spaces, one which allows trapped particle orbits and one which does not. The difference between the two is shown to be caused by the behavior of an effective radial potential (i.e. whether it has a kink or not). When varying the three parameters $(\alpha, \xi, \hat{P}_\theta)$ in the system, the behavior of the effective potential undergoes a critical transition.

The values of $(\alpha, \xi, \hat{P}_\theta)$ where the critical transition occurs yield an upper bound on the self-field parameter $2\omega_p^2/\omega_c^2 = \xi/\alpha^2$ for which trapped particle orbits exists. For an unbunched beam, the upper bound on the self-field parameter has been shown to be $2\omega_p^2/\omega_c^2 \leq 1$, which is precisely the Brillouin density limit. For a bunched beam, the maximum value of the self-field parameter is given in (10). The limit on the self-field parameter will always be less for the bunched beam than for the unbunched beam due to the higher local density of internal charges and induced surface charges, which contribute a higher electric field force.

The results reported in this paper are applicable to a relativistic charged-particle beam by a proper application of the Lorentz transformation from the laboratory frame to the frame of reference moving with the beam. Finally, it is anticipated that the results in this paper will provide a useful insight into the confinement of high-intensity bunched beams in linear accelerators as well as in high-power microwave sources such as klystrons.

ACKNOWLEDGEMENTS

This work was supported by the Air Force Office of Scientific Research, Grant No. F49620-97-1-0480 and Grant No. F49620-00-1-0007, and by the Department of Energy, Office of High Energy and Nuclear Physics, Grant No. DE-FG02-95ER-40919.

REFERENCES

1. Davidson, R.C., *Physics of Nonneutral Plasmas*, Addison-Wesley, Reading, Massachusetts, 1990.
2. Reiser, M., *Theory and Design of Charged Particle Beams*, John Wiley & Sons, New York, 1994.
3. Brillouin, L., *Phys. Rev.* **67**, 260 (1945).
4. See, for example, Chap. 1 and p. 545 of Ref. 1.
5. Sacherer, F.J., *IEEE Trans. Nucl. Sci.*, NS-18, 1105 (1971).
6. Barnard J.J. and Lund S.M., in *Proceedings of the Particle Accelerator Conference*, edited by M. Comyn (Institute of Electrical and Electronics Engineers, Piscataway, NJ, 1997), p. 1929.
7. Gluckstern, R.L., Fedotov, A.V., Kurennoy, S. and Ryne, R., *Phys. Rev. E* **58**, 4977 (1998).
8. Chen, C. and Pakter R., *Phys. Plasmas* **5**, 2203 (2000).
9. Chen, C. and Pakter, R. "Electron Beam Halo Formation in High-Power Klystron Amplifiers," IEEE Trans. Plasma Sci., in press (2000).
10. Chen and R. Pakter, "Halo Formation in Intense Electron Beams in High-Power Klystron Amplifiers," in *Intense Microwave Pulses VI*, edited by H. E. Brandt, SPIE Proc. 3702, 21 (1999).
11. Hess, M., Pakter, R., and Chen, C., "Green's function description of space charge in intense charged-particle beams," in *Proceedings of the Particle Accelerator Conference*, edited by A. Luccio and W. Mackay (Institute of Electrical and Electronics Engineers, Piscataway, NJ, 1999), p. 2752.
12. Hess, M. and Chen, C., *Phys. Plasmas*, to be published, (2000).

Studies of space-charge effects in ultrashort electron bunches

Gwenaël Fubiani,[1] Wim Leemans and Eric Esarey

Center for Beam Physics, Ernest Orlando Lawrence Berkeley National Laboratory, University of California, Berkeley CA 94720

Abstract. Laser-driven plasma-based accelerators are capable of producing ultrashort electron bunches in which the longitudinal size is much smaller than the transverse size. We present theoretical studies of the transport of such electron bunches in vacuum. Space charge forces acting on the bunch are calculated using an ellipsoidal bunch shape model. The effects of space charge forces and energy spread on longitudinal and transverse bunch properties are evaluated for various bunch lengths, energies and amount of charge.

INTRODUCTION

Plasma-based accelerators offer the possibility of providing compact, high energy electron accelerators [1]. Plasmas can sustain ultrahigh electric fields, thus providing for rapid acceleration. In addition, plasma-based accelerators can generate ultrashort electron bunches with a large amount of electrons per bunch. In a plasma, the wavelength of the accelerating field is the plasma wavelength, $\lambda_p = 2\pi c/\omega_p$, where $\omega_p = (4\pi n_e e^2/m_0)^{1/2}$ is the plasma frequency and n_e is the electron plasma density. In engineering units, the plasma wavelength is

$$\lambda_p[\mu m] \simeq 330(n_e[cm^{-3}])^{-1/2}.$$

For example, a laser wakefield accelerator (LWFA) in the standard regime typically has a density on the order of $n_e \simeq 10^{18}\ cm^{-3}$ and a plasma wavelength on the order of $\lambda_p \simeq 30\ \mu m$ (100 fs). If a mono-energetic electron bunch is injected into a wakefield such that it is accelerated while maintaining a small energy spread, then it is necessary that the bunch occupy a small fraction of the wake period, i.e., the bunch must be ultrashort, on the order of a few femtoseconds. Test particle simulations of the colliding pulse LWFA injector [2] - [4], in which two counterpropagating laser pulses are used to inject electrons from the background plasma directly into the wake, indicate the production of a trapped bunch ($\sim 10^7$ electrons) with a low

---
[1] Also at University of Paris XI, France

energy spread (< 1 %), low normalized emittance (~ 1 mm-mrad), and of ultra-short duration (~ 1 fs) may be possible. Such test particle simulations, however, neglected the space charge effects of the accelerated bunch.

Space charge effects can limit the amount of charge that can be transported in an ultrashort, tightly focused electron bunch, i.e., space charge can lead to a increase in both the longitudinal and transverse bunch dimensions. In a LWFA, space charge effects may not be of concern while the bunch is in the plasma wave, since the longitudinal and transverse fields of the wake are typically much greater than the space charge forces of the bunch. This is not the case, however, as the bunch exits the plasma into a vacuum region with no applied fields. In this case, space charge can lead to a rapid blow-up of the bunch. This paper will model the dynamics of ultrashort bunches in vacuum under the assumptions that (i) the spatial extent of the bunch remains ellipsoidal and (ii) the charge density within the bunch remains uniform.

INFINITE CYLINDRICAL BEAM

When a beam has a longitudinal size much larger than the transverse size, one can assume that the space charge force that occurs on the beam is only transverse. Since the longitudinal field is almost zero in this case, one can easily apply Gauss's law to calculate the transverse electric field. The radial electric field $E_{r'}$ in the beam frame (denoted by a prime) is given by

$$E_{r'} = \rho' r' / (2\epsilon_0), \qquad (1)$$

where ρ' the charge density in the beam frame (a frame moving with the average velocity of the beam electrons). The quantities in the lab frame are related to those in the beam frame by the Lorentz transforms [5],

$$\begin{aligned}\mathbf{E} &= \gamma(\mathbf{E}' - \boldsymbol{\beta} \times \mathbf{B}') - \gamma^2(\gamma+1)^{-1}\boldsymbol{\beta}(\boldsymbol{\beta} \cdot \mathbf{E}'), \\ \mathbf{B} &= \gamma(\mathbf{B}' + \boldsymbol{\beta} \times \mathbf{E}') - \gamma^2(\gamma+1)^{-1}\boldsymbol{\beta}(\boldsymbol{\beta} \cdot \mathbf{B}').\end{aligned} \qquad (2)$$

along with $E_r = \gamma E'_{r'}$, $B_\theta = \gamma \beta E'_{r'}$, $r' = r$, $z' = \gamma z$, and $\rho' = \rho/\gamma$, where $\gamma = (1-\beta^2)^{-1/2}$ is the relativistic factor, $\beta = v/c$ and v is the beam velocity (assumed to be along the z-axis). Thus, the radial Lorentz-Coulomb force in the lab frame is

$$F_r^{self} = q(E_r + \beta \times B_\theta) = q\gamma(1-\beta^2)E'_{r'} = \rho r/(2\gamma^2 \epsilon_0).$$

Newton's equation of motion can be easily computed under the assumptions $\beta \sim 1$, $\gamma \sim \gamma_0$ and $v_\theta(t=0) = 0$ (laminar flux) [6], *i.e.*,

$$\ddot{r} + \dot{\gamma}\dot{r}/\gamma = F_r^{self}/(\gamma m_0). \qquad (3)$$

where the dot denotes d/dt. Furthermore, assuming that the change in energy $\dot{\gamma}/\gamma$ remains small, and letting $s = ct$, we can approximate

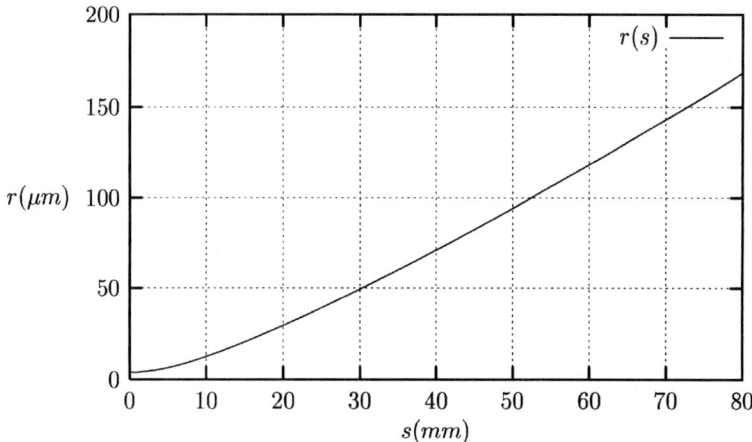

FIGURE 1. Transverse motion of a single electron with zero initial divergence. The motion is only due to space charge forces in the infinite beam limit.

$$\gamma m_0 d^2 r/ds^2 \simeq F_r^{self}/c^2. \tag{4}$$

Figure 1 shows the results of numerically solving Eq. (4), assuming a relatively long electron bunch with an average energy of $E_0 = 20 \ MeV$, a bunch radius of $\Delta r_0 = 4 \ \mu m$, a bunch length of $\Delta z_0 = 0.5 \ \mu m$ (the theory assumed an infinite beam, but it is necessary to define a bunch length for the calculation of an equivalent charge density), an angular spread of $\theta_0 = 0 \ mrad$, a total charge of $Q = 1 \ pC$ and a charge density $\rho = Q/(\pi \Delta r_0^2 \Delta z_0)$. Figure 1 is useful in that it shows the effects of space charge that one expects in the limit of a very long beam.

It is insightful to compare these results to those that are obtained from a more precise calculation that takes into consideration the finite duration of the bunch. The next section will discuss the space charge model for an ultrashort bunch, under the assumption of an ellipsoidal bunch shape. In this particular case one can calculate the potential U and the electric field everywhere at the surface and inside the ellipsoid. We will compare these two models and define the range where the simple case of an infinite beam can still adequately describe the bunch evolution.

TRANSVERSE AND LONGITUDINAL SPACE-CHARGE FORCES

The equations of motion that we use to model the effects of space charge on the transverse and longitudinal motion of a single beam electron are given by

$$\frac{d^2 r}{ds^2} = \frac{1}{\gamma_0 m_0 v_0^2} F_r^{self}, \tag{5}$$

$$\frac{d^2\sigma}{ds^2} = \frac{1}{\gamma_0^3 m_0 v_0^2} F_s^{self}, \qquad (6)$$

where F_r^{self} and F_s^{self} are the effective space charge forces in the transverse and longitudinal directions, respectively, $s = \int dt\, v_s + s_0$ is the longitudinal position of the electron with velocity v_s, s_0 is a constant, $\sigma = s - v_0 t_s(s)$ is the longitudinal distance between the electron and that of the synchronous particle located at the bunch center $\Delta s = v_0 t_s(s)$, v_0 is the average velocity of the bunch (assumed to be constant) and $t_s(s) = \int ds/v_s$. Here it is assumed that the electron is highly relativistic, $\beta \simeq 1$, with an energy near the average energy of the bunch, $\gamma \simeq \gamma_0$. Next we outline the derivations of Eqs. (5) and (6).

Transverse motion

Equation (3) describes in cylindrical coordinates the motion of a single electron of energy $\gamma m_0 c^2$ and charge $q = -e$. Assuming $v_\theta = 0$ (laminar beam), $\dot{\gamma}/\gamma$ small ($\gamma \simeq \gamma_0$, i.e., the electron energy is near the average energy of the bunch) and $\gamma_\perp \ll \gamma_\parallel \simeq \gamma_0$, the transverse equation of motion is given by

$$\ddot{r} = \frac{F_r^{self}}{\gamma_0 m_0}.$$

Furthermore, for the transverse motion, it is adequate to approximate $s \simeq v_0 t + s_0$. Hence,

$$\frac{d^2 r}{ds^2} = \frac{F_r^{self}}{\gamma_0 m_0 v_0^2}.$$

Longitudinal motion

We define ΔE_s as the energy spread induced by the longitudinal space charge force, i.e.,

$$\Delta E_s = \int ds\, F_s^{self}.$$

The energy spread of the electrons in the bunch is assumed small compared to the average energy $E_0 = \gamma_0 m_e c^2$ ($\gamma_0 \gg 1$) of the bunch, i.e., $|\Delta E_s|/E_0 \ll 1$. The variable $\eta = \Delta E_s/E_0$ is then appropriate to describe the longitudinal dynamics of the bunch [7], [8].

It is then straightforward to show that the spread in longitudinal momentum Δp_s due to the longitudinal space charge force is related to the variable η by

$$\frac{\Delta p_s}{p_0} = \frac{1}{\beta_0}\sqrt{(1+\eta)^2 - \frac{m_0 c^2}{E_0^2}} - 1 = f(\eta),$$

where $\beta_0 = v_0/c$ and $p_0 = \gamma_0 \beta_0 m_0 c$. Note that the derivatives of $f(\eta)$ with respect to η are given by

$$\frac{df(\eta)}{d\eta} = \frac{1}{\beta_0 \beta}, \quad \frac{df(0)}{d\eta} = \frac{1}{\beta_0^2}, \quad \frac{d^2 f(0)}{d\eta^2} = \frac{1}{\beta_0^4 \gamma_0^2}.$$

We now define the longitudinal position variable $\sigma = s - v_0 t_s(s)$ and find its time evolution by expanding $f(\eta) = \Delta p_s/p_0$ about $\eta = 0$. Note that

$$\frac{d\sigma}{ds} = 1 - v_0 \frac{dt_s}{ds} = 1 - \frac{v_0}{v_s} = 1 - \beta_0^2 \frac{df(\eta)}{d\eta}.$$

If $\eta \ll 1$, then $df(\eta)/d\eta \simeq df(0)/d\eta + \eta d^2 f(0)/d\eta^2$. This implies $d\sigma/ds \simeq \beta_0^{-2} \gamma_0^{-2} \eta$ and, hence,

$$\frac{d^2 \sigma}{ds^2} = \frac{1}{\gamma_0^3 m_0 v_0^2} \mathrm{F}_s^{self},$$

which is valid for relativistic bunches, $\gamma_0 \gg 1$, with small energy spreads $|\eta| = |\gamma - \gamma_0|/\gamma_0 \ll 1$.

ELLIPSOIDAL BEAM THEORY

We next calculate the electrostatic potential of the bunch in 3D by making a summation over all the charges inside the bunch. The model described below closely follows references [7]- [9]. The 3D bunch is described as an ellipsoid that maintains its ellipsoidal shape as it evolves. In the lab frame, the longitudinal bunch size is defined as σ_s along the longitudinal coordinate σ, and the transverse bunch sizes as σ_x and σ_y along the transverse coordinates x and y. In the frame moving with the bunch, the bunch boundary is assumed to be an ellipse given by

$$\frac{x'^2}{\sigma_x^2} + \frac{y'^2}{\sigma_y^2} + \frac{\sigma'^2}{\gamma_0^2 \sigma_s^2} = 1. \quad (7)$$

Furthermore, within this ellipse, the charge density of the bunch is assumed to be uniform.

In the interior of the ellipse, the electrostatic potential in the bunch frame (denoted by a prime) is a quadratic function of x', y' and σ',

$$U' = -Ax'^2 - By'^2 - C\sigma'^2 + D,$$

where A, B, C, and D are constants. The electric field in the bunch frame is $\mathbf{E}' = -\nabla U'$. Integrating over the ellipsoid volume, and using the expression of the total charge $Q = (4\pi/3)\sigma_x \sigma_y \sigma_s \rho$, gives

$$E'_x = \frac{3}{8\pi \epsilon_0} Q I_1 x' \quad (A = \frac{3}{16\pi \epsilon_0} Q I_1),$$

$$E'_y = \frac{3}{8\pi\epsilon_0} QI_2 y' \quad (B = \frac{3}{16\pi\epsilon_0} QI_2),$$

$$E'_s = \frac{3}{8\pi\epsilon_0} QI_3 \sigma' \quad (C = \frac{3}{16\pi\epsilon_0} QI_3),$$

where I_1, I_2 and I_3 are the elliptical integrals given by

$$I_1 = \int_0^\infty \frac{d\tau}{(\sigma_x^2 + \tau)\sqrt{\phi(\tau)}},$$

$$I_2 = \int_0^\infty \frac{d\tau}{(\sigma_y^2 + \tau)\sqrt{\phi(\tau)}},$$

$$I_3 = \int_0^\infty \frac{d\tau}{(\gamma_0^2 \sigma_s^2 + \tau)\sqrt{\phi(\tau)}},$$

with $\phi(\tau) = (\sigma_x^2 + \tau)(\sigma_y^2 + \tau)(\gamma_0^2 \sigma_s^2 + \tau)$.

The fields in the lab frame are given by Eq. (2), along with $E_x = \gamma_0 E'_x$, $E_y = \gamma_0 E'_y$, $E_s = E'_s$, $B_x = \beta_0 \gamma_0 E'_y$, $B_y = -\beta_0 \gamma_0 E'_x$, $B_s = 0$ (we assumed $\mathbf{B}' \sim \mathbf{0}$, i.e., the relative motion between the electrons inside the bunch is almost zero), $r' = r$, $y' = y$ and $\sigma' = \gamma_0 \sigma$.

We extract the Lorentz-Coulomb force $\mathbf{F}^{self} = q(\mathbf{E} + \mathbf{v} \times \mathbf{B}/c)$ from the fields,

$$F_x = \frac{3}{8\pi\epsilon_0} \frac{eQ}{\gamma_0} x \int_0^\infty \frac{d\tau}{(\sigma_x^2 + \tau)\sqrt{(\sigma_x^2 + \tau)(\sigma_z^2 + \tau)(\gamma_0^2 \sigma_s^2 + \tau)}},$$

$$F_y = \frac{3}{8\pi\epsilon_0} \frac{eQ}{\gamma_0} y \int_0^\infty \frac{d\tau}{(\sigma_z^2 + \tau)\sqrt{(\sigma_x^2 + \tau)(\sigma_z^2 + \tau)(\gamma_0^2 \sigma_s^2 + \tau)}},$$

$$F_s = \frac{3}{8\pi\epsilon_0} eQ\gamma_0 \sigma \int_0^\infty \frac{d\tau}{(\gamma_0^2 \sigma_s^2 + \tau)\sqrt{(\sigma_x^2 + \tau)(\sigma_z^2 + \tau)(\gamma_0^2 \sigma_s^2 + \tau)}}.$$

Assuming that the beam has a cylindrical symmetry, i.e., $\sigma_r \equiv \sigma_x = \sigma_y$, the forces are then given by,

$$F_r = \frac{3}{8\pi\epsilon_0} \frac{eQ}{\gamma_0} r I_r, \quad F_s = \frac{3}{8\pi\epsilon_0} eQ\gamma_0 \sigma I_s$$

where

$$I_r = \int_0^\infty d\tau (a+\tau)^{-2}(b+\tau)^{-1/2},$$

$$I_s = \int_0^\infty d\tau (a+\tau)^{-1}(b+\tau)^{-3/2},$$

with $a = \sigma_r^2$ and $b = \gamma_0^2 \sigma_s^2$. The integrals I_r and I_s can be calculated analytically,

$$I_r = \begin{cases} \dfrac{\pi}{2(a-b)^{3/2}} - \dfrac{b^{1/2}}{a(a-b)} - (a-b)^{-3/2} \tan^{-1}[b^{1/2}(a-b)^{-1/2}], & \text{for } a > b, \\ \dfrac{b^{1/2}}{a(b-a)} + \dfrac{1}{2}(b-a)^{-3/2} \ln[(\sqrt{b} - \sqrt{b-a})/(\sqrt{b} + \sqrt{b-a})], & \text{for } a < b, \end{cases}$$

$$I_s = \begin{cases} -\dfrac{\pi}{(a-b)^{3/2}} + \dfrac{2}{b^{1/2}(a-b)} + 2(a-b)^{-3/2}\tan^{-1}[b^{1/2}(a-b)^{-1/2}], & \text{for } a > b, \\ -\dfrac{2}{b^{1/2}(b-a)} - (b-a)^{-3/2}\ln[(\sqrt{b}-\sqrt{b-a})/(\sqrt{b}+\sqrt{b-a})], & \text{for } a < b, \end{cases}$$

and $I_r = I_s = (2/3)a^{-3/2}$ for $a = b$.

In the limit $a \ll b$ (i.e $\sigma_r \ll \gamma_0\sigma_s$), the above expressions can be expanded to yield

$$I_r = \frac{1}{b^{1/2}a} + b^{-3/2}[1 + (1/2)\ln(4b/a)][1 + O(a/b)],$$

$$I_s = b^{-3/2}[2 + \ln(4b/a)][1 + O(a/b)].$$

Notice that $I_s/I_r \sim O(a/b)$. Hence, to leading order in a/b, $I_r \sim 1/(\gamma_0\sigma_s\sigma_r^2)$ and $I_s \sim 0$ and we recover the forces on an infinite and cylindrical beam, i.e., $F_r = e\rho r/(2\gamma_0^2\epsilon_0)$ and $F_s \sim 0$, where the volume of the ellipsoid is $V_{ell} = 3/(4\pi\sigma_r^2\sigma_s)$.

ENVELOPE EQUATIONS

Space charge dominated

The trajectory of any particle within the bunch can be solved if the bunch radius and length (boundaries of the ellipse), σ_r and σ_s, are know as a function of propagation distance. Now under the condition of a space charge dominated beam, the forces on a particle scale as $F_r \sim r$ and $F_s \sim s$. Specifically, a particle at a larger radius will remain at a larger radius compaered to a particle at a smaller radius. We will estimate the evolution of the beam envelope by replacing r with σ_r and σ with σ_s in Eqs. (5) and (6), i.e.,

$$\sigma_r'' = \frac{K_r(\sigma_r, \sigma_s)}{\sigma_r}, \quad \sigma_s'' = \frac{K_s(\sigma_r, \sigma_s)}{\sigma_s}, \tag{8}$$

$$K_r = \frac{3}{8\pi\epsilon_0}\frac{eQ}{\gamma_0}I_r(\sigma_r, \sigma_s), \quad K_s = \frac{3}{8\pi\epsilon_0}eQ\gamma_0 I_s(\sigma_r, \sigma_s),$$

where the contribution of finite emittance to the envelope equations has been neglected.

Emittance dominated

It is interesting to compare the space charge dominated regime to the emittance dominated regime. In free space, neglecting space charge forces, the envelope equations for the electron bunch length and radius are given by

$$\sigma_r'' - \frac{\epsilon_r^2}{\sigma_r^3} = 0, \quad \sigma_s'' - \frac{\epsilon_s^2}{\sigma_s^3} = 0, \tag{9}$$

where ϵ_r and ϵ_s are the rms unnormalized emittances in the transverse and longitudinal direction, which are given by [10]

$$\epsilon_s = (\sigma_s^0/\gamma_0^2)\Delta p_s/p_0, \quad \epsilon_r = (\langle r^2\rangle\langle r'^2\rangle - \langle rr'\rangle^2)^{1/2},$$

where σ_s^0 is the initial bunch length, $\Delta p_s(\eta)/p_0$ is the longitudinal momentum spread, and the angular brackets denote an averaging over the particle distribution. The usual definitions for the normalized emittances are given by

$$\epsilon_{r,n} = \gamma_0\beta_0\epsilon_r, \quad \epsilon_{s,n} = \gamma_0^3\beta_0\epsilon_s$$

Equation (9) can be readily integrated to yield

$$\sigma_r^2(s) = (\sigma_r^0)^2 + 2\sigma_r^0\sigma_r'^0 s + [(\epsilon_r/\sigma_r^0)^2 + (\sigma_r'^0)^2]s^2,$$

$$\sigma_s^2(s) = (\sigma_s^0)^2 + 2\sigma_s^0\sigma_s'^0 s + [(\epsilon_s/\sigma_s^0)^2 + (\sigma_s'^0)^2]s^2,$$

where the initial conditions $\sigma(0) = \sigma^0$ and $\sigma'(0) = \sigma'^0$ have been assumed.

NUMERICAL SOLUTIONS AND COMPARISONS

Studies of ultrashort bunches

Next we will consider two cases: (i) electron bunches with low $\Delta E/E_0$ such as we expect to produce with colliding pulse LWFA injectors [2] - [4], and (ii) electron beams with 100% $\Delta E/E_0$ such as produced with self-modulated laser wakefield accelerators (SM-LWFA) [11] - [15].

Beam envelope growth in the colliding pulse LWFA regime

Typical bunch characteristics expected from a colliding pulse injector [2] - [4] are a total charge $Q \sim 1\ pC$, a low energy spread ($\eta \sim 1\%$), a transverse size of the order of the laser spot size ($\sigma_r \sim 5\ \mu m$) and an ultrashort bunch duration ($\Delta\tau \sim 1$ fs, i.e., $\sigma_s \sim 0.1 - 1\ \mu m$). Figures 2, 4 and 3 plot the bunch radius $\sigma_r(s)$ and length $\sigma_s(s)$ for the space charge dominated regime, Eq. (8), as a function of propagation distance and charge (at $s = 50\ mm$) assuming several electron bunch energies $E_0 = 20 - 40\ MeV$, an initial bunch radius $\sigma_r^0 = 4\ \mu m$, an initial bunch length $\sigma_s^0 = 0.5\ \mu m$, an initial divergence $\sigma_r'^0 = \sigma_s'^0 = 0$ (initially at the beam focus), and a total amount of charge from 1 to 100 pC. For comparison, we also plot results from the emittance dominated regime, Eq. (9), for a normalized transverse emittance of $\epsilon_{r,n} = 0.1, 0.5$ and 1.0 mm-mrad, and a normalized longitudinal emittance of $\epsilon_{s,n} = 6 \times 10^{-5}$ mm-mrad and $\epsilon_{s,n} = 6 \times 10^{-4}$ mm-mrad, which corresponds to an energy spread η of 1% and 10%. These figures clearly show that a fairly high energy electron bunch ($\sim 20\ MeV$) with a total amount of charge of several pC

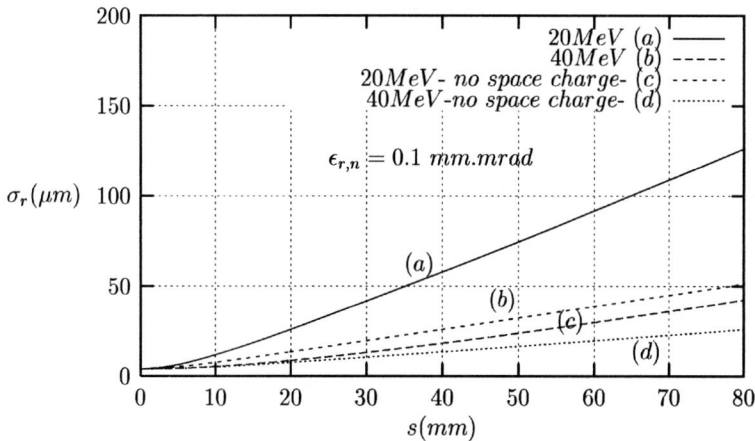

FIGURE 2. Transverse beam size versus propagation distance with space charge ($Q = 1$ pc) for a (a) 20 and (b) 40 MeV bunch. Curves (c) and (d) are for no space charge and normalized emittances of 0.1 mm-mrad

produced by colliding pulse injection can rapidly blow-up via space charge due to its very compact size if $\epsilon_{r,n} \sim 0.1$ mm-mrad. Below 30 pC per bunch the beam is emittance dominated when its value is of the order of $\epsilon_{r,n} = 1$ mm-mrad, which is a typical value obtained from numerical simulations, i.e., in this case one can neglect the effect of space charge. Above this value space charge must be considered and clearly participates in the beam growth. Around 50 pC both phenomenon must be

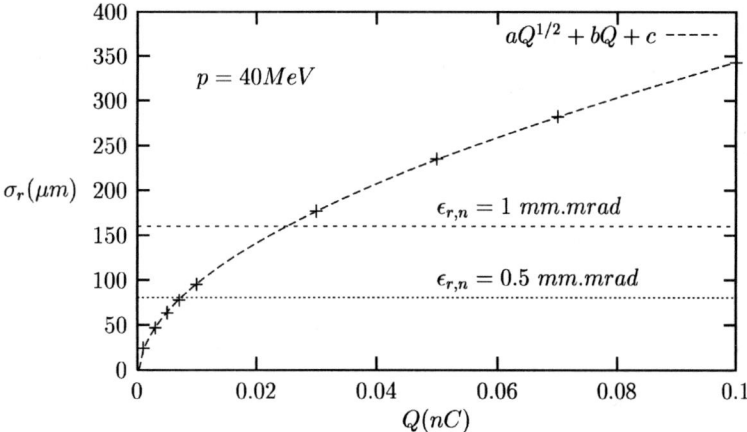

FIGURE 3. Transverse beam size (at $s = 50$ mm) versus charge with space charge for a 40 MeV bunch. The horizontal lines are for no space charge and normalized emittance of 0.5 and 1 mm-mrad.

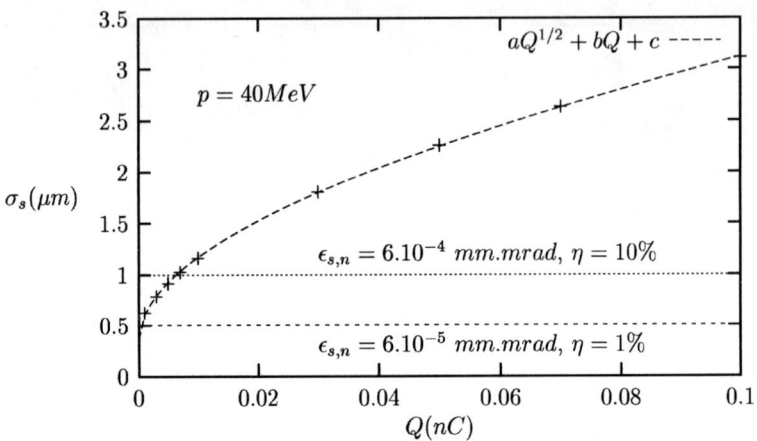

FIGURE 4. Longitudinal beam size (at $s = 50$ mm) versus charge with space charge for a 40 MeV bunch. The horizontal lines are for no space charge and normalized longitudinal emittance of 6×10^{-5} and 6×10^{-4} mm-mrad.

simultaneously studied and Eqs. (8) and (9) cannot be decoupled for the transverse motion, i.e., the coupled envelope equation is,

$$\sigma_r'' = \frac{K_r(\sigma_r, \sigma_s)}{\sigma_r} + \frac{\epsilon_r^2(\sigma_r, \sigma_s)}{\sigma_r^3}.$$

Conversely, one can consider that above a charge of 50 pC per bunch the beam is space charge dominated.

On the other hand from Fig. 4, one can see that the longitudinal motion for a 5 pC bunch is already in a space charge dominated regime if we assume an initial energy spread of $\eta = 1\%$. The longitudinal normalized emittance in this case is $\epsilon_{s,n} = 6 \times 10^{-5}$ mm-mrad.

Beam envelope growth in the SM-LWFA regime

In the SM-LWFA regime, the electron energy distribution is assumed to be similar to that observed experimentally [11] - [15], as shown in Fig.5, with a total charge per bunch on the order of 1 nC. To simulate the bunch size evolution, the method we use is to take the number of electrons, N_i, present in a small energy bin about the energy E_i, and to calculate the beam sizes $\sigma_{ri,si}$ for this group of electrons using Eq. (8) under the influence of the total bunch charge.

The overall bunch sizes $\sigma_{r,s}$ are estimated by summing the contribution of each group of electrons to the overall bunch profile, under the assumption that the

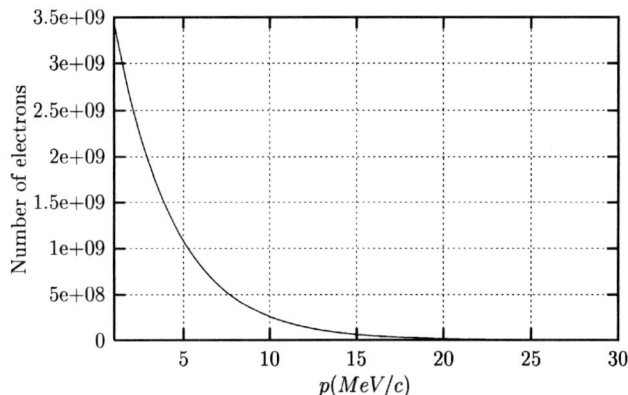

FIGURE 5. Momentum distribution model for an electron bunch produced by a SM-LWFA.

profiles for each group, as well as the overall profile, is flattop, i.e.,

$$\langle \sigma_r^2 \rangle = \sum_{i=0}^{n} N_i \sigma_{ri}^2 / \sum_{i=0}^{n} N_i, \quad \langle \sigma_s^2 \rangle = \sum_{i=0}^{n} N_i \sigma_{si}^2 / \sum_{i=0}^{n} N_i. \quad (10)$$

Using this model, it is found that the beam spot size blows-up rapidly due to the large amount of low energy electrons (energy spread per bunch η is close to 100%), which are more sensitive to space charge forces. Figure 6 shows the beam radius versus charge after $s = 50$ mm from the numerical solution of Eqs. (8) and (10), assuming $\sigma_r^0 = 5$ μm, $\sigma_s^0 = 10$ μm, $\sigma_s^{'0} = \sigma_r^{'0} = 0$, and neglecting the effect of

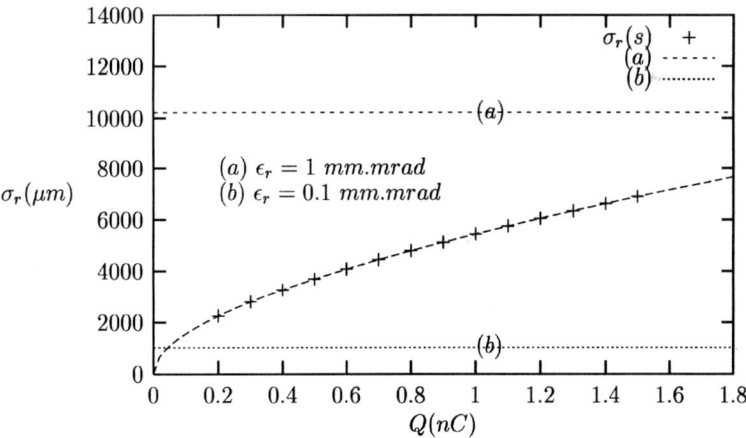

FIGURE 6. Transverse beam size (at $s = 50$ mm) as a function charge Q for a SM-LWFA bunch. The horizontal lines are for no space charge and unnormalized transverse emittances of 0.1 and 1 mm-mrad.

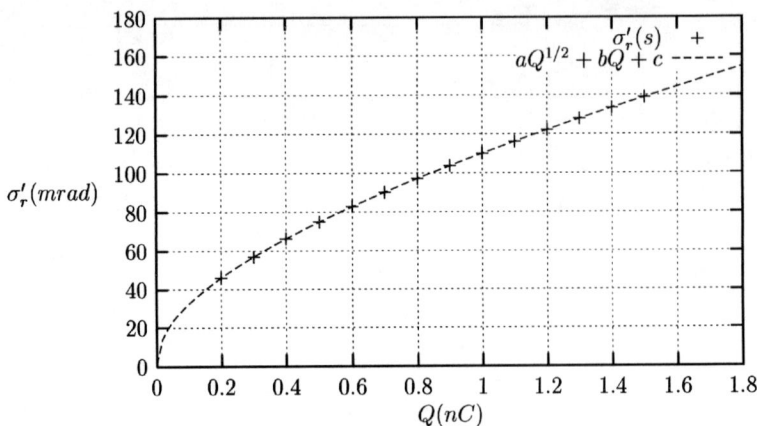

FIGURE 7. Beam divergence (at $s = 50$ mm) as a function of charge Q for a SM-LWFA bunch. Points are from a numerical calculation and the curve represents a fit of the form $aQ^{1/2} + bQ + c$, where a, b and c are constants.

emittance. For comparison with the emittance dominated regime, the solution of Eq. (9) is shown for an unnormalized transverse rms emittance $\epsilon_r = 1.0$ mm-mrad and $\epsilon_r = 0.1$ mm-mrad.

One can also study the average beam divergence $\langle \sigma_r'^2 \rangle^{1/2}$ and $\langle \sigma_s'^2 \rangle^{1/2}$, where

$$\langle \sigma_r'^2 \rangle = \sum_{i=0}^{n} N_i \sigma_{ri}'^2 / \sum_{i=0}^{n} N_i, \quad \langle \sigma_s'^2 \rangle = \sum_{i=0}^{n} N_i \sigma_{si}'^2 / \sum_{i=0}^{n} N_i.$$

Figure 7 shows the transverse beam divergence as a function of charge for the parameters of Fig. 6. The divergence $\langle \sigma_r'^2 \rangle^{1/2}$ is clearly dependent of the beam characteristics (σ_r^0, σ_s^0, Q, γ_0), but for fixed sizes $\langle \sigma_r'^2 \rangle^{1/2}$ is in good agreement with a fit of the form $aQ^{1/2} + bQ + c$, where a, b and c are constants, which can be used as an empirical scaling law.

CONCLUSION

Plasma-based accelerators offer the possibility of providing compact, high energy electron accelerators and are also capable of producing ultrashort electron bunches in which the longitudinal size is much smaller than the transverse size. Space charge effects are not of concern while the bunch is in the plasma wave, since the longitudinal and transverse fields of the wake are typically much greater than the space charge forces of the bunch, but space charge cannot be neglected when an electron bunch propagates in vacuum with no external fields, because of its very compact dimensions. The evolution of the bunch sizes under the influence of space charge has been considered with the assumptions that the beam shape remains ellipsoidal, and the charge density within the bunch is uniform. Space charge effects on bunches produced in the SM-LWFA and colliding pulse scheme have been examined computationally. Transverse and longitudinal beam growth and the normalized emittance growth depend strongly on energy, but even a fairly high energy electron bunch $\sim 40\ MeV$ can still be space charge dependent if it contains a high charge density.

The above analysis and simulations assumed that within the ellipsoidal bunch, the charge density was uniform. It is possible to perform a similar analysis that assumes the charge density is Gaussian, although this formulation is somewhat more difficult numerically. Previous studies [16] have found that the uniform charge model tends to under estimate the effects of space charge compared to the Gaussian charge distribution model, since a Gaussian distribution gives a higher charge density near the bunch center. Hence, the results presented in this paper present an lower bound on the effects of space charge on ultrashort bunches.

ACKNOWLEDGMENTS

The authors acknowledge useful discussions with David Bruhwiler and John Staples. This work was supported by the U.S. Department of Energy, Contract No. DE-AC-03-76SF0098.

REFERENCES

1. For a review see, E. Esarey et al., IEEE Trans. Plasma Sci. **PS-24**, 252 (1996).
2. E. Esarey et al., Phys. Rev. Lett. **79**, 2682 (1997).
3. C.B. Schroeder et al., Phys. Rev. E **59**, 6037 (1999).
4. E. Esarey et al., Phys. Plasmas **6**, 2262 (1999).
5. J.D. Jackson, *Classical Electrodynamics* (Wiley, 1975).
6. R.B. Miller, *An Introduction to the Physics of Intense Charged Particle Beams* (Plenum, 1985).
7. DESY Report, *DESY 88-013*, 1988.
8. DESY Report, *DESY 87-161*, 1987.
9. O.D. Kellogg, *Foundations of potential theory* (Dover, 1953).
10. M. Reiser, *Theory and design of charged particle beams* (Wiley, 1994).
11. A. Modena et al., Nature **377**, 606 (1995); D. Gordon et al., Phys. Rev. Lett. **80**, 2133 (1998).
12. K. Nakajima et al., Phys. Rev. Lett. **74**, 4428 (1995).
13. D. Umstadter et al., Science **273**, 472 (1996); R. Wagner et al., Phys. Rev. Lett. **78**, 3125 (1997).
14. A. Ting et al., Phys. Plasmas **4**, 1889 (1997); C.I. Moore et al., Phys. Rev. Lett. **79**, 3909 (1997).
15. W.P. Leemans, et al., these proceedings.
16. David Bruhwiler, private communication.

Low Emittance Electron Beam Formation with a 17 GHz RF Gun

W.J. Brown, S.E. Korbly, K.E. Kreischer, M.A. Shapiro, R.J. Temkin

Plasma Science and Fusion Center,
Massachusetts Institute of Technology, Cambridge, MA, 02139 USA

Abstract. We report on the design, construction, and initial operation of a new 1 1/2 cell 17 GHz RF gun and beamline. Emittance compensation is achieved with a 6.5 cm long, 0.5 T solenoid placed immediately after the RF Gun. The gun operates with 50 ns, up to 5 MW, pulses from a 17.13 GHz klystron amplifier built by Haimson Research Corp. Results of initial high power operation are reported. Cold test measurements of the azimuthal symmetry of the RF gun field profile along with results of 3D electromagnetic and beam dynamics simulations are also presented. For a bunch charge of 50 to 100 pC, initial bunch length of 1 ps, and beam energy of 1 MeV, a normalized rms emittance of less than 3 πmm-mrad has been measured after 35 cm of beam transport from the gun. This corresponds to a beam brightness of about 100 A/(mm-mrad)^2 at the gun exit. An rms energy spread of less than 2.5% has been previously measured and agrees well with simulations. Plans for increasing the beam energy and lowering the emittance are underway. Simulations predict that with a beam energy of 2 MeV produced by the gun, a normalized rms emittance of 0.5 π-mm-mrad can be achieved for 1 ps, 0.1 nC beam.

INTRODUCTION

The MIT 17 GHz photocathode RF gun is a 1.5 cell electron accelerating structure consisting of two coupled TM_{010} like cavities excited by side wall coupled microwaves from a WR-62 waveguide (Figure 1). The goal of the MIT 17 GHz RF gun experiment is to examine the advantages of operating an electron source at high frequency, and thereby produce an ultra-high quality electron beam capable of meeting the demands of future applications such as injectors for linear colliders or for accelerators for short wavelength free electron lasers [1], the stringent requirements of which (~100A peak current, and 1 πmm-mrad rms normalized emittance) have been unrealized by conventional electron sources. The scaling with RF frequency of the quality of the beam from an RF gun has previously been derived [2]. This study suggests that the emittance of the beam will scale inversely with RF frequency provided the charge and size of the beam are also scaled inversely with RF frequency (constant peak current) and the accelerating gradient is increased proportional to frequency. This implies a quadratic increase in the beam brightness with RF frequency can in principle be obtained. While technical limitations such as wall heating and breakdown may prevent the realization of this ideal scaling, it is believed that high frequency structures can still result in a significant improvement in beam brightness.

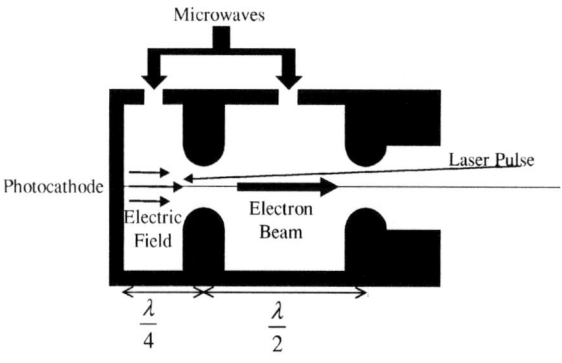

FIGURE 1. Schematic of the RF gun.

Initial experiments demonstrating beam production from a previous version of the 17 GHz RF gun have been completed at MIT and have been previously reported [3]. Results of beam measurements are listed in Table 1. This was the first RF gun experiment to operate above 3 GHz. In order to demonstrate an ultra-high brightness electron beam, a new emittance compensated RF gun and beamline with emittance diagnostics has been designed, built, and initial operation begun.

TABLE 1. Beam Measurement Results From Original Experiment.

Parameter	Measurement
Bunch Charge	0.1 nC
RF Injection Phase	10 - 40 °
Initial Bunch Length	1 ps (Laser Measurement)
Initial Bunch Radius	0.5 mm (Laser)
Cathode Electric Field	$\approx$ 200 MV/m
Beam Energy	1.05 MeV
Energy Spread (rms)	2.5% (at Spectrometer)

NEW GUN AND BEAMLINE DESIGN

A number of improvements have been made to the MIT 17 GHz RF gun experiment. Field profile measurements of the original gun revealed the field strength in the half cell to be about 20 percent stronger than that in the full cell. In order to optimize the acceleration efficiency and maximize the accelerating gradient, the new gun was equipped with tuners in both cells to allow for field balance optimization. The tuners consist of small plungers which retract from or fill up a small hole in both the half and full cells, but fall short of actually protruding into the cavity. This provides about 10 MHz of tunability without introducing protrusions into the cavity which could cause breakdown. To minimize the field asymmetries due to the coupling holes, it was insured that the coupling factor would be less than or equal to unity (i.e., the coupling holes would not be larger than necessary). The measured S_{11} curve from a cold test of this structure is shown in Fig. 2. It can be determined from the S_{11} curve that the coupling coefficient of the new gun is given by $\beta = 0.56$, and the ohmic quality factor is $Q = 1700$.

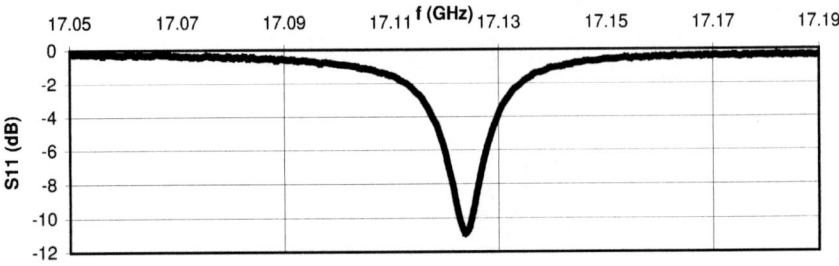

FIGURE 2. Measured S_{11} curve of the new 17 GHz RF gun.

Improvements have also been made in the beam transport and diagnostics. A schematic of the new beamline is shown in Fig. 3. Emittance compensation is performed with a 6.5 cm long, 5 kG peak field solenoid. The magnet consists of 4 double pancakes of 5x5 mm hollow core conductor, consisting of 80 total turns. In order to maximize the field inside the magnet and minimize it outside the magnet (i.e. at the cathode), the solenoid was encased in an iron yoke with an inside bore diameter of only 1.9 cm. The edge of the magnet is placed 2.0 cm from the gun cathode, resulting in a maximum magnetic field at the cathode of about 25 Gauss. The additional normalized emittance resulting from this magnetic field is only 0.04 πmm-mrad for a 0.5 mm radius beam.

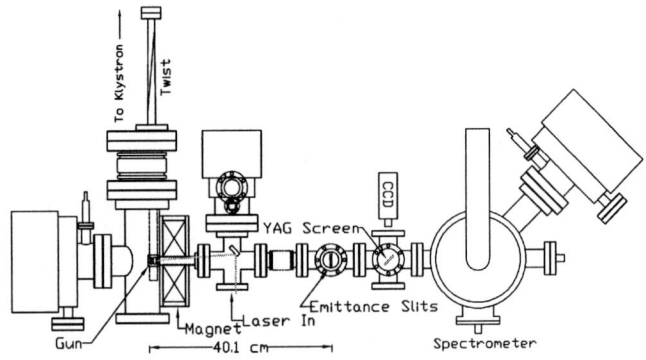

FIGURE 3. Schematic of New Beamline.

The emittance is measured by breaking the beam into individual beamlets using an emittance mask made from laser drilled 50 μm slits in a thin (0.125 mm) tantalum foil, and imaging the beamlets downstream of the slits using a scintillating YAG crystal and CCD camera in order to reconstruct the phase space. PARMELA simulations show that for a 0.1 nC, 1 ps, 2.4 MeV beam, a normalized rms emittance of 0.5 πmm-mrad can be produced after emittance compensation (Figure 4). For a 1 1/2 cell gun, peak accelerating fields of 350 MV/m will be required to obtain these parameters. In order to alleviate the necessity for such high fields, a 2 1/2 cell gun could be easily inserted in the beamline, bringing the needed peak field value down to 218 MV/m. The normalized rms brightness of the beam, defined by

$$B = \frac{2I_{peak}}{\varepsilon_n^2}, \tag{1}$$

could reach values of about 800 A/(πmm-mrad)^2 for these parameters. If this value is achieved, it would represent a record high value of beam brightness. Such a record value would allow for breakthroughs in accelerators designed for various applications including TeV colliders and free electron lasers.

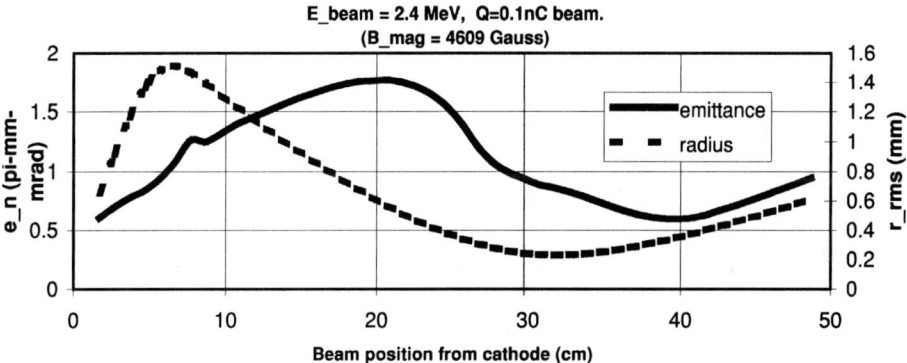

FIGURE 4. PARMELA Simulation of emittance compensation for 1 1/2 cell gun.

FIELD PROFILE MEASUREMENTS

In order to perform field profile measurements of the excited mode in the MIT RF gun, a "bead hang" method was developed. This is similar to "bead pull" measurements [4], but has a key difference in that the perturbing element is simply hung down into the RF cavity as opposed to being pulled all the way through it. The advantage of this method is that there is no need to have a hole in the cathode, allowing for the exact structure used in high power experiments to be measured. Ideally, the axial electric field profile in the gun can then be determined by mapping the perturbation in the resonant frequency of the excited mode as a function of the position of the bead, i.e.

$$|E(z)|^2 \propto \frac{\Delta f(z)}{\alpha_{bead}}.\qquad(2)$$

where α_{bead} is the electric polarizability of the bead.

In reality, the line used to hang the bead into the cavity also has a position dependent effect on the resonant frequency. This non-local perturbation, which can be expressed as

$$\Delta f_{line}(z) \propto \int^z |E(\xi)|^2 d\xi,\qquad(3)$$

where the integral is along the length of the support line up to the position of the bead, should be subtracted out of the measurement in order to obtain accurate results. This is especially necessary for smaller, high frequency structures like the MIT 17 GHz RF gun in which the line will have a more significant effect. The measured value for the electric field profile is then given by

$$|E(z)|^2 \propto \frac{(\Delta f(z) - \Delta f_{line}(z))}{\alpha_{bead}}.\qquad(4)$$

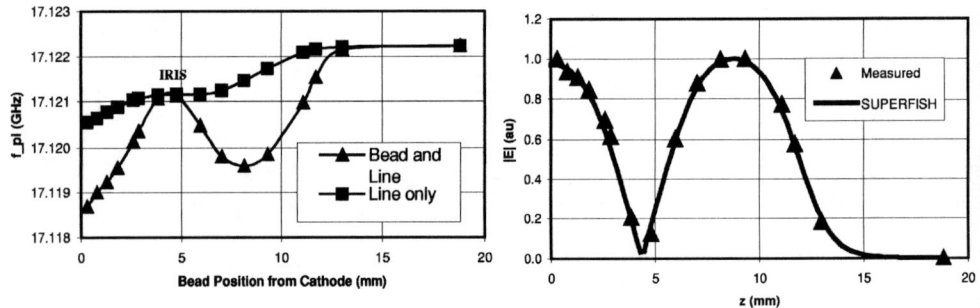

FIGURE 5. Measurement of the on axis field profile of the new RF gun.

Results of a bead hang measurement of the new cavity are shown in Figure 5. The bead used in the measurement was a 0.5 mm long, 0.2 mm diameter piece of copper wire. The support line used in the measurement was a short piece of 76 μm diameter fishing line (1.5 lb. tensile strength). As expected, the frequency perturbation due to the bead was zero at the position of the iris between the half and full cell.

The bead hang method was also employed to measure the azimuthal symmetry of the field profile in the RF gun (Figure 6). This was accomplished by placing a cylindrical metal wire halfway through the full cell of the gun and rotating the position around the gun axis at a constant radius. There is good agreement between the measurement and simulations of the RF gun obtained using the 3D electromagnetic code GDFIDL [5].

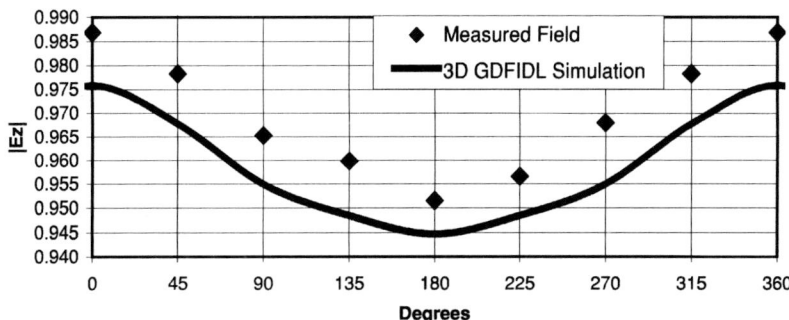

FIGURE 6. Azimuthal Symmetry measurement of RF Gun field profile at r = 1.6 mm. 0 degrees corresponds to the orientation closest to the coupling hole. Field values are normalized to the on axis field.

The field asymmetry seen in Figure 6 can result in an additional contribution to the emittance of the electron beam. In order to understand these effects, 3D beam dynamics simulations were performed using a single particle, no space charge, particle pushing code written at MIT. The code uses a 3D electromagnetic field map produced from the GDFIDL simulations to simulate the RF fields inside the gun, and pushes individual particles of various radial positions and RF injection phases to simulate the evolution of a beam envelope through the gun. The results of a simulation are seen in Figure 7, revealing the adverse effect the field asymmetries have on the beam emittance in the dimension perpendicular to the coupling holes.

The total emittance can be estimated by adding the RF induced emittance and the space charge induced emittance in quadrature (no correlation) to obtain a lower limit, and adding them linearly (unity correlation) to obtain an upper limit [6].

$$\sqrt{\varepsilon_{sc}^2 + \varepsilon_{rf}^2} \le \varepsilon_{n,rms} \le \varepsilon_{sc} + \varepsilon_{rf}. \quad (5)$$

From PARMELA simulations, the normalized rms emittance due to space charge for a 0.1 nC, 1 ps beam can be estimated to be 0.5 πmm-mrad. Combining this with an RF induced emittance due to field asymmetries of about 0.3 πmm-mrad, we can estimate a total normalized RMS emittance of about 0.6-0.8 πmm-mrad. Taking into consideration that the beam brightness goes as the inverse of the square of the emittance, it can be estimated that the obtainable beam brightness from an asymmetric gun is about half that obtainable from a field-symmetrized gun. A number of methods have been developed to symmetrize RF photoinjectors, including dual feed power coupling schemes, race-track geometries, and single feed power schemes symmetrized by passive coupling holes [7][8]. The MIT RF gun is currently being rebuilt with field symmetrization.

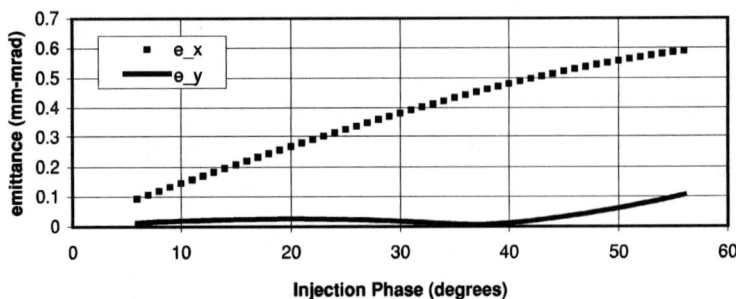

FIGURE 7. 3D simulation of the RF induced normalized rms emittance as a function of laser injection phase for an RF gun with a single sided power coupling scheme. The emittance in transverse dimension perpendicular to the coupling holes (x) is significantly larger than the emittance in the transverse dimension parallel to the coupling hole (y).

HIGH POWER OPERATION OF NEW BEAMLINE

The initial high power tests of the new experiment began in late 1999. The experiment utilizes a 17 GHz relativistic klystron amplifier constructed by Haimson Research Corporation [9] to provide a 50 ns to 1 µs pulse of up to 26 MW of microwave power. The klystron is driven by a 560 kV, 1 µs flattop modulator pulse. A 0.27 µperv Thomson CSF gun produces a space-charge limited electron beam at 560 kV with 95 A transmitted through the klystron. The amplifier chain includes a TWT amplifier to provide up to 5 W to the klystron. The klystron gain is approximately 67 dB. To date, up to 4 MW of incident power have been coupled into the cavity yielding accelerating gradients approaching 200 MV/m.

Laser and RF phase stability

The laser beam for the RF gun photocathode is generated by a Ti:Sapphire laser system, which produces 2 ps, 1.5 mJ, 2 mm diameter pulses at 800 nm after chirped pulse amplification. The pulse duration of 2 ps was verified using a single-shot autocorrelator. These pulses are frequency tripled to 10-20 µJ of UV, and then focused on the back wall of the copper cavity. In order to insure the photoemission is successful in producing an electron beam on every shot, a sophisticated timing scheme is required to force the synchronization of the laser arrival time to the RF phase within the gun. The 84 MHz Ti:Sapphire laser oscillator provides both the initial 2 ps, low energy, 800 nm laser pulse for input into a chirped pulse regenerative amplifier, as well as an initial low frequency RF signal (17GHz/204) which is filtered and used as input into a phase locked 17 GHz YIG oscillator. The YIG oscillator output is then used as the input into a TWT amplifier, which in turn is used as input to the klystron.

Figure 8 shows a measurement of the phase stability by means of observing the percentage of shots where an electron beam is observed with a Faraday cup as a function of RF phase shift. The width of this probability spectrum can be used to estimate the phase stability to be about ± 15 degrees. We are currently pursuing techniques for further reducing this jitter.

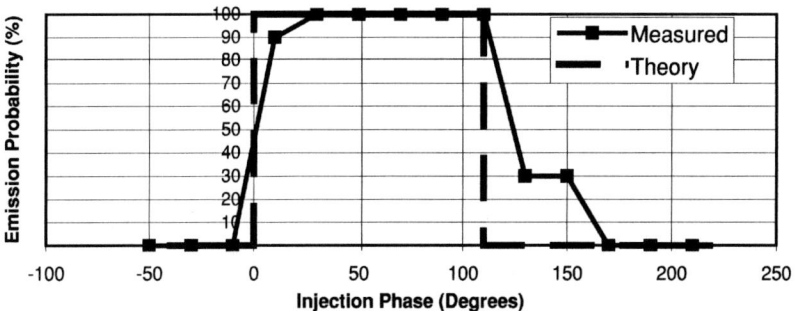

FIGURE 8. Probability of laser induced electron emission from the RF gun as a function of RF phase shifter setting. Comparison to that predicted from theory suggests a phase stability of about +- 15 degrees (or 2 ps).

Beam Measurements

The beam measurements performed on the new beamline have consisted of Faraday cup measurements of the beam bunch charge exiting the RF gun downstream of the laser injection chamber, transverse profile measurement with use of a YAG screen and CCD camera, and emittance measurements using a slit technique. With 15 µJ of UV incident on the gun cathode, and with about 200 MV/m accelerating gradients in the half cell, bunch charges up to 0.11 nC have been observed. This corresponds to a quantum efficiency for the copper cathode of about 3×10^{-5}.

Field enhancement of the laser induced electron emission was observed by varying the laser injection phase. According to photoemission theory, the work function of a material will be reduced in the presence of an electric field E, such that there is an effective work function given by

$$\Phi_{eff} = \Phi_0 - \sqrt{\frac{eE}{4\pi\varepsilon_0}} \quad . \tag{6}$$

This is known as the Schottky effect. The quantum efficiency goes as the square of the difference between the incident photon energy and the effective cathode work function, which leads to a predicted bunch charge vs. RF injection phase dependence given by [10]

$$Q_b \propto \left(E_v - \Phi_0 + \sqrt{\frac{eE\sin(\phi)}{4\pi\varepsilon_0}} \right)^2 , \tag{7}$$

where Q_b is the beam bunch charge, φ is the laser injection phase, E_v is the photon energy, and E is the peak electric field at the cathode. This scaling agrees well with the measured charge vs. laser injection phase data shown in Figure 9.

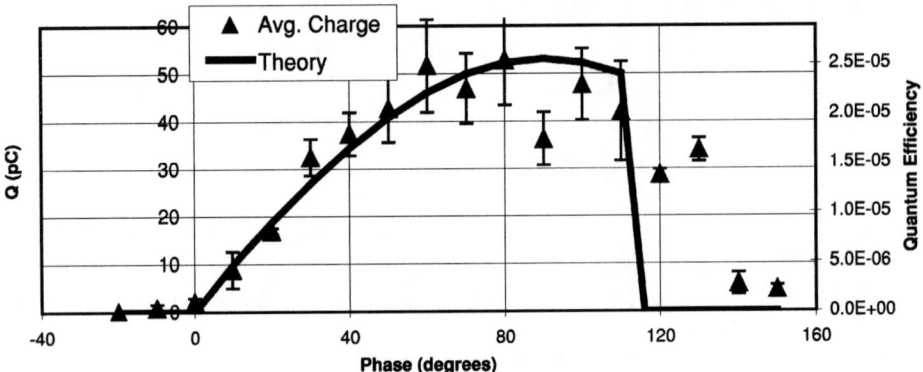

FIGURE 9. Average charge vs. laser injection phase. The increase of measured charge with increasing injection phase up to 90 degrees is consistent with the Schottky effect (i.e., the lowering of the cathode work function due to the presence of the electric field). The average laser energy for this scan was about 10 μJ, corresponding to a peak quantum efficiency, η, of about 2.5×10^{-5}.

Beam Imaging and Emittance Measurements

Transverse field profile measurements have been performed using a YAG crystal as a scintillator and storing the image using a CCD camera and frame grabber. The frame grabber was triggered with respect to the firing of the high voltage modulator in order to synchronize the acquired frame with the presence of the electron beam. Due to diffuse reflections of the laser light from the viewing mirror and the gun cathode, it was necessary to filter out the wavelengths associated with the laser (ie. 800nm, 400nm, and 266 nm). Figure 10 shows a typical transverse beam profile measurement for a beam near its waist after a soft focus at the position of the YAG screen (about 35 cm downstream of the emittance compensating solenoid). The profile is nearly Gaussian with an rms radius of about 1mm. The beam waist position and solenoid strength of 0.16 T suggests a beam energy of about 0.8 MeV, which is consistent with the energy calculated based on forward and reflected RF power measurements.

The emittance of the beam was measured using a slit technique. Individual beamlets are produced about 33 cm from the gun cathode by placing 50 μm slits in the beam path. The slits were laser drilled into 125 μm tantalum foil, which is thick enough to sufficiently scatter electrons with energies up to 2 MeV. The beamlets then travel through a drift space of about 13 cm before intersecting the YAG screen. By measuring the width of each beamlet at the YAG screen position as well as the relative intensity of each beamlet over the entire transverse dimension of the beam, the Twiss parameters can be determined and the geometric rms emittance can be calculated. Figure 11 shows some typical emittance measurements.

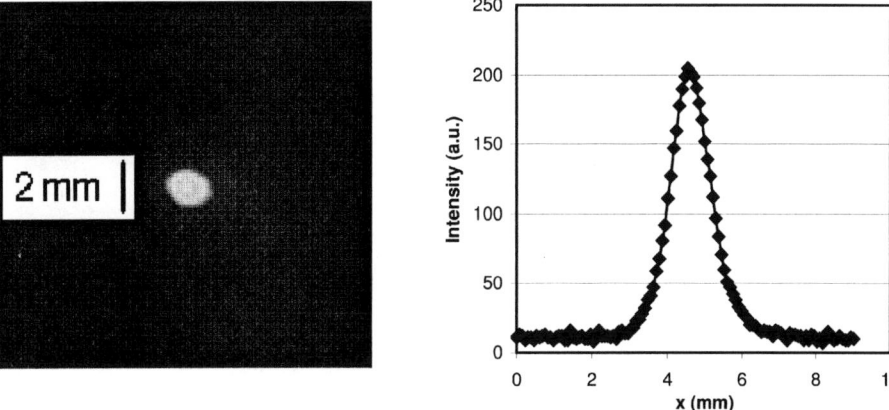

FIGURE 10. Image of 0.8 MeV electron beam (left) with Gaussian profile and 1 mm rms beam radius (right). The beam is imaged with a YAG crystal placed in the beam path.

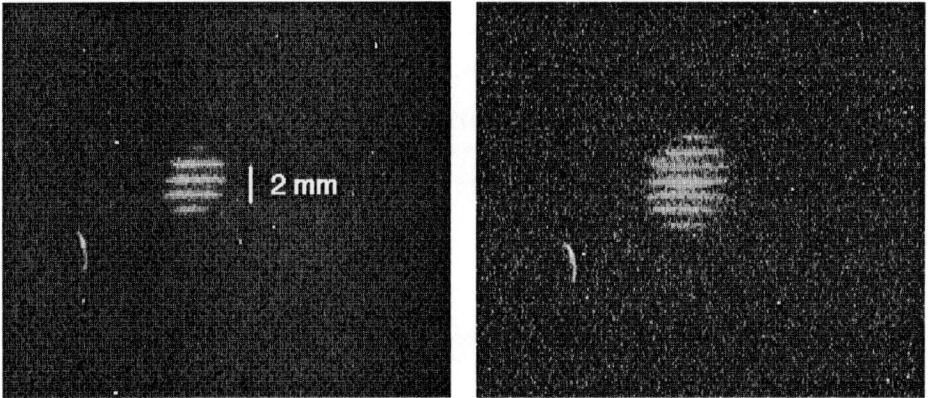

FIGURE 11. Images of individual beamlets produced by the slits after a 13 cm drift length. Left: 12 pC bunch charge with $\varepsilon = 0.9$ πmm-mrad. Right: 55 pC bunch charge with $\varepsilon = 1.4$ π mm-mrad. The relativistic factor $\beta\gamma$ is about 2.5.

The bunch charge for each emittance measurement is estimated by integrating over the intensity profile of the beamlets. A roughly linear dependence between emittance and charge is observed in the measurements. This is in good agreement with theory [6] as well as with PARMELA simulations. Figure 11 shows the measured dependence of emittance on charge with that predicted by PARMELA simulations of the acceleration process in the gun and transport through the emittance compensating solenoid and drift space to the location of the slits.

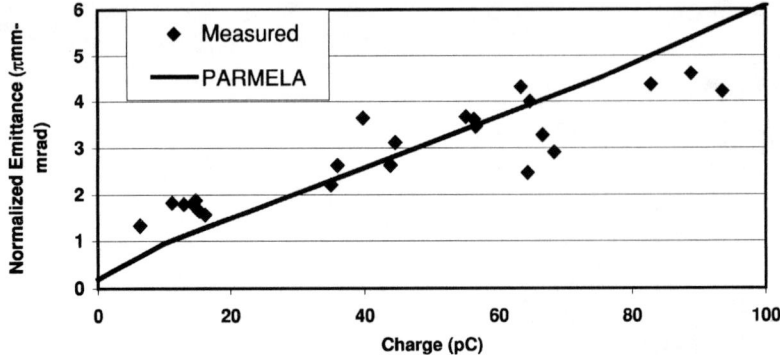

FIGURE 12. Dependence of normalized emittance on charge for a 1 MeV beam as determined from measurement (dots) and from PARMELA simulations (line).

For these beam energies (0.8-1.0 MeV), the PARMELA simulations suggest the emittance compensation is not effective, and there is actually a roughly 3 fold space charge induced emittance growth between the exit of the gun and the location of the slits. For the measured normalized emittance of $\varepsilon_n \approx 3\pi$mm-mrad for a 50 pC beam at the location of the slits, an emittance of about 1 πmm-mrad at the exit of the RF gun can be deduced, corresponding to a normalized rms brightness of

$$B_n = \frac{2I}{\varepsilon_n^2} \approx 100 \quad A/\pi mm-mrad^2 \quad . \tag{8}$$

IMPROVED OPERATION WITH A 2.4 CELL RF GUN

In order to produce a record high brightness of about 800 A/πmm-mrad, corresponding to $\varepsilon_n \approx 0.5$ πmm-mrad for a 100 A peak current beam, the energy of the electron beam must be increased to at least 2 MeV. In this case, the space charge forces are greatly reduced, and the longitudinal integrity of the beam is not degraded by large velocity spreads, allowing the emittance compensation to be effective (Figure 4). The primary impediment to obtaining these high energies is RF breakdown in the gun. For a 1 ½ cell gun, peak fields of about 350 MV/m must be obtained in order to reach this beam energy. However, breakdown in the gun is often observed for field gradients exceeding 200 MV/m, limiting our obtained beam energy to about 1 MeV. One solution to this problem is to build a gun with more cells in order to obtain higher beam energy for a given accelerating gradient.

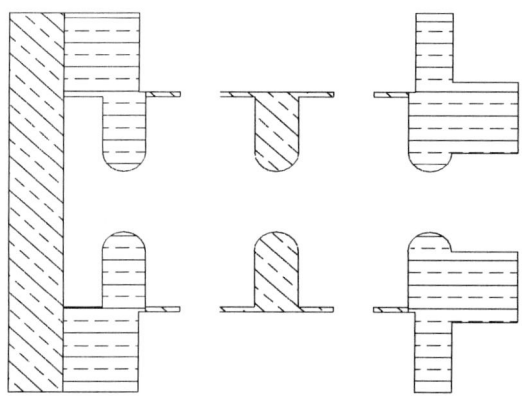

FIGURE 13. Schematic of a symmetrized 2.4 cell RF Gun. The shortened "half" cell provides for improved beam dynamics for the design gradient of 200 MV/m, while the opposing coupling holes provide dipole azimuthal field symmetry.

The 2.4 cell gun (Figure 13) is designed to produce a 2 MeV beam with peak accelerating fields of about 200 MV/m. To compensate for the relatively low normalized accelerating gradient [6], the "half" cell is shortened to 0.2 λ_0 from the 0.25 λ_0 of the 1 ½ cell design, where λ_0 is the vacuum RF wavelength. This is meant to improve beam capture, making the final beam energy less sensitive to laser injection phases and resulting in a significantly lower energy spread and improved longitudinal RF bunching than would be possible with a true half cell. Once space charge forces and emittance compensation are taken into account, simulations indicate a significantly lower emittance is obtainable with the 2.4 cell gun than with a 2.5 cell gun.

The 2.4 cell gun has sidewall coupling as previously described for the 1 ½ cell gun, but is also equipped with symmetrizing coupling holes to eliminate the dipole field asymmetry caused be the single waveguide coupling scheme (Figure 6) and the corresponding multipole field induced emittance (Figure 7). Figure 14 shows the result of an HFSS simulation for the 2.4 cell gun. The dipole field asymmetry is essentially eliminated, leading to very low RF induced emittance (< 0.1 π mm-mrad) for typical beam parameters.

The 2.4 cell RF gun is currently being constructed. It is believed that with this gun, a record high brightness beam is obtainable. PARMELA simulations of the beam transport with this gun (200 MV/m) yield results very similar to that shown in Figure 4 for the 1 ½ cell gun (350 MV/m), resulting in a normalized emittance of 0.5 πmm-mrad for a 100 pC, 1 ps electron beam, corresponding to a predicted beam brightness of 800 A/(πmm-mrad)^2.

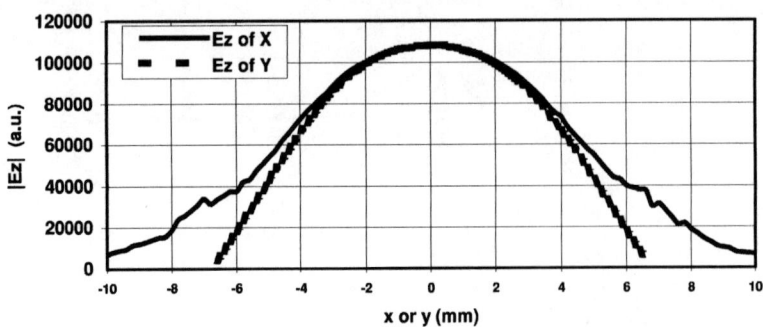

FIGURE 14. Results of an HFSS simulations showing the field symmetry of the symmetrized 2.4 cell gun. The dipole asymmetry is essentially eliminated, resulting in low RF induced emittance.

CONCLUSIONS

A tunable 17 GHz RF gun has been built and longitudinal and azimuthal field profile measurements have been performed by means of a "bead hang" method. The tunability of the new gun allowed for good field balance to be achieved between the half and full cells. The effect of the azimuthal asymmetry due to the coupling holes was also examined using a 3D particle dynamics code. A new beamline has also been built for improved operation and diagnostics, and initial operation begun. High power conditioning of the gun has yielded accelerating gradients of 200 MV/m, and electron beam emissions of over 0.1 nC have been measured corresponding to a quantum efficiency of 3×10^{-5} for the copper cathode. Schottky enhancement of photo-emission has been verified, and laser to RF phase stability of 2 ps has been measured. An rms normalized emittance of 3 π mm-mrad has been measured after 35 cm of beam transport from the gun for a 1 MeV, 50 pC beam. A 2.4 cell gun is currently under construction, and will be able to produce a 2 MeV beam with peak accelerating fields of 200 MV/m. With this gun, a record high brightness beam, corresponding to a normalized emittance of 0.5 πmm-mrad for a 100 pC, 1 ps electron beam should be obtainable.

REFERENCES

1. Carr R., Design of an X-ray free electron laser undulator AIP. American Institute of Physics Conference Proceedings **417** (1997) pp.29-34.
2. Rosenzweig J, Colby E., Charge and Wavelength Scaling of RF Photoinjector Designs, *Advanced Accelerator Concepts,* AIP Conf. Proc. **335** (1995) pp. 724-737.
3. Brown W. J., Trotz S., Kreischer K. E., Pedrozzi M., Shapiro M. A., Temkin R. J., Experimental and theoretical investigations of a 17 GHz RF gun, *Nucl. Instr. and Meth. A* **425** (1999) pp. 441-459.
4. S. Hanna, et. al. "Development of Characterization Techniques for X-Band Accelerator Structures", Proceedings of the 1997 Particle Accelerator Conference, vol.1, pp.539-41 (1998).
5. W. Bruns, "GdfidL: a finite difference program with reduced memory and CPU usage. Proceedings of the 1997 Particle Accelerator Conference (Cat. No.97CH36167). IEEE. Part vol.2, pp.2651-3 (1998).
6. Kim K., RF and Space-Charge Effects in Laser-Driven RF Electron Guns, *Nucl. Instr. and Meth. A* **275** (1989) pp. 201-218.
7. Palmer D.T., Wang X.J., Ben-Zvi I., Miller R.H., Beam Dynamics Enhancement due to Accelerating Field Symmetrization in the BNL/SLAC/UCLA 1.6 cell S-Band Photocathode RF Gun, *Proceedings of the 1997 Particle Accelerator Conference*, **3** (1998) pp.2846-8.
8. Haimson J., Mecklenburg B., Stowell G., "A Field Symmetrized Dual Feed 2 MeV RF Gun for a 17 GHz Electron Linear Accelerator", *Advanced Accelerator Concepts,* AIP Conf. Proc. **472** (1998) pp.653-667.
9. Haimson J. and Mecklenburg B., Initial Performance of a High Gain, High Efficiency 17 GHz Traveling Wave Relativistic Klystron for High Gradient Accelerator Research, *Pulsed RF Sources For Linear Collider*, AIP Conf. Proc. **337** (1995) pp.146-159.
10. Cardona M. and Ley L., *Photoemission in Solids*, Vol. 1, Springer, New York (1978) pp. 21-23.

Three-Dimensional Theory of Emittance in Compton Scattering

F.V. Hartemann,[a,b] A. Le Foll,[a,c] A.K. Kerman,[d] B. Rupp,[e] D.J. Gibson,[a,b] E.C. Landahl,[a,b] A.L. Troha,[a,b] N.C. Luhmann, Jr.,[b] and H.A. Baldis [a,b]

[a] *Institute for Laser Science & Applications, Lawrence Livermore National Lab., Livermore, CA 94550*
[b] *Department of Applied Science, University of California, Davis CA 95616*
[c] *Permanent Address: Ecole Polytechnique, 91128 Palaiseau, France*
[d] *Center for Theoretical Physics & Physics Department, Massachusetts Institute of Technology, Cambridge MA 02139*
[e] *Biology Division, Lawrence Livermore National Laboratory, Livermore CA 94550*

Abstract. A three-dimensional theory of Compton scattering is described, which accounts for the effects of the electron beam emittance and energy spread on the scattered x-ray brightness. The radiation scattered by an electron subjected to an arbitrary electromagnetic field in vacuum is derived in the linear regime, neglecting radiative corrections; it is found that each vacuum eigenmode gives rise to a single Doppler-shifted classical dipole excitation. This formalism is then applied to Compton scattering in a three-dimensional laser focus, and yields a complete description of the influence of the electron beam phase-space topology on the x-ray brightness; analytical expressions including the effects of emittance and energy spread are also obtained in the one-dimensional limit. With these results, the x-ray brightness generated by a 25 MeV electron beam is modeled, taking into account the beam emittance, energy spread, and the three-dimensional nature of the laser focus; its application to x-ray protein crystallography is outlined.

INTRODUCTION

Advances in ultrashort-pulse laser technology based on chirped-pulse amplification [1], and the development of high-brightness, relativistic electron sources [2] allow the design of compact, monochromatic, tunable, femtosecond x-ray sources using Compton scattering [3,4]. Such light sources will have a major impact on several fields of research, including the study of fast structural dynamics, advanced biomedical imaging, and protein crystallography; however, the quality of both the electron and laser beams is of paramount importance [5,6] in achieving the peak and average x-ray spectral brightness required for such applications. Here, we summarize a theoretical formalism capable of fully describing the three-dimensional (3D) nature of the interaction, and the influence of the electron and laser beam phase-space topologies upon the x-ray spectral brightness. Due to space limitations, most of the mathematical derivations have been omitted in this presentation; the details can be found in a forthcoming series of publications. We have striven, however, to emphasize salient facts; in particular, we address questions concerning the applicability of the Compton source to protein crystallography [7]. Analytical expressions of the x-ray spectral brightness, including emittance and energy spread, are obtained in the one-dimensional (1D) limit.

The aforementioned technical breakthroughs provide an opportunity to develop a new class of advanced x-ray sources, with characteristics similar to those of third-generation light sources [8], in a more compact and inexpensive package. This, offers the possibility of spin-offs of ultrashort-pulse laser technology for use in molecular biology, which is a rapidly growing field: the systematic study of protein structure and function will dominate biophysics in the first-half of the 21st Century; also, a new paradigm for drug design and synthesis has now emerged, using recombinant DNA technology and x-ray protein crystallography to produce new classes of drugs [9].

In protein crystallography, the 3D electron density of a molecule is obtained by mapping the diffraction-pattern intensity at different x-ray wavelengths, determining the phase by the Multi-wavelength Anomalous Diffraction (MAD) method [10], and Fourier-transforming back to the original molecular structure. The key characteristics of a useful x-ray source for protein crystallography are its small size, low divergence, reasonable transverse coherence, and high average spectral brightness; these requirements strongly impact the electron and laser beam quality, as discussed here.

RADIATION THEOREM

We first outline the following theorem (referred to henceforth as "Theorem I"): in the linear regime, where the 4-potential amplitude satisfies $eA/m_0 c \ll 1$, without radiative corrections [11-13], with the cutoff frequency $\omega \ll m_0 c^2 / \hbar$ in the electron frame, the spectral photon number density scattered by an electron interacting with an *arbitrary* electromagnetic field *in vacuo* is given by the momentum-space (k-space) distribution of the incident vector potential at the Doppler-shifted frequency:

$$\frac{d^2 N_x(k_\mu^s)}{d\omega_s d\Omega} = \frac{\alpha}{(2\pi)^4} \frac{1}{\gamma_0^2 \omega_s} \left| \mathbf{k}_s \times \int_{\mathbb{R}^3} \left[1 + \left(\frac{\mathbf{k}}{\kappa_s}\right) \mathbf{u}_0 \cdot \right] \tilde{\mathbf{A}} \left[\omega_s - \frac{\mathbf{u}_0}{\gamma_0} \cdot (\mathbf{k}_s - \mathbf{k}), \mathbf{k} \right] e^{i\mathbf{k}\cdot\mathbf{x}_0} d^3\mathbf{k} \right|^2, \quad (1)$$

where $k_\mu^s = (\omega_s, \mathbf{k}_s) = \omega_s(1, \hat{\mathbf{n}})$ is the 4-wavenumber of the wave scattered in the direction $\hat{\mathbf{n}}$, with frequency ω_s; α is the fine structure constant; $u_\mu^0 = (\gamma_0, \mathbf{u}_0)$ is the initial electron 4-velocity, $x_\mu^0 = (0, \mathbf{x}_0)$ is its initial 4-position, and the scattered light-cone variable is $\kappa_s = -u_0^\mu k_\mu^s = \gamma_0 \omega_s - \mathbf{u}_0 \cdot \mathbf{k}_s$. The term $[1 + (\mathbf{k}/\kappa_s)\mathbf{u}_0 \cdot]$ is treated as an operator acting on the Fourier transform of the spatial components of the 4-potential $A_\mu = (V, \mathbf{A})$,

$$\tilde{A}_\mu(k_\nu) = \frac{1}{\sqrt{2\pi}^4} \int_{\mathbb{R}^4} A_\mu(x^\nu) \exp(ik_\nu x^\nu) d^4 k_\nu. \quad (2)$$

The term $\exp(i\mathbf{k}\cdot\mathbf{x}_0)$ gives rise to the coherence factor [14,15]. This theorem will be applied to the specific case of Compton scattering [16-18] in a 3D Gaussian-elliptical focus [19,20]. The effects of the electron beam phase-space topology are included, in terms of energy spread and emittance [21]. Here, charge is measured in units of e, mass in units of m_0, time in units of the frequency, $\omega_0^{-1} = (ck_0)^{-1}$, and length is normalized to the corresponding reference wavelength, k_0^{-1}. Neglecting radiative corrections [11-13], the electron motion arises from the Lorentz force equation $d_\tau u_\mu = -(\partial_\mu A_\nu - \partial_\nu A_\mu)u^\nu$, where $u_\mu = d_\tau x_\mu$ is the electron 4-velocity, τ is the electron proper time, A_μ is the 4-potential of the electromagnetic field, and $\partial_\mu = (-\partial_t, \nabla)$ is the 4-gradient operator. The electromagnetic field considered here corresponds to a vacuum interaction; therefore, the 4-potential satisfies the wave equation, and can be expressed as a superposition of plane waves, as described in Eq. (2). Also, we choose to work in the Lorentz gauge, where $\partial_\mu A^\mu = 0$.

Covariant Linearization

Introducing the maximum amplitude of the 4-potential, A, we linearize the Lorentz force equation, stipulating that $A \ll 1$, which is satisfied in most experimental situations: e.g., for an ultrahigh-intensity laser focus, this corresponds to a maximum intensity $< 10^{17}$ W/cm^2 for visible wavelengths. Writing $u_\mu = u_\mu^0 + u_\mu^1 + u_\mu^2...$, where $u_\mu^n \propto A^n$, the Lorentz force equation yields, to first order, $d_\tau u_\mu^1 = -(\partial_\mu A_\nu - \partial_\nu A_\mu)u_0^\nu$. To solve this equation, Fourier transform the first-order perturbation of u_μ into k-space:

$$u_\mu^1(x^\nu) = \frac{1}{\sqrt{2\pi}^4}\int_{\mathbb{R}^4} d^4k_\nu\, \tilde{u}_\mu^1(k_\nu)\exp(ik_\nu x^\nu); \tag{3}$$

because of the orthogonality of complex exponentials, we have

$$\tilde{u}_\mu^1 ik_\nu\, d_\tau x^\nu = -u_0^\nu(ik_\mu \tilde{A}_\nu - ik_\nu \tilde{A}_\mu), \tag{4}$$

and $d_\tau x^\nu = u^\nu$ is approximated by u_0^ν in the linear theory presented here; one then finds that $\tilde{u}_\mu^1 \simeq \tilde{A}_\mu - k_\mu(\tilde{A}_\nu u_0^\nu / k_\nu u_0^\nu)$. Fourier transforming this result back into space-time yields

$$u_\mu(x^\nu) \simeq u_\mu^0 + \frac{1}{\sqrt{2\pi}^4}\int_{\mathbb{R}^4} d^4k_\nu \left[\tilde{A}_\mu(k_\nu) - k_\mu(\tilde{A}_\nu(k_\nu)u_0^\nu / k_\nu u_0^\nu)\right]\exp(ik_\nu x^\nu). \tag{5}$$

The linearization procedure used here is manifestly covariant.

Scattered Radiation

The electromagnetic radiation scattered by the accelerated charge is described by the number of photons radiated per unit frequency, per unit solid angle, which is determined by Fourier transforming the electron trajectory into k-space [22]:

$$\frac{d^2 N_x(k_\mu^s)}{d\omega_s d\Omega} = \frac{\alpha \omega_s}{4\pi^2} \left| \hat{\mathbf{n}} \times \int_{-\infty}^{+\infty} \mathbf{u}(\tau) \exp\left[-ik_\mu^s x^\mu(\tau)\right] d\tau \right|^2. \quad (6)$$

Here, the fact that $\hat{\mathbf{n}}$ is a unit vector is used to simplify the double cross product. The spatial component of the 4-velocity is replaced by the linearized solution, Eq. (5); with this, and using the Coulomb gauge, Eq. (6) is recast as

$$\frac{d^2 N_x(k_\mu^s)}{d\omega_s d\Omega} = \frac{\alpha}{(4\pi^2)^3} \frac{1}{\omega_s} \left| \mathbf{k}_s \times \int_{-\infty}^{+\infty} d\tau \int_{\mathbb{R}^4} d^4 k_\mu \left[\tilde{\mathbf{A}}(k_\mu) + \frac{\mathbf{k}}{\kappa} \mathbf{u}_0 \cdot \tilde{\mathbf{A}}(k_\mu) \right] e^{i(k_\mu - k_\mu^s) x^\mu(\tau)} \right|^2. \quad (7)$$

The electron 4-position is approximated by $x_\mu(\tau) \simeq x_\mu^0 + u_\mu^0 \tau$, which corresponds to the lowest-order convective term due to the ballistic component of the electron motion, and excludes harmonic production mechanisms [13,17,23,24]; this is valid for high Doppler-shift scattering, where the transverse-oscillation scale is given by $A\lambda_0/\gamma_0 \ll \lambda_0$. Using the light-cone variables, κ and κ_s, Eq. (7) becomes

$$\frac{d^2 N_x(k_\mu^s)}{d\omega_s d\Omega} = \frac{\alpha}{(4\pi^2)^3} \frac{1}{\omega_s} \left| \mathbf{k}_s \times \int_{\mathbb{R}^4} d^4 k_\mu \left[\tilde{\mathbf{A}} + \frac{\mathbf{k}}{\kappa} \mathbf{u}_0 \cdot \tilde{\mathbf{A}} \right] e^{i(k_\mu - k_\mu^s) x_0^\mu} \int_{-\infty}^{+\infty} d\tau \, e^{i(\kappa - \kappa_s)\tau} \right|^2. \quad (8)$$

The integral over proper time yields a δ-function: $\int_{-\infty}^{+\infty} d\tau \, e^{i(\kappa - \kappa_s)\tau} = 2\pi\delta(\kappa - \kappa_s)$. Using the change of variables $d^4 k_\mu = \left|\partial q_\nu / \partial k_\mu\right|^{-1} d^4 q_\nu$, where we introduce the 4-vector $q_\nu = [\kappa(k_\mu), \mathbf{k}] = (u_0^\mu k_\mu, \mathbf{k}) = (\gamma_0 \omega - \mathbf{u}_0 \cdot \mathbf{k}, \mathbf{k})$ in order to perform the integral over the δ-function, we find that the Doppler condition comes from the equality of $\kappa_s = -u_0^\mu k_\mu^s = \gamma_0 \omega_s - \mathbf{u}_0 \cdot \mathbf{k}_s$ and $\kappa = -u_0^\mu k_\mu = \gamma_0 \omega - \mathbf{u}_0 \cdot \mathbf{k}$: this yields the Compton scattering relation, $\omega_s = (\gamma_0 \omega - \mathbf{u}_0 \cdot \mathbf{k})/(\gamma_0 - \mathbf{u}_0 \cdot \hat{\mathbf{n}})$, in the limit where recoil is neglected. Finally, using the Jacobian of the transform, $\left|\partial q_\nu / \partial k_\mu\right|^{-1} = \gamma_0^{-1}$, we obtain the sought-after result, as expressed in Theorem I.

One-Dimensional Theory of Emittance in Compton Scattering

We consider the case of a linearly polarized plane wave with an arbitrary temporal profile: the 4-potential is $A_\mu(\phi) = \hat{\mathbf{x}} A_0 g(\phi) e^{-i\phi}$, where $\phi = -k_\mu^0 x^\mu$, and $k_\mu^0 = (1,0,0,1)$, for a wave propagating along the z-axis. Using the temporal Fourier transform of the pulse envelope, $\tilde{g}(\omega) = \int_{-\infty}^{+\infty} g(t) e^{-i\omega t} dt / \sqrt{2\pi}$, we have

$$\tilde{A}_\mu(k_\nu) = \hat{x}\sqrt{2\pi}^3 A_0 \delta(k_x)\delta(k_y)\delta(\omega - k_z)\tilde{g}(1-\omega), \qquad (9)$$

where $\delta(\omega - k_z)$ corresponds to the pulse propagation, and $\tilde{g}(1-\omega)$ is the spectrum of the pulse, centered around the normalized frequency $\omega_0 = 1$. Applying Theorem I, we find

$$\frac{d^2 N_x}{d\omega_s d\Omega} = \frac{\alpha}{2\pi} \omega_s \frac{A_0^2}{\kappa_0^2} \left| \hat{n} \times [\hat{x} + (u_{0x}/\kappa_0)\hat{z}] \right|^2 \tilde{g}^2 (1-[\kappa_0^s/\kappa_0]), \qquad (10)$$

where $\kappa_0 = \gamma_0 - \mathbf{u}_0 \cdot \hat{z} = \gamma_0 - u_{0z}$ and $\kappa_0^s = \omega_s(\gamma_0 - \mathbf{u}_0 \cdot \hat{n})$. Using the normalized Doppler-shifted frequency $\chi = \kappa_0^s / \kappa_0 = \omega_s(\gamma_0 - \mathbf{u}_0 \cdot \hat{n})/(\gamma_0 - u_{0z})$, and the differential scattering cross-section, $f = \left|[\hat{n} \times (\kappa_0 \hat{x}_0 + u_{0x}\hat{z})]/\kappa_0^2\right|^2$, this result can be recast as

$$\frac{d^2 N_x}{d\omega_s d\Omega} = \frac{\alpha}{2\pi} A_0^2 \omega_s f \tilde{g}^2(1-\chi). \qquad (11)$$

For a Gaussian pulse envelope and the interaction geometry shown in Fig. 1, Eq. (11) takes the form

$$\frac{4\pi}{\alpha A_0^2 \Delta\phi^2} S_0(\omega_s, \gamma, \theta, \varphi) = \omega_s e^{-\frac{\Delta\phi^2}{2}[\chi(\omega_s, \gamma_0, \theta, \varphi)-1]^2} \frac{[\gamma_0 \cos(\theta+\varphi) - u_0 \cos\theta]^2}{[\gamma_0 - u_0 \cos\varphi]^4}, \qquad (12)$$

where φ is the incidence angle between the initial electron velocity and the direction of propagation of the plane wave, and θ is the scattering angle, measured with respect to the initial electron velocity. Eq. (12) clearly shows that the scattering spectral density is proportional to the incident photon number density, as given by the laser intensity $A_0^2 \Delta\phi$, and that the cold spectral bandwidth of the x-rays is given by that of the incident laser pulse, $\Delta\phi^{-1}$. Eq. (12) also indicates that the peak intensity is radiated near the Doppler-shifted frequency, where $\chi(\omega_x, \gamma_0, \theta, \varphi) = 1$;

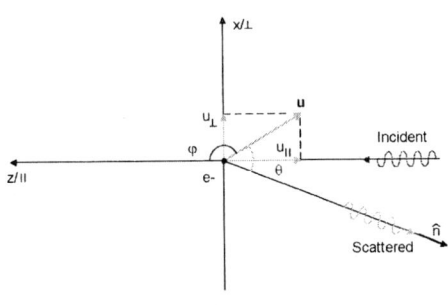

Fig. 1 Compton scattering schematic.

this yields $\hbar\omega_x(\gamma_0, \theta, \varphi) = \hbar\omega_0(\gamma_0 - u_0 \cos\varphi)/(\gamma_0 - u_0 \cos\theta)$. For a head-on collision

($\varphi = \pi$), the frequency radiated on-axis ($\theta = 0$) is the same as the free-electron laser (FEL) frequency for an electromagnetic wiggler: for ultra-relativistic (UR) electrons, we recover the well-known relation, $\omega_x = \gamma^2 (1+\beta)^2 \simeq 4\gamma^2$.

Energy Spread

We now model the influence of the electron beam phase-space topology for a linearly polarized plane wave. The "cold", 1D spectral brightness is given by Eq. (12). Since S_0 is a function of the electron initial energy, γ, scattering angle, θ, and incident angle, φ, we can perform incoherent summations over the electron initial energy and momentum distributions to study the effects of energy spread and emittance. For conciseness, the scattered frequency is now labeled ω, and the initial electron 4-velocity is labeled as $u_\mu^0 = (\gamma, \mathbf{u})$. For the beam energy spread, the "warm" beam brightness is given by

$$S_\gamma(\omega,\gamma_0,\Delta\gamma,\theta,\varphi) = \frac{1}{\sqrt{\pi}\Delta\gamma} \int_1^\infty S_0(\omega,\gamma,\theta,\varphi) \exp\left[-\left(\frac{\gamma-\gamma_0}{\Delta\gamma}\right)^2\right] d\gamma,$$

$$\simeq \frac{\omega\, f(\gamma_0,\theta,\varphi) \exp\left(\frac{u^2}{v} - w\right)}{\sqrt{1+\frac{1}{2}\left(\Delta\phi\frac{\Delta\gamma}{\gamma_0}\right)^2 \left[\frac{\omega\cos\varphi - \cos\theta}{\gamma_0^2(1-\cos\varphi)^2}\right]^2}}, \quad (13)$$

using a Gaussian distribution to model the beam longitudinal phase-space. The analytical expression is obtained by Taylor-expanding to second-order around the central electron energy, γ_0. The normalization constant is $\int_1^\infty e^{-[(\gamma-\gamma_0)/\Delta\gamma]^2} d\gamma \simeq \sqrt{\pi}\Delta\gamma$, valid for $\gamma_0 \gg 1$ and $\Delta\gamma/\gamma_0 \ll 1$. Here, $\Delta\gamma$ is the energy spread, $a = \frac{\omega}{\gamma_0^3} \frac{(\cos\varphi - \cos\theta)}{(1-\cos\varphi)^2}$, $b = \chi(\omega,\gamma_0,\theta,\varphi) - 1$, $u = \frac{1}{\Delta\gamma^2}\left[1 + \frac{a}{2}(\Delta\phi\Delta\gamma)^2\right]$, $v = \frac{\Delta\phi^2}{2}ab$,

and $w = \frac{\Delta\phi^2}{2}b^2$. Since v and w are both linear functions of b, which equals zero at the peak of the x-ray spectrum, the exponential equals one at $\omega = \omega_x$. Further, the factor $[\Delta\phi(\Delta\gamma/\gamma_0)]^2$ in the square root shows that the relative energy spread must be compared to the normalized laser pulse duration, which is equivalent to the number of electromagnetic wiggler periods; thus, if one increases the x-ray spectral brightness by lengthening the drive laser pulse, the condition on the electron beam energy spread becomes increasingly stringent. Fig. 2 illustrates the effects of energy spread, which symmetrically broaden the scattered x-ray spectrum and lower the peak intensity.

Emittance

We now look at the influence of the electron beam emittance:

$$S_\varepsilon(\omega,\gamma_0,\Delta\gamma,\theta,\varphi_0,\Delta\varphi) \simeq \frac{1}{\sqrt{\pi}\Delta\varphi} \int_0^{2\pi} S_\gamma(\omega,\gamma_0,\Delta\gamma,\theta-\delta,\varphi_0+\delta) e^{-(\delta/\Delta\varphi)^2} d\delta, \quad (14)$$

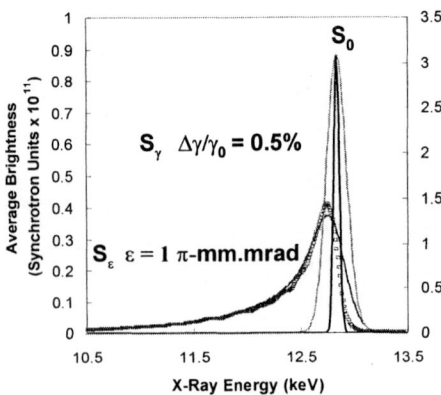

Fig. 2 X-ray spectral brightness.

where the spread of incidence angle is given in terms of the beam emittance, ε, and radius, r_b, by $\Delta\varphi = \varepsilon/\gamma_0 r_b$, and φ_0 is the mean incidence angle, defined by the laser and electron beams. Again, the normalization constant is given by $\int_0^{2\pi} e^{-(\delta/\Delta\varphi)^2} d\delta \simeq \sqrt{\pi}\Delta\varphi$, provided that $\Delta\varphi \ll 1$. In Eq. (14), we note the geometrical correction term, $\theta - \delta$, which comes from the fact that the scattering angle is measured with respect to the initial electron velocity. The effects of emittance are illustrated in Fig. 2, and are found to be independent of φ_0. Examining the on-axis x-ray spectral line, it is clear that emittance asymmetrically broadens the spectrum and decreases the peak spectral brightness; for near-head-on collisions, a low-energy tail develops because the maximum Doppler-shift equates to $\delta \simeq 0$: other electrons produce a smaller upshift, thus contributing to the lower energy photon population seen in Fig. 2.

Returning to the cold, 1D spectral brightness, the integral over a Gaussian distribution of incidence angle can be performed analytically, provided that the spectral density is approximated by the exponential of a bi-quadratic polynomial [25]:

$$\int_0^\infty e^{-\mu x^4 - 2vx^2} dx = \frac{1}{4}\sqrt{2v/\mu} \exp(v^2/2\mu) \overline{K}_{\frac{1}{4}}(v^2/2\mu), \quad (15)$$

where $\overline{K}_{\frac{1}{4}}$ is defined in terms of modified Bessel functions of fractional order:

$$\overline{K}_{\frac{1}{4}}(v^2/2\mu) = I_{-\frac{1}{4}}(v^2/2\mu) - \frac{v}{|v|} I_{\frac{1}{4}}(v^2/2\mu). \quad (16)$$

Since $\omega f(\gamma,\theta,\varphi)$ is a slow-varying function of the incidence angle, we seek an approximate expression for the cold spectral density of the form:

$$S_0(\omega,\gamma,\theta-\delta,\varphi+\delta) \simeq \omega f(\gamma,\theta,\varphi) e^{-\mu(\omega,\gamma,\theta,\varphi)\delta^4 - 2v(\omega,\gamma,\theta,\varphi)\delta^2 + \lambda(\omega,\gamma,\theta,\varphi)}. \quad (17)$$

The constant term is derived using $\delta = 0$: $\lambda(\omega,\gamma,\theta,\varphi) = -\Delta\phi^2[\chi(\omega,\gamma,\theta,\varphi)-1]^2/2$; the other coefficients are derived using $\cos\delta \simeq 1-(\delta^2/2!)+(\delta^4/4!)$, and $\sin\delta \simeq \delta-(\delta^3/3!)$. We then find that $\mu = [\Delta\phi^2/2](\mu_1+\mu_2)/(\gamma-u\cos\varphi)^4$, and $2\nu = [\Delta\phi^2/2](\nu_1-\nu_2)/(\gamma-u\cos\varphi)^4$, with

$$\mu_1 = (\gamma-u\cos\varphi)^2 \left\{ \begin{array}{l} \dfrac{u}{12}(\cos\varphi-\cos\theta)[\gamma(\omega-1)+u(\cos\varphi-\omega\cos\theta)] \\ -\dfrac{u^2}{3}(\omega\sin\theta+\sin\varphi)^2 + \dfrac{u^2}{4}(\cos\varphi-\omega\cos\theta)^2 \end{array} \right\},$$

$$\mu_2 = [\omega(\gamma-u\cos\theta)-\gamma+u\cos\varphi]^2 \left[\dfrac{u}{12}\cos\varphi(\gamma-u\cos\varphi) + \dfrac{u^2}{3}\sin^2\varphi - \dfrac{u^2}{4}\cos^2\varphi \right], \quad (18)$$

$$\nu_1 = (\gamma-u\cos\varphi)^2 \left\{ \begin{array}{l} u^2(\omega\sin\theta+\sin\varphi)^2 \\ -u(\cos\varphi-\omega\cos\theta)[\gamma(\omega-1)+u(\cos\varphi-\omega\cos\theta)] \end{array} \right\},$$

$$\nu_2 = [u\cos\varphi(\gamma-u\cos\varphi)+u^2\sin^2\varphi][\omega(\gamma-u\cos\theta)-(\gamma-u\cos\varphi)]^2.$$

This result is compared to a full 3D numerical simulation (squares) on Fig. 2; the agreement is quite good. To include the effects of energy spread and emittance, the analytical results given in Eqs. (15) and (18) are multiplied by the energy spread degradation factor, measured at the peak of the cold spectrum:

$$S_\varepsilon \simeq \dfrac{\omega_x f(\gamma_0,\theta,\varphi_0)}{\sqrt{1+\dfrac{1}{2}\left(\Delta\phi\dfrac{\Delta\gamma}{\gamma_0}\right)^2\left[\dfrac{\omega_x\cos\varphi_0-\cos\theta}{\gamma_0^2(1-\cos\varphi_0)^2}\right]^2}} \dfrac{\omega f(\gamma_0,\theta,\varphi_0)}{2\sqrt{\pi}\Delta\varphi}\sqrt{\dfrac{2\nu}{\mu}}e^{(\nu^2/2\mu)+\lambda}\overline{K}_{\frac{1}{4}}\left(\dfrac{\nu^2}{2\mu}\right). \quad (19)$$

Now, the combined effects of energy spread and emittance are studied by varying the bunch charge and modeling the behavior of the electron beam phase space as

$$\dfrac{\Delta\gamma(q)}{\gamma_0} \simeq \sqrt{\left[\dfrac{\gamma_0}{2}(\omega_{rf}\Delta\tau)^2\right]^2 + \left(\dfrac{e}{m_0 c^2}\dfrac{q}{2\pi\varepsilon_0 c\Delta\tau}\right)^2}, \quad (20)$$

where the first term is the spread due to the finite duration of the bunch in the rf accelerating bucket of frequency $\omega_{rf}/2\pi$, and the second term corresponds to space-charge; for the emittance, an empirical linear scaling with charge is chosen, with $\varepsilon(q) \simeq \sigma q$, and $\sigma = 1\pi$-mm·mrad/nC. This results in the brightness curve shown in Fig. 3, where the brightness first scales linearly with the charge, reaches a maximum near 0.5 nC, and starts degrading thereafter under the combined influences of energy spread and emittance. This optimum value of the charge is quite interesting as it nearly corresponds to the state-of-the-art for high-brightness photoinjectors.

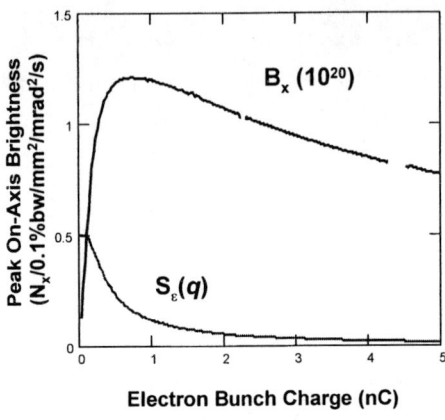

Fig. 3 Peak, on-axis brightness vs. charge.　　　Fig. 4 Brightness degradation with emittance.

Three-Dimensional Theory of Emittance in Compton Scattering

Theorem I is now applied to the case of a 3D laser focus. The transverse laser profile is specified at the focal plane, and propagated using the method discussed in Ref. 9, where the vector potential is derived from a generating function: $\mathbf{A} = \nabla \times \mathbf{G}$; in this manner, the Coulomb gauge condition, $\nabla \cdot \mathbf{A} = 0$, is automatically satisfied. For a linearly polarized Gaussian-elliptical focus, with focal waists w_{0x} and w_{0y}, and a monochromatic wave at the central frequency $\omega_0 = 1$, with a Gaussian envelope of duration $\omega_0 \Delta t = \Delta \phi$, the 4-potential is represented in k-space by

$$\tilde{A}_\mu(k_\nu) = \frac{\sqrt{\pi}}{2} A_0 w_{0x} w_{0y} \Delta t \exp\left\{-\left[\frac{w_{0x}k_x}{2}\right]^2 - \left[\frac{w_{0y}k_y}{2}\right]^2 - \left[\frac{\Delta t(\omega-1)}{2}\right]^2\right\} \qquad (21)$$
$$\times \delta\left(k_z - \sqrt{\omega^2 - k_x^2 - k_y^2}\right)[i(-\hat{\mathbf{x}}k_z + \hat{\mathbf{z}}k_x)],$$

where we recognize the $\mathbf{k}_\perp$-spectrum, the frequency spectrum, the propagator, $\delta(k_\mu k^\mu)$, and the curl operator, as expressed in momentum space.

Now the scattered radiation is determined by using Theorem I; to obtain an analytical result that can be exploited to include the phase-space topology of the electron beam interacting with the laser pulse using the method outlined above, the paraxial propagator formalism is used: the phase function $\delta\left(k_z - \sqrt{\omega^2 - k_x^2 - k_y^2}\right)$, is replaced by $\delta[k_z - \omega + (k_x^2/2k_0) + (k_y^2/2k_0)]$. The accuracy of the paraxial approximation has been studied in detail [19,20], and found to be valid over a wide range of parameters; however, in Compton scattering, this condition must be satisfied: $\kappa_0^2 w_{0x,y}^2 > \Delta \phi^2 \left[u_z(\kappa_s - \kappa_0) - u_{x,y}^2\right]$. This corresponds to the transition from a spectrum dominated by the spatial or temporal Gaussian profiles of the laser pulse, to a regime governed by diffraction, in the so-called "hourglass" axial profile of the laser focus. With this proviso, the integrals over the transverse wavenumber components converge, and can be performed analytically [25]:

$$\int_{-\infty}^{+\infty} e^{-(ax^2+2bx+c)} e^{i(px^2+2qx+r)} dx = \frac{\sqrt{\pi} \, e^{\frac{a(b^2-ac)-(aq^2-2bq+cp^2)}{a^2+p^2}} e^{i\left[\frac{1}{2}\arctan\left(\frac{p}{a}\right) - \frac{p(q^2-pr)-(b^2p-2abq+a^2r)}{a^2+p^2}\right]}}{\sqrt[4]{a^2+p^2}}, \quad (22)$$

and the fully 3D x-ray spectral brightness is now obtained as a function of the initial electron position and velocity, $\mathbf{x}_0$ and $\mathbf{u}_0$, as well as the laser parameters, and the scattered 4-wavenumber, $k_\mu^s = (\omega_s, \omega_s \hat{\mathbf{n}})$. Frequencies are normalized to the laser pulse central wavelength, $k_0 = \omega_0 = 1$, and axial positions are measured from the laser focal plane, lying at $z = 0$. The complete result is extremely complex, and was tracked analytically using Mathematica [26].

The cold (single electron) 3D brightness can be written as

$$\frac{\alpha A_0^2 \Delta\phi^2 w_{0x}^2 w_{0y}^2}{(4\pi)^3} S_0(k_\mu^s, x_\mu^0, u_\mu^0) = \frac{d^2 N_x}{d\omega d\Omega} = \frac{\alpha}{(4\pi)^3} \frac{\omega}{\kappa^2} A_0^2 w_{0x}^2 w_{0y}^2 \Delta t^2$$

$$\times \left| \hat{\mathbf{n}} \times \iint_{\mathbb{R}^2} (\alpha k_x^2 + \beta k_y^2 + \gamma k_x k_y + \delta k_x + \varepsilon k_y + \zeta) e^{-(ak_x^2 + bk_y^2 + ck_x k_y + dk_x + ek_y + f)} e^{i(pk_x^2 + qk_y^2 + rk_x + sk_y + t)} dk_x dk_y \right|^2.$$

(23)

These terms are functions of the laser parameters, the scattering 4-wavenumber, k_μ^s, and the initial electron position, x_μ^0, and velocity, u_μ^0; therefore, the 6-dimensional electron phase-space now appears explicitly in the 3D spectral density of the scattered x-rays. Finally, the incident and scattered light-cone variables are $\kappa = \gamma - u_z$ and $\kappa_s = \omega(\gamma - \hat{\mathbf{n}} \times \mathbf{u})$.

The advantage of a fully analytical treatment of the problem, afforded by the use of the paraxial propagator formalism, resides in the much shorter computing time required to map the radiation produced by each electron in the beam. A much simplified form of the general result is obtained by considering a centered electron, where $\mathbf{x}_0 = \mathbf{0}$: in this case,

$$\frac{d^2 N(k_\mu^s)}{d\omega_s d\Omega} = \frac{\alpha \omega_s A_0^2}{4\pi \kappa_0^2} \eta_x \eta_y \eta_t \left| \hat{\mathbf{n}} \times \mathbf{v}(u_\nu^0, x_\nu^0, k_\mu^s) \right|^2 \exp\left\{ -\frac{\Delta\phi^2(\chi-1)^2 + F(u_\nu^0, x_\nu^0, k_\mu^s)}{2\left[1 + \frac{\eta_x \Delta\phi^2 u_x^2}{\overline{w}_x^2} + \frac{\eta_y \Delta\phi^2 u_y^2}{\overline{w}_y^2}\right]} \right\}. \quad (24)$$

Here, we have defined $\overline{w}_{x,y} = \kappa_0 w_{0x,y}$, and the normalization constants

$$\eta_{x,y}(k_\mu^s) = \frac{\overline{w}_{x,y}^2}{\overline{w}_{x,y}^2 + \Delta\phi^2 u_z(\kappa_0 - \kappa_s)} \text{ and } \eta_t(k_\mu^s) = \left[\Delta\phi^{-2} + \eta_x\left(\frac{u_x}{\overline{w}_x}\right)^2 + \eta_y\left(\frac{u_y}{\overline{w}_y}\right)^2\right]^{-1} ; \quad (25)$$

we have also introduced the normalized Doppler-shifted frequency $\chi(u_\nu^0, k_\mu^s) = \kappa_s/\kappa_0$. Eq. (25) indicates that the minimum x-ray spectral width is given by the laser pulse duration; it also shows that 3D effects change the x-ray spectrum because of the convective terms due to the electron crossing the laser focus. The vector $\mathbf{v}(u_\nu^0, x_\nu^0, k_\mu^s)$, and the function $F(u_\nu^0, x_\nu^0, k_\mu^s)$ both depend on the initial position of the electron; when $x_\nu^0 = 0$, $F = 0$. The physics underlying these results can be easily interpreted: first, the 1D kernel $\frac{\alpha\omega_s A_0^2 \Delta\phi^2}{4\pi\kappa_0^2} e^{-\frac{\Delta\phi^2}{2}[\chi(\omega,\gamma,\theta,\varphi)-1]^2}$ always appears, indicating that the scattered radiation spectrum peaks near the Doppler-shifted frequency $\chi = 1$; furthermore, the scattering is proportional to the laser pulse energy $A_0^2\Delta\phi$, and the minimum spectral width is given by that of the laser, $\Delta\phi^{-1}$; second, the interaction geometry modifies the spectral width; in the case of a transverse interaction, wider bandwidths can be obtained for narrow focal spots, corresponding to the aforementioned shorter effective interaction time due to the convective motion of the electron through the focus; in addition, the spread of incidence angle $\Delta\varphi \simeq \pi\lambda_0/w_0$, corresponding to smaller spot sizes plays an important role when $\varphi_0 = \pi/2$, because of the $\cos\varphi_0$ behavior of the differential scattering cross-section; by contrast, when $\varphi_0 = \pi$, these corrections are quadratic and nearly negligible; third, in the case where the initial velocity is in the direction of polarization, the radiation observed along that direction results from the axial component of the 4-potential, which is a purely 3D effect; finally, we note that for larger values of the focal spot size, around $w_0 = 100$ μm, the spectra converge to the minimum spectral width of the laser regardless of the interaction geometry.

Three-Dimensional Compton Code

In order to fully exploit the results derived previously, we have developed a 3D code describing the radiation scattered by a distribution of N_e point charges having the same charge-to-mass ratio as electrons. The $6N_e$-dimensional phase space can be generated by randomly loading the particles in prescribed statistical distributions, or can be the output of an electron beam optics code, such as PARMELA.

In running the code initially, it was determined that good statistical convergence is obtained for $N_e \geq 3\times10^4$; generally, we have used 50,000 particles in the results presented here. Two different types of output data files are created by the code: spectral brightness measured at a prescribed scattering angle, or angular maps at a specified x-ray frequency; for angular maps, the code also integrates the flux over the map, in a small (typically 1 eV) x-ray photon energy interval. This is important for x-ray protein crystallography and other applications.

We have first systematically studied the effects of energy spread, emittance, electron beam focal spot size, bunch duration, and timing jitter. For the sake of clarity, we have chosen cylindrical focii. To distinguish between the various effects, we have varied one of the parameters listed above during each run, while keeping all other parameters equal to zero, and the radiation frequency $\chi = 1$; furthermore, to help comparing the various effects, we have normalized the brightness. The result for emittance is given in Fig. 4. The various degradation mechanisms studied yield similar curves, roughly scaling as $1/\sqrt{1+\xi^2}$, where ξ represents the degradation parameter, properly scaled; for example, for energy spread, $\xi \simeq \frac{\Delta\phi}{\sqrt{2}} \frac{\Delta\gamma}{\gamma_0} \frac{\omega}{\gamma_0^2} \frac{\cos\varphi - \cos\theta}{(1-\cos\varphi)^2}$. From an experimental point of view, the most stringent requirements are on emittance and energy spread. Finally, the optimum electron beam size does not correspond to a match with the laser mode; rather, the beam should be focused as tightly as possible within the laser focal spot. This can easily be understood, if we consider the laser pulse as an electromagnetic wiggler, which has maximal field strength, or photon density, on-axis. For high-energy beams, where emittance and space-charge effects are very small, and can therefore be focused over a few tens of nm, this is a potentially important result. We also note that this conclusion does not hold if the laser beam depletion becomes important.

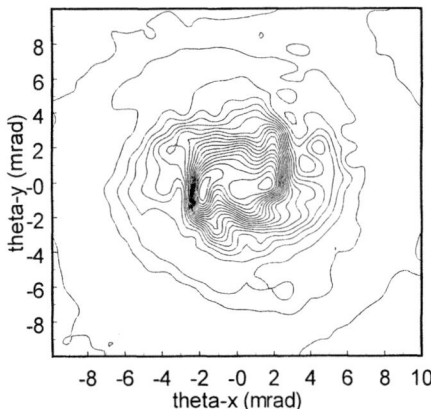

Fig. 5 Angular x-ray map, 1 eV spectral window.

The case of a realistic beam is presented in Figs. 2 and 5. The laser parameters are as follows: $\lambda_0 = 800$ nm, $w_0 = 10$ µm, $\Delta t = 200$ fs, and a pulse energy of 50 mJ. The electron bunch energy is 22.75 MeV, its charge is 0.5 nC, the relative energy spread is $\Delta\gamma/\gamma_0 = 0.5\%$, the beam normalized emittance is 1 π-mm·mrad, its duration is 1 ps, and the focal spot size matches the laser focal distribution. A repetition rate of 1 kHz is used to scale the average spectral brightness of the source; the maximum average brightness, $\langle B_x \rangle$, of 4.1×10^{10} photons/0.1%b.w./mm^2/mrad^2/s compares well with that produced by rotating anode tubes. The angle of incidence is 180°, and the spectra are observed on-axis; the angular maps are obtained at the spectral maximum of 12.77 keV. Note that quadrupole focusing effects are discernable on Fig. 5. Finally, the divergence is $\sqrt{\Delta\Omega} = \sqrt{\int_{4\pi}(d^2N_x/d\omega d\Omega)d\Omega/\widehat{B}_x} = 3.5$ mrad, where $\widehat{B}_x$ is the peak, on-axis brightness.

X-Ray Protein Crystallography

One important application identified and targeted for the 0.9 Å Compton x-ray source is protein crystallography, where recombinant DNA technology [27] is used to produce large quantities of a given protein by splicing the corresponding coding DNA sequence into the genetic material of a bacterium. Overexpression of large quantities of the protein is induced, and the protein is extracted, purified, and crystallized. The 3D electron density of the molecule is then reconstructed using the Fourier transform of the structure factor, which is proportional to the square root of the diffraction pattern intensity, and experimental phase information. The most accurate and prevalent method for obtaining experimental phases is the MAD phasing technique, where phase information is obtained from dispersive and anomalous differences between data sets collected at different wavelengths around the absorption edge of anomalously scatterering atoms introduced into in the protein [10].

This approach, first demonstrated at dedicated synchrotron beamlines, has proven extremely successful, for example, in designing *ab initio* inhibitor molecules targeting specific binding sites on proteins such as the HIV protease, thus providing a new paradigm for drug design and manufacturing [9]. As synchrotrons are large and expensive facilities, our goal is to develop compact, tunable, x-ray sources with nearly comparable characteristics at a fraction of the cost. Laser-driven Compton scattering allows one to use a much lower electron beam energy (tens of MeV's instead of GeV's), while retaining the tunability distinguishing synchrotron undulators from conventional x-ray sources.

We now compare the calculated radiation characteristics of the Compton source with the two main sources currently used for x-ray protein crystallography: rotating anode sources and synchrotrons. As one must match the x-ray beam divergence to the crystal mosaicity for high-resolution data collection, the 3.5 mrad half-angle of the x-ray cone produced in our simulations is a very encouraging result, as it matches the mosaicity of the best protein crystals, $> 0.2°$.

In terms of the spectral brightness requirement for the data acquisition time, a few preliminary comments are in order. First, the current average cost for the determination of the 3D structure of a protein is approximately $200k [28]; given the number of proteins coded for by human DNA, which is estimated to be of the order of 10^5, new technologies are clearly required to complete a genome-wide structural analysis of proteins (Structural Genomics) [28]. We also note that the average cost of a synchrotron beamline is $8M; smaller, less expensive sources could be extremely useful in terms of added flexibility, even at lower average spectral brightness. Second, the current amount of time required to isolate and amplify the target sequence, crystallize the corresponding protein, acquire the 3D x-ray diffraction pattern at different wavelengths to apply MAD, and post-process the data range between 1 and 6 months; therefore, an acquisition time of a few tens of hours is quite acceptable; in fact, a reasonable target would be to determine 50-100 structures/year on a compact device. Finally, extremely high flux, such as that produced by insertion devices on third-generation synchrotrons, frequently causes radiation damage in the biological material and loss of the crystal even in properly cryo-cooled specimens. Consequently, strong attenuation of the beam becomes necessary [29]. A first, simple comparison can be made by considering the average spectral brightness of different sources; the results are typically given in units of photons/0.1%b.w./mm^2/mrad^2/s. For modern rotating anode sources, given a 2.6 eV width for the K$_{\alpha 1}$ line of copper at 1.54051 Å, and up to 98% flux in that line, an average brightness around 10^9 is possible; synchrotron bends reach 10^{15} at 10 keV for a 6-8 GeV ring, while a 72-period, 3.3 cm-wavelength undulator produces 9.4 keV photons with a brightness of 4.8×10^{18} at the APS [8]; however, brightnesses $> 10^{12}$ cause radiation damage. The Compton source generates 12.77 keV photons with a brightness of 4.1×10^{10}; this compares quite favorably with x-ray tubes, especially in view of its tunability, which is indispensable for MAD.

A more detailed comparison between the Compton source and conventional rotating anode x-ray sources will now be given, where the protein diffraction acquisition time is evaluated in each case. We start from a known, state-of-the-art, rotating anode x-ray source with Osmic MaxFlux® confocal multi-layer fixed-focus optical system [30]: the flux through a 0.5 mm-diameter collimator is measured at 2.5×10^9 photons/mm^2/s [31]. The corresponding flux on the sample is 5×10^8 photons/s, and the total data acquisition time is 2.16×10^4 s, in high-resolution mode; including absorption and other losses, the total number of photons diffracted by the sample is 1.08×10^{13}. Depending on the resolution (the diffraction limit in the 2θ angle, expressed as the smallest observed d-spacing), 10^4-10^5 unique reflections are routinely collected for molecules weighing a few tens to a few hundred kDaltons (for example Lysozyme, space group $P4_3 2_1 2$, MW=14.5 kD, 19,500 unique reflections to a typically good resolution of 1.5 Å, 39,000 reflections for both Friedel wedges). The collection of these reflections requires a systematic, careful scanning of a large solid angle, with a sufficient integration time at each step to bring the signal-to-noise ratio to a desired value (usually $I/\sigma(I) = 2$ in the highest resolution shell; in our example 1.5 Å), which determines the data acquisition time. Using a typical number of 1.95×10^4 unique reflections (for Lysozyme), we see that the highest-resolution mode requires approximately 5.5×10^8 photons/reflection ($\sigma = 20.1$, which is excellent); further-more, the signal-to-noise ratio scales as the square root of this number. The Compton source produces an average flux of 2.5×10^7 photons/s in a 1 eV spectral interval, quite sufficient for MAD, that can be captured in a 3.5 mrad angle, which is much smaller than the 1° (17 mrad) acceptance angle of the aforementioned monochromator confocal mirrors [31]. For 10^8 photons/reflection ($\sigma = 18.4$), the acquisition time is estimated at 10 h / 10^4 reflections; lower-resolution calibration maps, with a decrease of the signal-to-noise ratio of one order of magnitude, could be acquired in a few minutes. These results are in line with our goal of 50-100 structures a year: a full 3-wavelengths MAD scan could be completed in 60 h. We also note that our results are preliminary, and that we are in the process of designing an optimized source for this application.

A unique characteristic of the Compton source is the very small size of the x-ray source, which essentially matches that of the laser focal spot size; for imaging applications, this translates into increased contrast and better resolution. This also has important implications for micro-crystals, which are more easily produced and often are of better quality (i.e., diffracting to higher resolution with smaller mosaicity) than larger ones. Cooling also becomes less problematic in smaller crystals due to the lower amount of heat generated and the more advantageous surface to volume ratios.

Finally, the laser-driven Compton source can produce much shorter x-ray flashes than those currently generated at synchrotrons: sub-100 fs pulses will be readily produced, in contrast with FWHM in the 35-100 ps range at ALS and 170 ps at APS [8].

ACKNOWLEDGEMENTS

This work was performed under the auspices of the U.S. Department of Energy by University of California Lawrence Livermore National Laboratory, through the Institute for Laser Science and Applications, under contract No. W-7405-ENG-48, and was partially supported by NIH Contract No. N01-CO-97113 and AFOSR MURI Grant No. F49620-99-1-0297. We also acknowledge challenging, yet considerate, discussions with W.P. Leemans.

REFERENCES

[1] G.A. Mourou, C.P.J. Barty, M.D. Perry, *Phys. Today* **51** (1), 22-28 (1998).
[2] *Advanced Accelerator Concepts, Eighth Workshop*, edited by W. Lawson, C. Bellamy, and D.F. Brosius, American Institute of Physics, Conference Proceedings No. 472, Woodbury, NY, 1999.
[3] R.W. Schoenlein, *et al.*, *Science* **274** (5285), 236-238 (1996).
[4] C. Bamber, *et al.*, *Phys. Rev.* **D60** (9), 092004/1-43 (1999).
[5] J. Drenth, *Principles of Protein X-Ray Crystallography*, 2nd Ed., Springer-Verlag, New York, NY, 1999.
[6] K.E. van Holde, W. Curtis Johnson, and P. Shing Ho, *Principles of Physical Biochemistry*, Prentice-Hall, Inc., Upper Saddle River, NJ, 1988, Chap. 6.
[7] W.P. Leemans, private communication (2000).
[8] D. Atwood, *Soft X-Rays and Extreme Ultraviolet Radiation*, Cambridge University Press, Cambridge, U.K., 1999.
[9] C. Gorman, *Time* **148** (29), 56-62 (1997).
[10] W.A. Hendrickson and C.M. Ogata, *Methods Enzymol*, **276**, 494–516 (1997).
[11] P.A.M. Dirac, *Proc. R. Soc. London, Ser.* **A167**, 148 (1938).
[12] F.V. Hartemann and A.K. Kerman, *Phys. Rev. Lett.*, **76**, 624-627 (1996).
[13] F.V. Hartemann, *Phys. Plasmas*, **5**, 2037-2047 (1998).
[14] M.J. Hogan, *et al.*, *Phys. Rev. Lett.*, **81**, 4867-4870 (1998).
[15] F.V. Hartemann, *Phys. Rev.* **E61**, 972-975 (2000).
[16] W. Greiner and J. Reinhardt, *Quantum Electrodynamics*, Springer-Verlag, Berlin, 1994, Chap. 3.7.
[17] F.V. Hartemann, *et al.*, *Phys. Rev.* **E54**, 2956-2962 (1996).
[18] S.K. Ride, E. Esarey, and M. Baine, *Phys. Rev.* **E52**, 5425-5442 (1995).
[19] F.V. Hartemann, *et al*, *Phys. Rev.* **E58**, 5001-5012 (1998).
[20] B. Quesnel and P. Mora, *Phys. Rev.* **E58**, 3719-3732 (1998).
[21] B.E. Carlsten, *Nucl. Instrum. Methods Phys. Res.* **A285**, 313-319 (1989).
[22] J.D. Jackson, *Classical Electrodynamics*, 2nd Ed., John Wiley and Sons, New York, NY, 1975, Chap. 14.
[23] E. Esarey, P. Sprangle, and J. Krall, *Phys. Rev.* **E52**, 5443-5453 (1995).
[24] E. Esarey, S.K. Ride, and P. Sprangle, *Phys. Rev.* **E48**, 3003-3021 (1993).
[25] I.S. Gradshteyn and I.M. Ryzhik, *Table of Integrals, Series, and Products*, 4th Ed., Academic Press, Orlando, FL, 1980, Equations 3.923, 3.924, and 3.323.3.
[26] S. Wolfram, *The Mathematica Book*, 3rd Ed., Wolfram Media, Champaign, IL, 1996.
[27] B.D. Hames, N.M. Hooper, and J.D. Houghton, *Biochemistry*, Springer, New York, NY, 1997, Section I.
[28] K. Garber, *Tech. Rev.* **103** (4), 46-56 (2000).
[29] W.P. Burmeister, *Acta Cryst.* **D56**, 328-341 (2000).
[30] M. Hart and L. Berman, *Acta Cryst.* **A54**, 850-858 (1998).
[31] L.C. Jiang, B. Verman, and K.D. Joensen, *J. Appl. Cryst.* **33** (1), 801-803 (2000).

A Ferroelectric Cathode Electron Gun for Use in High Power Microwave Sources

Y. Hayashi, X. Song, J. D. Ivers, D. Flechtner, J. A. Nation
and L. Schächter

*Laboratory of Plasma Studies and School of Electrical Engineering,
Cornell University, Ithaca, N.Y. 14853*

Abstract. A two-stage 500 kV, 200A, ferroelectric electron gun has been designed, fabricated and tested. We report on the operational characteristics of the gun including measurements of the beam dynamics. The optimum conditions for application of the trigger and its timing are also reported. Faraday cup measurement shows that the beam radius is 4.5 mm in good agreement with simulation. The gun is designed for use in traveling wave tube amplifiers (TWT) and testing of an X-band amplifier driven by the gun is reported. A peak output power of 5.9 MW has been observed from a single stage amplifier driven by a 100 A, 450 kV beam. This corresponds to energy converging efficiency of 13.1% and is the first observation of high power (~MW) microwave generation using the beam generated from a ferroelectric cathode.

INTRODUCTION

Ferroelectric cathodes have been developed in our laboratory over the last several years as a possible alternative to a thermionic cathode. Among the possible advantages of a ferroelectric cathode are room temperature operation, control of the emission by an external trigger pulse and most importantly high electron current density.

During the last decade the electron emission from a ferroelectric cathode has been investigated in several laboratories[1,2,3]. As part of this effort we have studied the use of ferroelectric cathodes for use in electron guns, with the development of a suitable source for high power microwave generation our principle application[4,5,6]. In this paper we describe our gun design and report on its performance in an X-band TWT amplifier. We have developed a suitable electron gun using a 2-stage compression scheme in which the beam is initially electrostatically compressed and subsequently magnetically compressed to its final radius of 4.5 mm. This arrangement allows us to obtain a sufficiently large beam current in matched beam with the desired radius.

A single stage X-band traveling wave tube amplifier has been designed and constructed for use with the gun. In these experiments microwave amplification of order 30 dB has been achieved and output powers of several MW have been observed. This is the first observation of microwave generation on the order of MW from a beam generated from a ferroelectric cathode.

In the following sections we present results extending our previously reported study of ferroelectric emission in a diode to electron beam generation at beam energies of up to 550 keV [5]. The adequacy of the beam quality has been determined by measurements of the beam emittance and directly by the observation of microwave amplification in X-band.

EXPERIMENTAL ARRANGEMENT

The experimental arrangement is shown schematically in Fig. 1. The cathode gun assembly is connected to the modulator described previously[5,6]. The gun geometry, which was designed using data from the aforementioned diode experiments, uses a planar cathode in a modified Pierce geometry. The emitting area can be made relatively small to produce the required beam current as a result of the high current density available from the ferroelectric cathode. In order to obtain a high electron current density the electric field on the ferroelectric cathode must be sufficiently high. On the other hand, in order to obtain the required electrostatic focussing, the emitting surface has to be recessed from the cathode-mounting surface. This causes the electric field on the cathode surface to be significantly reduced compared with that of our earlier work[5]. These constraints have lead us to select an operating current density at the cathode of ≥ 20 A/cm^2. The emission density is competitive with that available using thermionic cathodes.

Coupling these conflicting requirements has led us to a two-stage gun design, in order to satisfy both the requirements simultaneously. The first stage of the electron gun resembles a Pierce gun with the cathode located in the fringing magnetic field of the small diameter solenoid. In this region the beam is mainly compressed by the applied electric field. The geometry is chosen so that the beam compression follows the magnetic flux lines. The first stage of the gun provides a matched beam with relatively high current. However, the beam radius is still rather large and further compression is achieved in the second stage by adiabatically increasing the magnetic field within the drift tube.

An overview of the final geometry adopted is illustrated in Fig.1, in which the dimensions are all shown in millimeters.

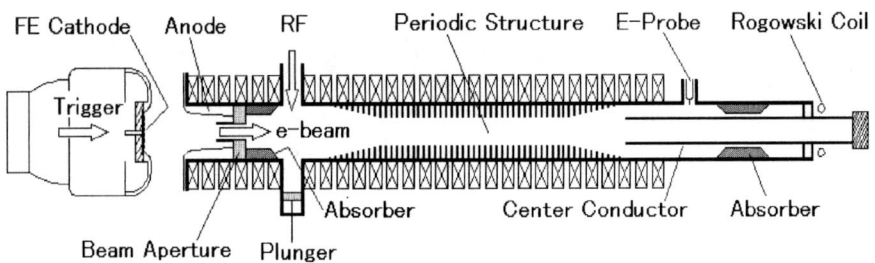

Figure 1: Experimental arrangement for the microwave amplification using an electron gun with a ferroelectric cathode.

Shown in Fig. 2 is an EGUN simulation result for the gun dynamics. In this figure z=0 is defined as the axial position of the anode plate. The magnetic fields at the flat

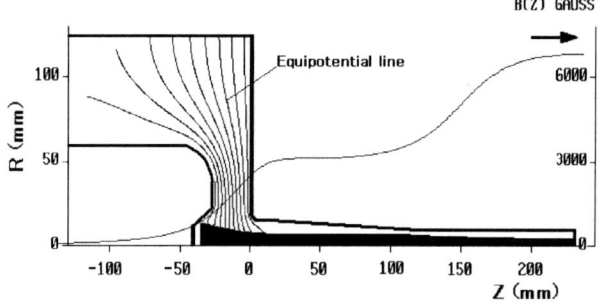

Figure 2. EGUN Simulation of Electron Gun Design

portion of first and second stages are 3.15 kG and 6.84 kG, respectively. The plot shows that the beam has been compressed from its initial radius of 15 mm down to 6.6 mm by the first stage compression. The second stage of the magnetic field coil, which starts at z = 120 mm, and is characterized by a slowly increasing magnetic field from 3.15 kG to 6.84 kG between 70 mm and 210 mm. In this region, the beam radius has been compressed from 6.6 mm down to 4.5 mm. Similar results for the beam dynamics have been found using the PIC code MAGIC. The closely identical results between these extreme cases (thermionic emission in EGUN and field emission in MAGIC) gives confidence that the ferroelectric emission will lie close to these results.

The ferroelectric cathode used in this experiment is a planar LZT-2 sample. The dimensions of the ferroelectric cathode are given in Table 1. The vacuum electric field at the emitting surface is 72 kV/cm. To avoid breakdown and emission from parts other than the emitting area, both the cathode and anode surfaces are relatively flat in comparison with the usual Pierce geometry. The maximum electric field on the cathode mount designed is 210 kV/cm, and is close to limit at which the stainless steel support will start to emit.

Table 1. Ferroelectric Cathode Parameters

Diameter	38.1 mm (1.5 in)
Thickness	1mm
Thickness of silver layer	10 μm
Width of the silver strip	0.4 mm
Emitting area	708 mm^2

The ferroelectric cathode is coated with a 0.2 mm silver grid on its front surface and a uniform silver layer on its back. The trigger pulse is adjusted to start slightly before the maximum of the gun voltage pulse, so that when a ~1.5 kV positive trigger pulse is imposed on the back of the ferroelectric cathode, a beam current pulse is obtained. The rise time of the beam current pulse, under appropriate trigger timing conditions, is much shorter than that of the secondary voltage.

The beam from the electron gun is used as an injector for a single stage X-band Traveling Wave Tube (TWT) amplifier. The disk loaded amplifier has 50 uniform irises and 11 tapered irises on each end. The inner and outer radii of the iris are 10 mm and 16.4 mm respectively, and the periodic spacing of the irises is 7mm. This corresponds to the $\pi/2$ synchronous mode at the operating voltage and at a wave frequency of 9.0 GHz. The input waveguide at the beginning of the TWT is connected to an X-band magnetron which provides ~10 kW power into the TWT. A reflector mounted on the end of the anode is located at a distance equal to a 3/4 wavelength of the TE_{11} mode, in order to reduce excitation of this mode.

At the end of the TWT, a coaxial mode converter/collector is used to separate the beam and amplified RF output [3]. The beam is dumped into the collector, in order to avoid re-acceleration of the beam electrons by the amplified RF. The amplified RF is efficiently mode converted from the TM_{01} mode of the slow wave structure into a TEM mode in the coaxial section. Measurement of the RF output power is made with an E field-probe mounted between the amplifier and the microwave absorber located at the end of the coaxial converter.

Experimental Results

Two sets of experiments have been conducted. The first is a study of the characteristics of the diode and beam dynamics. Optimum conditions for triggering the ferroelectric cathode, such as trigger voltage and timing, have been studied, and the longitudinal variation of the beam envelope has been measured with a 2-section Faraday cup. The second demonstrates the applicability of a beam generated by a ferroelectric cathode as an electron source for a microwave tube. Details of the experimental results will be described in this section.

Beam Generation

An important feature of ferroelectric cathodes is that the electron emission is initiated by a trigger signal, that may be used to control the timing of the current with respect to the voltage pulse.

The trigger timing and amplitude also may be used to control the current pulse shape and amplitude. However, the optimum conditions for applying the trigger, such as the timing and the magnitude, need to be determined. A series of experiment for testing the trigger conditions have been conducted. The trigger generator produces a pulse of 100 ns width and a voltage up to 3.2 kV. The trigger pulse is applied to the rear surface of the ferroelectric and is inductively decoupled from the main voltage pulse. Fig. 3 shows the waveforms for the diode voltage and beam current as a function of the trigger timing.

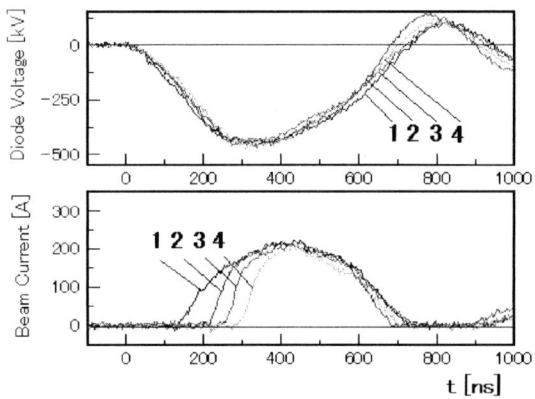

Figure 3. Gun Voltage and Beam current as a function of the timing of the trigger pulse.

Here t=0 is defined as the time when the diode voltage starts to rise. In the experiment, the peak diode voltage is kept at 450 kV and the applied trigger voltage is 2.1 kV. Varying the timing of the trigger pulse from 110 to 280 ns, the peak value of the beam current varies only between 200 to 210 A. However, changes of the pulse shape have been observed. The second curve, obtained by applying the trigger voltage at t = 200 ns, gives an approximately square pulse. The beam current is also observed for changing triggering voltages with a fixed trigger timing at 200 ns. Figure 4 shows the relation between the beam current and trigger voltage. The beam current is a rapid

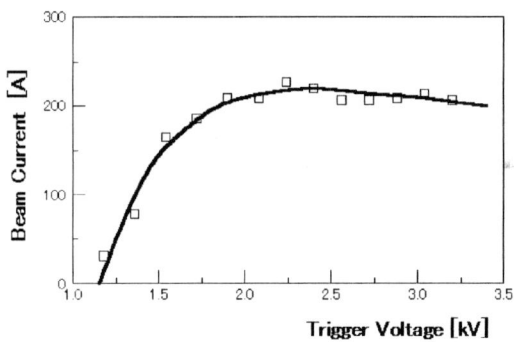

Figure 4. Beam Current as a function of the Trigger Pulse Amplitude.

function of the trigger voltage once the threshold has been exceeded and until the trigger voltage exceeds 2 kV, when the beam current becomes largely independent of the trigger amplitude. The working trigger voltage was chosen to be 1.9 kV.

A Faraday cup has been used to determine the beam radius and its axial variation. The cup consists of two coaxial parts, an inner section which measures the beam current through a central aperture and the main section that measures the total beam current. The radius of the aperture is 2.38 mm. Assuming the radial profile of current density is uniform, the ratio of the two currents and the radius of the aperture allow a determination of the beam radius. By moving the Faraday cup in longitudinal direction, the axial variation of the beam radius can be obtained. The results are shown in Fig.5. The magnetic compression in the 2nd stage is significant, the beam radius

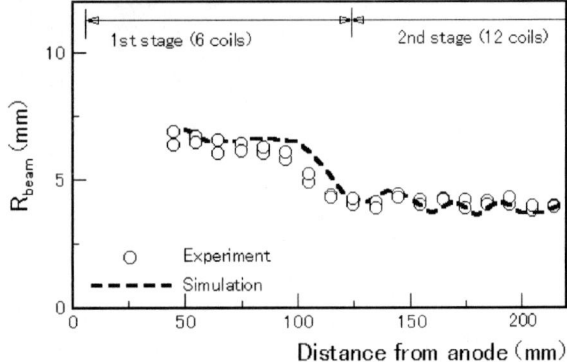

Figure 5. Beam Radius as a function of the axial position. The dashed line shows EGUN simulations using the measured magnetic field profile. Radial variations of 0.3 mm are almost one order of magnitude smaller than those found in ref 5.

being compressed from 6.5 mm to 4.1 mm. The difference between the measured beam radius and the design value as shown in Fig. 2 is caused by the difference of the magnetic field profile in two cases. The simulation result using actual magnetic field used is shown in Fig. 5. Good agreement with the experimental results has been found. The variation in the beam radius is less than 0.3 mm, and indicates that the beam is close to a matched condition.

Microwave Generation

The main purpose of this part of the experiment was to demonstrate that the electron beam generated from a ferroelectric cathode is suitable for high power microwave generation applications. The beam and TWT geometries have been indicated earlier. When the 200 A beam was used the TWT oscillated. As a result the beam current was limited by a 5 mm radius aperture to about 100 A. The remaining experiments were carried out using this beam. Fig. 6 shows the diode voltage, beam current, amplified microwave power and output from a heterodyne receiver in which part of the amplified signal is mixed with a reference signal from a local oscillator. The input microwave signal has a power of 18.7 kW at a frequency of 9.145 GHz. The peak of the output power is 5.9 MW, corresponding to a gain of 25 dB and an energy conversion efficiency of 13.1%. The heterodyne signal shows that the amplified signal is at a single frequency, which is equal to that of the input microwave signal.

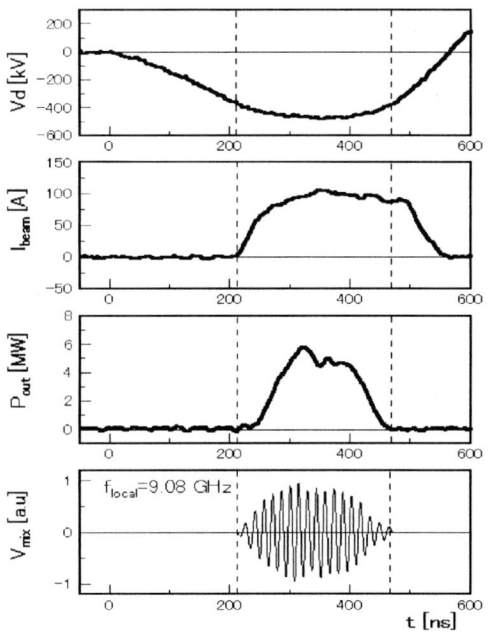

Figure 6. Waveforms from the TWT amplifier.

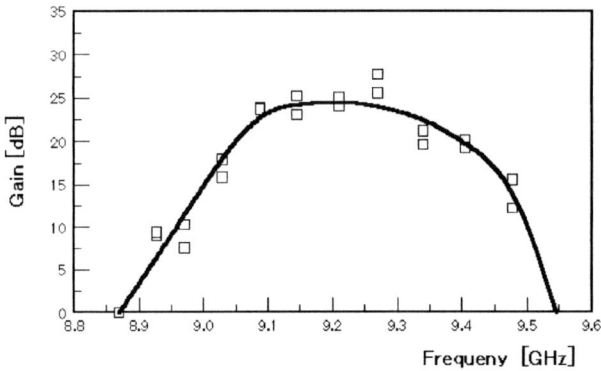

Figure 7. Frequency response of X-band TWT amplifier.

In Figure 7 we illustrate the frequency response of the single stage amplifier. Gain has been observed over a frequency range of nearly 0.7 GHz from 8.9 to 9.5 GHz. Note that the amplifier has not been, in any way, optimized for output power or efficiency. The purpose here was solely to demonstrate that the beam quality was adequate for use in high power microwave applications.

In a separate experiment an effort was made to estimate the beam emittance directly from a measurement of the beam expansion following transport through a seven slit emittance screen. In this experiment the beam radius was increased by a reduction in the confining magnetic field. In zero order this process will not change the emittance since the increase in beam radius is accompanied by a cooling of the beam in the adiabatic magnetic field expansion region. The subsequent transverse expansion of the beamlets due to the beam temperature was measured using radio-chromic film. The estimated normalized emittance was found to be 10 π mm.mrad. This figure is consistent with the observation of gain in the X-band. The gain observation is however a poor test of the beam quality as the amplifier gain is expected to not be reduced (in X-band amplifiers) until the emittance is much larger than the estimated value reported here. However, above results indicate that the ferroelectric cathodes can be used as an electron source for a high power X-band TWT and that microwave amplification of ~30 dB to MW levels is possible.

At present no life time tests have been carried out on ferroelectric emitters. Limitations in the cathode life may be severe if back sputtering occurs from the anode of the gun. A well designed beam channel and dump will be essential for ling life operation. The fact that ferroelectrics have been cycled through billions of switchings in memory chips offers an encouraging observation.

CONCLUSIONS

We have shown that ferroelectric cathodes may be used in electron guns to produce beams suitable for high power microwave applications. Peak powers of about 6 MW were measured in X band. This is the first reported observation of high power microwave sources using ferroelectric cathodes.

ACKNOWLEDGEMENTS

This work was supported by the US Department of Energy and the AFOSR under the MURI program. The partial support of the work by DoE through an SBIR sub-contract from FM Technologies is also acknowledged.

REFERENCES

[1] H. Gundel, H. Riege, E. J. N. Wilson, J. Handerek and K. Zioutas, "Fast Polarization Changes in Ferroelectrics and Their Application in Accelerator", Nucl. Instrum. Methods Phys. Res. A, Vol. 280, pp. 1-6, 1989.
[2] M. Okuyama, J.Asano and Y. Hamakawa, "Electron Emission from Lead-zirconate Ferroelectric Ceramic Induced by Pulse Electric Field", Jpn. J. Appl. Phys, Vol. 33, pp. 5506-5509, 1994.
[3] D. Shur and G. Rosenman, "Plasma-assisted Electron Emission from (Pb,La)(Zr,Ti)O3 Ceramic Cathode", J. Appl. Phys., Vol. 79, No. 7, pp. 3669-3674, Mar, 1993.
[4] C. B. Fleddermann and J. A. Nation, "Ferroelectric Sources and Their Application to Pulsed Power: a Review", IEEE Trans. On Plasma Science, Vol. 25, No. 2, pp. 212-220, April 1997.
[5] J. D. Ivers, D. Flechtner, Cz. Golkowski, G. Liu, J A Nation and L. Schächter, "Electron Beam Generation using a Ferroelectric Cathode", IEEE Trans. On Plasma Science, Vol. 27, No. 3, 1999.
[6] Y. Hayashi, J. D. Ivers, D. Flechtner, J. A. Nation, P. Wang, S. Banna and L. Schächter, "TWT Amplifier Using a Ferroelectric Cathode for Electron Beam Generation", Conference on Particle Accelerator 1999, pp. 3606.

Betatron Radiation from Electron Beams in Plasma Focusing Channels

E. Esarey, P. Catravas, and W.P. Leemans

Center for Beam Physics, Ernest Orlando Lawrence Berkeley National Laboratory, University of California, Berkeley CA 94720

Abstract. Spontaneous radiation emitted from an electron undergoing betatron motion is a plasma focusing channel is analyzed and applications to plasma wakefield accelerator experiments and to the ion channel laser (ICL) are discussed. Important similarities and differences between a free electron laser (FEL) and in an ICL are delineated. It is shown that the frequency of spontaneous radiation is a strong function of the betatron strength parameter a_β, which plays a similar role to that of the wiggler strength parameter in a conventional FEL. For $a_\beta \gtrsim 1$, radiation is emitted in numerous harmonics. Furthermore, a_β is proportional to the amplitude of the betatron orbit, which varies for every electron in the beam. This places serious limits on the possibility of realizing an ICL.

INTRODUCTION

The propagation of electrons beams through plasmas is relevant to a variety of advanced accelerator [1] and novel radiation source applications [2]. Presently two experiments are underway at the Stanford Linear Accelerator Center (SLAC) that explore the interaction of an intense electron bunch with a plasma. These are the E-150 plasma lens experiment [3] and the E-157 plasma wakefield experiment [4]. In the E-157 experiments, a 30 GeV electron beam of 2×10^{10} electrons in a 0.65 mm long bunch is propagated through a 1.4 m long lithium plasma with an electron density up to 2×10^{14} cm^{-3}. The electron bunch propagates through the plasma in the so-called blowout regime [5], i.e., the initial beam density is greater than the plasma density. In this regime, the head of the bunch expels the plasma electrons and leaves behind a nearly uniform ion channel. The bunch length and plasma density are chosen such that the blown-out plasma electrons come crashing back to the axis near the tail of the bunch, thus driving a very large axial electric field, on the order of several 100 MV/m, that can accelerate the electrons in the tail of the bunch.

One consequence of operating in the blowout regime is that the main body of the electron bunch resides in the nearly uniform ion channel, since the plasma electrons are blown out to approximately the plasma skin depth, $k_p^{-1} = c/\omega_p$, which is typically much greater than the bunch radius, where $\omega_p = (4\pi n_e e^2/m_e c^2)^{1/2}$ is the plasma frequency and n_e is the electron plasma density. Associated with the ion channel are very strong transverse fields, on the order of several thousand tesla per meter, that subsequently focus the body of the electron bunch. Since the initial beam radius $(50 - 100~\mu\text{m})$ is much greater that the matched beam radius $(\sim 5~\mu\text{m})$, the beam radius will undergo betatron oscillations as it propagates through the plasma [2], [3]. In the blowout regime, the radial space charge electric field is given [3] by $E_r = (em_e/c^2)k_p^2 r/2$. At the edge of the beam, $r = r_b$, this can be written in practical units as

$$E_r[\text{MV/m}] = 9.06 \times 10^{-15} n_e[\text{cm}^{-3}] r_b[\mu\text{m}]. \tag{1}$$

Likewise, in the blowout regime, the betatron wavelength is given by $\lambda_\beta = (2\gamma)^{1/2} \lambda_p$, where γ is the relativistic factor of the electron and $\lambda_p = 2\pi/k_p$ is the plasma wavelength, which can be written as

$$\lambda_p[\text{cm}] = 3.34 \times 10^6 (n_e[\text{cm}^{-3}])^{-1/2}. \tag{2}$$

Time integrated optical transition radiation has been used to study the transverse beam profile dynamics in the E-157 experiments [6]. Up to three betatron oscillations of the beam radius has been observed.

In addition to the blowout regime of the plasma wakefield accelerator, an accelerated electron bunch will experience transverse focusing forces in typical plasma-based accelerator configurations, such as the laser wakefield accelerator [1]. For example, in the laser wakefield accelerator that operates in the linear regime, the wakefield is often described by an electrostatic potential of the form $\Phi = \Phi_0 \exp(-r^2/r_p^2) \cos k_p(z - ct)$, where r_p is the radius of the wake and is proportional to radius of the drive beam. Notice that the axial electric field $E_z = -\partial\Phi/\partial z$ and the radial electric field $E_r = -\partial\Phi/\partial r$ are phased such that there exists a $\pi/2$ region of axial phase $k_p(z - ct)$ that is both accelerating and focusing. An electron residing off-axis will undergo radial betatron oscillations about the axis due to the transverse focusing force of the wakefield. The magnitude of the focusing field near the axis is given by $|E_r| = 2r\Phi_0/r_p^2$, assuming $\cos k_p(z - ct) = 1$, and the betatron wavelength is given by $\lambda_\beta = \pi r_p (2\gamma/\hat{\Phi}_0)^{1/2}$, where $\hat{\Phi}_0 = e\Phi_0/m_e c^2$ is the normalized amplitude of the wakefield. The density perturbation on axis associated with the wake is given by $\delta n_e/n_e = -\hat{\Phi}_0(1 + 4/k_p^2 r_p^2) \cos k_p(z - ct)$. Electron blowout near the axis occurs when $\hat{\Phi}_0 = k_p^2 r_p^2/4$, assuming $k_p r_p/2 < 1$.

As an electron undergoes betatron oscillations in a plasma focusing channel, it will emit synchrotron radiation [7]. In the limit of a small amplitude betatron oscillation, i.e., an electron displaced slightly from the axis, the wavelength of the synchrotron radiation is given by $\lambda = \lambda_\beta/2\gamma^2$, where λ_β is the wavelength of the betatron oscillation and γ is the relativistic factor of the electron. For plasma-based accelerators, this can easily be in the hard x-ray regime. For the E-157 experiment, $\lambda_\beta \sim 0.8$ m and $\gamma = 6 \times 10^4$, such that $\lambda \sim 0.1$ nm.

The betatron motion in a focusing channel also forms the basis of the ion channel laser [2]. In the ion channel, radiation at the resonant wavelength $\lambda = \lambda_\beta/2\gamma^2$ can feed back on the electron beam, leading to axial bunching of the beam, and coherent amplification of the radiation. The amplification process is analogous to that in a free electron laser, with the betatron motion analogous to the wiggler motion in a free electron laser (FEL) [8]. It has been suggested that the ICL mechanism can further enhance the spontaneous synchrotron radiation in the E-157 experiments [9], thus leading to partially coherent radiation near the 0.1 nm region. It is necessary that the details of the single particle synchrotron radiation in a plasma focusing channel be well understood, in order to access the prospects for the generation of self-amplified spontaneous emission (SASE) in an ICL.

In this article, spontaneous radiation emitted from an electron undergoing betatron motion in a plasma focusing channel is analyzed starting from basic principles. Application of these results to the E-157 experiment and to the ICL are examined. Important similarities and differences between SASE in an FEL and in an ICL are delineated. It is shown that the spontaneous radiation emitted along the axis of a plasma focusing channel from a single electron occurs near the resonant frequency $\omega_n = 2\gamma_{z0}^2 n\omega_\beta/(1 + a_\beta^2/2)^{1/2}$, where γ_{z0} is the initial gamma factor for the electron entering the channel, n is the harmonic number, $\omega_\beta = ck_\beta = 2\pi c/\lambda_\beta$ is the betatron frequency, $a_\beta = \gamma_{z0} k_\beta r_\beta$ is the betatron strength parameter, and r_β is the amplitude of the betatron orbit. The role of the betatron strength parameter a_β is analogous to that of the wiggler strength parameter a_w (or K_w) in FEL physics. In Ref. [2], the ICL was considered only in the limit $a_\beta^2 \ll 1$. When $a_\beta^2 \ll 1$, radiation is emitted primarily at the fundamental frequency $\omega = 2\gamma_{z0}^2 \omega_\beta$ and is independent of a_β. For $a_\beta \gtrsim 1$, however, radiation is emitted in numerous harmonics in which the resonant frequency is a strong function of a_β. This is the case in the E-157 experiments, in which $a_\beta \sim 2-50$. In an ideal FEL, the wiggler strength parameter a_w is a constant (a function of only the magnetic field of the wiggler) for all of the beam electrons. However, in an ICL, $a_\beta = \gamma_{z0} k_\beta r_\beta$ depends on both the electron energy γ_{z0} and the betatron amplitude r_β. Since r_β, and hence a_β, is different for every electron in a typical beam, this places serious limits on the possibility of realizing a SASE ICL.

ELECTRON MOTION IN PLASMA FOCUSING CHANNELS

The electron motion in a plasma focusing channel is governed by the relativistic Lorentz equation, which may be written in the form

$$d\mathbf{u}/dct = \nabla\hat{\Phi} \tag{3}$$

where $\hat{\Phi} = e\Phi/m_e c^2$ is the normalized electrostatic potential of the focusing channel, $\mathbf{u} = \mathbf{p}/m_e c = \gamma\beta$ is the normalized electron momentum, and $\gamma = (1+u^2)^{1/2} =$

$(1-\beta^2)^{-1/2}$ is the relativistic factor. Here only the transverse focusing force of the plasma is considered. Near the axis, $r^2 \ll r_0^2$, the space charge potential is assumed to have the form

$$\hat{\Phi} = \hat{\Phi}_0(1 - r^2/r_0^2), \tag{4}$$

such that the normalized radial electric field is $\hat{E}_r = -\partial\hat{\Phi}/\partial r = 2\hat{\Phi}_0 r/r_0^2$, where $\hat{\Phi}_0$ and r_0 are constants. The electrostatic potential is related to the electron plasma density by $\nabla^2\hat{\Phi} = k_p^2(n_e/n_0 - 1)$, where a uniform background of plasma ions of density n_0 is assumed. The maximum focusing field occurs in the when the plasma electrons are completely expelled (blown out) from the channel, $n_e = 0$. Notice that in the blowout regime, $\hat{E}_r = k_p^2 r/2$, hence, $\hat{\Phi}_0/r_0^2 \leq k_p^2/4$.

Equation (4) implies the existence of two constants of the motion, $du_z/dt = 0$ and $d(\gamma - \hat{\Phi})/dt = 0$. Inside the focusing channel, the electron orbits (assuming the electron orbit lies in the x-z plane) are given by

$$\tilde{\beta}_x \simeq k_\beta r_\beta \cos(k_\beta ct), \tag{5}$$

$$\tilde{x} \simeq r_\beta \sin(k_\beta ct), \tag{6}$$

$$\tilde{\beta}_z \simeq \beta_{z0}\left(1 - k_\beta^2 r_\beta^2/4\right) - \beta_{z0}(k_\beta^2 r_\beta^2/4)\cos(2k_\beta ct), \tag{7}$$

$$\tilde{z} \simeq z_0 + \beta_{z0}\left(1 - k_\beta^2 r_\beta^2/4\right)ct - \beta_{z0}(k_\beta r_\beta^2/8)\sin(2k_\beta ct), \tag{8}$$

where

$$k_\beta = (2\hat{\Phi}_0/\gamma_{z0} r_0^2)^{1/2} \tag{9}$$

is the betatron wavenumber, r_β is the constant amplitude of the betatron orbit, $u_z = u_{z0}$, $\gamma_{z0} = (1 + u_{z0}^2)^{1/2}$, $\beta_{z0} = u_{z0}/\gamma_{z0}$ and z_0 is a constant. Equations (5)-(8) are the leading order contributions to the orbits, assuming $k_\beta^2 r_\beta^2/2 \ll 1$. Notice that in the blowout regime $\hat{\Phi}_0 = k_p^2 r_0^2/4$ and the betatron wavenumber is given by $k_\beta = k_p/(2\gamma_{z0})^{1/2}$.

For an ensemble of particles comprising an electron beam, the evolution of the RMS beam radius r_b evolves via envelope equation [10]

$$d^2 r_b/dct^2 = \epsilon_n^2 \gamma^{-2} r_b^{-3} - k_\beta^2 r_b, \tag{10}$$

where $\epsilon_n \simeq \gamma r_b \theta_b$ is the normalized beam emittance, θ_b is the RMS beam angle, and $\gamma m_e c^2$ is the beam energy, where the effects of finite energy spread and space charge have been neglected. A matched beam occurs when $d^2 r_b/dct^2 = 0$, i.e., at a matched radius given by

$$r_{bm} = (\epsilon_n/\gamma k_\beta)^{1/2}, \tag{11}$$

at which point the expansion of the beam due to finite emittance is balanced by the focusing forces of the plasma channel. For parameters typical of the E-157 experiment ($\epsilon_n = 10$ mm-mrad, $\gamma = 6 \times 10^4$, and $\lambda_\beta = 0.82$ m), $r_{bm} = 4.7$ μm.

SYNCHROTRON RADIATION

The energy spectrum of the radiation emitted by a single electron in an arbitrary orbit $\tilde{\mathbf{r}}(t)$ and $\tilde{\beta}(t)$ can be calculated from the Lienard-Wiechert potentials [11],

$$\frac{d^2I}{d\omega d\Omega} = \frac{e^2\omega^2}{4\pi^2 c} \left| \int_{-T/2}^{T/2} dt \left[\mathbf{n} \times (\mathbf{n} \times \tilde{\beta})\right] \exp\left[i\omega(t - \mathbf{n}\cdot\tilde{\mathbf{r}}/c)\right] \right|^2, \quad (12)$$

where $d^2I/d\omega d\Omega$ is the energy radiated per frequency, ω, per solid angle, Ω, during the interaction time, T, and $\mathbf{n}$ is a unit vector pointing in the direction of observation. It is convenient to introduce the spherical coordinates (r, θ, ϕ) where $x = r\sin\theta\cos\phi$, $y = r\sin\theta\sin\phi$, $z = r\cos\theta$, and $\mathbf{n}$ is along the $\mathbf{e}_r$ direction. Using the betatron orbits given above, the radiation spectrum can be calculated with conventional techniques [12]. The resulting spectrum is given by

$$\frac{d^2I}{d\omega d\Omega} = \sum_{n=1}^{\infty} \frac{e^2 k^2}{4\pi^2 c} \left(\frac{\sin \bar{k}L/2}{\bar{k}}\right)^2$$

$$\cdot \left[C_x^2(1 - \sin^2\theta\cos^2\phi) + C_z^2\sin^2\theta - C_xC_z\sin 2\theta\cos\phi\right], \quad (13)$$

where

$$C_x = k_\beta r_\beta \sum_{m=-\infty}^{\infty} J_m(\alpha_z)\left[J_{n+2m-1}(\alpha_x) + J_{n+2m+1}(\alpha_x)\right], \quad (14)$$

$$C_z = \beta_{z0} \sum_{m=-\infty}^{\infty} J_m(\alpha_z)\{2(1 + k_\beta^2 r_\beta^2/4)J_{n+2m}(\alpha_x)$$

$$- (k_\beta^2 r_\beta^2/4)\left[J_{n+2m-2}(\alpha_x) + J_{n+2m+2}(\alpha_x)\right]\}, \quad (15)$$

$$\alpha_z = \frac{n(k/k_n)(a_\beta^2/4)\cos\theta}{\left[(1 + a_\beta^2/2)\cos\theta + 2\gamma_{z0}^2(1 - \cos\theta)\right]}, \quad (16)$$

$$\alpha_x = \frac{n(k/k_n)2\gamma_{z0}a_\beta\sin\theta\cos\phi}{\left[(1 + a_\beta^2/2)\cos\theta + 2\gamma_{z0}^2(1 - \cos\theta)\right]}, \quad (17)$$

$$\bar{k} = \alpha_0 k - nk_\beta, \quad (18)$$

$$\alpha_0 = 1 - \beta_{z0}(1 - k_\beta^2 r_\beta^2/4)\cos\theta, \quad (19)$$

and

$$a_\beta = \gamma_{z0} k_\beta r_\beta \quad (20)$$

is the betatron strength parameter. Here, $L = cT$ is the interaction length, $k = \omega/c$ is the radiation wavenumber, n is the harmonic number, and J_m are Bessel functions. For parameters typical of the E-157 experiment ($\gamma = 6 \times 10^4$ and $\lambda_\beta = 0.82$ m), $a_\beta = 45$ for $r_\beta = r_b = 100$ μm (typical unmatched beam radius) and $a_\beta = 2.1$ for $r_\beta = r_{bm} = 4.7$ μm (typical matched beam radius).

Provided $N_\beta \gg 1$, radiation is emitted in a series of harmonics and is confined in a narrow bandwidth about the resonant frequency of each harmonic, where $N_\beta = L/\lambda_\beta$ is the number of betatron periods that the electron undergoes. The frequency width of the radiation spectrum for a given harmonic is determined by the resonance function $R_n(k)$, where

$$R_n(k) = \left(\frac{\sin \bar{k}L/2}{\bar{k}L/2}\right)^2. \tag{21}$$

This function is sharply peaked about the resonant frequency, $\omega_n = ck_n$, given by $\bar{k} = 0$,

$$k_n = \frac{nk_\beta}{\alpha_0} \simeq \frac{2\gamma_{z0}^2 nk_\beta}{\left[(1+a_\beta^2/2)\cos\theta + 2\gamma_{z0}^2(1-\cos\theta)\right]}, \tag{22}$$

where $\gamma_{z0}^2 \gg 1$ was assumed. The width of the spectrum, $\Delta\omega$, about ω_n is given by $\Delta\omega/\omega_n = 1/nN_\beta$. Furthermore, $R_n(k) \to \Delta\omega_n \delta(\omega - \omega_n)$ as $N_\beta \to \infty$. The angular width $\Delta\theta_n$ within which can be found radiation with frequencies in $\Delta\omega$ about ω_n, for a single harmonic n, is given by $\Delta\theta_n \simeq (2\Delta\omega/M_0\omega_n)^{1/2}$. Typically, for frequencies of interest, the synchrotron radiation is confined to a cone angle $\theta^2 \ll 1$ and the resonant frequency can be approximated by

$$\omega_n \simeq nM_0 ck_\beta/(1 + M_0\theta^2/2), \tag{23}$$

where $M_0 = 2\gamma_{z0}^2/(1+a_\beta^2/2)$ is the relativistic Doppler upshift factor.

On-Axis Radiation

Of particular interest is the radiation emitted along the axis. Along the axis, $\theta = 0$, only the odd harmonics are finite, i.e., the even harmonics vanish. Setting $\theta = 0$ in the above expressions gives, for the nth odd harmonic, $\alpha_x = 0$, $\alpha_z = \alpha_n$, and

$$\frac{d^2 I_n(0)}{d\omega d\Omega} = 2e^2 \frac{\omega}{\omega_n} k_\beta N_\beta M_0^2 G_n F_n = \frac{4e^2}{c} \frac{\omega}{\omega_n} \frac{\gamma_{z0}^2 N_\beta^2 R_n F_n}{(1+a_\beta^2/2)}, \tag{24}$$

where

$$F_n(a_\beta) = n\alpha_n \left[J_{(n-1)/2}(\alpha_n) - J_{(n+1)/2}(\alpha_n)\right]^2 \tag{25}$$

is the harmonic amplitude function,

$$\alpha_n = \frac{n(\omega/\omega_n)a_\beta^2/4}{(1+a_\beta^2/2)}, \tag{26}$$

and

$$G_n(\omega) = \frac{R_n(k)}{\Delta\omega_n} = \frac{1}{\Delta\omega_n} \frac{\sin^2[\pi n N_\beta(\omega/\omega_n - 1)]}{[\pi n N_\beta(\omega/\omega_n - 1)]^2} \tag{27}$$

is the frequency spectrum function with the resonant frequency $\omega_n = nM_0ck_\beta$.

The energy radiated in the nth backscattered harmonic depends on the function $F_n(a_\beta)$, Eq. (25). For high harmonics, $n \gg 1$, F_n becomes significant when $a_\beta^2 \gg 1$. For $a_\beta^2 \ll 1$, only the fundamental, $n = 1$, is significant. Figure 1 shows a plot of $d^2I(0)/d\omega\delta\Omega$ versus $\omega/2\gamma_{z0}^2\omega_\beta$ for the first four harmonics ($n = 1, 2, 3$, and 4) with $N_\beta = 4$. The solid curve shows the radiation from a single electron with $a_\beta = \gamma_{z0}k_\beta r_\beta = 4$, indicating that radiation is emitted in well-defined harmonics. The dashed (dotted) curve shows the spectrum integrated over a flat-top (Gaussian) distribution of betatron amplitudes r_β with a rms value satisfying $a_{\beta,rms} = \gamma_{z0}k_\beta r_{\beta,rms} = 4$. Figure 2 shows a blow-up of Fig. 1. The effect of averaging over a distribution of electron orbits is clearly to smooth out the spectrum, since the frequency of the radiation emitted by a single electron is a strong function of a_β. Note that the fall-off on the right of the dashed (dotted) curve in Figs. 1 and 2 is artificial, since only the first four harmonic terms are included in this calculation of the spectrum.

An expression for the number of photons (N_n) radiated per electron can be obtained for photons in a narrow bandwidth near the resonant frequency by dividing Eq. (24) by the energy per photon ($\hbar\omega_n$). The total number of photons radiated per electron in the in the intrinsic bandwidth $\Delta\omega_n = \omega_n/nN_\beta$ about ω_n is given by integrating over this narrow frequency band and multiplying by the solid angle $2\pi(\Delta\theta^2/2)^{1/2}$, where $\Delta\theta \simeq (2\Delta\omega_n/M_0\omega_n)^{1/2}$, which gives

$$N_n \simeq 4\pi\alpha_f(\Delta\omega_n/\omega_n)(N_\beta/n)F_n(a_\beta), \qquad (28)$$

where α_f is the fine structure constant.

Ultra-Intense Behavior

For values of $a_\beta^2 \ll 1$, the scattered radiation will be narrowly peaked about the fundamental resonant frequency, ω_1, given by Eq. (22) with $n = 1$. As a_β approaches unity, scattered radiation will appear at harmonics of the resonant frequency as well, $\omega_n = n\omega_1$. When $a_\beta \gg 1$, high harmonic ($n \gg 1$) radiation is generated and the resulting synchrotron radiation spectrum consists of many closely spaced harmonics. Finite variations in the parameter $a_\beta = \gamma_{z0}k_\beta r_\beta$ within an electron beam can broaden the linewidth and cause the spectrum to overlap. Hence, in the ultra-intense limit, i.e., $a_\beta \gg 1$, the gross spectrum appears broadband, and a continuum of radiation is generated which extends out to a critical frequency, ω_c, beyond which the radiation intensity diminishes. The critical frequency can be written as $\omega_c = n_c M_0 \omega_\beta$, where n_c is the critical harmonic number. It is possible to calculate n_c by examining the radiation spectrum, Eqs. (13)-(15), in the ultra-intense limit, $a_\beta \gg 1$.

Asymptotic properties of the radiation spectrum for large harmonic numbers, $n \gg 1$, can be analyzed using conventional methods [12]. In particular, the asymp-

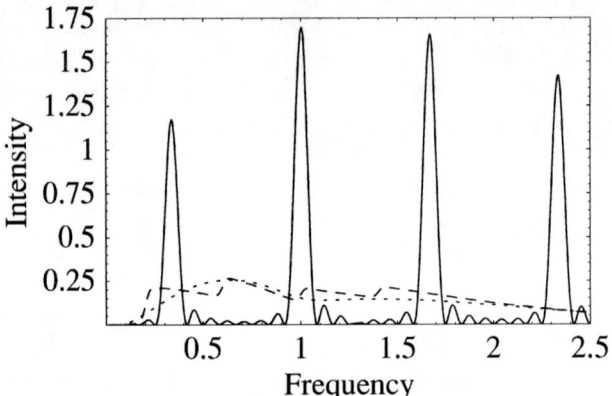

FIGURE 1. Normalized spectrum $d^2I(0)/d\omega\delta\Omega$ (arbitrary units) versus $\omega/2\gamma_{z0}^2\omega_\beta$ for the first four harmonics ($n = 1, 2, 3,$ and 4) with $N_\beta = 4$. The solid curve shows the radiation from a single electron with $a_\beta = \gamma_{z0}k_\beta r_\beta = 4$. The dashed (dotted) curve shows the spectrum integrated over a flat-top (Gaussian) distribution of betatron amplitudes r_β with $a_{\beta,rms} = \gamma_{z0}k_\beta r_{\beta,rms} = 4$.

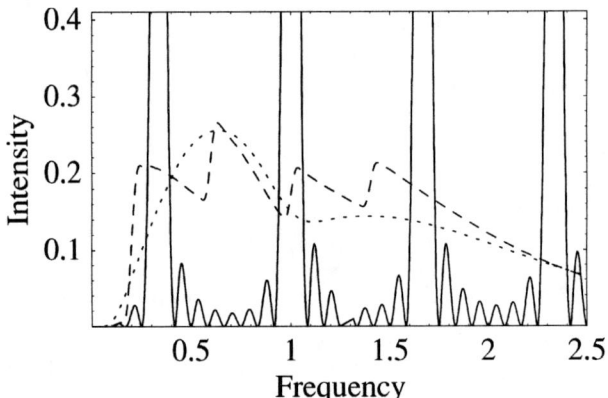

FIGURE 2. Blow-up of Fig. 1.

totic spectrum in the vertical direction ($\phi = \pi/2$) is given by

$$\frac{d^2I}{d\omega d\Omega} \simeq N_\beta \frac{6e^2}{\pi^2 c} \frac{\gamma_{z0}^2 \zeta^2}{(1+\gamma_{z0}^2\theta^2)} \left[\frac{\gamma_{z0}^2\theta^2}{(1+\gamma_{z0}^2\theta^2)} K_{1/3}^2(\zeta) + K_{2/3}^2(\zeta)\right], \tag{29}$$

where

$$\zeta = \frac{\omega}{\omega_c}(1+\gamma_{z0}^2\theta^2)^{3/2}, \tag{30}$$

$$\omega_c = n_c M_0 \omega_\beta \simeq 3a_\beta \gamma_{z0}^2 \omega_\beta \tag{31}$$

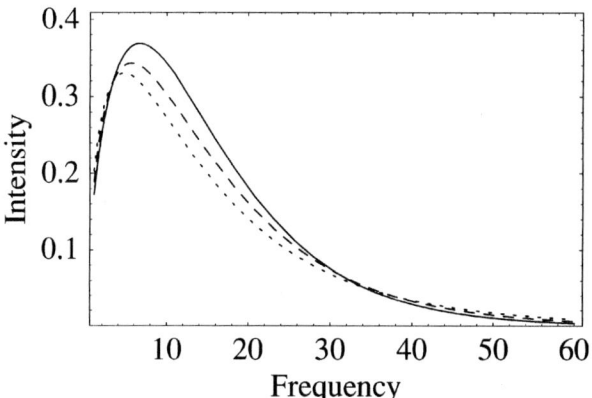

FIGURE 3. The function $Y(\xi) = \xi^2 K_{2/3}^2(\xi)$ versus $\omega/2\gamma_{z0}^2\omega_\beta$. The solid curve shows the radiation from a single electron with $a_\beta = \gamma_{z0}k_\beta r_\beta = 10$. The dashed (dotted) curve shows the spectrum integrated over a flat-top (Gaussian) distribution of betatron amplitudes r_β with $a_{\beta,rms} = \gamma_{z0}k_\beta r_{\beta,rms} = 10$.

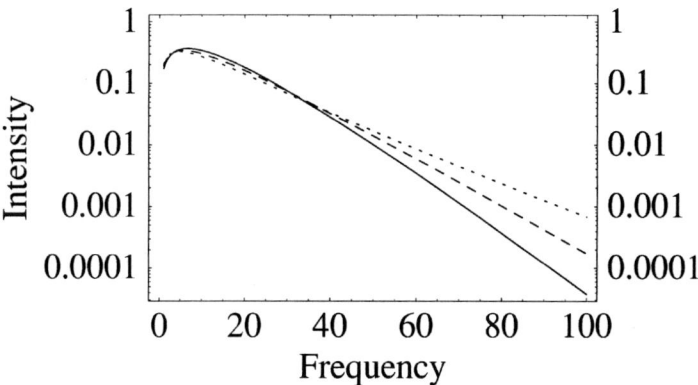

FIGURE 4. Figure 3 plotted on a log scale.

is the critical frequency,

$$n_c \simeq 3a_\beta^3/4 \tag{32}$$

is the critical harmonic number, and $M_0 \simeq 4\gamma_{z0}^2/a_\beta^2$. For E-157-like parameters ($\gamma = 6 \times 10^4$, $\lambda_\beta = 0.82$ m, and $a_\beta = 45$), $n_c \simeq 6.8 \times 10^4$ and $\lambda_c = 2\pi c/\omega_c \simeq 1.7 \times 10^{-12}$ m.

Along the axis $\theta = 0$, $d^2I(0)/d\omega d\Omega \sim \xi^2 K_{2/3}^2(\xi)$, where $\xi = \omega/\omega_c$. The function $Y(\xi) = \xi^2 K_{2/3}^2(\xi)$ is maximum at $\xi = 1/2$ and decreases rapidly for $\xi > 1$. The peak intensity occurs along the axis $\theta = 0$, at approximately the critical frequency,

$\omega \simeq \omega_c$, i.e., $n \simeq n_c = 3a_\beta^3/4$. Half the total power is radiated at frequencies $\omega < \omega_c/2$ and half at $\omega > \omega_c/2$. For harmonics below n_c ($\omega \ll \omega_c$), the radiation intensity increases as $(\omega/\omega_0)^{2/3}$, and above n_c ($\omega \gg \omega_c$), the radiation intensity decreases exponentially as $\exp(-2\omega/\omega_c)$. Furthermore, for $\omega \ll \omega_c$, the scattered radiation at a fixed frequency is confined to an angular spread $\Delta\theta = (\omega_c/\omega)^{1/3}/\gamma_{z0}$ about $\theta = 0$, whereas for $\omega > \omega_c$, $\Delta\theta = (\omega_c/3\omega)^{1/2}/\gamma_{z0}$. The average angular spread for the frequency integrated spectrum in the vertical direction ($\phi = \pi/2$) is is $\theta_v = \langle\theta^2\rangle^{1/2} \sim 1/\gamma_{z0}$. In the horizontal direction ($\phi = 0$), emission is confined to the angle $\theta_h \sim a_\beta/\gamma_{z0}$.

A plot of the function $Y(\xi) = \xi^2 K_{2/3}^2(\xi)$ versus $\omega/2\gamma_{z0}^2\omega_\beta$ is shown in Fig. 3 (linear scale) and Fig. 4 (log scale). The solid curve shows the radiation from a single electron with $a_\beta = \gamma_{z0}k_\beta r_\beta = 10$. The dashed (dotted) curve shows the spectrum integrated over a flat-top (Gaussian) distribution of betatron amplitudes r_β with a rms value satisfying $a_{\beta,rms} = \gamma_{z0}k_\beta r_{\beta,rms} = 10$. To calculate these averages, the quantity $\xi = \omega/\omega_c$ has been approximated by $\xi \simeq (\omega/2\gamma_{z0}^2\omega_\beta)(1 + 3a_\beta/2)^{-1}$, since the asymptotic form for the spectrum is not accurate when $a_\beta < 1$, i.e., Figs. 3 and 4 are inaccurate in the region $\omega/2\gamma_{z0}^2\omega_\beta \lesssim 1$.

RADIATED POWER AND ELECTRON ENERGY LOSS

The power radiated by a single electron, P_s, undergoing relativistic quiver motion in an intense laser field can be calculated from the relativistic Larmor formula [11]

$$P_s = (2e^2/3c)\gamma^2 \left[(d\mathbf{u}/dt)^2 - (d\gamma/dt)^2\right]. \tag{33}$$

Using the orbits described above, the power radiated by a single electron undergoing betatron motion is given by

$$\langle P_s \rangle \simeq r_e m_e c^3 \gamma_{z0}^2 k_\beta^2 a_\beta^2/3, \tag{34}$$

where an averaging was performed over the betatron period, $\gamma_{z0}^2 \gg 1$ was assumed, and $r_e = e^2/m_e c^2$ is the classical electron radius.

The rate at which a single electron loses energy due to radiating is given by $W'_{loss} = \langle P_s \rangle/c$, i.e.,

$$W'_{loss} \simeq r_e m_e c^2 \gamma_{z0}^2 k_\beta^2 a_\beta^2/3. \tag{35}$$

In the blowout regime, $k_\beta \sim n_0^{1/2}\gamma_{0z}^{-1/2}$, and the rate of energy loss scales as $W'_{loss} \sim n_0^2\gamma_{z0}^2 r_\beta^2$. In addition, if the betatron amplitude is equal to the matched beam radius $r_b = (\epsilon_n/\gamma_{z0}k_\beta)^{1/2}$, the energy loss scales as $W'_{loss} \sim \epsilon_n \gamma_{z0}^3 k_\beta^3 \sim \epsilon_n n_0^{3/2}\gamma_{z0}^{3/2}$. For example, in the blowout regime at a density $n_0 = 2 \times 10^{14}$ cm^{-3} ($\lambda_p = 0.24$ cm) and a beam energy of $\gamma_{z0} = 6 \times 10^4$, an electron with a betatron amplitude of $r_\beta = 100$ μm ($a_\beta = 45$) would lose energy at a rate of $W'_{loss} = 0.2$ MeV/m.

ION CHANNEL LASER

Under special conditions, e.g., sufficiently high electron beam quality, self-amplified spontaneous emission (SASE) can occur whereby the incoherent synchrotron radiation emitted by the electrons is amplified via the ion channel laser (ICL) mechanism [2]. In the ICL instability, the radiation beats with the betatron motion to create an axial $\mathbf{v} \times \mathbf{B}$ (i.e., ponderomotive) force that leads to bunching of the electron beam and growth of the radiation field. This can lead to large levels of semi-coherent or coherent radiation. In SASE, the incoherent, spontaneous radiation acts as a seed for the instability, in a manner analogous to the SASE mode of operation in a free electron laser (FEL) [8].

There are important differences between the ICL and FEL mechanisms, however, that limit the SASE mode of operation. For electrons undergoing betatron motion in a plasma focusing channel, the resonant frequency of the radiation emitted along the axis is given by $\omega = 2\gamma_{z0}^2 n \omega_\beta / (1 + a_\beta^2/2)$, as indicated by Eq. (23). For an FEL, the resonant frequency is $\omega = 2\gamma_{z0}^2 n \omega_w / (1 + a_w^2/2)$, where $\omega_w = c k_w = 2\pi c / \lambda_w$, λ_w is the wiggler wavelength, $a_w = eB_w/k_w m_e c^2$ is the wiggler strength, and B_w is the field amplitude of the wiggler magnet. In an ideal FEL, a_w is a constant since all the electrons experience the same value of B_w. This is contrast to the focusing channel, in which $a_\beta = \gamma_{z0} k_\beta r_\beta$ is a function of both the electron energy γ_{z0} and the radial position of the electrons via the betatron amplitude r_β. If a mono-energetic beam of finite radius is injected into a focusing channel (without any special tapering), electrons at different radii will have different betatron amplitudes r_β, different values of a_β, and hence different resonant frequencies.

Furthermore, for an ideal FEL with a planar wiggler of the form $\mathbf{B} = B_w \cos(k_w z) \mathbf{e}_x$, all of the beam electrons wiggle in the same plane with the same amplitude, i.e., $\mathbf{u}_\perp = a_w \cos(k_w z) \mathbf{e}_x$. Consequently, radiation emitted by all the electrons will have similar polarization. This is contrast to the focusing channel, in which the betatron motion, and hence the synchrotron radiation, will have a variety of polarizations in the x-y plane, depending on the position and angle of the electron as it enters the channel. Hence, to amplify radiation of a given frequency and polarization in a focusing channel, only those beam electrons with the proper values of γ_{z0} and r_β with be resonant with the radiation, and only a subset of these will have the proper polarization. This is contrast to an ideal FEL, in which all the electrons in a mono-energetic beam are resonant with the radiation field with the proper polarization.

It is straightforward to quantify some the conditions necessary for SASE to occur in a plasma focusing channel. In the following discussion, it is assumed that $k_\beta^2 r_\beta^2 \ll 1$. Consider an ideal mono-energetic electron beam of radius r_b injected into a focusing channel such that the beam centroid is along the z axis. A electron moving along the axis would have a betatron amplitude of $r_\beta = 0$, whereas an electron residing at the edge of the beam would have a betatron amplitude of $r_\beta = r_b$. For the beam to emit radiation along the axis with a narrow bandwidth $\Delta\omega/\omega \ll 1$, it is necessary that $a_\beta^2 \ll 1$ for all the electrons. This implies that the radiation wavelength satisfy $\lambda > \pi r_b/\gamma$. For a matched beam with a normalized emittance ϵ_n, the matched-beam radius is given by $r_{bm} = (\epsilon_n/\gamma k_\beta)^{1/2}$, and the condition $a_\beta^2 \ll 1$ implies

$$\lambda \gg \pi\epsilon_n/\gamma. \tag{36}$$

It is interesting to note the similarity of this condition with that usually required of a SASE FEL [8], $\lambda > 4\pi\epsilon_n/\gamma$.

The condition $\Delta\omega/\omega \ll 1$, however, is not sufficient for the SASE process to occur. A more stringent condition is that the normalized axial energy spread $\Delta\gamma_z/\gamma_z$ be small compared to the so-called Pierce or gain parameter ρ, i.e., $\Delta\gamma_z/\gamma_z \ll \rho$, where by analogy with an FEL,

$$\rho = \left[\frac{a_\beta k_{pb}}{4\gamma^{3/2}k_\beta}F_\Delta^2(a_\beta)\right]^{2/3}, \tag{37}$$

where $k_{pb} = 4\pi n_b e^2/m_e c^2$, n_b is the beam density, and

$$F_\Delta(a_\beta) = J_0\left(\frac{a_\beta^2/4}{1+a_\beta^2/2}\right) - J_1\left(\frac{a_\beta^2/4}{1+a_\beta^2/2}\right). \tag{38}$$

In terms of the beam current $I_b = ec\pi n_b r_b^2$, and evaluating the expression for ρ at $r_\beta = r_b$, gives

$$\rho = \left(I_b F_\Delta^4/4\gamma I_A\right)^{1/3}, \tag{39}$$

where $I_A = m_e c^3/e = 17$ kA. Using the equations of motion for an electron in a focusing channel, Eqs. (5)-(8), the normalized energy spread is given by $\Delta\gamma_z/\gamma_z \simeq a_\beta^2/4$, for a beam with a centroid along the axis. Hence, $\Delta\gamma_z/\gamma_z < \rho$ implies $a_\beta^2 < 4\rho$ or $\lambda > \pi r_\beta/(2\gamma\rho^{1/2})$. For a matched beam, this gives

$$\lambda > \frac{\pi\epsilon_n}{4\gamma\rho}. \tag{40}$$

This is considerably more stringent that the usual FEL constraint $\lambda > 4\pi\epsilon_n/\gamma$, since typically $\rho \ll 1$. For the parameters of the E-157 experiment, $\rho \simeq 5 \times 10^{-3}$.

In principle, it may be possible to tailor the energy distribution and radial profile of the beam such that a greater fraction of the beam electrons are in resonance with the radiation field. For example, consider a mono-energetic, very narrow beam of width Δr_b injected off-axis such that the centroid of the beam executes betatron oscillations of amplitude $r_\beta = r_{b0}$ with $r_{b0} \gg \Delta r_b$ [13]. In this case, all of the electrons in the beam would undergo approximately the same betatron orbit and would have approximately the same value for a_β, i.e., the spread in a_β is given by $\Delta a_\beta/a_\beta \simeq \Delta r_b/r_{b0}$. In this case the condition $\Delta\omega/\omega \ll 1$ implies $\Delta r_b/r_{b0} \ll (1+a_\beta^2/2)/a_\beta^2$, which in principle, could be easily satisfied. The more stringent condition, $\Delta\gamma_z/\gamma_z < \rho$, implies $\Delta r_b/r_{b0} < 2\rho/a_\beta^2$, which could be satisfied for sufficiently small values of a_β.

Even if the condition $\Delta\gamma_z/\gamma_z < \rho$ is satisfied, it is not clear that the SASE process would occur. In a conventional FEL, SASE requires that a number of conditions be satisfied (in addition to $\Delta\gamma_z/\gamma_z < \rho$) [8], i.e., $\epsilon_n < \gamma\lambda/4\pi$, $N_\beta\lambda_\beta \gg L_G$, $L_G < L_R$, and $N_\beta\lambda < L_e$, where N_β is the number of betatron oscillations, $L_g \simeq 0.046\lambda_\beta/\rho$ is the gain length, $L_R = \pi w_0^2/\lambda$ is the Rayleigh length of the radiation with spot size w_0, and L_e is the electron bunch length. Furthermore, for the case of an ICL driven by a narrow beam with a centroid undergoing betatron oscillations, it is likely that the gain (i.e., ρ) is reduced since the geometric overlap between the electron beam and the radiation is reduced, due to the betatron motion of the centroid. Such novel ICL configurations require a detailed analysis.

SUMMARY

Spontaneous radiation emitted from an electron undergoing betatron motion is a plasma focusing channel was analyzed starting from basic principles. Application of these results to the E-157 experiment and to the ICL were examined. Important similarities and differences between SASE in an FEL and in an ICL were delineated. In particular, the spontaneous radiation emitted along the axis of a plasma focusing channel from a single electron occurs near the resonant frequency $\omega_n = 2\gamma_{z0}^2 n\omega_\beta/(1 + a_\beta/2)^{1/2}$. The role of the betatron strength parameter a_β is analogous to that of the wiggler strength parameter a_w (or K_w) in FEL physics. In Ref. [2], the ICL was considered only in the limit $a_\beta^2 \ll 1$. When $a_\beta^2 \ll 1$, radiation is emitted primarily at the fundamental frequency $\omega = 2\gamma_{z0}^2\omega_\beta$ and is independent of a_β. For $a_\beta \gtrsim 1$, however, the resonant frequency is a strong function of a_β and radiation is emitted in numerous harmonics extending out to the critical harmonic number $n_c = 3a_\beta^3/4$. This is the case in the E-157 experiments, in which $a_\beta \sim 2-50$.

In an ideal FEL, the wiggler strength parameter a_w is a constant (a function of only the magnetic field of the wiggler) for all of the beam electrons. However, in an ICL, $a_\beta = \gamma_{z0}k_\beta r_\beta$ depends on both the electron energy γ_{z0} and the betatron amplitude r_β. Since r_β, and hence a_β, is different for every electron in a typical beam, this places serious limits on the possibility of realizing a SASE ICL. For an electron beam center about the axis with a radius r_b and $a_\beta(r_b) > 1$, the radiation from the beam is no longer emitted at discrete harmonics as it would be from a single electron with $r_\beta = r_b$. Rather, since $0 < a_\beta \lesssim a_\beta(r_b)$ for the electrons in the beam, the resulting radiation is in the form of a broad continuum as indicated by Figs. 1 and 2, even for the case of an initially mono-energetic beam. In the limit $a_\beta^2 \ll 1$, the radiation from the beam could be nearly monochromatic at the fundamental frequency. The condition $a_\beta^2 \ll 1$ implies $\lambda \gg \pi\epsilon_n/\gamma$ for a matched beam, which is similar to the criterion $\lambda > 4\pi\epsilon_n/\gamma$ often quoted for a SASE FEL. The condition $a_\beta^2 \ll 1$, however, is not sufficient to insure that the SASE ICL process will occur. A more stringent condition for the occurrence of SASE is on the axial energy spread of the beam within the focusing channel, i.e., $\Delta\gamma_z/\gamma_z \ll \rho$, where ρ is the effective Pierce (or gain) parameter. Again, since r_β varies across the beam, there exists a large energy spread $\Delta\gamma_z/\gamma_z \simeq a_\beta^2/4$. The condition $\Delta\gamma_z/\gamma_z \ll \rho$ implies $\lambda > \pi\epsilon_n/(4\gamma\rho)$ for a matched beam. Since typically $\rho \ll 1$, this restriction on the radiated wavelength $\lambda > \pi\epsilon_n/(4\gamma\rho)$ is much more stringent than that in a conventional SASE FEL. Furthermore, the betatron orbits in a typical beam in a focusing channel are not polarized in the same plane as they are in a conventional FEL. This also can reduce the gain in a SASE ICL. These arguments, however, assumed an untailored electron beam centered about the channel axis. It may be possible to relax this constraint on the radiated wavelength in a SASE ICL by appropriately tailoring the electron beam, for example, a narrow electron beam injected off-axis such that all of the beam electron execute approximately the same betatron orbit. Such novel ICL configurations require further analysis to access their viability.

ACKNOWLEDGMENTS

The authors acknowledge useful conversations with the participants of the working group on plasma wakefield accelerators and with the members of the E-157 collaboration. This work was supported by the Department of Energy under contract No. DE-AC-03-76SF0098.

REFERENCES

1. For a review see, E. Esarey et al., IEEE Trans. Plasma Sci. **24**, 252 (1996).
2. D.H. Whittum, A.M. Sessler, and J.M. Dawson, Phys. Rev. Lett. **64**, 2511 (1990); D.H. Whittum, Phys. Fluids B **4**, 730 (1992).
3. J. Ng et al., SLAC Preprint, SLAC-PUB-8501 (2000); SLAC E-150 web site, URL http://www.slac.stanford.edu/exp/e150.
4. M.J. Hogan et al., Phys. Plasmas **7**, 2241(2000); SLAC E-157 web site, URL http://www.slac.stanford.edu/grp/arb/e157.
5. J.B. Rosenzweig et al., Phys. Rev. A **44**, 6189 (1991).
6. P. Catravas et al., Proc. 1999 Particle Accelerator Conf., Ed. by A. Luccio and W. Mackay (IEEE, Piscataway NJ, 1999), pp. 2111-2113.
7. W.A. Barletta et al., Nuc. Instr. Meth. A **423**, 256 (1999).
8. *Handbook of Accelerator Physics and Engineering*, Edited by A.W. Chao and M. Tigner (World Scientific, Singapore, 1999).
9. S. Wang and C. Joshi, private communication.
10. M. Reiser, *Theory and Design of Charged Particle Beams* (Wiley, New York, 1994).
11. J.D. Jackson, *Classical Electrodynamics*, 2nd ed. (Wiley, New York, 1975), Chap. 14.
12. E. Esarey, S.K. Ride, and P. Sprangle, Phys. Rev. E **48**, 3003 (1993).
13. In collaboration with the working group on plasma wakefield accelerators, this workshop.

Commissioning and Measurements of the Neptune Photo-injector

S. G. Anderson, M. Loh, P. Musumeci, J. B. Rosenzweig, H. Suk, M. C. Thompson

*Department of Physics and Astronomy,
University of California, Los Angeles
405 Hilgard Ave., Los Angeles, CA 90095*

Abstract. The photo-injector for the Neptune Advanced Accelerator Laboratory is introduced. Its component parts, including the radio frequency gun, photo-cathode drive laser system, booster linac, RF system, chicane compressor, beam diagnostics, and control system are described. The injector is designed to produce high brightness, short pulse electron beams. Measurements of the photo-injector beams including quantum efficiency, emittance, pulse length, and pulse compression are presented.

THE NEPTUNE PHOTO-INJECTOR

The Neptune Advanced Accelerator Laboratory consists of two main components, the RF photo-injector and the high power, short pulse, two frequency Mars CO_2 laser. [1] The main goal of the lab is to accelerate a high quality, relativistic electron beam injected into a plasma beat wave accelerator (PBWA) to over 100 MeV, while preserving the phase space density of the injected beam. [2] The PBWA experiment can take two different forms, one where a beam of moderate charge and emittance (Q ≈ 1 nC, ε_n ≈ 5 mm mrad) covers more than one plasma wavelength, and the second where a shorter, low charge and emittance beam (Q ≈ 50 pC, ε_n < 0.1 mm mrad) is loaded into a single cycle of the PBWA.

Because of the range of injected beams required by the two phases of the PBWA experiment, the Neptune photo-injector is designed to produce an emittance compensated [3,4], optimized beam over an extent of different charges. This is done with a powerful method of charge scaling recently developed for photo-injectors. [5] This flexibility in the beam parameters the photo-injector can produce makes feasible many advanced accelerator experiments in the Neptune lab. These include free-electron laser (FEL) microbunching for injection into the PBWA, underdense plasma focusing [6-8], plasma wake-field acceleration (PWFA) [6,9], and inverse FEL acceleration [10]. In addition to these advanced acceleration experiments, studies of the high brightness beams themselves are underway at Neptune. These include the role of space-charge in emittance measurements, beam compressibility, and the process of emittance dilution in bends.

In the remainder of this section we describe the various component parts of the photo-injector and where applicable, their current performance. The components of the photo-injector include the RF gun, photo-cathode drive laser system, booster linac, rf system, chicane compressor, beam diagnostics, and control system. Figure 1 shows the layout of the photo-injector beamline.

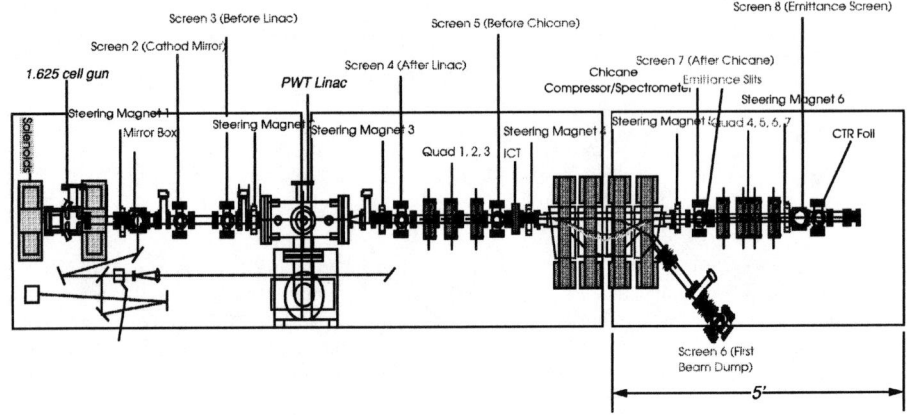

FIGURE 1. The Neptune Photo-injector beamline.

Accelerator Sections

The photo-injector has a split accelerator design consisting of a photo-cathode gun, a drift space, and a booster linac. The gun is a 1.625 cell π-mode standing wave cavity produced by a BNL-SLAC-UCLA collaboration. [11] The gun has been conditioned up to an input power of 6.5 MW, which corresponds to an on-axis peak field of 100 MV/m. The current operating power in the gun is somewhat lower than this (due to limited total output power of the rf system) and the nominal peak accelerating field is 85 MV/m. The original cathode of this gun was simply the OFHC copper backplane of the half-cell. More recently however, this backplane was replaced with one including a 1 cm diameter, 1 mm thick disk of single crystal copper (Cu_{100}). [12] The properties of the two cathodes will be discussed further in the measurements section below. The booster linac is a 7 and 2/2 cell π-mode standing wave structure. The linac design is that of a plane-wave transformer (PWT) which benefits from strong cell-to-cell coupling and large mode separation. [13] The linac has been conditioned up to 13 MW of input power and runs with a nominal peak accelerating field of 50 MV/m.

RF System

Low level RF starts with the 38.08 MHz output signal of the mode-locked laser oscillator (which is the first component of the photo-cathode drive laser). This signal is frequency multiplied by 75 to produce S-band RF. After passing though a phase shifter, which allows us to control the laser injection phase, the signal is amplified to approximately 700 Watts by a pulsed amplifier. The kW level RF is then used as the input to a SLAC XK-5 klystron. The klystron is pulsed by a modulator with a pulse length of 4 μsec. The modulator was designed to produce a flat-top pulse impedance matched to the klystron at high voltage when fired by an SCR-triggered thyratron. The current klystron typically makes 20 MW of RF power.

The RF power distribution system consists of SF_6 filled wave guide separating power manipulating elements. The first of these is a circulator, which protects the klystron from reflected power due to the impedance mismatch at the standing wave structures at the beginning and end of an RF pulse. The power is then split by a 4.77 dB divider sending one third to the gun. After the split high power attenuators control the power delivered to each accelerator. In addition the linac wave guide has a phase shifter to control the relative phase of the two structures. Figure 2 shows a schematic diagram of the Neptune RF system.

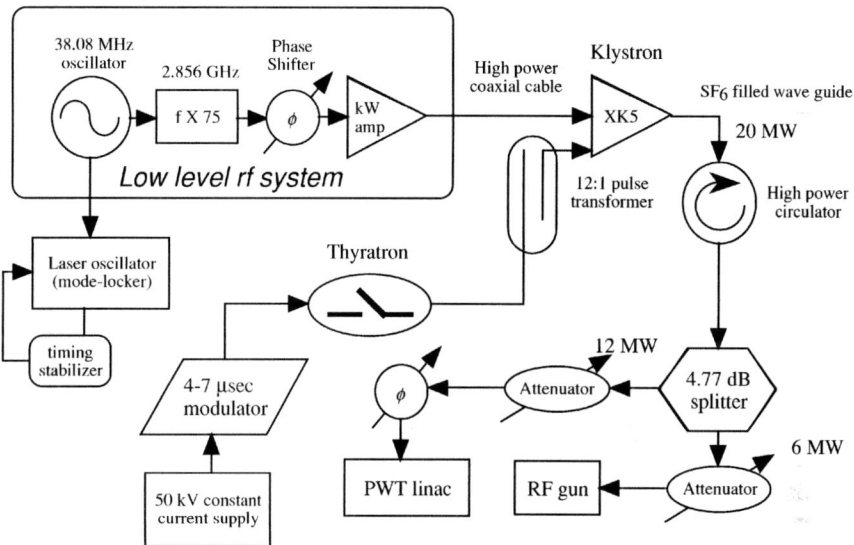

FIGURE 2. The Neptune RF System

Photo-cathode Drive Laser

The drive laser system begins with a 1064 nm mode-locked Nd:YAG laser oscillator which is matched into a 500 m long fiber to lengthen the pulse and yield a frequency chirp. The chirped pulse is then sent to a regenerative amplifier that increases the signal by a factor of one million. The chirp correlation is then removed and the pulse compressed by a grating pair. Adjustments to the grating pair allow control over the pulse length which is currently set at 6 psec (FWHM). At this point the pulse is frequency doubled by a BBO doubling crystal. The green laser light is then transported approximately 40 meters to the next BBO crystal which frequency doubles again to produce 266 nm light. The pulse energy in UV has been measured at

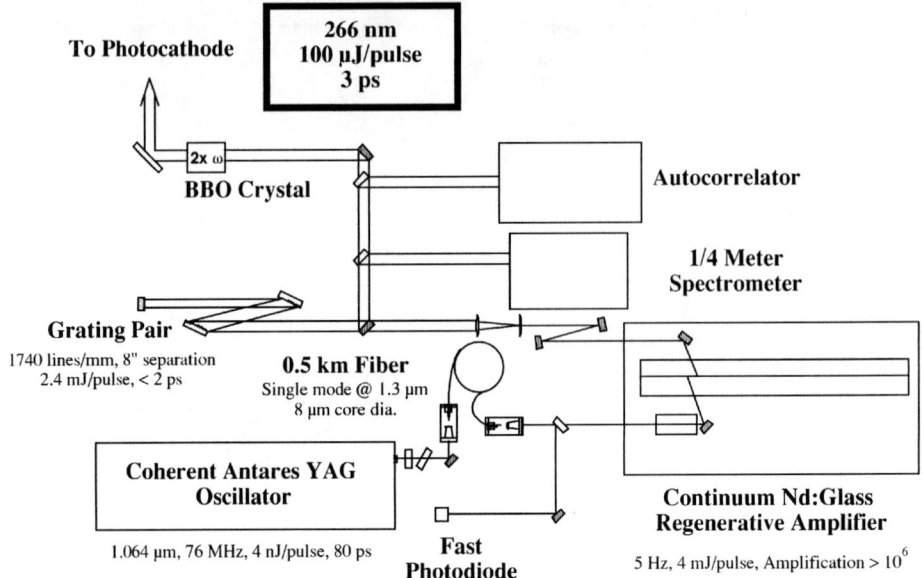

FIGURE 3. The Neptune Photo-cathode Drive Laser

130 µJ. Under nominal run conditions 100 µJ of UV energy is delivered to the cathode. Below is a schematic of the laser system.

Due to the long transport length, a vacuum transport system has been constructed to hold beam optics and to combat fluctuations in transverse position. To handle long time scale (≥ 10 sec) beam drift, a feedback system consisting of motorized mirror mounts and segmented photodiodes functions in the transport system. This system is computer automated by iteratively reading the position of a beacon laser with the photodiodes (beam position monitors) and adjusting the motorized mirrors accordingly.

Chicane Compressor

The compressor installed at Neptune was designed in part by scaling an L-band compressor designed for the TESLA Test Facility (TTF). [14] As shown in figure 4, it consists of four dipole magnets which can be configured either as a compressor or a spectrometer. In compressor mode a negative correlation in longitudinal phase space caused by running off-crest in the PWT is removed by the difference in path length of particles of different momentum. The problem of excessive vertical focusing in the chicane has been addressed by adjusting the initial and final edge angles to approximately equalize horizontal and vertical focusing in the device. By switching off the first two dipoles, the second two are used as a spectrometer. The chicane in spectrometer mode is used to measure the beam energy and energy spread. The typical beam energy at Neptune is 12 MeV with an energy spread of about 0.2%. The chicane is also being used to compress the beam and these measurements will be discussed in more detail below.

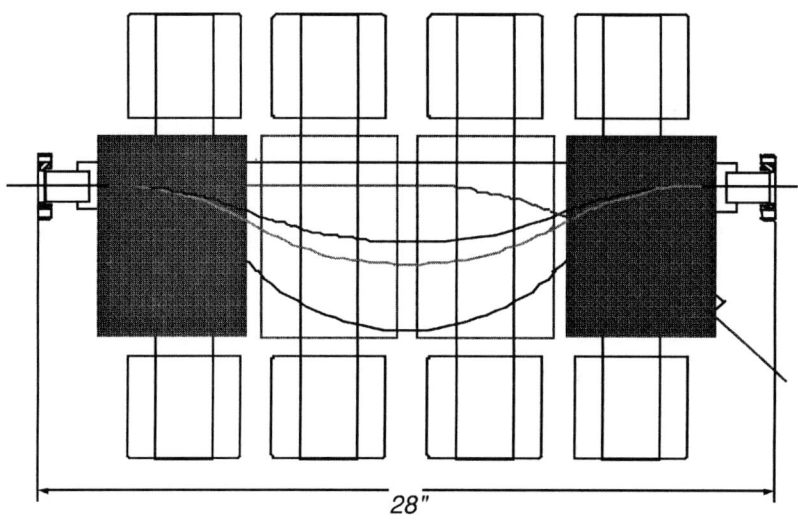

FIGURE 4. The Neptune Chicane Compressor

Beam Diagnostics

An important beam diagnosing tool is the pop-in view screen. At Neptune view screens are used for beam transport, spot size and profile measurement, and as an aid in the emittance compensation process. For this device phosphor is deposited on the downstream side of an aluminim foil mounted normal to the incident beam. A 45° mirror then directs light produced by the phospher out to a CCD camera. From there the video data is digitized by a computer and analysis is preformed on the image. In addition to phosphor, pop-in screens using YAG crystals, which offer higher resolution and better vacuum properties, are active at Neptune. The resolution limits of YAG crystals [15] are not a concern at Neptune because of the relatively low beam energy (12 MeV) and because in the photo-injector itself, the beam is never focused to a small spot.

Beam charge is measured primarily with an integrating current transformer (ICT). The ICT allows single shot, non-destructive charge measurements and has been used at Neptune to measure charges from 10 pC to over 1 nC. A charge of 600 pC is readily produced in the optimal accelerating phase of the RF gun. Measurements of charge and quantum efficiency (QE) will be detailed further below. At the end of the photo-injector beamline the beam is dumped into a Faraday cup, which was used to do initial charge measurements.

Additionally, beam diagnostics include a slit based emittance measurement system [16], and a bunch length measuremet technique using coherent transition radiation (CTR). [17] Both of these will be discussed in greater detail below.

Control System

The photo-injector control system begins with an Apple Macintosh computer. The computer has a video digitizing card which allows real time analysis such as dark current subtraction, spot size calculation, and emittance slit image analysis. Also, the computer is equipped with a GPIB interface which is used to import oscilloscope traces, and communicate with a GPIB controlled CAMAC crate. The CAMAC crate contains modules responsible for pop-in screen insertion, steering and quadrupole magnet control and read-back, chicane control, and rf attenuators and phase shifters.

BEAM MEASUREMENTS

In this section we discuss measurements made on the basic properties of the high brightness beams produced by the Neptune photo-injector. In particular, measurements of quantum efficiency, emittance, and pulse length are presented.

Quantum Efficiency

The number of electrons freed from the cathode surface per incident photon (the QE) is an important quantity for photo-injectors, both in the demands it puts on the drive laser and in its effect on the output beam quality (QE variations over the emitting surface of the cathode lead to variations in the beam density and can cause emittance growth [18]). For that reason, a number of studies of quantum efficiency and uniformity of emission of different materials are planned at Neptune. We present here the QE found for a single crystal copper (CU_{100}) cathode. [12]

The procedure used to measure the QE is straight forward. As mentioned above, an ICT was used to measure the beam charge. The ICT response was calibrated using a fast pulser and that calibration was in agreement with the factory specifications. The ICT signal was then fed into a fast, integrating ADC CAMAC module, which could be read by the computer. Knowing the calibrations of the ICT and the CAMAC module allowed the calculation of the charge on a single shot basis. As a further check, the ICT and Faraday cup signals were compared and found in agreement. UV energy was measured with a power meter placed behind a mirror. Because the amount of light transmitted through the mirror was very small, the power meter signal was amplified before being sent to a GPIB connected oscilloscope. The calibration between this signal and the UV energy was performed by measuring the energy directly with a second power meter. The relatively slow repetition rate of the photo-injector (1 Hz) allowed the computer to record both the charge and UV energy between shots and calculate the QE single-shot. The QE was obtained by varying the laser energy delivered to the cathode as shown in figure 5 below.

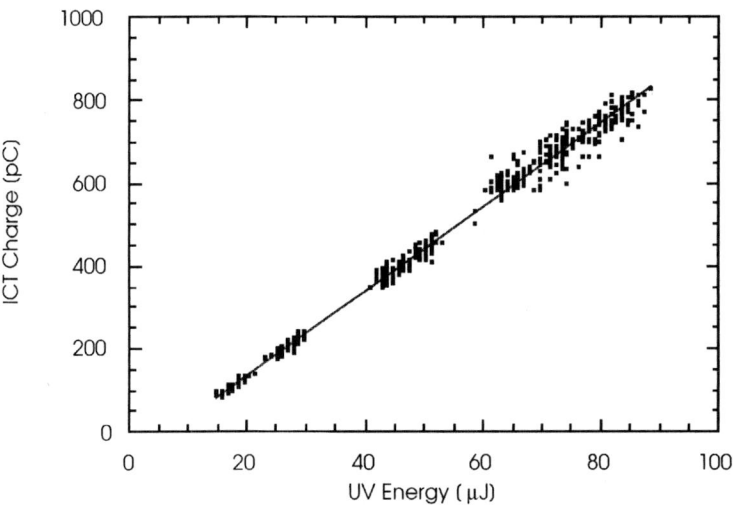

FIGURE 5. Plot of laser energy versus beam charge. The quantum efficiency is taken from the slope of a linear fit of the data.

We found the QE with a peak accelerating field of 85 MV/m and an injection phase of 45°, (corresponding to an electric field of about 60 MV/m at injection) to be 5×10^{-5}. This is a promising result because it is roughly twice the value found for an OFHC copper cathode operating under the same conditions. Measurements of the uniformity of emission, cathode cleaning with kV electrons [19], and the QE of different materials are planned for the near future.

Emittance

The space charge dominated behavior of the high brightness beams produced at Neptune can be seen through examination of the RMS envelope equation for a beam in a drift.

$$\sigma_x'' = \frac{\varepsilon_n^2}{\gamma^2 \sigma_x^3} + \frac{4I}{\gamma^3 I_0 (\sigma_x + \sigma_y)} \tag{1}$$

Here the ratio of the space charge to emittance terms determines the character of the electron beam.

$$R = \frac{2I\sigma_0^2}{I_0 \varepsilon_n^2} \tag{2}$$

For typical Neptune parameters $R \approx 20$, indicating a space charge dominated beam. Thus, any emittance measurement scheme based on beam propagation in a drift must take this ratio into account.

The slit based emittance measurement system is illustrated by figure 6. In this system the beam is collimated into narrow beamlets by a set of slits inserted into the beam path. The beamlets are then allowed to drift a given distance to a pop-in view screen. By integrating the digitized image over the vertical dimension (for horizontal emittance) an intensity profile is produced. This intensity profile is analyzed to determine the position, momentum, and momentum spread of each beamlet and thus, the emittance.

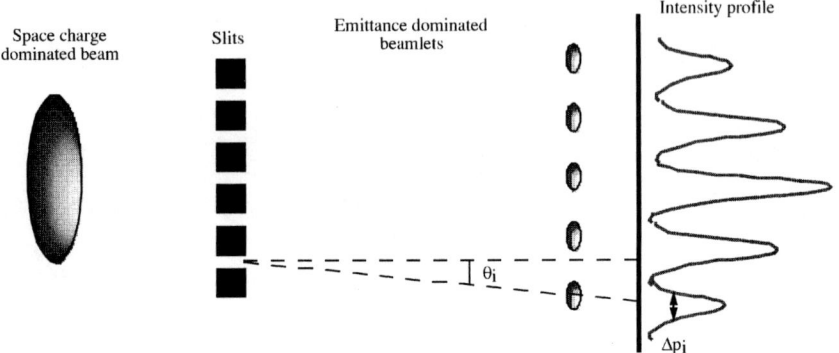

FIGURE 6. Cartoon illustrating the operation of a slit based emittance measurement system.

In addition, a crucial function of the slits is to produce *emittance* dominated beamlets. The ratio of space charge to emittance in the envelope equation for a beamlet is:

$$R_{beamlet} = \sqrt{\frac{2}{3\pi} \frac{I}{\mathcal{I}_0} \left(\frac{d}{\varepsilon_n}\right)^2} \qquad (3)$$

where d is the width of the slit. For the slits installed at Neptune, d = 50 μm and $R_{beamlet} \approx 0.04$. There are other issues that were considered in the design of the slit system including angular acceptance and interaction between beamlets. [16]

The single-shot nature of the slit emittance measurements allows real time calculation of the emittance via the analysis of the image intensity profile. This is done at Neptune with the computer running LabVIEW programs. An example of the computer program calculating emittance is shown in figure 7 below.

The horizontal, normalized emittance of the photo-injector has been measured using the slit system. We found typical normalize, RMS emittances of 5 mm mrad with a charge of 600 pC and pulse length of 3.6 ps (RMS). The complete set of nominal Neptune beam parameters is given in table 1 at the end of this report.

Planned beam dynamics studies using this technique are an examination of the emittance compensation process, and emittance growth in the compressor.

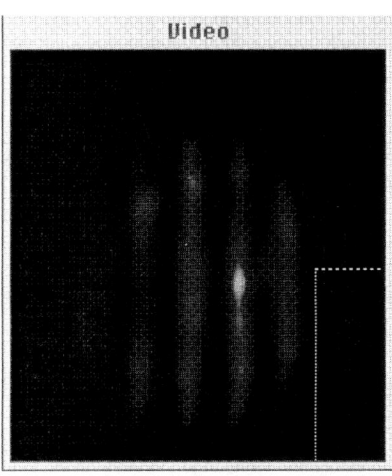
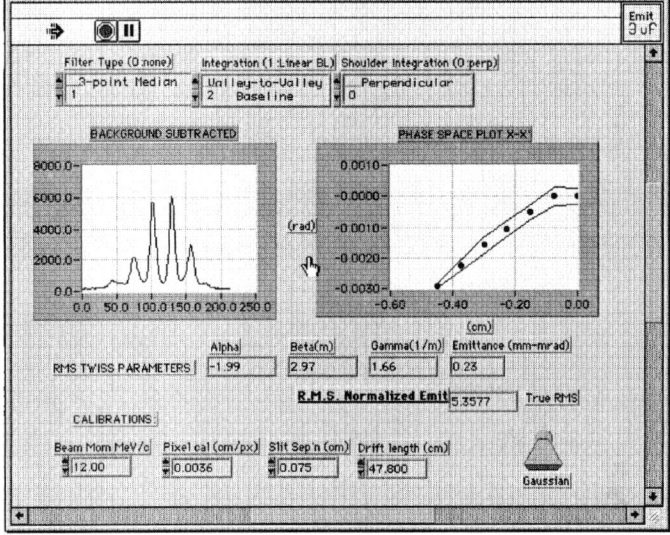

FIGURE 7. Emittance calculation software allows the measurement of emittance and other beam parameters such as charge in the same shot. The top image is a picture of the beam passing through the slits. The bottom image is the program calculating the emittance of the beam.

Pulse Length

Direct measurement of the pulse length of the electron beam is made with a device devoloped by Uwe Happek at the University of Georgia. [20] This device uses coherent transition radiation (CTR) to find the bunch longitudinal profile. The CTR is created by colliding the beam with a 45º foil. The radiation is directed out though a window into the device shown in figure 8. The device is a polarizing transmission (wire grating) Michelson interferometer.

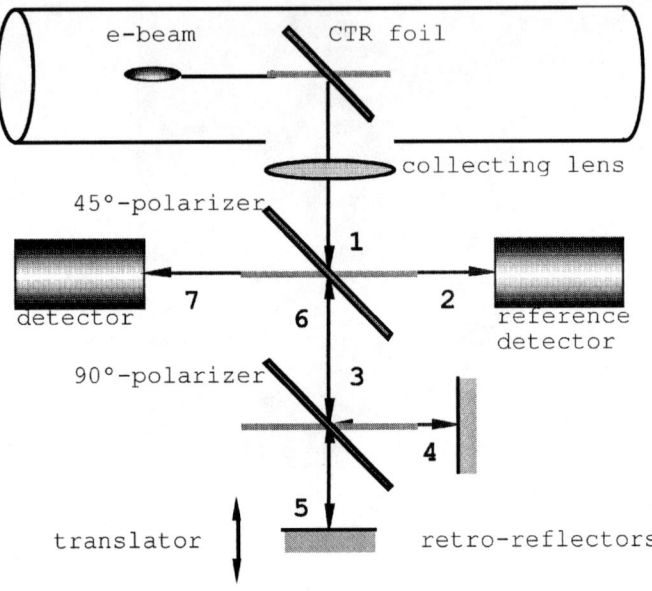

FIGURE 8. The CTR based pulse length measurement system.

Since the CTR created at the foil has the same temporal profile as the electron beam, the signal at the interferometer detector should be proportional to the autocorrelation of the beam density profile. [17] Thus, if the beam is gaussian longitudinally, the detector signal should also be a simple gaussian. The data, as shown in figure 9, does not show a simple form, but rather has multiple hills and valleys. This pattern was found in all cases the measurement was performed.

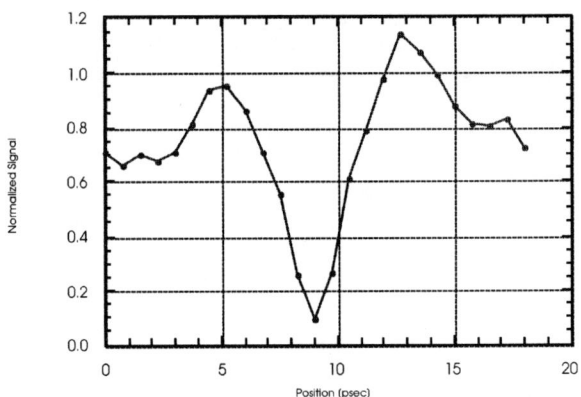

FIGURE 9. Raw data taken with the CTR interferometer.

The reason the raw data does not look like a simple autocorrelation of the beam profile is because the longer wavelength components of the CTR are lost in the device due to diffraction and finite apertures of the optics. Therefore, the data is the autocorrelation of the *filtered* beam distribution. The analysis of the time domain signal produced by the interferometer including this filtering effect was performed previously in reference 17. The key ideas of that analysis are presented now in order to obtain a pulse length from the measurements performed at Neptune.

The missing low frequency information in the autocorrelation can be modeled in the frequency domain by multiplying by a filtering function.

$$\tilde{\rho}_m(\omega) = \tilde{\rho}(\omega) g(\omega) \qquad (4)$$

We chose $g(\omega)$ to be

$$g(\omega) = 1 - e^{-\xi^2 \omega^2} \qquad (5)$$

This choice of filter function eases further analysis and is physically motivated. It is obtained by the aperturing of a diffraction-limited transverse Gaussian-mode photon beam of uniform initial frequency spectrum in the far field. With the filter function applied, the spectrum of the measured signal is

$$\tilde{s}(\omega) = |\tilde{\rho}_m(\omega)|^2 = |\tilde{\rho}(\omega)|^2 \left[1 - 2e^{-\xi^2 \omega^2} + e^{-2\xi^2 \omega^2} \right] \qquad (6)$$

By assuming a Gaussian beam profile we obtain an analytical expression for the signal in the time domain.

$$s(\tau) \propto \left[e^{-\frac{(\tau-\tau_0)^2}{4\sigma^2}} - \frac{2\sigma}{\sqrt{\sigma^2 + \xi^2}} e^{-\frac{(\tau-\tau_0)^2}{4(\sigma^2+\xi^2)}} + \frac{\sigma}{\sqrt{\sigma^2 + 2\xi^2}} e^{-\frac{(\tau-\tau_0)^2}{4(\sigma^2+2\xi^2)}} \right] \qquad (7)$$

Figure 10 shows the autocorrelation data for different compressor settings with this time domain fit function applied in each case. We see that the agreement between the time domain fit function and the data is good and that the compressor set to its design bend angle compresses the beam to a pulse length of 1.0 ps RMS (the uncompress pulse length was found to be 4 ps RMS).

The pulse length measurement scheme, in combination with the compressor and emittance slits, allows us to investigate the phenomenon of emittance growth in bends. Experiments in this area are planned for the near future and will be complimented by simulations using a three-dimensional code based on Lienard-Wiechart potentials. [21]

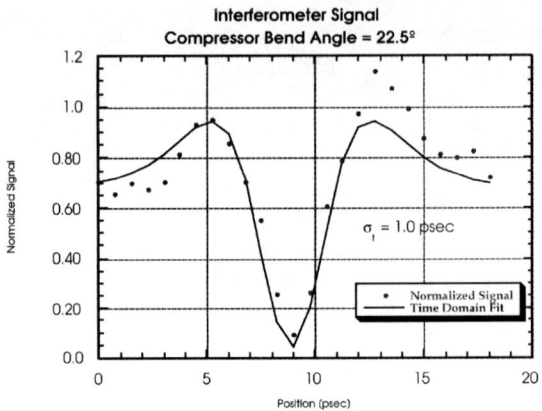

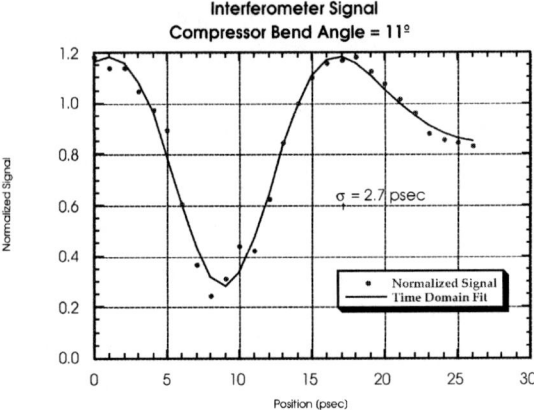

FIGURE 10. Interferometer data for two different compressor settings. In the first case the compressor magnets were set for the design bend angle of 22.5°. The pulse length calculated from the time domain fit is 1 ps (RMS). In the second case the bend angle was set to 11° and the calculated pulse length is 2.7 ps (RMS).

CONCLUSION

The photo-injector in the Neptune Advanced Accelerator Laboratory has been commissioned and the various properties of the beam have been measured. Table 1 summarizes the beam parameters. In addition to the PBWA experiment, the photo-injector will be utilized in a variety of advanced accelerator experiments as well as high brightness beam studies such as the process of emittance growth in bends.

TABLE 1. Neptune Photo-injector Parameters

Parameter	Value
Gun Power	6 MV
Gun Peak Field (E_z)	85 MV/m
PWT Power	12 MV
PWT Peak Field (E_z)	50 MV/m
Laser Energy (at 266 nm)	100 µJ
Laser Pulse Length (RMS)	3 ps
Beam Energy	12 MeV
Energy Spread (RMS)	0.2%
Pulse Length (RMS)	3.6 ps uncompressed
	1 ps compressed
Charge	600 pC typical
	up to 1 nC
Quantum Efficiency	5×10^{-5}
Emittance (normalized, RMS)	5 mm mrad at 600 pC

ACKNOWLEDGMENTS

The authors wish to thank Dennis Palmer for providing the single crystal copper cathode used at Neptune, as well as for numerous conversations and advice on the matter.

REFERENCES

1. C.E. Clayton et al., *Phys. Rev. Lett.* **54**, 2353 (1985); C.E. Clayton, et al., *Phys. Rev. Lett.* **70**, 37 (1993).
2. C.E. Clayton et al., *Nucl. Instrum. Methods A* **410**, 378 (1998).
3. B.E. Carlsten, *Nucl. Instr. Methods A* **285**, 313 (1989).
4. L. Serafini and J.B. Rosenzweig, *Phys. Rev. E* **55**, 7565 (1997).
5. J.B. Rosenzwieg and E. Colby, *Adv. Accelerator Concepts* **335**, 724 (AIP Conf. Proc., 1995).
6. J.B. Rosenzwieg et al., *Nucl. Instrum. Methods A* **410**, 532 (1998).
7. N. Barov et al., *Phys. Rev. Lett.* **80**, 81 (1998).
8. J.J. Su et al., *Phys. Rev. A*; S. Rojagapolan et al., *Phys. Rev. D*.
9. J.B. Rosenzweig et al., *Phys. Rev. A* **44**, 6189 (1991).
10. P. Musumeci and C. Pellegrini, these proceedings.
11. D.T. Palmer et al., *Proc. 1997 Particle Accelerator Conf.*, 2846 (IEEE, 1997).
12. D.T. Palmer et al., *Proc. The 2nd ICFA Advanced Accelerator Workshop on the Physics of High Brightness Beams*, to be published. (1999).
13. R. Zhang et al., *Proc. 1995 Particle Accelerator Conf.*, 1102 (IEEE, 1995).
14. E. Colby et al., *Proc. 1995 Particle Accelerator Conf.*, 1445 (IEEE, 1995).
15. A. Murokh et al., *Proc. The 2nd ICFA Advanced Accelerator Workshop on the Physics of High Brightness Beams*, to be published. (1999).
16. J.B. Rosenzweig et al., *Nucl. Instrum. Methods A* **341**, 379 (1994).
17. A. Murokh et al., *Nucl. Instrum. Methods A* **410**, 452 (1998).
18. S.G. Anderson and J.B. Rosenzweig, *Accepted for publication in Physical Review Special Topics - Accelerators and Beams* (July 2000).
19. M.C. Thompson et al., *Proc. The 2nd ICFA Advanced Accelerator Workshop on the Physics of High Brightness Beams*, to be published. (1999).
20. U. Happek, A.J. Sievers, E.B. Blum, *Phys. Rev. Lett.* **67**, 2962 (1991).
21. F. Ciocci et al., Nucl. Instrum. Methods A **393**, 434 (1997).

Summary of Japanese Advanced Accelerator Work

M.Uesaka

Nuclear Engineering Research Laboratory, University of Tokyo
Tokai-mura, Naka-gun, Ibaraki-ken, 319-1106, Japan

Abstract. Japanese advanced accelerator work is summarized with respect to compactness, high beam quality, short electron bunch and advanced application. GeV laser accelerator with the 300MeV microtron and 100TW 20fs Ti:Sapphire laser is under construction and the preliminary operation has started at JAERI-APR. Self-modulated wakefield accelerator in 40TW laser-plasma has yielded 35MeV electron (160GV/m) and 7MeV protons at ILE, Osaka Univ. ~1MeV electrons have been produced in moderately under-dense plasma with 1.8TW laser at ETL. Ultralow emittance of pico-mn has been achieved at KEK-ATF as a development forward to a linear collider. BNL/GUN-IV installed at Univ. Tokyo has generated 240fs (FWHM), 6πmm.mrad, 7nC bunches. Femtosecond streak camera, CTR interferometer, polychromator and fluctuation method are used for femtosecond electron bunch diagnostics. Time-resolved X-ray diffraction is going to visualize atomic motions in laser-irradiated GaAs monocrysral, which is one of the most promising applications of ultrashort beams.

INTRODUCTION

Advanced accelerator development has entered a new era recently in Japan. SPring8 as the third generation light source is achieving many remarkable results. Neutron and hadron science project is to start in 2001 under collaboration of JAERI and KEK. Thus, we have begun to aim at the development of the fourth generation light sources such as X-ray SASE-FEL and inverse Compton scattering hard X-ray source. They are closely related to a linear collider since they share many similar components. Laser plasma accelerator could be a promising candidate for the fifth generation accelerator. Large national project has not been authorized yet, but many small developments aiming at the above are under way. Here updated advanced accelerator work in Japan is summarized. The following abbreviations of the organizations are used in this paper.

NERL, U. Tokyo: Nuclear Engineering Research Laboratory, University of Tokyo
JAERI-APR   : Japan Atomic Energy Research Institute,
                 Advanced Photon Research Center
ILE, Osaka U. : Institute for Laser Engineering, Osaka University
ETL   : Electrotechnical Laboratory
KEK-ATF   : High Energy Accelerator Development Organization,
                 Advanced Test Facility

GeV LASER ACCELERATION RESEARCH AT JAERI-APR

Recently, there has been a great interest in ultrahigh field particle acceleration driven by ultraintense laser pulses in plasmas. Many world-wide experiments have shown 100 MeV electron acceleration and several tens MeV ion acceleration with ultrahigh gradient of ~100 GeV/m. In the self-modulated LWFA mechanism, however, the energy gain has been limitted up to ~100 Mev even for ~1 PW laser-plasma interactions. In last AAC98 there was an exciting topic on two LWFA experiments; "optimistic Japanese (JAERI) data" demonstrating more than 200 MeV acceleration with matched relativistic electron injection, "pessimistic French (Ecole Polytecnique) data" with mismatched injection. Those first-generation experiments suggest that the phase-matched injection and the stable optical guiding of intense laser pulses are inevitable breakthrough to accomplish high energy gain acceleration with high quality beam.

In JAERI-APR, since 1996 we have started "GeV laser wakefield acceleration project" aiming at more than 1 GeV energy gain with high quality beam as a second generation experiment[1,2]. We have developed the Cu-photocathode S-band RF gun as a high quality electron injecter associated with compact UV all solid-state lasers. We use a 150 MeV microtron for boosting the beam energy to make a beam emittance as small as possible. The new facility is depicted in Fig.1. For optical guiding, we have developed a cm-scale plasma waveguide using a fast Z-pinch capillary discharge plasma channel as shown in Fig.2[3]. We have succeeded in guiding a 2TW, 90fs high intensity laser pulse with $>10^{17}$ W/cm^2 over 2cm. In the capillary plasma waveguide driven by 40TW, 50fs laser pulses delivered from the 100TW laser system developed at JAERI-APR, properly phased electrons injected will be accelerated up to more than 1 GeV in a 2cm length.

On our recent status of the project, the commissioning of 25th turns microtron injected by the photocathode RF gun has been completed to deliver enough beam current for experiments. We plan to build a bunch slicing beam line consinsting of the undulator and the bunch slicing chicane for femtosecond bunch injection and femtosecond synchronization. The GeV laser acceleration expriment will be initiated after completion of overall laser transport system at JAERI-APR. The specifications of several proposed projects are summarized in Table.1.

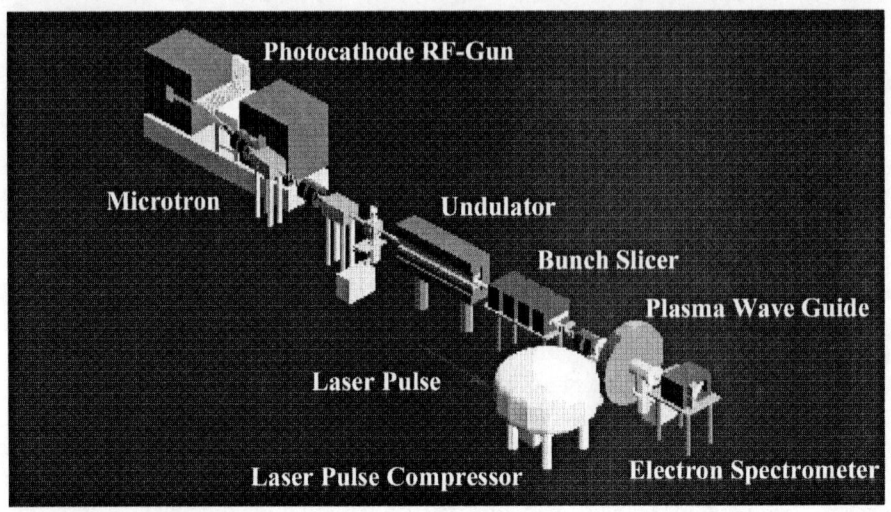

FIGURE 1. Laser Acceleration Test Facility at KAERI-APR

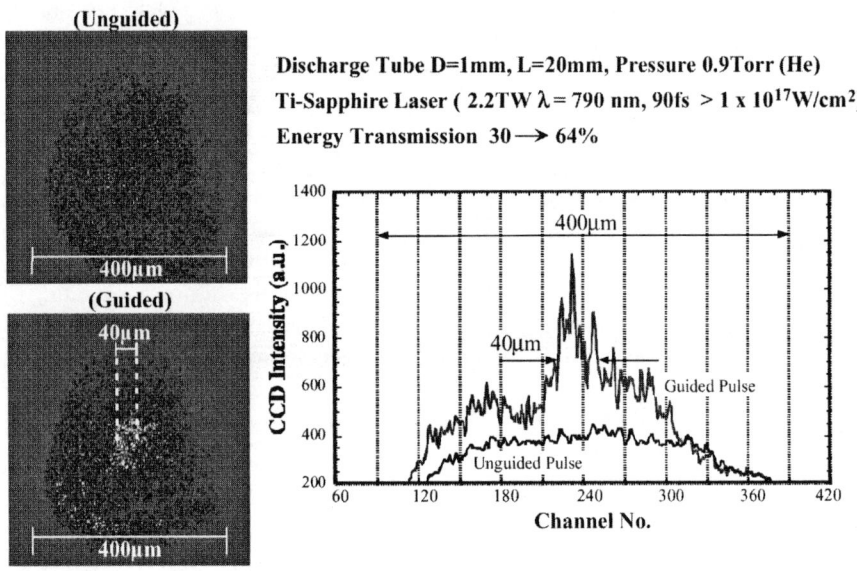

FIGURE 2. Terawatts laser pulses guided by the capillary discharge plasma

Mechanism	Standard LWFA	Channel-guided LWFA		Soliton Wake
Energy Gain	0.5 GeV	1 GeV	5 GeV	1 GeV
Particle Injector	150 MeV Electron Microtron			Tapered Capillary 100 keV proton
Laser Peak Power	100 TW	40 TW	20 TW	100 TW
Pulse Duration	20 fs	50 fs	100 fs	20 fs
Spot Radius	30 μm	20 μm	10 μm	10 μm
Laser Strength a_0	1.8	1.7	2.4	5
Plasma density	8.8×10^{18} cm^{-3}	1.4×10^{18} cm^{-3}	3.5×10^{17} cm^{-3}	$(1.74 \rightarrow 0.41) \times 10^{21}$ cm^{-3}
Accelerating gradient	1.9 GeV/cm	0.7 GeV/cm	0.55 GeV/cm	~ 3 GeV/mm
Diffraction length	11 mm	5 mm	1.2 mm	
Dephasing length	4 mm	55 mm	56 cm	
Optical guiding	No	1.5 cm Capillary	10 cm Capillary	< 1 cm Tapered Capillary
Number of particles accelerated	7×10^9	1.1×10^9	0.2×10^9	

TABLE 1. Test Parameter for laser acceleration

SELF-MODULATED WAKEFIELD EXCITATION IN 40TW LASER-PLASMA AND ELECTRON ACCELERATION

Illuminating a 40TW chirped pulse amplified laser on 0.01 critical density hydrogen plasma, we obtained a self-modulated wakefield in the relativistic region, whose amplitude is 24% of the average density, or the accelerating field is 160GV/m. The laser beam is relativistically self-focused, which in turn confined the plasma wave within 0.4mm diameter over 1mm axial length. The dephasing distance implies a electron acceleration to 48 MeV, which is consistent with the observation of the averaged energy 35 MeV. Protons of 7 MeV are observed. See ref[4].

The experimental set up is shown in Fig.3. The GEKKO MII short pulse laser system consists of a Ti:sapphire oscillator, a stretcher, a Ti:sapphire regenerative amplifier, a glass rod amplifier chain and a glass disk amplifier chain, and a four diffraction grating compressor. The system provides a 15cm diam. beam, which delivers an energy of 25J in 410 to 500fs at 1.053μm.

An off axial parabola mirror of f/3.8 focuses 30% energy into a spot size of 18μm. The focusable intensity is 2×10^{19} W/cm^2. The prepulse level is 2×10^{-4}. As a gas source, we used a electro-magnetically switched high speed gas puff. We have the exhausted H$_2$ gas of 2mm radius 2mm above the puff nozzle of 6mm aperture. We have obtained gas density of 10^{19} W/cm^3 over 4mm, which encourage us to accelerate electrons over few mm distance.

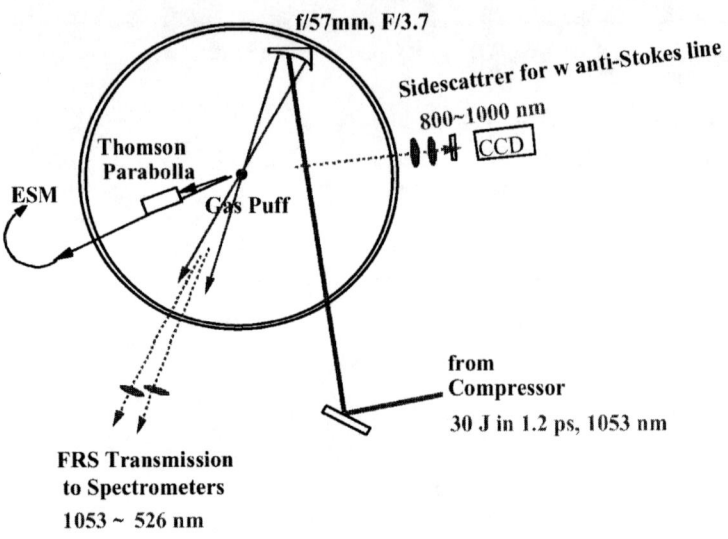

FIGURE 3. Experimental setup

Figure 4 shows a transmitted light spectrum around the second harmonics. Satellite peaks are Stokes and anti-Stokes and their harmonics laser associated with the Forward Raman and written as $\omega_s = 2\omega_0 \pm \omega_p$, $n = 1,2,$. Assuming that the second harmonic laser light is forward scattered by the plasma-wave driven grating, we can estimate the scattered power ratio P_1/P_0:

$$\frac{P_1}{P_0} = \left[\frac{p}{2}\frac{dn}{n_0}\frac{n_0}{n_c}\frac{L_z}{l_0}\right]^2 \left[\frac{\sin(DkL_z)}{DkL_z}\right]^2 \quad (1)$$

where L_z is given as 1mm from sidescattering image and $\Delta k = (1 \pm c/v_\phi)\omega_p/c = 3 \times 10^3 m^{-1}$. $P_1/P_0 = 0.008$ from Fig.4 yields then the density perturbation amplitude $\delta n/n_0 = 0.38$ to be 0.07. The harmonic analysis between Stokes lines, $(P_2/P_1)^{1/2}$, gives $\delta n/n_0$ without any assumption. Stokes scattering at the fundamental frequency $P_2/P_1 = 0.009$ provides $\delta n/n_0$ to be again 0.07. Harmonic analysis from three shots gives 0.24±0.14. Data gap between these ratios may be derived from the indefinite factors, such as L_z or phase mismatch. Or P_1 line sateration and broadening may rather be the reason. Spectral broadening of the line gives $\Delta \omega_p/\omega_p = 0.017-0.012$.

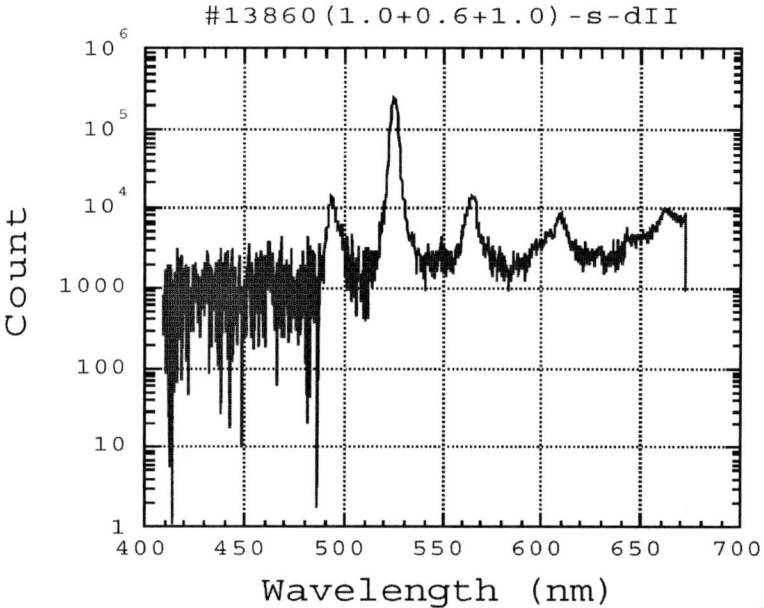

FIGURE 4. Forward scattered spectrum around the second harmonies

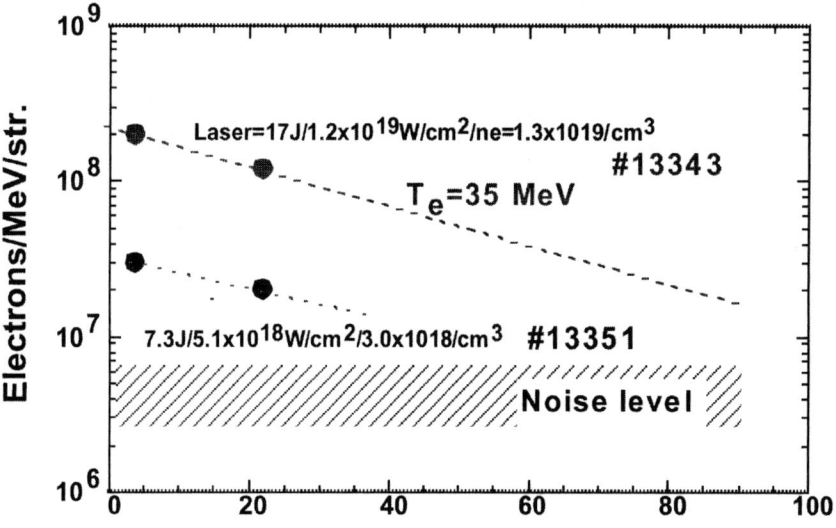

FIGURE 5. Electron emission from hydrogen gas into 43 degree

Figure 5 shows energy spectrum up to 22 MeV of the electrons emitted into 43 degree to the laser axis for the same shot. The detector is magnetically analyzed two channel Al filtered pilot-U scintillator. The dada indicate that the averaged energy is 35 MeV and electrons must be accelerated to 100 MeV.

We measured proton emission for #13347 to 45 degree using a Thomson parabola coupled to a CR-39 film, which is shown in Fig.6. The Thomson parabola was set 20cm from the target position to 45 degree from the laser axis. Since the CR-39 film was etched 16 hours, the upper detection limit is 8.8 MeV in this shot. The averaged ion energy seems to be 7 MeV.

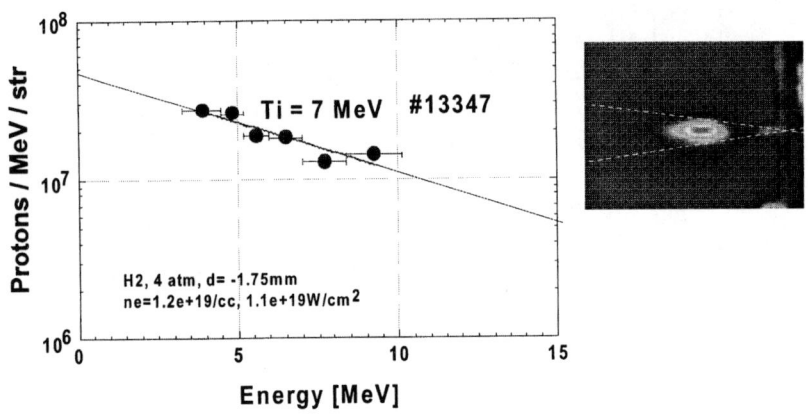

FIGURE 6. Proton 45 degree emission from hydrogen gas target

PRODUCTION OF MeV-ELECTRONS IN MODERATELY UNDER-DENSE PLASMAS

Experiments on high-intensity laser interaction with moderately under-dense plasmas ($n_e/n_c \approx 0.17$) are carried out at ETL[5]. A Ti:sappahire laser system provided the peak power of 1.8TW in a pulse duration (FWHM) of 100fs. The wavelength of the laser was 790nm. Laser pluses were focused on front edges of 2mm-dianeter supersonic gas jets which have flat-topped gas densities of $3.4 \times 10^{19} cm^{-3}$ at reservoir pressures of 800kPa, 800kA is approximately $5P_{cr}$ for a fully ionized He-plasma ($19P_{cr}$ for the 8+ argon-plasma). The vacuum intensity in a focal spot was $7 \times 10^{17} W/cm^2$.

We observed a transition of laser pulse propagation from a stable filament to unstable filaments in case of the high electron density as well as the high laser power by using a shadowgraphy technique. The time to the transition of the propagation varied by $P_L^{-0.4} P_{gas}^{-0.8}$, where P_{gas} is the gas density of the jet. This suggests the presence of the laser-hose instability, which has a growth tome of $P_L^{-1/2}$

$P_{gas}^{-3/4}$. We also observed high-energy electrons, which had maximum energy of at least 1 MeV. The dependence of the maximum electron energy on the laser power and the gas density was $P_L^{(0.7-1.4)} P_{gas}^{(1.8-2.3)}$. The direction of the electron beam was narrowly collimated (the half angle ≈ 3 - 4°), which scattered around optical axis shot by shot. Figure 7 shows the side scattering of the main beam and the shadow graph. Laser power and gas density dependence of the generated electron energy is shown in Fig.8

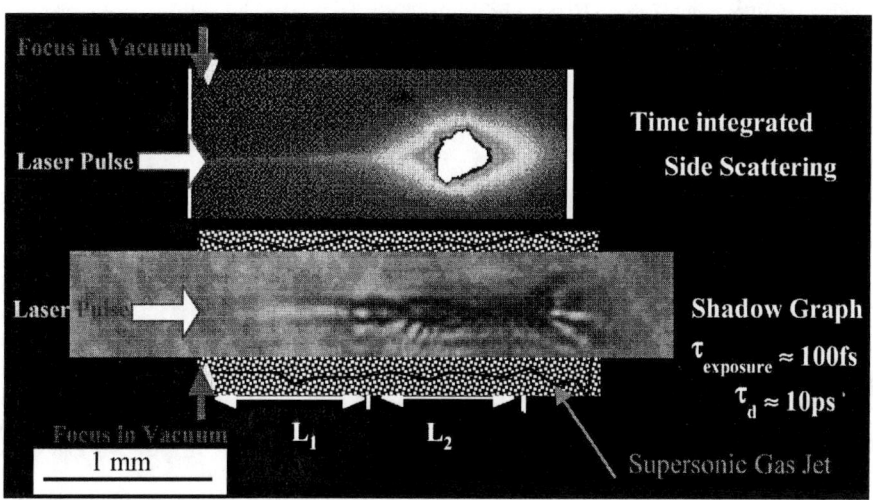

FIGURE 7. Side scattering of the main beam and the shadow graph

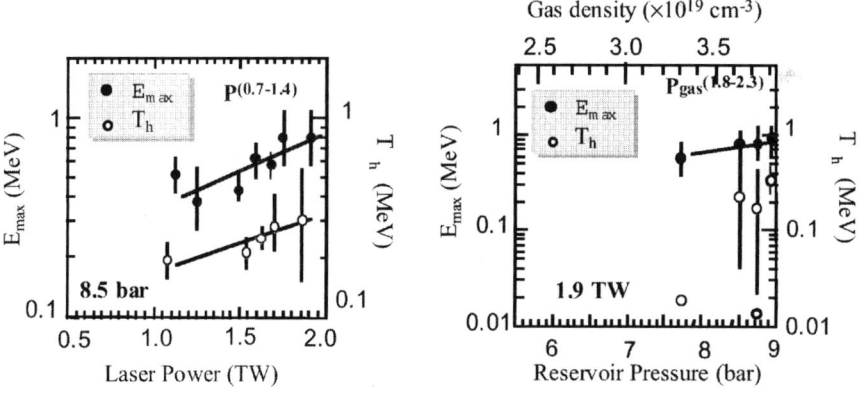

FIGURE 8. Laser power and gas density dependence of generated electron energy
It is hard to excite the Raman instability in the moderately under-dense plasmas, because the large difference of the phase velocity between the pump (1.1c) and

Raman scattered light (1.4c) leads to the very short dephasing length of 1m. If electrons are accelerated by the plasma wave, the maximum electric field limited by the wavebreaking and a trapping length of electrons are proportional to $n_e^{1/2}$ and $n_e^{-3/2}$, respectively. These relations imply that the maximum electron energy should be inversely proportional to the plasma density (n_e^{-1}), which dose not agrees with our expeiment.

One of the possible mechanisms to generate high-energy-electrons is the stochastic heating. In order to make clear the mechanism to the acceleration, we are carrying out the experiment by using new diagnostic equipment such as a high-sensitivity multi-electron spectrometer, and high-energy ion sensor.

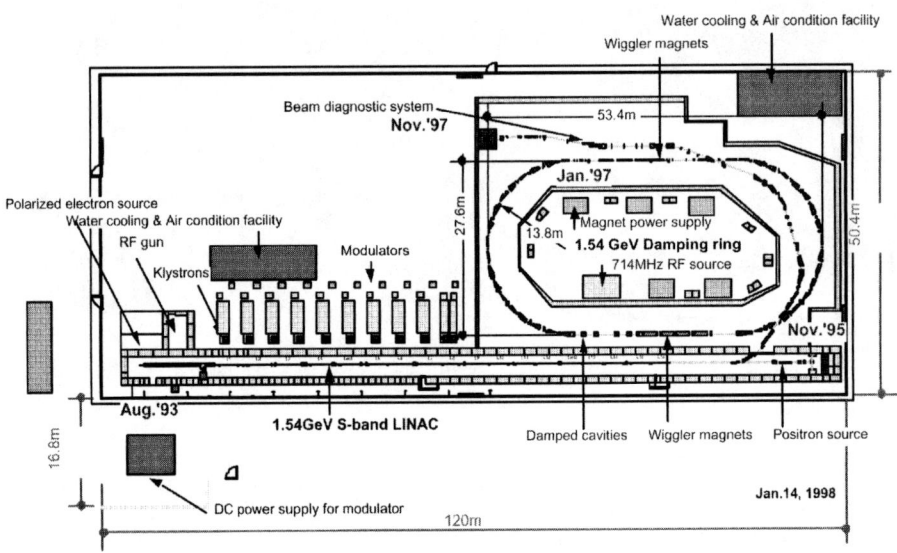

FIGURE 9. Overview of the KEK-ATF.

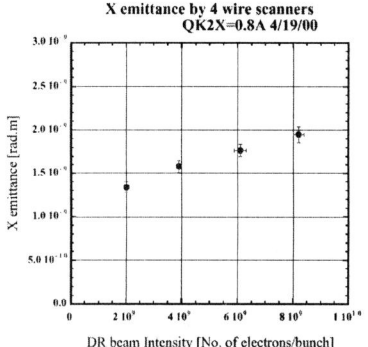
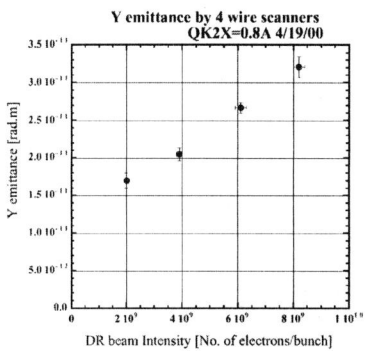
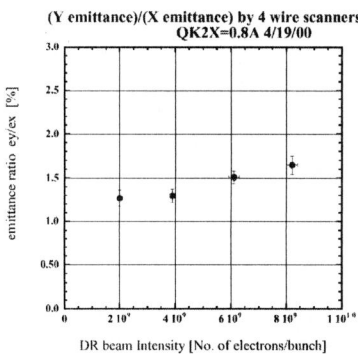

FIGURE 10. Recent results of emittance measurements using the wire scanners at the extraction line

ULTRA LOW EMITTANCE MEASUREMENT AT KEK-ATF[6]

The purpose of the ATF is to generate beams with very small transverse and longitudinal emittances as required for future linear colliders. The overview of the KEK-ATF is shown in Fig. 9. So far, a horizontal emittance of 1.4±0.3 nm was demonstrated by wire-scanner measurements in the extraction line and by a measurement of the horizontal spatial coherence using an SR interferometer. Also, a vertical emittance of 15±2.5 pm was measured by the 4 wire scanners in the extraction line as shown in Fig.10. Still much effort is needed in order to stably produce a beam with 10 pm vertical emittance at 1.3GeV. A large number of beam studies conducted in the ATF Linac since JFY 1995 proved the soundness of the ±ΔF multi-bunch beam loading compensation scheme. In general, multi-bunch operation in the linac was a success. Since JFY 1994 various new instrumentation has been developed for multi-bunch high-precision beam diagnostics, for example, synchrotron and optical transition radiation (SR and TDR) monitors with 3 nsec fast gate, a micron beam size monitor using the extraction line wire scanners for emittance measurements, an SR interferometer, and a laser wire beam size monitor in the ring. All of these are still being improved in order to establish beam control techniques adequate for the micron world.

ULTRASHOT BEAM RESEARCH AT NERL, U. TOKYO

A new Femtosecond Ultrafast Quantum Phenomena Research Facility has been installed at the laboratory in 1999 [7]. Here the upgraded femtosecond S-band twin linacs, the 12 TW 50 fs laser, the X-ray diffraction analysis devices, the X-ray CCD camera system, the X-ray photo-electron spectroscopy (XPS) device and the Fourier transform infra-red spectroscopy device (FTIR) have been installed as shown in Fig.11. In order to achieve subpicosecond linac-laser synchronization, we have designed and constructed a new system introducing the most advanced technologies as shown in the left-hand side of Fig.11. The design is based on the achievement at FELIX of FOM-Institute for Plasma Physics (Netherland) where the timing jitter between the electron and FEL pulse is 400fs (rms) [8]. The technologies are; the Kerr-lens-mode-locked Ti:Sapphire laser with the timing stabilizer at the 9th harmonics RF (Coherent, Synchro-lock), compact laser amplifiers, one stable 15 MW klystron RF power supplier (Mitsubishi, Modifed PV-3015), tempreture-controlled (within 1°C) laser clean room (CLASS 10,000) and vacuum laser transport line and so on. Timing jitters between the electron and laser pulses in the previous and new systems are evaluated in Table.2. K. Kobayashi et al., achieved 77 fs (rms) timing jitter at their mode-locked Ti:Sapphire laser [9]. Concerning the timing jitter at a laser oscillator, recent passive Ti:Sapphire passive mode-locker using the Kerr lens effect (for example, Coherent, Mira) is superior to conventional active Ti:Sapphire mode-lockers using an A/O crystal (for example, Spectra Physics, Tsunami). The nonlinear Kerr lens effect in the Ti:Sapphire crystal, governed by the van der Pol's equation, enables much faster feedback tuning of the resonance frequency of the laser pulses. The timing stabilizer operating at the 9th harmonics of the input RF of 79.3 MHz is also effective. This is because the spectral noise due to the timing jitter is enhanced by the square of the harmonic number, while that due to the power fluctuation is constant [10]. When we used two independent klystrons for the RF gun and accelerating tube, respectively, the electron bunch suffered from the mutual RF fluctuation between the two klystrons. In order to avoid this effect, we have chosen to use one klystron to feed RF into them. The electron beam jitter corresponds to the jitter of the electron bunch behind the chicane-type magnetic pulse compression due to fluctuations of RF power and phase, which are evaluated by the klystron performance data and PARMELA simulation.

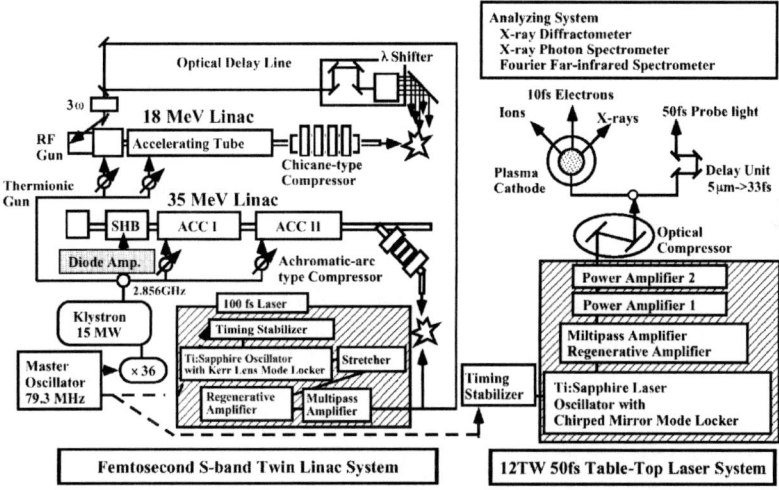

FIGURE 11. Femtosecond Ultrafast quantum Phenomena Research Facility

	Previous system (measured)		New system (design)
	Thermionic	Photocathode	Photocathode
RF linac (σ_{rf})	a few ps	a few ps	300fs
Mutual jitter between two klystrons (σ_{rf2})	a few ps	a few ps	~0fs
Laser (σ_{laser})	< 3ps	< 3ps	100fs
Laser mode-locker	Active by A/O		Passive by Kerr lens
Timing stabilizer	at the fundamental (79.3MHz)		at 9th harmonics (713.7MHz)
Total	3.7ps	3.5ps	320fs

TABLE 2. Measured and designed timing jitters [rms] in the previous and new systems

We measured and evaluated current precision of electron-laser synchronization by the femtosecond streak camera (Hamamatsu FESCA-200), where the time resolution is 200fs, as shown in Fig.12. We operated the timing, stablizer at the fundamental (79.33MHz) and 9th harmonics (714.97 MHz) modes. The light-hand side shows the synchronized femtosecond electron and laser pulses. We stacked the same data more than 100 times and measured their time differences for the two modes. The horizontal axis almost represents time in minute. We investigated that the slow drift attributed to the temperature change of the accelerating tube of the linac with not precise old water cooling system (<1°C). The jitter riding on the drift at 9th harmonics mode is ~2ps (p-p), which corresponds to ~330fs (rms), for a few minutes. This is because one data set was obtained in about a minute. This is close to the design value of 320fs. It is also clearly observed that the jitter at the 9th is superior compared to that at the fundamental. We are going to introduce a more precise water cooling system within 0.01° so as to establish 330fs synchronization for a longer time.

Laser photocathode RF gun (BNL/GUN-IV) is successfully under operation. Recently, we achieved 7nC/bunch, QE of 1.4×10^{-5} for 250 μJ 267nm irradiation [11]. Addition of a NEG getter vacuum pump remarkably improved the vacuum pressure (<5×10^{-9} torr during RF feeding) and the performance such as dark current (~10pC/5μs RF pulse) and laser cleaning of the Cu cathode. However, many small craters of tens μm are homogeneously observed at the cathode surface. This could attribute to electric dischage during the term of poor vacuum (>1×10^{-8} torr).

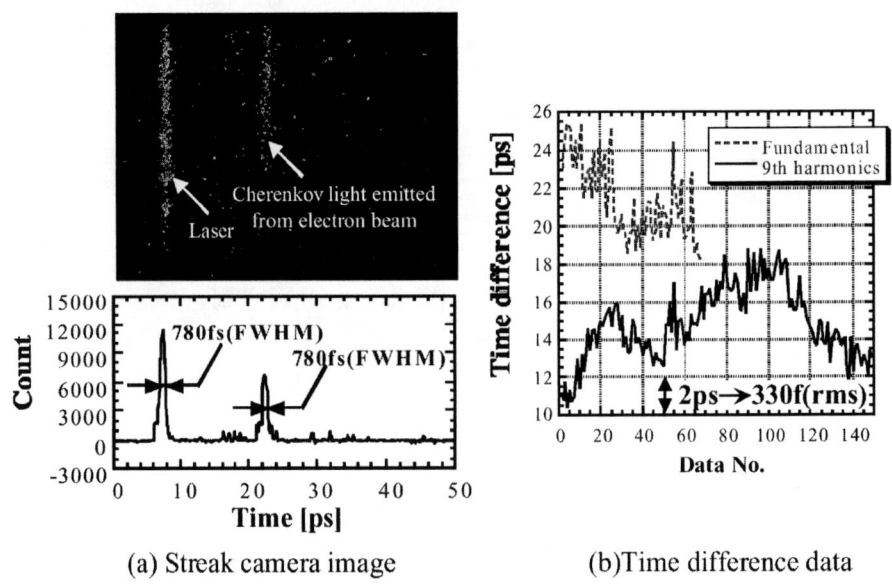

(a) Streak camera image   (b)Time difference data

FIGURE 12. Measured results of the linac-laser synchronaization

FEMTOSECOND ELECTRON BUNCH DIAGNOSTICS

The four major methodologies of the femtosecond streak camera [12], coherent transition radiation interferometer [13], polychromator [14] and fluctuation method [15] are used here. Advantages and limitations are summaried in Table.3. Up to 200fs (FWHM), the femtosecond streak camera is (FESCA-200, Hamamatsu) no doubt the best [12,13]. Beam dynamics including the jitters of intensity and timing can be watched by a single shot. Below 200fs, the combination and mutual checking among the rest three methods are necessary since neither of them has been yet established for such an ultrashort bunch diagnostics. Especially ~10fs MeV electron bunch is going to be generated in near future [16,17,18]. The highlights in these two years and the target in the coming two years are summarized in Fig.13. The coherent radiation interferometer looks the most promising because the sensitivity of optical sensor becomes higher for shorter bunches. This is because the autocorrelator is successfully used for fs lasers.

TENS FEMTOSECOND TIME-RESOLVED SYSTEM BY LASER PLASMA LINAC

Recently we have proceeded to the laser plasma linac to generate 10fs relativistic electron single bunch by a single 12TW 50fs laser pulse [18]. As the laser is more intense in a gas jet, the plasma wakefield changes from linear (sinusoidal) to nonlinear (sharper wavefront) and the election motion in plasma becomes from nonrelativistic to relativistic. Finally, beyond the critical value at the density-sloped part of the plasma, the wake wavebreaking occurs so that the wave energy is transferred to longitudinal electron momentum. Those electrons are accelerated and bunched by the wakefield at the flat-top. Numerical analysis by the PIC (Particle In Cell) –2D code shows that ~10MeV, ~10fs (FWHM) and ~3π mm.mrad(rms) electron single bunch with 10^{11} electrons is generated (refer[17,18]). We plan to verify this idea and generation of ~10fs ultrashort electron bunch experimentally in 2001. After we have measured the electron bunch, we plan to construct tens femtosecond time-resolved pump-and-probe analysis system using a laser beam splitter and optical delay line. Since the synchronization is done passively, the positioning precision of 5μm at the delay line corresponds to 33fs time delay without timing jitter. Updated results appears in ref [19].

	Measurement Source	Measurement Limitation (Reported)	Single-shot	In Vacuum	Non-Destructive
Femtosecond Streak Camera	Cherenkov Radiation	200 fs (~240 fs)	○	○ (Solid Radiator) × (Gas Radiator)	×
Coherent Radiation Interferometry 10ch Polychromator	CTR(Coherent Transition Radiation) CDR(Coherent Diffraction Raditation)	Unlimitted (~120 fs)	× ○	○	○ (CDR) × (CTR)
Fluctuation Method	Incoherent Radiations	Unlimitted (~ps)	○	○	○ (DR) × (TR)
Zero-phasing Method	Energy Spectrum of Electron Pulse	Unlimitted (~100 fs)	×	○	×
SMA Monitor	Induced Voltage from Electron Pulse	~ps (~ps)	○	○	○

TABLE 3. Characteristics of short electron bunch diagnostic methods

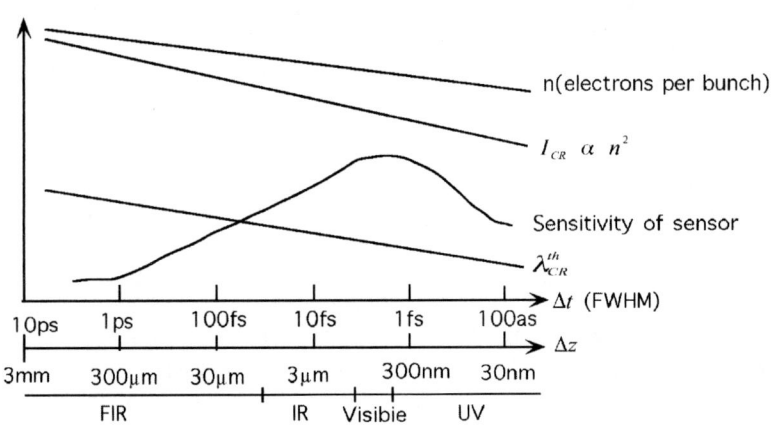

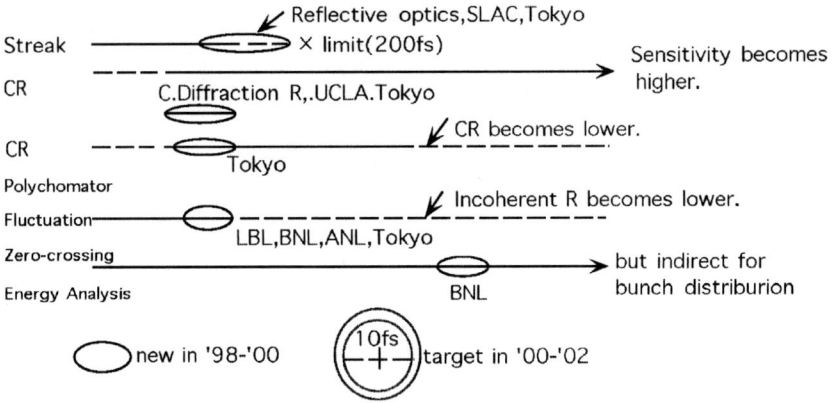

FIGURE 13. Progress and target in ultrashort electron bunch diagnostics

TIME-RESOLVED X-RAY DIFFRACTION TO VIVUALIZE ATOMIC MOTION

We are trying the pump-and-probe X-ray diffraction using an ultrashort X-ray pulse[20,21]. Here we try to visualize the atomic motion in laser-induced nonequilibrium thermal expansion via X-ray diffraction using the ultrashort X-ray pulse. We use the laser and X-ray pulses as a pump- and probe-pulses, respectively. At first, we generated the X-ray pulse by colliding an electron pulse to a Cu wire and use the characteristic components of $CuK\alpha_{1,2}$. We obtained the $CuK\alpha_{1,2}$ X-ray diffraction patterns from Si, GaAs, NaCl, KCl, CaF_2, BaF_2 clearly. Since the X-ray intensity is rather limited, we had to perform pump-and-probe data accumulation at more than 10^4 times and the sample surface was damaged during it. To enhance the X-ray intensity, we have shifted to laser plasma X-ray. The experimental configurarion is shown in Fig.14. X-ray diffraction pattern from a static GaAs microcrystal has been already obtained as shown Fig.15. The pump-and-probe analysis is now under progress. The computer code to visualize the atomic motion at thermal expansion and coherent acoustic phonon in the laser-irradiated GaAs monocrystal ((111) surface) has been developed. The shopshot at 150ps after laser-irradiation in shown in Fig.16, where we used our experimental results and those obtained by C. Rose-Petruck et al.[22].

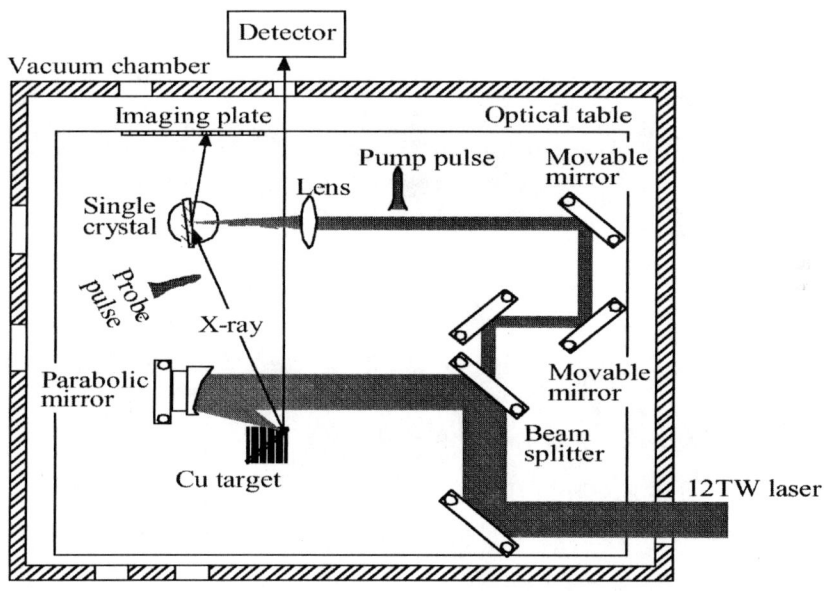

FIGURE 14. Experimental setup of the time-resolved X-ray diffraction

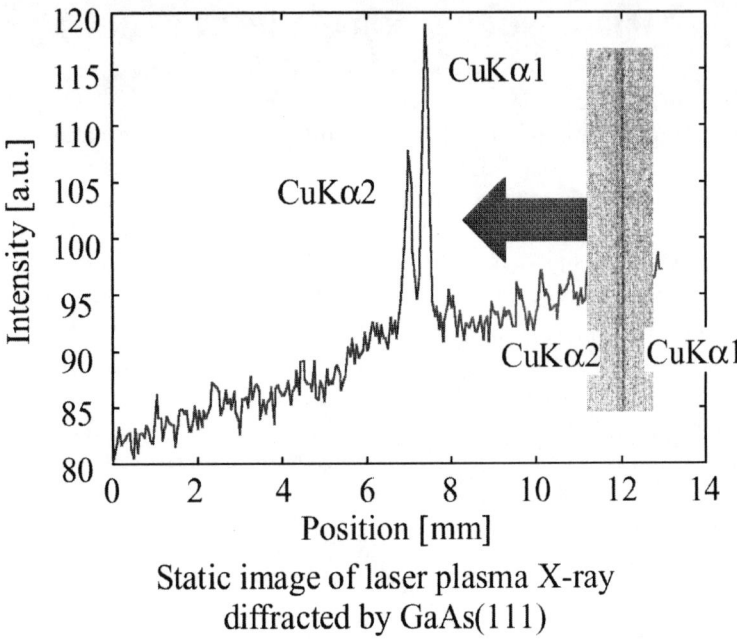

Static image of laser plasma X-ray
diffracted by GaAs(111)

FIGURE 15. X-ray diffraction pattern from a GaAs monocrystal

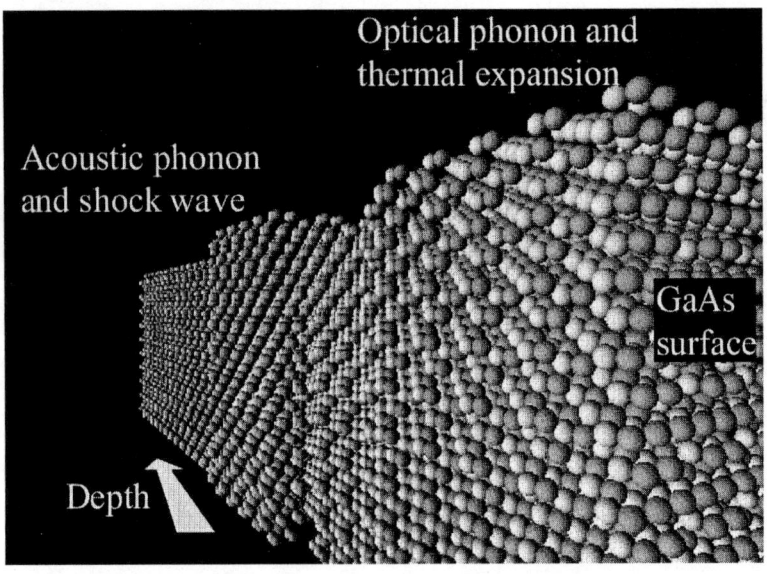

FIGURE 16. Snapshot of atomic motion GaAs monocrystal((111) surface) at 150ps after laser irradiation

CONCLUSION

Many remarkable achievements have been made in Japanese advanced accelerator work as described here. A new national partnership in this field to accelerate the research is now discussed in Japan. This activity is expected to be correlated with the same work in the world, aiming at future application to high energy physics, the fourth generation light source and industrial uses.

REFERENCE

[1] K. Nakajima, *Nucl. Instrum. Meth. A.*, Vol.410(3)(1998), pp.514-19.
[2] H. Dewa, et al., Nucl. Instrum. Meth. A., Vol.410(3) (1998), pp.357-363.
[3] T. Hosokai et al., Opitics Lett.,25(1)(2000), pp.10-12.
[4] Y. Kitagawa et al., In High-Power Laser in Energy engineering, Proc. SPIE, 3886 (2000),pp.105-112.
[5] K. Koyama, N. Saito, and M. Tanimoto, *Proceedings-SPIE* 3886 (1999), pp.76-80.
[6] J. Urakawa et al., Proc. of EPAC2000, in press.
[7] M.Uesaka et al.,Rdiat. Phys. Chem. (2000), in press.
[8] G.M.H. Knippels, et al., Optics Lett., Vol.23(22) (1998), pp.1754-1756.
[9] K. Kobayashi, T. Miura, Z. Zhang, and K. Torizuka, *SPIE*, 3616(1999), pp.156-164.
[10] D. von der Linde, *Appl. Phys.* B 39(1986), pp.201-217.
[11] T. Kobayashi ey al., Proc. of EPAC2000, in press.
[12] M. Uesaka, et al., *Nucl. Instrum. Methods A,* 406,371 (1998), pp.371-379.
[13] T. Watanabe, et al., *Nucl. Instrum. Methods A,* 437.1 (1999), pp.1-11.
[14] J. Sugahara et al., Proc. of PAC99, pp.2187-2189.
[15] P. Catravas et al., Phys. Rev. Lett., 82(26) (1999), pp.5261-5264.
[16] Umstadter, D., et al., *Phys. Rev. Lett.*, 79(12) (1996), pp.2073-2076.
[17] Esarey, E., et al., *Phys. Rev. Lett.*, 79(14) (1997), pp.2682-2685.
[18] N. Hafz et al., *Nucl. Instrum. Meth. A.*455 (2000),pp.148-154.
[19] N. Hafz et al., in this proceeding.
[20] H. Harano et al., J.Nucl. Materials (2000), in press.
[21] M. Uesaka et. al., *Nucl. Instrum. Meth. A.* 455 (2000),pp.90-98.
[22] C. Rose-Petruck et al., Nature, Vol.398.(1999).pp.310-312.

Plasma Focusing of High Energy Density Electron and Positron Beams

J.S.T. Ng[1], P. Chen[1], H.A. Baldis[2], P. Bolton[2], D. Cline[3], W. Craddock[1],
C. Crawford[4], F.J. Decker[1], R.C. Field[1], Y. Fukui[3], V. Kumar[3], M.J. Hogan[1],
R. Iverson[1], F. King[1], R.E. Kirby[1], T. Kotseroglou[1], K. Nakajima[5], R. Noble[4],
A. Ogata[6], P. Raimondi[1], D. Walz[1], A.W. Weidemann[7]

[1] *Stanford Linear Accelerator Center, Stanford, California, 94309*
[2] *Lawrence Livermore National Laboratory, Livermore, California, 94551*
[3] *University of California, Los Angeles, California, 90024*
[4] *Fermi National Accelerator Laboratory, Batavia, Illinois, 60510*
[5] *High Energy Accelerator Research Organization, Tsukuba, Ibaraki 305-0801*
[6] *Hiroshima University, Kagamiyama, Higashi-Hiroshima 739-8526*
[7] *University of Tennessee, Knoxville, Tennessee, 37996*

Abstract. We present results from the SLAC E-150 experiment on plasma focusing of high energy density electron and, for the first time, positron beams. We also present results on plasma lens-induced synchrotron radiation, longitudinal dynamics of plasma focusing, and laser- and beam-plasma interactions.

INTRODUCTION

The plasma lens was proposed as a final focusing mechanism to achieve high luminosity for linear colliders [1]. Previous experiments to test this concept had been carried out with low energy density electron beams [2]. The goals of the SLAC E-150 experiment are to study plasma focusing for high energy, high density particle beams in the regime relevant to linear colliders, to obtain better understanding of beam-plasma interactions, and to bench-mark computer codes for plasma lens designs. Such studies will help to develop compact and economical plasma lens designs suitable for collider experiments. In this paper, we present preliminary results obtained recently by the E-150 collaboration on plasma focusing of high energy density electron and positron beams.

TABLE 1. FFTB electron and positron beam parameters for this experiment.

Parameter	Value	Units
Bunch intensity	1.5×10^{10}	particles per pulse
Beam size	5 to 8 (X), 3 to 5 (Y)	μm
Bunch length	0.7	mm
Beam energy	29	GeV
Normalized emittance	3 to 5 $\times 10^{-5}$ (X), 0.3 to 0.6 $\times 10^{-5}$ (Y)	m-rad
Beam density	$\sim 7 \times 10^{16}$	cm^{-3}

EXPERIMENTAL SETUP

The experiment was carried out at the SLAC Final Focus Test Beam facility [3]. The experiment operated parasitically with the PEP-II B-factory; the high energy electron and positron beams were delivered to the FFTB at 1 - 10 Hz from the SLAC linac. The beam parameters are summarized in Table 1.

A layout of the beam line is shown in Figure 1. The beam size was measured using a wire scanner system. A carbon fiber 4 μm or 7 μm in diameter was placed downstream of the plasma lens, adjustable along the beam axis in a range of 8 to 30 mm from the center of the lens. The Bremsstrahlung photons were detected in a Cherenkov type detector located 35 m downstream of the lens. The variation in photon yield as the beam scans across the wire provided a measure of the transverse beam profile from which the beam size was determined. A set of ionization chambers interleaved with polyethylene blocks, located 33 m downstream of the lens, was used to monitor the synchrotron radiation emitted as a result of the strong deflection of the beam particles by the plasma lens. This detector provided an independent measure of the focusing strength. A streak camera, monitoring the Cherenkov radiation from an aerogel target installed in the electron beam line downstream, was used to measure the longitudinal plasma focusing dynamics.

To create the plasma lens, a short burst (800 μs duration) of neutral nitrogen

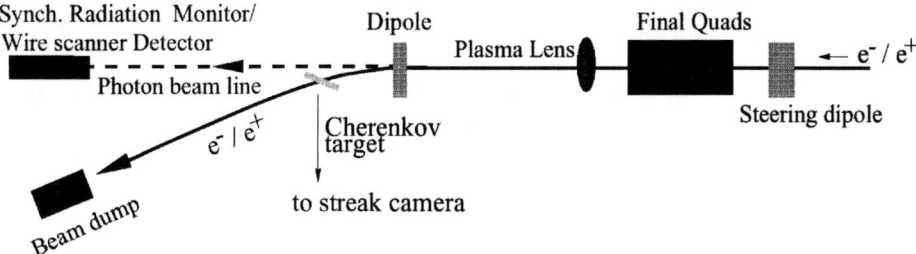

FIGURE 1. Layout of the plasma lens measurement setup.

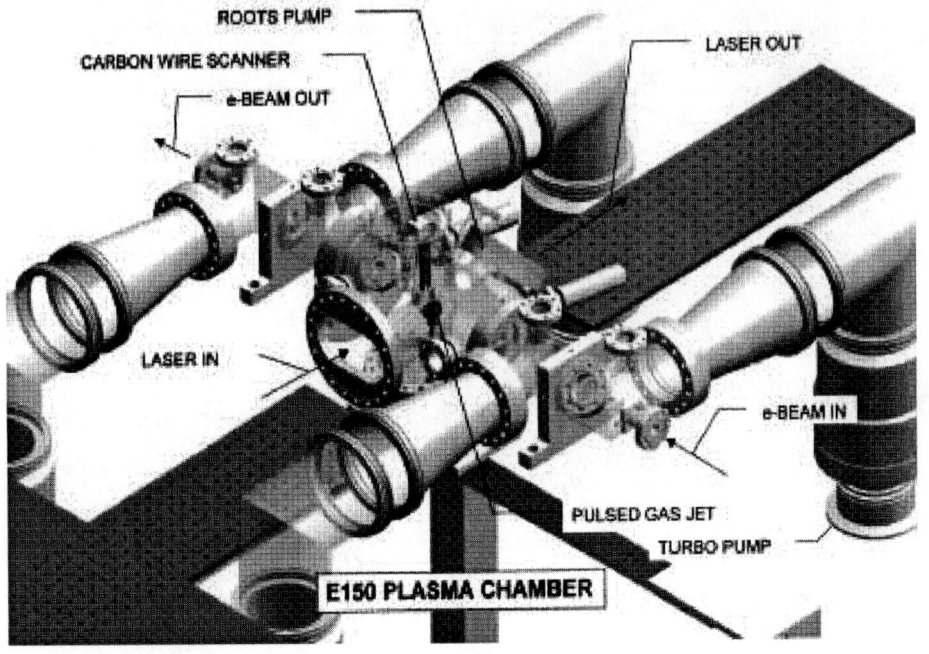

FIGURE 2. E-150 plasma chamber and differential pumping section.

or hydrogen gas, injected into the plasma chamber by a fast-pulsing nozzle, was ionized by a laser and/or the high energy beam. The solenoid valve operated at 0.5 - 2 Hz. Its 0.8 mm orifice was matched to a parabolic-shaped nozzle 5 mm long with an opening of 3 mm diameter. The transverse profile of the gas jet was obtained by scanning the high energy beam perpendicular to its axis. The gas jet diameter which determines the plasma lens thickness was found to be 3 mm over a distance of 1.4 to 3.4 mm from the nozzle exit. The neutral density was determined by interferometry to be 4×10^{18} cm^{-3} for N_2 and 5×10^{18} cm^{-3} for H_2 at a plenum pressure of 1000 psi. The injected gas was evacuated by a differential pumping system which made possible operation of the gas jet while maintaining ultra-high vacuum in the beam lines on either side of the chamber. A schematic drawing of the plasma chamber is shown in Figure 2.

RESULTS

The results presented below were obtained from data taken during the Fall of 1999 with electron beams and during the Spring of 2000 with positron beams. Measurements related to plasma focusing are presented first, followed by results on longitudinal focusing dynamics and laser- and beam-plasma interactions.

Plasma focusing

For a bunched relativistic beam traveling in vacuum, the Lorentz force induced by the collective electric and magnetic fields is nearly cancelled allowing it to propagate over kilometers without significant increase in its emittance. In response to the wakefield excited by the intruding beam, the initially uniform plasma electron distribution is perturbed in such a way as to neutralize the space charge of the beam and thereby cancel its radial electric field. For a positron beam, the plasma electrons are attracted into the beam volume thus neutralizing it; for an electron beam, the plasma electrons are expelled from the beam volume, leaving behind the less mobile positive ions which neutralize the beam. When the beam radius is much smaller than the plasma wavelength, the neutralization of the intruding beam current by the plasma return current is ineffective because of the small skin depth. This leaves the azimuthal magnetic field unbalanced which then "pinches" the beam, resulting in focusing. In this experiment, typical plasma densities are of the order of 10^{18} cm^{-3}, corresponding to a plasma wavelength of $\sim$30 μm which is indeed much larger than the incoming beam radius.

Beam self-ionization plasma focusing

A small fraction of the neutral gas molecules was ionized due to collisions with high energy beam particles. The secondary electrons from this impact ionization process were accelerated by the intense collective field in the beam, transverse to the direction of propagation, to further ionize the gas [4]. This beam self-ionization plasma was observed to focus the beam. That is, the head of the bunch was able to ionize the gas while the core and the tail of the bunch were focused. A more quantitative understanding requires detailed calculations which are not yet available for this experimental setup. The results for electron beams are shown in Figure 3. The wire scanner data were taken with the gas nozzle turned on and off at each beam position, as the beam scanned across the wire, so as to minimize the systematics due to changes in beam conditions when comparing beam sizes measured with and without plasma focusing. The beam envelope was measured by changing the downstream position of the wire scanner; a reduction by a factor two in the transverse beam spot area was observed.

The beam self-ionization efficiency was determined for nitrogen gas by comparing the plasma focusing effects in two data samples. One sample was taken with

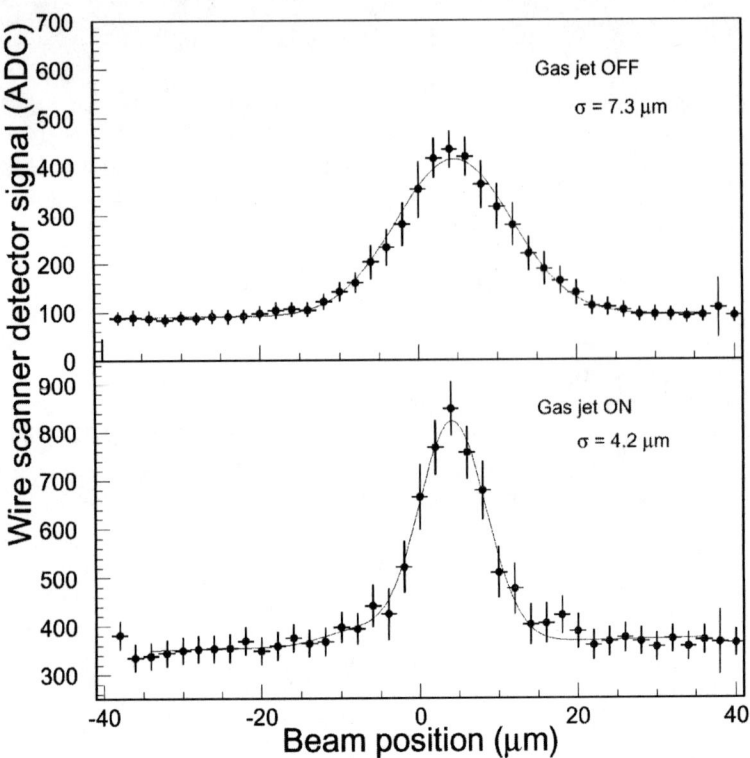

FIGURE 3. Wire scanner data showing focusing by a beam self-ionization plasma.

the plenum pressure at 220 psi with beam self-ionization and also ionization by a Terawatt laser; while the other sample at 550 psi was taken with beam self-ionization only. Based on the laser performance and optical field ionization data [5], the Terawatt laser's ionization efficiency was estimated to be 10%. The beam size reduction was observed to be the same for the two data samples at one location along the beam waist. Assuming the plasma densities to be the same in both data sets, we arrived at a beam ionization efficiency of 7%. Note that this argument is only valid for low ionization yields.

Laser avalanche ionization plasma focusing

The results on laser pre-ionization plasma focusing presented here were obtained using a turn-key infrared laser system. It delivered 1.5 Joules of energy per pulse of 10 ns FWHM while operating at 10 Hz at a wavelength of 1064 nm. The laser light was brought to a line focus at the gas jet; the plasma thus produced was approximately 0.5 mm thick as seen by the e^+/e^- beams.

With the relatively long infrared laser pulse, the pulse front was able to ionize a small fraction of the gas by multiple-photon absorption; the resulting secondary electrons were accelerated, transverse to the laser's incident direction, to further ionize the gas resulting in an avalanche growth in plasma density. This growth is similar to the beam self-ionization process discussed in the previous section. The ionization efficiency was determined by interferometry to be 45% in this case.

The results for laser (and beam) ionization plasma focusing of positron beams are shown in Figure 4. The measured transverse beam size is shown as a function of the distance (Z) between the plasma lens and the wire scanner along the beam axis; the convergence of the beam envelope towards a waist can be seen. The first plot shows the focusing for nitrogen gas at 800 psi, while the second shows the result for hydrogen gas at 1100 psi plenum pressure. A reduction in transverse beam size by a factor of two is observed with laser (and beam) induced ionization; focusing is also observed with beam-induced ionization alone.

Synchrotron radiation induced by plasma focusing

The synchrotron radiation induced by the strong deflection of beam particles inside the plasma lens provides an independent measure of the plasma focusing strength. It also provides a pulse by pulse diagnostic of the transient plasma lens effect. The critical energy of the emitted synchrotron radiation scales with the strength of the focusing; a monitor was designed to measure the penetration profile, which depends on the energy spectrum of the photon flux.

The photon flux detected in the synchrotron radiation monitor contained various background contributions. There was Bremsstrahlung radiation accompanying the beam due to off-axis beam particles interacting with the vacuum chamber wall or upstream collimators, and also synchrotron radiation (with a critical energy $\sim$250 keV) emitted when the spent beam was deflected towards the dump. This was measured with the gas jet turned off and the wire scanner removed from the beam. The contribution from beam-gas Bremsstrahlung, with the gas nozzle turned on, was determined by first measuring the Bremsstrahlung penetration profile from the carbon wire alone, then scaling this profile to the signal measured in the last depth section of the monitor with the beam-associated background removed. After these background contributions were subtracted, an excess in the photon signal beyond a depth of 3 radiation lengths was observed in the electron beam self-ionization plasma focusing data, for nitrogen gas at 550 psi plenum pressure. This excess was not observed when plasma focusing was weak (as determined from beam spot size measurements) for the case of lowered plenum pressure (to below 100 psi, for example), or for the case of hydrogen gas for which beam self-ionization was much less effective.

The penetration profile of the observed signal is consistent with synchrotron radiation with a critical energy of a few MeV according to Monte Carlo simulation. This value corresponds to a focusing gradient of the order of 10^6 T/m, as expected

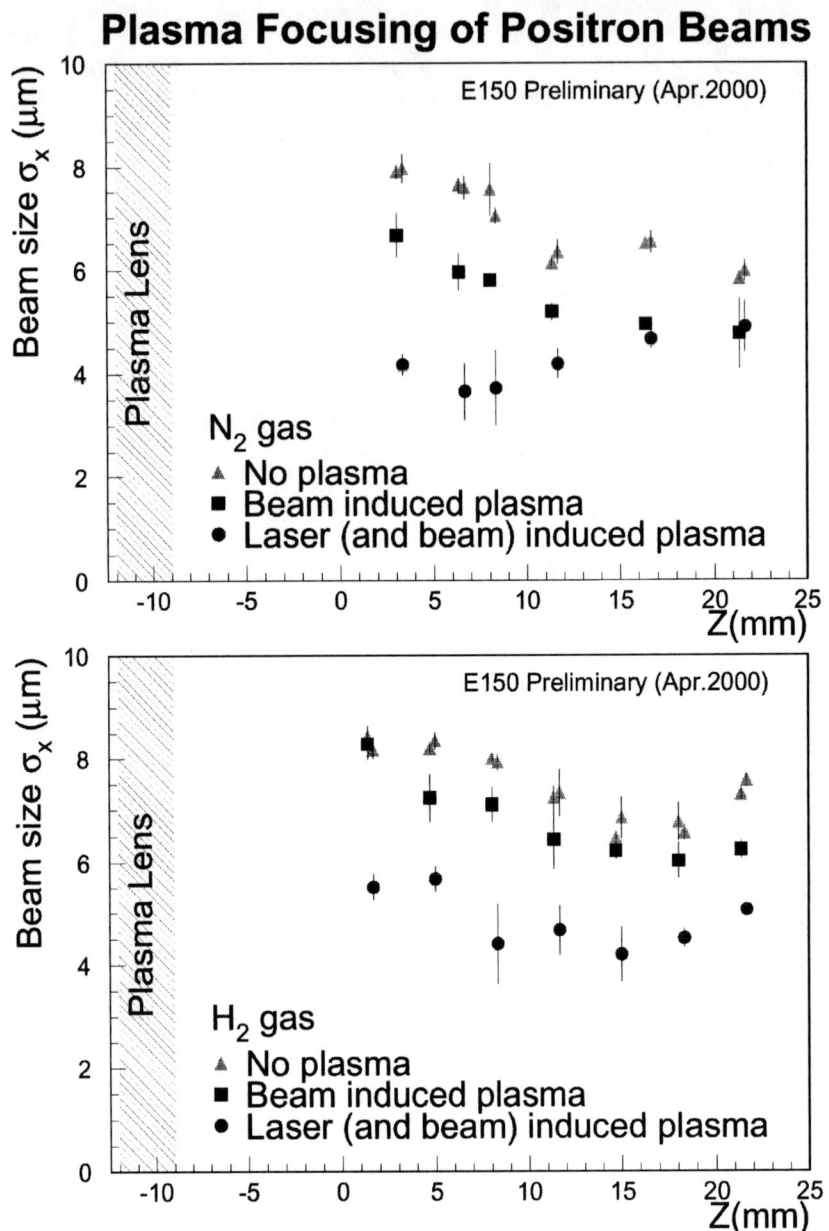

FIGURE 4. Evolution of the beam envelope, with and without plasma focusing, downstream of the plasma lens exit for two types of gases. The axis of the gas jet is at $Z = -10.5$ mm.

for the plasma lens parameters in this experiment.

Longitudinal focusing dynamics

A streak camera with pico-second time resolution was used to probe the variation in plasma focusing along the bunch. The Cherenkov light emitted as the beam passes through an aerogel target was imaged onto the streak camera. The vertical dimension of the beam was then measured in pico-second time slices along the bunch. The phase advance of the beam transport was such that focusing at the plasma lens becomes defocusing at the aerogel target. Therefore a larger beam spot at the Cherenkov target was expected because of the increased beam divergence due to plasma focusing.

The results from data taken with positron beams are shown in Figure 5, for nitrogen gas at two different pressures with beam self-ionization. The beam spot size in the vertical plane was measured with the gas nozzle turned on and off, and the difference is shown for time slices along the bunch. As expected, plasma focusing is significantly stronger at the longitudinal beam centroid. A detailed explanation of this result requires modeling of the beam self-ionization process which is currently under study.

Laser- and beam-plasma interactions

The laser avalanche ionization process has a finite build-up and decay time. By varying the laser pre-ionization timing before beam arrival at the plasma, different plasma densities can be probed by the high energy beam. The signal in the synchrotron radiation monitor is measured as a function of this advanced timing. Since this signal is predominantly synchrotron radiation induced by plasma lens focusing, the interaction between the high energy beam and the plasma at various stages of formation and decay is monitored. This also means that focusing is observed well into the "after-glow" regime of the laser-induced plasma.

The result for positron beams and nitrogen gas at 1050 psi is shown in Figure 6. Time zero corresponds to coincidence between the beam arrival time and the peak of the recombination signal detected in a photomultiplier. The ionization efficiency is measured at time zero also. The synchrotron radiation monitor signal from beam self-ionization plasma focusing is 300 counts (as measured with the laser turned off); the beam-only background signal level is 80 counts (as measured with the gas nozzle turned off).

A quantitative understanding of this delay curve requires detailed modeling of the laser- and beam-plasma interaction process, as well as gas dynamics. A qualitative explanation of this curve is given here. The initial rapid increase in synchrotron radiation monitor signal most likely is due to an increase in plasma focusing strength, peaking at 500 ns. The data shown in Figure 4 are collected with this time delay, when the plasma density has decayed to an optimal level with respect to the beam

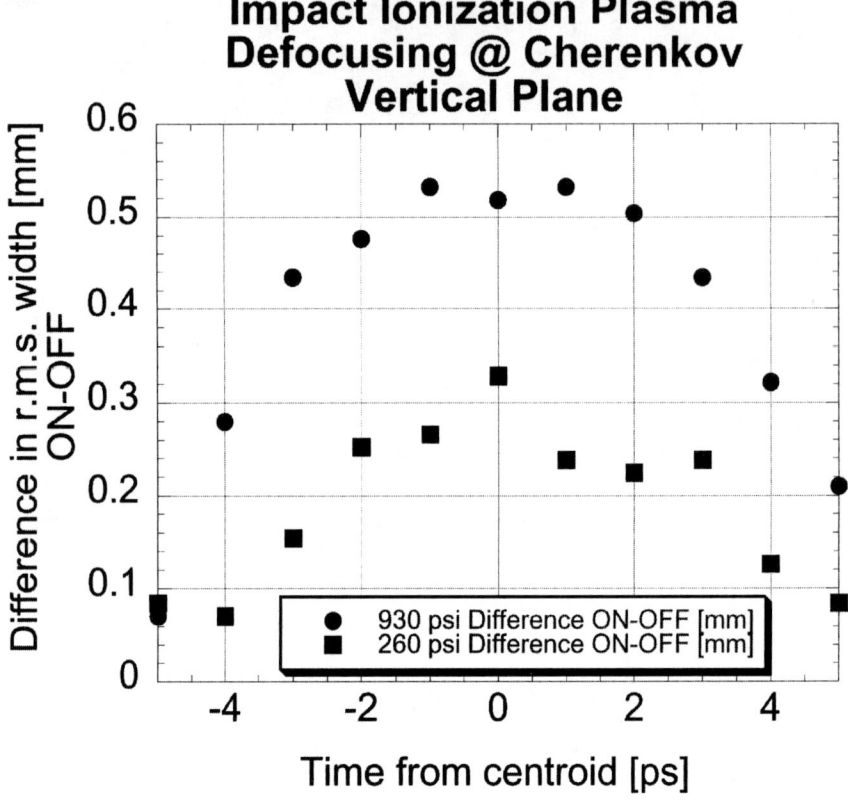

FIGURE 5. Streak camera diagnostics of plasma focusing longitudinal dynamics.

density. The rapid rise is followed by a sharp decrease to the background signal level of 80 counts. This is predominantly due to an almost complete expulsion of the neutral and ionized gas volume driven by a laser-induced shock wave. Each gas jet pulse, lasting more than 100 μs, supplied a continuous stream of neutral gas molecules into the vacuum chamber at sonic speed. Therefore, this abrupt reduction is also assisted by the motion of the plasma, and by its decay. The remainder of the curve starting at 1.5 μs shows a diffusion driven gas recovery phase. Initially, the gas rushed back to fill the void, giving rise to an increased gas density locally above ambient level at 8 μs. During this time, we observe a corresponding rise in beam-ionization plasma focusing signal, on a slow time scale determined by gas dynamics. This is followed by a decay back to local equilibrium level of 300 counts within the time interval 8 to 12 μs. Additional data have been collected since the Workshop, with varying plenum pressures for nitrogen and hydrogen gases. A detailed simulation study should provide further insight into this delay-correlated modulation of plasma focusing.

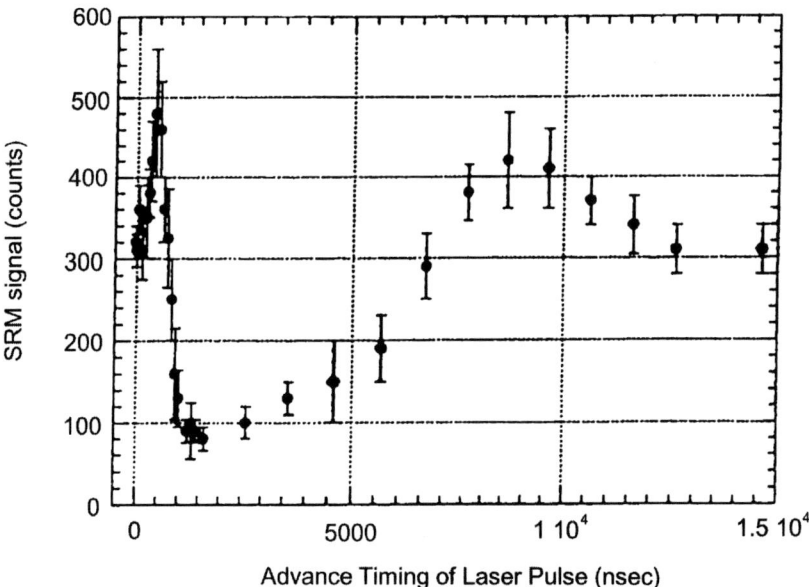

FIGURE 6. Synchrotron radiation monitor signal (without background subtraction) as a function of the advanced laser ionization timing with respect to beam arrival.

SUMMARY AND OUTLOOK

Results on plasma focusing of 29 GeV electron and, for the first time, positron beams have been presented. Beam self-ionization turned out to be an economical method for producing a plasma lens; reduction by a factor of two in the beam spot area was observed with this method. The plasma focusing strength was also measured independently by monitoring the synchrotron radiation emitted by particles focused by the lens. The infrared laser with a 10 ns long pulse also proved to be efficient in plasma production, resulting in the strong focusing of positron beams. The longitudinal focusing dynamics was diagnosed with a streak camera with pico-second time resolution and, as expected, the focusing was strongest at the longitudinal center of the bunch. The laser- and beam-plasma interaction was studied by varying the laser pre-ionization timing with respect to the beam arrival time; we observed a delay-correlated modulation of the plasma focusing in the "after-glow" regime.

Design studies for linear collider applications are just starting. The first issue to resolve is the effect of beam jitter on the achievable luminosity of plasma focused beams. Plasma lens parameters will also need to be optimized, which requires bench-marking of computer codes, as well as better understanding of the various plasma production processes. The experience gained in this experiment will serve as a basis for further engineering design studies for an eventual plasma lens application.

ACKNOWLEDGEMENT

This work was supported in part by the Department of Energy under contracts DE-AC02-76CH03000, DE-AC03-76SF00515, DE-FG03-92ER40695, DE-FG05-91ER40627, and the Univ. of California Lawrence Livermore National Laboratory, through the Institute for Laser Science and Applications, under contract No. W-7405-Eng-48; and by the US-Japan Program for Cooperation in High Energy Physics.

REFERENCES

1. P. Chen, Part. Accel., **20**, 171(1987).
2. J.B. Rosenzweig et al., Phys. Fluids B **2**, 1376(1990); H. Nakanishi et al., Phys. Rev. Lett, **66**, 1870(1991); G. Hairapetian et al., Phys. Rev. Lett. **72**, 2403(1994); R. Govil et al., Phys. Rev. Lett. **83**, 3202(1999).
3. V. Balakin et al., Phys. Rev. Lett. **74**, 2479(1995).
4. R.J. Briggs and S. Yu, LLNL Report UCID-19399, May 1982 (unpublished).
5. B. Chang et al., Phys. Rev. A **47**, No. 5, 4193(1993).

High-Brightness Electron Beam Production, Transport, and Measurement

Bruce E Carlsten

Los Alamos National Laboratory
Los Alamos, NM 87544

Abstract. This paper will review some of the recent developments in the area of high-brightness electron beams. Most of the paper will be devoted to emittance compensation concepts relating to high-brightness beam production in photoinjectors. Recent interpretation of emittance compensation and the residual emittance after compensation in terms of wave breaking has provided valuable insights into the process. Wave breaking also appears to have an influence on emittance growth and halo production in transport lines, which we will briefly discuss. There has been advances in understanding the emittance growth of bunches in circular motion, particularly related to the cancellation of longitudinal and transverse energy-independent forces. Finally, some non-intercepting bunch length and emittance diagnostic approaches will be reviewed.

INTRODUCTION

The following three sections review recent advances in emittance compensation, beam transport, and beam diagnostics, respectively. Due to the relatively large size of the high-brightness beam field due to its evolving maturity, a comprehensive review is not possible. Rather, the emphasis in this paper will be to review conceptual advances that have occurred recently that provide deeper fundamental understanding into some of the more critical aspects of high-brightness electron beam production, transport, and measurement.

A major insight in understanding emittance compensation was established with the introduction of using wave-breaking concepts to describe the final, residual emittance. These same concepts are applicable to a wide range of high-brightness electron beam dynamics, not just beam production from photoinjectors. Additionally, we can estimate the threshold initial emittance that leads to wave breaking and rapid thermalization in the first betatron period for high-brightness beams.

The next section will describe some aspects of high-brightness beam transport. The emphasis will primarily be on emittance preservation in bending systems. Recent work has demonstrated a remarkable cancellation between the energy-independent transverse and longitudinal forces. This cancellation is related to the cancellation of the centrifugal space-charge force with the beam's potential depression for DC beams. For periodic focusing channels, there is a halo and emittance growth mechanism related to the wave-breaking concepts of the earlier section.

In the final section, we will review some nonintercepting diagnostic developments. The transverse electric field from a short electron bunch has been measured via the phase shift of a laser beam in a crystal, which directly gives the wake field from that bunch, and potentially the bunch profile. Also recently, both the horizontal and vertical emittances have been measured with non-intercepting beam-position monitors. A final diagnostic which is very attractive is to use a chicane to rotate the beam's longitudinal phase space to directly obtain the bunch's longitudinal shape.

WAVE BREAKING AND EMITTANCE COMPENSATION

Emittance compensation has been used for over a decade to reduce the emittance from rf photoinjectors[1]. This emittance reduction is easy to understand in terms of coherent transverse plasma oscillations[2]. In Fig. 1 we see a typical plot of normalized emittance and beam size as a function of axial position[3].

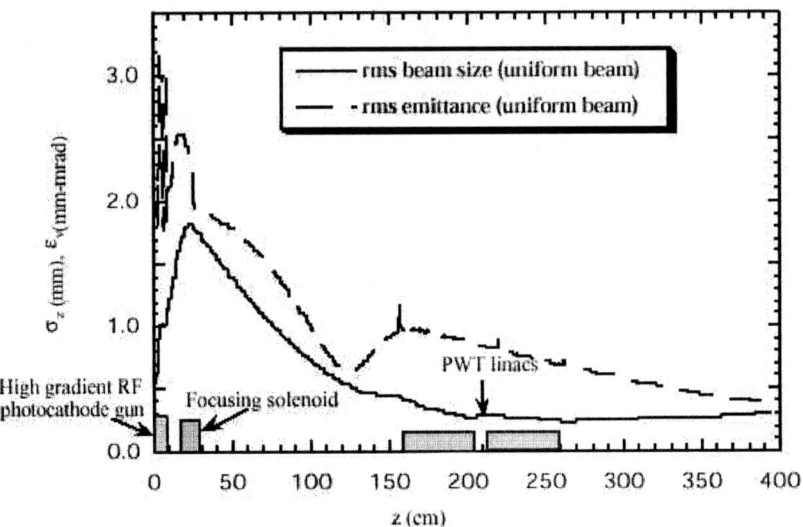

FIGURE 1. Simulation results of beam profile and emittance from a photoinjector design using an emittance-compensating solenoid and two PWT standing wave accelerator sections, from reference 3.

A nonuniform or bunched beam will undergo emittance oscillations due to the fact that the radial force is not simply proportional to radius everywhere. For a non accelerating beam, the radial equation of motion is:

$$\sigma'' + K\sigma - K_s/\sigma = 0 \quad (1)$$

where σ is the transverse coordinate of a particle in a slice of the beam at a normalized radius ζ, K is a normalized focusing and $K_s = 2I(r)/I_A\gamma^3\beta^3$ is the normalized space-charge force, where $I(\zeta)$ is the current within radius ζ, γ is the beam's relativistic mass factor, β is its velocity normalized to the speed of light, and the primes refer to an axial derivative. We will assume that the beam motion is very nearly laminar, and that ζ is a constant for each particle.

The equilibrium particle orbit is given by $\sigma_{eq} = \sqrt{K_s/K}$, and we expand the orbits about this equilibrium position:

$$\sigma(\zeta,z) = \sigma_{eq}(\zeta,z) + \delta(\zeta,z) \quad . \tag{2}$$

Then, to first order the radial equation of motion becomes

$$\delta''(\zeta) + 2K\delta(\zeta) = 0 \quad . \tag{3}$$

This simple expression tells us that all particles will oscillate at the same frequency (to lowest order), and that any emittance growth due to a spread in phase-space angles from these oscillations will vanish every half-oscillation period. Even though this oscillation period only depends on the external focusing force, we will still refer to these oscillations as coherent transverse plasma oscillations. In Fig. 2 (also from reference 3), we see the area in phase space defined by three particles initially at the same radius, but with three different equilibrium positions, due to different space-charge forces λ. Each of the particles have a circular orbit, which is traversed clockwise. Each orbit takes nearly the same time to complete, as seen in Eqn. (3). The triangle shown in Fig. 2 is the area containing all three particles after a quarter oscillation period. Note that after both a half and a full oscillation period that the triangle collapses to zero volume. If $x - x'$ phase space was plotted instead of $r - r'$, the area would include another triangle with both x and x' symmetry on the left side of the $x = 0$ axis, giving the phase-space area the appearance of a bow-tie.

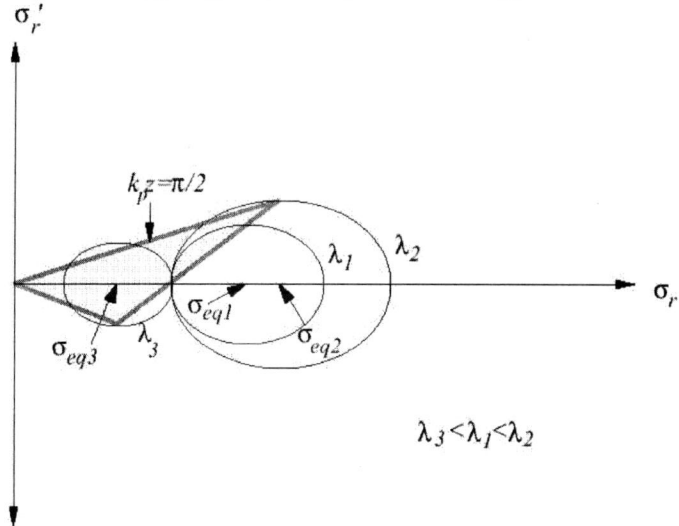

FIGURE 2. Formation of the "bow-tie" in phase space, outlined from three particles with the same initial positions, but with different equilibrium orbits (from reference 3).

Related emittance oscillations can be seen in many other areas of beam physics. The oscillations can be introduced by either radial or longitudinal nonlinearities. In Fig. 3, we plot emittance oscillations for a drifting 4 MeV, 4 kA, rms matched electron beam with zero initial emittance but with a nonuniform initial density distribution. The initial transverse kinetic energy of this beam is zero, but at the first emittance maximum, all the excess potential energy from the nonuniform density distribution has

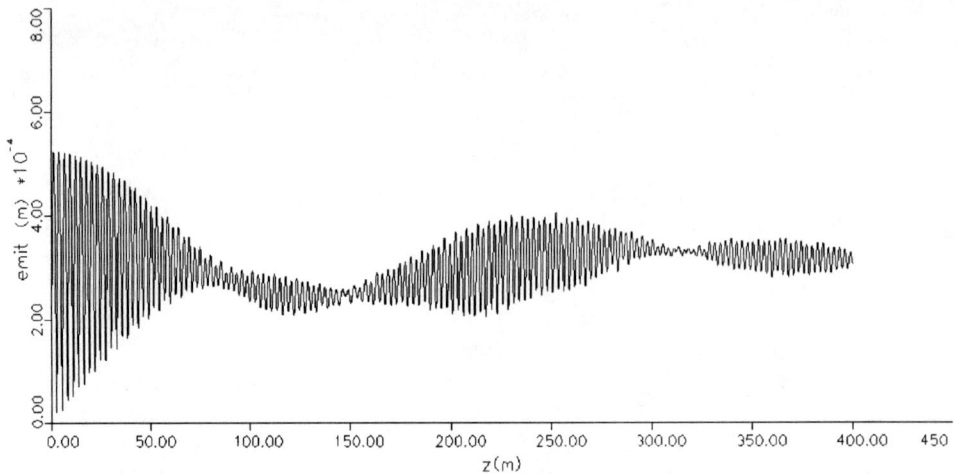

FIGURE 3. Emittance oscillations for a drifting 4-MeV, 4-kA electron beam.

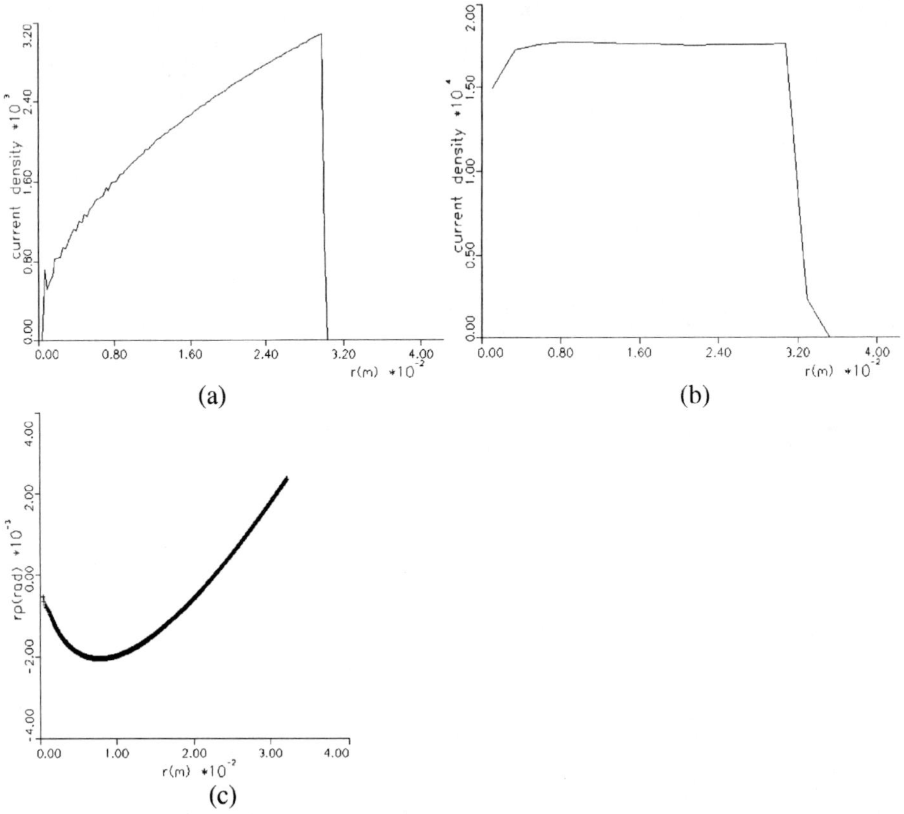

FIGURE 4. (a) Initial density distribution for the case shown in Fig. 3, (b) density profile at the first emittance maximum, and (c) phase space distribution at the first emittance maximum.

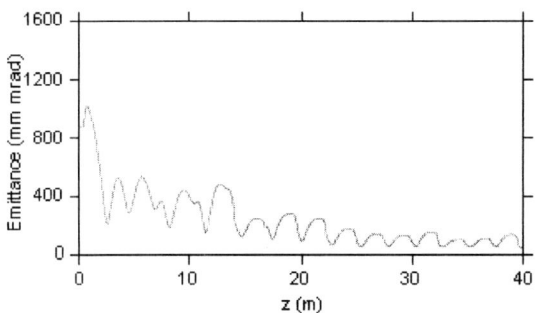

FIGURE 5. Normalized, 90% emittance evolution for 4-kA induction accelerator, showing emittance oscillations due to coherent transverse plasma oscillations.

been transferred into kinetic energy. Both the initial emittance maximum and the final emittance can be predicted from initial excess nonlinear field energy. For this case, the damping mechanism is from the spread of second-order frequencies. In general, though, thermalization results from wave breaking, where the $x - x'$ correlation in phase space becomes multivalued and the beam loses laminarity. Thermalization from wave breaking occurs relatively quickly, on the order of a half betatron period for mixing to occur.

In Fig. 4 we plot the initial density profile corresponding to the initial conditions of Fig. 3 (a), and the density profile (b) and the transverse phase space (c) at the first emittance maximum. These plots verify that the initial excess potential energy has been transferred into transverse kinetic energy. Note that the beam is rms matched, with a fixed point in phase space at the rms beam radius.

In Fig. 5 we see many of these emittance oscillations in a simulation of a 20-MeV, 4-kA induction linac, in which the pulse length is essentially DC. For the case of an rf photoinjector, there is typically only one emittance oscillation (and one-half of a transverse plasma oscillation) due to the rapid acceleration. For this simulation, nonlinear radial focusing in the diode and nonlinear radial space-charge forces induces the emittance oscillations.

With acceleration, the equations become more complex but the physics is essentially the same. The radial equation of motion is now:

$$\sigma'' + \sigma' \frac{\gamma'}{\gamma} + \sigma K \left(\frac{\gamma'}{\gamma \sin \phi} \right)^2 - \frac{K_s}{\sigma \gamma^3} = 0 \qquad (4)$$

where ϕ is the rf phase angle of the bunch. With a Cauchy transform (using $y = \ln \gamma)^2$ this becomes

$$\frac{d^2 \sigma}{dy^2} + \Omega^2 \sigma = \frac{S e^{-y}}{\sigma} \qquad (5)$$

where Ω has to do with focusing and the synchronous phase and $S(\zeta) = K_s(\zeta)/\gamma'^2$, which is a space-charge parameter. Reference 2 pointed out this equation has an exact solution

$$\sigma_{eq} = \sqrt{\frac{S}{\frac{1}{4}+\Omega^2}} \, e^{-y/2} \tag{6}$$

and the oscillations about this orbit (the invariant envelope) have (to first order) the same period. Although this solution has the same form as that for the coasting beam, there is a difference – the phase of the oscillations are not initially locked as before.

We can see this from the solutions of a trajectory perturbed from the equilibrium trajectory by an amount δ_0 (now we are thinking about slices that are radially uniform but differ in current longitudinally):

$$\sigma = \sqrt{\frac{S/\gamma}{\frac{1}{4}+\Omega^2}} + \delta_0 \cos(\omega \ln \gamma + \theta)$$

$$\sigma' = -\frac{\gamma'}{2}\sqrt{\frac{S/\gamma^3}{\frac{1}{4}+\Omega^2}} - \delta_0 \frac{\omega \gamma'}{\gamma} \sin(\omega \ln \gamma + \theta) \tag{7}$$

The phase of these oscillations is clearly more complicated, and one can achieve better alignment of the slices if there is some flexibility of the initial conditions at the cathode (some initial divergence or rf focusing)

So far we have been talking about *laminar* coherent transverse plasma oscillations. The beam can become nonlaminar either radially or longitudinally if the current density drops sufficiently low[4]. In an axial slice of the bunch, the outer electrons are not repelled enough by the space charge force (for rms matched focusing) if the beam density drops to a point somewhat less than ½ of the average density out to that point. If the outer particles are not repelled, they cross into the core of the beam, forming wave breaking. In Fig. 6(a) we see the beam's phase space exiting a diode for the case in the previous figure (4-kA and 3.5-MeV beam). Some amount of modest wave breaking is already seen at this location, due to nonlinear electrostatic and magnetic focusing. This simulation uses the diode simulation code EGUN[5], and assumes that the initial beam emittance (at the cathode) is zero. As the beam travels downstream in the accelerator, the particles which have wave broken (and thus have nonlaminar trajectories) pass through the core of the beam, forming a halo, as seen in Fig. 6(b). During the focusing part of each radial oscillation, additional particles at the edge of the beam's radial distribution wave break. Eventually, the wave broken particles oscillate about the 2:1 resonance, forming a ring in phase space around the laminar particles defining the core (shown in Fig. 6(c)).

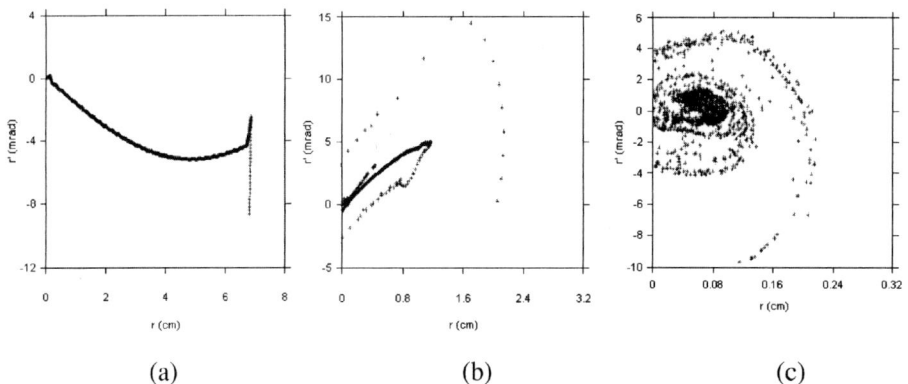

FIGURE 6. (a) Initial phase space distribution after the electron diode, (b) wave breaking particles passing through the beam core forming a halo early on, and (c) final evolved distribution with wave breaking particles spiraling about the 2:1 resonance.

At this point we know that emittance oscillations will occur if the beam either has an initial emittance or some initial density profile mismatch. If no wave breaking occurs, these oscillations will persist for very long times and distances (see Fig. 3 for example). However, if the beam wave breaks, the thermalization will occur much faster – essentially at the rate that the particles wave break. Note that the characteristic time scale for wave breaking is a radial compression, or about a ¼ betatron period. This time scale has been empirically described before[6]. Although we can not make general statements about the rate of wave breaking for particles in transport, we can easily estimate a characteristic emittance which will always lead to significant wave breaking in the first betatron period. Wave breaking of about ½ the particles will occur for a uniform density beam with an initial emittance exceeding[7]

$$\varepsilon_{w-b} = 4 r_e \sqrt{(I/I_A)/\gamma \beta} \quad . \tag{8}$$

By definition, high-brightness electron beams are under this limit, and do not undergo immediate, large-scale wave breaking. However, rapid phase-space mixing is common for ion beams (this emittance threshold is only 0.2 mm mrad for a 1-GeV, 100-mA, 1-mm radius hydrogen beam). This consideration explains why high-brightness electron beams do not thermalize like ion beams.

In an rf photoinjector, the radial nonlinearities are often less important than the time dependence of the current profile of the pulse. For this case, the ends of the electron bunch are not rms matched, and the emittance will oscillate from the mechanism shown in Fig. 2. The emittance thermalization will occur fairly rapidly as the head and tail slices with lower current will exhibit behavior essentially the same as if they wave broke relative to the slices near the center of the bunch. However, as long as the individual slices do not wave break, this emittance is still theoretically recoverable. Most importantly, for longitudinal current variations and no radial nonlinearities, there is no excess of nonlinear free energy which leads to a nonzero thermalized emittance. For the case of the rf photoinjector, if the phase of the emittance oscillations is properly matched, the emittance can theoretically vanish at the beam's application in the absence of wave breaking.

In Fig. 7 (from reference 3) we see phase-space plots from a simulation of an rf photoinjector, with no radial nonlinearities and with an initially radially uniform beam density, using the accelerator simulation code PARMELA[8]. The characteristic bow-tie shape is very evident in the $x-x'$ phase space shown in Fig. 7(a). In Fig. 7(b) we see the $x-z$ configuration space, which shows that the head and tail slices of the bunch have been overfocused (due to their lower space-charge forces). In Fig. 7(c) we see the $x-x'$ phase space of the distribution with $|\delta z|<0.1$ mm. Note that the core still has a bow-tie distribution, and the spread in the bow-tie angles is due to the second order dependence on the oscillation frequency about a given slice's equilibrium orbit. The effect of the head and tail particles in Fig. 7(a) show some small additional wave breaking characteristics.

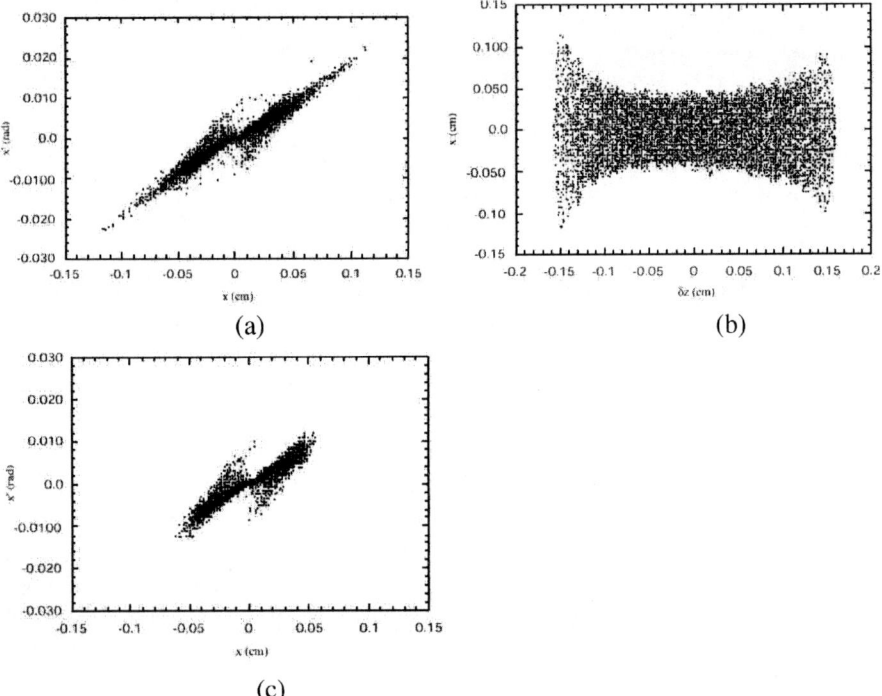

FIGURE 7. (a) Phase space for entire beam, (b) x-z correlation, (c) phase space for core of beam (from reference 3)

In Fig. 8, also from reference 3, we see the same plots for the same accelerator design, but with an initially nonuniform radial beam distribution. Now Figs. 8(a) and 8(c) are more similar, and demonstrate a subdistribution with strong wave breaking characteristics (the horizontal distribution perpendicular to the bow-tie distribution). The emittance of this case is four times larger than the previous case, with much less pronounced emittance oscillations (more rapid thermalization).

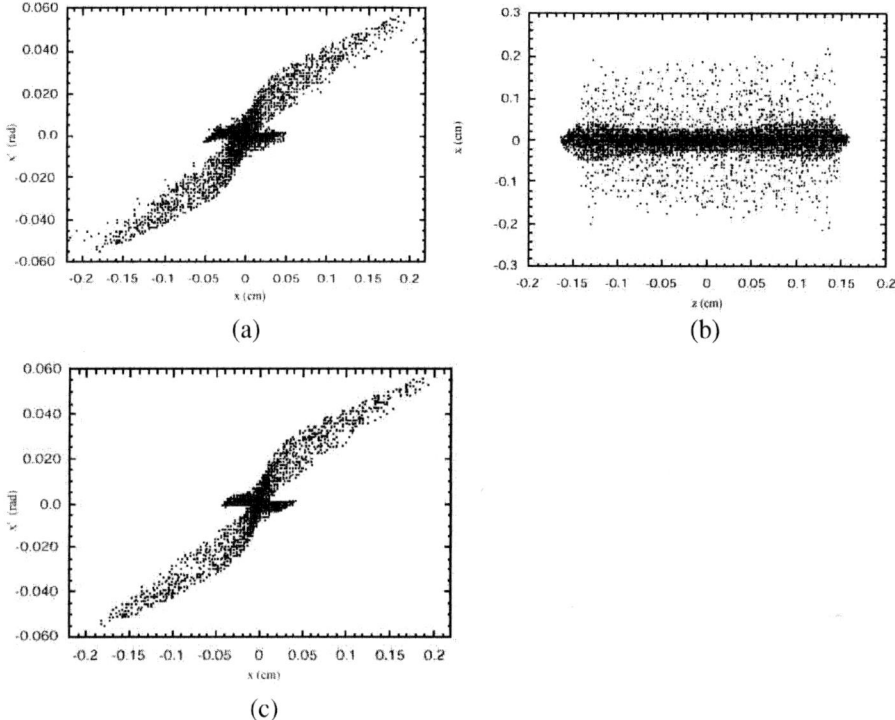

FIGURE 8. (a) Phase space for entire beam, (b) x-z correlation, (c) phase space for core of beam, for case showing radial wave breaking (from reference 3)

In practice, the rf emittance growth is not negligible, particularly at high frequencies. For many cases the rf emittance dominates, and is given by[9]

$$\varepsilon_{rf} = \frac{eE_o}{2\sqrt{2}m_o c^2} k^2 \sigma_x^2 \sigma_z^2 \quad . \tag{9}$$

This is probably the reason that L-band photoinjectors are competitive with higher frequency devices.

Timing stability for very high frequency rf photoinjector can be problematic[10]. The stability for W-band (90 GHz) photoinjectors is on the order of tens of fs, which is better than standard drive lasers can produce. One way around this is to use an essentially DC laser and bunch the electrons that get accelerated in an undulator[10]. Electrons from 65 degrees of phase get out of the gun (15 A, 30 pC in 2 ps at the cathode, and up to 70 A with 1.8 MeV energy at the exit). The energy spread is mostly linear, and the beam can be bunched to 600 A in the undulator, with about 1.5 mm mrad.

TRANSPORT ISSUES FOR HIGH-BRIGHTNESS ELECTRON BEAMS

We will review two recent developments in the transport of high-brightness electron beams. First, we will describe some recent work on halo production in periodically focused systems, where wave breaking can occur continually along the transport line The physics is different than for uniformly focused systems, and leads to a fuzzy separatrix. Second, we will review the emittance growth from coherent synchrotron radiation (CSR) effects in bends. This understanding is becoming quite mature, although there are still some loose ends.

Recent work[11] on periodic focusing channels show that the separatrix becomes fuzzy and particles can enter the 2:1 resonance that were initially in the core. Wang developed a new technique for stroboscopic plots for periodic focusing which reduces the flutter, and shows the separatrix becomes poorly defined. With varying focusing, particles can enter the 2:1 resonance (wave break) essentially anywhere along the beam line transport. Class I particles remain inside the core, Class II particles are outside the core but not in resonance with the core, Class III particles oscillate at ½ the core frequency (2:1 resonance), Class IV particles oscillate with large amplitudes, and Class V particles can be driven in and out of resonance by the periodic focusing and the flutter, an effect not found in uniform focusing cases. The action of the Class V particles can be described in terms of wave breaking from radial oscillations due to the periodic focusing.

There are several effects that can lead to beam brightness degradation for a beam traveling in a bend system. The first of these is if the optics are non-achromatic; the accelerator community knows how to avoid this and this is usually not a problem. There are also transverse space-charge forces which have a $1/\gamma^2$ scaling. There forces typically dominates at low energy and are also negligible at high energy. Finally, there are longitudinal space-charge forces which do not have the $1/\gamma^2$ scaling and the effects from these forces dominate at high energy. The resulting beam degradation can be exhitbited either as transverse emittance growth or energy spread. The energy independent longitudinal force has two parts, the coherent synchrotron radiation (CSR) which is a long range, radiative force, and the noninertial space-charge force (NISCF) which is short range and nonradiative.

The importance of these longitudinal forces is that they cannot be reduced by going to larger radius bends – the longitudinal forces (and resulting emittance growths) scales as $R^{1/3}$ and is independent of beam energy (at high enough energy)

The following derivation of the longitutinal space-charge forces in a bend clearly separates out the CSR and NISCF terms. The electric field from a line of charge is given by[12]:

$$E_\theta = -\frac{\partial A_\theta}{\partial t} - \frac{1}{r}\frac{\partial \phi}{\partial \theta} \tag{10}$$

where

$$\phi = \frac{1}{4\pi\varepsilon}\int_{\zeta_r}^{\zeta_f} \frac{\lambda}{r_{ret} - \vec{r}_{ret}\cdot\vec{u}_{ret}/c} d\zeta$$

$$A_\theta = \frac{1}{4\pi\varepsilon c}\int_{\zeta_r}^{\zeta_f} \frac{\lambda\beta\cos\zeta'}{r_{ret} - \vec{r}_{ret}\cdot\vec{u}_{ret}/c} d\zeta \quad . \tag{11}$$

Explicit calculation for a uniform density line of charge gives:

$$E_\theta = \frac{\lambda}{4\pi\varepsilon}\frac{1}{r_{ret} - \vec{r}_{ret}\cdot\vec{u}_{ret}/c}\left(\frac{1}{\gamma^2} - \beta^2\frac{x}{R} + \beta^2\frac{r}{R}(1-\cos(\zeta'))\right)\bigg|_{\zeta_r}^{\zeta_f} \tag{12}$$

where the retarded time is given by

$$\frac{R^2}{\beta^2}(\zeta'-\zeta)^2 = \rho^2 + 2R(R+x)(1-\cos\zeta') \quad . \tag{13}$$

We see three terms on the right-hand-side of Eqn. (12), which we can identify as: (1) the "usual" $1/\gamma^2$ space-charge term (but not the same as for straight-line motion), (2) the noninertial space-charge force (NISCF), and (3) the coherent synchrotron radiation force (CSR)

There is also an energy independent transverse force, the centrifugal space-charge force (CSCF), which cancels the effect from the beam's potential depression[13]. We can easily see this cancellation for the DC case, starting with the approximation $\vec{A} = \left(0, \frac{\beta}{c}\phi, 0\right)$, and defining $x = r - R$, where R is bend radius. For the DC case, the radial force is

$$F_r = e(E_r + v_z B_\theta - v_\theta B_z) = e\left(-\frac{1}{\gamma^2}\frac{d\phi}{dr} + \beta^2\frac{\phi}{R}\right) \quad . \tag{14}$$

The radial equation of motion is given by

$$\frac{d}{dt}\gamma\dot{r} = \frac{\gamma v^2}{r} - \frac{evB_{ext}}{m} + \frac{e}{m}\left(-\frac{1}{\gamma^2}\frac{d\phi}{dr} + \beta^2\frac{\phi}{R}\right) \quad . \tag{15}$$

We expand to lowest order in R about the equilibrium orbit, defined by $0 = \frac{\gamma_0 v^2}{r} - \frac{evB_{ext}}{m} + \frac{e}{m}\beta^2\frac{\phi(0)}{R}$ and let $\gamma_1 = \gamma - \gamma_0$, and the radial equation of motion becomes

$$\ddot{x} = -\frac{\dot{\gamma}_1}{\gamma}\dot{x} + \frac{v^2}{R}\left(\frac{\gamma_1}{\gamma} - \frac{x}{R}\right) + e\beta^2\frac{\phi(x)-\phi(0)}{\gamma mR} - \frac{e}{\gamma^3 m}\frac{\phi(0)}{R} \quad . \tag{16}$$

Now $\gamma_1 = -e\dfrac{\phi(x)-\phi(0)}{mc^2}$ so the curvature term cancels the potential depression term, leaving

$$\ddot{x} = \dfrac{v^2}{R^2}x - \dfrac{e}{\gamma m}\dfrac{\phi(0)}{R}\left(\beta_t^2 - \dfrac{1}{\gamma^2}\right), \qquad (17)$$

and the energy independent transverse space-charge term has vanished. The CSCF cancels the potential depression of the beam, leading to the surprising effect that the beam steers as if there was no potential depression of the beam. This effect has actually been experimentally verified[14].

Recently, R. Li[15] made important observation – for a bunched beam, the noninertial space charge force leads to a change in a particle's potential, which is canceled by the CSCF term! We can sketch this cancellation out (the derivation is similar to DC case, but now with time derivatives). The radial equation of motion is

$$\dfrac{d}{dt}\gamma m \dot{r} = e(\vec{E}+\vec{v}\times\vec{B})_r + \dfrac{\gamma m v_\theta^2}{r} \qquad (18)$$

Using $r = x + R$ and $\vec{E} = -\vec{\nabla}\phi - \dot{\vec{A}}$ we get

$$\gamma m \ddot{x} + \dot{\gamma} m \dot{x} = \dfrac{\gamma m v_\theta^2}{R+x} + e\dfrac{\partial}{\partial x}(-\phi - v_\theta A_\theta) - e\left(\dfrac{\partial}{\partial t}+\dfrac{v_\theta}{r}\dfrac{\partial}{\partial \theta}\right)A_r + e\dfrac{v_\theta A_\theta}{r} - ev_\theta B_{ext} \qquad (19)$$

and with $ev_\theta B_{ext} = \dfrac{\gamma_o m v_\theta^2}{R}$ this becomes ($\gamma = \gamma_o + \gamma_1$)

$$\ddot{x} = \dfrac{v_\theta^2}{R}\left(\dfrac{\gamma_1}{\gamma_o} - \dfrac{x}{R}\right) + \dfrac{e}{m\gamma}\left(\dfrac{\partial}{\partial x}(-\phi - v_\theta A_\theta) - \dfrac{d}{dt}A_r + \dfrac{v_\theta A_\theta}{r}\right). \qquad (20)$$

Using the azimuthal electric field,

$$\gamma_1 = \dfrac{e}{mc^2}\int\left(-\dfrac{1}{r}\dfrac{\partial}{\partial \theta}\phi - \dot{A}_\theta\right)ds = \dfrac{ev_\theta}{mc^2}\int\left(-\dfrac{1}{r}\dfrac{\partial}{\partial \theta}\phi - \dfrac{1}{v_\theta}\dfrac{\partial}{\partial t}\phi + \dfrac{1}{v_\theta}\dfrac{\partial}{\partial t}\phi - \dot{A}_\theta\right)dt$$

$$= \dfrac{ev_\theta}{mc^2}\int\left(-\dfrac{1}{v_\theta}\dfrac{d}{dt}\phi + \dfrac{1}{v_\theta}\dfrac{\partial}{\partial t}\phi - \dot{A}_\theta\right)dt . \qquad (21)$$

The radial equation of motion now becomes

$$\ddot{x} = -\dfrac{v_\theta^2 x}{R^2} + \dfrac{\beta_\theta^2 e}{\gamma_o Rm}\left(-\phi + \int(\dot{\phi}-v_\theta \dot{A}_\phi)dt\right)$$
$$+ \dfrac{e}{m\gamma}\left(\dfrac{\partial}{\partial x}(-\phi - v_\theta A_\theta) - \dfrac{d}{dt}A_r + \dfrac{v_\theta A_\theta}{r}\right) \qquad (22)$$

or (to lowest order)

$$\ddot{x} = -\dfrac{v_\theta^2 x}{R^2} + \dfrac{e}{\gamma_o Rm}\left[\left(-\beta_\theta^2 \phi + v_\theta A_\theta\right) + \beta_\theta^2\int(\dot{\phi}-v_\theta \dot{A}_\phi)dt + R\dfrac{\partial}{\partial x}((-\phi - v_\theta A_\theta))\right] \qquad (23)$$

where the first term is geometric, the second is the CSCF and NISCF cancellation, the third is the CSR term and the fourth is the centripetal space-charge force.

HIGH-BRIGHTNESS BEAM DIAGNOSTICS

The main challenges of diagnostics for high-brightness beams is to measure extremely small bunch lengths and to measure very low emittances. Both measurements provide difficulties, as even the process of measuring the beam can change it. In general, the more promising diagnostics are nonintercepting. In this section, we will review three relatively new diagnostic ideas. First, we will review progress on measuring the beam emittance using the quadrupole moments from beam-position monitors (BPMs). Emittances as low as 10 mm mrad have been measured successfully with this technique to date. Next, we will describe measurements made on the transverse electric field induced by the passing of a short electron bunch. This technique promises to measure both wake impedances of specific accelerator components and even the bunch current profile. Finally, we will describe an intercepting bunch length diagnostic that uses a chicane to rotate the longitudinal phase space.

Measuring the emittance using a BPM was suggested in 1983[16] and was first demonstrated in 1999[17]. The technique uses a single BPM, shown in Fig. 9, and two quadrupoles.

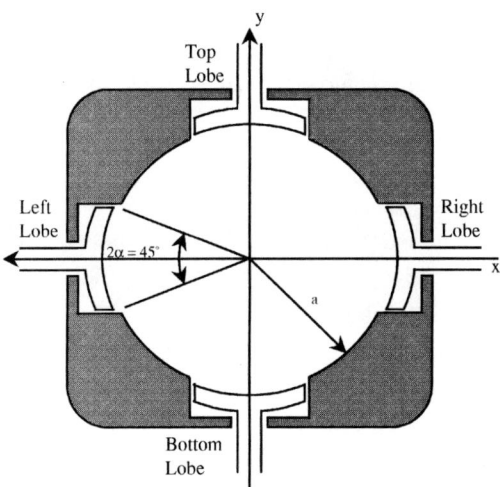

FIGURE 9. Beam-position monitor schematic.

The technique is based on the fact that a BPM is sensitive to the second moment of the beam's distribution. The difference between the horizontal and vertical second moments is given by

$$\langle x^2 \rangle - \langle y^2 \rangle + \bar{x}^2 - \bar{y}^2 = a^2 \frac{\alpha}{\sin 2\alpha} \frac{A_R + A_L - A_T - A_B}{A_R + A_L + A_T + A_B} \qquad (24)$$

where A_R, A_L, A_T and A_B are the signal amplitudes of the right, left, top and bottom electrodes of the BPM. The beam's phase space upstream of the quadrupole can be determined by inverting the relationship

$$\langle x^2 \rangle_{BPM} - \langle y^2 \rangle_{BPM} = (R_{11})^2 \langle x^2 \rangle_q + 2R_{11}R_{12} \langle xx' \rangle_q + (R_{12})^2 \langle x'^2 \rangle_q$$
$$- (R_{33})^2 \langle y^2 \rangle_q - 2R_{33}R_{34} \langle yy' \rangle_q - (R_{34})^2 \langle y'^2 \rangle_q, \quad (25)$$

where the BPM subscript refers to the BPM location and the q subscript to just upstream of the quadrupole. The constants R_{jk} are from the transfer matrix for the focusing channel between the upstream point and the BPM:

$$\bar{\bar{R}} = \begin{bmatrix} R_{11} & R_{12} & 0 & 0 \\ R_{21} & R_{22} & 0 & 0 \\ 0 & 0 & R_{33} & R_{34} \\ 0 & 0 & R_{43} & R_{44} \end{bmatrix}. \quad (26)$$

Tremendous progress has been made in using electro-optical effects to time-resolve the transverse electric field from a passing bunch of charge. One such experiment is taking place at Fermilab A LiTaO$_3$ electro-optic crystal is used to generate a phase shift in a laser beam due to the induced electric field from a passing electron bunch in a chamber. So far, the transverse field has been resolved with a time resolution of about 10 ps. Because the diagnostic is in a cross, the wake potential is actually measured; if the diagnostic were in a smooth pipe, the current profile of the bunch would be found.

The idea of the intercepting bunch-length diagnostic[19] is to pass the beam through a chicane followed by an off-phase accelerating structure with the following total transfer matrix and mapping from initial to final longitudinal phase space:

$$\begin{pmatrix} 1 & 0 \\ \alpha & 1 \end{pmatrix} \begin{pmatrix} 1 & \delta \\ 0 & 1 \end{pmatrix} \begin{pmatrix} t \\ E \end{pmatrix} = \begin{pmatrix} t + \delta E \\ \alpha t + (1+\alpha\delta)E \end{pmatrix}. \quad (27)$$

If $\alpha\delta = -1$, then the final energy of a particle only depends on its initial time, $E_{final} = \alpha t_{initial}$ and the bunch profile can be measured in a spectrometer. An experiment[20] measured the bunch current profile of an initially 19 MeV beam (accelerated to 27 MeV in the final accelerating structure) with 260 fs resolution. This technique is capable of a resolution of 16 fs for a beam of 100 MeV beam with a 10 MeV section at 10 GHz.

REFERENCES

[1] B. E. Carlsten, Particle Accelerators **49**, 27 (1995).
[2] L. Serafini and J. B. Rosenzweig, Phys. Rev. E **55**, 7565 (1997).
[3] S. G. Anderson and J. B. Rosenzweig, Phys. Rev. ST Accel. Beams **3**, 094201 (2000).
[4] O. A. Anderson, Particle Accelerators **21**, 197 (1987).
[5] W. B. Herrmannsfeldt, Linear Accelerator and Beam Optics Codes Workshop, San Diego, (1998).

[6] M. Reiser, *Theory and design of charged particle beams* (John Wiley and Sons, Inc., New York, 1994), Chap. 6
[7] B. E. Carlsten, Phys. Rev. E **60**, 2280 (1999).
[8] L. Young, Los Alamos National Laboratory, private communication.
[9] K.J. Kim, Nucl. Instrum. Meth. Phys. Res. **A275**, 210 (1989).
[10] D. T. Palmer and M. J. Hogan, 1999 Particle Accelerator Conference, IEEE Catalog Number 99CH36366, New York, 1997, (1999).
[11] T.S. F. Wang, Phys. Rev. E **61**, 855 (2000).
[12] W. Panofsky and M. Phillips, *Classical electricity and magnetism* (Addison-Wesley Publishing Company, Inc., Reading, MA, 1955)
[13] E. P. Lee, Particle Accelerators **25**, 241 (1990).
[14] P. Allison, D. C. Moir, and G. Sullivan, 1997 Particle Accelerator Conference, IEEE Catalog Number 97CH36167, Vancouver, 1144, (1997).
[15] R. Li, Thomas Jefferson National Accelerator Facility, private communication, 1999.
[16] R. Miller, J. Clendenin, M. James, and J. Sheppard, 12th International Conference on High Energy Accelerators, Fermilab, 602, (1983).
[17] S. J. Russell, Rev. Sci. Instrum. **70**, 1362 (1999).
[18] Michael Fitch, Fermilab, private communication, 2000.
[19] E. R. Crosson, K. W. Berryman, B. A. Richman, T. I Smith, and R. L. Swent, AIP Conference Proceedings **367**, 397 (1995).
[20] K. N. Ricci, T. I. Smith, and E. R. Crosson, Nucl. Meth. Instrum. Phys. Res. **A429**, II-61 (1999).

Aperture Effects in Intense Beams

S. Bernal, P.G. O'Shea, R. Kishek and M. Reiser

Institute for Plasma Research
University of Maryland
College Park, Maryland 20742

Abstract. An aperture affects an intense beam in two ways: first, it changes the relative contributions of thermal and space charge effects by changing the ratio of emittance to generalized beam perveance; second, the velocity space is altered depending on the aperture size relative to the beam size and also on the aperture location. It is observed that collimation of an expanding intense electron beam leads to the formation of an internal ring of charge that propagates towards the beam center. The effect depends on the presence of external (linear) focusing, but is independent of the nature of this focusing. Furthermore, the phenomenon is reproduced well in a simple model based on a K-V distribution for the bulk of the beam and special non-laminar orbits near the beam edge.

INTRODUCTION

Apertures are common in beam transport and beam generation systems, e.g. the anode aperture in a typical Pearson gun or a collimating aperture at the exit of the same gun. The study of the effects of apertures on beams has traditionally been limited to low-current, high energy beams where single-particle considerations are adequate, following a close analogy with ray optics [1]. Only recently have the effects of an aperture on the particle distribution in phase space been observed and studied in an intense beam [2]. The collimated phase space of the beam particles is "probed" with the addition of external focusing; the combined effects of an aperture and subsequent focusing lead to an initial "beam perturbation". In this paper, we concentrate our study on a simple model to represent the initial beam perturbation. As it turns out, the perturbation can be described as a type of "edge imaging" where non-linear space charge forces play a role and, counter-intuitively, net focusing is increased for higher beam currents.

The paper is organized as follows: in the next section we use the smooth approximation of periodic focusing to derive the effect of an aperture on the *tune depression*, a parameter commonly used to describe transport of intense beams. Then, we consider, in general terms, the effects on the *phase-space particle distribution*. The last section presents results of a simple model of the initial beam perturbation, with examples from experiments involving solenoid and quadrupole channels. We conclude with a smooth approximation treatment of the same model.

TUNE DEPRESSION

The transport of a beam of current I and constant radius a in a uniform focusing channel is characterized by the ratio of particle phase advances with and without space-charge, or tune depression TD. In terms of wave-numbers we can write [3]:

$$TD = \frac{k}{k_0} = \sqrt{1 - \frac{K}{k_0^2 a^2}} \qquad (1)$$

where $K = 2(I/I_0)(1/\beta^3\gamma^3)$ is the generalized beam perveance, I_0=17 kA, approximately, for electrons, and $\beta = v/c$, $\gamma = (1-\beta^2)^{-1/2}$ (c is the speed of light, and v is the particle's velocity.) Further, the effective (4×RMS) emittance, related to the natural divergence of the beam, must satisfy [3]

$$\varepsilon = ka^2 \qquad (2)$$

Equations [1] and [2] also represent the conditions of a *matched beam* in the *smooth approximation* of a periodic focusing channel [3].

An aperture at the entrance of the matching section will have an effect on K, and ϵ, which, together with the external focusing given by $k(z)$ will determine the "average" beam size a downstream in the periodic focusing lattice. Within the smooth approximation, where $k_0(z) = k_0$, constant, is substituted for $k(z)$, we can write

$$a = \frac{1}{k_0}\left[\frac{K}{2} + \sqrt{\varepsilon^2 k_0^2 + \left(\frac{K}{2}\right)^2}\right]^{1/2} \qquad (3)$$

Thus, the tune depression will depend on the ratio ϵ/K. It is generally assumed that the emittance scales linearly with the aperture size, so the ratio ϵ/K scales inversely with the aperture size, leading to a dependence of the tune depression on collimation. As an example, Figure 1 shows the tune depression and beam size vs. aperture radius for the University of Maryland Electron Ring [4], with a nominal full beam of 100 mA at 10 keV. The apertures are located on a rotating plate not far from the beam waist, near the anode of the electron gun.

PHASE-SPACE PARTICLE DISTRIBUTION

For a more detailed analysis of the effects of collimation, we assume that the beam just *upstream* of the aperture plane can be replaced by the equivalent K-V beam [5]. The phase-space particle distribution on the *downstream* plane of the aperture

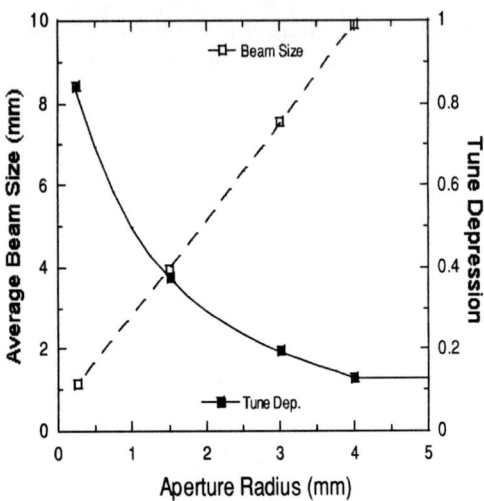

FIGURE 1. Effect of aperture size on tune depression and average beam radius in the University of Maryland Electron Ring. The smooth approximation is used with $\sigma_0 = 72^0$, zero-current phase advance.

is difficult to evaluate, however, because the K-V distribution is not separable, i.e., $f_{KV} \neq f(x,y)g(x',y')$, where primes denote "$d/dz$". The case of a one-dimensional K-V, though, can illustrate the effect of an aperture. In this case, the particles populate the *contour* of an ellipse in $x - x'$ space, so the action of an aperture (small compared to x_{max}) produces two sets of particles with velocities in opposite directions. Therefore, the initial uniform distribution is split and evolves so that 100% hollowness results downstream of the aperture. Of course, the situation is much more difficult to visualize in four-dimensional trace space (for a continuous beam), but we can infer that the collimated K-V is highly non-uniform in velocity space, if the aperture size is sufficiently small. This leads to a non-uniform distribution in space as the beam propagates in a focusing channel.

Simulations with the code WARP [2] confirm that the collimation of a K-V distribution and subsequent focusing lead to non-uniform density patterns. However, the patterns are very different from what is seen in experiments. A more realistic approach is to model the distribution on the *downstream* plane of the aperture as one with uniform density in space but non-uniform density in velocity. A semi-Gaussian (S-G) distribution with a uniform temperature profile is the simplest choice, with the temperature derived from the experimental beam parameters for the equivalent K-V distribution *after* collimation. As described in Ref. [2], simulations with this model are in fair agreement with experiment; a refined distribution with a non-uniform temperature profile yields better agreement.

Figure 2 shows a WARP simulation of the evolution of an (initially) S-G distribution in a uniform focusing channel. A bright ring appears inside the beam at about

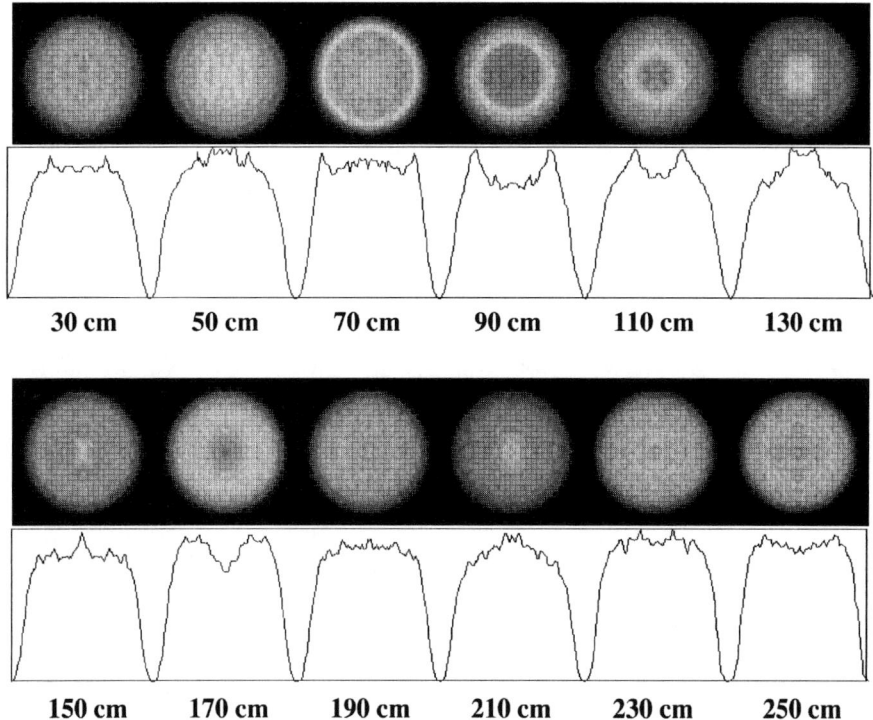

FIGURE 2. WARP simulation pictures of evolution of an initially (at $z = 0$ cm) semi-Gaussian distribution in a uniform focusing channel. The beam parameters correspond to those of Fig. 5(b) below: matched beam radius constant and equal to 8 mm, etc.

70 cm from the "aperture". The perturbation is followed by collective phenomena, seen in both experiments and simulations [2], whose study is outside the scope of this paper (see [6] for further discussion.) The rest of the paper is devoted to a simple model for the appearance of the ring in different focusing channels.

BEAM-EDGE PARTICLE TRACKING

The observed onset and initial evolution of the beam perturbation can be reproduced well with a simple calculation involving a K-V distribution for the bulk of the beam, and a special family of particle trajectories that start at the aperture near to and inside the beam edge, with relatively large initial slopes. The combined action of external focusing and the space charge repulsion from the bulk of the beam causes the edge particle trajectories to cross over inside the beam.

Solenoid Channels. We calculate the particle crossover in an axisymmetric (i.e. solenoidal) system by solving the following differential equations for the beam envelope $R(z)$ and the associated single-particle trajectory $r(z)$ (see, for example, Ref. [3]):

$$R''(z) + k_0^2(z)R(z) - \frac{K}{R(z)} - \frac{\varepsilon^2}{R^3(z)} = 0, \qquad (4)$$

$$r''(z) + k_0^2 r(z) = \begin{cases} \frac{K}{R^2(z)} r(z), & \text{if } r(z) \leq R(z), \\ \frac{K}{r(z)}, & \text{if } r(z) \geq R(z), \end{cases} \qquad (5)$$

where the different symbols have the same meanings as in Eqs. [1]-[2].

The results, using parameters from actual experiments with one and three solenoids, are shown in Figure 3. Three test trajectories have initial slopes larger

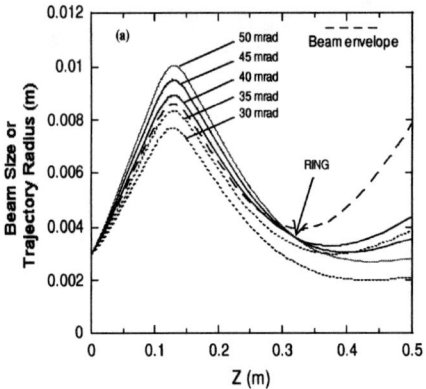
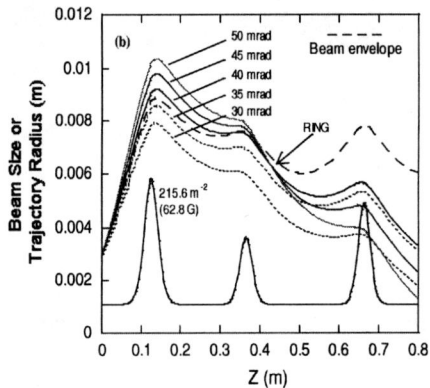

FIGURE 3: Particle tracking in solenoid channels: (a) single solenoid, and (b) three solenoids for envelope matching with $\sigma_0 = 85^0$.

than the initial beam slope (32 mrad); the motion observed can be contrasted with the approximately laminar trajectories calculated for particles starting at the same radial location but with slopes that are smaller than the initial beam slope. The choice of slope values is justified for purposes of illustration only: the population of crossover particles has initial slopes that are only in the immediate vicinity of 40 mrad, approximately.

The crossover points along the $z-axis$ in Fig. 3 agree well with the axial location of a phosphor screen (the main experiment diagnostics) where a bright ring first appears near the beam edge [2].

Quadrupole Channel. The beam-edge particle tracking calculations can be extended to quadrupole experiments; the results for a particular case are shown in Figure 4. The calculations are based on an extension of Eqs. [4]-[5] to non-axisymmetric beams [7]:

$$X''(z) + \kappa_x(z)X(z) - \frac{K}{X(z)+Y(z)} - \frac{\varepsilon_x^2}{X^3(z)} = 0 \tag{6a}$$

$$Y''(z) + \kappa_y(z)Y(z) - \frac{K}{X(z)+Y(z)} - \frac{\varepsilon_y^2}{Y^3(z)} = 0 \tag{6b}$$

$$x''(z) + \kappa_x(z)x(z) = \begin{cases} \frac{K}{X(z)[X(z)+Y(z)]}x(z), & if\ x(z) \leq X(z), \\ \frac{2K}{x(z)\left[x(z)+\sqrt{x^2(z)+Y^2(z)-X^2(z)}\right]}x(z), & if\ x(z) \geq X(z), \end{cases} \tag{7a}$$

$$y''(z) + \kappa_y(z)y(z) = \begin{cases} \frac{K}{Y(z)[X(z)+Y(z)]}y(z), & if\ y(z) \leq Y(z), \\ \frac{2K}{y(z)\left[y(z)+\sqrt{y^2(z)+X^2(z)-Y^2(z)}\right]}y(z), & if\ y(z) \geq Y(z), \end{cases} \tag{7b}$$

where $\kappa_y(z) = -\kappa_x(z)$, and it is assumed that $\epsilon_x = \epsilon_y$. The initial conditions for the solutions of the equations are the same as for the solenoid cases. To simplify the calculations, however, it is implicit in Eqs. [7a] that a particle starts on the horizontal plane ($y = 0$); similarly, Eq. [7b] is valid for a particle initially located on the vertical plane ($x = 0$).

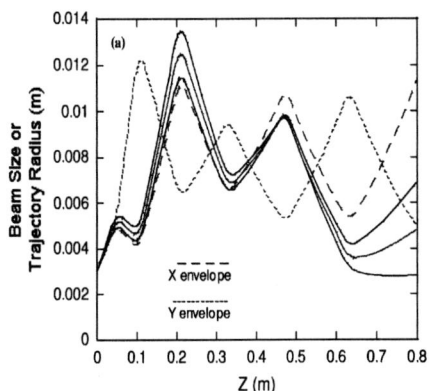
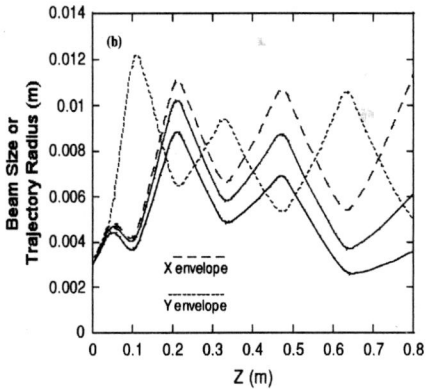

FIGURE 4: Particle tracking in a six-quadrupole matching experiment ($\sigma_0 = 85^0$, zero-current phase advance): (a) vertical motion of particles with initial slopes 40, 45 and 50 mrad, and (b) vertical motion of particles with initial slopes 30 and 35 mrad.

Uniform Focusing. Further insights can be gained if the ray tracing is applied to the uniform focusing case. It is found that the crossover distance is approximately equal to one-half the undepressed betatron wavelength, $\lambda_0/2$, as expected if external focusing is the dominant factor. Further, the crossover distance becomes increasingly smaller than $\lambda_0/2$ as the beam current (space charge) is larger, if the external focusing is adjusted to keep the beam size constant. These conclusions can be easily understood in terms of a linearized model for the motion of the beam edge particles. If we denote by a the (constant) beam radius and move the radial coordinate origin to the beam edge, we have from the second of Eqs. [5]:

$$r''(z) + k_0^2(r+a) - \frac{K}{(r+a)} = 0, \quad r \geq 0, \tag{8}$$

or

$$r''(z) + \left(k_0^2 + \frac{K}{a^2}\right)r(z) + \left(k_0^2 - \frac{K}{a^2}\right)a \approx 0, \quad r/a \ll 1. \tag{9}$$

Therefore, from Eq. [9], the effective half-wavelength of an edge-particle orbit is

$$\frac{\lambda}{2} \approx \frac{\pi}{\left(k_0^2 + \frac{K}{a^2}\right)^{1/2}} \to \frac{\lambda_0}{2\sqrt{2}}, \tag{10}$$

in the limit of laminar flow. As an example, if $K=0.001$ (4 keV, 17 mA electrons), $a=8$ mm, $k_0=4.1$ m^{-1}, we get, from Eq. [10], $\lambda/2 \approx 55$ cm. For comparison, calculations that employ Eqs. [4]-[5] yield a crossover at 62 cm from the aperture. These latter results are illustrated in Figure 5(b); two additional cases with lower and higher space charge are shown in Figs. 5(a) and (c), respectively. The focusing strength is adjusted to keep the beam size equal to 8 mm, with a 4×RMS emittance equal to 67 mm mrad in all three cases (see Eq. [3].) It is important to notice that the generalized beam perveance K increases by a factor of 16 from case (a) to (c), while focusing, as given by k_0^2, only increases by a factor of about 10. In other words, we encounter the seemingly contradictory situation, already expressed in Eq. [10], where additional space charge leads to increased focusing.

The ray tracing technique described provides a visual understanding of the onset of the beam perturbation; it displays the existence of a strong correlation between particle radial location and transverse velocity near the beam edge at the aperture, which in turn results from collimation of an expanding beam. The crossover point obtained in the calculations is relatively insensitive to the initial radial location of the test particles inside the beam, as long as this location and the chosen initial slopes are consistent with the available phase space. In other words, trajectories that start near and inside the beam edge, at the downstream plane of the aperture, can have relatively large initial slopes, while particles that start deeper inside the beam can only yield laminar-like motion.

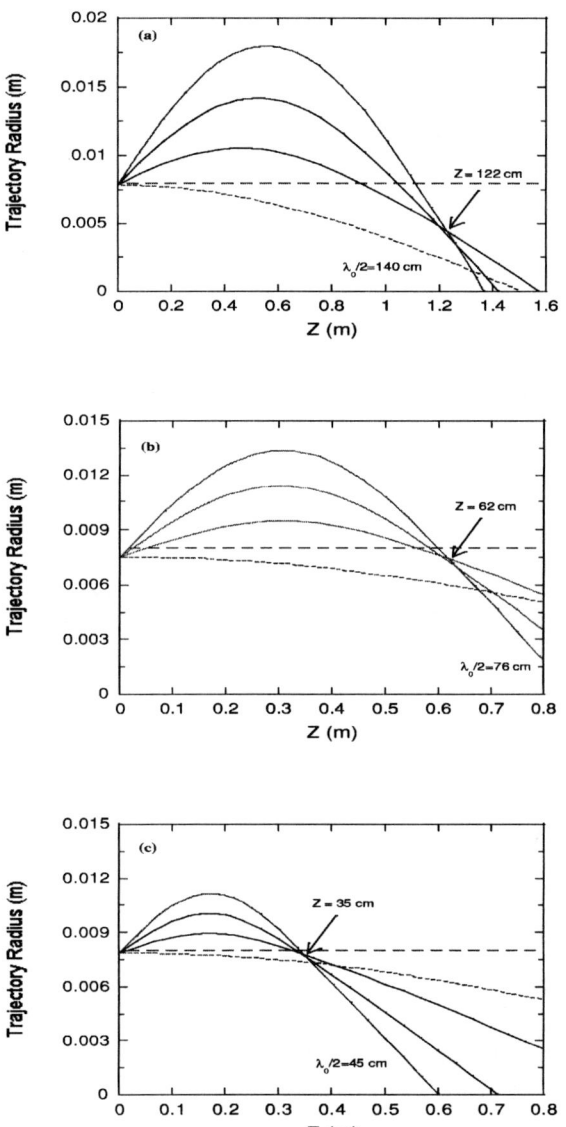

FIGURE 5: Tracking of beam-edge particles in a uniform focusing channel (see Eqs. [4] and [5]). The beam currents are: (a) 4.25 mA, (b) 17 mA, and (c) 51 mA. External focusing is such that $k_0^2 =$ (a) 5.04 m^{-2}, (b) 17.1 m^{-2}, and (c) 48.7 m^{-2}, to obtain a beam radius $a = 8$ mm (dashed line) in all three cases.

Naturally, the particle tracking model presented here is, unlike particle-in-cell simulations, not self-consistent. However, the close agreement with experiments in reproducing the location of the initial perturbation suggests that the initial beam-edge current is only a small fraction of the total beam current. Furthermore, it is conceivable that the model can be extended to predict the actual density of the ring perturbation if detailed knowledge of the initial phase space is available. Before this is attempted, however, additional studies are required for a detailed understanding of the effects of placing an aperture at a beam waist or at a beam envelope maximum. We can reason, though, that the effects are less pronounced than for an expanding beam, assuming a fixed aperture-size to beam-size ratio.

Finally, no analogue exists (to our knowledge) in geometrical optics to our "edge imaging". This type of imaging can include ring formation from particles emitted near e.g. the edge of a thermionic cathode, without the intervention of any aperture, but in the presence of external focusing. There is only a passing resemblance with the constructions of the geometrical theory of diffraction [8], where "diffracted" rays emanating from obstacle edges are used to construct diffraction fields.

REFERENCES

1. Hermann Wollnik, *Optics of Charged Particles*, New York: Academic Press, 1987, ch. 5.
2. S. Bernal, R.A. Kishek, M. Reiser and I. Haber, Phys. Rev. Lett. **82**, 4002 (1999).
3. Martin Reiser, *Theory and Design of Charged Particle Beams*, New York: Wiley, 1994, ch. 4.
4. M. Reiser, *et al.*, Fusion Eng. Des. **32-33**, 293 (1996). See also P.G. O'Shea *et al.*, these Conference Proceedings.
5. I. M. Kapchinskij and V. V. Vladimirskij, in Proceedings of the international Conference on High Energy Accelerators, Geneva, 1959 (CERN, Geneva, 1959), p. 274.
6. R. Kishek *et al.*, to be published.
7. Masanori Ikegami, Phys. Rev. E, **59**, 2330 (1999).
8. J.B. Keller, J. Opt. Soc. Amer. **52** 116 (1962).

Laser Acceleration of Protons from Thin Film Targets

K. Flippo,[†] S. Banerjee,[†] V. Yu. Bychenkov,[‡] S. Gu,[†] A. Maksimchuk,[†] G. Mourou,[†] K. Nemoto[+] and D. Umstadter[†]

[†]*Center for Ultrafast Optical Science, University of Michigan, Ann Arbor, MI 48109-2099, USA*
[‡]*P. N. Lebedev Physics Institute, Russian Academy of Science, Moscow 117924, Russia*
[+]*Central Research Institute of Electric Power Industry, 2-11-1, Iwado-kita, Komae-shi, Tokyo, 201-8511 JAPAN*

Abstract. A collimated beam of fast protons, with energies as high as 10 MeV and total number of 10^9, confined in a cone angle of $40° \pm 10°$ has been observed when a 10 TW laser with frequencies either ω_0 (corresponding to 1 µm) or $2\omega_0$ was focused to an intensity of a few times 10^{18} W/cm^2 on the surface of a thin film target. The protons, which originate from impurities on the front side of the target, are accelerated over a region extending into the target and exit out the backside in a direction normal to the target surface. Acceleration field gradients of ~10 GeV/cm are inferred. The maximum proton energy for $2\omega_0$ can be explained by the charge-separation electrostatic-field acceleration due to "vacuum heating." In other set of experiments when a deuterated polystyrene layer was deposited on a surface of a Mylar film and a ^{10}B sample was placed behind the target, we observed the production of ~ 10^5 atoms of positron active isotope ^{11}C from the nuclear reaction ^{10}B(d,n)^{11}C.

INTRODUCTION

The development of ultra-intense lasers using the Chirped Pulse Amplification technique has made intensities in excess of 10^{18} W/cm^2 routinely achievable. At these intensities an electron in a plasma, during a half laser cycle, can acquire energy that is equal to or greater then the electron's rest mass. Through the electrostatic field, caused by charge separation, the remaining ions can be accelerated to MeV energies. Since high-energy ions can participate in strong interactions, this opens a new realm dealing with the transformation of nuclei on a tabletop on a picosecond time scale.

Recently an interest has developed in ion acceleration by compact high-intensity subpicosecond lasers with potential applications for the initiation of nuclear reactions on a tabletop. Experiments now being carried out involve high-energy ions generated in the interaction of laser pulses with solid targets [1,2], gas jets [3,4], and clusters [5,6]. Most of the current research in this area is directed towards the development of a compact neutron source [7–10], while several other nuclear applications have recently been proposed: isotope production and initiation of fission reactions [11].

Critical for ion acceleration is the efficiency of laser-energy conversion into a high-energy electron component, since the latter through charge separation can produce the requisite strong electrostatic fields, E. Thermal expansion of a laser-driven plasma and ponderomotive electron expulsion constitute the most well-known examples of electrostatic-field production. While the former mechanism has been observed for many years [12], the latter one has only recently been observed in experiments with gas targets [3,4].

In this paper, we report on the direct observation of a high-energy ion beam accelerated by an electrostatic field [13] in the forward (laser) direction to an energy as high as ~ 10 MeV with the laser intensity of 6.10^{18} W/cm^2, and the observation of nuclear fusion reactions induced by high-energy deuterons.

ION ACCELERATION

In our experiments we observed that the protons, which appear to originate from impurities on the front side of the thin-film target, are accelerated over a region extending into the target and exit out the back side, in a direction normal to the target surface. The results are found to depend on a plasma scale length and for a high laser-contrast case ($2\omega_0$) c an be explained by the "vacuum heating" mechanism at a sharp interface due to the Brunel effect [14] or the $\mathbf{v} \times \mathbf{B}$ Lorentz force [15]. In previously reported experimental studies [1], the accelerated ions were found to propagate in the direction of plasma expansion, and the acceleration attributed to charge displacement by thermal expansion.

The experiments were performed using a 10 TW hybrid Ti:Sapphire/Nd:phosphate glass chirped pulse amplification (CPA) laser, which is able to deliver up to 4 J, in a 400 fs pulse at the fundamental wavelength of 1.053 µm with the intensity contrast ratio of ~10^4:1, as measured by the third order correlation technique, and 1 J at the second harmonic with the contrast improved to an estimated 10^7:1 by the frequency doubling in a 4-mm KDP crystal. The experiments with the green light were limited to 1 J, because of the nonlinear distortion of the laser wave front in the crystal. The laser beam was focused with an f/3 (f=16.5 cm) off-axis parabolic mirror on the surface of a thin foil targets. The maximum focused intensity was 6.10^{18} W/cm^2 for the fundamental frequency and 3.10^{18} W/cm^2 for $2\omega_0$.

Experiments at the Second Harmonic

The $2\omega_0$ laser light was focused on the surface of thin aluminum films with a thickness of 1.8 µm at 0° or 45° incidence angle with an off-axis parabolic mirror to 3.10^{18} W/cm^2.

The high-energy ion emissions were recorded by CR-39 plastic nuclear track detectors, which are able to record ions with energies ≥100 keV/nucleon as tracks on the surface of the detector, after being etched in sodium hydroxide solution.

To study the angular ion distribution we have used CR-39 covered with single thickness Mylar filters ranging from 2 to 26 µm. To determine the maximum energy and spectrum of the ions, the detectors were covered with steps of Mylar filters with thicknesses from 2 to 50 µm. To compare ion emission in the forward (through the thin-film target) and backward directions (the direction of plasma expansion) for 45° incidence angle, two CR-39 detectors, with sets of Mylar filters, were placed both in front and in back of the target.

It was found that the predominant component of the high-energy ion emission is protons. These protons originate from a thin layer on the target surface contaminated with hydrocarbon and/or water vapor. Such target contamination was observed in the late 1960s in laser-matter interaction experiments with nanosecond laser pulses as discussed in Ref. [12] and has been proven in the 1990s in experiments with subpicosecond high-intensity pulses [1,2]. In our experiments, we observed a high-energy proton beam, propagating inside a thin-film target and emerging through the rear surface as a beam with the angular divergence of about 40° ± 10° (Fig. 1a). The proton angular distribution is peaked in the center of the beam, where the maximum density of ion tracks was observed (Fig. 1a). The direction of the high-energy proton beam does not depend on the angle of incidence of laser radiation and is always normal to the target surface. From this fact one would conclude that the electric field

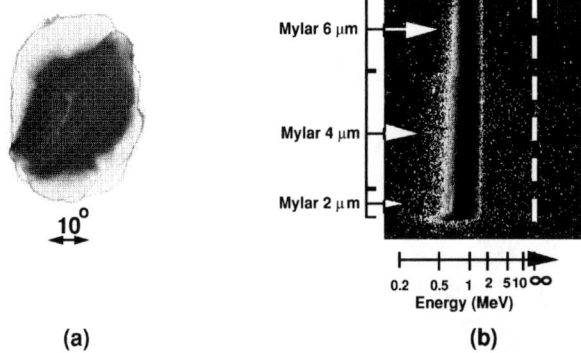

Figure 1. (a) A typical image of a proton beam observed in the forward (behind the target) direction in the interaction of 1.8-μm thick aluminum foil with a high-intensity laser. The proton beam passed through a 25 μm Mylar filter, which corresponds to energy above 1.2 MeV. The laser intensity on target is 2.10^{18} W/cm^2 at $2\omega_0$ and 0 degree angle of incidence; (b) Spectrogram of fast protons emitted in the forward direction and deflected in a dipole magnetic spectrometer. A dashed line shows the position of the slit image without the magnet. CR-39 was used as a detector and covered with three steps of Mylar filter of 2, 4 and 6-μm thickness, which corresponds to proton cut-off energies of 0.2, 0.3 and 0.5 MeV.

of the charge separation, accelerating protons inside the target, is directed along the target surface normal.

To verify that most of the energetic ions are indeed protons, we used a dipole magnetic spectrometer with a 200-m m slit in front of the magnet. CR-39 plastic with three steps of Mylar film in front was used as a detector. The Mylar filters had thicknesses of 2, 4, and 6 μm, which correspond to proton cutoff energies of 0.2, 0.3, and 0.5 MeV, respectively. Figure 1b presents a spectrogram of high-energy ions obtained with the dipole magnet. It shows a very sharp cutoff for particle energies of about 1.5 MeV, which was consistent with results on maximum proton energies as will be discussed below. On the left side of the spectrogram (Fig. 1b) by the arrows we have shown positions of above-mentioned cutoff energies.

These positions agree within ±10% of accuracy to the deflection of protons in the magnetic field, expressed by the formula: $x \approx LZebB/\sqrt{2\varepsilon_i M}$, where L is the distance between magnet and detector, $Z = 1$ for protons, e is the electron charge, b is the width of the magnet, B is the magnetic field strength, and ε_i and M are the ion energy and mass.

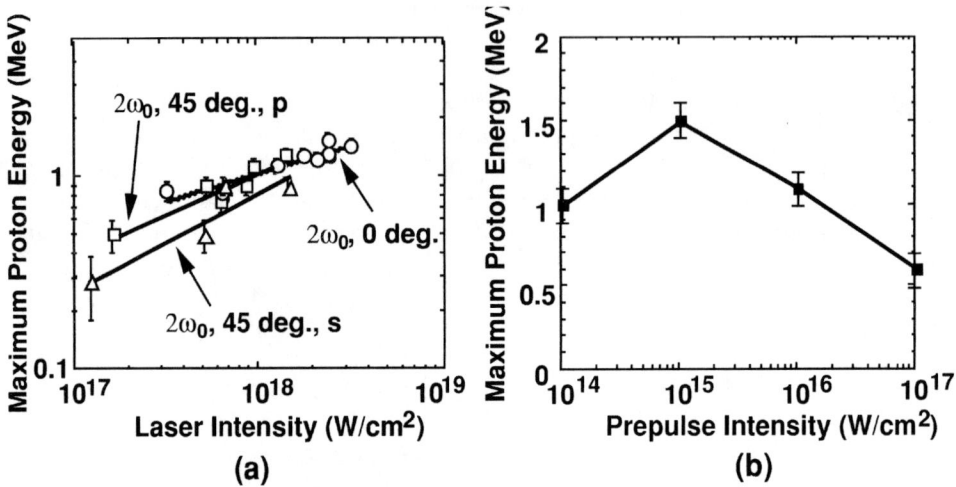

Figure 2. The maximum proton energy in the forward direction as a function of (a) laser intensity at $2\omega_0$ and for different conditions of illumination of 1.8-μm Al foil: circles — 0°; squares—45°, p —polarization; triangles —45°, s —polarization. Solid lines are the best power function fits to the experimental data. (b) The $2\omega_0$ prepulse intensity at 10^{18} W/cm2 main pulse intensity.

We studied the dependence of the maximum proton energy versus the laser intensity for different conditions of Al thin foil illumination. The highest proton energy of 1.6 MeV was observed for normal incidence at the maximum laser intensity of 3.10^{18} W/cm^2 (Fig. 2a). For 45° angle of incidence the highest intensity on a target was reduced by a factor of 2 due to increased spot size in the horizontal direction. Maximum observed proton energy, ε_i^{max}, was comparable for both cases for the same intensity and can be fitted as a function $\varepsilon_i^{max} \propto I^\alpha$, where α is between 0.3 and 0.4. Illumination with s polarization at 45° has produced protons with energies of 200–300 keV less than for p polarization. We also observed high-energy protons in the backward direction (in the direction of a plasma expansion) for 45° laser illumination, but their energy was twice lower than those in the forward direction. We changed the plasma gradient scale length by introducing a prepulse at $2\omega_0$, with the time delay of 50 ps in front of the main $2\omega_0$ pulse. We varied the prepulse intensity from 0.01% to 10% of main pulse intensity of 10^{18} W/cm^2. We found that there is an optimum in prepulse intensity of about 10^{15} W/cm^2 (Fig. 2b) for maximum proton energy production.

We estimated that at this intensity the scale length of the preformed plasma is about a few laser wavelength. The proton energy spectrum was measured by the foil attenuation method. Using steps of Mylar filters, with thicknesses differing by 2 μm provided proton energy resolution of about 100 keV. The proton spectrum (Fig. 3a) shows an exponential decay from energies of 400 keV to 1 MeV with a characteristic temperature of 230 keV (solid line). An interesting feature of the spectrum at higher energy is a plateau, which ends in a sharp energy cutoff at 1.5 MeV that is typical for the electrostatic mechanism of ion acceleration [11].

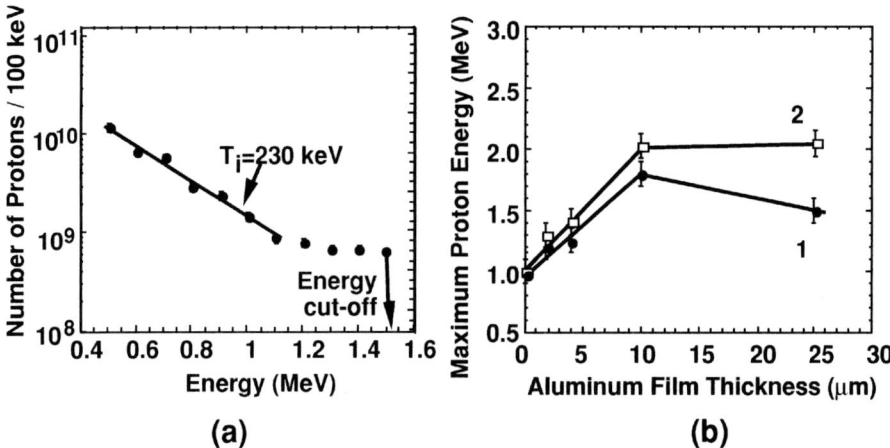

Figure 3. (a) Energy distribution of fast protons in the forward direction (circles), measured by attenuation of the beam in Mylar filters of different thicknesses for $2\omega_0$ illumination of 1.8-μm Al film with intensity of 2×10^{18} W/cm^2 at normal incidence. An arrow shows a sharp cutoff in proton energy on a spectral distribution below a detection threshold. The detection threshold was about 10^6 protons_100 keV. (b) Maximum energy of protons for laser intensity of 1.5×10^{18} W/cm^2 at $2\omega_0$ and 0° angle of incidence as a function of aluminum foil thickness—experiment (squares); ion losses are excluded (circles).

To determine the scale length of ion acceleration we varied the thickness of aluminum film from 0.1 to 25 μm. Circles on Fig. 3b are the experimental data points showing the dependence of maximum proton energy versus thickness l_0 of the film. For the thinnest target with $l_0 = 0.1$ μm, ε_i^{max} was about 1 MeV. We observed an increase in the maximum detected proton energy versus target thickness for $l_0 < 10$ μm. It reaches a peak at $l_0 \approx 10$ μm. At larger film thicknesses the proton energy decreases. This experiment demonstrates very efficient ion acceleration at a short spatial scale at the target front side where the predominant ion acceleration occurs and where ions gain their characteristic energy. For the maximum laser intensity we also have studied the dependence of maximum proton energy as a function of atomic number Z for 2 μm thick carbon, Mylar, molybdenum, and lead foils. We did not observe a significant difference in the maximum proton energies for different materials.

In spite of the absence of self-consistent theory of high-energy electron and ion generation in the laser-solid tar-get interactions, we may propose a heuristic pragmatic approach to get qualitative estimations. We will consider an interaction of a high-contrast laser pulse with an intensity $I > 10^{18}$ W/cm² at normal incidence in which the high-energy electrons with relativistic velocities $v \approx c$ can be produced. We assume when a high-intensity high-contrast laser pulse terminates at a target surface it produces a plasma with a size $l \approx \lambda/2$ [16], due to the longitudinal electron oscillations resulting from $\mathbf{v} \times \mathbf{B}$ oscillating Lorentz force. Near the target-vacuum surface the electrons are pushed in and out by the oscillating component of the ponderomotive force. Inside the target this force sharply vanishes. Twice in a laser period electrons reenter the target. Returning electrons are accelerated by the "vacuum" electric field and then deposit their energy inside the target. The electrons of this plasma are strongly heated by the laser light, penetrate inside the solid target with relativistic velocities, and constitute a low density ($n_e < n_c$) high-energy component of the entire electron population. They acquire an energy ε_e from the laser pulse which can be evaluated from energy balance: $n_e \varepsilon_e c \approx \eta I$, where n_e is the electron density of small size preformed plasma and coefficient η defines the efficiency of laser-energy conversion into the high-energy electrons. For laser intensities in the range of 10^{17} - $3 \cdot 10^{18}$ W/cm² we estimate η in a range from a few percent to 15%, in accordance with previous measurements [17] in similar experimental conditions.

The mean free path of high-energy electrons is very large—a 1 MeV free electron penetrates solid aluminum over a distance of ~1.5 µm. Nevertheless, in a plasma it is limited to a Debye length λ_{De} due to the charge-separation field, which for an electron density $n_e \sim 10^{19}$ -10^{20} cm⁻³ is comparable with the laser wavelength. High-energy electrons penetrating into a solid target (or even passing through it for the very thin foils) create an electrostatic field with $e\Phi \approx \varepsilon_e$, or $E \approx \varepsilon_e/\lambda_{De} \sim 10$ GeV/cm, which accelerates ions in the forward direction, and they in turn decelerate. An electrostatic field near the target surface has a bipolar structure with the more pronounced component accelerating ions in a forward direction. If the laser pulse duration is larger than the ion acceleration time in the layer of size λ_{De}, as is the case of our experiment, ions acquire an electrostatic energy $\varepsilon_i \approx Ze\Phi \approx Z\varepsilon_e$.

It has been known that the nonlinear regime of vacuum heating provides accelerated electrons with energies higher than predicted by the $\mathbf{v} \times \mathbf{B}$ force or P component of the laser field. This is related to the strong "heating" due to the self-intersection of electron orbits [18]. Instead of returning to the target in each cycle, many electrons remain on the front of a target and form a time-averaged density profile. In order to blow the electrons from the preformed plasma layer, the laser intensity must be enough to accelerate electrons up to an energy exceeding the Coulomb energy [19]: $\varepsilon_e \geq 2\pi e^2 n_e l R$, where R is the radius of focal spot, $R >> l$ ($R \cong 5$ µm). From this relation and energy balance, one can estimate the characteristic electron density $n_e \leq n_c$ $(a/2\pi)\sqrt{\eta 2\lambda/R}$ and energy $\varepsilon_e \geq \pi amc^2 \sqrt{\eta 2\lambda/R}$, where n_c is the critical density and $a = 0.85 \times 10^{-9} \lambda[\mu m]\sqrt{I[W/cm^2]}$ is the normalized amplitude of laser-field vector potential. Consequently, the maximum ion energy can be evaluated as $\varepsilon_i^{max} \geq Z\sqrt{\eta I R \lambda}$, where the laser intensity and spatial scales are measured in units of 10^{18} W/cm² and microns, respectively. For laser intensity of $3 \cdot 10^{18}$ W/cm² and for $\eta =$

10%, $\varepsilon_i^{max} \geq 1$ MeV, that is, in agreement with the experimental data (Fig. 2a). We may compare our estimate with an estimate based on the ponderomotive potential [20], which has been used for the description of ion acceleration in gas targets [3,11] $\varepsilon_i \cong Zmc^2 [\sqrt{1+a^2} -1]$. This formula predicts an ion energy an order of magnitude lower than observed in the present experiment at an intensity of 3.10^{18} W/cm². It is evident also that the thermal expansion of a skin-layer plasma cannot explain MeV ion generation because of low bulk electron temperature there. One may believe that stochastic electron heating due to parametric processes in the plasma corona can provide the high temperature component of electrons, T_h, and corresponding high thermal expansion velocity ~ $\sqrt{T_h/M}$, where M is the ion mass. However, at $2\omega_0$ illumination with a high laser-intensity contrast the micron spatial scale of a plasma corona prevents such processes.

We may explain also why ion energy cutoff depends on the foil thickness (Fig. 3b, curve 1). As follows from the estimations, the typical λ_{De} is about 1 μm so that for the foil thickness $l_0 \leq 1$ μm electrons will penetrate to the backside. Even for thicker targets ultrafast electrons (1 MeV), which cannot be stopped by charge separation, leave the target. Nevertheless, for the thick foils all electrons stop in the target and participate in generation of the electric field accelerating the ions. Because of this, one may expect an increase in the maximum ion energy with the foil thickness unless saturation occurs. However, ion energy losses increase with the foil thickness that should result in the optimum thickness, which according to Fig. 3b, curve 1, corresponds to l_0 ~ 10 μm. To demonstrate the effect of the saturation mentioned we plot also the similar dependence, $\varepsilon_i^{max}(l_0)$ where ion energy losses are excluded (Fig. 3b, curve 2).

Experiments at the Fundamental Frequency

In this section, we study the dependence of maximum proton energy as a function of laser intensity for the ω_0 illumination (Fig. 4). For $\lambda=1.053$ μm the maximum proton energy was found to vary linearly with the laser intensity and reached ~ 10 ± 1 MeV at 6.10^{18} W/cm².

Proton energy if given by the ponderomotive potential of standing electromagnetic wave [12] is only 0.7 MeV, which is much less than the observed maximum proton energy. An estimation of the proton energy from the mechanism of "vacuum heating" [5] is closer to the observed value but also underestimates it by a few times. However, for the intensity contrast of 5.10^5:1 it is questionable to apply the latter estimation because a preplasma very likely appears before the maximum laser energy reaches the target. Thus, one may attribute such a preplasma as a reason for enhanced electron generation and, hence, enhanced electrostatic field, which efficiently accelerates the ions. Electrons accelerated in an underdense preplasma (electron density n_e less than critical) up to the energy ε^{max} can penetrate inside a solid target to their Debye length, $\lambda_{De} \propto \sqrt{\varepsilon_e/n_e}$, and accelerate ions forward from the skin layer. Since the skin depth is shorter than λ_{De} a significant number of ions should have an energy equal to the electrostatic potential and the ion distribution function should demonstrate a plateau effect ("water bag" distribution) until energy cutoff $Z\varepsilon^{max}$. Such an ion distribution function has already been observed in previous experiments [13].

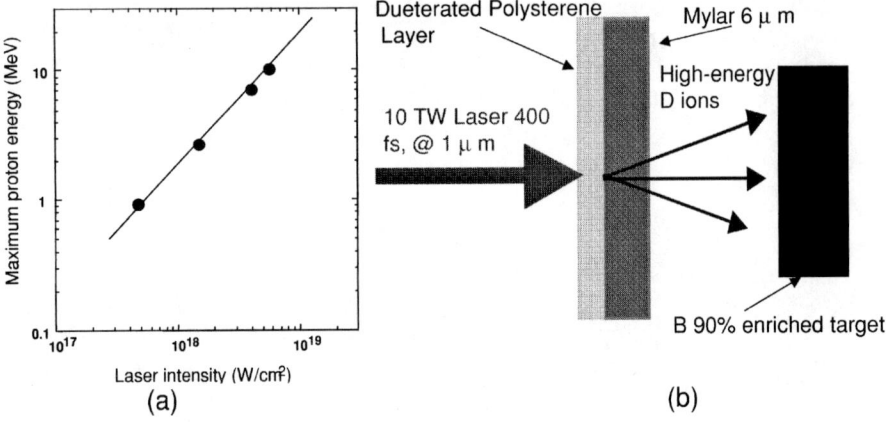

Figure. 4. (a) The maximum proton energy in the forward direction as a function of laser intensity ω_0, 45^0, p-polarization; (b) Schema for ^{10}B activation with high-energy deuterons in the reaction ^{10}B(d,n)^{11}C

NUCLEAR ACTIVATION

After demonstrating acceleration of a significant amount of ions to high-energies in the forward direction we performed an experiment on nuclear activation [21]. We deposited a thin layer of a deuterated plastic on a front side of a 6 μm thick Mylar foil. A 90% ^{10}B sample was positioned behind a thin film target at a distance of about 1 cm (Fig. 4b). A thin film target interacted with 1.053 μm light at the intensity of 6.10^{18} W/cm^2. We expected to produce a positron active isotope ^{11}C, which is a product of the reaction ^{10}B(d,n)^{11}C. A positron active isotope ^{11}C normally decays with the production of 2 photons with the energies of 511 keV each, propagating in an opposite direction. We have used a standard coincidence event system, which consist of two channels with NaI scintillators coupled to a photomultiplier tubes (PMT), amplifiers and a timing unit. The two channels send signals to a fast coincidence unit with a timing window of 50 ns followed by a counter unit. This allowed us to detect simultaneously produced photons with an efficiency measured to be in the range of 0.05-0.1. This was tested using a ^{22}Na radioactive source with a precisely known positron activity. The background noise was found to be 1-5 count/min and the radioactivity detection limit was ~ 10 pCi.

After activating the ^{10}B with the high-energy deuterons in a single shot we measured the radioactivity of produced ^{11}C. Fig. 5a shows this result. On the same graph we plotted a theoretical decay for positron active isotope ^{11}C with a half-life of 20 minutes. These results indicates that we produced ~ 10^5 atoms of ^{11}C, which corresponds to a radioactivity of ~2nCi directly after the initial activation. Another interesting feature of this study is the very strong power dependence of ^{11}C yield as a function of the laser intensity. By changing the laser intensity factor of 2 (from 3.10^{18} W/cm^2 to 6.10^{18} W/cm^2) we have increased ^{11}C yield by almost 2 orders of magnitude (Fig. 5b).

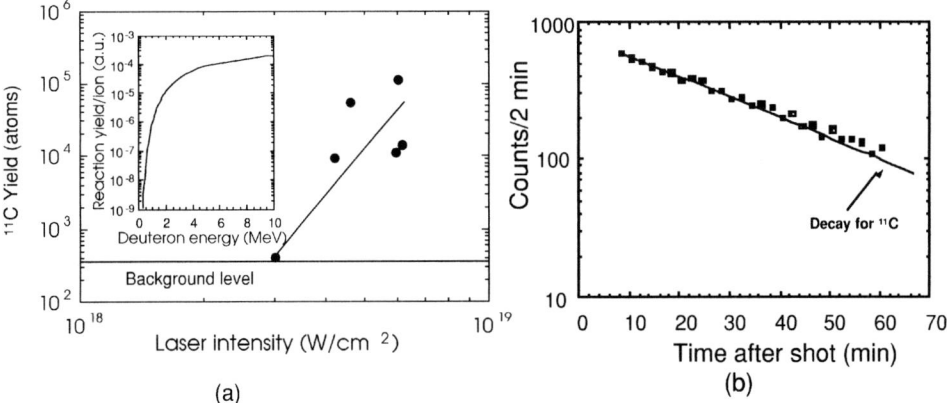

Figure 5. (a) Measured decay of radioactivity versus time for the reaction product ^{10}B(d,n)^{11}C (squares). Solid line is a theoretical decay of ^{11}C with a half-life of 20 minutes; (b) ^{11}C yield versus laser intensity for ω_0.

To verify that high-energy ions are accelerated from the front side of the target, that agrees with Ref. [22] and contradicts to Ref. [23], we irradiated a 6 μm thick Mylar target with a layer of a deuterated plastic on its back side and performed the same ^{10}B activation experiment described above [21]. In this case the activation signal was not above the background level. These results unequivocally indicate, that for the conditions of our experiment light ions are accelerated from the front side of the foil.

High-energy protons are also expected in the experiment with the layered deuterated target. They may participate in the ^{11}B(p,n)^{11}C reaction. The cross section of this reaction [24] has a higher threshold (above 3 MeV). Taking into account that the amount of ^{11}B in boron sample was only 10%, we expect ^{11}C yield for this reaction to be much less than that for deuterons. To verify that (p,n) reaction does not contribute sufficiently in ^{11}C yield we focused the laser at its highest intensity on a bare Mylar film behind which was a boron sample. No activation signal was observed.

SUMMARY

In summary we have demonstrated the production of a directed beam of high-energy protons in the interaction of high-intensity laser with a thin-film target. These ions are accelerated by the electrostatic field of charge separation due to hot-electron electrons generated on a target front surface and driven inside the target by the laser. A maximum proton energy of about 1 MeV has been observed for $2\omega_0$ illumination and scales with the laser intensity as $\varepsilon_i^{max} \sim I^{0.3}$. The maximum proton energy is about 10 MeV and approximately proportional to the laser intensity, I, for the fundamental. And in both cases the proton energy is many more times higher than estimated from the pondermotive potential. We have also demonstrated the production of $\sim 10^5$ atoms of a positron active isotope ^{11}C from the ^{10}B(d,n)^{11}C reaction. The activation results have shown that the generation of fast ions occurs on the front side of the target and the maximum proton energy agrees with an estimation of charge separation due to hot-electron generation by the vacuum heating effect for the $2\omega_0$ case. A beam of energetic protons might one day serve as a high-current proton injector or be used in basic science to study nuclear transformations on a picosecond time scale. A charge compensated high-energy deuteron beam can also be used as a fast ion ignitor for the direct and indirect drive inertial confinement fusion research.

ACKNOWLEDGMENTS

The High Energy Physics Division of the U.S. Department of Energy supported the Michigan participants and the National Science Foundation supported the laser facility. The work of V. Yu. B. was supported by the Russian Foundation for Basic. S. B. is thankful for the support of CUOS Fellowship Program and K. N. would like to thank S. Akita, S. Sasaki and T. Suzuki for general support.

REFERENCES

1. P. Fews et al., *Phys. Rev. Lett.* **73**, 1801 (1994); F. N. Beg et al., *Phys. Plasmas* **4**, 447 (1997).
2. P. E. Young et al., *Phys. Rev. Lett.* **76**, 3128 (1996).
3. G. S. Sarkisov et al., *JETP Lett.* **66**, 828 (1997); G. S. Sarkisov et al., *Phys. Rev. E* **59**, 7042 (1999).
4. K. Krushelnick et al., *Phys. Rev. Lett.* **83**, 737 (1999).
5. T. Ditmire et al., *Nature* (London) **386**, 54 (1997).
6. M. Lezius et al., *Phys. Rev. Lett.* **80**, 261 (1998).
7. P. A. Norreys et al., *Plasma Phys. Controlled Fusion* **40**, 175 (1998).
8. G. Pretzler et al., *Phys. Rev. E* **58**, 1165 (1998).
9. T. Ditmire et al., *Nature* (London) **398**, 489 (1999).
10. L. Disdier et al., *Phys. Rev. Lett.* **82**, 1454 (1999).
11. V. Yu. Bychenkov et al., *Sov. Phys. JETP* **88**, 1137 (1999).
12. S. J. Gitomer et al., *Phys. Fluids* **29**, 2679 (1986).
13. A. Maksimchuk, et al., *Phys. Rev. Lett.* **84**, 4108 (2000)
14. F. Brunel, *Phys. Rev. Lett.* **59**, 52 (1987).
15. W. L. Kruer and K. Estabrook, *Phys. Fluids* **28**, 430 (1985).
16. W. Yu et al., *Phys. Rev. E* **58**, 2456 (1998).
17. J. Yu et al., *Phys. Plasmas* **6**, 1318 (1999).
18. S. V. Bulanov et al., *Phys. Plasmas* **1**, 745 (1994).
19. T. Zh. Esirkerov et al., *JETP Lett.* **70**, 82 (1999).
20. S. C. Wilks, *Phys. Fluids B* **5**, 2603 (1993).
21. K. Nemoto et al. (to be published in *APL*)
22. E.L.Clark, et al., *Phys. Rev. Lett.* **84**, 670 (2000)
23. S.P. Hatchett, et al., *Phys. Plasmas* **7**, 2076 (2000)
24. Experimental Nuclear Reaction Data File, http://www.nndc.bnl.gov/nndc/exfor

Recent Advances in Electron and Positron Sources*

J.E. Clendenin

*Stanford Linear Accelerator Center
Stanford, CA 94309*

Abstract. Recent advances in electron and positron sources have resulted in new capabilities driven in most cases by the increasing demands of advanced accelerating systems. Electron sources for brighter beams and for high average-current beams are described. The status and remaining challenges for polarized electron beams are also discussed. For positron sources, recent activity in the development of polarized positron beams for future colliders is reviewed. Finally, a new proposal for combining laser cooling with beam polarization is presented.

ELECTRON SOURCES

Throughout the long history of electron sources, the standard and virtually universal configuration of the dc-biased gun has been a coaxial design with a ceramic high-voltage insulating tube providing the outside vacuum envelope and inside, supporting the cathode, a metal tube at atmospheric pressure as illustrated in Fig. 1(a). This design works well in practice, but the insulator is large, and the outside of the ceramic is subject to contamination from the atmosphere, which can lead to excessive leakage current. With the advent of polarized electron sources, various vacuum components associated with installing cathodes under vacuum are typically attached to the high-voltage flange, leading to an awkwardly large high-voltage assemblage [1]. These problems are solved for photocathode dc-biased guns by inverting the structure to have a smaller-diameter ceramic tube inside a larger metal vacuum chamber as shown in Fig. 1(b). The earliest such designs were developed in the early 1990s at SLAC [2] and independently at Novosibirsk [3]. The SLAC design was built and successfully tested at high voltage but has not yet been used to produce electrons. Variations on the basic idea include the double insulator design, Fig. 1(c), used at Amsterdam [4] and under development at Nagoya [5], and variations on the high-voltage connection, Fig. 1(d), as used at Mainz [6] and Bonn [7]. For a pulsed high-voltage gun, the Fig. 1(b) design looks most promising.

The past decade has seen the rapid development of photocathode rf guns [8]. These guns are especially well-suited as high-brightness sources. Since the emittance requirements of future colliders seems beyond the reach of any rf photoinjector design, the need for high-brightness rf guns comes mostly from free electron laser (FEL)

---

*Work supported by Department of Energy contract DE-AC03-76SF00515.

CP569, *Advanced Accelerator Concepts: Ninth Workshop,* edited by P. L. Colestock and S. Kelley
© 2001 American Institute of Physics 0-7354-0005-9/01/$18.00

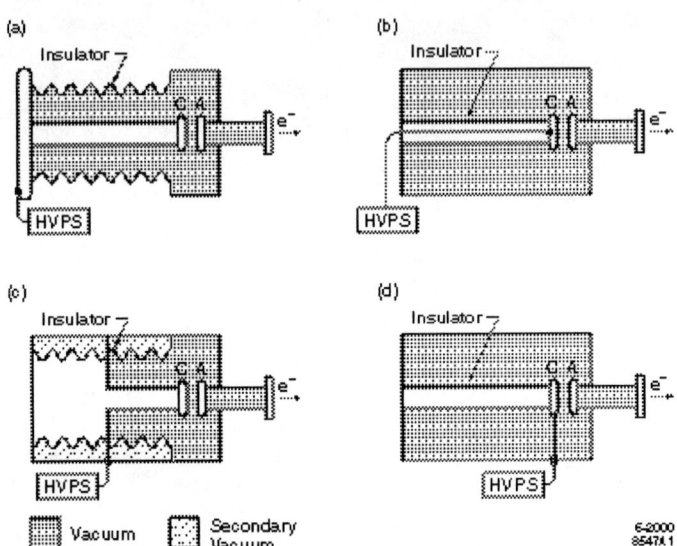

FIGURE 1. The inverted structure (IS) design and variations: (a) the conventional non-inverted design; (b) the original IS design; (c) the double insulator design; and (d) variation on connecting the high-voltage power supply (HVPS). For the IS design, the cathode (C) and anode (A) electrodes are permanent, while the photocathode (not separately shown) is removable.

developments. As an example, the Linac Coherent Light Source (LCLS) requires a 1-nC, 100-A beam from the photoinjector with a normalized rms transverse emittance, $\varepsilon_{n,rms}$, of 1 µm [9]. The emittance growth in an rf photoinjector is mostly correlated. Special techniques have been developed to reverse this growth and then lock-in the resulting emittance minimum just as the beam becomes relativistic. Experimentally the lowest emittance for an LCLS type beam is $\varepsilon_{n,rms}$~2 µm [10,11]. However, these measurements were done with a spatially uniform but temporally Gaussian charge distribution. Simulations using PARMELA, a multi-particle tracking code, indicate that if the temporal distribution is also uniform, $\varepsilon_{n,rms}$~1 µm should be achievable. Progress in exploring the relevant parameter space has been facilitated by the recent development of a semi-analytic code, HOMDYN [12]. Using newly discovered matching conditions [13], simulations now predict an emittance for an LCLS type beam of close to $\varepsilon_{n,rms}$~0.5 µm (thermal effects included) at 150 MeV [14]. See Fig. 2.

The possibility of achieving even higher brightness using a pulsed diode-structure photocathode gun with GV/m level fields was first reported in 1996 [15]. The high fields are achieved by using a pulsed voltage on the order of 1 MV across a gap of 0.5 to 1 mm. Very short voltage pulses on the order of nanoseconds are used to minimize breakdown and field emission. Ideally the laser pulse should be shorter than the voltage pulse. For an LCLS-type pulse, simulations predict an emittance of 0.4 µm measured 3.25 cm from the cathode [16]. Since space charge effects are still significant at 1 MeV, a design for matching this beam into an accelerating structure is needed to evaluate the

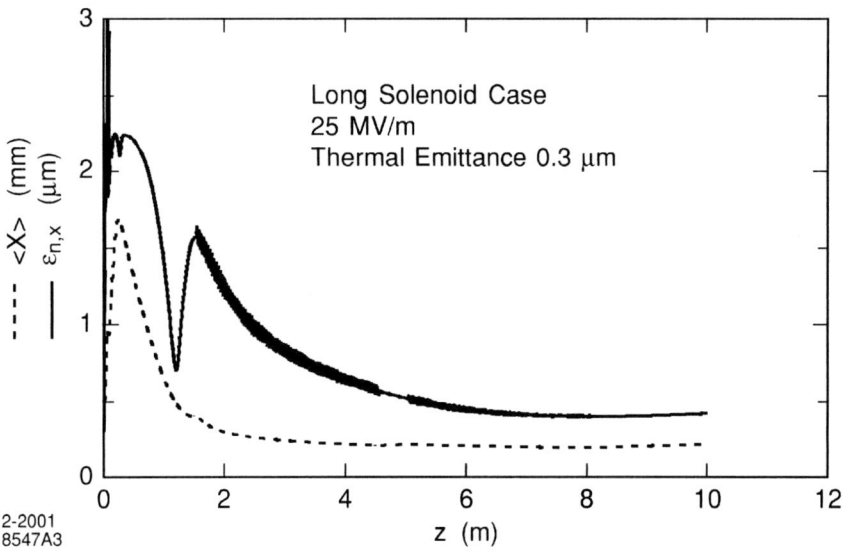

FIGURE 2. Transverse normalized rms emittance and beam size, computed using HOMDYN, for the LCLS photoinjector, which consists of a 1.6-cell S-band gun with cathode at z=0 and two 3-m TW sections beginning at z~1.5 m [14]. The peak field in the gun is 130 MV/m, while the sections, with a weak solenoid around the first, are operated at 25 MeV/m. Thermal emittance for a Cu cathode is included.

final emittance at high energy.

Photocathode guns using III-V semiconductor cathodes are now universally used as sources of polarized electron beams for accelerators. The successful operation of such beams at SLAC, Mainz, JLAB and elsewhere has demonstrated operating parameters (not all achieved at the same time) well matched to accelerator requirements. Some of these parameters are shown in Table 1.

There remain at least 3 challenges for future polarized electron sources: overcoming the cathode charge limit; increasing the polarization, P; and increasing the average current. 1) The maximum current density that can be extracted is limited by a surface barrier that dynamically grows when charge is temporarily trapped at the surface faster than it can recombine with holes. For a pulse train, such as required by most future collider designs, each pulse is influenced by the decaying surface barrier generated by the previous pulses. New cathode structures plus differential doping may solve this

TABLE 1. Operating Parameters Achieved for Polarized Electron Sources.

Parameter	Value	Where Achieved
Current Density, J	10 A cm^{-2}	SLAC
Average Current, I_{AVG}	5 mA	JLAB [17], GaAs, unpolarized
Polarization, P	80%	SLAC
Cathode 1/e Lifetime, τ	>1000 h	SLAC
Operating time per cathode	>5000 h	SLAC

problem [18]. 2) Higher polarization will improve the effective luminosity of any high-energy experiment which depends on polarized electrons. In addition, for a future collider, $P>95\%$ may be the only reasonable route to certain new physics. Most of the polarization loss in the cathode bulk can probably be eliminated, but losses in the band bending region may be unavoidable, limiting the maximum polarization to $P\sim90\%$. 3) Finally, the high average currents required by cw accelerators and some types of FELs result in a rapid loss of quantum efficiency (QE) due to ion bombardment at the cathode. Improving the vacuum near the cathode will minimize this effect.

Field emitter arrays, ferroelectrics, and secondary electron emitters have the potential to overcome the limitations found with photocathodes for producing high average currents [19]. At the present time, the latter is the most promising. It consists of an rf cavity equipped with a secondary emission surface at one end and a secondary emission grid at the beam exit. Startup electrons multiply rapidly during each rf cycle while simultaneously bunching. Steady state conditions are achieved within a few cycles for a pulse train of fixed charge and pulse length. While the pulse spacing is fixed by the rf frequency, the pulses are automatically synchronous with the rf. Proof of principle testing has been carried out at low charge, but simulations show that the charge per bunch can be up to 500 nC [20]. There appears to be some possibility of modulating the charge using a separate grid.

POSITRON SOURCES

Conventional positron sources for accelerators use a high-energy electron beam impinging on a high-Z material such as W to generate ~100 MeV γs by bremsstrahlung. The γs in turn create electron-positron pairs in the same material. Positrons exiting the target in the 2-20 MeV regime are confined by a magnetic field while being inserted into an rf accelerating field, bunched and accelerated to relativistic energies for transport to the main linac. Because the initial positron beam emittance is large, more damping is required than for an equivalent intensity electron beam. The NLC positron source design is essentially a scaled version of the SLC source.

It is highly desirable that the positron beam for a future collider be polarized [21]. For many types of experiments, the polarization of the two colliding beams combine to create in effect a single higher polarization. Thus a highly polarized electron beam colliding with a modestly polarized positron beam may be equivalent to the desired single beam polarization $P>95\%$. Another class of experiments is possible only with both beams polarized. At least 3 methods of producing polarized positron beams have been suggested for colliders: helical undulator; Compton scattering; and polarized electrons.

Circularly polarized γs in the required energy range and of sufficient intensity can be produced by passing a 150 GeV electron beam through a 150-m helical undulator [22]. The γs are directed to a thin conversion target placed downstream after the electron primary beam is bent away. This is the design chosen by TESLA [23] but to date rejected by NLC out of concern that the post-interaction beam is too disrupted, and that alternatively having the undulator in the main linac beamline will impose too great an operational restriction on the linac. Recently an interesting proposal has been made to operate the first part of the linac at double the normal rf repetition rate in order to accelerate positron-production electrons along with the main beam, then at the 150-point deflecting the positron-production electrons into a separate beamline having the undulator and target [24].

In the second method, circularly-polarized high-energy γs are produced by Compton backscattering of circularly-polarized photons by unpolarized high-energy electrons. Again the γs are directed to a thin conversion target. Such a scheme was first proposed for the JLC by Okugi et al. in 1996 [25]. Eighty-five CO_2 lasers, each producing 10 J per pulse (150 Hz) at the fundamental (10.6 μm) are required, one laser for each micropulse in the JLC pulse train. The cross section for Compton scattering is optimized by choosing a 6.7 GeV electron beam. A similar scheme has been proposed by Frisch (1997) that utilizes an Nd:glass laser (1.05 μm) and 1.7 GeV electrons [26]. In the latter case, in order to reduce the laser energy requirements, a resonant cavity is introduced to recycle the optical power, allowing the same optical pulse to interact with many electron bunches. The mirrors for the optical cavity are problematic because of the high energy in each laser pulse. A recent experiment at the Accelerator Test Facility (ATF) at KEK demonstrated that the production of positrons from a thin conversion target for which the γs were produced by scattering a 200 mJ Nd:YAG laser beam from the 1.26 GeV ATF electron beam was as expected [27].

The third method is a modification of the conventional scheme. If the incident electron beam is highly polarized, then both the high-energy end of the γ spectrum and of the resulting positrons will be polarized [28]. The problem here is that the yield is estimated to be 3 orders of magnitude below that required for colliders. One can imagine a number of ways to increase the total yield, including increasing the charge per pulse in the production beam, filling more rf buckets, using multiple sources, etc., so that in principle the yield might be forced to be adequate, but the practical aspects are daunting.

SIMULTANEOUS LASER DAMPING/POLARIZING

Potylitsin has recently proposed [29] that direct polarization of an unpolarized positron beam by Compton scattering [30] may be a more efficient source of polarized positrons than the methods discussed above.[1] For this case, the energy and polarization of the polarized positrons after N collisions with identical circularly polarized photons are (here only $\hbar = m_e = c = 1$) [29]:

$$\gamma_{(N)} = \frac{\gamma_0}{1+2\mu} \text{ and } \xi_{(N)} = \frac{\mu}{1+\mu}, \qquad (1)$$

where $\mu = \gamma_o \omega_o N = \frac{4}{3} \frac{A}{m_0 c^2} \gamma_o \left(\frac{r_e}{\sigma_{ph}}\right)^2$ and $\sigma_{ph}^2 = \frac{\lambda_o l_e}{8\pi}$. Here γ_0 and ω_0 are the initial positron and photon energies, A is the laser flash energy, r_e is the classical electron

---
[1] The process works equally well for electrons.

radius, and λ_o and l_e are the laser wavelength and positron bunch length respectively. For example, a 2 GeV positron bunch in a single interaction with a 25 J laser pulse (λ_o=1 µm) would be expected to result in a polarization of ~60% if l_e can be reduced to 0.2 mm [31]. For a collider such as the NLC with 95 microbunches per pulse train and 120 Hz repetition rate, an average laser power of 0.3 MW is required! However, since the positron beam must in any case be damped, let us review the requirements for laser damping.

A powerful laser can be used to damp an electron or positron beam by Compton scattering [32]. The requirement for a significant reduction in the initial transverse normalized emittance, ε_{no}, is that the electrons should lose a similar fraction of their initial energy, E_o, as a result of the Compton interaction:

$$\frac{\varepsilon_{no}}{\varepsilon_n} \cong \frac{E_o}{E} = 1 + \frac{64\pi^2 r_e^2 \gamma_o}{3m_o c^2 \lambda l_e} A, \qquad (2)$$

where $A[J] = \frac{25\lambda[\mu m] l_e[mm]}{E_o[GeV]} \left(\frac{E_o}{E} - 1\right)$. For E_o=2 GeV, A=10 J (at λ=1 µm) is required in a single pass to reduce the transverse emittance by a factor of 10 (again assuming l_e=0.2 mm), which is about the same laser requirement as for polarization.

Laser cooling can be combined with a storage ring [33] or a damping ring. The current design of the 1.98-GeV NLC damping ring [34] has a circumference of 297 m, so the rotation frequency, v_{rot}, is ~1.01 MHz. Three NLC pulse trains of 95 microbunches each are damped for 3 interpulse periods or ~25 ms. Therefore each microbunch passes a reference point in the ring ~2.5×10^4 times. The proposal here is to combine laser cooling with polarization in the NLC damping ring, substituting the laser interaction for the wiggler. By combining these functions, there might be a considerable cost savings. In addition, installation of the polarization function can in principle be delayed until sometime after the damping function is commissioned.

Combined with a damping ring, a modest Nd:glass laser system plus an optical resonator as suggested by Frisch for the case of Compton scattering from an unpolarized electron beam to produce γs can be envisioned as shown in Fig. 3. For maximum polarization, the total laser energy seen by each microbunch should be ~25 J or E_b=1 mJ per rotation. This is a reasonable energy per optical pulse for the resonator. The average laser energy (at 1 µm) required for each microbunch in the ring is $P_{b,avg} = E_b \times v_{rot}$~1 kW. The NLC ring rf is 712 MHz with spacing between microbunches of 2.8 ns (i.e., filling every other rf bucket). There is also a gap between each pulse train. With all 285 microbunches in the ring, the laser must operate at v_L=357 MHz, and thus the total average power is $P_{tot,avg}= E_b \times v_L$ =357 kW. The average power required of the injection laser is $P_{inj,avg}$=100 W (λ=1 µm) operating at 357 MHz, which is probably doable. Likewise the resonator gain of $G=P_{tot,avg}/P_{inj,avg}$=3.6×$10^3$ can probably be achieved. The principal uncertainty is that the stability of an optical resonator operating under these conditions is unknown. Also the ring design will have to accommodate 2 spin rotators, and the bunch length will have to be reduced, at least in the laser interaction region, to the order of 0.2 mm. Given these complications, a new ring design, optimized for both laser damping and polarization, should be considered.

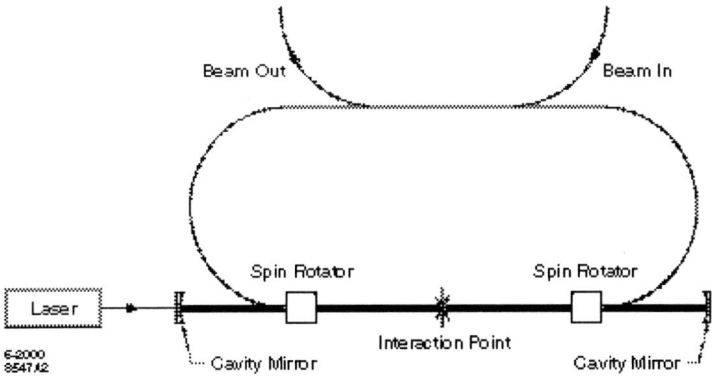

FIGURE 3. Conceptual layout of an NLC positron damping ring combining laser cooling and laser polarization.

ACKNOWLEDGMENTS

The author would like to thank J. Frisch, A. Kulikov, and D. Schultz (SLAC) for numerous useful conversations with respect to positron sources and also E. Bessenov (Lebedev) and A. Potylitsin (Tomsk) for sharing their insights into the possibilities and limitations for direct polarization by Compton scattering.

REFERENCES

1. Alley, R. et al., *Nucl. Instrum. and Meth. A* **365**, 1-27 (1995).
2. Breidenbach, M. et al., *Nucl. Instrum. and Meth. A* **350**, 1 (1994).
3. Gavrilov, N.G. et al., *Nucl Instrum. and Meth. A* **331**, ABS17 (1993).
4. Papadakis, N.H. et al., "Polarized Electrons at NIKHEF," in *Polarized Beams and Polarized Gas Targets*, edited by H.P. gen. Schieck and L. Sydow, World Scientific, Singapore, 1996, p. 323.
5. Nakanishi, T. et al., "Polarized Electron Source Development in Japan," in *Spin96 Proceedings*, edited by C.W. de Jager et al., World Scientific, Singapore, 1997, p. 712.
6. Aulenbacher, K. et al., *Nucl. Instrum. and Meth. A* **391**, 498 (1997).
7. Gowin, M. et al., "A 50kV Inverted Polarized Gun," in *Proceedings of Low Energy Polarized Electron Workshop (LE98)*, edited by Y.A. Mamaev et al., SPES-Lab-Publishing, St. Petersburg, Russia, 1998, p. 115.
8. Clendenin, J.E., "RF Photoinjectors," in *Proceedings of the XVII International Linear Accelerator Conference*, edited by C. Hill and M. Vretenar, CERN, Geneva, CH, 1996, p. 298.
9. Cornacchia, M., "The LCLS X-Ray FEL at SLAC," in *Free-Electron Laser Challenges II*, edited by H.E. Bennett and D.H. Dowell, SPIE **3614**, Bellingham,WA, 1999, p. 109.
10. Babzien, M. et al., *Phys. Rev. E* **57**, 6093 (1998).
11. Gierman, S., *Streak Camera Enhanced Quadrupole Scan Technique for Characterizing the Temporal Dependence of the Trace Space Distribution of a Photoinjector Electron Beam*, a Ph.D dissertation, University of California, San Diego, 1999, ch. 6.
12. Ferrario, M. et al., *Part. Acc.* **52**, 1 (1996).

13 Ferrario, M. et al., "HOMDYN Study for the LCLS RF Photo-Injector," contributed to the *2nd ICFA Advanced Accelerator Workshop on The Physics of High Brightness Beams*, Los Angeles, November 9-12, 1999.
14 Ferrario, M., INFN-LNF, private communication, 2000.
15 Srinivasan-Rao, T. and Smedley, J., "Table Top, Pulsed, Relativistic Electron Gun with GV/m Gradient," and F. Villa, "Acceleration of Kiloampere Current at 2.65 GV/m," in *Advanced Accelerator Concepts Seventh Workshop*, edited by S. Chattopadhyay et al., AIP Conference Proceedings 398, New York, 1997, pp. 730 and 739 respectively.
16 Srinivasan-Rao, T. et al, "Simulation, Generation, and Characterization of High Brightness Electron Source at 1 GV/m Gradient," in *Proceedings of the 1999 Particle Accelerator Conference*, edited by A. Luccio and W. MacKay, IEEE Operations Center, Piscataway, NJ, 1999, p. 75.
17 Bohn, C.L. et al., "Performance of the Accelerator Driver of Jefferson Laboratory's Free-Electron Laser," in *Proceedings of the 1999 Particle Accelerator Conference*, edited by A. Luccio and W. MacKay, IEEE Operations Center, Piscataway, NJ, 1999, p. 2450.
18 Togawa, K. et al., *Nucl. Instrum. and Meth. A* **414**, 431 (1998).
19 Nation, J.A. et al., *Proc. of the IEEE* **87**, 865 (1999).
20 Len, L.K. and F.M. Mako, "Self-Bunching Electron Guns," in *Proceedings of the 1999 Particle Accelerator Conference*, edited by A. Luccio and W. MacKay, IEEE Operations Center, Piscataway, NJ, 1999, p. 70.
21 Subashiev, A.V. and Clendenin, J.E., "Polarized Electron Beams with P≥90%, Will It Be Possible?," Preprint SLAC-PUB-8312, 2000; and Clendenin, J.E., *Int. J. Mod. Phys. A* **13**, 2507 (1998). See also Omori, T., "A Polarized Positron Beam for Linear Colliders," KEK Preprint 98-237, 1999.
22 Mikhailichenko, A.A., "Use of Undulators at High Energy to Produce Polarized Positrons and Electrons," *in Proceedings of the Workshop on New Kinds of Positron Sources for Linear Colliders*, edited by J. Clendenin and R. Nixon, SLAC-R-502, Stanford, 1997, p. 229.
23 "Conceptual Design of a 500 GeV e+e- Linear Collider with Integrated X-ray Laser Facility," edited by R. Brinkmann et al., DESY 1997-048/ECFA 1997-182.
24 Frisch, J., SLAC, private communication, 2000.
25 Okugi, T. et al., *Jpn. J. Appl. Phys.* **35**, 3677 (1996).
26 Frisch, J., "Design Considerations for a Compton Backscattering Positron Source," *in Proceedings of the Workshop on New Kinds of Positron Sources for Linear Colliders*, edited by J. Clendenin and R. Nixon, SLAC-R-502, Stanford, 1997, p. 125.
27 Dobashi, K. et al., *Nucl. Instrum. and Meth. A* **437**, 169 (1999).
28 Bessonov, E.G. and Mikhailichenko, A.A., "A Method of Polarized Positron Beam Production," in *Proceedings of the 5th European Particle Accelerator Conference*, edited by S. Myers et al., IoP Publishing, Bristol, UK, 1996, p. 1516; Potylitsin, A.P., *Nucl. Instrum. and Meth. A* **398**, 395 (1997).
29 Potylitsyn, A., "Single-Pass Laser Polarization of Ultrarelativistic Positrons," Preprint arXiv:physics/0001004, 2000.
30 Polarization of electrons in a storage ring using circularly polarized photons was proposed in Yu. A. Bashmakov, E.G. Bessonov, and Ya. A. Vazdik, *Pis'ma Zh. Tekh. Fiz.* **1**, 520 (1975), English translation *Sov. Tech. Phys. Lett* **1**, 239 (1975). See also comments on this proposal in Ya. S. Derbenev, A.M. Kondratenko and E.L. Saldin, *Nucl. Instrum. and Meth.* **165**, 201 (1979).
31 Processes not accounted for in reference 29 may affect the final polarization value. A. Potylitsyn, Tomsk, private communication, 2000.
32 Telnov, V., *Phys. Rev. Lett.* **78**, 4757 (1997).
33 Huang, Z. and Ruth, R., "Radiation Damping and Quantum Excitation in a Focusing-Dominated Storage Ring," in *Quantum Aspects of Beam Dynamics*, edited by P. Chen, World Scientific, Singapore, 1999, p. 34.
34 Corlett, J.N. et al., "The Next Linear Collider Damping Ring Complex," in *Proceedings of the 1999 Particle Accelerator Conference*, edited by A. Luccio and W. MacKay, IEEE Operations Center, Piscataway, NJ, 1999, p. 3429.

Development of High-Brightness Laser Synchrotron Source at BNL ATF

I.V. Pogorelsky[a], I. Ben-Zvi[a], T. Hirose[b], S. Kashiwagi[c], K. Kusche[a],
T. Kumita[b], T. Omori[d], V. Yakimenko[a], K. Yokoya[d], J. Urakawa[d],
M. Washio[c]

[a] *Accelerator Test Facility, Brookhaven National Laboratory, 725C, Upton, NY 11973, USA*
[b] *Physics Department, Tokyo Metropolitan University, Japan*
[c] *Waseda University, Japan*
[d] *KEK, Japan*

Abstract. The counter-propagating picosecond CO_2 laser pulse and picosecond or femtosecond electron bunch produced by the photocathode linac shape the optimum configuration for the ultra-bright relativistic Thomson scattering x-ray source. Using these components, a proof-of-principle experiment at the ATF reached the x-ray photon yield of 8×10^6 measured in the 1.8-2.3 Å window per 3.5 ps pulse. Upon completion of the ongoing upgrade of the ATF laser and linac, 300 fs x-ray pulses with intensity of 5×10^{22} photon/second will be obtained. This will provide also an opportunity for fundamental study of the nonlinear Thomson scattering at the laser strength parameter close to 1. Finally, we discuss the feasibility of a high repetition rate Thomson source and how plasma channels facilitate this design.

1. INTRODUCTION

The concept of the laser synchrotron source is based on Thomson scattering between laser photons and relativistic electrons. The cut-off wavelength in relativistic Thomson scattering depends upon the angle between the electron and laser beam ϕ,

$$\lambda_x = \frac{\lambda}{2\gamma^2}\left(\frac{1+a^2/2}{1+\cos\phi^2}\right), \qquad (1)$$

where λ is the laser wavelength, γ is the electron Lorentz factor, and $a \equiv \frac{eE}{m\omega c} \approx 0.85\times10^{-9}\lambda[\mu m]I^{1/2}[W/cm^2]$ is the normalized laser vector-potential. Compare Eq.(1) with a similar expression for synchrotron radiation produced from the wiggler in a conventional synchrotron source:

$$\lambda_x = \frac{\lambda_w}{2\gamma^2}\left(1+K^2/2\right), \qquad (2)$$

where λ_w is the wiggler period and K is the normalized strength parameter of the wiggler magnetic field. We see that the laser beam behaves like a wiggler of the ultra-short period. This introduced the terminology of a laser synchrotron source (LSS) [1]. Because the laser wavelength is 1000 times shorter than the wiggler period, only moderate electron energy is needed to produce hard x-rays. This allows using relatively compact linear accelerators instead of GeV synchrotrons. Another distinctive feature of LSS is that the present day technology enables production of femtosecond x-ray pulses. Due to these capabilities, the LSS promises to progress to a new generation of light-source facilities to support ultra-fast x-ray research in Basic Energy Science. First attempts to use the femtosecond LSS in ultra-fast structural dynamics studies look promising [2,3].

Thomson gamma sources can benefit High Energy Physics as well when utilized in the future e^--e^+ and γ-γ colliders. An example of such an application is a Japan-US collaborative project on the polarized positron source for the Japan Linear Collider (JLC) [4]. This projected source will be based on the e^--e^+ pair creation using Thomson-scattered polarized gamma rays [5].

Enticed by these prospects, the BNL ATF initiated proof-of-principle LSS tests based on a combination of a high-brightness linac and a high-power laser. The ATF LSS experiment combines a photocathode RF linac and a picosecond CO_2 laser. Selection of such components is not occasional and is based on a systematic approach to optimize the LSS towards the maximum photon yield [6,7]. To compare with other previously demonstrated and proposed LSS driven by ultra-fast solid state lasers [8-11], the CO_2 laser driver offers advantages reviewed in [7]. In particular, the CO_2 laser beam, having wavelength $\lambda=10$ μm, carries 10 times more photons than a solid state laser ($\lambda=1$ μm) of the same power. That implies a proportionally higher x-ray yield for the LSS. Another advantage is the capability of the gas laser technology to high repetition rate and high average power.

When maximizing the intensity of the prospective LSS one shall choose the right interaction geometry. Options are defined by the set of available components. If we have a femtosecond laser and a conventional picosecond or longer electron bunch then, in order to get an x-ray pulse as short as the laser pulse, we choose the 90^0-interaction. Note however, that at $\phi=90^0$ the x-ray pulse length is defined not just by the laser pulse duration but by the transverse time through the electron and laser focus that is about 300 fs at the typical beam size of $\sigma\approx 50$ μm [9].

In the $\phi=180^0$ configuration, the x-ray pulse duration $\tau_x = \tau_b + \tau_L/4\gamma^2$ is defined primarily by the electron bunch length τ_b and, as one can see, is very slightly affected by the laser pulse length τ_L. With the already demonstrated 200 fs electron bunches from the RF linac [12] and the recent idea of a highly-relativistic chirped bunch compression to 10-20 fs [13], the counter-propagation geometry can achieve the absolutely shortest x-ray pulses. Note that the 180^0-LSS is capable of producing femtosecond x-ray pulses using picosecond and even nanosecond laser pulses (for nanosecond pulses, channeling is required [14]).

Designing a high-yield LSS, we choose backscattering ($\phi=180^0$) also as the most efficient interaction geometry. The time interval when the counter-propagating focused laser and electron beams stay in interaction is normally $\pi r_L / \lambda$ times (where r_L is the laser beam radius) longer than in the 90^0 geometry prompting correspondingly higher numbers of scattered photons. To obtain this ratio, we take a proportion between the Rayleigh range $z_0 = \pi r_L^2 / \lambda$, which defines the interaction length in the backscattering configuration, to the laser beam radius that is important for the 90^0 configuration.

Combining our findings we decide that, in order to built the ultra-bright fs x-ray source based on relativistic Thompson scattering, we better choose a picosecond CO_2 laser and femtosecond e-beam in counter-propagating configuration.

2. PROOF OF PRINCIPLE LSS EXPERIMENT

Validating this conclusion, we set the proof-of-principle experiment schematically shown on Fig. 1. The 60 MeV electron beam is focused in the middle of the interaction cell. Other electron beam parameters are: bunch charge 0.5 nC, energy spread 0.15%, normalized emittance $\varepsilon_n=2$ mm mrad, bunch duration 3.5 ps, focus spot size $\sigma_b=32$ μm. Recollimated after the interaction cell, the electron beam is deflected by the dipole magnet that separates it from the backscattered Thomson x-rays.

The 0.6 GW, 180 ps CO_2 laser pulses are sent in a head-on collision with the e-beam. In order to focus the laser beam tightly, the 15 cm focal length parabolic copper mirror is placed on the way of the e-beam. Naturally, the mirror has a hole (5 mm in diameter) drilled to transmit both the electrons and the backscattered x-rays.

To avoid laser energy loss and mirror ablation at the central hole, the initially Gaussian laser beam profile is telescoped into the "donut" shaped beam using a pair of axicon lenses. We demonstrated that a parabolic mirror with a hole focuses the "donut" beam to the diffraction limit. On burn patterns made by the laser on fax paper and shown in Fig.2 we see how the axial hole in the beam fills with radiation at the focus.

For precise measurement of the electron and laser spot sizes the following technique is used. First we measure the e-beam spot size in the interaction region by a wire scan. Then, the laser spot size is determined by measuring the Thompson x-ray signal via transverse scanning of the electron beam across the interaction point. Results of these scans indicate that the laser focus closely matches the electron beam size with $\sigma_L=\sigma_b=32$ μm.

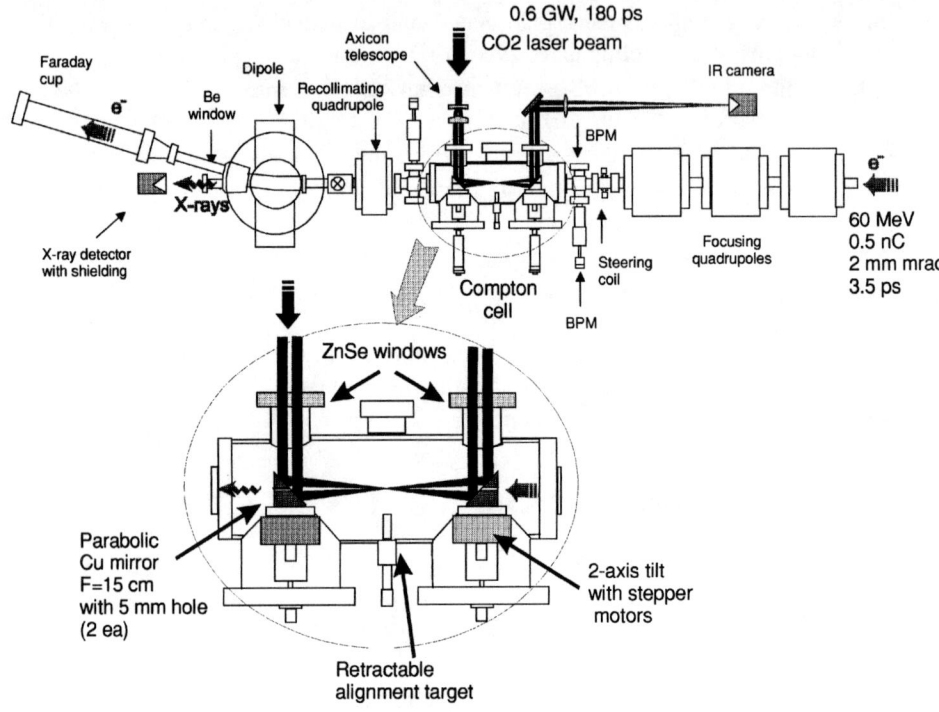

Figure 1. Principle diagram of the CO_2 LSS experiment.

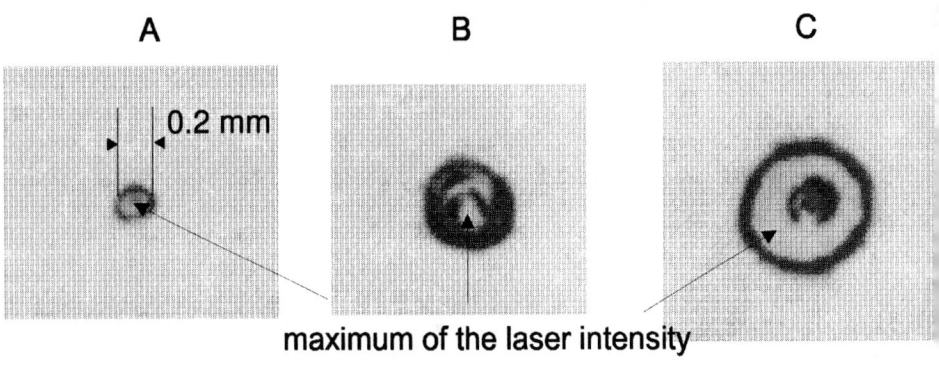

Figure 2. Burn patterns produced by the "donut"-shaped CO_2 laser beam on the thermal paper placed at the focal plane (image A) and at the distance 3.5 mm (image B) and 7 mm (image C) from the focus.

We know that the temporal shape of the electron bunch normally defines the backscattered x-ray pulse. The bunch shape was measured by adjusting the rf phase in the second linac section to produce a linear energy chirp to the electron bunch. A collimating slit in the dispersive region of the monochromator positioned after the linac filters out a narrow slice of the bunch that is measured by a Faraday cup. The result of these measurements is 3.5 ps FWHM as is shown in Fig.3.

The same Fig. 3 shows the time envelope of the laser pulse. A narrow shaded area corresponds to the portion of the laser pulse that fits into the focal waist length. The rest of the laser energy was wasted. As is discussed below, for the next experiment the laser pulse will be compressed to 25-30 ps for a better match to the focal waist length. But even under the present conditions a strong Thomson x-ray signal was detected by the Si diode placed in front of the output Be window. Calibration of the Si detector shows that this signal corresponds to 3×10^6 photon/pulse.

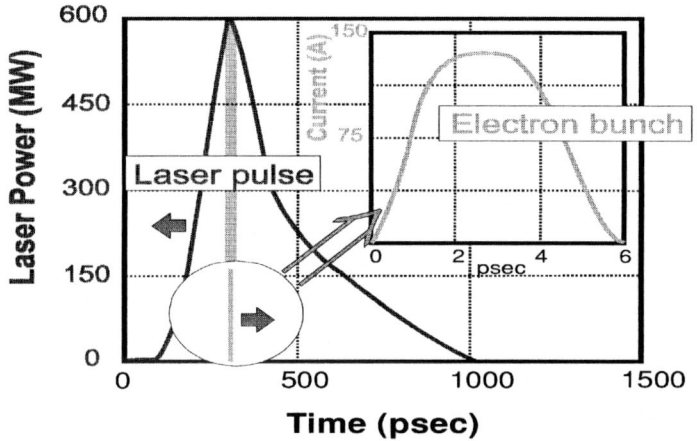

Figure 3. The 3.5 ps electron bunch counter-propagating with the 180 ps laser pulse radiates 3.5 ps x-ray pulse, shaded is a portion of the laser pulse that is utilized for the x-ray production (fits into the laser waist length);

The measured photon flux was compared with the theoretically expected value for the conditions of the experiment. Feeding the measured parameters of the electron and laser beams into the Monte-Carlo code CAIN [15] we calculated 2.9×10^7 photon/pulse integrated over the entire spectrum. However, when we introduced correction on the angular acceptance of the detector and spectral transmission of the Be window and the air on the way to the detector we obtained the same number as the measured one. This proves that the experiment indeed produced the record x-ray yield and intensity ever achieved via laser Thomson scattering on relativistic electron beams. Parameters of the ATF proof-of-principle LSS experiment are compiled in the first column of Table 1.

Diagram in Fig. 4 shows brightness of the ATF LSS together with previous results obtained in LBL and NRL. One of the success factors is the λ-proportional x-ray yield

$$N_x \propto \frac{E_L Q \lambda}{\sigma_{L,b}^2}, \tag{3}$$

where E_L is the portion of the laser energy within the time interval $4z_0/c$ that actually participates in interaction with the electron bunch and Q is the electron bunch charge. Other important ingredients are the quality of the ATF e-beam and the tight focusing of both beams (high Q and low σ_b and σ_L).

3. UPCOMING NONLINEAR COMPTON SCATTERING EXPERIMENT

In the next ATF experiment, the beam of 10^{16} W/cm² peak intensity from the upgraded CO_2 laser will be brought into head-on collision with 70-MeV electrons. At such intensity, there is a high probability for an electron to absorb several laser photons before emitting a single photon of higher energy. In addition to harmonics, the relativistic transverse oscillation of an electron inside the laser beam leads also to a shift in the effective mass of the electron to be seen in the spectrum of radiated photons.

TABLE 1. ATF CO_2 LSS experimental results and near-future design parameters

PARAMETER	1-st stage (1999)	2-nd stage (2000)
CO_2 LASER		
Pulse Length [ps]	180	30
Pulse Energy [J]	0.2	30
Peak Power [GW]	0.6	1000
RMS Radius at Focus [μm]	32	32
Waist Length [mm]	2.5	2.5
ELECTRON BEAM		
Electron Energy [MeV]	60	60 (70)
Bunch Duration FWHM [ps]	3.5	3.5
Bunch Charge [nC]	0.5	0.5
Normalized Emittance [mm mrad]	2	2
Momentum Spread [%]	0.15	0.15
RMS Radius at Focus [μm]	32	32
X-RAYS		
Peak Wavelength [Å]	1.8	2.6 (1.8)
Pulse Duration [ps]	3.5	3.5
Photons per Pulse (total spectrum)	2.9×10^7	1.3×10^{10}
Photons per Second (total spectrum)	8×10^{18}	4×10^{21}
Spectral Bandwidth [%]	0.5	~10
Peak Brightness [photon/sec mm² mrad²]	2.5×10^{18}	1.3×10^{21}

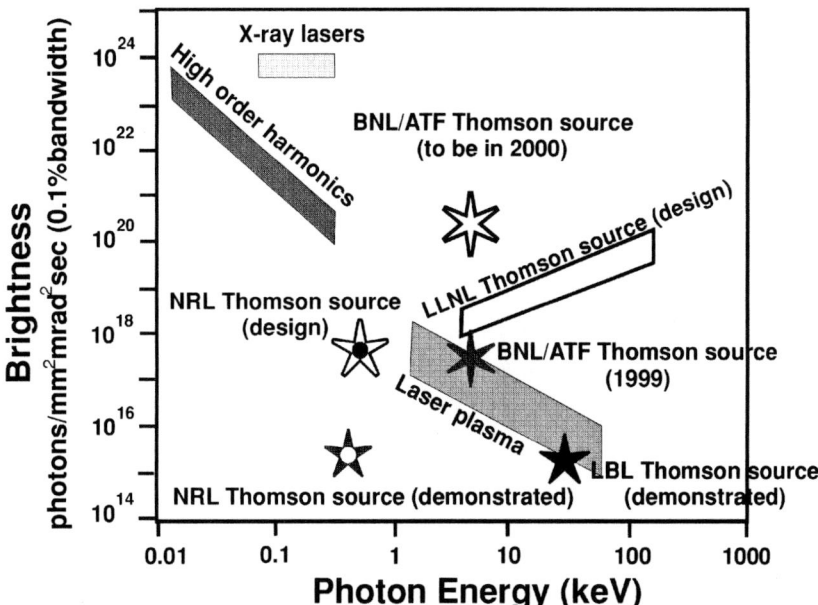

Figure 4. Peak brightness of existing and projected picosecond and femtosecond laser-driven x-ray sources

Although nonlinear relativistic Thomson scattering was studied theoretically since 1912 [16,17], just a few photons of the second harmonic radiation have been observed at a very week laser field strength parameter $a=0.01$ [18]. This did not allow verification of mass-shift effect.

By focusing the 1 TW CO_2 laser beam into the $\sigma_L=30$ µm spot we will achieve the laser strength parameter $a=0.75$. Such a laser field produces highly relativistic quiver motion in the counter-propagating electron beam. According to Eq.(1) this results in a notable downshift in the high-energy peak in the simulated Thomson spectrum illustrated by Fig. 5 (right). Simultaneously, we shall observe a significant expansion of the x-ray spectrum towards the hard x-ray region due to harmonics (see Fig. 5 (left)). Fig.6 maps angular distribution of the harmonics. The most remarkable difference from the fundamental is a minimum intensity along the axis. The next ATF Thomson scattering experiment shall confirm this feature as well. The design parameters of the nonlinear Thomson scattering experiment are compiled in the second column of Table 1.

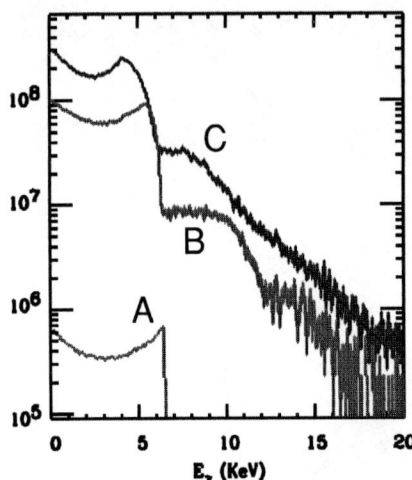 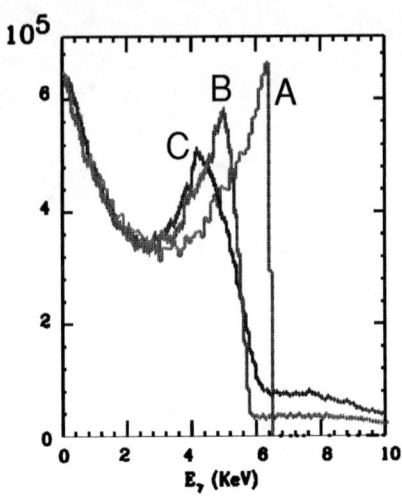

Figure 5. Simulated photon density (photon/keV) spectra in logarithmic scale (left) and normalized linear scale (right);

curve A:	E_L=0.2 J,	τ_L=180 ps,	a=0.02,	normalized scale coefficient ×1
curve B:	E_L=10 J,	τ_L=30 ps,	a=0.43,	normalized scale coefficient ×200
curve C:	E_L=30 J,	τ_L=30 ps,	a=0.75,	normalized scale coefficient ×600

4. PROSPECTIVE ULTRA-BRIGHT LASER SYNCHROTRON SOURCE IN PLASMA CHANNEL

A relatively high spectral bandwidth indicated in the second column of Table 1 for the relativistically strong laser beam is due to the dependence of the x-ray wavelength upon the laser strength parameter a according to Eq. (1). The nonuniform transverse and longitudinal profile of the laser pulse results in the x-ray spectrum smearing. Keeping the nonlinear term in the relative energy spread below 0.1% requires $a \leq 0.03$. This imposes a fundamental restriction on the x-ray spectral brightness $B = \dfrac{N_x \gamma^2}{4(\pi \sigma_b)^2 \tau_b \Delta \omega_x}$. Indeed, when the nonlinear term dominates the spectral bandwidth, the increase in N_x, that is proportional to the laser intensity, is compensated by the equivalent increase in the x-ray spectral bandwidth.

In order to abate the constraint discussed above, it was proposed to confine the laser-electron interaction region in a plasma channel extended over several centimeters [19]. If we find a channel with low optical losses, the enhancement of the x-ray yield may be equal to the number of Rayleigh lengths within the channel length.

Permitting the use of nanosecond laser pulses instead of picosecond pulses, the channeling approach opens a possibility to utilize the existing technology of high-repetition rate TEA (transverse electrical discharge, atmospheric pressure) CO_2 lasers, thus, opening new prospects for high average intensity and brightness LSS. High repetition rate, multi-kilowatt average power and ~1 ns, pulse duration have been demonstrated with TEA lasers [20].

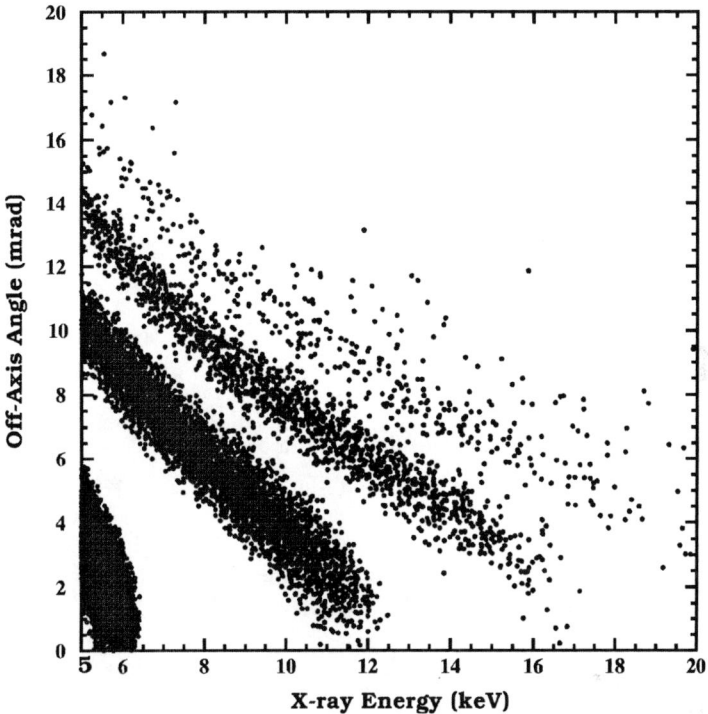

Figure 6. Spectral and angular distributions in nonlinear Thomson scattering simulated for the conditions: E_L=10 J, τ_L=30 ps, σ_L=32 μm, a=0.43 (note that the x-ray energy scale starts from 5 keV that is close to the air and Be window transmission cut-off)

There are a number of proposed schemes and experimental demonstrations for channeling of high intensity laser pulses. These include: laser guiding in micro-capillary tubes [21,22], plasma channels produced by electric discharge in the dielectric capillary tube [23], and by laser breakdown in the gas [24,25]. All the plasma channel experiments have been performed using picosecond solid state lasers. However, the condition for plasma channeling $\Delta n_e [cm^{-3}] = 10^{20}/r_L^2 [\mu m]$ (where Δn_e

shows the increase of the plasma density from the laser beam axis to the point of $r=r_L$) is not sensitive to the laser wavelength.

In collaboration with Hebrew University, Jerusalem the ATF initiated a study of the capillary discharge [23] for channeling of the CO_2 laser pulse. In this scheme, the laser beam is confined in the plasma core 10 times smaller than the physical dimensions of the capillary tube. That allows for an interaction between the laser and electron beam and for extraction of the Thomson scattered x-rays without obstruction by the walls.

The plasma channel set-up will be installed into the electron beamline as is shown in Fig.7. The goal of this experiment will be demonstration of the efficient generation of picosecond and femtosecond high-brightness x-rays using subnanosecond laser pulses.

Finally, let us review other factors that affect the spectral bandwidth when the nonlinear effect is not an issue. The calculations are done for the proof-of-principle ATF LSS experiment described above.

The fractional bandwidth of the Thomson radiation depends upon the total number of laser periods that fit into the electron-laser interaction interval. With demonstrated laser spot radius r_L=45 μm, this time interval is 15 ps and the number of periods is 500. This results in $\Delta\omega'_x/\omega_x$ =0.2%.

The x-ray bandwidth is also related to the momentum spread or "temperature" of the e-beam. For the ATF e-beam, the temperature smearing in the x-ray spectrum is $\Delta\omega''_x/\omega_x = 2\Delta\gamma/\gamma$=0.3%.

A finite angular divergence α of the electron beam also disperses the x-ray angular distribution; $\alpha = \dfrac{\varepsilon_n}{\gamma\sigma_b}$ =0.5 mrad results in $\Delta\omega'''_x/\omega_x = (\varepsilon_n/\sigma_b)^2$ =0.4% spectrum smearing.

The combined total bandwidth is equal to 0.5% for the conditions of the ATF proof-of-principle experiment. This confirms the possibility of achieving ultra-high peak spectral brightness using the CO_2 laser driven LSS.

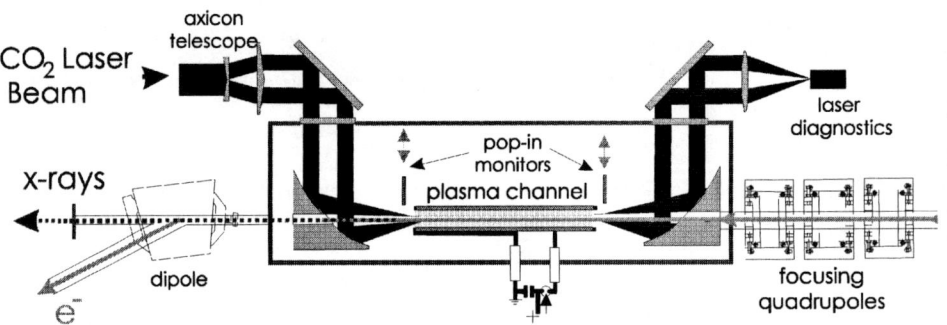

Figure 7. Prospective Thomson scattering experiment in plasma channel

ACKNOWLEDGMENTS

Authors wish to thank all collaborators on the US-Japan Compton experiment who contributed at different stages of the experiment preparation and in particular A. Tsunemi, K. Dobashi, T. Muto, T. Kobuki, R. Kuroda, T. Okugi, Z. Segalov, P. Siddons, Y. Liu, P. He, and D. Cline, Optoel company for designing the interaction cell to our specifications, W. Kimura and K. McDonald for providing essential optical components and diagnostics, R. Tatchyn for useful discussions, and J. Skaritka and R. Harrington for technical assistance. This study is supported by the U.S. Dept. of Energy under the Contract No. DE-AC02-98CH10886 and Japan/U.S. cooperation in the field of high energy physics.

REFERENCES

1. P. Sprangle, A. Ting, E. Esarey, A. Fisher, *J. Appl. Phys.* **72**, 5032 (1992)
2. A. H. Chin et al., *Phys. Rev. Lett.* **83**, 336 (1999)
3. M. Uesaka, H. Kotaki, K. Nakajima, H. Harano, K. Kinoshita, T. Watanabe, T. Ueda, K. Yoshii, M. Kando, H. Dewa, S. Kondo, and F. Sakai, "Generation and Application of Femtosecond X-ray Pulse", *these Proceedings*.
4. BNL Proposal "Study of Compton Scattering of Picosecond Electron and CO_2 Laser Beams to Prototype the Polarized Positron Source for Japan Linear Collider", http://www.nsls.bnl.gov/AccTest/experiments/Compton/compton.htm
5. T. Okugi, Y. Kurihara, M. Chiba, A. Endo, R. Hamatsu, T. Hirose, T. Kumita, T. Omori, Y. Takeuchi, and M. Yosioka, *Jpn. J. Appl. Phys.* **35**, 3677 (1996)
6. I.V. Pogorelsky, I. Ben-Zvi, *Particle Accel. Conf. 97*, Vancouver, B.C., Canada, May 12-16, 1997, http://www.triumf.ca/pac97/papers/pdf/6V040
7. I.V. Pogorelsky, *Nucl. Instrum. and Methods in Phys. Res. A* **411**, 172 (1998)
8. A. Ting, R. Fischer, A. Fisher, C.I. Moore, B. Hafizi, R. Elton, K. Krushelnick, R. Burris, S. Jakel, K. Evans, J.N. Weaver, P. Sprangle, M. Baine, S. Ride, *Nucl. Instrum. and Methods in Phys. Res. A* **375**, ABS 68 (1996)
9. R.W. Schoenlein, W.P. Leemans, A.H. Chin, P. Volbeyn, T.E. Glover, P. Balling, M. Zolotarev, K.-J. Kim, S. Chattopadhyay, C.V. Shank, *Science* **274**, 236 (1996)
10. K. Nakajima, *Proceedings of LASERS'97*, New-Orleans, LA, December 15-19, 1997, STS Press, McLean, VA, 778 (1998)
11. T. Cowan, "Intense laser-electron interaction research at LLNL", *Nucl. Instrum. & Methods in Phys. Res. A*, to be published.

12. M. Uesaka, K. Kinoshita, T. Watanabe, T. Ueda, K. Yoshii, H. Harano, J. Sugahara, K. Nakajima, A. Ogata, F. Sakai, H. Dewa, M. Kando, H. Kotaki, S. Kondo, *8th Workshop on Advanced Accelerator Concepts*, July 1998, Baltimore, MD, *AIP* **472**, 908 (1999)

13. X.J. Wang, "Producing and Measuring Small Electron Bunches", Proceedings *of Particle Accel. Conf. 99,* New York, NY, April 1999

14. I.V. Pogorelsky, I. Ben-Zvi, X.J. Wang, T. Hirose, "Femtosecond laser synchrotron sources based on Compton scattering in plasma channels", *Nucl. Instrum. & Methods in Phys. Res. A*, to be published.

15. The manual of code CAIN, ftp://lcdev.kek.jp/pub/Yokoya/manual-cain21e.ps.zip

16. G.A. Scott, Electromagnetic Radiation. Cambridge University Press, 1912

17. E. Esarey, S.K. Ride, P. Sprangle, *Phys. Rev. E*, **48**, 3003 (1993)

18. T.J. Englert and E.A. Rinehart, *Phys. Rev. A*, **28**, 1539 (1983)

19. I.V. Pogorelsky, I. Ben-Zvi, X.J. Wang, T. Hirose, "Femtosecond laser synchrotron sources based on Compton scattering in plasma channels", *Nucl. Instrum. & Methods in Phys. Res. A*, to be published.

20. "Handbook of Molecular Lasers", ed. P.K. Cheo, Marcel Dekker, Inc, New-York, 1987

21. Y Jackel, R. Burris, J. Grun, A. Ting, C. Manka, K. Evans, and J. Kosakovskii, *Opt. Lett.* **20**, 1086 (1995)

22. F. Dorchies, J.R. Marques, B. Cros, G. Matthieussent, C. Courtois, T. Velikorossov, P. Audebert, J.P. Geindre, S. Rebibo, G. Hamoniaux, and F. Amiranoff, *Phys. Rev. Lett.* **82,** 4655(1998)

23. Y. Ehrlich, C. Cohen, A. Zigler, J. Krall, P. Sprangle, and E. Esarey, *Phys. Rev. Lett.* **77**, 4186 (1996)

24. C.D. Durfee III & H.M. Milchberg, *Phys. Rev. Lett.* **71**, 2409 (1993)

25. E.W. Gaul, S.P. Le Blank, and M.C. Downer, *8th Workshop on Advanced Accelerator Concepts*, July 1998, Baltimore, MD, *AIP* **472**, 377(1999)

Muon Cooling - Emittance Exchange [1]

Zohreh Parsa

Brookhaven National Laboratory
Physics Department 510 A, Upton, NY 11973-5000, USA

Abstract. Muon Cooling is the key factor in building of a Muon collider, (to a less degree) Muon storage ring, and a Neutrino Factory. Muon colliders potential to provide a probe for fundamental particle physics is very interesting, but may take a considerable time to realize, as much more work and study is needed. Utilizing high intensity Muon sources - Neutrino Factories, and other intermediate steps are very important and will greatly expand our abilities and confidence in the credibility of high energy muon colliders. To obtain the needed collider luminosity, the phase-space volume must be greatly reduced within the muon life time. The Ionization cooling is the preferred method used to compress the phase space and reduce the emittance to obtain high luminosity muon beams. We note that, the ionization losses results not only in damping, but also heating. The use of alternating solenoid lattices has been proposed, where the emittance are large. We present an overview of the cooling and discuss formalism, solenoid magnets and some beam dynamics.

INTRODUCTION

Alternating solenoid lattices has been proposed as desirable for use in the earlier cooling stages of Muon Colliders, where the emittances are large. Since the minimum $\beta_\perp$'s must decrease in order to obtain smaller transverse emittances as the muon beam travels down the cooling channel. This can be done by increasing the focusing fields and/or decreasing the muon momenta, where the current carrying lithium lenses may be used (to get a stronger radial focusing and to minimize the final emittance) for the last few cooling stages. The use of 'bent solenoids' may provide the required dispersion for the momentum measurement. Where the off-momentum muons are displaced vertically by an amount:

$$\Delta y \approx \frac{P}{eB_s}\frac{\Delta P}{P}\theta_{\text{bend}}, \qquad (1)$$

---

[1] Supported by U.S. Department of Energy Contract Number DE-AC02-98CH10886.
[§] E-mail: parsa@bnl.gov

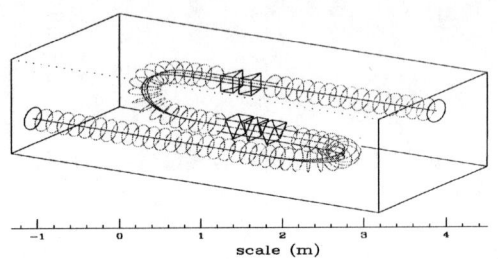

FIGURE 1. Example of bent solenoids and Wedges - for emittance exchange. May be used for muon collider longitudinal cooling

where B_s is the field of the bent solenoid and θ_{bend} is the bend angle. That eq. (1) describes the deflection of the 'guiding ray' (or axis) of the helical muon trajectory and not the trajectory itself. Also, the muon's momentum cannot be determined simply by measuring the height of its trajectory at the entrance and exit of a bent solenoid. Rather, the height of the guiding ray must be reconstructed at both places, which requires precise measurements of the helical trajectories. The momentum resolution of a bent solenoid spectrometer is given by:

$$\frac{\sigma_{D_P}}{P} \approx \frac{1}{\theta_{\text{bend}}} \frac{P}{eB_s} \frac{\sigma_{D_x}}{L^{5/2}} \sqrt{\frac{720}{n}}, \tag{2}$$

L is the length of the trajectory observed (before and after the bend) with n samples/m, each with transverse spatial resolution σ_{D_x}. Where the momentum resolution improves with increasing magnetic field. But, to reduce the cost of the solenoid channel one should try to use a relatively low magnetic field. The momentum resolution is improved by increasing θ_{bend} and by increasing L.

In Fig.1, the bending of the solenoid produce the dispersion required for the longitudinal to transverse emittance exchange. Where after one bend and one set of wedges the beam cross-section is asymmetric then the symmetry is restored by going through the second bend and wedge system (which is rotated by 90 degrees w.r.t. the first). [1].

FORMULATION AND MAPS FOR SOLENOIDS

The canonical equations in 2n-Dimensional phase space (e.g. 6 Dim., in our calculation) can be expressed as $\frac{d\psi_i}{dt} = [\psi_i, H]$, $i = 1, 2, \ldots 2n$, and in terms of the Lie transformations as

$$\frac{d\psi_i}{dt} = - : H : \psi_i, \qquad i = 1, 2, \ldots 2n \tag{3}$$

Where the Lie operator $(: H :)$ is generated by the Hamiltonian, (H), and Lie transformation, $M = e^{-t:H:}$, could generate the solution to Eq. (3) as $\psi_i = M\psi_i(0)$, where ψ_i is the value of $\psi_i(t)$ at $t > 0$ and $\psi_i(0)$ is the initial trajectory. The interest is to find solutions to equations of motion which differ slightly from the reference orbit Thus, one can choose the canonical variables, from the values for the reference trajectory (for small deviations) and Taylor expand the Hamiltonian (H) about the design trajectory $H = H_2 + H_3 + \ldots$. Where H_n is a homogeneous polynomial of degree n in the canonical variables. After transformations to the normalized dimensionless variables, one can obtain the effective Hamiltonian H^{New}, expressed as

$$H^{\text{New}} = F_2 + F_3 + F_4 \ldots . \qquad (4)$$

Thus the particle trajectory $\vec{\psi} = (X, P_X, Y, P_Y, \tau, P_\tau)$ through a beamline element of length L can be described by $\psi_i^f = -:H^{\text{New}}:\psi_i$, $i = 1, 2, \ldots 2n$. The exact symplectic map that generates the particle trajectory through that element is $M = e^{-L:H^{\text{New}}:}$, where M describes the particle behavior through the element of length L. Using the factorization and expanding H^{New} as in Eq. (4), results in

$$M = e^{-L:H^{\text{New}}:} = e^{:f_2:}e^{:f_3:}e^{:f_4:}\ldots , \qquad (5)$$

(e.g., for a map through 3rd order we need to include terms of f_2, f_3, and f_4).

To illustrate the above formalism, consider the evolution of the motion of particles in an external electromagnetic field described by the Hamiltonian

$$H = [m^2c^4 + c^2((p_x - qA_x)^2 + (p_y - qA_y)^2 + (p_z - qA_z)^2)]^{1/2} + e\phi(x, y, z; t)$$

where m and q are the rest mass and charge of the particle, A and ϕ are the vector and scalar potentials such that $\vec{B} = \nabla \times \vec{A}, \vec{E} = -\nabla\phi - \nabla\vec{A}/\partial t$.

Making a canonical transformation from H to H_1 and changing the independent variable from time t to z (for convenience) for a particle in magnetic field (e.g. of solenoid) results in
$p_z = [(p_z - qA_x)^2 + (p_y - qA_y)^2 + p_t^2/c^2 - m^2c^2]^{1/2}$.

Where $H = -p_t$, $H_1 = -p_z$ and $t = (z/v_{0z})$ the time as a function of z. We next make a canonical transformation from H_1 to H^{New}, with a dimensionless deviation variables (for convenience), $X = x/l$, $Y = y/l$, $\tau = c/l(t - z/v_{0z})$, $P_x = p_x/p_0$, $P_y = p_y/p_0$, $P_\tau = (p_t - p_{0t})/p_0c$, where l is a length scale (taken as 1 m in our analysis), with $\mathbf{P} = \vec{P}_x + \vec{P}_y$ and $\mathbf{Q} = \vec{X} + \vec{Y}$ defined as two dimensional vectors [6], p_0 and p_0c are momentum and energy scales. Where p_0 is the design momentum, v_{0z} is the velocity on the design orbit and p_{0t} is a value of p_t on the design orbit ($p_{0t} = \sqrt{m^2c^4 + p_0^2c^2}$) (reminding that design orbit for the solenoid is along the z-axis). Thus, expanding the new Hamiltonian Eq. (4) leads to:

$$F_2 = \frac{P_\tau^2}{(2\beta^2\gamma^2)} - \frac{1}{2}B_0(\vec{Q} \times \vec{P}) \cdot \hat{z} + \frac{1}{8}B_0^2Q^2 + \frac{P^2}{2} \qquad (6)$$

$$F_3 = \frac{P_\tau^3}{(2\beta^3\gamma^2)} - \frac{P_\tau}{2\beta}B_0(\vec{Q} \times \vec{P}) \cdot \hat{z}$$

$$+\frac{P_\tau}{8\beta}(B_0^2 Q^2 + 4P^2) \tag{7}$$

$$F_4 = \frac{P_\tau^4(5-\beta^2)}{8\beta^4\gamma^2} + \frac{P_\tau^2 Q^2 B_0^2(3-\beta^2)}{16\beta^2}$$

$$-\frac{P_\tau^2}{2}(\vec{Q}\times\vec{P})\cdot\hat{z}\frac{B_0(3-\beta^2)}{2\beta^2}$$

$$+\frac{P_\tau^2}{2}\frac{P^2(3-\beta^2)}{2\beta^2} + \frac{Q^4}{16}(B_0^4 - 4B_0 B_2)/8$$

$$+\frac{Q^2}{4}\frac{P^2 3 B_0^2}{4} + \frac{Q^2}{4}(\vec{Q}\times\vec{P})\cdot\hat{z}(B_2 - B_0^3)/4$$

$$-\frac{1}{8}(\vec{P}\cdot\vec{Q})^2 B_0 - \frac{P^2}{4}(\vec{Q}\times\vec{P})\cdot\hat{z}B_0 + \frac{P^4}{8} \tag{8}$$

Following the Hamiltonian flow generated by:

$$H^{\text{New}} = F_2 + F_3 \ldots$$

from some initial ψ_0 to a final ψ_f coordinates we can calculate the transfer map M (Eq. (5)) for the solenoid. Where F_2, F_3, and F_4 would lead to the 1st, 2nd, and 3rd order maps. The effects of which can be seen from Eqs. (6–8). For example, the 2nd order effects due to solenoid transfer maps are purely chromatic aberrations Eq. (7). In addition, we note the third order geometric aberrations Eq. (8). As shown by Eqs. (6–8), the coupling between X, Y planes produced by a solenoid is rotation about the z-axis which is a consequence of rotational invariance of the Hamiltonian H^{New}, due to axial symmetry of the solenoid field. For beam simulations, M can be calculated to any order using numerical integration techniques such as Runge-Kutta method depending on the computer memory and space available [6].

In obtaining Eq. 8, the correlations were neglected (as e.g. in the Status Report see [1]), e.g. $\langle xP_x\rangle = 0$), and the relation $\langle x^2\rangle = \epsilon\beta_\perp = \frac{\epsilon_n \beta_\perp}{\gamma\beta}$ was used, which can not be assumed if the correlations are properly taken into account. Thus, if $\langle xP_x\rangle \neq 0$ then transverse cooling to be expressed as

$$\frac{d\epsilon_n}{ds} = \frac{1}{\beta^2}\frac{dE_\mu}{ds}\frac{\epsilon_n}{E_\mu} + \frac{1}{\beta^3}\frac{\langle x^2\rangle(0.014 GeV)^2}{2\epsilon_n E_\mu m_\mu L_R} + \ldots, \tag{9}$$

As in Fig. 1, by introducing a transverse variation in the wedge (absorber) density or thickness, where there is dispersion (i.e. the transverse position is energy dependent), the energy spread, and the longitudinal emittance can be reduced. As we noted earlier, from theoretical point of view, a situation with ionization cooling completely corresponds to a situation with radiation cooling whose theory is well developed. For some standard "hierarchy" of methods for analyzing such systems see e.g. Ref. [5].

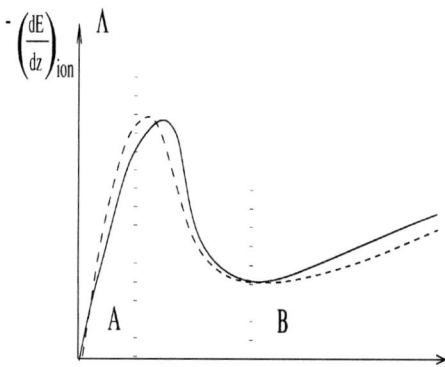

FIGURE 2. Schematic of the dependence of ionization losses on momenta.

IONIZATION COOLING

In ionization method, muons passing through a material medium lose momentum and energy through ionization interactions in transvers and longitudinal directions. The normalized emittance is reduced due to transvers energy losses. The curve in Fig. (2) shows the dependence of ionization losses on momenta. Damping rates (decrements) of individual particles in the absence of wedges (natural damping rate) are defined by the following formula:

$$\lambda_\perp = -\frac{dE}{dz}_{ion} \frac{1}{2\beta^2 \gamma mc^2}$$
$$\lambda_\parallel = -\frac{1}{z}\frac{d}{dp}\left[\left(\frac{dE}{dz}\right)_{ion} \frac{1}{v}\right]$$
(10)

Where $\lambda_\perp$ and $\lambda_\parallel$ are natural transverse and longitudinal damping respectively. Here $\left(\frac{dE}{dz}\right)_{ion}$ is the ionization losses of energy, m is the muon mass, β, γ are relativistic parameters, p, v are momentum and longitudinal velocity of muons being cooled. It was established, that the sum of all increments is invariant of the cooling system: $\Lambda = 2\lambda_\perp + \lambda_\parallel$. This curve is also plotted in Fig. (2) (as the dotted line). In Fig. (2) we see that there are two natural regions for cooling: region A ("frictional cooling") and region B ("ionization cooling" for intermediate and high energies). Frictional Cooling is convenient only for cold (low energy) muons (e.g. Kinetic energy 10 to 150 KeV), and therefore it is difficult to use for high energy muon source, (in addition to big noises due to coulomb scattering etc.). Classical Ionization Cooling is usable for kinetic energy range of 30 to 100 MeV. Which due to absence of "natural" longitudinal cooling it is necessary to use "wedges" for which R & D is needed. A proposal for such studies is being considered [1].

COOLING FOR A NEUTRINO FACTORY

In a Neutrino Factory, a proton driver of moderate energy (< 50 GeV) and high average power,(e.g., 1-4 MW), similar to that required for a muon collider, but

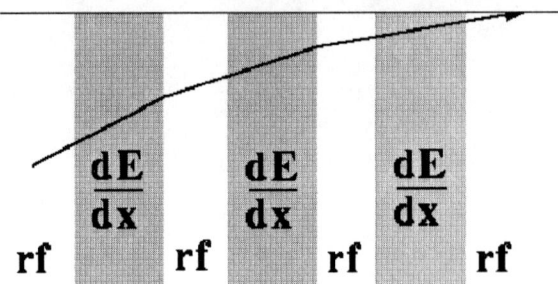

FIGURE 3. Schematic of Ionization Cooling concept (Ionization takes away momentum, and the RF acceleration puts momentum back along the z-axis, resulting in a Transverse Cooling).

with a less stringent requirements on the charge per bunch and power is needed. This is followed by a target and a pion-muons capture system. A longitudinal phase rotation is performed to reduce the muon energy spread at the expense of spreading it out over a longer time interval. The phase rotation system may be designed to correlate the muon polarization with time, allowing control of the relative intensity of muon and anti-electron neutrinos. Some cooling may be needed, to reduce phase space, about a factor of 50 in six dimensions. This is much smaller than the factor of 10^6 needed for a muon collider. Production is followed by fast muon acceleration to 50 GeV (for example), in a system of linac and two recirculating linear accelerators (RLA's), which may be identical to that for a first stage of muon collider such as a Higgs Factory.

Figure 3 shows a schematic of Ionization Cooling concept. Ionization cooling that has been proposed involves passing the beam through an absorber in which the muons lose transverse- and longitudinal-momentum by ionization loss (dE/dx). The longitudinal momentum is then restored by coherent re-acceleration, leaving a net loss of transverse momentum (transverse cooling). The process is repeated many times to achieve a large cooling factor. The beam energy spread can also be reduced using ionization cooling by introducing a transverse variation in the absorber density or thickness (e.g. a wedge) at a location where there is dispersion (the transverse position is energy dependent). Theoretical studies have shown that, assuming realistic parameters for the cooling hardware, ionization cooling can be expected to reduce the phase-space volume occupied by the initial muon beam by a factor of $10^5 - 10^6$. Ionization cooling is a new technique that has not yet been demonstrated. Special hardware needs to be developed to perform transverse and longitudinal cooling. It is recognized that understanding the feasibility of constructing an ionization cooling channel that can cool the initial muon beams by factors of $10^5 - 10^6$ is on the critical path to the overall feasibility of the muon collider concept.

MUON COOLING "MERIT FACTOR"

Luminosity of collider L is defined by the following expression:

$$L \sim \frac{N^2 f}{g_x g_y} = \frac{N^2 f}{\epsilon_\perp^f \cdot \beta_\perp^f} \tag{11}$$

Where N = a number of muons per bunch, f = mean repetition frequency of collisions, $\epsilon_\perp^f$ = emittance at collision point and $\beta_\perp^f$ = β-function at collision point. Usually $\beta_\perp^f$ is limited by condition: $\beta_\perp^f \geq \sigma_z^f$ where σ_z^f is a longitudinal bunch size. Let us assume, that: 1) $\frac{\Delta p_t}{p}$ is known (monochromatic experiments); 2) we can redistribute emittances inside a given six-dimensional phase volume. Then, taking into account losses in the cooling system, we can rewrite Eq. (11) in the following form:

$$L \sim \frac{N_0^2 \exp\left(-\frac{2}{cT_0}\int_0^z \frac{dz}{\gamma(z)}\right) D^2}{\sqrt{V_6^N \cdot \epsilon_\parallel^f}} \cdot \left(\frac{\Delta p}{p}\right)_\parallel^f \tag{12}$$

Here "N_0" is a number of particles at an entrance of the cooling system, "exp" describes muon decay, "D" describes muon losses in cooling section, and "V_6^N" is an invariant six-dimensional phase volume of muon beam.

Thus we can introduce "merit factor" which describes a quality of muon cooling system. We obtain

$$R = \frac{D^2 \exp\left[-\frac{2}{cT_0}\int_0^z \frac{dz}{\gamma(z)}\right]}{\sqrt{V_6^N}} \tag{13}$$

Note that, the dependence on V_6^N may be stronger. With account of all the circumstances, we can write

$$R \sim (V_6^N)^\alpha \tag{14}$$

with α in interval (0.5; 2/3). For more info. see references.

MUCOOL

The MUCOOL collaboration has been formed to pursue the development of a muon ionization cooling channel for a high luminosity muon collider. During the presentation we discussed the MUCOOL cooling studies and the Feasibility I studies for Fermilab side specific muon storage based neutrino facility, which we will not include here due to space limitation. We refer the reader to the MUCOOL - the muon ionization cooling experimental R&D page [1]. The examples presented include some of the scenarios being explored by our Neutrino Factory and Muon Collider Collaboration [12].

REFERENCES

1. C.N. Ankenbrandt etal., *Ionization Cooling Research and Development Program for a High Luminosity Muon Collider*, FNAL-P904 (April 15, 1998), MUCOOL Notes http://www-mucool.fnal.gov/notes; http://www.fnal.gov/projects/muon collider/cool/
2. Ankenbrandt et al. (Muon Collider Collaboration), Phys. Rev. ST Acc. Beams 2, 081001 (99).
3. Z. Parsa, ed., *Beam Stability and Nonlinear Dynamics*, AIP-Press CP **405**, 1997.
4. K. McDonald TN/98-17 (98), http://puhep1.princeton.edu/mumu/nuphys/ J. Norem, Private Comm.
5. Z. Parsa, Ionization Cooling and Muon Dynamics, in *Physics Potential & Development of Muon-Muon Colliders*, Ed. D. Cline, AIP CP 441, pp 289-294 (1998).
6. E.g., used in L. Gluckstern, F. Neri, G. Raingarajan, A. Dragt, Univ. of Maryland Note (1988) and MARYLIE; W. Lysenko, M. Overley, AIP CP 177, (1988). Z. Parsa, Proc. of IEEE 0-7803-1203,p. 509 (1993).
7. Z. Parsa, Muon Dynamics and Ionization Cooling at Muon Colliders, Procd. of EPAC98, Stockholm, Sweden, Vol 2, pp.1055- (1998).
8. *Neutrino Factory Feasibility Studies at Fermilab*: http://www.fnal.gov/projects/muon_collider/nu_factory/
9. The Neutrino Factory and Muon Collider Collaboration http://www.cap.bnl.gov/mumu/
10. Z. Parsa, *Muon Storage Rings - Neutrino Factories*, in *Next Generation Nucleon decay and Neutrino detector (99)*, ed. M. Diwan and C. Jung, AIP CP 533, pp. 181-195, N.Y. (2000).
11. Z. Parsa, *Intense Muon Beams and Neutrino Factories* in *MUMU99*, ed. D. Cline (2000).
12. [1]- [10] and refrences therein.

Modeling Beam-Driven and Laser-Driven Plasma Wakefield Accelerators with XOOPIC

David L. Bruhwiler,[a] Rodolfo Giacone,[b] John R. Cary,[a,b] John P. Verboncoeur,[c] Peter Mardahl,[c] Eric Esarey[d] and Wim Leemans[d]

[a] *Tech-X Corporation, 5541 Central Avenue, Suite 135, Boulder CO 80301, USA*
[b] *University of Colorado at Boulder, Physics Department, Boulder CO 80309-0390, USA*
[c] *University of California Berkeley, EECS Department, Berkeley CA 94720-1770, USA*
[d] *Lawrence Berkeley National Laboratory, University of California, Berkeley CA 94720, USA*

Abstract. We present 2-D particle-in-cell simulations of both beam-driven and laser-driven plasma wakefield accelerators, using the object-oriented code XOOPIC, which is time explicit, fully electromagnetic, and capable of running on massively parallel supercomputers. Simulations of laser-driven wakefields with low ($\sim 10^{16}$ W/cm^2) and high ($\sim 10^{18}$ W/cm^2) peak intensity laser pulses are conducted in slab geometry, showing agreement with theory. Simulations of the E-157 beam wakefield experiment at the Stanford Linear Accelerator Center, in which a 30 GeV electron beam passes through 1 m of preionized lithium plasma, are conducted in cylindrical geometry, obtaining good agreement with previous work. We briefly describe some of the more significant modifications to XOOPIC required by this work, and summarize the issues relevant to modeling electron-neutral collisions in a particle-in-cell code.

INTRODUCTION

The quest to understand the fundamental nature of matter requires ever higher energy particle collisions, which in turn leads to ever larger and more expensive particle accelerators. Plasma-based accelerators can sustain electron plasma waves (EPW) with longitudinal electric fields on the order of the nonrelativistic wave breaking field, $E_0 = cm_e\omega_p/e$, where $\omega_p = (4\pi n_e e^2/m_e)^{1/2}$ is the plasma frequency at an electron density n_e (see Ref. 1 for a review). For $n_e = 10^{18}$ cm^{-3}, the electric field is $E_0 \cong 100$ GV/m, with a phase velocity close to the speed of light. Laser plasma accelerators have demonstrated accelerating gradients of 100 GV/m -- several orders of magnitude higher than for conventional structures -- providing hope for reaching new energy regimes. Such large amplitude plasma wakefields can also be driven by intense relativistic particle beams.

Laser-Driven Plasma Wakefield Acceleration (LWFA)

Research in laser plasma acceleration is very active, with many innovative concepts being explored through theory[2-9] and experiment.[10-16] In the "standard" laser wakefield accelerator (LWFA) concept,[1] a single short (<1ps), ultrahigh intensity (>10^{18}

W/cm^2) laser pulse injected into an underdense plasma excites an EPW behind the pulse. The plasma wake is excited by the ponderomotive force created by rapid oscillations of the electromagnetic field. The wakefield amplitude is maximum when the laser pulse length L is approximately equal to the plasma wavelength L=λ_p, where λ_p=$2\pi c/\omega_p$. A correctly placed trailing electron bunch can be accelerated by the axial electric field and focused by the transverse electric field of the plasma wake.

Both 2-D and 3-D LWFA simulations are extremely demanding computationally, due to multiple time and space scales. The multiple scales arise, because the laser radiation field and the transverse electron oscillations evolve on a short time scale -- governed by the laser frequency ω -- with a correspondingly short wavelength, while the longitudinal plasma dynamics and consequent particle acceleration evolve on a much longer time scale -- governed by the electron plasma frequency ω_p -- and longer wavelength. Depending on the density of the plasma, the ratio ω/ω_p can vary from order unity to as high as 100. Thus, simulation codes for these problems must be parallelizable, so they can run on massively parallel processors (MPP), and they must also implement a "moving window" algorithm to follow the laser pulse over distances long compared to the pulse length.

A particle-in-cell (PIC) treatment of laser plasma acceleration[17-20] provides the most detailed simulation of the relevant physics, but is generally constrained to follow the short time scale evolution of the laser pulse, and thus is the most computationally expensive approach. Fluid treatments[21-23] are computationally more efficient, especially models that average over the faster time scales, and are less noisy than PIC, but cannot model the dynamics of accelerated electrons. A third approach[24] uses a PIC treatment of time-averaged equations, along with the use of the quasistatic approximation[25] and sometimes other assumptions. Quasistatic approximations impose the assumption that there is a single forward-propagating laser pulse, thus ruling out certain instabilities, as well as all accelerating concepts involving multiple laser pulses that are incident from various angles.

Beam-Driven Plasma Wakefield Acceleration (PWFA)

Beam-driven plasma wakefield accelerators (PWFA)[26,27] are also capable of providing dramatic accelerating gradients, and thus may lead to a next-generation of smaller, cheaper high-energy accelerators. The acceleration mechanism in the PWFA is analogous to that in the LWFA, only in the PWFA the EPW is excited by the space charge force of the drive bunch, as opposed to the ponderomotive force of the laser pulse. It has been proposed[28] to use the PWFA concept as a means of doubling the beam energy of the Stanford Linear Accelerator Center (SLAC) Linear Collider (SLC) in a distance of only seven meters. This so called "afterburner" would possibly enable detection of the Higgs particle.

A PWFA experiment, referred to as the E-157 experiment,[29-33] aimed at demonstrating accelerating gradients on the order of 1 GeV/m is currently underway at SLAC. In this experiment, a 30 GeV electron bunch is injected into a 1 to 1.5 m long plasma column with density on the order of 2-3 x 10^{14} cm^{-3}. E-157 operates in the "blow-out" regime of the PWFA, meaning the number density of the electron bunch is greater than the plasma density, so that all of the plasma electrons are expelled from the axis in the vicinity of the electron bunch. The EPW generated by the electron bunch is expected to accelerate electrons in the tail of the bunch to higher energies. The plasma afterburner concept is a scaled up version of E-157, which will operate at much higher plasma density, thus requiring a much short duration electron bunch, which will generate an EPW with much stronger longitudinal fields.

In E-157, the laser-ionized lithium plasma density is roughly 10% of the neutral lithium density, $n_0 \sim 2 \times 10^{15}$ cm^{-3}. One proposal for an SLC afterburner[28] requires a plasma density two orders of magnitude larger, corresponding to a neutral lithium density of $n_0 \sim 2 \times 10^{17}$ cm^{-3}. At such high densities, the effects of electron-neutral collisions could modify the physics of the EPW.

The XOOPIC Particle-in-Cell Code

The standard PIC scheme[34] solves the equations representing a coupled system of charged particles and fields. The particles are followed in a continuum space, while the fields are computed on a mesh. First, forces due to the electric and magnetic fields are used to advance the velocities of the particles, and subsequently the velocity is used to advance the position. Particle boundary conditions such as emission and absorption are then applied. If collisions with a neutral background gas are included, the velocities are updated to reflect elastic and inelastic collisions. Next, the particle positions and velocities are used to compute the charge density and current density on the mesh. The charge density and current density provide the source terms for the integration of the field equations (Poisson equation in the electrostatic limit, Maxwell's equations in the electromagnetic limit) on the mesh. The fields resulting from the integration are then interpolated to particle locations to provide the force on the particles.

The XOOPIC (X11-based object oriented particle-in-cell)[35] code started as a pioneering effort to apply object oriented techniques to plasma simulation codes. XOOPIC is written in C++, and includes the XGrafix[36] user interface. Applications have ranged from high pressure discharges to relativistic microwave devices. XOOPIC, along with the rest of the suite of plasma device codes developed at University of California at Berkeley, is in use by over one thousand researchers worldwide (including students), with over 70 journal publications and hundreds of conference publications over the last seven years.

XOOPIC models two spatial dimensions in both Cartesian (x,y) and cylindrical (r,z) geometry, including all three velocity components, with both electrostatic and electromagnetic models available. All three components of both the electric and the magnetic fields are modeled, but there is no spatial variation along the ignored coordinate.

XOOPIC uses the message passing interface (MPI)[37] to take advantage of massively parallel, symmetric multiprocessor and distributed architectures, and has demonstrated linear speed up with 16 processors on the Cray T3E. A 3-D version is now under development.[38] This new code is designed around the C++ architecture of the 2-D XOOPIC code. The architecture is extended in four important areas: the advisor, the particle algorithms, the field algorithms and the boundary conditions.

The code presently supports a non-uniform orthogonal mesh and arbitrary placement of most boundary conditions on that mesh. Static magnetic fields can be added analytically using the equation evaluator, or read from an external file. A number of different charge and current weighting algorithms are available, as well as Poisson and Langdon-Marder divergence corrections for non-conservative current weighting schemes. The code includes a fully relativistic model for inertial particles, as well as a Boltzmann model for inertia-less electrons. Particles and fields can each run on independently subcycled time steps, improving computational efficiency. A temporal filtering scheme reduces high frequency noise, and a spatial digital filtering algorithm reduces short wavelength noise.

XOOPIC also includes volumetric and surface plasma injection, including thermionic and field emission models. Particle statistics can be collected at arbitrary surfaces, and field and particle data can be averaged over arbitrary volumes and surfaces. A Monte Carlo collision (MCC) technique[39] allows multiple background gases at arbitrary partial pressures. The features described are all adjustable from the input file, using MKS (or arbitrary) units for input parameters.

MODIFYING XOOPIC FOR USE IN PLASMA-BASED ACCELERATOR SIMULATIONS

Previous to the work presented here, XOOPIC had never been used for high-energy particle accelerator applications, but had been used extensively to model microwave devices, plasma diodes, plasma display panels, and other low-energy systems, usually in single precision. The authors have modified and enhanced XOOPIC so that it can be used to model high-energy plasma-based accelerators in 2-D Cartesian or cylindrical geometry, in double precision, on the massively parallel Cray T3E. The excellent object-oriented architecture of XOOPIC made it possible to complete this task in a relatively short time.

Development of a Moving Window Algorithm

Plasma-based accelerators are too large to simulate the entire device, and it is only the small region in the vicinity of the particle beam or laser pulse that must be modeled. Because this beam or pulse is moving at the speed of light, it is possible to implement a "moving window" algorithm, such that the simulation follows the small region of interest and ignores the rest of the device.

There are two fundamental approaches to implementing a moving window. One is to move the mathematical mesh along with the particles, and give the background and walls a velocity relative to the mesh. A 3-D moving-window algorithm for cylindrical geometry was implemented in this manner in the ELBA code.[40] The other approach is to keep the mesh stationary with respect to the background, create new particles and fields on the leading edge, shift existing particles and fields to neighboring mesh points, and discard any particles and fields on the trailing edge. The second approach is used for XOOPIC (and also for the OSIRIS code[44]), because it required no modifications to the basic field solve and particle push, and because it eliminates numerous other complications.

For a moving window which is following a group of particles moving to the right, new analytic fields (typically all zero) are introduced into the rightmost row of mesh points, and the fields in the rightmost row of mesh points are copied to the row immediately to the left, and so on. When this shift in the fields takes place, all the particles must also be shifted. At this time, any particles in the leftmost row of cells (for a rightward moving window) are discarded, for they have left the moving window. New particles may be introduced in the rightmost row of cells, if required.

Boundary conditions present no difficulty, if the moving window travels at the speed of light. In the case of a rightward-moving window, disturbances at the leftmost boundary cannot propagate into the moving window, because all electromagnetic waves are constrained to move with a velocity less than or equal to the speed of light. Similarly, incoming fields on the right hand side are not affected by the contents of the moving window to the left, so fields here may be safely specified analytically in a simple way.

Combining parallel operation and the moving window leads to some additional complication. Whenever a shift in fields takes place (usually every time step, or every few time steps), the shifted fields and particles must be passed to the downstream computational region. The moving window in XOOPIC uses MPI to pass particles and fields across computational boundaries, showing linear scaling up to 16 processors on the Cray T3E.

Adding a New Electromagnetic Pulse Launcher

We designed and implemented within XOOPIC the ability to launch a linearly polarized electromagnetic pulse, with a Gaussian time profile and a Gaussian spatial profile along one transverse dimension. The other transverse dimension is along the ignored coordinate, so there is no variation in that direction. The peak intensity, the wavelength and the pulse length can all be specified from the input file, as can the initial divergence or convergence of the pulse.

Implementation within the code was fairly straightforward. We created a new derived class, PortGauss, which inherits from the previously existing boundary class, Port. Due to the benefits of dynamic polymorphism, a PortGauss instance (or object) can be used anywhere in the code as a substitute for the old Port object.

Generalization of the Particle Beam Emitters

The beam emission boundary conditions in XOOPIC have been extended to handle more general cases. Spatial dependence has been added to both the BeamEmitter algorithm, which emits particles of a specified computational particle weight as well as the VarWeightBeamEmitter algorithm, which emits particles of variable weights. Particle weight is defined as the ratio of the charge of a computational particle to that of a physical particle, $w=q_c/q_p$. The particle weight in the VarWeightBeamEmitter has also been generalized to have both spatial and temporal dependence, and the weight can be adjusted automatically to emit a fixed number of particles per time step.

The previously existing emitter models in XOOPIC emitted a specified time-dependent current, I(t), which could be specified from the input file. Only uniform current density was possible. Furthermore, the VarWeightBeamEmitter only allowed for variation of the weight of particles based on the radial origin of emission for particles, $w(r)=w_{max}r/r_{max}$.

LASER-DRIVEN PLASMA ACCELERATOR SIMULATIONS

We present two simulations of the standard LWFA, one driven by a low (5.5×10^{16} W/cm^2) and the other a high (3×10^{18} W/cm^2) peak intensity laser pulse, both in slab geometry. These simulations have relevance to ongoing LWFA experiments at the l'OASIS laboratory of Lawrence Berkeley National Laboratory.[41,42] To understand the detailed particle trapping mechanisms in these experiments, PIC simulations will be performed with parameters similar to the examples shown in this section.

These results demonstrate the capabilities of XOOPIC. Previously, XOOPIC has been used to model the effects of colliding laser pulses.[43]

Modeling the Wakefield Generated by a Low Intensity Laser Pulse

We first consider the plasma wakefield generated by a low intensity laser pulse. The electron plasma density is $n_e=3\times10^{19}$ cm^{-3}, which corresponds to an EPW wavelength of $\lambda_p = 6$ µm $= 6.2$ c/ω_p and a plasma frequency of $\omega_p=3.1\times10^{14}$ rad/s.

The laser pulse is linearly polarized, with a transverse Gaussian profile. The minimum laser spot size is 5 µm = 5.2 c/ω_p, and the Rayleigh length is $\lambda_R = 97$ µm = 100 $c/\omega_p = 16$ λ_p. In order to maximize the EPW amplitude, the laser pulse length is chosen to be of order λ_p, with a full width at half maximum $\tau_{fwhm} = 6.7$ fs $= 2$ µm $= \lambda_p/3$. The peak laser intensity is $I_L=5.5\times10^{16}$ W/cm^2, corresponding to a dimensionless amplitude $a_0=0.2$, and the laser wavelength is $\lambda=1$ µm = 1 c/ω_p.

Figure 1 shows a surface plot of the longitudinal electric field E_x over the mesh. The length of the simulation region is $L_x = 30$ µm $= 31$ c/ω_p in the x (longitudinal) direction and $L_y = 50$ µm $= 52$ c/ω_p in the y (transverse) direction. The simulation uses 7 macro-particles per cell to represent the plasma, and the initial plasma is cold.

The plasma wake can be clearly seen behind the laser pulse. The plasma wake is linear, with a peak gradient of $E_x \sim 5.5$ GV/m ~ 0.01 E_0. The peak field of the EPW is significantly smaller than the peak longitudinal field of the laser pulse.

The Wakefield of a High-Intensity Laser Pulse

We now consider the plasma wakefield generated by a high intensity laser pulse. All the physical and simulation parameters are the same as for the low intensity pulse of the previous subsection, but the peak intensity is now $I_L=3\times10^{18}$ W/cm^2, corresponding to a dimensionless amplitude $a_0=1.5$.

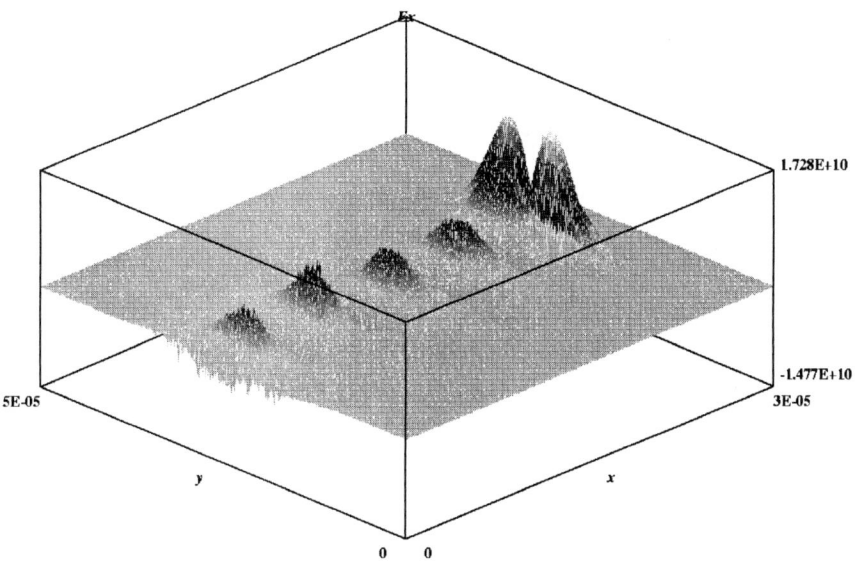

FIGURE 1. Surface plot of the longitudinal electric field generated by the 5.5×10^{16} W/cm^2 ($a_0=0.2$) laser pulse (large peaks to the right) and the resulting plasma wake (smaller peaks, left and center).

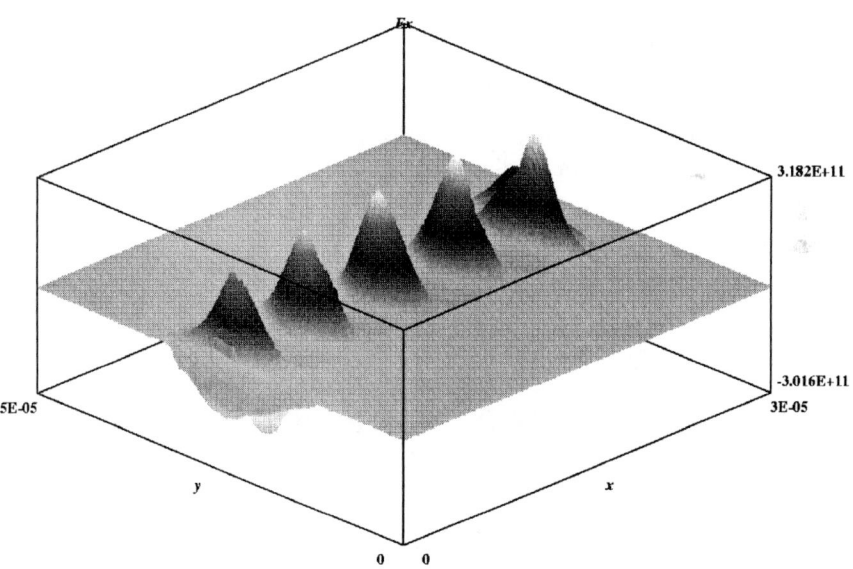

FIGURE 2. Surface plot of the longitudinal electric field generated by the 3×10^{18} W/cm^2 ($a_0=1.5$) laser pulse (smaller, partially hidden peaks to the far right) and the resulting plasma wake (larger peaks).

Figure 2 shows a surface plot of the longitudinal electric field E_x over the mesh for the high-intensity case. The EPW can be clearly seen behind the laser pulse, but in this case the wake is nonlinear, and has a peak gradient of $E_x \sim 300$ GV/m $\sim 0.56\ E_0$. The wake amplitude is found to increase linearly with the peak laser intensity (as the square of the dimensionless amplitude) when $I_L < 3 \times 10^{18}$ W/cm^2, in agreement with theory.

BEAM-DRIVEN PLASMA ACCELERATOR SIMULATIONS

Here we present some XOOPIC simulations of plasma wakefield acceleration. Our simulations of the E-157 PWFA experiment at SLAC show good agreement with results obtained previously using the OSIRIS[44] code. We also discuss the important issue of electron-neutral collisions.

Modeling the SLAC E-157 Experiment

We have modeled E-157 with XOOPIC and found agreement with previous work.[31,32] The simulation region, in 2-D cylindrical geometry, is 0.9 mm in r by 5.4 mm in z, with the corresponding number of grid points $n_r=32$ and $n_z=192$, for a total of 6144 cells. With 4 macro-particles per cell representing the plasma electrons, there are 24,576 plasma particles. The 30 GeV electron beam is represented by 9 macro-particles per cell, and the beam covers 8 by 64 grids (initially) for 4608 beam particles. The grid size is dz=dr=28 μm. The time step, chosen to satisfy the Courant condition, is dt=.5*dz/c=4.69x10^{-14} s. Thus, it requires 71,400 time steps to propagate the beam through the 1 m lithium plasma.

The plasma density is taken to be 2.1×10^{14} cm^{-3}, which implies an electron plasma frequency of $\omega_p = 8.2 \times 10^{11}$ rad/s. Thus, $\omega_p*dt=0.04$ and the electron plasma frequency is being resolved, which is required for stability in a time-explicit PIC code. The lithium plasma is assumed to be cold, but very little numerical heating is observed, because the moving window algorithm "sweeps" the electrons through at the speed of light.

Figure 3 shows the initial 30 GeV beam in cylindrical coordinates, with r on the vertical axis and z on the horizontal axis, and dimensions in m. Figure 4 shows the plasma wake. The crossing of particle trajectories in the wake indicates highly non-laminar flow, which cannot be modeled with a fluid code. The structure of the wake is independent of the beam radius. Figure 5 shows the accelerating field generated by the wake in V/m. With higher resolution, the peak field on axis is greater than 1 GV/m.

The peak accelerating field overlaps the tail of the beam. Figure 6 shows the resulting acceleration of beam particles after 1m of propagation through the lithium plasma. The vertical axis is $p_z=\gamma v_z$ in units of m/s. These results agree well with previously published work.[31,32]

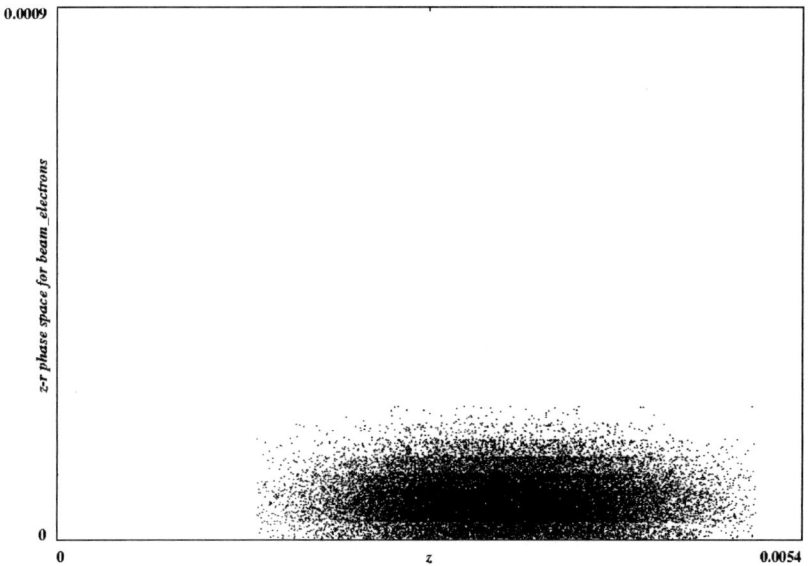

FIGURE 3. Initial distribution of the 30 GeV beam.

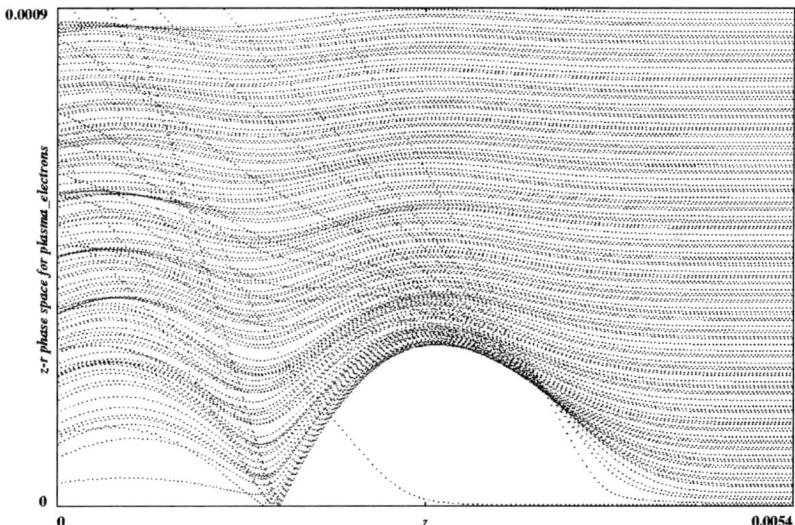

FIGURE 4. Plasma wake.

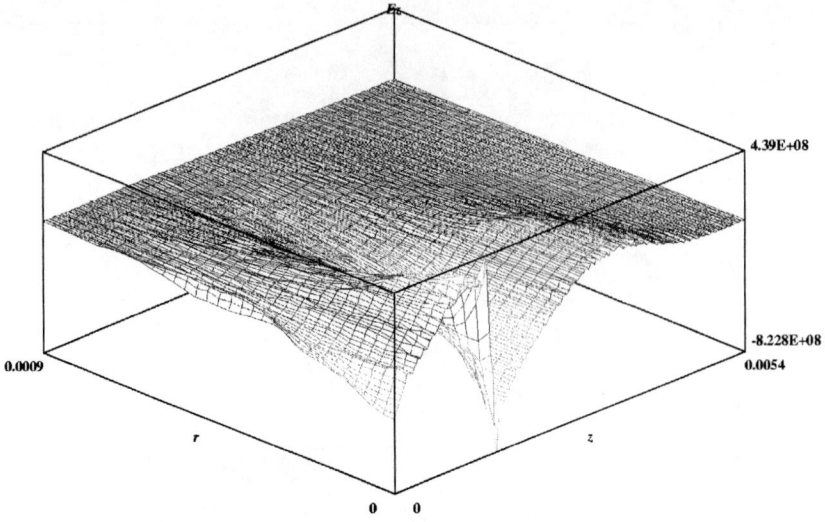

FIGURE 5. Longitudinal electric field.

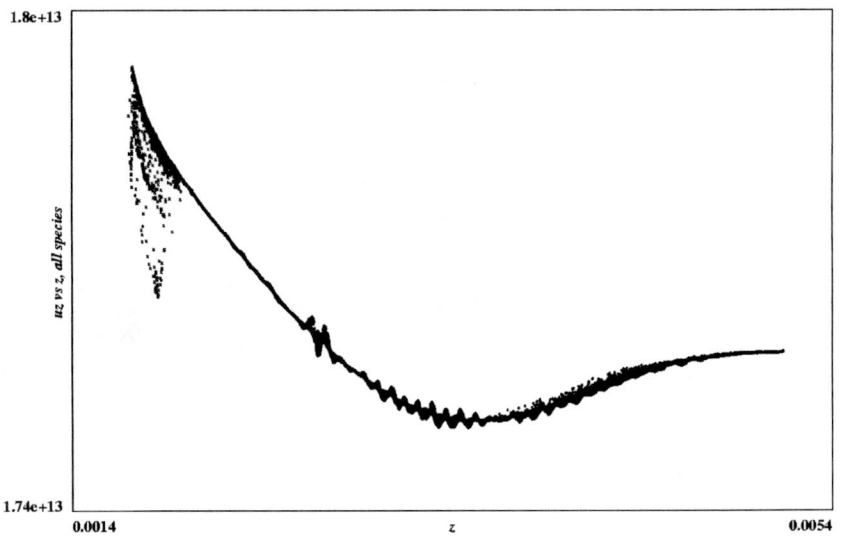

FIGURE 6. z-P_z phase space of beam at 1 m.

Modeling Electron-Neutral Collisions

XOOPIC uses the null collision method[39] for MCC treatment of electron-impact excitation and ionization and for electron-neutral elastic scattering. MCC models for Ar, Ne, He and H have been used for some time, and we recently added an ionization model for Li, using cross sections from the literature.[45] However, the cross section and scattering models assume the impact energy is nonrelativistic.

The bulk of the plasma electrons in the wake are nonrelativistic for E-157, but a significant fraction are not. Modeling collisional effects involving the drive beam must, of course, be fully relativistic. Electron-neutral collision cross sections $\sigma(E)$ fall from their maximum like $ln(E)/E$ for impact energies E<200KeV.[46,47] Relativistic effects break this scaling, leading to a minimum[48,49] in $\sigma(E)$ for E~1 MeV, followed by logarithmic growth, which eventually saturates at a density dependent energy (the Fermi plateau).[49-52]

A simple fitting function for impact ionization cross sections has been developed,[53] using the ionization energy and two adjustable parameters, approximately capturing both low-energy and relativistic behavior (but not the Fermi plateau). The fitting parameters are determined largely by data for impact energies near 1 MeV, which has been published for a number of gasses[48] (although not lithium). Fitting functions have also been developed for the energy distribution of the secondary electrons,[54-57] and there is some applicable theory in certain regimes.[58-60] Several works discuss the angular distribution for elastic and inelastic scattering in relativistic and nonrelativistic regimes.[50,56,57,60-62] We are presently compiling new and comprehensive parametric models for impact ionization and elastic scattering for a wide range of energies, to be published elsewhere.

Figure 7 shows parametric fits to electron-impact cross-sections for lithium, for an energy range of 0.1 eV < E < 10 GeV. The solid line is for ionization of neutral lithium (ejecting the outer electron from the 2s shell). The dashed line is for elastic scattering. The fitting parameters for ionization were determined from Ref. 45 (low energy) and Ref. 63 (high energy). The fitting parameters for elastic scattering were determined from Ref. 64 (low energy) and Ref. 63 (high energy). We note that the cross section for elastic scattering is orders of magnitude larger than the ionization cross section at low energies.

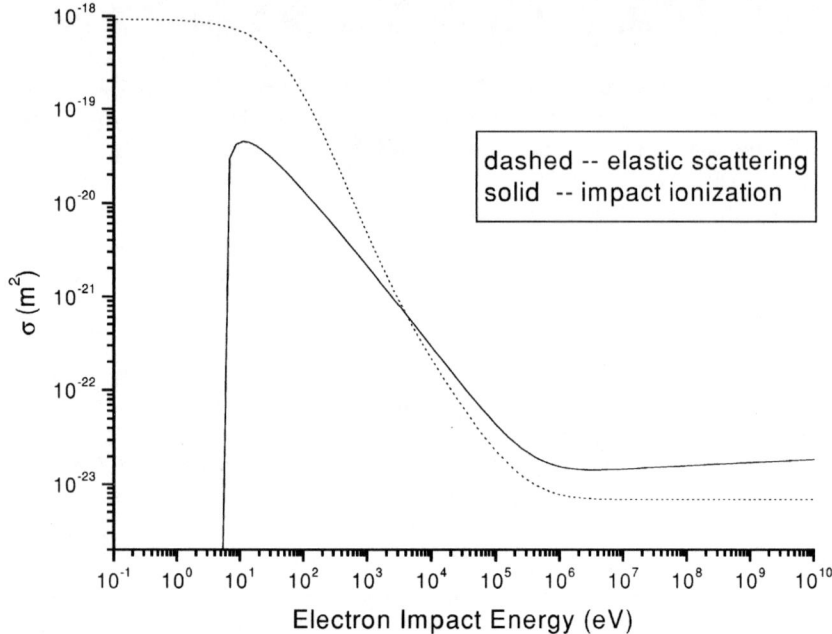

FIGURE 7. Total cross sections (in m^2) for electron collisions with neutral lithium, as a function of impact energy (1 m^2 = 10^4 cm^2 = 10^{28} barns). The dashed line is for elastic scattering and the solid line is for impact ionization. These lines are parametric fits to previously published results.

We have conducted some simulations using a nonrelativistic impact ionization model for lithium.[65] The neutral lithium background density is assumed to be ten times the plasma density, which corresponds to 2×10^{15} cm^{-3} for E-157. Impact ionization is found to be negligible for E-157. We also modeled the case where the lithium density is larger by a factor of 100 than the density for E-157 (parameters similar to those for the recently proposed afterburner concept[28]). Simulations indicate that impact ionization is not completely negligible for this higher density, but that the plasma wake and corresponding accelerating fields are not significantly modified. However, the electron-neutral scattering cross section is much larger, and this scattering might disrupt the wake, even for neutral lithium densities of order 10^{17} cm^{-3}. A detailed study of these effects is the subject of a future publication.

SUMMARY

The code XOOPIC has been modified to enable simulations of plasma based accelerators on massively parallel platforms. Modifications include the development of a moving window algorithm, adding a new electromagnetic pulse launcher, generalization of the particle beam emitters, and further optimization to allow efficient use on parallel platforms. As examples of the utility of XOOPIC, simulations of the standard

LWFA with both low and high intensity laser pulses were performed, and the results were in agreement with the theoretically predicted wake amplitudes. In addition, simulations of the PWFA were preformed. Simulations for the parameters of the E-157 experiment were found to be in agreement with previous studies. Simulations using the nonrelativistic impact ionization model for lithium indicated insignificant effects for the parameters of E-157. However, electron-neutral scattering could possibly alter the wake for neutral lithium densities of order 10^{17} cm^{-3} or higher.

ACKNOWLEDGMENTS

The authors gratefully acknowledge many helpful and interesting conversations with C.K. Birdsall, P. Chen, R. Hemker, R. Hubbard, C. Joshi, T. Katsouleas, W. Mori, J. Ng and F. Tsung. This work is supported by the U.S. Department of Energy, under Contract No.'s DE-FG03-99ER82903, DE-FG03-95ER40926 and DE-AC03-76SF00098, and by Tech-X Corporation. This research used resources of the National Energy Research Scientific Computing Center, which is supported by the Office of Science of the U.S. Department of Energy under Contract No. DE-AC03-76SF00098.

REFERENCES

1. Esarey, E., et al., IEEE Trans. Plasma Science **24**, 252 (1996).
2. Umstadter, D., et al., Phys. Rev. Lett. **76**, 2073 (1996).
3. Hemker, R.G., et al., Phys. Rev. Lett. **57**, 5920 (1998).
4. Esarey, E., Hubbard, R.F., Leemans, W.P., et al., Phys. Rev. Lett. **79**, 2682 (1997).
5. Schroeder, C.B., et al., Phys. Rev. E **59**, 6037 (1999).
6. Esarey, E., Schroeder, C.B., Leemans, W.P., and Hafizi, B., Phys. Plasmas **6**, 2262 (1999).
7. Moore, C.I., et al., Phys. Rev. Lett. **82**, 1688 (1999).
8. Duda, B.J., Hemker, R.G., Tzeng, K.C., and Mori, W.B., Phys. Rev. Lett. **83**, 1978 (1999).
9. Andreev, N.E., et al., Phys. Rev. Special Topics -- Accelerators and Beams **3**, 021301 (2000).
10. Modena, A., Najmudin, Z., Dangor, A.E., *et al.*, Nature **377**, 606 (1995).
11. Nakajima, K., Fisher, D., Kawakubo, T., *et al.*, Phys. Rev. Lett. **74**, 4428 (1995).
12. Coverdale, C., Darrow, C.B., Decker, C.D., *et al.*, Phys. Rev. Lett. **74**, 4659 (1995).
13. Wagner, R., Chen, S.Y., Maksimchuk, A., and Umstadter, D., Phys. Rev. Lett. **78**, 3125 (1997).
14. Ting, A., Moore, C.I., Krushelnick, K., *et al.*, Phys. Plasmas **4**, 1889 (1997).
15. Gordon, D., Tzeng, K.C., Clayton, C.E., *et al.*, Phys. Rev. Lett. **80**, 2133 (1998).
16. Volfbeyn, P. , Esarey, E., and Leemans, W.P., Phys. Plasmas **6**, 2269 (1999).
17. Decker, C.D., Mori, W.B., and Katsouleas, T., Phys. Rev. E **50**, R3338 (1994).
18. Decker, C.D., Mori, W.B., Tzeng, K.C., and Katsouleas, T., Phys. Plasmas **3**, 2047 (1996).
19. Bulanov, S.V., et al., Phys. Fluids B **4**, 1935 (1992).
20. Bulanov, S.V., Pegoraro, F., and Pukhov, A.M., Phys. Rev. Lett. **74**, 710 (1995).
21. Andreev, N.E., et al., JETP Letters **55**, 571 (1992).
22. Krall, J., Ting, A., Esarey, E., and Sprangle, P., Phys. Rev. E **48**, 2157 (1993).
23. Antonsen, T.M., and Mora, P., Phys. Fluids B **5**, 1440 (1993).
24. Mora, P., and Antonsen, T.M., Phys. Plasmas **4**, 217 (1997).
25. Sprangle, P., Esarey, E., and Ting, A., Phys. Rev. Lett. **64**, 2011 (1990).
26. Chen, P., Dawson, J., Huff, R. and Katsouleas, T., Phys. Rev. Lett. **54**, 693 (1985).
27. Ruth, R., Chao, A., Morton, P., and Wilson, P., Part. Accel. **17**, 171 (1985).

28. Katsouleas, T., *et al.*, these proceedings.
29. Assmann, R. *et al.*, in *Proc. 1999 Particle Accelerator Conference* (IEEE, NY, 1999), p. 130.
30. Muggli, P., et al., in *Proc. 1999 Particle Accelerator Conference* (IEEE, NY, 1999), pp. 3651-3653.
31. Lee, S., et al., Phys. Rev. E **61**, 7014 (2000)
32. Hogan, M.J., et al., Phys. Plasmas **7**, 2241-2248 (2000).
33. SLAC E-157 web site, URL http://www.slac.stanford.edu/grp/arb/e157
34. Birdsall, C.K., and Langdon, A.B., *Plasma Physics via Computer Simulation*, New York: McGraw-Hill, 1985.
35. Verboncoeur, J.P., Langdon, A.B., and Gladd, N.T., Comp. Phys. Comm. **87**, 199-211 (1995).
36. Vahedi, V., Verboncoeur, J.P., and Birdsall, C.K., "XGrafix: an X-Windows Environment for Real-Time Interactive Simulations," Proc. 14th Conf. on the Numerical Simulation of Plasmas, 1991.
37. The MPI home page at URL http://www.mcs.anl.gov/mpi .
38. Mardahl, P., and Verboncoeur, J.P., IEEE Intl. Conf. Plasma Science, (Monterey, CA, June 1999).
39. Vahedi, V., and Surendra, M., Computer Phys. Comm. **87**, 179-198 (1995).
40. Joyce, G., Krall, J., and Slinker, S., Laser and Particle Beams **12**, 273-282 (1994).
41. Leemans, W.P., *et al.*, Phys. Plasmas **5**, 1615 (1998).
42. Leemans, W.P., *et al.*, these proceedings.
43. Giacone, R.E., Cary, J.R., Bruhwiler, D.L., Mardahl, P., and Verboncoeur, J.P., *Proc. Seventh European Particle Accelerator Conference*, (Vienna, June, 2000), pp. 907-909.
44. Hemker, R.G., et al., in *Proc. 1999 Particle Accelerator Conf.* (IEEE, NY, 1999), pp. 3672-3674.
45. Younger, S.M., Journal of Quantum Spectroscopy and Radiative Transfer **26**, 329 (1981).
46. Younger, S.M., and Märk, T.D., in *Electron Impact Ionization*, edited by T.D. Märk and G.H. Dunn, Vienna: Springer-Verlag, 1985, pp. 24-41.
47. Brown, S.C., *Basic Data of Plasma Physics, The Fundamental Data on Electrical Discharges in Gases*, New York: American Institute of Physics, 1994.
48. Rieke, F.F., and Prepejchal, W., Phys. Rev. A **6**, 1507 (1972).
49. Jackson, J.D., *Classical Electrodynamics*, 2nd Edition, New York: Wiley, 1975, pp. 618-653.
50. Fermi, E., Phys. Rev. **57**, 485-493, 1940.
51. Rossi, B., *High-Energy Particles*, New York: Prentice Hall, 1952, pp. 27-29.
52. Cobb, J.H., Allison, W.W.M., and Bunch, J.N., Nucl. Instrum. and Meth. **133**, 315-323 (1976).
53. Reiser, M., *Theory and Design of Charged Particle Beams*, New York: Wiley, 1994, pp. 273-278.
54. Opal, C.B., Peterson, W.K., and Beatty, E.C., J. Chem. Phys. **55**, 4100 (1971).
55. Slinker, S.P., Taylor, R.D., and Ali, A.W., J. Appl. Phys. **63**, 1 (1988).
56. Rudd, M.E., Phys. Rev. A **44**, 1644 (1991).
57. Rudd, M.E., *et al.*, Phys. Rev. A **47**, 1866 (1993).
58. Moeller, C., Annalen der Physik 14, 531-585 (1932).
59. Bethe, H.A., and Ashkin, J., in *Experimental Nuclear Physics*, Vol. 1, edited by E. Segrè, New York: John Wiley & Sons, 1953, pp. 166-357.
60. Kalinovskii, A.N., Mokhov, N.V., and Nikitin, Y.P., *Passage of High-Energy Particles through Matter*, New York: American Institute of Physics, 1989, pp. 29-54.
61. Mott, N.F., and Massey, H.S.W., *The Theory of Atomic Collisions*, 2nd Edition, Oxford: Oxford University Press, 1949.
62. Goldstein, H., *Classical Mechanics*, 2nd Edition, Reading: Addison-Wesley, 1980, pp. 309-320.
63. Perkins, S.T., Cullen, D.E., and Seltzer, S.M., Lawrence Livermore National Laboratory report UCRL-50400, Vol. **31** (1991).
64. Bray, I., Fursa, D.V., and McCarthy, I.E., Phys. Rev. A **47**, 1101-1110 (1993).
65. Bruhwiler, D.L., Cary, J.R., Verboncoeur, J.P., Mardahl, P., and Giacone, R., *Proc. Seventh European Particle Accelerator Conference*, (Vienna, June, 2000), pp. 877-879.

Transformer Ratio Enhancement Using A Ramped Bunch Train In A Collinear Wakefield Accelerator

J.G. Power, W. Gai, A. Kanareykin[†]

High Energy Physics Division, Argonne National Laboratory, Argonne, IL 60439, USA
[†]*St. Petersburg Electrical Engineering University, 5 Prof. Popov St., St. Petersburg 197376, Russia*

Abstract. We present a practical method for achieving a transformer ratio (R) greater than 2 with any collinear wakefield accelerator – i.e. with either plasma or structure based wakefield accelerators. It is known that the transformer ratio cannot generally be greater than 2 for a symmetric drive bunch in a collinear wakefield accelerator. However, using a ramped bunch train (RBT) where a train of n electron drive bunches, with increasing ('ramping') charge, one can achieve $R = 2n$ after the bunch train. We believe this method is feasible from an engineering standpoint using existing technology and an experiment to be preformed at the Argonne Wakefield Accelerator (AWA) is planned.

INTRODUCTION

In general, the wakefield theorem [1] restricts the maximum accelerating field behind the drive bunch in a wakefield accelerator to be less than twice the maximum retarding field inside the drive bunch thus limiting the efficiency which can be obtained. One of the concepts central to the physics of wakefield acceleration is the transformer ratio, R, defined as R = (Maximum energy gain behind the bunch)/(Maximum energy loss inside the drive bunch). For the case of a collinear drive and witness beam geometry device, R is less than 2 except in a few special cases. For the purposes of this paper we only consider one regime where the wakefield theorem does not apply, namely, the use of an axially asymmetric charge distribution in the drive bunch. We can understand the RBT method by invoking linear superposition and expansion of $W(z)$, into the normal modes of the accelerator - i.e. $W(z) = \sum_l W_l \cos(k_l z)$

Several schemes have been proposed to obtain $R > 2$ in collinear wakefield accelerators, but no experimental results have been obtained due to the inherent difficulties of these experiments. One of the more promising schemes [2] sends a single drive bunch, with an asymmetric axial current distribution (Fig. 1a) through a collinear wakefield accelerator. Simulations show that R can be much greater than 2 for the triangular (ramped) bunch distribution as seen in the figure. Notice that most of the particles in the drive bunch experience the same decelerating wakefield, W^-, but the accelerating wakefield behind the bunch, W^+, is much larger. Using a similar idea, a second scheme tailors the profile of a train of drive bunches [3] into a triangular ramp (see dotted line in Fig. 1b) to produce $R > 2$. In this later scheme, the individual bunches in the train are symmetric (e.g. gaussian) separated by a distance d. The charge is then ramped up such that the first bunch in the train has the lowest charge and the last bunch the highest. From the figure we see that all four drive bunches in the drive train experience the same maximum decelerating field W^- just like in the case of a single triangular ramped bunch. Thus, the fundamental condition for both of these schemes is that the trailing particles (bunches) in the drive bunch (train) are positioned in the accelerating phase of the leading particles (bunches) so that all the driving particles experience the same maximum decelerating field.

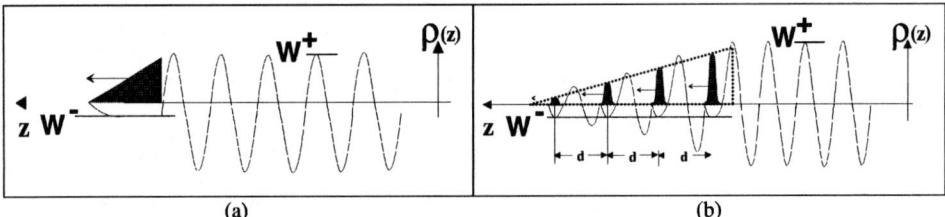

FIGURE 1. Two schemes that have been proposed to generate $R = W^+/W^- \gg 2$. The height of the shaded area, $\rho(z)$, represents the total amount of charge in the bunch at location z while the solid, sine-like, line is the amplitude of the wakefield driven by the beam. (a) A single drive bunch with a triangular axial current distribution moving to the left. (b) A train of gaussian drive bunches with an overall triangular pattern of the train (see dotted line) moving to the left.

The difficulty with the schemes that propose to use asymmetric axial current distribution to achieve $R > 2$, arises from the lack of suitable techniques to tailor the axial distribution of the drive beam. For example, for nearly 10 years it has been known that beam dynamics codes (such as PARMELA) predict improved beam quality from *rf* photoinjectors when driven by a 'flat top' axial current distribution. Despite this knowledge, no one has successfully produced a 'flat top' although several experimenters are getting close. Since the 'flat top' pulse is most likely easier to generate than the 'triangular top' pulse it may be some time before this method is used to obtain $R > 2$.

In this paper we consider the later method, here termed, the 'ramped' bunch train (RBT) method of transformer ratio enhancement. Since the AWA facility has already generated a 'flat' bunch train [4] it should be easy to generate a RBT. How the RBT is generated and the difficulties we are likely to encounter will be discussed in a later section.

We begin by reviewing the concept of transformer ratio for a single, symmetric beam in a collinear wakefield accelerator and studying the trade off between the acceleration gradient and the transformer ratio. We then examine the ramped bunch train method and present an algorithm for choosing the spacing of the bunch and charge of the different bunches in the train. Finally, we describe a proof of principle experiment where we propose to send a train of 4 electron bunches, with a ramped charge distribution through a dielectric lined waveguide.

TRANSFORMER RATIO AND ACCELERATION GRADIENT FOR A SINGLE BUNCH

In this section we examine the dependence of the transformer ratio (R_0) and the peak acceleration gradient (W^+) of a single bunch - where the subscript '0' denotes a single bunch. As will be seen in the next section, the transformer ratio enhancement of the ramped bunch train is maximum when $R_0 = 2$.

Numerical simulations of a single drive bunch passing through a collinear wakefield accelerator are presented. The particular collinear accelerator we use for our simulations is a dielectric lined cylindrical structure with inner radius a, outer radius b and dielectric constant ε as shown in Fig. 2. Although we use a dielectric wakefield accelerator (DWFA) for analysis the results we derive in this section are general to dielectric structures, metallic structures, and plasmas. In the last section of this paper we propose an experiment for using a RBT to drive both a DWFA and a plasma wakefield accelerator (PWFA).

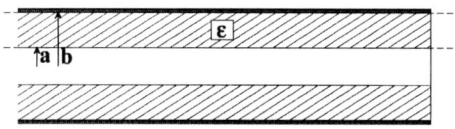

FIGURE 2. The dielectric wakefield accelerator. A hollow dielectric (ε) cylinder of inner radius a and outer radius b covered by a copper jacket. The drive bunch passes through the vacuum hole of radius a.

Using the analytic theory of Rosing and Gai [5] we numerically simulate the longitudinal wakefield in a dielectric structure excited by an axial current distribution. We consider a dielectric structure of inner radius $a = 2.5$ mm, $b = 2.8$ mm, dielectric constant $\varepsilon = 38$ excited by a single drive bunch with a gaussian axial current distribution of width $\sigma = 1.5$ mm. From the simulation results (plotted in Fig. 3) we obtain the wavelength of the fundamental accelerating mode ($\lambda_0 = 7.89$ mm) the maximum value of the accelerating field behind the bunch ($W^+ = 0.54$ MeV/m/nC) and the minimum value of the decelerating field within the bunch ($W^- = 0.28$ MeV/m/nC). From this we calculate $R_0 = W^+/W^- = 1.9$ for a normalized bunch length of $\sigma/\lambda = 0.19$.

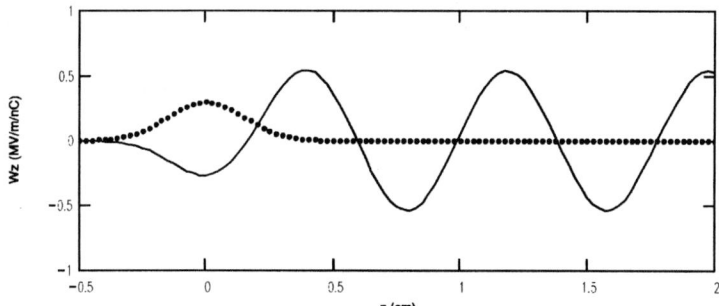

FIGURE 3. Longitudinal wakefield (solid line) excited by a gaussian bunch (dotted line) with $Q = 1$ nC and $\sigma = 1.5$ mm. The fundamental wavelength $\lambda_0 = 7.89$ mm.

In general, the transformer ratio, R_0, and the peak acceleration gradient behind the bunch, W^+, are functions of the normalized bunch length (σ/λ). Keeping the dielectric structure fixed, thus fixing λ, we can vary σ in our simulations. We now plot the dependence of R_0 and normalized W^+ as a function of the normalized bunch length, σ/λ, in Fig. 4. (W^+ has been normalized to W^+,max - the maximum peak acceleration gradient for a given bunch length σ.)

As one would expect, the simulations (plotted in Fig. 4) show that the peak acceleration gradient W^+ increases as the bunch length decreases. A detailed study shows that the maximum value of the peak acceleration gradient, W^+, max is obtained for values of $\sigma = \lambda/100$ and below. This means that once the bunch length is below $\lambda/100$ no appreciable increases to W^+ occur.

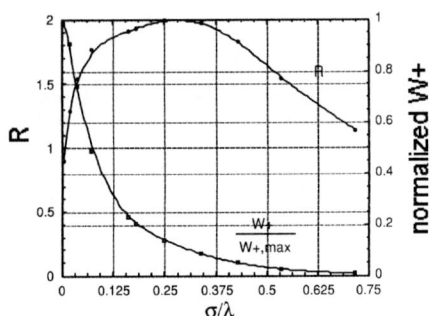

FIGURE 4. Transformer ratio, R, and peak accelerating field, W^+ as a function of the normalized bunch length.

The behavior of R_0 as a function of σ/λ is more complicated than the monotonically decreasing behavior of W^+. From Fig. 4 we see that R_0 is peaked near bunch length $\lambda/4$, while dropping off towards one for bunch lengths either side of $\lambda/4$. (As an aside, it is well known that the transformer ratio is equal to 2 for a delta function even though the asymptotic value for Fig. 4 is 0.75. This isn't surprising since the function R is simply discontinuous for zero bunch length - i.e. a delta function.)

We now see that there is a tradeoff between R_0 and gradient W^+. Maximum gradient is obtained by driving the collinear accelerator with a very short bunch while maximum transformer ratio is obtained when $\sigma=\lambda/4$. The optimal point of operation is arguably near $\lambda/20$, but since we are interested in maximizing the transformer ratio, we choose σ near $\lambda/4$, so that $R_0 = 2$.

ENHANCED TRANSFOMER RATIO WITH A RAMPED BUNCH TRAIN

In this section we describe a method [3] for enhancing R beyond 2 in a collinear wakefield accelerator. We call this the ramped bunch train (RBT) method and show that it works by simple linear superposition of the fields from a train of drive bunches using a clever arrangement of drive bunch spacing and charge. In addition to the work of [3] we add a new condition for enhancing R - the symmetric, single wake.

Flat Bunch Train Wakefields

In the typical 'flat' bunch train experiments, such as those at CLIC and AWA a train of equal intensity bunches ($Q_0 = Q_1 = ...$), separated by one fundamental wavelength ($d = \lambda_0$) is used to drive the collinear wakefield accelerator. This separation of $d = \lambda_0$ means that the second bunch is placed in the deceleration phase of the first bunch so that the self-wake of the second bunch and the decelerating field left behind by the first bunch reinforce each other. Consider the example of a two bunch, 'flat' bunch train. Using the single bunch wakefield (charge $Q_0 = 1$ nC) of Fig. 3, we calculate the net wakefield generated by both the single bunch and a second bunch, of equal charge ($Q_1 = 1$ nC) located at $d = \lambda_0 = 7.89$ mm. Since the single (first, $n=0$) bunch of the above example had $W_0^+ = 0.54$ MeV/m/nC and $W_0^- = 0.28$ MeV/m/nC, by linear superposition, we obviously have the peak 'net' wakefield left behind the second ($n=1$) bunch, $W_{1,\,net}^+ = W_0^+ + W_1^+ = 2W_0^+ = 1.08$ MeV/m/nC and 'net' minimum decelerating field within the bunch train, $W_{1,\,net}^- = W_0^- + W_1^- = 3W_0^- = 0.84$ MeV/m/nC, located within the second bunch. Thus our peak acceleration gradient has doubled, but our minimum decelerating field within bunch has tripled resulting in a 2/3 reduction in the transformer ratio to 1.3. In general, for a flat bunch train, the net fields after the nth bunch are $W_{n,\,net}^+ = (n+1)W_0^+$ and $W_{n,\,net}^- = W_0^- + nW_0^+ = $ and thus $R_n = (n+1)/(n+1/R)$ which goes to $R_n = 1$ for large n.

Ramped Bunch Train Wakefields

In the RBT method the second bunch is placed in the accelerating phase of the first bunch (Fig. 5) or at distance $d = 1.5\lambda_0$. The effect of this is to cancel out some of the self-wakefield of the second bunch with the accelerating field of the first bunch so that the second bunch experiences a reduced deceleration. With the second bunch located at the accelerating phase of the first bunch, the charge of the second bunch (Q_1) is increased until the decelerating field it experiences is equal to that of the first bunch (i.e. W_0^-). For example, let the first ($n=0$) bunch have charge Q_0 and produce wakefields W_0^- and W_0^+ such that $W_0^+ = R_0 W_0^-$. If the second ($n=1$) bunch is located at $d = 1.5\lambda_0$ and has charge $Q_1 = kQ_0$, then the net decelerating field it experiences ($W_{1,\,net}^-$) is the sum of the accelerating field of the first bunch (W_0^+) minus its self-wake ($W_1^- = kW_0^-$) or $W_{1,\,net}^- = (R_0-k)W_0^-$. Now, if $W_{1,\,net}^-$ is to be equal to W_0^- then we must have $(R_0-k) W_0^- = -W_0^-$. or $k = R_0 + 1$. If $R_0 = 2$ then $k=3$ and we have $Q_1 = 3Q_0$ and $W_1^+ = 3W_0^+$. To calculate the 'net' peak accelerating field left behind the second bunch we simply subtract the wakefield produced by the first bunch (W_0^+) of charge Q_0 and from the wakefield produced by the second bunch ($W_1^+ = 3W_0^+$) of charge $3Q_0$ and we have $W_{1,\,net}^+ = 3W_0^+ - W_0^+ = 2W_0^+$. Finally, we can *calculate* $R_1 = W_{1,\,net}^+ / W_{1,\,net}^- = 2W_0^+ / W_0^- = 2 R_0$.

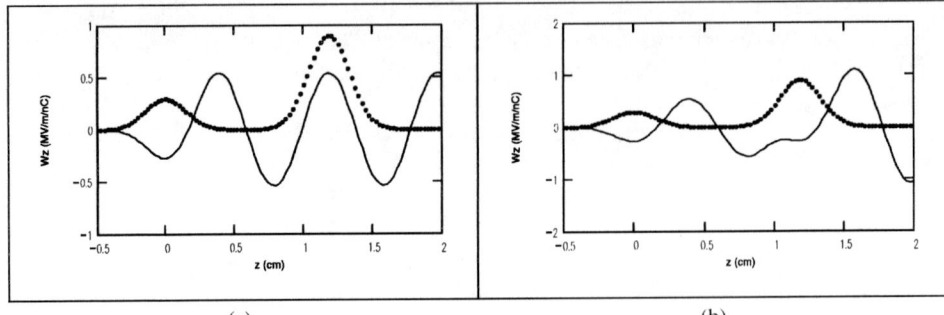

FIGURE 5. A 'two-bunch' train. (a) The location and relative magnitude of the 2nd bunch (dotted line) is shown relative to the location of the first bunch (dotted line) and its wake (solid line). (b) The same as in (a), but the contribution to the wakefield from the second bunch is included.

RBT Algorithm For Transformer Ratio Enhancement

The previous analysis is generalized to a train of N drive bunches. Given the transformer ratio after the first ($n=0$) bunch (R_0) then the maximum transformer ratio that can be achieved after the nth bunch (R_n) is,

$$R_n = (n+1)R_0 \qquad (n = 0,1,2,...N-1) \qquad (1)$$

This maximum enhancement of the transformer ratio can only be achieved if the following conditions are satisfied. The separation between the bunches (see 'd' in Fig. 1b) is $1/2$ integer or,

$$d = (m + 1/2)\lambda_0 \qquad (m = 1, 2, ...,N-1) \qquad (2)$$

where λ_0 is the fundamental wavelength of the structure. The charge ratio of the individual bunches within the train increases according to,

$$Q_n = Q_0 \, [nR_0 + 1] \qquad (n=0,1,2,...N-1) \qquad (3)$$

where Q_0 is the charge in the first drive bunch. In addition to the requirements of spacing (Eqn. 2) and relative charge ratio (Eqn. 3) for the bunch train, there are also two requirements on the single bunch for obtaining maximal transformer ratio enhancement. The first condition is in addition to the work of [3]. The self-wake generated by the single bunch must be symmetric with respect to the center of the bunch. In other words, for a gaussian distribution, centered at $z = 0$, the self-wake within the bunch must satisfy, $W_0(-z) = W_0(+z)$. If this 'symmetric, single wake' condition is not met, the self-wakefields of the trailing bunch will be phase shifted from the accelerating bucket of the leading bunch resulting in only a partial cancellation of the fields. This can be corrected by changing the spacing ('d' in Fig. 1) of the bunches within the train so that full cancellation is obtained, but since it is easier to generate a bunch train of equal spacing we choose to satisfy the above condition. Numerical simulations show that the 'symmetric, single wake' condition is met when the bunch length satisfies

$$\sigma/\lambda_0 = 0.2 \qquad (4)$$

For a given R_0, the transformer ratio is maximized if the conditions specified in Eqn. 2, Eqn. 3, and Eqn. 4 are satisfied. This means that if $R_0 = 1$, then the fastest R_n can increase is $R_0 = 1$, $R_1 = 2$, $R_2 = 3$, $R_3 = 4$, etc. However, if $R_0 = 2$, then R_n could increase as $R_0 = 2$, $R_1 = 4$, $R_2 = 6$, $R_3 = 8$, etc. Therefore, since our goal is to maximize R_n we desire $R_0 = 2$ and therefore, bunch length satisfies,

$$\sigma/\lambda_0 = 0.25 \qquad (5)$$

as shown in the previous section. Although the last two conditions cannot be satisfied simultaneously, an intermediate value is sufficient for our experiments.

DIELECTRIC AND PLASMA EXPERIMENTS

In this section we outline plans for two experiments to be performed at the AWA. First we describe in some detail a DWFA experiment and finish by sketching a PWFA experiment.

Transformer Ratio Enhancement Using A RBT To Excite A DWFA

The available bunch length and finite group velocity considerations drove the choice of the structure for the DWFA experiment at the AWA facility. The upgraded AWA facility [6] will be able to produce a 40 nC beam with 1 mm of charge. For our experiment we use a conservative bunch length of $\sigma = 1.5$ mm. Based on the RBT algorithm we want $\sigma/\lambda_0 = 0.2$ and thus choose the dielectric structure of the previous section with $\lambda_0 = 7.89$ mm (the wavelength of our previous examples). If we locate the second bunch in the first acceleration bucket (i.e. m = 1 of the previous section) after the first bunch we have $d = (m+1/2) \lambda_0 = (1.5)* 7.89$ mm $= 11.84$ mm. We plot this example in Fig. 5a to show the position of the second bunch relative to the first bunch - its wakefield is not taken into account in the figure. The second consideration, finite group velocity, means that high ε is best, which can be understood as follows. Since the group velocity of the *rf* packet $\neq 0$, one must make sure that the length of the tube must be long enough so that the wakefields of the four drive bunches overlap. If the dielectric constant is high ($\varepsilon = 38$) then the group velocity is low; $\beta_g \approx c/\varepsilon = 0.026c$, where c is the speed of light. By the time the 4th drive bunch enters the tube, the *rf* packet from the first bunch has traveled a distance $\beta_g 3d$ where 3d is the separation between the first and last bunch. Thus L only need to be greater than $\beta_g 3d$ or about 2 cm.

If we had an electron source that could produce bunches separated by 11.84 mm we could do the experiment as it is described. However, the last constraint we must consider for our design is due to the AWA facility. The AWA *rf* frequency is based on 1.3 GHz and therefore we can only produce drive beams separated by $\lambda_{rf} = 23$ cm. Since the dielectric structure of the previous section has $\lambda_0 = 0.8$ cm, then we must operate at m^{th} harmonic of 23/0.8 or $m = 28^{th}$ harmonic.

Using the above DWFA with $\lambda_0 = 0.8$ cm, we construct a bunch train according to the algorithm given above. We choose $\sigma/\lambda_0 = 0.2$, $d = (28 + 1/2)\ 0.8$ cm $= 23$ cm, and $Q_n = Q_0 = 3$, $Q_1 = 9$, $Q_2 = 15$, and $Q_3 = 21$. The resultant wakefield excited by this RBT is shown in Fig. 6. A close examination of Fig. 6 shows that each bunch experiences the same decelerating wakefield $W_n^- = -0.8$ MeV/m/nC, while the peak accelerating field behind each bunch W^+_n increases as $W^+_0 = 1.63$, $W^+_1 = 3.26$, $W^+_2 = 4.90$, and $W^+_3 = 6.54$ all in units of MeV/m/nC. This results in a transformer ratio behind each bunch as, $R_0 = 2$, $R_1 = 4$, $R_2 = 6$, $R_3 = 8$ or an overall transformer ratio for this experiment of 8.

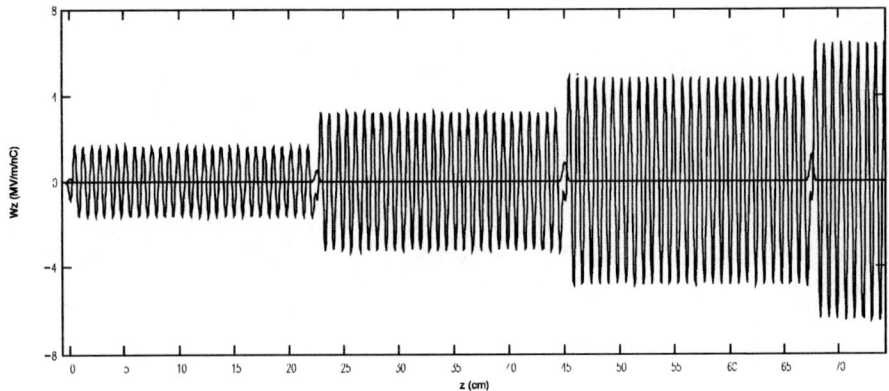

FIGURE 6. The resultant wakefield produced in a DWFA by a train of 4 electron bunches of bunch length 1.5 mm and charge magnitude $Q_0 = 3$, $Q_1 = 9$, $Q_2 = 15$, and $Q_3 = 21$. The Transformer Ratio $R = 8$ for this example.

Transformer Ratio Enhancement Using a RBT To Excite A PWFA

As a second example we briefly consider using a RBT to excite plasma wakefields. For a plasma density of $n_0 = 10^{14}$ /cm^3 we have a plasma wavelength of $\lambda_p = 0.334$ cm. For the PWFA experiment we choose a beam density of $n_b = 10^{14}$ /cm^3, bunch length $\sigma_z = 0.055$ cm, beam energy of 15 MeV, and normalized emittance of 40 mm mrad.

A bunch train of N = 5 bunches can be designed by using the above RBT algorithm. Given the fundamental wavelength of $\lambda_p = 0.334$ cm we choose d = $1.5\lambda_p \sim 0.5$ cm as the bunch separation. (In this example we are using a slightly non-optimized case where d = $1.45\lambda_p$ since we are using $\sigma_z/\lambda_p = 0.165$ and the condition of Eqn. (4) is not exactly satisfied.) Given $Q_0 = 5$ nC we have, $Q_1 = 15$, $Q_2 = 25$, $Q_3 = 35$, and $Q_4 = 45$. Lastly, we expect $R_0 \sim 2$ since $\sigma_z/\lambda_p = 0.165$, is near the optimal value of 0.2.

The peak plasma wakefield amplitude excited by a single bunch can be calculated by using the linear theory factor [7] in which the transverse ($\sigma_r^2 = \beta\varepsilon$) and longitudinal ($\sigma_z$) beam sizes, but not the gaussian dependency are included. Although we are not operating in the linear regime, this scaling law is consistent with detailed PIC simulations and should be a good indicator of wakefield amplitude. The amplitude is given by,

$$E_0 = 100\sqrt{n_0} \frac{n_b k_p \sigma_z}{n_0 \left(1 + \frac{1}{k_p^2 \sigma_r^2}\right)} \qquad (6)$$

where k_p is the plasma wave number and E_0 has units of V/m. If the beta function (β) is 10 cm we have $E_0 = 3.398*10^8$ V/m. The wakefield produced by a gaussian bunch of the above parameters is calculated (to first order) by convoluting the Green function solution of the plasma over a gaussian bunch distribution and scaling it by the linear theory factor, E_0. Finally, the wakefield excited by the entire bunch train is calculated (see Fig. 6) by invoking linear superposition.

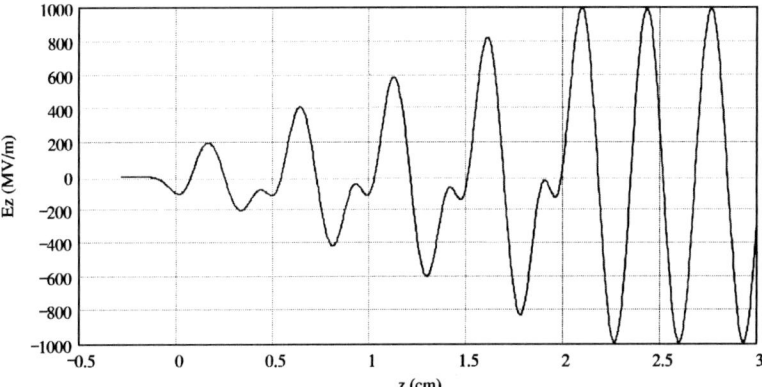

FIGURE 7. The resultant wakefield produced in a PWFA by a train of 5 electron bunches of bunch length 0.55 mm and charge magnitude $Q_0 = 5$, $Q_1 = 15$, $Q_2 = 25$, $Q_3 = 35$, $Q_3 = 45$. The Transformer Ratio $R = 10$ and the acceleration gradient is 1GV/m for this example.

The maximum accelerating fields after the bunches (W^+_n) are seen to ramp as, as $W^+_0 = 200$, $W^+_1 = 400$, $W^+_2 = 600$, $W^+_3 = 800$, and $W^+_4 = 1000$, all in units of MeV/m/nC while the decelerating fields are all equal to $W_n \sim -100$ MeV/m/nC. This results in a transformer ratio behind each bunch as, $R_0 = 2$, $R_1 = 4$, $R_2 = 6$, $R_3 = 8$, $R_4 = 10$ or an overall transformer ratio for this experiment of 10.

Tail Of The Distribution

If one examines Fig. 6 and Fig. 7 closely he will notice that the tails of the drive bunches spills over into the adjacent decelerating bucket. This means that the minimum decelerating field within the drive bunch train is larger than previously stated and the transformer ratio is therefore lower. To this there are two responses. First, the amount of charge contained in adjacent bucket is extremely small. For the example of the PWFA with $\sigma_z/\lambda_p = 0.165$ the total amount of charge in the adjacent decelerating bucket is < 1%. Second, just to show that this is a real effect, one could design an experiment using a multimoded DWFA [4] that has decelerating buckets spread very far apart so that no charge ends up in the adjacent bucket.

EXPERIMENTAL SETUP

In this section we describe how the RBT experiment can be done. We first describe how the bunch train is generated and the associated difficulties with transporting the RBT and finish with a discussion of how we will measure the transformer ratio.

Generation And Transport Of The RBT

A RBT of 4 electron bunches is made by optically splitting a single laser pulse into 4 separate pulses (in much the way as was done for the flat bunch train (FBT) experiments [4]) and sending these pulses into the AWA *rf* drive photoinjector [6]. The optical splitter for the RBT [8] differs from the FBT experiment in that splitters used in this case are not 50/50 ones but are chosen to produce the desired charge ratio. The distance between bunches is adjusted optically by moving mirrors on translation stages in the delay line. Initially, the distance between bunches is crudely measured with a streak camera, using the 10 ns sweep rate, thus giving us a timing resolution of ~10 ps between bunches. Final bunch spacing must be done during the experiment by making the deceleration of trailing bunches equal to the deceleration of the lead bunches after they emerge from the dielectric structure. Although it is easy to generate the laser pulse train with the correct intensity using the optical splitter, it may be difficult to transport the electron bunches down the accelerator. This is because the electron charge of the first bunch is so much less than that of the last bunch, that it will not be possible to run the accelerator system at an optimized machine tune for either bunch. (For example, $Q_0 = 3$ nC & $Q_3 = 21$ nC in the DWFA experiment.)

Measurement Of R

To measure the transformer ratio R, we must infer the deceleration gradient, W^-, experienced by the four drive beams by measuring their energy loss after emerging from the dielectric structure of Fig. 2. Since all drive beams will experience the same decelerating field, we will only be able to measure the combination of the four beams with our spectrometer (Fig. 8).

The energy will be measured with the spectrometer shown in Fig. 8. The energy measurement system has a resolution of 0.2% with the tungsten slit set to 300 μm. To complete the measurement of the transformer ratio, one must also know W^+ behind each bunch – i.e. one must map out the wakefield left behind the drive bunch train with a witness beam. For a length of tube $L = \frac{1}{2}$ m, we expect the drive bunches to only lose 0.4 MeV. Thus the drive beam will exit the structure with 15.2 MeV – 0.4 MeV = 14.8 MeV. The witness beam will enter the structure with 4 MeV and receive a maximum acceleration of 3 MeV thus exiting with energy of 7 MeV.

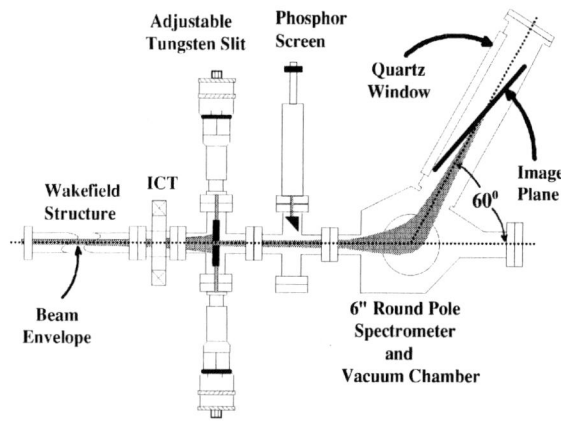

FIGURE 8. Energy Measurement System. The decelerated drive bunch train passes through a tungsten slit of adjustable width, which is imaged through a 60^0 dipole (diameter = 6") onto a phosphor screen located at the image plane. The phosphor screen is viewed with an intensified camera through the quartz window.

SUMMARY

We have described how to use the RBT method to achieve a transformer ratio (R) greater than 2 in any collinear wakefield accelerator – i.e. in either plasma or structure based wakefield accelerators. We presented an algorithm for designing a RBT experiment and have outlined two experiments for obtaining $R \gg 2$ in both a DWFA and a PWFA. We believe this method is feasible from an engineering standpoint using existing technology. An experiment to be preformed at the AWA facility to measure a $R \gg 2$ in a collinear DWFA is planned for the near future. Achieving $R \gg 2$ could have important implications for the future development of any collinear wakefield accelerator.

REFERENCES

[1] P.B. Wilson, Proc. of the 13th SLAC Summer Inst. on Particle Physics, SLAC Report No. 296, p. 273, E. Brennan ed., (1983)
[2] K.L. Bane, P. Chen, P.B. Wilson, IEEE Trans. Nucl. Sci. 32 3524 (1985)
[3] P. Schutt, T. Weiland, V.M. Tsakanov, On the Wake Field Acceleration using a Sequence of Driving Bunches., Nor Amberb Conf. Proc., Armenia, (1989)
[4] J.G. Power, M.E. Conde, W. Gai, A. Kanareyken, R. Konecny, and P. Schoessow, these Proceedings
[5] M. Rosing, W. Gai, Phys. Rev. D Vol. 42, No. 5, p. 1829, (1990
[6] M. E. Conde, W. Gai, R. Konecny, X. Li, J. G. Power, and P. Schoessow Phys. Rev. ST Accel. Beams **1**, 041302
[7] T. Katsouleas and W.B. Mori, private communication
[8] J.G. Power, WF-198, http://gate.hep.anl.gov/awa/awa/pubs.htm

Particle Beam Stability in the Hollow Plasma Channel Wake Field Accelerator

Carl B. Schroeder[1] and Jonathan S. Wurtele

Department of Physics
University of California
Berkeley, California 94720

Abstract. The electromagnetic wake field response of a hollow plasma channel to a driver (laser or charged particle beam) of arbitrary velocity is derived. The dispersion and loss factors of excited fundamental and higher order azimuthal modes are computed. Growth rates for beam breakup instabilities are calculated and beam transport is studied. External focusing is shown to provide a method of controlling transverse instabilities. For parameters of interest for high gradient plasma-based accelerators, it is shown that the most severe limitation to the interaction length of a single accelerator stage based on the hollow plasma channel structure is the transverse instability of the particle beam.

INTRODUCTION

Plasma-based accelerators have the ability to sustain extremely large accelerating gradients, with possible high-energy physics applications [1,2]. Diffraction is a severe limitation for laser-driven plasma-based accelerators. Therefore, any successful design of a plasma-based accelerator which utilizes a laser driver must include some form of optical guiding. Two schemes for optical guiding are currently being explored for overcoming diffraction: relativistic self-focusing [3] and plasma channel guiding [4].

Relativistic self-focusing relies on the energy dependence of the plasma frequency to modify the plasma index of refraction. Relativistic self-focusing of long laser pulses (i.e., pulse lengths much longer than the plasma wavelength) suffer from Raman forward and sidescatter instabilities [5]. These instabilities lead to break up of the pulse into small pulses of order the plasma wavelength and therefore limit the propagation distance of the laser pulse. For short laser pulses (i.e., pulse lengths of order the plasma wavelength), such as those used in the standard laser wake field accelerator, relativistic self-focusing is substantially reduced. This is due to the generation of a plasma density perturbation by the ponderomotive force of

---

[1] present address: University of California, Los Angeles, California 90095

the laser. For short pulses, the plasma frequency decrease from relativistic effects is balanced by this density perturbation [4]. Consequently, the index of refraction will have no transverse variation, and the plasma cannot optically guide a short laser pulse.

Plasma channel guiding provides optical guiding through the use of a plasma channel that has a higher plasma density outside the channel than inside the channel giving the plasma channel an index of refraction which decreases from the channel axis. A fixed plasma channel is analogous to an optical fiber and its guiding properties can be similarly analyzed. Plasma channels can be used to guide short pulses and have been studied analytically using axisymmetric models for a parabolic plasma density variation [4] and for hollow plasma channels [6].

Calculations show that a hollow plasma channel, in addition to optically guiding the laser pulse, supports a plasma wave with attractive properties for particle acceleration. The driver excites a surface mode in the plasma which extends into the channel. Unlike in a homogeneous plasma or parabolic channel, the transverse profile of the driver is decoupled from the transverse profile of the accelerating mode. Therefore, for a relativistic driver, the accelerating gradient of the fundamental mode is uniform and the focusing fields are linear [6]. In addition, the excited fields in a hollow plasma channel are fully electromagnetic, unlike the electrostatic fields excited in a homogeneous plasma. These properties make a hollow plasma channel well-suited as a structure for both particle beam and laser-driven wake field accelerators.

Since the original demonstration of the guiding of a low-intensity laser pulse in a plasma channel at the University of Maryland [7], several research groups have been examining experimental methods of plasma channel formation and guiding of high-intensity lasers [8–11]. Methods of forming a plasma channel include: inverse bremsstrahlung heating of the plasma by a precursor laser pulse resulting in hydrodynamic expansion and channel formation [8,9] and discharge ionization of a preformed capillary tube [10,11].

In this paper an externally formed hollow plasma channel is characterized as an accelerating structure, independent of the structure excitation mechanism (laser or particle beam). The results provide the basic scalings for the plasma channel accelerator, including current limiting higher-order mode couplings. Higher-order moments of the drive pulse distribution, present due to drive pulse shape or misalignment, will excite higher-order modes in addition to the fundamental (accelerating) mode. These higher-order modes can cause beam instabilities, limiting the propagation distance and, therefore, the energy gain. In this paper we examine instabilities resulting from the beam-plasma coupling.

MODE STRUCTURE OF THE HOLLOW PLASMA CHANNEL

Our derivation of the excited fields in the hollow plasma channel is similar to that used for conventional metallic structures. The driver (beam current or laser pulse) is assumed to be unaltered during the excitation of the channel. The hollow plasma channel is modeled as having an equilibrium electron plasma density $n_e(r) = n_o\Theta(r - r_w)$, where Θ is the Heaviside step function, r_w is the radius of the channel wall, and n_o is the number density of the plasma outside the channel. The ion plasma density is assumed to be equal to the equilibrium electron plasma density. The ions are also assumed to remain motionless since the drive pulse duration is taken to be much shorter than the response time of the ions.

For a hollow plasma channel, $\nabla n_e = 0$ in all regions, and the wave equation for the electric field $\vec{E}$, obtained from the Maxwell equations and the linearized cold collisionless fluid equations modeling the plasma response, is

$$\left(c^2 \nabla^2 - \frac{\partial^2}{\partial t^2} - \omega_p^2\right) \vec{E} = 4\pi \frac{\partial}{\partial t} \vec{J}_{\text{ext}} + 4\pi c^2 \nabla \rho_{\text{ext}}, \tag{1}$$

where ω_p is the plasma frequency. Here $\vec{J}_{\text{ext}}$ and ρ_{ext} are the external drive current and charge densities respectively. Note that if ∇n_e is nonvanishing, then resonant absorption is possible in the plasma channel walls and the excited fields can mode convert into an electrostatic Langmuir wave. The electromagnetic fields in the channel decay as the electrostatic Langmuir wave in the plasma wall is excited. This leads to an effective quality factor of this hollow plasma structure [12] and a corresponding limit on the number of bunches that can be accelerated.

The source terms in the wave equation are determined by the external driver. For beam-driven excitation of the plasma channel, $\vec{J}_{\text{ext}} = \vec{v}_b \rho_b$, where ρ_b is the beam charge density and $\vec{v}_b$ is the beam velocity. For excitation by a laser pulse, the current source is driven, to lowest-order, by the ponderomotive force of the laser pulse envelope (i.e., the gradient of the radiation pressure). To second-order in $a < 1$, the current source driven by the ponderomotive force is

$$4\pi \frac{\partial}{\partial t} \vec{J}_{\text{ext}} = \frac{m_e c^2}{e} \omega_p^2 \nabla \frac{a^2}{2}, \tag{2}$$

where $a = e|\vec{A}|/m_e c^2$ is the normalized vector potential of the laser.

For the linear analysis presented in this paper to be valid, surface plasma density perturbations should be small compared to the channel radius. This implies a laser pulse driver must satisfy $a^2 \ll 1$ (assuming the laser spot size is of order the channel radius $r_0 \sim r_w$ and the laser pulse duration is of order the plasma period $\omega_p \tau_L \sim 1$), and a particle drive beam must satisfy $N_b \ll (\omega_p/c) r_w^2 / r_e$, where $r_e = e^2/(m_e c^2)$ is the classical electron radius and N_b is the number of electrons per bunch. For example, if $r_w = 20$ μm and $n_0 = 7 \times 10^{16}$ cm^{-3}, then the linear theory is valid for beams with $N_b \ll 7 \times 10^9$ electrons.

The nonevolving driver propagates axially with group velocity near the speed of light $c\beta \simeq c$. Thus, we make the "frozen-field" approximation: the modes are functions only of the co-moving coordinate $\tau = t - z/(\beta c)$. The fields are decomposed into discrete azimuthal modes with mode index m and a Fourier transform in the co-moving coordinate τ is performed such that solutions are of the form $\exp(-i\omega_m\tau + im\theta)$, with the mode frequencies ω_m. The boundary conditions across the channel wall are: continuity of the electric and magnetic field components $\epsilon_m \vec{E} \cdot \hat{r}$, $\vec{E} \times \hat{r}$, and $\vec{B}$, where $\epsilon_m = 1 - \omega_p^2(r)/\omega_m^2$ is the dielectric function of the hollow plasma structure.

To study the excited channel modes synchronous with the driver, let $\vec{E} = \hat{A}_m \vec{e}_m(r,\theta) \exp(-i\omega_m\tau)$ and $\vec{B} = \hat{A}_m \vec{b}_m(r,\theta) \exp(-i\omega_m\tau)$, where $\hat{A}_m$ are constants determined by the excitation mechanism. With these definitions, the equation for the plasma wave electric field behind the drive pulse is

$$\left[c^2 \nabla_\perp^2 - \omega_m^2(\beta^{-2} - \epsilon_m)\right]\vec{e}_m = 0, \tag{3}$$

where $\nabla_\perp^2$ is the transverse Laplacian. Note that only purely electromagnetic modes (i.e., $\nabla \cdot \vec{E} = 0$) exist in the channel, and since there are no linear surface currents, the continuity of $\nabla \times \vec{E}$ requires that the mode in the plasma also satisfies $\nabla \cdot \vec{E} = 0$.

For the fundamental mode $m = 0$, the solutions to the homogeneous wave equation Eq. (3) are

$$e_{0z} = \frac{ik_1 c\beta}{\omega_0} \frac{I_0(k_1 r)}{I_1(k_1 r_w)}, \tag{4}$$

$$e_{0r} = \frac{I_1(k_1 r)}{I_1(k_1 r_w)}, \tag{5}$$

$$b_{0\theta} = \beta \frac{I_1(k_1 r)}{I_1(k_1 r_w)} \tag{6}$$

in the channel $r < r_w$, and

$$e_{0z} = -\frac{ik_2 c\beta}{\omega_0 \epsilon_0} \frac{K_0(k_2 r)}{K_1(k_2 r_w)}, \tag{7}$$

$$e_{0r} = \frac{1}{\epsilon_0} \frac{K_1(k_2 r)}{K_1(k_2 r_w)}, \tag{8}$$

$$b_{0\theta} = \beta \frac{K_1(k_2 r)}{K_1(k_2 r_w)} \tag{9}$$

in the plasma $r > r_w$, where I_m and K_m are m^{th}-order modified Bessel functions of the second kind. Here $k_1 = (\omega_m/c)(\beta^{-2}-1)^{1/2}$ and $k_2 = (\omega_m/c)(\beta^{-2}-1+\omega_p^2/\omega_m^2)^{1/2}$. Note that in the limit of an ultra-relativistic driver ($\beta \to 1$), $k_1 \simeq 0$ and $k_2 \simeq \omega_p/c$. The fundamental mode $m = 0$ frequency (eigenvalue equation) is

$$\omega_0 = \omega_p \Omega_0 = \omega_p \left[1 + \frac{k_2 I_1(k_1 r_w) K_0(k_2 r_w)}{k_1 I_0(k_1 r_w) K_1(k_2 r_w)}\right]^{-1/2}, \tag{10}$$

where $\Omega_m = \omega_m/\omega_p$ is the normalized frequency of the mth mode.

The higher-order modes $m > 0$ of the excited plasma wave are

$$e_{mz} = \frac{k_1 c\beta}{\omega_m} \frac{I_m(k_1 r)}{I_m(k_1 r_w)} f(m\theta), \text{ and} \tag{11}$$

$$b_{mz} = \frac{k_1 c\beta}{\omega_m} \Upsilon \frac{I_m(k_1 r)}{I_m(k_1 r_w)} g(m\theta) \tag{12}$$

in the channel $r < r_w$, and

$$e_{mz} = \frac{k_2 c\beta}{\omega_m} \frac{K_m(k_2 r)}{K_m(k_2 r_w)} f(m\theta), \text{ and} \tag{13}$$

$$b_{mz} = \frac{k_2 c\beta}{\omega_m} \Upsilon \frac{K_m(k_2 r)}{K_m(k_2 r_w)} g(m\theta) \tag{14}$$

in the plasma $r > r_w$, where $f(m\theta) = \cos(m\theta)$ and $g(m\theta) = -\sin(m\theta)$ for even modes or $f(m\theta) = \sin(m\theta)$ and $g(m\theta) = \cos(m\theta)$ for odd modes. In Eqs. (12) and (14), Υ is

$$\Upsilon = \frac{2m}{\beta} \left(\frac{k_2^2 - k_1^2}{r_w^2 k_1^2 k_2^2} \right) \left[\frac{I_{m+1}(k_1 r_w) + I_{m-1}(k_1 r_w)}{k_1 r_w I_m(k_1 r_w)} + \frac{K_{m+1}(k_2 r_w) + K_{m-1}(k_2 r_w)}{k_2 r_w K_m(k_2 r_w)} \right]^{-1}. \tag{15}$$

The transverse fields for the higher-order modes can be computed directly from the axial components Eqs. (11)-(14) using the relations

$$\vec{e}_{m\perp} = \frac{ic\beta^2}{\omega_m(1 - \epsilon_m \beta^2)} \left[\hat{z} \times \nabla_\perp b_{mz} - \beta^{-1} \nabla_\perp e_{mz} \right] \tag{16}$$

$$\vec{b}_{m\perp} = \frac{-ic\beta^2}{\omega_m(1 - \epsilon_m \beta^2)} \left[\epsilon_m \hat{z} \times \nabla_\perp e_{mz} + \beta^{-1} \nabla_\perp b_{mz} \right]. \tag{17}$$

The eigenvalue equation for the higher-order modes is

$$\frac{4m^2}{\beta^2} \left(\frac{k_2^2 - k_1^2}{r_w^2 k_1^2 k_2^2} \right)^2 = \left[\frac{I_{m+1}(k_1 r_w) + I_{m-1}(k_1 r_w)}{k_1 r_w I_m(k_1 r_w)} + \frac{K_{m+1}(k_2 r_w) + K_{m-1}(k_2 r_w)}{k_2 r_w K_m(k_2 r_w)} \right] \times \left\{ \frac{I_{m+1}(k_1 r_w) + I_{m-1}(k_1 r_w)}{k_1 r_w I_m(k_1 r_w)} + \frac{\epsilon_m \left[K_{m+1}(k_2 r_w) + K_{m-1}(k_2 r_w) \right]}{k_2 r_w K_m(k_2 r_w)} \right\}. \tag{18}$$

The solutions of Eq. (18) provide the higher-order mode frequencies ω_m.

Ultra-Relativistic Limit

In the ultra-relativistic limit ($\beta \to 1$), the linearly excited mode frequencies of the hollow plasma channel (Eqs. (10) and (18)) become [13]

$$\omega_m = \omega_p \Omega_m = \omega_p \left[\frac{(1+\delta_{m0})(m+1)K_{m+1}(R)}{2(m+1)K_{m+1}(R) + RK_m(R)} \right]^{1/2} , \qquad (19)$$

where $R = \omega_p r_w / c$ is the normalized channel radius and $\delta_{m0} = 1$ for $m = 0$ and zero otherwise.

The forces on a beam due to the excited fields have attractive properties for particle acceleration. The excited fundamental mode fields in the channel (Eqs. (4)-(6)) provide the axial and transverse forces

$$F_z \simeq -eA_0 \cos(\omega_0 \tau) , \qquad (20)$$

$$F_r \simeq eA_0 (1 - \beta_z \beta) \frac{\omega_0}{2} r \sin(\omega_0 \tau) , \qquad (21)$$

where A_0 is a constant determined by the excitation mechanism and $c\beta_z$ is the axial velocity of a witness charged particle beam. Inside the channel, in the ultra-relativistic limit, the axial (accelerating) field is uniform with respect to transverse position as indicated by Eq. (20). Therefore electrons at different radii gain energy at the same rate, minimizing the energy spread due to the transverse extent of the beam. The transverse fields are linear with respect to the radial position as indicated by Eq. (21), which implies the root-mean-squared transverse normalized emittance will be conserved for any beam slice. Note that the focusing due to the excited fundamental mode fields is typically small in the ultra-relativistic limit (i.e., $|F_r/F_z| \sim 1 - \beta_z\beta \ll 1$). In addition, there is a $\pi/4$ phase region where the fundamental channel mode both focuses transversely and accelerates longitudinally. These properties make the hollow plasma channel well-suited for an accelerating structure independent of excitation mechanism.

LOSS FACTOR

The interaction of the beam with the accelerator environment can be quantified by a calculation of the loss factors. The loss factor per unit length κ relates the accelerating gradient to the energy stored per unit length in the structure U by $\kappa = E_z^2/4U$. The loss factor is purely geometrical and is independent of the excitation mechanism. Since the loss factor is independent of the means of energy deposition, it is a figure of merit for comparisons of accelerating structures.

The loss factor per unit length for the mth mode [13] is

$$\kappa_m = \frac{\omega_p^2}{c^2} \left[\frac{K_m(R)}{RK_{m+1}(R)} \right] \left[1 + \frac{RK_m(R)}{2(m+1)K_{m+1}(R)} \right]^{-1} , \qquad (22)$$

where the axial electric fields of the higher-order modes have been evaluated at the channel radius. The energy stored in the plasma structure U_m is equal to the energy deposited by the driver. For an ultra-relativistic charge q at a radius r_b, with $r_b < r_w$, the total energy deposited in the plasma structure by the charge can be written as

$$U = \sum_m U_m = \sum_m \kappa_m (r_b/r_w)^{2m} q^2 . \tag{23}$$

For comparison, the fundamental mode of a scaled disk-loaded copper SLC structure [14] has a loss factor of $\kappa_0(\text{SLC}) \approx 2.1 \times 10^3 \lambda^{-2}(\text{cm})$ V/(pC m), while the fundamental mode loss factor in a hollow plasma channel is $\kappa_0 = 3.6 \times 10^3 \lambda_0^{-2}(\text{cm}) K_0(R)/[RK_1(R)]$V/(pC m) where $\lambda_0 = \Omega_0^{-1} 2\pi c/\omega_p$ is the accelerating wavelength. For a normalized channel radius of $R = 1$, the fundamental mode loss factor is $\kappa_0 = 2.5 \times 10^3 \lambda_0^{-2}(\text{cm})$ V /(pC m), somewhat larger than the conventional resonantly-excited conducting structure, which implies stronger beam loading and smaller stored energy per unit length for a given accelerating gradient. Note that a larger loss factor results not only in greater energy deposition, but also in larger wake field excitation. The latter could result in instabilities and lead to greater energy spread.

PARTICLE BEAM DYNAMICS IN A HOLLOW PLASMA CHANNEL

The longitudinal and transverse forces on an ultra-relativistic beam due to its interaction with the plasma can be calculated from the convolution of the charge distribution of the beam with the wake fields $\vec{W} = \vec{E} + \hat{z} \times \vec{B}$ produced by all proceeding charges. The wake fields excited inside the hollow plasma channel by a ultra-relativistic point charge q, passing through the channel at radius r_b, with $r_b < r_w$, and azimuthal angle $\theta = 0$, are

$$\vec{W}_\parallel = -q \sum_m \hat{W}_{\parallel m}(\tau) r^m r_b^m \cos(m\theta) \hat{z} , \tag{24}$$

$$\vec{W}_\perp = q \sum_m \hat{W}_{\perp m}(\tau) r^{m-1} r_b^m [\hat{r} \cos(m\theta) - \hat{\theta} \sin(m\theta)] , \tag{25}$$

with the wake functions,

$$\hat{W}_{\parallel m}(\tau) = \frac{2\kappa_m}{r_w^{2m}} \cos[\Omega_m \omega_p \tau] , \tag{26}$$

$$\hat{W}_{\perp m}(\tau) = \frac{2m\kappa_m c}{r_w^{2m} \Omega_m \omega_p} \sin[\Omega_m \omega_p \tau] . \tag{27}$$

The mode frequencies, Ω_m, and loss factors, κ_m, are given by Eqs. (19) and (22), respectively. The Laplace transform of the wake functions Eqs. (26) and (27) yields the impedance of the plasma structure. These point charge wake fields (Eqs. (24) and (25)) can be used as Green functions to compute the longitudinal and transverse forces produced by an arbitrary beam charge distribution, ρ_b. The longitudinal wake fields tend to cause energy variation within a bunch, and the transverse wake fields cause beam breakup instabilities. In the case that the charge is near the axis of the channel, $r_b \ll r_w$, the longitudinal wake field is dominated by the fundamental ($m = 0$) mode and the transverse wake field is dominated by the dipole ($m = 1$) mode.

Longitudinal Effects

The longitudinal wake fields will cause the energy spread σ_γ within the beam to grow. Energy spread constraints will therefore limit the beam current. The energy change of an ultra-relativistic electron bunch propagating along the axis of the hollow plasma channel is described by the equation

$$\frac{\partial \gamma}{\partial z} = cI_o^{-1}E(\tau) - \int_0^\tau cI_o^{-1}I(\zeta)\hat{W}_{\|0}(\tau - \zeta)d\zeta , \qquad (28)$$

where $\hat{W}_{\|0}$ is the longitudinal fundamental mode wake function given by Eq. (26). Here, $E(\tau) = A_0 \cos(\omega_0 \tau - \varphi_{\text{inj}})$ is the accelerating gradient, with A_0 the peak axial electric field of an excited plasma wave (created by a drive pulse) and φ_{inj} the injection phase of the head of the bunch with respect to the plasma wave. For a delta function bunch $I(\tau) = q\delta(\tau)$ (i.e., a bunch much shorter than the period of the accelerating field $\omega_p \tau_b \ll 1$), one finds $\sigma_\gamma/\gamma \approx q\hat{W}_{\|0}(0)/[2E(0)] = q\kappa_0/E(0)$. For illustration, if an energy spread of order 0.1% is required in a plasma structure with $r_w = 20$ μm, $R = 1$, and an accelerating gradient of 10 GV/m, then the beam-induced gradient should be held to $2\kappa_0 q \approx 20$ MV/m. The single-bunch charge q is then limited to 0.9 pC or 5×10^6 particles. In principle, the energy spread within a single bunch can be minimized and the charge limits increased by shaping the charge distribution of the bunch [15], although this may be difficult to achieve in practice.

Transverse Instabilities

It is well-known in accelerator physics that interaction of the beam with the structure geometry can result in transverse instabilities coupled to the off-axis displacement of the beam centroid, or beam breakup instabilities [16]. This section discusses beam breakup instabilities in a hollow plasma channel.

Consider the effect of a small displacement of the beam centroid $X(z,\tau)$ in the transverse direction. The transverse displacement is expressed as a function of two variables: the propagation distance, z, and the distance from the head of the beam, τ. The variable $\tau = t - z/v_b$ indexes beam slices where $v_b \simeq c$ is the axial beam velocity. The beam extends from $\tau = 0$ (the beam head) to $\tau = T_b$ (the beam tail). Beam electrons remain approximately at a fixed τ, as they advance in z along the length of the accelerator.

From the Lorentz force equation, assuming the beam is monoenergetic, the evolution of the transverse displacement of the beam due to the dipole transverse wake field is

$$\left[\frac{\partial}{\partial z}\gamma(z)\frac{\partial}{\partial z} + \gamma(z)k_\beta^2(z)\right] X(z,\tau) = \int_0^\tau cI(\zeta)I_o^{-1}\hat{W}_{\perp 1}(\tau - \zeta)X(z,\zeta)d\zeta, \qquad (29)$$

where $I(\tau)$ is the beam current and $I_o = m_e c^3/e \approx 17$ kA is the Alfvén constant. The transverse dipole wake function $\hat{W}_{\perp 1}$ is given by Eq. (27) with $m = 1$ and determines the Lorentz force on an electron at τ as it arrives at z due to the fields generated by the beam segment at $\zeta < \tau$. The right-hand side of Eq. (29) is the cumulative force due to the transverse dipole wake fields of the proceeding charges in the beam. The transverse focusing force in the channel from a plasma wave (created by a drive pulse) and from any external magnets can be described, in the linear approximation, by the betatron wavenumber $k_\beta(z)$. This model [i.e., Eq. (29)] is valid in the ultra-relativistic limit of the beam velocity, where phase slippage between particles in the bunch is small. Equation (29) can be solved in a variety of limits to study the single-bunch beam breakup instability.

Single-Bunch Beam Breakup

Assuming the beam density, which is proportional to the beam current, remains constant, Eq. (29) can be rewritten as

$$\left[\frac{\partial}{\partial z}\gamma(z)\frac{\partial}{\partial z} + \gamma(z)k_\beta^2(z)\right] X(z,\tau) = \int_0^\tau d\zeta G(\tau - \zeta)X(z,\zeta), \qquad (30)$$

where the Green function G is given by the excited wake field,

$$G(\tau) = c\frac{I}{I_o}\hat{W}_{\perp 1}(\tau) = \left(\frac{I}{I_o}\frac{2\kappa_1\omega_p}{R^2\Omega_1}\right)\sin(\omega_1\tau) = G_1\sin(\omega_1\tau). \qquad (31)$$

In the limit of a short bunch (i.e., $\omega_p T_b \ll 1$), the wake field response is approximately linear $G(\tau) \simeq (G_1\omega_1)\tau$.

If the growth length of the instability is much less than k_β^{-1} (i.e., the weak-focusing regime), then the term due to transverse focusing on the left-hand side of Eq. (30) can be neglected. This will typically be valid for ultra-relativistic beams propagating in hollow plasma channels without external focusing since the transverse focusing forces in the channel due to the excited fundamental mode fields will be small in the ultra-relativistic limit, as indicated by Eq. (21) with $(1 - \beta_z\beta) \simeq 0$.

Equation (30) can be solved following the method in Ref. [17] (i.e, by assuming that the growth of the transverse beam displacement and the change in energy to be slow on the scale of the accelerating field). In this approximation Eq. (30) has the solution

$$X(z,\tau) = X_0 \left(\frac{\gamma_0}{\gamma}\right)^{1/4} \frac{1}{2\pi i} \int_{-i\infty}^{i\infty} ds \frac{1}{s} \exp\left[s\tau \pm \frac{\sqrt{G_1\omega_1}}{s} \int_0^z dz_1 \gamma^{-1/2}(z_1)\right], \quad (32)$$

where X_0 is the initial transverse displacement of the beam. Applying the method of steepest descent to the integral in Eq. (32), one finds

$$X(z,\tau) \approx X_0 \left(\frac{\gamma_0}{\gamma}\right)^{1/4} \frac{\exp(\Lambda_w)}{\sqrt{8\pi\Lambda_w}}, \quad (33)$$

with the exponent

$$\Lambda_w = 2\left[\left(G_1\omega_1\tau^2\right)^{1/2} \int_0^z dz_1 \gamma^{-1/2}(z_1)\right]^{1/2}. \quad (34)$$

For an ultra-relativistic beam, longitudinal slippage is negligible the beam slices remain at fixed τ. With fixed τ and $\omega_p T_b \ll 1$, the axial force Eq. (20) is constant along the beam and the energy growth is linear. Thus, $\gamma = \gamma_0 + gz$, where γ_0 is the initial energy and g is the constant acceleration gradient. The exponent Λ_w becomes

$$\Lambda_w = 2^{3/2} \left(\frac{G_1\omega_1\tau^2}{g^2}\right)^{1/4} \left(\gamma^{1/2} - \gamma_0^{1/2}\right)^{1/2}. \quad (35)$$

Asymptotically (for large z), the exponent Eq. (35) is

$$\Lambda_w \to 2^{3/2} \left(\frac{G_1\omega_1\tau^2 z}{g}\right)^{1/4} = \left(\frac{z}{L_w}\right)^{1/4}, \quad (36)$$

with the characteristic growth length

$$L_w = g\left(2^6 G_1\omega_1\tau^2\right)^{-1} = 2^{-7}\frac{I_o}{I}\frac{gR^2}{\kappa_1(\omega_p\tau)^2}. \quad (37)$$

This growth length will impose an upper bound on the accelerator length for a given $I\tau_b^2$ product. For example, in a plasma channel with plasma wavelength of 125 µm, channel radius of 20 µm, and accelerating gradient of 10 GV/m, a 3 fs beam with a charge of 1 pC will have an instability growth length of $L_w \approx 5$ mm. As Eq. (37) indicates, the instability growth length can be increased by increasing R, which, in turn, will lower the loss factor of the structure (assuming a fixed plasma density).

The asymptotic growth of the transverse displacement of the beam centroid can also be determined for a particle beam which is traveling through an unexcited

($g = 0$) hollow plasma channel (i.e., a coasting beam or drive beam). In this case, the exponent from Eq. (35) becomes

$$\Lambda_w = 2\left(\frac{G_1\omega_1\tau^2}{\gamma_0}\right)^{1/4} z^{1/2} = \left(\frac{z}{L_{wu}}\right)^{1/2}, \tag{38}$$

with the characteristic growth length

$$L_{wu} = \frac{1}{4\tau}\left(\frac{\gamma_0}{G_1\omega_1}\right)^{1/2} = 2^{-5/2}\left[\frac{I_o}{I}\frac{\gamma_0 R^2}{\kappa_1(\omega_p\tau)^2}\right]^{1/2}. \tag{39}$$

As one can see from comparison of the growth lengths Eqs. (37) and (39), acceleration tends to reduce the influence of transverse beam displacements.

Single-Bunch Beam Breakup with External Focusing

The transverse focusing in the hollow plasma channel provided by the accelerating (fundamental) wake field is weak for relativistic beams. For high-energy applications one may prefer to provide external focusing to operate in the strong-focusing regime (i.e, k_β^{-1} is much less than the instability growth length). As shown below, external focusing (e.g., magnetic quadrupole lens) can substantially reduce the asymptotic growth of transverse beam displacements.

The asymptotic growth of the transverse centroid displacement of an accelerated and strongly focused beam can be determined by applying an eikonal approximation to Eq. (30), assuming the growth will be slow on the scale of a betatron period. Consider the slowly varying amplitude of the transverse centroid displacement $\chi(z,\tau)$ such that

$$X(z,\tau) = \left(\frac{\gamma_0\dot\theta_0}{\gamma\dot\theta_\beta}\right)^{1/2} \chi(z,\tau)\exp[i\theta_\beta(z)], \tag{40}$$

with the betatron phase

$$\theta_\beta(z) = \int_0^z dz_1 k_\beta(z_1) \tag{41}$$

and $\dot\theta_0 = \dot\theta_\beta(z=0)$. Substituting Eq. (40) into Eq. (30), assuming the eikonal approximation such that $|\dot\gamma| \ll |\gamma\dot\theta_\beta|$ and $|\dot\chi| \ll |\chi\dot\theta_\beta|$, and taking a Laplace transform in τ yields

$$\frac{\partial}{\partial z}\tilde\chi(z,s) = \frac{\tilde G}{2i\gamma\dot\theta_\beta}\tilde\chi(z,s), \tag{42}$$

which has the solution

$$\tilde{\chi}(z,s) = \tilde{\chi}_0 \exp\left[\frac{\tilde{G}(s)}{2i}\int_0^z \frac{dz_1}{\gamma(z_1)\dot{\theta}_\beta(z_1)}\right], \qquad (43)$$

where $\tilde{\chi}_0 = \tilde{\chi}(z=0,s)$. Inverting the Laplace transform, the solution for the amplitude of the beam centroid is

$$\chi(z,\tau) = \frac{X_0}{2\pi i}\int_{-i\infty}^{i\infty} \frac{ds}{s} \exp\left[s\tau + \frac{\tilde{G}(s)}{2i}\int_0^z \frac{dz_1}{\gamma(z_1)\dot{\theta}_\beta(z_1)}\right], \qquad (44)$$

where the initial condition $\chi(z=0,\tau) = X_0\Theta(\tau)$ is assumed. The integral in Eq. (44) may be computed approximately by the method of steepest descent. Using this method, one finds the transverse beam displacement is

$$X(z,\tau) \approx X_0 \frac{3^{1/4}}{2^{3/2}\pi^{1/2}}\left(\frac{\gamma_0\dot{\theta}_0}{\gamma\dot{\theta}_\beta}\right)^{1/2} \frac{\exp(\Lambda_s)}{\Lambda_s^{1/2}}\cos\left(\theta_\beta - \frac{\Lambda_s}{\sqrt{3}} + \frac{\pi}{12}\right), \qquad (45)$$

with the exponent

$$\Lambda_s = \frac{3^{3/2}}{4}\left[G_1\omega_1\tau^2 \int_0^z \frac{dz_1}{\gamma(z_1)\dot{\theta}_\beta(z_1)}\right]^{1/3}. \qquad (46)$$

In deriving Eq. (45), a short bunch, $\omega_p T_b \ll 1$, was assumed such that $G(\tau) \simeq G_1\omega_1\tau$.

Considering linear energy growth $\gamma(z) = \gamma_0 + gz$ and assuming the betatron wavenumber has an energy dependence such that $k_\beta(z) = \dot{\theta}_\beta = \dot{\theta}_0(\gamma_0/\gamma)^\alpha$, the transverse beam displacement of a short bunch becomes

$$\frac{X(z,\tau)}{X_0} \approx \frac{3^{1/4}}{2^{3/2}\pi^{1/2}}\left(\frac{\gamma_0}{\gamma}\right)^{(1-\alpha)/2} \frac{\exp(\Lambda_s)}{\Lambda_s^{1/2}}\cos\left(\theta_\beta - \frac{\Lambda_s}{3^{1/2}} + \frac{\pi}{12}\right), \qquad (47)$$

with the betatron phase

$$\theta_\beta = \frac{\gamma_0^\alpha k_0}{g(1-\alpha)}\left(\gamma^{1-\alpha} - \gamma_0^{1-\alpha}\right) \qquad (48)$$

and exponent

$$\Lambda_s = \frac{3^{3/2}}{2^{5/3}}\left[\frac{I}{I_o}\frac{\kappa_1(\omega_p\tau)^2}{\alpha g\gamma_0^\alpha k_0 R^2}(\gamma^\alpha - \gamma_0^\alpha)\right]^{1/3}, \qquad (49)$$

where $k_0 = k_\beta(z=0)$ is the initial betatron wavenumber at injection. Asymptotically, $\Lambda_s \to (z/L_s)^{\alpha/3}$, with the instability growth length

$$L_s = \frac{2^{5/\alpha}}{3^{9/2\alpha}}\left(\frac{I_o}{I}\right)^{1/\alpha}\left[\frac{\alpha g^{1-\alpha}\gamma_0^\alpha k_0 R^2}{\kappa_1(\omega_p\tau)^2}\right]^{1/\alpha}. \qquad (50)$$

For example, if $\alpha = 1/2$ (e.g., a magnetic quadrupole lens), then the growth rate scales as $L_s \propto (I/I_o)^{-2}(\omega_p\tau)^{-4}$, which is a more favorable scaling than Eq. (37).

For a coasting or drive beam [i.e., no acceleration ($g = 0$)] traveling through an unexcited hollow plasma channel structure with strong external focusing, $\gamma(z) = \gamma_0$, $\theta(z) = k_\beta z$, and

$$\Lambda_s = \frac{3^{3/2}}{2^{5/3}} \left(\frac{G_1 \omega_1 \tau^2 z}{2 k_\beta \gamma_0} \right)^{1/3} = \left(\frac{z}{L_{su}} \right)^{1/3}, \tag{51}$$

where the characteristic growth length is

$$L_{su} = \frac{2^6 k_\beta \gamma_0}{3^{9/2} G_1 \omega_1 \tau^2} = \frac{2^5}{3^{9/2}} \frac{I_o}{I} \frac{k_\beta \gamma_0 R^2}{\kappa_1 (\omega_p \tau)^2}. \tag{52}$$

CONCLUSIONS

This paper characterizes the hollow plasma channel in terms of fundamental accelerator parameters: the mode frequencies (eigenvalues) and the loss factors (eigenfunctions) of the electromagnetic channel modes excited by a driver (laser or particle beam) of arbitrary velocity. With these results, one can quantify the performance of a high-energy machine based on this plasma structure. In order to reach TeV-energies, such a plasma-based accelerator would consist of many stages. With optical guiding provided by the plasma channel, the length of a single stage based on a hollow plasma channel structure would be fundamentally limited by the shorter of the dephasing length and the driver depletion length. In practice, the length of a plasma-based accelerator may be limited by beam-plasma instabilities.

Plasmas provide strong coupling for acceleration of particle beams, quantified in the loss factors. At the same time this strong coupling extends to strong deflection and breakup of the beam. In this paper, the stability of a particle beam propagating in a hollow plasma channel was examined. We addressed the coupling of the dipole wake field excited in the plasma channel to the transverse displacement of the beam. Single-bunch beam breakup was analyzed for accelerating and coasting beam propagation in a hollow plasma channel in the weak-focusing and the strong-focusing regimes. These results show that the most favorable scalings are achieved for beams propagating in an excited hollow plasma channel in the strong-focusing regime.

With diffraction overcome by a plasma channel, the most severe limitation to the length of a single accelerator stage based on the hollow plasma channel structure is the transverse stability of the particle beam. That is, for typical parameters of plasma-based accelerator experiments the transverse beam breakup growth length will be shorter than both the dephasing lengths or the driver depletion lengths). As it does in the study of high-energy accelerators based on conventional rf technology, the transverse instabilities will set limits on allowable jitter and alignment tolerances. Using the analysis presented in this paper, such limits can be meaningfully estimated for the hollow plasma structure.

ACKNOWLEDGEMENTS

The authors thank Prof. David H. Whittum for useful discussions. This work was supported by the US Department of Energy, Division of High-Energy Physics.

REFERENCES

1. Wurtele, J., *Phys. Fluids B* **5**, 2363 (1993).
2. Esarey, E., Sprangle, P., Krall, J., and Ting, A., *IEEE Trans. Plasma Sci.* **PS-24**, 252 (1996).
3. Sun, G.-Z., Ott, E., Lee, Y. C., and Guzdar, P. *Phys. Fluids* **30**, 526 (1987).
4. Sprangle P., Esarey, E., Krall, J., and Joyce, G. *Phys. Rev. Lett.* **69**, 2200 (1992).
5. Antonsen, T. M., and Mora, P., *Phys. Fluids B* **5**, 1440 (1993).
6. Chiou, T. C., et. al., *Phys. Plasmas* **2**, 310 (1995).
7. Durfee, C. G., and Milchberg, H., *Phys. Rev. Lett.* **71**, 2409 (1993).
8. Durfee, C. G., Lynch, J., and Milchberg, H., *Phys. Rev. E* **51**, 2368 (1995).
9. Volfbeyn, P., Esarey, E., and Leemans, W. P., *Phys. Plasmas* **6**, 2269 (1999).
10. Ehrlich, Y., et. al., *Phys. Rev. Lett.* **77**, 4186 (1996).
11. Kaganovich, D., et. al., *Phys. Rev. E* **59**, R4769 (1999).
12. Shvets, G., Wurtele, J. S., Chiou, T. C., and Katsouleas, T. C., *IEEE Trans. Plasma Sci.* **PS-24**, 351 (1996).
13. Schroeder, C. B., Whittum, D. H., and Wurtele, J. S., *Phys. Rev. Lett.* **82**, 1177 (1999).
14. Wilson, P., *Laser Acceleration of Particles*, New York: AIP Press, 1985, pp. 560-597.
15. Loew, G. A., and Wang, J. W., *IEEE Trans. Nucl. Sci.* **NS-32**, 3228 (1985).
16. Lau, Y. Y., *Phys. Rev. Lett.* **63**, 1141 (1989).
17. Whittum, D. H., et. al., *Phys. Rev. A* **46**, 6684 (1992).

Trapping of Background Plasma Electrons in a Beam-Driven Plasma Wake Field Using a Downward Density Transition

H. Suk, N. Barov, J. B. Rosenzweig and E. Esarey[*]

Department of Physics and Astronomy, University of California, Los Angeles, CA 90095
[*]*Center for Beam Physics, Lawrence Berkeley National Laboratory, Berkeley, CA 94720*

Abstract. Trapping of background plasma electrons by a beam-driven plasma wake field is studied as a new self-injection method. In this scheme, a short electron beam pulse is sent through an underdense plasma with a downward density transition and some background plasma electrons are trapped by the strong wake field due to the sudden increase of the wake wave wavelength at the density transition. Two-dimensional PIC (Particle-In-Cell) simulations show that a significant amount of plasma electrons are trapped and accelerated to a higher energy than the driving beam energy. Furthermore, the trapped-beam quality is fairly good. In this paper, the 2-D simulation results, dynamics of the trapped beam and the driving beam, and the proposed experiment for the UCLA Neptune Laboratory are described.

INTRODUCTION

In recent years, extensive studies have been performed to explore advanced accelerators based on a beam-driven plasma wake-field accelerator (PWFA) [1,2] or a laser wake field accelerator (LWFA) [3]. In both cases, injection is an important issue. In the LWFA, the plasma wavelength is in the 10 to 100 μm range for typical cases and a much shorter electron beam than the plasma wavelength should be injected to have a small energy spread. Injecting such a short external beam into the acceleration phase of the wake field by using lasers requires a femto-second timing accuracy which is beyond the current technology. Hence, to avoid such timing difficulties a few self-injection mechanisms were studied. They include the self-modulated laser wake-field acceleration (SMLWFA) [4], laser injection methods [5,6] and self-trapping of plasma electrons using wavebreaking of laser-induced wake wave in an inhomogeneous plasma density [7]. For the beam-driven PWFA, the plasma wavelength is rather larger ($\lambda_p \sim$ a few mm for typical cases) so that the timing accuracy requirement is mitigated, but still a separate external beam should be generated to be injected.

To avoid such complexities, a new self-injection method for the PWFA was recently proposed by the authors of this paper [8]. In this scheme, a short ($\sigma_z \sim c/\omega_p$) electron beam pulse is sent through an underdense plasma (plasma density n_0 < beam density n_b) with a sharp downward plasma density transition. When the beam passes through the sharp density transition, the wavelength of the plasma wake field increases suddenly so that some background plasma electrons are injected into the acceleration phase of the wake field. The injected particles are trapped and accelerated to a high energy by the wake field. In this scheme, the *trapped electron beam* is focused in a nearly linear field of in a plasma-electron-free ion channel and the trapped beam does not have collisions with plasma electrons. Thus a good beam quality is expected and our simulation results verified that the new injection method can generate fairly good quality beams.

In this paper, we show the 2-D PIC (Particle-In-Cell) simulation results for trapping and acceleration of the background plasma electrons. They also include emittance and energy spread simulation results. In addition, transverse dynamics of the trapped beam, collision effects of the trapped beam with the background ions, and dynamics of the driving beam are investigated. Finally we propose an experiment for the UCLA Neptune Advanced Accelerator Laboratory to verify the trapping mechanism experimentally.

2-D PIC SIMULATIONS

In order to show trapping of the background plasma electrons, we performed 2-D simulations with the MAGIC code [9]. The plasma and driving beam parameters for the simulations are summarized in Table I. In the simulation, the driving beam has a bi-Gaussian density profile of $n_b(r,z) = n_{b,0} e^{-r^2/2\sigma_r^2} e^{-\xi^2/2\sigma_z^2}$ ($\xi = z - v_b t$) and it propagates through an underdense plasma ($n_{b,0} = 2.4 n_0' = 3.4 n_0''$) with a downward density transition ($n_0'' = 0.7 n_0'$).

TABLE 1. Plasma and driving beam parameters for simulations.

Plasma densities	$n_0' = 5 \times 10^{13} cm^{-3}$ for z<1 cm
	$n_0'' = 3.5 \times 10^{13} cm^{-3}$ for z>1 cm
Plasma electron temperature (kT_e)	3 eV
Driving beam energy (E_b)	50 MeV
Beam density ($n_{b,0}$)	$10^{14} cm^{-3}$
Beam size (σ_r, σ_z)	$k_p'' \sigma_z = 1$
	$k_p'' \sigma_r = 0.56$

Figure 1 shows the phase space plot (r vs. z) of the background plasma electrons when the driving beam center is located at 1.76 cm and 13.76 cm, respectively. As shown in the plot, plasma electrons are almost completely expelled from the beam path due to the space-charge force of the driving beam and a plasma-electron-free ion cavity is formed behind the driving beam. In the plot, it is shown that some plasma electrons are injected into the acceleration phase of the wake field when the plasma wake wave passes the density transition (located at z=1 cm). The trapped particles are transversely focused in the plasma-electron-free ion cavity due to the plasma lens effect [10] and they execute betatron oscillations with a wavelength of $\lambda_\beta = (2\pi\gamma\beta/n_0 r_e)^{1/2}$, where r_e is the classical electron radius, γ and β are the relativistic factor and normalized velocity of the trapped particles. In the bottom plot of Fig. 1, the trapped beam is tightly focused after $\lambda_\beta/4$ and separated from other plasma electrons. The trapped beam is so short (beam length $<<1/k_p$) that the trapped beam does not develop the electron hose instability [11] and other instabilities during acceleration. However, if the trapped beam propagates farther in the ion channel, it begins to be affected by some slipped electrons from the driving beam, which is caused by depletion of the driving beam energy. This will be discussed later in detail.

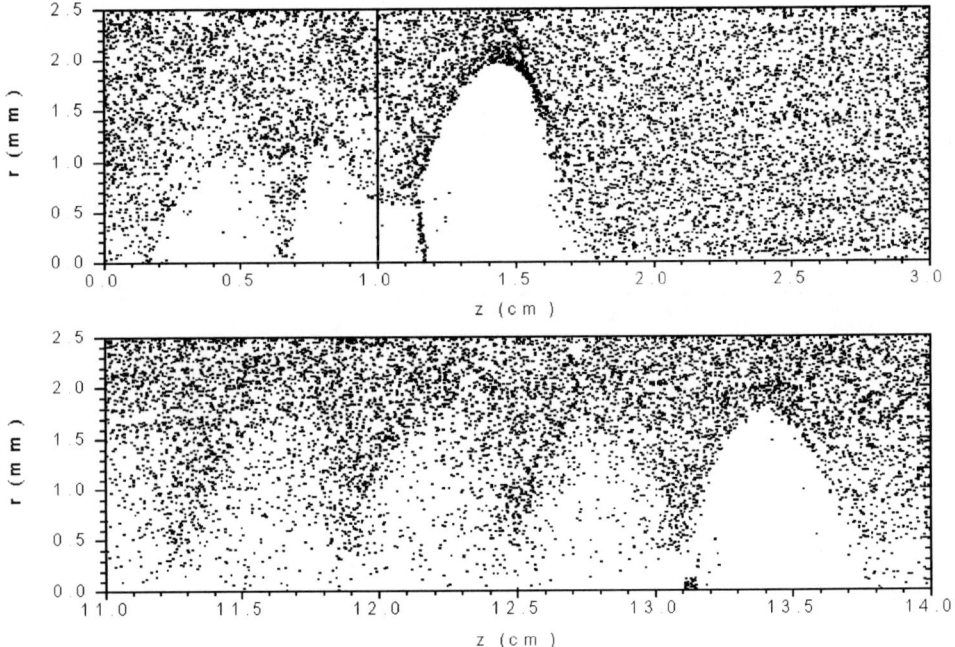

FIGURE 1. Phase space plot (r vs. z) of the background plasma electrons.

Figure 2 shows the phase space plot for momentum of the trapped particles in Fig. 1. The trapped plasma electrons are observed to be rapidly accelerated to a relativistic energy and bunched in the first electron-free ion cavity. In the figure, note that a small amount of plasma electrons are also trapped in other rarefied cavities, but this trapping is due to wavebreaking which is completely different from the trapping mechanism in the first rarefied cavity. The figure shows that a lot more particles are trapped and accelerated in the first rarefied cavity than in other cavities. In this simulation, the total charge of the trapped particles in the first rarefied cavity is about 850 pC and the pulse duration is about 1 ps.

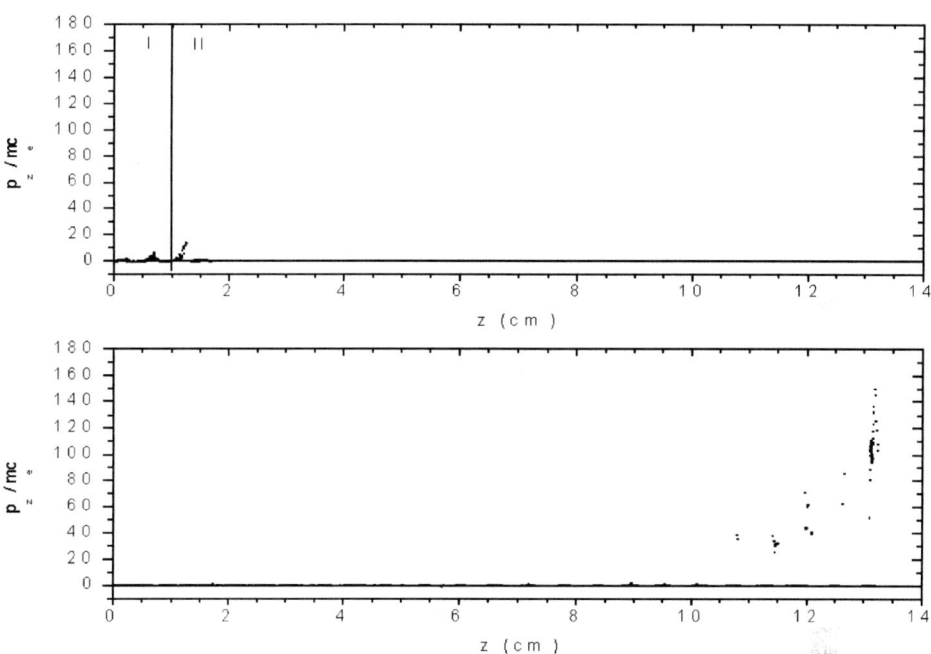

FIGURE 1. Phase space plot (p_z vs. z) of the trapped plasma electrons in Fig. 1.

The trapped particles are almost linearly accelerated with a gradient of $dE/dz \cong 430$ MeV/m in the beginning, but the energy is gradually saturated and then it begins to decrease. This is shown in Fig. 3. The saturation in the trapped beam energy is caused by depletion of the driving beam energy and this happens after the driving beam propagates a distance of $d = E_b / eE_0$, where e is the electron charge, E_b is the beam energy, and E_0 is the longitudinal electric field in the plasma wave. Figure 3 also shows the energy spread of the trapped beam during acceleration. The trapped particles have a certain energy distribution along the longitudinal direction, with significant contributions from some stray particles in the head and tail parts of the beam. Hence, the particles in the head and tail parts should be cut off to have a smaller energy spread and emittance. Figure 3 shows the energy spread during acceleration for three different cases (20 %, 50 % and 90 % of the trapped particles). In the bottom plot of Fig. 2, for example, the rms energy spread $\Delta E/E_{rms}$ is 1.7 % for 20 % (Q=170 pC) of the trapped particles. In this case, the normalized rms emittance $\varepsilon_{n,rms}$ is estimated to be about 1 mm-mrad. Thus, the beam quality after cutting off the stray particles is fairly good.

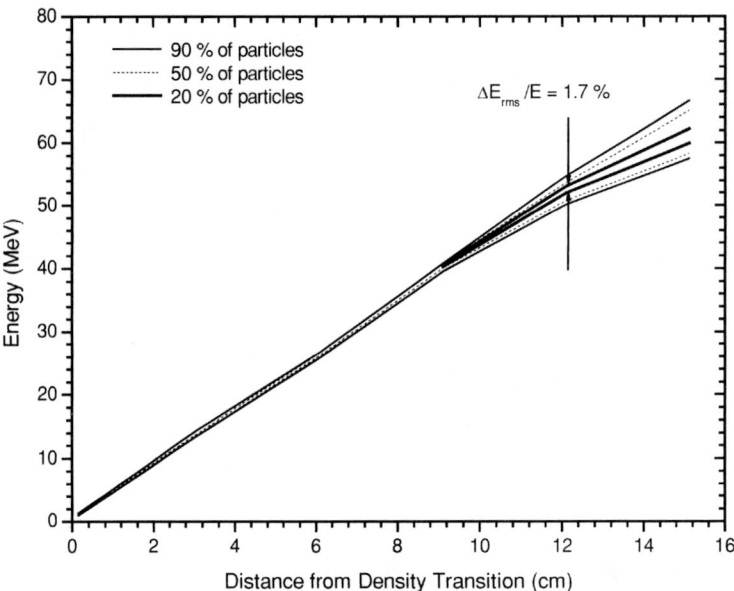

FIGURE 3. Energy of the trapped plasma electrons for three different cases (90 %, 50 % and 20 % of the trapped particles).

RADIUS OF THE TRAPPED BEAM

When the trapped beam propagates in the wake field, they are accelerated longitudinally and focused transversely at the same time. During acceleration the trapped beam radius R is governed by [12]

$$R'' + \frac{\gamma'}{\beta^2 \gamma} R' + \frac{\gamma''}{2\beta^2 \gamma} R + K_0 (\gamma^2 f_e - 1) \frac{1}{R} - \frac{\varepsilon_n^2}{\beta^2 \gamma^2} \frac{1}{R^3} = 0, \qquad (1)$$

where R'' and R' indicate d^2R/dz^2 and dR/dz, respectively, f_e is the neutralization factor given by $f_e = n_0/n_{t,b}$ ($n_{t,b}$ = trapped beam density), and ε_n is the normalized effective emittance. In the equation, K_0 is the generalized beam perveance defined by $2I/I_0(\beta\gamma)^3$, where I is the current and I_0 is the characteristic current given by $I_0 = ec/r_e$. As the trapped beam is accelerated, its radius changes. However, the radius gradually reaches an equilibrium (for a matched beam) as the energy of the trapped beam is saturated. In this case, the equilibrium radius $\overline{R}$ is obtained from

$$K_0(\gamma^2 f_e - 1)\frac{1}{\overline{R}} = \frac{\varepsilon_n^2}{\beta^2\gamma^2}\frac{1}{\overline{R}^3}, \qquad (2)$$

which leads to

$$\overline{R} = \left(\frac{\frac{r_e I}{e\beta c} + \sqrt{\left(\frac{r_e I}{e\beta c}\right)^2 + 2\pi r_e \gamma^3 n_0 \varepsilon_n^2}}{2\pi r_e \gamma^2 n_0}\right)^{1/2}. \qquad (3)$$

If the trapped beam is emittance-dominated, i.e., if $\varepsilon_n^2/\beta^2\gamma 2\overline{R}^3 \gg K_0/\overline{R}$, Eq. (3) reduces to the well-known result $\overline{R} = (\varepsilon_n^2/2\pi n_0 r_e)^{1/4}$, while in the case of space-charge-dominated beams, it reduces to $\overline{R} = (I/\pi c n_0 e\beta\gamma^2)^{1/2}$. If the beam is not matched, the envelope of the trapped beam will oscillate around the equilibrium radius $\overline{R}$.

EMITTANCE AND ENERGY SPREAD CHANGE DUE TO COLLISIONS OF THE TRAPPED PARTICLES WITH THE BACKGROUND IONS

When the trapped beam is accelerated in wake field, its emittance changes. This can be caused by several physical mechanisms, but multiple scattering of the trapped electrons with the background plasma ions and adiabatic damping during acceleration may be dominant. In the presence of uniform focusing the scattering with the background ions leads to an emittance growth that is linearly proportional to the longitudinal distance, while the adiabatic damping effect reduces the emittance. Hence, the overall emittance change is the sum of these two effects and it is given by [13]

$$\left(\frac{d\varepsilon}{dz}\right)_{tot} = \left(\frac{d\varepsilon}{dz}\right)_{scatt} + \left(\frac{d\varepsilon}{dz}\right)_{acc}$$

(4)

$$= \frac{1}{\beta^4 \gamma^2} \left[\frac{2\lambda_\beta \rho}{\alpha_f L_0} \left(\frac{m_e}{m_{ion}}\right)^2 - \beta \varepsilon_n \gamma' \right],$$

where ρ is the ion mass density, α_f is the fine structure constant defined by $\alpha_f = mcr_e / \hbar$ ($\hbar$ = Planck constant) and L_0 is the radiation length defined by $d\gamma/dz = \rho(\gamma-1)/L_0$. In the case of high acceleration gradient, which is typical in the plasma wake field acceleration, calculations show that the adiabatic damping effect is several orders of magnitude larger than the emittance growth effect from the multiple scattering. Thus the emittance growth effect due to scatterings can be neglected.

Since the scattering effect does not lead to a significant emittance growth, it can be conjectured that the scattering effect will not increase the energy spread noticeably, either. To support this argument, we can calculate the ion cross section σ_c. The nuclear radius r_n of an ion (with Z protons) for head-on collisions is given by $r_n = 0.57 Z^{1/3} r_e$ [13], and based on this the mean free path $1/n_0 \sigma_c$ can be calculated and shown to be several orders of magnitude larger than the typical plasma length (0.1 ~ 1 m) for the PWFA. Hence, the collision effect on the energy-spread growth can be completely neglected.

DYNAMICS OF THE DRIVING BEAM

Transverse dynamics of the driving beam for the previous simulation example is shown in Fig. 4. It shows that the head part of the beam expands due to the beam emittance, but particles in the middle and tail parts of the beam execute betatron oscillations in the rarefied ion cavity. In the beginning, they move coherently, i.e., they move as a whole. As the beam propagates further, however, the so-called phase mixing gradually occurs and the beam reaches a nearly equilibrium state. This state, which may be good for plasma particle acceleration, can not last long since the beam loses its energy continuously to excite the plasma wave and some beam particles eventually begin to slide to the acceleration phase of the wake field, as shown in Fig. 5. In this case, the driving beam electrons and trapped beam electrons are mixed together and they interact. This causes a rapid quality (represented by emittance and energy spread) deterioration of the trapped electron beam. Thus the trapped beam must be ejected before this happens. If the driving beam propagates further, more particles flow into the acceleration phase of the wake field, but soon these particles also lose their energy and have a slippage to the next period of the plasma wake field.

Eventually a kind of a longitudinal beam break-up occurs and this deforms the plasma wave form seriously.

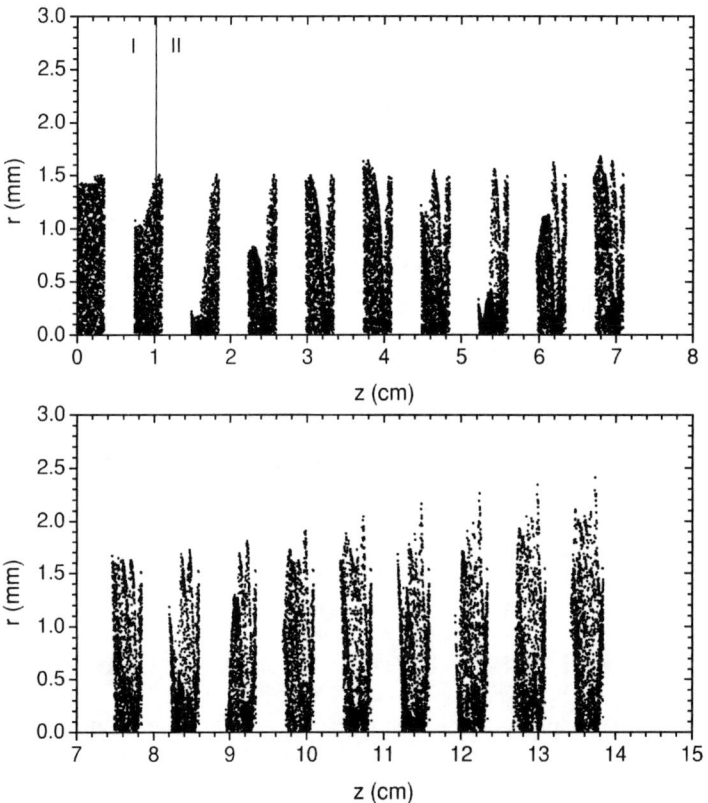

FIGURE 4. Trajectory of the driving beam electrons in the plasma.

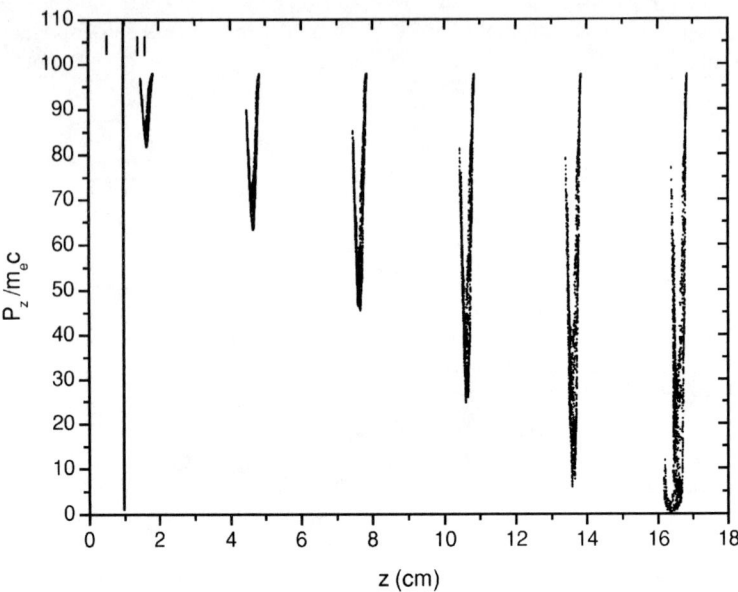

FIGURE 5. Momentum change of the driving beam electrons.

PROPOSED EXPERIMENT FOR THE NEPTUNE LABORATORY

In order to verify the trapping mechanism described above, we propose a proof-of-principle experiment for the UCLA Neptune Advanced Accelerator Laboratory. There are two plasma source options for this experiment. One is to use the existing argon discharge plasma source which was originally developed and tested for the UCLA underdense plasma lens experiment [14]. For the trapping experiment, minor modifications can be made to have a density transition in the plasma. This can be done by installing a mesh at the entrance of the beam-plasma interaction chamber. In this configuration, however, it may be difficult to have a sharp density transition due to plasma diffusion. In this case, another option can be considered. In this option, a rectangular (for example, 5mm×10cm) high-power UV laser beam is illuminated on a lithium oven source transversely and a semi-transparent UV filter is placed to have an intensity step. Thus a lithium plasma column with a sharp density transition can be made. In this configuration, a high power UV laser is required and it is in the UCLA electrical engineering group already. A small (~10 cm long) lithium oven is also needed, but it can be built easily [15]. The Neptune beamline gives an electron beam of about 16 MeV and this is high enough to do the proof-of-principle experiment. However, a high driving-beam charge requirement (>>1 nC) for the experiment will be challenging.

SUMMARY

A new self-injection method using a density transition for the PWFA was studied. The 2-D simulation results show that some plasma electrons are trapped and accelerated to a higher energy than the driving beam energy. Trapped beam quality for the simulation example was obtained and shown to be fairly good (Q =170 pC, $\Delta E_{rms}/E$ =1.7 %, $\varepsilon_{n,rms}$ =1 mm-mrad). Collisions of ions and the trapped beam was investigated and its effect on the beam emittance and energy spread growth was shown to be negligibly small. Finally two different options for a proof-of-principle experiment was proposed for the UCLA Neptune Laboratory.

ACKNOWLEDGMENTS

One of the authors (H.S.) would like to thank Dr. Chris Clayton at UCLA for helpful discussions on the rectangular UV beam method. Eric Esarey was supported by the U.S. Department of Energy, Contract No. DE-AC-03-76SF0098.

REFERENCES

1. P. Chen et al., *Phys. Rev. Lett.* **54**, 693 (1985).
2. J.B. Rosenzweig et al., *Phys. Rev. A* **44**, R6189 (1993).
3. T. Tajima and J. Dawson, *Phys. Rev. Lett.* **43**, 267 (1979).
4. For a review, see E. Esarey et al., *IEEE Trans. Plasma Sci.* **24**, 252 (1996).
5. D. Umstadter et al., *Phys. Rev. Lett.* **76**, 2073 (1996).
6. E. Esarey et al., *Phys. Rev. Lett.* **79**, 2682 (1997).
7. S. Bulanov et al., *Phys. Rev. E* **58**, R5257 (1998).
8. H. Suk et al., submitted to *Phys. Rev. Lett.* for publication (2000).
9. B. Goplen et al., *Comput. Phys. Commun.* **87**, 54 (1995).
10. T. Katsouleas, *Phys. Rev. A* **33**, 2056 (1986).
11. D. Whittum et al., *Phys. Rev. Lett.* **67** (1991).
12. M. Reiser, *Theory and Design of Charged Particle Beams*, John Wiley & Sons, New York, Chap. 4 (1994).
13. J.D. Lawson, *The Physics of Charged-Particle Beams*, Clarendon Press, Oxford, Chap. 5 (1988).
14. H. Suk et al., *IEEE Trans. Plasma Sci.* **28**, 271 (2000).
15. P. Muggli et al., *IEEE Trans. Plasma Sci.* **27**, 791 (1999).

Development of Multiple Beam Guns for High Power RF Sources

Lawrence Ives, George Miram

Calabazas Creek Research, Inc., 20937 Comer Drive, Saratoga, CA 95070-3753, USA

Abstract. The next generation of accelerators will require RF sources producing many megawatts of RF power. Typically, such sources operate at voltages exceeding several hundred kV, requiring pulse modulators and significant amounts of x-ray shielding. In addition, electrical circuits must be designed to handle the high electrical stresses imposed by such high voltages. Multiple beam guns allow for dramatic reduction in the operating voltage while still allowing generation of high RF power levels. These devices have been developed for power levels up to several megawatts; however, significantly higher RF power levels will be required for new accelerators. This paper describes a program to extend the power level of multiple beam devices above 50 MW using doubly convergent electron guns with confined flow focusing.

INTRODUCTION

The next generation of high energy accelerators will require RF sources producing output power levels in the range of 50 - 150 MW at frequencies from 700 MHz to 11 GHz and higher. Traditional klystrons are being developed to provide this power; however, they typically operate with beam voltages of several hundred kV. The proposed 75 MW klystron for the Next Linear Collider (NLC), for example, will require beam voltage of 490 kV with a beam current on the order of 250 A, assuming operating efficiency of 60% [1]. One of the major cost drivers for these devices is the high operating voltage. Not only are the power supply costs significant, but the high voltage leads to increased circuit length, higher radiation hazards, larger insulating ceramics, and increased problems with high voltage breakdown.

One way to avoid high operating voltages is to use a klystron with a multiple beam electron gun to raise the effective perveance. In the multiple beam gun (MBG), the cathode emits a number of 'beamlets' that traverse the tube in separate beam tunnels. This reduces space charge forces that drive the voltage requirement. The perveance of each beamlet can be lower than would otherwise be necessary, leading to increased efficiency and greater bandwidth.

This paper describes an approach to multiple beam gun design where the magnetic field strongly controls the electrons in a confined flow configuration. This allows determination of cathode placement based on area convergence, magnetic field compression, beam location, and cathode loading.

The current generation of MBGs typically employs Brillouin focusing. With Brillouin focusing, the forces on the electrons due to the magnetic field are equal to the space charge forces in the beam, tending to cause divergence. If the electrons enter the magnetic field with no radial motion, then the beam will be transported with a constant beam diameter. In a typical implementation, as shown in Figure 1, electrons are emitted from the cathode in a region where the magnetic field is essentially zero, and the beam is electrostatically focused to a point within the beam tunnel. Iron is used to prevent the leakage of magnetic flux into the cathode-anode region. The geometry is designed so that the magnitude of the magnetic field achieves the Brillouin value at approximately the minimum beam diameter resulting from electrostatic focusing. If precise matching is achieved, the beam will be transported through the device with no change in beam diameter.

In practice, however, it is difficult to achieve the precise conditions for Brillouin flow. Any error in achieving these condition results in electron beams where the beam envelope oscillates about the equilibrium position, and the beam is said to "scallop." Even if the precise conditions are achieved, the balance is disrupted when the beam undergoes electron bunching within an electronic circuit, such as a klystron or traveling wave tube.

These problems are reduced if the magnetic flux is allowed to penetrate through the electron emitter or cathode. This allows a stronger magnetic field that will avoid scalloping and provide additional control of the electron beam in the presence of electron bunching. This is referred to as confined flow focusing. In a typical configuration, as shown in Figure 2, the opening in the iron polepiece is increased to allow magnetic flux to penetrate into the cathode-anode region so that electrons are emitted along the appropriate flux line. Additional magnet coils may also be used to provide precise control in this region. For most applications, magnetic fields of 2-3 times the Brillouin value are usually sufficient for beam transport through the RF circuit.

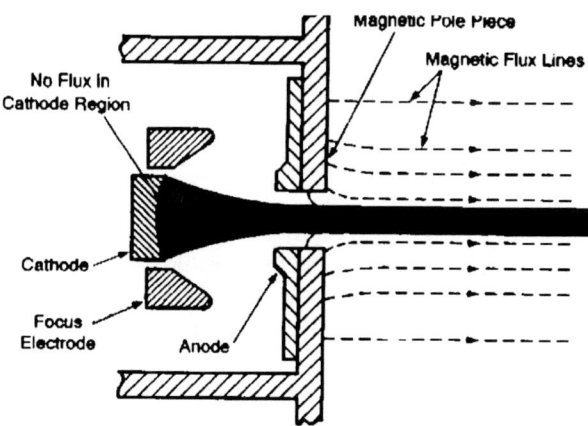

Figure 1. Gun and polepiece configuration for launching Brillouin flow. Reprinted with permission from *Principles of Traveling Wave Tubes* by A. S. Gilmour, Jr., Artech House, Inc., Norwood, MA, USA, www.artechhouse.com, 1994.

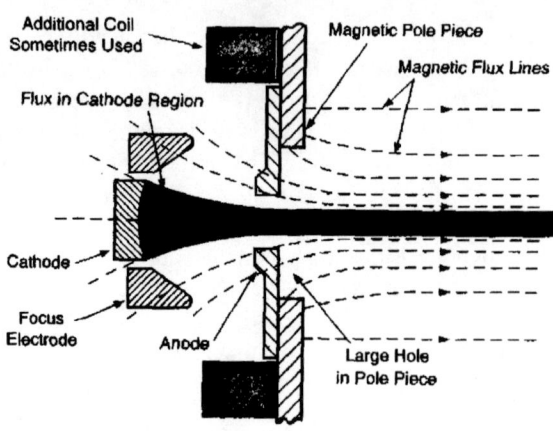

Figure 2. Magnetic field for launching confined (immersed) flow. Reprinted with permission from *Principles of Traveling Wave Tubes* by A. S. Gilmour, Jr., Artech House, Inc., Norwood, MA, USA, www.artechhouse.com, 1994.

Confined flow focusing for multiple beam guns, however, presents several challenges. For those beams whose center is radially positioned away from the device axis, the flux penetrating through an opening in the iron pole piece is no longer symmetric with respect to the center of the emitter. The presence of radial shear in the magnetic field at the cathode will distort the electron beam and prevent adequate transmission through the RF circuit. It is, therefore, necessary to understand the precise configuration of the asymmetry and develop techniques for either removing the asymmetry or modifying the electron emitter to provide precise beam focusing through the region of interest.

An additional complication arises for doubly convergent MBGs. Figure 3 demonstrates the difference between singly and doubly convergent MBGs. Singly convergent MBGs can be employed at lower frequencies where the cavity radius can be sufficiently large, or for lower power devices where cathode current densities are not a major issue. For higher frequency devices at high power levels, for example, an X-Band klystron over 50 MW, the doubly convergent design is required. This allows the cathode size to be large enough to keep cathode emission densities within acceptable limits while providing the electron beam at the proper radius for the RF cavities.

A key issue for doubly convergent MBGs is the rotation of the beamlets as they are compressed toward the axis of the tube. If the electrons are emitted in a nonzero magnetic field, they will possess canonical angular momentum. In addition, v x B forces may impose a rotational drift. Part of the electric field will be a resultant of space charge forces within the beamlet and will depend on beam voltage and perveance. It will be desirable to minimize this motion and account for the remainder in the anode and drift tunnel design.

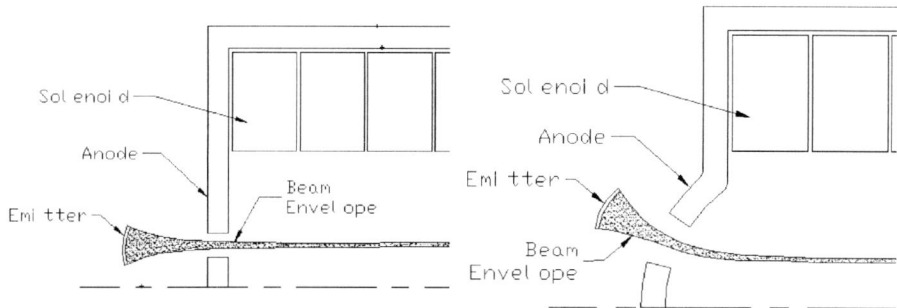

Figure 3. Configuration on the left is a singly convergent electron gun. A doubly convergent gun is schematically shown on the right.

A number of analytical tools are being used to understand the beam dynamics and to model three-dimensional magnetic and electric fields. A new 3D beam optics program called Beam Optics Analysis (BOA), which is currently in development, will allow complete modeling of the charged particle beam from the gun to the spent beam collector. The current suite of design tools includes a 2D version of BOA, the finite difference charged particle code UGUN, the magnetostatic-electrostatic analysis programs Maxwell 2D and Maxwell 3D, and the 3D finite difference program MAFIA.

Research to date has examined both singly and doubly convergent multiple beam guns. For the singly convergent gun, the approach is to analyze the fields and particle dynamics for an on-axis electron beam and for an identical beam displaced in radius. The differences are determined and quantified in order to understand the impact of the radial displacement. Techniques are then explored to compensate for these changes so that the beam dynamics and field of the off-axis gun duplicate those of the on-axis gun. If the field conditions for both guns are identical, the initial assumption is that the beam transport characteristics will also be essentially identical, resulting in confined flow transport through the RF circuit.

In actuality, the principle challenge involves only the magnetics design. The symmetry of the electric fields is easily maintained, at least for singly convergent guns, by maintaining the configuration of cathode, focus electrode, and anode about the axis of each individual beam.

Significant progress has also been realized in the development of doubly convergent MBGs. The challenge is similar to that in the singly convergent case, that is, to properly shape the magnetic field to guide the electrons along the intended path. Research to date incorporates primarily 2D modeling. What the 2D development does, in effect, is model a series of concentric, cylindrical beams. A key question to be resolved was the difficulty in producing a magnetic field configuration with perpendicular flux lines emitting from the curved cathode surface to minimize beam scalloping and maintain high beam quality.

A design configuration was developed incorporating strategically placed iron to accomplish this purpose. As will be shown later, configurations were obtained producing perpendicular magnetic fields in the vicinity of the emission surface. The configurations developed represent a novel, innovative approach to beamlet control in a highly convergent, multiple-beam, electron gun.

Another important issue relates to the magnitude of the magnetic field required to bind the electrons to the flux lines. Based on results obtained by Stanford Linear Accelerator Center (SLAC), it is expected that a magnetic field that is 2-3 times Brillouin will be sufficient to bind the electrons to the flux lines in the gun region. Once the beamlets enter their individual beam tunnels, forces from adjacent beams will be eliminated, and the electrons will continue to follow the magnetic field through the tube. The field will also be sufficient to bind the electrons to the flux lines in the cavity gaps of the RF circuit.

A schematic diagram of a 19-emitter gun is shown in Figure 4. Currently no analytical tools for 3D gun trajectory simulations are available; however, Calabazas Creek Research, Inc. (CCR) working with Ansoft Corporation, developed Beam Optics Analysis (BOA), a 2D gun particle simulation code using finite element, unstructured, adaptive meshing. BOA was constructed using object-oriented programming and modern data structures to facilitate the introduction of 3D elements and full 3D capability. CCR is currently funded through a U.S. DOE Small Business Innovative Research program to implement the 3D elements and emission algorithms (Grant Number DE-FG03–00ER92966). When available, the 3D code will be used to determine the effects of magnetic shear on the shape and stability of the beamlets as well as azimuthal drifting, or corkscrewing, of the beamlets.

The current research program will result in a complete methodology for the design of multiple beam guns, including magnetics design, beam simulations, thermal analysis procedures, and mechanical design techniques for ensuring proper alignment of the cathode with the magnetic circuit. Test models of multiple beam guns will be built and tested to verify the simulation predictions and extend the technology in a series of experiments. The simplest gun that demonstrates the basic concepts will be built first and tested and compared with simulations. After these tests are complete, the gun will be incorporated into a beamstick, which will simulate the transmission through the entire circuit into the collector.

Figure 4. Possible emitter configuration for 19 beam gun.

The device will be targeted toward linear accelerator applications, though other uses are envisioned. The commercial potential of such a device is significant if it is reliable, meets the requirements of RF circuit designers, and can be manufactured at reasonable cost. Such a device would be applicable for all high-peak power RF sources for charged particle accelerators. The NLC will require thousands of klystrons and the Muon Collider will require several hundred. If the multiple-beam gun can be developed for these applications, particularly if it can incorporate permanent magnet focusing, the commercial potential is quite large, and it will dramatically decrease the cost of new accelerator systems.

In particular, a multiple beam klystron operating below 200 kV would eliminate the requirement for a pulse forming network and pulse transformer and allow use of switch mode power supplies. A 140 kV switch mode power supply is currently being developed and tested by Diversified Technologies, Inc., under a Phase II DOE SBIR program, and it is anticipated that this switch could be extended to 200 kV. A single 200 kV supply could be used to power a series of klystrons using a relatively simple design. The recharge current for each klystron would be approximately 0.168 A, so the supply could be sized to provide the current demands of the klystrons connected to the supply. This consists simply of stacking power supply modules together within the main supply.

A key issue for large accelerators is the overall efficiency of operation. With the pulse modulator, the power expended during the rise and fall time of the pulse is lost. By some estimates, this is approximately 20-30% of the total power. This rise and fall time are virtually eliminated with the solid state power supplies.

The doubly convergent cathode design investigated in this program incorporates a modulating/ focus grid surrounding each individual beamlet. This allows increased separation between the cathode and anode to reduce field gradients. This also allows the cathode to be maintained at full beam voltage with the beam modulated on and off using the grid. Simulations predict that a voltage swing of approximately 15 kV will be required to go from beam-off to beam-on operation. This might eliminate the need for a high voltage switch to pulse the cathode and would result in additional simplification in the power supply and improved efficiency of operation.

One must also consider the additional problems incurred when dealing with very high voltage components, including cabling, breakdown, interlocks, protection, etc. Another important consideration is the difference in x-ray generation between 500 kV electrons and 200 kV electrons. The shielding requirements are not linear, and the higher voltage will require significantly more shielding than the 200 kV approach.

SINGLY CONVERGENT MULTIPLE BEAM GUNS

The research program investigated an approach to multiple beam gun development where the magnetic field strongly controls the electrons in a confined flow configuration using a singly convergent configuration. The magnetic flux profile required to transport each beamlet through the circuit is determined consistent with the desired convergence and perveance. Once the magnetic field profile is shaped as required, the cathode is optimized to fit the magnetic profile in the gun region.

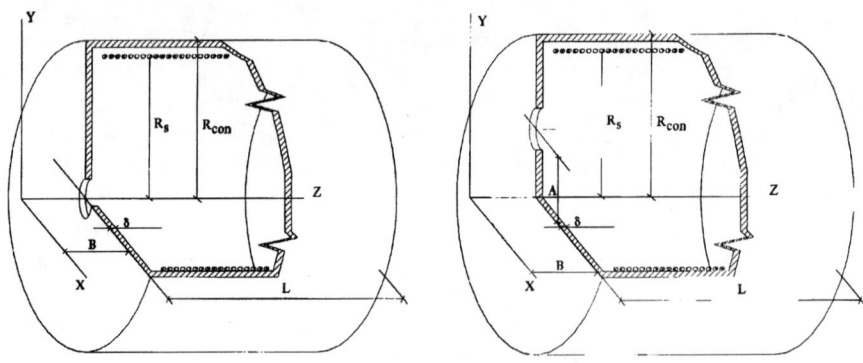

Figure 5. Magnetic circuit for axial (A) and off-axis (B) electron beams. Dimensions are A= 50 mm, B=60 mm, Delta=5 mm, L=180 mm, R_s=100 mm, R_{con}=130 mm. The hole diameter is 27.4 mm.

Research on singly convergent multiple beam electrons guns examined an S-Band application requiring an output power level around 1 MW. The program investigated the magnetic field configurations produced by the magnetic circuits shown in Figure 5. Configuration A represents a typical, axial device where an opening in the iron pole piece is used to provide the required flux leakage for confined flow focusing of an axial gun. In configuration B, the opening is radially displaced as required for an off-axis electron beam. For the S-Band application investigated, this displacement was 5.0 cm.

Figure 6 shows contours of magnetic field for the two configurations. Note that there is axial symmetry about the cathode axis for the axial configuration and that there is a variation for the off-axis configuration. As expected, the fall-off in the magnetic field is less pronounced toward the device axis.

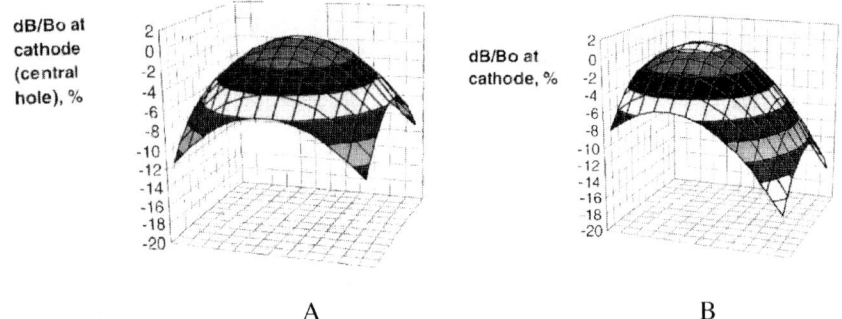

Figure 6. Three-dimensional plots showing variation of magnetic field about the center of each beam location. Plot A is for the beam centered on the axis and plot B is for the beam centered about the off-axis beam tunnel.

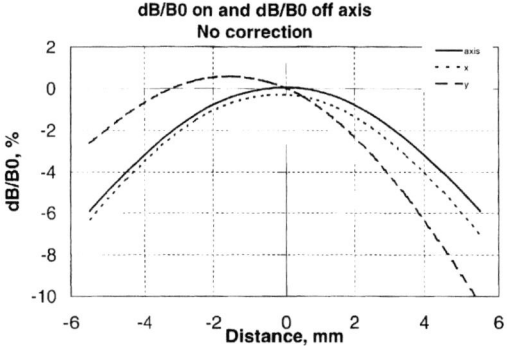

Figure 7. Variation of magnetic field along the X and Y axes, centered along the respective beam axes for the axial beam and the off-axis beam

This is further demonstrated by x-y plots of the magnetic field centered on the axis of each beam tunnel, as shown in Figure 7. This figure plots the radial variation in magnetic field as a percentage of the field at the cathode. The solid line represents the field variation for the axial case, and the two dashed lines represent the variations in the X and Y directions for the off-axis case. The x-direction corresponds to the azimuthal component at the center of the cathode and the y-direction corresponds to the radial direction. For the off-axis case, negative values of distance are in the direction of the device axis, where, as expected, the field falls off less rapidly.

Once the magnitude of the magnetic asymmetry was quantified, techniques were developed to correct the asymmetry of the off-axis cathode. A configuration consisting of strategically placed iron and an external coil were designed to reduce the asymmetries, as shown in Figure 8. Note that the field is symmetric about the center of the cathode; however, the field falls off more quickly with perpendicular distance.

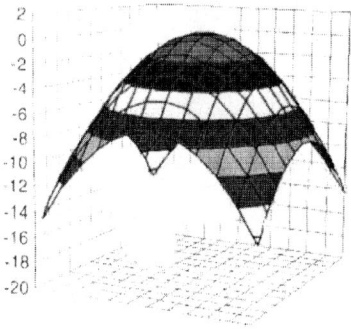

Figure 8. Variation of magnetic field about beam center for off-axis beam for new design.

This is further demonstrated by Figure 9 which also indicates that there is a slight variation between the field shape in the x (azimuthal) direction and the y (radial) direction), though the variation is less than 1% for a 1 centimeter diameter cathode.

Efforts are now in progress to correct the remaining differences using specially shaped iron structures around the gun solenoid. The next step will be to implement azimuthal variations in the iron structure of the gun coil to reduce the difference in field profile in the x and y directions at the location of the off-axis cathode.

Doubly Convergent Multiple Beam Guns

While the singly convergent multiple beam gun is adequate for the 1 MW S-Band application, it can not be used for power levels required by the current and next generation of high-energy accelerators. As the required beam power increases, the current emission density at the cathode begins to exceed levels appropriate for long life. Current oxide cathode technology is such that emission densities in excess of 25 A/cm^2 will only provide approximately one thousand hours of operation before barium depletion causes failure of the cathode.

Doubly convergent multiple beam guns are required for RF sources producing more than 50 MW at frequencies at X-Band or higher [2]. The proposed configuration would employ a ring circuit where the beam–RF interaction occurs in one or more concentric rings about the axis of the tube. The ring circuit is driven by mechanical considerations related to the number of individual beams that can pass through the interaction gap at 11.424 GHz. The ring circuit allows greater spacing for separation of the individual beamlets and increases the mechanical integrity.

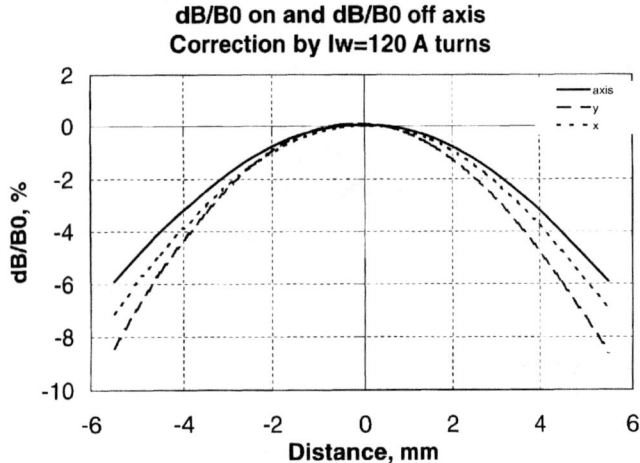

Figure 9. Variation of magnetic field along the X and Y axes, centered along the respective beam axes for the axial beam and the off-axis beam with the new design

Figure 10 shows an electrostatic simulation of a doubly convergent multiple beam gun appropriate for a 75 MW, X-Band klystron operating at 150 kV. As this 2D simulation demonstrates, an acceptable electrostatic design can be achieved for concentric beams. The more significant challenge was development of a 2D magnetic field that provided flux lines that were perpendicular to the cathode surfaces and properly shaped to focus the beams through the concentric beam tunnels. Figure 11 shows a magnetic field configuration that provides for area convergence as well as focusing of individual beamlets. Iron structure is required in the vicinity of the emitter to properly shape the field. A mechanical configuration was designed such that the temperature of all magnetic materials remained below the Curie point while using modular construction techniques.

The next phase of the research will focus on modeling the magnetic field configuration in three dimensions using MAFIA and Maxwell 3D. The goal will be to obtain magnetic field configurations where the contours of magnetic field are parallel to the cathode emission surfaces and the gradient of the field contours are everywhere parallel to the desired electron beam trajectories. The field structures must properly focus the electron beamlets into their respective beam tunnels. The gradients must be similar to those shown in Figure 11 but applied to three-dimensional cathodes, like those shown in Figure 4, rather than the current 2D cylindrical cathodes.

Once an acceptable electrostatic and magnetostatic field configuration is obtained, particle simulations can be initiated, assuming a 3D charged particle code is available. Some preliminary simulations can be performed using MAFIA; however, the size of the problem will quickly become unmanageable for a regular mesh program. Key questions for the 3D analysis relate to possible rotation of the electrons about the device axis as well as about the local axis of each beamlet. Small rotation rates may be manageable in a practical RF device, but large rotation rates may require more innovative devices than are currently contemplated.

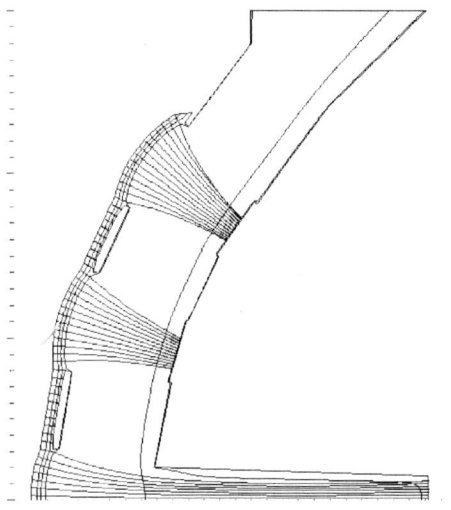

Figure 10. Xgun simulation of multiple beam gun at 150 kV.

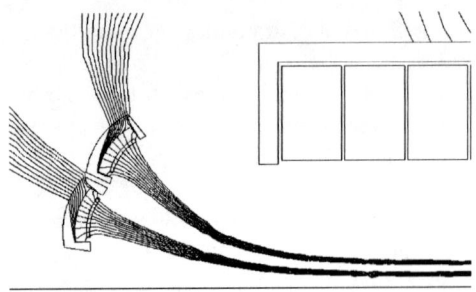

Figure 11. Magnetic field configuration for two circumferential rows of doubly convergent electron beamlets.

Field Emission Cathodes

Another area of research directly relevant to multiple beam devices is the development of high emission density cathodes. Recall that the doubly convergent design was required to keep cathode emission densities below currently acceptable levels. For the current generation of thermionic cathodes, 20-25 A/cm^2 appears to be the upper limit for long lifetime (> 1000 hours) cathodes. Field emission cathodes, however, are capable of several hundred A/cm^2. If multiple beam guns with field emission cathodes can be developed, the requirement for double convergence is eliminated and much simpler configurations, such as those described above for singly convergent guns, could be applicable for high power devices.

Field emission cathodes have other advantages. The emission current is determined by the gate voltage and can be varied by more than 5 orders of magnitude with a voltage change on the order of 100 volts. This provides unprecedented control for switching of the electron beam providing high efficiency operation. Another advantage is that the devices are manufactured using conventional solid state etching techniques. The cathodes could be produced in high volumes at very low cost.

Principle disadvantage is their fragility. Even a small arc through the cathode will completely destroy the device. This is a serious problem, since arcs are common during initial processing of microwave tubes. Advances in this area are being made, however. Northrop-Grumman recently reported operation of a traveling wave tube that incorporated a field emission cathode. The tube went through typical processing, including high temperature bakeout, and operated successfully for many hours before an arc in an unprotected grid circuit destroyed the tube [3]. While the voltage and current were considerably less than would be required in a high power device, this was the first demonstration that field emission cathodes could be implemented in typical vacuum tube.

This research program will investigate techniques for processing field emission cathodes up to the high voltage and currents required for 50-100 MW RF devices. If promising techniques can be developed, prototype guns will be built and tested at the high voltages and currents required.

SUMMARY

The development of high power multiple beam guns will require addressing many challenging issues. A key element for success will be development of advanced modeling tools applicable to complicated 3D structures. It will also require innovative solutions to generation of the complex electrostatic and magnetostatic fields required for proper focusing of the individual beamlets through the RF circuit. Parallel development of high emission density cathodes offers the hope that simpler solutions can be incorporated for generating the high beam currents required for the next generation of RF sources.

ACKNOWLEDGEMENTS

This research was initially funded by a Small Business Innovative Research (SBIR) program from the U.S. Department of Energy (DE-FG97ER82341). Subsequent funding was obtained from the U.S. Air Force (Prime contract number F34601-96-C-072C, subcontract E3X-JS4292). Current research is funded by U.S. DOE SBIR grant number DE-FG03-00ER82964. Funding is also provided by a U.S. DOE SBIR program to develop a 3D finite element charged particle analysis program with adaptive meshing (DE-FG03–00ER82966).

REFERENCES

1. Robert M. Phillips and Daryl W. Sprehn, "High-Power Klystrons for the Next Linear Collider," Proceed. of the IEEE, Vol. 87, No. 5, May 1999.
2. Development of Multiple Beam Guns for High Power RF Applications, Final Report of Phase I SBIR contract Number DE-FG03-97ER82341, Calabazas Creek Research, Inc., May 1998
3. Dave Whaley, Northrop-Grumman Corporation, private communications, March 2000.

Symmetric And Asymmetric Mode Interaction in High-Power Traveling Wave Amplifier

Pingshan Wang, Zhou Xu, Chris. Grabowski, John A. Nation, Samer Banna and Levi Schächter

School of Electrical Engineering and Laboratory of Plasma Studies,
Cornell University, Ithaca, NY 14853

Abstract. High power microwave TWT amplifier operation has been studied for use in electron accelerators. The performance of the amplifiers has been marred, in some cases, by pulse shortening of the microwave signal, possibly due to hybrid HEM_{11} mode interaction with the beam. In this paper we describe experiments which investigate high power operation and the effects of HEM modes on the amplifier performance. We report the high output powers (>50MW) with efficient (>54%) amplification of microwaves in an X-band traveling wave amplifier, and present preliminary data showing operation at a few megawatt output levels in Ka band. In some experiments peak power levels in X-band exceeding 120 MW were measured at an efficiency of 47%. The excitation of the asymmetric hybrid electro-magnetic mode was monitored carefully, but does not seem to have a critical impact on the main interaction process in spite of the fact that its dispersion curve almost overlaps that of the symmetric interacting mode. Theoretical analysis of the interaction in a tapered traveling wave structure indicates that, even if the amount of power in the asymmetric modes at the input of the structure is comparable to that in the symmetric mode, the asymmetric modes cause no power reduction in the symmetric mode. For the case of off axis beams the TM_{01} output power may drop by about 30% and the power in the hybrid mode reach about one third of that in the symmetric mode. In order to avoid hybrid mode excitation it is necessary to suppress the reflections from both ends of the output structure several dB below the gain level of the asymmetric mode.

INTRODUCTION

Spectrum improvement has been one of the major concerns in the development of relativistic TWT amplifiers. Previously reported results included side-bands observed at all power levels with the output spectrum extending over ~300 MHz, which at the highest output levels carried up to ~50% of the power[1]. The side-bands were attributed to reflections within the tube. In the experiments described in this paper we make extensive use of slow wave structure tapers to minimize reflections, and use SiC severs and absorbers to further reduce reflected signals. We demonstrate that the use of appropriately distributed severs and absorbers eliminates the sideband formation without limiting the high beam power to microwave conversion efficiencies. Particle-in-cell (PIC) simulations[2] using the Magic Code[3] have shown that a combination of tapering of the iris aperture with each successive period until the desired aperture is reached, together with extensive use of absorbers, can lower the reflection level dramatically. This technique is useful in TWT amplifiers with ~40-50 dB gain over the instantaneous bandwidth of the amplifier.

We further report in this paper a series of experiments designed to examine the importance of hybrid modes in high output microwave power (~100 MW) X-band traveling wave tube amplifiers. Two amplifiers were designed and tested such that in one case the passbands of the lowest order TM and HEM modes almost overlapped, while in the other the modes were well separated. Hybrid modes have been shown to be important in particle accelerators and may be responsible for the microwave pulse shortening observed in some high power microwave sources[4]. In these experiments the hybrid modes were found to be unimportant and the microwave pulse duration was equal to that of the primary pulse power even at output power levels of 120 MW.

High beam to microwave conversion efficiency is another driving issue in TWT investigations. In the transit-time isolation experiment reported above the efficiency was ~ 20%[5]. Two of the possible reasons for low conversion efficiency are the re-absorption of microwave energy by the 'used' electrons and inefficient beam bunching. Coaxial extraction[6] and traveling wave compression[7] concepts have been proposed to solve these two problems. In the first of these we use a coaxial beam dump to decouple the microwaves from the electrons and to simultaneously convert the TM_{01} mode into a TEM mode. The elimination of the microwave re-absorption by the beam electrons increases the amplifier efficiency and has been demonstrated using PIC simulations and in experiments at low efficiency (~20%)[6]. The second efficiency enhancement technique relies on the use of a traveling wave structure with a cold phase velocity considerably above the value required by the synchronous operation condition to produce efficient beam bunching. This concept, which has been developed using PIC simulations[7], showed that a high phase velocity bunching section followed by a lower phase velocity output section can then be used to efficiently extract the energy from a tightly bunched electron beam. However, the off synchronous 'tuning' decreases the growth rate within the structure, increases its length, and enhances the possible growth of hybrid modes. These modes can efficiently drive the electrons to the tube wall and cause beam break-up instability (BBU) problems[8-9]. These effects may nullify the improved TWT amplification process and cause microwave pulse shortening.

Following this, we summarize recent simulation results based on a model reported elsewhere[10-12]. These reveal that as long the beam is well aligned with the structure, excitation of the hybrid modes is small even if we assume that the power at the input is four times that in the symmetric mode.

In the last section of this study the extension of the work to Ka-band amplifier operation is discussed and early results showing gains of ~30 dB and power levels of a few megawatts are reported from a single stage amplifier.

EXPERIMENTAL ARRANGEMENT

We show schematically the experimental arrangement for the amplifiers in Figure 1. An 80 nsec duration pulsed electron beam having an energy of 700 kV and a current of ~200 A in a 7 mm diameter beam is guided by a 1m long, 0.5 T, axial magnetic field through the structures. The beam voltage and current are flat to better than ± 4% and 2% respectively. The input microwave signal to the amplifier is

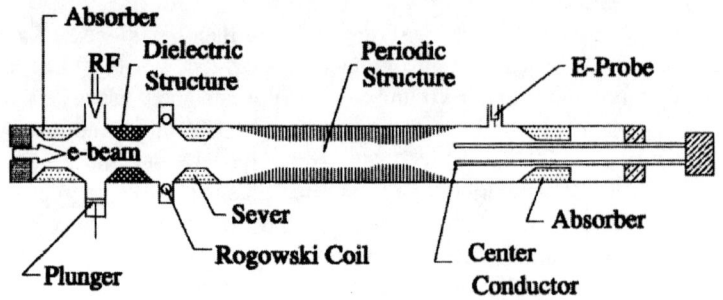

FIGURE 1. Experimental Schematic for X-Band Amplifier.

provided by a magnetron coupled to the amplifier through a sidearm and ranges from a few hundred watts up to about 30 kW. The base pressure in the system is ~10^{-6} Torr. Two amplifier configurations (A and B respectively) have been tested.

A detailed description of the amplifier design and parameters has been presented previously in Applied Physics letters[13] Briefly we note that each amplifier has two separate stages. The first stage is a low gain dielectric amplifier that is used to buffer the main amplifier section from the input. It is followed by a non-uniform second stage in which the first part of the amplifier has a high cold wave phase velocity designed to maximize beam bunching and the second part a low phase velocity section to optimize the rf conversion.

Table 1. Design Parameters for the two amplifiers.

Section		Structure A.	Structure B.
Input Taper		17.13—11.75 mm in 10 cells	14.8 – 8.0 mm in 14 cells
Buncher	R_{ext}	17.13 mm	14.8 mm
	R_{int}	11.75 mm	8.0 mm
	# Cells	45	24
	β_{ph} at 9.0 GHz	1.05	1.05
	β_{ph} at 9.48 GHz	0.94	0.78
	Spatial Growth Rate	1.43 dB/cm at 9.0 GHz	3.3 dB/cm at 9.0 GHz
	Interaction Impedance	49 Ohms	608 Ohms
Transition Taper	R_{ext}	17.13 mm	14.8-15.2 mm in 4 cells
	R_{int}	11.75 – 10.75 mm in 4 cells	8.0 mm
Output Section	R_{ext}	17.13 mm	15.2 mm
	R_{int}	10.75 mm	8mm
	# Cells	7	7
	β_{ph} at 9.0 GHz	0.84	0.84
	β_{ph} at 9.48 GHz	0.72	0.6
	Spatial Growth Rate	1.41 dB/cm at 9.0 GHz	2.79 dB/cm at 9.0 GHz
	Interaction Impedance	29 Ohms	224 Ohms
Output Taper		10.75--17.13 mm in 14 cells	8 –15.2 mm in 14 cells

The relevant design figures for the two amplifiers used are given in table 1 and the dispersion relations for the main amplifier stages are shown in Figure 2. The dispersion curves include the axi-symmetric TM_{01} mode and the lowest branch of the dipole like HEM_{11} mode. The input and output to the second stage consists of a tapered disk loaded guide designed to keep reflections to a minimum. In both amplifiers the periodic length of the disk-loaded sections is 7.5 mm, of which 1.5 mm comprises the disk. The periodicity corresponds to phase advance per cell of $\sim\pi/2$ at 9 GHz.

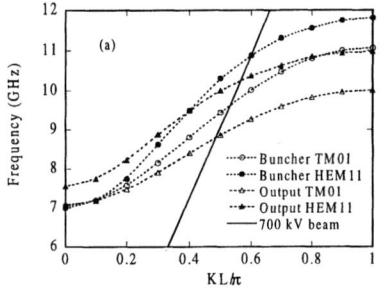 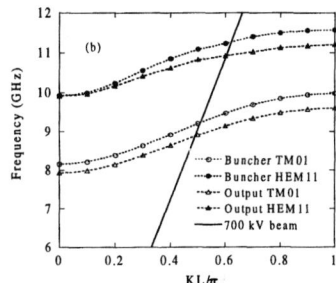

FIGURE 2. Dispersion relations for the two structures tested. In both cases the TM and HEM modes are shown for the buncher and output sections. The beam line is also indicated.

The dielectric and disk loaded sections are separated by a sever consisting of 4 orthogonal SiC fins joined to the outside wall of the tube and tapering in from ~17.13mm to 11.1 mm radius. Each fin has a length of 11.5 cm and a thickness of 3 mm. The single pass attenuation of the sever is greater than 17.5 dB in the TM mode, and its power reflection coefficient is less than 1%. Similarly designed absorbers are located at the input of the amplifier and after the output taper. The sever and absorbers are essential for good operation of the amplifier, since it is important to limit feedback from the amplifier output to the input for both symmetric and asymmetric modes.

The microwave signal from the magnetron is coupled into the cylindrical waveguide immediately following the front-end absorber. Some tuning is available from a second rectangular guide arm located directly opposite to the input arm. The tuner is used to minimize coupling into the TE_{11} /TM_{11} modes of the circular waveguide.

In some of the experiments to be described in the following sections a coaxial output configuration, consisting of a coaxial hollow metallic tube with 10.6 mm outer diameter, was used to de-couple the electron beam from the amplified microwave signal. The coaxial transition also functions as an efficient mode converter from the symmetric TM mode in the circular tube to a TEM mode in the coaxial section. The inner of the coaxial conductor also serves as a beam dump for the electrons. The transition is typically designed to operate correctly close to the end of the output section and before the wave has speeded up in the tapers. For the complete frequency range used the power reflection coefficient is less than 3%.

Measurements of the microwave amplitude are made immediately following the dielectric amplifier and after the disk loaded amplifier output taper. Measurements are made within the coaxial guide, when the inner conductor is in place. The microwave signal is measured using previously calibrated electric field (E field) probes located in the walls of the guide. The detected signal is split into two parts with one going to a crystal detector to determine the power envelope and the other going to a double balanced mixer where the frequency of the amplified wave is checked. The IF signal is also checked to determine the spectral content of the amplified pulse.

EXPERIMENTAL RESULTS

In this section of the paper we describe the experimental results obtained for the two amplifiers studied. In both configurations the same dielectric-loaded waveguide was used as the input buffer amplifier. The role of the first stage is to absorb the injected microwave signal and to pre-modulate the beam. We require a low gain for this stage because the reflection from the input coupler junction is relatively large and it is easy, if the gain is too large, to develop side bands from the reflections. Following the dielectric section we have a sever. A SiC absorber is also located at the output end of the disk-loaded amplifier. Both the sever and the absorber have low reflection coefficients and allow amplifier gains of order 40-50 dB. Note that the sever effectively eliminates the amplified signal from the first stage but allows propagation and further bunching of the electron beam prior to the tapered entrance to the disk loaded sections.

High Power And High Efficiency Operation

Figure 3 shows a typical microwave output pulse superimposed with beam data taken from amplifier A at a location following the output taper in the mode-converter section.

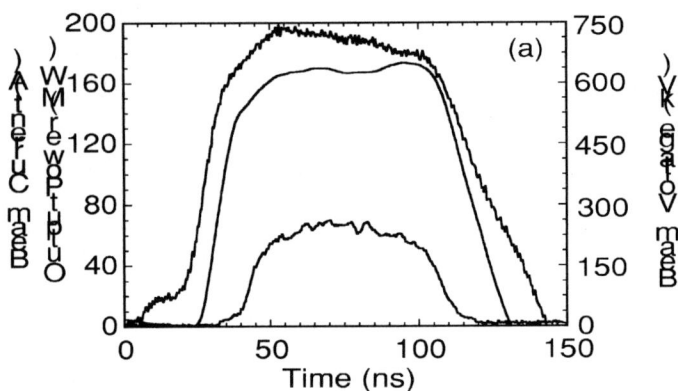

FIGURE 3. Experimental Waveforms showing the Microwave Output Power and the Beam Current and Voltage Pulses

Note that the microwave power signal follows the power envelope of the beam and that there is no evidence of pulse shortening. The data shown corresponds to an output power 78 MW. The microwave conversion efficiency during the 'flat top' of the pulse was 54 %. Allowing for the actual rise and fall time of the beam pulse, and for which amplification did not occur, this data yielded a 43 % energy conversion efficiency. With a longer pulse duration and/or better rise time the power and energy conversion efficiencies should become closer to the measured power conversion efficiency. In the absence of the output coupler the best performance of the amplifiers showed output powers of 63 MW (for A) and 58 MW (for B) from the 700 keV beam. Experiments have also been carried out with a 300A beam current and at a beam energy of 800 keV using amplifier A. Output signals similar to those shown in Figure 3 were obtained at power levels of up to 120 MW corresponding to an efficiency of 47%. These results were obtained in the absence of the coaxial beam dump. Higher power levels and efficiencies would be expected with the output converter in place.

Amplifier Gain and Frequency Response

The output power of amplifier A, measured without the coaxial mode converter shows a well-defined linear operational range for input signal levels of less than 10 kW. The measured gain at 9.478 GHz is about 35 dB. The signal saturates at approximately 60 MW. Less detailed data was obtained for structure B where the gain was approximately 43 dB at 9.145 GHz. In Figure 4 we show plots of the frequency response of the amplifiers indicating that the 3 dB bandwidth is at least 300 MHz with that for

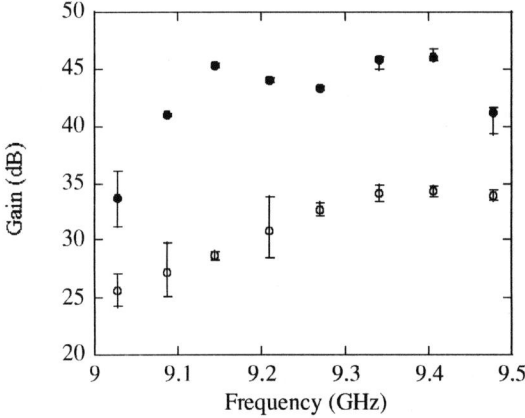

FIGURE 4. Frequency Response for the two amplifiers. The solid dots represent the data for amplifier B and the open dots for amplifier A. Typical data scatter is indicated by the error bars.

amplifier A being, as expected, greater than that for amplifier B. It is noted that above 9.3 GHz the fluctuation level in the output of amplifier B increased significantly. This is expected due to an increase in the output reflection coefficient of the output section above 9.2 GHz. In all other cases however the amplifier outputs are smooth and the measured FFT signals are well behaved and similar to those illustrated in reference 13.

Hybrid Mode Excitation

As indicated above there was no evidence of shortening of the microwave pulse in the experiments. Note that in amplifier A the hybrid mode dispersion relation is almost superimposed on the TM mode, whereas in amplifier B the modes are well separated – see Figure 2. We have in addition carried out a detailed check of the output at frequencies other than the input magnetron frequency. In these experiments, using amplifier A, we used a high pass filter to eliminate the amplified signal at the input wave frequency and searched using the power envelope detector and a double-balanced mixer for signals at other frequencies, especially at the expected hybrid mode interaction frequency. Some microwave power was detected between 9.4 and 9.8GHz with a peak power level about 25-30 dB lower than that of the amplified signal. As may be checked in Figure 2(a), amplification in the TM_{01} mode at the frequency, 9.6 GHz corresponds to synchronous operation with the electrons. Synchronous interactions in the HEM_{11} mode would occur around 10.5 GHz. No signal has been detected at this frequency at power levels down to ~40 dB lower than that of the amplified TM_{01} signal. Similar results were obtained using the second amplifier.

We have attempted to deliberately enhance the probability of hybrid mode excitation by use of an asymmetric beam. In this case the beam cross section was made approximately elliptical with a two to one variation in the major and minor radii. The asymmetric beam was propagated through the tube in two orthogonal orientations. In neither case did we detect any hybrid mode coupling. This result was reinforced by checking the detected signal in two orthogonal planes. The experimental results show an azimuthal symmetric field distribution, another indicator of the absence of HEM_{11} mode. The asymmetric HEM_{11} mode signal level was insignificant even when the input wave excitation was preferentially made using a TE_{11} or TM_{11} mode. Finally we propagated a regular pencil beam off axis so that one part of its outer extremity propagated along the guide axis. Once again there was no evidence of significant growth of the hybrid mode. Further reinforcement of these results was obtained from simulations that rely on a theoretical model [10-12].

Theoretical models developed in the past for the description of the interaction in a traveling wave structures, focused exclusively on the symmetric mode. Recently, in parallel with the experimental studies reported here, we developed [10-12] a theoretical model that takes into consideration asymmetric modes. In principle, it calculates the full 3D dynamics of the particles and 3D variations of the electromagnetic field though we assume that the interaction affects the field variation only in one direction – parallel to the main motion of the electrons. Details of the theoretical results may be found in a paper submitted for publication in IEEE Trans on Plasma Science. Other than the symmetric mode it is assumed that there might be two additional asymmetric modes in the system: one (HEM_{11}) can be excited by the beam in the buncher which will oscillate at 10.866GHz and a second which may be excited by the beam in the output section which will oscillate at 10.227GHz.

Following an extensive series of simulations we draw the following conclusions:

a. If the input hybrid mode power to the amplifier is small then there is virtually no growth in the hybrid mode in the amplifier.
b. If there is an input hybrid mode power of the order of tens of kilowatts then a reduction of order ten megawatts is found in the TM mode power if the beam is off axis. This is accompanied by an increase in the hybrid mode power to about one megawatt corresponding to at least 10dB amplification. Note that in the experiments reported above, the suppression for the reflections in the TM_{01} mode are better than 40dB and although we do not know exactly the suppression of the hybrid mode reflections, even if it is 20dB, then it is still more than sufficient to suppress the 10dB gain associated with the asymmetric modes.
c. For an hybrid mode input of the same order of magnitude as that for the TM_{01} beam modulation (~200kW) there is still no significant effect on the interaction if the beam is on axis. However, when the beam is off axis we observed a significant reduction of the TM mode output power. The power in the symmetric mode dropped below 40MW and the power of the first hybrid mode increased by more than 10dB to about 3MW. A significant impact of the second hybrid mode was observed when the beam is off axis and when the power at the input was about a factor of four larger than the power in the TM_{01}. The power in this case increases by almost 15dB. The reduction in the power in the TM_{01} mode is similar to that in the previous case.
d. Finally, for an off axis beam, if both hybrid modes are present at the amplifier input and they have a total power of about 800 kW, compared to the TM input of about 200 kW, then the TM mode amplification drops substantially to about 25 dB while the hybrid mode power increases again by about ten dB. For an on axis case there is no significant effect on the TM mode amplification.

It is evident that if the electron beam is well aligned, then the excitation of hybrid modes is very unlikely if the overall reflection coefficient is several dB larger than the gain. Therefore the presence of the SiC absorbers as well as the tapers is crucial for the suppression of the hybrid modes as well as eliminating the reflections in the TM_{01} mode.

Ka-Band Amplifier

Based on the successful operation of a relatively broadband X-band amplifier we have initiated work on a Ka band device. The results summarized in the following section have been obtained using a device similar to that used at X-band scaled for operation at 35 GHz. It consists of a buffer dielectric amplifier followed by a disk loaded amplifier. The latter has utilized 70 to 90 uniform cells with ten tapered cells at each end. In these experiments the phase advance per cell is designed to be 90 degrees and the operation is synchronous with the electrons. There was no effort made to optimize the rf conversion efficiency or to maximize the output power. Severs are used as in the X band amplifiers. The passband of the TM01 mode extended from about 24 to 40 GHz and was substantially overlapped by the lower hybrid HEM11 mode. The beam current was approximately 100 A at an electron energy of 850-900 keV. The input microwave signal came from a magnetron and was about 2.7 kW at a frequency of 35.18 GHz.

Based on MAGIC simulations the dielectric amplifier produced a small ~1.75 % modulation of the beam for an input 5 kW microwave signal. The gain in this stage was about 10 dB. The sever separating the dielectric amplifier from the disk loaded section absorbs the microwave signal and, as in our X-band experiments, the modulated beam allows for reconstruction of the rf signal in the disk loaded section. Simulation showed peak amplified powers of 12.6 MW (at a beam current of 140 A.) at the end of the 100 cell disk loaded section. Only a fraction of this power is expected to be available at the output, since some of the wave energy is returned to the electrons in the tapered output. Note that the coaxial output section was not used in this device. The overall gain for the system was predicted to be about 31.5 dB.

In the experiments carried out the beam current was only 100 A and the input microwave power was a factor of two smaller than the level used in the simulations. The experimentally detected power level at the end of the dielectric stage was in the range of 40-80 kW and comparable to that predicted by the simulations. At the end of the output taper of the composite amplifier, peak powers of about 2.0 ± 0.2 MW were detected corresponding to an overall gain of about 28.4 dB. Bearing in mind the differences between the simulation parameters and those used experimentally this agreement is satisfactory. Note that the overall system efficiency is only slightly in excess of 2 % under these conditions. This however is as expected for the parameter ranges examined to date.

Though the power levels and the efficiency are modest, the results do show that the TWT amplifier is operating at gains close to those predicted by numerical simulations. The microwave pulse envelopes are not as flat as desired, but this feature may be partially attributed to the fact that the current amplitude varies by 20 % during the pulse. Further increases in power levels can be obtained by introducing a coaxial transition at the end of the disk-loaded structure and by increasing the current level to ~200 A. Simulation results for a 200 A beam injected into the amplifier predict that power levels of up to 28 MW can then be obtained at an efficiency of 16 %. Further design improvements suggest, in simulation at least, that overall efficiencies of 40 % are achievable.

Discussion of Results and Conclusions

The data presented above showed the efficient operation of two different X-band TWT's using non-uniform slow wave structures, in which traveling wave bunching sections have cold phase velocities well above that determined by the synchronous condition. The experimentally observed amplifier efficiency came relatively close to the PIC code prediction of 58% . The achieved power and efficiency level was comparable to that obtained in recent high power klystron experiments[14,15]. The beam bunching in the amplifiers develops in the traveling wave bunching sections and the power extraction occurs mainly in the short traveling wave output sections. For a given beam the interaction is determined by the wave phase velocity in the slow wave structures. The two amplifiers have different cell dimensions, and therefore different interaction impedances, passbands and lengths. However the cold wave phase velocities and phase-advance per cell at 9.0 GHz for TM_{01} mode were kept constant at $\pi/2$ for both amplifiers. As a result the experimental performance of the amplifiers was similar.

Operation of the buncher in a regime where the cold phase-velocity is high, entails lower gain and consequently requires a longer interaction distance to achieve a given degree of beam bunching. In long uniform structures, the probability of excitation of the BBU instability, associated with the growth of non-azimuthal symmetric HEM_{11} modes, is increased. Its effect would be to drive the beam to the wall [8-9]. The two structures examined have different gain characteristics and dispersion relations. Specifically their gains differ by about 10 dB and the HEM_{11} mode overlaps the TM_{01} mode for structure A and is separate from it in structure B. Structure A is also much longer than structure B. We have not found any experimental evidence of excitation of HEM modes in either structure, as indicated by the absence of wave excitation at the expected HEM mode interaction frequencies, and by the propagation of the injected beam current throughout the whole structure without significant loss. The measured frequency-spectrum of the amplifiers approximates that of a square pulse modulated with a single frequency signal, and its ~20-30 MHz width is approximately equal to that expected for the pulse power duration. The measured single frequency spectrum is indicative of the absence of reflections and the excitation of spurious modes. The three severs and the transition sections contributed to the effective suppression of reflections inside the structures. It is observed in the experiment that the first absorber is essential for stable performance of the dielectric modulation stage and the overall amplifier performance. Structure B has higher gain and higher reflections above 9.2 GHz. As a result the relevant spectral width is somewhat broader. Further minimization of reflections could be achieved by exponential tapering of the cells' inner radii within the transition sections. In the experiments linear tapering was adopted for experimental convenience. It is noteworthy that these spectrum improvement methods do not impose any pulse duration limitations or sacrifice any of the power handling ability of the structures. The successful operation of the amplifiers over the available magnetron frequency range demonstrates that these methods also enable broad-band operation.

We have also successfully demonstrated amplifier operation at 35 GHz with output power levels of a few Megawatts and amplifier gains of 28 dB. These numbers are consistent with code predictions and suggest that efficient TWT amplifier operation should be achievable within Ka-band.

ACKNOWLEDGEMENTS

We wish to acknowledge early contributions made to this work by S. Naqvi and the continued engineering support provided by J. Ivers and S. Wright. The work was supported by DOE and the AFOSR.

REFERENCES

[1] D. Shiffler D., J.D. Ivers, G.S. Kerslick, J.A. Nation and L. Schächter; J. Appl. Phys. **70**(1), 106-113(1991).
See also L. Schächter, J.A. Nation and D. Shiffler; J. Appl. Phys. **70**(1), 114-124(1991).
[2] S. Naqvi, PhD dissertation, Cornell University, 1996
[3] B. Goplen, L. Ludeking, D. Smithe, and G. Warren, Comput Phys. Commun., 87, 55 (1995)
[4] G. Westenskow, J. Boyd, T. Houck, D. Rogers, R. Ryne, Proc. Of 1991 IEEE Particle Accelerator Conference 91CH3038-7 p.646-648 (1991).
[5] E. Kuang, T.J. Davis, G.S. Kerslick, J.A. Nation and L. Schächter; Phys. Rev. Lett. 71(16) 2666 (1993).
[6] S.A. Naqvi, G.S. Kerslick, J.A. Nation and L. Schächter; Appl. Phys. Lett. 69, p. 1550(1996).
[7] S.A Naqvi, J.A. Nation, L. Schächter and Q. Wang; IEEE Trans. Plasma Science - Special Issue, Vol.26(3) p. 840-5(1998).
[8] J.Haimson and B.Mecklenburg, SPIE vol.1629, Intense microwave and Particle Beams III , p. 209 (1992).
[9] R.H. Helm and G.A. Loew, Chapter B.1.4, Linear Accelerator, Eds. P. Lapostolle and A. Septier, pp. 173-221, North Holland Publishing Co., Amsterdam, 1970.
[10] S. Banna, L. Schächter, J. A. Nation, and P Wang; PAC, March 1999, p.3609
[11] S. Banna, J.A. Nation, L. Schächter and P. Wang, Physical Review E – accepted for publication.
[12] S. Banna, J.A. Nation, L. Schächter and P. Wang; IEEE Trans. Plasma Science - Special Issue. Accepted for publication.
[13] P. Wang, Zhou Xu, J.D. Ivers, J.A. Nation, S.A. Naqvi, and L. Schachter; Appl. Phys. Lett., 75, 2506 (1999)
[14] High Energy-Density Microwaves, Edt. R.M. Phillips, AIP Conference Proceedings 474 (1998).
[15] G. Caryotakis, IEEE Trans. on Plasma Science, 22(5), 683-691, 1994.

Development of a 10 MW, 91 GHz Gyroklystron for Accelerator Applications

R. Lawrence Ives*, Wes Lawson[@], Jeff M. Neilson*, Michael Read*

*Calabazas Creek Research, Inc., 20937 Comer Drive, Saratoga, CA 95070, USA
[@]Institute for Plasma Research, University of Maryland, College Park, MD 20742-3285, USA

Abstract. A 10 MW, 91 GHz gyroklystron is under development for W-Band accelerator applications. The device will generate 1.5 microsecond pulses at 120 Hz and will be provided to Stanford Linear Accelerator Center for testing of W-Band accelerator components and subsystems. A magnetron injection gun operating at 500 kV will provide a 55 amp beam for interaction in a 5 cavity circuit. The output will be in a hybrid TE_{01}/TE_{02} mode that can be converted to a more suitable mode at the accelerator. The device is expected to operate with efficiency close to 40% with a gain of 55 dB. A depressed collector will be implemented to allow improvement in the total efficiency to more than 50%.

INTRODUCTION

To expand the experimental frontier of high-energy physics, an international effort is underway to design advanced generations of linear electron-positron colliders with anticipated center-of-mass energies of 0.5 TeV and beyond. While conventional state-of-the-art klystrons at 11.424 GHz may be suitable for the 0.5 TeV energy level [1], the expected performance requirements for RF drivers of linear colliders with energies above 1 TeV are well beyond the state-of-the-art in amplifier technology. Many believe, for example, that a 5 TeV collider extrapolated from current RF-based structures will require a drive frequency somewhere between Ka-Band (e.g. 35 GHz) and W-Band (e.g. 91 GHz) [2].

Several novel source concepts are under development in the United States to meet the anticipated driver requirements. The schemes are usually divided into two classes: discrete sources, which typically utilize long pulses and pulse compression, and two-beam designs, which must produce hundreds of megawatts of power for time durations somewhat longer than the accelerator fill time (depending on the length of the bunch train). Potential discrete sources include klystrons [3], gyroklystrons [4], traveling-wave tubes [5], magnicons [6], CARMs [7], and free-electron lasers [8]. The principal two-beam schemes under consideration involve relativistic klystrons [9] and free-electron lasers [10].

To date, none of the devices mentioned above have demonstrated all the necessary driver requirements for advanced colliders. Of these requirements, the most difficult two appear to be the ability to produce the required peak power and the ability to produce the necessary efficiency. To keep the operating costs of the linear collider manageable, it is expected that the minimum acceptable wall plug-to-accelerator input energy conversion efficiencies for RF drivers will be 40% [11]. Included in the calculation are losses in the power supply, magnet supplies, beam transport, RF interaction, pulse compression, and microwave transport. Because the best RF interaction efficiencies are typically near 50%, there is little room for loss in the other systems. Consequently, this efficiency requirement may well lead to the use of DC supplies, gridded or modulated cathodes, permanent (PPM or solenoidal) or superconducting magnets, and energy recovery (e. g. depressed collector) systems. While these technologies are fairly well developed for low-power, CW, linear-beam tubes, they have not been generally applied to high-power, short-pulse systems. Furthermore, depressed collectors have only recently been applied to gyrotron and other rotating-beam devices [12-15]. These requirements on efficiency and peak power, along with the other requirements for high gain, stability, tube lifetime, average power, phase noise, etc., combine to make the driver design a formidable problem.

First-harmonic gyrotrons have proven to be reliable, efficient, high power sources of microwave and millimeter wave radiation. Gyrotron oscillators have, for example, produced average powers of about 1 MW at frequencies above 100 GHz by operating with highly overmoded cavities [16]. First-harmonic gyroklystron amplifiers have produced over 70 kW of peak power near 94 GHz with a duty factor of about 10% by using moderately overmoded (TE_{011}) cavities and a relatively low beam voltage of 70 kV [17]. This result represents more than an order of magnitude increase over the present capabilities of linear tubes; further increases are possible by moving to the beam voltages that are typically utilized in RF drivers for accelerators. For example, state-of-the-art values of peak power density have already been demonstrated in experiments at the University of Maryland, where a two-cavity, second-harmonic gyroklystron produced 32 MW at 19.7 GHz [4].

Magnetron injection guns (MIGs) were used to generate the rotating beams in all the experiments mentioned above. Simple design rules developed for MIGs indicate that under some circumstances they can be scaled such that the power produced by the MIG decreases only in proportion to the frequency of a device [18]. Because the beam power available is often the limiting factor in low average power gyroklystrons, the gyroklystron power can be subsequently scaled in proportion to the wavelength. This idea has been tested a number of times. A recent design study performed by the group at the University of Maryland showed that the 19.7 GHz result could be scaled to 95 GHz with a simulated output power of about 7.5 MW [19]. The theoretical study included the design of a double-anode MIG and a three-cavity, second-harmonic gyroklystron circuit. The drive cavity operated at the first harmonic (47.5 GHz) in the TE_{011} mode, and the buncher and output cavities operated at 95 GHz in the TE_{021} mode.

Unfortunately, there were a number of problems with the design reported in the literature. The main problem from the standpoint of an RF driver was that the efficiency of 34% was lower than desired. A second problem was that the tube was not zero drive stable, because the output cavity was unstable to the TE_{221} mode near 91 GHz. The simulated beam quality for the MIG was good, but the average cathode loading was near 10 A/cm^2. Furthermore, no designs were given for the input and output waveguide systems. Nevertheless, it appeared that the design had potential for improvement and was a good starting point for a design for the 91 GHz driver for accelerators.

The research program described here indicated that a complete W-Band, second-harmonic gyroklystron system that suffers from none of the defects mentioned above and meets the expected requirements for an RF driver can be designed. A complete description of the proposed design will be provided later in this paper. The design builds on work performed by the Naval Research Laboratory in development of a 94 GHz gyroklystron for radar applications and extends that technology to the high peak powers required for accelerator applications. The power level for the proposed gyroklystron is 10 MW with a calculated efficiency of 37.4%. While the efficiency is lower than the desired 40%, it should be practical to increase the total efficiency to at least 50% with a depressed collector.

The design frequency is 91.392 GHz, which is 32 times the current SLAC frequency and a frequency at which accelerator hardware is already being developed. Values for the beam and magnetic field parameters are given in Table 1. The beam voltage of 500 kV is well within the capability of current modulator technology, and the perveance of 0.14 µPervs is quite low. The nominal beam radius is consistent with operating in the TE_{021} mode in the second-harmonic cavities. The average ratio of perpendicular to parallel velocity of 1.6 is typical for gyroklystron designs. The required magnetic field is quite low for superconducting technology. Given the recent advances that have resulted in low-cost, closed cyro-system superconducting magnets, it should be possible to procure a low-maintenance, low-power-consumption magnet to generate the required field.

Table 1. Gyroklystron Parameters

Parameter	Design Value
Beam Voltage (kV)	500
Beam Current (A)	55
Average beam radius (mm)	1.649
Average velocity ratio	1.6
Drive Frequency (GHz)	45.696
Magnetic Field (kG)	27.6
Output Power (MW)	10.25
Electronic Efficiency	37.4
Gain (dB)	55

All design parameters fall within operating ranges that have been demonstrated with similar devices and development and assembly of a prototype is now under way. A layout drawing of the gyroklystron is shown in Figure 1.

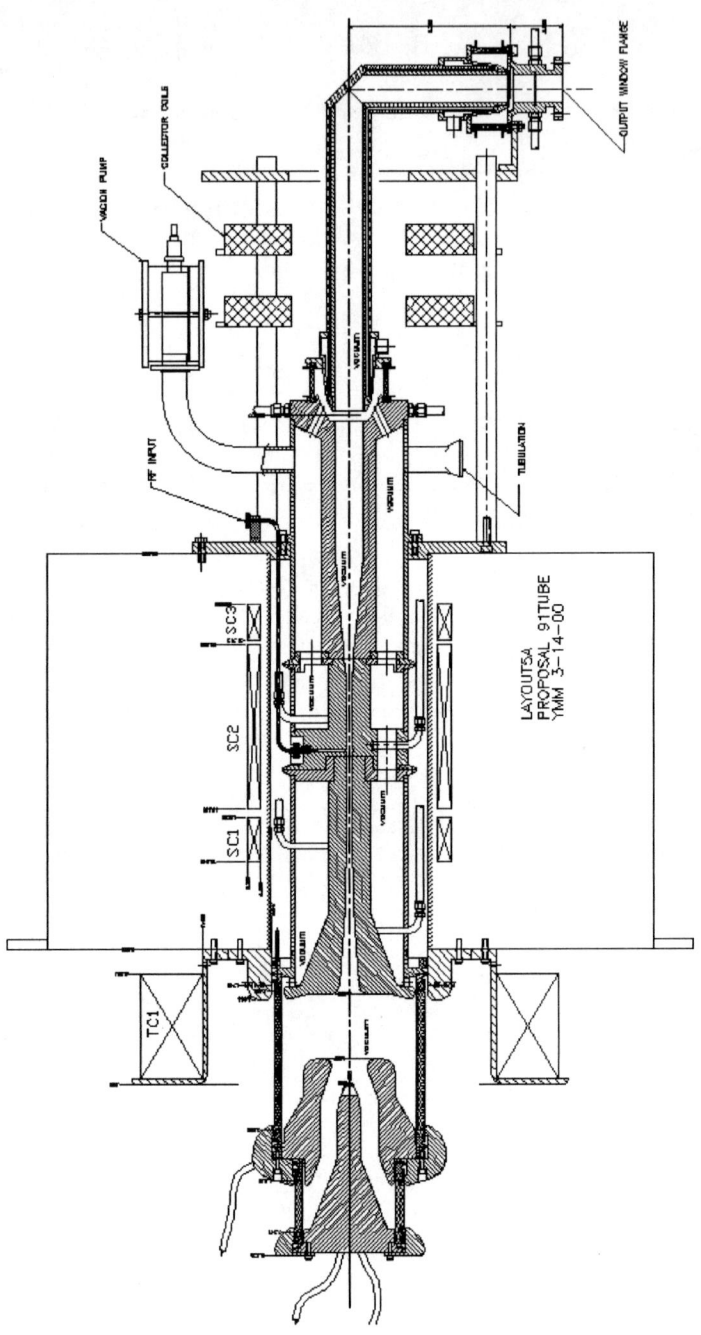

FIGURE 1. Layout drawing of 10 MW, 91 GHz, second-harmonic gyroklystron

GYROKLYSTRON DESIGN

The initial design study analyzed all device subsystems in detail and determined that acceptable designs could be achieved that met the performance requirements while not exceeding accepted limitations on voltage gradients or thermal and mechanical stresses. The sections that follow provide additional detail on all major subsystems and components.

Magnetron Injection Gun

A double-anode MIG that will provide an electron beam with the required specifications for the 91.392 GHz gyroklystron amplifier system was designed. The design is nominally scaled from a 20 GHz MIG that was built by Communications and Power Industries, Inc. (CPI) and ran successfully for many years at the University of Maryland.

The layout for the MIG is shown in Figure 2. The nominal beam power was 27.5 MW and the magnetic compression was 38.4. The control anode voltage was adjusted to produce the required average velocity ratio. The electrodes for the MIG were generated from a sequence of line and arc segments. The magnetic field profile was generated by a set of superconducting coils complemented by a normal-conducting bucking coil located near the control anode - main anode gap.

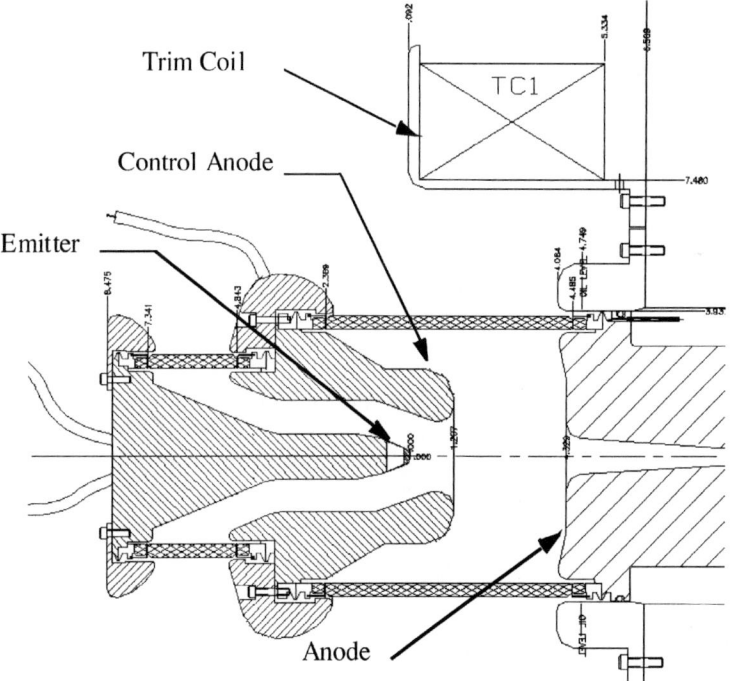

FIGURE 2. Layout drawing magnetron injection gun

Simulations were performed using EGUN. The required velocity ratio of 1.6 was achieved with an axial velocity spread of 3.4% with an average beam radius in the circuit of 1.65 mm. The perpendicular velocity spread was 1.3%, and the beam is approximately half of the space charge limited current. The beam radius provides for a beam clearance in the drift tunnels of 0.014 inches, which is typical for W-Band gyroklystrons.

Detail studies were performed to insure that the device would perform within specifications over the full range of operating parameters. The velocity spread stays below 10% for currents from 40 A to 80 A. The corresponding change in the control voltage necessary to keep the desired velocity ratio goes from 51.5 kV to 56.5 kV. At the nominal current, the velocity ratio can be varied from 1 to 2 by changing the control voltage from 47 kV to 56 kV with the corresponding value of velocity spread remaining below 5.5%. Predicted efficiency exceeds 34% for spreads up to 7%.

Of primary concern were electric field gradients in the electron gun, and several design iterations were required to achieve acceptable values. Maximum gradient occurs along the outside surface of the control voltage electrode and is approximately 170 kV/cm. Peak field gradient along the high voltage ceramic is approximately 30 kV/cm.

RF Circuit

Electrical Design

A schematic of the circuit layout is shown in Figure 3. Five cavities were used to achieve a large signal gain in excess of 55 dB. The input cavity interacts at the first harmonic in the TE_{011} mode; all other cavities interact near the second harmonic in the TE_{021} mode. The nominal drive frequency is 45.696 GHz. The cavities are stagger-tuned to increase the efficiency and bandwidth. The drift tube radius is 0.3175 cm between all cavities. The TE_{01} mode at the drive frequency and the TE_{02} mode at the operating frequency are cut off in the drift tubes by 25% and 15%, respectively.

The walls of the first four cavities are formed by abrupt radial transitions. Mode conversion in the three intermediate cavities from the TE_{02} mode to the TE_{01} mode was minimized by adjusting the cavity length to provide destructive interference. This was required because the TE_{01} mode was not cut off in the drift regions at 91 GHz.

The quality factors were chosen as a trade-off between the desire to increase circuit gain and the need to ensure that no spurious oscillations can exist in the cavities. All modes with azimuthal indices from zero to four in the frequency range from 40 GHz to 110 GHz were investigated. The only potentially troublesome mode was the TE_{211} mode; however, using the techniques described by Castle et al [20], the quality factor for that mode is expected to be less than 40 and therefore stable.

The output cavity geometry is more complex. Adiabatic radial wall transitions were used to define the cavity for two reasons. First, the transitions reduce the maximum wall loading in the output cavity. The wall loading along the outer wall of the output cavity at the nominal design point is indicated in Figure 4. Second, the adiabatic

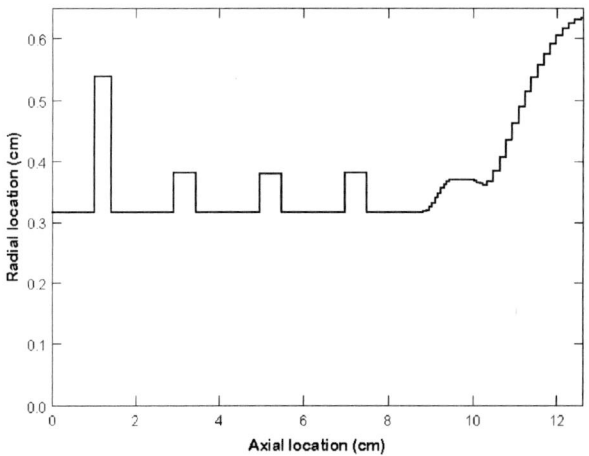

Figure 3. Schematic diagram of the five cavity circuit.

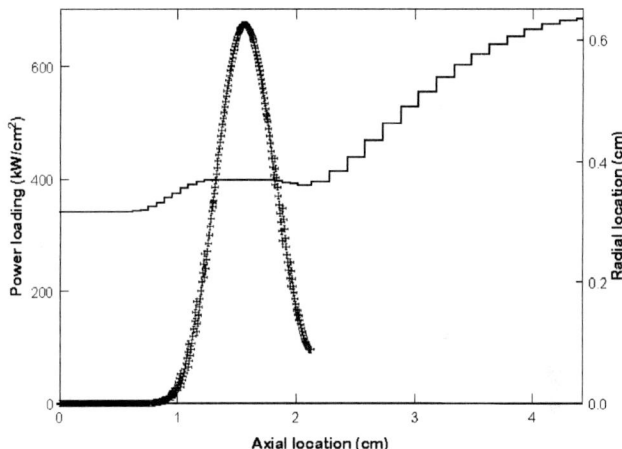

Figure 4. Cavity wall loading superimposed on the output cavity geometry used for RF simulations. The step structure of the geometry is an artifice of the simulation program.

transitions allow for greater flexibility is selecting the cavity length, and this was essential to maximize conversion efficiency. The output cavity has a diffractive iris to minimize mode conversion. The other tapers followed simple sinusoidal variations. The simulated mode purity of the output signal was over 99.8%. The fractional power that travels into the drift region toward the penultimate cavity was less than .022% of the power that enters the output waveguide.

The ideal zero spread efficiency of the circuit is near 40%, and the predicted efficiency exceeds 34% for spreads up to 7%; after that the efficiency drops rapidly. The large signal gain also drops rapidly with spread. The system is optimized at an input power of 24.4 W and an efficiency of 37.4%. The output power drops rapidly below an input power of 16 W. This is a consequence of the nonlinearity of the second harmonic interaction.

The bandwidth for the initial design was about 0.078%, which is consistent with other second harmonic tubes. The design is currently being investigated to achieve additional bandwidth at the expense of gain; however, a maximum required input power of 50 W has been selected for the program.

Thermal Design

While the average power density in the output cavity is low, the peak power density in the output cavity exceeds 600 kW/cm^2. A finite element thermomechanical analysis was performed to investigate the impact of this power density on the copper walls of the cavity. In particular, it was of interest to determine if shock heating on the cavity walls could lead to permanent distortion.

A 2D, axisymmetric, finite element model was generated using the geometry and heat flux data for ideal oxygen free (OFE) copper. A wall thickness of approximately 0.5 cm was used with 20° C water cooling. The power pulse used was a square wave of 1 microsecond at a rate of 120 Hz. The heat flux was increased by a factor of 1.25 for Glidcop models. Steady state thermal results (using the average power level) were used as the initial conditions for transient analyses. The thermal results were used as the loading for linear and nonlinear thermal stress analyses.

The steady state temperatures are quite low and should not present any problems for cooling. The average temperature increase is less than 15° C and would not be a problem for a continuous wave (CW) device; however, there are concerns about thermal stresses due to the high thermal gradients generated by the pulsed operation. The transient thermal response shows that the peak temperature increase at the end of the 1 microsecond pulse is approximately 125° C. The analysis showed that there are some plastic strains on the OFE cavity; that is, the material is stressed beyond its yield point.

The analysis was repeated using Glidcop Al-15 instead of OFE copper. Peak temperature increase on the cavity wall was approximately 39° C under steady state conditions and 152°C at the end of each pulse. The peak Von Mises stresses were approximately 1.53 kpsi under steady state conditions and 37.7 kpsi at the end of each pulse. Work hardened glidcop can withstand stresses in excess of 50 kpsi before plastic deformation will occur. The analysis did not show any plastic deformation, and, therefore, the cavity will be constructed from Glidcop AL-15

These results are consistent with research performed by David P. Pritzkau et al. at SLAC [21]. That research reached a preliminary conclusion that temperature increases of 100° C were acceptable for OFE copper under controlled conditions, but may not be acceptable for RF heating applications. The predicted temperature increase for the output cavity manufactured using OFE copper was 105° C. Dr. Pritzkau's research also indicated that glidcop should be able to tolerate temperature rises of 240° C before reaching a damage threshold, much more that the predicted temperature increase for the glidcop output cavity of 152° C. These results support the decision to build the output cavity from glidcop.

Input Cavity

The input cavity operates in the circular TE_{011} mode at the fundamental frequency. Input power to excite the cavity is supplied through a rectangular TE_{10} waveguide. One way of coupling the input rectangular TE_{10} mode with the circular TE_{011} mode is the 'wrap-around' mode-converter (WAMC). This device consists of a coaxial cavity excited by the rectangular TE_{10} mode. The coax cavity encircles the cylindrical TE_{011} cavity and is coupled via radial slots. The resonant mode in the coax cavity is usually chosen to be the $TE_{N,1,1}$ where N is the number of radial coupling guides

The azimuthal mode sets that can be excited by the WAMC are the L*N azimuthal modes where L=0,1,2,.. and N is the number of slots. For example, if there are four slots, the first two azimuthal mode sets that can be excited are the TE_{0n} and TE_{4n} modes. Thus the coax cavity mode must be chosen such that the number of slots is greater than the cutoff frequency in the cavity of the second azimuthal mode set. The design of the radial cavity and coupling section for the proposed gyroklystron is shown in Figure 5.

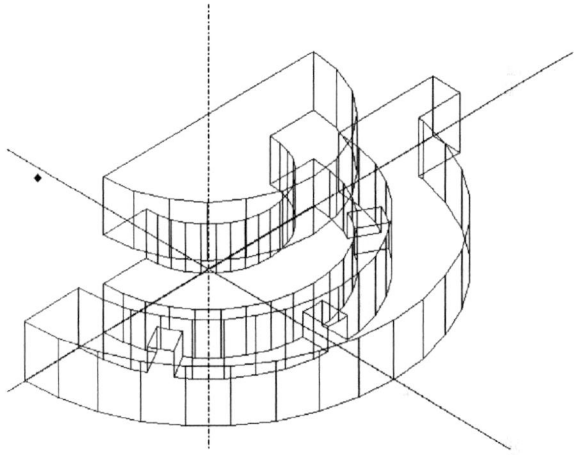

Figure 5. Simulation model of the input coupler

For the cavity dimensions (radius=0.5396 cm) the TE_{4n} modes are cut off at the cavity resonant frequency of 45.6 GHz, so a four slot converter could be used. However, it was found that to get the resonant frequency of the $TE_{4,1,1}$ coax cavity to match the cylindrical cavity resonant frequency, the width of the coax section had to be not much larger than the cutoff width of the rectangular TE_{10} input mode. This led to the use of the coax $TE_{6,1,1}$ resonant mode instead, as the coax cavity width for resonance at 45.6 GHz was significantly above the cutoff width of the rectangular TE_{10} mode.

The design procedure for obtaining the desired loaded Q of 350 is as follows. First the cavity drift tube lengths were adjusted to give an unloaded Q of 700. (The drift tube radius was increased to the cavity radius at the end of the drift tube section.) The proper cavity Q was verified using Calabazas Creek Research's mode matching program *CASCADE*. This design was then input into Ansoft Corporation's High Frequency Structure Simulator *(HFSS)* along with the $TE_{6,1,1}$ coax cavity with six radial coupling sections. A scattering matrix calculation of this assembly was then performed from 45 to 46 GHz. The scattering matrix generated by *HFSS* was then imported into *CASCADE*, where a matching iris in the rectangular input waveguide was cascaded with the input scattering matrix from *HFSS*. Using the optimizer in *CASCADE*, the iris dimensions were adjusted for critical coupling. The frequency was then swept over the resonant frequency band, and the Q calculated by observing the frequency width between the 3 dB points in the transmission calculation. This process was repeated multiple times while the width of the radial coupling guides was varied until the desired loaded Q of 350 was achieved.

Collector

In the device now being developed, the collector will serve as the output waveguide for the RF circuit and the beam collection region for the spent electron beam. A schematic of the current collector design is shown in Figure 6. An uptaper from the RF circuit to the diameter of the collector incorporates a mode converter that generates a hybrid TE_{01}/TE_{02} mode that will be described later. This mode allows incorporation of gaps in the waveguide wall without loss of RF power. The collector design incorporates gaps between the body of the gyroklystron and the output window to allow for voltage depression of the collector. This will allow partial recovery of energy remaining in the electron beam following interaction in the circuit, thereby increasing the overall efficiency.

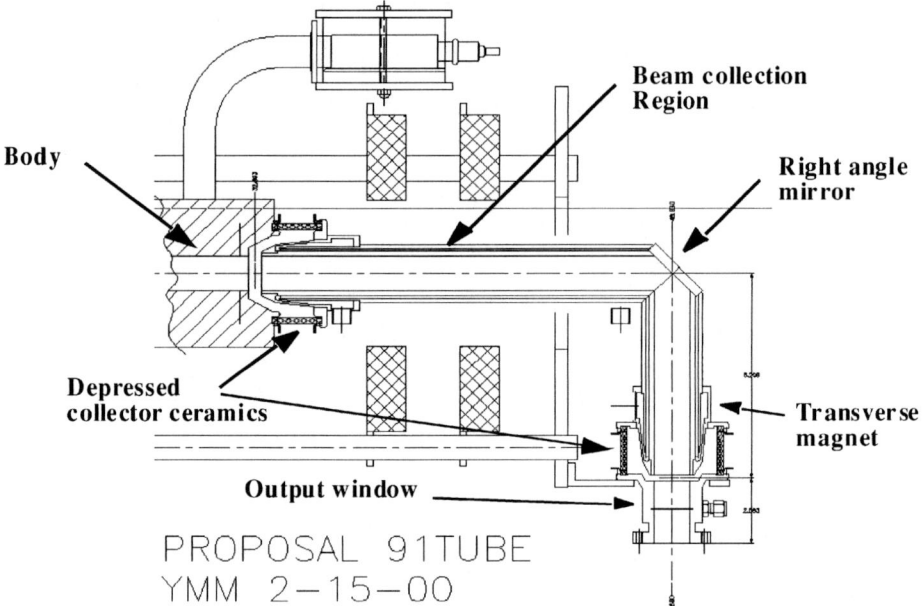

Figure 6. Schematic of gyroklystron collector, including output window.

The trajectories of the spent electron beam were modeled using EGUN. The peak power density is approximately 250 kW/cm^2. Using the predicted power density profile, the temperature increase can be approximated from two limits. The first is made assuming no penetration of the electrons into the surface, i.e., that there is simply a flux of heat onto the surface. For a one microsecond pulse, the temperature rise in copper will be about 80° C. This is an acceptable rise. Further, it is an upper bound, since 500 keV electrons striking normal to the surface will penetrate copper by about 0.03 cm. Their heat will thus be distributed into a volume of the collector wall. Assuming that all of the electrons strike the surface at 90°, the energy is deposited uniformly along their paths; and assuming that the heating is instantaneous, the temperature rise in the penetration volume will be about 2.5° C. While none of these assumptions is correct, even with a safety factor of 10, the calculation indicates that the heating will be small.

The power averaged over many pulses will be quite small. At a pulse repetition frequency (PRF) of 180 Hz with a one microsecond pulse width, the average power will be 45 W/cm^2. This can be cooled with minimal effort.

A recognized difficulty with this type of collector is that of reflected primary electrons or secondary electrons impacting the output window (which is mounted at the end of the waveguide). As is commonly done in high power gyrotron oscillators, these can be suppressed with a cross field magnet in between the collector and window. The magnetic field must be such that the cyclotron length is significantly less than the length over which the field is effective. A field of 1.3 kG, which can be produced by a permanent magnet, will result in a cyclotron radius of 2 cm, and a cyclotron length of 4 cm. The use of two magnets will probably be adequate.

The TE_{01}/TE_{02} mode combination can also be used in conjunction with a specially shaped mirror to redirect the RF beam 90° from the original direction. The mirror is located where the two RF modes are correctly phased for location of a waveguide wall gap. The design of the waveguide bend was developed in an SBIR program funded by the U.S. Department of Energy (DE-FG03-97ER82343). That program is currently in Phase II and is tasked with developing high power waveguide components for the Next Linear Collider.

With this design, significant additional protection is afforded to the output window from direct impact of reflected electrons. Also, the output window is now in a vertical position, which provides better protection from arcing due to debris on the window surface.

Output Waveguide and Window

Output mode converter

It is desirable to allow for a break in the output waveguide so a depressed collector can be added to the tube for increasing efficiency. To obtain adequate clearance for preventing breakdown across the gap, several cm per 100 kV of voltage depression will be required. The output power from the circuit is in the TE_{02} mode, which has reasonably good transmission across a gap. The gap loss versus length for the TE_{02}

mode can be calculated using a radial mode-matching code. Using this code, a calculation of the transmission for the TE_{02} mode across a gap at 1 cm and 2 cm radius was performed. For the case where the gap radius is 2 cm, the transmission is somewhat low (95%). A gap length of 2 cm would probably be the minimum gap length. The transmission can be improved significantly, however, by using a hybrid $TE_{01/02}$ mode mix. A combination of 80% TE_{01} and 20% TE_{02} will result in cancellation of the axial component of the magnetic field on the wall. This cancellation significantly decreases the diffraction losses of the hybrid mode. The hybrid mode has significantly increased transmission over the single mode (TE_{02}) transmission. At a radius of 2 cm, the transmission is still greater than 99.9% at a gap length of 4 cm. Using this hybrid mode will allow a gap to be placed in the output guide of sufficient length to allow significant collector depression.

A converter is required to produce the hybrid mixture from the cavity output mode (TE_{02}). In addition, a final conversion to 100% TE_{01} mode must be allowed if conversion to other modes such as a rectangular TE_{10} is required for the tube application. This requirement places an upper bound on the gap radius; mode control becomes problematic as the radius becomes larger. Also, as the radius becomes larger, the mode converter ripple amplitude becomes increasingly smaller and the length becomes larger, which makes construction more sensitive to machining tolerances.

Several designs were generated to convert the TE_{02} mode from the cavity output (radius of 0.635 cm) to the hybrid mode mixture. The first design consisted of a constant average radius converter with one period. It achieved a combined mode purity of 99.9% at a radius of 0.635 cm and a length of 2.5 cm. A taper from the output of the converter to 1.75 cm with a length of 20 cm was required to achieve a combined mode purity of 99.0%. A second design using an 'unfolded' converter was performed. The best result was achieved (length and mode purity) using a 1.5 period converter. The combined mode purity and length of this design were 99.7% and 19.7 cm, respectively. The profile of this converter is shown in Figure 7.

As previously mentioned, the $TE_{01/02}$ mode combination achieves a phase combination that allows gaps in the waveguide walls approximately every 20 cm. The combination can be converted back to a pure TE_{01} mode after the last collector ceramic gap or after the RF has exited through the output window. As mentioned in the collector section, the hybrid mode allows incorporation of gaps or 90 degree bends approximately every 20 cm. This provides unprecedented flexibility in transporting the RF power to the accelerator with low ohmic losses.

Output Window

The window will consist of a single ceramic disk. It is currently proposed to use Al-995 in the 3.50 cm diameter output waveguide. Because the average power is low, only mechanical stresses due to the differential pressures across the window are of concern. The initial design consists of a ceramic disk with a thickness of 3/2 of a guide wavelength at the operating frequency. Figure 8 shows the transmission for the TE_{01} and TE_{02} modes. Also included is the transmission for the TE_{22} mode, which is the

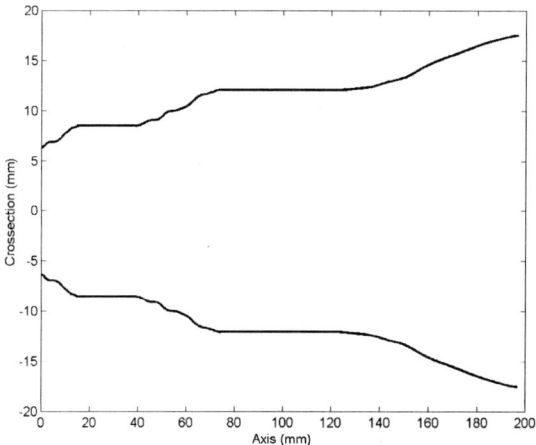

Figure 7. Wall profile for "unfolded" mode converter to generate hybrid mode mixture from cavity output taper.

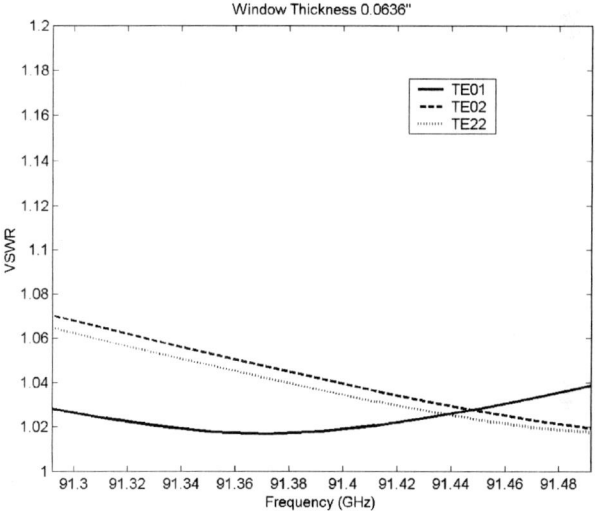

Figure 8. Output window VSWR for output mode and nearest competing modes in the output cavity.

principal competing mode in the output cavity. It is important to have good transmission of competing modes to avoid increasing the quality factor, which lowers the starting current, for competing modes. The average power deposition in the window is only 0.07 watts, so thermal issues were not addressed in the initial study. Only mechanical stresses due to pressure loading were considered. The tensile strength of Al-995 is 20-35 kpsi. With a load of 3 atmospheres (vacuum inside tube and waveguide pressurized to two atmospheres), the maximum tensile stress for the 0.0636 inch thick window ceramic is 6.17 kpsi. This is less than one third of the material's tensile strength.

The output window design appears straightforward, although performance is very sensitive to the ceramic thickness. This is not unexpected and is within typical machining tolerances for these devices.

Magnet

The requirements for the magnetic field are typical for a gyrotron, and are summarized in Table 2. For the 2.7 Tesla peak field, a niobium-titanium superconducting magnet is appropriate. The current density limit for that field is about 35,000 A/cm^2. The magnet bore will be vertical. For this configuration, and without active cooling, clearances required between the air surfaces of the dewar and the coils are: bottom, 3.81 cm; bore diameter, 2.54 cm; top, 6.35 cm.

The coil locations, sizes and current densities were determined iteratively with the gun and cavity designs. The field that produced both a low spread in the beam and a high efficiency is shown in Figure 9. The coil configuration that produced this field is shown in Figure 10.

Table 2. Magnetic field parameters

Peak Field	~2.7 Tesla
Cathode Field	~680 Gauss
Uniform field length	~6 cm
Magnet warm bore inner diameter	20 cm

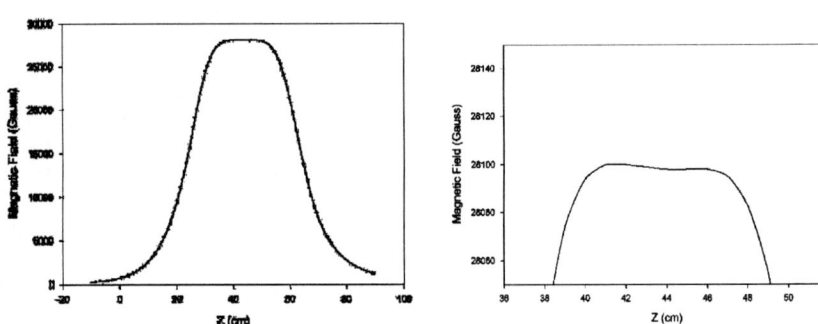

Figure 9. Magnetic field for the gyroklystron. Left view is the total field. Right view is enlargement of the magnetic field for the circuit region.

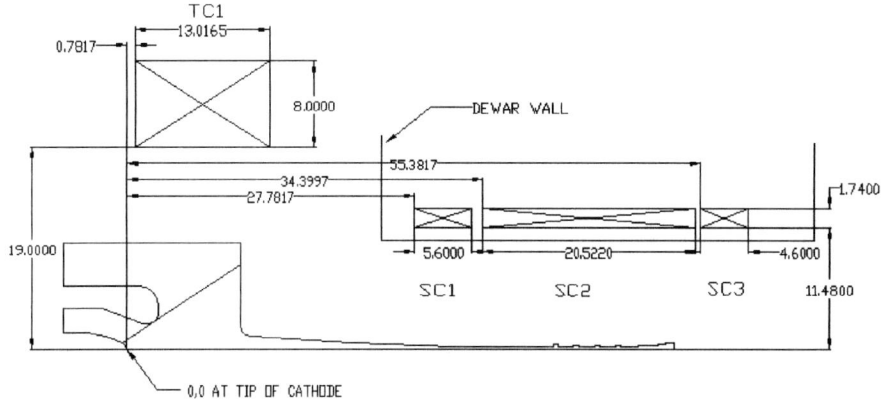

Figure 10. Magnet coils and dewar specifications. All dimensions are in centimeters.

PROGRAM SCHEDULE

The first prototype test is scheduled for late summer in 2001. The gyroklystron will be assembled using the facilities of Communications and Power Industries, Inc. in Palo Alto, CA, and the tube will be tested at Stanford Linear Accelerator Center. A rebuild of the tube is included in the program, which is scheduled for completion in June 2002.

ACKNOWLEDGEMENTS

This program is funded by a Small Business Innovative Research Grant from the U.S. Department of Energy. Communications and Power Industries, Inc are providing facilities for assembly.

REFERENCE

1. G. Caryotakis, "Development of X-Band Klystrons at SLAC", SLAC-PUB-7548, May 1997, Presented at the *1997 Particle Accelerator Conference*, Vancouver, May 12-16, 1997
2. Perry Wilson, "RF Power Sources for 5-15 TeV Linear Colliders," *Proc. 3rd* Workshop on Pulsed RF Sources for Linear Colliders, April, 1996, KEK Proceedings 97-1, p. 9.
3. H. Bohlen, F. Friedlander, E. Lien, "HOM-Klystrons," *Proc. 3rd* Workshop on Pulsed RF Sources for Linear Colliders, April, 1996, KEK Proceedings 97-1, p. 148.
4. V. Granatstein and W. Lawson, "Gyro-amplifiers as Candidate RF-Drivers for TeV Linear Colliders," *IEEE Trans. Plasma Sci.* **24**, 648 (1996).
5. D. Shiffler, J. A. Nation, and G. S. Kerslick, IEEE Trans. Plasma Sci., **18**, 546 (1990).
6. E. V. Kozyrev, et al., "The Latest Experience with 7 GHz Pulsed Magnicon Amplifier," *Proc. 3rd* Workshop on Pulsed RF Sources for Linear Colliders, April, 1996, KEK Proceedings 97-1, p. 234.
7. B. Danly, in *Proc. 1993 Int. Workshop on Pulsed RF Sources for Linear Colliders,* 251 (1993).
8. M. E. Conde and G. Bekefi, *IEEE Trans. Plasma Sci.,* **20**, 240 (1992).

9. G. Westenskow, in *Proc. 1993 Int. Workshop on Pulsed RF Sources for Linear Colliders*, 197 (1993).
10. T. J. Orzechowski, *et al.*, *Phys. Rev. Lett.*, **57**, 2172 (1986).
11. The NLC Design Group: *Zeroth Order Design Report for the Next Linear Collider*, SLAC Report 474, SLAC, Stanford, CA (1996).
12. K. Sakamoto, et al., "Major Improvement of Gyrotron Efficiency with Beam Energy Recovery," *Phys. Rev. Lett.*, **73** (1994), pp. 3532-3535.
13. Piosczyk, B., et al., "Progress Report on the 165 GHz Coaxial Cavity Gyrotron," 24th Intern. Conf. on Infrared and Millimeter-Waves, Monterey, CA September 1999, TU-A9.
14. M. Thumm, et al., "1 MW, 140 GHz, CW Gyrotron for Wendelstein 7-X", 24th Intern. Conf. on Infrared and Millimeter-Waves, Monterey, CA September 1999, W-2.
15. M. Kuntze, et al., "140 GHz Gyrotron with 2.1 MW Output Power," 24th Intern. Conf. on Infrared and Millimeter-Waves, Monterey, CA September 1999, W-46.
16. K. Hayashi, et al., "Developments of High Power Gyrotrons," *Proc. 3rd* Workshop on Pulsed RF Sources for Linear Colliders, April, 1996, KEK Proceedings 97-1, p. 243.
17. M. Blank, et al., "Experimental Investigation of W-Band gyroklystron Amplifiers," *IEEE Trans. Plasma Sci.* **26**, 409 (1998).
18. W. Lawson, "Magnetron Injection Gun Scaling," IEEE Trans. Plasma Sci., vol. 10 (1988), p. 290.
19. M. R. Arjona and W. Lawson, "Design of a 7 MW, 95 GHz, Three Cavity Gyroklystron," *IEEE Trans. Plasma Sci.* **27**, 438 (1999).
20. M. Castle, et al., "An Overmoded Coaxial Buncher Cavity for a 100 -MW Gyroklystron," IEEE Microwave and Guided Wave Letters, Vol. 8, No. 9, September 1998.
21. David P. Pritzkau, Gordon B. Bowden, Al Menegat, Robert H. Siemann, "Possible High Power Limitations from RF Pulse Heating," SLAC-PUB-8013, Nov. 1998

X-Band Dielectric Loaded Traveling-wave Acceleration Structure

P. Zou, W. Gai, R. Konecny, X. Sun, and T. Wong*

Argonne National Laboratory, Argonne, IL 60439, USA
*Also Illinois Institute of Technology, Chicago, IL 60616, USA

Abstract. We report on the construction, numerical modeling and experimental testing of a traveling-wave acceleration structure based on a dielectric-lined circular waveguide. This type of structure has similar acceleration properties to disk-loaded metal slow wave structures but with some distinct advantages in terms of simplicity of fabrication, suppression of parasitic wakefield effects, and having no dark current. Efficient coupling of external RF power to the cylindrical dielectric waveguide is a technical challenge, particularly to structures loaded with very high dielectric constant materials. We have designed and constructed an 11.4GHz structure loaded with ceramic with dielectric constant of 20, to be powered by an external RF power source. High efficiency RF coupling has been achieved using a combination of a tapered dielectric end section and a carefully adjusted coupling slot. Bench tests using a network analyzer have demonstrated a power coupling efficiency in excess of 95% with bandwidth of 30 MHz, and vacuum tests have also shown that this dielectric loaded structure can be operated in an ultra-high vacuum environment. Thus, this work provides a necessary basis for construction of an accelerator using this kind of structure. We have also simulated the parameters of this structure using MAFIA. Within the limits of the approximations used, the results are in agreement with the bench measurements.

1. INTRODUCTION

The proposed use of RF driven dielectric based structures for particle acceleration can be traced to the early 1950's [1]. Since then, numerous studies have examined the use of dielectric materials in accelerating structures [2, 3]. The advantages of using dielectrics are discussed in the references above and are summarized here:

- Simplicity of fabrication – the device is little more than a tube of dielectric surrounded by a conducting cylinder. This may be a great advantage for high frequency (10 GHz) structures over conventional structures where extremely tight fabrication tolerances are required.
- Comparable impedance to conventional metal disk-loaded structures – for a given power flow, the maximum acceleration field gradient that can be established in the dielectric loaded structure is close to that of the conventional metal disk-loaded acceleration structure.
- No dark current – because there are no free electrons in the dielectric ceramic, dark current cannot be formed inside the dielectric loaded structure. This is good for from the standpoint of eliminating loading of the structure due to dark current from field emission.

- Simple reduction of coupled bunch effects – it has been demonstrated experimentally that it is relatively straightforward to build deflection mode damping into dielectric structures so that very large attenuation (≈250 dB/m) of all but TM_{0n} modes can be obtained [4].
- Acceleration field is the maximum field in the structure – unlike the conventional iris loaded acceleration structure, in which the electric field at the iris is about twice the electric field on axis.

However, there are some potential problems of applying the dielectric loaded structure to particle acceleration, such as:
- Dielectric breakdown
- Joule heating
- Absorption of gases in the dielectric materials
- Dimensional tolerances.

At present, little information on these problems is available. Whether they are fatal or not should be answered through experiments conducted at high RF power level. In the past, the losses of dielectric materials were high, and they were easy to break down under high electric field intensities. The recent development of high dielectric constant ($\varepsilon \sim 20 - 40$), low loss dielectric materials ($Q \sim 10,000 - 40,000$) has resulted in a serious reexamination of dielectric structures as acceleration devices [5].

Another practical problem to be addressed when building a dielectric accelerator occurs because the outer diameter of the dielectric lined waveguide is much smaller than the width of rectangular waveguide that couples the external RF power to the device. Therefore, impedance matching becomes a difficult task. There is also no previous work known to us in this area to provide guidance on coupling design. We found that by using a combination of side coupled slots and tapering the dielectric section near the coupling slots, one can efficiently couple the RF power from a rectangular waveguide to the circular dielectric waveguide. We have designed and constructed a prototype 11.4 GHz dielectric loaded accelerator (Figure 1) to study RF coupling techniques and electrical properties of the dielectric under high vacuum and high power RF fields. We found that by careful tuning of the RF coupling slots and, more importantly, by tapering the inner radius of the dielectric tube near the coupling slot, one can obtain a coupling coefficient greater than 95%. We have also verified this coupling scheme using a numerical simulation and report the results in section III.

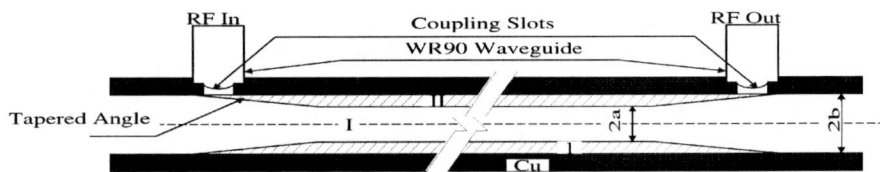

FIGURE 1. Dielectric loaded traveling wave accelerator prototype used for the network analyzer measurements. The length of the dielectric is 25 cm and consists of 1-inch-long segments. Regions 0: vacuum; 1: dielectric.

2. CONSTRUCTION AND TESTING OF AN X-BAND DIELECTRIC ACCELERATOR

2.1 Construction and Bench Testing

We have developed a design for an X-band structure (11.4GHz) using the parameters given in Table 1. The theoretical basis of the dielectric loaded traveling-wave accelerator have been published elsewhere [6]. The choice of X-band for the test dielectric structure permits comparison with the expected performance of RF structures designed for the Next Linear Collider (NLC) [7], since this technology represents the current state of the art in conventional metallic accelerating structures. High power X-band klystron RF sources are presently available at SLAC [7]; high power tests of these prototype dielectric devices will eventually be carried out there.

The dielectric material is an MgCaTi compound that has a dielectric constant of 20 and can be readily obtained from commercial vendors. The desired group velocity for the NLC structure design is in the $0.03c \sim 0.05c$ range [7]. The dielectric loaded structure has a comparable shunt impedance and group velocity to the conventional X-band structure. The rectangular waveguide used for RF coupling to the device is WR90. A complete solution for the dispersion curve of the dielectric structure has been derived elsewhere [6, 8, 9]; the dispersion curves of our device are shown in Figure 2 for the TM_{01} and HEM_{11} modes. We would like to point out that one of the interesting characteristics of this structure is that the frequency of the HEM_{11} mode (first deflection mode) is lower than that of the acceleration mode. Because the deflection force is a function of $sin(kz)$, this implies very different and improved conditions for the single bunch BBU instability compared to conventional structures where the HEM_{11} is always higher in frequency than the accelerating TM_{01} mode.

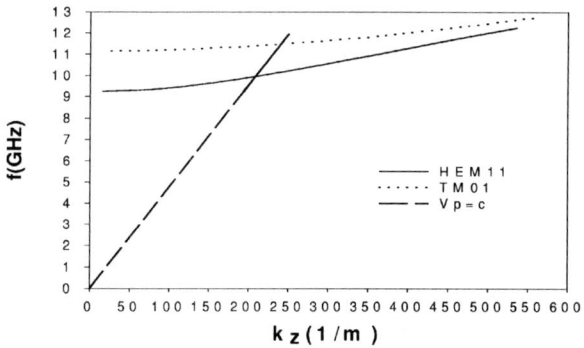

FIGURE 2. Calculated dispersion curve of the dielectric structure with the parameters of Table I

The RF coupling scheme we use here is similar to the side coupled method used for conventional disk-washer RF cavities. Impedance matching of the coupling slots is more difficult in the high ε dielectric case because the outer radius of the dielectric tube is much smaller than the waveguide.

A 25-cm long prototype structure was constructed with the parameters given in Table 1. The dielectric materials were obtained from Trans Tech [5]. The device is a constant impedance structure with shunt impedance of 70 MΩ/m. Our goal is to obtain maximum RF transmission through the two coupling slots. We found that high efficiency coupling can be achieved by tapering the inner diameter of the dielectric tube. This tapered section serves as an impedance transformer for impedance matching. For the structure described here the taper angle was chosen to be 8°. While no other angles were tested, it is not expected that the results will depend strongly on the taper angle. The detailed configuration of the tapered dielectric structure and coupling slots are shown in Figure 1.

TABLE 1. Dimensions and physical properties of the 11.4GHz dielectric loaded accelerator

Coefficient	Value
Material	MgCaTi
Dielectric Constant ε	20
Loss tangent δ	10^{-4}
Taper angle	8°
Inner radius a	0.3cm
Outer radius b	0.456cm
Frequency of TM_{01} mode	11.42GHz
Frequency of HEM_{11} mode	9.96GHz
Group velocity v_g	0.057c
Attenuation	4dB/m
Power required (10MV/m gradient)	2.6MW

Optimal coupling was obtained by adjusting the coupling slot dimensions and monitoring the S-parameters using an HP8510C network analyzer until no further improvement was observed. Plots of the measured S-parameters vs. frequency for the optimized coupling case are shown in figures 3. The maximum transmission coefficient S_{21} is -1.7 dB at 11.421 GHz. The reflection coefficient S_{11} at this frequency < -20dB. The optimized coupling slot dimensions are 4.7 mm (axial) × 5.69 mm (transverse). The calculated attenuation of the RF in the waveguide is 4 dB/m using the dielectric loss factor of 10^{-4} supplied by the manufacturer and the nominal conductivity of copper, and is in good agreement with our measurements. One interesting feature of high dielectric constant loaded waveguides is that the attenuation is dominated by the copper wall losses rather than dielectric losses. We conclude that further reduction of the loss tangent of the dielectric material will not improve the device performance unless the outer wall is replaced by a superconductor.

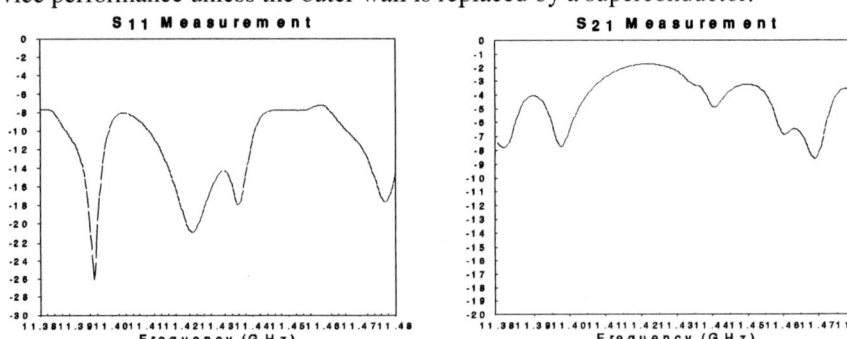

FIGURE 3. S-parameter measurement results

2.2 Vacuum Test

We have conducted vacuum test on the dielectric loaded accelerator prototype to verify that it can be operated in an ultra-high vacuum environment. After vacuum baking-out at 150°C, 10^{-9} torr was achieved and there was no change in dielectric properties. There was only one ion pump in the vacuum testing stand, while two ion pumps are mounted at the both ends of the dielectric loaded accelerator section in the actual high power experiment setup. Therefore, better vacuum can be guaranteed for the experiment.

2.3 High Power Experiment

The eventual goal of a high power test of this dielectric accelerator section is to investigate fundamental issues such as RF breakdown limits, Joule heating, and vacuum properties of dielectric loaded structure under high power RF. With 100 MW X-band RF power available at SLAC, we can test this structure at a 60 MV/m gradient, comparable to that planned for the NLC structures [7].

The high power test setup is shown in figure 4. If RF breakdown happens in the dielectric loaded waveguide, it can be observed from the viewports. The whole structure in figure 4 was shipped to SLAC for installation in June 2000. However, some damage during shipping was found when it was unpacked. The transmission coefficient from port to port was changed seriously. Mechanical improvement and readjustment on this dielectric loaded accelerator is underway. The high power test is now scheduled at the end of August 2000.

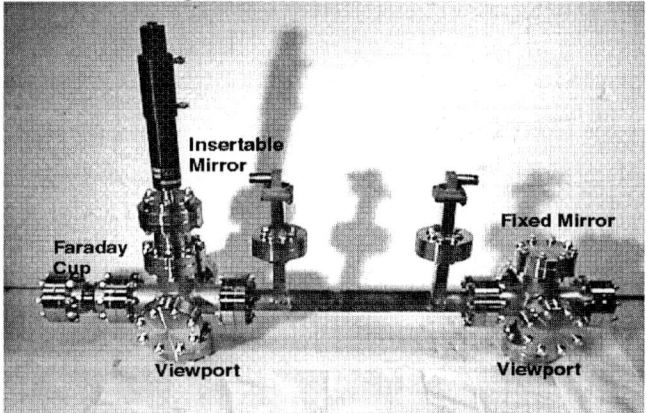

FIGURE 4. High power experiment setup

3. NUMERICAL SIMULATIONS OF THE COUPLING PORT USING MAFIA

In order to further investigate the coupling method developed empirically, we have used the MAFIA code suite [10] to perform a full 3-D time domain electromagnetic simulation of the junction of the WR90 waveguide and dielectric loaded waveguide with the parameters described in the last section. Due to the relatively small size of the coupling slot, special attention has to be given to the finite-difference mesh size in the neighborhood of the slot to faithfully reproduce the actual device geometry.

Figure 5 shows the calculated S-parameter as a function of frequency. The optimally coupled frequency in the simulation occurs at a slightly different frequency (11.45 GHz) from the measurements. Since the calculation does not include dielectric and wall losses, the computed transmission parameter S_{21} is approximately -0.4 dB as shown in Figure 5, compared to -1.7 dB obtained in the actual measurement (Figure 3) in which wall losses dominate. Therefore, the simulation results are in reasonable agreement with the bench measurements.

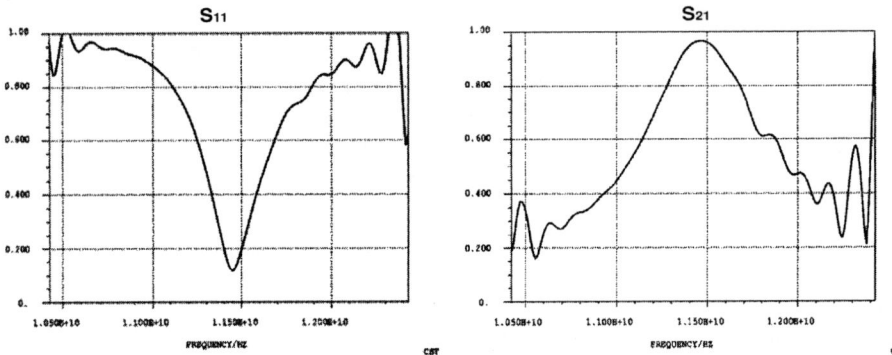

FIGURE 5. S-parameter from MAFIA simulation

In order to verify the mode coupled into the dielectric tube is indeed predominantly TM_{0n}, we have obtained the electric field pattern of the propagating wave at 11.45GHz in MAFIA simulation [10] as shown in Figure 6. We can see the dominant component is TM-like mode and there are small amount of HEM mode components. The hybrid mode component may be caused by the asymmetric excitation configuration using one coupling slot on the wall of the dielectric loaded waveguide. Using symmetric coupling configuration can reduce the hybrid mode components. We also did not attempt to include hybrid mode suppression in this simulation, which would effectively filter out all HEM components in the accelerating tube.

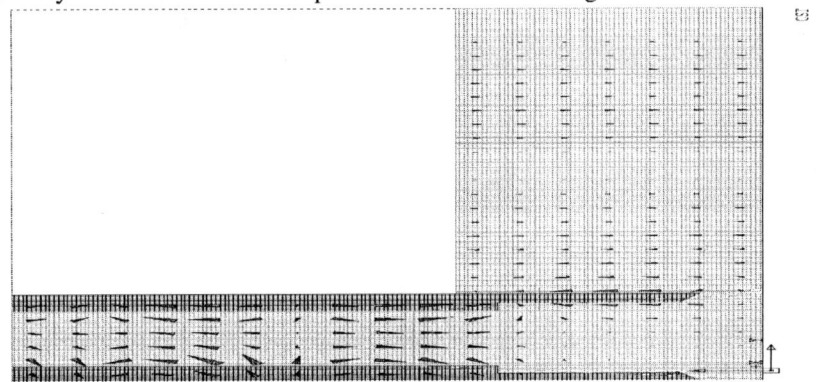

Figure 6. Calculated electric field pattern at the waveguide junction

4. SUMMARY

We have constructed and studied an 11.4GHz dielectric loaded structure for particle acceleration. Careful engineering considerations were implemented. We have achieved efficient RF coupling from port to port. A demonstration accelerator for a high power test has been constructed and successfully completed vacuum testing. Our goal is to achieve 50-100 MV/m gradients so the structure can be used as a viable alternative to conventional accelerating structures. We also simulated the parameters of this structure and the results are in reasonable agreement with the bench test results. Some fundamental issues concerning high power breakdown and Joule heating of dielectric loaded structure will be answered by the upcoming high power experiments.

ACKNOWLEDGMENTS

This work is supported by DOE, High Energy Physics Division under contract No. W-31-109-ENG-38. We would like to thank Dr. Juwen Wang at SLAC for his help during the construction and installation.

REFERENCES

1. Flesher, G. and Cohn, G., *AIEE Transactions*, **70,** 1951, pp. 887-893.
2. Zhang, T-B., Hirshfield, J., Marshall, T., and Hafizi, B., *Physical Review E*, **56,** 1997, pp. 4647.
3. Gai, W., Konecny, R., and Simpson, J., in *Proceedings of 1997 Particle Accelerator Conference*, Vancouver BC, May 1997, pp. 636-638.
4. Chjonacki, E., Gai, W., Ho, C., Konecny, R., Mtingwa, S., Norem, J., Rosing, M., Schoessow, P., Simpson, J., *J. Applied Physics* **69**, 1991, pp. 6257.
5. *Trans-Tech Inc., Catalog*, No. 5520 Adamstown, MD 21710.
6. Zou, P., Gai, W., Konecny, R., Sun, X., Wong, T., and Kanareykin, A., *Review of Scientific Instruments*, **71**, No. **6**, 2000, pp. 2301-2304.
7. SLAC report 474, 1996.
8. Ng, K-Y., *Physical Review* **D42**, 1990, pp. 1819-1828.
9. Rosing, M., and Gai, W., *Physical Review* **D42**, 1990, pp. 1829-1834.
10. MAFIA Version 4.0, Gesellschaft fur Computer-Simulationstechnik, Lauteschlagerstrabe 38, D-64289, Darmstadt, Germany.

RTA Beam Dynamics Experiments: Limiting Cumulative Transverse Instability Growth In A Linear Periodic System*

Tim Houck, Steve Lidia**, and Glen Westenskow

Lawrence Livermore National Laboratory, 7000 East Avenue, L-645
Livermore, CA 94550-9234
** Lawrence Berkeley National Laboratory, 1 Cyclotron Road, Mailstop 47-112
Berkeley, CA 94720

Abstract. A critical issue for a Two-Beam accelerator based upon extended relativistic klystrons is controlling the cumulative dipole instability growth We describe a theoretical scheme to reduce the growth from an exponential to a more manageable linear rate, and a new experiment to test this concept. The experiment utilizes a 1-MeV, 600-Amp, 200-ns electron beam and a short beamline of periodically spaced RF dipole pillbox cavities and solenoid magnets for transport. Descriptions of the RTA injector and the planned beamline are presented, followed by theoretical studies of the beam transport and dipole mode growth.

INTRODUCTION

A Lawrence Livermore National Laboratory (LLNL) and Lawrence Berkeley National Laboratory (LBNL) collaboration is studying the application of induction accelerator technology to the generation of microwave power. We refer to this scheme of power generation as the Relativistic Klystron Two-Beam Accelerator (RK-TBA) [1]. This scheme is considered a TBA approach as the extraction of microwave power is distributed along a drive beam parallel to the high-energy RF linear accelerator. The RK designation indicates that the power is generated by the interaction of the relativistic modulated drive beam with resonant structures similar to those used in a conventional klystron.

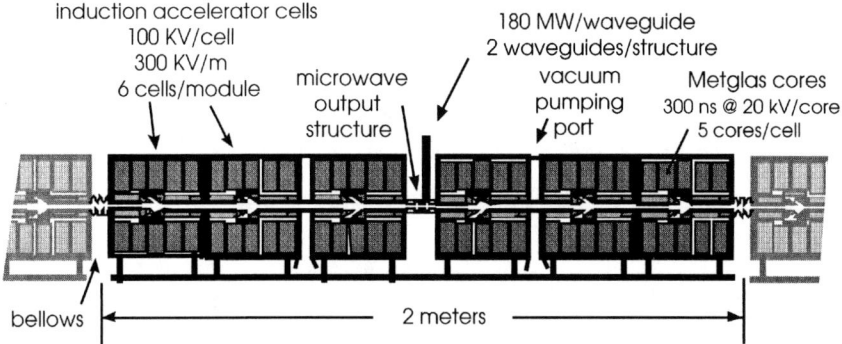

FIGURE 1. Illustration of a repeating module that comprises the extraction section of a RK.

The primary advantage of TBA concepts is that the conversion of drive beam power to microwave power can be highly efficient (> 90%). This efficiency is realized by distributing the power extraction over an extended length. The interest in RK-TBA's is that induction accelerators are efficient at producing very high power electron beams. Present induction accelerators operate at currents of several kiloamperes and accelerate the beam to 10's of MeV for beam power of 100's GW [2]. The induction accelerator can realize improved efficiency at converting wall plug power into beam power by replacing the standard electromagnet solenoids for beam transport with permanent magnets. Even higher efficiency can be attained if the induction cells are used as high-voltage step up transformers driven by a relative low voltage (~ 20 kV) pulsed power system. Present designs of a RK-TBA predict efficiency of about 40% in conversion of wall plug power into induction beam power [3], or a total wall plug to microwave power efficiency of about 36%.

The main section of an RK where the microwave power is generated is comprised of many repeating modules as illustrated in Figure 1. Within each module, the induction cells replace the energy extracted from the electron beam by the microwave output structure. The efficiency of this process – extraction and reacceleration – is nearly 100%. Not shown in Figure 1 are the beam generation and modulation sections and the final beam dump. Fixed energy losses in those processes have to be included in calculating the total beam energy to microwave conversion efficiency. Thus, it is imperative that the RK have many of the efficient extraction and reacceleration cycles to reduce the relative value of fixed losses with respect to the total energy transferred to the beam.

Several proof-of-concept experiments have been performed to demonstrate the viability of the RK-TBA concept. These experiments have shown the generation of collider-scale drive beam in induction linacs, production of high-quality, high-power microwaves from standing- and traveling-wave structures driven by induction accelerator beams, and multiple reacceleration and extraction cycles [4, 5]. As will be described below, we are continuing to perform experiments to study specific physics and technology issues while constructing a prototype relativistic klystron.

RTA FACILITY STATUS

The RTA Facility at LBNL was established to study issues related to RK-TBA designs. The principle effort is the construction of a prototype RK to serve as a test bed for physics, engineering, and cost studies. We have completed the first major component of the prototype RK, the commissioning of the induction gun [6]. A schematic of the gun is shown in Figure 2. The induction cores are individually driven at 15 kV as a demonstration of the type of pulsed power system proposed for a full scale RK. The 3.5" M-type dispenser cathode and electrodes are designed to produce a normalized edge emittance of 300 • mm-mr. Currently we are performing some minor refurbishment to the gun and upgrades to the pulsed power system. Once this work is finished, we will re-characterize the beam and then begin a series of beam dynamics experiments (described below).

A second effort involves parameter and optimization studies of possible RK-TBA designs. We have considered microwave frequencies from X-band to Ka-band [7] for powering colliders with center-of-mass energies up to 5-TeV. In support of the higher frequency designs, we participated in an experiment where an induction beam was modulated at 35 GHz using a FEL and then transported through a resonant cavity to generate power [8]. This experiment was performed at CESTA with collaborators and support from the University of Bordeaux, LBNL, LLNL, and CERN. Additional experiments are planned.

We are also continuing to develop relevant induction accelerator technology in collaboration with other induction accelerator programs at LLNL, LBNL, and Los Alamos National Laboratory (LANL) as well as with small businesses [9]. A recent example is the testing of a High Gradient Insulator (HGI) [10] on LLNL's ETA-II accelerator. The HGI is a multilayered construction of alternating insulator and conductor material that has shown great ability in holding off electrical fields. The HGI has also been used in induction cell designs to produce lower impedance. The test placed an HGI in an actual induction cell (modified ETA-II cell) under electrical field stress in excess of 100 kV/cm during accelerator operations with a 2-kA beam. The successful demonstration of the insulator under realistic conditions was a major milestone towards the incorporation of HGI technology into operating accelerators.

An important issue for the RK-TBA is the conversion of wall plug to microwave power efficiency. Most of the components of an RK can be independently tested to determine efficiency, cost, and lifetime – e.g. pulsed power components and induction cells. However, the number of extraction cavities determines the overall efficiency of the RK, and this number is limited by beam dynamics. We expect the prototype RK to be able to address all the relevant beam dynamics issues. However, prior to the completion of the prototype, we intend to do a series of smaller proof of concept experiments to study specific beam dynamics issues.

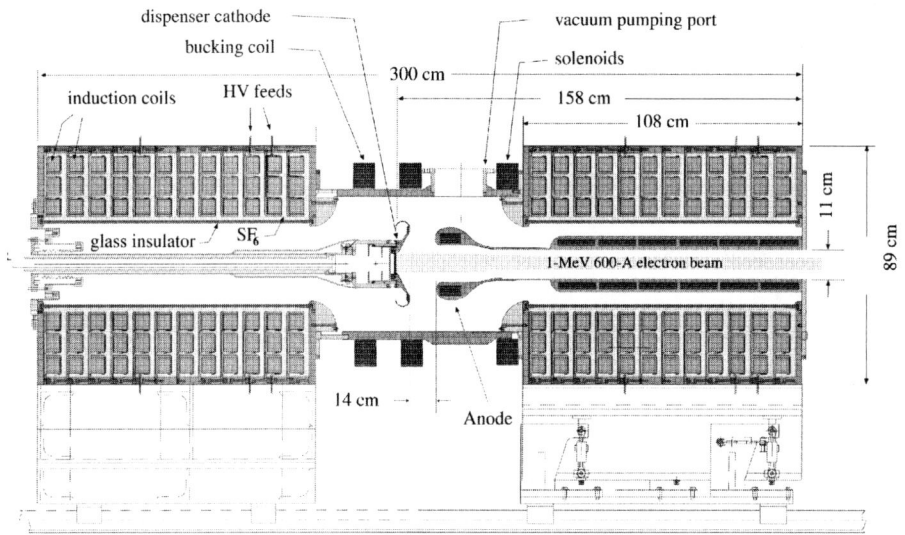

FIGURE 2. Schematic of the RTA induction electron gun.

BEAM DYNAMICS ISSUES

The ultimate efficiency of a RK is determined by the induction beam dynamics – i.e. the number of extraction structures that the beam can transit. We have identified three critical areas of beam dynamics that must be understood. The first involves maintaining the longitudinal modulation of the beam or "RF bucket" structure. In the drifts between output structures, space charge forces will cause the beam to lengthen in phase space, i.e., "debunch". If this effect is not corrected, the RF current (Fourier component of the beam at the modulation frequency) will decrease resulting in a decrease in the microwave power that can be extracted. Inductively detuning the output structures, similar to the penultimate cavity in conventional klystrons, can counter the space charge forces. The requirement for long-term longitudinal stability is reestablishing the initial longitudinal charge distribution at the end of a synchrotron period. Computer simulations have shown that with proper detuning, the RF current can be maintained over the 150 output structures envisioned for a full scale RK-TBA.

The other issues involve transverse instabilities. The beam will excite dipole modes in the induction cell accelerating gaps as well as in the resonant output structures. The induction cell accelerating gaps can be severely damped with RF absorbers for all resonant modes since the applied voltage pulse is quasi-static compared to the resonant frequencies. In addition, the natural energy spread over the RF bucket contributes to phase mixed, or Landau, damping. The combination of RF absorbers and energy spread is expected to maintain the transverse instability due to the dipole modes in the accelerating gaps at acceptable levels.

The resonant output structures present a more difficult transverse instability issue. The fundamental mode must couple sufficiently with the beam to extract the required energy. Various techniques exist to damp higher order modes in both output and accelerating structures. However, the permanent magnet focusing system envisioned for an RK-TBA allows the application of a new technique that we refer to as the Betatron Node Scheme.

Transverse beam instability theory is well developed and the exponential growth predicted is supported by experiment. However, the standard theoretical approach assumes that the discrete cavities interacting with the beam are closely spaced compared to the betatron wavelength due to the focusing system. Our design for an RK-TBA system requires strong focusing to maintain the required beam radius and a constant average energy over each extraction/reacceleration cycle. This combination leads to spacing between output structures of one betatron wavelength and the basic assumption of the standard theoretical approach does not hold.

An alternative approach to studying the transverse instability uses transfer matrices [11]. Assuming a monoenergetic beam and a thin cavity, Equations (1) through (3) indicate the salient parts of this theory. Equation (1) represents the transverse momentum change an electron receives passing through the cavity. R is an integral operator that accounts for the part of the beam that has already passed through the cavity. The first matrix on the RHS of Equation (2) is then the transfer matrix for the beam going through the cavity. For a sufficiently thin cavity, the transverse position does not change. Only the momentum is affected. The second matrix represents the betatron motion of the electrons between cells where θ is the phase advance. Thus, Equation (2) advances the position and momentum of electrons from the exit of on cavity to the exit of the following cavity. By repeatedly multiplying the two transfer matrices, the position and momentum at the exit of any cavity can be related to the initial conditions. For the situation where θ is constant for all sections and $\theta \ll 1$, the series of matrix multiplications can be shown to yield the same expected exponential growth as the more standard approach.

$$\Delta p_\perp = Rx \qquad (1)$$

$$\begin{bmatrix} x \\ p_\perp \end{bmatrix}_{n+1\,\mathrm{exit}} = \begin{bmatrix} 1 & 0 \\ R & 1 \end{bmatrix} \begin{bmatrix} \cos\theta & \sin\theta/\omega \\ \omega\sin\theta & \cos\theta \end{bmatrix} \begin{bmatrix} x \\ p_\perp \end{bmatrix}_{n\,\mathrm{exit}} \qquad (2)$$

For the case where $\theta = 2\pi$ (or any integral multiple of π), the matrix multiplication is greatly simplified. The betatron motion returns the electrons to the original transverse position and momentum (oppositely directed for odd multiples of π). The multiplication involves only the matrix describing the effect of the cavity, and, as shown in Equation (3), this leads to a linear growth in the transverse instability.

$$\begin{bmatrix} x \\ p_\perp \end{bmatrix}_{n+1\,\mathrm{exit}} = \begin{bmatrix} 1 & 0 \\ R & 1 \end{bmatrix}^n \begin{bmatrix} x \\ p_\perp \end{bmatrix}_{\mathrm{exit}} = \begin{bmatrix} 1 & 0 \\ nR & 1 \end{bmatrix} \begin{bmatrix} x \\ p_\perp \end{bmatrix}_{\mathrm{exit}} \qquad (3)$$

There are many nonideal factors in a realistic accelerator including cavities of finite thickness and variation in phase advance due to energy and/or focusing errors. Parameter studies through computer simulations indicate that the transverse instability is significantly reduced for systems with reasonable variations in parameters. We intend to experimentally test the validity and robustness of the Betatron Node Scheme.

BETATRON NODE SCHEME EXPERIMENT

The basic elements involved in a test of the Betatron Node Scheme are: a set of devices that generate a localized transverse impedance, a tunable focusing and transport system, and diagnostics to measure the BBU mode signal on the beam as a function of time and distance along the beamline. A schematic for a possible beamline is shown in Figure 3. The localized impedances are generated in simple pillbox cavities, tuned so that the TM_{110} mode frequency matches the modulation of the beam; a series of solenoid magnets provide tunable focusing; and rf ("B-dot") loops (rf BPM's) placed between cavities provide a means of collecting the dipole mode signal carried by the beam. We have built several sections of this beamline, using off-the-shelf components wherever possible. Each section is one betatron wavelength long and is comprised of one pillbox cavity, a pumping port, a diagnostic, and three solenoids. A photograph of a single section is shown in Figure 4.

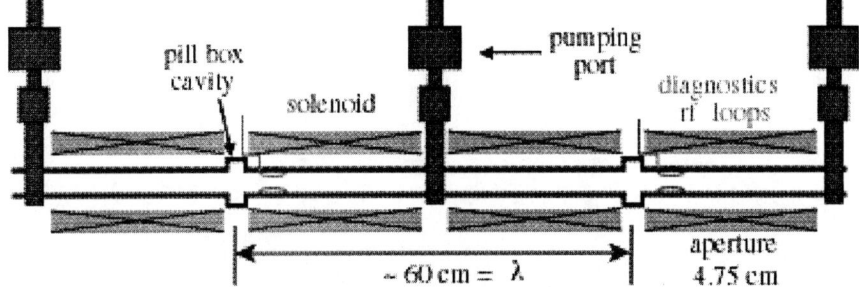

FIGURE 3. Schematic of the minimum beamline configuration required to demonstrate one period of the Betatron Node Scheme.

The designed solenoidal focusing field and expected beam envelope is indicated in Figure 5. The ten solenoids shown prior to the exit of the injector can be identified in the schematic of the electron gun (Figure 2). The next four solenoids are used to transport the beam through a transport region consisting of an inline gate valve and pumping station, and to match the beam into the repeating sections of the Betatron Node Scheme experiment. We will start with only two sections to verify transport calculations and diagnostic performance. Additional sections (up to a total of ten) and an initial driven tickler cavity will be added over the following months.

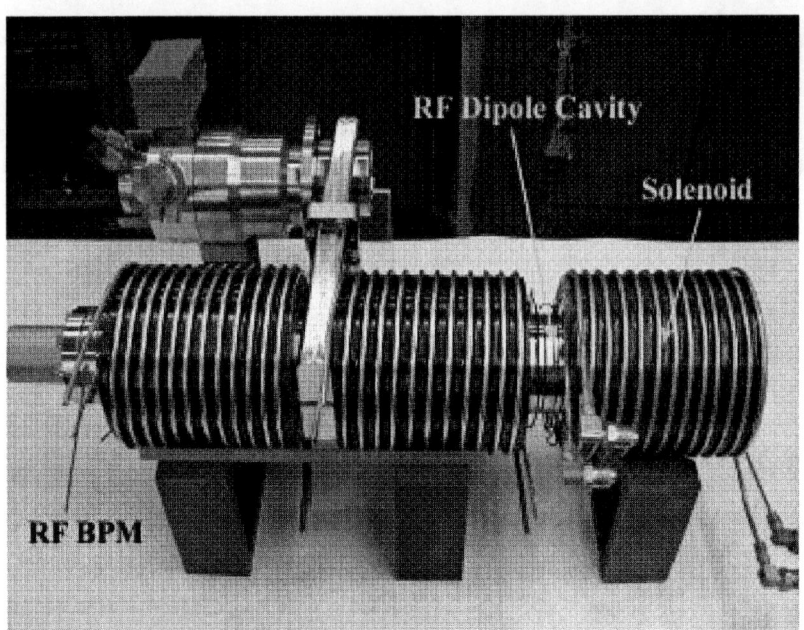

FIGURE 4. Realization of the minimum beamline configuration in hardware.

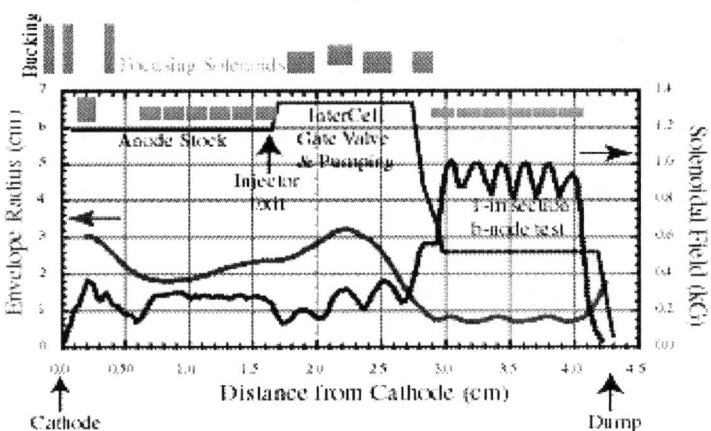

FIGURE 5. Solenoid field and beam envelope as a function of distance along beamline.

Computer simulations of the increase in power measured by the RF diagnostics at the dipole mode frequency are shown in Figure 6. Variations of ± 10 % in the solenoidal field (betatron phase advance) from the optimum should produce several orders of magnitude increase in measured mode power after only a few sections. The graphs in Figure 6 indicate the maximum power expected during the main body of the beam ("flat top"). The temporal power variation during the pulse (not shown) is predicted to have different characteristics between under- and over-focused scenarios.

In addition to the variation in power with focusing field, the change in temporal characteristics will be a stringent test of the simulations.

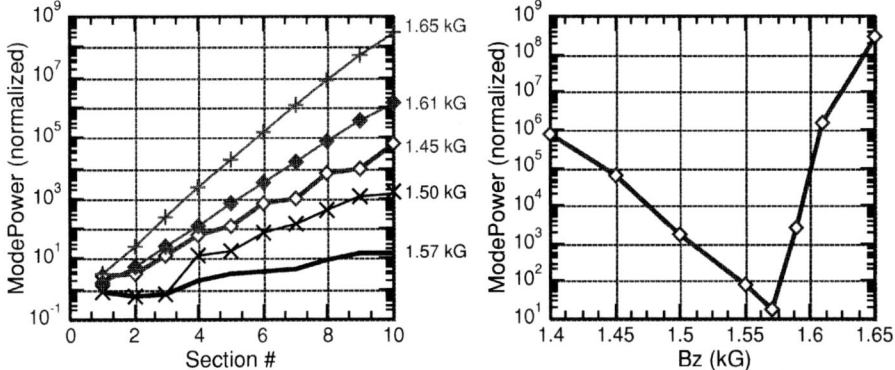

FIGURE 6. Computer predictions of BBU mode growth as a function of the number of sections (cavities) the beam has passed through for five values of the average solenoid field (left graph), and as a function of the average solenoid field (betatron phase advance) after 10 sections (right graph).

SUMMARY

The long-term goal of the RTA Facility is to build a prototype relativistic klystron that has all the major components required for a RK suitable for collider applications. The prototype would serve as a test bed for examining physics, engineering, and cost issues. The first major component, the 1-MeV, 600-A, induction electron gun, of the prototype has been completed and commissioned. Before continuing with the next section of the prototype, we intend to perform a series of beam dynamics experiments. In particular, we will demonstrate the effectiveness of the Betatron Node Scheme. We are also continuing to study and optimize collider designs based on the RK-TBA scheme.

ACKNOWLEDGEMENTS

We thank Swapan Chattopadhyay, George Caporaso, Kem Robinson, and Simon Yu for their support and guidance. Dave Vanecek and Wayne Greenway provided invaluable mechanical engineering and technical services. John Corlett and Bob Rimmer designed the RF diagnostics. This work was performed under the auspices of the U.S. Department of Energy by University of California Lawrence Berkeley Livermore National Laboratory under contract No AC03-76SF00098 and Lawrence Livermore National Laboratory under contract No. W-7405-Eng-48.

REFERENCES

1. Sessler, A.M. and Yu, S.S., *Phys. Rev. Letters* **54**, 889 (1987).
2. Burns, M.J., et al., "DARHT Accelerator Update And Plans For Initial Operation", in *Proceedings of IEEE 1999 Part. Accel. Conf.*, NY, 1999, pp. 617–621.

3. Houck, T.L. (ed.), et al., "Appendix A: A RF Power Source Upgrade to the NLC Based on the Relativistic Klystron Two-Beam Accelerator Concept," in the *Zeroth Order Design Report for the Next Linear Collider*, Stanford, Stanford Linear Accelerator Center, May 1996.

4. Westenskow, G.A. and Houck, T.L., "Relativistic Klystron Two-Beam Accelerator," *IEEE Trans. Plasma Sci.*, **22**, pp. 424–436 (1994).

5. Westenskow, G.A. and Houck, T.L., "Results of the Reacceleration Experiment: Experimental Study of the Relativistic Klystron Two-Beam Accelerator," in *Proceedings of the 10th Int'l Conference on High Power Particle Beams*, San Diego, CA (1994).

6. Lidia, S.M., et al., "Initial Commissioning Results of the RTA Injector", in *Proceedings of IEEE 1999 Part. Accel. Conf.*, NY, 1999, pp. 3390–3392.

7. Lidia, S.M., et al., "RK-TBA Studies in Ka-Band", in *Proceedings of XIX Intl. Linear Accel. Conf.*, IL, 1998, pp. 97–99.

8. Lefevre, T., et al., *Phys. Rev. Letters* **84**, 1188–1191 (2000).

9. Godlove, T.F., "Induction Modulator for a Relativistic-Klystron Two-Beam Accelerator, Final Technical Report", *FM Technolgies, Inc.*, Report No. 96-05-01, May 1996.

10. Sampayan, S.E., et al., "High-Performance Insulator Structures For Accelerator Applications", in *Proceedings of IEEE 1997 Part. Accel. Conf.*, Vancouver, BC, 1997, pp. 1308–1310.

11. Neil, V.K., Hall, L.S., and Cooper, R.K., "Further Theoretical Studies of the Beam Breakup Instability", *Particle Accelerators* **9**, pp. 213–222 (1979).

CARM-Klystron Amplifier for Accelerator Applications

Steven H. Gold and Arne W. Fliflet

*Beam Physics Branch, Plasma Physics Division
Naval Research Laboratory
Washington, DC 20375-5346*

Abstract. We consider the possibility of a cyclotron-autoresonance-maser (CARM) klystron configuration for accelerator applications as an alternative to the gyroklystron amplifier. The potential advantages, compared to gyroklystrons, include: 1) comparable efficiencies at lower values of the electron beam pitch ratio α, which should improve the beam quality and make the device substantially more stable against the excitation of parasitic mode, 2) operation far from cutoff, which should reduce the fields at cavity walls, allowing higher power operation, and 3) operation at lower magnetic fields for the same cyclotron harmonic number. However, there are two significant issues associated with the design of efficient, high-power CARMs. First, because of the higher value of k_z, compared to gyroklystrons, CARMs are substantially more sensitive to parallel velocity spread (pitch-angle spread). Second, conventional cavities support a variety of near-cutoff modes, which can compete with the CARM interaction. Therefore, one must consider either Bragg resonators or quasioptical cavity configurations.

INTRODUCTION

The klystron amplifier has been the foundation of linear accelerator rf systems for many years, and seems likely to retain this position at least through X-band systems. However, its limitations in scaling to high frequencies (e.g., Ka-band and above) has motivated research into a variety of other tube technologies, including magnicons, gyroklystrons, gyroharmonic converters, etc. In this paper, we examine the possibility of developing a cyclotron-autoresonance-maser (CARM) amplifier for accelerator applications as an alternative to the gyroklystron.

The most common type of cyclotron resonance maser is the gyrotron, a device that (by definition) operates at a frequency ω near $s\Omega_c$ and in the regime $k_z \ll \omega/c$, where s is the harmonic number and Ω_c is the relativistic cyclotron frequency, $eB/\gamma m$. A variety of gyrotron configurations have been developed, including gyromonotron oscillators, gyro-backward-wave oscillators (gyroBWOs), gyroklystron amplifiers, gyro-traveling-wave amplifiers (gyroTWAs), and gyrotwystron amplifiers. Accelerators require a stable, narrow-bandwidth amplifier, and for this application, the gyroklystron has proved to be the most suitable gyrotron amplifier configuration. Gyroklystron amplifiers for accelerator applications are presently under development at frequencies from X-band to W-band. Each gyrotron oscillator or amplifier configuration, except for the gyro-BWO, has a Doppler-upshifted CARM analog that operates at higher k_z, and at a frequency of approximately $s\Omega_c/(1-h\beta_z)$ where $h=k_zc/\omega$, and β_z is the electron parallel velocity normalized to the speed of light. CARMs offer potentially higher efficiency than gyrotrons, and operate effectively at lower values of the velocity pitch ratio α, but require electron beams with very low parallel velocity spreads to achieve the predicted high efficiencies. Moreover, CARMs are subject to serious competition from near-cutoff gyrotron and gyroBWO modes, and therefore effective CARM cavities must select for high-k_z modes, and discriminate against near-cutoff modes. For this purpose, Bragg resonators configurations have typically been used. However, even experiments employing Bragg resonators often suffer from competition with near-cutoff modes as well as other non-Bragg modes, such as output window reflection modes. The two CARM configurations that have been investigated experimentally are the CARM oscillator and the CARM traveling-wave amplifier (CARM TWA). The combination of beam quality problems and mode competition have severely limited most of the previously reported CARM experiments, resulting either in oscillator experiments dominated by spurious modes or with relatively low efficiency (<10%), or in traveling-wave amplifier experiments with high-gain, but relatively low efficiency. The recent history of CARM experiments is summarized in Ref. [1] and references therein.

The best example of a CARM oscillator substantially exceeding these results is the work of Bratman et al., who achieved 13 MW at 26% efficiency in a 38-GHz CARM oscillator using a Bragg resonator cavity and a 100-A, 500-keV beam from a field-emission cathode [2]. In this device, the transverse momentum required to drive the CARM interaction was produced by a half-period wiggler magnet, and great care was taken to minimize the parallel velocity spread of the electron beam. (There have also been recent experiments examining devices operating with coupled near-cutoff and Doppler-shifted modes [3,4]. However, these configurations have employed large-orbit helical beams generated from linear beams that are spun up using kicker or wiggler magnets, rather than the small-orbit configurations produced by magnetron injection guns (MIGs). Large-orbit (i.e., axicentered orbit) gyrodevices, with coherent helical electron trajectories, are a different class of device than those driven by thermionic MIGs, and this configuration imposes a selection rule on the interaction, that the sth harmonic interaction will only couple to a TE_{sm} or TM_{sm} modes.)

Nusinovich et al. [5] have considered the effect of velocity spread on a CARM twystron configuration. In the CARM twystron, a beam prebunched in a drive cavity generates output power in a traveling-wave output section. In the output section, the radiation power grows in a single pass without feedback, which can increase the interaction length substantially compared to a device with a short output cavity. Nusinovich's study suggests that a CARM twystron would have a very great sensitivity to parallel velocity spread, making an efficient, high-power device extremely difficult to achieve. The purpose of this study is to explore comparable limits for a CARM-klystron configuration.

In this paper, we consider the possibility of an efficient CARM-klystron device for accelerator applications. Such a device could in principle match or exceed the efficiency of a comparable gyroklystron device at much lower values of the electron beam pitch ratio α. The lower value of α would make the entire device substantially more stable. One of the greatest difficulties in building an efficient high-power gyroklystron amplifier for accelerator applications is the problem of producing a high quality, high-α beam. Typically, efficient, high-power gyroklystron designs require α~1.5. However, with inadequate beam quality, electrons may begin to reflect before this value of α is achieved. Moreover, the maximum value of beam α is also frequently constrained by the onset of oscillations in the gun or beam tunnels. (A major issue in gyroklystron design and engineering is to substantially load drift sections to prevent oscillation without simultaneously loading the cavities whose fields leak into those drift spaces.)

One more motivation for this study is the existence of a high power MIG that was designed for a somewhat different CARM oscillator experiment (see Ref. [6]). New simulations suggest that this gun could produce a 500 kV, 100-200 A electron beam with sufficient beam quality for an interesting CARM amplifier at 34 GHz. Including the Doppler-shift effect, this would require magnetic fields of ~10-12 kG, and, for high efficiency, α~0.7-1.0.

THEORETICAL MODEL

The model that was used to analyze the gyroklystron efficiency was derived from the cold-beam CARM theory presented in Ref. [7]. In the present work, that formulation was generalized to account for a spread in beam velocities subject to the constraint of zero energy spread. The modified equations of motion for an electron beam interacting with a co-propagating wave of the form

$$A_i \propto F_s g^+(z) e^{-i(\omega t - k_z z + \xi)}$$

are given by

$$\frac{d\bar{p}_{ti}}{d\zeta} = -\frac{\bar{p}_{ti}^{s-1}}{2\bar{p}_{zi}} F_s g^+(\zeta) \cos(\vartheta_i + \xi)$$

$$\frac{d\vartheta_i}{d\zeta} = \frac{1}{\bar{p}_{zi}} \left[\Delta_i - u_i - b\hat{q}_i + \frac{s}{2} F_s \bar{p}_{ti}^{s-2} g^+(\zeta) \sin(\vartheta_i + \xi) \right]$$

$$\frac{d\hat{q}_i}{d\zeta} = -\frac{\bar{p}_{ti}^s}{2\bar{p}_{zi}} F_s \frac{dg^+(\zeta)}{d\zeta} \sin(\vartheta_i + \xi)$$

where the normalized magnitudes of the transverse and axial momenta of the ith electron are given by

$$\bar{p}_{ti} = \frac{p_{ti}}{\gamma_0 m_0 c \bar{\beta}_{t0}}$$

$$\bar{p}_{zi} = \frac{p_{zi}}{\gamma_0 m_0 c \bar{\beta}_{z0}}$$

where m_0 is the electron rest mass, γ_0 is the relativistic mass ratio, c is the speed of light in vacuum, and

$$\bar{\beta}_{t0} = \bar{v}_{t0}/c$$

$$\bar{\beta}_{z0} = \bar{v}_{z0}/c$$

are the average values of the beam transverse and axial velocity divided by c, respectively. The distribution of axial velocities is taken to be Gaussian, truncated after two standard deviations. The normalized electron energy change during the interaction is given by:

$$u_i = \frac{2}{\bar{\beta}_{t0}^2}\left(1 - \frac{\bar{\beta}_{z0}}{\beta_{ph}}\right)\left(\frac{\gamma_0 - \gamma}{\gamma_0}\right)$$

At any point in the interaction the momenta are related to the change in energy according to:

$$\bar{p}_{ti} = \sqrt{\bar{p}_{t0i}^2 - u_i}$$

$$\bar{p}_{zi} = \bar{p}_{z0i} - b u_i$$

where the parameter b characterizes the coupling between the energy and the axial momentum, and is given by

$$b = \frac{\bar{\beta}_{t0}^2}{2\bar{\beta}_{z0}(1 - \bar{\beta}_{z0}/\beta_{ph})}$$

where β_{ph} is the wave phase velocity, ϑ_i is the phase shift of the electron orbit relative to the interacting wave, and Δ_i is the normalized kinematic phase shift during the interaction given by

$$\Delta_i = \bar{\Delta} + \frac{\bar{p}_{z0i} - 1}{b - \frac{1-\gamma_0^{-2}}{b(1+\bar{\alpha}^2)}\left(b + \frac{\bar{\alpha}^2}{2}\right)^2}$$

where

$$\overline{\Delta} = \frac{2(1-\overline{\beta}_{z0}/\beta_{ph})}{\overline{\beta}_{t0}^2(1-\beta_{ph}^{-2})}\left(1-\frac{\overline{\beta}_{z0}}{\beta_{ph}}-\frac{s\Omega}{\omega}\right)$$

is the average detuning parameter of the beam, s is the harmonic number, ω is the wave angular frequency, and

$$\Omega = \frac{eB_0}{\gamma_0 m_0}$$

is the relativistic cyclotron frequency prior the interaction. The wave envelope function is assumed to have a Gaussian form:

$$g^+(\zeta) = \exp\left[-4\left(\frac{2\zeta}{\zeta_{out}}-1\right)^2\right]$$

where the interaction occurs between $\zeta=0$ and $\zeta=\zeta_{out}$. At the cavity input the phase parameter is given by

$$\vartheta_i(\zeta=0) = \vartheta_0 + q_b \sin\vartheta_0$$

where q_b is the bunching parameter and ϑ_0 is uniformly distributed between 0 and 2π. The electronic efficiency is given by:

$$\eta = \frac{\overline{\beta}_{t0}^2 \langle u_i(\zeta_{out})\rangle_i}{2\left(1-\frac{\overline{\beta}_{z0}}{\beta_{ph}}\right)(1-\gamma_0^{-1})}$$

where the angular brackets denote an average over entrance phases and velocities.

RESULTS

Figure 1 shows the results of calculations of efficiency versus parallel velocity spread for CARM klystron output cavities operating at two typical values of the parameter b (0.25, 0.4), and similar results for devices with two different values of α and with small Doppler shift ($b=0.1$), i.e., in the gyroklystron regime. The bunching parameter q_b was taken to be 2 in all cases corresponding to strong bunching. The interaction length was optimized for efficiency, typically $\zeta_{out}\sim 8$. The velocity spread parameter corresponds to one standard deviation of the axial velocity spread. It is noteworthy that for zero parallel velocity spread, the CARM result with $\alpha=0.82$ has a higher efficiency than the gyroklystron with $\alpha=1.5$, and the CARM result with $\alpha=0.71$ has a higher efficiency than the gyroklystron result with $\alpha=1$. However, the two CARM curves fall rapidly with increasing parallel velocity spread, while the gyroklystron results are relatively insensitive to parallel velocity spread. It appears that the CARM klystron has potential advantages for parallel velocity spreads less than 2%, while the gyroklystron has clear advantages for spreads greater than 5%. For the interval between 2% and 4%, it appears that the two devices may be competitive, depending on details of the design, such as beam α, cavity length, and the value of the b parameter.

CONCLUSIONS

Based on this preliminary study of a CARM output cavity driven by a prebunched electron beam, it appears that the CARM-klystron configuration may be of interest for developing high-frequency (e.g. Ka-band or W-band) amplifiers for accelerator applications. These simulations indicate that the potential interaction efficiency of the CARM klystron can equal that of the more common gyroklystron configuration, while using an electron beam with significantly lower transverse momentum. However, the value of the axial velocity spread is critical. Moreover, this limited study has not examined the effects of velocity spread on the bunching induced in a high-gain multicavity configuration, which would be expected to increase the deleterious effects of parallel velocity spread. In addition, we have not considered how to design high-k_z cavities while avoiding the excitation of near-cutoff modes. These issues will be the subject of future work.

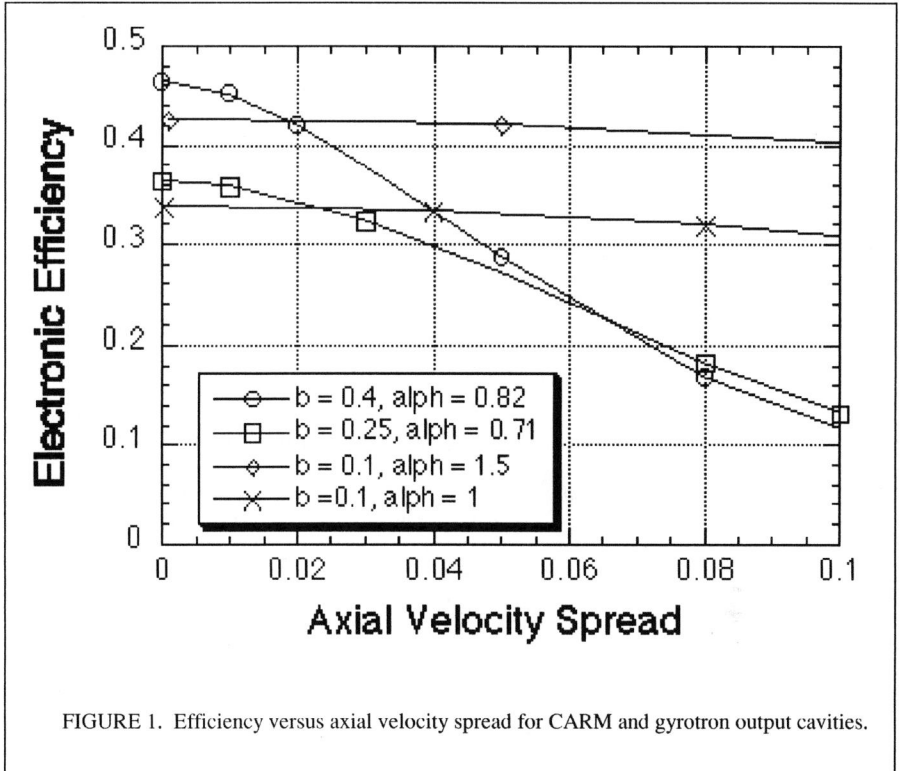

FIGURE 1. Efficiency versus axial velocity spread for CARM and gyrotron output cavities.

ACKNOWLEDGMENTS

This work was supported by the US Department of Energy and the Office of Naval Research.

REFERENCES

1. Gold, S.H. and Nusinovich, G.S., *Rev. Sci. Instrum.* **68**, 3945-3974 (1997).
2. Bratman, V.L., Denisov, G.G., Kol'chugin, B.D., Samsonov, S.V., and Volkov, A.B., *Phys. Rev. Lett.* **75**, 3102-3105 (1995).
3. Denisov, G.G., Bratman, V.L., Cross, A.W., He, W., Phelps, A.D.R., Ronald, K., Samsonov, S.V., and Whyte, C.G., *Phys. Rev. Lett.* **81**, 5680-5683 (1998).
4. Savilov, A.V., Bratman, V.L., Phelps, A.D.R., and Samsonov, S.V., *Nucl. Instrum. Methods Phys. Res. A* **445**, 230-235 (2000).
5. Nusinovich, G.S., Latham, P.E., and Li, H., *IEEE Trans. Plasma Sci.* **22**, 796-803 (1994).
6. McCowan, R.B., Pendleton, R.A., and Fliflet, A.W., *IEEE Trans. Electron Devices* **39**, 1763-1767 (1992).
7. Fliflet, A.W., *Int. J. Electronics* **61**, 1049-1080 (1986).

Multi-Moded Passive RF Pulse Compression Development at SLAC

Christopher D. Nantista and Sami G. Tantawi[1]

Stanford Linear Accelerator Center, P.O. Box 4349, Stanford, CA 94309, U.S.A.

Abstract. The design of a pulse compressing power distribution system for the Next Linear Collider has evolved significantly in the past few years. This system allows the combined power of several klystrons to be directed, by means of drive phase manipulation, to different accelerator feeds at different times during each pulse. The desire to reduce the amount of required low-loss, circular-waveguide delay line has led to multi-moded schemes [1], in which different modes are propagated through the same waveguide to different destinations. We will present current plans for a system utilizing two modes, in which manipulations are done primarily in overmoded rectangular guide. We will describe several novel, passive waveguide components developed for this system. Because these must carry up to 600 MW pulsed rf power, features that invite breakdown, such as coupling slots, irises, and septa, are avoided, and h-plane symmetry is exploited to allow the use of overheight waveguide. This X-band pulse compressor is scalable to higher frequencies.

INTRODUCTION

The NLC (Next Linear Collider) [2] is a proposed X-band, linac-based, e^+e^--colliding high-energy physics facility designed to provide center-of-mass energies in the range 0.5–1 TeV. To achieve high accelerating gradients in such linacs, it has become common to employ rf pulse compression to match the output capabilities of sources (longer than needed pulses) to the accelerator structure input requirements (higher than directly producible peak power). In the course of R&D for the high-power rf system of the NLC, the original SLED-II [3] pulse compression technique was replaced early on by an inherently more efficient scheme of effective pulse compression called DLDS (Delay Line Distribution System) [4]. This eliminated the need for resonant rf storage and utilized the time-of-flight of the electron/positron beam to reduce the required length of delay line compared to an earlier, similar proposal called BPC (Binary Pulse Compression) [5]. To further reduce the amount of bulky, low-loss waveguide required, the concept of multi-moding was introduced [1], replacing simple DLDS with MDLDS (Multi-moded DLDS).

In a DLDS system, groups of klystrons are made to deliver their combined power, during consecutive time-bin divisions of their full operating pulse, to distantly separated accelerator feeds. Power direction is accomplished via phase shifts in the rf drives. With a proper passive, matched circuit, four sources, for example, can be made to feed four feeds through the four orthogonal combinations of phases. The first

---

[1] Also with the Communications and Electronics Department, Cairo University, Giza, Egypt.

time-bin is shipped furthest upstream; the last is delivered locally. The shortened propagation distance of the rf combines with the later arrival of the beam to allow time for the structures powered by each consecutive feed to be filled. With MDLDS, we eliminate the need for a separate delay line to transport the rf to each feed. Power is delivered to two or more feeds through the same waveguide by using different modes as the carriers and inserting specially designed extractors to direct each mode to its proper destination.

In a previous paper [1], we described plans for a three-mode system, utilizing the TE_{01}° mode and both polarizations of the TE_{12}° mode. Component designs for that system involved coupling slots and irises, which had the potential to occasion rf breakdown limitations. As our system must be able to carry up to 600 MW, we have more recently concentrated our efforts on a more conservative approach in which we manipulate the rf in moderately overmoded, rectangular waveguide. Transitioning to and from the highly overmoded circular delay line waveguide will be done through special mode-order-preserving tapers. Current component designs exploit planar symmetry, which allows for more facile mode manipulation and for the use of arbitrarily tall waveguide to limit field levels. Our goal has been to keep the electric field below ~40 MV/m while at the same time aiming for compactness to minimize ohmic losses.

Figure 1 gives a schematic of a module of our MDLDS, whose components will be described below. Eight 75 MW, 11.424 GHz klystrons are combined in pairs, yielding four independently phased sources. The combined 600 MW, 1.5 µs pulse is phased in four 375 ns time bins. The feeds are numbered to show the order in which they receive power. Each feed delivers a nominal 200 MW to each of three structures. A number of interleaved modules power the intervening structures.

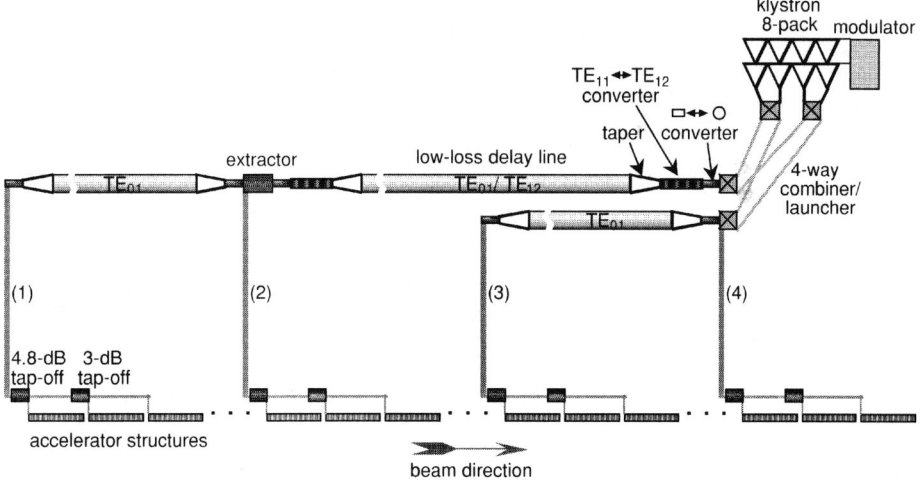

FIGURE 1. Schematic of the dual-moded MDLDS with four accelerator feeds.

CIRCULAR-TO-RECTANGULAR TAPER CONVERTER

For low-loss transmission of rf power, the ideal waveguide mode is the circular TE_{01}° mode, for which the attenuation falls off at large radius as a^{-3}. Much experience has been gained in the past decade in transporting hundreds of megawatts at 11.424 GHz in this mode [6]. At a diameter of 4.75 inches, the next most efficient mode is TE_{12}°. Recent transmission experiments, in which no coupling between cross-polarizations was detected, demonstrated the viability of this mode [7]. One can generate TE_{12}° at a smaller diameter from TE_{11}° via a Marié mode converter with wall undulations designed to pass TE_{01}° unperturbed. The challenge then becomes to launch TE_{01}° and TE_{11}° into the same waveguide and to extract one while passing the other.

The components to be described below perform these functions with the rectangular TE_{10} and TE_{20} modes. We can use them in our system without sacrificing the benefits of circular waveguide delay lines if we can transition between the two cross-sections in such a way that a one-to-one correspondence is achieved for the respective operating modes. An adiabatic cross-section taper naturally converts TE_{10} to TE_{11}°. TE_{20}, however, tends to produce a combination of the circular modes TE_{21}° and TE_{01}°. The cross-section deformation must be done in two or more properly designed and spaced taper sections to yield finally a pure TE_{01}° wave [8]. A taper-mode transducer design is being finalized that will accomplish this in the space of a few inches without compromising the TE_{11}° conversion. A preliminary design is illustrated in Figure 2 with electric field plots from an HP HFSS [9] simulation. (Since the mode evolution along the taper is order preserving, and the TE_{21}° cutoff falls below that of TE_{01}°, it is technically the higher cutoff polarization, or TE_{02} mode, which produces TE_{01}°. However, a slight rectangular taper in the dimension along which the field varies, necessary to match into the following components, will correct this nomenclature.) The ending diameter is 1.5 inches. After the Marié converter the transition to 4.75 inch guide will be made through a taper designed by Calabazas Creek Research, Inc. [10] to preserve both TE_{01}° and TE_{12}°.

A number of rectangular to circular conversions are needed in parts of our system that carry a single operating mode. For these, a more compact design might be used which needn't be optimized for TE_{10}/TE_{11}° transmission.

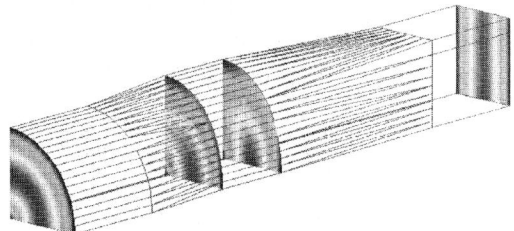

FIGURE 2. Taper Converter ¼ geometry with field patterns illustrating conversion between circular TE_{01}° and rectangular TE_{20}. An incident circular TE_{11}° wave would be converted to rectangular TE_{10}.

CROSS POTENT SUPERHYBRID/LAUNCHER

The move in our pulse compression work toward rectangular waveguide components, in which planar symmetry is exploited to allow arbitrary height, began with the design of planar hybrids [11] to replace magic T's in the ASTA (Accelerator Structure Test Area) SLED-II system. These had been exhibiting rf breakdown problems above 200 MW, particularly at the mouth of the E-plane port. One novel hybrid design has an "H" geometry. Its central guide is wide enough to support two TE modes, and, at its junctions, triangular wall protrusions yield essentially double mitred bends. Prototypes have been built and successfully operated at peak power levels approaching 500 MW.

If two such "magic H" hybrids, with ports half the width of the central guide, are placed side-by-side and their common wall removed, the resulting oversized ports have the same cross-section as the central guide. If these are split again with T's at the proper distance, the symmetry is completed, and an eight port device in the shape of a cross potent (cross with a cross bar at each extremity) results [12]. This "cross potent superhybrid" can be used to combine power from four input ports into any one of four output ports, by proper phasing. Opposite pairs of cross arms are isolated. A prototype has been built and its scattering parameters measured with a network analyzer, with very satisfactory results.

Of course, one can leave off the T split on one or more of the arms, substituting posts for matching, and consider the TE_{10} and TE_{20} modes as the orthogonal outputs or inputs for such arms. This cross potent launcher configuration and its function are illustrated in the HP HFSS simulation field plots of Figure 3. The wide waveguide width is 1.442 inches, and the height can be chosen to accommodate the taper converter. Combined with a taper converter and Marié converter, it will allow us to launch the desired modes into our circular waveguide delay lines from four independently phased sources.

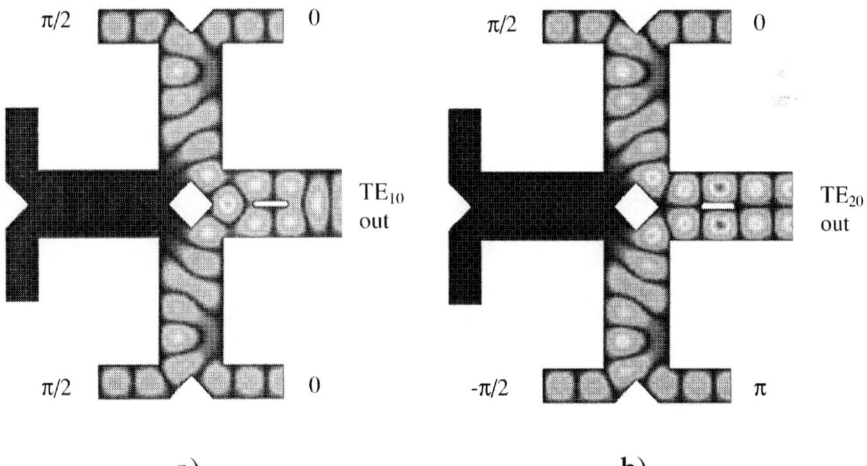

FIGURE 3. Cross Potent Launcher with simulated electric field plots illustrating launching a) TE_{10} and b) TE_{20} in the right overmoded rectangular port with the indicated relative phases for four equal amplitude inputs. Alternate phasings of the inputs sends the power to either of the left ports.

EXTRACTOR

One system feed spacing before the end of a dual-moded delay line, one mode is extracted and fed into the accelerator. Following the proper tapers and converters, the extractor begins in the same rectangular dimensions as the launcher output. A 45° H-plane bend with an inner-wall radius-of-curvature of 1.055 inches first mixes the two rectangular guide modes, converting either input into an equal combination of TE_{10} and TE_{20}. A short straight section is used to achieve the proper relative phase. Then a doubly matched T split, at which the TE_{10} field adds constructively to one lobe of TE_{20} and destructively to the other, sends all the power one way for a given extractor input mode and all the power the other way for the other input mode. Since the two input modes result in combinations with opposite relative phases, they excite opposite ports at the split. Again, we illustrate the geometry and function with field plots in Figure 4. Since it corresponds to the circular delay line mode with the greater attenuation, the TE_{10} mode is selected for extraction.

Single-moded 45° H-plane bends orient the extraction port waveguide perpendicular to the delay line and the through port waveguide parallel to it, albeit offset. The latter is then tapered to full width and sent through a dogleg or jog converter, described below, which simultaneously brings the port back in line with the delay line axis and restores the TE_{20} mode. An identical mode converter can be appended to the extraction port, so that, through taper converters, the power will be relaunched in either the delay line or the accelerator feed in the more efficient circular TE_{01}° mode.

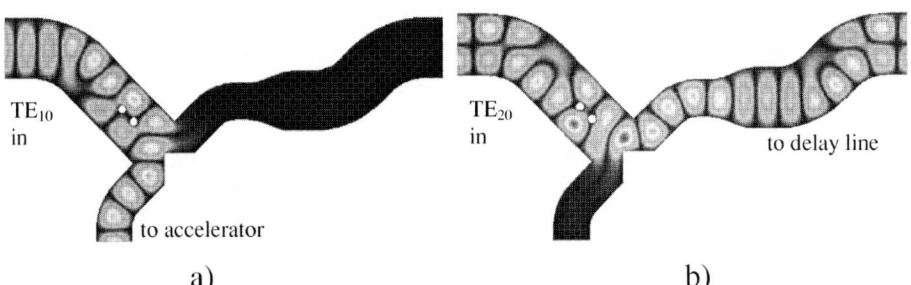

FIGURE 4. Extractor with simulated electric field plots illustrating a) extraction of the TE_{10} mode and b) passing the TE_{20} mode. For the latter, a "jog converter" attached to the through port after a width taper restores the mode and brings the port back in line with the delay line.

BENDS AND OTHER COMPONENTS

The physical layout of our rf system will require bends which, because the system is overmoded, are not completely trivial. Even at places where a single mode is used, power levels generally do not allow us to reduce the cross-section back to single-moded waveguide. We now describe plans for negotiating such bends and otherwise manipulating the rectangular modes.

Overmoded H-plane Bend

We have designed a 90° H-plane bend in the 1.442 inch-wide waveguide mentioned above. Just as we found a radius-of-curvature that gave 50% conversion at 45° for the extractor, one expects that certain bending radii will bring power coupled along the bend completely back to the entering mode. We find that, for this cross-section at our operating frequency, a bend with a radius-of-curvature from the inner wall of 1.409 inches transmits the two operating modes purely from input to output, as shown in Figure 5. As with the above planar components, other propagating modes, including TM modes, are not coupled in these bends. Since reflections are negligible and the coupling only between TE_{10} and TE_{20}, the solution automatically works for both modes.

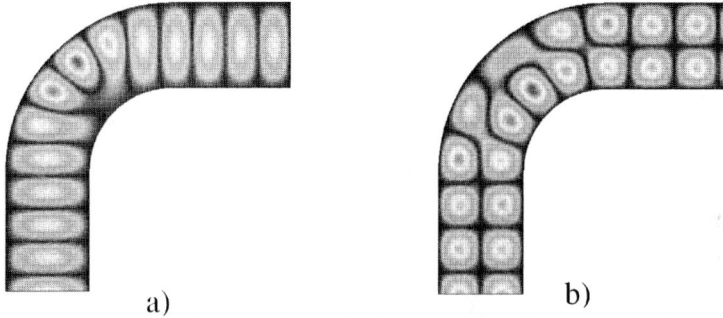

FIGURE 5. Overmoded H-plane bend in 1.442" waveguide with simulated electric field plots illustrating a) TE_{10} mode transmission and b) TE_{20} mode transmission.

Jog Converter

Following a 45° bend like that at the beginning of the extractor, which leaves power entering in either operating mode in an equal mixture of the two, a second such bend may be located so as to either return the coupled power to the original mode or transfer the remaining power from it. This second bend can be in the same or the opposite sense, resulting in a 90° bend or a jog.

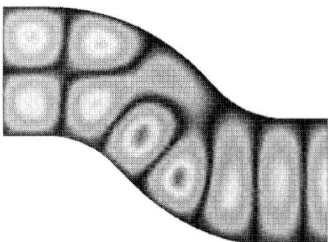

FIGURE 6. Jog converter geometry with simulated electric field plots illustrating conversion from TE_{20} to TE_{10} (left to right) or from TE_{10} to TE_{20} (right to left).

The "jog converter" is a compact mode transducer, consisting of two oppositely oriented 45° bends, separated by a very short phasing section, which gives complete conversion between TE_{10} and TE_{20}. It works in either direction, for either input mode.

This simple device is shown in Figure 6. It is used at several points in our rf system plans. It can be combined with a rectangular waveguide taper and a rectangular-to-circular taper converter of the type described above, to form a novel TE_{10} to TE_{01}° launcher.

Overmoded E-plane Bend and Height Taper

There are points in our rf system where components are required for which planar symmetry cannot be maintained. Two rectangular waveguide components which fall into this category are E-plane bends and height tapers.

Our power-combining waveguide configuration requires some E-plane bends, to get into the proper plane. It is desireable, for power handling and matching into other components, that these be in full overmoded height. Fortunately, they needn't be in our full overmoded width; thus, only TE_{10} must be transmitted. This mode is coupled strongly to TM_{11} in such a bend. At a 0.900"×1.210" cross-section, an inner (bottom) wall radius-of-curvature of 2.30 inches returns a pure TE_{10} mode at 90°, as seen in Figure 6a). The surface field is enhanced within the bend, but we only require it to carry ¼ of the total power.

While the above bend and parts of other components propagate a single TE_{n0} mode, it has been assumed that the rectangular guide components described above will be built with an overmoded height (allow propagation of modes with non-zero second index). At points in the rf system, specifically at the structure inputs, it will be necessary to transition to true single-moded waveguide. Figure 7b) shows a preliminary double-stepped height taper design shown in going from 0.900"×0.400" (WR90) to 0.900"×1.210" in less than an inch with perfect transmission. However, even with well-rounded edges, the field enhancement gives ~70 MV/m for 200 MW. This must be reduced when both these nonplanar components are redesigned to accommodate the height of the final taper converter design.

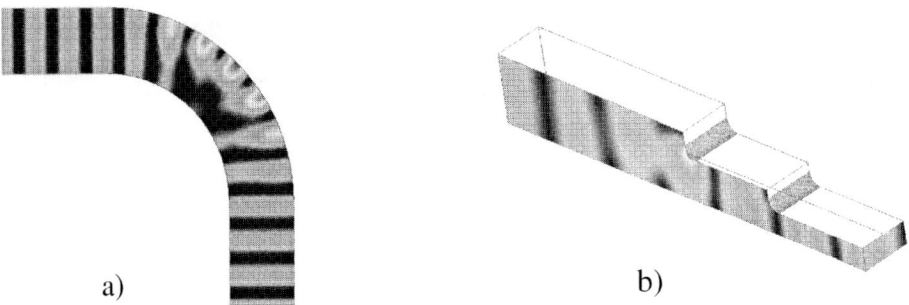

FIGURE 7. a) overmoded E-plane bend and b) quarter geometry of a height taper with simulated electric field plots for TE_{10} in 0.9"×1.210" waveguide. The midplane field in each shows enhancement and local TM_{11} coupling.

TAP-OFFS

Each feed from our MDLDS will deliver ~600 MW to a set of three consecutive accelerator structures. After the waveguide turns to run parallel to the accelerator, first one third of the power will need to be removed for the first structure, and then one half of the remaining power will need to be removed for the second structure before the feed terminates in the third. We plan to do this as well in overmoded planar rectangular waveguide components.

One idea is to simply peel off a lobe from the appropriate TE_{n0} mode for each tap-off. The TE_{30} mode can be generated in widened waveguide from TE_{10} or TE_{20} by a planar converter. One third of the waveguide could then be interrupted by a mitred bend, which leaves two thirds of the field pattern and of the power to form a TE_{20} wave in the continuing guide. At the second structure, one half of a TE_{20} wave is similarly diverted. This concept for a 4.77-dB and a 3.01-dB power divider is illustrated in Figure 8, though the designs shown need slight refinement.

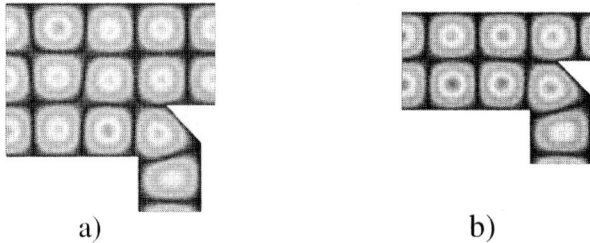

a)   b)

FIGURE 8. Tap-offs utilizing the idea of achieving fractional power division by deflecting a lobe of a TE_{n0} field pattern. This requires converting to TE_{30} before the first structure input.

To isolate the structures in case of rf breakdown, it may be preferable to use directional couplers instead. A hybrid of the "magic H" type on which the cross potent launcher is based can serve for the second tap-off. This is illustrated in Figure 9. For the first tap-off, the hybrid design can be modified to give the proper 1/3-2/3 split. This is simply a matter of adjusting the differential phase length of the coupling section while maintaining the match. Any reflected power from the structures would then travel back through the waveguide system or into high power loads on the fourth ports of the directional couplers.

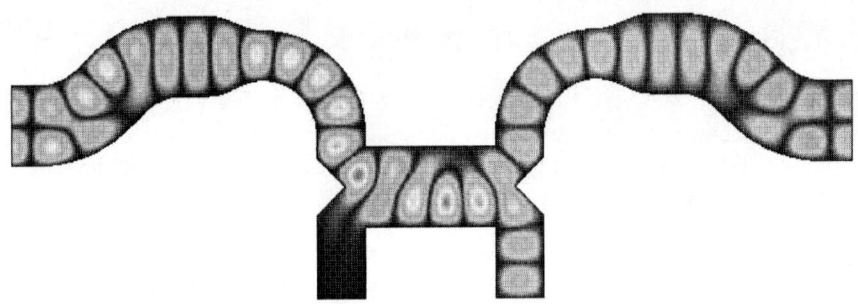

FIGURE 9. A 3-dB directional coupler tap-off with simulated field plots. Power flow is from left to right. Again jog converters are used to give the proper mode. A second design with reduced coupling can serve for the 1/3 power tap-off.

As the structure inputs will be spaced about two meters apart, it is worthwhile to convert back to the circular $TE_{01}^°$ mode in between them. Taper converters combined with jog converters will get us to and from the proper rectangular mode.

CONCLUSIONS

We have described a dual-moded MDLDS rf waveguide system for the NLC and several novel components designed to accomplish the various required functions. The relatively open geometry and exploitation of planar symmetry in our design motif allows us to keep the peak surface fields at reasonable levels for 600 MW at X-band. The use of highly-overmoded, circular waveguide is maintained for low-attenuation in delay lines and power transport, while rectangular waveguide is used in moderately overmoded components for the flexibility and simplicity of design provided by the separation of dimensions. The current two-mode scheme reduces the amount of delay line by 1/3 from what a simple DLDS would require. The designs described here might be suitably scaled for operation at 30 GHz or higher to serve in a pulse compression system for a higher frequency linear collider or for any other rf system with high power and small bandwidth requirements. Their simplicity may recommend them for use at low power as well.

ACKNOWLEDGMENTS

We gratefully acknowledge the contributions, feedback, and discussions of this development provided by Norman Kroll, Perry Wilson, Karen Fant, and all who participate in our regular rf power handling meetings at SLAC.

This work is supported by Department of Energy contract DE-AC03-76SF00515.

REFERENCES

1. S.G. Tantawi, et al., "A Multi-Moded RF Delay Line Distribution System for the Next Linear Collider," proc. of the Advanced Accelerator Concepts Workshop, Baltimore, MD, July 5-11, 1998, pp. 967-974.
2. The NLC Design Group, Zeroth-Order Design Report for the Next Linear Collider, LBNL-PUB-5424, SLAC Report 474, and UCRL-ID 124161, May 1996.
3. C. Nantista et al., "High-Power RF Pulse Compression With SLED-II at SLAC," presented at the IEEE Particle Accelerator Conference, Washington, D.C., May 17-20, 1993; SLAC-PUB-6145.
4. H. Mizuno and Y. Otake, "A New RF Power Distribution System for X Band Linac Equivalent to an RF Pulse Compression Scheme of Factor 2^N," presented at the 17th International Linac Conference (LINAC 94), Tsukuba Japan, August 21-26, 1994.
5. T.L. Lavine et al., "High-Power Radio-Fequency Binary Pulse-Compression Experiment at SLAC," presented at the IEEE Particle Accelerator Conference, San Francisco, CA, May 6-9, 1991; SLAC-PUB-5451.
6. S.G. Tantawi, et al., "The Generation of 400-MW RF Pulses at X Band Using Resonant Delay Lines," IEEE Trans. Microwave Theory Tech., vol. 47, no. 12, pp. 2539-2546, Dec. 1999; SLAC-PUB-8074.
7. Sami G. Tantawi, et al., "Evaluation of the TE_{12} Mode in Circular Waveguide for Low-Loss, High-Power RF Transmission," Phys. Rev. ST Accel. Beams, vol.3, 2000.
8. S.G. Tantawi, et al., "RF Components Using Over-Moded Rectangular Waveguides for the Next Linear Collider Multi-Moded Delay Line RF Distribution System," presented at the 18th Particle Accelerator Conference, New York, NY, March 29-April 2,1999.
9. HP High Frequency Structure Simulator, Version 5.4, Copyright 1996-1999 Hewlett-Packard Co.
10. Calabazas Creek Research, Inc., 20937 Comer Dr., Saratoga, CA 95070.
11. C.D. Nantista, et al., "Planar Waveguide Hybrids for Very High Power RF," presented at the 1999 Particle Accelerator Conference, New York, NY, March 29-April 2, 1999; SLAC-PUB-8142.
12. C.D. Nantista and S.G. Tantawi, "A Compact, Planar, Eight-Port Waveguide Power Divider/Combiner: The Cross Potent Superhybrid," submitted to IEEE Microwave and Guided Wave Letters.

Design and Fabrication of a 94 GHz Klystron

G. Scheitrum[1], G. Caryotakis[1], A. Haase[1], L. Song[2],
B. Arfin[3], Y. Cheng[4], B. Shew[4], B. James[5]

[1]*Stanford Linear Accelerator Center, Menlo Park, CA 94025*
[2]*University of California at Davis, Davis, CA 95616*
[3]*Arfin Associates, San Carlos, CA 94070*
[4]*Synchrotron Radiation Research Center, Hsinchu, Taiwan*

Abstract. The design and fabrication issues arising from the development of a 94 GHz, 100kW klystron are discussed. The klystron beam is focused using periodic permanent magnets that are mounted in the vacuum envelope. Several circuit fabrication processes were considered before the "LIGA" lithographic process was chosen. LIGA has the dimensional tolerances and surface finish required for klystron operation at 94 GHz.

INTRODUCTION

The W-band (94 GHz) klystron being developed at SLAC is designed to address multiple applications in accelerators, radars, and communications. The device has been called the 'klystrino' at SLAC. The goal of the program is to produce a 100kW peak, 1kW average power modular klystron in a relatively small package. The klystrino can be stacked in a multi-klystrino module to produce relatively high peak and average power in a small package. Figure 1 shows an example of a four klystrino module that should produce 400 kW peak power and 4 kW average power in a 4" x 4" x 14" package

The major design issues encountered in the W-band klystron development are; beam formation and focusing, simulation of the three dimensional circuit, fabrication of the circuit, circuit efficiency and heat transfer in the output cavity. Each of these topics will be addressed in turn.

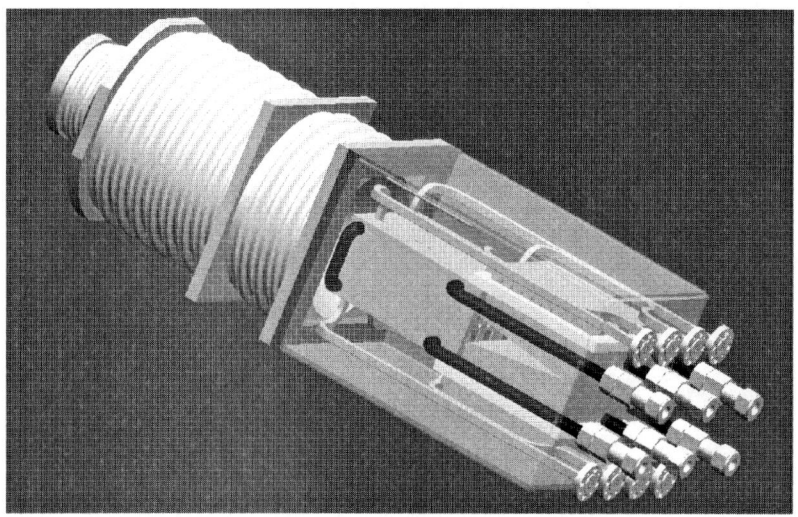

FIGURE 1. Solid model of a four-klystrino module.

DESIGN AND FABRICATION ISSUES

Electron Gun and PPM Focusing

The electron gun design was constrained by the beam voltage and current, final beam radius, and maximum acceptable cathode current density. The design parameters for the electron beam are based on the goal of 100 kW peak output power. Assuming an efficiency of 40%, the beam power must be greater than 250 kW. A 110 kV, 2.4 A beam was chosen to keep the perveance low for high efficiency and the current low to reduce the required magnetic field. The design perveance of the klystrino is 0.066 µP. The beam tunnel diameter of 0.8 mm was chosen to keep the radial propagation constant γa close to 1.

The 110 kV beam voltage is high for an individual device with this power output but the benefits outweigh the issues associated with high voltage insulators, power supplies and modulators. The high beam velocity leads to a longer output circuit with more surface area to dissipate RF circuit losses. Equally as important, the high beam velocity leads to a larger plasma wavelength making the periodic permanent magnet (PPM) focusing much easier. The ratio of plasma wavelength to magnet period (λ_p/L) is a standard measure of the ease of focusing an electron beam. Most PPM focused tubes have a λ_p/L around 2 or 3. This tube has a λ_p/L of 5, which is similar to the PPM design for the 75 MW NLC klystron at SLAC that had >99% transmission at 450 kV.

The cathode loading for the klystrino was limited to 15 A/cm². This was deemed an acceptable compromise to keep the area convergence under 100. The cathode diameter is fixed at 4.5 mm by the beam diameter, beam current and cathode loading yielding an area convergence of 81. The cathode, anode and focus electrode designs were completed by modeling the electron trajectories using DEMEOS, a 2.5 D deformable mesh gun code.

Because the pole pieces and magnets for the PPM focusing were designed to be placed inside the vacuum envelope, the magnetic focusing cannot be adjusted once the tube is exhausted. In order to verify that the magnetic circuit would adequately focus the beam without adjustment, a beamstick was designed and tested. The cross-section of the beamstick is shown in Fig. 2. A simple copper tube is used to pinch off the tube and it also serves as the collector. The magnetic circuit is mounted to one side of the anode plate and the gun and ceramic insulator are attached to the other side. Several ports are provided for thermocouples and optical viewports.

The completed beamstick is shown in Fig. 3. Since SLAC did not have a 110 kV power supply and modulator, the beamstick tests were done using a 5045 SLAC klystron modulator. The 5045 klystron was used to provide a high current load and the klystrino was operated in parallel. This provided a stable cathode voltage for testing.

The tests were limited to monitoring beam current, beam transmission, and circuit temperatures. Beam transmission is critical because it limits the duty that the klystrino can handle. The beam transmission results matched the simulations with 95% to 98% transmission at cathode voltages up to 70 kV. Breakdown occurring in the high voltage feedthru prevented testing at higher voltages and eventually punctured the ceramic feedthru. A new insulator is being constructed to enable testing at 110 kV. The perveance of the gun was a little higher than the design value but it is expected to drop as the beam velocity increases. The thermocouples showed no change with beam on or off which gave additional evidence that the beam interception was minimal.

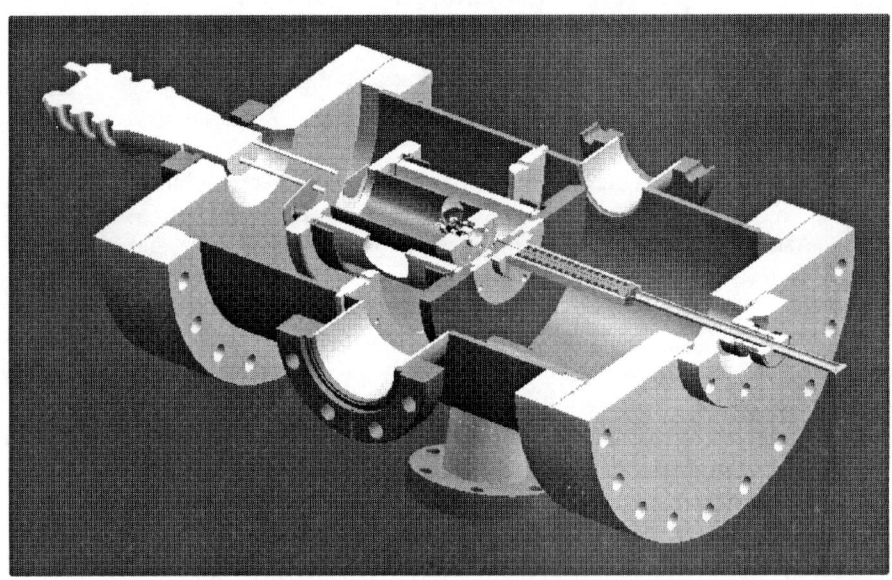

FIGURE 2. Cross-section of the 94 GHz klystron beamstick.

FIGURE 3. Photograph of beamstick under test.

Circuit Fabrication

At 94 GHz, circuit dimensions for linear beam tubes such as the klystron are on the order of one to two millimeters. The rate of change in cavity resonant frequency versus cavity dimensions can be as high as 35 MHz/micron. This implies that dimensional tolerances must be on the order of ±1 micron in order to achieve the desired resonant frequencies without requiring an external tuner.

In order to accurately fabricate the circuit without exorbitant expense, only electric discharge machining (EDM) and LIGA were evaluated as possible fabrication methods. EDM uses an electric discharge in a dielectric fluid to remove material. This leaves the finished surface very rough. Several cavities at 91 GHz and 94 GHz were made using EDM and the Q's of the test cavities ranged from 400 to 800. Etching or electropolishing of the finished surface brought the Q's up to 1000 to 1200 but these processes remove roughly three microns of material which would change the resonant frequency of the cavities by more than 200 MHz. These issues eliminated EDM as a potential microfabrication tool for the 94 GHz klystron.

The other alternative, LIGA, is a lithographic process developed in Germany. The acronym stands for LIthographie, Galvanoformung, Abformung (lithography, electroplating, molding). LIGA is capable of producing structures with thicknesses up to a few millimeters and maintaining dimensional accuracy of better than ±1 micron. As in integrated circuit manufacturing, a mask is made with the circuit features in place. A substrate is coated with a photoresist material and then exposed through the mask. The exposed photoresist is etched away and the substrate is grown to occupy the area of the etched photoresist. In LIGA, mask is usually gold and short-wavelength synchrotron light is used to expose the photoresist. The photoresist is polymethylmethacrylate (PMMA) and it can be up to two millimeters thick. LIGA is a planar process and accurately defines only the two lateral dimensions of the finished part. The height of the part is defined by a planarization step. The planarization step is usually done either by lapping or diamond flycutting in order to maintain comparable dimensional accuracy.

The following figures describe the LIGA process as it applies to the 94 GHz klystrino circuit fabrication. Figure 4 schematically shows the processes used in LIGA fabrication. The molding step is not used in prototyping but would be used for high volume fabrication of LIGA structures.

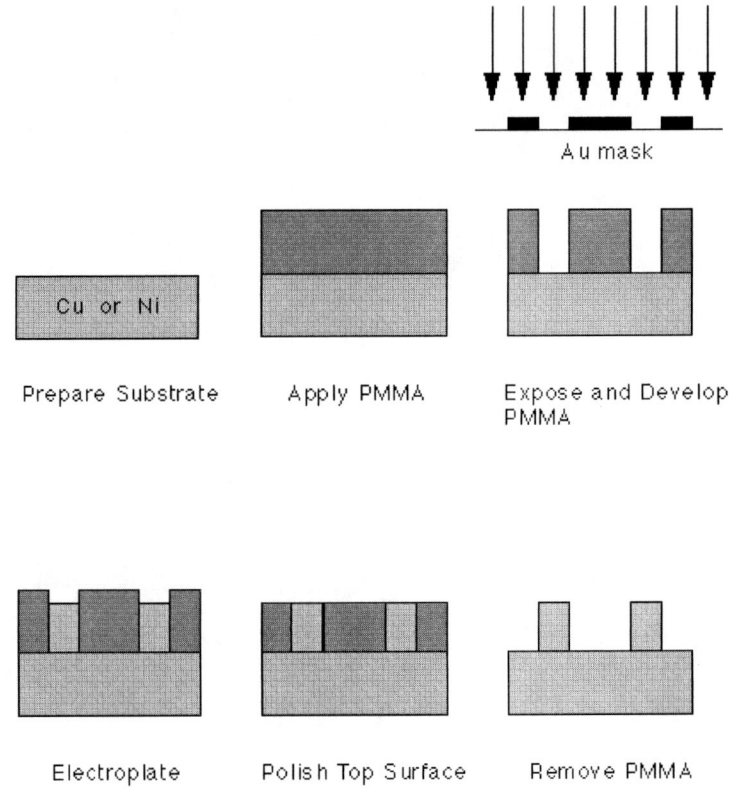

The LIGA Process

(Lithographie, Galvanoformung Abformung)

FIGURE 4. Schematic of the lithography and electrodeposition processes used in LIGA

Figure 5 is an exploded view of one half of the klystrino circuit including the magnetic circuit and cooling passages.

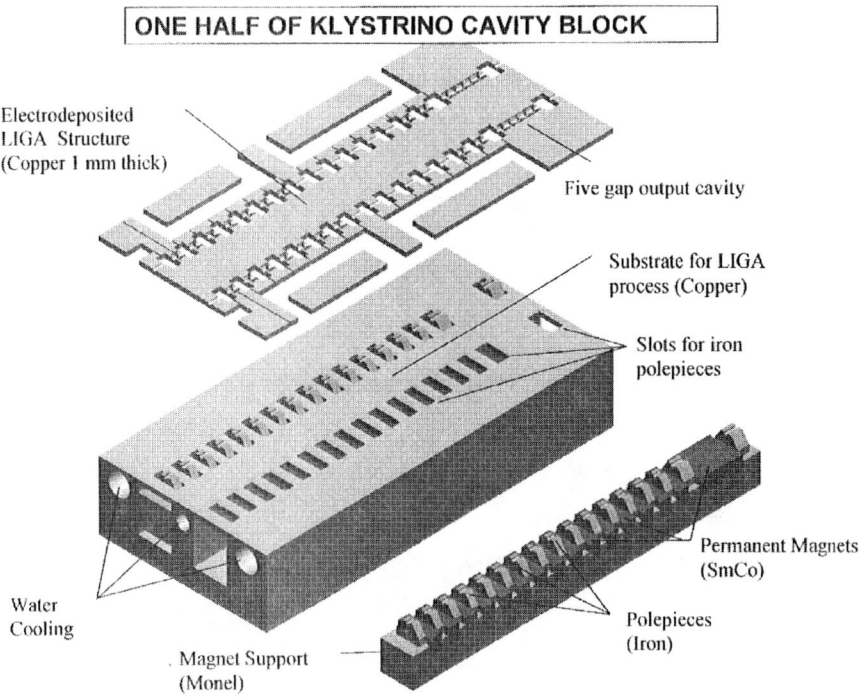

FIGURE 5. Exploded view of one half of a two-klystrino circuit. The large slots accommodate the pole piece halves and are slotted to improve vacuum pumping

Figure 6 is a photograph of the etched PMMA from the first LIGA substrate. The mask, exposure and etching were done at the Synchrotron Radiation Research Center in Taiwan. The SLAC letters were added to the LIGA mask to evaluate the minimum radius for LIGA features. The letters in the photograph are about 200 microns high. The next step in the process is to electrodeposit copper until it covers the PMMA and then lap to the desired final height. After lapping, the remaining PMMA is etched away and the EDM features (beam tunnel, coupling slots) are machined into the parts.

FIGURE 6. Photograph of exposed and etched PMMA prior to electrodeposition

FIGURE 7. Photograph of finished LIGA structure showing top and bottom halves of a three-cavity klystron circuit with a three-gap output cavity. The holes on the left half were used to evaluate the tuning of the cavities resonant frequencies by deforming the cavity walls.

Figure 7 shows the two halves of a completed LIGA structure. The right half shows two copies of the same circuit. Each circuit has three cavities, two rectangular cavities and one three-gap reentrant output cavity. The cavities, the two upper waveguide sections and the four alignment holes were made with LIGA. The beam tunnel, lower waveguide, and the coupling slots in the output cavity were done using EDM. Pins in the four alignment holes insure accurate registration of the two circuit halves. Since no current crosses the joint between circuit halves, the joint does not degrade the cavities' intrinsic Q's and the parts do not need to be brazed or diffusion bonded together. The measured intrinsic Q's for the LIGA cavities are around 1400 to 1500 which is close to the theoretical value for a rectangular copper cavity.

The AUTOCAD drawing file used to create the mask for the final klystrino design is shown in Fig. 8. It contains the geometry for four six-cavity klystrinos in a 60 mm square. The output section is a five gap extended interaction cavity. The five gaps increase the interaction impedance and reduce the risk of RF breakdown in the 0.5 mm output gaps. The additional slots in one of the klystrino circuits mark the locations of the pole pieces for the magnetic circuit.

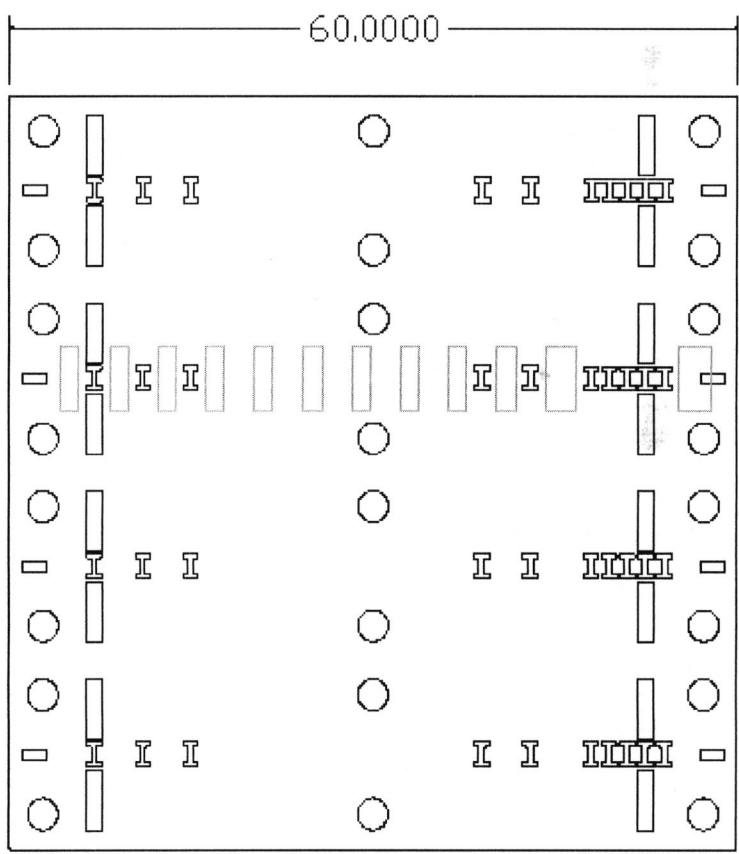

FIGURE 8. AUTOCAD drawing used to generate the gold mask for LIGA exposure.

The final LIGA figures show the four klystrino LIGA substrate after exposure and etching. The gold surfaces are the LIGA mask that was bonded to the PMMA before exposure to the synchrotron light source at SRRC. The base of the LIGA structure is etched aluminum, which is used to provide a strong bond to the PMMA.

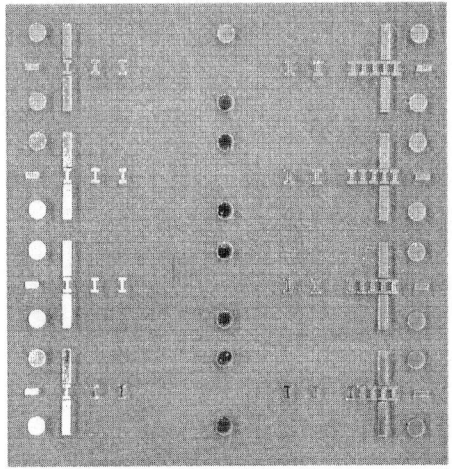 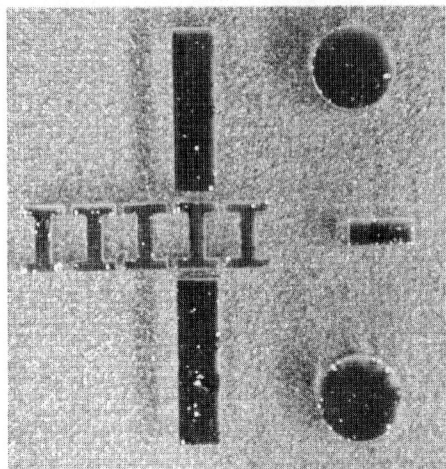

FIGURE 9. The left photograph shows the whole 60 mm square LIGA structure. The image on the right is an enlarged image of the output section showing the five reentrant cavities, two waveguides, two alignment holes and the alignment bar for the beam tunnel. The beam tunnel and the coupling slots will be added later with EDM.

Thermal Analysis

Given the small surface area of the klystrino circuit, it is critical to minimize heat input to the circuit. Both the RF circuit losses and beam interception must be kept very low to avoid damage to the circuit. The RF losses are reduced by making the circuit efficiency as high as possible. This is difficult at millimeter wave frequencies since the skin depth is very small and the surface resistivity is roughly three times that at X-band.

In order to evaluate the capability of the output circuit to handle the heat load, CPI ran several 3-D ANSYS thermal simulations for both instantaneous and average temperature. The heat inputs were based on an 80% circuit efficiency with the heat input distributed exponentially along the output section and 1% beam interception at full energy distributed uniformly on the output section. Figure 10 shows the ANSYS results without beam interception. The peak temperature rise during the pulse is 80°C, which is acceptable as far as pulsed heating is concerned. The highest average temperature at 1% duty is 180°C without beam interception and 240°C with 1% intercepted in the output section. These results indicate that the design is capable of handling the RF losses and some interception and still support 1% duty.

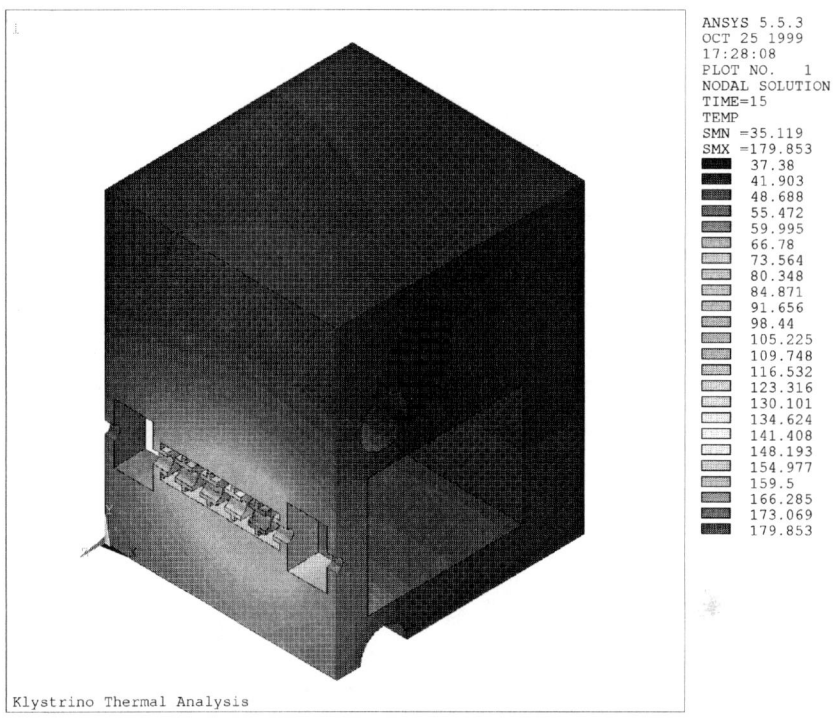

Figure 10. Three-dimensional ANSYS simulation of heat input in klystrino output section.

Computer Modeling

The rectangular cavity shapes dictated by LIGA should be modeled with fully 3-D computer simulations; however, this requires much more time and/or computational resources than currently available. In order to adequately model the klystrino it was necessary to develop an equivalent 2-D model for the 3-D klystrino circuit. Figure 11 shows the output of a 3-D MAFIA simulation of the zero mode for the five-gap output section. From this model, the resonant frequency, R/Q, Q_0, and cavity coupling were determined. An equivalent, 2-D, cylindrically-symmetric model was made that had the same characteristics as the actual 3-D structure. The 2-D, electromagnetic PIC code, MAGIC was then used to model the beam-circuit interaction.

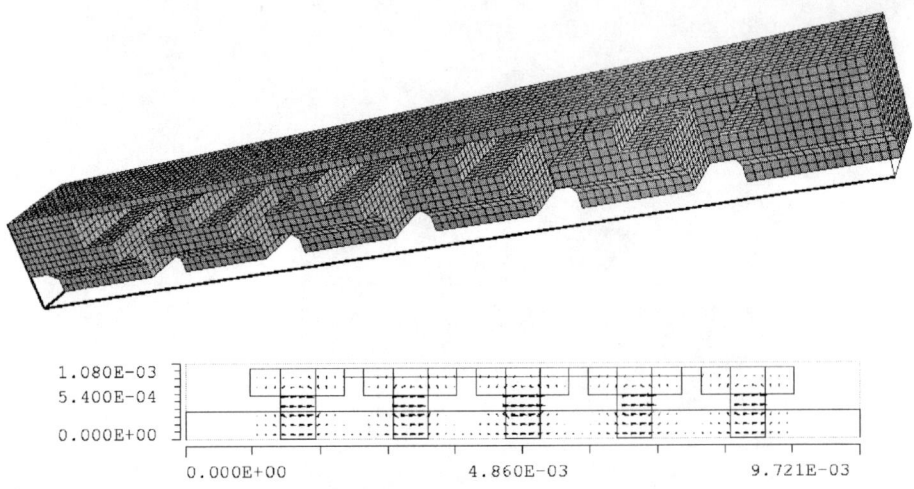

FIGURE 11. MAFIA model of 3-D klystrino output section and field pattern for the zero mode.

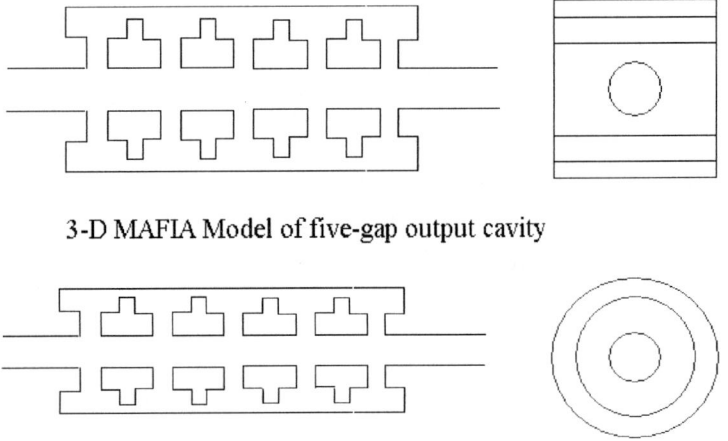

FIGURE 12. Equivalent 3-D and 2-D models for the five-gap klystrino output cavity. Note that the equivalent coupling slot in the 2-D model is cylindrically symmetric and therefore not physically realizable.

The MAGIC simulation for the output section of the klystrino requires about 25 hours to converge on a 450 MHz Pentium III processor. In order to model the complete six-cavity tube from input cavity to output cavity requires about 100 hours of computer time. Faster processors and/or multiprocessors should significantly reduce this time. Figure 13 shows an output simulation using a prebunched beam with the equivalent RF current ($I_1/I_0=1.8$) from a disk model JAPANDSK simulation. The predicted output power is roughly 120 kW. Simulations of the complete tube produce an I_1/I_0 of 1.6 and an output power of 100 kW. Further work on the input and gain cavity frequencies should produce the same RF current as in the disk model.

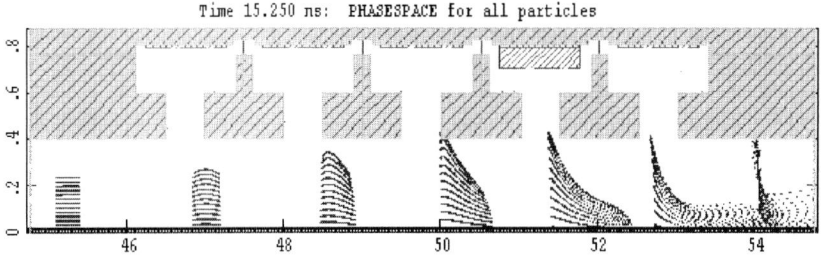

FIGURE 13. MAGIC simulation of five-gap klystrino output. Hatched sections at the outer wall of the cavity are conductance regions that simulate the power that goes into wall losses and into the output coupler.

The beam interception seen in the figure is a result of operating with the same magnetic field as in the gain section of the klystrino. The output section actually has a single magnet that is roughly 1.5 times the Brillouin field compared to 1.1 times Brillouin in the gain section.

In the 2-D simulations it is apparent that the beam has slowed sufficiently in the last gap that the voltage and current are in quadrature and no power is generated in the last gap. Work is ongoing to develop a phase tapered output section that should significantly increase the output power by keeping the voltage and current in phase throughout the five-gap output.

CONCLUSIONS

The 94 GHz klystron design has several areas where significant attention has been directed in order to insure the device will work as desired. All potential problem areas have been addressed. The beam transmission is excellent with no adjustments required to the PPM magnets. LIGA fabrication has produced cavities with the required dimensional accuracy and sufficiently high Q's to minimize RF circuit losses. The ANSYS thermal analysis indicates that the pulse heating is within acceptable levels and the average power handling will support operation at 1% duty. The modeling of the klystrino has predicted 100 kW output power with realistic input parameters.

The six-cavity LIGA substrate shown in Fig. 9 is currently being electrodeposited. The parts will be diamond flycut to the correct height and then the beam tunnel and coupling slots will be added. The two halves will be fastened together and waveguides and coolant pipes will be added to the assembly. Once the magnets and pole pieces are inserted, the completed RF circuit will be inserted in the beamstick vacuum envelope. The tube should be exhausted and available for test in December 2000.

ACKNOWLEDGEMENTS

Funding for this effort was provided by both the Air Force Office of Scientific Research MURI program – grant# F49620-95-1-0253 and an Air Force Research Laboratory grant from Phillips Lab. This work was also supported by Department of Energy contract DE-AC03-76SF00515. The support of these three programs is greatly appreciated. The support provided by the SLAC ARDB group with the W-band cold test measurements is greatly appreciated, especially the help given by Dr. E. Colby and Dr. D. Palmer.

Comparison of Discrete Klystron Produced RF to Two-Beam Produced RF for Large Accelerator Systems

Rainer Pitthan [1]
CERN and SLAC

Abstract. We compare here some technical aspects, and with it the cost, of constructing a 500 GeV center of mass Linear Collider with either Discrete Klystron or with Two-Beam (relativistic Klystron) technology using X-band for the main linac. A comparison concept is applied to CLIC and NLC technologies, but not to a particular CLIC or NLC design. The methodology created can be extended to higher c.m.s. energies, if the reader so desires.

I INTRODUCTION

One of the critical issues for the success of future electron linear colliders is the efficient and reliable generation of the RF power needed for high-gradient acceleration. Different schemes have been proposed to do that, mainly the use of discrete Klystrons, as already used for SLAC and the SLC, and Two-Beam (relativistic Klystron) technologies as in the CLIC proposal.

This talk is somewhat of a hybrid, because the organizers asked me to talk about the "CLIC concept and its technical details" and "thoughts (or my study) on the advantages and disadvantages of Two-Beam acceleration vs. discrete sources". Up to now research into Two-Beam production (relativistic Klystron) of RF has been mainly geared toward the use of high frequency RF. Research into discrete Klystron technology has been mainly directed toward lower RF frequencies. Since it is the total **system** difficulties and cost which determine the overall cost of a large accelerator, it is possible to directly compare global ramifications of discrete Klystron (as in NLC X-band at 11.4 GHz) vs. Two-Beam (as in CLIC at 30 GHz) schemes.

The combinations of various possibilities are too numerous to examine all here, so a choice has to be made for meaningful comparison. What do we compare, how do we compare? Are there any rules or models to compare seemingly incomparable technical components? And which arbitrary choices do we have to make to be able to compare?

The emphasis here is on technical comparisons, but costs obviously enter the

---

[1] Work partially supported by CERN and partially by the Department of Energy, contract DE-AC03-76SF00515.

picture. To put things in perspective, Figure 1 shows the reality of costing in R&D. The probability to find that things are easier than anticipated is smaller than the opposite. What is commonly called the contingency can be huge. The lesson to be learned from this figure is that to put a contingency of 25% and more on such estimates is well justified.

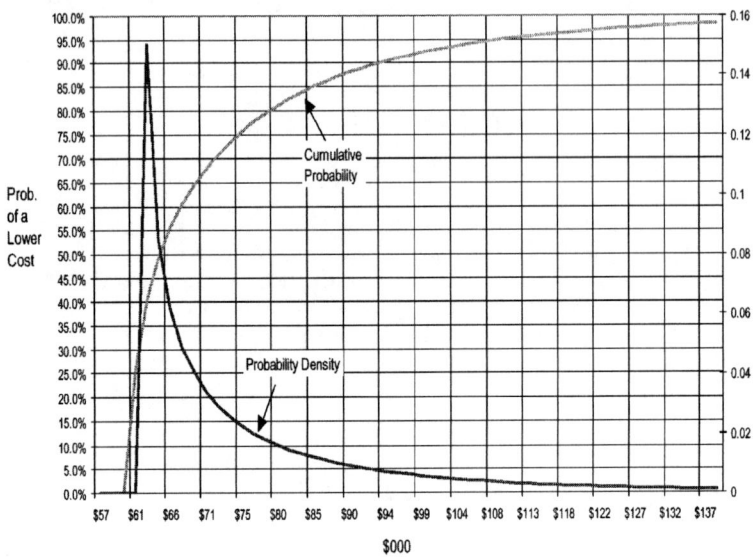

FIGURE 1. An example of a cost contingency for R&D and Hi-Tech items (Figure from of J. Cornuelle, SLAC). The highest probability, what one could naively call the price or the cost, of an item is determined to be $63500. There is a small probability it could be cheaper (easier to produce) and a large probability, with long tails, that it will cost more.

The discrete Klystron reference technology is relatively easy to define: it used to be SLAC S-band, but with 10 years of development at SLAC of X-band Klystrons and acceleration structures [1], and the NLCTA in operation [2], X-band is now the dominant technology, just as the SLAC S-band was before.

The Two-Beam reference technology is more difficult to define. There are many combinations of (low energy, high current) Drive Beam and (high energy, low current) Main Beam RF-wavelengths possible, because the drive beam, when created, can produce in deceleration units any RF frequency desired.

As driver frequencies L-band (937, 952 or 1428 MHz), UHF-band (476 MHz) and induction linacs have been investigated [1]. In the NLC ZDR [1] the Livermore Induction Linac technology was identified as a probable and possible energy upgrade path [3], but see also Ref. [4] for more recent information.

---

[1] These frequencies are not always integer multiples of a common frequency because of the slight difference between European and US S-band standard frequency. Also, many of the considerations were never formally published - they are only available in seminar or colloquia notes, and personal communications.

Proposals have been made for the main linac to use X-band (11.4), 2X-band (22.8) and Ka-band (30 GHz). More recently, because of the uncertainty of the validity of the "higher gradients at higher frequency" argument, thinking has again been focused on X-band. Nevertheless, in recent years CLIC made a giant conceptual leap forward, so that now its concept of RF production with a fully-loaded low-frequency linac at room temperature [5], even if not the precise implemention, is also regarded as a valid contender for production of a drive beam.

The inherent flexibility is one of the great conceptual strengths of the Two-Beam approach. This flexibility extends beyond the choice of frequency for the high energy main linac (=30 GHz in the case of CLIC). Once the drive beam exists, it can be used to power RF equipment which otherwise could not be powered. This includes provisions for beam loading compensation (RF ramps) [6], harmonic acceleration (higher mode cavities) [7], RF Quadrupoles for the Final Focus [8], or RF Quadrupoles to do BNS damping while avoiding the introduction of a large energy spread into the beam [9], and other possible applications and refinements.

So as requested, this paper will first describe the CLIC technology but then will focus on an adaptation of the CLIC approach for RF production to an X-band-based main accelerator.

This CLIC scheme, developed over many years at CERN, has indeed some intriguing possibilities as mentioned above. However, its main tenets, namely effective production of a high-current drive beam, deceleration without beam break-up, and the reliable existence of ≈150 MeV/m gradients at 30 GHz, have still to be proven and will not be known until at least until 2005 when the 3rd Clic Test Facility (CTF3) will be fully implemented [10]. But through the continuous test of concepts and material in its Test Facilities (CTF1 and CTF2) it has progressed to a point where it is seriously considered by CERN as a successor to post-LHC time [11].

We want to re-emphasize the main weakness encountered in past comparisons of large accelerator systems: often the focus is on one detail, say RF production, while the overall costs are really dominated by **system costs**. But the system costs are identical for most parts of a Collider, no matter what the source of the RF, so the relative total cost differences are bound to be not very large.

II PROBLEMS OF PULSED LINEAR ACCELERATORS

A Problem #1: The Gradient

As soon as linear accelerators had been invented, the quest began for higher gradients to get to higher beam energies. It was clear in principle how to get there (apart from just making it longer): higher RF power and/or higher frequency. The SLAC Blue Book ([12], Table 6-2) in the 60's has a very detailed comparison of the relative virtues of different frequency bands (L, S, and X were compared). The advantage of going to X-band were clear even then: higher gradients at higher frequencies through higher permissible field strength. But in 1960 there were no power sources available in X-band which were strong enough. The peak power available then for S-band

Klystrons was 25MW; 2.5MW was "assumed" to be "possible" at X-band. So the compromise choice for SLAC was S-band [12].

SLAC first increased the beam energy achieved through increasing the peak RF power through the SLED scheme [13]. For the higher frequency path it took SLAC a longer time of concentrated development effort to build an X-band Klystron with 50MW power. The people who designed NLC knew about the (supposed) advantages of even higher frequencies with respect to the achievable gradients. But again, there was no proven power source available.

B Problem #2: The Power Source.

The main thrust behind the CLIC effort [14] is to develop an appropriate power source to reach 30 GHz (Ka-band), and with it the promised higher gradients. The main topic of this paper is to develop a concept for comparison between Discrete Klystron and Two-Beam technology. Since the CLIC Power Source [5] basically is also based on Klystrons (L-band, which here creates the drive beam), the comparison must compare Klystrons in different frequency bands.

CLIC is now optimized for 3 TeV with 30 GHz technology, but the methods are regarded as useful at other frequency bands, for example X-band [15] which for NLC has been optimized to the 1 TeV range.

To summarize, similar to the S-band decision for SLAC in the 60's, X-band for NLC today is a compromise between theoretical expectations at higher frequency and the availability of power sources today. As mentioned above, the high gradients (150 MeV/m) held out at 30GHz may be possible to achieve, but have not been convincingly proven to be usable in long structures.

In the following comparison we choose X-band for the main Linac and compare X-band RF production with discrete Klystrons and Two-Beams. The generic machine we will call P(rototype)LC. It is designed for 500 GeV cms but with conventional facilities costed to allow an upgrade to 1.0 TeV. This upgrade does not assume higher gradients than assumed for the 500 GeV case. Doing so keeps the road open to get to 1.5 TeV in an upgrade, if higher gradients can be achieved at a later date. As nomenclature we use DKPLC (discrete Klystron) and TBPLC (Two-Beam).

Such a comparison is useful. Already the NLC ZDR has investigated the path to higher energy using Two-Beam technology. It is believed from today's point of view that Two-Beam technology is a possible route to follow for energy upgrades of X-band based accelerators, provided that higher gradients can be achieved and tolerated by the structures.

C A way out: Pulse Compression

Despite many differences in technical details, there are common concepts for all linear accelerators: energy storage and pulse compression. The average power is (relatively) low, otherwise the wall plug power could not be provided, but through energy storage and pulse compression one gets (nanoseconds) of high luminosity at high pulse power and high beam energy.

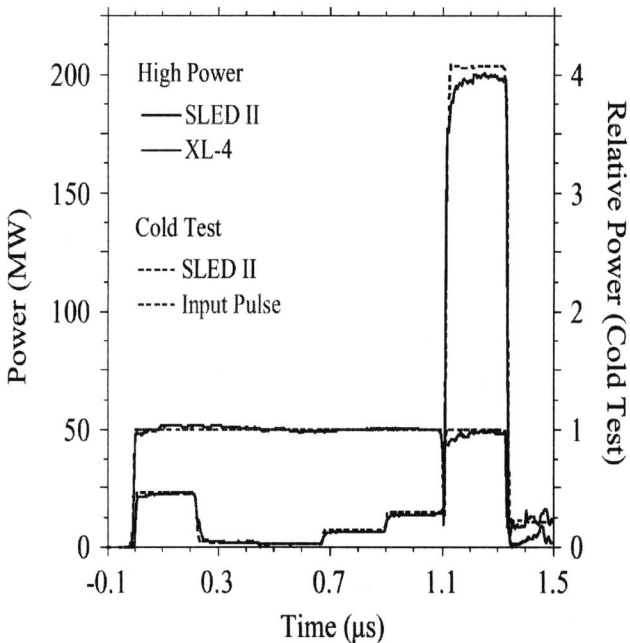

FIGURE 2. Results of pulse compression using the SLED II method for NLCTA, reference [17]. Note, how spectacularly square the pulse stays through the compression process. This is quite an improvement over previous SLED schemes.

Originally SLAC reached 20 GeV maximum beam energy [12] with 3 μsec long RF pulses. The original energy storage was done in capacitors in pulse forming networks (PFN's). Thyratrons were the switches which discharged the capacitors into Klystrons which powered the accelerator's disk loaded waveguide. This compression was done before the pulse reached the Klystron. At low power there is nothing in this scheme which could not be built in anyone's garage: it is the high power which matters.

To reach higher beam energy, shorter pulses with higher peak power are needed. These pulses can not be created efficiently in a direct way, because the rise and fall times of the Modulators become a useless large fraction of the beam. Also, the Klystron peak power would be too large. Consequently pulse compression **after** the Klystron is used: the front parts of a pulse are delayed and recombined with later parts. This results in a shorter pulse, but with higher peak power.

SLAC used this SLED (SLAC Energy Doubler) idea to store and compress RF pulses [13]. With 800 nsec RF pulses 50 GeV were reached for the SLC [16]. Whether power is stored in superconducting cavities (TESLA), in Delay Lines (NLCTA), or in RF cavities as in the old SLED (or the Japanese C-band), most schemes to store and compress pulses are variation of the original idea. Figure 2 shows a measurement of such compression for NLCTA [17].

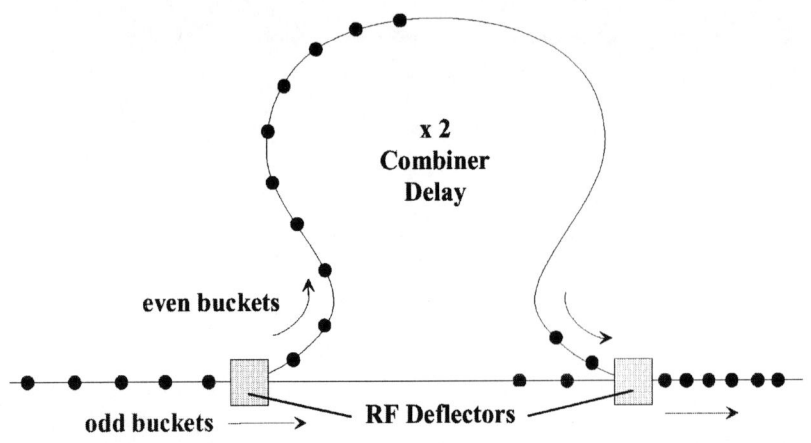

FIGURE 3. The delay loop is the first element in the chain of CLIC combiner rings. The delay loop works without a kicker, because the original 130 nsec bunch trains are alternatively put into odd and even buckets, so that a subharmonic RF deflector can be used. Although the delay loop only increases the current by a factor of 2, it is a very important element of the compression chain, by creating a gap long enough so that in later combiner rings a kicker can kick the accumulated and stacked bunch trains out.

III THE CLIC WAY

CLIC approaches the energy storage and pulse compression task differently, adapted in an innovative manner to the needs of a Two-Beam scheme of RF production. They started originally with a single superconducting RF produced drive beam, which needed to be re-accelerated many times [18]. This has been replaced by a multi-drive-beam scheme, all produced by the same room-temperature linac [14].

The most significant difference is the way CLIC proposes to do compression by frequency multiplication of electron bunches. Figure 3 shows the first element in the chain of 3 rings to get the high current needed for extraction of RF power out of the extraction units. These recent changes have increased the theoretical range of the system to c.m.s. energies of 3 to 5 TeV, based on an assumed high gradient (100 to 200 MeV/m) for acceleration at 30 GHz [5].

We will not describe here the whole scheme with all its details, but in a nutshell, the CLIC approach needs a current of about 250 A in the drive beam trains for each main accelerator. This high current is needed to extract a peak RF power of $\approx$400-500MW needed for acceleration at 30 GHz. Such a high current can not be created in an existing linac, so it is made by making 32 separate but contiguous 130 nsec long trains (which add up to 4.6 μsec total length) with an 937 MHz RF linac, filling every second bucket and combining them in a delay loop (Figure 3) and 2 combiner rings (Figure 4). The average current in each original train is planned to be 7.6 A.

For 3 TeV cms CLIC needs drive trains with a total duration of 92 μsec. For a lower c.m.s. energy, in the CLIC scheme, one uses simply a smaller number of 4.6 μsec

X4 PULSE COMPRESSION AND FREQUENCY MULTIPLICATION USING A COMBINER RING

R.CORSINI - J.P.DELAHAYE / CERN

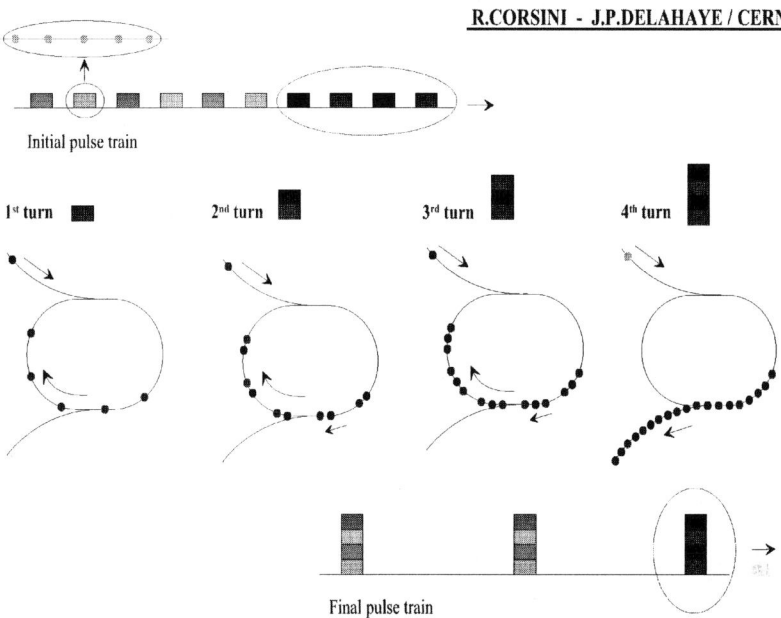

FIGURE 4. A combiner ring increases the density of bunches in a train by a factor of 4 by injecting 4 separate trains (already doubled after the delay loop) and then ejecting the combined new train in the time gap created by the delay loop. In the CLIC case this frequency multiplication by 4 is repeated a second time. For the X-band case, depending if one starts from either 952 or 1428 MHz, multiplication by either 12 or 8, respectively, would be appropriate. Or one could also choose 16 or 24; neither of these multiplication numbers is "magic", but there is a natural limit. This limit is reached when the distance between bunches becomes equal to the wavelength of the RF one wants to extract.

trains. Although the total number of L-band Klystrons is not reduced, the pulse width of each one is. Consequently energy costs, and the Klystron and Modulator costs, will scale linearly with the cms energy, see Equation (1), below.

The bunches in these original 32 trains have a distance of 64 cm between each other. (In an X-band equivalent scheme starting from 1428 MHz (1/2 of S-band frequency) [15] they would start with 43 cm distance from each other.) These 32 trains in the CLIC case are stacked in a series of delay loops and combiner rings such that one drive train of 130 nsec length results, which has 32 times the original current.

In other words, the bunches have now a distance from each other of 2 cm, well matched in time structure to the 30 GHz extraction units. Conceptually the reduction in bunch spacing, called frequency multiplication, is equivalent to pulse compression in as much as the bunch train intensity (current) is increased. In the X-band case, if starting from 952 or 1428 MHz L-band, the compression has to be optimized in a way that high enough final currents are achieved, in order to make the RF extraction from the deceleration units efficient. The development of bunch characteristics for

Parameter	Name	Initial	After Delay	After 1st Ring	After 2nd Ring	Unit
Pulse Length	τ_P	~92	~92	~92	~92	µs
Trains/Pulse	N_T	1	352	88	22	
Train Length	τ_T	92	0.130	0.130	0.130	µs
Bunch Separation	Δ_B	64	32	8	2	cm
Train Periodicity	Δ_T	-	0.26	1.04	4.16	µs
Pulse Current	I_P	7.6	15.2	61	244	A

FIGURE 5. Development of the bunch characteristics as they progress along the compression system for the 30GHz case, taken from reference [17]. The final compression is 32. For the X-band case at 952 MHz drive beam frequency, the frequency we will use below for costing, the final compression was chosen to go up to 24.

the CLIC case is collected in the table in Figure 5 [19].

One of the advantages of bunch compression via frequency multiplication is the efficiency. Direct RF pulse compression always has some losses; frequency multiplication does not, at least in theory.

One requirement for successful frequency multiplication is that the rings are isochronous, i.e., that the bunches preserve their separation. As a first step toward showing the validity of frequency multiplication CLIC recently has shown [20] that isochronicity is preserved in the LEP electron-positron accumulator ring EPA with simple modification of the strength of the Quadrupoles, without any hardware modification. In 50 turns no lengthening of the bunches could be observed.

After creation in the combiner rings, the combined, high current (250A), 130 nsec drive beam trains are now being sent separately to the main linac in a counter flow pattern. With the correct timing they are bent around in 180^0 achromats in the tunnel and sent through deceleration units immediately next to the main linac acceleration structures (Figure 6). The 30 GHz RF is extracted in the decelerators and is being used to accelerate the main beam. At the end of each sector, the energy depleted drive train (at about 100 MeV) is kicked into a local dump, and a new drive train takes over.

One of the interesting advantages of this system is that all high power installation can be centralized. There are no Klystrons to be supplied and cooled in a long tunnel.

IV   COMPARISON

A   Identical Systems

Let us first identify the elements of any Collider which are the same and will cost the same. Then we will pick a model to compare the RF system for both the cases described above.

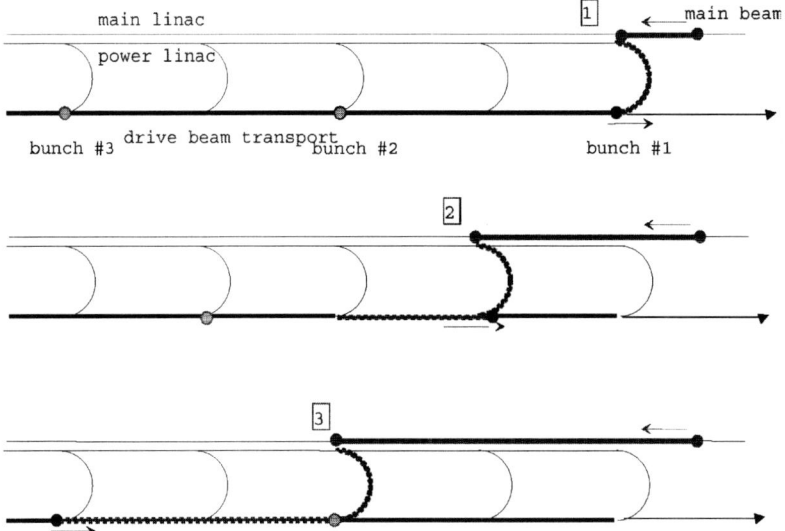

FIGURE 6. A snapshot of three 130 nsec drive trains as they move along the drive beam transport line (bottom line) at an energy of *approx* GeV. They are kicked into an 180^0 achromat which brings them around to the "power linac" (deceleration units), where the RF is extracted. The RF is fed to the main linac where it accelerates the main beam. At the end of each sector the energy depleted drive train (at about 100 MeV) is kicked into a local dump, and a new drive train takes over.

- The tunnels. [1]

- Other civil engineering, like electricity distribution, water and waste.

- The cooling and power systems.

- The injectors, positron production, and damping rings.

- The main beam line systems of the main linac and its alignment and control system.

- The beam delivery, collimation, and the final focus.

B How to Compare the RF Producing System?

This leaves a comparison of the RF production. While here we only look at the initial cost, there is an important difference in the cost of a main beam energy upgrade, depending on how it is done.

---

[1] The TBPLC tunnels are slightly longer because of the combiner rings. This expense has been proposed to be partly off-set in the TBPLC, by parasitically using the drive linac for Injector purposes [15]. On the other hand, this is not completely parasitically because then the drive linac needs to have a longer pulse then is otherwise would need. So we just leave it at the actual length.

- If the upgrade is done by extending the active length of the linac, starting from existing systems and keeping the gradient constant, the TBPLC needs just longer pulses in the (existing) L-band Modulators and Klystrons, but DKPLC needs more additional Modulators and Klystrons. This naturally raises the question, can the L-band Klystrons (and Modulators) be upgraded at a fraction of the new cost, or at great savings, or does one need completely new ones, at great cost?

- If the energy upgrade is done by raising the gradient in the X-band acceleration structure, the DKPLC needs to upgrade the (square of the) voltage by upgrading every modulator-klystron package. The TBPLC can upgrade by a combination of more current, more Klystron power on the existing structures, and/or additional Klystrons and accelerating structures. This can be tricky and the original optimization of the drive linac has to be done with great care to save as much as possible from the fully loaded condition in the upgrade.

Inverse learning curve @ 90%

avg. power/kW   39   100
freq/MHz   2856   952
data: $150k for the last one of many

		S-band (39kW)			L-band (100 kW)		
Lot Size	cumu- lative	cost	diff. cost	avg cost	cost	diff. cost	avg cost
256	512	150	38400	150		43	10940	43
128	256	167	21333	167		47	6078	47
64	128	185	11852	185		53	3377	53
32	64	206	6584	206		59	1876	59
16	32	229	3658	229		65	1042	65
8	16	254	2032	254		72	579	72
4	8	282	1129	282		80	322	81
2	4	314	627	314		89	179	90
1	2	348	348	348		99	99	99
1	1	387	387			110	110	
total			86400				24600	
average cost			169				48	

remember: power x frequency ^{2}
cost factor: 100/39/9 =   0.28

FIGURE 7. Normalized cost estimate (for a 100kW L-band average power) Klystron derived from the SLAC data of the actual design and purchasing experience of many 39kW S-band Klystrons. Equation (1) is used as a model.

The elements one has to compare are the Klystrons. They are used in a different way for RF production in the main linac (X-band) and in the drive linac (L-band) Klystrons. In order to remove as much arbitrariness from the process as possible, we need a formalized model to estimate cost between different wavelength Klystrons.

A Klystron Figure of Merit (f.o.m. = cost, or difficulty) scaling law, widely used in the Klystron Industry, is:

$$cost = average power \ x \ frequency^2 \qquad (1)$$

Inverse learning curve @ 90%

avg. power/kW		1200			100		
freq/MHz		476			952		
data:	$270k for #2-8 from Bfactory *						

band		UHF (1200kW)			L-band (100kW)		
Lot Size	cumu-lative	cost	diff. cost	avg cost	cost	diff. cost	avg cost
256	512	143	36733	143	48	12244	48
128	256	159	20407	159	53	6802	53
64	128	177	11337	177	59	3779	59
32	64	197	6299	197	66	2100	66
16	32	219	3499	219	73	1166	73
8	16	243	1944	243	81	648	81
4	8	270	1080	270	90	360	90
2	4	300	600	300	100	200	100
1	2	333	333	333	111	111	111
1	1	370	370		123	123	
	total		82600			27500	
	average cost		161			54	

remember: power x frequency ^{2}
cost factor: $100/1200*4 = 0.33$
* Note however: $470k for #10,11 from B-factory

FIGURE 8. Normalized cost estimate (for a 100kW average power L-band Klystron) derived from the actual design and purchasing experience of a modest number of 1200kW UHF Klystrons. The note on the bottom is a reminder that prices in the bidding process can easily be different up to a factor of two. This should put a damper on expecting too accurate a number from the fortuitous agreement between Figures 7 and 8. A range of two for bids in non-standard hi-tech or R&D items is also the experience of CERN. A consistency check between UHF and S-band Klystrons gives agreement within 16%.

We will use this model in the following.

Looking at Modulator costs it is found that they track closely the cost of the associated Klystrons. For this study we assume they are identical.

Some argument has been made of better energy efficiency of one design vs. another. We find that the total power needed for RF Production and cooling is nearly identical for the same center-of-mass energies. The efficiencies for wall plug to beam power are typically just below the 10% level. Consequently, much of the wall plug power related infrastructure needed is the same.

On the next level of scrutiny, we find that all finesse of using different L-band starting frequencies, different combiner ratios, and different pre-acceleration schemes, does not make much difference in parameters and cost.

One question which has to be re-addressed by both the TB-community and the DK-community is the number of Klystrons needed as stand-by (see Reference [21]). These "spare" Klystrons are needed when a Klystron fails, to ensure that the energy of the drive linacs (TBPLC) and the main linacs (DKPLC) can be kept constant to the 10^{-3} level as needed. For SLC this overhead was about 6%; LEP needed 10% and more [22]; the NLC ZDR [1] assumed 3%. [1].

This stand-by Klystron power needs to be mechanically connected to the accelerator sections all the time, to be available to go on-line in seconds. This creates problems with the present schemes of coupling together 2, 4, or even 8 Klystrons. This topic needs more research and thought, but we could take 6% as a usable number in the final cost summary for the DKPLC in analogy to the SLC experience. This number would double to 12% for the TBPLC because of the fully loaded condition of the drive linac and even 24% if 2 Klystrons are coupled together.

To compare appropriate items we will focus now on an X-band main linac and develop parametric differences between Two-Beam and Discrete Klystron RF production. Because of continuous progress this is a moving target. For NLC, e.g., the number of Klystrons needed has gone down by a large factor in recent years with the advent of more powerful Klystrons with a longer pulse length [23]. While a few years ago it was many thousands, it is now only in the many 100s.

We know that buying large quantities of anything reduces the unit price. But there is no experience with R&D in, and procurement of, a large number of L-band Klystrons. For estimating their cost we use the methodology of the DOE Cost Estimating Guide [24], "Effects of Doubling Production". Instead of the learning curve described in Ref. [24] we use a reverse learning curve.

The use of the reverse learning curve is appropriate here, because good data for large quantities, including past bids, exist from SLAC's 5045 S-band Klystron. In addition we can estimate the cost from below in frequency using the SLAC B-factory Klystron experience, although from a smaller number of units.

Figures 7 and 8 show the cost of L-band Klystrons derived from the classical SLAC S-band Klystron and from the B-factory UHF Klystron. For our further considerations we use the average, $51k for a L-band Klystron normalized to 100kW.

V SUMMARY

The cost have been estimated for a 500 GeV cms X-band Collider using Discrete Klystron and Two-Beam RF production technology. The summary from TB simulations using 952 MHz Klystrons for the drive beam is shown in Figure 9 [2].

The Klystron cost have been given above in US$, but in Figure 10 the cost are in € (Euro). The original cost research at CERN [21], which was used to create Figure 10, was done in Swiss Francs (CHF) and Euros (€). For conversion into US$ we used a rate which is believed to reflect the purchasing power, that is 1€ ≈ 1US$ ≈ 1.5CHF.

The cost of X-band RF equipment has been calculated as if only half the main linac tunnel would be filled for 0.5 TeV cms. But all infrastructure needed to reach

---

[1] One should realize that unlike SLC, the LEP Klystrons were not accessible during operation. On the other hand, global corrections in RF power are easier with CW Klystrons when one Klystron fails. So probably any number between 5 and 10% is realistic.

[2] I am thanking Roberto Corsini for the use of his "magic" spread sheet.

Main Beam		
	CM Energy (TeV)	0.54
	Maximum Gradient (MeV/m)	68
	Actual Gradient (MeV/m)	62
	2 Linac Length (km)	10.9
	Repetition Frequency (Hz)	120
	Pulse Length (nsec)	263
	Number of Bunches	95
	Charge per Bunch (10^9)	9
	HE Beam Total Energy (KJ)	37
Drive Beam		
	Number of Drive Beams per Linac	4
	Rf Pulse Total Energy (KJ)	122
	Rf Pulse Length (nsec)	375
	Deceleration Section Length (m)	1350
	DriveBeam Pulse Length (Microsec)	36
	Total Drive Beam Energy (KJ)	175
	Drive Beam Energy (GeV)	0.98
	Drive Beam Current (A)	4.9
	Frequency of DBA (MHz)	952
	Active Length of DBA (m)	296
	Structure Length (m)	2.97
	Klystron Power per Structure (MW)	49.4
	Delay Line Length (m) x2	112.5
	1st Combiner Length (m) x3	225
	2nd Combiner Length (m) x4	675
	Frequency Multiplication (2x3x4)	24
Total		
	Number of 50 MW Klystrons (for 2 Linacs)	200
	Wall Plug Power (MW)	82
	Total RF Efficiency (%)	39
	Wall to Beam Efficiency (%)	9.7

FIGURE 9. Summary of technical parameters for a TBPLC based on 952 MHz for the L-band frequency. The optimization was done for a gradient of 62 MeV/m in the X-band linac, and to always keep the power needed for the drive linac Klystrons below 50 MW. If the TB-RF system is optimized in a way to match the planned NLC RF power per 1.8 m section (170MW corresponding to ≈68 MeV/m), the L-band Klystron count goes up to 226.

1.0 TeV has been costed, assuming a main linac tunnel length of 22 km. It was also assumed that a total wall plug power capability of 200 MW was installed.

With this length of 22 km a full complement of acceleration structures and Klystrons could reach 1.0 TeV at a gradient of 62 MeV/m. This gradient seems to be within reach from the NLCTA experience. Also, with this tunnel length and a gradient of 93 MeV/m a cms energy of 1.5 TeV could be reached. All calculations assume a fill factor of 80% and an energy overhead for BNS and other items of 8%.

A PLC built strictly only to reach 0.5 GeV, without the infrastructure in tunnel, power, water, and some other minor items, to get to 1.0 TeV, would be ≈ 500M€ cheaper.

The L-band Modulators and Klystrons for TBPLC have been costed with a 36 μsec pulse and 216 kW average power to be able to reach 0.5 TeV with 11 km of main linacs. These numbers would double to 72 μsec pulse and 432 kW to double

System	TBPLC	DKPLC	Remarks	f(E)
				c = constant
Conventional Facilities				E = proportional
Tunnels (incl. IR's)	850	800	25k€ / m	c+E
Power, Cooling, Water, Waste etc.	300	300		c+E
	1150	1100		
Injector Systems				
Damping Rings	200	200		c
pre-linacs(L,S,C,X)	100	100		c
other Systems	200	200		c
	500	500		
Main Linac				
RF	0	500	1M€ / GeV, no standby	E
BL Systems	300	300	Structures, Quads, Movers, ...	E
other Systems	200	200	Installation, Integration	c+E
	500	1000		
Drive Beam				
L-band Linacs	220	0	100x2 Kly, no standby	E
Frequency Multiplication	70	0	delay line + 2 combiner rings	c
Transports, Turnarounds, Dumps	30	0	10% of Linac BL Systems	E
decelerators	80	0	25% of Linac BL Systems	E
	400	0		
Control System	200	200		c+E
Beam Delivery	200	200	incl. IRs, no Detector	c+E
Services @ 20%	590	600	Tech. Support, Pre-ops	c+E
	3540	3600	in M€	

FIGURE 10. Summary of costs with the assumptions as described in the text. It is clear that this estimate is crude. It is also clear that items which have never been built and operated, like fully loaded linacs, combiner rings and decelerators, carry a larger contingencies than standard equipment. No contingency is assigned here.

the main linacs lengths to 22 km for 1 TeV.

It is clear from Figure 10 and the text that the raw cost of producing RF using a Two-Beam scheme could be lower than the direct Discrete Klystron scheme. But the major component cost in both systems go up proportional to energy. For the same (low) gradient of 62 MeV/m, there is no clear advantage of one vs. the other.

However, there are some conceptual advantages (and also a whole host of as of yet unproven assumptions). We mentioned already the possibility to have the high power installation at a central place and not distributed in the tunnel. So even if Two-Beam would be **more** expensive (which it is not), it has certain advantages in flexibility, which should give pause for thought.

One concrete example is the CLIC simulation [6] which shows that with a staggered RF ramp an energy spread of only $\Delta E/E = 5 \cdot 10^{-4}$ can be reached. The energy spread produced in linear colliders operated in a classical way is large, $\Delta E/E = \approx 10^{-2}$. This large value creates a whole range of beam dynamics problems one could easily do without.

Acknowledgments

The foundation for this analysis were laid while on Sabbatical in the CLIC Study Group at CERN in 1999. The work was continued more recently at a much lower intensity at SLAC. I want to thank John Cornuelle, Roberto Corsini, Franz-Josef Decker, John Irwin, Chris Pearson, Ray Larsen, Peter Pearce, Louis Rinolfi, Heinz Schwarz, Igor Syratchev, Juwen Wang, among many, for enlightenment and education on many of the details covered here. And last but not least, I want to thank Steven Gold, Gregory Nusinovich and David Sutter, for inviting me to the AAC workshop.

REFERENCES

1. C. Adolphsen et al., "Zeroth Order Design Report for the Next Linear Collider", *SLAC-Report* **474** (1996).
2. R.D. Ruth et al., SLAC-PUB-7532, "Results from the SLAC NLC Test Accelerator", 17th IEEE Particle Accelerator Conference (PAC 97), Vancouver, Canada, 1997.
3. The Livermore effort with Induction Linac Technology is documented in Appendix A in: C. Adolphsen et al., "Zeroth Order Design Report for the Next Linear Collider," *SLAC-Report* **474** (1996).
4. G.A. Westenskow, T.L. Houck et al., "Progress on the Relativistic Klystron Two-Beam Accelerator prototype", in *Advanced Accelerator Concepts: Eighth Workshop, Baltimore 1998*, edited by C. Bellamy, D. Brosius, p. 983, AIP Conference Proceedings: 472.
5. H.H. Braun, R. Corsini, (ed.), et al., "The CLIC Power Source: a Novel Scheme of Two-Beam Acceleration for Electron-Positron Linear Colliders". CERN-99-06 (Yellow Report), Sept. 1999. 206pp.
6. R. Corsini, J.-P. Delahaye and I. Syratchev, "CLIC Main Linac Beam-Loading Compensation by Drive Beam Phase Modulation", CLIC-Note-408, CERN, 1999.
7. David H. Whittum, Xintian E. Lin, and Frank Zimmermann, "Principles of Harmonic Acceleration for a Prescribed Beam Profile", ARDB Technical Note 166, SLAC, 1998.
8. S. Fartoukh and J.B. Jeanneret, "A Proposal to use Microwave Quadrupoles to Shorten the Beam Delivery Section of CLIC", CLIC-Note-423, CERN, 1999.
9. Rainer Pitthan, "Using Octupoles for Background Control in Linear Colliders - an Exploratory Conceptual Study", CLIC-Note-418, CERN, 1999.
10. Rainer Pitthan, "Impact Report for CTF3", A Clic Internal Report, CERN, 1999.
11. I. Wilson, "1999 Activities Report for the CLIC Study Group", CLIC-Note-427.
12. "The Stanford Two-Mile Accelerator", R.B. Neal, ed., W.A. Benjamin, 1968.
13. Z.D. Farkas et al., "SLED: a Method of Doubling SLAC's Energy", 9th Int. Conf. on High Energy Accelerators, SLAC, Stanford, Calif., 1974, p. 576, QCD183:I5:1974.
14. R. Bossart, et al., "The CLIC Study of a Multi-TeV e+e- Linear Collider", 1999 Particle Accelerator Conference, New York, USA.
15. John Irwin, Ron Ruth, private communication.
16. SLAC Linear Collider Design Handbook, Chapter 2, R. Erickson, ed., SLAC 1984
17. S.G. Tantawi et al., "The Next Linear Collider Test Accelerator's RF Pulse Compression and Transmission Systems", SLAC-PUB-7247, Presented at 5th European Particle Accelerator Conference (EPAC 96), Sitges, Spain, 1996.

18. Roberto Corsini, "The Drive Beam Generation for Two-Beam Accelerators", in Advanced Accelerator Concepts: Seventh Workshop, Lake Tahoe 1996, edited by S. Chattopadhyay, J. McCullough, P. Dahl, p. 126, AIP Conference Proceedings: 335; and CERN-CLIC-Note 330, 1997. An update can be found in CLIC-Note 367.
19. "Proposals for Future CLIC Studies and a new CLIC Test Facility (CTF3)", CLIC-Note 402, 1999, to be found under
http://cern.web.cern.ch/CERN/Divisions/PS/CLIC/Publications/1999.html
20. R. Corsini, J.P. Potier, L. Rinolfi, T. Risselada, "Isochronous Optics and Related Measurement in EPA", CLIC-Note-440, 7th European Particle Accelerator Conference, 2000, Vienna, Austria.
21. Rainer Pitthan, "Impact Report for the CLIC Power Source", A CLIC Internal Report, CERN, 2000.
22. Gunther Geschonke, private communication.
23. D. Sprehn et al., "X-Band Klystron Development at the Stanford Linear Accelerator Center", SLAC-PUB-8346, Presented at Intense Microwave Pulses VII at SPIE 2000 AeroSense Symposium, Orlando, Florida, 2000.
24. DOE G 430.1.1, "Cost Estimating Guide", Chapter 21. All chapters can be found under: http://peak.lanl.gov:1776/pdfs/doe/doetext/neword/430/g4301-1toc.html#RELATED

Two-Channel Active High-Power X-Band Pulse Compressor

A. L. Vikharev [1,2], A. M. Gorbachev [1,2], O. A. Ivanov [1,2], V. A. Isaev [2], V. A. Koldanov [2], S. V. Kuzikov [2], A. G. Litvak [2], M. I. Petelin [2], J. L. Hirshfield [1,3], and O. A. Nezhevenko [1]

[1]*Omega-P, Inc. 202008 Yale Station, New Haven, CT 06520-2008*
[2]*Institute of Applied Physics, Nizhny Novgorod, 603600 Russia*
[3]*Department of Physics, Yale University, POB 8120, New Haven, CT 06520-8120*

Abstract. A two-channel active pulse compressor has been developed that is able to provide output pulses of at least 100 MW peak power with pulse duration of 100 nsec at X-band, with a power gain of 12-15 and with an energy efficiency of 60%. This paper describes the design of the compressor and the driving generator-compressor microwave circuit. Each channel of the compressor is connected to the driving generator and the load via a novel 3-dB quasi-optical coupler. Variations in phase of compressed output pulses from this active pulse compressor were measured. The moderate-power tests of a prototype design of such a compressor using 100 kW-level microwaves demonstrated coherent addition of the compressed pulses from each of the compressor channels. The paper also describes design of a modified output reflector, with which the two-channel active pulse compressor can produce output pulses with a peak power of at least 500 MW and a power gain 12-15.

INTRODUCTION

It is widely accepted that microwave pulse compression is required for a future high-gradient electron-positron collider in order to achieve the high pulse power. For this purpose two methods are being developed—passive and active pulse compression. From our point of view active pulse compressors are rather promising due to achieving higher power gain than do passive compressors. But up to now there was no active pulse compressor developed capable of providing output pulses of power at least 100 MW at X-band. Furthermore, there was no active compressor capable of having output pulses with acceptable phase stability and capable of good reproducibility of shape and amplitude of the compressed pulses.

Omega-P, Inc. and Institute of Applied Physics teams managed to develop a novel active microwave compressor [1,2] with characteristics suitable for accelerator use. The compressor, named the active Bragg compressor (ABC), utilizes an oversized cylindrical resonator operating in the "breakdown-proof" TE_{01n} mode (with $n \gg 1$), with an input Bragg reflector and an output reflector with electrically controlled gas discharge switches. In this one-channel ABC the microwave radiation of the generator is coupled into the cavity through the input

Bragg reflector and is coupled out through the output reflector based on the step-wise widening of a circular waveguide. This stepped widening section comprises a cylindrical resonator containing a quartz ring-shaped gas discharge tube. The discharge tube is used to switch in about 10 ns from the regime of energy storage to the regime of energy extraction.

This paper describes design of the improved active Bragg compressor—two-channel ABC with combined input-output reflector. Each channel of the compressor is connected to the driving generator and the load via a novel 3-dB quasi-optical coupler. Two-channel ABC makes it possible to isolate the generator from the reflected microwave power at the initial stage of energy storage, to improve ABC efficiency [3], and to make the ABC design simpler.

DESIGN OF TWO-CHANNEL ACTIVE BRAGG COMPRESSOR

The scheme of the compressor is shown in Fig. 1. The compressor consists of two identical channels. Each of the compressor channels has mode converters ($TE_{01} \to TE_{11}^0$, $TE_{11}^0 \to TE_{01}^0$) connected with smooth tapered transitions and a resonator formed by an input-output reflector, a section of a cylindrical oversized waveguide and a cone-shaped reflector. The central part of the cavity for storing microwave energy is a section of an oversized 1-m long waveguide 80 mm in diameter, which is equipped with a tapered 400 mm long transition to a narrower waveguide. The diameter of the latter waveguide was 55 mm, and the TE_{01}^0 mode is the only propagating mode of all the axially symmetric ones.

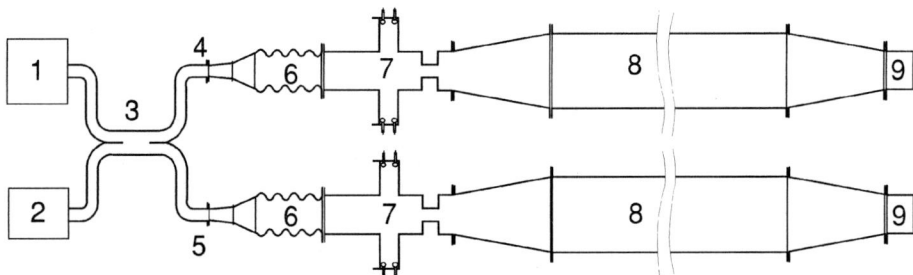

FIGURE 1. Schematic diagram of the two-channel active Bragg compressor operating in the TE_{01} mode: 1 - driving generator, 2 - load, 3 - 3dB directional coupler, 4 - first channel of ABC, 5 - second channel of ABC, 6 - TE_{01}-mode converter, 7 - input-output electrically controlled reflector, 8 - storage cavity, 9 - reflector.

The input-output reflector consists of active and passive sections. The active section is based on step-wise widening of a circular waveguide. This stepped widening section comprises a cylindrical TE_{031} mode resonator containing the quartz ring-shaped two gas-discharge tubes. The diameter of the wider waveguide is 140 mm and it can excite axially symmetric modes with high-order radial indices. The passive section is a waveguide with an over-critical narrowing. This combined input-output reflector makes it possible to reduce intensity of the electric field in the region of gas discharge tubes in the active section. By changing dimensions of the over-critical waveguide narrowing in the passive section one can change the transmission coefficient and thus control the amplitude and duration of the compressed pulse.

In two-channel ABC 3-dB directional coupler is one of the main elements of the compressor. This coupler makes it possible not only to decouple the generator from the reflected microwave power at the initial stage of energy storage but also allows one to operate both channels into one load.

In the frame of the Project for realization of a two-channel active Bragg compressor we manufactured three two-channel ABC prototypes. One is an evacuated version made of copper. Two of them are non-vacuum versions made of aluminum capable of operating at frequencies of 11.4 GHz and 9.4 GHz, respectively. These prototypes were made for testing ABC characteristics at different level of microwave power.

LOW-POWER TEST MEASUREMENTS

An important point for successful use of active microwave compressors in accelerators is the issue of phase correlation between the input and compressed pulses. Phase characteristics of the output radiation of non-vacuum ABC prototype at 11.4 GHz were measured experimentally by Mach-Zender interferometric scheme, shown in Fig.2.

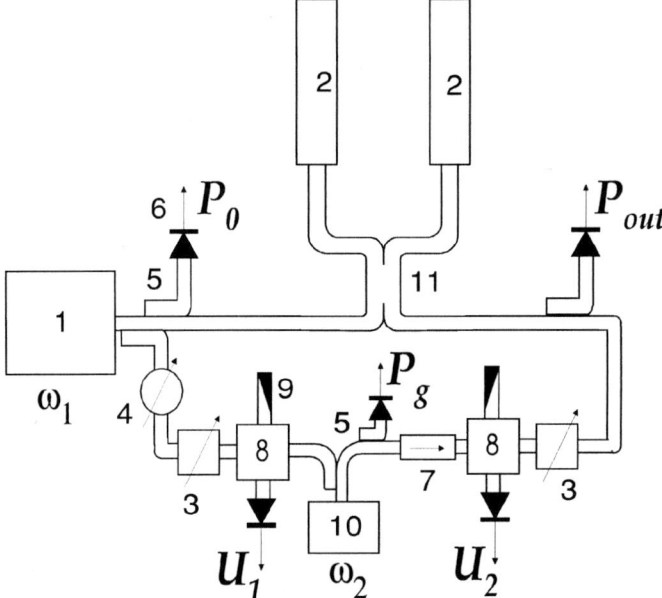

FIGURE 2. Schematic diagram of the experimental set-up for measurements of phase characteristics: 1 - microwave generator, 2 - two-channel ABC, 3 - attenuator, 4 - phase shifter, 5 - directional coupler, 6 - detector, 7-feirrite isolator, 8-magic-T, 9-matched load, 10 - diagnostic microwave generator, 11 - 3 dB directional coupler.

An interferometer with heterodyne-type frequency conversion was used as well as two microwave generators with their frequencies ω_1 and ω_2 that differed by the value of the intermediate frequency, $\Omega = \omega_1 - \omega_2$. Microwaves were mixed using a waveguide magic-T joint, both in the reference and in the measuring line. The phase of the output microwave radiation of the compressor was measured by comparing the signals at the intermediate frequency of the both detectors. The first detector yields the signal with its reference phase Ωt

$$U_1 \propto P_g + P_o + 2\sqrt{P_g P_o} \cos[\Omega t]. \quad (1)$$

The second detector receives the radiation from the compressor output, thus yielding the signal with the measured phase

$$U_2 \propto P_g + P_{out} + 2\sqrt{P_g P_{out}} \cos[\Omega t + \Delta\varphi(t) + \varphi_0]. \quad (2)$$

Time dependence of phase difference $\Delta\varphi(t)$ between the input and output radiation and the envelope of the compressed pulse are shown in Fig.3.

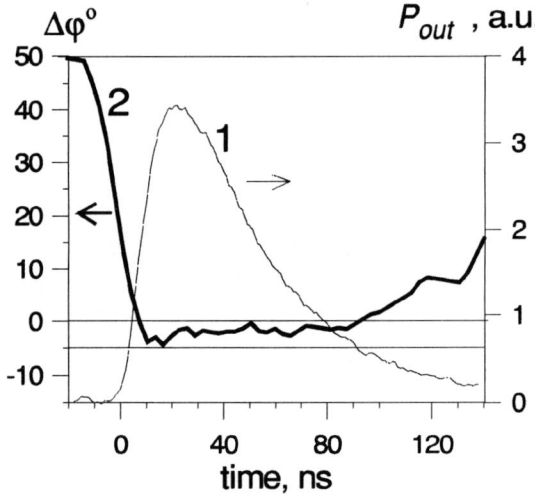

FIGURE 3. Time dependence of the power of the compressed pulse (1) and of phase mismatch (2) between the output and input pulses.

As seen from the figures, after the input-output reflector switches, the radiation phase at the compressor output changes rapidly to ~30-60° and then tunes back smoothly. It is seen that the phase of microwave radiation stays within the 5° variation range for a major part of duration of the compressed output pulse. One must note good reproducibility of the shape and amplitude of the compressed microwave pulse of the two-channel ABC from pulse to pulse.

Operation of the non-vacuum two-channel ABC prototype at 11.4 GHz was also tested using experimental scheme, shown in Fig. 1. In this experiment we tested the superposition of microwave radiation from two channels of ABC. Oscillograms of compressed pulses after the each channel and the composite pulse are shown in Fig. 4. The composite pulse is identical to the individual pulses from the channels, except for a factor-of-4 increase in amplitude. It should be noted that parameters of the composite pulse demonstrated excellent reproducibility which proves that time variations of radiation phase in each of the channels are synchronous. Moreover, the absence of noticeable difference in the shape of the composite pulse from the shapes of individual compressed pulses makes it possible to state that variations in phase of microwave radiation in each compressed pulse are identical. Thus this experiment demonstrated coherent addition of the compressed pulses from each of the compressor channels.

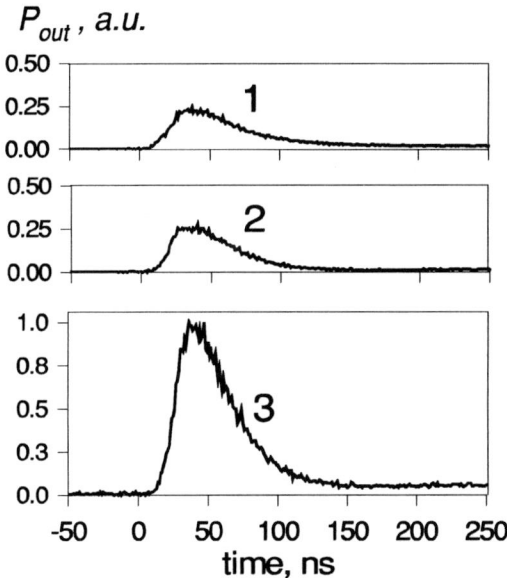

FIGURE 4. Oscillograms of compressed pulses of the two-channel ABC: 1- output pulse of ABC at operation with attenuator inserted in the second channel, 2- output pulse of ABC at operation with attenuator inserted in the first channel, 3- composite output pulse of ABC at operation without attenuators.

TEST USING 100 KW-LEVEL MICROWAVES

The testing of non-vacuum two-channel ABC prototype using 100 kW-level microwaves was performed at 9.4 GHz. In this experiment we used the same scheme shown in Fig. 1 but with a 160 kW magnetron as the driving generator. It was found that the two-channel ABC is able to provide a power gain of over 20. The oscillograms of the input and compressed pulses are shown in Fig. 5. Experimental results are listed in Table 1.

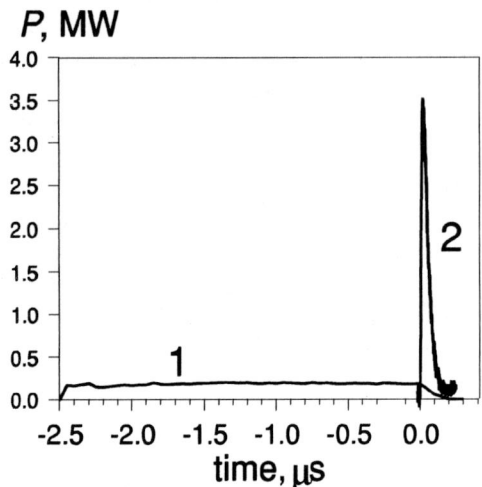

FIGURE 5. Oscillograms of input (1) pulse with 160 kW peak power and compressed (2) pulse with 3.4 MW peak power.

TABLE 1. Experimental results of 100kW-power test

Parameters	Values
Frequency	9.4 GHz
Operating mode	TE_{01}
Resonator Q-factor	$5 \cdot 10^4$
Input pulse duration	2.5 μs
Power gain	21
Output pulse duration	52 ns
Input power	160 kW
Output power	3.4 MW
Efficiency	44%

For testing of two-channel ABC prototype we used 3-dB directional coupler made from single-mode waveguide. Such a coupler cannot be used at 100 MW level of power. Therefore we developed a novel 3-dB quasi-optical coupler for operation of vacuum ABC prototype with at least 100 MW peak power.

QUASI-OPTICAL 3DB DIRECTIONAL COUPLER AT HIGH POWER OPERATION

The basic idea of the novel quasi-optical coupler is to use the effect of image multiplication in oversized rectangular waveguide [4]. In such a waveguide an asymmetrically injected wave beam is divided, at a certain distance, into two beams of equal amplitudes, but phase shifted by 90°; see Fig.6. On the basis of this phenomena we developed a 3-dB coupler which is being manufactured now.

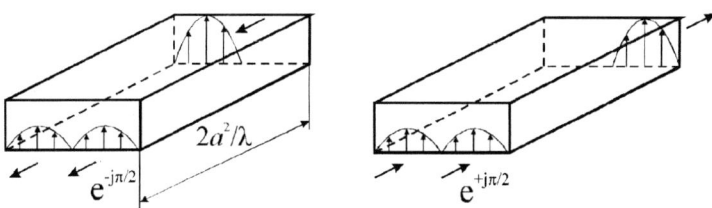

FIGURE 6. Quasi-optical 3dB directional coupler.

MODIFIED OUTPUT REFLECTOR

The input-output reflector of existing ABC has a number of unique features. It make it possible to control the power gain of the compressor, as well as the shape and duration of the compressed pulses. Besides, it allows for obtaining good phase characteristics of the compressed pulse. Experiments performed demonstrated that ABC with such an output reflector makes it possible to obtain output pulses with only a slight variation of phase with time.

However, it must be noted that compressed microwave pulses at levels well above 100 MW are required for Next Linear Collider (NLC). At a power gain 11-12, ABC (as now configured) cannot handle output pulses of 500-700 MW. The principal limitations to output power above a level of about 100 MW in the existing ABC arise from strong electromagnetic fields in the output reflector that lead to self-breakdown in the plasma switch.

Operation of the existing ABC in the TE_{01} mode was modeled numerically by the FDTD method [5]. In numerical calculations, parameters and dimensions of the compressor were chosen to be close to those of the manufactured compressor. Calculations showed that at the 100 MW power of the output pulse the maximum electric field at the discharge tube in the combined output reflector will be 25 kV/cm. Thus, the studies performed showed that it is necessary to develop an output reflector for the ABC possessing significantly weaker fields in the region of the tubes while retaining the other positive features of combined output reflector.

To obtain high-power pulses using ABC we suggest a modified output reflector. Schematic diagram of ABC with modified plasma switch and frequency characteristics of plasma switch are shown in Fig. 7. In the regime of energy storage the plasma switch has the reflection coefficient close to unity in a wide frequency band. When the gas-discharge tubes are broken down the total reflection coefficient decreases resonance-wise, as in Fig. 7b.

Operation of the modified switch was modeled numerically. The calculations were performed for a storage cavity 1 m long with a loaded Q-factor of 2.5×10^4. Numerical calculations showed that at the 100 MW pulsed power output level, the maximum electric field at the tubes in the compressor will be 8 kV/cm. Thus the calculation performed showed principal possibility of building a plasma switch that would provide a high power gain of the compressor combined with high stability of phase of the compressed pulse. Intensity of the electric field in the vicinity of gas-discharge tubes in such a switch is three times lower, as compared to the fields in the combined output reflector of the existing ABC.

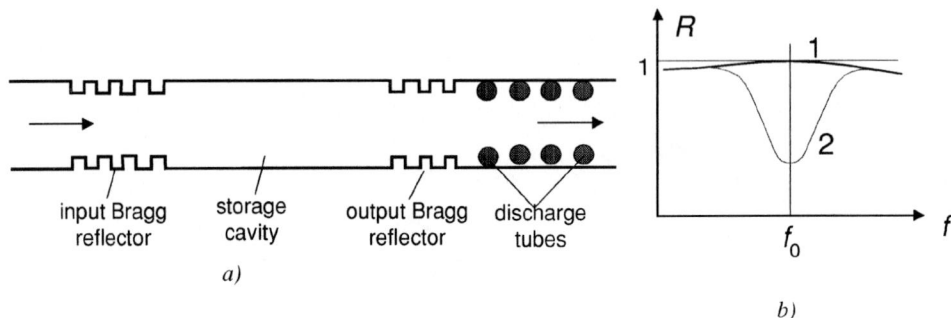

FIGURE 7. Schematic diagram of ABC with modified plasma switch (a) and frequency characteristics of plasma switch (b): 1- in the regime of energy storage, 2- in the regime of energy extraction.

ANTICIPATED PARAMETERS OF ABC

The numerical calculations were performed with the assumption that the loaded Q-factor of the storage cavity of ABC with a modified plasma switch equals the Q-factor of the cavity in the existing ABC. Calculations and measurements have shown the efficiency of pulse compression in the existing ABC is 50-60% [2]. The loaded Q-factor, namely $Q=2.5\times10^4$, is determined by coupling losses and is optimal when pumping pulses with their duration of 1µs are used. In the existing ABC the efficiency of pulse compression is limited due to energy losses in the dielectric tube in the active section of the combined output reflector. These losses reduce the ohmic Q-factor, thus affecting efficiency of the device [6]. In ABC with the modified plasma switch the tubes have less influence on the Q-factor (and efficiency) since the field at the tubes is much less intense. That is why for the ABC with a modified plasma switch (as shown by estimates) the efficiency of pulse compression will be 60-65%. Thus, design parameters for the active Bragg compressor (ABC) with a modified plasma switch could be as listed in Table 2.

TABLE 2. Parameters of ABC with modified plasma switch

Parameters	Values
Ohmic cavity Q	110,000
Loaded cavity Q	25,000
Input pulse width	1-1.5 μs
Output pulse width	0.1 μs
Power gain	10-15
Input power	50 MW
Output power	500-600 MW
Efficiency	60-65%

CONCLUSION

The ABC concept has strong potential for application to NLC. The 100 MW level of power in compressed pulses in the existing ABC is achieved by using oversized low-loss TE_{01} mode circular waveguide as an energy storage cavity with Bragg reflectors. The 100 MW limitation is connected with occurrence of a self-breakdown in the plasma switch used. If the breakdown characteristics of the output reflector are improved by reducing the electric field in the vicinity of the plasma switch, the existing ABC will be able to sustain more powerful compressed pulses.

Preliminary computations demonstrated the basic possibility of designing and building a modified plasma switch that would provide:
1) compressed pulses with high power gain (in the range 10-15);
2) high stability of the phase of the compressed pulse (the output radiation phase within the 0.5 degree variation range);
3) 3-fold reduction of electric field as compared to the existing plasma switch, corresponding to a 5-10-fold increase of ABC output power;
4) high energy efficiency of pulse compression, in the range 60-65%.

Thus, this new-generation ABC will allow full use of magnicon and/or klystron outputs with power in the range 50-60 MW to obtain output pulses of the order of 500 MW and higher as required for NLC.

ACKNOWLEDGMENTS

This work was supported by the US Department of Energy—Division of High Energy Physics under grant DE-FG02-98-ER 82630, and by Russian Ministry of Science and Technologies.

REFERENCES

1. Vikharev, A.L., Gorbachev, A.M., Ivanov, O.A., Isaev, V.A., Kuzikov, S.V., Kolysko, A.L., Litvak, A.G., Petelin, M.I., and Hirshfield, J.L., "Active microwave pulse compressors employing oversized resonators and distributed plasma switches" in *Advanced Accelerator Concepts*, Eighth Workshop, edited by W.Lawson, C.Bellamy, D.F.Brosius, ., AIP Conference Proceedings 472, New York: American Institute of Physics, 1998, pp. 975-982.

2. Vikharev, A.L., Gorbachev, A.M., Ivanov, O.A., Isaev, V.A., Kuzikov, S.V., Kolysko, A.L., Litvak, A.G., Petelin, M.I., Hirshfield, J.L., Nezhevenko, O.A., and Gold, S.H., "100 MW active X-band pulse compressor" in *Proceedings of the 1999 Particle Accelerator Conference*, edited by A.Luccio and W.MacKay, IEEE Conference Proceedings 1-5, 1999, pp. 1474-1476.
3. Vikharev, A.L., Gorbachev, A.M., Ivanov, O.A., Isaev, V.A., Kuzikov, S.V., Petelin, M.I., and Hirshfield, J.L., "Active pulse compression" in *Proceeding of Strong Microwaves in Plasmas Workshop*, edited by A.G.Litvak, Nizhny Novgorod: IAP, 2000, pp. 896-914.
4. Denisov, G.G., Kuzikov, S.V., "Microwave systems based on controllable interference of paraxial wavebeams in oversized waveguides" in *Proceeding of Strong Microwaves in Plasmas Workshop*, edited by A.G.Litvak, Nizhny Novgorod: IAP, 2000, pp. 618-624.
5. Yee, K.S., IEEE Trans. Antennas Propagat. **AP-14**, 302 (1966).
6. Alvarez, R.A., Rev. Sci. Instruments **57**, no10, 2481 (1986).

Some Thoughts about Millimeter-Wave Drivers for Future Linear Colliders

Gregory S. Nusinovich

Institute for Plasma Research
University of Maryland
College Park, Maryland 20742

Abstract. In this paper, an attempt is made to overview some problems important for the development of high-power millimeter-wave drivers for future linear colliders. Since the microwave pulse duration required at high frequencies is much shorter than at low ones, two options seem possible. The first one is to develop 'moderate' power level, long-pulse tubes based on relatively reliable technology and then greatly compress these microwave pulses. The second one is to operate at much higher voltages and to directly generate very high-power pulses of the required length. Besides discussing pros and cons of these options, an overview of the methods of mode selection in oversized microwave circuits required for producing multi-megawatt power at millimeter wavelengths is presented. Also the issue of thermal limitations caused by microwave losses in circuit walls is discussed, and some scaling laws for the maximum power and pulse duration are given.

INTRODUCTION

This paper should be considered as an introduction to the topics that are the focus of the Working Group on "Millimeter-Wave Sources" of the present AAC Workshop. In order to review some facts from the history of the development of millimeter-wave (mm-wave) sources, let us start from the paper entitled "Millimeter Waves" published by J. Pierce 50 years ago [1]. First of all, it makes sense to pay attention to the editorial comment that states that "he (the author) thinks that the world needs smaller and better vacuum tubes, but his considered opinion is that it is easier to write about them than to make them."

The chart shown in Fig. 1, reproduced from Ref. 1, shows the level of average power as a function of wavelength that existed 50 years ago. At millimeter wavelengths, this power is at the level of 10 mW to 1W. Also, Pierce lists the 'headaches' associated with the development of mm-wave tubes. Those are: difficulties of manufacturing of circuit elements, high current density in small cathodes (which may limit the lifetime of tubes), Ohmic losses, and heat dissipation problems.

In our days, 50 years later, the power level of mm-wave gyrotron oscillators operating at frequencies higher than 100 GHz, in very long pulses (1 second and above) is about 1 MW [2]. Also, the average power of W-band gyroklystrons developed at NRL for radar applications exceeds 10 kW [3] (duty-cycle is about 10%).

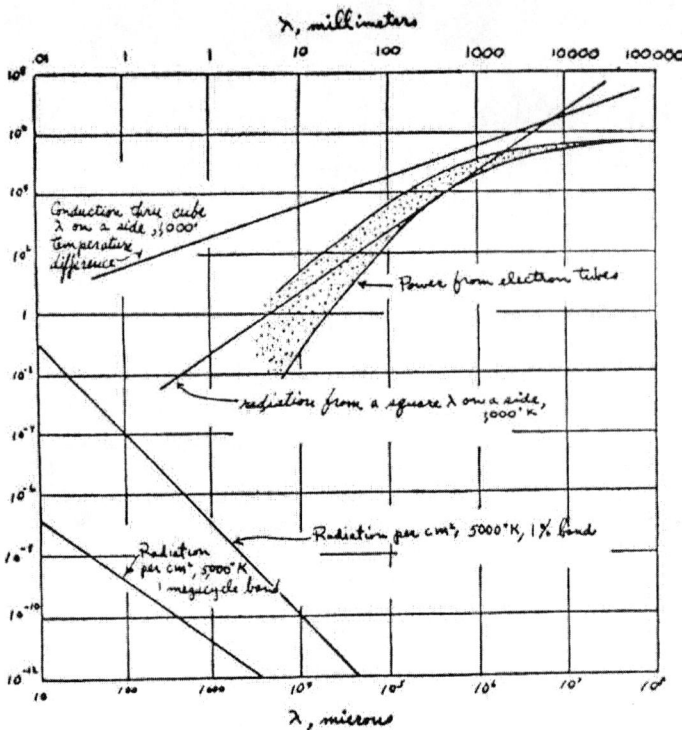

Fig.1. Historical chart "average power (in Watts) vs. wavelength (in millimeters, at the top)" from Ref. 1.

Averaging over the 50 years time scale, this shows that the progress in the development of mm-wave tubes leads to an increase in their average power by roughly an order of magnitude every 10 years. We can also make another conclusion from these data. Let us take such typical parameters for W-band sources as 5 MW power and 1 µs pulse duration (which are discussed in Ref. 4, which is devoted to the development of mm-wave drivers for future linear colliders) and assume a 200 Hz repetition frequency. This combination corresponds to 1 kW average power. So we may conclude that, in terms of the average power, we already have sources of the required level. Obviously, the issue is not average but peak power level.

Peak Power

As follows from Ref. 1, 50 years ago the peak power of pulsed magnetrons reached the 30-40 kW level at 'long' millimeters and the 1-2 kW level at 'short' millimeters. Presently, Ka-band (36 GHz) traveling-wave-tube amplifiers produce 100 MW power in short pulses (less than 10 nsec) [5]. At higher frequencies (140 GHz) the Livermore free-electron laser (FEL), driven by a 6 MeV, 2.5 kA electron beam, produced up to 2 GW in 20-nsec pulses [6]. A detailed overview of the present status of mm-wave sources is the subject of a paper by B. Danly at this Workshop [7]. Therefore we will not repeat it here. Instead of this, let us discuss the most important issues in the development of mm-wave drivers.

Pulse Duration

As is known, the filling time of existing accelerating structures scales with the wavelength as $\lambda^{3/2}$. This means that, if we choose as a starting point a 200 ns pulse duration for operation in the X-band (NLC case), the corresponding pulse duration required for operation in the W-band will be about 10 nsec. Such a short pulse duration makes two options in the development of mm-wave systems possible:

Option 1 is to develop relatively long-pulse (µs pulse duration) mm-wave tubes and then greatly compress these pulses. Such tubes and relevant supplies can be based on quite reliable technology since, first, the required operating voltage can be about 500 kV only, and second, for generating µsec pulses, the thermionic cathodes can be used which often may operate with a long lifetime. However, this option implies the use of pulse compressors with very high compression ratios. In this case, clearly, passive pulse compressors will be out of the game, because their efficiency is high only at moderate (3-5 times) compression ratios [8]. The applicability of active pulse compressors will depend on the progress in their development. The present status of active pulse compressors is the subject of papers by S. Tantawi [9] and A. L. Vikharev [10] at this Workshop.

Option 2 is to develop very high power (GW level) mm-wave sources generating short pulses of the required duration. This option does not need any pulse compressors which is a clear advantage. However, the required technology is less matured and can be less convenient. For instance, the operating voltage should be at the multi-MV level. Also it is not clear whether thermionic cathodes should be used. Their technology is well developed; however the cathode loading (on the order of 10 A/cm^2) is relatively small so the cathode area can be quite large. This may require high beam compression ratios in the process of electron beam transport from the gun region to the circuit. On the contrary, field emission cathodes can produce much higher current densities (on the order of kA/cm^2), so there will be no need of high beam compression. At the same time, these cathodes operate less reliably even in short pulses, so the reliability and lifetime can be an issue for such cathodes. Note that there are known examples of successful operation of microwave tubes employing the field-emission cathodes for quite a long time. For instance, in the nanosecond-gigawatt radar (NAGIRA) built by the Russians for Marconi, X-band backward-wave oscillators are used which are driven by short-pulse electron beams generated by field-emission cathodes [11] (see also Ref. 12).

As a two-and-a-half option also the two-beam accelerator concept can be mentioned. However, this concept is out of the scope of this paper, which is focused on 'discrete' tubes.

Very High Power

To realize very high power levels (~1 GW) of microwave radiation, one should operate at very high voltages (> 1 MV) and in high-order modes of the microwave circuit. Operation at very high voltages is, certainly, associated with various technical difficulties and complications. Nevertheless, it is possible in principle and has been demonstrated in many experiments. At the same time, it is obvious that just the voltage increase by itself increases the microwave power density in the interaction region, which may cause some problems associated with either RF breakdown or the dissipation of too much microwave power in the circuit walls. To solve these problems one should enlarge the interaction volume V, which implies operation in high-order modes.

Operation in high-order modes can be accompanied by the excitation of various parasitic modes, because, in general, the density of the mode spectrum increases with the V/λ^3 ratio. To avoid mode competition, various methods of mode selection can be used which are well developed not only for lasers but also for mm-wave tubes (see, e.g., Ref. 13).

MODE SELECTION

The methods of mode selection can be divided into two categories, electrodynamic selection and electron selection (see, e.g., Ref. 14). Let us consider these two categories separately.

Electrodynamic Selection

Electrodynamic selection is based on the development of highly-selective oversized cavities and/or waveguides. At millimeter wavelengths it is preferable to use methods based on the difference in diffractive losses for different modes, i.e., to develop open selective structures (see, e.g., Ref. 15). As known, in closed cavities the density of the mode spectrum scales as

$$\frac{\Delta\omega}{\omega} \propto \frac{\lambda^3}{V}.$$

When slightly irregular open waveguides excited at frequencies close to cutoff are used as cavities (see Fig. 2a), the modes with one axial variation have much smaller diffractive losses than other modes [16]. So, in such open cavities, the axial mode selection is provided, and therefore the spectral density of high-Q modes scales as

$$\frac{\Delta\omega}{\omega} \propto \frac{\lambda^2}{S_\perp},$$

where $S_\perp$ is a waveguide cross-sectional area. To explain the reason for the difference in diffractive losses, let us recall that a diffractive Q-factor of a section of waveguide is proportional to $\omega L/v_{gr}$, where L is the section length and v_{gr} is the wave group velocity. In the case of fast waves, $v_{gr} = c^2/v_{ph}$, where the phase velocity is $v_{ph} = \omega/k_z$ and the axial wavenumber k_z is proportional to the number of axial variation of the wave q, $k_z \sim q\pi/L$. From these simple formulas, it immediately follows that Q_{dif} is inversely proportional to q. Note that when the wave reflection from waveguide ends is considered more accurately, this yields an even stronger dependence, $Q_{dif} \sim 1/q^2$ [see Ref. 16].

Some methods of mode selection in transverse indices are based on the use of coaxial cavities as shown in Fig. 2b [17]. When a coaxial insert has a constant radius the selection is due to the difference in the effect of the insert on the cutoff frequencies of different modes. When the ratio of outer to inner wall radius is chosen properly, this may lead to rarefaction of the mode spectrum in the vicinity of the operating mode. The use of tapered inserts is even more advantageous, because a proper tapering of the insert may provide discrimination in diffractive Q-factors against unwanted modes [17,15,18]. For instance, a down-tapered coaxial insert pushes the rays of modes with a large number of radial variations out of the cavity.

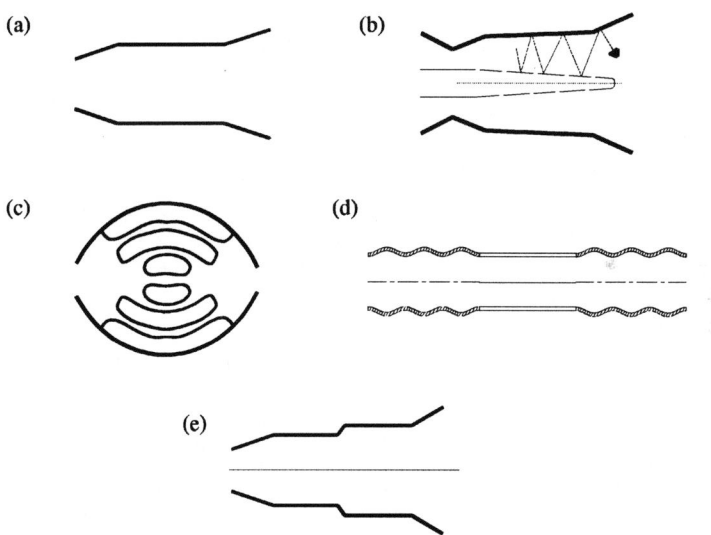

Fig.2. Open, highly selective cavities: a) open slightly irregular waveguide used as a cavity in a majority of gyrodevices, b) coaxial cavity, c) two-mirror cavity, d) Bragg resonator, e) step-profile cavity.

At the same time, the field of whispering gallery modes having a large number of azimuthal variations but a small radial index can be perturbed only slightly by such a coax when the radius of the coax is smaller than the caustic radius of the modes. Recall that gyrotron oscillators using coaxial cavities successfully operate even at such high-order modes as $TE_{28,16}$ [19] (here 28 and 16 are the azimuthal and radial indices, respectively).

It is also possible to use two-mirror quasioptical cavities shown in Fig. 2c that are commonly used in lasers. When the mirror dimensions are chosen properly, the only modes having a high diffractive Q-factor are the modes with one transverse variation of the field at the mirror. The density of the spectrum of such modes scales as

$$\frac{\Delta\omega}{\omega} = \frac{\lambda}{2L}$$

where L is the distance between mirrors.

A high degree of mode selection can be realized in Bragg resonators [20]. In such resonators (as shown in Fig.2d) the wall contains a spiral corrugation, i.e., the wall radius is

$$R = R_0 + R_1 \cos(2\pi z / d + m_B \phi)$$

[where $R_1 \ll R_0$, d is the corrugation period, and m_B is the azimuthal index of the corrugation]. Therefore, only two counter propagating waves whose axial wavenumbers $k_{z1,2}$ and azimuthal indices $m_{1,2}$ obey Bragg resonance conditions:

$$k_{z1} + k_{z2} \approx 2\pi/d, \quad m_B = \pm(m_1 - m_2)$$

form here a high-Q mode, while all other waves propagating with a large group velocity are scattered by the wall. Recall that Bragg resonators were used in many experiments with cyclotron autoresonance masers (CARMs), including a 26% efficient, 13 MW Ka-band CARM [21].

Another kind of highly selective structure is a step-profile cavity, as shown in Fig. 2e (see, e.g., Ref 15 and references therein). In each part of such a cavity, a specific partial mode can be excited. It is possible to choose the radii of these two waveguide sections in such a way that two partial modes having the same azimuthal but different radial indices will have the same frequency. Then just these two partial modes will form a normal mode of the whole structure, while other modes will be localized in one of the two waveguide sections, thus having a shorter length for interacting with electrons.

Electron Selection

Electron selection is based on the choice of beam parameters that provide selective excitation and stable efficient operation in the desired mode of the overmoded circuit.

The methods of electron selection can be divided into linear and nonlinear ones (see, e.g., Ref. 13).

Linear methods are focused on providing a difference in the starting currents for the operating mode and parasites. Recall that the starting current is inversely proportional not only to the cavity Q-factor but also to the coupling impedance of the beam to the mode and to the imaginary part of the susceptibility of the beam to the mode. The latter, in terms of the equivalent circuit representation that is widely used in the theory of klystrons, is often called the beam conductance.

The coupling impedance depends on the beam location and on the transverse structure of the mode field. The mode selection due to the difference in coupling impedances of various modes can be especially efficient when a device operates with fast waves whose fields have a number of transverse variations. Then, one can place a beam (or several beams) in a position where the field of the desired mode is strong, while the fields of parasites are weak, and thus strongly discriminate against unwanted modes. This statement is illustrated by Fig. 3, which shows the radial dependence of coupling impedances for TE_{01} and TE_{31} modes in a gyrotron coaxial cavity.

Note that another version of linear electron selection is also possible. This version is based on introducing an additional electron beam, which does not interact with the operating mode but serves as a passive load for parasites. In the case of gyrodevices operating at cyclotron harmonics, this can be a beam of electrons moving linearly along the magnetic force lines. As is known (see, e.g., Ref. 23), the interaction at cyclotron harmonics ($s>1$, where s is harmonic number) is possible due to the transverse inhomogeneity of the RF field, since the field rotating synchronously with gyrating electrons has a structure of an s-th order multipole. Correspondingly, electrons moving linearly will not interact with an RF field at cyclotron harmonics; however, they can interact at the fundamental cyclotron resonance with an RF field having a dipole structure. As is known, under condition of the normal Doppler effect, an initially linear electron beam can only absorb the microwave radiation. Therefore

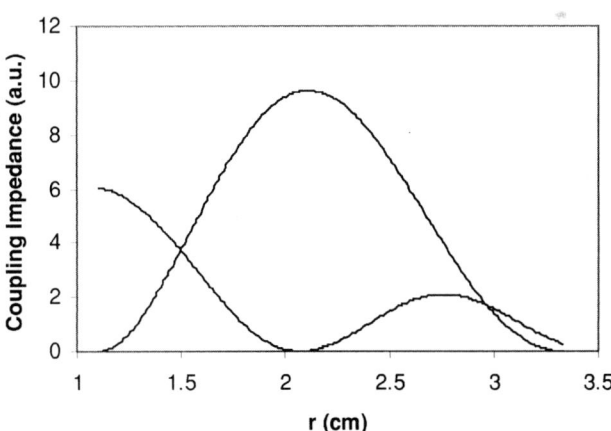

Fig.3. Coupling impedance of electrons to the TE01 and TE31 modes in a gyrotron coaxial cavity with outer and inner radii equal to 3.325 cm and 1.1 cm, respectively, for the case of fundamental cyclotron resonance.

the interaction of such a beam with parasitic modes at the fundamental resonance will increase their starting currents as shown in Ref. 24.

Another method of linear selection, known as a cyclotron-resonance selection [25], was used in RF sources of Cherenkov radiation. This method is based on the fact that, in a certain range of guiding magnetic field values, the Cherenkov resonance condition for the first space harmonic of the operating wave

$$\omega \approx (k_{z0} + 2\pi/d)v_z$$

(d is the SWS period) can be satisfied simultaneously with the condition of the cyclotron resonance for the zero space harmonic

$$\omega - k_{z0}v_z \approx \Omega.$$

(Here k_{z0} is the axial wavenumber of the zero harmonic and Ω is the cyclotron frequency.) Recall that the zero harmonic can be a fast wave having several radial variations in the transverse structure. Therefore, an annular electron beam can be located at such a position that its coupling impedance to the zero harmonic of the operating wave is negligibly small while the cyclotron interaction with parasitic modes can be substantial. As discussed above, such an interaction will lead to the absorption of the RF field of these modes, which increases their starting currents. As reported in Ref. 25, by using this method it was possible to generate 1.4 GW power in an X-band backward-wave oscillator operating in the TM02-mode.

Recall that in the sources of Cherenkov and/or cyclotron radiation, the above-mentioned susceptibility depends on the detuning of the corresponding resonance (ω-kzvz in Cherenkov devices and ω-kzvz-Ω in the sources of cyclotron radiation). Therefore, by varying the operating voltage and (in the case of cyclotron devices) also the magnetic field, one can maximize the imaginary part of this susceptibility for the operating mode, thus minimizing its starting current, while for other modes with different frequencies this detuning of the resonance will be nonoptimal.

In general, this susceptibility also depends on the amplitude of the RF field. This means, first, that the detuning which minimizes the starting current of the device is not optimal for its efficiency; and second, that as the amplitude of the RF field increases, the chance to excite different modes varies. This is already a nonlinear effect, which is the next topic of discussion.

Nonlinear methods of mode selection are based on the effect of nonlinear interaction between modes in the large-signal regimes. These methods are especially important for highly overmoded devices, in which previously described methods are insufficient, and in which a number of modes can be excited when the device is turned on. Possible nonlinear effects caused by mode interaction in such multimode devices were recently reviewed in Ref. 13. Therefore, without going into details, let us only mention that very often a mode excited first can suppress other modes. This effect of

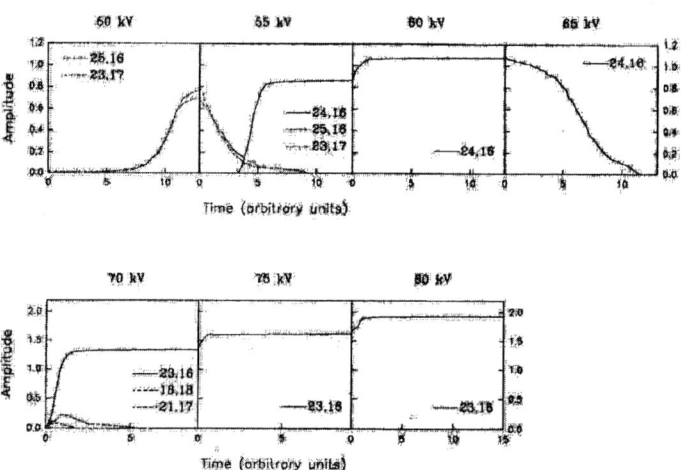

Fig.4. Start-up scenario for a 1 MW 280 GHz coaxial gyrotron operating at the TE$_{23,16}$ mode (from Ref. 18, © 1994 IEEE). (It is assumed that the voltage rise time is much larger than the cavity fill time)

mode suppression allows one to develop a start-up scenario in which a desired mode can be excited prior to other modes, and then, in spite of the presence of parasites in the spectrum, the desired mode can be driven to the regime of highly efficient operation while simultaneously suppressing the parasites. In some cases, like the one shown in Fig. 4, several modes can be excited during the voltage rise. However, this process may culminate in the onset of efficient oscillations of the desired mode.

Note that the mode suppression effect is also used in so-called mode priming. This means that a driver initially excites the desired mode. Then this mode, which starts from the level determined by the available driver, grows much faster than parasites, which start from the noise level, and thus suppresses them. Of course, nonlinear methods of mode selection do not provide amplifiers with zero-drive stability.

THERMAL LIMITATIONS

As mentioned above, operation in high-order modes of oversized structures is important, partly, because it provides a solution to some problems associated with the dissipation of microwave power in the circuit walls due to ohmic losses. The main problem here is single pulse heating, because the duty-cycle of μsec amplifiers operating with a repetition frequency of about 200 Hz is very low.

In order to analyze this problem, let us consider the pulse heating of a simple cylindrical cavity open in the axial direction. The temperature rise in the cavity wall, ΔT, follows from a known equation

$$c_t \rho V (\Delta T) = W_{ohm}$$

Here c_t, ρ, V, and W_{ohm} are specific heat, density, volume and the energy of the ohmic losses. The ohmic loss energy relates to the radiated microwave energy $W_{RF}=P_{RF}\tau$ as

$$W_{ohm} = \frac{Q_{dif}}{Q_{ohm}} W_{RF}$$

where the ohmic Q-factor, Q_{ohm}, can be estimated as the ratio of the cavity radius R to the skin depth

$$\delta \sim \sqrt{\lambda c/\sigma}$$

(here σ is the wall conductivity). The volume V in which this energy is deposited is the area of the wall surface $S=2\pi RL$ multiplied by the depth of the layer. This depth, in principle, is determined by the skin depth δ and by distance of heat propagation during the pulse. The skin depth in copper for mm-wave radiation does not exceed 1μm. According to the heat transfer equation, the heat propagation distance, l_h, for copper scales with the pulse duration τ as

$$l_h \approx 1.1\sqrt{\tau}$$

where l_h is given in cm and τ in seconds. For microsecond pulses, this yields l_h on the order of 10 μm. So we may conclude that heat propagation plays the dominant role in determining the depth of the microwave pulse power deposition. In general, as can easily be shown, for copper this statement is correct when the pulse duration (in seconds) and the frequency (in Hz) obey the condition $f\tau > 150$.

By using the simple formulas given above, one can easily find that the maximum pulse duration, restricted by the maximum allowable temperature rise $(\Delta T)_{max}$, scales as

$$\tau_{max} \sim v^4 \lambda^5 \left[(\Delta T)_{max} \frac{L/\lambda}{Q_{dif} P_{RF}} \right]^2 \quad (1)$$

Here $v=2\pi R/\lambda_{cut} \approx 2\pi R/\lambda$ is the eigennumber of the mode which we assume to be excited near cutoff. This scaling shows the importance of operation in high-order modes. Certainly, the resulting scaling is greatly influenced by the dependence of the radiated power on transverse dimensions. If we assume that the radiated power level does not depend on the mode, then Eq. (1) yields $\tau_{max} \sim (R/\lambda)^4$, which is a very strong dependence. However, typically P_{RF} increases with the ratio R/λ, which mitigates this scaling. In particular, when $P_{RF} \sim (R/\lambda)^2$ (which implies a constant power density in the interaction space), τ_{max} does not depend on R/λ at all. The same is valid for the scaling of τ_{max} with the wavelength. Although it is given in Eq. (1) that $\tau_{max} \sim \lambda^5$, when the power is restricted by ohmic losses, P_{RF} is proportional to $\lambda^{5/2}$, so that τ_{max} does not depend on the wavelength. Note that Q_{dif} in Eq. (1) is proportional to $(L/\lambda)^2$. So, when the L/λ ratio is constant, the ratio $(L/\lambda)/Q_{dif}$ in Eq. (1) is constant as well.

One of the possible ways to mitigate the restrictions caused by pulse heating is to operate with traveling waves having larger group velocities. As mentioned previously, the diffractive Q-factor is proportional to $\omega L/v_{gr}$. Therefore the increase in v_{gr} decreases Q_{dif}, and correspondingly, decreases W_{ohm}, which is proportional to Q_{dif}.

As an example, let us consider an output waveguide of the gyrotwystron, a device which consists of an input cavity and output waveguide separated by a drift space. The power flow at the waveguide output is proportional to

$$P_{out} \propto A_{out}^2 R^2 v_{gr},$$

or inversely, the amplitude of the outgoing wave is

$$A_{out} \propto R^{-1} P_{out}^{1/2} v_{gr}^{-1/2}.$$

The energy extracted from one electron is proportional to $eA_{out}L$ and for highly efficient regimes this value should be on the order of the initial kinetic energy of the electrons, $mc^2(\gamma_0-1)$. This means that the interaction length should scale inversely proportional to A_{out}, and therefore the L/λ ratio scales as

$$\frac{L}{\lambda} \sim \frac{v}{P_{out}^{1/2} v_{gr}^{-1/2}}$$

At the same time the diffractive Q scales as

$$Q_{dif} \sim \frac{\omega L}{v_{gr}} \sim \frac{L}{\lambda}\frac{1}{v_{gr}} \sim \frac{v}{P_{out}^{1/2} v_{gr}^{1/2}}$$

Since, as was shown above,

$$W_{ohm} = \frac{Q_{dif}}{Q_{ohm}} P_{out} \tau \propto (\Delta T)\sqrt{\tau}$$

this yields

$$\Delta T \propto \sqrt{P_{out}\tau/v_{gr}}$$

(Of course, to make correct estimates from these formulas, corresponding units and normalization coefficients should be used.)

ELECTRON BEAM LIMITATIONS

As mentioned above, 50 years ago it was already realized that wavelength shortening leads to a miniaturization of the tube that implies the use of small cathodes. For generating a high enough current from small areas of emitters, the cathode loading should be increased. In the case of thermionic cathodes, this may shorten the lifetime of the tubes.

An alternative solution is to leave the emitter area fixed but to increase the beam compression, which means to combine emitters of a large diameter with the circuits of small dimensions. Limitations on high compression of electron beams produced by Pierce guns were recently discussed in detail in Ref. 26. Therefore, addressing readers to Ref. 26, let us mention only that area compression ratios up to 2500 seem to be realistic.

In the case of magnetron injection guns (MIGs) used in gyrodevices, the wavelength shortening leads to quite specific modifications of cathode dimensions. Recall that in MIGs, a thin annular electron beam is generated by a ring emitter and the gun is located in a region of weak external magnetic field (see Fig. 5). Then the beam propagates toward the interaction space located in a region of maximum magnetic field. In the process of this propagation, the beam undergoes adiabatic magnetic compression, which can be characterized by the ratio $\alpha_B = B_0/B_C$ (here B_0 and B_C are the values of magnetic field in the interaction and cathode regions, respectively). By using the adiabatic theory of such beams [27,28], one can readily find that this compression ratio (for a beam with a fixed strength of the electric field at the cathode surface and a fixed orbital electron velocity in the interaction region) scales with the wavelength as $\alpha_B \propto \lambda^{-2/3}$. At the same time, from Busch's theorem it follows that $\alpha_B = (R_e/R_g)^2$ where R_e and R_g are the emitter radius and the radius of electron guiding centers in the interaction space, respectively. Therefore, in the case of operation in a given mode (when $R_g \sim \lambda$), the emitter radius scales as $R_e \sim \lambda^{2/3}$. This means that we can keep the beam current I_b and the cathode loading j constant only if the width of the emitter, l_e, scales with wavelength as $l_e \sim \lambda^{-2/3}$.

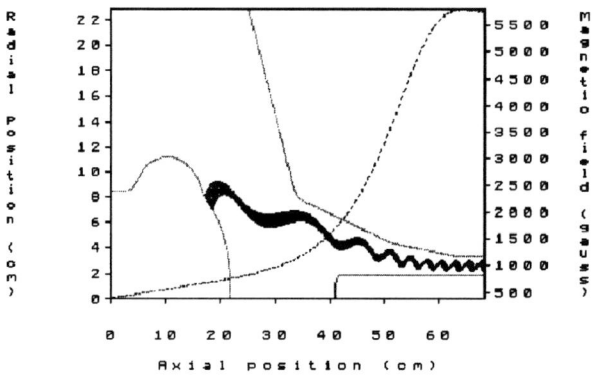

Fig.5. MIG with electron beam.

Such an increase in the emitter width as the wavelength is shortened enhances the spread in electron guiding center radii $(\Delta R_g) \sim \lambda^{-1/3}$; i.e., it is possible only until the spread reaches the maximum value $(\Delta R_g)_{max}$, which should be a substantial fraction of the wavelength. (Typically, it is assumed that $(\Delta R_g)_{max} = (1/6 - 1/4)\lambda$). Larger values of (ΔR_g)'s lead to efficiency degradation due to the radial inhomogeneity of the RF field. After (ΔR_g) reaches its maximum, the emitter width scales with the wavelength as $l_e \sim \lambda^{2/3}$. Correspondingly, the total beam current for the gun with a fixed cathode loading scales as $I_b \sim \lambda^{4/3}$. (Note that, if we assume that the cathode loading is a fixed percentage of the space-charge limited current, this yields $j \sim \lambda^{-1/3}$, which results in the beam current scaling $I_b \sim \lambda$ given in Ref. 28.)

One should also bear in mind that, as the beam compression increases, the beam density increases as well, so stronger magnetic fields could be required for providing the beam transport. This increase in beam density also makes the role of beam self-fields more important. Also recall that the wavelength shortening leads to smaller tolerances in the circuit fabrication and smaller clearances between the beam and the circuit walls, which makes possible misalignments of tube elements more dangerous.

SUMMARY

Summarizing this brief overview it seems possible to make two statements:

1. Development of millimeter-wave sources is already a quite mature field of activity.

2. Development of millimeter-wave drivers for future linear colliders is a new and very exciting field with a lot of challenges for scientists and engineers.

ACKNOWLEDGMENTS

This work was sponsored by the Division of High Energy Physics of the Department of Energy. The author wishes to thank S.Gold for useful discussions.

REFERENCES

1. Pierce, J. R., Physics Today, Nov. 1950, p. 24.
2. Felch, K. L. et al., Proc. IEEE **87**, 752 (1999).
3. Blank, M. et al., Phys. Plasmas **6**, 4405 (1999).
4. Whittum, D. H., Plenary Talk at the 22nd Int. Conf. on Infrared and Millimeter Waves, Wintergreen, VA, July 1997, Conf. Digest 54.
5. Abubakirov, E. B. et al., Sov. Phys. Tech. Phys. **35**, 1341 (1990).
6. Allen, S. L. et al., Phys. Rev. Lett. **72**, 1348 (1994).
7. Danly, B. G., "High-Power Millimeter Wave Amplifier Technology," this Workshop.
8. Wilson, P. B., in *Applications of High-Power Microwaves*, edited by A. V. Gaponov-Grekhov and V. L. Granatstein, Artech House, Boston, 1994, Ch. 7.
9. Tantawi, S., "Multi-Megawatt Semiconductor Microwave Switches," this Workshop.
10. Vikharev, A. L. et al., "Two-Channel Active High-Power X-Band Pulse Compresosr," this Workshop.
11. Clunie, D. et al., in *Strong Microwaves in Plasmas*, edited by A. G. Litvak, Inst. Appl. Phys., Nizhny Novgorod, 1996, v. 2, p. 886.
12. Gold, S. H. and Nusinovich, G. S., Rev. Sci. Instrum. **68**, 3945 (1997).
13. Nusinovich, G. S., IEEE-PS **27**, 313 (1999).
14. Petelin, M. I., Int. J. Electron. **67**, 137 (1989).
15. Gaponov, A. V. et al., Int. J. Electron. **51**, 277 (1981).
16. Vlasov, S. N. et al., Radiophys. Quantum Electron. **12**, 972 (1969).
17. Vlasov, S. N. et al., Radio Eng. Electron Phys. **21** [7], 96 (1976).
18. Nusinovich, G. S. et al., IEEE-ED **41**, 433 (1994).
19. Piosczyk, B. et al., IEEE-PS **26**, 393 (1998).
20. Denisov, G. G. and Reznikov, M. G., Radiophys. Quantum Electron. **25**, 407 (1982).
21. Bratman, V. L. et al., Phys. Rev. Lett. **75**, 3102 (1995).
22. Main, W. et al., IEEE-PS **22**, 566 (1994).
23. Flyagin, V. A. et al., IEEE-MTT **25**, 514 (1977).
24. Zapevalov, V. E. and Tsimring, Sh. E., Radiophys. Quantum Electron. **33**, 954 (1990).
25. Abubakirov, E. B. et al., Sov. Tech. Phys. Lett. **9**, 230 (1983).
26. Yakovlev, V. P. and Nezhevenko, O. A., "High Energy Density Microwaves," Pajaro Dunas, CA, 1998, edited by R. M. Phillips, AIP Conf. Proc. 474, Woodbury, NY, p. 316 (1999).
27. Gol'denberg, A. L. and Petelin, M. I., Radiophys. Quantum Electron **16**, 106 (1973).
28. Lawson, W., IEEE-PS **16**, 290 (1988).

10-MW, W-Band RF Source for Advanced Accelerator Research

J. L. Hirshfield,[*,†] O. A. Nezhevenko,[†] Changbiao Wang,[*]
V. P. Yakovlev,[†] A. A. Bogdashov,[¶] V. L. Bratman,[¶] A. V. Chirkov,[¶]
G. G. Denisov,[¶] A. N. Kuftin,[¶] S. V. Samsonov,[¶] and A. V. Savilov[¶]

Beam Physics Laboratory, Yale University, 272 Whitney Ave., New Haven, CT 06511
†Omega-P, Inc., 345 Whitney Ave., New Haven, CT 06511
¶Institute of Applied Physics, Russian Academy of Sciences, 603600 Nizhny Novgorod, Russia

Abstract. A conceptual design is presented for a W-band RF source that should be suitable for testing advanced accelerator structures and related components. The source is an 8th-gyroharmonic converter, in which 28 MW of X-band power at 11.424 GHz is used to energize and spin up an injected 500 kV, 40 beam in a TE$_{111}$ cavity; and in which over 10 MW of W-band power at 91.392 GHz is extracted from the beam in a TE$_{811}$ output cavity. A mode converter is employed to provide a Gaussian output beam.

INTRODUCTION

Efforts have been directed at design and fabrication of accelerating structures to operate at W-band (91.4 GHz) because of the expectation of achieving an acceleration gradient ~1 GeV/m, based on empirical scaling [1]. This gradient allows a 5 TeV electron/positron collider to be built within a length of 5 km, not including the final focus region. It is this dramatic reduction in size of a future multi-TeV collider that has provided much of the stimulus for the W-band work. But to test the susceptibility of accelerating structures and components at high-power to rf breakdown and fatigue, it is apparent that a high-power W-band source will be required.

No megawatt-level W-band source exists for this task. Some moderate power sources at W-band are presently available, and others are under design. Notable are the W-band gyro-klystron amplifiers developed at Naval Research Laboratory (NRL) [2] and at Institute of Applied Physics in Nizhny Novgorod, Russia [3]. These devices currently deliver peak output powers of over 100 kW and over 200 kW, respectively. Design of a W-band multi-beam klystron that embodies several 100 kW "klystrinos" is also currently underway [4]. Furthermore, a preliminary design has been published for a 7.5 MW, W-band three-cavity second-harmonic gyro-klystron [5]. No other W-band amplifier has been designed heretofore with a peak power of more than 10 MW.

CONCEPTUAL DESIGN

A design for an 8th-harmonic frequency multiplier is described here that is predicted to have 40% efficiency for power conversion from 11.424 GHz up to 91.392 GHz. A conceptual drawing of the heart of the device is shown in Fig. 1. Computations described below show a peak output power of 11.3 MW at W-band, for an input of 28 MW at X-band. This converter could be driven from an X-band SLAC klystron [6], or from the Omega-P/NRL 60-MW, X-band magnicon [7]. Either driver would allow one to obtain a 1-3 μsec W-band output pulse with a repetition rate determined by the available modulator. The 11.3 MW output power level is not an absolute upper limit, but is set in the present design by the beam current (40 A) from the available modulator at NRL. Features of the tube include:

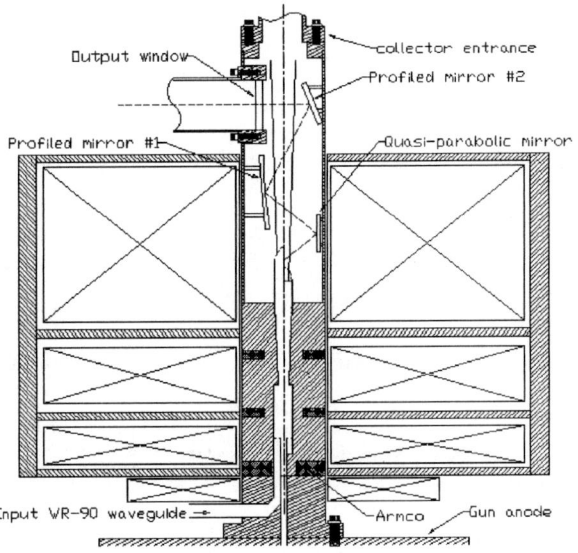

Figure 1. Conceptual design of W-band source.
Not shown are the gun at bottom, and the beam collector at top.

a. Two WR-90 input waveguides, with an H-plane miter-bend to feed a coupling aperture in the bottom face of the TE_{111} input drive cavity. The second input waveguide is disposed at 90° with respect to the first, and is not seen in the drawing. Opposite each input waveguide is an aperture in the cavity wall with a tuning section to symmetrize fields in the drive cavity. Each waveguide is to carry 14 MW of 11.424 GHz input power from the rf driver, a level well below the waveguide breakdown limit.

b. A four-coil room-temperature magnet structure, with three carefully tailored Armco rings that help produce the steeply contoured axial magnetic field required for achievement of high efficiency in this device. Such a steep contour may be difficult to produce with a cryomagnet of the same room-temperature inner bore diameter as this coil system (70 mm). The entire magnet structure can be raised up over the top of the tube without breaking vacuum, to allow bakeout and other adjustments. The lowest coil is to provide the field needed to match the beam emerging from the gun anode and pole piece as it flows through the long beam tunnel to the input cavity.

c. An input drive cavity that supports a rotating TE_{111} mode at 11.424 GHz, as fed by the two input waveguides phased in time-quadrature. Cavity radius and length are 8.10 mm and 40.9 mm. The ratio of ohmic-to-beam-loaded quality factors for this cavity is 136, indicating that the efficiency for imparting rf drive power to the beam is 99.2%. Peak surface rf electric field in this cavity at rated drive power is 250 kV/cm, far below the breakdown limit at 11.4 GHz. A short cut-off section is inserted between the drive cavity and the output cavity, with radius 4.6 mm and length 12 mm.

d. A TE_{811} mode output cavity, with radius and length 5.07 mm and 14.7 mm, ending in a gently up-tapered output waveguide. Output power is 11.3 MW at 91.392 GHz, with only 0.85 MW in spurious modes at higher harmonics. Lower harmonic spurious modes are suppressed by careful circuit design. Peak surface rf electric field in this cavity is 727 kV/cm, safely below the breakdown limit at 91.4 GHz.

e. A quasi-optical mode converter for producing a Gaussian output wave beam, consisting of a helical cut in the output waveguide, a quasi-parabolic mirror, two profiled mirrors, and an output window. In addition to producing the low-transmission-loss Gaussian beam, the mode converter separates the wave beam from the spent electron beam, which passes to the collector above. Spurious modes are not focused by the mode converter.

Further detail on each of these elements is given below.

The modulator that could be used to power the W-band tube is rated at 500 kV, 250 A, in 1.1 μsec pulses at a 10 Hz rep rate. This modulator now powers the 11.4 GHz magnicon that can furnish rf drive power for the W-band tube, so that the driver and converter will be operated in parallel on the same modulator. Since the magnicon requires 210 A to run the tube at rated output and gun voltage, only 40 A will be left over for the W-band tube.

Fig. 2 shows the optimum result found for design of the gun, using the code Super-SAM [8]. The beam is seen to be in nearly ideal laminar flow. Maximum electric field on the focus electrode is found to be 195 kV/cm, which is well below the 230-420 kV/cm fields that are sustained by SLAC X-band klystron gun focus electrodes [7]. Cathode loading of 8 A/cm^2 is well within an acceptable range for dispenser cathodes. Geometric emittance is below the irreducible thermal emittance value of 6π mm-mrad; 500:1 electrostatic compression produces a beam with a diameter of 1.1 mm, and subsequent magnetic area compression of about 5:1 in the gradually-increasing field of the input cavity gives a beam diameter computed to be about 0.5 mm. But to provide a margin of safety, simulations of converter performance shown below are for a beam with diameter of 0.8 mm, corresponding to an overall area compression of ~1000:1.

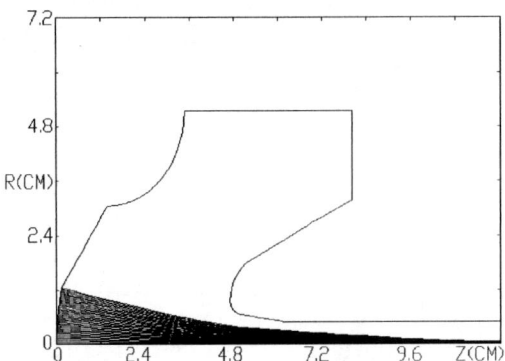

Figure 2. Optimum conceptual design of electron gun for W-band source.

The mechanical design for this gun will be very similar to a gun already delivered from Budker INP for the 34 GHz magnicon now under development by Omega-P, Inc. The only significant design difference between the gun for the W-band source and the 34 GHz gun is the cathode diameter, the corresponding focus electrode changes, and small modifications in anode shape. The existing gun is designed for 200 A with a cathode diameter of 50 mm, while the new gun will be designed for only 40 A with a cathode diameter of 25 mm. The low perveance value of 0.12×10^{-6} A-$V^{-3/2}$ for the new gun will allow a long focal length into the 11.4 GHz drive cavity, perhaps allowing room between the gun and the first iron pole piece for a gate valve.

Designs were optimized of three linked sub-systems of the W-band source, namely magnet, drive cavity, output cavity and mode converter. Fig. 1 shows these elements together. The highest tolerable 11.4 GHz input power level was found to be 28 MW for the available beam current of 40 A; the beam is accelerated in the drive cavity by 700 keV to produce a 1.2 MeV beam at the entrance of the output cavity. For rf drive power above 28 MW, beam energy spread increased and output power decreased. The output cavity, with radius of 5.07 mm and length of 14.7 mm, and a tapered output waveguide, has a diffraction Q of about 400. Suppression of harmonics below the eighth was achieved in the design shown here because the cavity radius is such that these are cutoff. To study the device in detail, a self-consistent system of equations for the electromagnetic fields and electron motions was applied, using time-tested interaction codes.

Fig. 3 shows the optimum magnetic field profile, the structure of the two-cavity/up-tapered output system, and resulting electron trajectories. The power that emerges from the cavity and passes along the following up-taper is shown in Fig. 4; it is seen to be about 11.3 MW. Some power (totaling about 0.85 MW) is radiated into other harmonics, also shown in Fig. 4. But these harmonics are not expected to be focused into a Gaussian output beam by the output converter optics that are designed for the TE$_{81}$ mode.

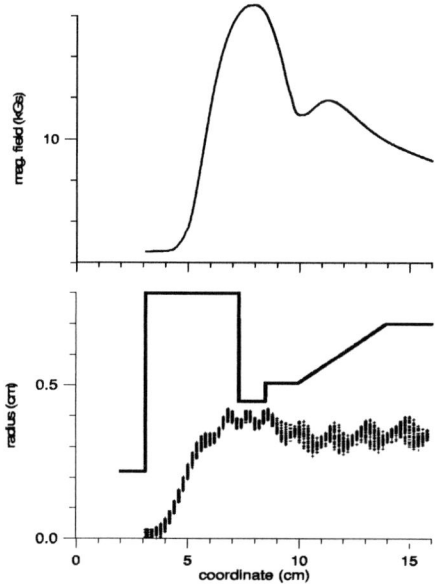

Figure 3. Electron-wave interaction region, showing (top) magnetic field profile; and (bottom) rf structure profile and radii of electron trajectories.

The magnet system is shown in Fig. 5; it has been designed to provide the required magnetic field profile, as shown in full in Fig. 6. The room-temperature coil system consists of three axially symmetric solenoids with independent current control. The solenoids are to be wound with copper wire of rectangular cross-section, and are to be equipped with a two-stage oil-water cooling system. The primary cooling circuit (whose coolant is transformer oil) consists of a heat exchanger and a pump. The secondary circuit of the heat exchanger is cooled by chilled water. Influence of the first coil near the gun, introduced to match the beam into the fields at the input cavity, is neglected in studying the electron dynamics in the interaction regions, but its influence upon fields in the interaction region should be small. Design parameters for the main coils include the following:

coil # 1: $j = 2.15$ A/mm^2, $I = 37$ A (or 306 A), $P = 0.21$ kW.
coil # 2: $j = 18.0$ A/mm^2, $I = 306$ A, $P = 24.9$ kW.
coil # 3: $j = 6.43$ A/mm^2, $I = 109$ A (or 306 A), $P = 7.2$ kW.
coolant flow rate (transformer oil): 3.6 m^3/h, 3-4 atm.
water flow rate for oil cooling: 4 m^3/h, 2 atm.

The total power required is seen to be 32.3 kW, but this figure depends on coil temperature and thus on oil-cooling water flow rates, and may vary within ±10%.

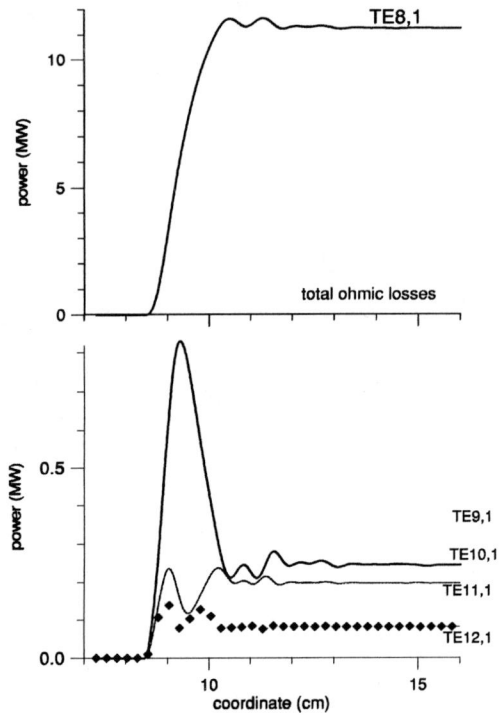

Figure 4. Output power of operating mode and ohmic losses (top); and power in parasitic modes (bottom); as functions of distance along the device.

To convert the $TE_{8,1}$ 91.4 GHz output mode into a nearly Gaussian wave beam that can be transported to a load with negligible loss, a quasi-optical mode converter has been designed. This mode converter consists of a special waveguide cut, a quasi-parabolic mirror, and two profiled turning mirrors, as shown in Fig. 7. The quasi-parabolic mirror transforms the radiation into a paraxial wave beam. The following shaped mirrors maximize the Gaussian mode content in the beam. The shaped mirrors are shown in Fig. 8. The output structure ends with 40 mm diameter window. The mode converter is supported within a 70 mm diameter pipe, so it is possible to hoist the magnet off the W-source so as to bake out the source or to make other adjustments. Fig. 9 shows the intensity and phase variations of the output Gaussian wave beam at the output window.

Collector design is essentially the same as that of a collector already designed and built by Omega-P, Inc. for the 34 GHz magnicon [7].

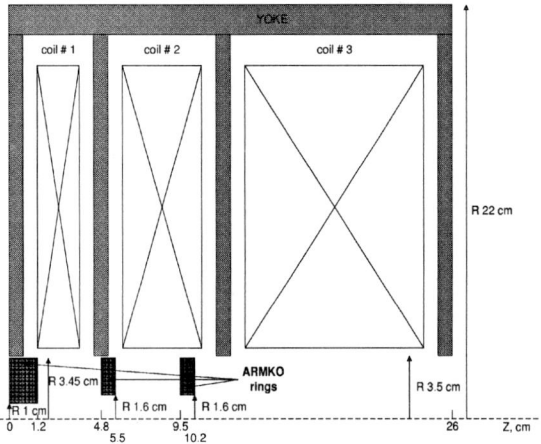

Figure 5. Schematic of the room-temperature magnet system. Shown are the dimensions of the three coils, and of the Armco rings and iron yoke structure. This structure, mounted vertically, is designed to allow removal from the tube without breaking vacuum.

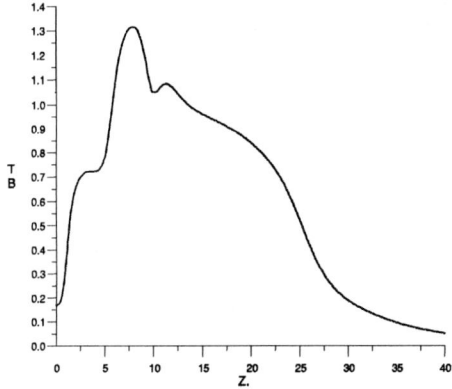

Figure 6. Computed guide magnetic field profile for the structure shown in Fig. 5.

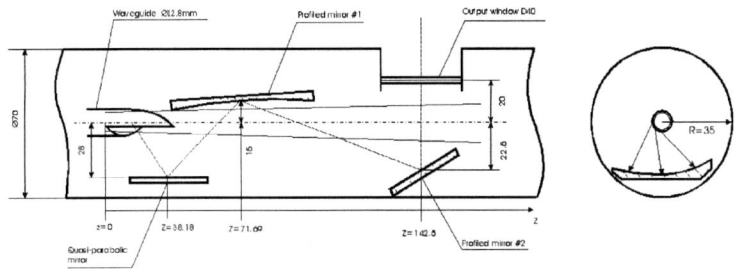

Figure 7. Quasi-optical output structure to convert TE_{81} mode to Gaussian wave beam.

771

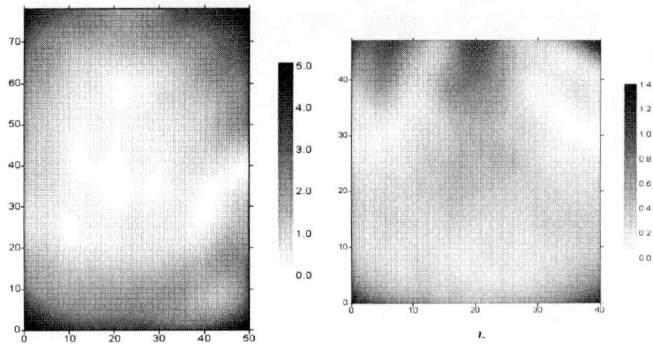

Figure 8. Synthesized mirrors for the quasi-opical mode converter. The first mirror (*a*) is 50×77.9×5.46 mm; the second mirror (*b*) is 40×46.7×1.48 mm.

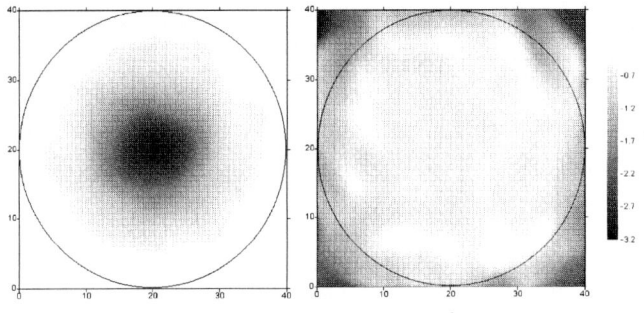

Figure 9. Intensity (*a*), and phase (*b*) patterns at the output window. Gaussian content: $\eta_a = 99.15\%$, $\eta_{a,\varphi} = 98.56\%$; Gaussian parameters: $a_x = 8.26$ mm, $a_y = 7.99$ mm, $S_x = -2.68 \cdot 10^3$ mm, $S_y = 4.66 \cdot 10^3$ mm; Diffraction losses: $\approx 8\%$.

A summary of the main parameters embodied in the 10-MW W-band source design is given in Table I.

For higher applied rf power, higher beam current is needed, and higher W-band output can be obtained. Indeed, the W-band source described above is designed for a 40 A beam since that is the highest current that can be derived from the NRL modulator operating in parallel with the 11.4 GHz magnicon. Drive power at 11.4 GHz of over 50 MW is expected to be available, but no more than 28 MW of this can be used because of the current limitation. But there is no fundamental limitation to achievement of higher W-band output from this type of source. Study of the performance of the 8th gyroharmonic W-band source reveals, for example, that operation at 80 A should be possible with only minor modifications of the device. Of course, a new gun—similar to that described above—would have to be built with

beam voltage	480 kV
beam current	40 A
gun perveance	0.12×10^{-6} A-V$^{-3/2}$
cathode diameter	25 mm
maximum cathode loading	8 A/cm^2
beam area compression	1000:1
maximum E-field in gun	195 kV/cm
rms beam thermal emittance	6π mm-mrad
rms beam geometrical emittance	0.5π mm-mrad
11.4 GHz drive power	28 MW
91.4 GHz output power	11.3 MW
Output mode purity	$\eta_a = 99.15\%$, $\eta_{a,\varphi} = 98.56\%$
Output beam radii	$a_x = 8.26$ mm, $a_y = 7.99$ mm
power in spurious harmonics*	0.85 MW
peak magnetic field	13.1 kG
total magnet power	32.3 kW

Table I. Main parameters for proposed 10-MW, W-band source.
*Spurious modes are not focused by output mode converter.

larger cathode and with suitably modified focus electrode and anode. Furthermore, the output cavity would have to be shortened slightly to reduce the diffraction Q to below its present value of 400. The 80 A beam would have a Brillouin diameter of about 0.8 mm, and would thus pass the beam tunnel and interact well in the existing drive cavity. With twice the present drive power, or 56 MW at 11.4 GHz, it is estimated that the W-band output power would be greater than 20 MW.

SUMMARY

The 10-MW W-band source described here embodies several attractive features:
- A high-convergence diode electron gun that is similar to one already built for Omega-P 34 GHz magnicon is employed, so no significant new gun development is needed.
- A high-power beam collector that is similar to one built for the Omega-P 34 GHz magnicon is to be employed, so no significant new collector development is needed.
- An existing X-band magnicon will serve as the rf driver.
- Modulator requirements are to be satisfied by adding a second socket to the existing NRL modulator that powers the Omega-P/NRL X-band magnicon, thereby operating both the driver and the harmonic converter in parallel from the same power supply.
- The required magnetic field is to be provided by room-temperature (non-cryogenic) coils.

- The W-band tube design allows removal, without breaking vacuum, of surrounding magnet structure for high-temperature bakeout and other external adjustments.
- W-band output is in the form of a Gaussian wave beam, suitable for low-loss transmission to a wide class of loads, such as test accelerator structures and associated components.
- For an 80 A beam and 56 MW of 11.4 GHz drive power, it is estimated that a W-band output of about 20 MW could be obtained from this device.

ACKNOWLEDGMENT

This work was sponsored by the US Department of Energy.

REFERENCES

[1] P. J. Chou et al, *Advanced Accelerator Concepts, AIP Conf. Proc.* 398, New York, Am. Inst. Phys., (1996) pp. 501-517; D. T. Palmer, at *AAC'98*, Baltimore, 1998; P. B. Wilson, *SLAC-Pub-7449, April 1997*; R. B. Palmer, "Pulsed rf sources for linear colliders," *AIP Conf. Proc.* 337, New York, Am. Inst. Phys. (1994) pp. 1-15.

[2] M. Blank et al, *Phys. Rev. Lett.* 79, 4485 (1997).

[3] E. V. Zasypkin et al, *IEEE Trans. Plasma Sci.* 24, 666 (1996).

[4] G. Caryotakis et al, "High power W-band klystrons," in *High Energy Density Microwaves*, R. M. Phillips, ed., CP474, Am. Inst. Phys. (1999) pp. 59-73.

[5] M. R. Arjona and W. Lawson, to be pub. In *Proc. EPAC'98*, Stockholm (1998); see also "Design of a 10 MW, 91.392 GHz Gyroklystron for Advanced Accelerators," by W. Lawson, R. L. Ives, J. Neilsen, M. E. Read, *abstract submitted to EPAC2000*.

[6] E. Wright et al, "Design of a 50-MW-Klystron at X-band," *Proc. RF'94*, AIP Conf. Proc. 337, pp. 58-66 (Am.Inst. Phys., New York 1995).

[7] O. Nezhevenko, *Phys. Plasmas* 7, 2224 (2000).

[8] D.G. Myakishev, M.A. Tiunov and V.P. Yakovlev, "Code SuperSAM for Calculation of Electron Guns with High Beam Area Convergence", XV-th Int. Conf. on High Energy Accelerators, 1992, Hamburg. *Int. J. Mod. Phys.* A (proc. Suppl.) 2B (1993), v-2, pp.915-917.

Design of Flat-Field, High-Aspect Ratio RF Structures

D. Yu and A. V. Smirnov

DULY Research Inc., Rancho Palos Verdes, CA 90275

Abstract. Recent work at DULY Research on electron planar devices in flat field rf structures is presented. This new class of rf structures is useful for mm-wave power generation in sheet-beam klystrons, for generation of flat beams from a round beam, and acceleration of flat beams in advanced linacs and future colliders. Several approaches of flat-field designs with effective aspect ratio ~10 are characterized here.. Comparisons are made using the concept of a generalized "flat" impedance. High group velocity ~0.1 is achieved in a modified muffin-tin structure with side dielectric slab.

INTRODUCTION

Efficient, high-gradient acceleration schemes are essential for future linear colliders and advanced commercial electron accelerators. Many of the rf structure based schemes operate at high frequencies (Ka to W bands) in order to achieve high gradients and high wall-plug-to-acceleration efficiency. Examples of several linear collider designs scaled in frequency and energy from the NLC were worked out a few years ago by Perry Wilson et al [1]. To meet these rf power requirements and gradients, parameters of a sheet beam klystron (SBK) can be easily scaled to higher frequencies from an X-band structure designed [2,3] by DULY Research (see Table 1). At present, SLAC continues this R&D work to pursue the SBK at W-band.

Because of the smaller size of the accelerating structure with increasing operating frequency, conventional shop fabrication techniques are replaced by modern microfabrication techniques, such as EDM and LIGA. Planar structures are more suitable than traditional axially symmetric structures if the microfabrication techniques are used [4]. In principle, planar rf structures can also be used to accelerate a high-aspect ratio flat beam of particles. A flat beam with low asymmetric emittances can be accelerated over long distances without further emittance compensating the damping ring. This is partly because of the lower charge density associated with a flat beam which mitigates the space charge effect and in part due to reduced wake fields associated with planar structures. For future electron-positron linear colliders a flat beam in the final focus region has an additional benefit of reducing background events caused by beamstrahlung of round beam, thus providing cleaner experimental data from the electron positron collisions [4]. In order to be able to accelerate high-aspect ratio flat beam in a planar rf structure the structure should be designed to have a constant longitudinal electric field in the maximum transverse dimension (say, horizontal) of the accelerating gap. Such structures designed for high aspect-ratio beams and having 'flat' impedance profile in a maximum transverse cross-section are referred to here as flat-field structures.

TABLE 1. Sheet-Beam Klystron Parameters.

Frequency, GHz	11.42	34.27	91.39	91.39
Voltage, keV	400	450	500	560
Cathod Loading, A/cm^2	12	12	12	12
Compression Ratio	10	15	25	28
Beam Thickness, mm	11.78	6050	2.965	18.8
Current density, A/mm^2	19.2	4.05	1.00	1.21
Microperveance, per square, A/(V$^{3/2}$ mm^2)	0.076	0.013	0.03	0.03
Beam Width, cm	16	16	16	18
Beam Current, A	768	432	288	363
Efficiency, %	65	70	70	70
Peak Power, MW	200	136	101	142
Pulse Compression Ratio	1	8	32	32
Pulse Compression Efficiency, %	100	80	70	70
Input Peak Power, MW/m	200	871	2258	3186

CHARACTERIZATION OF FLAT-FIELD STRUCTURES

In designing flat-field structures for planar accelerators it is necessary to introduce a special figure of merit to allow comparison of rf structures with different aspect ratios of the aperture S, aperture dimensions, operating frequencies $f=2\pi c/\lambda$ and phase advance per cell θ (for periodic structures). In this way we define here the following parameter:

$$R_g/Q = S \cdot (R/Q) \cdot (a/\lambda),$$

where Q is standard Q-factor, $2a$ is vertical gap and S is effective aspect ratio.

We refer here to this parameter R_g/Q as generalized 'flat' impedance, which reduces to the standard definition of impedance $R=R(0,0)$ at the center of cavity, $R(x,y)$ is standard shunt impedance $R(x,y) = \left| \int E_z e^{i\frac{k}{\beta}z} dz \right|^2 / P$, where β is the particle velocity divided by the speed of light, and P is cavity power losses.

Note, effective aspect ratio S is defined as dimension where the longitudinal impedance is essentially flat divided by another (vertical) dimension of the aperture. The flatness can be characterized by dimensionless parameter $F_R(x) = \dfrac{a^2}{2R} \dfrac{\partial^2 R}{\partial x^2}$. We define some rf geometry design as "flat-field structure" if for some area X takes place $F_R(x) \ll 1$ on this area $x \subset X$.

Impedance for flat structures described above is not the only generalization which is possible and other definitions can be introduced. For example, such values as group velocity β_{gr} that plays an important role in designing waveguide structures, and flatness F_R are not included into this figure of merit explicitly. Nevertheless for convenience in evaluating different geometries we will use this generalized impedance R_g/Q along with group velocity as the basic criteria.

FLAT-FIELD STRUCTURE DESIGNS

The main problem in designing effective flat-field structures is to modify boundary conditions in such a way to provide field flatness along with high enough R_g/Q value. Besides, there are certain manufacturing requirements for the geometry. As a rule for common microfabrication processes the structure should be etched with a single X-ray mask resulting in a uniform cavity depth and arbitrary cavity shape. It is very difficult to satisfy all of these requirements and we consider here a number of approaches and geometries that can be of potential interest for current and future designs.

A barbell cavity designed earlier [3] by DULY Research for an X-band sheet beam klystron can serve as a starting point for the design of a flat-field cavity. Although the barbell design cannot be used directly in a 1-step LIGA fabricated planar structure because the "bells" are larger than the space available between contiguous accelerating cells, the design in principle may be adapted for other configurations. A simple variation is to turn the bells sideways so they have the same depth as the cavity. This design may be applicable to devices such as SBK in which the cavities are not contiguous, but not for devices such as traveling wave accelerating structure. Our approach is to "stretch out" the fields by imposing appropriate boundary modifications outside the transverse region where field flatness is required. (In the barbell, this is accomplished by the "bells" at either end, which have a cutoff frequency very close to the resonant frequency of the "bar" to which they are attached.) Feasibility of this idea to obtain high-aspect ratio flat-field cavity with increased inductive or capacitive load at the sides is demonstrated in the following for both standing wave and traveling wave structures.

Some of the applications require tapered (constant gradient, CG) and biperiodic (or even three periodic) coupled SW flat-field structures that in turn require development of essentially new geometries.

Basic approaches to the planar accelerating cells proposed and examined here are the following:
1. Side metal wall of a rectangular cavity modified by dielectric, allowing an electric field to have a non-vanishing component parallel to the metal boundary separated or covered by the dielectric.
2. Side slots for magnetic coupling with resonance elements providing field flatness.
3. Profiling (transverse tapering) of the iris (vane) thickness to match the field uniformity along with side openings.

Some results for the designed novel configurations are summarized in Table 2 for $\beta=1$ and $\beta_{ph}=1$ (for TW structures). Computations and design optimization were made using GdfidL [5] and MAFIA [6] codes.

TABLE 2. Flat-Field High-Aspect-Ratio Planar Structures – Cavities And Waveguides.

Type #	Brief description	f, GHz	Q	r/Q, kΩ/m	R/Q, Ω	θ	β_{gr}	a/λ	R_g/Q	S	ε	Figs and refs.
C1	original barbell cavity	11.67	6145	0.617	18.13	N/A	N/A	0.1	20.9	11.5	N/A	[3]
C2	T-end barbell cavity	11.745	6145	0.714	21.02	N/A	N/A	0.1	25.2	12	N/A	Fig. 1
C3	simple pillbox+dielectr.	11.71	4465	0.49	14.45	N/A	N/A	0.1	16.6	11.5	4.92	Fig. 2
M1	side slot+rod	11.78	6050	2.965	18.8	$\pi/2$	0.06	0.1	21.6	11.5	N/A	
M1	side slot+rod	12.12	5480	2.834	23.4	$2\pi/3$	0.05	0.1	29.3	11.5	N/A	Fig. 3
M1	side slot+rod	11.81	6030	2.047	25.9	π	0	0.1	31.1	12	N/A	
M2	side slot+rod+T-end iris	12.15	5753	2.76	22.7	$2\pi/3$	0.074	0.1	25.9	11.4	N/A	Fig. 4
M2	side slot+rod+T-end iris	11.43	5789	3.835	32.5	$2\pi/3$	0.02	0.06	37.1	19	N/A	
M3	profiled iris thickness	12.4	4529	3.08	18.7	$\pi/2$	0.068	0.1	16.7	8.94	N/A	Fig. 5
M4	side diel. long strip	12.19	4896	2.89	17.7	$\pi/2$	0.1	0.1	22.4	12.6	12.8	Fig. 6
M5	snake-wise asymmetry	12.08	6074	1.732	21.5	π	0	0.114	21.7	8.84	N/A	Figs. 7, 8

Although the computations were done for nearly the same frequency, the geometries designed can be easily scaled without recalculations for any other frequency. The only exception would be those designs containing dielectric: scaling may require additional computations to take into account non-linear properties of material (losses, permeability).

Initially we designed two new variants of flat-field single-cell cavities.

The first variant (C1 in Table 2) is a simple modification of the original barbell cavity to make it more compatible with microfabrication process. The structure view is shown in Figure 1 with accelerating field profile along wide horizontal. Field flatness depends very strongly on length of the side cavity and can be adjusted to be nearly perfect as it is seen from Figure 1. The next structure (C2) that is not shown is simple: this is a rectangular pillbox cavity partially filled with dielectric from the sides. Proper option of dielectric depth and/or permeability results in flattening of field profile as depicted in Figure 2, obtained for $\varepsilon=4.92$ that is very close to quartz. Although this design is very simple and can be characterized analytically it is not suitable when high levels of rf power are required. High overvoltage in the dielectric can cause electric discharge and dielectric breakdown. Besides, Q-factor is unavoidably lower in designs having dielectric (by ~30% in our case).

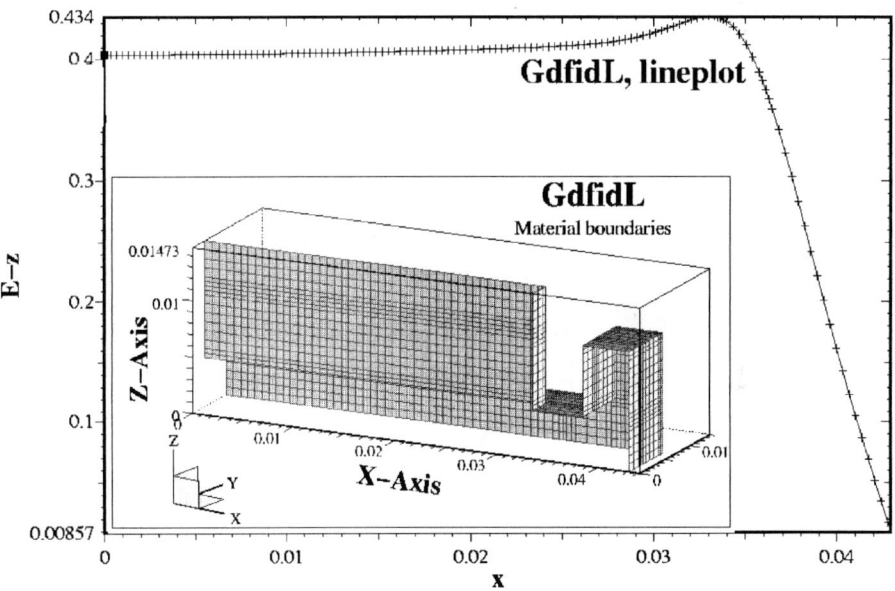

FIGURE 1. Longitudinal field profile across horizontal dimension for flat-field cavity of uniform depth (type C1). Inset shows one-eighth part of the cavity with beam pipe attached.

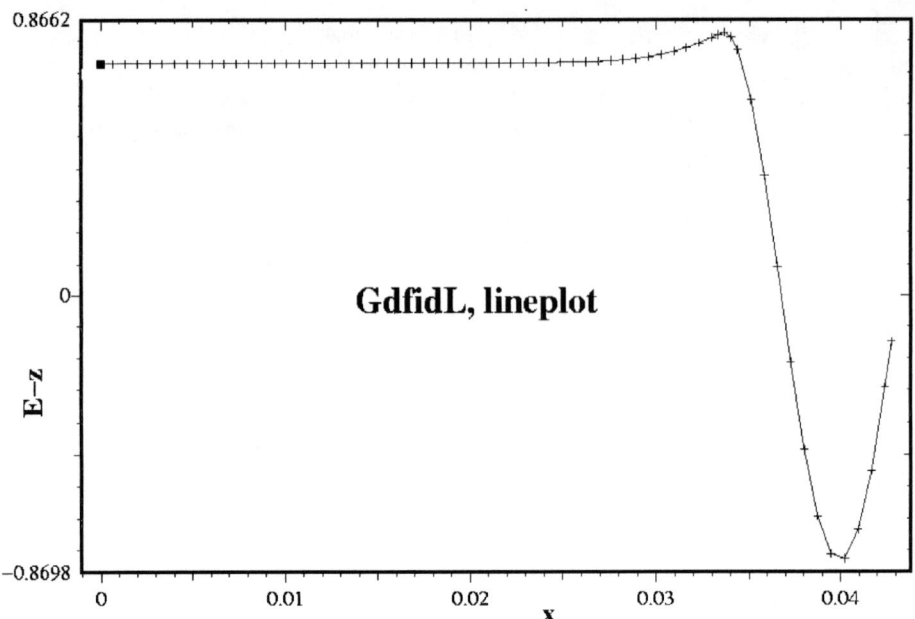

FIGURE 2. Longitudinal field profile for pillbox cavity with beam pipe and side space filled by dielectric (type C2).

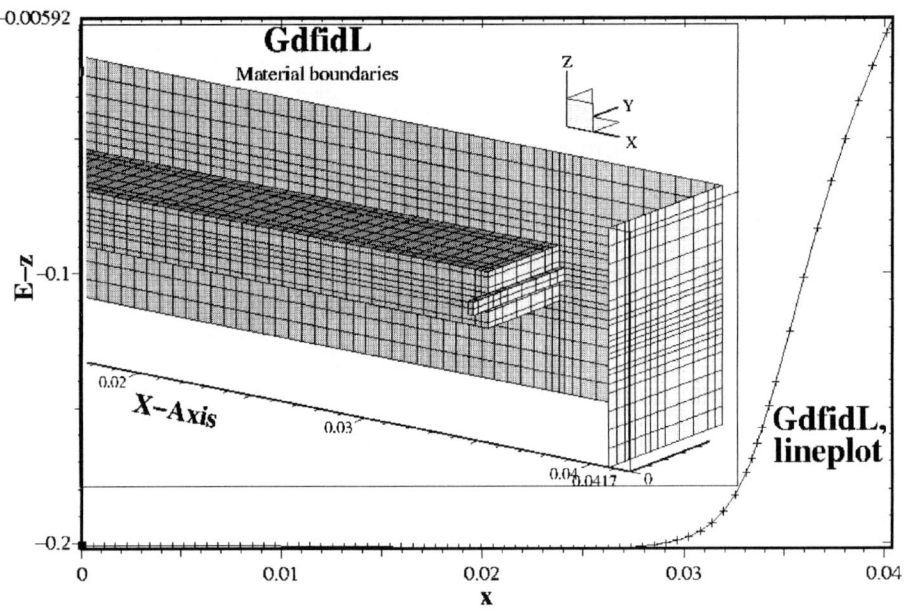

FIGURE 3. Field profile for muffin-tin $2\pi/3$ TW structure with side slot and short rod (type M1). The inset shows part of quarter $2\pi/3$-cell.

The next set of designs is developed for multi-cell resonance planar structures operating at speed of light phase velocity $\beta_{ph}=1$.

For muffin-tin structures with side vertical slots resulting in magnetic overcoupling and short compensating rod at the ends of irises (variants M1 in Table 2, see inset in Figure 3) one can provide good flatness by optimization of the rod height. However, presence of such an insertion element as a rod of intermediate height is difficult for single-step fabrication. One can facilitate the fabrication by making the rod have the same height as the minimum vertical gap: in this case it is easier to insert continuous rods throughout the structure. Since continuous rods overcompensate the magnetic coupling in previous design, we can recover equilibrium responsible for the flatness by shaping the ends of irises. Such a structure is shown in the inset of Figure 4 (M2 variants in Table 2). It has irises with T-ends and continuous rods across the structure. Field profile for the structure is optimized numerically (Figure 4). At low frequencies, the rods can be used as adjustment elements to correct flatness mechanically in real structures of the types M1 and M2.

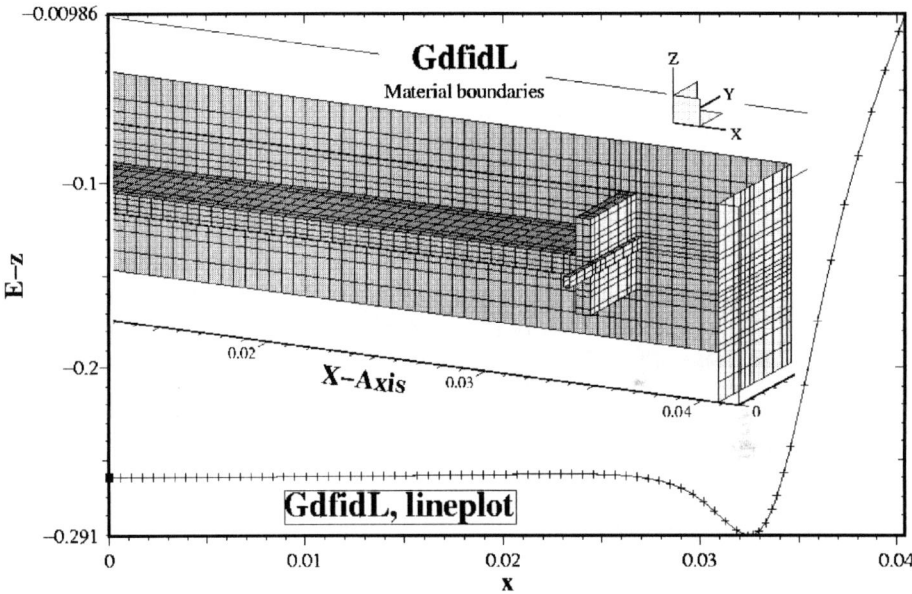

FIGURE 4. Field profile for muffin-tin $2\pi/3$ TW structure with T-end shaped irises and rods going throughout the structure across the gap (type M2). The inset shows part of quarter $2\pi/3$-cell.

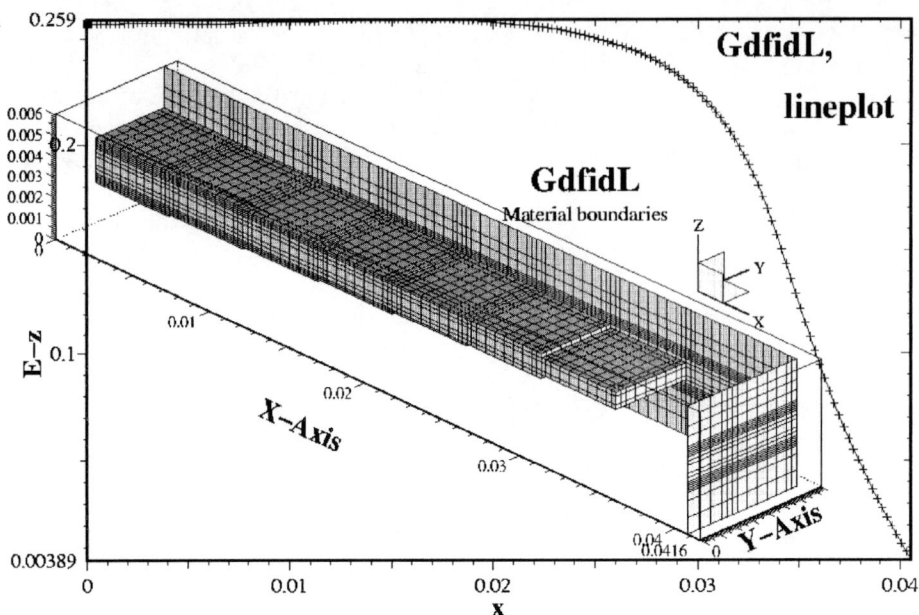

FIGURE 5. Field profile for muffin-tin $\pi/2$ TW structure with iris having profiled thickness (type M3). The inset shoes quarter part of flat-field muffin-tin $\pi/2$-cell.

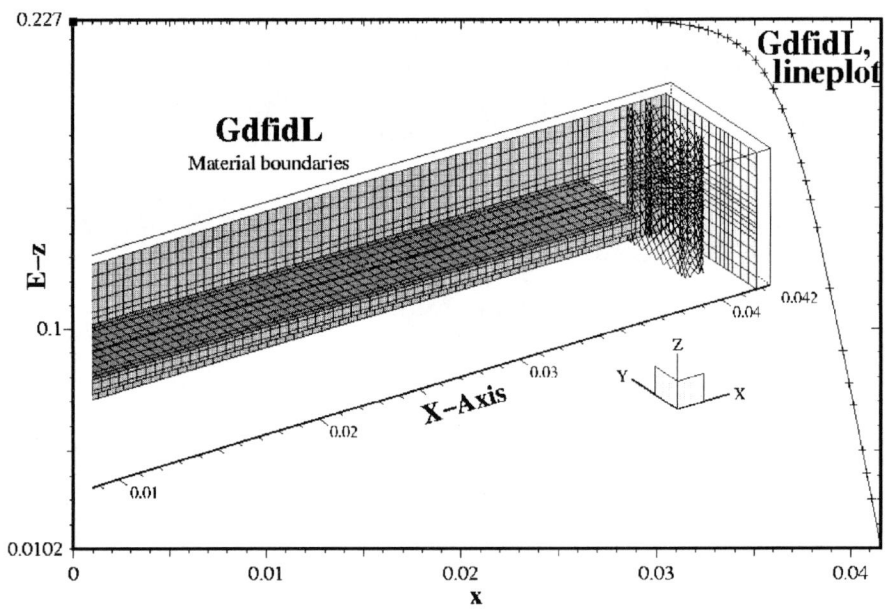

FIGURE 6. Field profile for muffin-tin $\pi/2$ TW structure with side dielectric slab (type M4). The inset shows part of quarter $\pi/2$-cell.

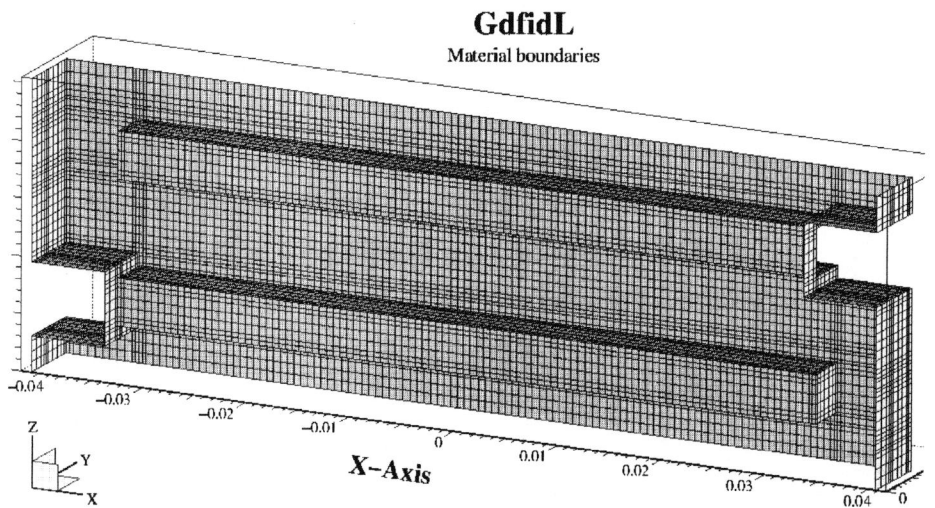

FIGURE 7. Half of two π-cells of flat-field asymmetric muffin-tin cavity (type M5).

At high frequencies the rods hardly can be fabricated with high enough accuracy within available technology. A more suitable design for microfabrication process could be the next type of structure M3 having profiled irises (with tapered thickness, see inset in Figure 5). The tapered iris design has an effective aspect ratio which is less than that for previous structures (see Figure 5, Table 2). The reason for noticeable reduction of computed Q-factor for this configuration (see Table 2) is the step-wise numerical approximation for the profiled iris.

A long dielectric slab inserted into side space parallel to side wall of a modified traveling wave muffin-tin structure can be rather effective. We optimized this configuration (type M4) shown in the inset of Figure 6 using a longitudinally continuous dielectric strip made from SiC material. One can see from Figure 6 high quality flatness in the region of interest and decreased field in location of dielectric material. The flatness can be adjusted effectively by variation of the horizontal distance separating side metal wall and dielectric.

So far we dealt with symmetrical structures with respect to vertical YOZ plane. Next we turn our attention to asymmetrical structure as illustrated in Figure 7 (type M5). For standard circular structures such an asymmetry was used earlier for classical biperiodic SW structures, but in our case it gives flatness provided by longitudinally interchanging magnetic and electrical side coupling. One can notice some small asymmetry of field profile (see Figure 8) that is not serious while taking into account compensation effect in terms of shunt impedance of two adjacent $\pi/2$ cells. Current design is characterized in the last row in Table 2 as having zero group velocity and π operating mode.

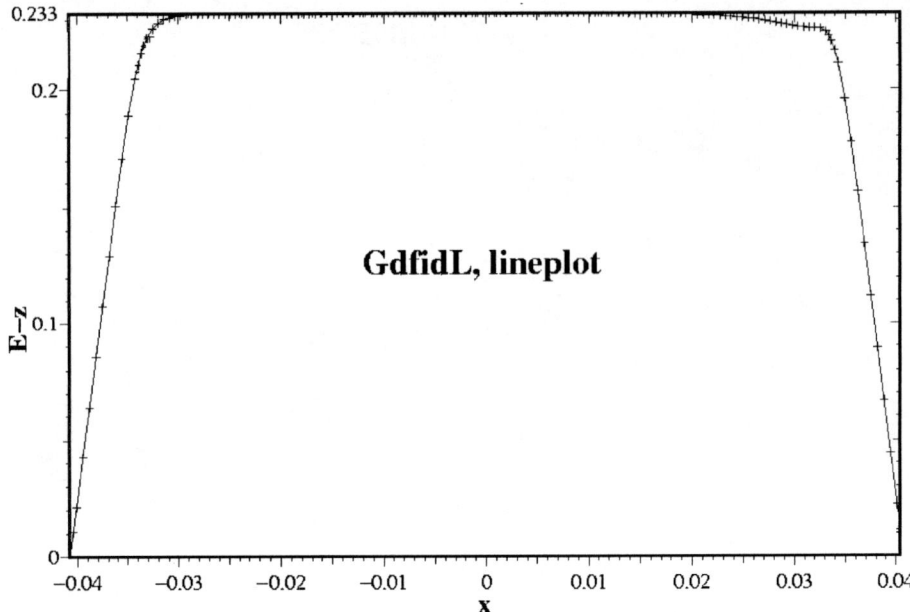

FIGURE 8. Field profile between irises for M5 type structure.

CONCLUSION

We considered a number of different designs of flat-field structures which are of potential interest. From Table 2 one can conclude that the most effective flat-field TW structure is that using rods: highest value of generalized impedance at a given aperture. RF breakdown and manufacturing problems remain to be solved. The second type of TW structures is muffin tin structure with side slot supplied by longitudinal side dielectric slab: it gives highest group velocity $\beta_{gr}=0.1$. However, Q-factor is ~20% less and dielectric material stability at high microwave power and microfabrication problems are not addressed yet.

It is important to note that since flat-field structures have additional degrees of freedom and are characterized by new flatness parameters (unlike circular structures) the practical design should be tolerant of manufacturing errors in terms of flatness. Computationally this stability can be expressed in terms of the number of iterations to achieve the flatness required and keeping phase synchronism ($\beta_{ph}=1$ in our case). In this way we noticed that the most stable design is M4 (with dielectric) and the most unstable are M5 and M3 types. For TW structures with uniform thickness (M1, M2, M5) the stability is defined basically by group velocity, whereas for structures with non-uniform iris thickness (M3) the flatness stability is extremely critical to deflections from ideal iris profile.

ACKNOWLEDGMENTS

This work is supported by DOE SBIR No. DE-FG03-96ER82213.

REFERENCES

1. P. Chen, J. Irwin, T.W. Markiewicz, T. Raubenheimer, R. Ruth, P. Wilson, D. Yeremian, in Proc. of *the 1996 DPF/DPB Summer Study on New Directions for High-Energy Physics*, Snowmass '96, p. 360.
2. D. Yu, J. Kim, P. Wilson, in *AIP Conf. Proc. 279, Advanced Accelerator Concepts Workshop*, Port Jefferson, Long Island, NY, June 14-20, 1992, p. 85.
3. D. Yu, P. Wilson, in *Proc. of the 1993 Particle Accelerator Conf.*, Washington D.C., May 1993, p. 2681.
4. D. Yu, S. Ben-Menahem, P. Wilson, R. Miller, R. Ruth, A. Nassiri, in *AIP Conference Proceedings, Advanced Accelerator Concepts*, Fontana, WI, 1994, Edited by P. Schoessow, (AIP 335), p. 800.
5. W. Bruns, in Proceedings of Particle Accelerator Conference (PAC`97), Vancouver, B.C., Canada 12-16 May 1997, p. 2651
6. R. Klatt, F. Krawczyk, U. Laustroer, E. Lawinsky, S. G. Wipf, T. Weiland, T. Barts, M. J. Browman, R. K. Cooper, G. Rodenz, H. K. Stokes, B. Steffen, *MAFIA User Guide*, MAFIA Collaboration, 1998.

34 GHz Pulsed Magnicon Project

O.A. Nezhevenko[1], M.A. LaPointe[1], S.V. Schelkunoff[1], V.P. Yakovlev[1],
J.L. Hirshfield[1,2], E.V. Kozyrev[2,3], G.I. Kuznetsov[3], B.Z. Persov[3]; A. Fix[4].

[1] *Omega-P, Inc., 345 Whitney Avenue, New Haven, CT 06511*
[2] *Department of Physics, Yale University, POB 8120, New Haven, CT 06520-8120*
[3] *Budker Institute of Nuclear Physics, Novosibirsk, 630090 Russia*
[4] *Institute of Applied Physics, Nizhny Novgorod, 603600 Russia*

Abstract. A high efficiency, high power magnicon amplifier at 34.272 GHz has been designed as a radiation source to drive a multi-TeV electron-positron linear collider. Simulations predict a peak output power of 45 MW in a 1.5 microsecond pulse with an efficiency of 45% and a gain of 55 dB. The amplifier is a frequency tripler, or third harmonic amplifier, in that the output frequency of 34.272 GHz is three times the input drive frequency of 11.424 GHz. Thus the rotating TM_{110} modes in the drive cavity, 3 gain cavities and 2 penultimate cavities are resonant near 11.424 GHz; and the rotating TM_{310} mode in the output cavity is resonant at 34.272 GHz. A 500 kV, 215 A high area compression electron gun will provide an electron beam with a diameter less then 1 mm. A superconducting solenoid magnet will provide a magnetic field of 13 kG in the deflection system and 23 kG in the output cavity. Simulation results for the operation of the entire magnicon amplifier (gun, magnetic system, RF system and collector) will be given, and the status described of critical hardware components.

1. INTRODUCTION

In order to achieve high gradients and consequently to keep a future multi-Tev linear collider length within reasonable bounds one has to operate in the millimeter wavelength domain. For example, a 1.0 TeV c.m. collider at 11.424 GHz is expected to have a loaded gradient of 64 MV/m, a length of 17.7 km, and a wall-plug power of 180 MW [1]. In contrast, a 34.272 GHz upgrade to NLC would have a loaded gradient of 189 MV/m, so that each 1.0 TeV of additional c.m. energy would require an extra length of about 6 km and extra wall-plug power of about 90 MW, once a 34.272 GHz amplifier with 45% efficiency becomes available. In the case of two-beam linear collider (e.g. CLIC [2]) it is also desirable to have a high power RF source in order to test accelerating structures and RF components, and to determine limits in breakdown and metal fatigue (see, e.g. [3]).

The magnicon is a RF source, based on a circular deflection of electron beam, whose main features are high power and high efficiency [4,5]. These properties make the tube especially attractive for accelerator applications. Furthermore, since RF cavities in the magnicon are significantly larger than in a klystron at the same operating frequency, magnicons can be designed for higher peak and average power.

The first magnicon developed at Budker INP was a fundamental harmonic amplifier at 915 MHz; it operated with an efficiency of 73% using a 300 kV, 12A electron beam

[6]. The measured output power was 2.6 MW and pulse width was 30 μsec. In experimental tests also at Budker INP [7], a second harmonic magnicon amplifier operating at 7 GHz achieved an output power of 55 MW in a 1.1 μsec pulse and a repetition rate of 3 Hz, with a gain of 72 dB and efficiency of 56%. This device is driven at 3.5 GHz and uses 430 kV, 230 A beam with an area compression ratio of about 2300:1 [8]. The demonstrated performance of the second harmonic magnicon operation strongly indicates that a third harmonic magnicon amplifier could be a viable high power source at high frequencies for a future linear colliders.

2. THIRD HARMONIC MAGNICON AMPLIFIER

In scaling magnicon amplifiers to higher frequencies (consequently, smaller physical dimensions), a few design problems arise at high power due to the limitations imposed by cathode loading, breakdown field and pulse heating of the cavity walls. The concept of a third harmonic magnicon amplifier is introduced to overcome these limitations [9].

In general, a magnicon (as a klystron) consists of four major components, namely: electron gun, magnet, RF system and beam collector. The Pierce electron gun injects a 500 kV, 215 A beam with a diameter of 0.8 mm into a chain of cavities forming the RF system. The deflection system consists of a drive cavity, three gain cavities and two "penultimate" cavities (working in "angle summing" mode [10]). The external magnetic field provides both beam focussing and interaction between the electrons and the RF fields in the cavities. The electron beam is radially deflected by the RF magnetic fields of a rotating TM_{110} mode in the deflection system. The scanning beam rotates at the frequency of the drive signal (11.424 GHz), then enters the output cavity and emits radiation at three times the drive frequency (34.272 GHz) by interacting with the TM_{310} mode

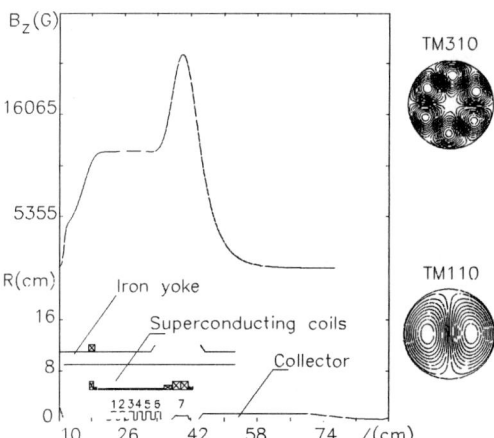

FIGURE 1. Required axial magnetic field profile (top); superconducting coil and iron yoke layout (bottom). Cavity chain and collector are also shown. Inserts at the right show RF field patterns for cavities #1-6 of deflecting system (TM_{110} mode at 11.424 GHz), and for the output cavity (TM_{310} mode at 34.272 GHz).

Fig. 1 shows the required magnetic field profile (top) and the superconducting coil configuration and iron yoke geometry to achieve this profile (bottom). For effective deflection, the magnetic field in the deflection system should be such that $\Omega/\omega \sim 1.5$, where Ω is the cyclotron frequency and ω is the drive frequency. In the output cavity, however, for efficient extraction of energy, the magnetic field should be chosen such that $\Omega/3\omega \sim 0.9$ [9].

Fig. 2 show results of steady-state computations for the axial evolution of radial orbit displacement and particle energy. The beam particles are seen to lose a substantial part of their energy in the output cavity. Beam dynamics for the finite

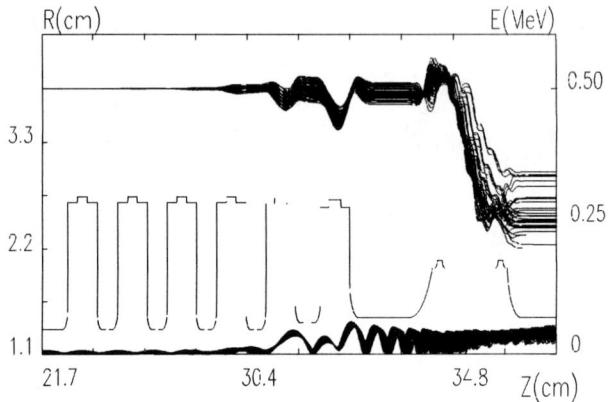

FIGURE 2. Computed steady-state evolution with z of radial orbit displacement (bottom) and particle energies (top). Also shown are the cavity outlines.

thickness have been optimized using realistic magnetic fields and realistic cavity geometries [9]. The resulting design parameters of this amplifier are given in Table 1.

TABLE 1. 34.3 GHz magnicon prameters.	
Operating frequency, GHz	34.272
Power, MW	44-48
Pulse duration, µs	1.5
Repetition rate, Hz	10
Efficiency, %	41-45
Drive frequency, GHz	11.424
Drive power, W	150
Gain, dB	54
Beam voltage, kV	500
Beam current, A	215
Beam diameter, mm	0.8-1.0
Magnetic field, deflecting cavities, kG	13.0
Magnetic field, output cavity, kG	22.5

3. THE TUBE DESIGN

The complete design of 34.3 GHz magnicon is presented in Fig. 3.

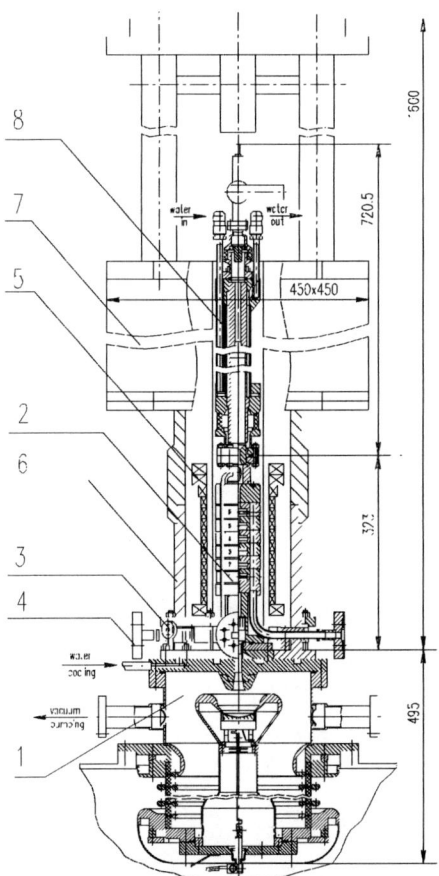

FIGURE 3. 34.3 GHz magnicon amplifier tube: 1-electron gun, 2-RF system, 3-output waveguide (WR28), 4-WR90 waveguide, 5-superconducting coils, 6-iron yoke, 7- cryostat, 8-beam collector. The dimensions are in mm.

1. Electron gun.

The gun design [9,11] calls for a cathode current density of 12 A/cm^2, and a maximum surface electric field strength of 238 kV/cm on the focus electrode. Beam compression in this gun is only partially electrostatic (500:1). Higher electrostatic compression would lead to a higher electric field at the focus electrode, and would require a magnetic field of about 13 kG at the edge of the pole piece, leading to undesirable saturation in the iron [11]. Thus a magnetic compression of about 2:1 occurs as the beam passes through the hole in the pole piece into a ~5 kG field, and a further factor of 3:1 occurs adiabatically as the magnetic field gradually rises up to 13 kG. The resulting compression ratio of 3000:1 is comparable to the 2300:1 compression ratio for the 7 GHz magnicon [7,8]. It is found in this design that 95% of the current is within a diameter of 0.8-mm [11]. In order to decrease the beam halo the width of thermal gap is made to be very small. Note, that for a small thermal gap it is not necessary to use negative bias of focus electrode to prevent emission from the cylindrical part of the cathode edge, as was done for the gun for 11.4 GHz magnicon [12]

At present, the gun is delivered. Before delivering it was tested up to 100 kV. In Fig. 4 measured and calculated dependences of the beam current versus voltage are shown. The figure shows that the gun current and, thus, perveance is very close to design value.

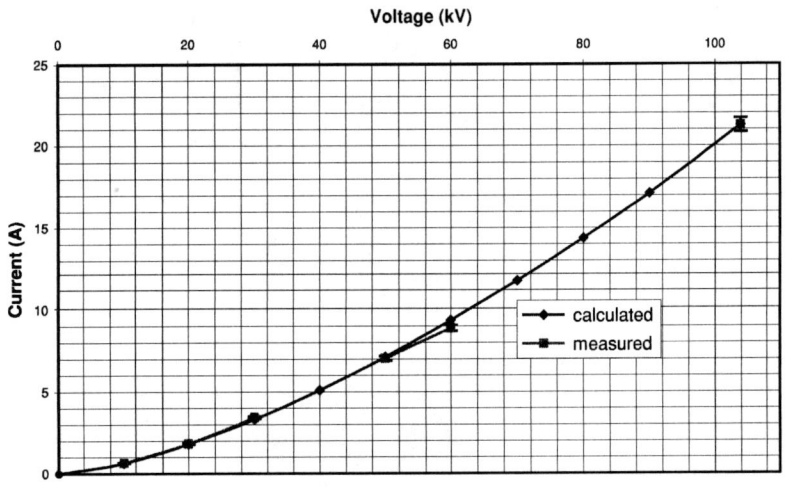

FIGURE 4. The gun current versus voltage.

Modulator and pulse transformer have been tested up to 500 kV, and now the gun is in preparation for full voltage tests, which are scheduled for September, 2000.

2. RF system.

The RF system (see Fig. 3) consists of seven cavities: one drive (#1) and three gain cavities (#2-4), two "penultimate" cavities (#5-6), and one output cavity (#7). The shapes and dimensions of the cavities are chosen to avoid monotron self-excitation of axisymmetric modes, and harmonic frequency modes [9]. All cavities of the deflection system are about 1.25 cm long and their diameters are about 3.0 cm. In Fig. 5 there is a drawing of the RF system.

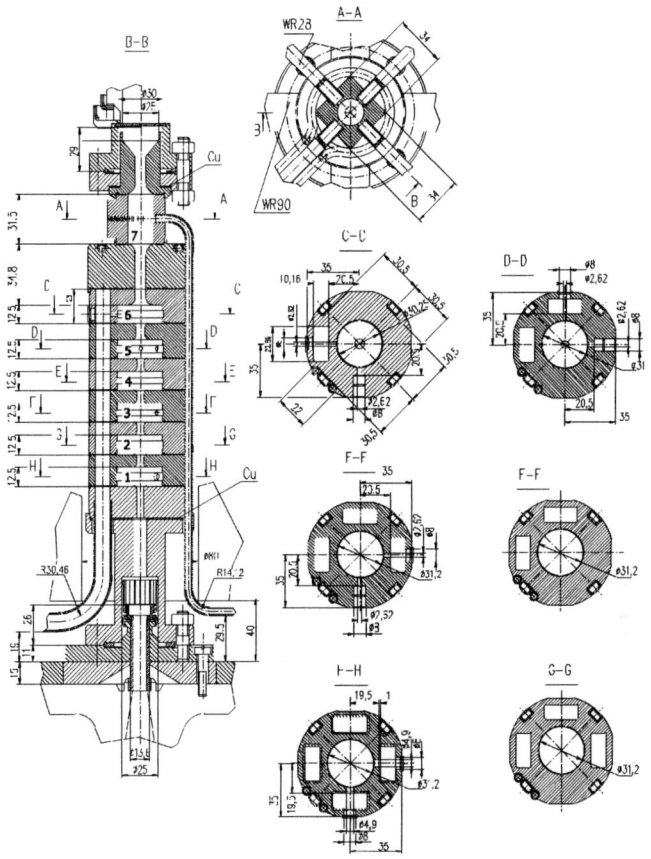

FIGURE 5. RF system of 34.3 GHz magnicon amplifier. The dimensions are in mm.

The two "penultimate" cavities are used in order to reduce the RF field in deflecting system. For double-gap penultimate cavity [4], it was found that the 0-dipole mode (as opposed to the desired π-dipole mode) may self-excite for large deflection angles of the electrons. This effect would severely limit the pulse width to 0.4 μsec or less. The solution to the problem is to use two decoupled "penultimate" cavities working in the "angle summing" mode [10]. Proper gap spacing and tuning of these cavities has been shown to deflect the beam effectively. This approach was proven experimentally in the deflecting system of X-band magnicon [13,14].

There are four WR90 waveguides built in the body of deflecting system (see Figs.3 and 5). One of them is for drive cavity, and the rest are for measurements in the cavities #3, 5 and 6. These waveguides will also be used for pumping.

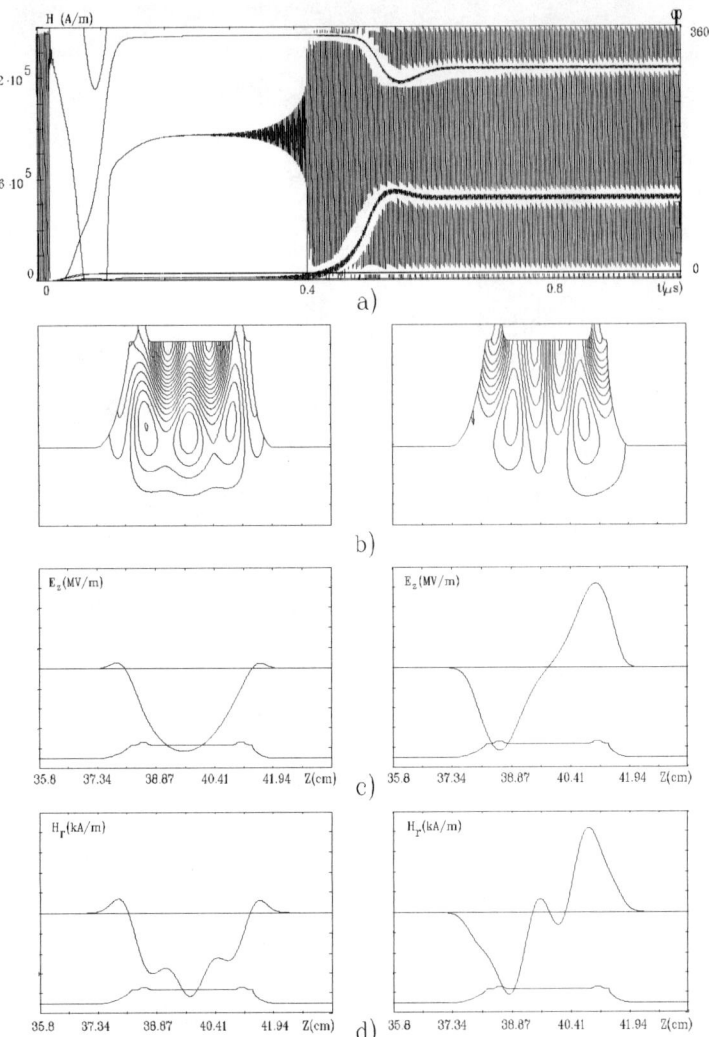

FIGURE 6. An example of the parasitic mode self-excitation in the magnicon cavity (a): transient process for the operational mode (top) and the parasitic mode (bottom). Phase slippage of the parasitic mode is also shown. Field map (b), electric field (c), and magnetic field (d) along the cavity for operational (left) and parasitic (right) modes

It was found that in the output cavity there is a parasitic TE_{313} mode, which has resonance frequency close to frequency of the operating mode TM_{310}. When the magnicon operates with resonance load (i.e., for accelerating structure), the loaded quality factor of this mode may be high enough, and this mode may self-excite. In the Fig. 6a there is an example of the parasitic mode self-excitation. In the Figs 6b-6d field pattern and longitudinal field distribution for both modes are shown. Finally, the length of the output cavity (3.15 cm) and its shape were optimized to achieve not only maximum efficiency and acceptable surface electric fields, but also absence of parasitic oscillations mentioned above [9]. The diameter of the output cavity is about 1.75 cm. Power will be extracted by a set of four WR28 waveguides with an azimuthal separation $\Delta\theta = \pi/2$ that couple to both field polarizations [9]. One of them is shown in Fig. 3. The RF system is made as a brazed monoblock (see Fig. 5) that allows baking up to 400° C.

At present, cold tests are completed, and manufacturing has begun. The RF system is expected to be ready to the end of 2000.

1. Magnet

A superconducting solenoid (see Fig.3) provides a magnetic field of 13 kG in the deflection system and 23 kG in the output cavity. The magnet coils consist of three independently driven sections for adjustment of the magnetic field profile. Each section has a persistent switch. The coils are placed in a liquid helium cryostat with a vertical room temperature bore of 80 mm in diameter. The cryostat holds about 25 l of liquid helium. In order to decrease the liquid helium evaporation rate, radiation shields in the cryostat are cooled down to 30° K by a cryo refrigerator. The storage time of liquid helium is about 30 days.

The magnet is scheduled for delivery in August 2000.

2. Collector

The collector design is shown in Fig. 7. To prevent the collector damage caused by possible beam ion focusing during the tube conditioning (or operation without drive signal), there is a 100mm long, 10mm diameter pipe placed after the collector taper. In Fig. 8a,b electron trajectories in the beam collector for deflected and undeflected beam are shown. One can see that there is no beam current in the end of the pipe in either case. But if ion focusing takes place, beam electrons may reach the end of the pipe and will be intercepted by a molybdenum target placed in the end of the pipe (see Fig.7). This target is insulated from the collector. It gives the ability to detect electron current on the target. The pumping port placed after the molybdenum cylinder, provides a good vacuum in the collector, decreasing the probability of the ion focusing as well.

The beam collector is ready for installation.

4. CONCLUSION

It is expected that the most critical parts will be ready to the end of 2000. Tube tests are scheduled to begin in early 2001.

This research is supported by US DoE.

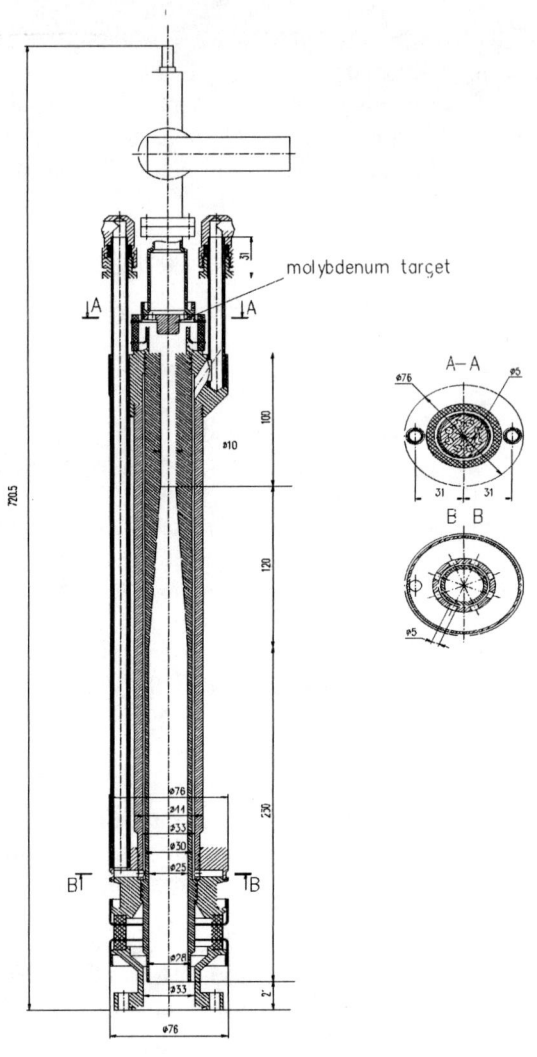

FIGURE 7. The beam collector. The dimensions are in mm.

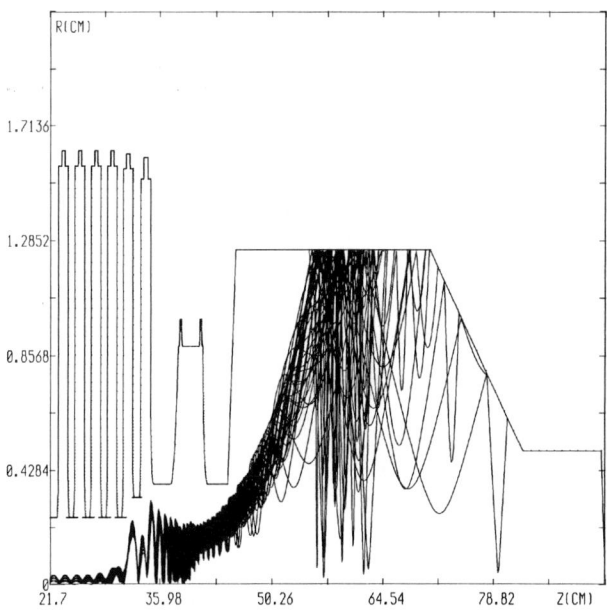

FIGURE 8a. Particle trajectories in the beam collector for full power operating condition.

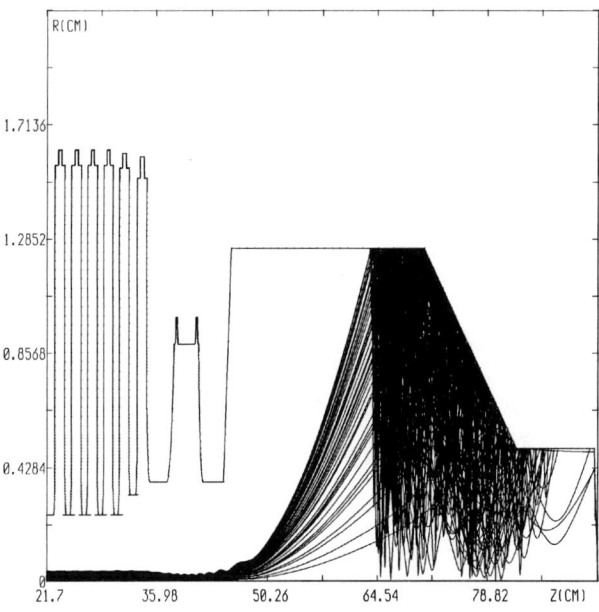

FIGURE 8b. Particle trajectories in the beam collector in the absence of drive power.

REFERENCES

1. P.B. Wilson, SLAC-PUB-7449, April 1997.
2. "Proposal for Future CLIC Studies and a New CLIC Test Facility (CTF3)", CLIC Study Team, *CERN/PS 99-047 (LP), CLIC Note 402,* Geneva, Switzerland, July 1999.
3. O.A. Nezhevenko, Proceedings of 1997 Particle Accelerator Conference, Vancouver, 1997,(IEEE, Piscataway, NJ, 1998), p. 3013.
4. O.A. Nezhevenko, *IEEE Trans. Plasma Sc.*, vol.22 pp. 765-772, October 1994.
5. O.A. Nezhevenko, *Physics of Plasmas*, vol.7 pp. 2224-2231, May 2000.
6. M.M. Karliner, E.V. Kozyrev, I.G. Makarov, O.A. Nezhevenko, G.N. Ostreiko, B.Z. Persov, and G. V. Serdobinstev, Instrum. Methods Phys. Res. A 269, 459 (1988).
7. E.V. Kozyrev, O.A. Nezhevenko, A.A. Nikiforov, G.N. Ostreiko, S.V. Shchelkunoff, G.V. Serdobinstev, V.V. Tarnetsky, V.P. Yakovlev, and I.A. Zapryagaev, *AIP Conference Proc. 474*, p. 187.
8. Yu.V. Baryshev, I.V. Kazarezov, E.B. Kozyrev, G.I. Kuznetsov, I.G. Makarov, O.A. Nezhevenko, B.Z. Persov, M.A. Tiunov, V.P. Yakovlev, and I.A. Zapryagaev, Nuclear Inst. and Methods in Phys. Research A 340 (1994), pp. 241-258.
9. O.A. Nezhevenko, V.P. Yakovlev, A.K. Ganguly, and J.L. Hirshfield, *AIP Conference Proc. 474*, p. 195.
10. O.A. Nezhevenko, V.P. Yakovlev, *IEEE Trans. of Plasma Sci.*, 8th Special Issue on High Power Microwave Generation, to be published.
11. V.P. Yakovlev, O.A. Nezhevenko, *AIP Conference Proc. 474*, p. 316.
12. V.P. Yakovlev, O.A. Nezhevenko, R.B. True, PAC97, Vancouver, 1997, p. 3153.
13. S.H. Gold, A.W. Fliflet, A.K. Kinkead, B. Hafizi, O.A.Nezhevenko, V.P. Yakovlev, J.L. Hirshfield, R. True, *Phys. Plasmas* 4 (5), May 1997, pp. 1900-1906.
14. O.Nezhevenko, V.Yakovlev, J.Hirshfield, E.Kozyrev, S.Gold, A.Fliflet, A.Kinkead, "X-Band Pulsed Magnicon for Next Linear Collider", Bull. Am. Phys. Soc. 44, No 7, p. 252 (1999)

3-D Space Charge Simulations of High-Power Magnicon Amplifiers

V.P. Yakovlev, O. V. Danilov, B. Hafizi, and O. A. Nezhevenko

Omega-P, Inc., 345 Whitney Avenue, New Haven, CT 06511

Abstract. One of the most attractive candidates for the RF source for future particle accelerator applications is the magnicon - a microwave amplifier with circular deflection of an electron beam. The magnicon has demonstrated both high power and high efficiency in decimeter and centimeter wavelength domains. The models and codes developed for magnicon simulations include realistic RF field distributions and DC magnetic field profile, as well as finite beam size. The results of these codes can simulate performance of the entire tube, and are in excellent agreement with the experimental results for a moderate beam perveance and power level up to tens of MW. However, taking into account space charge effects could be extremely important when one proceeds to still higher power and perveance. This paper describes a physical model and a 3-D code developed to simulate electron beam dynamics in magnicons, including space charge effects. Preliminary simulations that show the influence of beam space charge on the performance of a high-power magnicon are presented.

INTRODUCTION

Continued progress in high energy and nuclear physics demands a new generation of particle accelerators, which in turn require new high-power, high efficiency RF sources. One attractive candidate for this purpose is the magnicon – a microwave amplifier with circular deflection of an electron beam [1]. Magnicons have demonstrated high power at high efficiency, good phase stability and are relatively insensitive to variations in the load [1,2]. Magnicons provide the best parameters in the decimeter, centimeter, and millimeter wavelength domains, which are used in accelerators [2-5]. These characteristics render the magnicon not only attractive for collider applications, but also for a variety of other accelerator applications. Experimental investigations have confirmed the unique properties of magnicon. The first magnicon experiment at the Budker Institute of Nuclear Physics (INP) yielded an efficiency of 73% at 915 MHz with output power of 2.6 MW [2]. A second harmonic pulsed magnicon amplifier at 7 GHz has demonstrated 56% efficiency with 55 MW output power and pulse duration of 1 μs [3]. These encouraging results prompted Omega-P, in close collaboration with the Naval Research Laboratory (NRL), to undertake design, construction and evaluation of an 11.424 GHz, 60 MW pulsed magnicon [4]. Presently this magnicon is being tested. Additionally, Omega-P is building a third harmonic magnicon at 34 GHz [5].

Modern colliders require simultaneous operation of hundreds or thousands of high power, high efficiency microwave amplifiers in the decimeter and centimeter wave

CP569, *Advanced Accelerator Concepts: Ninth Workshop,* edited by P. L. Colestock and S. Kelley
© 2001 American Institute of Physics 0-7354-0005-9/01/$18.00

range. Needless to say, the vast capital outlays and operating expenses associated with the RF system can be prohibitive. There is, therefore, a tremendous incentive for reducing the cost of the RF system. A factor-of-two increase in the output power of the RF tube will cut down the number of RF sources by a similar number--clearly a huge saving in costs. For example, for the NLC project the power output of each 11.4 GHz RF source is 75 MW [6]. Thus, for NLC applications a 150 MW magnicon can replace two klystrons, reducing the number of tubes from about 6600 to about 3300. Another example of possible application of the magnicon is in the long pulse, high power, decimeter wavelength domain. An example is the two-beam accelerator project CLIC [7] in which RF tubes feed a drive beam that in turn excites the main linac. Use of a magnicons having 60% efficiency instead of proposed conventional klystrons having 45-47 % efficiency will reduce the wall plug power by 80-100 MW. Recent developments in multi-TeV linear colliders show that it is necessary to increase the accelerating gradient and thus to employ sources in the millimeter wavelength domain. This will require further optimization of millimeter wave range magnicons (e.g., a 34 GHz magnicon [5]) in order to increase their power and efficiency.

As the power and frequency are raised the influence of space charge effects on the beam dynamics in magnicon becomes significant. At higher power and frequency it is necessary to develop new methods and computer codes for simulations of the beam dynamics in magnicons allowing for interaction with both RF and space charge fields.

The magnicon is a conceptually new RF source (see Figure 1, which displays a schematic of an 11.4 GHz frequency-doubling magnicon amplifier [4]) and has a number of novel characteristics that require special treatment in numerical simulations.

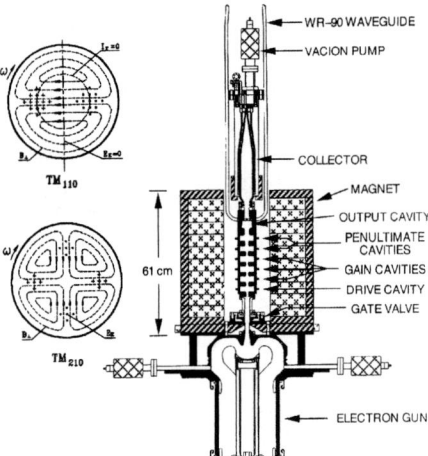

FIGURE 1. Schematic of 11.4 GHz magnicon amplifier. Inserts at the left show RF field pattern for deflecting cavities (TM_{110} mode) and for the output cavity (TM_{210} mode).

In magnicon, rotating TM_{110} modes are used to spin up a pencil electron beam to high transverse momentum in a series of deflection cavities, the first externally driven, and a synchronously rotating TM_{n10} mode of output cavity is used to extract the transverse momentum as microwave power at the drive frequency or one of its harmonics. The electron dynamics in the magnicon are essentially three dimensional in nature (see Figure 2). Further, for high efficiency operation it is essential to employ a small beam radius r_b compared to the operating wavelength λ: $r_b/\lambda = 0.01\text{-}0.03$ [1-5]. As is well known, in the presence of an axial magnetic field the minimum electron beam

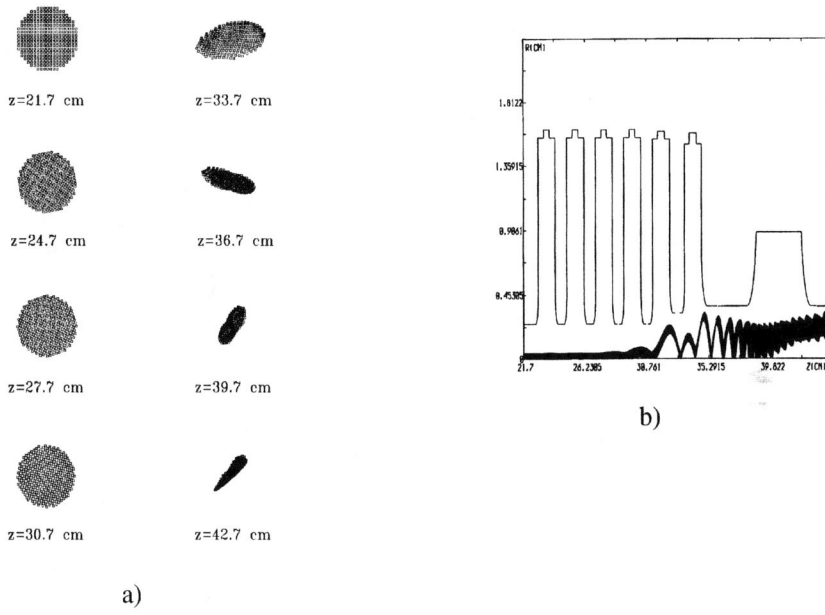

FIGURE 2. a) Charge distributions in the beam transverse cross-section for different longitudinal coordinates for 34GHz magnicon [5]; b) Beam trajectories in $(r\text{-}z)$ plane for 34 GHz magnicon.

size corresponds to the Brillouin radius. Techniques for the generation and transport of electron beams near the Brillouin limit have been developed and demonstrated in Ref. 8. For parameters of current interest the Brillouin radius is indeed small compared to the RF wavelength. For a near-Brillouin beam space charge forces may have a significant impact on electron dynamics and hence on tube operation. Moreover, it has been amply demonstrated--both numerically and experimentally— that the fringing fields associated with beam tunnels and cavity openings and roundings affect the beam dynamics and the efficiency [9]. Thus it is imperative to include accurate realistic modeling of the cavities, beam tunnels and fringing fields in simulation codes. In addition, it is necessary to have the means to investigate transient processes and non-linear regimes of magnicon operation.

At the present time a number of codes are available for designing magnicons and modeling their operation [10-14]. These codes permit the evaluation of the magnicon as a whole and include realistic RF fields (and, thus, effects of fringing fields), the realistic static magnetic field and finite electron beam size. These codes are invaluable in understanding the experiments and refining the designs. The existing codes model the tube by advancing slices of electron beams through the cavities in the presence of the axial magnetic field, and self-consistently evolving the RF fields. The time-dependent codes model transient processes and parasitic modes that compete and affect the efficiency of the desired mode. The results of these codes can simulate performance of the entire tube, and are in excellent agreement with the experimental results for a moderate beam perveance and power level up to tens of MW [2,3]. However, magnicon codes do not include space charge for deflected beam simulations.

Three-dimensional space charge simulation of RF sources (and accelerators) is highly advanced both in terms of physical models and methodology [15-22]. Based on this, many sophisticated and user-friendly codes are available for investigation and design purposes. However, due to the peculiar nature of the interaction and geometry, as noted above, these codes cannot be utilized for the magnicon amplifier. For the codes based on direct integration of the Maxwell equations combined with a particle-in-cell (PIC) method for electron dynamics, a large number of mesh points are required in order to adequately model the region around a pencil beam. Also integration of the field and electron variables over many RF cycles is necessary, because the duration of the transient processes is much longer than the oscillation period. In addition, the narrowness of the resonance, which is typical of RF amplifiers for accelerator applications, implies that for adequate description of the resonant component of the field high precision must be maintained throughout the integration.

Note that 2D models, where the charge distribution in the beam cross section is used to estimate transverse space charge forces, cannot be utilized because the charge distribution in a magnicon is neither axisymmetric nor elliptic and hence simple analytical formulae for the transverse field are not valid. In addition, self-fields depend on boundary conditions. Changing the beam cross section not only modifies the beam transverse dynamics but also longitudinal dynamics. Due to varying shielding radii, the average beam energy in the beam tunnels differs from that in the cavities. Fine-scale beam characteristics that affect the tube efficiency, such as pitch angle, phase angle and energy spreads, depend on the extent of shielding. For high beam perveance these effects may be important.

Since none of the available codes is adequate for simulations of the magnicon including space charge effects, new fully 3D space charge models and codes should be developed.

GENERAL CODE DESCRIPTION

The basic idea underlying the physical model is that the fields can be separated into two parts. The first part is resonant (i.e., oscillates primarily near a multiple of the operating frequency $n\omega_b$, where n is a natural number) with a slowly varying envelope. The remainder is nonresonant and represents the self-fields. In a magnicon the transverse dimension of the electron beam is small compared with the wavelength and thus retardation effects may be neglected near the source, i.e., the electron beam.

That is, the instantaneous location of the electrons determines the self-fields inside the beam. At every time step the self electric field is simply given by the solution of the Poisson equation for which the source term (charge density) is given by the distribution of the electrons at that instant. Similarly, the self-magnetic field is determined by the instantaneous current distribution and its evaluation can also be simplified. It should be recalled that the self-magnetic transverse field force for a relativistic beam tends to cancel the self electric field force to within a factor $1/\gamma^2$, where γ is the relativistic factor. This model is exact in the absence of electron beam rotation. However, in a magnicon the beam rotates as a whole around the cavity axis. Due to this rotation additional, azimuthal eddy current electric fields are induced. In practice these fields are small since the rotation velocity of the electron beam is small compared to the speed of light.

Due to the stroboscopic nature of the electron beam motion in the rotating RF fields the dynamics of any single beam slice are to similar to any other since electrons undergo identical motion but for a displacement in phase. Thus the instantaneous distribution of charge can be calculated from the motion of a single slice. Further, a two-dimensional subdivision of the electron beam can be usefully employed, in lieu of a three dimensional one, allowing considerable savings in the required computer memory and time. Knowing the charge distribution the self-field can be determined and the total field is given by adding this to the RF field. In given RF fields the self-consistent beam dynamics may be obtained by iteration. Transient processes and steady state RF field amplitudes may be determined in the slow time scale approximation.

Numerical realization of the physical model described above proceeds as follows. The Poisson equation for the electrostatic field is solved by a finite element method (FEM) [23]. The FEM method is based on a Galerkin [24] expansion of the potential in terms of basis functions, and makes use of rectangular prisms as the basic element, from the 'serendipity' family. Quadratic shape functions are employed, with 20 nodes, and an isoparametric transformation of the global coordinates to local coordinates is utilized. Quadratic elements give more accurate description of both the problem geometry and the solution. This transforms the differential equation into a matrix equation that is inverted using direct methods associated with LU forms [25], where L and U refer to lower and upper triangular, respectively. Note, that in the present case the matrix is symmetric and the matrix L can be used to obtain the solution. Since the geometry of the magnicon is invariant, so is the matrix form and the inversion need be made once and for all. The potential distribution at each mesh point can then be simply constructed for each stage of the space charge iteration process. If the number of mesh nodes is denoted by N it is simple to show that the ratio of the time required to prepare the LU form to the time required to solve the equation using the given LU form scales as $N^{5/6}$. It should be noted that there are a number of general-purpose library routines available for the solution of the Poisson equation. However, these routines are not optimized for the case of interest here. It should also be remarked that to simplify the process of electron beam trajectory calculation a rectangular mesh is employed inside the beam region.

A magnicon amplifier consists of a series of deflection cavities and beam tunnels, and an output cavity. A key point of the solution method is to break up the entire device into a number of independent, subdivisions. This leads to drastic savings in computation time and turns the code into a versatile, practical tool. To properly subdivide the amplifier it is necessary to introduce imaginary boundaries separating the subdivisions. These boundaries are inserted near the middle of the beam tunnels and the appropriate boundary condition on the fields is of the Neumann type (i.e., prescribed by the derivative normal to the boundary). Introduction of the imaginary boundaries will perturb the beam dynamics insignificantly for the following reasons.

i) The self-electric field is predominantly transverse in the beam tunnels due to the absence of bunching.

ii) The perturbed contribution to the small longitudinal component of the static self-electric field resulting from the introduction of the imaginary boundaries is evanescent due to the shielding inside the beam tunnels.

Thus the problem breaks up into a set of independent subdivisions wherein the Poisson equation is to be solved.

The code is organized as follows. A slice of particles is advanced in the fixed field of the first subdivision. After the slice reaches the end of the subdivision the charge density in the subdivision is determined. This density is used on the right hand side of Poisson's equation for calculation of the self-field. Since the particle trajectories are independent in the fixed fields, the code is organized such that the trajectories are determined in parallel on a multiprocessor computer. After the slice reaches the end of the first subdivision, the Poisson equation is solved, in multi-tasking parallel fashion, for the first cavity, while at the same time a slice is advanced in the second subdivision. This process can be continued until the slice reaches the end of the output cavity or the beam collector. Moreover, the determination of the potential from the Poisson equation can be readily vectorized. The reason for this is that the matrix to be inverted has a band structure with regular entries and the bandwidth is uniform. Clearly the expedient of dividing the amplifier into independent subdivisions, along with the vectorizability of the matrix, leads to many-fold savings in CPU time on the multiprocessor computer. Realistic RF fields and DC magnetic fields are taken from the RF and magnetostatic codes, as in existing magnicon codes.

Stability of the steady state solution is ascertained by making use of Lyapunov's method, writing the slow-time-scale equations for the RF amplitude u_j and phase φ_j in the j-th subdivision as follows

$$\frac{du_j}{dt} = -\frac{u_j}{\tau_j}\left(1 - \frac{\Re P_j}{P_{jd}}\right),$$

$$\frac{d\varphi_j}{dt} = \delta\omega_j + \frac{\Im P_j}{W_j},$$

where P_j is the complex beam power loss in the cavity, P_{jd} is the power dissipated in the walls or in the load, τ_j is the time constant, $\delta\omega_j$ is the difference between the cavity resonance frequency and the operating frequency and W_j is the stored energy. After linearization near a stationary (fixed) point, derivatives of the right hand sides of these equations are numerically determined by solving the steady state problem with perturbed amplitude and phase. The eigenvalues of the linearized system determine the instability growth rates and boundary. This method is tested in the existing magnicon code [10].

Transient processes are analyzed by using the slow-time-scale approximation, taking into account excitation of the drive cavity by an external RF source [10]. Time integration is combined with space charge iterations to determine the correct steady state self-consistent solution—if any--during transient processes, or to identify parasitic mode self-excitation.

SIMULATION RESULTS

Based on the model, outlined above, the code prototype for space charge simulation was written. In Fig. 3 the process of space-charge iteration to convergence is shown for 3D simulation of a Brillouin beam propagating in a cavity with adjoining beam tunnels. The 3D finite-element mesh is shown schematically in the Fig. 4. The beam and cavity parameters of the 11.4 GHz magnicon have been employed [4]. The relative amplitude of scalloping is plotted versus iteration number. The mesh consists of 35×35×30 elements. Observe that for the self-consistent solution (e.g., at the 11 th iteration) a relatively small beam scalloping persists because of the focusing properties of the beam tunnel openings in the presence of space charge: the shielding radius changes when the beam electrons enter or leave the cavity (see also Fig. 5c). Note that for this case the ratio of the Brillouin radius (0.65 mm) to the cavity radius (32 mm) is about 0.02 [4].

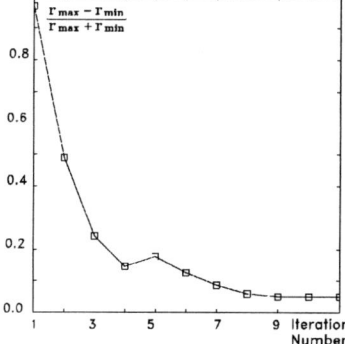

FIGURE 3. Iteration process for 3D simulation of a Brillouin beam. Ratio of the beam scalloping radius to the average beam radius versus iteration number.

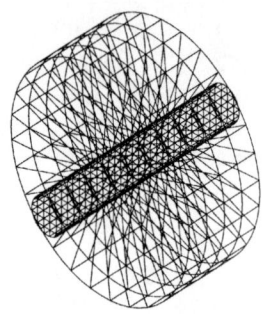

FIGURE 4. 3D mesh for deflecting cavity

In Fig. 5 the trajectories in the (r-z) plane for the beam propagating in a cavity with beam tunnels (considered above) are shown for different mesh node numbers.

The scalloping amplitude for the same beam propagating in a uniform but otherwise identical tunnel is computed to be about 1.5%, using the same mesh in the area of the beam propagation. This very small scalloping amplitude demonstrates the high precision of the potential and electric field calculations— since for a Brillouin beam the forces due to self electric field, external static magnetic field and self magnetic field exactly balance centrifugal forces.

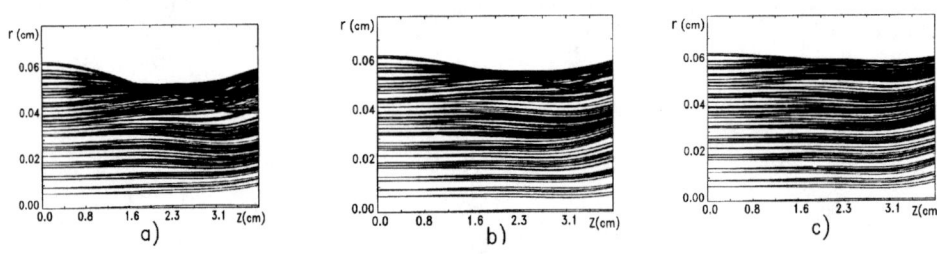

FIGURE 5. Beam trajectories in (r-z) plane for three meshes: a) 25×25×x30; b) 30×30×30; c) 35×35×30

In Fig. 6 the result of a 3D simulation of a deflected beam in 11.4 GHz magnicon [4] is displayed. The trajectory of every 10th macro-particle is shown. Figure 7 shows the computed equipotential contours obtained from the passage of a beam through a cavity structure in Fig. 6. The beam enters the cavity symmetrically through the input beam tunnel and the equipotential contours are observed to be correspondingly symmetrical, as indicated in Fig. 7 a). On passing through the cavity the beam is deflected (see Fig. 6) and thus the equipotential contours in the output beam tunnel are concentrated towards the boundary.

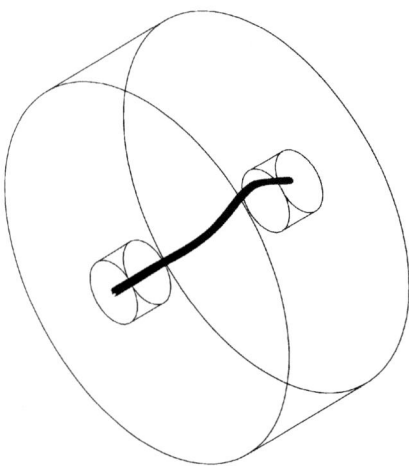

FIGURE 6. Representative trajectories from 3D simulations of deflected beam.

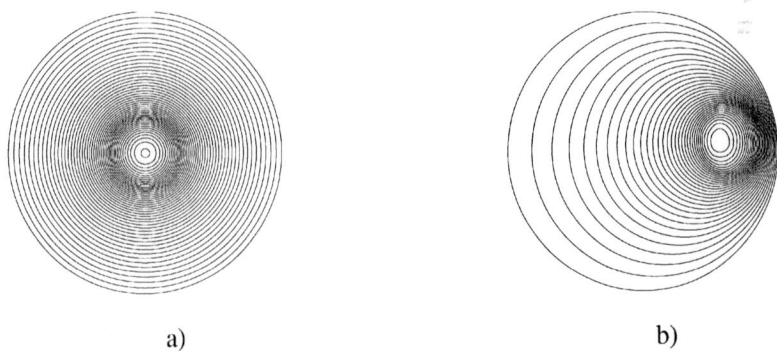

a)   b)

FIGURE 7. Equipotential contours due to space charge set up by a beam passing through a cavity. In the input beam tunnel leading to the cavity the equipotentials are concentric with the wall as shown in a) while the equiptentials are displaced towards the wall of the output beam tunnel due to deflection of the beam on passing through the cavity, as shown in b).

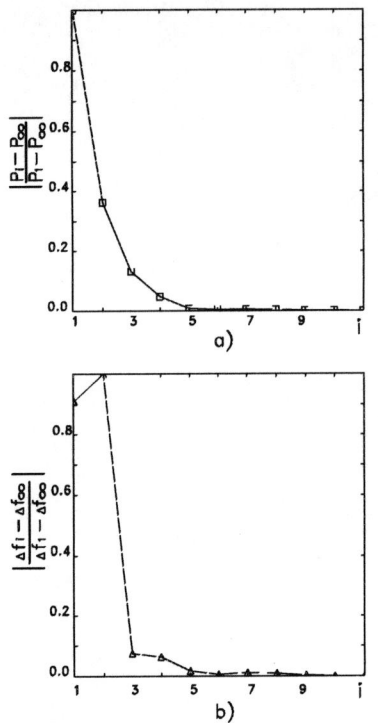
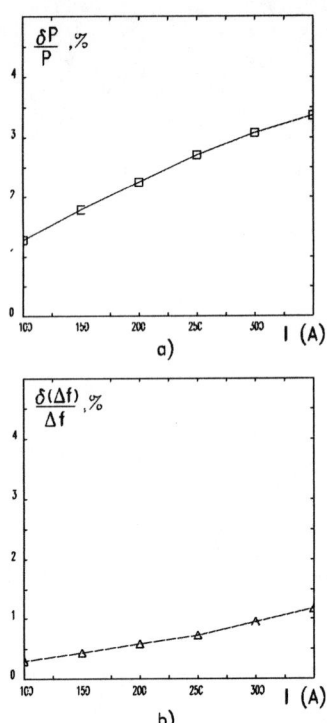

FIGURE 8. The beam power loss P and cavity detuning Δf due the beam versus iteration number i. P_1 and Δf_1 are initial values (no space charge). P_∞ and Δf_∞ correspond to final solution with space charge.

FIGURE 9. The relative changes of the beam power loss and cavity detuning by the beam caused by space charge versus the beam current.

In Fig. 8 beam power loss P and cavity detuning Δf due to the beam are shown versus number of iterations of space charge effects. One can see that convergence is achieved in 6-7 iterations. In Fig. 9 changes of a cavity detuning and beam power loss caused by space charge are shown versus beam current for a 11.4 GHz magnicon [4]. The beam voltage is 500 kV. Note that for beam current up to 400 A the influence of space charge on these quantities is still not very significant. Of course, space charge may have a significant effect on beam energy spread in the penultimate cavity and thus, on tube efficiency.

The computations described here were performed on the NERSC 32×100 MHz Cray J90 (Killeen) computer without any attempt at optimization or parallelization. The variant that employs a mesh having 25×25×30 elements, consumes ~9,000 sec of CPU time. Rough estimates indicates that extensive code vectorization and multitasking, optimization of the data file access (Killeen has comparatively slow HD access) and using the 24×300 MHz SV1 (Franklin) or the 20×300 MHz SV1 (Seymour and Bhaskara) computers in batch mode will lead to at least a ten-fold reduction in run time. That is, the run-time for simulation of an entire full-scale tube should not exceed 1-2 hours.

SUMMARY

A physical model is developed for 3D space charge simulations of a high power, high efficiency magnicon amplifier. Making use of a number of simplifying approximations that are relevant to the magnicon configuration, a fast and efficient simulation code has been formulated. The code prototype has been generated and preliminary calculations have been made to investigate space charge effects on the beam – cavity interaction in a single deflection cavity. The developed model has been tested and shown to provide an accurate description of the beam dynamics in magnicon amplifier.

ACKNOWLEDGMENTS

This work is supported in part by a DoE Small Business Innovative Research (SBIR) grant to Omega-P, Inc. The authors wish to thank Prof. J.L. Hirshfield for useful discussions.

REFERENCES

1. O.A. Nezhevenko, "Gyrocons and Magnicons: Microwave Generators with Circular Deflection of the Electron Beam," *IEEE Trans. Plasma Sc.*, vol.22 pp. 765-772, October 1994.
2. M.M. Karliner, E.V. Kozyrev, I.G. Makarov, O.A. Nezhevenko, G.N. Ostreiko, B.Z. Persov, and G.V. Serdobintsev, " The Magnicon-An Advanced Version of the Gyrocon," *Nucl. Instrum. Methods. Phys. Res.*, vol. A269, pp. 459-473, 1988.
3. E.V. Kozyrev, O.A. Nezhevenko, A.A. Nikiforov, G.N. Ostreiko, G.V. Serdobintsev, S.V. Schelkunoff, V.V. Tarnetsky, V.P. Yakovlev, I.A. Zapryagaev, "Present Status on Budker INP 7 GHz Pulsed Magnicon," *AIP Conference Proc. 474*, (American Institute of Physics, Melville, NY, 1999), p. 187.
4. S.H. Gold, A.W. Fliflet, A.K. Kinkead, B. Hafizi, O.A.Nezhevenko, V.P. Yakovlev, J.L. Hirshfield, R. True, "X-Band magnicon amplifier for the next linear collider," *Phys. Plasmas* vol. 4 1997, pp. 1900-1906.
5. O.A. Nezhevenko, V.P. Yakovlev, A.K. Ganguly, and J.L. Hirshfield, "High Power Pulsed Magnicon at 34 GHz", *AIP Conference Proc. 474*, (American Institute of Physics, Melville, NY, 1999), p. 195.
6. T.O. Raubenheimer and K. Yokoya, "Proposed ILC Parameters", *SLAC MEMO*, 06-05-98. *Deposited into the NTIS database.*
7. "Proposal for Future CLIC Studies and a New CLIC Test Facility (CTF3)", CLIC Study Team, *CERN/PS 99-047 (LP), CLIC Note 402*, Geneva, Switzerland, July 1999. *Deposited into the NTIS database.*
8. Yu.V. Baryshev, I.V. Kazarezov, E.B. Kozyrev, G.I. Kuznetsov, I.G. Makarov, O.A. Nezhevenko, B.Z. Persov, M.A. Tiunov, V.P. Yakovlev, and I.A. Zapryagaev. "A 100 MW electron source with extremly high beam area compression", Nuclear Inst. and Methods in Phys. Research A 340 (1994), pp. 241-258.
9. O.Nezhevenko, I.Kazarezov, E. Kozyrev, G. Kuznetsov, I. Makarov, A. Nikiforov, B. Persov, G. Serdobintsev, M.Tiunov, V. Yakovlev, I. Zapryagaev., *Proceedings of 1993 IEEE Particle Accelerator Conference*, Washington D.C., 1993, (IEEE, Piscataway, NJ, 1994), p. 2650.
10. V. Yakovlev, O. Danilov, O. Nezhevenko, V. Tarnetsky, *Proceedings of 1995 Particle Accelerator Conference*, Dallas, 1995 (JEEE, Piscataway, NJ, 1996), p.1569.

11. D. Myakishev, V. Yakovlev, "The new possibilities of SuperLANS Code," *Proceedings of 1995 Particle Accelerator Conference and International Conference on High Energy Accelerators*, Dallas, 1995, pp.2348-2350.
12. M.A. Tiunov, B.M. Fomel, V.P. Yakovlev, *Proceedings of XIII International Conference on High Energy Accelerators*, Novosibisk, 1986 (Publishing House "Nauka" Siberian Division, Novosibirsk, 1987), v.1, p. 353.
13. D.G. Myakishev, M.A. Tiunov and V.P. Yakovlev, " Code SuperSAM for Calculation of Electron Guns with High Beam Area Convergence", XV-th International Conference on High Energy Accelerators, 1992, Hamburg. Int. J. Mod. Phys. A (proc. Suppl.) 2B (1993), v-2, pp.915-917.
14. B. Hafizi and S.H. Gold, "Optimization Studies of Magnicon Efficiency," Phys. Fluids, vol. 2, 1995, p. 902.
15. B. Goplen, L. Ludeking, D. Smithe and G. Warren, "User –friendly configurable MAGIC for electromagnetic PIC calculations," Computer Physics Communications 87 (1995), pp. 54-86.
16. B. Goplen, J. MacDonald and G. Warren, SOS Reference Manual, MRC/WDC-R-190 (Mission Research Corporation. March 1989).
17. D.B. Seidel, M.L. Kiefer, R.S. Coats, T.D. Pointon, J.P. Quintez and W.A. Jonson, Multitasking the 3-D electromagnetic PIC code QUICKSILVER, *Proceedings of 13th Conference on Numerical Simulations of Plasma*, Santa Fe, NM, September, 1989. Paper IM6.
18. V.Tarakanov, private communication.
19. J.W. Eastwood, W. Arter, N.J. Brealey, R.W. Hokney, "Body-fitted electromagnetic PIC software for use on parallel computers", Computer Physics Communications 87 (1995), pp. 155-178.
20. A. Novokhatsky and A. Mosnier, "Wakefields of short bunches in the canal covered with thin dielectric layer", *Proceedings of 1997 Particle Accelerator Conference*, Vancouver, 1997 (JEEE, Piscataway, NJ, 1998), pp.1661-1663.
21. Guy Le Meur, Francois Touze, "PRIAM/ANTIGONE: a 2D/3D Package for Accelerator Design", *Proceedings of the European Particle Accelerator Conference*, London, 1994, (World Scientific, Singapore, 1995), pp. 1321-1323.
22. M. Botton, T.M. Antonsen, B. Levush, K.T. Nguyen, A.N. Vlasov, "MAGY: A Time-Dependent Code for Simulation of Slow and Fast Microwave Sources", IEEE Transactions on Plasma Science, Vol. 26, No 3, June 1998. pp.882-892.
23. O.C. Zienkiewicz and R.L. Taylor, *Finite element method*, (McGraw-Hill, London, 1989).
24. A.J. Baker, *Finite element computational fluid mechanics*, (Hemisphere, Washington DC 1983).
25. J.H. Wilkinson, *The algebraic eigenvalue problem*, (Clarendon Press, Oxford, 1988).

Designs of Three-Cavity Frequency Quadrupling Coaxial Gyroklystrons

I. Yovchev, G. S. Nusinovich, W. Lawson, V. L. Granatstein, and E. S. Gouveia

Institute for Plasma Research and Electrical Engineering Department
University of Maryland, College Park, MD 20742

Abstract. Two designs of Ka-band 34.2 GHz three-cavity coaxial gyroklystrons that might be applied to driving high gradient linear accelerators are presented. The input, buncher and output cavities operate in the TE_{011}, TE_{021} and TE_{041} modes, respectively. The buncher cavity doubles the signal frequency and the output cavity quadruples it. The electron beam current of 540 A and voltage of 460 kV are used in both designs. The simulations show that for the first design, a maximum efficiency of 15%, a maximum gain of 31 dB and a bandwidth of 1 MHz can be realized for an orbital-to-axial velocity ratio at the entrance to the microwave circuit of $\alpha_0 = 1.6$. The efficiency and gain for the second design are 16.3% and 42.1 dB, respectively, at $\alpha_0 = 1.3$; but the bandwidth is reduced to 600 kHz. For both designs the output cavity is zero-drive unstable.

INTRODUCTION

The gyroklystron is one of the main candidates for driving the future TeV-scale linear colliders [1]. Recent experiments with a three cavity 1-1-1 coaxial gyroklystron performed at the University of Maryland showed the viability of this device; viz., high peak microwave power of 80 MW at frequency of 8.6 GHz was obtained [2]. The sequence "1-1-1" means that the experimentally studied gyroklystron has 3 cavities (input, buncher and output), each of them operating at the fundamental cyclotron resonance (i.e. with the cyclotron harmonic number equal to one).

For TeV-scale linear colliders it is desirable to increase the radiation frequency in order to sustain higher accelerating gradients. In gyroklystrons, higher frequency can be realized either by operating with stronger magnetic fields, or with a given magnetic field by designing the output cavity for operation at higher harmonics of the electron cyclotron frequency. Furthermore, study of frequency multiplying gyroklystrons is important since relatively inexpensive low-frequency (solid-state) drivers can be utilized for driving such tubes.

Recently, experiments with a frequency-doubling, two-cavity gyroklystron were successfully performed at the University of Maryland resulting in 32 MW of peak power at 19.76 GHz with efficiency near 28.6% and a gain of 27.2 dB [3]. Further simulation studies have been done on frequency doubling three-cavity 1-2-2 and four-cavity 1-2-2-2 gyroklystron circuits [4], and experimental studies on these designs are on-going. The efficiency and gain for the 1-2-2 circuit are 32% and 49 dB, respectively, whereas for the 1-2-2-2 design – 35% and 60 dB. Both designs were performed with the following beam parameters: current of 540 A, voltage of 460 kV, orbital-to-axial velocity ratio of 1.5, and a parallel velocity spread of 6.4%.

The present paper reports on the results from the numerical study of two three-cavity frequency-quadrupling 1-2-4 designs. The simulations have been performed based on the presently available set-up in the gyroklystron laboratory at the University of Maryland: magnetron-injection gun (beam current and voltage of 540 A and 460 kV, respectively) and set of solenoidal magnets (pancake coils, creating the guide magnetic field). The first design was performed for the existing geometry of the coils, while for the second one we decreased the distances between some of them, effectively altering the magnetic field profile.

OUTPUT CAVITY DESIGN

The first two cavities of both 1-2-4 gyroklystron designs (Table 1) are the same as in the previous frequency-doubling designs (three-cavity 1-2-2 and four cavity 1-2-2-2 systems) [4]. We have performed the design of the output cavity (OC), which operates in the TE_{041} mode using the code CASCADE [5]. By optimizing the profile of the output

Cavity or section	No.	Inner Radius (cm)	Outer Radius (cm)	Length (cm)
Inlet	1	1.825	3.325	5.000
Input	2	1.100	3.325	2.286
Drift 1	3	1.825	3.325	5.525
Buncher	4	1.610	3.520	1.691
Drift 2	5	1.825	3.325	7.664
Sinusoidal transition	6	–	–	0.600
Output	7	1.693	3.457	2.9
Iris:				
a) quarter circle	8	1.717	3.433	0.025
b) sinusoidal transition	9	–	–	1.200
Outlet	10	1.650	3.500	1.000

Table 1. Dimensions of first 1-2-4 circuit design

iris we were able to achieve the following parameters: left-to-right power ratio of – 24.87 dB, mode purity of 99.87%, and power carried by the TE_{041} at the exit of the OC (i.e. behind the iris) of 99.76%. The OC for these simulations was modeled as having sinusoidal transitions at both ends - i.e. transition from the second drift into OC (Fig.

1), and transition from OC to the last section of the system (No. 10 in Fig.1). The output cavity's iris consists of two parts: a quarter circle and the second sinusoidal transition (indicated by No. 8 and No. 9, respectively, in Fig. 1).

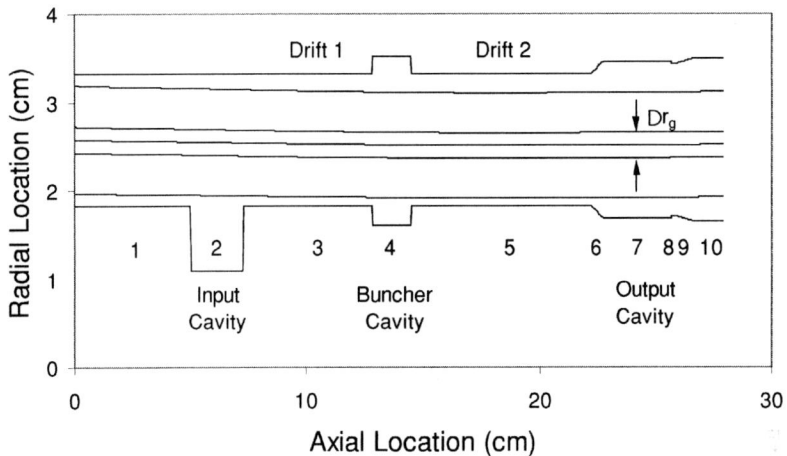

Figure 1. Longitudinal view of beam spread within 1-2-4 circuit

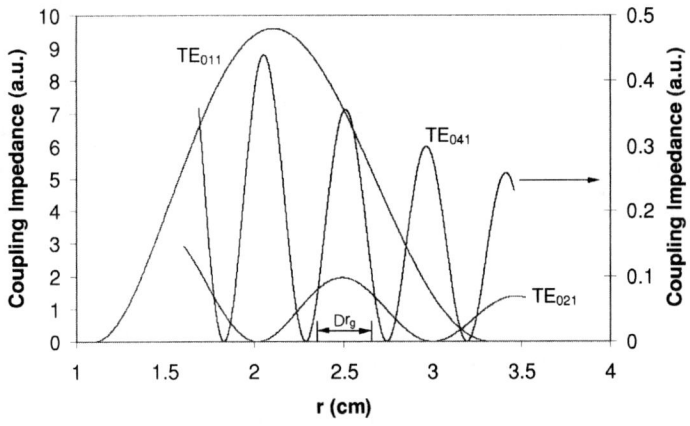

Figure 2. Coupling impedance of various modes

For such an iris design it is enough to vary only the sinusoidal transition length in order to change the OC Q-factor. To better explain this let us recall that the axial structure of the field in the output cavity (f) shown in Fig. 1 can be described by the inhomogeneous string equation [6]

$$\frac{d^2 f}{dz^2} + k_z^2(z) f = 0$$

where $k_z^2 = (w/c)^2 - k_\perp^2(z)$ with the transverse wavenumber $k_\perp = n_{mp}/R(z)$ determined by the eigennumber n_{mp} of the operating TE_{mp} wave and the variable radius of a structure wall $R(z)$. This equation has the same form as the one-dimensional stationary Schrödinger equation. This means that an output cutoff iris can be treated as the potential barrier for the field localized in such a well whose potential is described by $k_\perp^2(z)$, and the diffractive output radiation is proportional to the probability of wave passage through the barrier. Now it is obvious that the variation in the sinusoidal transition length of the iris changes the width of this barrier, and therefore affects the diffractive Q-factor.

A very important issue for the design of short wavelength gyroklystrons is the influence of the spread in the electron guiding center radius Δr_g. Electron gun simulations show that our existing magnetron injection gun generates an electron beam with the spread $\Delta r_g \cong 0.3$ cm at the entrance to the interaction region. For an output cavity operating in the TE_{041} mode there exist 4 peaks in the radial distribution of the coupling coefficient between the electron beam and the mode. Taking into account that these four peaks are located at the distance between the outer and inner OC radii of about 1.8 cm (see Table 1) and assuming roughly that each peak is approximately a half-cutoff wavelength wide - i.e. its width is about 0.45 cm - it is clear that Δr_g is of the same order as the peaks' width (Fig. 2). Note that in the previous designs of the lower frequency fundamental and frequency-doubling gyroklystrons [4,7], the spread in guiding center radii was not taken into account because the peaks' widths in these cases are considerably wider than Δr_g. This situation is illustrated in Fig. 2, which shows the radial dependencies of coupling impedances of the TE_{01}, TE_{02} and TE_{04} modes which can be excited at, respectively, the fundamental, second and fourth harmonic in the output cavity.

FIRST DESIGN OF 1-2-4 CIRCUIT

The cylindrically symmetric coaxial geometry of the circuit is shown in Fig. 1 and the dimensions are tabulated in Table 1. The resonant frequencies, Q-factors and operating modes of the cavities are given in Table 2, while the beam parameters are presented in Table 3.

	Resonant frequency (GHz)	Q-factor	Operating mode
Input cavity	8.567	30	TE_{011}
Buncher cavity	17.136	80	TE_{021}
Output cavity	34.267	1300	TE_{041}

Table 2. Parameters of cavities in 1-2-4 circuit

Average velocity ratio	1.6
Axial velocity spread (%)	7.2
Beam voltage (kV)	460
Beam current (A)	540

Table 3. Beam Parameters

The five curves located between the tube walls in Fig. 1 represent the electron beam. The central curve corresponds to the central trajectory of the electron guiding center. The two curves located symmetrically around the central one and close to it depict the maximum and minimum guiding center radii, therefore the difference represents the spread in the guiding center radius. Other two lines (the most inner and the most outer ones) designate the inner and outer extrema of the beam. Hence the difference represents the total width of the beam, taking into account the electron Larmor radii.

A larger scale of circuit geometry along with the pancake coils and the magnetic field created by them is shown in Fig. 3. Seven pancake coils form a structure, which can be divided into 3 groups driven by independent currents – first group (the first two coils), second group (the middle three ones), and third group (the last two coils). The values of the coil currents (and therefore the profile of the magnetic field) are optimized using MAGYKL and EGUN codes, so as to produce the highest efficiency of a given circuit. These currents (in Ampere-turns) through the three groups of coils are I_1 = 43500 A.t, I_2 = 38050 A.t, and I_3 = 46500 A.t. The corresponding variation of the magnetic field along the circuit axis is small – about 0.27 kG (minimum value of 5.07 kG at the entrance to the system and maximum value of 5.34 kG between the buncher cavity (BC) and the output cavity (OC)). This determines the small bending of the beam in the gyroklystron microwave circuit shown in Fig.1.

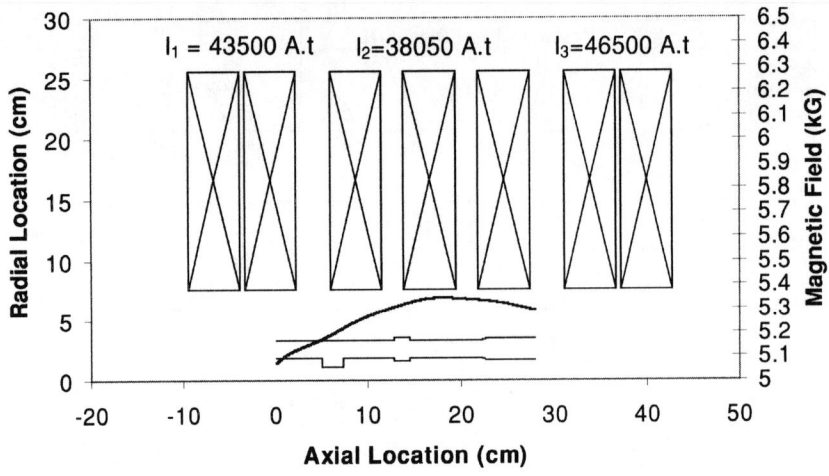

Figure 3. Schematic of entire circuit and magnetic field profile.

When the pitch-ratio α_0 (orbital-to-axial velocity ratio) at the entrance to the system is 1.6 (see Table III), the α-values in the middle of the IC, BC, and OC are 1.67, 1.75, and 1.76, respectively. For these values of α, the start oscillation curves for each cavity were calculated, using the code QPB [8]. The results are shown in Fig. 4.

The solid circles in Fig. 4 designate the operating points of each cavity. Their position along the abscissa coordinate specify the magnetic fields in the middle of the corresponding cavity, while the ordinate indicates the Q-factor of the cavity: $B_1 = 5.19$ kG with Q=30 (for the IC), $B_2 = 5.32$ kG with Q=80 (for the BC), and $B_3 = 5.33$ kG with Q=1300 (for the OC). From Fig. 4 it follows that the Q-factors at which the IC and the BC operate are lower than the corresponding start oscillation Q-factor. For the output cavity, however, this is not valid, which means that the last cavity is zero-drive unstable. The OC is zero-drive unstable with respect also to other modes – with azimuthal index m=0 (TE_{011}, TE_{021}, TE_{022} and TE_{031}) and with m=1 (TE_{111}, TE_{121}, TE_{122}, TE_{131}, TE_{132}, and TE_{141}). Modes with azimuthal index m > 1 have not been checked for stability.

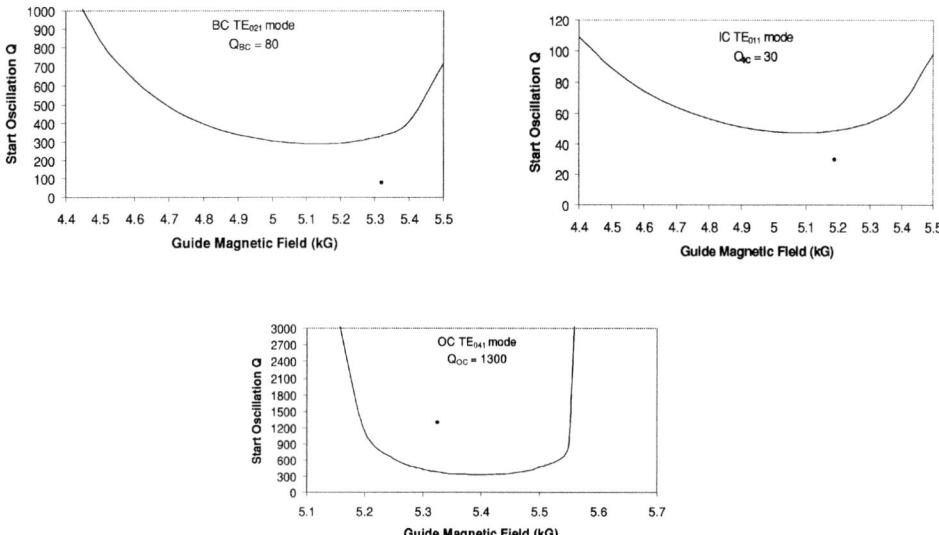

Figure 4. Start oscillation curves for each cavity in circuit

Previous studies of stability in zero-drive unstable gyroklystrons have shown that an initially excited operating mode can suppress possible parasites [7].

The calculation of the efficiency was performed using the large signal code MAGYKL [9,10]. To take into account the spread in the guiding center radii, a new subroutine was embedded into this code.

The dependence of the efficiency η and gain G on the input power P_{in} is presented in Fig. 5 for drive frequency f_{dr} of 8.568 GHz. The maximum efficiency of about 15% with corresponding gain of 29.5 dB is realized when P_{in} = 39 kW. The maximum gain, in excess of 31 dB, is reached when P_{in} = 27 kW. Note that at $P_{in} \leq 27$ kW it was impossible to reach the specified convergence with respect to the complex amplitude of the electromagnetic field in the OC, using the code MAGYKL. This result can be interpreted as being due to the generation of self-oscillations in the zero-drive unstable output cavity at low P_{in}'s.

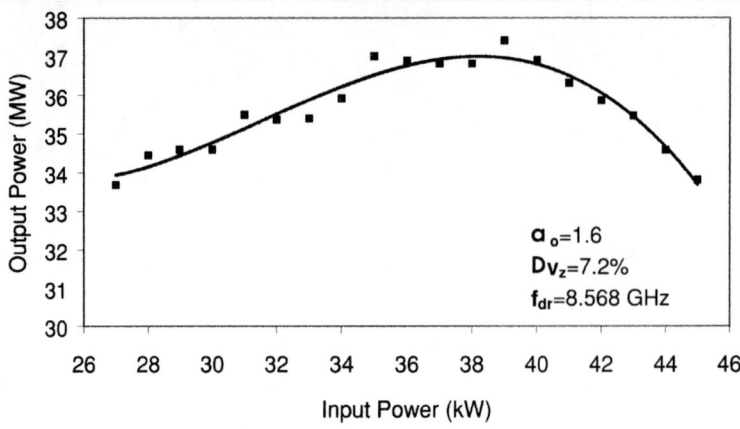

Figure 5. Variation of the efficiency and gain with respect to input power

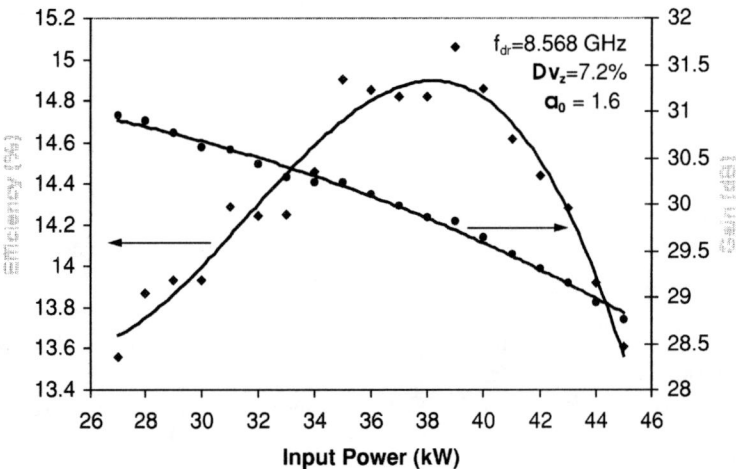

Figure 6. Output power as a function of input power

Fig. 6 displays the output power P_{out} vs. P_{in} for f_{dr} = 8.568 GHz, and it indicates that the maximum output power is about 37 MW.

The drive curve for P_{in} = 39 kW is shown in Fig. 7. Data for P_{out} in the cases f_{dr} ⊟8.5679 GHz and f_{dr} ⊛ 8.5689 GHz are absent because of the poor convergence of the

OC complex field amplitude at these frequencies. Nevertheless, a conclusion can be drawn that the bandwidth is on the order of 1 MHz (0.012%). When f_{dr} = 8.5682 GHz,

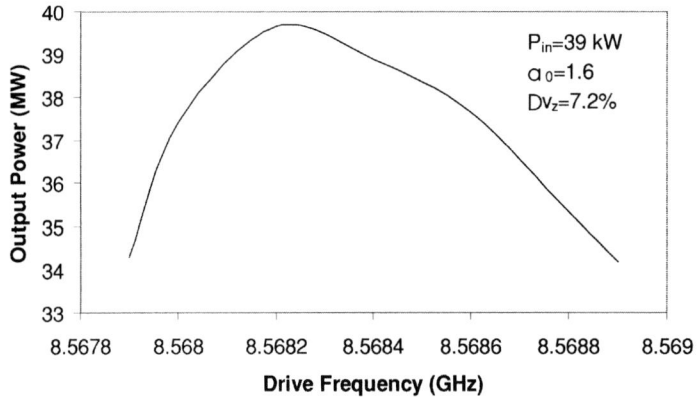

Figure 7. Output power as a function of the frequency of the driver

the output power (~40 MW) is higher than in the case f_{dr} = 8.568 GHz, for which Fig. 6 was calculated.

A positive feature of the first 1-2-4 gyroklystron design is that the guide magnetic field changes, as mentioned above, along the system only slightly and therefore the beam trajectory is almost a straight line. This enables us to locate the beam approximately in the middle between the inner and the outer drift tube radii, thus preventing the beam from interception with the drift tube walls. However, the large values of α along the system could lead to the mirroring of decelerated electrons.

SECOND DESIGN OF 1-2-4 CIRCUIT

In an attempt to reduce the possibility for electron mirroring, a second 1-2-4 gyroklystron design has been performed, in which we decreased α_0 and the magnetic field in the IC and BC. On the other hand, it was necessary to preserve the magnetic field in the OC close to ~5.3 kG in order to realize strong interaction between the beam and the TE_{041} mode in the OC (see Fig. 4).

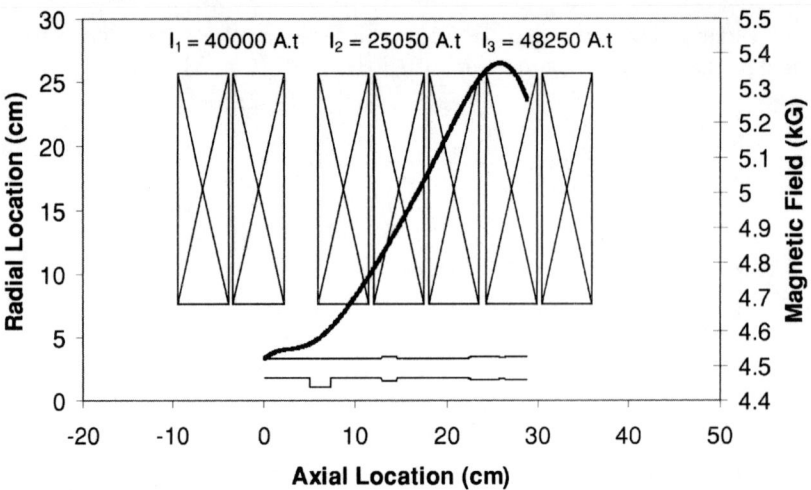

Figure 8. Schematic of entire circuit and magnetic field profile

To do this, the currents I_1 and I_2 have been decreased, while the current I_3 was increased. In order to enhance the influence of the current I_3 upon the magnetic field in the output cavity, it was decided to put the third coil set closer to this cavity. The configuration of the pancake coils for the second 1-2-4 design and their magnetic field along the system are presented in Fig. 8. The changes in solenoids' locations are the following: 1) the distance between coils in the second set was decreased from 2.41 cm (for the first design) to 0.51 cm and 2) the distance between the second and the third set of coils was reduced from 3.68 cm (first design) to 0.80 cm. The optimum currents I_1, I_2, and I_3 are now 40000 A.t, 25050 A.t, and 48250 A.t, respectively, and the magnetic fields in the middle of the IC, BC and OC are $B_1 = 4.58$ kG, $B_2 = 4.86$ kG, and $B_3 = 5.35$ kG, respectively. While B_3 is approximately the same as in the previously discussed design, the values of B_1 and B_2 for this second design are considerably lower than the corresponding values for the first.

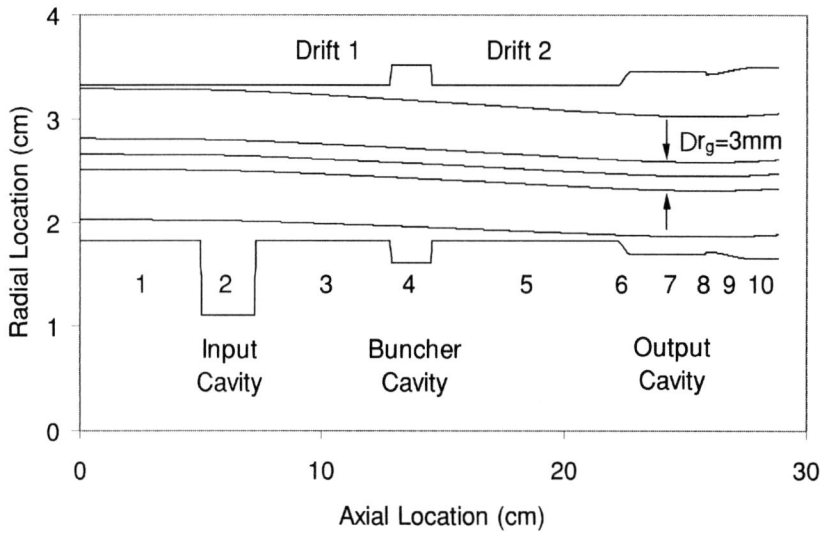

Figure 9. Longitudinal view of beam spread within the second 1-2-4 circuit.

The circuit geometry is shown in Fig. 9. Because of the larger variation of the magnetic field along the system – 0.85 kG (with the minimum value of 4.52 kG at the entrance to the system and the maximum of 5.37 kG at the end of the OC) in comparison with the first design (0.27 kG), the bending of electron beam trajectory is stronger. At the entrance to the circuit and immediately before the OC, the beam propagates very close to the drift tube walls (the minimum clearance is about 0.3 mm). The dimensions of the second 1-2-4 gyroklystron design are tabulated in Table 4.

The only differences in comparison with the first design are the lengthening of the OC (from 2.9 cm to 3.035 cm) and of the iris' sinusoidal transition (from 1.2 cm to 2.0 cm), along with a small variation in the output cavity inner and outer radii.

The α-values in the IC and BC are 1.324 and 1.440, respectively, i.e. substantially smaller than the corresponding α-values for the first design, which reduces the possibility of electron mirroring. On the other hand, α's in the middle of the OC for both designs are very close, compare 1.76 in first design with 1.70 in the second one.

The beam parameters are shown in Table 5 and the resonant frequencies, Q-factors and operating modes are presented in Table 6. The average velocity ratio at the entrance to the system (1.3) is considerably smaller than that for the first design (1.6);

the axial velocity spread (4.9%), which scales with α, is also smaller (compare with 7.2% for the first design), whereas the Q-factors of all cavities are higher.

Cavity or section	No.	Inner Radius (cm)	Outer Radius (cm)	Length (cm)
Inlet	1	1.825	3.325	5.000
Input	2	1.100	3.325	2.286
Drift 1	3	1.825	3.325	5.525
Buncher	4	1.610	3.520	1.691
Drift 2	5	1.825	3.325	7.664
Sinusoidal transition	6	–	–	0.600
Output	7	1.694	3.456	3.035
Iris:				
a) quarter circle	8	1.717	3.433	0.025
b) sinusoidal transition	9	–	–	2.000
Outlet	10	1.650	3.500	1.000

Table 4. Dimensions of second 1-2-4 circuit design

Average velocity ratio	1.3
Axial velocity spread (%)	4.9
Beam voltage (kV)	460
Beam current (A)	540

Table 5. Beam Parameters

	Resonant frequency (GHz)	Q-factor	Operating mode
Input cavity	8.5695	125	TE_{011}
Buncher cavity	17.138	620	TE_{021}
Output cavity	34.264	2000	TE_{041}

Table 6. Parameters of cavities in second 1-2-4 circuit

The start oscillation curves for each cavity are plotted in Fig. 10. The Q-factors for the IC and BC are chosen to be very close to the start oscillation curves, but these cavities are still zero-drive stable, whereas the output cavity is zero-drive unstable. The efficiency dependence on drive power at $f_{dr} = 8.568$ GHz is shown in Fig. 11.

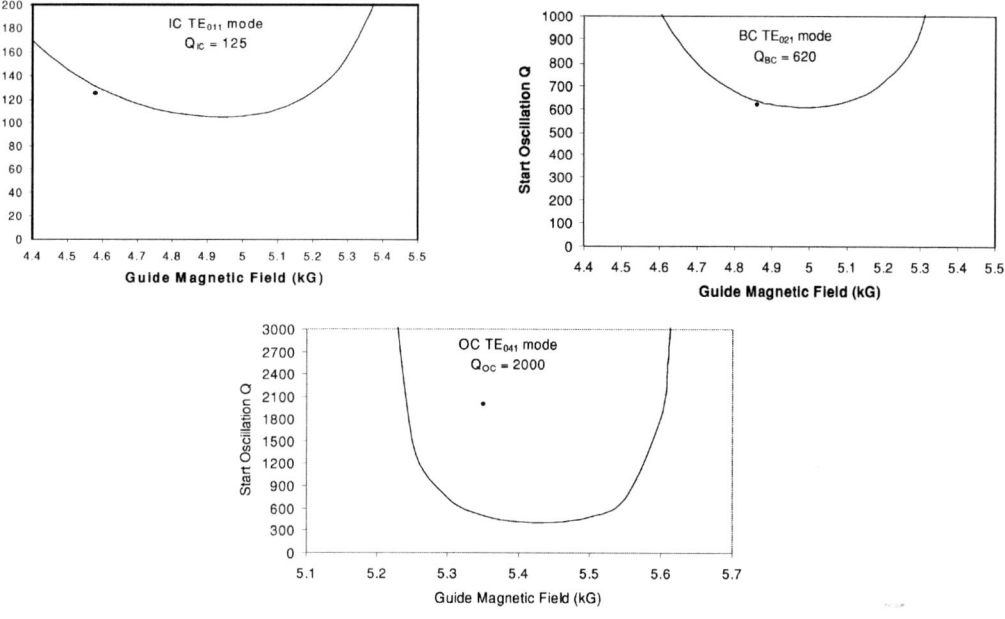

Figure 10. Start oscillation curves for each cavity in circuit

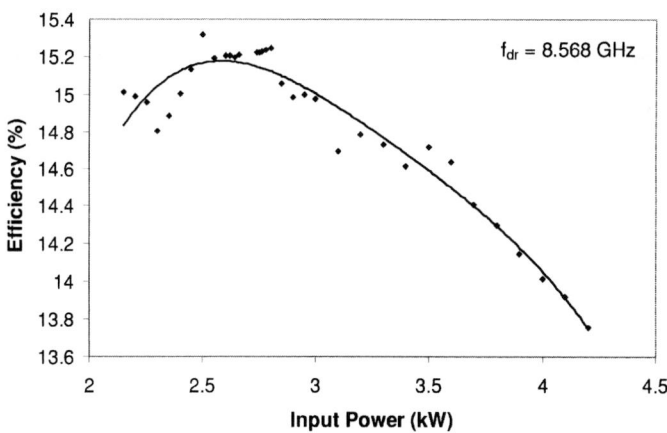

Figure 11. Variation of the efficiency with respect to input power

The maximum efficiency of 15.2% occurs at P_{in} = 2.5 kW. This value of P_{in} is more than one order of magnitude smaller than in the first design where η is maximum at P_{in} = 39 kW, which is caused by the higher Q-factor values of the IC and BC. In Fig. 12, the efficiency dependence on drive frequency is presented for P_{in} =2.5 kW. When f_{dr} = 8.5676 GHz, η = 16.3%, which corresponds to P_{out} = 40.5 MW; correspondingly the gain is 42.1 dB. These values for the efficiency, output power and gain are larger than corresponding values for the first design. However, the frequency bandwidth Δf_{dr} is 0.6

MHz (relative bandwidth of 0.007%), i.e. narrower than for the first design (1 MHz). The bandwidth of 0.6 MHz is close to the Fourier limit of 1.5 μsec pulses we will work with.

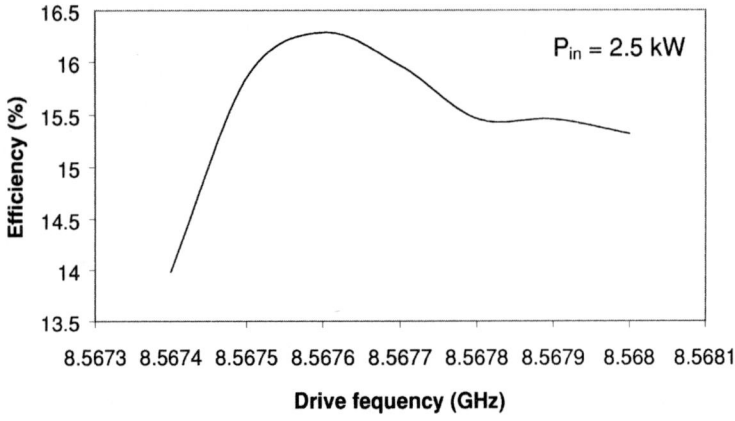

Figure 12. Efficiency as a function of the frequency of the driver

As in the case of the first design, the efficiency data in Fig. 11 for the region P_{in} < 2.1 kW and in Fig. 12 for the regions defined by f_{dr} < 8.5674 GHz and f_{dr} > 8.5680 GHz are not available due to the self-oscillations of the output cavity.

CONCLUSIONS

From the simulation study presented in this paper it follows that the three-cavity, frequency-quadrupling coaxial gyroklystron, using the existing magnetron injection gun, driver and pancake coils, could generate microwave power in excess of 40 MW at 34.3 GHz. The interaction efficiency is in the range of 15-16% and the gain is over 40 dB.

However, these preliminary results indicate that in general fundamental or frequency-doubling gyroklystrons are better candidates for drivers of future electron-positron linear colliders. From the comparison between previous simulation results [4]

and the present ones we can deduce that the frequency-doubling gyroklystrons have the following advantages over the 1-2-4 circuit:

– Higher efficiency (32% or 35% for 1-2-2 and 1-2-2-2 schemes, respectively, vs. 15-16% for the 1-2-4 scheme);

– Zero-drive stable output cavity;

– Approximately one order of magnitude wider bandwidth (0.09% for the 1-2-2-2 scheme vs. 0.007% – 0.012% for the 1-2-4 scheme).

Also recall that the requirement of larger magnetic fields leads to an effective reduction in beam width which allows fundamental and second-harmonic designs with narrower drift sections, hence minimizing possible cross-talk between cavities as well as self-oscillations in drift regions.

In spite of these disadvantages, the studied frequency-quadrupling scheme has two positive characteristics as well: 1) The gain of over 40 dB obtained at the second 1-2-4 design is high enough and could be improved further by insertion of an additional buncher cavity; 2) The wavelength shortening at a fixed magnetic field is beneficial for accelerator needs. Recall that the number of amplifiers, required for linear colliders is approximately proportional to the parameter $\mathbf{a}=\lambda^2/(P\tau)$ [11]. Here λ, P and τ are respectively the wavelength, the power, and the duration of the amplifier output pulse. The best results from previous simulations showed that the frequency doubling 1-2-2 gyroklystron could deliver 80 MW of output power at 17 GHz [4], while our 1-2-4 design predicts 40 MW at 34 GHz. Therefore, the parameter $\mathbf{a}$ for the 1-2-4 scheme would be half in comparison with the same parameter for the 1-2-2 scheme. Also, the required magnetic field would be reduced by a factor of two in the 1-2-4 scheme. These positive characteristics of the 1-2-4 circuit lead to the conclusion that it is worthwhile continuing the simulation work, aiming to improve the design of the frequency-quadrupling schemes and further testing them experimentally.

ACKNOWLEDGEMENTS

This work was supported by the Department of Energy, Division of High Energy Physics, under Contract DE-FG02-94ER40855.

REFERENCES

1. Granatstein, V. L. and Lawson, W., *IEEE Trans. Plasma Sci.* **24**, 648 (1996).
2. Lawson, W., *et al., Phys. Rev. Lett.* **81**, 3030 (1998).
3. Matthews, H. W., *et al., IEEE Trans. Plasma Sci.* **22**, 825 (1994).
4. Yovchev, I., *et al., IEEE Trans. Plasma Sci.* **28** (2000), in press.
5. Lawson, W. and Latham, P. E., *IEEE Trans. –MTT* **40**, 1973 (1992)

6. Vlasov, S. N., *et al.*, *Radiophys. And Quantum Electron.* **12**, 972 (1969).
7. Saraph, G. P., *et al.*, *IEEE Trans. Plasma Sci.* **24**, 671 (1996).
8. Latham, P. E., *et al.*, *Phys. Rev. A* **45**, 1197 (1992).
9. Latham, P. E., *et al.*, *IEEE Trans. Plasma Sci.* **22**, 804 (1994).
10. Blank, M., *et al.*, *IEEE Trans. Plasma Sci.* **26**, 409 (1998).
11. Wilson, P. B., in *Applications of High Power Microwaves*, Gaponov-Grekhov, A. V. and Granatstein, V. L., Eds., Norwood, MA: Artech House, 229 (1994).

Optically Driven Emitter Of Neutrinos For Testing Of Neutrino Oscillations

Andrew V. Pakhomov and Yoshiyuki Takahashi

Department of Physics, University of Alabama in Huntsville, Huntsville, Alabama 35899

Abstract. This paper proposes a new concept for generating controlled, high-flux pulses of neutrinos. Laser-induced generation of relativistic protons, followed by meson production and decay, provides the neutrino source. By conservative estimate, the source will yield nanosecond-range pulses of muon neutrinos, with fluxes of ~ 10^{19} ν_μ s^{-1}sr^{-1} and energies of ~ 20 - 40 MeV. This source is proposed for use in ν_e appearance and ν_μ disappearance tests. The expected lower limit for assessable ν_e/ν_μ mass difference is at least 10^{-3} eV^2. Concept feasibility depends upon further progress in high-energy lasers; the process assumes a driving laser with pulse energy ~ 8 kJ, providing an irradiance of ~ 4×10^{23} W/cm^2.

INTRODUCTION

The purpose of this paper is to describe a new concept for a pulsed neutrino source: an Optically Driven Emitter of Neutrinos (ODEN). The source is based on laser wakefield acceleration at relativistic regime, leading in turn to the emission of a relativistic proton pulse. The protons generate charged pions, which decay into muon - neutrino pairs. Although these processes can also be induced using accelerators, the fact that the whole chain of transformations is triggered and seeded by a high-energy pulsed laser permits new control options. Significant advantages of the new source include (1) fine temporal neutrino pulse-widths, limited by pion lifetime (~ 40 ns due to time dilation); (2) the triggering of the detector by the same laser pulse; and (3), when it will become possible, fixed repetition rates set by the laser. This permits the efficient temporal filtering of background noise and residual electron neutrino contaminations from muon decays, by simply fast time-gating the detector. The flux of generated muon neutrinos at the source is expected to be as high as ~10^{19} ν_μ s^{-1}sr^{-1}, with an energy of ~ 20 MeV or higher. (The requirements for the generation of GeV-neutrinos will be discussed below). The ODEN offers new experimental opportunities for addressing the neutrino deficiency phenomena, via neutrino appearance/disappearance tests limited to $\Delta m^2 \geq 10^{-3}$ eV^2.

BASIC PRINCIPLES

As was established recently, metal foils emit relativistic protons when exposed to irradiances of ~ 10^{18} - 10^{20} W/cm^2, from focused sub-picosecond laser pulses. This phenomenon was independently reported by groups of researchers from the Rutherford Appleton Laboratory (RAL, United Kingdom) [1], the University of Michigan (UMich) [2] and the Lawrence Livermore National Laboratory (LLNL) [3]. Their data is summarized in Table I. The strongest emissions (~ 3×10^{13} protons per pulse, accelerated by ponderomotive forces to kinetic energies of ~ 50.0 MeV) were observed by the LLNL group. This phenomenon of proton emission provides the core principle of ODEN.

TABLE 1. Laser Irradiation and Proton Emission Characteristics

Laser Wavelength, μm	Pulse width, ps	Pulse Energy, J	Intensity, W/cm²	Mean Proton Energy, MeV	Number of Protons per Pulse	Protons/Laser Pulse Energy Ratio	Reference
1.053	0.4	1	3×10^{18}	1.6	$\geq 10^9$	$\geq 2.4\times10^{-4}$	2
1.053	0.9 - 1.2	50	5×10^{19}	~10	10^{12}	0.06	1
1.053	0.45	500	4×10^{20}	29	3×10^{13}	0.2	4
~ 1.0	0.04	8,000	4×10^{23}	1000	10^{13}	0.2	This paper

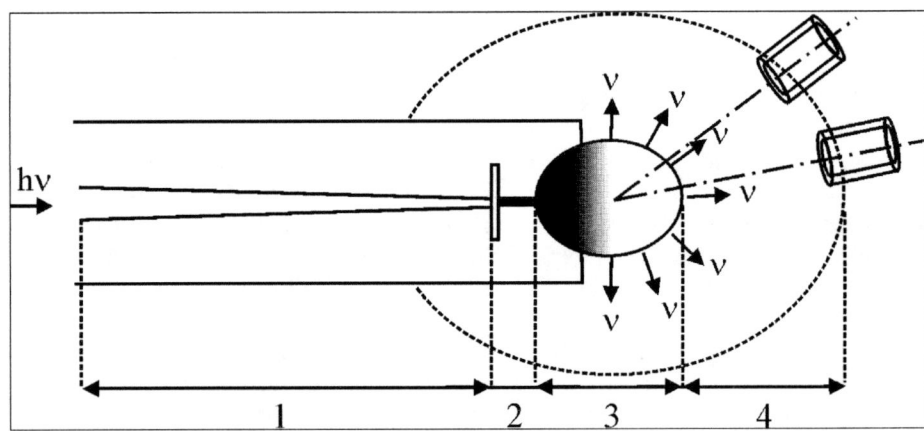

FIGURE 1. Principal schematics of ODEN, see the text for details.

One possible configuration of ODEN is shown in Figure 1. The figure is partitioned into four zones: 1) a laser beam zone, 2) a proton beam zone, 3) a proton beam stop / pion decay zone and 4) a detector zone (several possible detector positions are shown).

In Zone 1, a petawatt laser beam is focused onto a solid target (1-2 interface). A relativistic proton pulse, emitted from the opposite (unexposed) side of the target, is directed toward a beam stop (Zone 2, and 2-3 interface). Within the beam stop the proton flux is transformed into a charged pion flux by the following inclusive reaction:

$$p + p \rightarrow \pi^{\pm} + X \qquad (1)$$

The pion pulse emerging from the proton stop (2-3 interface) decays by the reaction:

$$\pi^{\pm} \to \mu^{\pm} + v_{\mu}(\bar{v}_{\mu}) \qquad (2)$$

which yields a first pulse of muon neutrinos. Reaction (2) is followed by muon decays:

$$\mu^{\pm} \to e^{\pm} + \bar{v}_{\mu}(v_{\mu}) + v_{e}(\bar{v}_{e}) \qquad (3)$$

Upon complete decay of all parents by reactions (2,3) ODEN will generate neutrinos of v_e/v_μ flavor ratio one-to-two. The pion pulse, emerging at the (2-3) interface, is stopped at the target within Zone 3. Neutrinos generated within Zone 3 are radially emitted toward detector(s) (Zone 4). The detector zone could be composed, for example (see Figure 1), of several cylindrical tanks with longitudinal axes directed radially outward from the geometric center of Zone 3. Calculation of the detector fiducial volume and selection of the detector liquid require a more detailed feasibility study (currently underway). However, several important features of the detector can be noted here. Detector photomultiplier tubes must be triggered by laser pulses in order to detect Cherenkov cones within a fixed time window. The duration of this window should match the width of the pulse of neutrinos from reaction (2). The fiducial length of the detector, then, will be about 12 meters. The window will permit temporal exclusion of neutrinos produced by reaction (3), thus eliminating contamination by electron flavor. Background neutrinos can also be excluded, based on their non-radial directionality and times of interaction. Finally, the detection system will permit distinguishing neutrino flavors, should disappearance tests be feasible. (This will be the case, provided the ODEN can generate neutrinos with energies exceeding 1 GeV.) To veto contamination by cosmic muons, the ODEN should be located underground or underwater.

NEUTRINO ENERGY AND FLUX

The electron energy in the wakefield plasma (E) can be assumed proportional to the square root of the irradiance (intensity), I [4]:

$$E = mc^2 \sqrt{1 + 2U_p/mc^2}, \qquad (4)$$

where m is the electron mass and U_p is the non-relativistic ponderomotive potential:

$$U_p = (9.33 \times 10^{-14} s)I\lambda^2 \qquad (5)$$

λ (μm) is the wavelength. From (4,5) the mean proton energy can be assumed proportional to the field, or to the square root of irradiance; i.e. $E_p \propto I^\alpha$, where $\alpha =$

0.5 . This simple relationship is clearly corroborated by the data of RAL and LLNL from Table I: the square root of the intensity ratio is 0.35, and the proton energy ratio is 0.36. The research group from UMich reported α to be in the 0.3 - 0.4 range [2], although their irradiation conditions are essentially below the intensities of RAL and LLNL.

The LLNL collaboration achieved a field of ~ 10^{14} V/m [4]. Using this as a reference value, to generate 1.0-GeV protons would require a field on the order of 2×10^{15} V/m, corresponding to 4×10^{23} W/cm^2 irradiance. As will be shown below, 1.0-GeV protons are at about the minimum energy level required for ODEN development. An irradiance of 4×10^{23} W/cm^2 has not yet been reached; however, such irradiances seem feasible for the near future. For example, even the recently dismantled DOE Nova laser could be linearly upgraded to these levels of irradiance, by a scaling up the compression of the pulse and extending the amplifier chain. Expected progress in the development of beam compression and amplification techniques will raise the irradiances to 10^{24}-W/cm^2 levels; even 10^{29}-W/cm^2 levels are feasible. Some of these concepts are based on chirped pulse amplification (CPA) [5], like, for example, large-aperture, open multipass amplifiers proposed by Fork, et al. [6]. Some advanced concepts are going beyond CPA. For example, Shvets et al. proposed the principle of superradiant amplification, which can allow to amplify intense ultrashort pulses without initial decompression in a plasma medium, i.e. rod overheating is not an issue [7,8].

Next, to allow estimating the proton yield per pulse, assume that 4×10^{23} W/cm^2 irradiances and mean proton energies E_p ~ 1.0 GeV are actually achieved. Suppose further a hypothetical laser producing 8 kJ pulses. (This assumption is not unrealistic, given that a single amplifier chain of Nova was able to generate 4 kJ pulses [9].) For a wavelength of ~ 1.0 μm, and a final recompressed beam of 1.0 meter in diameter, the diffraction-limited spot formed by a parabolic mirror with a 4-m focal length will have a diameter of ~ 8.0 μm. The required irradiance (4×10^{23} W/cm^2) can then be achieved by a single pulse of ~ 40 fs width. Thus, for generation of 1.0-GeV protons the driving laser must generate 8-kJ, 40-fs pulses. Using a protons/laser-pulse-energy ratio of 0.2 (Table I), one can expect the generation of 10^{13} 1.0-GeV protons per pulse. The corresponding irradiation and proton emission characteristics are summarized in the Table I.

One can now estimate the yield of charged pions, generated by reaction (1) at the 3-4 interface of Figure 1. A logical way to obtain this estimate would be to conduct a Monte-Carlo simulation for all possible channels, provided one had time profiles of proton-beam energy- and number densities, and a blueprint of the ODEN. Neither of these conditions currently exists; for the present it must suffice to use the available Monte-Carlo simulation data, for inelastic scattering of proton beams [10]. From that data, for scattering events due to 1.0-GeV protons, one expects about a 25% yield of each type of charged pion (with a slight excess of π^+), with mean energies of ~ 100 MeV. Therefore, one expects a pulse of ~10^{13} 1.0-GeV scattered protons to generate 5×10^{12} charged pions (neglecting kaon production and multi-meson events, due to the lack of initial energy.)

The decay by reaction (2) of charged pions with mean energy 100 MeV will yield an initial pulse of 5×10^{12} muon neutrinos (antineutrinos) with mean energy ~ 20 MeV (see, for example, Ref. 11). In the laboratory frame the decay of 100-MeV pions will be dilated to ~ 45 ns. For in-flight decay a ~ 10-m long shaft (Zone 3) will be required.

Perhaps the most attractive feature of the ODEN is its ability to trigger the detector and separate the products of reactions (2,3) in the time domain. Placing the muon stop 10^2 - 10^3 m away from the interface (2-3) will separate the muon and pion decays temporally (30 - 300 ns) and spatially. The composition of the neutrino pulse formed from pion decay is quite similar to that of atmospheric neutrinos. In both cases the neutrino energies are generally less than the rest energies of the parent pions (as was mentioned above), and correct estimation of neutrino fluxes requires three-dimensional calculations (see Ref. 11,12). Assuming a uniform 4π-steradian angular spread of neutrinos, and a distance-to-detector sufficient to neglect the linear size of the pion decay zone (~ 10 m), one obtains an "effective" emission of ~ 10^{19} v_μ $s^{-1}sr^{-1}$, or 4×10^{11} v_μ sr^{-1} per pulse. A detector placed 100 m away from the pion decay zone will experience a flux of 10^{11} v_μ $s^{-1}cm^{-2}$. A shorter source-to-detector distance or a higher irradiance at the laser target will produce correspondingly higher fluxes.

It is worth noting that optically triggering the detector offers a clear advantage over any other accelerating system. For example, Ref. 12 described temporal synchronization of an RF-driven linear accelerator with a femtosecond-pulsed laser, to study transverse profiles of 50 MeV electron beams. The repetition rates of linac and laser oscillators were adjusted through a phase-locked loop, while the arrival times of laser pulses at the interaction point were synchronized with electron bunches by optical delay line, with 1-2 ps jitter [13]. However, this technique is limited to electron energies within a few hundred MeV. Beyond this the use of transition radiation for optical triggering (as employed in Ref. [13]) becomes impractical [14]. In contrast, ODEN is driven by a single oscillator, so that synchronization is not required; a beam pick-up can be used for setting an optical triggering line.

TESTS ADDRESSING NEUTRINO OSCILLATIONS

The oscillation probability function (OPF) for two-flavor oscillations $v_x \rightarrow v_y$ can be expressed as follows:

$$P_{xy} = \sin^2 2\theta \sin^2\left(\pi \frac{L}{L_o}\right) = \sin^2 2\theta \sin^2\left(\frac{1.27 \Delta m^2 L}{E_v}\right) \quad (6)$$

where θ is the mixing angle, Δm^2 is the mass difference (eV^2), L is the distance (m), L_o is an oscillation length, and E_v is the neutrino energy (MeV). The oscillation length L_o can be defined as a spatial period of the OPF in linear scales of the Δm^2 - $\sin^2 2\theta$ plane (see Figure 2). Using this definition, the OPF is symmetrical with respect to the median plane of the Δm^2 range. One can now find L_o using (6) for a given decimal range of Δm^2 (where the range for $\sin^2 2\theta$ is 0 - 1). This value of L_o can be used as the source-to-detector distance. The following sections will discuss the conditions under which the ODEN can be used for testing $v_\mu \rightarrow v_e$ and $v_\mu \rightarrow v_\tau$ oscillations.

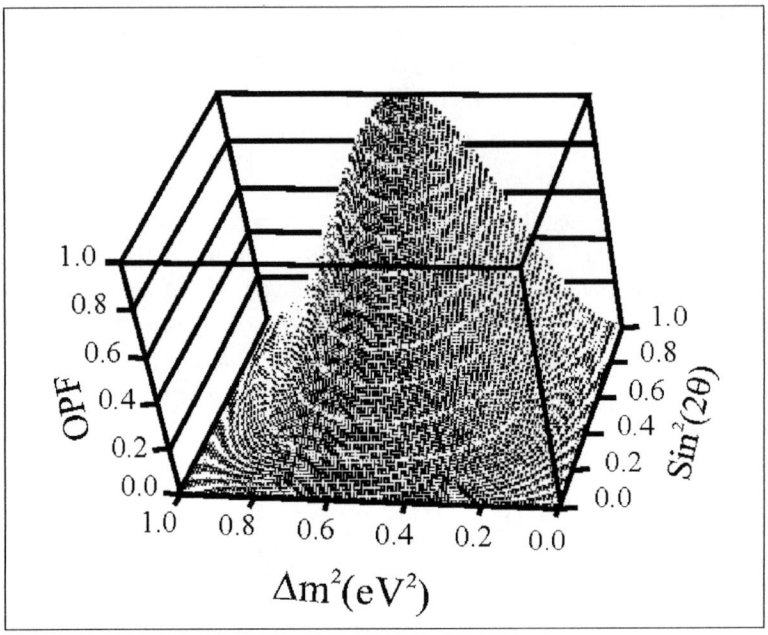

FIGURE 2. Single period of OPF (vertical axis) vs. Δm^2 and $\sin^2(2\theta)$. White contours on the surface mark 0.1 step of OPF.

$v_\mu \rightarrow v_e$ Oscillations

The $v_\mu \rightarrow v_e$ oscillations are the only type of oscillations which can be addressed at the marginal E_v value of ~ 20 MeV. The ODEN can be used to probe for v_e appearance phenomena. The clear advantage of the ODEN will be its ability to provide a pure muon neutrino pulse, one that is uncontaminated by electron flavor.

Studies of v_e appearance were conducted previously by the Liquid Scintillator Neutrino Detector (LSND) at Los Alamos Meson Physics Facility (LAMPF) [15]. In those studies, the LSND was placed 30 m away from a source providing a total flux of $(3.75\pm0.26)\times10^{13}$ v_μ cm^{-2}, accumulated discontinuously over a five-month period [15]. Assuming conservatively that the LAMPF operated for 1% of this period, this flux would correspond to 3×10^8 v_μ cm^{-2}s^{-1}. The ODEN on the other hand, emitting 10^{19} v_μ s^{-1}sr^{-1}, will provide a flux of 10^{12} v_μ s^{-1}sr^{-1} to a detector placed 30 m away.

The OPF maxima at Δm^2 mid-ranges of 5×10^{-1} eV^2, 5×10^{-2} eV^2 and 5×10^{-3} eV^2 correspond respectively to L_o = 500 m, 5.0 km, and 50.0 km. Detectors at these distances will experience neutrino fluxes of 5×10^9, 5×10^7, and 5×10^5 v_μ cm^{-2}s^{-1}, respectively. These fluxes are comparable to those of the LAMPF linac.

$\nu_\mu \rightarrow \nu_\tau$ Oscillations

The threshold restriction on charged current interactions rules out conducting ν_μ disappearance tests at $E_\nu \sim 20$ MeV. However, a 100-fold increase in irradiance produces a corresponding 100-fold increase in E_ν (~ 2.0 GeV). At this energy level, ν_μ disappearance tests become feasible. The Super-Kamiokande collaboration suggested, for $\nu_\mu \rightarrow \nu_\tau$ oscillations of atmospheric neutrinos, a Δm^2 range of $10^{-3} - 10^{-1}$ eV^2 [16]. Corresponding values of L_o lie between 50 and 500 km. Therefore, placement of the first detector near the ODEN, with collinear placement of the second about 100 miles away, permits performance of K2K-type tests with fluxes of 4×10^2 ν_μ cm^{-2}s^{-1} through the second detector. Experimental verification of the OPF via assessment of ν_μ disappearance on the second detector could then provide a practical proof of the oscillation hypothesis.

CONCLUSIONS

This brief paper has provided a first sketch of the concept of the ODEN. The authors realize that they have left many important matters unaddressed; a few of these deserve some mention. First, a more detailed picture is needed of neutrino generation characteristics. This paper presents only "back of envelope" estimates of neutrino energies and fluxes; follow-on feasibility studies should include a more detailed assessment. Further concept development is tightly tied both to progress in high-power pulsed lasers, and to improved understanding of relativistic proton beam-formation phenomena. Progress in these fields will permit assembling a detailed picture of neutrino spectra and densities, and developing statistics on neutrino oscillation tests. This will lead in turn to hardware profiles needed for ODEN development.

ACKNOWLEDGMENTS

Authors would like to express their deep gratitude to Dr. Toshiki Tajima (UT Austin / LLNL) for helpful discussions and encouragement. The precious help of Dr. R. David Hampton (UAH) on preparation of the final version of the paper is greatly appreciated.

REFERENCES

1. Clark, E.L., Krushelnick, K., Davies, J.R., et al., *Phys. Rev. Letters* **84**, 670-673 (2000).
2. Maksimchuk, A., Gu, S., Flippo, et al., *Phys. Rev. Letters* **84**, 4108-4111 (2000).
3. Snavely, R.A. et al., *Bull. Am. Phys. Soc.* **44**, 228 (1999).
4. Cowan, T.E., Hunt, A.W., Phillips, T.W. et al., *Phys. Rev. Letters* **84**, 903-906 (2000).
5. Perry, M.D., and Mourou, G., *Science* **264**, 917-923 (1994).
6. Fork, R.L., Cole, S.T., Gamble, L.J., et al., *Optics Express* **5**, 273-301 (1999).

7. Shvets, G., Fisch, N.J., Pukhov, A., *et al.*, *Phys. Rev. Letters* **81**, 4879-4882 (1998).
8. Malkin, V.M., Shvets, G., and Fisch, N.J., *Phys. Rev. Letters* **84**, 1208-1211 (2000).
9. Schappert, G.T., Caldwell, S.E., Hsing, W.W., *et al.*, LA-13355-PR Progress Report 1995-1996, 78-82 (1996).
10. Metropolis, N., Bivins, R., Storm, M. *et al.*, *Phys. Rev.* **110**, 204-219 (1958).
11. Gaisser, T.K., *Cosmic Rays and Particle Physics*, Cambridge University Press, Cambridge, 1990, p. 42.
12. Lee, H. and Bludman, S.A., *Phys. Rev. D* **37**, 122-125 (1988).
13. Leemans, W.P., Schoenlein, R.W., Volfbeyn, P., *et al.*, *Phys. Rev. Letters* **77**, 4182-4185 (1996).
14. Wartski, L., Roland, S., Lasalle, J., *et al.*, *J. Appl. Phys.* **46**, 3644-3653 (1975).
15. Athanassopoulos, C., Auerbach, L.B., Bauer, D.A., *et al., Phys. Rev. Letters* **75**, 2650-2653 (1995).
16. Fukuda, Y., Hayakawa, T., Ichihara, E., *et al Phys. Rev. Letters* **82**, 2644-2648 (1999).

Multi-Stage, High-Gradient, Cyclotron Resonance Proton Accelerator Concept

J. L. Hirshfield,[*,†] Changbiao Wang,[*] and Robert Symons[¶]

[*]*Beam Physics Laboratory, Yale University, 272 Whitney Ave., New Haven, CT 06511*
[†]*Omega-P, Inc., 345 Whitney Ave., New Haven, CT 06511*
[¶]*Litton Electron Devices Division, 960 Industrial Road, San Carlos, CA 94070-4194*

Abstract. Simulations are presented that show the possibility of high-gradient, high-efficiency cyclotron-resonance acceleration of protons that drift through a cascade of cavities in a uniform axial magnetic field. Means to maintain phase synchronism with a pulsed proton beam are discussed, with resonance frequencies for successive cavities in the cascade lowered by a fixed frequency interval. In an illustrative example, acceleration of a 114 mA (average current) beam by nearly 118 MeV is predicted for a cascade of two cavities at 100 MHz and 94 MHz in a 6.7 T magnetic field, with an efficiency of over 75% and an average acceleration gradient of over 20 MeV/m. Extension of this concept to ~12 stages could allow design of an accelerator to generate a 1.0 GeV, high-power proton beam.

INTRODUCTION

A preliminary design study has been conducted for a novel high current, high gradient, high efficiency, multi-stage proton cavity cyclotron accelerator. As will be shown below, this concept uses available technology to provide energy gains of over 50 MeV/stage, at an acceleration gradient exceeding 20 MV/m in room temperature cavities, with currents of over 100 mA and efficiency over 75%. Acceleration is provided *via* cyclotron resonance, so a strong static magnetic field is required. An innovation in design is described to minimize the diameters of the cavities and the corresponding diameters of the cryomagnets.

Recent developments in high-intensity proton accelerators, and new applications of high-intensity proton beams, have been reviewed by W. T. Weng [1]. It is possible that the high-gradient proton accelerator concept described in this paper could lead to machines for some of the same applications as existing proton linacs [2], but without need for the substantial real estate associated with conventional acceleration gradients of only ~2 MeV/m. These applications include: injectors into synchrotrons for acceleration to multi-TeV energies for fundamental particle physics research; basic studies in condensed-matter physics, materials science, chemistry, polymer science, and structural biology; radiography; materials characterization; medical isotope production; nondestructive evaluation; tomographic surveillance; burning of plutonium, energy amplifier, burning of commercial nuclear waste, and production of tritium. Staged cavity cyclotrons might also be used for acceleration of muons, and an example of this is given to illustrate.

A considerable body of material has been published on the design and operation of *electron* accelerators that invoke cyclotron resonance. These include a 400 keV, 12 MW cyclotron autoresonance accelerator (*CARA*) [3]; a 1.2 MeV, 35 MW *CARA* [4]; and a 150 MeV laser-driven *CARA* (*LACARA*) [5]. *CARA* is a microwave-driven electron accelerator and *LACARA* is laser-driven. In contrast, the proton accelerator described here would be a VHF-band rf machine, operating at frequencies in the range 40-120 MHz for acceleration up to about 0.5-1.0 GeV. *CARA* and *LACARA* operate with non-uniform magnetic fields having spatial profiles designed to preserve cyclotron resonance along the acceleration path, while the proton accelerator described here employs a uniform magnetic field in which slippage in and out of resonance phase occurs in each stage. The magnetic field in the proton accelerator must be uniform across all stages, since—as protons drift from one stage to the next—an increasing field would lead to loss of axial momentum and stalling, while a decreasing field would lead to an unmanageable increase in orbit radius. Successive cavity stages of the proton accelerator operate at successively-lower rf frequencies, to maintain approximate resonance as the proton mass increases: acceleration from 10 keV to 1.0 GeV requires a frequency reduction between the first and last stages of approximately a factor-of-two. This diminution in frequency is opposite to the temporally-increasing frequency variation typical for synchrotrons [6], where the magnetic field also increases. Electron cyclotron accelerators in laboratory tests have be shown to be capable of operation with rf-to-beam power efficiencies of over 90% [3]. High efficiency is also possible in proton cavity cyclotron accelerators, as will be shown below. These superficial comparisons suggest that the accelerator described here is qualitatively different from existing proton machines, most notably in its high acceleration gradient and high efficiency. However, it must be stressed that results shown here are preliminary and are not optimized with respect to any particular system parameter. Clearly, additional design computations, and construction and operation of a prototype, should be carried out to show that the presently-perceived virtues of the concept can be realized in practice.

Results are presented in this paper of a preliminary computational study to illustrate the new concept for proton acceleration. Computations have been carried out for a two-stage system that seems to embody much of the relevant physics; extension to a large number of stages is straightforward. The rf structure for computations described here is shown in Fig. 1. It consists of back-to-back TE_{111} rotating mode room temperature cavities immersed in a strong uniform axial magnetic field. Protons are injected at 10 keV directly from an ion source. The cavities are driven with rf power at different frequencies, with the resonant frequency of the second cavity lower than that of the first. Phase slip of up to ±3-4% occurs in each cavity during acceleration.

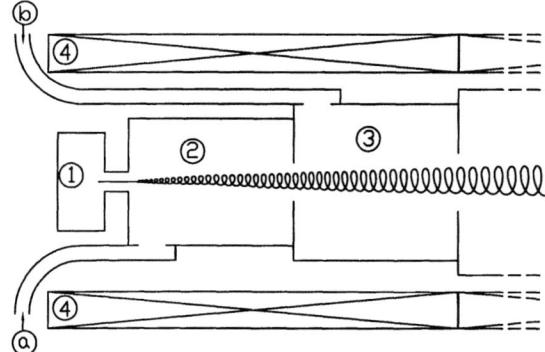

Figure 1. Sketch of two stages in a multi-stage high-gradient cavity proton accelerator. (1) ion source; (2) 100 MHz cavity; (3) 94 MHz cavity; (4) solenoid coil. (a) and (b) are input feeds for the cavities. A proton orbit is also sketched.

RESULTS OF COMPUTATION

For the illustrative example to be presented here, the first cavity is driven with 10 MW of rf power at 100 MHz, the second with 7.7 MW at 94 MHz. The unloaded and beam-loaded quality factors for the first cavity are Q_o = 100,000 and Q_L = 30,000; while for the second cavity they are Q_o = 100,000 and Q_L = 17,000. These values imply that 70% of the incident rf power is absorbed by the proton beam in the first cavity, and 83% in the second cavity; the beam power after the second stage is 13.4 MW. A uniform magnetic field of 67.0 kG threads both cavities. The injected proton beam energy is 10 keV, the final proton energy is 114.0 MeV and the proton current is 117.6 mA. The beam is taken—for purposes of this illustration—to have zero initial emittance and zero initial energy spread. Sixteen computational particles to simulate the beam are injected at time intervals of 1.25 nsec, corresponding to rf phase intervals of $\pi/4$ over two cycles at 100 MHz and to a pulse width of 20 nsec; the injected particles have zero initial radial coordinate. The histories of average energy gain and axial velocity variation along the first cavity are shown in Fig. 2, for three values of axial guide magnetic field B_z = 66.8, 67.0, and 67.2 kG. This cavity has a radius of 110 cm. The energy gain at the end of the cavity (z = 249 cm) is maximum for 67.0 kG, where the decrease in axial velocity within the cavity is not as severe as for the 67.2 kG case. Further increase in B_z is found to lead to a reversal of the sign of axial velocity, i.e. to particle reflection. This stalling effect is attributable to a ponderomotive axial force, which evidently depends on the precise details of the proton orbit. For B_z = 67.0 kG, a net energy gain $\overline{U} - \overline{U}_o$ = 59.5 MeV (γ =1.063) is found during passage through the cavity, where $\overline{U}$ and $\overline{U}_o$ are the ensemble average final and initial proton energies. The small diminution in particle energy for z > 200 cm is attributable to excessive phase slip, since the cyclotron frequency of the protons has fallen to below 94% of the rf frequency at this stage. The average acceleration gradient in the first cavity is 23.8 MeV/m. With a beam current of 117.6 mA, the efficiency of the first cavity is 70%. The strong axial acceleration gradient is possible since protons make a large number of gyrations, and follow a long path moving nearly parallel to the rotating rf electric field. For this example, the protons execute about 48 turns in the first cavity, and reach a final gyration radius of about 17 cm. This rapid,

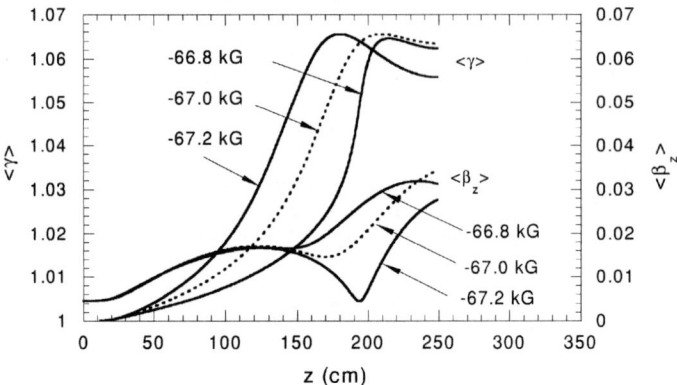

Figure 2. Computed variations of mean proton energy, in units of $\langle\gamma\rangle = 1 + \overline{U}[\text{MeV}]/938$, and mean axial velocity $\langle\beta_z\rangle = \overline{v}_z/c$, as functions of axial coordinate z within the first cavity. Examples are for $I = 117.6$ mA and for other parameters as described in text, and for three values of B-field.

efficient cyclotron resonance acceleration of protons in a TE_{111} cavity with a uniform magnetic field is reminiscent of similar results reported for electrons by Jory and Trivelpiece [7], who showed evidence of acceleration by 100's of keV.

Fig. 3 shows the energy gain and axial velocity for two cavities operated in tandem. The second cavity, operating at 94 GHz, has a radius of 110 cm and a length of 302 cm. The relative phase difference between fields in the first and second cavities

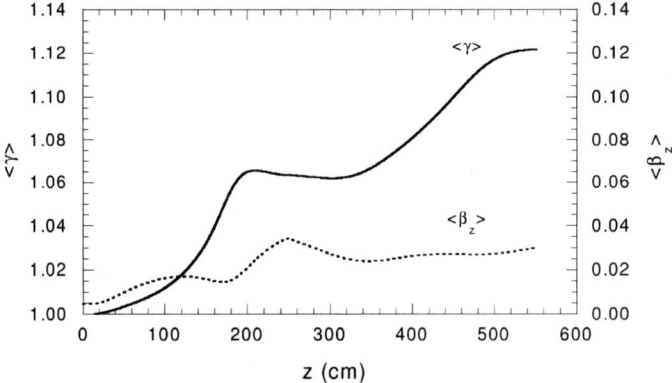

Figure 3. Energy gain for protons in traversing two cavities. Final proton energy is 114 MeV, beam current is 117 mA, average acceleration gradient is 20.7 MeV/m, and rf-to-beam efficiency is 75.6%.

(reckoned at the initial time) is set at 0.70π. This phase difference allows gyrating protons to enter the second cavity with their velocity vectors aligned nearly parallel to the rotating rf electric field; phase synchronism is discussed in the next section of this paper. From Fig. 3, it is seen that the energy gain in the two cavities together reaches 113.96 MeV ($\gamma = 1.1215$), while the axial velocity remains sensibly constant throughout the second cavity. The beam-loaded Q (17,000) and rf drive power (7.7 MW) were adjusted to accommodate the same current (117.6 mA) as in the first cavity; this adjustment required several steps of computational iteration. The protons execute about 43 turns in the second cavity, and reach a final gyration radius of about 22 cm. The average acceleration gradient for both cavities is 20.7 MeV/m. Figs. 4 and 5 show projections in the transverse (x-y) and longitudinal (x-z) planes of the orbit of a single proton during the course of its acceleration.

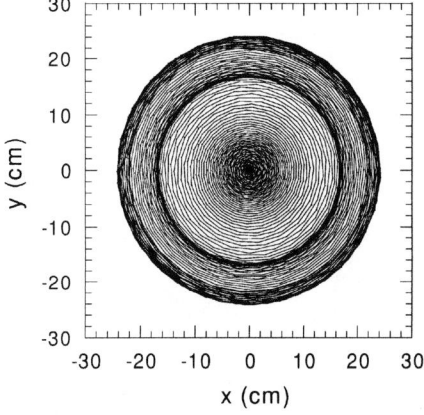

Figure 4. Projection in the transverse plane of the orbit of a proton undergoing acceleration, as in Fig. 3.

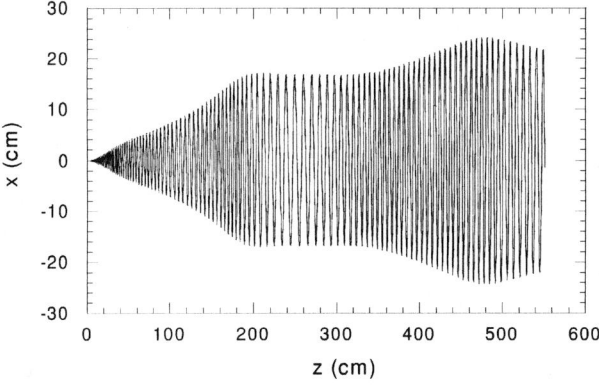

Figure 5. Projection in a longitudinal plane of the orbit of a proton undergoing acceleration as in Figs. 3 and 4. The proton executes 91 turns during acceleration. At $z = 550$ cm, $\beta_\perp / \beta_z \approx 15$.

The same principle that is shown in the above example for acceleration of protons can also be applied to acceleration of other charged species, namely electrons [7], muons, or heavy ions. In view of the current strong interest in muon accelerators [8], it may be instructive to provide an example of muon acceleration at cyclotron resonance using cavities in a strong uniform magnetic field. Fig. 6 shows an example for two cavities in a uniform 67.0 kG B-field, for parameters as follows.

1st cavity: $f = 850$ MHz, $P = 10$ MW, $Q_0 = 40{,}000$, $Q_L = 20{,}000$, $R = 13$ cm, $L = 29$ cm;
2nd cavity: $f = 700$ MHz, $P = 4.0$ MW, $Q_0 = 40{,}000$, $Q_L = 10{,}000$, $R = 15$ cm, $L = 39$ cm.

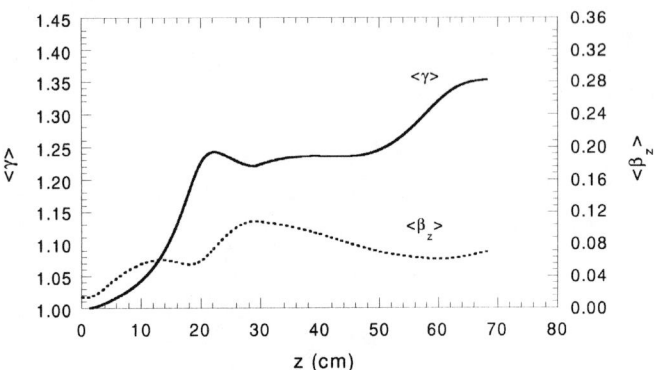

Figure 6. Normalized mean energy $\langle \gamma \rangle = 1 + \overline{U}[\text{MeV}]/105.7$ and axial velocity $\langle \beta_z \rangle$ for muons in the two-cavity cyclotron accelerator described above.

Acceleration in the first cavity is from 10 keV to 23.24 MeV, and in the second cavity to 37.1 MeV. The beam current is 215 mA, maximum orbit radius is 3.8 cm, average acceleration gradient is 54.4 MeV/m, and overall efficiency is 57%. These values compete favorably with conventional muon linacs.

PHASE SYNCHRONISM CONSIDERATIONS

In the cavity cascade concept described here, protons drift from one TE_{111} cavity to the next, but successive cavities must have lower resonance frequencies in order to effect cumulative acceleration, since the imposed axial magnetic field is uniform and the effective proton mass is increasing. Thus, the issues of phase synchronism and stability must be examined carefully. For a single narrow bunch of protons, it is not difficult to imagine acceleration through a cascade of cavities, provided the phases for fields in each cavity are properly adjusted. Specifically, as the proton bunch arrives at each cavity, if the orientation of the electric vector of the rotating TE_{111} mode is parallel to the proton momentum, maximum acceleration is afforded. However, uniform acceleration of a train of proton bunches can occur only if the phases of disparate frequencies in successive cavities are judiciously sequenced, to insure that all bunches have identical histories as they progress through the cascade. This can be arranged if the cavity frequencies decrease in equal increments. For example, suppose the frequency decrease between cavities is 5 MHz, and the cavity frequencies are 100, 95, 90, 85, 80...50 MHz. Also suppose that the initial phases of the fields in each cavity are such as to provide optimized cumulative acceleration to the first proton bunch. If successive bunches are injected at time intervals of $(5\ \text{MHz})^{-1} = 200$ nsec, then the fields seen by each bunch are identical to those seen by the first bunch. This is so since, after each 200 nsec interval, fields in the respective cavities will have advanced by precisely 20, 19, 18, 17, 16...10 cycles, and will thus reconstruct the sequence seen by the first bunch. Since precise reconstruction only occurs at 200 nsec intervals (in this example), protons in a finite width bunch experience slightly different acceleration histories, leading to a finite energy spread for the bunch. However, careful choice of the median phase difference between successive cavities can minimize this spread. Moreover, phase focusing can also occur. In these regards, the cavity cascade has features in common with a conventional rf linac.

Examples of the effects of finite proton bunch width are shown in Fig. 7. Within the first cavity, acceleration is independent of the time of injection. But, due to the phase

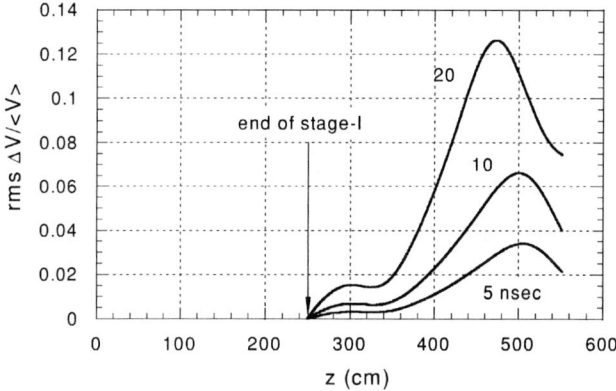

Figure 7. Influence of finite bunch width on beam rms energy spread, for parameters described in text. The examples shown correspond to duty factors of 3%, 6%, and 12%.

dependence of acceleration in the second cavity, energy spread increases with pulse width. For this example, parameters for the first cavity at 100 MHz are as in Figs. 2-5. In the second cavity at 94 MHz, parameters are also as in Figs. 2-5, except for small variations in Q_L and final average beam energy. For the 5, 10, and 20 nsec examples these are 13,200, 13, 600, and 17,000; and 116.2, 116.0, and 114.0 MeV, respectively. The relative initial phase difference between fields in the two cavities is 0.70π. For the 5 nsec case, the final energy spread is seen to be about 2%.

Fig. 8 illustrates the influence of relative phase difference upon beam energy spread. Here, for a 5 nsec pulse width (3% duty factor), acceleration history and evolution of beam rms energy spread are plotted for three values of relative phase shift, namely 0.65π, 0.70π, and 0.75π. It is seen that a final rms energy spread of about 0.7% is found for the 0.75π case, lower by nearly a factor-of-3 than the 0.70π case. Here, the second cavity Q_L and final average beam energies are 16,500, 13,200, and 11,600; and 114.2, 116.2 and 117.3 MeV, respectively. The facts that energy spread decreases significantly after $z \approx 500$ cm, and that minimum spread accompanies maximum final energy, strongly suggest that longitudinal phase focusing is occurring, a phenomenon that is sensitive to small changes in relative phase between cavities.

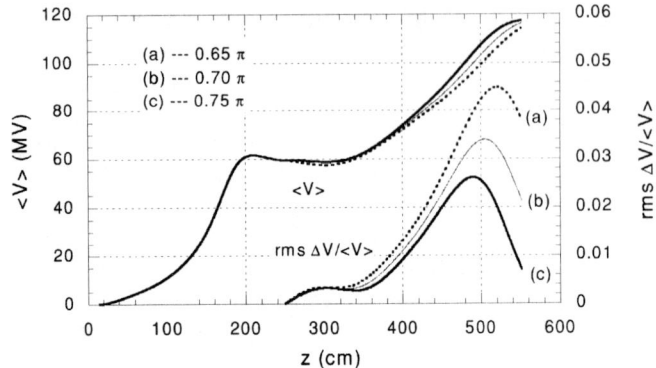

Figure 8. Acceleration history and evolution of rms energy spread in a two-cavity proton accelerator for three values of relative initial phase between fields in the two cavities. Note the significant diminution in energy spread for case (c) towards the end of the trajectory.

It is also instructive to illustrate the bunch shape during acceleration. Figs. 4 and 5 show the orbit for a typical proton, but the instantaneous distribution of charge for a finite-length bunch does not lie along this curve, since orbits of successive protons are rotated in the x-y plane at the rf frequency. To illustrate, Figs. 9(a) and 9(b) show the

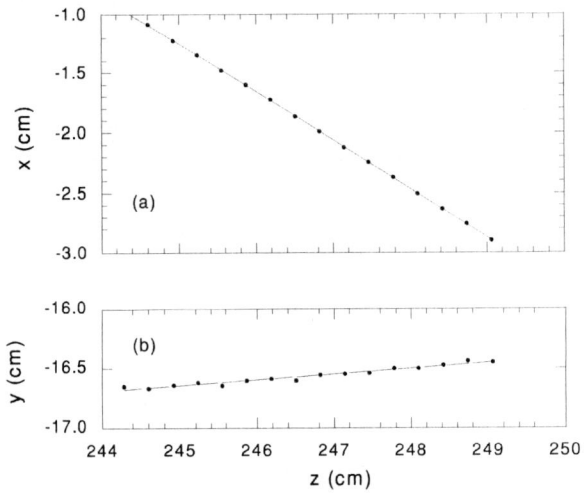

Figure 9. Loci in x-z and y-z for protons in a 5 nsec bunch, at the end of the first cavity. The bunch is seen to be a straight line of charge that gyrates as a whole about the cavity axis.

x-z and y-z loci for 16 protons injected on axis within a 5 nsec long bunch, at the instant (671.5 nsec after injection) that the head of the bunch reaches the end of the first cavity at $z = 249.06$ cm. The particles are seen to lie along a nearly straight line ~4.8 cm long, with a deviation from linearity of less than 0.4 mm. These particle loci can be contrasted with the trace in x-z for the first particle during its final 4.8 cm of travel, which is a half-cycle of oscillation with radius ~17 cm, as shown in Fig. 5. During acceleration, the proton bunch advances in z and rotates about z nearly as a straight rigid object. The small deviations from linearity arise from phase slip between proton momenta and the rf electric field, and from small energy differences during acceleration between the head and tail of the bunch. The near-uniformity of the axial charge distribution within such a long bunch should mitigate against longitudinal instability.

CAVITY DESIGN CONSIDERATIONS

The 100 MHz and 94 MHz TE_{111} cavities for the example of the first two stages of the proton accelerator discussed above have diameters of 220 cm, yet the maximum proton orbit diameters are only 34 and 44 cm. At least in these first stages, most of the cavity volume is not traversed by the proton beam, but is permeated with magnetic flux lines from the surrounding coils. The 67 kG cryomagnet would need a room-temperature bore diameter of perhaps 240 cm (to allow room for the rf feeds, as sketched in Fig. 1); while this is probably within the present state-of-the-art, it would be highly desirable if a means were found to reduce this bore diameter. One such means to achieve this is to employ thick radial vanes in the cavity that provide capacitive loading and thereby reduce the cutoff frequency for the desired dipole modes. When four symmetric vanes are used, the structure resembles that for a radio-frequency quadrupole, except that it is the two degenerate dipole modes that are of interest, rather than the quadrupole modes. To obtain a rotating ("circularly polarized") field, the two dipole modes must be excited in time-quadrature. A simple example of such a structure has been analyzed using the HFSS structure simulation code; the structure outline and computed E-field flux lines are shown in Fig. 10.

For the structure shown, with an outer diameter of 130 cm, a ridge width of 15 cm, and a central gap between opposing ridges of 30 cm, the cutoff frequency for the dipole mode is found to be 73.7 MHz, while the cutoff frequency for the quadrupole mode is 78.97 MHz. Thus a section of structure 222 cm in length would have a dipole resonance frequency of 100 MHz, and a quadrupole resonance frequency of 104 MHz. Operation with Q_L of the order of 1,000-10,000 should thus be possible purely in the dipole mode, without significant coupling by the beam to the quadrupole mode. This idealized example is shown to illustrate the possibility of devising an all-metal structure for the cavities that will have outer diameters significantly smaller than for simple TE_{111} cylindrical cavities. Clearly, it is required to refine the analysis of such structures, including optimizing the shape of the vanes, rounding of sharp corners to reduce surface field strengths, and provision of input coupling for excitation of both degenerate dipole modes in time quadrature.

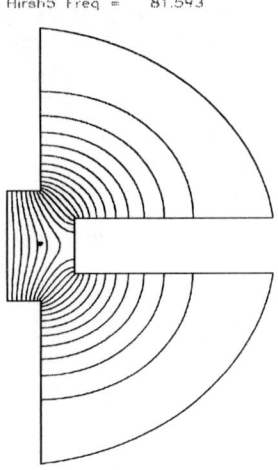

Figure 10. Example of cross-section of 4-vaned cavity structure for proton cyclotron accelerator. Note that only one-half of the structure is shown, after cutting along the vertical axis of symmetry. Electric field lines shown for the dipole mode are seen to be nearly uniform near the axis.

CONCLUSIONS

Preliminary computations indicate that a cavity cascade can be designed and built for cyclotron resonance acceleration of an intense proton beam to energies in the range of 1 GeV. This new concept is characterized by high acceleration gradient for room-temperature cavities (~20 MV/m), high rf-to-beam efficiency (~75%), and high average beam currents (~100 mA). A design innovation is introduced to minimize the diameters of the cavities, and the diameters of the associated cryomagnets.

ACKNOWLEDGMENT

This research was sponsored in part by US Department of Energy.

REFERENCES

[1] W. T. Weng, "Ultra-high intensity proton accelerators and their applications," *Proc. 1997 Particle Accelerator Conf. (Vancouver)* vol. 1, pp. 42-46 (IEEE, 1998).

[2] See, for example, J-M. Lagniel, S. Joly, J-L. Lemaire, and A. C. Mueller, *Proc. 1997 Particle Accelerator Conf. (Vancouver)* vol. 1, pp. 1120-1122 (IEEE, 1998).

[3] J. L. Hirshfield, M. A. LaPointe, A. K. Ganguly, R. B. Yoder, and C. Wang, "Multi-megawatt cyclotron autoresonance accelerator," *Phys. Plasmas* **3**, 2163 (1996); see also "Experimental demonstration of high efficiency electron cyclotron autoresonance acceleration," *Phys. Rev. Lett.* **76**, 2718 (1996).

[4] M. A. LaPointe, C. Wang, and J. L. Hirshfield, "Cyclotron autoresonance accelerator for electron beam dry scrubbing of flue gases," in CP475 *Applications of Accelerators in Research and Industry*, J. L. Duggan & I. L. Morgan, eds., pp. 945-948 (Am. Inst. Phys., 1999).

[5] J. L. Hirshfield and C. Wang, "Laser-driven cyclotron autoresonance accelerator," *Proc. 1999 Particle Accelerator Conf. (New York)* vol. 5, pp. 3630-3632 (IEEE, 1999); "Laser-driven electron cyclotron autoresonance accelerator with production of an optically chopped electron beam," *Phys. Rev. E* **61**, pp. 7252-7255 (2000).

[6] See, for example, E. J. N. Wilson, "Synchrotrons and storage rings," in *Handbook of accelerator physics & engineering*, A. W. Chao & M. Tigner, eds. (World Scientific, Singapore, 1999), pp. 42-44.

[7] H. R. Jory and A. W. Trivelpiece, "Charged-particle motion in large amplitude electromagnetic fields," *J. Appl. Phys.* **39**, pp. 3053-3060 (1968).

[8] Y. Zhao, R. Palmer, R. Fernow, J. Gallardo, and H. Kirk, "A normal conducting accelerator for a muon collider demonstration machine," *Proc. 1997 Particle Accelerator Conf. (Vancouver)* vol. 1, pp. 408-410 (IEEE, 1998).

W-band Accelerator Study in KEK

Xiongwei Zhu, Kazuhisa Nakajima

High Energy Accelerator Research Organization
1-1 Oho ,Tsukuba , Ibaraki ,305 , Japan

Abstract. In this paper, we summarize the W-band accelerator study in KEK. We present a design study on W-Band photocathode RF gun which is capable of generating and accelerating 300pC electron bunch. The design system is made up of 91.392GHz photocathode RF gun and 91.392GHz travelling wave linac cells. Based on the numerical simulation using SUPERFISH and PARMELA and the conventional RF linac scaling law, the design will produce 300pC at 1.74MeV with bunch length 0.72ps and normalized tranverse emittance 0.55mm mrad. We study the beam dynamics in high frequency and high gradient; due to the high gradient, the ponderomotive effect plays an important role in beam dynamics; we found the ponderomotive effect still exist with only the fundamental space harmonics (synchrotron mode)due to the coupling of the transverse and longitudinal motion.

1. INTRODUCTION

Recently, high gradient, high frequency accelerators of mm-scale are of interest in advanced accelerator research now[1]. Research on short wavelength, high gradient, RF driven acceleration has concentrated on 90GHz which is in the range of W-Band (75GHz-110GHz), it involes the understanding of millimeter wave source, high power operation of millimeter wave, beam dynamics in high frequency and high gradient, and technologies for fabrication and measurement of millimeter accelerator.

The scaling of RF accelerator with respect to RF frequency f is as follows
Shunt impedance $r \sim f^{1/2}$,
Accelerating gradient $G \sim f$,
Peak power required per meter $\sim f^{1/2}$.
In this paper, we use the above scaling law and codes (SUPERFISH, PARMELA) to design the W-Band photoinjector, and study the beam dynamics in high frequency and high gradient.

2. DESIGN OF W-BAND PHOTOINJECTOR

The design system is made up of photocathode RF cavity, emittance compensation drift section, and 60 $2\pi/3$ travelling wave cells. We use the conventional RF linac scaling law to scale down the BNL S-Band photocathode RF cavity[2,3] and S-Band SLAC $2\pi/3$ travelling wave cells. The RF cavity length is 0.46cm, the drift length 2.3cm and the linac section length 6.55cm as shown in Fig.1, the drift length is decided in order to get good match between the photoinjector and linac cells.

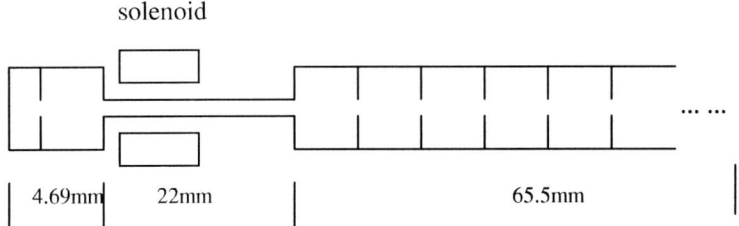

Fig. 1 Setup of the design system

Table 1 and Table 2 give the structure parameters of the RF cavity and linac cell are calculated using SUPERFISH, the shunt impedance and quality factor agree with the scaling law well.

Table 1 The gun cavity design parameters

Inner radius of the cell(mm)	1.3
Radius of the iris(mm)	0.39
Width of the iris(mm)	0.689
Length of the cavity(mm)	4.69
π mode frequency(GHz)	91.392
Shunt impedance(MΩ/m)	269.67
Quality factor	2806

Table 2 The linac cell design parameters

Inner radius of the cell(mm)	1.303
Radius of the iris(mm)	0.409
Width of the iris(mm)	0.182
Length of the cavity(mm)	1.092
$2\pi/3$ mode frequency(GHz)	91.392
Shunt impedance(MΩ/m)	284.5
Quality factor	2435

3. SIMULATION OF W-BAND LINAC

We use PARMELA to study the beam dynamics of our system, the RF cavity works in π mode, while the Linac cells work in $2\pi/3$ mode. Table 3 gives the typical parameters of the photoinjector.

Table 3 Photoinjector parameters

solenoid peak field[T]	5.8
gradient[GV/m]	1
charge[pC]	300
bunch length[ps]	0.72
emittance[mm mrad]	0.55
energy[MeV]	1.74
energy spread	7.2%

Because of the high frequency (small structure), it is difficult to get high charge bunch, the maximum bunch charge we get in the simulation is 300pC, above this value, the beam break up fast with the space charge effect. Fig.2 shows the r.m.s. emittance relation with bunch charge under the condition B=5.8T, and the average RF cavity gradient=1GV/m. Single bunch charge in W-Band linac is constrained by the beam break up. Zimmermann[6] has estimated the longitudinal and transverse wake fields in a W-Band (91GHz) accelerating structure. The transverse wake field is almost completely determined by the structure geometry (iris radius). For a 60pC charge, and a/$\lambda \geq$0.18, the transverse beam break up is negligible.

In order to get low emittance bunch, we use POISSON to design the emittance compensation solenoid whose peak field is as high as 5.8T. We have simulated the emittance dependance on the solenoid field shown in Fig.3 under the condition Q=300pC, average gradient=1GV/m.

In our case, the peak electric field in the half and the full cell of the RF cavity is not the same, the electric field amplitude has effect on RF focusing and defocusing which play a role in the emittance evolution of the beam. Fig.4 shows the beam r.m.s emittance evolution with the photoinjector longitudinal position and Fig.5 shows the beam energy spectrum and phase spectrum at the exit of the photoinjector for bunch charge Q=300pC .

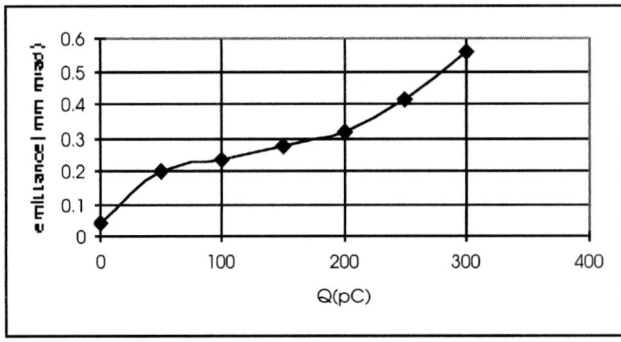

Fig. 2 The r.m.s emittance dependance on the bunch charge

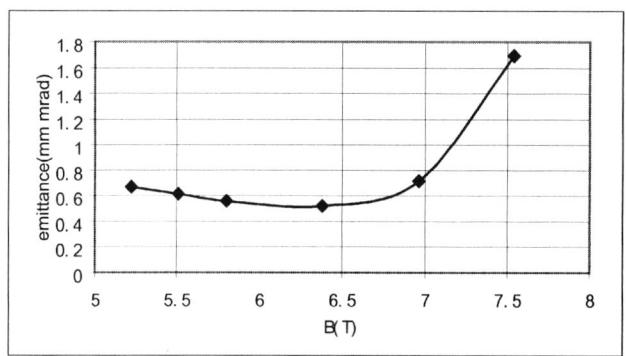

Fig. 3 The r.m.s emittance dependance on the peak magnetic field

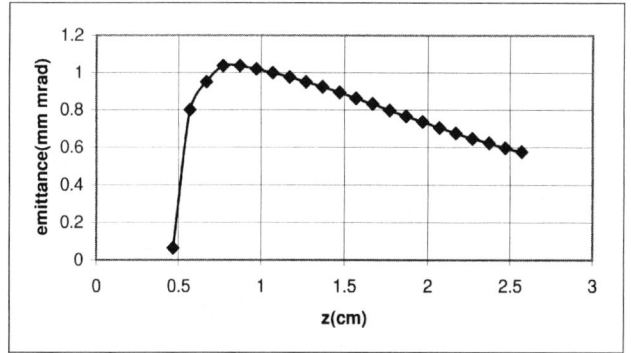

Fig. 4 The r.m.s emittance evolution with the longitudinal distance

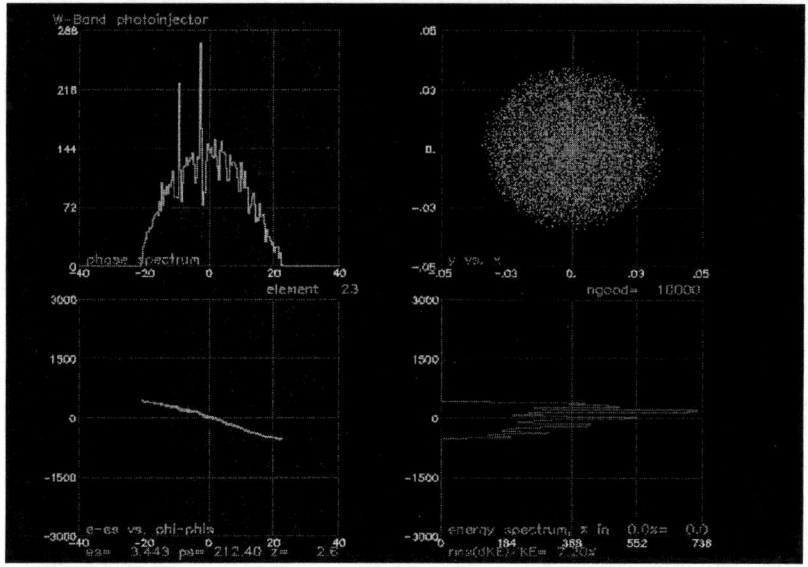

Fig. 5 The bunch energy and phase spectrum at the exit of photoinjector

Up to now, there are few works on the beam dynamics in high frequency and high gradient case. Due to high electric field, the RF focusing and defocusing will become more important. We use 60 linac cells operated at the gradient 1GeV/m to study the beam dynamics in W-band, the cells operated at the fundamental $2\pi/3$ mode. In the simulation, SUPERFISH is used to calculate the RF field coefficients, the high space harmonics were not included. We observe the beam envelope osillation, the periodical time of this kind of osillation approximately like the square of the electric field amplitude approximately, and is almost independent of the bunch charge. Hence, we can conclude that there exists the transverse ponderomotive effect of the fundamental mode. According to the rf ponderomotive focusing theory developed by Hartman and Rosenzweig[4,5], there will be no such effect if only fundamental travelling mode available. If the coupling of the longitudinal and transverse motion is considered, i.e. there still exists the pondermotive effect for the case of fundamental travelling mode.

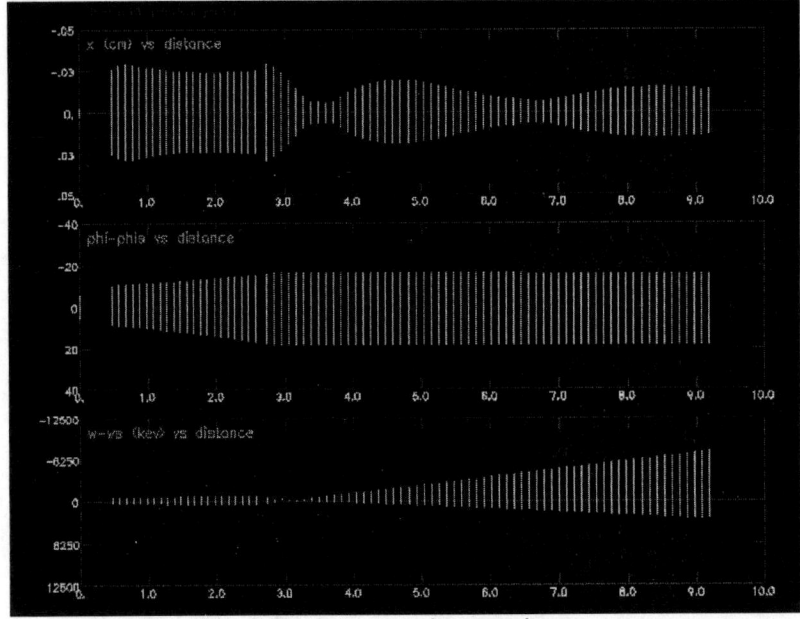

Fig. 6 The beam envelope motion

4. CONCLUSION

The conventional RF linac scaling law works well in W-Band. The pondermotive effect becomes obvious due to the high gradient. There will exist the pondermotive effect in the case of fundamental travelling mode due to the coupling of the longitudinal and transverse motion.

ACKNOWLEDGEMENT

This work is supported by the High Energy Accelerator Research Organization (KEK) and the Japan Society for the Promotion of Science(JSPS). Thanks to C.D.Barnes and Igor Stamenin for the useful disscussion about this work.

REFERENCES

[1] R.H.Siemann, Advanced Electron Linacs (1997).
[2] D.Palmer, the next generation photoinjector, Ph.D Dissertation,Stanford University .
[3] W.Gai, Nuclear Instruments and Methods, A410 (1998)431.
[4] J.Rosenzweig, L.Serafini, Transverse particle motion in radio frequency linear accelerators, Physical Review E,Vol.49,No.2 (1994)1599.
[5] S.C.Hartman, J.Rosenzweig, pondermotive focusing in asisymmetric rf linacs, Physical Review E,Vol.47, No.3 (1993) 2031.
[6] F.Zimmermann, D.H.Whittum, C.K.Ng, Wake fields in a mm-wave Linac, SLAC-PUB-7899 (1998).

Principle of Alternating Gradient Acceleration

Ming Xie
Lawrence Berkeley National Laboratory
Berkeley, CA94720, USA

Abstract.
Principle of alternating gradient acceleration is proposed to overcome one of the major limitations on laser acceleration: phase slippage due to the difference between phase velocity of acceleration wave and longitudinal speed of accelerated particle. According to the principle, net acceleration can be achieved in various ways even under continuous phase slippage. Single particle dynamics of both longitudinal and transverse motions are investigated under alternating gradient acceleration.

I INTRODUCTION

In all known accelerators, acceleration gradient is kept positive along the entire passage of accelerated particles. This is so for a good reason, of course, as synchronous acceleration or all-positive-gradient acceleration can be achieved so easily that there is no need to do it otherwise. However, situation can be drastically different when acceleration wavelength is scaled down, where phase slippage becomes a critical problem limiting net acceleration.

For direct field laser acceleration, major difficulties have been encountered when trying to maintain synchronous acceleration. In all schemes proposed so far, the approach to overcome phase slippage is to either place structures in the near field of intense laser and particle beams or load gases directly in the passage of the beams, in an effort to either terminate the interaction or slow down the acceleration wave. As a result, hefty price has to be paid by sacrificing acceleration gradient and beam current, and in doing so the very attractiveness of the laser acceleration schemes is severely compromised. Phase slippage is also a major limitation for ponderomotive driven acceleration in all laser-plasma-based schemes.

In this report, Principle of Alternating Gradient Acceleration (PAGA) is proposed to overcome the phase slippage problem. According to the principle, net acceleration can be achieved in various ways even under continuous phase slippage. Several schemes based on the principle have been proposed to achieve high average gradient for both direct field and ponderomotive driven accelerations [1–5].

Therefore, synchronous acceleration, a concept so deeply rooted in the conventional accelerator physics and practice, is neither a necessary condition nor worthy of striving for in order to achieve high gradient laser acceleration. It is now the time to cross over a major conceptual threshold.

The dramatic change of operation principle from synchronous acceleration to alternating gradient acceleration may have profound impact on the beam dynamics we used to know. The main purpose of this report, therefore, is to establish a new foundation for both longitudinal and transverse dynamics, and reveal some unique characteristics of beam dynamics under the new principle.

Taking a perspective in a broader scale, PAGA is proposed as one of the two pillars for a new landscape of laser accelerations. The other pillar, Concept of Oversized Open Waveguide (COOW), is required to overcome other two major limitations on laser acceleration: diffraction of laser field and structure damage by high power laser. By supporting each other in harmony, the two pillars together support a great variety of architectures that are capable of overcoming simultaneously the three major limitations and hence to provide high gradient acceleration over much extended length in single stage with durable solid-state acceleration structures. Readers are referred to my recent articles [1–7] for a more complete exposition of the entire framework.

II THE PRINCIPLE

Longitudinal field of a traveling wave seen by a charged particle moving on an orbit defined by $t(z)$ is of the form

$$\mathcal{E}[z, t(z)] = E(z) \cos \psi(z) . \tag{1}$$

Assuming phase velocity of the wave and longitudinal speed of the particle are different, the particle is accelerated over a distance L_a corresponding to a π phase slippage, and then decelerated over L_d correspondingly to another π phase slippage. Respectively, the energy gain and loss are

$$\Delta W_a = \int_0^{L_a} E(z) \cos \psi(z) dz = q E_a L_a T_a ,$$

$$\Delta W_d = \int_{L_a}^{L_{2\pi}} E(z) \cos \psi(z) dz = -q E_d L_d T_d ,$$

where $L_{2\pi} = L_a + L_d$, $E_a(E_d)$ is the magnitude of peak acceleration (deceleration) field, and $T_a(T_d)$ is a transit factor over the distance $L_a(L_d)$ satisfying $0 < T_a \leq 1$ ($0 < T_d \leq 1$). The average gradient over $L_{2\pi}$ is then

$$G_{2\pi} = \frac{\Delta W_a + \Delta W_d}{L_a + L_d} = G_a \frac{[1 - (\frac{E_d}{E_a})(\frac{L_d}{L_a})(\frac{T_d}{T_a})]}{1 + \frac{L_d}{L_a}} ,$$

where $G_a = \Delta W_a/L_a > 0$ is the average gradient over L_a. Thus, the condition requiring net energy gain over a 2π phase slippage follows

$$G_{2\pi} > 0 \quad \Longrightarrow \quad \left(\frac{E_d}{E_a}\right)\left(\frac{L_d}{L_a}\right)\left(\frac{T_d}{T_a}\right) < 1 . \tag{2}$$

Equation (2) is a general statement of the principle of alternating gradient acceleration, applicable to all types of accelerations under continuous phase slippage. It also reveals the variety of approaches in achieving net acceleration.

III LONGITUDINAL DYNAMICS

We now discuss single particle dynamics of longitudinal motion in the field of Eq.(1) under continuous phase slippage. In particular, we consider direct field acceleration by an electromagnetic wave confined in a waveguide. For electron with $q = -e$, energy equation can be written as

$$\frac{d\gamma}{dz} = -ka\cos\psi , \tag{3}$$

where

$$a = \frac{eE(z)\lambda}{2\pi mc^2} , \quad \psi = \omega t - \int_0^z k_z(s)ds .$$

The electron phase may be separated into two parts, $\psi = \psi_l + \psi_\gamma$, where ψ_l depends on accelerator lattice

$$\psi_l = \int_0^z ds[k - k_z(s)] ,$$

and ψ_γ depends on particle's energy and initial phase

$$\psi_\gamma = \int_0^z dsk\left(\frac{1}{\beta(s)} - 1\right) + \psi_0 , \quad \psi_0 = \omega t_0 .$$

We are interested in the relativistic regime satisfying

$$\left|\frac{d\psi_l}{dz}\right| \gg \left|\frac{d\psi_\gamma}{dz}\right| .$$

Consequently, we may define a lattice period, $L_{2\pi}$, according to the condition for a 2π phase slippage by

$$|\Delta\psi_l(\Delta z = L_{2\pi})| = 2\pi .$$

In addition, we introduce $a(z) = a_s(z)f_l(z)$, where $a_s(z)$ is assumed to vary slowly in the scale $L_{2\pi}$, $f_l(z)$ and $k_z(z)$ are periodic functions with period $L_{2\pi}$ and $0 < f_l(z) \leq 1$.

Taking average of Eq.(3) over the fast scale $L_{2\pi}$, one gets

$$\frac{d\bar{\gamma}}{dz} = -ka_s(C_l \cos\psi_\gamma - S_l \sin\psi_\gamma) ,$$

where $\bar{\gamma} = <\gamma>$, $C_l = <f_l \cos\psi_l>$, $S_l = <f_l \sin\psi_l>$ are averaged quantities. Obviously, net energy exchange is possible if $C_l \neq 0$ or $S_l \neq 0$. For convenience we may set the lattice such that $C_l = 0$ and $S_l > 0$. Hence, longitudinal equations of motion in slow scale become

$$\frac{d\bar{\gamma}}{dz} = ka_s S_l \sin\psi_\gamma , \qquad \frac{d\psi_\gamma}{dz} = \frac{k}{2\bar{\gamma}^2} . \qquad (4)$$

It is important to note that Eq.(4) is casted into the same form as the usual equations of motion for a linac [8]. As a result, our entire wealth of knowledge about synchronous acceleration can be directly applied to the case of alternating gradient acceleration. For example, period of small amplitude synchrotron oscillation follows directly from Eq.(4)

$$\lambda_s = \frac{\bar{\gamma}^{3/2} \lambda}{\sqrt{a_s S_l \cos\psi_\gamma}} . \qquad (5)$$

IV   TRANSVERSE DYNAMICS

Transverse motion of an electron under direct field alternating gradient acceleration by a TM mode can be derived from the radial equation of motion

$$\frac{d(\gamma r')}{dz} = -\frac{e}{mc^2}(E_r - \beta c\mu_0 H_\phi) , \qquad (6)$$

where $r' = dr/dz$. The transverse field components can be related to the acceleration field, $E_z(r, z, t) = J_0(k_\perp r)\mathcal{E}(z, t)$, in paraxial approximation by [8]

$$E_r(r, z, t) = r \left.\frac{\partial E_r}{\partial r}\right|_{r=0} = -\frac{r}{2\epsilon}\frac{\partial(\epsilon\mathcal{E})}{\partial z} ,$$

$$H_\phi(r, z, t) = r \left.\frac{\partial H_\phi}{\partial r}\right|_{r=0} = \frac{\epsilon r}{2}\frac{\partial \mathcal{E}}{\partial t} ,$$

where $\epsilon = \epsilon_0 \epsilon_1$, $\epsilon_1 = \nu_1^2$, $\nu_1 = 1 + \delta\nu_1$. Hence from Eq.(6)

$$\frac{d(\gamma r')}{dz} = \frac{er}{2mc^2}\left[\frac{\partial\mathcal{E}}{\partial z} + \frac{\beta\epsilon_1}{c}\frac{\partial\mathcal{E}}{\partial t} + \left(\frac{1}{\epsilon_1}\frac{\partial\epsilon_1}{\partial z}\right)\mathcal{E}\right] . \qquad (7)$$

Upon introducing reduced variable $Q = \sqrt{\bar{\gamma}} r$ [9], Eq.(7) can be converted into the form of Hill's equation

$$Q'' + K_r Q = 0 , \qquad (8)$$

where

$$K_r = \frac{k^2}{\gamma}\left(\frac{a^2 \cos^2\psi}{4\gamma} - \frac{a\sin\psi}{2\gamma^2} + a\delta\nu_1 \sin\psi - ab\cos\psi\right),$$

$$b = \frac{\lambda}{2\pi}\frac{\partial \delta\nu_1}{\partial z} .$$

There are four small parameters: $1/\gamma \ll 1$, $a \ll 1$, $|b| \ll 1$ and $|\delta\nu_1| \ll 1$. To compare the relative magnitude of b and $\delta\nu_1$, it is convenient to introduce a profile function

$$f_d(z) = \frac{1}{1+e^{-(z+L_d/2)/\Delta}} - \frac{1}{1+e^{-(z-L_d/2)/\Delta}} , \qquad (9)$$

where L_d is the width of the profile, Δ is the width of rising or falling edge, and $0 \leq f_d \leq 1$. Note, with $\Delta/L_d \ll 1$, f_d approaches a square profile. Using Eq.(9), we have

$$\left|\frac{b}{\delta\nu_1}\right|_{max} = \frac{\lambda}{2\pi}\left|\frac{\partial \ln(f_d)}{\partial z}\right|_{max} = \frac{\lambda}{2\pi\Delta} .$$

Thus, for highly relativistic particle in a medium loaded structure with slow rising and falling edges satisfying $(\lambda/2\pi\Delta) \ll 1$, K_r is dominated by one term

$$K_r = \frac{k^2 a \delta\nu_1 \sin\psi}{\gamma} .$$

To realize alternating gradient acceleration, we consider a specific scheme [1,2] in which an oversized open waveguide supporting TM_{01} mode is periodically loaded along z direction with plasma layers, each of width L_d and separated by a vacuum of distance L_a. The plasma density profile can be specified by $n(z) = n_0 f_d(z)$, corresponding to a reduction in the index of refraction $\delta\nu_1 = -f_d(z)/2\gamma_p^2$, where $\gamma_p = \omega/\omega_p \gg 1$ and ω_p is the plasma frequency associated with the peak density n_0. In this scheme, $E_a = E_d = E$, $T_a = T_d = 2/\pi$, $f_l = 1$, and

$$S_l = \frac{2}{\pi}\left(\frac{1-\xi}{1+\xi}\right) ,$$

where $\xi = L_d/L_a$, which varies in the range $\{0-1\}$, and under the condition $\Delta/L_d \ll 1$, f_d is approximated by a square profile, shifted in z from Eq.(9)

$$f_d(z) = \begin{cases} 0 & : \quad 0 \leq z \leq L_a \\ 1 & : \quad L_a < z \leq L_{2\pi} \end{cases} . \qquad (10)$$

When considering dynamics in the fast scale, we may neglect slow synchrotron motion by setting $\psi_\gamma = \psi_0$, and furthermore, $\psi_0 = \pi/2$ for maximum acceleration, thus

$$\frac{d\gamma}{dz} = ka\sin\psi_l \,, \tag{11}$$

$$K_r(z) = -\kappa f_d(z)\cos\psi_l(z) \,, \tag{12}$$

where $\kappa = k^2 a/2\gamma\gamma_p^2$. Noting that $k_z = k\nu_1(1 - 1/2\gamma_g^2)$ [1,2,4,5], where γ_g is a quantity determined by the waveguide, the lattice phase ψ_l can be expressed as

$$\psi_l(z) = \left(\frac{k}{2\gamma_g^2}\right)z + \left(\frac{k}{2\gamma_p^2}\right)\int_0^z f_d(s)ds \,. \tag{13}$$

Thus by definition, we obtain from Eq.(13), using Eq.(10)

$$L_a = \gamma_g^2 \lambda \,, \quad L_d = \frac{\gamma_p^2 \lambda}{1 + (\gamma_p/\gamma_g)^2} \,.$$

Finally, the lattice phase ψ_l can be reduced to

$$\psi_l(z) = \begin{cases} \frac{\pi}{L_a} z & : \; 0 \leq z \leq L_a \\ \frac{\pi}{L_d}(z - L_a) + \pi & : \; L_a \leq z \leq L_{2\pi} \end{cases} \,. \tag{14}$$

Indeed, with Eq.(14), we observe $C_l = 0$.

Given Eqs.(8,10,12,14), an alternating gradient focusing lattice is fully specified. Applying the smooth approximation [10] to the lattice, the β-function is found to be

$$\beta_t = \left(\frac{\gamma m c^2}{G_{2\pi}}\right) f_\xi \,, \tag{15}$$

where

$$G_{2\pi} = \frac{d(\bar{\gamma}mc^2)}{dz} = eES_l \,,$$

$$f_\xi = \frac{4}{\pi\sqrt{2\xi(1+\xi) - 16\xi^2/\pi^2}} \,.$$

It is noted that because of the scaling $\beta_t \sim \gamma$, electron beam size remains constant during acceleration for the AG focusing lattice discussed here. Thus, the acceptable value for normalized rms beam emittance can be determined by the condition

$$\epsilon_n \leq \frac{G_{2\pi}\sigma_{max}^2}{mc^2 f_\xi} \,, \tag{16}$$

where σ_{max} is the maximum rms beam size limited by considerations of waveguide aperture and beam dynamics.

V EXAMPLE

An example of alternating gradient acceleration is given in table 1 for highly relativistic electron satisfying the conditions $(\gamma/\gamma_g)^2 \gg 1$ and $(\gamma/\gamma_p)^2 \gg 1$. The example is taken from [2], although the acceleration scheme was also presented in [1]. In this example, direct field acceleration by TM_{01} mode is taken place in a capillary waveguide periodically loaded with plasma layers. Taking $\sigma_{max}/R = 0.1$, it then follows from Eq.(16) that $\epsilon_n \leq 0.25$mm-mrad. For electron energy of 1 GeV, Eq.(5) gives $\lambda_s = 4.9$m assuming $\cos\psi_\gamma = 1$ for maximum longitudinal focusing, and Eq.(15) gives $\beta_t = 3.1$m.

Table 1. Direct Field Alternating Gradient Acceleration

λ [μm]	1	P_0 [TW]	10	E_a [GV/m]	1.9
R/λ	200	$n_0[10^{17}/\text{cm}^3]$	1.1	E_s [GV/m]	1.5
ν_2	1.5	L_{attn} [m]	5.3	$G_{2\pi}$ [GeV/m]	1
γ_g	328	L_a [cm]	10.8	ΔW_a [MeV]	127
γ_p	100	L_d [cm]	0.91	ΔW_d [MeV]	-11
S_l	0.53	f_ξ	3.1	a_s [10^{-3}]	0.59

VI CONCLUSIONS

The analysis of single particle dynamics reveals no show-stopper for alternating gradient acceleration. In fact, the longitudinal dynamics in the slow scale is shown to be identical to that for synchronous acceleration. On transverse dynamics, the AG focusing inherent from the direct field acceleration is found to be surprisingly favorable for its weaker focusing strength, resulting in much relaxed alignment tolerance and smaller emittance growth. Although the dynamics presented here is focused on direct field acceleration, the general dynamical characteristics under alternating gradient acceleration are expected to hold for ponderomotive driven acceleration as well. Effects of plasma wakefields driven by laser and particle beams will be treated elsewhere. This work was supported by the Director, Office of Science, Office of High Energy and Nuclear Physics, High Energy Physics Division, of the U.S. Department of Energy under contract No.DE-AC03-76SF00098.

REFERENCES

1. M. Xie, PAC99 Conf. Proc., 3678 (1999).
2. M. Xie, LBNL-42783, 1999.
3. M. Xie, LBNL-42784, 1999.
4. M. Xie, AIP Conf. Proc. **472** for AAC Workshop, 701 (1998).
5. M. Xie, EPAC98 Conf. Proc., 830 (1998).
6. M. Xie, LBNL-40558, 1997.

7. M. Xie, PAC97 Conf. Proc., 660 (1997).
8. R. Helm and R. Miller, in *Linear Accelerators*, edited by P. Lapostolle and A. Septier (North-Holland, 1969), p115.
9. M. Reiser, *Theory and Design of Charged Particle Beams*, (Wiley, 1994), p77.
10. K. Symon et al., Physical Review, **103**, 1837 (1956).

The Status of Ionization Cooling Tests for Muon Colliders and Neutrino Factories

David B. Cline

Center for Advanced Accelerators
Department of Physics and Astronomy, Box 951547
University of California, Los Angeles, CA 90095-1547 USA

Abstract. The concept of ionization cooling is about 30 years old, but it has never been tested directly. Friction cooling could be useful if new modes of accelerators are used. We discuss the cooling needs for muon colliders at neutrino factories. We then mention a long-term program, MUCOOL, and near-term tests of the cooling principle. We describe an experiment in a muon beam at TRIUMF now, MUSCAT, that is related to this issue and some ideas for a modest cooling experiment.

INTRODUCTION

The modern development of a $\mu^+\mu^-$ collider requires intense ionization cooling.[1-6] The current ideas for a neutrino factory also require cooling, albeit less than the $\mu^+\mu^-$ collider. There are two forms of ionization cooling: (A) frictional and (B) higher energy cooling. The first process has been demonstrated in various experiments, while the second has not yet been demonstrated experimentally. This second form of cooling requires a delicate balance of energy loss (dE/dx) and heating (multiple and plural scattering). Furthermore, depending on which side of the Bragg peak that one uses, the longitudinal beam energy can be heated or "cooled."

In this report, we discuss the prospects for new measurement of the multiple scattering and some ideas for the demonstration of cooling.

FRICTION COOLING REVISITED

The only form of ionization cooling that has been demonstrated experimentally is friction cooling. Both μ^+ and μ^- have been cooled by this method. Furthermore, very cold beams can be produced (see Refs. 3-5 for some discussions of friction cooling in the context of a $\mu^+\mu^-$ collider). At the Catalina Neutrino Factory-$\mu\mu$ Collider Collaboration meeting in May 2000, a group of physicists from Columbia and UCLA discussed a new study of the use of this cooling method mainly for the $\mu^+\mu^-$ collider option. In this case, there is a trade off between the very cold $\mu^\pm$ beams that can be produced and the $\mu^\pm$ intensity and high gradient accelerator. A schematic of the proposed friction-cooled $\mu^\pm$ source is shown in Fig. 1. A Higgs factory requires very cold beams.

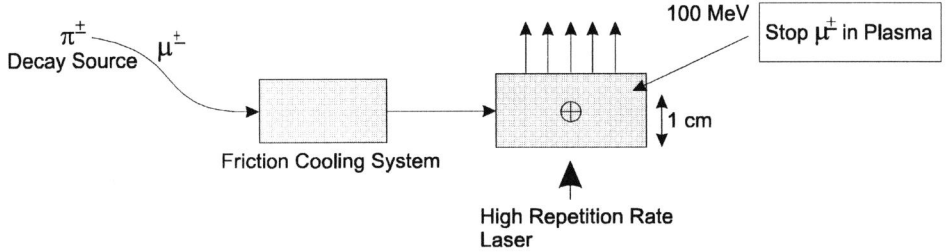

FIGURE 1. Schematic of a possible friction cooling experiment in a plasma to achieve high accelerating gradient.

THE TRIUMF MUON SCATTERING EXPERIMENT, MUSCAT

In order to carry out an experimental study of multiple scattering by muons in low-z targets, a group (MUSCAT) has been formed to do the experiment at TRIUMF.[8] Figure 2 shows a very old set of direct measurements of multiple scattering in the appropriate γ region.[9] Figure 3 shows the MUSCAT detector at TRIUMF.

The first run of the MUSCAT experiment this Spring and Summer was very successful. The UCLA team provided the fast timing system that was used to cleanly separate the $\mu^\pm$ particles from the rest by time of flight. In this run, low-z targets beyond hydrogen were studied. In the Spring of 2001, the next MUSCAT run will take place with the use of liquid hydrogen targets. Once the data is fully analyzed, we will have a complete picture of the "heating" components of the muon cooling problem.

IDEAS FOR A MODEST COOLING EXPERIMENT

The earliest ideas for a $\mu^+\mu^-$ collider discussed the possibility of ionization cooling.[10] Many people are now thinking about a modest ionization cooling experiment, including the author, K. MacDonald, the MUSCAT team, and others. Figure 4 shows the schematic of one such project by P. Gruber of CERN.

The MUCOOL program (at FNAL and part of the Neutrino Factory-$\mu\mu$ Collider Collaboration goals) has encountered difficulties and is unlikely to use a real muon beam to cool muons. Thus, the modest cooling projects discussed here could be the only near-term experimental test of ionization cooling.

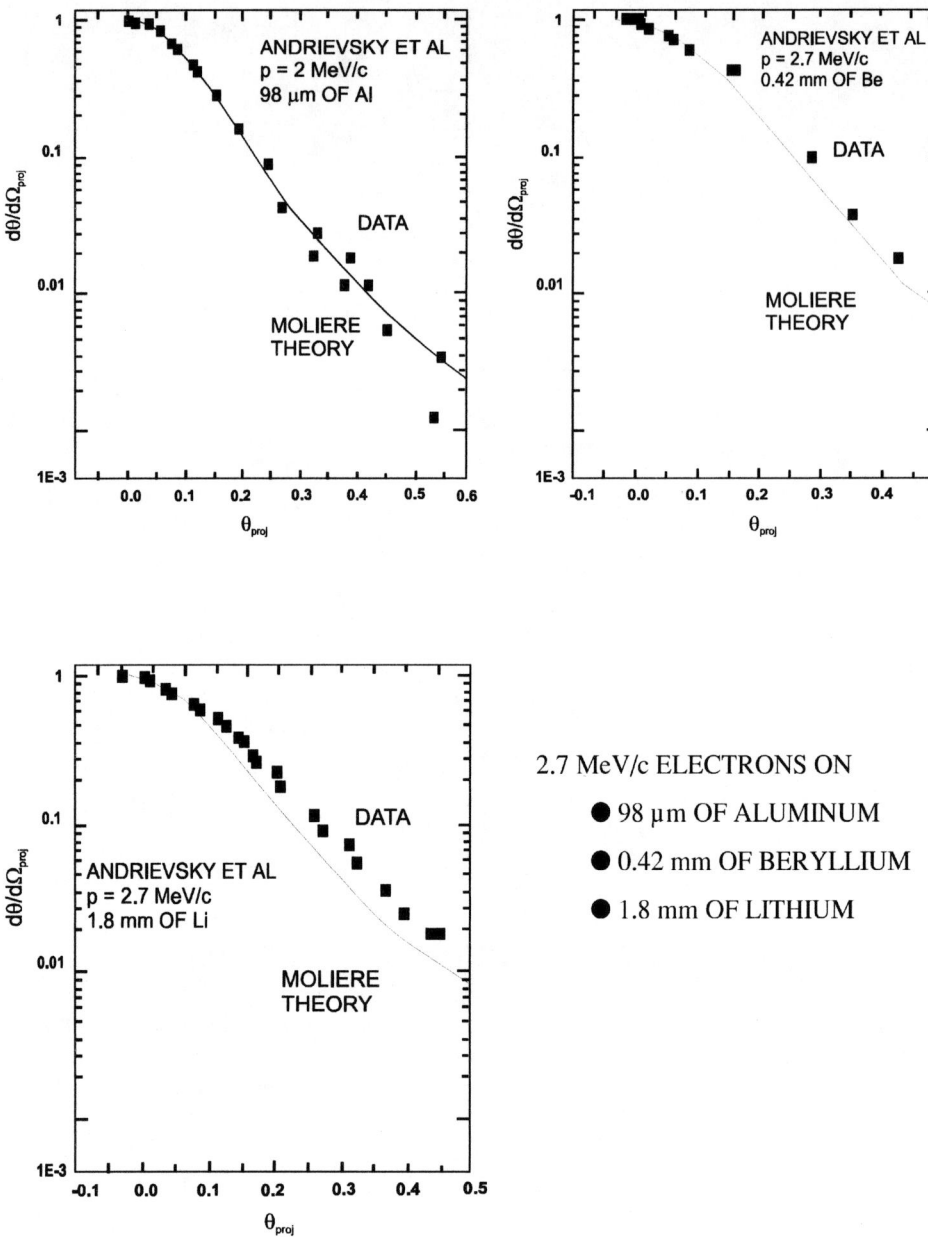

FIGURE 2. Current electron multiple scattering data in the correct energy range for a $\mu^{\pm}$ collider/neutrino factory.

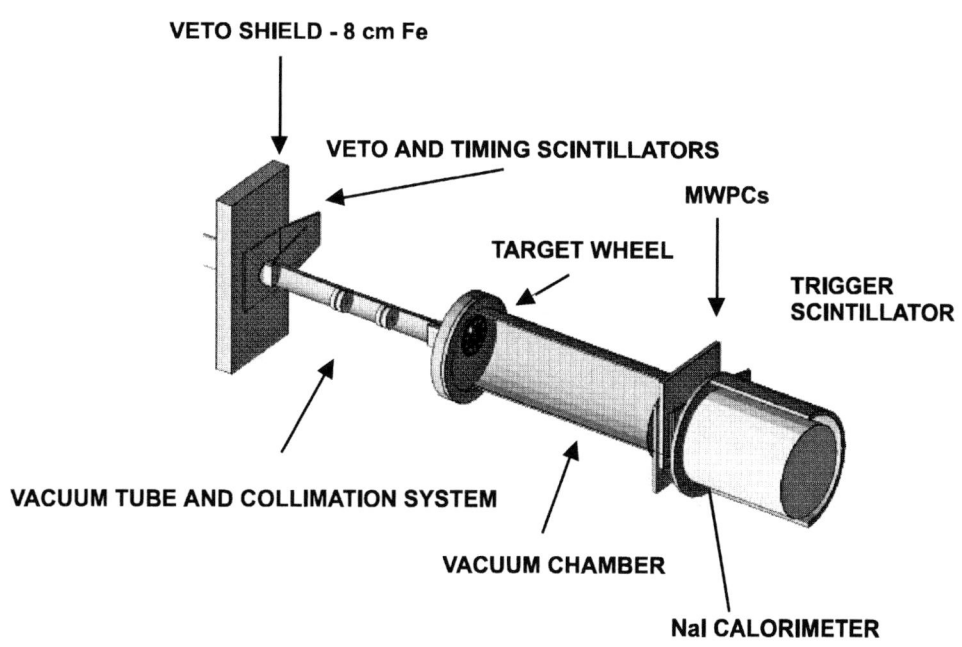

FIGURE 3. MUSCAT detector at TRIUMF.[8]

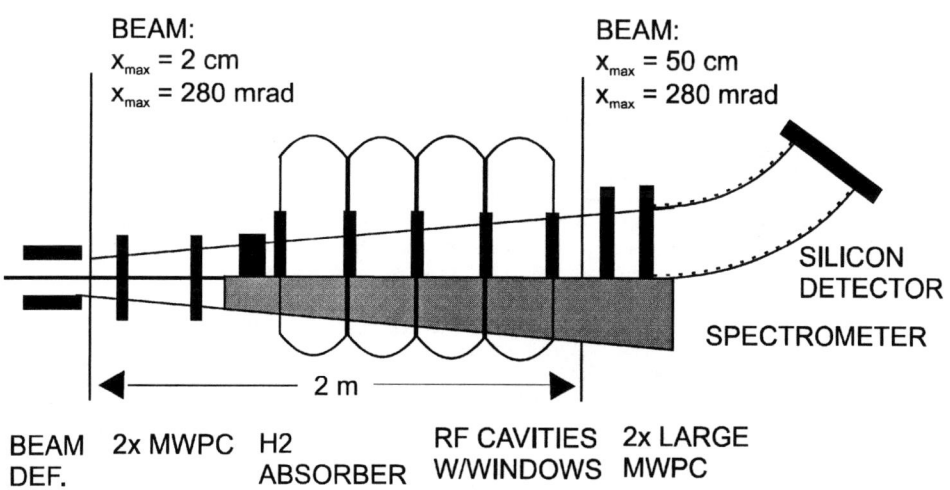

FIGURE 4. Schematic of an ionization cooling demonstration experiment.[11]

SUMMARY

Ionization cooling is an extremely important concept that has only been tested at very low energies after 30 years or more of discussion. A first step is to measure directly the effects of multiple scattering in low-z materials. The MUSCAT experiment is making a good start on this problem. It is essential to move to the next step soon to test ionization cooling at higher energies. While MUCOOL was to do this initially, it looks like a more modest cooling demonstration will be required in the near future.

REFERENCES

1. Cline, D., *Nucl. Instrum. Methods* **A350**, 24 (1994).
2. See collection of papers from the First Workshop on the Physics Potential and Development of $\mu^+\mu^-$ Colliders (Napa, CA, 1992), in *Nucl. Instrum. Methods* **A350**, 24-56 (1994).
3. *Physics Potential and Development of $\mu^+\mu^-$ Colliders* (Proc., 2nd Wksp., Sausalito, CA, 1994), edited by D. B. Cline, AIP Conference Proceedings 352, New York, 1996.
4. *Physics Potential & Development of $\mu^+\mu^-$ Colliders* (Proc. 3rd Intl. Conf., San Francisco, Dec. 1995), edited by D. B. Cline, *Nucl. Phys. B* (PS), **51A** (1996).
5. *Beam Dynamics and Technology Issues for $\mu^+\mu^-$ Colliders* (Proc., 9th Advanced ICFA Beam Dynamics Workshop, Montauk, LI, NY, Oct. 1995), edited by J. C. Gallardo, AIP Conference Proceedings 372, New York, 1996.
6. *Future High Energy Colliders* (Proc., Symp., Santa Barbara, CA, Oct. 1996), edited by Z. Parsa, AIP Conference Proceedings 397, New York, 1997.
7. Cline, D.B., "A Higgs Factory $\mu^+\mu^-$ Collider," in *New Directions for High-Energy Physics* (Proc., 1996 DPF/DPB Summer Study, Snowmass, CO, June-July, 1996), edited by D. G. Cassel, L. Trindle Gennari, and R. H. Siemann, IEEE, New York, 1997, pp. 593-597.
8. Attwood, D., Bell, P., Benveniste, S., et al., "MUSCAT - the Muon Scattering Experiment," presented at the 2nd Int'l Wksp. on Neutrino Factories based on Muon Storage Rings, NUFACT 2000 (Monterey, CA, May 2000) and to be published in NIMPR-A.
9. Andrievsky, J., et al., *J. Phys.* **6**, 278 (1942).
10. Early references for $\mu\mu$ colliders are: Perevedentsev, E. A. and Skrinsky, A. N., in *Proc., 12th Int. Conf. on High Energy Accelerators* (Madison, WI, 1983), editors R. T. Cole and R. Donaldson, p. 485; also Neuffer, D., *Part. Accel.*, **14**, 75 (1984); also Neuffer, D., in *Advanced Accelerator Concepts* (Proc., Symp., Madison, WI, 1986), edited by F. E. Mills, AIP Conference Proceedings 156, New York, 1987, p. 201.
11. Gruber, P. (CERN) private communication (2000).

Wake-Amplification by a Solid-State Active Medium

L. Schächter[1], E. Colby[2] and R.H. Siemann[2]

[1]*Department of Electrical Engineering, Technion, Haifa 32000, ISRAEL*
[2]*SLAC, Stanford University, MS-7, 2575 Sand-Hill Road, Menlo Park, CA 94025*

Abstract. A relativistic bunch moving in the vicinity of a dielectric slab excites Cerenkov radiation. The spectrum of this radiation depends on the dielectric coefficient, the energy of the bunch and its geometric shape. If the dielectric slab is active and the bunch excites the proper frequency, the wake will be amplified. We determine the dependence of the spatial growth of the bunch on the geometric and optical parameters of the system. The wake propagation in the active medium is analyzed and compared with the spontaneous radiation generated and amplified in the medium.

INTRODUCTION

In all the acceleration schemes suggested in the past two decades, the initial energy is either stored in the driving electron beam or laser pulse[1-2]. In recent years we have shown that there are some important advantages if the energy required for acceleration is stored in the background medium [3-6]. Firstly, this type of interaction mechanism virtually eliminates the constraint imposed in the other schemes on the driving electron beam or laser pulse that must carry all the necessary energy. Secondly, the longitudinal electric field is inherent in this acceleration scheme without the difficulties associated with focusing i.e., Rayleigh length in laser driven schemes. In order to illustrate the concept in the simplest possible way, we have used in general a background gas characterized by

$$\varepsilon(\omega) = \varepsilon_r + \frac{\omega_p^2}{\omega_0^2 - \omega^2 + 2j\omega\omega_1} \quad (1)$$

where $\omega_p^2 = (N_1 - N_2)(2\omega_0 \mu_{12}^2)/\hbar\varepsilon_0$, ω_0 is the resonance angular frequency, ω_1 is the resonance width and ω_p defines "plasma" frequency in terms of the microscopic parameters of the constituents; $N_1[N_2]$ is the density of atoms in which the bound electron is in the low [high] state; μ_{12} is the dipole-moment of the transition from one energy state to the other. Clearly this quantity is negative if the population is inverted ($N_2 > N_1$). In a medium described by Eq.(1) the reaction-field of the medium to the motion of a relativistic point-charge on the particle itself is

$$E_z = \frac{q}{4\pi\varepsilon_0\varepsilon_r} \frac{2\omega_p^2}{c^2\varepsilon_r} = \frac{q}{4\pi\varepsilon_0\varepsilon_r} \frac{\omega_0}{c} \frac{4\mu_{12}^2(N_1-N_2)}{c\hbar\varepsilon_0\varepsilon_r} \qquad (2)$$

and in case of population inversion this corresponds to an accelerating field. As an example consider a resonance at 10μm, a dipole-moment $\mu_{12}= (1.6\times10^{-19}) \times (2\times10^{-10})$, a point charge that consists of 10^7 electrons and a large population inversion $N_2-N_1 = 10^{25} m^{-3}$. The accelerating gradient in such a case is of the order of 1.3GV/m – ignoring saturation in the process.

There is a practical difficulty associated with this scheme. Since the gradient is proportional to the number of electrons in the bunch, the latter needs to be large in order to have a gradient that is competitive with other mechanisms. In addition, even if we knew how to generate such a large number of electrons in a pulse that is much smaller than the resonance wavelength, there still is the problem of the space-charge that tends to spread its spatial distribution. For these reasons we suggested [3] to separate the bunch into two: a trigger and a witness bunch. The triggering bunch consists of a relatively small number of electrons (say 10^4-10^5) depending on the geometry of the structure and the energy of the electrons or a train of such micro-pulses that form one macro-pulse. This macro-pulse triggers the medium in the sense that it generates Cerenkov radiation. The latter is amplified by the active medium. This amplified wake has a phase velocity that is identical with the velocity of the trigger bunch. The witness bunch that trails the trigger bunch, is accelerated by this amplified wake. In fact, the accelerated bunch should be also split into a train of micro-bunches in order to avoid de-bunching. Moreover, the witness macro-bunch may be located into the saturation region in order to maintain the energy variation from one micro-bunch to another as small as possible.

WAKE IN A UNIFORM MEDIUM

Let us now show a simple evaluation of such a wake in space and in time. Using the background gas model (1) but this time the gas and the electromagnetic field are confined by a circular waveguide of radius R. The wake a point-charge generates is given by

$$E_z(r,z,t) = \frac{-q}{4\pi\varepsilon_0 R^2} \sum_{s=1}^{\infty} \left[\frac{2}{J_1(p_s)}\right]^2 J_0(p_s \frac{r}{R}) \left\{ \frac{1}{2\pi} \int_{-\infty}^{\infty} d\omega \frac{j\omega \left[\varepsilon(\omega)-\beta^{-2}\right] e^{j\omega(t-z/v)}}{\varepsilon(\omega) \left[\omega^2 \left[\varepsilon(\omega)-\beta^{-2}\right] - \omega_{c,s}^2 \right]} \right\} \qquad (3)$$

where it has been assumed that the particle is on axis, $\omega_{c,s}=cp_s/R$ are the cut-off angular frequencies and p_s are the zeros of $J_0(\xi)$. In order to demonstrate the growing character of the eigen-mode associated with this wake we can consider the poles exactly at the Cerenkov condition $(\varepsilon_r-\beta^{-2} = 0)$; a more general solution is also possible but it is more complex and it was presented elsewhere[4]. The relevant poles in such a case are given by

$$\omega^2 = \omega_0^2 \frac{\omega_{c,s}^2}{\omega_{c,s}^2 + \omega_p^2} \quad (4)$$

implying that only the modes that satisfy $(\omega_{c,s})^2 < -(\omega_p)^2$, grow in space. According to this result, we should be able to control the number of growing modes with the population inversion. In this specific configuration it is possible to ensure a single mode operation if $(\omega_{c,1})^2 < -(\omega_p)^2 < (\omega_{c,2})^2$. The wake can be evaluated analytically to read

$$E_z(\tau) = \frac{q}{4\pi\varepsilon_0\varepsilon_r R^2} \sum_{s=1}^{\infty} \left[\frac{2}{J_1(p_s)}\right]^2 \frac{\omega_p^2}{\omega_p^2 + \omega_{c,s}^2 + \varepsilon_r\omega_0^2}$$

$$\times \left\{\cos\left[\sqrt{\omega_0^2 + \omega_p^2/\varepsilon_r}\,\tau\right] + \frac{\varepsilon_r\omega_0^2}{\omega_p^2 + \omega_{c,s}^2}\cos\left[\sqrt{\frac{\omega_0^2\omega_{c,s}^2}{\omega_{c,s}^2 + \omega_p^2}}\,\tau\right]\right\}h(\tau) \quad (5)$$

and it can be shown that the first term in the curled brackets is associated with the dielectric function [$\varepsilon(\omega)=0$] and the second corresponds to the hybrid eigen-modes of the system [$\omega^2(\varepsilon(\omega)-\beta^{-2})-\omega_{c,s}^2=0$]. The first term corresponds to an accelerating force whereas the second can be either decelerating or accelerating according to the value of ω_p relative to $\omega_{c,s}$. In any event, as indicated above, only if $(\omega_{c,1})^2 < -(\omega_p)^2$ there will be a mode that grows in space. It is interesting to note that the trigger bunch experiences net acceleration if $(\omega_{c,1})^2 > -(\omega_p)^2$ however, if $(\omega_{c,1})^2 < -(\omega_p)^2 < (\omega_{c,2})^2$ then the first mode (the one that grows) tends to decelerate it, whereas all the other modes accelerate it.

The background gas model has its inherent advantage that the results can be presented analytically at a reasonable degree of complexity. In addition, injecting the bunch in an inverted gas has the advantage that there is no exponential decay of the evanescent field as it happens in vacuum. However, the energy density that can be stored in a gas is significantly smaller than in solid-state material. Also the probability of breakdown is significantly higher in a gas medium comparing to vacuum based systems. A solid-state medium on the other hand has the advantage that it can withstand much higher electric fields before breakdown, and it may store much more energy in a given volume. Its drawback consists of the exponential decay of the field from the triggering bunch. If the bunch is at a height h above a surface, then in zero order the field varies according to $e^{-\omega h/c\beta\gamma}$. At 1μm wavelength and for a "mildly" relativistic particle ($\gamma=2-3$) the particle needs to be within less than one wavelength from the surface in order for the structure to experience its field. This problem is practically eliminated if the triggering bunch is ultra-relativistic (say 30GeV).

WAKE IN A SOLID-STATE MEDIUM

With the increase in the energy density and reduction of wavelength associated with solid-state active medium there remains the question of confining all the energy in a single mode. Since as the energy density increases, the wavelength decreases and thus for a given geometry the number of modes increases. Yet single mode operation is crucial for this scheme to have a reasonable efficiency. In what follows we shall present a design of a single mode solid-state system and we shall conclude with a proposed experiment that will measure amplified Cerenkov radiation

In order to ensure single-mode operation, the present design relies on two periodic dielectric mirrors that confine the resonant wavelength and allow other frequencies to leave the system – see left frame in Figure 1. On top of each dielectric mirror, there is a thin layer of active dielectric that stores the energy required for acceleration. In what follows we shall calculate the mode(s) excitation by a bunch as it moves along this structure. In order to simplify the model we shall assume a circular symmetric system –see right frame of Figure 1. The circumference of the active layer is similar to the width of the two active layers combined; the thickness is much smaller than the radius of the dielectric loop. At resonance, the reflection from the dielectric mirror is assumed to be unity ($\rho = e^{-j\psi}$).

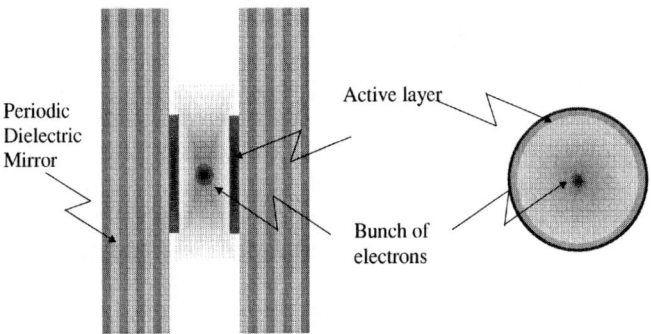

FIGURE 1. Cross-section (left) view of a Wake-Amplified Accelerator. A couple of dielectric mirrors confine the radiation to the vacuum and active medium regions. The bunch excites a wake that is amplified by the medium. The wake moves at the speed of the trigger bunch thus it can accelerate a trailing bunch. A symmetric simplified model (right) is used for the analysis. The inner radius of the active layer is R and its thickness is $\Delta R \ll R$.

In order to describe an arbitrary symmetric bunch we shall consider a thin charge-loop of radius r_c moving in the center of the structure. The longitudinal component of the electric field that is due to reflection from the boundary is given by

$$E_z(r,\tau) = \frac{q}{4\pi\varepsilon_0 R^2} \frac{2}{2\pi} \int_{-\infty}^{\infty} d\theta (j\theta) e^{j(\theta\tau c\gamma\beta/R)} F_0(\theta\frac{r}{R}) I_0(\theta\frac{r_c}{R}) \frac{K_0(\theta) + \alpha K_1(\theta)}{I_0(\theta) - \alpha I_1(\theta)} \quad (6)$$

$$\alpha = \frac{\gamma\beta}{\varepsilon_a}\sqrt{\varepsilon_a - \beta^{-2}} \frac{\sin(\phi) + \zeta\cos(\phi)}{\cos(\phi) - \zeta\sin(\phi)}, \quad \zeta = -\tan(\chi/2)\frac{\varepsilon_a}{\varepsilon_m}\frac{\sqrt{\varepsilon_m - \beta^{-2}}}{\sqrt{\varepsilon_a - \beta^{-2}}}$$

where $\phi = \theta\gamma\beta\frac{\Delta R}{R}\sqrt{\varepsilon_a - \beta^{-2}}$, ε_a is the dielectric coefficient of the active medium, ε_m is the dielectric coefficient of the mirror's first layer, the phase χ is the total phase shift associated with the reflection from the dielectric mirror $\chi = \pi/2 - \psi + 2\theta\gamma\beta\sqrt{\varepsilon_a - \beta^{-2}}(R+\Delta R)/R$ and

$$F_0\left(\theta\frac{r}{R}\right) \equiv \begin{cases} I_0\left(\theta\frac{r}{R}\right)K_0\left(\theta\frac{r_c}{R}\right) & r < r_c \\ K_0\left(\theta\frac{r}{R}\right)I_0\left(\theta\frac{r_c}{R}\right) & r > r_c \end{cases} \quad (7)$$

is the transverse Green's function. In order to evaluate the expression for E_z presented above, it is necessary to analyze its poles. These are determined by the zeros of the denominator

$$D(\omega,\varepsilon_a) \equiv I_0(\theta) - \alpha(\theta)\, I_1(\theta) = 0. \quad (8)$$

When the active layer is not pumped its dielectric coefficient is denoted by ε_r [equivalent to taking $\omega_p = 0$ in Eq.(1)] and we shall assume that the eigen-mode that is the closest to the resonance of the medium oscillates at ω_0 (its phase velocity is v) or explicitly,

$$D(\omega_0, \varepsilon_r) = 0. \quad (9)$$

This dispersion relation may have an infinite number of solutions due to the periodic character of the function. However for $\omega_0 R/c \gg 1$ and $\omega_0 R/c\gamma \ll 1$, there are two simple cases that can be expressed analytically. In one case the dielectric mirror sets a phase such that $\chi = \pi$ whereas in the other $\chi = 2\pi$. The corresponding thickness of the active layer is

$$\Delta R = \frac{\lambda}{4}\frac{1}{\sqrt{\varepsilon_r - 1}} \quad \text{and} \quad \Delta R = \frac{\lambda}{2}\frac{1}{\sqrt{\varepsilon_r - 1}}. \quad (10)$$

Note that according to its definition [prior to Eq.(7)], χ itself is a function of ΔR however, we tacitly assume that the dielectric mirror can be designed such that at the frequency of interest the conditions $\chi = \pi$ or $\chi = 2\pi$ are satisfied. Clearly an accurate exact design will have to take into consideration the exact characteristics of the dielectric mirror. Any practical value of ΔR will be between the two values presented above. After establishing the conditions for single mode operation, we next shall determine the effect of the active medium on the propagation of this mode.

The pump introduces an "increment" in the dielectric coefficient as well as a small shift in the frequency therefore the "dispersion" function can be expanded in a Taylor series with respect to these two variations:

$$D(\omega,\varepsilon_a) \approx D(\omega_0,\varepsilon_r) + (\omega^2 - \omega_0^2)D_\omega + \frac{\omega_p^2}{\omega_0^2 - \omega^2 + 2j\omega\omega_1}D_\varepsilon,$$

$$D_\omega \equiv \left[\frac{d}{d\omega^2}D(\omega,\varepsilon_r)\right]_{\omega=\omega_0} = \frac{1}{2\omega_0}\left[\frac{d}{d\omega}D(\omega,\varepsilon_r)\right]_{\omega=\omega_0}, \quad (11)$$

$$D_\varepsilon \equiv \left[\frac{d}{d\varepsilon_a}D(\omega_0,\varepsilon_a)\right]_{\varepsilon_a=\varepsilon_r}.$$

According to our definition in (9) the first term in the first row of (11) is zero and the main contribution to the integral is from the vicinity of the resonance as revealed by the next expression for the longitudinal field

$$E_z = \frac{q}{4\pi\varepsilon_0 R^2} \frac{2}{2\pi} \frac{F_0(\theta_0 r/R)I_0(\theta_0 r_c/R)}{\theta_0 I_1(\theta_0)D_\theta} \int_{-\infty}^{\infty} d\theta \frac{j\theta e^{j\theta\tau c\gamma\beta/R}}{\theta^2 - \theta_0^2 + \frac{\theta_p^2 D_\varepsilon/D_\theta}{\theta_0^2 - \theta^2 + 2j\theta\theta_1}}$$
(12)

$$\approx \frac{q}{4\pi\varepsilon_0 R^2} \frac{2F_0(\theta_0 r/R)I_0(\theta_0 r_c/R)}{\theta_0 I_1(\theta_0)D_\theta} \cos(\omega_0 \tau)\cosh(\Omega\tau)h(\tau)$$

In this expression $\theta_p^2 \equiv (\omega_p R/c\gamma\beta)^2$, $D_\theta \equiv D_\omega(c\gamma\beta/R)^2$ and $\Omega \equiv (|\omega_p^2|/4\omega_1)(D_\varepsilon/\omega_0^2 D_\omega)$; the second line presents the evaluation of the integral for an inverted medium ($\omega_p^2 < 0$). For a relativistic particle $\theta_0 \ll 1$, the form-factor of the growth rate was estimated to be $D_\varepsilon/(\omega_0^2 D_\omega) \approx 1/(\varepsilon_r - 1)$ thus it is dependent only on the dielectric coefficient of the active medium. Consequently the the growth rate is

$$\frac{|\omega_p^2|}{4\omega_1\sqrt{\varepsilon_r - 1}}.$$
(13)

Further simplification of the result in Eq. (12) is possible by assuming that the trigger bunch and the accelerated bunch have the same radius R_b and the electrons are distributed uniformly both radially and longitudinally. Averaging over all particles in the bunch we obtain a form-factor defined by

$$F(u) \equiv \frac{2}{u^2}\int_0^u dx\, x \frac{2}{u^2}\int_0^u dy\, y\, I_0(y)F_0(x,y) \approx -\ln(u)$$
(14)

where the last term is valid only for relativistic particles $u \equiv \theta_0 R_b/R \ll 1$; in the framework of the same approximation $\omega_0^2 D_\omega \approx -(\pi/4)(\omega_0 R/c)\sqrt{\varepsilon_r - 1}/\varepsilon_r$. Typical bunches available today are much longer than 1-10μm therefore the average field generated by a bunch is zero due to the random phase. However, each electron does generate a certain amount of power at the resonant frequency thus the total "noise" at the particle location at ω_0 is given by

$$P_0 = \frac{e^2 N_e c}{4\pi\varepsilon_0 R^2} \frac{-\ln(\theta_0 R_b/R)}{\frac{\pi}{4}\left(\frac{\omega_0}{c}R\right)\frac{1}{\varepsilon_r}\sqrt{\varepsilon_r - 1}}$$
(15)

This power is amplified by the medium.

AMPLIFICATION

From the perspective of a potential experiment, there are two main questions that need to be addressed: what is the actual amplification of the Cerenkov radiation? And how it compares with the amplified spontaneous emission? Regarding the Cerenkov radiaton generated, consider a bunch of $N_e=2\times10^{10}$ electrons of radius $R_b=75\mu m$; the material in this case is assumed to have a resonance at $\lambda=1\mu m$ and the geometry is designed such that at $\varepsilon_r=3$ there is only one mode in the active slab i.e. $\Delta R=0.3\mu m$. With these parameters, the amount of Cerenkov radiation generated by the bunch is $P_0= 2\mu W$; the *average* (in space) energy flux of the Cerenkov radiation at this wavelength, assuming $R=1mm$, is approximately $I_c=0.6\times10^{-4}$ W/cm^2; this value is somewhat higher (by a factor of ε_r) in the dielectric region but we shall ignore this change at this point.

The linear gain in an active medium is in general characterized by three parameters: the interaction cross-section (σ) that for Nd:YAG is 4.9×10^{-19}cm^2, the population inversion density (n_{pop}) and the interaction length (d); explicitly the linear gain is $\exp(\sigma n_{pop}d)$. For Nd:YAG the inverted population density may be as high as 2.5×10^{19}cm^3 therefore the gain can be as high as 100dB/cm. Note that the energy stored in the medium is $\hbar\omega n_{pop}\Delta R\, 2\pi R\, d = 1$mJ consequently if we assume that all the energy is "drained" during the $\tau_p=2$psec pulse, then the peak power we may expect is 0.5GW. Let us pursue a more rigurous analysis of the output signal.

The input energy flux is assumed to be gaussian i.e. $I_{in}(\tau)=I_c\exp[-(\tau/\tau_p)^2]$ where $\tau=t-z/v$ is just the delay time. Since the bunch is ultra-relativistic we may adopt the formulation used for description of the propagation of fast light pulses taking into consideration the impact of the laser field on the medium – see for example Ref. 7. The average (in space) number of inverted atoms per unit area is $N_{tot}(\tau) = d\, n_{pop}$; this is the total number per unit area of inverted atoms that participate in the interaction process. Consequently, the energy flux at the output, $I_{out}(\tau)$, is given by

$$I_{out}(\tau) = I_{in}(\tau)e^{\sigma N_{tot}(\tau)}. \qquad (16)$$

Energy conservation for a system in which the lower level empties rapidly reads

$$\frac{d}{d\tau}N_{tot}(\tau)=-\frac{1}{\hbar\omega}[I_{out}(\tau)-I_{in}(\tau)] \qquad (17)$$

where it was assumed that the pulse is short comparing to the relaxation times and therefore these terms were ignored. Substituting (16) in (17) we obtain an explicit expression for the gain

$$I_{out}(\tau) = \frac{I_{in}(\tau)}{1-(1-e^{-\sigma N_0})e^{-U_{in}(\tau)/U_{sat}}} \qquad (18)$$

where N_0 is the population inversion in equilibrium prior to the pulse entrance, $U_{in}(\tau) = \int_{-\infty}^{\tau} dt\, I_{in}(t)$ is the input energy per unit area of the medium and $U_{sat} = \hbar\omega/\sigma$ is the saturation energy per unit area - saturation fluence. For the parameters of interest

$N_0=2.5\times10^{20}$ and $U_{sat}=0.25J/cm^2$ therefore the maximum fluence associated with input pulse ($U_{in,max}=\tau_p I_c = 10^{-16}J/cm^2$) is negligible comparing to the saturation fluence hence

$$I_{out}(\tau) \approx \frac{I_{in}(\tau)e^{\sigma N_0}}{1+e^{\sigma N_0}U_{in}(\tau)/U_{sat}}. \tag{19}$$

This last expression indicates that as long as the gain is smaller than 150dB the saturation effect is negligible however for $\sigma=4.9\times10^{-19}cm^2$ and $N_0=2.5\times10^{20}cm^{-2}$ is more than three times as large therefore saturation is expected to play an important role (except if the slab length is reduced from 10cm to 2.8cm). In fact in zero order we can evaluate the output power at saturation to be

$$I_{out,max} \approx \frac{U_{sat}}{\tau_p\sqrt{\pi/2}} \tag{20}$$

and for the parameters previously defined it equals $0.14TW/cm^2$ corresponding to a total power of 2.5MW (comparing to the 0.5GW which was the rough estimate based on the total energy stored). The top frame in Figure 2 illustrates the output power for three different lengths (d=3.0, 2.8 and 2.6cm) of the active medium. For the parameters chosen, a total gain of almost 120dB amplification is obtained. In the case that d=3cm there is clear indication of saturation in the sense that the front of the Cerenkov pulse saturates the medium and the amplification of the peak pulse is significantly smaller than that of the front. As a result, the peak of the output pulse occurs prior to the peak input pulse. In order to further illustrate the saturation effect we show in the right frame of Figure 2 the change of the population inversion density during the pulse duration. For a 3cm long medium there is a change of about 6% in the population inversion but for all the other cases, the change is negligible. Before we address the second question raised at the beginning of this section, an important distinction has to be made: the spatial growth rate developed in (13) differs from the growth rate of a TEM mode that was used in this section. The relation between the two is $g_{TM01} = g_{TEM}(\varepsilon_r/\varepsilon_r-1)^{0.5}$ – recall that $g_{TEM} = \sigma N_0$.

In the analysis presented above the spontaneous radiation was ignored and the question is to what extent this radiation may affect our measurement. Ignoring at this stage saturation in the medium and in the absence of an external signal the intensity of the radiation in *steady-state* is $I(z)=A\exp(gz)+B\exp(-gz)$. The first term represents the amplification of the spontaneous radiation in the positive direction whereas the second term represents the intensity of the field in the opposite direction. In order to determine the two coefficients (A and B) two conditions are imposed: the total energy flux from the two ends equals the pump flux i.e. $I(d)-I(0)=I(d)+|I(0)|=I_p$. In addition, assuming that both ends are identical $I(d)+I(0)=I(d)-|I(0)|=0$. Consequently,

$$I(z) = \frac{I_p}{2}\frac{\sinh[2g(z-d/2)]}{\sinh(gd)}. \tag{21}$$

This expression reflects the fact that when ignoring the energy that converts into heat as well as saturation, the power from the pump splits into two and leaves the system, maintaining a certain level of population inversion. *Saturation* changes the distribution but not the power at the two output ends. In other words, saturation affects the distribution of the population inversion but in equilibrium conditions, the energy flux at the two output ends will be $I_p/2$ at each end.

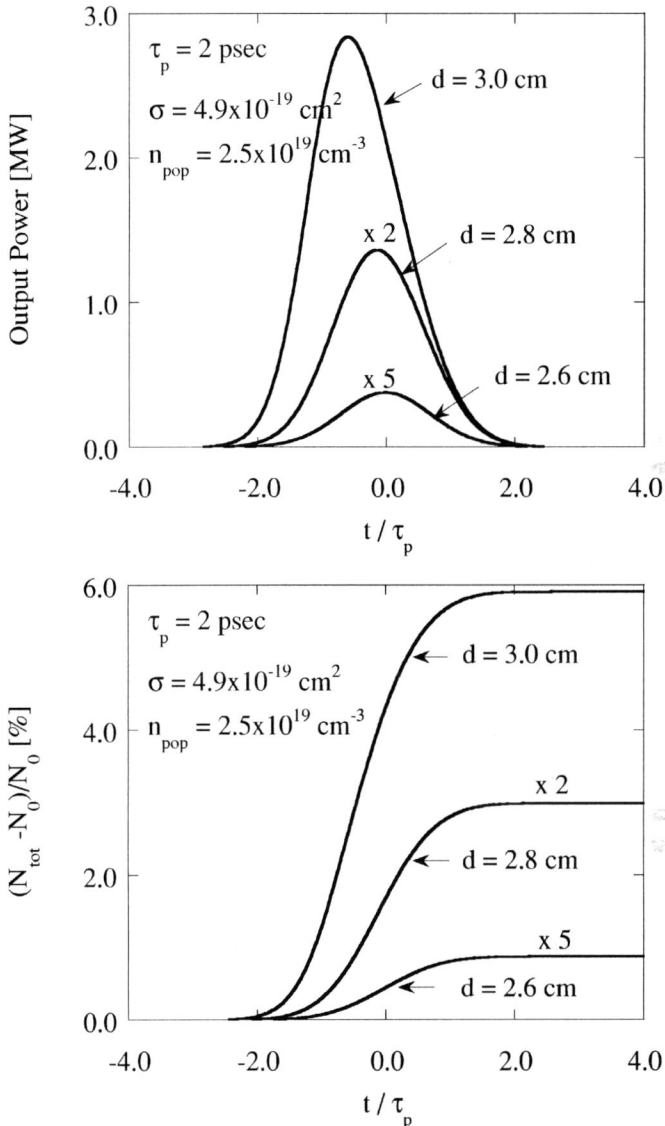

FIGURE 2. Top: Intensity of the output pulse for three different lengths of the active medium. Bottom: the relative change in the population inversion as the Cerenkov pulse propagates and is being amplified by the medium.

In the center of the dielectric medium the intensity of the radiation field vanishes due to symmetry: whatever the gain and the saturation are, the power flux from left to right equals that from right to left. Consequently the population inversion density in the center of the slab remains unaffected – $n_{pop,0}$. Since the system is in steady-state, energy conservation implies that the energy flux at one output end is $I_p/2$, therefore the population inversion density is

$$n_{pop}(z=d) = \frac{n_{pop,0}}{1+I_p/2I_{sat}} \qquad (22)$$

where $I_{sat} = \hbar\omega/\sigma\tau_{eff}$ is the saturation energy flux and for the values presented above I_{sat}=1kW/cm^2 (τ_{eff}=2msec for typical solid-state medium). If the pump feeds 2mW in the medium the equivalent intensity is I_p=0.1kW/cm^2 and the maximum change in the population inversion due to the amplification of the spontaneous radiation is 5%. Although this is comparable to the effect of the amplified Cerenkov signal on the population inversion density (see Figure 2), the effect of the spontaneous radiation on the amplification of the Cerenkov pulse is negligible. It is important to point out that (21) and the estimates that follow, as well as (22), are an *upper limit* to the emitted spontaneous radiation. The actual emission is by at least one order of magnitude smaller than the pump intensity [8].

In conclusion, a solid-state active medium amplifies a Cerenkov signal generated by a 2psec long bunch in the same medium. Altough the amplified spontaneous emission may affect the population inversion, the impact on the amplified Cerenkov pulse is small. An amplification of 120dB seems feasible with a 3cm long Nd:YAD active medium with population inversion of n_{pop}=2.5x10^{19}cm^{-3} and collision cross-section σ=4.9x10^{-19} cm^2. The corresponding signal-to-noise ration is large (+90dB) for a pump input power of 1mW. The wake depletes only a small (<1%) fraction of the energy stored in the medium.

This work was supported by the United States Department of Energy.

REFERENCES

1. *Advanced Accelerator Concepts*- 1996, edited by S. Chattaopadhyay, AIP Conference Proceedings 398, New York, American Institute of Physics
2. *Advanced Accelerator Concepts*- 1998, edited by W. Lawson, AIP Conference Proceedings 472, New York, American Institute of Physics
3. L. Schächter, *Phys. Lett. A*., **205**, p. 355 (1995).
4. L. Schächter, *Phys. Rev. E.*, **53**, p. 6427 (1996).
5. L. Schächter, *Phys. Rev. Lett.*, **83**, p.92 (1999).
6. L. Schächter, *Phys. Rev. E*, July (2000).
7. "Lasers" by A.E. Siegman, University Science Book, Mill Valley, California 1986 pp. 363-373.
8. "Solid-StateLaser Engineering" by W. Koechner, 3rd edition, Springer-Verlag, Berlin, 1992, p. 182.

Resonant Absorption Instability: Acceleration and Radiation Amplification

Levi Schächter

*Department of Electrical Engineering
Technion – Israel Institute of Technology
Haifa 32000, ISRAEL*

Abstract. A new type of instability is demonstrated. It occurs when a space-change wave that oscillates at a frequency close to the resonance of the medium in which the electron beam propagates. The space-charge wave grows in space according to the loss associated with the resonance namely, it is directly related to the attenuation of a pure electromagnetic mode in the medium. The spatial growth is proportional to the square root of the current density and for a given current, it is inverse proportional to the normalized momentum to the power of 1.5 i.e. $(\gamma\beta)^{-3/2}$. We briefly discuss possible application to particles acceleration and millimeter wave amplification.

The next linear collider (NLC) requires gradients in excess of 100 MV/m in order to achieve energies of 0.5TeV (c.m.) or higher. According to the operating frequency, the various acceleration approaches that are being pursued by groups around the world, can be divided into two categories, microwave [1-8] and laser based schemes [9-12]. The need for a high gradient on one hand and power levels below the breakdown threshold on the other hand, push the operating frequency upward. From the perspective of microwave regime i.e. 2<f[GHz]<30, the difficulty with operation at higher frequencies is associated with the small transverse dimensions necessary to maintain a single-mode in the system. This geometric constraint imposes stringent limitations on the amount of power that an electron beam may carry into the system and thus on the feasibility of most microwave or millimeter wave sources. We suggest here to combine two well known effects: resistive wall instability and resonant absorption, in order to amplify radiation or accelerate particles. In other words, we use *microscopic* cavities i.e. atoms or molecules, for the generation of an instability that causes space-charge waves to grow in space. To a large extent, this paradigm is free of the geometrical constraints and limitations mentioned previously however it is not free of limitations since collisions, breakdown and dark current may be a serious impediment. The method presented here is an additional interaction mechanism to the one presented in the past [13-16] where internal resonances of the medium play an important role in the interaction process. The main difference between the present one and the earlier publications is that here the resonant medium is lossy whereas in the other publications the medium was assumed to be active i.e. energy was stored in the medium.

Consider a background gas or a mixture of gases that consists of atoms/molecules characterized by a series of resonances. For sake of simplicity we shall assume that the constituent atom has a single dominant resonance close to the frequency where the system is excited; all the other resonances in this region are represented by a frequency independent dielectric coefficient ε_r

$$\varepsilon(\omega) = \varepsilon_r + \frac{\omega_p^2}{\omega_0^2 - \omega^2 + 2j\omega\omega_1} \tag{1}$$

where ω_0 is the resonance angular frequency, ω_1 is the resonance width and ω_p is the plasma frequency. A beam of electrons is injected in this medium. Its average velocity is v and its average density is n_b. For simplicity we shall ignore the scattering process of these electrons with the constituents of the resonant medium and the possible ionization of the medium by the fronf of the beam. The constitutive relation of the beam, assuming steady state ($e^{j\omega t}$) and variation in the z-direction according to e^{-jkz}, reads

$$J_z(\omega, k) = -j\omega\varepsilon_0 \frac{\omega_{p,b}^2}{(\omega - kv)^2} E_z(\omega, k) \tag{2}$$

where $\omega_{p,b} = (e^2 n_b/m\varepsilon_0\gamma^3)^{1/2}$ is the relativistic plasma frequency of the beam; $\gamma = (1-\beta^2)^{-1/2}$ and $\beta = v/c$. In this expression it was also tacitly assumed that: the beam propagates in the z-direction, the motion of the electrons is always confined to the z-direction and their energy spread is negligible. With these assumptions in mind, the dispersion relation of the electromagnetic and space-charge waves ignoring any transverse variations reads

$$\left[\varepsilon(\omega)\frac{\omega^2}{c^2} - k^2\right]\left[1 - \frac{\omega_{p,b}^2}{(\omega - kv)^2} \frac{1}{\varepsilon(\omega)}\right] = 0. \tag{3}$$

The first term, $k^2 = \varepsilon(\omega)\omega^2/c^2$, represents the pure *electromagnetic* (TEM) mode therefore the decay associated with the imaginary part of the wave number is

$$\mathrm{Im}(k_{EM}) = \frac{\omega}{c}\mathrm{Im}\left[\sqrt{\varepsilon(\omega)}\right]. \tag{4}$$

In a similar way, the wave numbers associated with the *space-charge* waves are given by $k_{SC} = \omega/v \pm \omega_{p,b}/v\sqrt{\varepsilon(\omega)}$ and the corresponding imaginary part reads

$$\mathrm{Im}(k_{SC}) = \frac{\omega_{p,b}}{v}\mathrm{Im}\left[\frac{1}{\sqrt{\varepsilon(\omega)}}\right]. \tag{5}$$

This result indicates that the imaginary part of the dielectric function is responsible to the spatial growth of the space-charge wave. In other words, the same mechanism that causes the opaqueness of the medium, is responsible to the spatial growth of the space-charge wave. The mechanism is similar to the so-called resistive wall instability that occurs when space-charge waves propagate in the vicinity of a metallic wall of finite conductivity. However there are several significant differences: the "conductivity" in our case is narrow-band and the beam propagates through the medium rather than in the vicinity of the medium. In addition, this mechanism has a particular appeal in the millimeter, sub-millimeter or even in the optical range where resonant absorption is a significant effect.

The spatial growth-rate of the space-charge mode is smaller from the spatial decay rate of the electromagnetic mode by the ratio of the plasma frequency of the beam and the operating (resonant) frequency. In addition, for a given current, this ratio is inversely proportional to the momentum to the power of 3/2 implying that there must be an optimal momentum since at very low momentum (corresponding to a few hundreds eV) the process will diminish due to ionization of the medium.

In order to envision the potential of this mechanism let us consider the analysis in Ref.17, that indicates that peak absorption coefficient $\alpha = 2\,\text{Im}(k_{SC})$ of a pure rotational line at room temperature is given by

$$\alpha_{max}[m] = 1.4 \times 10^{-12} \mu^2 \omega_0^3 / \omega_1$$

where μ is the dipole moment of the atom or molecule expressed in Debye, the frequencies are expressed in GHz; the typical values of μ vary between 1-4 Debye (e.g. $\mu_{Ammonia} = 1.44$ Debye, $\mu_{water} = 1.84$). The line-width depends on the characteristics of the molecule as well as the mechanism of energy loss such as collisions between molecules. Typically, the higher the pressure, the shorter the mean free path and therefore the larger the line-width (ω_1) - in fact, for sufficiently high pressure the line-width is proportional to the pressure. Accordingly, the line-width may vary quite dramatically from a few kHz in the case of the 24GHz resonance of Ammonia [18] to MHz and higher [17]. In the example to follow we shall assume $\omega_1/2\pi = 10$MHz and $\mu = 2.5$ Debye and operation at 125 GHz thus

$$\alpha_{max} = 6.72\text{cm}^{-1}[\text{Im}(k_{EM}) = 3.36\text{cm}^{-1}].$$

A relativistic beam ($\gamma \gg 1$) is injected in this gas and for a beam plasma frequency of 2GHz, the growth rate of the system is 0.054cm^{-1}. This corresponds to 0.47dB/cm thus we may expect a 50dB gain in about 120cm of interaction length. Consequently, assuming an initial modulation that corresponds to 100W at the input, after 110cm a total power of 10MW can be expected. The same concept can be applied for acceleration of particles. Consider an electron pulse that is 4nsec long and an initial modulation that corresponds to a 0.15MV/m gradient. Following the same arguments of the above, the growth of the space-charge wave along the electron pulse is 50dB therefore the gradient towards the end of the bunch corresponds to 45MV/m. These two examples illustrate the potential of behind the resonant absorption instability. Before we proceed however it is important to emphasize that in this calculation we assumed an equal excitation of both space-charge waves. In both cases we ignored in the process the possibility of ionization of the medium by the intense microwave radiation and non-linear effects.

The result presented above can be extended to the case of finite cross-section beam. For this purpose it is necessary to determine the wave-number k by solving the following dispersion relation

$$\frac{\Lambda R_b I_1(\Lambda R_b)}{I_0(\Lambda R_b)} = \frac{\Gamma R_b K_1(\Gamma R_b)}{K_0(\Gamma R_b)} \tag{6}$$

where $\Gamma^2 = k^2 - \varepsilon(\omega)\omega^2/c^2$ and $\Lambda^2 = \Gamma^2\left[1 - \omega_{p,b}^2/\varepsilon(\omega)(\omega-kv)^2\right]$. The solution can be expressed in terms of the solution in (5) and plasma filling-factor namely, $\omega_{p,b}^2 \rightarrow \omega_{p,b}^2 FF_{ex}$. An approximate solution for the latter may be determined by dividing the area of the beam and the effective area of the space-charge wave using the energy flux (S_z) as the weighting function $A_{wave} \equiv \left|\int_0^\infty drr(\pi r^2)S_z(r)\right|/\left|\int_0^\infty drr\, S_z(r)\right|$ and $A_{beam} \equiv \pi R_b^2$ hence $FF_{flux} = A_{beam}/A_{wave}$. A crude estimate of this filling-factor can be evaluated by assuming vacuum outside the beam in which case $FF_{approx} = 1 - e^{-\omega R_b/2c\gamma\beta}$. All three filling-factors are shown in Figure 1. In what follows we shall use the FF_{approx} that is a reasonable approximation as indicated in Figure 1. Based on this explicit (but approximate) expression for the filling factor, the normalized spatial growth-rate reads

$$\text{Im}(k_{SC}) = \frac{1}{R_b}\sqrt{\frac{e\eta I}{\pi mc^2}\frac{1}{(\gamma\beta)^3}\left[1 - e^{-\frac{\omega R_b}{2c\gamma\beta}}\right]\text{Im}\left[\frac{1}{\sqrt{\varepsilon(\omega)}}\right]} \quad (7)$$

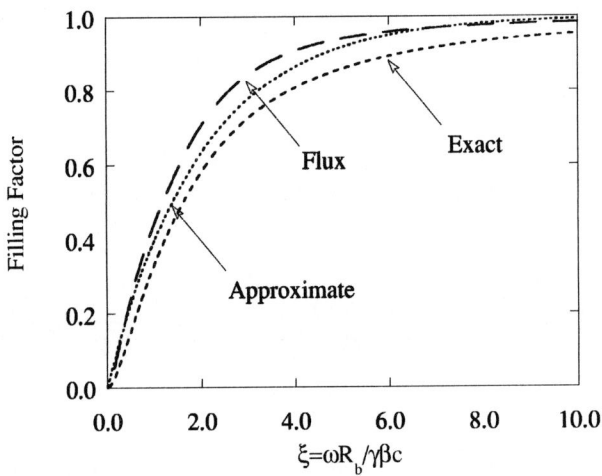

FIGURE 1. Three approximations for the filling factor: FF_{exact} is the exact expression assuming ξ is a real quantity; FF_{flux} is the ratio of the area of the beam and the effective area of the space-charge wave and the third, FF_{approx} is an approximate expression based on the former.

Figure~2 illustrates the real and imaginary part of a typical resonant material. For comparison, we illustrate the imaginary part of $\sqrt{\varepsilon(\omega)}$ that represents the spatial decay of the electromagnetic mode (TEM) as well as the imaginary part of the reciprocal of this quantity, $\text{Im}[1/\sqrt{\varepsilon(\omega)}]$, as a function of the normalized frequency, ω/ω_0; here $\omega_p/\omega_0=0.2$ and $\omega_1/\omega_0=0.03$. Note that the maximum spatial growth-rate of the medium is similar to the minimum of the real part of the dielectric coefficient.

In order to envision the excitation and propagation of a space-charge wave in a resonant medium consider a beam of radius R_b that is modulated by a high-frequency field (e.g. millimeter or sub-millimeter wave). This field propagates transverse to the electron beam and it intersects the latter at z=0 thus the beam experiences an electric field $E_z(r=0, z=0, t) = E\cos(\Omega\tau)e^{-(t/\tau)^2}$ where Ω is the source's angular frequency and τ is the width of the pulse. Assuming that the velocity-modulation at the input (z=0) is zero, the longitudinal electric field on axis (r=0) is given by

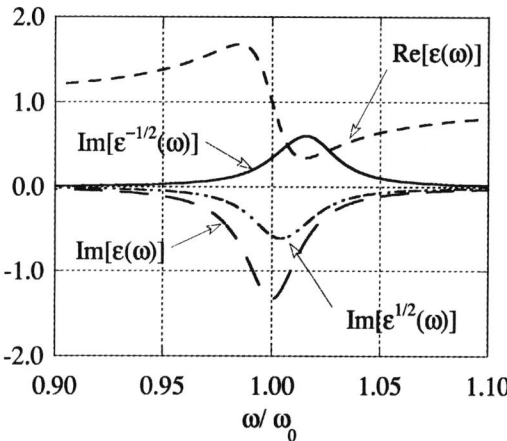

FIGURE 2. The real and imaginary part of a typical resonant material and, for comparison, we illustrate the imaginary part of $\sqrt{\varepsilon(\omega)}$ that represents the spatial decay of the electromagnetic mode (TEM) and the imaginary part of the reciprocal of this quantity, $\text{Im}[1/\sqrt{\varepsilon(\omega)}]$, as a function of the normalized frequency, ω/ω_0; here we used $\omega_p/\omega_0=0.2$ and $\omega_1/\omega_0=0.03$.

$$E_z(r=0,z,t) = \frac{E\tau}{\sqrt{\pi}} \int_0^\infty d\omega\, e^{-(\omega-\Omega)^2\tau^2} \text{Re}\left[e^{j\omega(t-z/v)} \cos(K_{SC}(\omega)z)\right]. \quad (8)$$

This result may be further simplified if we assume a *long pulse* such that its duration (τ) is much longer than the reciprocal of the line width i.e. $\omega_1\tau \gg 1$. In addition, it is assumed that there are many periods in this pulse ($\Omega\tau \gg 1$) and that the analysis is limited to the solution far away from the excitation point i.e., $\text{Im}[k_{SC}(\Omega)] z > 1$. In this case the Gaussian function in Eq.(8) is sharp and the main contribution to the integral is from the close vicinity of its peak value i.e. $\omega=\Omega$ hence

$$E_z(r=0,z,t) = \frac{E}{2}\cos[\Omega(t-z/v)]\cos[\text{Re}(k_{SC}(\Omega))z]e^{\text{Im}(k_{SC}(\Omega))z} \quad . \qquad (9)$$

A second regime of interest occurs for a *short pulse* duration. In this case the pulse is short on the scale of the reciprocal of the line-width ($\omega_1\tau \ll 1$). An analytic expression can be developed by taking advantage of the maximum that Im(k_{SC}) has a maximum as a function of the frequency (see Figures~2 and 3) and use the steepest descent path method for evaluation of the integral. For this purpose we assume that ω_m is the angular frequency for which the Im(k_{SC}) is maximum however we shall assume that it equals Ω consequently,

$$E_z(r=0,z,t) = E\sqrt{\frac{\tau^2}{2z\ddot{k}}}\cos[\Omega(t-z/v)]\cos[\text{Re}(k_{SC}(\Omega))z]e^{\text{Im}(k_{SC}(\Omega))z} \qquad (10)$$

as in the previous case, here we assumed that $\Omega\tau \gg 1$; $\ddot{k} \equiv d^2k/d\omega^2$ at the maximum. Clearly in both cases the field grows exponentially in space.

The analysis so far was performed subject to the assumption that the wave is azimuthally symmetric. While this is a valid assumption with regards to what is required, it is not necessarily what we are able to excite and in many cases non-symmetric modes develop. For a simple comparison of the interaction of symmetric and asymmetric modes we shall assume that both the beam and the medium are confined by a circular waveguide of radius R. Consequently, the boundary conditions impose the transverse variation of the field and the dispersion relation reads

$$\left[\varepsilon(\omega)\frac{\omega^2}{c^2} - k^2\right]\left[1 - \frac{\omega_{p,b}^2}{(\omega-kv)^2}\frac{1}{\varepsilon(\omega)}\right] = \frac{p_{n,s}^2}{R^2} \qquad (11)$$

where $p_{n,s}$ are the zeros of the Bessel function of the first kind and n'th order i.e., $J_n(p_{n,s})=0$. The solution of this dispersion relation looks very similar to that presented in Figure 2 and for comparison Figure 3 illustrates the maximum value of Im(k_{SC})R for a given beam energy. Two facts are evident: firstly, there is little difference between the lowest symmetric and asymmetric modes. Secondly, for large values of γ, the spatial growth rate drops as $\gamma^{3/2}$ as the simple analytic model predicts. It is important to indicate that the frequency where this maximum occurs is practically identical – see Figure 4.

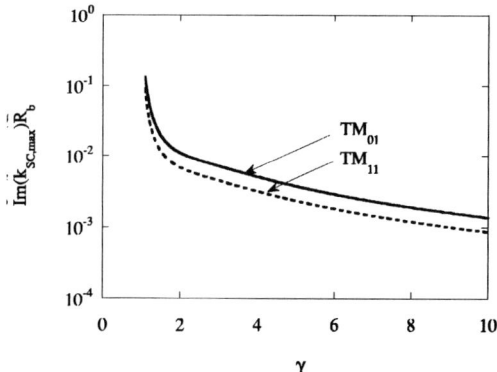

FIGURE 3. The maximum spatial growth-rate as a function of beam's energy for both lowest symmetric and asymmetric modes. There is relatively small difference between the two therefore special care will be required for the suppression of asymmetric modes.

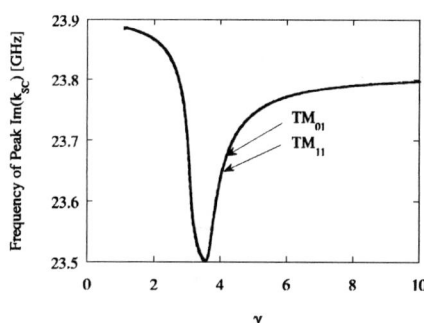

FIGURE 4. Frequency where maximum Im(k_{SC}) occurs.

In conclusion, the resonant absorption in a gas causes a spatial growth of a space-charge wave that propagates along an electron beam that propagates in that medium. This instability can be used for conversion of kinetic energy from the beam and the amplification of high-frequency radiation when regular macroscopic cavities impose constraints that are too stringent for a realistic application. Moreover, this mechanism may used for actual acceleration of particles.

This study was supported by the United States Department of Energy.

REFERENCES

1. R. Ruth, Proceedings of the Third Workshop on Pulsed RF Sources for Linear Colliders (RF-96) Kanagawa, Japan, 1996 (KEK Proceedings 97-1) p. 54; G. Caryotakis, Proceedings of the Third Workshop on Pulsed RF Sources for Linear Colliders (RF-96) Kanagawa, Japan, 1996 (KEK Proceedings 97-1) p. 72; S.G. Tantawi et. al., in Advanced Accelerator Concepts, Lake Tahoe, CA, 1996, (AIP Woodbury NY 398), p. 805; N. Kroll et. al., in Advanced Accelerator Concepts, Lake Tahoe, CA, 1996, (AIP Woodbury NY 398), p. 455.
2. L. Schächter and J.A. Nation, Phys. Rev. E, **57**, pp.7176 (1998); Lawson W. et. al., in Proceedings of the Third Workshop on Pulsed RF Sources for Linear Colliders (RF-96), Kanagawa, Japan 1996 ,(KEK Proceedings 97-1), p. 225.
3. H. Henke, in Advanced Accelerator Concepts, Lake Tahoe, CA, 1996, (AIP Woodbury NY 398), p. 485.
4. A. M. Sessler and S.S. Yu, Phys. Rev. Lett., **54**, 889 (1987).
5. R. Corsini, in Advanced Accelerator Concepts, Lake Tahoe, CA, 1996, (AIP Woodbury NY 398) p. 126 .
6. S. Lidia et. al.; in Advanced Accelerator Concepts, Lake Tahoe, CA, 1996, (AIP Woodbury NY 398), p. 842 .
7. Gai W. et. al., Phys. Rev. Lett. **61**, 2756 (1988).
8. P. Wilson, in High Brightness Accelerators (NATO ASI Series Vol. 178, Plenum Press, New York, 1988), p.129.
9. E. Esarey et. al., IEEE - Plasma Science **24**, 252 (1996).
10. J. R. Fontana and R.H. Pantell; J. Appl. Phys. **54**, 4285 (1983). See also W.D. Kimura et.al. Phys. Rev. Lett., **74**, 546 (1995).
11. R.B. Palmer, in Laser Acceleration of Particles, Los Alamos, 1982 (AIP, Woodbury NY, 91) p. 179. See also N.M. Kroll, in Laser Acceleration of Particles, Malibu, 1985 (AIP Woodbury NY 130), p. 253.
12. E.D. Courant et. al., Phys. Rev. A. **32**, 2813 (1985).
13. L. Schächter, *Phys. Lett. A* ., **205**, p. 355 (1995).
14. L. Schächter, *Phys. Rev. E.*, **53**, p. 6427 (1996).
15. L. Schächter, *Phys. Rev. Lett.*, **83**, p.92 (1999).
16. L. Schächter, *Phys. Rev. E*, July (2000).
17. "Microwave Spectroscopy" by W. Gorgy, W.V. Smith and R.F. Trambarulo, Dover Publications Inc., New York 1966; p.95-6, 191-2.
18. J.P. Gordon, H.J. Zeiger and C.H. Townes; Molecular Microwave Oscillator and New Hyperfine Structure in the Microwave Spectrum of NH_3, Phys. Rev. **95**, 282L (1954).

Table Top Accelerator with Extremely Bright Beam

A. Mikhailichenko

Cornell University, Wilson Laboratory, Ithaca, NY 14850

Abstract. We considered here an accelerator having microstructures excited by laser radiation. The gradient achievable ~1GeV/m and invariant emittance of the beam 10^{-9} cm rad with 10^6 particles per pulse and repetition rate ~kHz allows numerous applications for this type of accelerator. We discuss here some possible utilization of such a linac.

1. INTRODUCTION

Technology, developed in a framework of high-energy linear collider [1] might have a side product such as a compact high gradient accelerator. This accelerator has the number of the particles about 10^6/pulse and, hence, low average power, but it has extremely small emittance of the order $\gamma\varepsilon \cong 10^{-9} cm \cdot rad$. The envelope function one can expect here be of the order of a centimeter or smaller, so the angular divergence defined by opening angle of radiation $\sim 1/\gamma$ in both directions mostly, not by the angular spread in the beam. The number of the photons, generated by Compton scattering of the electron beam per second, goes here to $\dot{N}_\gamma \cong 10^8 s^{-1}$. So the brightness of the beam for gammas, having energy around 160 *keV*, can rich for 100 *MeV* primary *electron* beam the value about

$$B \cong \frac{\dot{N}_\gamma}{(\gamma\varepsilon) \cdot (\beta/\gamma)\gamma^{-2}} = \frac{\gamma^3 \cdot \dot{N}_\gamma}{(\gamma\varepsilon) \cdot \beta} \cong 2 \cdot 10^{24} \, photons/cm^2/rad^2/\sec.$$

This accelerator can work with ions and protons too. The sources of these particles could be small and adequate to the tabletop size. Sequentially, this accelerator can have numerous applications in instrumentation, technology and medicine, such as high-energy ion implanter, the photon/proton source, precise beam therapy and so on. The main task for this accelerator, however, is a proof of principle, what could be used for large-scale projects, such as a multi $-TeV$ linear collider.

2. SOURCE OF ELECTRONS/IONS/PROTONS

The source of particles must be adequate to the tabletop size. Obviously it is not possible to keep a damping ring as injector, what is the case for 2×200 GeV machine for example, [2]. Basically the sources with minimal emittance exist. There were developed for scanning electron microscopy an ion implantation [2]. The microscopes have a few nanometer resolutions, what means that the electron beam, generated by the source of microscope, at the object plane has the same *nanometer* size [3]. The sources for these scanning microscopes are working in a DC regime, so in principle one can consider a bunching process for a proper time structure arrangements.

General problem with electron source for microstructures excited by laser radiation is similar to one for scanning electron microscope: the size of the beam must be as small as possible. One circumstance in relief is that the beam used in laser driven linac is of μm scale, rather than nanometer one. If we suggest, that the size of passing holes δ of the microstructure need to be fabricated with a fraction of the accelerating wavelength value λ_{ac}, then we can write $\sim \delta \cong 0.2 \cdot \lambda_{ac}$, where factor 0.2 reflects the fractional size of the passing hole. Taking into account that this must have a safety margins as big as ten sigma of the beam, one can estimate $\sigma \cong \sqrt{(\gamma \varepsilon) \cdot \beta / \gamma / \pi} \cong 0.1 \cdot \delta = 0.02 \cdot \lambda_{ac}$. So the emittance required must be not bigger, than $\gamma \varepsilon \le 4\pi \cdot 10^{-4} \lambda_{ac}^2 \cdot \gamma / \beta$. One can see from here that situation worsens for low energy. That is why the source is a general problem for tabletop accelerator. With *RF focusing* one can obtain $\beta \cong 5mm$, so this yields for $\lambda_{ac} \cong 1\mu m$ an emittance required $\sim \gamma \varepsilon \le 8\pi \cdot 10^{-12} \cong 2.4 \cdot 10^{-11} cm \cdot rad$. An attempt to have such a small emittance makes considerations about possible limitations for emittance arising from the quantum mechanical phenomena actual [1]. Let us discuss the possibility to satisfy such emittance requirements. First of all the injection sections could have less safety margins, arranging some scrapping the beam. This will require adequate cooling of first sections of the source. So if we refuse ten sigma margins at first sections, leaving only one sigma per side, the emittance allowable becomes a hundred times bigger $\gamma \varepsilon \le 2.4 \cdot 10^{-9} cm \cdot rad$ and this becomes within practical possibility, as it will be shown lower. Other possibility to squeeze the bunch – to move further in reduction of envelope function. In principle, the energy of the particle from injector is in a MeV scale. So, the alternative phase focusing together with appropriate profile of the passing holes which are focusing (for example in x –direction as shown Fig.3), could rise a possibility for a envelope function even as fraction of a millimeter.

For the accelerator beam source we are suggesting to use the type of source similar to what are in use for scanning microscopes, namely an emission from the tips, [3], see Fig.1. At the tip the electric field strength could reach V / ρ, where V is a voltage applied between the tip and anode and ρ is the radius of the tip. In this case the divergence of electrical lines defined by the height of the tip at distances bigger, than the radius. If, however, the other small height tip with height about its radius attached to the first tip, then the radius of the top tip will define the divergence.

This gives typical divergence about 1/100 *rad*. Basically this is nothing else, but formation of the best potential profile with the help of material. Similar to what is done for macroscopic electron guns. Now it is possible to obtain the tips with nanometer-level radius routinely. So the emittance of the electron beam, one could expect from a single super-tip goes to

$$\gamma\varepsilon \cong 10^{-7} cm \cdot 10^{-2} rad = 10^{-9} cm \cdot rad.$$

This is an invariant emittance, as the electrons are nonrelativistic. This is an impressive number. The tips described above routinely working with DC currents ~100 nA. The last number gives the total charge ~10^{-7} Q/s. In its turn, this corresponds to $N \cong Q/e \cong 6.25 \cdot 10^{11}$ electrons per second. The longitudinal emittance defined by the temperature of the tip. In any case the longitudinal energy spread is less, than the work function eU_W.

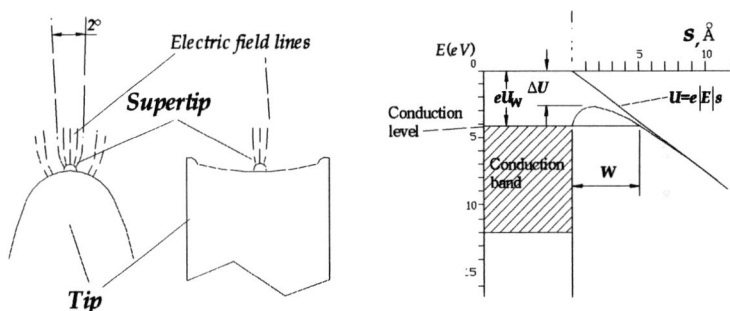

FIGURE 1: To the physics of electron source based on cold emission from the tip. At the left two types of tips are represented. At the right the pure W-vacuum surface shown. The field strength shown at the right is ~ $4V/5A° = 8 \cdot 10^7 V/cm$ as example.

For practical reasons the emitting surface installed into the end wall of the structure similar to the main one. The profile of the first cell optimized so that the maximal electric field strength reaches its maximum at the central region. The materials with low out work could be used here as well. These arrangements are shown in Fig.2. In principle, this figure is similar to the ones represented elsewhere for RF guns. The difference is, first, that the accelerating structure is a micrometer size scale, and second, there is no special laser radiation illuminating the cathode. In our case tremendous electric field makes electron emission possible due to Schottky regime of operation.

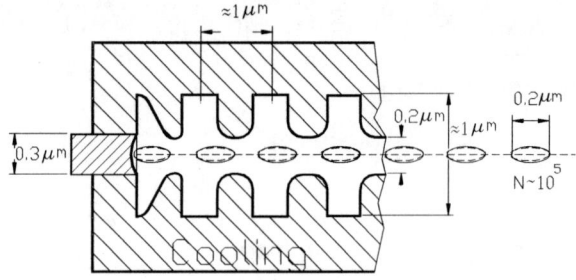

FIGURE 2: Schematic view of micro-RF gun. All elements illuminated by the same laser source swept. High electric field strength reached at the emitting surface carries out electrons into RF bucket.

The area of emitting region shown on Fig.2 is about $S \cong 10^{-9} cm^2$. Time duration of the pulse is $\tau \cong l_{bunch}/c \cong 6 \cdot 10^{16} s$ what is about the time of electron plasma reaction in metal. Absolute limit for the number of electrons defined by the velocity of electrons at the Fermi surface, what is $N_e \cong 6 \cdot 10^{23} \cdot 6 \cdot 10^{16} \cdot 10^{-3} \cdot 3 \cdot 10^{10} \cdot 10^{-9} \cong 10^7$. Thus this is far above our needs, what is $N \cong 10^5$ electrons only. Peak current is $J \cong eN/\tau \cong 25A$. The current density, however, goes to $j \cong J/S \cong 2.5 \cdot 10^{10} A/cm^2$, what is high, but taking into account that the time duration is going only $\tau \cong 6 \cdot 10^{-16} s$, that possibly makes this current density safe. In any case experiments need to be done to investigate the limiting current in short pulses.

Other possibility to make an adequately small and bright source of electrons, on the basis ponderomotiv repulsion of electrons from plasma described during this Workshop [2].

3. ACCELERATING STUCTURE

Described in [1], [2] as a basis, Fig. 3. Once again we would like to stress here that any other type of accelerating structure could be used with TLF method as well.

The foxhole type accelerating structure has advantages in coupling with the outer space, easy pumping, and could be treated as open for high order modes, generated by the bunch. When tuned, the structure has inductive coupling with outer space. Passing slits (having dimension δ on Fig.3) might have different shapes for focusing in both directions.

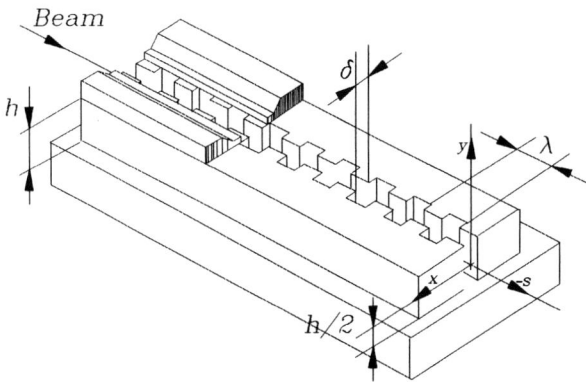

FIGURE 3: The foxhole type accelerating structure fragment. Height h is $h \cong \lambda_W/2$, where λ_W –is a wavelength of laser radiation inside the cell. Height of the structure (y-direction is $\cong \lambda_W/2$, what gives inductive coupling of each cell with outer space.

4. LASER UNDULATOR

Let us consider here one example of how this accelerator can be used for generation of X–rays. We are suggesting here use of undulator, arranged with the standing waves of the laser radiation from the same laser source, which feeds the accelerating structure. With other terminology (language), the photon generation in undulator could be treated as Compton back-scattering as well. Advantages of such an undulator/wiggler are the following. First of all, period of this undulator could be made of the order of the laser wavelength. Second –is the absence of any material boundary close to the beam. Third–is a possibility to have a helical/elliptical polarization easily exchangeable. Extremely low period, of the order of the laser wavelength, allows using a relatively low energy beam for X–rays generation. Disadvantage of such device might be a low deflection parameter, or K factor, $K = eH\lambda_u/2\pi mc^2$, where λ_u – is the undulator period, H–is magnetic field value.

As the radiation is orthogonal to the beam's trajectory, the maximal frequency is the same as defined for ordinary undulator. The difference occurs, however, when the energy of the quanta radiated becomes comparable with the energy of the particle. Let us make estimations however.

First of all, for the undulator radiation the number of photons, radiated at first harmonic by every particle, could be written as the following

$$N_\gamma \cong 4\pi\alpha \frac{L}{\lambda_u} \frac{K^2}{1+K^2} \approx Lr_0^2 \frac{H^2}{\hbar\omega}, \qquad (4.1)$$

where L –is the length of undulator, $\omega = c/\lambda_u$. One can describe the process of radiation in term of scattering process also. Here the number of the secondary quantas, radiated by electron, passing the region filled with photons, having volume density n_γ, could be calculated in term of effective length as

$$N_\gamma \cong L\sigma_\gamma n_\gamma \approx Lr_0^2 \frac{H^2}{\hbar\omega}, \tag{4.2}$$

i.e. the same value.

The principal scheme of the undulator is represented in Fig. 4. Ideologically it is similar to so-called laser wire, developed in SLAC. We however would like to stress the significant difference is that in our case the size of the laser waist in longitudinal direction is significant (a cm level).

For a primary laser radiation with a wavelength around micrometer, $\hbar\omega_0 \cong 1eV$, energy of electron beam $E \cong 100 MeV$, scattered photon will have energy about $\hbar\omega' \cong mc^2\gamma \cdot x/(1+x)$, with x –parameter $x \cong 2\gamma^2\hbar\omega_0/mc^2\gamma \cong 1.6 \cdot 10^{-3}$ (factor 2 instead of 4 appears due to orthogonal illumination). So $\hbar\omega' \cong 160 keV$.

Let us estimate the photon flux. For the laser flash with, say $16mJ$, the number of primary photons goes to $N_\gamma \cong 10^{17}$. Effective volume of interaction for $100ps$-laser flash could be estimated as $V \cong 3cm \times 5\mu m \times 5\mu m \cong 7.5 \cdot 10^{-7} cm^3$. Here we suggested that the waist of the laser radiation be in synchronism with electron bunch. So the photon density goes to $n_\gamma \cong N_\gamma/V \cong 1.3 \cdot 10^{23} cm^{-3}$ and length of interaction goes to

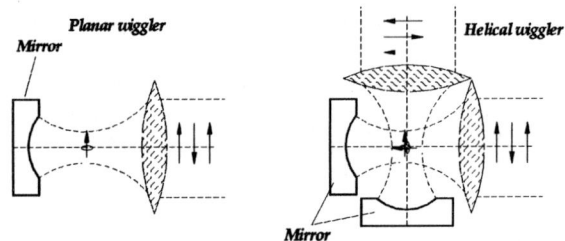

FIGURE 4: The wiggler arranged with standing wave of the laser beam. At the left the planar wiggler is represented. Combination of two planar wigglers gives a helical wiggler if the phase shift between the rays is $\pi/2$. Circular polarization could be obtained also if the laser beam has *circular* polarization without a mirror. Any other polarization is possible if the phase delay between the rays arranged. Elliptical polarization is also possible if appropriate attenuation is arranged in one ray path.

$l_{eff}^{-1} \cong n_\gamma \sigma_T = n_\gamma \frac{8\pi}{3} r_0^2 \cong 1.3 \cdot 10^{23} \cdot 6.65 \cdot 10^{-25} = 0.086 \, cm^{-1}$, so $l_{eff} \cong 11.5 cm$. The last means, that each electron radiates in average 0.3 photons per pass. This brings the photon number to $N_\gamma \cong 3 \cdot 10^4 /bunch/pass$. Suggesting repetition rate $1kHz$ and number of bunches 10 one can expect the photon flux $\dot{N}_\gamma \cong 3 \cdot 10^8 s^{-1}$.

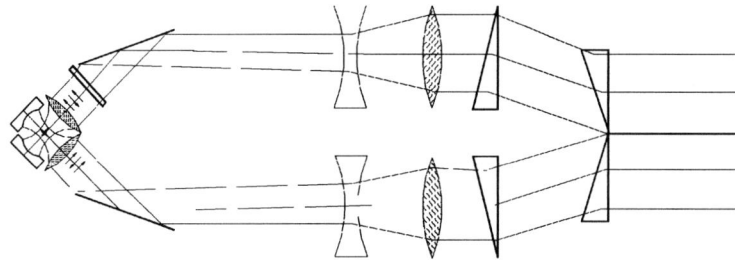

FIGURE 5: The way to feed the two rays from the same source.

The bunch number defined by illumination time for every point of acceleration structure. The last number defined by the size of the laser spot on the surface of the structure, what in it's turn defined by the number of resolved spots associated with the sweeping device [1]. For a micrometer level of accelerating wavelength the number of resolved spots could be expected of the order of 100. So the longitudinal size of the laser spot on the surface will be $3cm/100 \sim 300\mu m = 300\lambda_{ac}$. So, roughly 300 cells will be illuminated simultaneously. Taking into account that the only central part of this spot is useful for acceleration, on can suggest, that about 50-100 bunches could be accelerated as a train. We suggested that the quality factor of each cell is low, and does not tolerating critically to the raise time of the field in each cell.

5. SCHEME

Possible arrangements of the tabletop accelerator represented in Fig.6. Sections with accelerating structures feed by single sweeping device are about 3 cm long. Vacuumed cover for the beam part is shown in dashed lines. This cover has windows transparent for laser radiation. The design is also possible, when the sweeping devices placed in a vacuumed volume also.

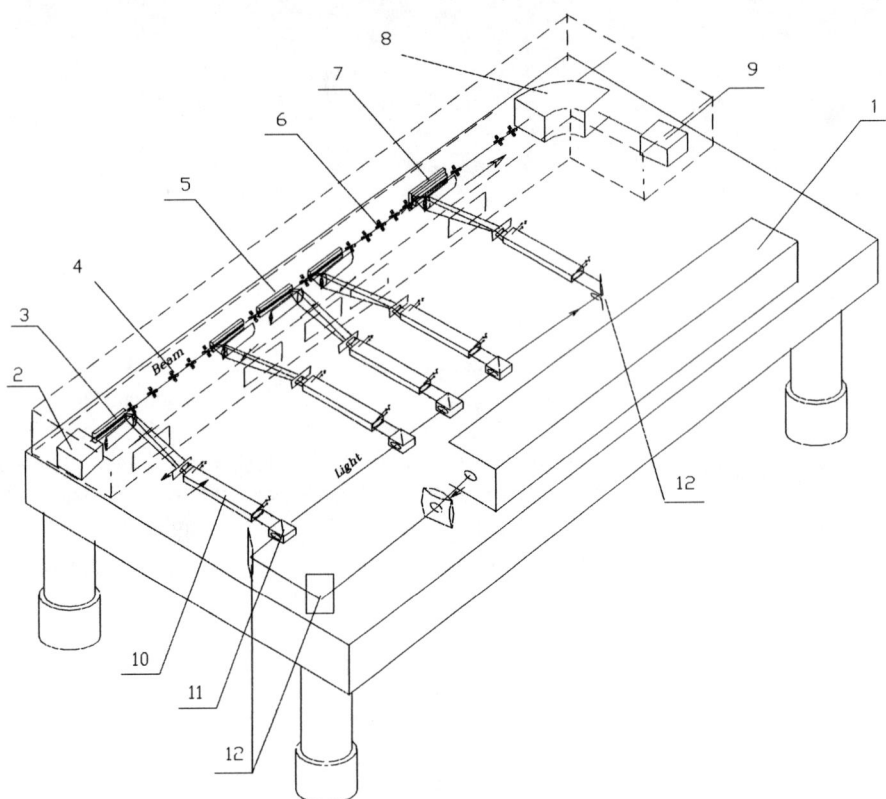

FIGURE 6: 100 MeV accelerator with a Compton wiggler concept. All elements installed on a platform. Light means laser beam. Other comments are in the text. Vacuumed cover for the beam part is shown in dashed lines. 1 –is a laser, 2–is source of particles, including micro-tip and movers, 3–RF prebuncher, 4–ias a space for buncher (if necessary) and beam longitudinal structure formation, 5–is main acceleration modules, 6–focusing elements, 7–a region for laser wiggler, 8–bending magnet, 9–is beams dump, 10–is a sweeping device, 11–is a splitting device, 12–is a mirror.

TABLE 1. Parameters of the laser-driven tabletop linac.

Wavelength	$\lambda_{ac} \cong 1\mu m$
Energy of $e^{\pm}$ beam	100 MeV
Active linac length	10 cm
Main linac gradient	1.0 GeV/m
Bunch population	10^6
No. of bunches/pulse	10 (≤ 100)*
Laser flash duty	100 ps
Laser flash energy	5 mJ**
Repetition rate	160 Hz
Average laser power	~0.8 W
Average beam power	26 mW
Bunch length	0.1 μm
$\gamma\varepsilon_x / \gamma\varepsilon_y$	$\approx 10^{-8}/10^{-8}\, cm\cdot rad$
Length of section/Module	3 cm
Wall plug power	3.5 kW

*–Maximal possible number.
**–Power for laser undulator is not included here.

6. CONCLUSION

The work required for demonstration of operationability of this linac could be very well staged, with clear results after each stage completed. One stage includes fabrication and cold testing of accelerating structures. Other stage includes development and test of the sweeping device. The low emittance solid-state electron/proton or ion source is another stage for this job.

The general problem of low emittance source could be resolved successfully, as we could conclude this from positive experience with scanning electron microscopes, for example.

In conclusion I would like to thank K. Finkelstein, who attracted my interest to the TableTop Accelerator problematic.

7. REFERENCES

[1] A. Mikhailichenko, *Particle Acceleration in Microstructures Excited by Laser Radiation*, CLNS 00/1662, Cornell 2000.
[2] A. Mikhailichenko, *Parameters of $2\times 200 GeV$ Linear Collider with Microstructures Excited by Laser Radiation*, this Workshop.
[3] L.Reimer, *Scanning Electron Microscopy*, Springer Series in Optical Science Vol. 45, ISSN 0342-4111, 1998.
[4] G. Stupakov, M. Zolotorev, this Workshop. A private communication from M. Zolotorev.

Muon Sources –
ν Factory to $\mu^{\pm}$ Colliders[1]

Zohreh Parsa

Brookhaven National Laboratory
Physics Department 510 A, Upton, NY 11973-5000, USA

Abstract. Employing intense muon sources to carry out forefront low energy research, such as the search for muon - number non–conservation, or for the purpose of providing intense high energy neutrino beams (ν factory) represents very interesting possibilities. If successful, such efforts would significantly advance the state of muon technology and provides intermediate steps in technologies required for a future high energy muon collider complex. High intensity muon: production, capture, cooling, acceleration and multiturn muon storage rings are some of the key technology issues that needs more studies and development. A muon collider require basically same number of muons as for the muon storage ring Neutrino Factory, but would require more cooling, and simultaneus capture of both $\pm\mu$. We present an overview of Muon Sources - Neutrino Factories, example of a muon storage ring at BNL, and possible upgrades to a full Muon Collider.

INTRODUCTION

A full high energy muon collider may take considerable time to realize. However, intermediate steps in its direction are possible and could help facilitate the process. Building a muon storage ring for the purpose of providing intense high energy neutrino beams is particularly exciting. Such neutrino factories could have their own world class research program, with neutrino oscillation studies as the primary focus. High intensity muon experiments, neutrino factories, and other intermediate steps toward the muon collider are extremely important. They will greatly expand our abilities and build confidence in the credibility of high energy muon colliders.

Many of the recent, exciting results in neutrino physics have been obtained by non-accelerator experiments, although the neutrino mass and mixing parameters appear to require a new generation of accelerator based experiments. For this, an intense source of well-collimated neutrinos is needed.

Excitement is high in the accelerator physics community because Atmospheric-neutrino results suggest that the long-baseline accelerator experiments such as MINOS [4], K2K [3], and NGS [5] should also find neutrino oscillations. Further, the

---

[1] Supported by US Department of Energy contract DE-AC02-98CH10886.
[†] E-mail: parsa@bnl.gov

LSND experiment that was conducted at a short-baseline accelerator facility, can be confirmed by future accelerator experiments such as MiniBooNE [6], ORLanD [7], and CERN P311 [8]. Moreover, physics associated with some interpretations of the solar-neutrino deficit may be accessible to studies in accelerator-based experiments, if neutrino-beam fluxes can be improved by 1-2 orders of magnitude.

To obtain a factor of 100 improvement in neutrino flux, the best prospect appears to be neutrino-beams derived from a muon-storage-ring, rather than from direct pion decays. However, such an approach requires considerable development before it can be realized in the laboratory.

In the following sections we present schematics of a Neutrino Factory Facility concept (based on various muon storage rings), its components and a possible upgrade to a full muon collider. Figures and the examples described are based on some of the scenarios being explored by our Neutrino Factory and Muon Collider Collaboration, [10].

NEUTRINO FACTORY - FACILITY

A neutrino factory based on a muon storage ring is a challenging extension of present accelerator technology. Conventionally, neutrino beams employ a proton beam on a target to generate pions, which are focused and allowed to decay into neutrinos and, muons [4]. The muons are stopped in the shielding, while the muon-neutrinos are directed toward the detector. In a neutrino factory, pions are made the same way and allowed to decay, but it is the decay muons that are captured and used. The initial neutrinos from pion decay are discarded, or used in a parasitic low-energy neutrino experiment. But the muons are accelerated and allowed to decay in a storage ring with long straight sections. It is the neutrinos from the decaying muons (both muon-neutrinos and anti-electron-neutrinos) that are directed to a detector.

In a Neutrino Factory, a proton driver of moderate energy (< 50 GeV) and high average power,(e.g., 1-4 MW), similar to that required for a muon collider, but with a less stringent requirements on the charge per bunch and power is needed. This is followed by a target and a pion-muons capture system. A longitudinal phase rotation is performed to reduce the muon energy spread at the expense of spreading it out over a longer time interval. The phase rotation system may be designed to correlate the muon polarization with time, allowing control of the relative intensity of muon and anti-electron neutrinos. Some cooling may be needed, to reduce phase space, about a factor of 50 in six dimensions. This is much smaller than the factor of 10^6 needed for a muon collider. Production is followed by fast muon acceleration to 50 GeV (for example), in a system of linac and two recirculating linear accelerators (RLA's), which may be identical to that for a first stage of muon collider such as a Higgs Factory. A muon-storage ring with long straight sections could point to one or more distant neutrino detectors for oscillation studies, and to one or more near detectors for high intensity scattering studies.

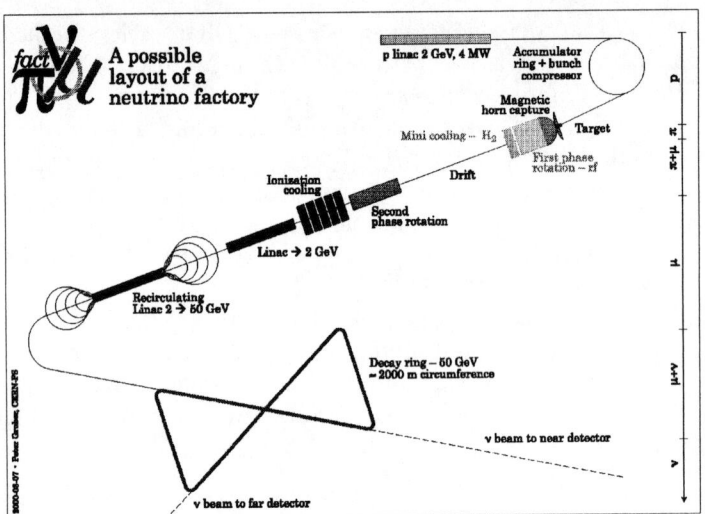

FIGURE 1. A schematic concept of a Neutrino Factory Facility based on a muon storage lattice for CERN [15].

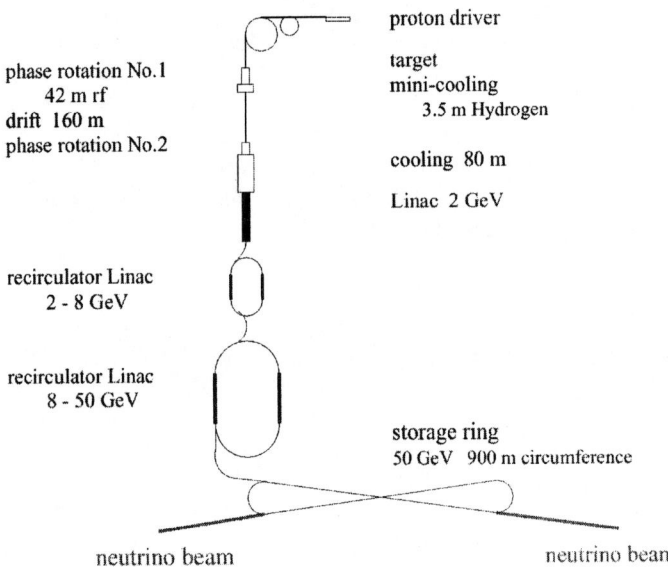

FIGURE 2. A schematic concept of Neutrino Factory Facility based on a bowtie muon storage lattice.

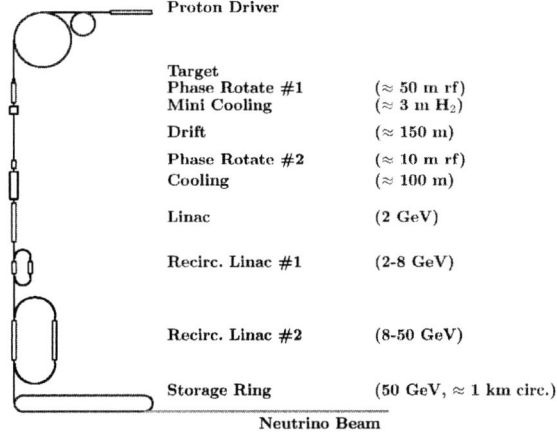

FIGURE 3. Overview of a Neutrino Factory Concept, with a Racetrack Muon - Storage Ring

A planar bowtie - shaped ring (illustrated in Figure 2) can be designed and oriented to send neutrino beams to any two detector sites. Since, there is no net bending, the polarization may be preserved. (A disadvantage of the Bowtie - shaped ring is that it may need extra bending. Since there is geometry constrains on the ratio of short to long straight sections, the ring circumference may increase.) With the ring in a tilted plane, both long straight sections would point down into the earth, such that neutrinos can be directed into two very distant detectors. Triangular-shaped storage rings also have this advantage.

Figure 3, illustrates components of a Neutrino Factory based on a racetrack - shaped muon storage lattice [10].

Figure 2 and Figures 3 show examples of the scenarios being explored by our Collaboration, [10].

In the following sections, a description and simulation of target through cooling-channel and a bowetie-shaped muon storage lattice will be discussed.

FRONT-END SYSTEM

The number of pions per proton produced with an optimized system varies linearly with the proton energy, Thus, the number of pions, and the number of muons into which they decay, is essentially proportional to the proton beam power.

Table 1 presents possible parameters for proton drivers at BNL and FNAL. The target requirements are very similar to those for the muon collider, except the instantaneous shock heating is somewhat less because protons are distributed in a larger number of bunches. In the scheme presented here, it is assumed that the liquid mercury jet solution is used. The capture solenoid is likely to be the same as described in the muon collider status report [9]. Figure 4, shows the pion production target, solenoidal capture, decay channel and beginning of phase rotation. At the

TABLE 1. Example of parameters for various Proton driver scenarios at BNL and FNAL.

	BNL_1	BNL_2	$FNAL_1$	$FNAL_2$
Energy [GeV]	24	24	16	16
Power [MW]	1	4	1	4
Rep. Rate [Hz]	2.5	5	15	15
p's/fill	10^{14}	$2\,10^{14}$	$2.5\,10^{13}$	10^{14}
Bunches	6	6	4	4
Circumference [m]	807	807	474	474
Bunch spacing [m]	135	135	118	118
σ_t [nsec]	1	1	1	1

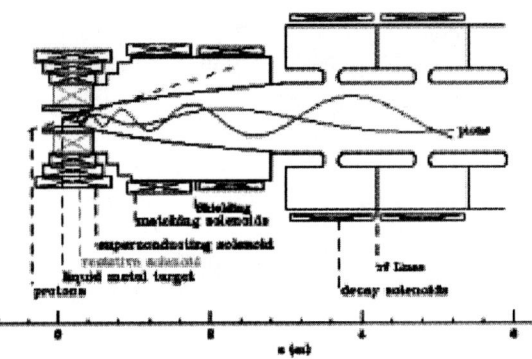

FIGURE 4. A Schematic of Targetry, Pion Capture, and beginning of Phase Rotation.

end of this first phase rotation stage, the bunch length increases by about a factor of 6 and the energy spread decreases by the same amount. Whether this first stage of phase rotation can be eliminated is being investigated. An example of designs and simulations being explored by NFMCC is illustrated in Figure 5 (shows schematics of a Muon source front-end compenents), Fig. 6 (Particle composition in the target-to-linac channel), and in Fig. 7 (the muon emittance variation in the target-to-linac channel).

The challenges of further acceleration and storage of the muon beam will be substantially easier if we reduce the transverse phase area of the beam by an additional factor of 10. This may not be accomplished in a single step of ionization cooling, but involves alternating ionization cooling and rf acceleration, all in a magnetic channel. The acceleration from $\sim$ 100 MeV to e.g., $\sim$ 50 GeV may be accomplished in recirculating linacs with superconducting rf cavities, after which muons are injected into a muon storage ring. The desire for multiply directed neutrino beams with very small angular divergence may require a more novel design for the storage ring, with a plane that is far from horizontal. The R&D needs for a muon collider are very similar, but with additional challenges in cooling and storage ring design. At

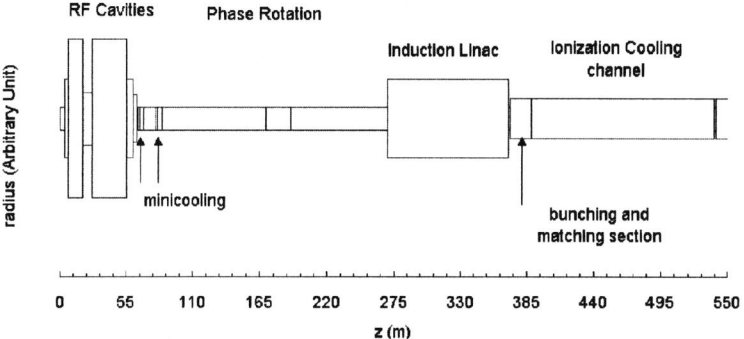

FIGURE 5. Schematics of the Muon Source from Target to Linac.

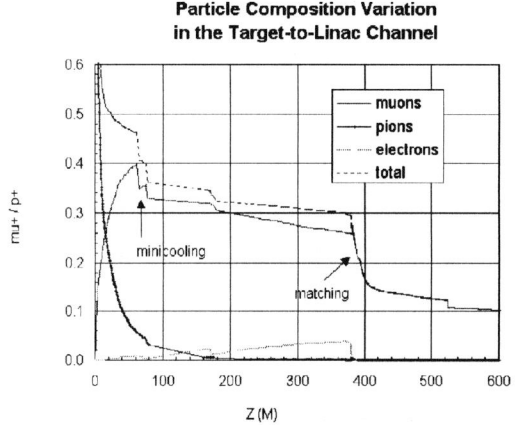

FIGURE 6. Particle Composition from Target to Linac.

least four orders of magnitude more cooling (including continual exchange between transverse and longitudinal emittance) are required for a muon collider than a neutrino factory. Also, a different ring is needed to maximize collider luminosity than simply to hold the muons while they decay.

Ionization cooling that has been proposed involves passing the beam through an absorber in which the muons lose transverse- and longitudinal-momentum by ionization loss (dE/dx). The longitudinal momentum is then restored by coherent re-acceleration, leaving a net loss of transverse momentum (transverse cooling). The process is repeated many times to achieve a large cooling factor. The beam energy spread can also be reduced using ionization cooling by introducing a transverse variation in the absorber density or thickness (e.g. a wedge) at a location where there is dispersion (the transverse position is energy dependent). Theoretical studies have shown that, assuming realistic parameters for the cooling hardware, ionization cooling can be expected to reduce the phase-space volume occupied by

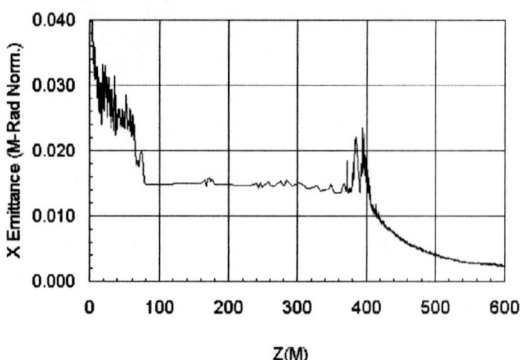

FIGURE 7. Muon emittance variation in Target to Linac channel.

the initial muon beam by a factor of $10^5 - 10^6$. Ionization cooling is a new technique that has not yet been demonstrated. Special hardware needs to be developed to perform transverse and longitudinal cooling. It is recognized that understanding the feasibility of constructing an ionization cooling channel that can cool the initial muon beams by factors of $10^5 - 10^6$ is on the critical path to the overall feasibility of the muon collider concept.

STORAGE RINGS

Figure 3 illustrated a racetrack - shaped configuration, with two long straight sections. and Figure 2 a bowtie-shaped ring. The planar ring can be designed and oriented to send neutrino beams to any two detector directions and with bypass(es) that could be added, to send beams to additional detector sites. In the bowtie-shaped lattice design, the lattice has two long-straight sections, two short-straight sections and two arcs. A racetrack muon storage - ring can be configured to deliver one neutrino beam to an arbitrary detector site. Bowtie - shaped, triangle shaped rings can be configured to deliver neutrino beams to two arbitrarily selected detector sites. This can be done by appropriate choice of, 1) the ring plane, 2) the orientation of the ring in that plane and 3) the angle at the crossing point between the two long straight sections. By inclusion of bypasses, additional detector sites may be accessible from a single muon storage-ring source.

A bypass would lie in a plane that includes the original long straight section (but differs from that of the ring), and begin and end on one of the long straight sections. Its magnets would be powered when one desires to send the muons along the deformed bypass path rather than along the normal straight path. In such a bypass, dipoles would produce a roughly triangular path in the bypass plane, one of whose sides would point to the desired detector. The two necessary degrees of

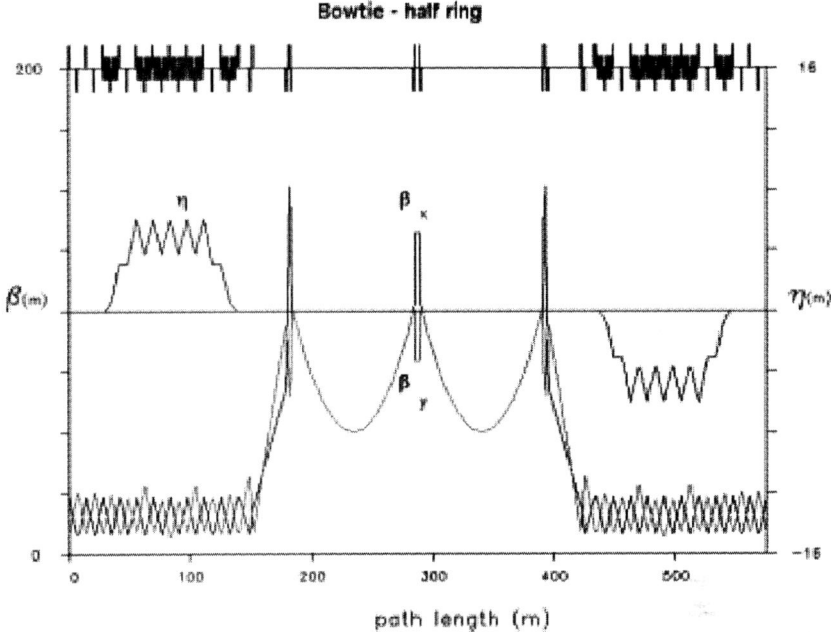

FIGURE 8. Example of Lattice Functions for Bowtie-shaped Half Ring.

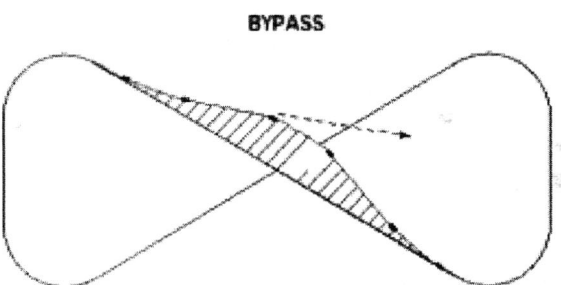

FIGURE 9. Lattice functions for Bowtie-shaped Ring with Bypass. The arrow illustrates direction of a neutrino beam to additional detector site(s) via the Bypass.

freedom are provided by the angle between the bypass and ring planes and by the magnitude of the deflection given by the bypass dipoles. To suppress the dispersion pairs of dipoles should be placed 180 deg apart, in FODO cells.

MUON STORAGE RING BASED NEUTRINO SOURCE AT BNL?

As is known, the BNL-AGS proton beam parameters are very suited for use as a source for muon storage ring based neutrino factory and muon collider. Table 2 illustrates basic BNL-AGS proton beam properties.

With a muon storage ring - neutrino source at BNL (Figure 10), detectors at Fermilab or Soudan, Minnesota (1715 km), become very interesting possibilities. The feasibility of constructing and operating such a muon-storage-ring based Neutrino-Factory, including geotechnical questions related to building non-planar storage rings (e.g. for BNL-fermilab; at 8° angle for BNL-Soudan, and 31° angle for BNL-Gran Sasso) along with the design of the muon capture, cooling, acceleration, and storage ring for such a facility is being explored by our growing Neutrino Factory and Muon Collider Collaboration (NFMCC), but requires additional studies for a BNL site specific example.

Conventionally, neutrino beams employ a proton beam on a target to generate pions, which are focused and allowed to decay into neutrinos and, muons [4]. The muons are stopped in the shielding, while the muon-neutrinos are directed toward the detector. In a neutrino factory, pions are made the same way and allowed to decay, but it is the decay muons that are captured and used. The initial neutrinos from pion decay are discarded, or used in a parasitic low-energy neutrino experiment. But the muons are accelerated and allowed to decay in a storage ring with long straight sections. It is the neutrinos from the decaying muons (both muon-neutrinos and anti-electron-neutrinos) that are directed to a detector.

Figure 10 shows schematics of space angles [19] and baselines for example of a muon storage neutrino source at BNL, with detectors (placed at Fermilab; Soudan; Minnesota (1715 km); or Gran Sasso, Italy (6527 km)) at various global locations.

MUON COLLIDER

Fig. 11 shows a schematic of a muon collider components [9]. A high intensity proton source is bunch compressed and focused on a heavy metal target. The pions

TABLE 2. BNL- AGS Proton Beam Properties

Parameters	BNL-AGS	Muon Collider
Proton Energy [GeV]	24	16 - 24
Proton/Bunch	1.6×10^{13}	5×10^{13}
Bunch No.	6	2
Proton/cycle	1.0×10^{14}	1.0×10^{14}
Bunch Length [μs]	2.2	1
Bunch spacing [ns]	440	1000

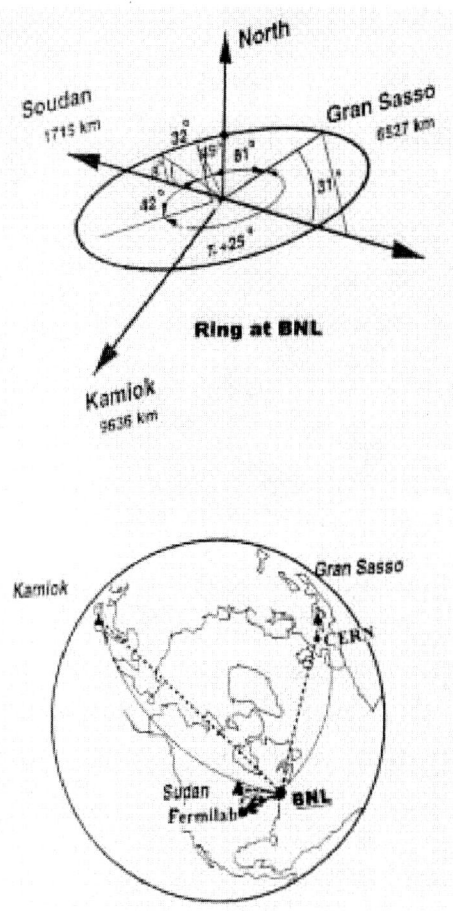

FIGURE 10. Shows space angles and baselines for a Muon - Storage Ring at BNL and possible detector sites (at Fermilab, Sudan, CERN, Kamioka and Gran Saso).

generated are captured by a high field solenoid and transferred to a solenoidal decay channel within a low frequency linac. The linac reduces, by phase rotation the momentum spread of the pions and of the muons into which they decay. Subsequently, the muons are cooled by a sequence of ionization cooling stages. Each stage consists of energy loss, acceleration, and emittance exchange by energy absorbing wedges in the presence of dispersion. Once they are cooled the muons must be rapidly accelerated to avoid decay losses. This can be done in recirculating accelerators (as at CEBAF) or in fast pulsed synchrotrons. Muon collisions occur in a separate high field collider storage ring with a single very low beta insertion.

It is expected that the first stage, proton driver would be 16 to 30 GeV; but

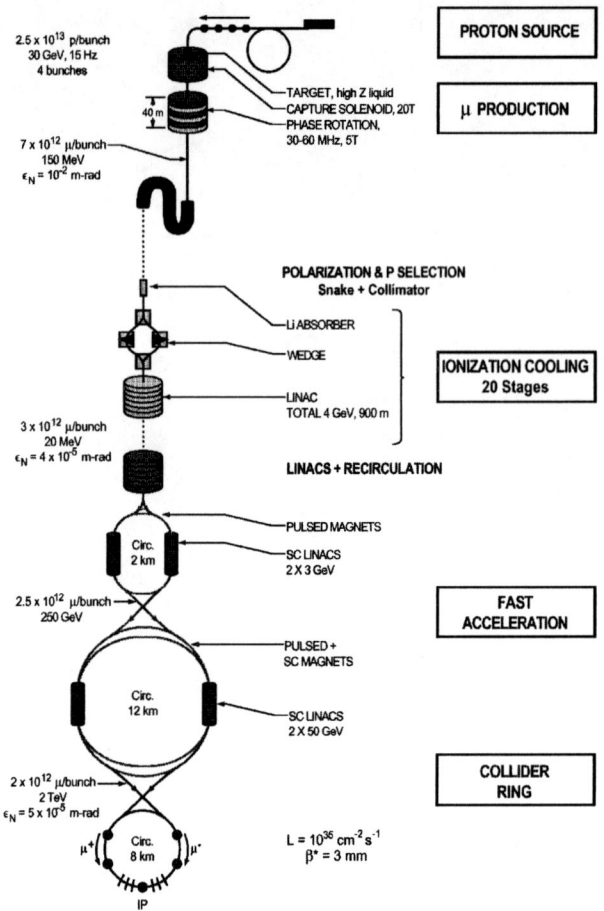

FIGURE 11. Schematic of a 4 TeV Muon Collider.

would be much faster pulsed, keeping the number of protons per pulse the same or smaller than the AGS, which is about 6×10^{13} protons per pulse and with some upgrade to about 10^{14} protons per pulse.

Roughly one expect to get 1 muon/proton on target which would give luminosity between 10^{34} to 10^{35} the envisioned muon collider. Although the accelerating component is large, the other components can fit within it and the whole machine is compact enough to fit on existing Brookhaven or Fermilab sites. For more information on the Muon Collider and parameters under study, see e.g. [9], [22] – [26].

Table 3 shows examples of the parameters of potential muon colliders at 100 GeV, 500 GeV and 4 TeV center of mass energy. The 100 GeV collider would be interesting for the study of the lowest mass Higgs. The 4 TeV collider should be in the

TABLE 3. Parameters of $\mu^+\mu^-$ collider Rings.

Energy (C.M.) TeV	4	0.5	0.1
Beam Energy TeV	2	0.25	0.05
Beam γ	19,000	2,400	473
Rep. rate Hz	15	2.5	15
p Energy GeV	30	24	16
p/pulse	10^{14}	10^{14}	5×10^{13}
μ/bunch	2×10^{12}	4×10^{12}	4×10^{12}
Bunches/sign	2	1	1
Beam Power MW	38	0.7	1.0
ϵ_N π mm-mrad	50	90	195
Bending Field T	9	9	
Circumference km	8	1.3	0.3
Ave. ring field B T	6	5	3.5
Effective turns	900	800	450
β^* mm	3	8	9
IP beam size μm	2.8	17	187
β_{max} km	200-400	10-20	1.5
Lumin. $cm^{-2}s^{-1}$	10^{35}	10^{33}	2×10^{31}

energy range of most of the heavy Higgs in the minimal SUSY model (if that is the correct theory).

SUMMARY

Building a muon storage ring for the purpose of providing intense high energy neutrino beams is particularly exciting. Such neutrino factories could have their own world class research program, with neutrino oscillation studies as the primary focus (see e.g. Indeed, if very high intensities, $\sim 10^{21} \frac{\nu}{year}$, are attained and nature has been kind in her neutrino mass and mixing parameters, one could envision a complete exploration of the 3×3 neutrino mixing matrix and even the detection of CP violation in the oscillation phenomena. If a neutrino factory is successfully accomplished, it would provide a major advancement. Its ambitious goals would test essentially all aspects of the muon collider concept, muon production, collection, cooling and acceleration. Furthermore, if properly coordinated, the neutrino factory complex might be suitably expanded into the First Muon Collider, perhaps a Higgs factory with center of mass energy ~ 100 GeV.

High intensity muon experiments, neutrino factories, and other intermediate steps toward the muon collider are extremely important. They will greatly expand our abilities and build confidence in the credibility of high energy muon colliders. many would prefer, but remember, Rome was not built in a day.

At BNL, a 20 GeV muon storage ring intense muon (neutrino) source would be very interesting but expensive? An alternative source of intense muons are the conventional Horn Beams which seems to be not only competitive with the lower energy muon storage rings but also at a lower cost. For example, with the same number of proton (p) on target and same size (kTon) detector the BNL – AGS 1 $GeV \nu_\mu^{peak}$ Horn $\simeq$ $10 GeV$ Muon Storage Ring (statistically if L/E is fixed). Upgraded Horn facility is potentially powerful. Further R&D on $6 \times 10^{14} p/sec$ driver and target at BNL are important for both the muon storage ring and Horn. For more info. see references, [2] - [30].

REFERENCES

1. See e.g., MECO Presentation in *CIPANP2000 Proceedings* ed. Z. Parsa, W. Marciano, [25].
2. The Neutrino Factory & Muon Collider Collaboration http://www.cap.bnl.gov/mumu/
3. http://neutrino.kek.jp/~melissa/K2K/K2K2./html
   http://www.awa.tohoku.ac.jp/html/KamLAND/
4. See, e.g., http://www.hep.anl.gov/NDK/Hypertext/minos_tdr.html
5. http://www.cern.ch/NGS/ngs99.pdf
6. The MiniBooNE project: http://www.neutrino.lanl.gov/BooNE
7. The Oak Ridge Large ν Detector http://www.orau.org/orland/
8. Search for $\nu_\mu \to \nu_e$ Oscillations at CERN PS,
   http://chorus01.cern.ch/~pzucchel/loi/
9. C.M. Ankenbrandt *et al.*, *Status of muon collider research and development and future plans*, Phys. Rev. ST Accel. Beams **2**, 081001 (1999), and refrences therein.
10. See e.g., [9] - [29]; and references therein.
11. Z. Parsa, *Muon Storage Rings - Neutrino Factories*, in *Next Generation Nucleon decay and Neutrino detector (99)*, ed. M. Diwan and C. Jung, AIP CP 533, pp. 181-195, N.Y. (2000).
12. Z. Parsa, *Intense Muon Beams and Neutrino Factories* in AIP CP542, ed. D.Cline (2000).
13. K.T. McDonald, ed. for the Neutrino Factory and Muon Collider Collaboration, physics/9911009, 6 Nov, 1999, and references therein; ibid, Private comm.
14. http://lyoninfo.in2p3.fr/nufact99/ R.B. Palmer, C. Johnson, E Keil, BNL-66971,
15. http://muonstoragerings.web.cern.ch/ ref for a schematic of CERN Nu-factory layout figure.
16. R.B. Palmer, Draft Paramets of a Neutrino Factory, MUC0046.
17. C. Kim, simulation of the target to linac, (99).
18. A. Garren, Private Comm.: Bowtie-lattice with code SYNCH.
19. Y. Fukui's Drawing was modified for Fig. 1; also see AIP CP542, ed. D. Cline.
20. S. Geer, Phys. Rev. D **57**, 6989 (1998), hep-ph/9712290;
21. A. Blondel, http://alephwww.cern.ch/~bdl/muon/nufacpol.ps
22. D. B. Cline (ed.), *Physics Potential & Development of $\mu^+\mu^-$ Colliders* AIP CP **352** (1996).
23. The $\mu^+\mu^-$ Collider Collaboration, $\mu^+\mu^-$ *Collider Feasibility Study*,
    BNL-52503, FERMILAB-Conf-96/092, LBNL-38946 (July 1996);
24. Z. Parsa, ed., *Future High Energy Colliders*, AIP CP **397**, NY (1997).
25. Z. Parsa, New High Intensity Muon sources and Flavor Changing Neutral Currents, World scientific Publishing, pp 147-153 (1998).
26. Z. Parsa, Lasers and Future High Energy Collliders, STS-Press, pp 823-830 (1997).
27. Z. Parsa, Polarization and Luminosity requirements for the First Muon Collider, Procd. of AAC98, AIP-Press, NY.(1998).
28. Z. Parsa, Muon Dynamics and Ionization Cooling at Muon Colliders, Procd. of EPAC98, Stockholm, Sweden, Vol 2, pp.1055-.
29. *Neutrino Factory Feasibility Studies at Fermilab*:
    http://www.fnal.gov/projects/muon_collider/nu_factory/
30. N. Mokhov, Carbon and Mercury Targets in 20-T Solenoid with Matching, MUC0061

AN ULTRA-HIGH GRADIENT PLASMA WAKEFIELD BOOSTER[1]

P. Chen [1], S. Cheshkov [2], R. Ruth [1], T. Tajima [2,3]

[1] *Stanford Linear Accelerator Center, Stanford University, CA 94309*
[2] *Department of Physics, University of Texas, Austin, TX 78712*
[3] *Lawrence Livermore National Laboratory, Livermore, CA 94550*

Abstract. We present a Plasma Wakefield Acceleration (PWFA) scheme that can in principle provide an acceleration gradient above 100 GeV/m, based on a reasonable modification of the existing SLAC beam parameters. We also study a possible up-grade of the Stanford Linear Collider (SLC) to hundreds of GeV center-of-mass energy using such a PWFA as a booster. The emittance degradation of the accelerated beams by the plasma wakefield focus is relatively small due to a uniform transverse distribution of the driving beam and the single stage acceleration.

INTRODUCTION

Since the introduction of the plasma accelerator concepts [1,2], there has been substantial progress both experimentally and theoretically that further advances the schemes [3]. Nevertheless, a macroscopic demonstration of a high gradient plasma acceleration with reasonable accelerated-beam quality, is still lacking. In the case of Laser Wakefield Accelerator (LWFA) [1], very high acceleration gradients have been observed. But the challenge has been to overcome the laser Rayleigh divergence in the plasma so as to extend the distance of acceleration. The propagation of the laser in a hollow plasma channel appears to be a promising idea [4]. On the other hand, while the electron-beam driven Plasma Wakefield Accelerator (PWFA) [2] can indeed be staged in macroscopic scale [5], the expected acceleration gradient tends to be lower than that in the LWFA scheme unless the driving beam pulse is shaped in either the linear [6,7] or the nonlinear [8,9] regime to optimize the *transformer ratio* [6,7].

In this paper we present our study of PWFA parameters based on a reasonable extension of existing beam conditions at the Stanford Linear Accelerator Center (SLAC). We invoke the scheme of a multi-stage bunch compression that would both

---

[1] Work supported by the Department of Energy, contracts DE–AC03–76SF00515 and DE–W–7405–ENG–48.

compress and shape the 50 GeV SLAC beam to tens of micrometers in length. Such high density shaped beams can then excite plasma wakefields that would provide acceleration gradients of more than 100 GeV/m. We also study the beam dynamics of the trailing accelerated beam, the associated beam-beam interaction effects and the luminosity deliverable. Specifically a rough design is presented for a high energy linear collider built upon adding a PWFA "booster" to the Stanford Linear Collider (SLC). We demonstrate that the collider operation at several hundred GeV center-of-mass energy is possible. It is interesting to note that such a "Plasma Booster" was actually proposed when the PWFA concept [10] was first introduced.

PLASMA WAKEFIELDS

Our main motivation is to find a physically realizable parameter set for a linear collider application of the PWFA scheme. The fundamental principles have already been laid down when the concept was originally introduced [10,2], and studied in some detail [11,12]. For concreteness, we invoke the existing SLAC beam parameters as our starting point. Several conditions must be satisfied to use the SLAC beam as a driver. A number of assumptions are made.

First, we assume that the SLAC beams can be "bunch compressed" to a much shorter length. This is essentially the rotation of the beam in its longitudinal phase space, where the adiabatically damped relative energy spread, $\delta p/p$, is exchanged with the length of the bunch. We second assume that during bunch rotation one is able to shape the beam into an asymmetric head-to-tail density distribution for large transformer ratios, applicable to the linear regime of plasma perturbation, or a uniform distribution from head to tail for the application to the nonlinear regime. Finally, to minimize transverse focusing of the accelerated beams, we assume that the driving beam is also transversely shaped into a uniform distribution. This can in principle be accomplished by applying proper octupole magnetic fields in the beam line.

The general expressions for the longitudinal and transverse plasma wakefields are [13]

$$W_\parallel = -\frac{4\pi e n_b}{k_p^2} \partial_\zeta Z(\zeta) R(r) , \qquad (1)$$

$$W_\perp = -\frac{4\pi e n_b}{k_p^2} Z(\zeta) \partial_r R(r) , \qquad (2)$$

where

$$Z(\zeta) = k_p \int_\zeta^\infty d\zeta' \rho(\zeta') \sin k_p(\zeta' - \zeta) , \qquad (3)$$

and $\rho(\zeta)$ is the normalized longitudinal density distribution of the driving bunch. As we assume a uniform transverse distribution, the function $R(r)$ is

$$R(r) = \begin{cases} 1 - k_p a K_1(k_p a) I_0(k_p r), & r < a, \\ k_p a I_1(k_p a) K_0(k_p r), & r > a. \end{cases} \quad (4)$$

Here $\zeta = z - ct$ is the beam comoving coordinate, $k_p = \sqrt{4\pi r_e n_p}$ is the plasma wave number, $r_e = e^2/mc^2$ is the classical electron radius, n_p is the ambient plasma density and n_b is the beam density. K_i's and I_i's are the modified Bessel functions.

For $k_p a \gg 1$ and $r/a \ll 1$ we get

$$W_\| = -\frac{4\pi e n_b}{k_p^2} Z'(\zeta) \left(1 - k_p a \sqrt{\frac{\pi}{2k_p a}} e^{-k_p a}\right), \quad (5)$$

$$W_\perp = \frac{4\pi e n_b}{k_p^2} Z(\zeta) \frac{\sqrt{2\pi}}{4} k_p^{5/2} a^{1/2} e^{-k_p a} r. \quad (6)$$

We see that the transverse wakefield is exponentially suppressed, whereas the longitudinal wakefield is slightly reduced by the form factor $F(k_p a)$ given by

$$F(k_p a) \equiv \left(1 - k_p a \sqrt{\frac{\pi}{2k_p a}} e^{-k_p a}\right) \approx 1. \quad (7)$$

There is a quarter-wavelength region in the wake with simultaneous acceleration and focusing, which is the phase suitable for placing the accelerating beam.

LONGITUDINAL BUNCH SHAPING

The wakefield acceleration gradient is sensitive to the longitudinal bunch shape of the driving beam. In order to search for the optimal acceleration gradient, we shall consider three representative cases of longitudinal bunch shaping that ranges from Case A: a parabola beam; Case B: a "doorstep", or optimized, beam; and Case C: a "flat-top" (uniform density) beam in the nonlinear beam-plasma interaction regime. The first two cases are in the linear regime, where the plasma density is sufficiently higher than that of the beam. In Case A we intend to study the plasma wakefield generated by an unshaped, high energy beam, which is typically in Gaussian distribution. Since mathematically the parabolic density distribution is found to be easier to handle analytically than the Gaussian distribution [12], while the characteristics of the excited plasma wakefields are essentially the same, we shall invoke the parabolic, instead of the Gaussian, distribution, for Case A.

The idea is to use the existing SLC beam with compression and appropriate profile shaping as a driver for the plasma wakefield based accelerator setup in a high density plasma (or gas). This beam is characterized by an energy of 48 GeV, number of particles $N = 2 \times 10^{10}$, normalized emittance $\epsilon_n = 3 \times 10^{-5}$m and $\sigma_z = 700\mu$m. However, to produce tens to hundreds of GeV/m field we need to further shorten the bunch. Such shortening can be achieved by several bunch compression stages utilizing rotation in the longitudinal phase space of the beam.

Assuming that the beam energy injected from the Damping Ring into the LINAC is 1.2 GeV, one can in principle achieve a total reduction of the bunch length by a factor of 40 when the beam reaches its final energy of 48 GeV. With the initial bunch length at $\sigma_z = 700\mu$m, the final bunch length would be $\sigma_z = 17.5\mu$m. Of course, when such a Gaussian bunch is further shaped, the total length of the beam will be different from this *rms* value.

Our goal is to achieve an acceleration gradient of the order of a 100 GeV/m. For this purpose we choose a high plasma density. The actual plasma densities in the following three different cases will be determined by different constraints. Secondly, we wish to maximally reduce the transverse wakefield, i.e., we insist that $k_p a \gg 1$. Thirdly, we want to accelerate the particles over a distance of the order of 1m to achieve significant final energy. When all three constraints are put together, we find it a reasonable compromise to choose $a = 2\sigma_r = 20\mu$m, and the betatron wavelength of the beam becomes $\beta = \gamma \sigma_r^2/\epsilon_n \approx 30$ cm. These should allow us to meet the above requirements.

Case A: Parabola (Linear Regime)

We first examine the wakefield generated by an unshaped beam. As we have explained, it is mathematically simpler to work with a parabolic, instead of a Gaussian, distribution. In this approach, we take the half-length of the parabola, b, to be the *rms* value of the corresponding Gaussian distribution, i.e., $b = \sigma_z$.

The longitudinal beam density profile is given by

$$\rho(\zeta) \equiv \frac{n_b(\zeta)}{n_b} = (1 - \zeta^2/b^2) \, , \, -b \leq \zeta \leq b \tag{8}$$

where n_b is the peak density of the driving beam determined from

$$N = n_b \int_0^a \int_{-b}^b 2\pi r \, dr d\zeta (1 - \zeta^2/b^2) = \frac{4}{3} n_b \pi a^2 b \rightarrow n_b = \frac{3N}{4\pi a^2 b}. \tag{9}$$

In this case the longitudinal wakefield on axis behind the beam is

$$\begin{aligned}W_\| &= 4\pi e n_b F(k_p a) \int_{-b}^b d\zeta'(1 - \zeta'^2/b^2) \cos k_p(\zeta' - \zeta) \\ &= -\frac{16\pi e n_b}{k_p^2 b}[\cos k_p b - \frac{1}{k_p b}\sin k_p b] F(k_p a) \cos k_p \zeta \, .\end{aligned} \tag{10}$$

At locations where $k_p \zeta = 2n\pi$, the longitudinal wakefield $W_\|$ reaches maxima. We define the maximum value of $|eW_\||$ as the acceleration gradient G.

We want to optimize the acceleration gradient behind the driving beam (with fixed beam parameters) by matching the bunch length with a properly chosen plasma density. This can be determined by demanding $\delta G/\delta(k_p b) = 0$. In our case, this results in a choice $k_p b \approx 2$. Unfortunately this solution would correspond to too large a beam-to-plasma density ratio, $\alpha = n_b/n_p \approx 2/3$, which clearly violates the assumption of linear plasma perturbation. As a compromise (but not much), we choose $k_p b = \pi$ so as to increase the plasma density and reduce α_0. For the given $b = 17.5 \mu m$, we find $n_p = 9.2 \times 10^{17} cm^{-3}$. The density ratio is now reduced to $\alpha = 1/4 \ll 1$. With $a = 20 \mu m$, we have $k_p a = 3.6$, and thus $F(k_p a) = 0.93$. Then the acceleration gradient on the axis is

$$G = \frac{16\pi e^2 n_b}{k_p^2 b}[\cos k_p b - \frac{1}{k_p b}\sin k_p b]F(k_p a) \approx 28 \text{GeV/m} . \qquad (11)$$

We see that this acceleration gradient, though substantial, falls short of achieving the 100 GeV/m goal.

Case B: "Doorstep" (Linear Regime)

The distribution for a "doorstep" bunch is defined as

$$n_b(\zeta) = \alpha_0 n_p \begin{cases} 1, & 0 < \zeta < \frac{\pi}{2k_p}, \\ 1 + k_p(\zeta - \frac{\pi}{2k_p}), & \zeta > \frac{\pi}{2k_p}. \end{cases} \qquad (12)$$

Then the wake potential inside the bunch is

$$Z^- = \alpha_0 \begin{cases} 1 - \cos k_p \zeta, & 0 < \zeta < \frac{\pi}{2k_p}, \\ 1 + k_p(\zeta - \frac{\pi}{2k_p}), & \zeta > \frac{\pi}{2k_p}. \end{cases} \qquad (13)$$

To find the wake potential Z^+ behind the bunch, we start with the general expression

$$Z^+ = C_1 \cos k_p \zeta + C_2 \sin k_p \zeta . \qquad (14)$$

Matching the boundary conditions at the end of the bunch $\zeta = b$, $Z^-(\zeta = b) = Z^+(\zeta = b)$ and $Z^{-'}(\zeta = b) = Z^{+'}(\zeta = b)$, we get

$$C_1 = \alpha_0 \left\{ -\sin k_p b + \cos k_p b \left[1 + k_p b(1 - \frac{\pi}{2k_p b}) \right] \right\}, \qquad (15)$$

$$C_2 = \frac{\alpha_0}{\cos k_p b} \left\{ 1 - \sin^2 k_p b + \sin k_p b \cos k_p b \left[1 + k_p b(1 - \frac{\pi}{2k_p b}) \right] \right\} \qquad (16)$$

An interesting special case is when $k_p b = n\pi$. Then

$$C_1 = \alpha_0 \left[1 + n\pi(1 - \frac{1}{2n}) \right], \qquad (17)$$

$$C_2 = \alpha_0 . \qquad (18)$$

Therefore the transformer ratio becomes

$$R = \frac{W^+_{\|max}}{W^-_{\|max}} = 1 + n\pi\left(1 - \frac{1}{2n}\right). \tag{19}$$

The maximum acceleration gradient is

$$G = \alpha_0 \left[1 + n\pi\left(1 - \frac{1}{2n}\right)\right] k_p mc^2 F(k_p a). \tag{20}$$

Specifically, let us take $k_p b = 8\pi$. With the total length of the beam assumed to be $b = 2\sigma_z = 35\mu m$, this corresponds to a plasma density of $n_p = 7.3 \times 10^{18} \text{cm}^{-3}$. To ensure self-consistency in the linear approximation of the plasma perturbation, from Eq.(12) we require that the ratio of the maximum beam density and the end of the bunch to that of the plasma be much smaller than unity. For definiteness, we set

$$\alpha = 0.5 = \frac{n_b(\zeta = b)}{n_p} = \alpha_0 \left[1 + k_p b - \frac{\pi}{2}\right], \tag{21}$$

and this fixes the parameter $\alpha_0 = 0.02$. Inserting these values into Eq.(20), we find

$$G \approx 180 \, \text{GeV/m}. \tag{22}$$

Note that this gradient is reasonably smaller than the so-called wavebreaking limit, 270 GeV/m, at the given plasma density.

Case C: "Flat-Top" (Nonlinear Regime)

Now we examine the case of a "flat-top" bunch in the nonlinear regime. By this we mean that the longitudinal density distribution of the beam is uniform from head to tail. Such a bunch distribution, though not optimized, can also provide a transformer ratio larger than 2, if the plasma density is matched to be exactly twice that of the beam [8]. One can also shape the beam in a more sophisticated manner to further optimize R [9], similar to that in the linear regime [6,7]. But for the sake of simplicity, we will consider the uniform distribution only.

Since the density of the uniform beam is $n_b = N/\pi a^2 b \approx 4.5 \times 10^{17} \text{cm}^{-3}$, the matched plasma density, i.e., with $\alpha = n_b/n_p = 0.5$, is $n_p = 2n_b \approx 9 \times 10^{17} \text{cm}^{-3}$, which gives $2k_p b = 6.3 = \tau_f$. Using the results in [8], the maximum accelerating gradient is

$$G = \sqrt{\chi - 1} k_p mc^2 F(k_p a), \tag{23}$$

where χ is the solution to

$$\tau_f = \sqrt{\chi}\sqrt{\chi - 1} + \ln\left[\sqrt{\chi - 1} + \sqrt{\chi}\right]. \tag{24}$$

Eq.(24) leads to $\chi \approx 5$ and $G \approx 166$ GeV/m. The transformer ratio in this case is simply

$$R = \sqrt{T_f} \sim 2.5 \,. \tag{25}$$

Comparing Case B and Case C, we note that operating in the nonlinear regime has the advantage that it achieves the similar level of acceleration gradient without necessarily invoking a much higher plasma density. The price to pay, however, is that the transformer ratio in the nonlinear case is much smaller. This means with $R = 2.5$, the driving beam with initial energy of 48 GeV cannot sustain more than $L \sim 0.72$m in acceleration length.

The parameters discussed above are listed in Table 1.

TABLE 1. Various bunch shapes

Cases	A: Parabola	B: Doorstep	C: Flat-Top
Beam Parameters			
E [GeV]	48	48	48
N [10^{10}]	2	2	2
ϵ_n [10^{-5} mrad]	3	3	3
σ_z [μm]	17.5	17.5	17.5
b [μm]	17.5	35	35
σ_r [μm]	10	10	10
a [μm]	20	20	20
n_b [10^{17} cm^{-3}]	2.4	6.4	4.5
Plasma Parameters			
n_p [10^{17} cm^{-3}]	9	72	9
α	0.25	0.5	0.5
k_p [cm^{-1}]	1800	5000	1800
$k_p a$	3.6	10	3.6
$k_p b$	π	8π	2π
G [GeV/m]	28	180	167
R	2	24.5	2.5
β [cm]	32	32	32

BEAM DYNAMICS AND BEAM-BEAM INTERACTION ISSUES

Beam Dynamics Issues

The emittance of the trailing (accelerated) bunch degrades due to the strong wakefield focusing (combined with structure errors) and binary collisions in the background plasma. To have a reasonable luminosity we need to start with a high quality beam and to deliver it to the collision point. The trailing bunch needs to be very short for two reasons - the shortness of the driver (PWFA) wavelength and to avoid big losses at the interaction point (IP) ([18]). At present it is not clear if a portion of the driver can be used as a trailing bunch due to the stringent requirements to its quality and parameters. If we use the "doorstep" scenario with $n_p = 7.2 \cdot 10^{18} \text{cm}^{-3}$ plasma density, it gives $\lambda_p = 13\mu$, which makes the useful accelerating period about 3 microns. We can calculate the accelerated bunch betatron length using ("flat top" driver)

$$\beta = \left(\frac{\gamma mc^2}{G \sin \Psi} \sqrt{\frac{8}{\pi k_p a}} \frac{e^{k_p a}}{k_p} \right)^{1/2} . \qquad (26)$$

Taking the initial beam energy $\gamma mc^2 = 1$ GeV, $G \approx 180$ GeV/m we obtain $\beta_i \approx 1\text{cm}/\sqrt{\sin \Psi}$. If $\sigma_z = 0.3\mu$m then $k_p \sigma_z \approx 0.15$ so we need to take the phase at least as $\Psi = 0.6$ which gives $\beta_i \approx 1.5$cm. It means that even in a single stage design we need alignment control [14–16] in the submicron range to preserve the emittance of the accelerated beam (assuming initial normalized emittance of 2 μm, see Table 2).

Following [17], we now consider the emittance growth rate from the multiple scattering in the plasma:

$$\frac{d\epsilon_n}{dz} = \frac{\gamma \beta}{2} \frac{d}{dz} <\Theta^2>_p , \qquad (27)$$

where

$$\frac{d}{dz} <\Theta^2>_p = \pi n_p \left(\frac{4r_e^2}{\gamma^2} \right) \ln \left(\frac{\lambda_D}{R_0} \right) , \qquad (28)$$

where $R_0 = 0.7 \times 10^{-13}$cm is the effective radius of the proton and λ_D is the Debye length ($\lambda_D = k_p^{-1}\sqrt{kT/mc^2}$). Assuming $\gamma = \gamma_i + \gamma' z$ we integrate and obtain

$$\Delta \epsilon_n = \frac{4\pi n_p r_e^2}{\gamma'} \ln \left(\frac{\lambda_D}{R_0} \right)(\beta_f - \beta_i) , \qquad (29)$$

where β_f and β_i are the final and the initial betatron lengths, respectively. We used the fact that $\beta \propto \sqrt{\gamma}$. In our design the acceleration gradient is 180 GeV/m which corresponds to $\gamma' \approx 3.5 \cdot 10^5$ m^{-1}. In [17] the electron temperature is chosen to be 5 eV , in our case it might be much higher but the result is very insensitive to this value. In particular for $kT = 5$ eV and $n_p \approx 7.2 \cdot 10^{18}$ cm^{-3} we obtain:

$$\Delta \epsilon_n \text{ [m]} \approx 3.2 \cdot 10^{-8} (\beta_f - \beta_i) \text{ [m]} , \qquad (30)$$

which is clearly negligible compared to the initial emittance.

TABLE 2. Trailing beam parameters at the IP for 500 GeV, $\mathcal{L}_g = 10^{33}$cm^{-2}s^{-1} e^+e^- linear collider.

P_b(kW)	$N(10^9)$	f_c(kHz)	$\epsilon_n(\mu m)$	$\beta_x^*(\mu m)$	σ_x(nm)	$\sigma_z(\mu m)$	Υ	$\frac{\mathcal{L}(498<W_{cm}<500 \text{GeV})}{\mathcal{L}_g}$
250	5	0.6	2	30	11	0.30	393	0.4
790	1.6	6	2	30	11	0.30	124	0.6
2500	0.5	60	2	30	11	0.30	39	0.7

Beam-Beam Interaction Issues

In the doorstep case the linear calculation gives $G \approx 180$GeV/m. The required center of mass energy of 500 GeV can be achieved in a single stage (for each arm) with a length of about 1 m. The luminosity requirement we impose is $\mathcal{L}_g = 10^{33}$cm^{-2}s^{-1} which was considered in [20]. If we are to work with the SLAC driver with repetition rate of 600 Hz, the number of particles in the accelerated bunch needs to be $5 \cdot 10^9$ which would cause severe beam loading issues. To study the luminosity distribution at the IP we use K. Yokoya's QED code "CAIN" [19]. The Fig. 1 shows three cases which differ by their repetition frequency at IP and the number of particles in the beam but have the same geometrical luminosity. The corresponding beam parameters: beam power P_b, particle number N, repetition frequency f_c, normalized emittance ϵ_n, IP betatron length β_x^*, IP beam transverse size σ_x, bunch length σ_z, the beamstrahlung parameter Υ, and the relative luminosity in the last simulation bin, are listed in Table 2. Clearly, high repetition frequency is necessary to reduce the number of particles required and to improve the differential luminosity (to achieve a higher peak at the design 500 GeV center of mass energy). See Table 1 and Fig. 1. These results indicate the importance of studying multibunch loading in a single shot created wakefield.

LIMITATIONS AND UNRESOLVED ISSUES

The primary beam is self focused (by its own wakefield), and to avoid the emittance growth due to the phase space mixing it should be matched to the focusing. However, the focusing is different in the head/tail of the bunch, so there is concern associated with this.

The primary beam experiences various instabilities: transverse two-stream [21,7], Weibel [22], electron-hose [23]. These are summarized in [3] and need investigation for the discussed scenario.

Stability of the trailing bunch also needs to be studied. It has smaller number of particles but very high density, so the loading issues might be very important.

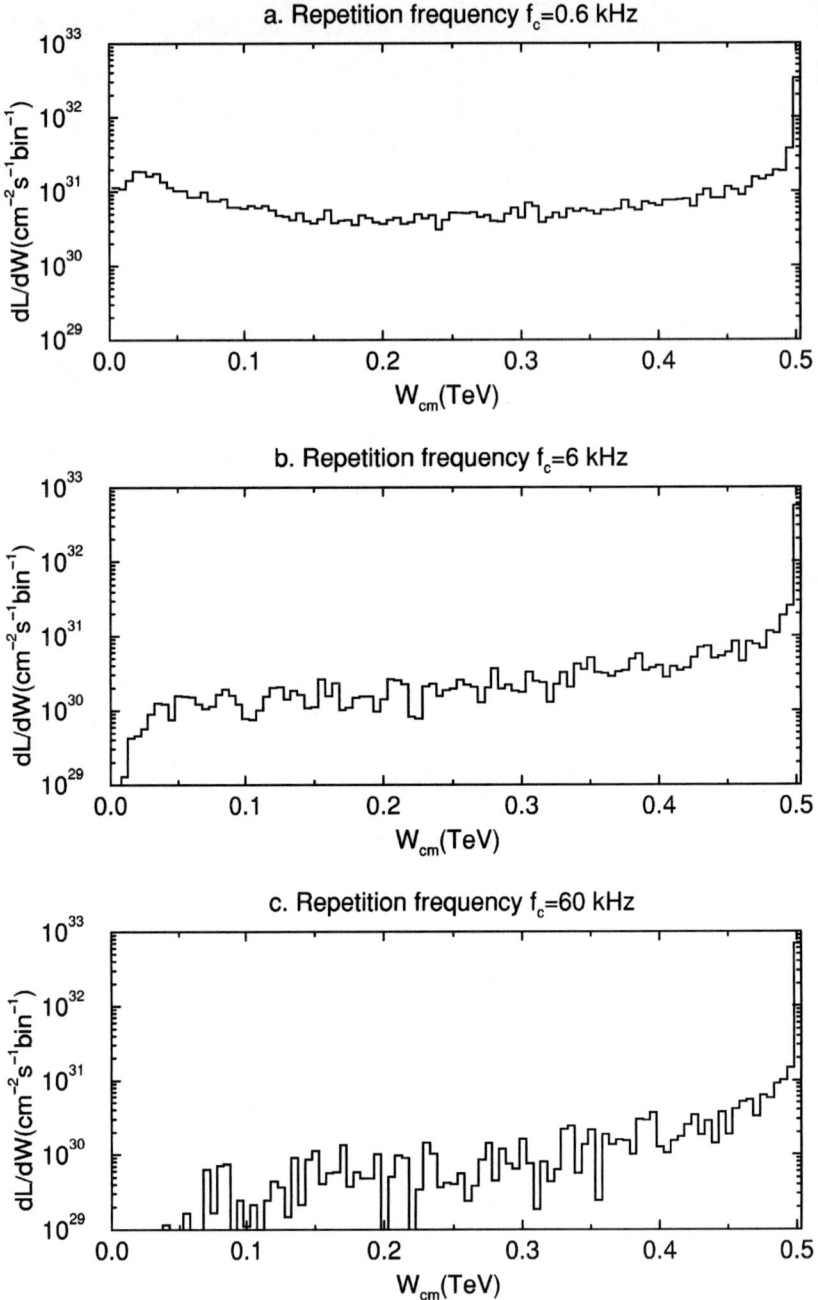

FIGURE 1. Differential luminosity at repetition frequency $f = 0.6$, 6, and 60 kHz, respectively. The geometrical luminosity $\mathcal{L}_g$ was chosen 10^{33} cm^{-2}s^{-1}

REFERENCES

1. T. Tajima and J. M. Dawson, *Phys. Rev. Lett.* **43**, 267 (1979).
2. P. Chen, J. Dawson, R. Huff and T. Katsouleas, *Phys. Rev. Lett.* **54**, 693 (1985).
3. E. Esarey, P. Sprangle, J. Krall and A. Ting. *IEEE Trans. Plasma Science* **24**, No. 2, 252 (1996).
4. T. Chiou, T. Katsouleas, C. Decker, W. Mori, J. Wurtele, G. Shvets and J. Su, *Phys. Plasmas* **2**, 310 (1995).
5. M. J. Hogan et al., *Physics of Plasmas* **7**, 2241 (2000).
6. K. Bane, P. Chen and P. Wilson, *IEEE Trans. Nucl. Sci.* **NS-32**, 3524 (1985).
7. P. Chen, J. Su, J. Dawson, K. Bane and P. Wilson, *Phys. Rev. Lett.* **56**, 1252 (1986).
8. J. Rosenzweig, *Phys. Rev. Lett.* **58**, 555 (1987).
9. Y. Yan and H. Chen *Phys. Rev. A* **38**, 1490 (1988).
10. P. Chen, R. W. Huff, and J. M. Dawson, "A Plasma Booster for Linear Accelerators", UCLA Report PPG-802, 1984.
11. R. Ruth, A. Chao, P. Morton and P. Wilson, *Part. Acc.* **17**, 171 (1985).
12. R. D. Ruth and P. Chen, "Plasma Accelerators", in *Supersymmetry*, SLAC Report No. 296 (1986).
13. Pisin Chen, *Part. Accelerators* **20**, 171 (1987).
14. Tajima, T., Cheshkov, S., Horton, W., and Yokoya, K., in *Advanced Accelerator Concepts 8*, edited by W. Lawson, (AIP, New York, 1999), p.153.
15. Cheshkov, S., Tajima, T., Horton, W., and Yokoya, K., in *Advanced Accelerator Concepts 8*, edited by W. Lawson, (AIP, New York 1999), p.343.
16. Cheshkov, S., Tajima, T., Horton, W., and Yokoya, K. *Phys. Rev. ST Accel. Beams* **3**, 071301 (2000).
17. B. Montague, and W. Schnell, in *Laser Acceleration of Particles 2* (AIP, New York, 1985) p.146 (1985).
18. Xie, M., Tajima, T., Yokoya, K., and Chattopadhyay, S., in *Advanced Accelerator Concepts 7*, edited by S. Chattopadhyay, (AIP, New York, 1997), p.233.
19. K. Yokoya, *CAIN21b*.
    http://jlcux1.kek.jp/subg/ir/Program-e.html
20. S. Cheshkov, and T. Tajima, *Int. J. Mod. Phys. A* **15**, 2555 (2000).
21. T. Katsouleas, *Phys. Rev. A* **33**, 2056 (1986).
22. R. Keinings, and M. Jones, *Phys. Fluids* **30**, 252 (1987).
23. D. Whittum, W. Sharp, S. Yu, M. Lampe, and G. Joyce, *Phys. Rev. Lett.* **67**, 991 (1991).

AUTHOR INDEX

A

Alexeev, I., 231
Anderson, S. G., 391, 487
Antonsen, Jr., T. M., 242
Archambault, L., 136
Arfin, B., 712

B

Babzien, M., 146
Baldis, H. A., 450, 518
Banerjee, S., 553
Banna, S., 652
Barat, K. L., 136
Barov, N., 630
Ben-Zvi, I., 146, 571
Bernal, S., 405, 544
Bogdashov, A. A., 765
Bolton, P., 518
Bratman, V. L., 765
Breitling, F., 163
Brown, W. J., 436
Bruhwiler, D. L., 591
Brussaard, G. J. H., 136
Bychenkov, V. Y., 553

C

Campbell, L. P., 146
Carlsten, B. E., 529
Cary, J. R., 591
Caryotakis, G., 712
Catravas, P. E., 136, 473
Chattopadhyay, S., 136
Chen, C., 415
Chen, P., 183, 518, 903
Cheng, Y., 712
Cheshkov, S., 163, 177, 903
Chirkov, A. V., 765
Chiu, C., 163
Clendenin, J. E., 563
Cline, D. B., 146, 518, 858
Colby, E. R., 47, 863
Conde, M. E., 61, 258, 287

Cowan, T. E., 391
Craddock, W., 518
Crane, J. K., 391
Crawford, C., 518

D

Danilov, O. V., 797
Dawson, J. M., 3
Decker, F. J., 518
Denisov, G. G., 765
Dickinson, M. R., 136
Dilley, C. E., 146
DiMaggio, S., 136
Ditmire, T., 391
Donahue, R., 136
Downer, M. C., 35, 105, 177

E

Esarey, E., 136, 154, 214, 423, 473, 591, 630

F

Fan, J., 231
Fang, J.-M., 316
Field, R. C., 518
Fix, A., 786
Flechtner, D., 465
Fliflet, A. W., 695
Flippo, K., 553
Floyd, J., 136
Fubiani, G., 136, 423
Fukui, Y., 518

G

Gai, W., 47, 258, 287, 605, 679
Gallardo, J. C., 146
Gaul, E. W., 105, 177
Geddes, C. G. R., 136
Giacone, R., 591

Gibson, D. J., 450
Godlove, T., 405
Gold, S. H., 65, 695
Goloviznin, V. V., 223
Gorbachev, A. M., 741
Gottschalk, S. C., 146
Gouveia, E. S., 809
Grabowski, C., 652
Granatstein, V. L., 809
Grigsby, F. B., 177
Gu, S., 553
Gueye, P., 127

H

Haase, A., 712
Haber, I., 405
Hafizi, B., 242, 797
Hafz, N., 122
Haldemann, P., 405
Han, P., 127
Hartemann, F. V., 450
Hayashi, Y., 465
He, P., 146
Hemker, R., 122
Hess, M., 415
Hirose, T., 571
Hirshfield, J. L., 294, 305, 316, 326, 741, 765, 786, 833
Hogan, M. J., 518
Hosokai, T., 112
Houck, T., 686
Hubbard, R. F., 242

I

Isaev, V. A., 741
Ivanov, O. A., 741
Ivers, J. D., 465
Iverson, R., 518
Ives, L., 640
Ives, R. L., 663

J

James, B., 712
Joshi, C., 26, 85, 340

K

Kamp, L. P. J., 223
Kanareyken, A., 258, 605
Kanazawa, S., 112
Kando, M., 112
Kang, Y. W., 335
Kashiwagi, S., 571
Katsouleas, T., 61
Kehne, D., 405
Keppel, C., 127
Kerman, A. K., 450
Kim, K. Y., 231
Kimura, W. D., 146
King, F., 518
Kirby, R. E., 518
Kishek, R. A., 405, 544
Koga, J. K., 97, 122
Koldanov, V. A., 741
Kondo, S., 112
Konecny, R., 258, 287, 679
Korbly, S. E., 436
Kotaki, H., 112
Kotseroglou, T., 518
Kozyrev, E. V., 786
Kreischer, K. E., 436
Kuftin, A. N., 765
Kumar, V., 518
Kumita, T., 571
Kusche, K. P., 146, 571
Kuzikov, S. V., 741
Kuznetsov, G. I., 786

L

Landahl, E. C., 450
Langhoff, H., 105
LaPointe, M. A., 786
Lawson, W., 663, 809
Le Blanc, S. P., 105, 177
Leemans, W. P., 136, 154, 214, 423, 473, 591
Le Foll, A., 450
Le Sage, G. P., 391
Li, H., 405
Li, X., 204
Lidia, S., 686
Litvak, A. G., 741
Liu, Y., 146

Loh, M., 487
Luhmann, Jr., N. C., 450

M

Maksimchuk, A., 553
Mardahl, P., 591
Margolin, L. Y., 231
Marshall, T. C., 294, 316
Matlis, N. H., 105
Mikhailichenko, A., 365, 881
Milchberg, H. M., 231
Miram, G., 640
Mori, W. B., 35
Mourou, G., 553
Musumeci, P., 249, 487

N

Nakagawa, K., 97
Nakajima, K., 97, 112, 518, 844
Nantista, C. D., 702
Nation, J. A., 465, 652
Neilson, J. M., 663
Nemoto, K., 553
Newsham, D., 274
Nezhevenko, O. A., 741, 765, 786, 797
Ng, J. S. T., 518
Noble, R., 518
Nusinovich, G. S., 65, 751, 809

O

Ogata, A., 518
Omori, T., 571
O'Shea, P. G., 405, 544

P

Pakhomov, A. V., 825
Palmer, D. T., 57
Pantell, R. H., 146
Park, S.-Y., 305, 316
Parra, E., 231
Parsa, Z., 583, 890
Pellegrini, C., 249

Peñano, J. R., 242
Persov, B. Z., 786
Petelin, M. I., 741
Pitthan, R., 725
Pogorelsky, I. V., 146, 571
Power, J. G., 258, 287, 605
Pruessner, M., 405
Pyatnitskii, L. N., 231

Q

Quimby, D. C., 146

R

Raimondi, P., 518
Read, M., 663
Reiser, M., 405, 544
Reitsma, A. J. W., 223
Rodgers, D., 136
Rosenzweig, J. B., 33, 374, 391, 487, 630
Rundquist, A. R., 105, 177
Rupp, B., 450
Ruth, R., 903

S

Saleh, N., 127
Samsonov, S. V., 765
Savilov, A. V., 765
Schächter, L., 342, 465, 652, 863, 873
Scheitrum, G., 712
Schelkunoff, S. V., 786
Schep, T. J., 223
Schoessow, P., 258, 287
Schroeder, C. B., 214, 616
Shadwick, B. A., 136, 154, 214
Shapiro, M. A., 436
Shew, B., 712
Short, R., 136
Shvets, G., 195, 204
Siemann, R. H., 863
Skaritka, J., 146
Smirnov, A. V., 775
Smith, A., 136
Song, L., 712

Song, X., 465
Spitkovsky, A., 183
Sprangle, P., 242
Steinhauer, L. C., 146, 266
Suk, H., 487, 630
Sun, X., 287, 679
Symons, R., 833

T

Tajima, T., 23, 77, 163, 177, 903
Takahashi, Y., 825
Tantawi, S. G., 702
Tarkenton, G. M., 154
Temkin, R. J., 436
Thompson, M. C., 374, 487
Ting, A., 242
Troha, A. L., 450

U

Uesaka, M., 122, 500
Umstadter, D., 127, 553
Urakawa, J., 571

V

van Steenbergen, A., 146
van Tilborg, J., 136
Verboncoeur, J. P., 591
Vikharev, A. L., 741
Virgo, M., 405

W

Walz, D., 518
Wang, C., 326, 765, 833
Wang, J. G., 57

Wang, P., 652
Washio, M., 571
Weidemann, A. W., 518
Westenskow, G., 686
Wong, E., 136
Wong, T., 679
Wurtele, J. S., 136, 214, 616

X

Xie, M., 850
Xu, Z., 652

Y

Yakimenko, V., 146, 571
Yakovlev, V. P., 765, 786, 797
Yanovsky, V., 127
Yoder, R. B., 294
Yokoya, K., 571
Yokoyama, T., 112
Yovchev, I., 809
Yu, D., 274, 775
Yun, V., 405

Z

Zeng, J., 274
Zgadzaj, R., 105
Zhang, W., 405
Zhu, X., 844
Zigler, A., 242
Zou, P., 287, 679

Previous Proceedings in the Series of Advanced Accelerator Concepts

	Year	Held in	Publisher	ISBN
8th	1998	Baltimore, Maryland	AIP Conf. Proceedings vol. 472	1-56396-889-4
7th	1996	Lake Tahoe, California	AIP Conf. Proceedings vol. 398	1-56396-697-2
6th	1994	Fontana, Wisconsin	AIP Conf. Proceedings vol. 335	1-56396-476-7
5th	1992	Port Jefferson, New York	AIP Conf. Proceedings vol. 279	1-56396-191-1

Other Related Titles from AIP Conference Proceedings

546 Beam Instrumentation Workshop 2000: Ninth Workshop
Edited by Kenneth D. Jacobs and R. Coles Sibley III, December 2000, 1-56396-975-0

542 Physics Potential and Development of Muon Colliders and Neutrino Factories: Fifth International Conference
Edited by David B. Cline, November 2000, 1-56396-970-X

521 Synchrotron Radiation Instrumentation: Eleventh US National Conference
Edited by Piero Pianetta, John Arthur, and Sean Brennan, May 2000, 1-56396-941-6

520 Bates 25: Celebrating 25 Years of Beam to Experiment
Edited by T. W. Donnelly and W. Turchinetz, June 2000, 1-56396-949-1

512 Nuclear Physics at Storage Rings: Fourth International Conference: STORI99
Edited by Hans-Otto Meyer and Peter Schwandt, June 2000, 1-56396-928-9

480 Space Charge Dominated Beam Physics for Heavy Ion Fusion
Edited by Yuri K. Batygin, June 1999, 1-56396-860-6

475 Applications of Accelerators in Research and Industry: Proceedings of the Fifteenth International Conference
Edited by J. L. Duggan and I. L. Morgan, June 1999, 2 vol. hard cover set, 1-56396-825-8

474 High Energy Density Microwaves
Edited by Robert M. Phillips, May 1999, 1-56396-796-0

473 Heavy Ion Accelerator Technology: Eighth International Conference
Edited by Kenneth W. Shepard, April 1999, 1-56396-806-1

To learn more about these titles, or the AIP Conference Proceedings Series, please visit the webpage **http://www.aip.org/catalog/aboutconf.html**